Lecture Notes in Computer Science 8783

Commenced Publication in 1973
Founding and Former Series Editors:
Gerhard Goos, Juris Hartmanis, and Jan van Leeuwen

Sherman S.M. Chow   Jan Camenisch
Lucas C.K. Hui   Siu Ming Yiu (Eds.)

Information Security

17th International Conference, ISC 2014
Hong Kong, China, October 12-14, 2014
Proceedings

 Springer

Volume Editors

Sherman S.M. Chow
Chinese University of Hong Kong, Department of Information Engineering
Ho Sin Hang Engineering Building, Sha Tin, N.T., Hong Kong, China
E-mail: smchow@ie.cuhk.edu.hk

Jan Camenisch
IBM Research Zurich
Säumerstrasse 4, 8803 Rüschlikon, Switzerland
E-mail: jca@zurich.ibm.com

Lucas C.K. Hui
The University of Hong Kong, Department of Computer Science
Chow Yei Ching Building, Pokfulam Road, Hong Kong, China,
E-mail: hui@cs.hku.hk

Siu Ming Yiu
The University of Hong Kong, Department of Computer Science
Chow Yei Ching Building, Pokfulam Road, Hong Kong, China
E-mail: smyiu@cs.hku.hk

ISSN 0302-9743   e-ISSN 1611-3349
ISBN 978-3-319-13256-3   e-ISBN 978-3-319-13257-0
DOI 10.1007/978-3-319-13257-0
Springer Cham Heidelberg New York Dordrecht London

Library of Congress Control Number: 2014953917

LNCS Sublibrary: SL 4 – Security and Cryptology

Typesetting: Camera-ready by author, data conversion by Scientific Publishing Services, Chennai, India

Printed on acid-free paper

Springer is part of Springer Science+Business Media (www.springer.com)

Preface

The Information Security Conference (ISC) is an annual international conference dedicated to research on the theory and applications of information security. It started as a workshop in 1997, changed to a conference in 2001, and has been held on five different continents. ISC 2014, the 17th in the series, was the first conference in this series to be held in Hong Kong, during October 11–13, 2014. The conference was organized by The University of Hong Kong.

There were 106 submissions from 30 different countries. Each submission was reviewed by around three Program Committee members on average (could be four or five if the submission was co-authored by a Program Committee member), and the review process was done in a "double-blind" fashion. The committee decided to accept 20 full papers and 17 short papers (one short paper was later withdrawn). The program also included two invited talks given by Prof. Shengli Liu on "Public-Key Encryption with Provable Security: Challenges and Approaches" and Prof. Ahmad-Reza Sadeghi on "Gone with the Gadgets: The Continuing Arms Race of Return-Oriented Programming Attacks and Defenses."

We would like to thank all the people who contributed to the success of ISC 2014. First, we would like to thank all authors for submitting their works to us. We deeply thank the 56 Program Committee members (from 18 different countries) as well as the 89 external reviewers for their volunteer work of reading and discussing the submissions.

We are grateful to the ISC Steering Committee, and Prof. Masahiro Mambo in particular, for their advice and support. We thank the Information Security and Forensics Society, which provides an excellent platform for user registration using Paypal and credit card; the general co-chairs Dr. Lucas C.K. Hui and Dr. S.M. Yiu, and the local organizing staff, especially Ms. Catherine Chan, for their unlimited support for ISC. This conference would not have been successful without their great assistance.

Last but not least, we would like to thank EasyChair for providing a user-friendly system we used for both the review phase and the camera-ready phase of the papers, and Springer for providing a meticulous service for the timely production of the proceedings.

October 2014

Jan Camenisch
Sherman S.M. Chow

Organization

ISC 2014 was organized by the Center for Information Security and Cryptography, Department of Computer Science, The University of Hong Kong, and Department of Information Engineering, The Chinese University of Hong Kong.

Steering Committee

Sherman S.M. Chow	Chinese University of Hong Kong, Hong Kong SAR
Ed Dawson	Queensland University of Technology, Australia
Javier López	University of Málaga, Spain
Masahiro Mambo	Kanazawa University, Japan
Eiji Okamoto	University of Tsukuba, Japan
Bhavani Thuraisingham	University of Texas at Dallas, USA
Susanne Wetzel	Stevens Institute of Technology, USA
Yuliang Zheng	University of North Carolina at Charlotte, USA

Program Co-chairs

Jan Camenisch	IBM Research Zürich, Switzerland
Sherman S.M. Chow	Chinese University of Hong Kong, Hong Kong SAR

General Co-chairs

Lucas C.K. Hui	University of Hong Kong, Hong Kong SAR
Siu-Ming Yiu	University of Hong Kong, Hong Kong SAR

Local Organizing Committee

Catherine K.W. Chan	University of Hong Kong, Hong Kong SAR
T.W. Chim	University of Hong Kong, Hong Kong SAR
Leo C.Y. Yeung	University of Hong Kong, Hong Kong SAR
Russell W.F. Lai	Chinese University of Hong Kong, Hong Kong SAR
Yongjun Zhao	Chinese University of Hong Kong, Hong Kong SAR
Tao Zhang	Chinese University of Hong Kong, Hong Kong SAR

Sponsoring Institution

Information Security and Forensics Society, Hong Kong SAR

Program Committee

Gail-Joon Ahn	Arizona State University, USA
Claudio Agostino Ardagna	Università degli Studi di Milano, Italy
Endre Bangerter	Bern University of Applied Sciences, Switzerland
Kevin Bauer	MIT Lincoln Laboratory, USA
Filipe Beato	KU Leuven and iMinds, Belgium
Elisa Bertino	Purdue University, USA
Alex Biryukov	University of Luxembourg, Luxembourg
Marina Blanton	University of Notre Dame, USA
Joseph Bonneau	Princeton University, USA
Rainer Böhme	University of Münster, Germany
Jan Camenisch	IBM Research Zürich, Switzerland
Bogdan Carbunar	Florida International University, USA
Hao Chen	University of California at Davis, USA
Sherman S.M. Chow	Chinese University of Hong Kong, Hong Kong SAR
Roberto Di Pietro	Bell Labs, France
Josep Domingo-Ferrer	Universitat Rovira i Virgili, Catalonia
Maria Dubovitskaya	IBM Research Zürich, Switzerland
Manuel Egele	Boston University, USA
Carol Fung	Virginia Commonwealth University, USA
Thomas Groß	University of Newcastle upon Tyne, UK
Christian Hanser	Graz University of Technology, Austria
Amir Herzberg	Bar Ilan University, Israel, and TU Darmstadt, Germany
Susan Hohenberger	Johns Hopkins University, USA
Nicholas Hopper	University of Minnesota, USA
Yan Huang	Indiana University Bloomington, USA
Pan Hui	Hong Kong University of Science and Technology, Hong Kong SAR
Stefan Katzenbeisser	Technische Universität Darmstadt, Germany
Angelos Keromytis	Columbia University, USA
Patrick P.C. Lee	Chinese University of Hong Kong, Hong Kong SAR
Feng-Hao Liu	University of Maryland, USA
Joseph K. Liu	Institute for Infocomm Research, Singapore
Xiapu Luo	Hong Kong Polytechnic University, Hong Kong SAR

Leonardo A. Martucci	Karlstad University, Sweden
Fabio Massacci	University of Trento, Italy
Yi Mu	University of Wollongong, Australia
Miyako Ohkubo	National Institute of Information and Communications Technology (NICT), Japan
Charalampos Papamanthou	University of Maryland at College Park, USA
Josef Pieprzyk	Queensland University of Technology, Australia
Bart Preneel	KU Leuven and iMinds, Belgium
Kui Ren	State University of New York at Buffalo, USA
Pierangela Samarati	Università degli Studi di Milano, Italy
Jörg Schwenk	Ruhr University Bochum, Germany
Haya Shulman	Technische Universität Darmstadt, Germany
Claudio Soriente	ETH Zürich, Switzerland
Jinyuan Sun	University of Tennessee at Knoxville, USA
Jakub Szefer	Yale University, USA
Mahesh Tripunitara	University of Waterloo, Canada
Carmela Troncoso	Gradiant, Spain
Wen-Guey Tzeng	National Chiao Tung University, Taiwan
Christian Wachsmann	Intel Collaborative Research Institute for Secure Computing, Germany
Susanne Wetzel	Stevens Institute of Technology, USA
Shouhuai Xu	University of Texas at San Antonio, USA
Tsz Hon Yuen	Huawei, Singapore
Moti Yung	Google and Columbia University, USA
Kehuan Zhang	Chinese University of Hong Kong, Hong Kong SAR
Hong-Sheng Zhou	Virginia Commonwealth University, USA
Jianying Zhou	Institute for Infocomm Research, Singapore
John Zic	Commonwealth Scientific and Industrial Research Organisation (CSIRO), Australia

External Reviewers

Aslan, Mahmoud	Davi, Lucas	Farràs, Oriol
Baldimtsi, Foteini	de La Piedra, Antonio	Fischer, Simon
Beckerle, Matthias	De Ryck, Philippe	Föerg, Fabian
Blanco-Justicia, Alberto	Defreez, Daniel	Goldfeder, Steven
Bleikertz, Sören	Derbez, Patrick	Guo, Fuchun
Brasser, Ferdinand	Derler, David	Henricksen, Matt
Büscher, Niklas	Diao, Wenrui	Henry, Ryan
Chang, Ee-Chien	Ding, Jintai	Heuser, Stephan
Copos, Bogdan	Emura, Keita	Hoshino, Fumitaka
Cui, Hui	Enderlein, Robert	Jayaram Masti, Ramya
Daud, Malik Imran	Englehardt, Steven	Junod, Pascal

Kelley, James
Kemerlis, Vasileios P.
Khorsandroo, Sajad
Khovratovich, Dmitry
Kontaxis, Georgios
Korff, Stefan
Lai, Russell W.F.
Li, Nan
Liang, Kaitai
Liu, Xiangyu
Luhn, Sebastian
Luykx, Atul
Mao, Ming
Marforio, Claudio
Melara, Marcela
Mendel, Florian
Meng, Weizhi
Meyer, Bernd
Mladenov, Vladislav

Neven, Gregory
Oren, Yossef
Oya, Simon
Papadopoulos, Dimitrios
Pappas, Vasilis
Perrin, Léo
Polychronakis, Michalis
Pulls, Tobias
Rial, Alfredo
Romero-Tris, Cristina
Russello, Giovanni
S.P.T., Krishnan
Saleh, Moustafa
Seo, Anes Seung-Hyun
Seo, Jae Hong
Seth, Karn
Skowyra, Richard
Slamanig, Daniel
Song, Yingbo

Soria-Comas, Jordi
Tang, Qiang
Tews, Erik
Trieu Phong, Le
Velichkov, Vesselin
Vercauteren, Fréderik
Visegrady, Tamas
Wang, Xiao
Wei, Lei
Xu, Jia
Yan, Fei
Zhang, Bingsheng
Zhang, Cong
Zhang, Tao
Zhang, Yuan
Zhao, Yongjun
Zhou, Yajin
Zhou, Zhe

Invited Talks (Abstracts)

Gone with the Gadgets:
The Continuing Arms Race of Return-Oriented Programming Attacks and Defenses

Ahmad-Reza Sadeghi

Technische Universität Darmstadt, Germany

Abstract. Code reuse attacks, such as Return-oriented Programming (ROP), are elegant techniques that exploit software program vulnerabilities to redirect program control logic within applications. In contrast to code injection attacks, ROP maliciously combines short instruction sequences (gadgets) residing in shared libraries and the application's executable. Even after three decades the security vulnerabilities enabling these attacks still persist, and the cat and mouse game plays on. To date, ROP has been applied to a broad range of architectures (including Intel x86, SPARC, Atmel AVR, ARM and PowerPC). As a consequence, a variety of defenses have been proposed over the last few years, most prominently code randomization (ASLR) techniques and control-flow integrity (CFI). Particularly, constructing practical CFI schemes has recently become a hot topic of research, as it promises the most general mitigation methodology.

In this talk, we go on a journey through the evolution of return-oriented programming (ROP) attacks and defenses. We investigate the security of various defense technologies proposed so far with particular focus on control-flow integrity (CFI). We bring the bad news that currently no practical solution against ROP attacks exists: We show how to bypass all recently proposed "practical" CFI solutions, including Microsoft's defense tool EMET. Finally, we discuss new research directions to defend against code reuse attacks, including our current work on hardware-assisted fine-grained control-flow integrity.

Public-Key Encryption with Provable Security: Challenges and Approaches

Shengli Liu

Shanghai Jiao Tong University, China

Abstract. The golden standard security notion for public key encryption is IND-CCA2 security. However, new attacking techniques impose new security requirements, and IND-CCA2 security might not suffice for some specific scenarios. In this talk, we introduce Selective-Opening Attacks, Key-Leakage Attacks, Key-Dependent Security, Key-Related Attacks and Randomness-Related Attacks. We show how to formalize those attacks to get new security models and set up stronger security notions. We also present the available techniques used to achieve those new security notions.

Table of Contents

Public-Key Encryption

Authentication

Symmetric Key Cryptography

Zero-Knowledge Proofs and Arguments

Outsourced and Multi-party Computations

Implementation

Information Leakage

Firewall and Forensics

Web Security

Android Security

Short Papers

Fully Secure Self-Updatable Encryption in Prime Order Bilinear Groups

Pratish Datta, Ratna Dutta, and Sourav Mukhopadhyay

Department of Mathematics
Indian Institute of Technology Kharagpur
Kharagpur-721302, India
{pratishdatta,ratna,sourav}@maths.iitkgp.ernet.in

Abstract. In CRYPTO 2012, Sahai et al. raised the concern that in a cloud control system revocation of past keys should also be accompanied by updation of previously generated ciphertexts in order to prevent unread ciphertexts from being read by revoked users. Self-updatable encryption (SUE), introduced by Lee et al. in ASIACRYPT 2013, is a newly developed cryptographic primitive that realizes ciphertext update as an inbuilt functionality and thus improves the efficiency of key revocation and time evolution in cloud management. In SUE, a user can decrypt a ciphertext associated with a specific time if and only if the user possesses a private key corresponding to either the same time as that of the ciphertext or some future time. Furthermore, a ciphertext attached to a certain time can be updated to a new one attached to a future time using only public information. The SUE schemes available in the literature are either (a) fully secure but developed in a composite order bilinear group setting under highly non-standard assumptions or (b) designed in prime order bilinear groups but only selectively secure. This paper presents the *first fully secure* SUE scheme in *prime order bilinear groups* under *standard assumptions*, namely, the Decisional Linear and the Decisional Bilinear Diffie-Hellman assumptions. As pointed out by Freeman (EUROCRYPT 2010) and Lewko (EUROCRYPT 2012), the communication and storage, as well as, computational efficiency of prime order bilinear groups are much higher compared to that of composite order bilinear groups with an equivalent level of security. Consequently, our SUE scheme is highly cost-effective than the existing fully secure SUE.

Keywords: public-key encryption, self-updatable encryption, ciphertext update, prime order bilinear groups, cloud storage.

1 Introduction

Cloud storage is gaining popularity very rapidly in recent years due to its lower cost of service, easier data management facility and, most importantly, the accessibility of stored data through Internet from any geographic location. However, since these databases are often filled with oversensitive information, these are prime targets of attackers and security breaches in such systems are not uncommon, especially by insiders of the organizations maintaining the cloud servers.

S.S.M. Chow et al. (Eds.): ISC 2014, LNCS 8783, pp. 1–18, 2014.
© Springer International Publishing Switzerland 2014

In particular, access control is one of the greatest concerns, i.e., the sensitive data items have to be protected from any illegal access, whether it comes from outsiders or even from insiders without proper access rights. Additionally, organizations storing extremely sensitive data to an external cloud server might not want to give the server any access to their information at all. Similar problems can easily arise when dealing with centralized storage within an organization, where different users in different departments have access to varying levels of sensitive data.

One possible approach for this problem is to use attribute-based encryption (ABE) that provides cryptographically enhanced access control functionality in encrypted data [10]. However, in a cloud storage data access is not static. To deal with the change of users' credentials that takes place over time, revocable ABE (R-ABE) [1] has been suggested in which a user's private key can be revoked. In R-ABE, a revoked user is restricted from learning any partial information about the messages encrypted when the ciphertext is created after the time of revocation. Sahai et al. [13] highlighted the fact that R-ABE alone does not suffice in managing dynamic credentials for cloud storage. In fact, R-ABE cannot prevent a revoked user from accessing ciphertexts that were created before the revocation, since the old private key of the revoked user is enough to decrypt those ciphertexts. Thus a complete solution has to support not only the revocation functionality but also the ciphertext update functionality such that a ciphertext at any arbitrary time can be updated to a new ciphertext at a future time by the cloud server just using publicly available information and thereby making previously stored data inaccessible to revoked users.

Self-updatable encryption (*SUE*), introduced by Lee et al. [8], is a newly developed cryptographic primitive that realizes ciphertext update as an inbuilt functionality and thus improves the efficiency of key revocation and time evolution in cloud management. In SUE, a ciphertext and a private key are associated with specific times. A user who has a private key with time T' can decrypt a ciphertext with time T if and only if $T \leq T'$. Additionally, a ciphertext attached to a particular time can be updated to a new ciphertext attached to a future time just using public values.

SUE is related to the notion of forward secure encryption (FSE) [3]. In FSE, a private key associated with a time is evolved to another private key with the immediate subsequent time, and then the past private key is erased. Forward security ensures that an exposed private key with a time cannot be used to decrypt a past ciphertext. In addition to satisfying a similar objective, SUE offers more flexibility in the sense that SUE supports one-to-many communication and thus is suitable for constructing more complex cryptographic primitives by combining with ABE or predicate encryption (PE) [8]. We may view SUE as the dual concept of FSE since the role of private keys and that of the ciphertexts are reversed.

Being an encryption scheme with enhanced time control mechanism, SUE can be applied in various other cryptographic primitives such as time-release encryption [12] and key-insulated encryption [4] to provide better security guarantees.

It is worth observing that there are simple inefficient ways (without the independence guarantee) to realize SUE assuming a vanilla identity based encryption (IBE) scheme [14]. For instance, we can view time periods as identities. An SUE ciphertext for a particular time T is constructed as the collection of IBE ciphertexts for the message corresponding to all times T' in the range $[T, T_{max}]$, wwhere T_{max} is the maximum time bound in the system. An SUE private key attached to a specific time T'' consists of IBE private keys for all time periods upto T''. The ciphertext update procedure simply deletes the IBE ciphertext for the least time period contained in an SUE ciphertext. However, in this approach the ciphertext and key sizes depend polynomially on T_{max}. Constructing an SUE scheme in which the ciphertext and private key sizes are *polylogarithmic* in terms of the maximum time period in the system is not straightforward. In particular, this requires managing the time structure in some suitable manner and design the decryption policy, as well as, the ciphertext update functionality accordingly.

Related Works: Lee et al. [8] constructed a fully secure SUE scheme in composite order bilinear group under non-standard assumptions. To the best of our knowledge, this scheme is so far the only SUE scheme in the literature that achieves full security. However, as pointed out by Freeman [5] and Lewko [9], group operations and pairing computations are prohibitively slow in composite order bilinear groups as compared to prime order bilinear groups with an equivalent level of security. Although, SUE in prime order bilinear groups are proposed in [8], [7] under standard assumptions those are only selectively secure. We note that the work of Sahai et al. [13] can be viewed as a restricted form of SUE but their constructions are highly customized for ABE. In particular, they first added ciphertext delegation to ABE, and then, represented time as a set of attributes, and by doing so reduced ciphertext update to ciphertext delegation. Also in their system all legitimate users are given keys corresponding to the same time and they can decrypt ciphertexts attached to the same time period as their keys. Further, their systems have composite order bilinear group setting.

Our Contribution: We present the *first* fully secure cost effective SUE scheme in prime order bilinear groups under standard assumptions. As in [8], [7], in order to make the ciphertext and private key sizes of our SUE scheme polylogarithmic in terms of the maximum time period supported by the system, we employ a full binary tree to efficiently manage the time structure in the system. We assign a time value to each node of the full binary tree. The ciphertext associated with a particular time consists of a set of components corresponding to some specific nodes in the tree. When a user with a private key associated with a particular time T' attempts to decrypt a ciphertext attached to a certain time T, it can encounter a node associated with a component in the target ciphertext on the path from the root node to the node corresponding to the time T' if and only if $T \leq T'$ and hence the decryption is successful if and only if $T \leq T'$. To design our SUE scheme, we start from the hierarchical identity based encryption (HIBE) scheme of Waters [14] and exchange the private key structure with the ciphertext structure of this HIBE scheme. This is not a trivial task. Unlike [8], [7], the decryption algorithm of our SUE scheme does not involve repeated application of the ciphertext update algorithm to convert a ciphertext associated with a previous time to a ciphertext corresponding to the time associated with the private key used for decryption, resulting in a much faster decryption.

Specifically, the salient features of our SUE scheme lie in the following two aspects:

– *Firstly*, our scheme is proven fully secure under two well-studied assumptions in prime order bilinear groups, namely, the Decisional Linear assumption and the Decisional Bilinear Diffie-Hellman assumption. We employ the dual system encryption methodology introduced by Lewko and Waters [11], [14] for proving security of our SUE scheme. In a dual system, there are two kinds of keys and ciphertexts: normal and semi-functional. Normal keys and ciphertexts are used in the real system, while their semi-functional counter parts are only invoked in the proof. These objects must be constructed satisfying certain relationships. Two crucial properties of the underlying bilinear group that are exploited in proofs employing the dual system encryption technique are *canceling* [5] and *parameter hiding* [9]. These features are achieved naturally when the underlying bilinear group is of composite order by the orthogonality property of the subgroups and the Chinese Remainder theorem. Replicating these properties in prime order groups is quite challenging [5], [9], which makes the construction of SUE in prime order bilinear groups rather difficult. To overcome the difficulty, we construct the SUE ciphertexts and private keys in such a way that we can associate additional random spaces to the semi-functional form of one that is orthogonal to the normal form of the other and vice versa. This approach of achieving the required relations between the normal ciphertexts and private keys with their semi-functional counter parts is quite different from the generic tools developed in [5], [9] and is one of the major contribution of this work.

– *Secondly*, our SUE scheme outperforms the existing fully secure SUE scheme [8] in terms of both communication and computation. In our SUE construction, a ciphertext consists of at most $5 \log T_{max} + 7$ group elements together with $2 \log T_{max}$ elements of $\mathbb{Z}_p$ and a private key involves at most $\log T_{max} + 7$ group elements together with $\log T_{max}$ elements of $\mathbb{Z}_p$, where T_{max} is the maximum time period in the system and p is the order of the bilinear group. At a first glance, these values seem worse than the corresponding values of $3 \log T_{max} + 2$ and $\log T_{max} + 2$ group elements respectively for [8]. However, the important point to observe here is that the order of a composite order bilinear group used in [8] must be at least 1024 bits in order to prevent factorization of the group order, whereas, the size of a prime order bilinear group used in our design that provides an equivalent level of security is only 160 bits [5] which is more than six times smaller. Thus, when compared in terms of bit length, both the ciphertext and key sizes of our SUE scheme are roughly one-third of the corresponding sizes for [8]. The prime order bilinear group has a similar advantage over composite order group in terms of computation as well [5], [9]. For this reason, all the algorithms in our scheme are much faster than [8]. For instance, the decryption algorithm of our SUE scheme is roughly 25 times faster than that of [8] and the ciphertext update algorithm is more than 3 times faster than that of [8].

2 Preliminaries

A function ϵ is *negligible* if, for every integer c, there exists an integer K such that for all $k > K$, $|\epsilon(k)| < 1/k^c$. A problem is said to be *computationally hard* (or *intractable*) if there exists no probabilistic polynomial time (PPT) algorithm that solves it with non-negligible probability (in the size of the input or the security parameter).

2.1 Self-Updatable Encryption

Syntax of Self-Updatable Encryption: A self-updatable encryption (SUE) scheme consists of the following PPT algorithms:

SUE.Setup$(1^\lambda, T_{max})$: The key generation center takes as input a security parameter 1^λ and the maximum time T_{max}. It publishes public parameters PP and generates a master secret key MK for itself.

SUE.GenKey$(T', \mathsf{MK}, \mathsf{PP})$: On input a time T', the master secret key MK together with the public parameters PP, the key generation center outputs a private key $\mathsf{SK}_{T'}$ for a user.

SUE.Encrypt(T, PP): Taking as input a time T and the public parameters PP, the encryptor creates a ciphertext header CH_T and a session key EK. It sends CH_T to the ciphertext storage and keeps EK secret to itself.

SUE.ReEncrypt$(\mathsf{CH}_T, T+1, \mathsf{PP})$: On input a ciphertext header CH_T for a time T, the next time $T+1$ and the public parameters PP, the semi-trusted third party, employed for managing users' access on ciphertext by updating the time component, publishes a ciphertext header for time $T+1$ and a partial session key that will be combined with the session key of CH_T to produce a new session key. The re-encryption procedure may be realized by sequentially executing the following two algorithms although it is not mandatary.

> **SUE.UpdateCT**$(\mathsf{CH}_T, T+1, \mathsf{PP})$: The semi-trusted third party managing users' access on ciphertexts takes as input a ciphertext header CH_T for a time T, a next time $T+1$ together with the public parameters PP, and it outputs an updated ciphertext header CH_{T+1}.

> **SUE.RandCT**$(\mathsf{CH}_T, \mathsf{PP})$: Taking input a ciphertext header CH_T for a time T and the public parameters PP, the semi-trusted ciphertext access managing third party outputs a re-randomized ciphertext header $\widetilde{\mathsf{CH}}_T$ and a partial session key $\widetilde{\mathsf{EK}}$ that will be combined with the session key EK of CH_T to produce a re-randomized session key.

SUE.Decrypt($\mathsf{CH}_T, \mathsf{SK}_{T'}, \mathsf{PP}$): A user takes as input a ciphertext header CH_T, its private key $\mathsf{SK}_{T'}$ together with the public parameters PP, and computes either a session key EK or the distinguished symbol $\perp$.

Correctness: The correctness property of SUE is defined as follows: For all PP, MK generated by **SUE.Setup**, all T, T', any $\mathsf{SK}_{T'}$ output by **SUE.GenKey**, and any $\mathsf{CH}_T, \mathsf{EK}$ returned by **SUE.Encrypt** or **SUE.ReEncrypt**, it is required that:

- If $T \leq T'$, then **SUE.Decrypt**($\mathsf{CH}_T, \mathsf{SK}_{T'}, \mathsf{PP}$) = EK.
- If $T > T'$, then **SUE.Decrypt**($\mathsf{CH}_T, \mathsf{SK}_{T'}, \mathsf{PP}$) = $\perp$ with all but negligible probability.

Security: The security property for SUE schemes is defined in terms of the indistinguishability under a chosen plaintext attack (IND-CPA) by means of the the following game between a challenger $\mathcal{B}$ and a PPT adversary $\mathcal{A}$:

Setup: $\mathcal{B}$ runs **SUE.Setup**($1^\lambda, T_{max}$) to generate the public parameters PP and the master secret key MK, and it gives PP to $\mathcal{A}$.

Query 1: $\mathcal{A}$ may adaptively request a polynomial number of private keys for times $T'_1, \ldots, T'_{q'} \in [0, T_{max}]$, and $\mathcal{B}$ gives the corresponding private keys $\mathsf{SK}_{T'_1}, \ldots, \mathsf{SK}_{T'_{q'}}$ to $\mathcal{A}$ by executing **SUE.GenKey**($T'_i, \mathsf{MK}, \mathsf{PP}$).

Challenge: $\mathcal{A}$ outputs a challenge time $T^* \in [0, T_{max}]$ subject to the following restriction: For all times $\{T'_i\}$ of private key queries, it is required that $T'_i < T^*$. $\mathcal{B}$ chooses a random bit $\beta \in \{0, 1\}$ and computes a ciphertext header CH_{T^*} and a session key EK^* by performing **SUE.Encrypt**(T^*, PP). If $\beta = 0$, then it gives CH_{T^*} and EK^* to $\mathcal{A}$. Otherwise, it gives CH_{T^*} and a random session key to $\mathcal{A}$.

Query 2: $\mathcal{A}$ may continue to request private keys for additional times $T'_{q'+1}, \ldots,$ $T'_q \in [0, T_{max}]$ subject to the same restriction as before, and $\mathcal{B}$ gives the corresponding private keys to $\mathcal{A}$.

Guess: Finally $\mathcal{A}$ outputs a bit β'.

The advantage of $\mathcal{A}$ is defined as $\mathsf{Adv}_{\mathcal{A}}^{\mathrm{SUE}}(\lambda) = |\Pr[\beta = \beta'] - 1/2|$ where the probability is taken over all the randomness of the game.

Definition 1 (Security of SUE). *An SUE scheme is fully secure under a chosen plaintext attack* (CPA) *if for all* PPT *adversaries $\mathcal{A}$, the advantage of $\mathcal{A}$ in the above game, i.e., $\mathsf{Adv}_{\mathcal{A}}^{\mathrm{SUE}}(\lambda)$ is negligible in the security parameter λ.*

2.2 Bilinear Groups of Prime Order and Complexity Assumptions

Let $\mathbb{G}$ and $\mathbb{G}_T$ be two multiplicative cyclic groups of prime order p. Let g be a generator of $\mathbb{G}$ and $e : \mathbb{G} \times \mathbb{G} \to \mathbb{G}_T$ be a bilinear map. The bilinear map e has the following properties: (a) *Bilinearity*: For all $u, v \in \mathbb{G}$ and $a, b \in \mathbb{Z}_p$, we have $e(u^a, v^b) = e(u, v)^{ab}$ and (b) *Non-degeneracy*: $e(g, g) \neq 1$. We say that $\mathcal{G} = (p, \mathbb{G}, \mathbb{G}_T, e)$ is a bilinear group if the group operation in $\mathbb{G}$ and the bilinear map $e : \mathbb{G} \times \mathbb{G} \to \mathbb{G}_T$ are both efficiently computable.

Definition 2. [**Decisional Bilinear Diffie-Hellman (DBDH)**]: *Given* $(\mathcal{D}_1 = (g, g^{c_1}, g^{c_2}, g^{c_3}), z) \in \mathbb{G}^4 \times \mathbb{G}_T$ *as input for random generator* $g \in \mathbb{G}$ *and random exponents* $c_1, c_2, c_3 \in \mathbb{Z}_p$, *the* DBDH *assumption holds if it is computationally hard to decide whether* $z = e(g, g)^{c_1 c_2 c_3}$. *Formally, we define the advantage of an algorithm* $\mathcal{B}$ *in solving the* DBDH *problem as follows:*

$$\mathsf{Adv}_{\mathcal{B}}^{\mathrm{DBDH}}(\lambda) = |\mathsf{Pr}[\mathcal{B}(\mathcal{D}_1, z = e(g, g)^{c_1 c_2 c_3}) = 1] - \mathsf{Pr}[\mathcal{B}(\mathcal{D}_1, z \text{ random in } \mathbb{G}_T) = 1]|$$

The DBDH *assumption holds if* $\mathsf{Adv}_{\mathcal{B}}^{\mathrm{DBDH}}(\lambda)$ *is negligible for any* PPT $\mathcal{B}$.

Definition 3. [**Decisional Linear (DLIN)**]: *Given* $(\mathcal{D}_2 = (g, f, \nu, g^{c_1}, f^{c_2}), z) \in \mathbb{G}^6$ *as input for random generators* $g, f, \nu \in \mathbb{G}$ *and random exponents* $c_1, c_2 \in \mathbb{Z}_p$, *the* DLIN *assumption holds if it is computationally hard to decide whether* $z = \nu^{c_1 + c_2}$. *Formally, the advantage of an algorithm* $\mathcal{B}$ *in solving the* DLIN *problem is defined as follows:*

$$\mathsf{Adv}_{\mathcal{B}}^{\mathrm{DLIN}}(\lambda) = |\mathsf{Pr}[\mathcal{B}(\mathcal{D}_2, z = \nu^{c_1 + c_2}) = 1] - \mathsf{Pr}[\mathcal{B}(\mathcal{D}_2, z \text{ random in } \mathbb{G}) = 1]|$$

The DLIN *assumption holds if* $\mathsf{Adv}_{\mathcal{B}}^{\mathrm{DLIN}}(\lambda)$ *is negligible for any* PPT $\mathcal{B}$.

3 Our Self-Updatable Encryption

In this section we present our fully secure SUE scheme in prime order bilinear group.

3.1 Managing the Time Structure

As in [8], [7], we use a full binary tree $\mathcal{T}$ to represent time in our SUE scheme by assigning time periods to all tree nodes. In the full binary tree $\mathcal{T}$, each node (internal node or leaf node) is assigned a unique time value by using pre-order tree traversal that recursively visits the root node, the left subtree, and the right subtree; i.e., the root node of $\mathcal{T}$ is associated with 0 time value and the right most leaf node of $\mathcal{T}$ is associated with $2^{d_{max}+1} - 2$ time value where d_{max} is the maximum depth of the tree. We fix some notation at this point.

- ν_T = The node associated with time T.
- **Parent**(ν_T) = The parent node of ν_T in the tree.
- **Path**(ν_T) = The set of path nodes from the root node to the node ν_T.
- **RightSibling**(**Path**(ν_T)) = The set of right sibling nodes of **Path**(ν_T).
- **TimeNodes**$(\nu_T) = \{\nu_T\} \cup$ **RightSibling**(**Path**(ν_T))\ **Path**(**Parent**(ν_T)).

We mention that, if a node on **Path**(ν_T) is itself a right sibling of some node in the tree $\mathcal{T}$, then we include it in **RightSibling**(**Path**(ν_T)).

We instantiate the above notations using Figure 1. Let us consider the node ν_{13}, the node associated with the time $T = 13$ according to pre-order traversal. From the figure we have, **Parent**$(\nu_{13}) = \nu_{12}$, **Path**$(\nu_{13}) = \{\nu_0, \nu_8, \nu_{12}, \nu_{13}\}$,

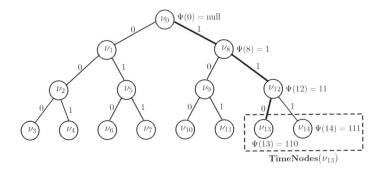

Fig. 1. The tree-based time structure for SUE

RightSibling(Path(ν_{13})) $= \{\nu_8, \nu_{12}, \nu_{14}\}$, and **TimeNodes($\nu_{13}$)** $= \{\nu_{13}, \nu_{14}\}$ (marked with dotted rectangle).

Intuitively, if we consider all subtrees corresponding to all the times that are greater than and equal to the time T according to pre-order traversal, then **TimeNodes(ν_T)** contains the root of each such subtree. More precisely, we have the following lemma, the proof of which is straightforward.

Lemma 1. *TimeNodes(ν_T) $\cap$ Path($\nu_{T'}$) $\neq \varnothing$ if and only if $T \leq T'$.*

To use the structure of HIBE [14], we associate each node with a unique label string $L \in \{0, 1\}^*$. The label of each node in the tree is assigned as follows: Each edge in the tree is assigned with label 0 or 1 depending on whether the edge connects a node to its left or right child node respectively. The label L of a node ν_T is defined as the bit string obtained by reading all the labels of edges in **Path(ν_T)**. We assign a special empty string 'null' to the root node as a label. Note that the length of the label string associated with a node increases with the depth of the node. For example, in Figure 1, the label of ν_8 is 1, that of ν_{12} is 11, ν_{13} is 110 etc. For a label string $L \in \{0, 1\}^*$ of length d, we define the following notations:

- $L[i]$ = the ith bit of L.
- $L|_i$ = the prefix of L with i-bit length, where $i = 0$ means that $L|_i$ is the empty string.
- $L \| L'$ = concatenation of the string L and another string $L' \in \{0, 1\}^*$.
- $L^{(j)} = L|_{d-j} \| 1$, $1 \leq j \leq d$. $L^{(0)}$ is defined to be the label string L itself.
- $d^{(j)}$ = the length of $L^{(j)}$, $0 \leq j \leq d$. Note that $d^{(0)} = d$.

A simple observation would guarantee the validity of the following lemma.

Lemma 2. *If a node ν_T has label string L of length d, then $\{L^{(j)} | 1 \leq j \leq d\}$ contains the label strings of all the nodes in RightSibling(Path(ν_T)). Further, if $L^{(j)} = L|_{d-j+1}$, for $1 < j \leq d$, then $L^{(j)}$ falls in Path(Parent(ν_T)) and hence they are excluded from TimeNodes(ν_T). On the other hand, if $L^{(j)} \neq L|_{d-j+1}$, $1 \leq j \leq d$, then $L^{(j)}$ does not correspond to a node in Path(Parent(ν_T)), rather, it is contained in TimeNodes(ν_T).*

For example, if $L = 110$, then $L[1] = 1$, $L[2] = 1$, $L[3] = 0$, and $L|_1 = 1$, $L|_2 = 11$, $L|_3 = 110$. In Figure 1, ν_{13} has label string $L = 110$ and the set $\{L^{(1)}, L^{(2)}, L^{(3)}\} = \{111, 11, 1\}$ consists of the label strings of $\{\nu_{14}, \nu_{12}, \nu_8\} = $ **RightSibling(Path**(ν_{13})). Also note that $L^{(2)} = 11 = L|_{3-2+1}$ corresponds to the node ν_{12} which is in **Path(Parent**(ν_{13})) and similarly for $L^{(3)}$, whereas $L^{(1)} = 111 \neq 110 = L|_{3-1+1}$ and the node with label string $L^{(1)}$, i.e., ν_{14} is not contained in **Path(Parent**(ν_{13})), rather, it is included in **TimeNodes**(ν_{13}).

We define a mapping Ψ that maps time T corresponding to a tree node ν_T to a unique label L associated with ν_T. Specifically, we define $\Psi(T) = L$ where the node ν_T, according to pre-order traversal, has label L assigned using the method discussed above. For instance, in Figure 1, $\Psi(8) = 1$, $\Psi(12) = 11$, $\Psi(13) = 110$ etc. The SUE ciphertext header for time T consists of a ciphertext component for each node in **TimeNodes**(ν_T).

3.2 Construction

SUE.Setup($1^\lambda, T_{max}$): The key generation center takes as input a security parameter 1^λ and a maximum time T_{max} where $T_{max} = 2^{d_{max}+1} - 2$. It proceeds as follows:

- It considers a full binary tree of depth d_{max} so that d_{max} is the maximum length of label strings associated with the nodes in the tree.
- It generates a bilinear group $\mathcal{G} = (p, \mathbb{G}, \mathbb{G}_T, e)$ of prime order p where $e : \mathbb{G} \times \mathbb{G} \to \mathbb{G}_T$ is the bilinear map.
- It chooses random generators $g, v, v_1, v_2, w, u_1, \ldots, u_{d_{max}}, h_1, \ldots, h_{d_{max}} \in \mathbb{G}$ and exponents $a_1, a_2, b, \alpha \in \mathbb{Z}_p$. Let $\tau_1 = vv_1^{a_1}$, $\tau_2 = vv_2^{a_2}$.
- It publishes the public parameters

$$\mathsf{PP} = (\mathbb{G}, \mathbb{G}_T, e, g^b, g^{a_1}, g^{a_2}, g^{ba_1}, g^{ba_2}, \tau_1, \tau_2, \tau_1^b, \tau_2^b, w^{a_1}, \{u_i^{a_1}, h_i^{a_1}\}_{i=1}^{d_{max}}, \Omega),$$

where $\Omega = e(g, g)^{\alpha a_1 b}$, and sets the master secret key

$$\mathsf{MK} = (g, g^\alpha, g^{\alpha a_1}, v, v_1, v_2, w, \{u_i, h_i\}_{i=1}^{d_{max}})$$

for itself.

SUE.GenKey($T', \mathsf{MK}, \mathsf{PP}$): Taking as input a time T', the master secret key MK and the public parameters PP, the key generation center generates a secret key $\mathsf{SK}_{T'}$ for a user as follows:

- It computes the label string $L' = \Psi(T') \in \{0, 1\}^n$ (say).
- It selects random $s_1, s_2, z_1, z_2, \mathsf{ktag}_1, \ldots, \mathsf{ktag}_n \in \mathbb{Z}_p$ and sets $s = s_1 + s_2$.
- It outputs the private key $\mathsf{SK}_{T'}$ that implicitly includes T' as

$$\mathsf{SK}_{T'} = \begin{pmatrix} D_1 = g^{\alpha a_1} v^s, D_2 = g^{-\alpha} v_1^s g^{z_1}, D_3 = (g^b)^{-z_1}, D_4 = v_2^s g^{z_2}, \\ D_5 = (g^b)^{-z_2}, D_6 = (g^b)^{s_2}, D_7 = g^{s_1}, \\ \{K_i = (u_i^{L'|_i} w^{\mathsf{ktag}_i} h_i)^{s_1}\}_{i=1}^n, \{\mathsf{ktag}_i\}_{i=1}^n \end{pmatrix} \quad (1)$$

where for any $z \in \mathbb{G}$ and any label string $L \in \{0,1\}^*$, $z^L = y \in \mathbb{G}$ such that

$$\log_z y = \sum_{i=1}^{|L|} L[i] 2^{i-1} \pmod{p}, \ |L| \text{ being the length of } L.$$

SUE.Encrypt(T, PP): Taking as input a time T and the public parameters PP, the encryptor prepares a ciphertext header consisting of components corresponding to all the nodes in **TimeNodes**(ν_T) and a session key by performing the following operations:

1. It first computes $L^{(0)} = \psi(T) \in \{0,1\}^d$ (say). It generates the ciphertext component $\mathsf{CH}^{(0)}$ corresponding to the node ν_T as follows:
   - It selects random $\mu_1, \ldots, \mu_d, r_1, r_2, \mathsf{ctag}_1, \ldots, \mathsf{ctag}_d \in \mathbb{Z}_p$ and sets $t = \sum_{i=1}^{d} \mu_i$.
   - It sets

$$\mathsf{CH}^{(0)} = \begin{pmatrix} C_1 = (g^b)^{r_1+r_2}, C_2 = (g^{ba_1})^{r_1}, C_3 = (g^{a_1})^{r_1}, C_4 = (g^{ba_2})^{r_2}, \\ C_5 = (g^{a_2})^{r_2}, C_6 = \tau_1^{r_1} \tau_2^{r_2}, C_7 = (\tau_1^b)^{r_1}(\tau_2^b)^{r_2}(w^{a_1})^{-t}, \\ \{E_{i,1} = [(u_i^{a_1})^{L^{(0)}|_i}(w^{a_1})^{\mathsf{ctag}_i} h_i^{a_1}]^{\mu_i}, E_{i,2} = (g^{a_1})^{\mu_i}\}_{i=1}^d, \\ \{\mathsf{ctag}_i\}_{i=1}^d \end{pmatrix} \tag{2}$$

2. For $1 \le j \le d$, it sets $L^{(j)} = L^{(0)}|_{d-j} \| 1$, i.e., it sets $L^{(j)}$ to be the label of the nodes belonging to **RightSibling(Path**$(\nu_T))$, and proceeds the following steps:

   Case(a): $(L^{(j)} = L^{(0)}|_{d-j+1})$ In this case, $L^{(j)}$ is the label of a node in **Path**(ν_T). The encryptor sets $\mathsf{CH}^{(j)}$ as an empty one since it either corresponds to the node ν_T $(j = 1)$, for which the ciphertext component has already been computed, or it corresponds to a node in **Path(Parent** $(\nu_T))$ which is not in **TimeNodes**(ν_T) (as explained in Figure 1) and hence for which the ciphertext component is not needed.

   Case(b): $(L^{(j)} \ne L^{(0)}|_{d-j+1})$ In this case, $L^{(j)}$ is the label of a node in **TimeNodes**(ν_T). The encryptor obtains $\mathsf{CH}^{(j)}$ as follows:
   - It chooses fresh random $\mu'_{d-j+1}, \mathsf{ctag}'_{d-j+1} \in \mathbb{Z}_p$, sets $\mu'_i = \mu_i$, $\mathsf{ctag}'_i = \mathsf{ctag}_i$ for $1 \le i \le d-j$ and defines $t' = \sum_{i=1}^{d-j+1} \mu'_i$.
   - It computes

$$\mathsf{CH}^{(j)} = \begin{pmatrix} C'_1 = (g^b)^{r_1+r_2}, C'_2 = (g^{ba_1})^{r_1}, C'_3 = (g^{a_1})^{r_1}, C'_4 = (g^{ba_2})^{r_2}, \\ C'_5 = (g^{a_2})^{r_2}, C'_6 = \tau_1^{r_1} \tau_2^{r_2}, C'_7 = (\tau_1^b)^{r_1}(\tau_2^b)^{r_2}(w^{a_1})^{-t'}, \\ \{E'_{i,1} = [(u_i^{a_1})^{L^{(j)}|_i}(w^{a_1})^{\mathsf{ctag}'_i} h_i^{a_1}]^{\mu'_i}, E'_{i,2} = (g^{a_1})^{\mu'_i}\}_{i=1}^{d-j+1}, \\ \{\mathsf{ctag}'_i\}_{i=1}^{d-j+1} \end{pmatrix} \tag{3}$$

Note that $L^{(j)}$ and $L^{(0)}$ have the same prefix string of length $d - j$.

– It also prunes redundant elements, namely, $C'_1, \ldots, C'_6, \{E'_{i,1}, E'_{i,2}\}_{i=1}^{d-j}$, $\{ctag'_i\}_{i=1}^{d-j}$ from $CH^{(j)}$ which are already contained in $CH^{(0)}$.
3. It removes all empty $CH^{(j)}$'s and sets $CH_T = (CH^{(0)}, \ldots, CH^{(d')})$ for some $d' < d$ that consists of nonempty $CH^{(j)}$'s.
4. It sends the ciphertext header CH_T that implicitly includes T to the ciphertext storage and keeps a session key as $EK = \Omega^{r_2}$ private to itself. Note that $CH^{(j)}$'s are ordered according to pre-order traversal.

SUE.UpdateCT$(CH_T, T+1, PP)$: On input a ciphertext header $CH_T = (CH^{(0)},$ $\ldots, CH^{(d')})$ for a time T, a next time $T+1$ and the public parameters PP, the semi-trusted third party, employed for managing users' access on ciphertexts by updating a ciphertext header associated with some time to a ciphertext header corresponding to a future time, performs the following steps, where $L^{(j)}$ is the label of $CH^{(j)}$.
Case(a): ($L^{(0)} = \Psi(T)$ is the label of an internal node in the tree)

– It first obtains $CH^{L^{(0)}\|0}$ and $CH^{L^{(0)}\|1}$ as follows:
  • Let $CH^{(0)} = (C_1, \ldots, C_7, \{E_{i,1}, E_{i,2}\}_{i=1}^d, \{ctag_i\}_{i=1}^d)$.
    It selects a random $\mu_{d+1}, ctag_{d+1} \in \mathbb{Z}_p$.
  • It computes

$$CH^{L^{(0)}\|0} = \begin{pmatrix} C_1^\dagger = C_1, C_2^\dagger = C_2, C_3^\dagger = C_3, C_4^\dagger = C_4, \\ C_5^\dagger = C_5, C_6^\dagger = C_6, C_7^\dagger = C_7(w^{a_1})^{-\mu_{d+1}}, \\ \{E_{i,1}^\dagger = E_{i,1}, E_{i,2}^\dagger = E_{i,2}\}_{i=1}^d, \\ E_{d+1,1}^\dagger = [(u_{d+1}^{a_1})^{L^{(0)}\|0}(w^{a_1})^{ctag_{d+1}} h_{d+1}^{a_1}]^{\mu_{d+1}}, \\ E_{d+1,2}^\dagger = (g^{a_1})^{\mu_{d+1}}, \{ctag_i^\dagger = ctag_i\}_{i=1}^{d+1} \end{pmatrix}$$

  • In a similar fashion, $CH^{L^{(0)}\|1}$ is computed from $CH^{(0)}$ by selecting fresh random $\mu'_{d+1}, ctag'_{d+1} \in \mathbb{Z}_p$.
  Note that in this case, $L^{(0)}\|0$ corresponds to the node ν_{T+1} and $L^{(0)}\|1$ corresponds to a node in **TimeNodes**(ν_{T+1}) according to pre-order traversal. See Figure 1 for a pictorial understanding.
– It also prunes redundant elements in $CH^{L^{(0)}\|1}$ which are contained in $CH^{L^{(0)}\|0}$ in the same way as in **SUE.Encrypt**.
– It outputs an updated ciphertext header $CH_{T+1} = (CH^{\dagger(0)} = CH^{L^{(0)}\|0}, CH^{\dagger(1)}$ $= CH^{L^{(0)}\|1}, \ldots, CH^{\dagger(d'+1)} = CH^{(d')})$. Note that all the nodes contained in **TimeNodes**(ν_T) except the node ν_T itself are also in **TimeNodes**(ν_{T+1}) due to pre-order traversal as can be seen from Figure 1.

Case(b): ($L^{(0)} = \Psi(T)$ is the label of a leaf node) It copies common elements in $CH^{(0)}$ to $CH^{(1)}$ and simply removes $CH^{(0)}$, since $CH^{(1)}$ is the ciphertext component corresponding to ν_{T+1} in this case and the node ν_T corresponding to the ciphertext component $CH^{(0)}$ is excluded from **TimeNodes**(ν_{T+1}) by pre-order traversal. It outputs an updated ciphertext header as $CH_{T+1} = (CH^{\dagger(0)} =$

$\mathsf{CH}^{(1)}, \ldots, \mathsf{CH}^{\dagger(d'-1)} = \mathsf{CH}^{(d')}$). Observe that in this case also, the nodes included in **TimeNodes**(ν_T) except ν_T itself are also included in **TimeNodes**(ν_{T+1}). See Figure 1 for an instantiation.

SUE.RandCT$(\mathsf{CH}_T, \mathsf{PP})$: The semi-trusted third party managing the users' access on ciphertexts takes as input a ciphertext header $\mathsf{CH}_T = (\mathsf{CH}^{(0)}, \ldots, \mathsf{CH}^{(d')})$ for a time T, and public parameters PP. Let $L^{(j)}$ be the label of $\mathsf{CH}^{(j)}$ and $d^{(j)}$ be the length of the label $L^{(j)}$. It proceeds as follows:

1. It first obtains a re-randomized ciphertext component $\widetilde{\mathsf{CH}}^{(0)}$ from $\mathsf{CH}^{(0)} = (C_1, \ldots, C_7, \{E_{i,1}, E_{i,2}\}_{i=1}^{d^{(0)}}, \{\mathsf{ctag}_i\}_{i=1}^{d^{(0)}})$ as follows:
   - It chooses random exponents $\mu'_1, \ldots, \mu'_{d^{(0)}}, r'_1, r'_2 \in \mathbb{Z}_p$ and sets $t' = \sum_{i=1}^{d^{(0)}} \mu'_i$.
   - It obtains

   $$\widetilde{\mathsf{CH}}^{(0)} = \begin{pmatrix} \widetilde{C}_1 = C_1(g^b)^{r'_1+r'_2}, \widetilde{C}_2 = C_2(g^{ba_1})^{r'_1}, \widetilde{C}_3 = C_3(g^{a_1})^{r'_1}, \\ \widetilde{C}_4 = C_4(g^{ba_2})^{r'_2}, \widetilde{C}_5 = C_5(g^{a_2})^{r'_2}, \\ \widetilde{C}_6 = C_6(\tau_1^{r'_1}\tau_2^{r'_2}), \widetilde{C}_7 = C_7(\tau_1^b)^{r'_1}(\tau_2^b)^{r'_2}(w^{a_1})^{-t'}, \\ \{\widetilde{E}_{i,1} = E_{i,1}[(u_i^{a_1})^{L^{(0)}|i}(w^{a_1})^{\mathsf{ctag}_i}h_i^{a_1}]^{\mu'_i}\}_{i=1}^{d^{(0)}}, \\ \{\widetilde{E}_{i,2} = E_{i,2}(g^{a_1})^{\mu'_i}\}_{i=1}^{d^{(0)}}, \\ \{\widetilde{\mathsf{ctag}}_i = \mathsf{ctag}_i\}_{i=1}^{d^{(0)}} \end{pmatrix}$$

2. For $1 \le j \le d'$, it computes $\widetilde{\mathsf{CH}}^{(j)}$ from pruned $\mathsf{CH}^{(j)} = (C'_7, \{E'_{d^{(j)},1}, E'_{d^{(j)},2}\}, \{\mathsf{ctag}'_{d^{(j)}}\})$ as follows:

   - It selects random $\mu''_{d^{(j)}} \in \mathbb{Z}_p$ and defines $t'' = \sum_{i=1}^{d^{(j)}-1} \mu'_i + \mu''_{d^{(j)}}$. Note that $L^{(j)}$ has the same prefix string of length $d^{(j)} - 1$ as $L^{(0)}$.
   - It computes

   $$\widetilde{\mathsf{CH}}^{(j)} = \begin{pmatrix} \widetilde{C}'_7 = C'_7(\tau_1^b)^{r'_1}(\tau_2^b)^{r'_2}(w^{a_1})^{-t''}, \\ \{\widetilde{E}'_{d^{(j)},1} = E'_{d^{(j)},1}[(u_{d^{(j)}}^{a_1})^{L^{(j)}}(w^{a_1})^{\mathsf{ctag}'_{d^{(j)}}}h_{d^{(j)}}^{a_1}]^{\mu''_{d^{(j)}}}\}, \\ \{\widetilde{E}'_{d^{(j)},2} = E'_{d^{(j)},2}(g^{a_1})^{\mu''_{d^{(j)}}}\}, \{\widetilde{\mathsf{ctag}}'_{d^{(j)}} = \mathsf{ctag}'_{d^{(j)}}\} \end{pmatrix}$$

   Note that the other components of $\widetilde{\mathsf{CH}}^{(j)}$ have been pruned.

3. It publishes a re-randomized ciphertext header as $\widetilde{\mathsf{CH}}_T = (\widetilde{\mathsf{CH}}^{(0)}, \ldots, \widetilde{\mathsf{CH}}^{(d')})$ and a partial session key as $\widetilde{\mathsf{EK}} = \Omega^{r'_2}$ that will be multiplied with the session key EK of CH_T to produce a re-randomized session key.

SUE.Decrypt$(\mathsf{CH}_T, \mathsf{SK}_{T'}, \mathsf{PP})$: A user takes as input a ciphertext header $\mathsf{CH}_T = (\mathsf{CH}^{(0)}, \ldots, \mathsf{CH}^{(d')})$, its own private key $\mathsf{SK}_{T'} = (D_1, \ldots, D_7, \{K_i\}_{i=1}^n, \{\mathsf{ktag}_i\}_{i=1}^n)$ such that $L' = \Psi(T') \in \{0,1\}^n$ together with the public parameters PP.

1. If the user can find $\mathsf{CH}^{(j)} = (C_1, \ldots, C_7, \{E_{i,1}, E_{i,2}\}_{i=1}^{d^{(j)}}, \{\mathsf{ctag}_i\}_{i=1}^{d^{(j)}})$ from CH_T such that its corresponding label string $L^{(j)} \in \{0,1\}^{d^{(j)}}$ is a prefix of L' and for $1 \le i \le d^{(j)}$, it holds that $\mathsf{ctag}_i \ne \mathsf{ktag}_i$, then the user retrieves EK by the following computation:

$$A_1 = \prod_{i=1}^{5} e(C_i, D_i), \quad A_2 = e(C_6, D_6)e(C_7, D_7), \quad A_3 = A_1/A_2,$$

$$A_4 = \prod_{i=1}^{d^{(j)}} \left[\frac{e(E_{i,1}, D_7)}{e(E_{i,2}, K_i)} \right]^{\frac{1}{\mathsf{ctag}_i - \mathsf{ktag}_i}} \quad \text{and finally, } \mathsf{EK} = A_3/A_4.$$

2. Otherwise, the user obtains $\bot$.

3.3 Correctness

The SUE ciphertext header CH_T of a time T consists of the ciphertext components $\mathsf{CH}^{(0)}, \ldots, \mathsf{CH}^{(d')}$ each of which is associated with a node belonging to **TimeNodes**(ν_T). If the SUE private key $\mathsf{SK}_{T'}$ for a time T' satisfies $T \le T'$, then **TimeNodes**$(\nu_T) \cap$ **Path**$(\nu_{T'}) = \{\nu_{\widehat{T}}\}$, where $T \le \widehat{T} \le T'$, for a unique node $\nu_{\widehat{T}}$, by the property of pre-order tree traversal as stated in Lemma 1. Let $\mathsf{CH}^{(j)}$ be the ciphertext component that is associated with the node $\nu_{\widehat{T}}$. Now as $\nu_{\widehat{T}} \in$ **Path**$(\nu_{T'})$, the label string $L^{(j)}$ of $\nu_{\widehat{T}}$ is a prefix of the label string $L' = \Psi(T')$. Assume that $L^{(j)}$ and L' are of length $d^{(j)}$ and n respectively. According to the scheme description, we have $\mathsf{SK}_{T'}$ and $\mathsf{CH}^{(j)}$ are of the forms as defined in equations (1) and (2) or (3) respectively. Thus we obtain

$$A_1 = \prod_{i=1}^{5} e(C_i, D_i) = e(g,g)^{\alpha a_1 b r_2} e(v,g)^{b(r_1+r_2)s} e(v_1, g)^{b a_1 r_1 s} e(v_2, g)^{b a_2 r_2 s}$$

$$A_2 = e(C_6, D_6)e(C_7, D_7) = e(v,g)^{b(r_1+r_2)s} e(v_1,g)^{b a_1 r_1 s} e(v_2 g)^{b a_2 r_2 s} e(g,w)^{-a_1 t s_1}$$

$$A_3 = A_1/A_2 = e(g,g)^{\alpha a_1 b r_2} e(g,w)^{a_1 t s_1}$$

$$A_4 = \prod_{i=1}^{d^{(j)}} \left[\frac{e(E_{i,1}, D_7)}{e(E_{i,2}, K_i)} \right]^{\frac{1}{\mathsf{ctag}_i - \mathsf{ktag}_i}} = e(g,w)^{a_1 s_1 \sum\limits_{i=1}^{d^{(j)}} \mu_i} \quad \text{if } \mathsf{ctag}_i \ne \mathsf{ktag}_i, 1 \le i \le d^{(j)}$$

$$\mathsf{EK} = A_3/A_4 = e(g,g)^{\alpha a_1 b r_2}, \text{ since } t = \sum_{i=1}^{d^{(j)}} \mu_i$$

Further, one can easily verify that **SUE.UpdateCT** outputs a valid ciphertext header and **SUE.RandCT** re-randomizes the input ciphertext header. These procedures are sequentially executed to implement **SUE.ReEncrypt**, whereby any semi-trusted third party, managing users' access on ciphertexts, can update a ciphertext header attached to a certain time to a new ciphertext header attached to a future time using only public information.

4   Security Analysis

Theorem 1. *For any* PPT CPA *adversary $\mathcal{A}$ for the SUE scheme, introduced in Section 3, we have* $\mathsf{Adv}_{\mathcal{A}}^{\mathrm{SUE}}(\lambda) \leq \mathsf{Adv}_{\mathcal{B}_1}^{\mathrm{DLIN}}(\lambda) + q\mathsf{Adv}_{\mathcal{B}_2}^{\mathrm{DLIN}}(\lambda) + \mathsf{Adv}_{\mathcal{B}_3}^{\mathrm{DBDH}}(\lambda)$, *where* $\mathsf{Adv}_{\mathcal{B}_i}^{\mathrm{DLIN}}(\lambda)$, *for* $i \in \{1, 2\}$, *denotes the advantage of an algorithm $\mathcal{B}_i$ in solving the DLIN problem;* $\mathsf{Adv}_{\mathcal{B}_3}^{\mathrm{DBDH}}(\lambda)$ *denotes that of an algorithm $\mathcal{B}_3$ in solving the DBDH problem; and q is the maximum number of private key queries made by $\mathcal{A}$. Thus, the SUE scheme of Section 3 is fully secure under* CPA *if the DBDH and DLIN assumptions hold.*

Proof Outline: To prove the security of our SUE scheme, we use the dual system encryption technique of Lewko and Waters [11], [14]. In dual system encryption, ciphertexts and private keys can be normal or semi-functional type. The normal type and the semi-functional type are indistinguishable and a semi-functional ciphertext cannot be decrypted by a semi-functional private key. The whole security proof consists of hybrid games that change the normal challenge ciphertext and the normal private keys to the semi-functional challenge ciphertext and semi-functional private keys respectively. In the final game, the adversary given the semi-functional private keys and the semi-functional challenge ciphertext cannot distinguish a proper session key from a random session key.

We define the semi-functional type of ciphertexts and private keys as follows:

SUE.GenKeySF: This algorithm generates a semi-functional private key $\widehat{\mathsf{SK}}_{T'}$ for a time T' as follows:

- It first computes the label string $L' = \Psi(T') \in \{0, 1\}^n$ (say).
- Next, it creates a normal private key

$$\mathsf{SK}_{T'} = \begin{pmatrix} D_1 = g^{\alpha a_1} v^s, D_2 = g^{-\alpha} v_1^s g^{z_1}, D_3 = (g^b)^{-z_1}, D_4 = v_2^s g^{z_2}, \\ D_5 = (g^b)^{-z_2}, D_6 = (g^b)^{s_2}, D_7 = g^{s_1}, \\ \{K_i = (u_i^{L'|_i} w^{\mathsf{ktag}_i} h_i)^{s_1}\}_{i=1}^n, \{\mathsf{ktag}_i\}_{i=1}^n \end{pmatrix}$$

for the time T' under random exponents $s_1, s_2, z_1, z_2 \in \mathbb{Z}_p$ using the master key MK along with the public parameters PP by running the algorithm **SUE.GenKey**$(T', \mathsf{MK}, \mathsf{PP})$.
- Then, it chooses a random $\gamma \in \mathbb{Z}_p$.
- It sets

$$\widehat{\mathsf{SK}}_{T'} = \begin{pmatrix} \widehat{D}_1 = D_1 g^{-a_1 a_2 \gamma}, \widehat{D}_2 = D_2 g^{a_2 \gamma}, \widehat{D}_3 = D_3, \widehat{D}_4 = D_4 g^{a_1 \gamma}, \\ \widehat{D}_5 = D_5, \widehat{D}_6 = D_6, \widehat{D}_7 = D_7, \\ \{\widehat{K}_i = K_i\}_{i=1}^n, \{\widehat{\mathsf{ktag}_i} = \mathsf{ktag}_i\}_{i=1}^n \end{pmatrix}$$

- It outputs
$$\widehat{\mathsf{SK}}_{T'} = (\widehat{D}_1, \ldots, \widehat{D}_7, \{\widehat{K}_i\}_{i=1}^n, \{\widehat{\mathsf{ktag}_i}\}_{i=1}^n).$$

SUE.EncryptSF: This algorithm computes a semi-functional ciphertext header $\widehat{\mathsf{CH}}_T$ for a time T as follows:

- It first computes the label string $L^{(0)} = \Psi(T) \in \{0,1\}^d$ (say).
- Next, it obtains a normal ciphertext header $\mathsf{CH}_T = (\mathsf{CH}^{(0)}, \ldots, \mathsf{CH}^{(d')})$ for some $d' < d$ that consists of nonempty $\mathsf{CH}^{(j)}$, $1 \leq j \leq d$, and a session key EK for the time T by running $\mathbf{SUE.Encrypt}(T, \mathsf{PP})$, where $L^{(j)}$ is the label string of $\mathsf{CH}^{(j)}$ and $d^{(j)}$ is the length of $L^{(j)}$. Note that $d^{(0)} = d$.
- Then, it modifies the components of CH_T as follows:
  1. Let
  $$\mathsf{CH}^{(0)} = \begin{pmatrix} C_1 = (g^b)^{r_1+r_2}, C_2 = (g^{ba_1})^{r_1}, C_3 = (g^{a_1})^{r_1}, C_4 = (g^{ba_2})^{r_2}, \\ C_5 = (g^{a_2})^{r_2}, C_6 = \tau_1^{r_1}\tau_2^{r_2}, C_7 = (\tau_1^b)^{r_1}(\tau_2^b)^{r_2}(w^{a_1})^{-t}, \\ \{E_{i,1} = [(u_i^{a_1})^{L^{(0)}|_i}(w^{a_1})^{\mathsf{ctag}_i}h_i^{a_1}]^{\mu_i}, E_{i,2} = (g^{a_1})^{\mu_i}\}_{i=1}^{d^{(0)}}, \\ \{\mathsf{ctag}_i\}_{i=1}^{d^{(0)}} \end{pmatrix}$$

  - It chooses a random $x \in \mathbb{Z}_p$.
  - It sets
  $$\widehat{\mathsf{CH}}^{(0)} = \begin{pmatrix} \widehat{C}_1 = C_1, \widehat{C}_2 = C_2, \widehat{C}_3 = C_3, \widehat{C}_4 = C_4 g^{ba_2 x}, \\ \widehat{C}_5 = C_5 g^{a_2 x}, \widehat{C}_6 = C_6 v_2^{a_2 x}, \widehat{C}_7 = C_7 v_2^{a_2 bx}, \\ \{\widehat{E}_{i,1} = E_{i,1}, \widehat{E}_{i,2} = E_{i,2}\}_{i=1}^{d^{(0)}}, \{\widehat{\mathsf{ctag}}_i = \mathsf{ctag}_i\}_{i=1}^{d^{(0)}} \end{pmatrix}$$

  2. For $1 \leq j \leq d^{(0)}$, it proceeds as follows: If $\mathsf{CH}^{(j)}$ is not present in CH_T, then it sets $\widehat{\mathsf{CH}}^{(j)}$ as an empty one. Otherwise, let
  $$\mathsf{CH}^{(j)} = \begin{pmatrix} C_7' = (\tau_1^b)^{r_1}(\tau_2^b)^{r_2}(w^{a_1})^{-t'}, \\ \{E'_{d^{(j)},1} = [(u_{d^{(j)}}^{a_1})^{L^{(j)}}(w^{a_1})^{\mathsf{ctag}'_{d^{(j)}}}h_{d^{(j)}}^{a_1}]^{\mu'_{d^{(j)}}}, \\ E'_{d^{(j)},2} = (g^{a_1})^{\mu'_{d^{(j)}}}\}, \{\mathsf{ctag}'_{d^{(j)}}\} \end{pmatrix}$$

  It creates the semi-functional ciphertext component $\widehat{\mathsf{CH}}^{(j)}$ as
  $$\widehat{\mathsf{CH}}^{(j)} = \begin{pmatrix} \widehat{C}_7' = C_7' v_2^{a_2 bx}, \{\widehat{E}'_{d^{(j)},1} = E'_{d^{(j)},1}, \widehat{E}'_{d^{(j)},2} = E'_{d^{(j)},2}\}, \\ \{\widehat{\mathsf{ctag}}'_{d^{(j)}} = \mathsf{ctag}'_{d^{(j)}}\} \end{pmatrix}$$

  Observe that $C_1', C_2', C_3', C_4', C_5', C_6', \{E'_{i,1}, E'_{i,2}\}_{i=1}^{d^{(j)}-1}, \{\mathsf{ctag}'_i\}_{i=1}^{d^{(j)}-1}$ were pruned in the normal ciphertext component $\mathsf{CH}^{(j)}$ and so are in $\widehat{\mathsf{CH}}^{(j)}$.
  3. It removes all empty $\widehat{\mathsf{CH}}^{(j)}$ and sets $\widehat{\mathsf{CH}}_T = (\widehat{\mathsf{CH}}^{(0)}, \ldots, \widehat{\mathsf{CH}}^{(d')})$.
  4. It outputs a semi-functional ciphertext header as $\widehat{\mathsf{CH}}_T$ and a session key as $\widehat{\mathsf{EK}} = \mathsf{EK} = \Omega^{r_2}$.

The details of the proof of Theorem 1 can be found in the full version.

5   Efficiency

Table 1 and 2 present the comparison between our fully secure SUE scheme and that of [8], the only SUE scheme with full security available in the literature,

Table 1. Communication and Storage Comparison

SUE	$|\mathcal{G}|$	Complexity Assumptions	$|$PP$|$	$|$MK$|$	$|$SK$_T|$	$|$CH$_T|$
[8]	n (composite)	Variants of SD, Composite DH	$4k + 2$ in $\mathbb{G}$, 1 in $\mathbb{G}_T$	1 in $\mathbb{G}$, 1 in $\mathbb{Z}_n$	$k + 2$ in $\mathbb{G}$	$3k + 2$ in $\mathbb{G}$
Ours	p (prime)	DLIN, DBDH	$2k + 10$ in $\mathbb{G}$, 1 in $\mathbb{G}_T$	$2k + 7$ in $\mathbb{G}$	$k + 7$ in $\mathbb{G}$, k in $\mathbb{Z}_p$	$5k + 7$ in $\mathbb{G}$, $2k$ in $\mathbb{Z}_p$

Here, $k = \log T_{max}$ where T_{max} is the maximum time in the system.

Table 2. Computation Comparison

SUE	$|\mathcal{G}|$	SUE.Setup	SUE.GenKey	SUE.Encrypt	SUE.ReEncrypt	SUE.Decrypt
[8]	n (composite)	1 in $\mathbb{G}_T$; 1	$2k + 3$ in $\mathbb{G}$	$4k + 2$ in $\mathbb{G}$, 1 in $\mathbb{G}_T$	$4k + 6$ in $\mathbb{G}$, 1 in $\mathbb{G}_T$	$2k - 2$ in $\mathbb{G}$; $k + 2$
Ours	p (prime)	$2k + 12$ in $\mathbb{G}$, 1 in $\mathbb{G}_T$; 1	$3k + 9$ in $\mathbb{G}$	$9k + 10$ in $\mathbb{G}$, 1 in $\mathbb{G}_T$	$9k + 20$ in $\mathbb{G}$, 1 in $\mathbb{G}_T$	k in $\mathbb{G}$; $2k + 7$

Here, $k = \log T_{max}$ where T_{max} is the maximum time in the system.
In this table, 'x; y' signifies 'x many exponentiations and y many pairings'.

in terms of communication together with storage and computation respectively. The other previously known SUE schemes, viz., the prime order constructions of [8], [7] are only selectively secure and hence we do not consider them for efficiency comparison. Let us first concentrate on the communication and storage efficiency. Just by looking at Table 1, it may appear that the communication efficiency of our scheme is worse than that of [8]. However, our scheme is built on prime order bilinear group as opposed to [8] which is built on composite order bilinear group and this difference in the group order results in a complete turn-around of the situation. As noted in [5], the only known instantiations of composite order bilinear group use elliptic curves (or more generally, abelian varieties) over finite fields. Since the elliptic curve group order must be infeasible to factor, it must be at least 1024 bits. On the other hand, the size of a prime order elliptic curve group that provides an equivalent level of security is only 160 bits which is more than six times smaller. Thus, when compared in terms of bit length, we can readily infer from Table 1 that both the ciphertext and key sizes in our SUE scheme are roughly one-third of the corresponding sizes in [8]. Our public parameters contain almost half the number of group elements in the public parameters of [8]. Regarding the master key size of our SUE scheme, note that, although it is not constant, it grows moderately with the maximum time bound in the system. For instance, if the time periods are taken to be months, then the maximum time bound $T_{max} = 2^{17}$ is more than 850 years, and in this case our master key would involve only 41 group elements.

Table 2 displays the computational cost of our SUE scheme in comparison with that of [8]. Here also our scheme is benefited from the use of prime order bilinear group. Due to the excessive bit length of the group order, group operations and pairing computations are prohibitively slow on composite order elliptic curves [5], [9]. In particular, an exponentiation is at least six times slower and a pairing computation is roughly 50 times slower on a 1024 bit composite order elliptic curve than the corresponding operations on a comparable prime order curve [5]. In this light, we can rapidly obtain from Table 2 that the **SUE.GenKey** algorithm of our SUE scheme is almost 4 times faster, the **SUE.Encrypt** and the **SUE.ReEncrypt** algorithms are almost 3 times faster, as well as, the **SUE.Decrypt** algorithm is roughly 25 times faster than the respective algorithms in [8]. Finally, following [6], if we assume that in prime order elliptic curve groups, a pairing is equivalent to six exponentiations, then our **SUE.Setup** algorithm would have lower computational complexity compared to that of [8] even for $T_{max} = 2^{200}$. However, we note that in last few years there has been significant improvement in the efficiency of pairing computation. Despite this we may conclude that our **SUE.Setup** algorithm would still have lower complexity than [8] for sufficiently large value of T_{max}.

Remark 1. Note that, Freeman [5] and Lewko [9] have developed abstract frameworks to simulate the key properties of bilinear groups of composite order that are required to construct secure pairing based cryptosystems in prime order groups. Their tools are not generic scheme conversions in the sense that they do not transform any composite order cryptographic protocol to a prime order protocol. On the contrary, their techniques serve in replicating certain tricks in the security proofs of independent prime order protocol designs in a manner analogous to that of their composite order variants. In the security proof of our SUE scheme, we have incorporated the required features from the underlying prime order bilinear groups in a different fashion. Further, we have developed a *tag based approach* to overcome certain subtility which they addressed using *nominal semi-functionality* [9].

Remark 2. As mentioned earlier in the paper, the work of Sahai et al. [13] can be viewed as a restricted SUE. However, Lee et al. [8] have already shown that their SUE scheme is more efficient than [13] in terms of both communication and storage, as well as, computation. Thus, our scheme having better performance guarantee compared to [8], naturally outperforms [13]. On the other hand, although achieving a similar goal, SUE should not be confused with the notion of updatable encryption introduced by Boneh et al. [2]. An updatable encryption scheme supports the functionality of re-encrypting a certain data encrypted under some symmetric key to a fresh encryption of the same data under another key. On the contrary, SUE associates ciphertexts with time and not with keys and thus it has more flexibility. Furthermore, in case of updatable encryption re-encrypting a ciphertext requires a re-encryption key derived from the current encryption key and the target one, whereas, in SUE ciphertexts are publicly updatable without requiring any re-encryption key.

6   Conclusion

In this paper, we have designed an SUE scheme in prime order bilinear groups and proved its full CPA security under standard assumptions, namely, the DLIN and the DBDH assumptions. To the best of our knowledge, our SUE scheme is the *first* to achieve full security in prime order bilinear group setting. We have employed the dual system encryption technique of Lewko and Waters [11], [14]. In our SUE scheme, the master key size is not constant. Also the ciphertext and key sizes are not very short because the tag values could not be compressed. Developing an SUE scheme that overcomes these difficulties could be an interesting future work. Note that, an SUE scheme cannot be secure against adaptive chosen ciphertext attack (CCA2) due to the re-randomizability guarantee of SUE ciphertexts. However, one may attempt to analyse the possible relaxations of CCA2 security for SUE and design an SUE scheme, secure in that relaxed CCA2 security model, in prime order bilinear groups under standard assumptions.

References

1. Boldyreva, A., Goyal, V., Kumar, V.: Identity-based encryption with efficient revocation. In: Proceedings of the 15th ACM Conference on Computer and Communications Security, pp. 417–426. ACM (2008)
2. Boneh, D., Lewi, K., Montgomery, H., Raghunathan, A.: Key homomorphic PRFs and their applications. In: Canetti, R., Garay, J.A. (eds.) CRYPTO 2013, Part I. LNCS, vol. 8042, pp. 410–428. Springer, Heidelberg (2013)
3. Canetti, R., Halevi, S., Katz, J.: A forward-secure public-key encryption scheme. In: Biham, E. (ed.) EUROCRYPT 2003. LNCS, vol. 2656, pp. 255–271. Springer, Heidelberg (2003)
4. Dodis, Y., Katz, J., Xu, S., Yung, M.: Key-insulated public key cryptosystems. In: Knudsen, L.R. (ed.) EUROCRYPT 2002. LNCS, vol. 2332, pp. 65–82. Springer, Heidelberg (2002)
5. Freeman, D.M.: Converting pairing-based cryptosystems from composite-order groups to prime-order groups. In: Gilbert, H. (ed.) EUROCRYPT 2010. LNCS, vol. 6110, pp. 44–61. Springer, Heidelberg (2010)
6. Hohenberger, S.: Advances in signatures, encryption, and e-cash from bilinear groups. Ph.D. thesis, Citeseer (2006)
7. Lee, K.: Self-updatable encryption with short public parameters and its extensions, http://eprint.iacr.org/2014/231.pdf
8. Lee, K., Choi, S.G., Lee, D.H., Park, J.H., Yung, M.: Self-updatable encryption: Time constrained access control with hidden attributes and better efficiency. In: Sako, K., Sarkar, P. (eds.) ASIACRYPT 2013, Part I. LNCS, vol. 8269, pp. 235–254. Springer, Heidelberg (2013)
9. Lewko, A.: Tools for simulating features of composite order bilinear groups in the prime order setting. In: Pointcheval, D., Johansson, T. (eds.) EUROCRYPT 2012. LNCS, vol. 7237, pp. 318–335. Springer, Heidelberg (2012)
10. Lewko, A., Okamoto, T., Sahai, A., Takashima, K., Waters, B.: Fully secure functional encryption: Attribute-based encryption and (hierarchical) inner product encryption. In: Gilbert, H. (ed.) EUROCRYPT 2010. LNCS, vol. 6110, pp. 62–91. Springer, Heidelberg (2010)
11. Lewko, A., Waters, B.: New techniques for dual system encryption and fully secure HIBE with short ciphertexts. In: Micciancio, D. (ed.) TCC 2010. LNCS, vol. 5978, pp. 455–479. Springer, Heidelberg (2010)
12. Paterson, K.G., Quaglia, E.A.: Time-specific encryption. In: Garay, J.A., De Prisco, R. (eds.) SCN 2010. LNCS, vol. 6280, pp. 1–16. Springer, Heidelberg (2010)
13. Sahai, A., Seyalioglu, H., Waters, B.: Dynamic credentials and ciphertext delegation for attribute-based encryption. In: Safavi-Naini, R., Canetti, R. (eds.) CRYPTO 2012. LNCS, vol. 7417, pp. 199–217. Springer, Heidelberg (2012)
14. Waters, B.: Dual system encryption: Realizing fully secure IBE and HIBE under simple assumptions. In: Halevi, S. (ed.) CRYPTO 2009. LNCS, vol. 5677, pp. 619–636. Springer, Heidelberg (2009)

Related-Key Security for Hybrid Encryption[*]

Xianhui Lu[1,2], Bao Li[1,2], and Dingding Jia[1,2]

[1] Data Assurance and Communication Security Research Center,
Chinese Academy of Sciences, Beijing, China
[2] State Key Laboratory of Information Security,
Institute of Information Engineering,
Chinese Academy of Sciences, Beijing, 100093, China
{xhlu,lb,ddj}@is.ac.cn

Abstract. We prove that, for a KEM/Tag-DEM (Key Encapsulation Mechanism/ Tag Data Encapsulation Mechanism) hybrid encryption scheme, if the adaptive chosen ciphertext secure KEM part has the properties of key malleability and key fingerprint and the Tag-DEM part is a one-time secure tag authenticated encryption, then the hybrid encryption is seucure against related key attacks (RKA). We show that several classical KEM schemes satisfy these two properties.

Keywords: public key encryption, related-key attack, hybrid encryption.

1 Introduction

The traditional model for the security of public key encryption schemes assumes that cryptographic devices are black-boxes and the private keys are completely hidden and protected from the attackers. For example, in the definition of adaptive chosen ciphertext attacks (IND-CCA2), the adversary can only communicate with the challenger through the encryption oracle and the decryption oracle, and can not get any internal information of the cryptographic devices. However, real attacks demonstrate that the adversary may get or modify the private key by using physical side-channels [26,7,9,18].

Related-key attacks (RKAs) were first proposed in [25,6] as a cryptanalysis tool for block ciphers. Real attacks [7,9], in which the attacker can modify the keys stored in the memory, turned this theoretical analysis model for block ciphers to a practical attack model for all kinds of cryptographic primitives such as public key encryption (PKE), identity based encryption (IBE), digital signature and so on. The theoretical definition of RKA security was first proposed by Bellare and Kohno [4], who treated the case of PRFs (PseudoRandom Functions) and PRPs (PseudoRandom Permutations). Research then expanded to other primitives [1,3,16,17].

[*] Supported by the National Basic Research Program of China (973 project, No.2013CB338002), the National Nature Science Foundation of China (No.61070171, No.61272534).

S.S.M. Chow et al. (Eds.): ISC 2014, LNCS 8783, pp. 19–32, 2014.

The RKA security of public key encryption schemes was first considered by Bellare, Cash and Miller [3]. They showed how to leverage the RKA security of block ciphers to provide RKA security for high-level primitives including SE (symmetric encryption), PKE, IBE and digital signature. They also showed that IBE is an enabling primitive in the RKA domain: achieving RKA secure IBE schemes yields RKA secure CCA-PKE (chosen ciphertext secure PKE) and digital signature schemes. Their main idea is to protect the secret key of a high level primitive with a RKA secure PRG (PseudoRandom Generator) which can be constructed from a RKA secure PRF. Since affine functions and polynomial functions contain constant functions, RKA secure PRFs for these functions can not exist. That is, the framework of [3] can only get RKA security for linear functions. To overcome the linear barrier, Bellare et al. [5] proposed a framework enabling the constructions of RKA secure IBE schemes for affine functions and polynomial functions of bounded degree. To go beyond the algebraic barrier and achieve RKA security for arbitrary key relations, Damgård et al. [15] proposed the bounded tamper resilience model, in which the number of tampering queries the adversary is allowed to ask for is restricted.

Wee [31] firstly proposed direct constructions of RKA secure public key encryption schemes. Wee showed that the Cramer-Shoup CCA secure constructions [12,13] do not satisfy the property of finger-printing. To achieve this property, Wee turned to the "all-but-one extraction" paradigm [8]. Dingding Jia et al. [22] showed that the Cramer-Shoup paradigm satisfy a similar property as finger-printing and proposed two RKA secure public key encryption schemes based on the Cramer-Shoup paradigm.

1.1   Our Contribution

We focus on how to uniformly enhance an IND-CCA2 secure PKE schemes to CC-RKA (adaptive chosen ciphertext related key attack) security. Specifically, for the KEM/Tag-DEM framework [11], we prove that CC-RKA secure hybrid encryption schemes can be constructed from an IND-CCA2 or IND-CCCA [19](Constrained Chosen Ciphertext Attacks) secure KEM and a one-time secure tag authenticated encryption. In addition, we require that the KEM scheme has the properties of key-malleable and key-fingerprint. We show that several classical IND-CCA2 or IND-CCCA secure KEM schemes satisfy these two properties. Thus we get efficient RKA secure hybrid encryption schemes.

In the construction of RKA secure schemes, key malleability is a useful and widely used property [2,3,5] (also name as "key homomorphism" in [31]). Key malleability means that the decryption of a ciphertext C using a secret key $\phi(sk)$, where ϕ denotes a function, equals the decryption of some other ciphertext C' using the original secret key sk. If the adversary can not find a (ϕ, C) pair such that C' equals the challenge ciphertext C^*, then the key malleability property reduces the CC-RKA security to the IND-CCA2 security. To prevent the adversary from getting such pairs, Wee [31] combined a tag-based CCA secure scheme with a one-time signature scheme, where the tag is derived from the verification key of the one-time signature scheme. In addition, Wee required that C and C'

share the same tag. If C and C' are valid ciphertexts and share the same tag, then the one-time signature scheme tells us that $C - C'$. To achieve CC-RKA security, Wee required that C^* is an invalid ciphertext under any $\phi(sk) \neq sk$. As a result, the adversary can not find a (ϕ, C) pair such that $C = C' = C^*$ is a valid ciphertext under $\phi(sk)$. This property is another useful property named as "key fingerprint" in [2] and "finger-printing" in [31].

Instead of adding a one-time signature scheme to a tag-based CCA secure PKE scheme as in[31], we show that the AE-OT (one-time secure authenticated encryption) secure DEM part in the hybrid scheme itself is a good choice to prevent the adversary from finding a valid (ϕ, C) pair such that $C' = C^*$, here C, C', C^* denote the ciphertext of the KEM part. That is, the AE-OT secure DEM part can provide an integrity authentication service for the KEM part. When $C \neq C' = C^*$, the AE-OT property guarantees that the adversary can not construct a valid DEM part. More formally, we prove that the KEM/Tag-DEM hybrid encryption is CC-RKA secure if the IND-CCA2 (or IND-CCCA) secure KEM part has the properties of key malleability and key fingerprint and the Tag-DEM part is a one-time secure tag authenticated encryption.

Compared with Wee's construction [31], our construction can get CC-RKA secure public key encryption schemes from the "all-but-one extraction" paradigm and the Cramer-Shoup paradigm uniformly, while Wee's construction can only get CC-RKA secure public key encryption schemes from the "all-but-one extraction" paradigm.

1.2 Outline

In section 2 we review the definition of related key attacks, key encapsulation mechanism and data encapsulation mechanism with tag. In section 3 we propose our new construction. In section 4 we show that several classical IND-CCA2 secure KEM schemes satisfy key malleability and key fingerprint. Finally we give the conclusion in section 5.

2 Definitions

If S is a finite set, $s \xleftarrow{R} S$ denotes that s is sampled from the uniform distribution on S. If A is a probabilistic algorithm and x an input, then $A(x)$ denotes the output distribution of A on input x. Thus, we write $y \leftarrow A(x)$ to denote of running algorithm A on input x and assigning the output to the variable y.

2.1 Related Key Attacks

We follow the definition of related key attacks from [31]. A public key encryption scheme is secure against adaptive chosen ciphertext related key attacks (CC-RKA) if the advantage of any adversary in the following game is negligible in the security parameter k.

1. The challenger runs the key generation algorithm $(PK, SK) \leftarrow \text{KeyGen}(1^k)$ and sends the public key PK to the adversary.
2. The adversary makes a sequence of calls to the related key decryption oracle RKA.Dec$(\cdot, \cdot)$ with (ϕ, C). Here $\phi \in \Phi$, Φ is a class of related-key deriving functions, C is a ciphertext. The challenger decrypts the ciphertext C using $\phi(SK)$ and sends the result to the adversary.
3. The adversary queries the encryption oracle with (m_0, m_1). The challenger computes:

$$b \xleftarrow{R} \{0, 1\}, C^* \leftarrow Enc_{PK}(m_b)$$

and responds with C^*.
4. The adversary queries the related key decryption oracle continuously with (ϕ, C). The challenger acts just as in step 2. The only restriction is that the adversary can not query the related key decryption oracle with (ϕ, C) that $\phi(sk) = sk$ and $C = C^*$.
5. Finally, the adversary outputs a guess b'.

The adversary's advantage in the above game is defined as $\text{Adv}_{A,\Phi}^{\text{rka}}(k) = |\Pr[b' = b] - 1/2|$. An encryption scheme is Φ-CC-RKA secure if for all PPT adversary the advantage $\text{Adv}_{A,\Phi}^{\text{rka}}(k)$ is a negligible function of k.

2.2 Key Encapsulation Mechanism

A key encapsulation mechanism consists of the following algorithms:

- KEM.KG(1^k): A probabilistic polynomial-time key generation algorithm takes as input a security parameter (1^k) and outputs a public key PK and a private key SK. We write $(PK, SK) \leftarrow \text{KEM.KG}(1^k)$
- KEM.E(PK): A probabilistic polynomial-time encapsulation algorithm takes as input the public key PK, and outputs a pair (K, ψ), where $K \in K_D$(K_D is the key space) is a key and ψ is a ciphertext. We write $(K, \psi) \leftarrow \text{KEM.E}(PK)$
- KEM.D(SK, ψ): A decapsulation algorithm takes as input a ciphertext ψ and the private key SK. It returns a key K. We write $K \leftarrow \text{KEM.D}(SK, \psi)$.

A KEM scheme is secure against adaptive chosen ciphertext attacks if the advantage of any adversary A in the following game is negligible in the security parameter k.

1. The adversary queries a key generation oracle. The key generation oracle computes $(PK, SK) \leftarrow \text{KEM.KG}(1^k)$ and responds with PK.
2. The adversary makes a sequence of calls to the decapsulation oracle. For each query the adversary submits a ciphertext ψ, and the decapsulation oracle responds with KEM.D(SK, ψ).
3. The adversary queries an encapsulation oracle. The encapsulation oracle computes:

$$b \xleftarrow{R} \{0, 1\}, (K_1, \psi^*) \leftarrow \text{KEM.E}(PK), K_0 \xleftarrow{R} K_D,$$

and responds with (K_b, ψ^*).

4. The adversary makes a sequence of calls to the decapsulation oracle. For each query the adversary submits a ciphertext ψ, and the decapsulation oracle responds with KEM.D(SK, ψ). The only restriction is that the adversary can not request the decapsulation of ψ^*.
5. Finally, the adversary outputs a guess b'.

Let $\Pr[\mathcal{A}_{suc}]$ be the probability that the adversary $\mathcal{A}$ succeeds in the game above. The adversary's advantage in the above game is

$$\text{Adv}_{\mathcal{A}}^{\text{cca}}(k) = |\Pr[\mathcal{A}_{suc}] - 1/2| = |\Pr[b' = b] - 1/2|.$$

If a KEM is secure against adaptive chosen ciphertext attacks defined in the above game, we say it is IND-CCA2 secure.

Hofheinz and Kiltz [19] proposed a relaxed notion of IND-CCA2 named as "constrained chosen ciphertext security" (IND-CCCA). In the definition of IND-CCCA the adversary is allowed to make a decapsulation query if it already has some priori knowledge of the decapsulated key K. That is, the adversary need to provide an efficiently computable boolean predicate $pred : K \rightarrow \{0, 1\}$. To construct a predicate $pred(K)$ that evaluates to 1, the adversary has to have a high priori knowledge about the decapsulated session key K. The formal definition of IND-CCCA is similar to that of IND-CCA2, while the only difference is that the adversary provides a $(\psi, pred)$ pair in the decapsulation query, and the challenger verifies whether $pred(K) = 1$ or not. If $pred(K) = 1$ then K is returned, and $\perp$ otherwise. The adversary's advantage is defined as

$$\text{Adv}_{\mathcal{A}}^{\text{ccca}}(k) = |\Pr[\mathcal{A}_{suc}] - 1/2| = |\Pr[b' = b] - 1/2|.$$

Φ-**Key malleability.** We say that a KEM has the property of Φ-key malleability if there is a PPT algorithm T such that for all $\phi \in \Phi$, PK, SK and ψ:

$$\text{KEM.D}(\phi(SK), \psi) = \text{KEM.D}(SK, T(PK, \phi, \psi)).$$

Φ-**Key fingerprint**. For all PK, SK and ψ, we say that a KEM has the property of Φ-key fingerprint if for any PPT adversary the probability to find a function $\phi \in \Phi$ such that $\phi(SK) \neq SK$ and $T(PK, \phi, \psi) = \psi$ is a negligible value ϵ_{kf}.

2.3 Data Encapsulation Mechanism with Tag

A data encapsulation mechanism with tag consists of two algorithms:

– Tag-DEM.E(K, m, t): The encryption algorithm takes as inputs a key K, a message m, a tag t and outputs a ciphertext χ. We write $\chi \leftarrow$ Tag-DEM.E (K, m, t)
– Tag-DEM.D(K, χ, t): The decryption algorithm takes as inputs a key K, a ciphertext χ, a tag t and outputs a message m or the rejection symbol $\perp$. We write $m \leftarrow$ Tag-DEM.D(K, χ, t)

We require that for all $K \in \{0,1\}^{l_e}$ (l_e denotes the length of K), $m \in \{0,1\}^*$ and $t \in \{0,1\}^*$,we have:

$$\text{Tag-DEM.D}(K, \text{Tag-DEM.E}(K, m, t), t) = m.$$

A Tag-DEM scheme is IND-OT (indistinguishability against one-time attacks) secure if the advantage of any PPT adversary $\mathcal{A}$ in the following game is negligible in the security parameter k:

1. The challenger randomly generates an appropriately sized key K and a tag t^*.
2. The adversary $\mathcal{A}$ queries the encryption oracle with two messages m_0 and m_1) such that $|m_0| = |m_1|$. The challenger computes

$$b \xleftarrow{R} \{0,1\}, \chi^* \leftarrow \text{Tag-DEM.E}(K, m_b, t^*)$$

and responds with χ^* and t^*.
3. Finally, $\mathcal{A}$ outputs a guess b' .

The advantage of $\mathcal{A}$ is defined as $\text{Adv}_{\mathcal{A}}^{\text{ind-ot}}(k) = |\Pr[b = b'] - 1/2|$. We say that the Tag-DEM is one-time secure in the sense of indistinguishability if $\text{Adv}_{\mathcal{A}}^{\text{ind-ot}}(k)$ is negligible.

A Tag-DEM scheme is INT-OT (one-time secure in the sense of ciphertext integrity) secure if the advantage of any PPT adversary $\mathcal{A}$ in the following game is negligible in the security parameter k:

1. The challenger randomly generates an appropriately sized key K and a tag t^*.
2. The adversary $\mathcal{A}$ queries the encryption oracle with a message m. The challenger computes

$$\chi^* \leftarrow \text{Tag-DEM.E}(K, m, t^*)$$

and responds with χ^* and t^*.
3. Finally, the adversary $\mathcal{A}$ outputs a ciphertext χ and a tag t such that $\text{Tag-DEM.D}(K, \chi, t) \neq \perp$.

The advantage of $\mathcal{A}$ is defined as

$$\text{Adv}_{\mathcal{A}}^{\text{int-ot}}(k) = \Pr[(\chi, t) \neq (\chi^*, t^*)].$$

We say that the Tag-DEM is one-time secure in the sense of ciphertext integrity if $\text{Adv}_{\mathcal{A}}^{\text{int-ot}}(k)$ is negligible.

A Tag-DEM is one-time secure in the sense of tag authenticated encryption (Tag-AE-OT) iff it is IND-OT secure and INT-OT secure. Similar to AE-OT secure ciphers in [24], Tag-AE-OT secure ciphers can also be constructed from a SE (symmetric encryption) scheme and a MAC (message authentication code) scheme. The only difference is that the MAC scheme takes the ciphertext of the SE scheme and the tag t as inputs.

3   Hybrid Encryption Against Related Key Attacks

In this section we prove that the KEM/Tag-DEM hybrid encryption [11] is Φ-CC-RKA secure if the IND-CCA2 (or IND-CCCA) secure KEM part has the properties of Φ-key malleability and Φ-key fingerprint and the Tag-DEM part is a one-time secure tag authenticated encryption. The KEM/Tag-DEM framework for hybrid encryption can be described as follows.

- KeyGen(1^k): The key generation algorithm is the same as that of the KEM scheme.
$$(PK, SK) \leftarrow \text{KEM.KG}(1^k)$$
- Encrypt(PK, m): The encryption algorithm works as follows:
$$(K, \psi) \leftarrow \text{KEM.E}(PK), \chi \leftarrow \text{Tag-DEM.E}(K, m, \psi), C \leftarrow (\psi, \chi)$$
- Decrypt(SK, C): The decryption algorithm works as follows:
$$K \leftarrow \text{KEM.D}(SK, \psi), m \leftarrow \text{Tag-DEM.D}(K, \chi, \psi)$$

Before formal proof, we give a direct understanding of the CC-RKA security of the KEM/Tag-DEM hybrid encryption scheme. Intuitively, for an IND-CCA2 secure public key encryption scheme, if the private key is completely protected, the ciphertext is non-malleable. That is, the adversary can not construct a ciphertext based a valid ciphertext C^*. However, in the CC-RKA model, the ciphertext may be malleable since the adversary can modify the private key. In the KEM/Tag-DEM framework the Tag-DEM part provides the integrity authentication service to the KEM part. The INT-OT security of the Tag-DEM scheme guarantees that the adversary can not extend an existing ciphertext to get a new valid ciphertext.

First we prove the CC-RKA security of the hybrid encryption scheme when the KEM part is IND-CCA2 secure.

Theorem 1. *If the KEM part is IND-CCA2 secure and has the properties of Φ-key malleability and Φ-key fingerprint, the Tag-DEM part is Tag-AE-OT secure, then the hybrid encryption above is Φ-CC-RKA secure.*

Proof. Suppose that an adversary $\mathcal{A}$ can break the Φ-CC-RKA security of the hybrid encryption. To prove the theorem, we construct an adversary $\mathcal{B}$ to break the IND-CCA2 security of the KEM scheme. The construction of $\mathcal{B}$ is described as follows.

Setup: The adversary $\mathcal{B}$ gets the public key PK from the challenger and sends it to the adversary $\mathcal{A}$.

Decryption oracle1: When $\mathcal{A}$ queries the related key decryption oracle with (ϕ, C), where $C = (\psi, \chi), \phi \in \Phi$, the adversary $\mathcal{B}$ computes m as follows and returns it to $\mathcal{A}$.

$$\psi' \leftarrow T(PK, \phi, \psi), K \leftarrow D_{kem}(\psi'), m \leftarrow \text{Tag-DEM.D}(K, \chi, \psi).$$

Here $D_{kem}(\cdot)$ denotes the decapsulation oracle of the KEM scheme, T is the transform function according to the Φ-key malleability property. According to the Φ-key malleability property, we have

$$K = D_{kem}(\psi') = \text{KEM.D}(SK, \psi') = \text{KEM.D}(\phi(SK), \psi).$$

Thus the adversary $\mathcal{B}$ simulates the related key decryption oracle perfectly in this step.

Challenge: The adversary $\mathcal{A}$ queries the encryption oracle with two messages m_0 and m_1. The adversary $\mathcal{B}$ computes as follows.

$$(K^*, \psi^*) \leftarrow E_{kem}(PK), b \xleftarrow{R} \{0, 1\},$$

$$\chi^* \leftarrow \text{Tag-DEM.E}(K^*, m_b, \psi^*), C^* \leftarrow (\psi^*, \chi^*).$$

Here $E_{kem}(PK)$ is the encryption oracle of the KEM scheme, K^* is randomly chosen from the key space or equals to $\text{KEM.D}(SK, \psi^*)$. The adversary $\mathcal{B}$ sends C^* the $\mathcal{A}$.

Decryption oracle2: When $\mathcal{A}$ queries the decryption oracle with (ϕ, C) continuously, $\mathcal{B}$ computes $\psi' \leftarrow T(PK, \phi, \psi)$ and acts as follows.

- Case 1: $\psi' \neq \psi^*$. The adversary $\mathcal{B}$ computes m as follows and returns it to $\mathcal{A}$.

$$K \leftarrow D_{kem}(\psi'), m \leftarrow \text{Tag-DEM.D}(K, \chi, \psi).$$

- Case 2: $\psi \neq \psi' = \psi^*$. The adversary $\mathcal{B}$ returns a rejection symbol $\perp$. According to the INT-OT security of Tag-DEM, $\text{Tag-DEM.D}(K^*, \chi, \psi) = \perp$ except with the probability of $\text{Adv}_{\mathcal{A}}^{\text{int-ot}}$.
- Case 3: $\psi = \psi' = \psi^*$ and $\chi = \chi^*$. The adversary $\mathcal{B}$ returns a rejection symbol $\perp$. Since the adversary $\mathcal{A}$ can not query $(\phi, (\psi, \chi))$ such that $(\psi, \chi) = (\psi^*, \chi^*)$ and $\phi(SK) = SK$, we have that $\phi(SK) \neq SK$. According to the Φ-key fingerprint property, the probability that $\mathcal{A}$ can find a function $\phi \in \Phi$ such that $\phi(SK) \neq SK$ and $T(PK, \phi, \psi) = \psi$ is a negligible value ϵ_{kf}.
- Case 4: $\psi = \psi' = \psi^*$ and $\chi \neq \chi^*$. The adversary $\mathcal{B}$ returns a rejection symbol $\perp$. According to the INT-OT security of Tag-DEM, $\text{Tag-DEM.D}(K^*, \chi, \psi) = \perp$ except with the probability of $\text{Adv}_{\mathcal{A}}^{\text{int-ot}}$.

According to the four cases above, we have that $\mathcal{B}$ simulates the related key decryption oracle in this step perfectly except with the probability of $\text{Adv}_{\mathcal{A}}^{\text{int-ot}} + \epsilon_{\text{kf}}$.

Guess: Finally when $\mathcal{A}$ outputs b', $\mathcal{B}$ outputs 1 if $b' = b$ and 0 otherwise.

To compute the advantage of $\mathcal{B}$ in breaking the IND-CCA2 security of the KEM scheme we first consider the probability that $\mathcal{B}$ succeeds in guessing $K^* = K_1$ or $K^* = K_0$. According to the algorithm above, we have:

$$\Pr[\mathcal{B}_{suc}] = \Pr[b' = b | K^* = K_1] \Pr[K^* = K_1] + \Pr[b' \neq b | K^* = K_0] \Pr[K^* = K_0], \tag{1}$$

where $\Pr[\mathcal{B}_{suc}]$ denotes the probability that $\mathcal{B}$ succeeds in guessing $K^* = K_1$ or $K^* = K_0$.

According to the definition of the IND-CCA2 security for KEM, we have that:

$$\Pr[K^* = K_1 = \text{KEM.D}(SK, \psi^*)] = \Pr[K^* = K_0] = 1/2. \qquad (2)$$

If $K^* = K_1 = \text{KEM.D}(SK, \psi^*)$, we have that $\mathcal{B}$ simulates the Φ-CC-RKA challenger perfectly except with the probability of $\text{Adv}_{\mathcal{A}}^{\text{int-ot}} + \epsilon_{\text{kf}}$. Thus we have:

$$\Pr[b = b'|K^* = K_1] \geq 1/2 + \text{Adv}_{\mathcal{A},\Phi}^{\text{rka}} - (\text{Adv}_{\mathcal{A}}^{\text{int-ot}} + \epsilon_{\text{kf}}). \qquad (3)$$

If $K^* = K_0$, since K_0 is randomly chosen from the key space we have that K^* is independent from the point view of $\mathcal{A}$ except the probability of $\text{Adv}_{\mathcal{A}}^{\text{int-ot}} + \epsilon_{\text{kf}}$. Thus we have:

$$\Pr[b \neq b'|K^* = K_0] \geq 1/2 - \text{Adv}_{\mathcal{A}}^{\text{ind-ot}} - (\text{Adv}_{\mathcal{A}}^{\text{int-ot}} + \epsilon_{\text{kf}}). \qquad (4)$$

From equations (1), (2) and (3) we have that:

$$\begin{aligned}
\Pr[\mathcal{B}_{suc}] &= \Pr[b' = b|K^* = K_1]\Pr[K^* = K_1] + \\
&\quad \Pr[b' \neq b|K^* = K_0]\Pr[K^* = K_0] \\
&\geq (1/2 - \text{Adv}_{\mathcal{A}}^{\text{ind-ot}} - (\text{Adv}_{\mathcal{A}}^{\text{int-ot}} + \epsilon_{\text{kf}}))1/2 + \\
&\quad (1/2 + \text{Adv}_{\mathcal{A},\Phi}^{\text{rka}} - (\text{Adv}_{\mathcal{A}}^{\text{int-ot}} + \epsilon_{\text{kf}}))1/2 \\
&= 1/2(\text{Adv}_{\mathcal{A},\Phi}^{\text{rka}} - \text{Adv}_{\mathcal{A}}^{\text{ind-ot}}) + 1/2 - (\text{Adv}_{\mathcal{A}}^{\text{int-ot}} + \epsilon_{\text{kf}})
\end{aligned} \qquad (5)$$

Finally we can get the advantage of $\mathcal{B}$ as follows:

$$\begin{aligned}
\text{Adv}_{\mathcal{B}}^{\text{cca}} &= |\Pr[\mathcal{B}_{suc}] - 1/2| \\
&\geq \tfrac{1}{2}\text{Adv}_{\mathcal{A},\Phi}^{\text{rka}} - \tfrac{1}{2}\text{Adv}_{\mathcal{A}}^{\text{ind-ot}} - \text{Adv}_{\mathcal{A}}^{\text{int-ot}} - \epsilon_{\text{kf}}.
\end{aligned} \qquad (6)$$

This completes the proof of theorem 1. $\square$

Now we prove the CC-RKA security of the hybrid encryption scheme when the KEM scheme is IND-CCCA secure.

Theorem 2. *If the KEM part is IND-CCCA secure and has the properties of Φ-key malleability and Φ-key fingerprint, the Tag-DEM part is Tag-AE-OT secure, then the hybrid encryption above is Φ-CC-RKA secure.*

The proof of theorem 2 is similar to that of theorem 1. The only difference is that in the decryption oracle $\mathcal{B}$ needs to provide a boolean predicate function *pred* when querying the decapsulation oracle of the IND-CCCA secure KEM challenger. Just as in [19], we can use the ciphertext of the Tag-DEM scheme as the boolean predicate function. That is $pred_\chi(K) = 0$ if $\text{Tag-DEM.D}(K, \chi, \psi) = \perp$ and $pred_\chi(K) = 1$ otherwise.

4   Instantiations

In this section we show that several classical KEM schemes have the properties of key malleability and key fingerprint. Specifically, we consider the KEM schemes from the Cramer-Shoup paradigm and the "all-but-one extraction" paradigm.

4.1   KEM Schemes from the Cramer-Shoup Paradigm

We show that the IND-CCCA secure KEM scheme proposed in [24] has the properties of key malleability and key fingerprint. First we review the scheme as follows.

- KeyGen(1^k): Assume that G is a group of order q where q is large prime number.

$$(g_1, g_2) \xleftarrow{R} G, (x_1, x_2) \xleftarrow{R} Z_q^*, h \leftarrow g_1^{x_1} g_2^{x_2},$$

$$PK \leftarrow (g_1, g_2, h, \mathrm{H}), SK \leftarrow (x_1, x_2),$$

  where H is a 4-wise independent hash function.
- Encapsulation(PK):

$$r \xleftarrow{R} Z_q^*, u_1 \leftarrow g_1^r, u_2 \leftarrow g_2^r, K \leftarrow \mathrm{H}(h^r), \psi \leftarrow (u_1, u_2).$$

- Decapsulation(SK, ψ): $K \leftarrow \mathrm{H}(u_1^{x_1} u_2^{x_2})$.

Now show that the KEM above satisfies $\Phi^\times$-key malleability and $\Phi^\times$-key fingerprint, where $\phi^\times(s) = as \mod q$, $\phi^\times \in \Phi^\times, a \in Z_q$.

$\Phi^\times$**-Key malleability**. For $PK = (g_1, g_2, h, H), SK = (x_1, x_2), \psi = (u_1, u_2)$ and $\phi^\times(x_1, x_2) = (a_1 x_1, a_2 x_2)$, the transform function T is defined as:

$$T(PK, \phi^\times, \psi) = (u_1^{a_1}, u_2^{a_2}).$$

The correctness of T can be verified as follows:

$$\mathrm{KEM.D}(\phi^\times(SK), \psi) = \mathrm{H}(u_1^{a_1 x_1} u_2^{a_2 x_2}).$$

$$\mathrm{KEM.D}(SK, T(PK, \phi^\times, \psi)) = \mathrm{KEM.D}(SK, (u_1^{a_1}, u_2^{a_2})) = \mathrm{H}(u_1^{a_1 x_1} u_2^{a_2 x_2}).$$

$\Phi^\times$**-Key fingerprint**. For $PK = (g_1, g_2, h, H), SK = (x_1, x_2), \psi = (u_1, u_2)$ and $\phi^\times(x_1, x_2) = (a_1 x_1, a_2 x_2)$, if $(a_1 x_1, a_2 x_2) \neq (x_1, x_2)$ we have that

$$T(PK, \phi^\times, \psi) = (u_1^{a_1}, u_2^{a_2}) \neq (u_1, u_2).$$

Thus the KEM above satisfies $\Phi^\times$-key fingerprint.

There are several KEM schemes from the Cramer-Shoup paradigm [14,27,24,20]. It is easy to verify that KEM scheme in [20] also satisfies these two properties, while the KEM schemes in [14] and [27] are not $\Phi^\times$-key malleability.

4.2   KEM Schemes from the All-But-One Extraction Paradigm

We show that the IND-CCA2 secure KEM scheme proposed in [21] satisfies key malleability and key fingerprint. First we review the scheme as follows.

- KeyGen: The key generation algorithm chooses uniformly at random a Blum integer $N = PQ = (2p+1)(2q+1)$, where P, Q, p, q are prime numbers, then computes:

$$g \xleftarrow{R} \mathrm{QR}_N, x \xleftarrow{R} [(N-1)/4], X \leftarrow g^{x 2^{l_K + l_H}},$$

$$PK \leftarrow (N, g, X), SK \leftarrow x,$$

where $\mathrm{H} : QR_N \rightarrow \{0,1\}^{l_H}$ is a TCR (Target Collision Resistant) hash function, l_H is the bit length of the output value of H, l_K is the bit length of the encapsulated key K.

– Encapsulation:

$$\mu \stackrel{R}{\leftarrow} [(N-1)/4], R \leftarrow g^{\mu 2^{l_K + l_H}}, t \leftarrow \mathrm{H}(R), S \leftarrow \left| (g^t X)^\mu \right|,$$

$$K \leftarrow \mathrm{BBS}_N(g^{\mu 2^{l_H}}),$$

where $\mathrm{BBS}_N(s) = \mathrm{LSB}(s), \cdots, \mathrm{LSB}(s^{2^{l_K-1}})$, $\mathrm{LSB}(s)$ denotes the least significant bit of s.

– Decapsulation: Given a ciphertext (R, S) and PK, the decapsulation algorithm verifies $R \in Z_N^*, S \in Z_N^* \cap [(N-1)/2]$, then computes:

$$t \leftarrow \mathrm{H}(R),$$

$$\text{if } \left(\frac{S}{R^x}\right)^{2^{l_K + l_H}} = R^t \text{ then computes}$$

$$2^\gamma = \gcd(t, 2^{l_K + l_H}) = \alpha t + \beta 2^{l_K + l_H},$$

$$\text{returns } K \leftarrow \mathrm{BBS}_N \left(\left((SR^{-x})^\alpha R^\beta \right)^{2^{l_H - \gamma}} \right),$$

else returns the rejection symbol $\perp$.

Now we show that the KEM scheme above has the properties of Φ^+-key malleability and Φ^+-key fingerprint, where $\phi^+(s) = s + a$, $\phi \in \Phi, a \in Z_N$.

Φ^+-**Key malleability.** For $PK = (N, g, X), SK = x, \phi^+(x) = x + a$, and $\psi = (R, S) = (g^{\mu 2^{l_K + l_H}}, \left| \left(g^t X g^{a 2^{l_K + l_H}} \right)^\mu \right|)$ the transform function T is defined as:

$$T(PK, \phi^+, \psi) = (R, SR^{-a}) = (g^{\mu 2^{l_K + l_H}}, \left| (g^t X)^\mu \right|).$$

The correctness of T can be verified as follows:

$$\mathrm{KEM.D}(\phi^+(SK), \psi) = \mathrm{BBS}_N(g^{\mu 2^{l_H}}).$$

$$\mathrm{KEM.D}(SK, T(PK, \phi^+, \psi)) = \mathrm{KEM.D}(SK, (R, SR^{-a})) = \mathrm{BBS}_N(g^{\mu 2^{l_H}}).$$

Φ^+-**Key fingerprint.** For $PK = (N, g, X), SK = (x), \phi^+(x) = x + a$, and $\psi = (R, S) = (g^{\mu 2^{l_K + l_H}}, \left| (g^t X)^\mu \right|)$ if $x + a \neq x$ and $T(PK, \phi^+, \psi) = (R, SR^{-a}) = (R, S)$ we have $R^{-a} = 1 \mod N$. Since $a \in Z_N, R \in QR_N$ we have $pq \leq a = \delta pq \leq 4pq$, where $\delta \in \{1, 2, 3, 4\}$. Thus we can find pq from the equation $pq = a/\delta \approx \frac{N-1}{4}$ and then factor N. So the adversary can not find such ϕ^+ except with negligible probability.

There are several KEM schemes from the all-but-one extraction paradigm [10,23,21,30,29,28]. It is easy to verify that KEM schemes in [30,29,28] also satisfy these two properties, while the KEM schemes in [10,23] do not satisfy Φ^+-key malleability.

5   Conclusion

We proved that the KEM/Tag-DEM hybrid encryption is Φ-CC-RKA secure if the IND-CCA2 (or IND-CCCA) secure KEM part satisfies Φ-key malleability and Φ-key fingerprint and the Tag-DEM part is a one-time secure tag authenticated encryption. We showed that several KEM schemes satisfy these two properties. Thus we can get efficient CC-RKA secure hybrid encryption schemes.

Compared with Wee's construction [31] we do not need the one-time signature. In addition, we can get CC-RKA secure public key encryption schemes from the "all-but-one extraction" paradigm and the Cramer-Shoup paradigm uniformly. While Wee's framework can only get CC-RKA secure public key encryption schemes from the "all-but-one extraction" paradigm.

References

1. Applebaum, B., Harnik, D., Ishai, Y.: Semantic security under related-key attacks and applications. In: ICS, pp. 45–60 (2011)
2. Bellare, M., Cash, D.: Pseudorandom functions and permutations provably secure against related-key attacks. In: Rabin, T. (ed.) CRYPTO 2010. LNCS, vol. 6223, pp. 666–684. Springer, Heidelberg (2010)
3. Bellare, M., Cash, D., Miller, R.: Cryptography secure against related-key attacks and tampering. In: Lee, D.H., Wang, X. (eds.) ASIACRYPT 2011. LNCS, vol. 7073, pp. 486–503. Springer, Heidelberg (2011)
4. Bellare, M., Kohno, T.: A theoretical treatment of related-key attacks: RKA-PRPs, RKA-PRFs, and applications. In: Biham, E. (ed.) EUROCRYPT 2003. LNCS, vol. 2656, pp. 491–506. Springer, Heidelberg (2003)
5. Bellare, M., Paterson, K.G., Thomson, S.: RKA security beyond the linear barrier: IBE, encryption and signatures. In: Wang, X., Sako, K. (eds.) ASIACRYPT 2012. LNCS, vol. 7658, pp. 331–348. Springer, Heidelberg (2012)
6. Biham, E.: New types of cryptoanalytic attacks using related keys (extended abstract). In: Helleseth, T. (ed.) EUROCRYPT 1993. LNCS, vol. 765, pp. 398–409. Springer, Heidelberg (1994)
7. Biham, E., Shamir, A.: Differential fault analysis of secret key cryptosystems. In: Kaliski Jr., B.S. (ed.) CRYPTO 1997. LNCS, vol. 1294, pp. 513–525. Springer, Heidelberg (1997)
8. Boneh, D., Boyen, X.: Efficient selective-ID secure identity-based encryption without random oracles. In: Cachin, C., Camenisch, J.L. (eds.) EUROCRYPT 2004. LNCS, vol. 3027, pp. 223–238. Springer, Heidelberg (2004)
9. Boneh, D., DeMillo, R.A., Lipton, R.J.: On the importance of checking cryptographic protocols for faults (extended abstract). In: Fumy, W. (ed.) EUROCRYPT 1997. LNCS, vol. 1233, pp. 37–51. Springer, Heidelberg (1997)
10. Boyen, X., Mei, Q., Waters, B.: Direct chosen ciphertext security from identity-based techniques. In: ACM Conference on Computer and Communications Security, pp. 320–329. ACM (2005)

11. Chen, Y., Dong, Q.: RCCA security for KEM+DEM style hybrid encryptions. In: Kutyłowski, M., Yung, M. (eds.) Inscrypt 2012. LNCS, vol. 7763, pp. 102–121. Springer, Heidelberg (2013)

12. Cramer, R., Shoup, V.: A practical public key cryptosystem provably secure against adaptive chosen ciphertext attack. In: Krawczyk, H. (ed.) CRYPTO 1998. LNCS, vol. 1462, pp. 13–25. Springer, Heidelberg (1998)

13. Cramer, R., Shoup, V.: Universal hash proofs and a paradigm for adaptive chosen ciphertext secure public-key encryption. In: Knudsen, L.R. (ed.) EUROCRYPT 2002. LNCS, vol. 2332, pp. 45–64. Springer, Heidelberg (2002)

14. Cramer, R., Shoup, V.: Design and analysis of practical public-key encryption schemes secure against adaptive chosen ciphertext attack. SIAM J. Comput. 33, 167–226 (2004), http://dl.acm.org/citation.cfm?id=953065.964243

15. Damgård, I., Faust, S., Mukherjee, P., Venturi, D.: Bounded tamper resilience: How to go beyond the algebraic barrier. In: Sako, K., Sarkar, P. (eds.) ASIACRYPT 2013, Part II. LNCS, vol. 8270, pp. 140–160. Springer, Heidelberg (2013)

16. Goldenberg, D., Liskov, M.: On related-secret pseudorandomness. In: Micciancio, D. (ed.) TCC 2010. LNCS, vol. 5978, pp. 255–272. Springer, Heidelberg (2010)

17. Goyal, V., O'Neill, A., Rao, V.: Correlated-input secure hash functions. In: Ishai, Y. (ed.) TCC 2011. LNCS, vol. 6597, pp. 182–200. Springer, Heidelberg (2011)

18. Halderman, J.A., Schoen, S.D., Heninger, N., Clarkson, W., Paul, W., Calandrino, J.A., Feldman, A.J., Appelbaum, J., Felten, E.W.: Lest we remember: Cold boot attacks on encryption keys. In: USENIX Security Symposium, pp. 45–60 (2008)

19. Hofheinz, D., Kiltz, E.: Secure hybrid encryption from weakened key encapsulation. In: Menezes, A. (ed.) CRYPTO 2007. LNCS, vol. 4622, pp. 553–571. Springer, Heidelberg (2007)

20. Hofheinz, D., Kiltz, E.: The group of signed quadratic residues and applications. In: Halevi, S. (ed.) CRYPTO 2009. LNCS, vol. 5677, pp. 637–653. Springer, Heidelberg (2009)

21. Hofheinz, D., Kiltz, E.: Practical chosen ciphertext secure encryption from factoring. In: Joux, A. (ed.) EUROCRYPT 2009. LNCS, vol. 5479, pp. 313–332. Springer, Heidelberg (2009)

22. Jia, D., Lu, X., Li, B., Mei, Q.: RKA secure PKE based on the DDH and HR assumptions. In: Susilo, W., Reyhanitabar, R. (eds.) ProvSec 2013. LNCS, vol. 8209, pp. 271–287. Springer, Heidelberg (2013)

23. Kiltz, E.: Chosen-ciphertext secure key-encapsulation based on gap hashed diffie-hellman. In: Okamoto, T., Wang, X. (eds.) PKC 2007. LNCS, vol. 4450, pp. 282–297. Springer, Heidelberg (2007)

24. Kiltz, E., Pietrzak, K., Stam, M., Yung, M.: A new randomness extraction paradigm for hybrid encryption. In: Joux, A. (ed.) EUROCRYPT 2009. LNCS, vol. 5479, pp. 590–609. Springer, Heidelberg (2009)

25. Knudsen, L.R.: Cryptanalysis of loki 91. In: Seberry, J., Zheng, Y. (eds.) AUSCRYPT 1992. LNCS, vol. 718, pp. 196–208. Springer, Heidelberg (1993), http://dx.doi.org/10.1007/3-540-57220-1_62

26. Kocher, P.C.: Timing attacks on implementations of diffie-hellman, RSA, DSS, and other systems. In: Koblitz, N. (ed.) CRYPTO 1996. LNCS, vol. 1109, pp. 104–113. Springer, Heidelberg (1996)

27. Kurosawa, K., Desmedt, Y.: A new paradigm of hybrid encryption scheme. In: Franklin, M. (ed.) CRYPTO 2004. LNCS, vol. 3152, pp. 426–442. Springer, Heidelberg (2004)

28. Lu, X., Li, B., Liu, Y.: How to remove the exponent GCD in HK09. In: Susilo, W., Reyhanitabar, R. (eds.) ProvSec 2013. LNCS, vol. 8209, pp. 239–248. Springer, Heidelberg (2013)
29. Lu, X., Li, B., Mei, Q., Liu, Y.: Improved efficiency of chosen ciphertext secure encryption from factoring. In: Ryan, M.D., Smyth, B., Wang, G. (eds.) ISPEC 2012. LNCS, vol. 7232, pp. 34–45. Springer, Heidelberg (2012)
30. Lu, X., Li, B., Mei, Q., Liu, Y.: Improved tradeoff between encapsulation and decapsulation of HK09. In: Wu, C.-K., Yung, M., Lin, D. (eds.) Inscrypt 2011. LNCS, vol. 7537, pp. 131–141. Springer, Heidelberg (2012)
31. Wee, H.: Public key encryption against related key attacks. In: Fischlin, M., Buchmann, J., Manulis, M. (eds.) PKC 2012. LNCS, vol. 7293, pp. 262–279. Springer, Heidelberg (2012)

ARBRA: Anonymous Reputation-Based Revocation with Efficient Authentication

Li Xi, Jianxiong Shao, Kang Yang, and Dengguo Feng

Trusted Computing and Information Assurance Laboratory
Institute of Software, Chinese Academy of Sciences, China
{xili,shaojianxiong,yangkang,feng}@tca.iscas.ac.cn

Abstract. Service providers (SPs) that allow anonymous access need to protect their services against misbehaving users. Several schemes are proposed to achieve anonymous revocation without a trusted third party, thus protecting users' privacy. They either have linear computational complexity in the size of the blacklist (EPID, BLAC, BLACR), or require all misbehaviors being identified in a time window (PEREA, PERM).

In ESORICS 2012, Yu *et al* propose an efficient scheme called $PE(AR)^2$ which does not require the SPs to review sessions in a timely manner. However, we find there are security problems in $PE(AR)^2$. We propose ARBRA, a reputation-based revocation system for which the SPs can assign positive or negative scores to anonymous sessions and block the users whose reputations are not high enough. ARBRA allows the SPs to ramp up penalties for repeated misbehaviors from the same user and does not require the SPs to judge misbehaviors within a time window. Our benchmark shows that ARBRA has the best performance on the SP side among existing schemes and is also efficient on the user side even if the misbehavior list contains one million entries.

1 Introduction

Internet services that allow anonymous access are desirable as they protect users' privacy. Without anonymity, users can be tracked by intrusive advertising companies, or may hesitate to honestly express their opinions under an oppressive government. However, full anonymity can be abused as users can misbehave without getting blocked. Thus accountability and anonymity must be balanced, service providers (SPs) should be able to revoke misbehaving users to protect their services. To supports anonymous blacklisting, an intuitive approach is to introduce a trusted third party (TTP) who can deanonymize or link users' accesses. There are many TTP-based revocation schemes, such as schemes based on group signatures [1,2], schemes based on accumulators [3,4,5] and Nymble-like systems [6,7]. However, having a trusted third party can be dangerous as a user can never be sure that his activity remains anonymous.

To revoke misbehaving users while protecting users' privacy, several TTP-free schemes are proposed, including BLAC [8], EPID [9], PEREA [10], FAUST [11], BLACR [12], PERM [13], $PE(AR)^2$ [14]. These schemes allow the server to block

S.S.M. Chow et al. (Eds.): ISC 2014, LNCS 8783, pp. 33–53, 2014.

misbehaving users without a TTP, thus users' identities can not be deanonymized and their authentications can not be linked. Though being the state of the art in anonymous revocation, these schemes have problems for real-world deployment.

BLAC [8] and EPID [9] require a user to prove in zero-knowledge that his credential is not associated with any ticket on the blacklist, and the computational complexity of authentication is linear with the size of the blacklist which makes it impractical for SPs with large blacklists. BLACR [12] introduces the *reputation-based revocation* and improves the authentication time by proposing the express lane authentication. Nevertheless BLACR can only support a misbehavior list with a few thousands of entries. Ryan Henry *et al* propose batch proof technique [15] to improve the efficiency of BLAC but their approach can not be applied to BLACR-weighted.

Realizing the importance of reducing the computational complexity at the SP, several revocation-window-based schemes are proposed, including PEREA [10] and PERM [13]. PEREA and PERM require a user's misbehavior being caught before the user authenticates K more times. The authentication time of PEREA and PERM at the SP side has linear dependence on K, so K can not be too large. PERM improves on PEREA to support reputation-based revocation by adding memory and reduces the authentication time at the user side by introducing the judgement pointer. At the SP, PERM is slower than PEREA. [13] shows that for PERM with $K = 10$, an SP equipped with an 8-core server can support about 10 authentications per minute, which is not enough for heavy-loaded SPs such as popular internet forums with thousands of posts per minute.

A problem of these revocation-window-based schemes is the SP has to review transactions in a serial and timely manner, i.e., *all misbehaviors* must be judged within a time window. However, while we can expect most misbehaviors can be detected in a short period, some dispute cases need more time to judge, especially in large internet forums with many sessions and posts. Note that a user can only authenticate K times within the time window, so simply increasing the time window would lead to usability constraint.

On the user side, PEREA and PERM both require rate-limiting. However, at some circumstances, users may want to quickly "re-anonymize" their sessions, especially when doing some privacy-sensitive stuffs.

FAUST [11] reduces the computational complexity by using whitelist, but only supports limited functionality and can not resist collusion attack.

In ESORICS 2012, Yu *et al* propose PE(AR)^2 [14], though it only supports *positive* reputation and "*1-strike-out*" policy (a user would get revoked if he misbehaves once), it does not require the SP to judge sessions serially, thus the SP can have better flexibility and more time in judging dispute cases. Unfortunately, we find there are security problems in PE(AR)^2 which make it unusable.

Note that existing reputation-based revocation schemes are still not efficient enough for heavy-loaded SPs with hundreds or thousands of sessions per minute and the other efficient schemes only support limited functionalities or suffer from security problems. A scheme that supports reputation-based revocation and is efficient at the SP would be a significant improvement by making reputation-based revocation highly practical for the real world. Thus we propose ARBRA.

Our Contribution. We first point out the security problems in PE(AR)^2, then we propose a new scheme called ARBRA (Anonymous Reputation-Based Revocation with efficient Authentication) that has the following advantages:

1. *Reputation-based anonymous revocation.* The SP can assign positive and negative scores under multiple categories to sessions, e.g., popular video: +5 in category content, offensive words: -2 in category comment. Then the SP can use boolean combinations of reputation policies to block users that do not meet the requirement, e.g., reputation in category content should > -5 AND reputation in category comment should > -10.
2. *Weighted reputation.* ARBRA allows the SP to penalize repeated misbehavior, e.g., double the severity of a user's second misbehavior. The only existing scheme which has this functionality is BLACR. In BLACR, the SP punishes repeated misbehaviors with regard to each category, for example, if a user first misbehaves in category comment and then misbehaves in category content, he would not get punished for *repeated* misbehavior. ARBRA allows the SP to punish repeated misbehavior without regarding categories.
3. *No rate-limiting and serial judgement.* Unlike PEREA and PERM, ARBRA does not need rate-limiting and serial judgement, thus is more desirable to both users and SPs. The users can quickly re-anonymize sessions many times when necessary. The SPs do not have to judge sessions in a serial manner.
4. *Much more efficient at the server.* At the SP side, the computational overhead is independent of reputation list size and is more efficient than existing schemes. Our benchmark shows when $K = 80$, for ARBRA an 8-core SP can handle 1653 authentications per minute while PERM can only handle 43 authentications per minute.
5. *Efficient at the user.* We propose practical techniques which significantly improve the efficiency at the user side. Even with a misbehavior list which contains one million entries, for ARBRA it takes 13.6 seconds at the user to authenticate which we believe is acceptable.

Paper Outline. We point out the security problems of PE(AR)^2 in §2 and give an approach overview of ARBRA in §3. We present the construction of ARBRA in §4 and analyze the efficiency of ARBRA in §5. We discuss various issues in §6 and conclude in §7.

2   The Security Problem of PE(AR)^2

In the Auth protocol of PE(AR)^2, a user uses the non-membership witness (NMW) [4] of the product value of his tickets to prove all the tickets belong to him are not in the blacklist. Regarding reputation, the Redeem protocol allows a user to reclaim the positive scores assigned to his tickets. Details of PE(AR)^2 can be found in [14]. Unfortunately, we find there are security problems in PE(AR)^2 which make it unusable.

Incorrect Reputation. In $PE(AR)^2$, the user can only use the Redeem protocol to get positive scores assigned to his tickets. To achieve unlinkability, T_{old} which is the product of the tickets he wants to redeem should not be revealed. So the user has to prove in zero-knowledge that T_{old} actually belongs to his ticket set T. In the Redeem protocol [14, §4.2], the user generates the zero-knowledge proof of knowledge (ZKPoK) $PK_3\{(t_{i+1}, T, T_{old}, T_{new}, \sigma, s, s', \mathbb{S})$: $T_{old}|T \wedge s' = \sum_{s_j \in \mathbb{S}} s_j \wedge_{(t_j, s_j, \sigma_j) \in \mathbb{S}} 1 = \text{Verify}_{\text{CL}}((t_j, s_j), \sigma_j, \text{pk}_{\text{CL}}) \wedge 1 = \text{Verify}_{\text{CL}}((T, t_{i+1}, s), \sigma, \text{pk}_{\text{Acc}})\}$ to proves that $T_{old}|T$. Unfortunately, this is not enough, as $T_{old}|T$ does not necessarily mean T_{old} are *actually used* by this user.

First, two users A and B can conspire in protocol Reg or Redeem, so that there is one ticket P that is in both T_A and T_B, where T_A is A's ticket set and T_B is B's ticket set. Notice that this conspiracy is easy to achieve, as T_A and T_B are never revealed and there is no check of whether there is *collision* between ticket sets in the Reg protocol and Redeem protocol. So if A uses ticket P to do something good and gains a positive score, both A and B can redeem this ticket to get a higher reputation. Second, in the redeem protocol, A can always generate T_{new} which contains P as there is no check whether T_{new} contains used tickets and T_{new} is never revealed, thus A can redeem P as many times as he wishes.

Misuse of CL Signature. A key idea in $PE(AR)^2$ is using the product value $T = \prod_{i=1}^{K} t_i$ of all tickets $\{t_1, \cdots, t_K\}$ in non-membership witness (NMW) generation. Given an accumulator acc and let $X = \prod_{t \in acc} t$, the NMW of T actually proves $gcd(X, T) = 1$, thus proving that all the K tickets in T are not accumulated, as $\{t_i\}_{i=1}^{K}$ are all prime numbers. $PE(AR)^2$ requires the SP to generate a CL signature for T. As shown in [14, §4.2], to avoid collision, each prime ticket is at least $l_t = 166$ bits, and K is recommended to be 1000. Thus T is $l_T = 166000$ bits. It means the SP has to generate a 166002-bit prime number e in the CL signature, which is obviously impractical. In $PE(AR)^2$, this problem is overlooked, as it only requires the length of the prime number $l_e > l_t + 2 = 168$ in [14, §4.2].

3   ARBRA: Overview

3.1   Approach Overview

ARBRA adopts the idea in $PE(AR)^2$ that use the product value of tickets to generate NMW, but utilizes a completely different construction that not only avoids the security problem of $PE(AR)^2$ but also significantly improves functionality.

In ARBRA, users register with an SP and obtain credentials from the SP. Then the users can authenticate anonymously with the SP. Each authenticated session is bound to a session identifier (ticket) and would be tied to one particular action such as a page edit or a video uploading. The SP assigns scores to reviewed sessions under categories and maintains a reputation list, each entry of which contains the ticket of a reviewed session, the corresponding scores and a CL signature that binds the ticket with the scores. In addition, the SP also maintains a misbehavior list as an accumulator which contains the tickets that are judged

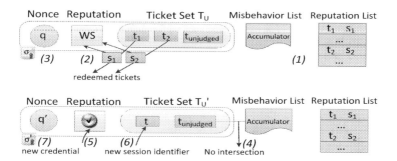

Fig. 1. (1) The user obtains the misbehavior list and reputation list from the SP, then (2) redeems her tickets $\{t_i\}_{i=1}^n$ in the newly-added reputation list and adds the corresponding (weighted) scores $\{s_i\}_{i=1}^n$ to her reputation using her credential σ and ZKPoK. (3) She proves the freshness of her credential using q, and (4) proves all the unredeemed tickets are not in the misbehavior list using NMW. (5) She proves her reputation satisfies the policy and (6) presents a ticket t as the session identifier. (7) She gets a new credential σ' on her new ticket set and other information from the SP.

as misbehaviors. Figuer 1 briefly shows how a user authenticates in ARBRA, which will be explained below.

Ticket Set. Instead of using tickets that are already fixed in registration and redeem protocol like PE(AR)^2, in ARBRA, a user maintains a value T_U which is the product value of the tickets that are *used by her but not redeemed*. For simplicity, we will not distinguish the ticket set between the product value T_U, as tickets are all prime numbers. When authenticating, the user redeems tickets $\{t_i\}_{i=1}^n$ which are the tickets that belong to her in the newly-added reputation list, and reclaims the scores assigned to these tickets. The user has to prove $\{t_i\}_{i=1}^n$ belong to T_U. Then she presents a new ticket t as the session identifier. After the authentication, the redeemed tickets are removed from T_U and the new ticket t is added to T_U: $T_U \leftarrow t \cdot T_U/T_r$, where $T_r = \prod_{i=1}^n t_i$. Note that SP will check if t has been used before, thus the ticket set collision between different users can be avoided. ARBRA guarantees that a user can only redeem the tickets that are *used by her* previously and each ticket can be redeemed only *once*.

Shorter Prime Tickets. In PEREA and PE(AR)^2, the prime tickets have to be large enough to avoid collision, for example the tickets should be 230 bits so that the probability of collision is about 2^{-112}[10]. It is necessary in PEREA and PE(AR)^2 as the prime ticket is used to prove the freshness of the credential in case of credential replay. In ARBRA the freshness of the credential is proved by using a nonce q bound in credential. After the freshness is proved, the user can submit a short prime ticket, if this ticket has been used (happens with small probability, for example, 2^{-20}), then the user can choose another one. The reason why do not let the SP assign a prime ticket is that the user maintains a ticket pool which will be explained afterwards.

The ticket length l_t in ARBRA is shorter than PEREA and PE(AR)[2]: $l_t = 56$ will suffice as there are more than 2^{50} prime tickets in this set[1]. Given a ticket set S_{used} which contains 2^{30} used prime tickets (considering there have been about $2^{29.4}$ edits to English Wikipedia), the probability that a random generated prime falls into S_{used} is 2^{-20}.

The length of T_U $l_T = K \cdot l_t$, where K is the number of tickets in T_U, thus the length of prime numbers in CL signatures can still be large when K grows. To support bigger K, our second trick is splitting T_U into n l_m-bit message blocks $\{T_i\}_{i=0}^{n-1}$ which satisfies $T_U = \sum_{i=0}^{n-1} 2^{i \cdot l_m} T_i$. Typically $l_m = 224$.

Our improvement has two advantages. First, the length l_e of the prime number e in a CL signature will be much shorter as $l_e > l_m + 2$, which means better efficiency on the SP side. Second, the update of NMW on the user side will be more efficient, as the main computation overhead is $O(\Delta_{acc} \cdot l_t)$ exponentiation, where Δ_{acc} is the number of newly-added tickets in the accumulator. Given $K = 40$, updating NMW in ARBRA is about 160 times faster than PEREA (note that PEREA has to update K NMWs).

Positive and Negative Scores. While users are motivated to redeem positive scores, they certainly would not voluntarily reduce their reputations. So the SP has to mandate the user to reclaim all the negative scores that belong to him. This is the main reason of the inefficiency of schemes such as EPID, BLAC and BLACR. The user not only has to retrieve the entries belong to him in the blacklist, but also has to prove that *all the other entries* do not belong to him.

In the authentication protocol of ARBRA, the user first redeems the tickets $\{t_i\}_{i=1}^n$ in the reputation list using zero-knowledge proof of knowledge (ZKPoK) protocol, then utilizes the NMW of T_U/T_r, where $T_r = \prod_{i=1}^n t_i$, to prove that all the other tickets in T_U are not in the misbehavior list. Thus the SP will be sure that the user has redeemed all the negative-score tickets that belong to him. In the authentication protocol, the only leaked information about the user is the number of the tickets that he redeems this time, however the compromise of privacy is minimum as this session can not be linked to the user's past sessions.

Moreover, we provide an efficient re-login protocol below. Sensitive users can always redeem their tickets using the auth protocol, then *re-anonymize* their sessions using the re-login protocol as the re-anonymized sessions reveal nothing.

Efficient Re-Login. Consider a user who only has a few non-negative tickets to redeem, he wants a quick login this time and wants to redeem the tickets later. To facilitate these users and to reduce the computational complexity on the SP side, we propose a re-login protocol. In re-login protocol, the user does not redeem any ticket and he just has to prove that all the tickets in T_U are not in the misbehavior list, thus proving his reputation does not decrease. Notice that a user gets a new CL signature only if his reputation satisfies the policy at the last authentication. So there is no need for the user to prove that his reputation satisfies the policy again (the policy is used to block users whose reputations are

---

[1] For $x > 598$, there are more than $\frac{x}{ln(x)}(1 + \frac{0.992}{ln(x)})$ primes in $[1, x]$[16].

not high enough). Using these optimizations, the re-login protocol significantly reduces the computational complexity on the SP side.

Weighted Reputation. SPs such as WikiPedia want to penalize repeated misbehavior. For example, if the score of offensive words is -2, the SP may want to double it if it is the second misbehavior of a user, thus disincentivizing repeated misbehaviors. The SP can do the same thing to reward good behaviors. If s is the score assigned to a ticket t, t is a user's N_tth bad behavior and the adjusting factor for N_tth bad behavior is δ_{N_t}, then the weighted score of t is $\delta_{N_t} \cdot s$. So far as we know, the only existing scheme which has this functionality is BLACR. In BLACR, the SP can only punish repeated misbehavior with regard to each category, for example, if a user first misbehaves in category comment, then misbehaves in category content, he would not get punished for *repeated* misbehavior. ARBRA allows the SP to punish misbehaviors without regarding categories.

Achieving this functionality is non-trivial as the reputation list contains positive-score tickets, negative-score tickets and zero-score tickets. The efficiency will be significantly reduced if we simply use the OR-proof [17]. We also have to mandate users to redeem tickets in correct order. In ARBRA each entry in the reputation list contains a serial number which indicates the order of the ticket in the list. When redeeming tickets, the user has to prove that the ticket to be redeemed has a bigger serial number than previous redeemed ones.

Calculate NMW Using a Predefined Ticket Pool. In $PE(AR)^2$ the ticket set is generated in reg and redeem protocol and does not change in auth protocol. When the ticket set T_U remains unchanged, the user just has to update the NMW of T_U for each authentication. However, we show that this pre-fixed ticket set leads to serious security problem. As in ARBRA a user maintains a ticket set T_U which contains the tickets that are used but not redeemed, i.e., the ticket set changes for each authentication, two new problems arise. First, before authentication the user has a NMW w_U of T_U against old misbehavior list BL, during the authentication the user redeems $\{t_i\}_{i=1}^n$. Let $T_r = \prod_{i=1}^n t_i$. To authenticate, the user has to generate NMW for T_U/T_r against new misbehavior list BL' efficiently. Second, after every authentication $T_U' \leftarrow t \cdot T_U/T_r$, thus the user has to generate NMW for T_U'. As T_U' can not be revealed, the user has to generate the NMW himself, which will be inefficient, especially when there are millions of entries in the accumulator. To make ARBRA practical, an efficient NMW generation algorithm is very important.

We show given two numbers T_1 and T_2, $T_1|T_2$, together with the NMW w_2 of T_2 against an accumulator acc, how NMW w_1 of T_1 against acc can be calculated efficiently. We denote this novel algorithm as $\mathtt{nmw}_{\mathtt{pool}}(T_1, T_2, w_2, acc)$. For the first problem, a user can first calculates NMW of T_U/T_r against BL using $w' \leftarrow \mathtt{nmw}_{\mathtt{pool}}(T_U/T_r, T_U, w_U, \mathtt{BL})$, then updates w' to w using BL'/BL.

For the second problem, in a time period, for example, one year, a user generates a ticket pool P_{tk} which contains n_P tickets he wants to use in this period, typically $n_P = 3000$. An active user can generate more tickets. The user calculates $T_P = \prod_{t \in P_{tk}} t$ and maintains the NMW of T_P against the accumulator (as shown in §4.1, generating and updating NMW for T_P is almost as efficient as for a single ticket). After each authentication, the NMW of T_U can be calculated efficiently from the NMW of T_P as $T_U|T_P$.

3.2 Security Goals

We present the informal descriptions of desired security properties below, the formal definition and the sketched proof of security of ARBRA can be found in the appendix.

Authenticity. In ARBRA, SPs accept authentications only from registered users who satisfy the authenticity policy.

Revocation auditability. Users should be able to check whether they are revoked before trying to authenticate.

Anonymity. In the re-login protocol of ARBRA, all that an SP can infer about an authenticating user is that the user satisfies the authenticity policy. In the authentication protocol of ARBRA, the only additional information that an SP can get is the number of the tickets that the user redeems this time.

Non-frameability. No coalition of third parties should be able to get an honest user who satisfies the authenticity policy revoked from an honest SP.

4 ARBRA: Construction

4.1 Building Blocks

In this subsection, we first introduce the various cryptographic primitives that are used in ARBRA. Then we present a novel algorithm $\texttt{nmw}_{\texttt{pool}}(T_1, T_2, w_2, acc)$ and show how this algorithm can be used to reduce the computational complexity at the user.

ZKPoK Protocols. We use the notion introduced by Camenisch and Stadler [18], for example, $PK\{(x) : y = g^x\}$ denotes a Zero-Knowledge Proof-of-Knowledge (ZKPoK) protocol that proves the prover knows a number x which satisfies $y = g^x$.

Commitment Scheme. The commitment scheme used in ARBRA is developed by Fujishaki and Okamoto [19] and improved by Damgard and Fujishaki [20]. A commitment on m is denoted by $\mathbb{C}(m)$.

CL Signature. CL signature [21] is a signature scheme that has two useful protocols: 1) generating signature on blocks of committed values. We denote this protocol as $\texttt{cl}_{\texttt{sign}}(\mathbb{C}(m_1), ..., \mathbb{C}(m_n))$. After this protocol, the user gets a CL signature on $(m_1, ... m_n)$; 2) proving knowledge of a signature σ on committed values m. We denote this protocol as $PK\{(m, \sigma) : C = \mathbb{C}(m) \wedge \texttt{cl}_{\texttt{verif}}(\sigma, m) = 1\}$, where $\texttt{cl}_{\texttt{verif}}(\sigma, m) = 1/0$ if σ is a valid/invalid CL signature on m.

Signature-Based Range Proof. We use the signature-based range proof proposed by Camenisch et.al [22] in ARBRA. This range proof is efficient when the interval is not too large.

Dynamically Univeral Accumulator (DUA). We mainly focus on the functionality of DUA [4] needed by ARBRA, i.e., the non-membership witness (NMW). Detailed introduction of DUA can be found in appendix. Given an accumulator acc, the corresponding accumulator value is $\mathsf{V} = g^{\prod_{t \in acc} t}$, where g is the exponential base. $w = \mathtt{nmw_{gen}}(x, acc)$ denotes the algorithm to generate NMW w for x against acc *without the private key*. $w' = \mathtt{nmw_{update}}(x, w, \triangle_{acc}, \mathsf{V})$ denotes the algorithm to update NMW for x where $\triangle_{acc}$ are newly-accumulated values. $\mathtt{nmw_{verif}}(x, w, \mathsf{V}) = 1/0$ if w is a valid/invalid NMW for x against acc.

The Algorithm. $w_1 \leftarrow \mathtt{nmw_{pool}}(T_1, T_2, w_2, \mathsf{V})$ Given two numbers T_1 and T_2, $T_1 | T_2$, let $\tilde{T} = T_2/T_1$. V is the accumulator value of an accumulator acc: $\mathsf{V} = g^U$, where g is the exponential base and $U = \prod_{t \in acc} t$. The NMW of T_2 against the accumulator acc is $w_2 = (a, d)$, $a < T_2$. So there exist b such that $aU + bT_2 = 1$ and $g^{-b} = d$ (recall the vulnerability of DUA [23] and the fix in [14, §4.1]).

We want to calculate the NMW w_1 for T_1. Intuitively we want to find $(a', g^{-b'})$, such that $a'U + b'T_1 = 1$, $0 < a' < T_1$. As $aU + b\tilde{T}T_1 = 1$, we calculate $a' = a$ mod T_1, and find r such that $a = a' + rT_1$. Note that $r < \tilde{T}$ as $a < T_2$. Thus $b' = b\tilde{T} + rU$, $d' = g^{-b'} = g^{-rU}d^{\tilde{T}} = \mathsf{V}^{-r}d^{\tilde{T}}$. The NMW w_1 of T_1 is (a', d'). Calculating w_1 from w_2 requires one exponential of size r and one exponential of size $\tilde{T}$, $r < \tilde{T} < T_2$.

Now we show how algorithm $w_1 = \mathtt{nmw_{pool}}(T_1, T_2, w_2, \mathsf{V})$ can be used to reduce the computational burden at the user. In ARBRA, a user generates a ticket pool P_{tk} which contains n_P random prime tickets he wants to use, and maintains the NMW w_P of $T_P = \prod_{t \in P_{tk}} t$ against the misbehavior list BL. The accumulator value of BL is $\mathsf{V_{BL}}$. After a successful authentication the user obtains a new ticket set T_U, w_U which is the NMW of T_U can be computed efficiently from w_P as $T_U | T_P$: $w_U \leftarrow \mathtt{nmw_{pool}}(T_U, T_P, w_P, \mathsf{V_{BL}})$.

Let l_t be the length of the tickets, $\triangle_{tk}$ be the number of the newly-added tickets in the accumulator each day, updating the NMW w_P for T_P each day needs exponential of size $\triangle_{tk} \cdot l_t$, generating the NMW of T_U from w_P needs exponential of size $2 \cdot n_P \cdot l_t$. Our benchmark shows when $n_P = 2000, \triangle_{tk} = 2000, l_t = 56$, it takes about 4.5 seconds to update w_P and calculate NMW for T_U from w_P using $\mathtt{nmw_{pool}}$. For comparison, given $\triangle_{tk} = 2000$ each day, we can assume there are about one million tickets in the accumulator, calculating a NMW for T_U against this accumulator using $\mathtt{nmw_{gen}}$ takes about 700 seconds.

4.2   Subprotocols

The following subprotocols are used in ARBRA.

$\mathbb{P}_{\mathtt{cred}}(C_T)$. $\mathbb{P}_{cred}$ allows a prover Alice to prove that she has a CL signature on message blocks $\{T_i\}_{i=0}^{n-1}$ and the value T_U committed in C_T satisfies $T_U = \sum_{i=0}^{n-1} 2^{i \cdot l_m} T_i$, where l_m is the length of T_i. As described above, we split the large number T_U into n l_m-bit messages $\{T_i\}_{i=0}^{n-1}$ to allows for smaller primes e in CL signature. Alice generates $PK\{(\{T_i\}_{i=0}^{n-1}) : C_T = \mathbb{C}(T_U) \wedge T_U = \sum_{i=0}^{n-1} 2^{i \cdot l_m} T_i \wedge \mathtt{cl_{verif}}(\sigma, \{T_i\}_{i=0}^{n-1}) = 1\}$, and sends it to the verifier who verifies the proof.

$\mathbb{P}_{\texttt{rdm}}(\mathfrak{L}, \mathfrak{D}, C_{T,0}, C_{WS,0}, C_{GB,0}, C_{\Theta,0}, C_T, C_R, C_{GB}, C_{\Theta})$. Given a reputation list $\mathfrak{L}$, the tickets in $\mathfrak{L}$ that belong to Alice are $\{t_i\}_{i=1}^n$. For each ticket t_i, the corresponding entry in the reputation list $\mathfrak{L}$ is: $(t_i, \{S_{j,i}\}_{j=1}^m, b_{G,i}, b_{B,i}, \theta_i, \sigma_i)$, where $\{S_{j,i}\}_{j=1}^m$ are the scores in m categories, $(b_{G,i}, b_{B,i}) = (1,0)/(0,1)/(0,0)$ if t_i is a good/bad/zero-score behavior, θ_i is the serial number of t_i on the list, $\theta_i > \theta_{i-1}$, σ_i is a CL signature on these messages. The final weighted reputations of Alice depends on $\mathfrak{L}$, the set of adjusting factors $\mathfrak{D}$ and a set of initial values, including the initial weighted reputation and the number of good/bad behaviors that Alice has already done.

The initial values are denoted by $(T_0, \{WS_{j,0}\}_{j=1}^m, N_{G,0}, N_{B,0}, \theta_0)$. T_0 is the product value of the tickets that are used but not redeemed, $\{WS_{i,0}\}_{i=1}^m$ are weighted reputations, $N_{G/B,0}$ is the number of good/bad behaviors that Alice has already done, θ_0 is the initial serial number. $C_{T,0} = \mathbb{C}(T_0), C_{WS,0} = \mathbb{C}(\{WS_{j,0}\}_{j=1}^m), C_{GB,0} = \mathbb{C}(N_{G,0}, N_{B,0}), C_{\Theta,0} = \mathbb{C}(\theta_0)$.

$\mathbb{P}_{\texttt{rdm}}$ allows Alice to redeem $\{t_i\}_{i=1}^n$ and convince the SP that

- The number committed in C_T is $T_U = T_0/T_{rdm}$, where $T_{rdm} = \prod_{i=1}^n t_i$.
- The numbers committed in C_R are weighted reputations $\{WS_j\}_{j=1}^m$ and $\{WS_j = \sum_{i=0}^n WS_{j,i}\}_{j=1}^m$, where $WS_{j,i}$ is the weighted score of t_i in jth category (recall $\{WS_{j,0}\}_{j=1}^m$ are the initial weighted reputations).
- The numbers committed in C_{GB} are (N_G, N_B) which are the numbers of (good, bad) behaviors, i.e., $N_{G/B} = N_{G/B,0} + \sum_{i=1}^n b_{G/B,i}$.
- The number committed in C_Θ is θ_n and $\theta_i > \theta_{i-1}$ for $1 \le i \le n$.

This protocol proceeds as follows:

1) For each t_i, Alice checks whether t_i is the $N_{G/B,i}$th good/bad behavior and finds the corresponding adjusting factor δ_i using $\mathfrak{D}$. Alice calculates the weighted scores $\{WS_{j,i} = \delta_i \cdot S_{j,i}\}_{j=1}^m$ and $T_i = T_{i-1}/t_i$. Alice produces the auxiliary commitments: $C_{S,i} = \mathbb{C}(S_{1,i}, \cdots, S_{m,i}), C_{GB,i} = \mathbb{C}(b_{G,i}, b_{B,i}), C_{WS,i} = \mathbb{C}(WS_{1,i}, ..., WS_{m,i}), C_{\Theta,i} = \mathbb{C}(\theta_i), C_{T,i} = \mathbb{C}(T_i)$.

2) Alice generates the following proof Π, sends the auxiliary commitments and Π to the SP.

$$\Pi = SPK \left\{ \begin{array}{c} (\{t_i, \{S_{j,i}\}_{j=1}^m, b_{G,i}, b_{B,i}, N_{G,i}, N_{B,i}, \theta_i, \{r_{l,i}\}_{l=1}^6\}_{i=1}^n) : \\ \left(\begin{array}{c} \texttt{cl}_{\texttt{verif}}(\sigma_i, t_i, \{S_{j,i}\}_{j=1}^m, b_{G,i}, b_{B,i}, \theta_i) = 1 \wedge \\ C_{S,i} = h^{r_{1,i}} \prod_{j=1}^m g_j^{S_{j,i}} \wedge C_{GB,i} = g_1^{b_{G,i}} g_2^{b_{B,i}} h^{r_{2,i}} \wedge \\ \prod_{k=0}^i C_{GB,k} = g_1^{N_{G,i}} g_2^{N_{B,i}} h^{r_{3,i}} \wedge \\ \texttt{cl}_{\texttt{verif}}(\sigma_{d_i}, N_{G,i}, N_{B,i}, b_{G,i}, b_{B,i}, \delta_i) = 1 \wedge \\ C_{WS,i} = C_{S,i}^{\delta_i} \cdot h^{r_{4,i}} \wedge C_{T,i-1} = C_{T,i}^{t_i} h^{r_{5,i}} \wedge \\ C_{\Theta,i} = g_1^{\theta_i} h^{r_{6,i}} \wedge \theta_i > \theta_{i-1} \end{array} \right)^n \right\}_{i=1}^n \quad (M)$$

The above ZKPoK proves that for each t_i,

1. The values $N_{G/B,i}$ which are committed in $\prod_{k=0}^i C_{GB,k}$ equal $N_{G/B,0} + \sum_{k=1}^i b_{G/B,k}$, i.e., $N_{G/B,i}$ is the number of the good/bad behaviors Alice has done up to t_i. This is due to the homomorhpic property of commitments.

2. The values committed in $C_{S,i}$ are the scores of t_i and the values committed in $C_{WS,i}$ are the scores of t_i multiplying with correct adjusting factor δ_i. The correctness of δ_i is explained below.
3. The value T_i which is committed in $C_{T,i}$ equals to T_{i-1}/t_i. The serial number θ_i is bigger than θ_{i-1}.

Recall that each session would be tied to one particular action such as a page edit or a video uploading, thus a session would not be both good and bad, e.g., good in category content but bad in category comment. Instead of using the OR proof [17] to prove the session is a good/bad/zero-score behavior and δ_i is the correct adjusting factor, we use the following trick to achieve better efficiency.

The SP generates CL signature σ_{d_i} to bind the adjusting factor δ_i with $N_{G,i}$, $N_{B,i}$, $b_{G,i}$, $b_{B,i}$ together. If $(b_{G,i}, b_{B,i}) = (1,0)/(0,1)$, then δ_i equal the the adjusting factor of the $N_{G,i}$th good behavior/$N_{B,i}$th bad behavior. If $b_{G,i} = b_{B,i} = 0$ then $\delta_i = 0$. The SP has to generate $3 \cdot M_G \cdot M_B$ CL signatures, where $M_{G/B}$ is the maximum number of one user's good/bad behaviors. We assume $M_G = 2^9$ and the number of allowed misbehaviors $M_B = 2^5$, thus the SP has to generate $3 \cdot 2^{14}$ signatures which is acceptable.

3) The verifier verifies the proof Π and calculates $C_R = \prod_{i=0}^n C_{WS,i}$, $C_{GB} = \prod_{i=0}^n C_{GB,i}$, due to the homomorhpic property of commitments, the verifier is convinced that the values committed in C_R are the weighted reputations and the values committed in C_{GB} is the numbers of good/bad behaviors. Let $C_T = C_{T,n}$, $C_\Theta = C_{\Theta,n}$, the verifier is convinced that the value committed in C_T is T_0/T_{rdm}, the value committed in C_Θ is θ_n.

$\mathbb{P}_{\text{pol}}(\text{Pol}, C_R)$. This protocol is used to prove the reputations $\{R_j\}_{j=1}^m$ committed in $C_R = \mathbb{C}(R_1, \cdots, R_m) = h^r \prod_{j=1}^m g_j^{R_j}$ satisfy the authentication policy Pol. The authentication policy Pol is of the form $\vee_{k=1}^J (\wedge_{j=1}^m P_{kj})$, i.e., a combination of conjunctive clauses. Each clause P_{kj} is of the form $(\geqslant \eta_{kj})$ which means the user's reputation in category j should be equal or higher than η_{kj}. P_{kj} can also be $\perp$ which means the kth subpolicy $\wedge_{j=1}^m P_{kj}$ does not involve the category j. The prover generates $PK\{(\{R_j\}_{j=1}^m, r) : \vee_{k=1}^J (C_R = h^r \prod_{j=1}^m g_j^{R_j} \wedge_{j=1}^m (R_j \geqslant \eta_{kj}))\}$, and sends it to the verifier who verifies the proof.

4.3   Construction

Setup. The parameters $param = (l_t, K, l_r, l_m, l_n, l, l_s, l_e, l_\phi, l_H)$ where l_t (56) is the ticket length and K is the maximum number of a user's tickets that are used but not redeemed, typically $K = 80$. l_r is the length of the random nonce used to prove the freshness of the credential, $l_r = 224$ can guarantee probability of collision is about 2^{-112} due to the birthday paradox. l_m (224) is length of message blocks in CL signature (we split a $K \cdot l_t$-bit number into $K/4$ l_m-bit messages to allow for smaller e). l_n (2048) is the size of the RSA modulus, l (160) is a security parameter of CL signature, l_s and l_e are sizes of the components s and e in the CL signature. For its security we require $l_s > l_n + l_m + l$, $l_e > l_m + 2$. l_ϕ (80) is the security parameter controlling the statistical zero-knowledge property, l_H (160) is the output length of the hash function used for the Fiat-Shamir heuristic.

Given *param*, the SP generates a l_n-bit safe-prime product $N = pq$ as a RSA modulus, where p and q are random safe primes. The secret key is $\phi(N)$. Let m be the number of categories, the SP chooses $\{g_i\}_{i=1}^{K/4+m+3}, h, g_{acc} \in_R \mathbb{QR}_N$, where g_{acc} is the exponentiation base used in accumulator. The SP chooses a l_t-bit prime $\hat{t}$ which is used to fill a user's ticket set during registration. Let J be the number of subpolicies, the SP chooses a policy Pol of the form $\vee_{k=1}^{J}(\wedge_{j=1}^{m} P_{kj})$.

The SP generates the set of adjusting factors $\mathfrak{D} = \mathfrak{D}_G \cup \mathfrak{D}_B$. $\mathfrak{D}_{G/B} = \{i, \delta_{G/B,i}\}_{i=1}^{M_{G/B}}$, $\delta_{G/B,i}$ is the adjusting factor for ith good/bad behavior, $M_G = 2^9$ is the maximum number of a user's good behaviors, $M_B = 2^5$ is the allowed maximum number of a user's bad behaviors. The SP also publishes $3 \cdot M_G \cdot M_B$ CL signatures on tuples $(N_G, N_B, b_G, b_B, b_G \cdot \delta_{G,N_G} + b_B \cdot \delta_{B,N_B})$, where $(b_G, b_B) = (1,0)/(0,1)/(0,0)$ corresponds to good/bad/zero-score behavior.

SP initials the reputation list $\mathtt{RL} = \varnothing$ and the misbehavior list $\mathtt{BL} = \varnothing$, the corresponding accumulator value is $\mathtt{V} = g_{acc}$. The SP publishes $(N, \{g_i\}_{i=1}^{K/4+m+3}, h, g_{acc}, \hat{t}, \mathtt{Pol}, \mathfrak{D}, \mathtt{RL}, \mathtt{BL})$ as the public parameter, the private key is $\phi(N)$.

Registration. 1. A user Alice generates a ticket pool P_{tk} which contains $\hat{t}$ and $\tilde{n}$ random l_t-bit prime tickets, typically $\tilde{n} = 3000$. Let $T_P = \prod_{t \in P_{tk}} t$. Alice then generates a l_r-bit random nonce q. let $Q = (\{T_i\}_{i=0}^{K/4-1}, q, \{R_j = 0\}_{j=1}^{m}, N_G = 0, N_B = 0, \theta = 0)$, where $\sum_{i=0}^{K/4-1} 2^{i \cdot l_m} T_i = \hat{t}$, $\{R_j\}_{j=1}^{m}$ denote the initial reputations in m categories, N_G/N_B denote the initial numbers of good/bad behaviors and θ denotes the initial serial number. Alice commits Q as $C_Q = \mathbb{C}(Q)$ and sends C_Q together with $PK\{(q) : C_Q = \mathbb{C}(\{T_i\}_{i=0}^{K/4-1}, q, \{0\}_{m+3})\}$ to the SP.

2.If the ZPKoK is valid, the SP sends the current blacklist $\mathtt{BL}$ and the corresponding accumulator value $\mathtt{V}$ to Alice, and runs the CL signature generation protocol $\mathtt{cl_{sign}}(\mathbb{C}(Q))$ with Alice, thus Alice gets a CL signature σ on Q.

3. Alice computes $w_P = \mathtt{nmw_{gen}}(T_P, \mathtt{BL})$ and $w_U = \mathtt{nmw_{pool}}(\hat{t}, T_P, w_P, \mathtt{V})$. Alice stores $cred = (T_U = \hat{t}, T_P, \sigma, w_U, w_P, q, \{R_j\}_{j=1}^{m}, N_G, N_B, \theta, \mathtt{BL}, \mathtt{V})$.

Authentication. Alice is in possession of $cred = (T_U, T_P, \sigma, w_U, w_P, q, \{R_j\}_{j=1}^{m}, N_G, N_B, \theta, \mathtt{BL}, \mathtt{V})$. Alice authenticates with the SP anonymously as follows.

1. Alice gets the latest reputation list $\overline{\mathtt{RL}}$, the latest misbehavior list $\overline{\mathtt{BL}}$ and the corresponding accumulator value $\overline{\mathtt{V}}$ from the SP. Let $\triangle_{\mathtt{BL}} = \overline{\mathtt{BL}} \setminus \mathtt{BL}$, which means $\triangle_{\mathtt{BL}}$ is the set of newly added negative-score tickets.

2. Let $\{t_i\}_{i=1}^{n} \subset \overline{\mathtt{RL}}$ are the tickets to be redeemed, for each t_i, Alice calculates the weighted scores $\{WS_{j,i} = \delta_i \cdot S_{j,i}\}_{j=1}^{m}$ base on $\mathfrak{D}$, where δ_i is the adjusting factor for t_i and $\{S_{j,i}\}_{j=1}^{m}$ are t_i's scores. Alice computes $(\{\overline{R}_j\}_{j=1}^{m}, \overline{T}_U, \overline{N}_G, \overline{N}_B, \overline{\theta})$. $\{\overline{R}_j = R_j + \sum_{i=1}^{n} WS_{j,i}\}_{j=1}^{m}$ are her latest reputations. $\overline{T}_U = T_U/T_{rdm}$ where $T_{rdm} = \prod_{i=1}^{n} t_i$. $\overline{N}_G/\overline{N}_B$ is the latest number of her good/bad behaviors. $\overline{\theta}$ is the largest serial number of $\{t_i\}_{i=1}^{n}$.

   She checks if $\{\overline{R}_j\}_{j=1}^{m}$ satisfies Pol, if not, aborts. Otherwise, she calculates $\overline{w}_U$ which is the NMW of $\overline{T}_U$ against $\overline{\mathtt{BL}}$ as follows. First she computes NMW of $\overline{T}_U$ against BL: $\overline{w}'_U = \mathtt{nmw_{pool}}(\overline{T}_U, T_U, w_U, \mathtt{V})$, then updates $\overline{w}'_U$ to $\overline{w}_U = \mathtt{nmw_{update}}(\overline{T}_U, \overline{w}'_U, \triangle_{\mathtt{BL}}, \mathtt{V})$.

3. Alice generates the following commitments: $C_{\overline{R}} = \mathbb{C}(\overline{R}_1, \ldots, \overline{R}_m)$, $C_{\overline{T}} = \mathbb{C}(\overline{T}_U)$, $C_{\overline{GB}} = \mathbb{C}(\overline{N}_G, \overline{N}_B)$, $C_{\Theta} = \mathbb{C}(\overline{\theta})$. Alice sends the nonce q and these commitments to the SP. She then generates a proof $\prod$ to convince the SP that these commitments are correctly formed and her latest reputations satisfy the policy Pol as follows:

   - Execute protocol $\mathbb{P}_{\mathtt{cred}}$ to convince the SP that Alice has a CL signature on $(T_U(\mathrm{split}), q, \{R_i\}_{i=1}^m, N_G, N_B, \theta)$ and these values are committed in $C_T = \mathbb{C}(T_U)$, $C_R = \mathbb{C}(\{R_i\}_{i=1}^m)$, $C_{GB} = \mathbb{C}(N_G, N_B)$, $C_{\Theta} = \mathbb{C}(\theta)$.
   - Execute protocol $\mathbb{P}_{\mathtt{rdm}}(\overline{\mathtt{RL}}, \mathfrak{D}, C_T, C_R, C_{GB}, C_{\Theta}, C_{\overline{T}}, C_{\overline{R}}, C_{\overline{GB}}, C_{\overline{\Theta}})$ to redeem tickets $\{t_i\}_{i=1}^n$ and convince the SP that $(C_{\overline{T}}, C_{\overline{R}}, C_{\overline{GB}}, C_{\overline{\Theta}})$ are correctly formed.
   - Execute protocol $PK\{(\overline{T}_U, \overline{w}_U) : C_{\overline{T}} = \mathbb{C}(\overline{T}_U) \wedge 1 = \mathtt{nmw}_{\mathtt{verif}}(\overline{T}_U, \overline{w}_U, \overline{V})\}$ to convince the SP that all the unredeemed tickets are not in $\overline{\mathtt{BL}}$.
   - Execute protocol $\mathbb{P}_{\mathtt{PoL}}(C_{\overline{R}}, \mathtt{Pol})$ to convince the SP that the reputations committed in $C_{\overline{R}}$ satisfy the policy Pol.

4. The SP proceeds only if q is fresh and $\prod$ is valid. Alice sends the SP a ticket $t \in P_{tk}$ as the session identifier. If t has already been used, then Alice *deletes* t from P_{tk}, calculates $T_P \leftarrow T_P/t$, and sends another ticket in P_{tk} to the SP.

5. Alice generates a new nonce q' and commits it in $C_{q'} = \mathbb{C}(q')$. She splits $t \cdot \overline{T}_U$ into n l_m-bit parts $\{T_i\}_{i=0}^{n-1}$, $n = K/4$. Alice commits $\{T_i\}_{i=0}^{n-1}$ into $C_{split} = \mathbb{C}(\{T_i\}_{i=0}^{n-1})$ and executes $PK\{(\{T_i\}_{i=0}^{n-1}, t \cdot \overline{T}_U) : t \cdot \overline{T}_U = \sum_{i=0}^{n-1} 2^{i \cdot l_m} T_i\}$ with SP. Then Alice executes the protocol $\mathtt{sign}_{\mathtt{CL}}(C_{split}, C_{q'}, C_{\overline{R}}, C_{\overline{GB}}, C_{\overline{\Theta}})$ with the SP to get a CL signature $\overline{\sigma}$ on $(\{T_i\}_{i=0}^{n-1}, q', \overline{R}_1, \cdots, \overline{R}_m, \overline{N}_G, \overline{N}_B, \overline{\theta})$.

6. After a successful authentication, Alice *deletes* the tickets that are redeemed from the ticket pool: $\overline{T}_P = T_P/T_{rdm}$, generates the NMW $\overline{w}_P$ of $\overline{T}_p$ against $\overline{\mathtt{BL}}$ as follows: $\overline{w}'_P = \mathtt{nmw}_{\mathtt{pool}}(\overline{T}_P, T_P, w_P, \mathtt{V})$, $\overline{w}_P = \mathtt{nmw}_{\mathtt{update}}(\overline{T}_P, \overline{w}'_P, \Delta_{\mathtt{BL}}, \mathtt{V})$. Alice then calculates $\overline{w}_U = \mathtt{nmw}_{\mathtt{pool}}(t \cdot \overline{T}_U, \overline{T}_P, \overline{w}_P, \overline{V})$. Alice stores the credential $cred = (t \cdot \overline{T}_U, \overline{T}_P, \overline{\sigma}, \overline{w}_U, \overline{w}_P, q', \{\overline{R}_i\}_{i=1}^m, \overline{N}_G, \overline{N}_B, \overline{\theta}, \overline{\mathtt{BL}}, \overline{V})$.

Re-Login. Basically, re-login is the same as authentication except that Alice does not redeem any tickets and does not need to prove that her reputation satisfies the policy. Alice first executes protocol $\mathbb{P}_{\mathtt{cred}}$ to assure the SP that Alice has a CL signature on (split) T_U and uses the NMW to prove no ticket in T_U is in the misbehavior list, thus her reputation would not decrease. After a successful re-login, Alice gets a new CL signature from the SP as in the auth protocol.

List Management. For each reviewed session t, the SP assigns scores $\{S_j\}_{j=1}^m$ to t under m categories and generates two bits $(b_G, b_B) = (1, 0)/(0, 1)/(0, 0)$ if t is good/bad/zero-score behavior. Let n be the number of entries on the current reputation list RL, let $\theta = n + 1$. The SP generates a CL signature σ on $(t, \{S_j\}_{j=1}^m, b_G, b_B, \theta)$ and adds $(t, \{S_j\}_{j=1}^m, b_G, b_B, \theta, \sigma)$ into the RL, updates $n \leftarrow n+1$. If $b_B = 1$, the SP also adds t into the misbehavior list BL, and updates the corresponding accumulator value $V \leftarrow V^t$.

5   Performance Evaluation

We evaluate the performance of ARBRA both analytically and quantitatively, and compare it to PEREA, BLACR and PERM as they are the closest schemes that support reputation-based revocation. Our bechmark shows that ARBRA has the best performance at the SP side among these schemes. At the user side, ARBRA is also efficient even if the misbehavior list contains one million entries.

5.1   Data Transfer

The communication costs (downlink and uplink) for ARBRA are given in Figure 2(a). In the following analysis, we assume the number of categories $m = 5$, the number of sub-policies $J = 5$, each score is 5 bits, the reputation of the user in each category is 10 bits. Suppose there are 20000 anonymous authentications per day (it corresponds to 1/5 edits on wikipedia are anonymous which is assumed by Au et al [13]), for ARBRA the downloaded data per day is about 11MB which is acceptable in nowadays. In average, the number of redeemed tickets in each login is one, as each login only adds one ticket. For $K = 80$, the average size of the uploaded proof generated by users in each login is less than 87KB.

5.2   Computation

We now analyze the computational performance of ARBRA. Figure 2(a) outlines the major operations for ARBRA. EN1 denotes an exponentiation of a random base with a l_t-bit exponent modulos N, and EN2 denotes an exponentiation of a fixed base with l_n-bit exponent modulo N using precomputation. EN1 is used for updating the NMW. ZKPoK involving a l_x-bit x at the user needs an exponentiation of $(l_x + l_\phi + l_H) = (l_x + 240)$ bits $= \frac{l_x+240}{2048}$EN2. Note that the SP can calculate $l_x + 240 \mod \phi(N)$ if $l_x + 240 > \phi(N)$.

For ARBRA, recall that the average number of redeemed tickets in each login is one. We assume 80% logins are made using the re-login protocol (assume there are 10% non-zero tickets and a user will use re-login protocol until she has non-zero tickets to redeem). Thus the average computation complexity of one login at the SP is $(47 + \frac{232}{2048}K + \frac{990}{2048}m + \frac{6}{5}Jm)$EN2.

The benchmark is obtained on a Lenovo T4990d pc, equipped with a Intel i7-3770M CPU and 8GB RAM. Timing of EN1 and EN2 are obtained using the MIRACL library running in Windows 8. For pairing-based operations used in BLACR and PERM, the benchmark is obtained using Pairing-Based Cryptography (PBC) library, type D pairing with $|p| = 224$, running in Ubuntu 12.04 on the same machine. Note that the computations are highly parallelizable at the SP such as verifications of commitments and signatures. Thus the speed increases when more cores are involved. We assume that SPs are equipped with 8-core servers in our performance analysis. For users, we still consider that only one core is involved.

Downlink	Uplink (in average)
$(4796 + 5m)\Delta_{RL}$ bits	$133148+232K+745m+$ $21501Jm$ bits
Authentication at SP	Authentication at User
$[17 + \frac{232}{2048}K + \frac{500}{2048}m +$ $(30 + \frac{490}{2048}m)\delta_{rdm} +$ $6Jm]$EN2	$(\Delta_{BL} + 2K)$EN1$+[17+$ $\frac{680}{2048}K + \frac{500}{2048}m + (36 +$ $\frac{56}{2048}K + \frac{490}{2048}m)\delta_{rdm} +$ $6Jm]$EN2
Re-login at SP	Re-login at User
$(17 + \frac{232}{2048}K + \frac{500}{2048}m)$ EN2	$(\Delta_{BL}+2K)$EN1$+(17+$ $\frac{680}{2048}K + \frac{500}{2048}m)$EN2

(a) Complexity analysis of ARBRA. $\Delta_{RL/BL}$ is the number of entries in the newly-added reputation/misbehavior list, δ_{rdm} is the number of redeemed tickets, J/m is the number of sub-policies/categories

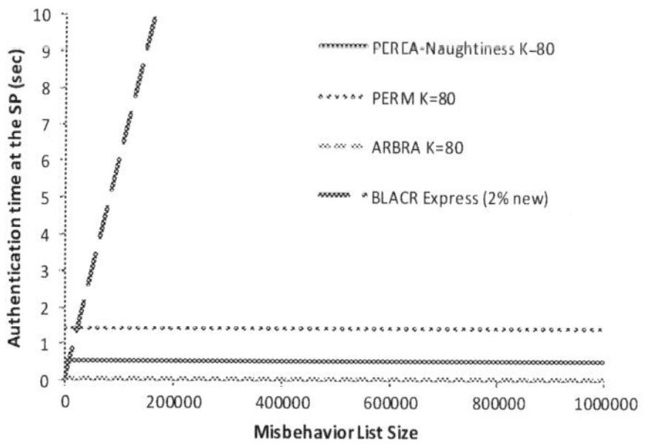

(b) Authentication time at the SP

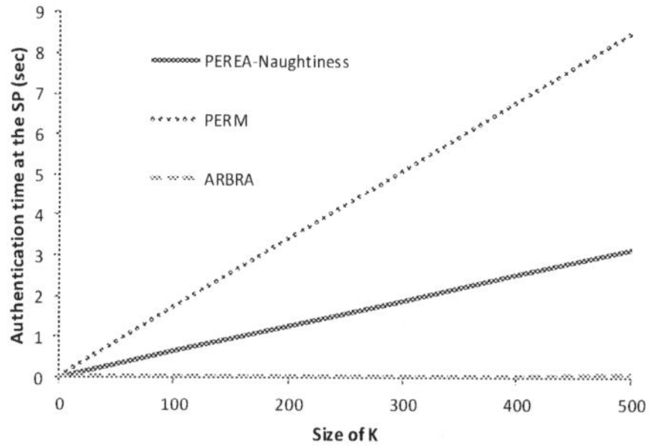

(c) Authentication time at the SP as K increases

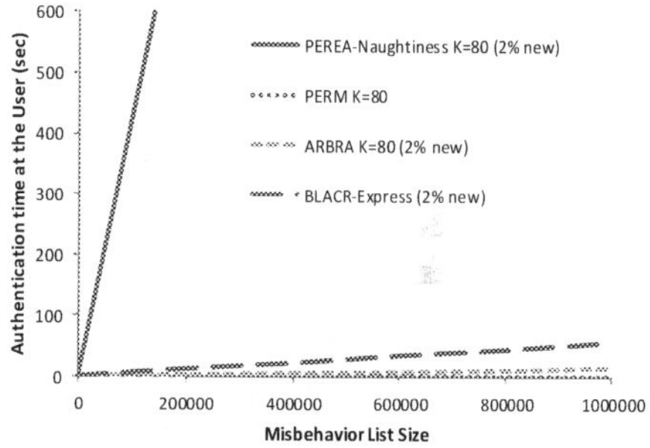

(d) Authentication time at the User

Fig. 2. Complexity analysis of ARBRA and the benchmark for authentication

Performance at the SP. Figure 2(b) shows the performance of authentication of PEREA, BLACR, PERM and ARBRA at the SP. On the SP side, the authentication times of PEREA, PERM and ARBRA are independent of the length of misbehavior list. When $K = 80$, for PEREA the SP can supports 118 logins/minute and for PERM the SP can support 43 logins/minute. For ARBRA, the SP can support 1653 logins/minute which is much better than PEREA and PERM. When the misbehavior list contains one million entries, for the BLACR-Express (with 2% new entries), the SP can only support 1 login/minute.

Figure 2(c) shows the authentication times of PEREA, PERM and ARBRA at the SP as K increases. The authentication time of ARBRA increases much slower than that of PEREA and PERM. When $K = 500$, for ARBRA the SP can support 1071 logins/minute while for PEREA the SP can support 19 logins/minute. For PERM, the SP can only support 7 logins/minute. Note that for PEREA and PERM, a user can only authenticate K times in a revocation window. If a heavy-loaded SP wants to have longer time to review sessions, then K has to be increased to avoid usability constraint.

Performance at the User. Figure 2(d) shows the performance of authentication of these schemes at the user. PEREA's main drawback is that a user has to update K NMWs before authentication thus is slow at the user. For ARBRA, a user only has to update one NMW of T_U and the length of prime ticket is only 1/4 of the length in PEREA, thus the efficiency is significantly improved. When the misbehavior list contains one million entries and assume there are 2% new entries since the last authentication, PEREA ($K = 80$) takes over 70 minutes which is unacceptable. ARBRA (K=80) only needs 13.6 seconds. BLACR-Express needs 56 seconds and PERM (K=80) with precomputation takes about 1 second. ARBRA is slower than PERM at the user but is still efficient enough to be acceptable, moreover if a user can updates the NMW frequently (e.g. once a day) and has 0.5% new entries when authenticating, it only takes 3.8 seconds at the user.

Note that the bottleneck is *on the SP side*, ARBRA has the best performance at the SP among existing schemes. At the user, ARBRA is also efficient enough to be acceptable even if the misbehavior list contains one million entries.

6 Discussion

Efficient Authentication. In PEREA, PERM and ARBRA, a user maintains a buffer which contains some tickets. In PEREA and PERM this buffer is a ticket queue while in ARBRA is a ticket set. As a ticket queue has the "first in first out" (FIFO) property, which means a ticket used first must be removed first, the SP has to review all tickets serially. By using the ticket set, ARBRA allows the SP to have more time in judging disputed cases as there is no FIFO property. More importantly, ARBRA uses *one* NMW of the product value of ticket set to prove all the unredeemed tickets are not in the misbehavior list, thus guarantees the user has redeemed all the negative-score tickets. PEREA and PERM have to prove each ticket in the ticket queue has the right score one by one. Thus ARBRA achieve better efficiency at the SP than PEREA and PERM.

Timing Attack. In ARBRA, the time needed to update the NMW is linear with the size of newly added misbehavior list. Thus if an SP is able to know the time of updating NMW, he may be able to deduce when the user authenticated last time. There are several approaches to prevent this timing attack. First every user should run a daemon process that automatically updates NMW. Thus the time of updating NMW only relates to when the *NMW is updated last time*. Second, all users are required to download the latest misbehavior list anonymously and update NMWs *before* trying to authenticate. By separating NMW updating and authentication, the relation between the time of NMW updating and the authentication can be hidden, given there are many list downloading and authentications in a short time period.

7 Conclusion

Anonymous revocation schemes without TTP are desirable as they protect users' privacy. Recent works such as BLACR and PERM present *reputation-based revocation*. Though being the state of the art of anonymous revocation, the efficiency of these schemes at the SP is not satisfying. We present ARBRA, an anonymous reputation-based revocation scheme which is much more efficient than existing schemes on the SP side. ARBRA allows the SPs to ramp up the punishment of multiple misbehaviors from the same user, thus disincentivize repeated misbehaviors. On the user side, ARBRA is also efficient enough to be acceptable even if the misbehavior list contains one million entries.

References

1. Boneh, D., Boyen, X., Shacham, H.: Short group signatures. In: Franklin, M. (ed.) CRYPTO 2004. LNCS, vol. 3152, pp. 41–55. Springer, Heidelberg (2004)
2. Boneh, D., Shacham, H.: Group signatures with verifier-local revocation. In: Proceedings of ACM CCS 2004, pp. 168–177. ACM (2004)
3. Camenisch, J., Lysyanskaya, A.: Dynamic accumulators and application to efficient revocation of anonymous credentials. In: Yung, M. (ed.) CRYPTO 2002. LNCS, vol. 2442, pp. 61–76. Springer, Heidelberg (2002)
4. Li, J., Li, N., Xue, R.: Universal accumulators with efficient nonmembership proofs. In: Katz, J., Yung, M. (eds.) ACNS 2007. LNCS, vol. 4521, pp. 253–269. Springer, Heidelberg (2007)
5. Nguyen, L.: Accumulators from bilinear pairings and applications. In: Menezes, A. (ed.) CT-RSA 2005. LNCS, vol. 3376, pp. 275–292. Springer, Heidelberg (2005)
6. Tsang, P.P., Kapadia, A., Cornelius, C., Smith, S.W.: Nymble: Blocking misbehaving users in anonymizing networks. IEEE Transactions on Dependable and Secure Computing 8(2), 256–269 (2011)
7. Henry, R., Henry, K., Goldberg, I.: Making a nymbler nymble using VERBS. In: Atallah, M.J., Hopper, N.J. (eds.) PETS 2010. LNCS, vol. 6205, pp. 111–129. Springer, Heidelberg (2010)
8. Tsang, P.P., Au, M.H., Kapadia, A., Smith, S.W.: Blac: Revoking repeatedly misbehaving anonymous users without relying on ttps. ACM Transactions on Information and System Security (TISSEC) 13(4), 39 (2010)
9. Brickell, E., Li, J.: Enhanced privacy id: A direct anonymous attestation scheme with enhanced revocation capabilities. In: Proceedings of WPES 2007, pp. 21–30. ACM (2007)
10. Au, M.H., Tsang, P.P., Kapadia, A.: Perea: Practical ttp-free revocation of repeatedly misbehaving anonymous users. ACM Transactions on Information and System Security (TISSEC) 14(4), 29 (2011)
11. Lofgren, P., Hopper, N.: Faust: efficient, ttp-free abuse prevention by anonymous whitelisting. In: Proceedings of WPES 2011, pp. 125–130. ACM (2011)
12. Au, M.H., Kapadia, A., Susilo, W.: Blacr: Ttp-free blacklistable anonymous credentials with reputation. In: Proceedings of NDSS 2012, San Diego, CA, USA (2012)
13. Au, M.H., Kapadia, A.: Perm: Practical reputation-based blacklisting without ttps. In: Proceedings of ACM CCS 2012, pp. 929–940. ACM (2012)
14. Yu, K.Y., Yuen, T.H., Chow, S.S.M., Yiu, S.M., Hui, L.C.K.: PE(AR)^2: Privacy-enhanced anonymous authentication with reputation and revocation. In: Foresti, S., Yung, M., Martinelli, F. (eds.) ESORICS 2012. LNCS, vol. 7459, pp. 679–696. Springer, Heidelberg (2012)
15. Henry, R., Goldberg, I.: Thinking inside the blac box: Smarter protocols for faster anonymous blacklisting. In: Proceedings of WPES 2013, pp. 71–82. ACM (2013)

16. Dusart, P.: The k th prime is greater than k (ln k+ ln ln k-1) for k 2. Mathematics of Computation, 411–415 (1999)
17. Damgard, I.: On σ-protocols. Lecture notes for CPT (2002)
18. Camenisch, J., Stadler, M.: Efficient group signature schemes for large groups. In: Kaliski Jr., B.S. (ed.) CRYPTO 1997. LNCS, vol. 1294, pp. 410–424. Springer, Heidelberg (1997)
19. Fujisaki, E., Okamoto, T.: Statistical zero knowledge protocols to prove modular polynomial relations. In: Kaliski Jr., B.S. (ed.) CRYPTO 1997. LNCS, vol. 1294, pp. 16–30. Springer, Heidelberg (1997)
20. Damgård, I.B., Fujisaki, E.: A statistically-hiding integer commitment scheme based on groups with hidden order. In: Zheng, Y. (ed.) ASIACRYPT 2002. LNCS, vol. 2501, pp. 125–142. Springer, Heidelberg (2002)
21. Camenisch, J., Lysyanskaya, A.: A signature scheme with efficient protocols. In: Cimato, S., Galdi, C., Persiano, G. (eds.) SCN 2002. LNCS, vol. 2576, pp. 268–289. Springer, Heidelberg (2003)
22. Camenisch, J., Chaabouni, R., Shelat, A.: Efficient protocols for set membership and range proofs. In: Pieprzyk, J. (ed.) ASIACRYPT 2008. LNCS, vol. 5350, pp. 234–252. Springer, Heidelberg (2008)
23. Peng, K., Bao, F.: Vulnerability of a non-membership proof scheme. In: SECRYPT 2010, pp. 1–4. IEEE (2010)

A   Dynamically Universal Accumulator

ARBRA makes use of dynamically universal accumulator (DUA). DUA has a vulnerability as described in [23], fortunately this vulnerability is easy to fix without reducing much efficiency as shown in [14]. We summarize the correct DUA as follows and we mainly focus on the non-membership witness (NMW).

DUA Setup: The setup algorithm takes the security parameter 1^k as input, chooses a k-bit safe-prime product N and $g \in_R QR_N$. It outputs the accumulator public key $PK_{acc} = (N, g)$, the private key $SK_{acc} = \phi(N)$, and sets up the initial accumulator $acc = \varnothing$, the initial ticket product $U = 1$, the initial accumulator value $V = g$.

DUA Accumulating Tickets: The accumulating algorithm $\tilde{V} = \mathtt{accu}(V, S_t)$ is used to accumulate tickets $S_t = \{t_i\}_{i=1}^n$ into an accumulator. It takes the original accumulator value V and S_t as input, outputs the updated accumulator value $\tilde{V} = V^{\prod_{t \in S_t} t}$.

DUA NMW Generation without Secret Key: The NMW generation algorithm $w \leftarrow \mathtt{nmw_{gen}}(T, acc)$ is used to generate a NMW w for a ticket T against acc using public key. This algorithm proceeds as follows:

1. Compute the ticket product $U = \prod_{t \in acc} t$.
2. Using Euclidean algorithm to find a and b, such that $aU + bT = 1, 0 < a < T$.
3. The NMW w is $(a, g^{-b} \bmod N)$.

DUA NMW Update: The NMW update algorithm $\hat{w} \leftarrow \mathtt{nmw_{update}}(T, w, S_t, V)$ is used to update the NMW $w = (a, d)$ for T. The newly accumulated tickets are S_t, $gcd(T, \prod_{t \in S_t} t) = 1$. This algorithm proceeds as follows:

1. Compute the product of the new added accumulated tickets $T_{add} = \prod_{t \in S_t} t$.
2. Using Euclidean algorithm to find a_0 and r_0, such that $a_0 T_{add} + r_0 T = 1, 0 < a_0 < T$. Compute $\hat{a} = a_0 a \bmod T$.

3. Calculate r such that $\hat{a}T_{add} = a + rT$, the updated NMW $\hat{w} = (\hat{a}, d\mathsf{V}^r \bmod N)$.

DUA NMW Proof-of-Knowledge: Given an element $h \in QR_n$ such that $log_g h$ is unknown to the prover, a number T, a random value r and the NMW (a, d), the prover can use the ZKPoK $PK\{(T, r, a, d) : C_1 = g^T h^r \wedge \mathsf{V}^a = d^T g\}$ to prove that $gcd(T, U) = 1$.

Notice the above ZPKoK actually proves that $gcd(T, U) = 1$, thus we can use the NMW of the product value $T = \prod_{t_i \in S_t} t_i$ to prove that all t_i in S_t are not in the accumulator if t_i are all prime numbers. This technique is proposed in PE(AR)^2 and adopted in ARBRA.

B   Security Analysis

We use an ideal-world/real-world model to prove security of ARBRA. In the real world, the players run cryptographic protocol with each other while in the ideal world the players send all their inputs to and receive all their outputs from an ideal trusted party T. T achieves the functionality that ARBRA is supposed to realize. Note that communications with T are not anonymous. In both worlds, dishonest user are controlled by an adversary A. There is an environment $\mathcal{E}$ that provides the players with inputs and arbitrarily interacts with A. ARBRA is secure if for every A and $\mathcal{E}$, there exists a simulator S controlling the same parties in the ideal world as A does in the real world such that $\mathcal{E}$ can not distinguish whether it is interacting with A in the real world or interacting with S in the ideal world. We consider a static model in which the number of honest and dishonest players are fixed before the system setup. ARBRA supports the following functionalities:

1. *Setup* The systems starts when $\mathcal{E}$ fixes the number of honest and dishonest users and SPs and the authentication policy Pol.
   - Real World. The SP generates a key pair (PK, SK) and the authentication policy Pol. PK and Pol are available to all players.
   - Ideal World. The trusted party T initializes a set U, which is used to store the registration and authentication information of all users.
2. *Registration* $\mathcal{E}$ instructs user i to register with the SP.
   - Real World. User i sends a request for registration to the SP. The user and the SP output individually the outcome of this event to $\mathcal{E}$. If SP is honest, he will reject a request from a user that has already registered.
   - Ideal World. User i sends a request to T who will inform the SP that user i wants to register and whether this user has registered before. The SP returns accept/reject to T who return it to the user. If it is user i's first successful registration, T stores the registration status of user $(i, T_U = \phi)$ in U. The user and SP output individually the outcome to $\mathcal{E}$.
3. *Authentication* $\mathcal{E}$ instructs user i to authenticate with the SP.
   - Real World. User i executes the auth protocol with the SP. The user and the SP output individually the outcome and the session identifier to $\mathcal{E}$.
   - Ideal World. User i sends a message Auth to T, who informs SP an anonymous user wants to authenticate. The SP responds to T with the reputation list RL, blacklist BL. T forwards RL, BL and whether user i satisfies the policy to user i. The user i chooses whether to proceed or not. If user i proceeds, i sends a session identifier t and the tickets T_r he wants to redeem this time to

T. T checks whether T_r belongs to i using U and whether user i's reputation satisfies Pol after tickets T_r are redeemed, then informs the SP whether this anonymous user satisfies Pol and the number of tickets he wants to redeem this time. The SP responds with accept or reject to T, who relays the answer to the user. If the response is accept, T updates $(i, T_U) \leftarrow (i, (T_U \cup t)/T_r)$ if (i, T_U) is in the set U. The user and the SP output individually the outcome of this event and the session identifier to $\mathcal{E}$.

4. *Re-login* $\mathcal{E}$ instructs user i to re-login with the SP.
   - Real World. User i executes the re-login protocol with the SP. The user and the SP output individually the outcome and the session identifier to $\mathcal{E}$.
   - Ideal World. User i sends a message Re-login to T, who informs SP an anonymous user wants to re-login. The SP responds with the reputation list RL, misbehavior list BL. T forwards RL, BL and whether user i can re-login to user i. The user i chooses whether to proceed or not. If user i proceeds, i sends a session identifier t to T. T informs the SP whether this anonymous user can re-login. The SP responds with accept or reject to T, who relays the answer to the user. If the response is accept, T updates $(i, T_U) \leftarrow (i, T_U \cup t)$ if (i, T_U) is in the set U. The user and the SP output individually the outcome of this event and the transaction identifier to $\mathcal{E}$.

5. *List Management.* $\mathcal{E}$ instructs the SP to assign scores $(s_1, ..., s_m)$ and two bits (b_M, b_B) to a session identifier t.
   - Real World. If t is not a transaction identifier that SP outputs accept or if $b_M = b_B = 1$ an honest SP aborts. Else SP adds $(s_1, ..., s_m, b_M, b_B, t)$ to the reputation list, if $b_B = 1$ SP also adds t to the misbehavior list. SP outputs the outcome of this event to $\mathcal{E}$.
   - Ideal World. SP forward the request to T, who checks that t is a session identifier that SP outputs accept. T also checks whether $b_M = b_B = 1$, then T informs SP whether the check is successful. If the check succeeds, the SP adds $(s_1, ..., s_m, b_M, b_B, t)$ to the reputation list, if $b_B = 1$ SP also adds t to the misbehavior list. SP outputs the outcome of this event to $\mathcal{E}$.

The ideal-world ARBRA provides all the security properties and functionalities we want. In the Authentication protocol, T only informs the SP that an anonymous user wants to authenticate and the number of the tickets she wants to redeem this time. In the re-login protocol, the SP only knows an anonymous user wants to re-login and nothing else. Thus anonymity is guaranteed. T checks whether the user satisfies policy in the auth and re-login protocol, thus authenticity and non-frameability is satisfied. T informs users whether they satisfy policy before authentication, thus revocation auditability is guaranteed.

Definition 1. *Let $\mathbf{Real}_{\mathcal{E},\mathcal{A}}(\lambda)$ (resp. $\mathbf{Ideal}_{\mathcal{E},\mathcal{S}}(\lambda)$) be the probability that $\mathcal{E}$ output 1 running in the real world (resp. ideal world) with adversary $\mathcal{A}$ (resp. $\mathcal{S}$ with access to $\mathcal{A}$). ARBRA is secure if for any PPT algorithm $\mathcal{E}$, $\mathcal{A}$, $|\mathbf{Real}_{\mathcal{E},\mathcal{A}}(\lambda) - \mathbf{Ideal}_{\mathcal{E},\mathcal{S}}(\lambda)| = \mathtt{negl}(\lambda)$*

Proof (sketch): We will briefly show how $\mathcal{S}$ can be constructed thus $\mathcal{E}$ can not distinguish interacting with S in the ideal world from interacting with A in the real world
Case 1: the SP is not corrupted.

   Setup: representing honest SP to $\mathcal{A}$. $\mathcal{S}$ generates the key pair (PK, SK) and the authentication policy Pol.

   Registration: representing a corrupt user i to T/ the honest SP to $\mathcal{A}$. $\mathcal{A}$ register with $\mathcal{S}$ (as the SP). $\mathcal{S}$ utilizes the zero-knowledge extractor to extract the nonce q and store (i, q) into a set U_q, q will be used to identify the user i.

Authentication: representing the corrupt user i to T/ the honest SP to $\mathcal{A}$. $\mathcal{A}$ runs the authenticate protocol with $\mathcal{S}$ (as the honest SP). Note that $\mathcal{A}$ may use any corrupted user's credential, thus $\mathcal{S}$ has to use U_q and the nonce q $\mathcal{A}$ presents in the protocol to locate the right corrupted user i to interact with T. At the end of authenticate $\mathcal{A}$ will get a new credential on a new nonce q', $\mathcal{S}$ utilizes the zero-knowledge extractor to extract the nonce q' and updates the ith entry in U_q to (i, q').

Re-login: representing the corrupt user i to T/ the honest SP to $\mathcal{A}$. Basically, $\mathcal{S}$ acts the same as in authentication.

Note that due to the soundness of PK and the unforgeability of CL signature, the probability that the location or the extractor fails is negligible.

Case 2: the SP is corrupted.

Setup: representing honest users to $\mathcal{A}$. $\mathcal{S}$ receives the public key PK and the authentication policy Pol from $\mathcal{A}$.

Registration: representing the corrupt SP to T/ the honest user to $\mathcal{A}$. T tell $\mathcal{S}$ that an anonymous user wants to register, then $\mathcal{S}$ utilizes the zero-knowledge simulator to simulate the proof of knowledge and gets the corresponding credential from $\mathcal{A}$.

Authentication: representing the corrupt SP to T/ the honest user to $\mathcal{A}$. T tell $\mathcal{S}$ that an anonymous user wants to authenticate and the number of the tickets he wants to redeem, then $\mathcal{S}$ utilizes the zero-knowledge simulator to simulate the proof of knowledge to authenticate with $\mathcal{A}$.

Re-login: representing the corrupt user i to T/ the honest SP to $\mathcal{A}$. T tell $\mathcal{S}$ that an anonymous user wants to re-login, then $\mathcal{S}$ utilizes the zero-knowledge simulator to simulate the proof of knowledge to authenticate with $\mathcal{A}$.

Attribute-Based Signatures
for Circuits from Multilinear Maps*

Fei Tang[1,2,3], Hongda Li[1,2], and Bei Liang[1,2,3]

[1] State Key Laboratory of Information Security, Institute of Information Engineering
of Chinese Academy of Sciences, Beijing, China
[2] Data Assurance and Communication Security Research Center of Chinese Academy
of Sciences, Beijing, China
[3] University of Chinese Academy Sciences, Beijing, China
tangfei127@163.com, {lihongda,liangbei}@iie.ac.cn

Abstract. In this paper, we construct an Attribute-Based Signature
(ABS) scheme for general circuits from multilinear maps. Our scheme
is inspired by Garg et al.'s Attribute-Based Encryption (ABE) scheme
at CRYPTO 2013. We prove selective unforgeability of our scheme in
the standard model under the Multilinear Computational Diffie-Hellman
(MCDH) assumption. The privacy security of our scheme is perfect.

Keywords: Attribute-based signatures, multilinear maps, general circuits.

1 Introduction

Attribute-Based Signatures (ABS) allow a user to sign a message with fine-
grained control over identifying information. Similar to the notion of attribute-
based encryption [27,11], we may define two variants for ABS: Key-Policy ABS
and Signature-Policy ABS. In a Key-Policy ABS system, a secret key sk_f (issued
by a trusty authority) is associated with a boolean (policy) function f chosen
from some class of admissible functions $\mathcal{F}$. The holder of sk_f then can sign a
message m on behalf of an attribute set x if and only if $f(x) = 1$. In a Signature-
Policy ABS system, the secret key is associated with an attribute set x and the
signature is associated with a policy function f. Key-policy and signature-policy
ABS systems are useful in different contexts.

Attribute-based signatures provide anonymous signing paradigm, that is, a
valid signature proves that the signer possesses an appropriate policy function
f (resp. attribute set x) without reveling any further information about the
signer. This requirement is very similar to signature variants like group signatures [6], ring signatures [25], functional signatures [2], and policy-based signatures [1]. The common theme of all these signature primitives is that they allow

---

* This research is supported by the National Natural Science Foundation of China
(Grant No. 60970139) and the Strategic Priority Program of Chinese Academy of
Sciences (Grant No. XDA06010702).

the signer fine-grained controlled over how much of her personal information is revealed by the signature. Attribute-based signatures have natural applications in many systems where users' capabilities depend on possibly complex policy function (resp. attribute set), such as private access control, anonymous credentials, trust negotiations, distributed access control mechanisms for ad hoc networks, attribute-based messaging, and so on. (See [21] for more descriptions about the applications of the attribute-based signatures.)

Many attempts have been made to realize the notion of ABS. Maji et al. [20] described the first ABS scheme which supports predicates having AND, OR, and threshold gates, but the security of their scheme was only proven in the generic group model. Maji et al. [21] then gave an ABS scheme by using NIZK. Scheme in [18] is proven secure under the standard computational Diffie-Hellman assumption, but it considers only (n, n)-threshold, where n is the number of attributes purported in the signature. Shahandashti et al. [26] then extended [18]'s scheme to that supports (k, n)-threshold. Li et al. [17] designed an efficient ABS scheme only for single level threshold predicates of the form k-of-n among a certain set of attributes. Kumar et al. [15] then designed an ABS scheme for bounded multi-level threshold circuits. In [23], Okamoto and Takashima designed a fully secure ABS scheme built upon dual pairing vector spaces [22] and used proof techniques from functional encryption [19]. In [13], Herranz et al. constructed a short ABS for threshold predicates by using Groth-Sahai proof system [12]. Okamoto et al. [24] and Kaafarani et al. [16] considered a variant of attribute-based signatures called decentralized ABS. Interestingly, there is no paper considers the direct realization of the key-policy ABS system yet.[1]

The notion of multilinear maps was first postulated by Boneh and Silverberg [3] where they discussed potential applications such as non-interactive n-party Diffie-Hellman key exchange. Unfortunately, they also showed that it might be difficult or not possible to find useful multilinear forms within the realm of algebraic geometry. However, Garg, Gentry, and Halevi [9] announced a surprising result. They used ideal lattices to design a candidate mechanism that would approximate or be the moral equivalent of multilinear maps. Then, Coron, Lepoint, and Tibouchi [7] described a different construction that works over the integers instead of ideal lattices. Some subsequent works showed that the multilinear maps are a powerful cryptographic primitive, some successful examples including programmable hash functions [8], full domain hash and identity-based aggregate signatures [14], private outsourcing of polynomial evaluation and matrix multiplication [28], broadcast encryption [5], attribute-based encryption [10], and constrained pseudorandom functions (PRF) [4].

More specifically, in [10], Garg, Gentry, Halevi, Sahai, and Waters designed an ABE scheme for general circuits from multilinear maps. Whereafter, Boneh and Waters [4] utilized their technique to design a constrained pseudorandom functions where the accepting set for a key can be described by a polynomial size

---

[1] We have notice that Bellare and Fuchsbauer [1] recently showed that how to transform a policy-based signature scheme to a key-policy ABS scheme.

circuit. Security of Garg et al.'s ABE and Boneh et al.'s constrained PRF both are based on the Multilinear Decisional Diffie-Hellman (MDDH) assumption.

In this paper, we also follow Garg et al.'s technique [10]. However, our goal is to design an attribute-based signature scheme. In our ABS system, the policy function f can be described by a polynomial size circuit. We prove selective unforgeability of our scheme in the standard model under the Multilinear Computational Diffie-Hellman (MCDH) assumption, and the privacy security of our scheme is perfect. We construct an ABS system of the key-policy variety where signature descriptors are an n-tuple x of boolean variables and keys are associated with boolean circuits (denoted by f) of a max depth ℓ.

2   Attribute-Based Signatures for Circuits

In this section we give the definitions and security models for attribute-based signature system for circuits. In the beginning, we give some notations of circuit. The following circuit notation is same as that in [10] (in subsections 2.2 and 2.4).

2.1   Circuit Notation

There is a folklore transformation that uses De Morgan's rule to transform any general boolean circuit into an equivalent monotone boolean circuit, with negation gates only allowed at the inputs. Therefore, we can focus our attention on monotone circuits. Note that inputs to the circuit correspond to boolean variables x_i, and we can simply introduce explicit separate attributes corresponding to $x_i = 0$ and $x_i = 1$. Because of this simple transformation, in this paper we only consider ABS for monotone circuits.

For our application we restrict our consideration to certain classes of boolean circuits. First, our circuits will have a single output gate. Next, we will consider layered circuits. In a layered circuit a gate at depth j will receive both of its inputs from wires at depth $j - 1$. Finally, we will restrict ourselves to monotonic circuits where gates are either AND or OR gates of two inputs.

Our circuit will be a five-tuple $f = (n, q, A, B, \mathtt{GateType})$. We let n be the number of inputs and q be the number of gates. We define inputs $= \{1, \ldots, n\}$ (shorthand as $[n]$), Wire $= \{1, \ldots, n + q\}$, and Gates $= \{n + 1, \ldots, n + q\}$. The wire $n + q$ is the designated output wire. $A :$ Gates $\rightarrow$ Wires/outputwire is a function where $A(w)$ identifies w's first incoming wire, and $B(w)$ means w's seconding incoming wire. Finally, $\mathtt{GateType} :$ Gates $\rightarrow \{\mathrm{AND}, \mathrm{OR}\}$ is a function that identifies a gate as either an AND or OR gate.

We require that $w > A(w) > B(w)$. We also define a function $\mathtt{depth}(w)$ where if $w \in$ inputs then $\mathtt{depth}(w) = 1$ and in general $\mathtt{depth}(w)$ of wire w is equal to the shortest path to an input wire plus 1. Since our circuit is layered we require that for all $w \in$ Gates that if $\mathtt{depth}(w) = j$ then $\mathtt{depth}(A(w)) = \mathtt{depth}(B(w)) = j-1$.

We will abuse notation and let $f(x)$ be the evaluation of the circuit f on input $x \in \{0, 1\}^n$. In addition, we let $f_w(x)$ be the value of wire w of the circuit on input x.

2.2   Definitions for ABS for Circuits

We now give a formal definition of our attribute-based signatures for circuits. Formally, a key-policy ABS scheme for circuits consists of the following four PPT algorithms:

- **Setup**$(1^\lambda, n, \ell)$: The setup algorithm takes as input the security parameter λ, the length n of input and depth ℓ of the circuits. It outputs the public parameters pp and a master key msk. For ease of notation on the reader, we suppress repeated pp arguments that are provided to every algorithms. For example, we will write **KeyGen**(msk, f) instead of **KeyGen**(pp, msk, f).
- **KeyGen**$(msk, f = (n, q, A, B, \texttt{GateType}))$: The key generation algorithm takes as input the master key msk and a description of a circuit f. It outputs a secret key sk_f.
- **Sign**$(sk_f, x \in \{0,1\}^n, m)$: The signing algorithm takes as input the secret key sk_f, a description of an attribute set $x \in \{0,1\}^n$, and a message m. It outputs a signature σ if $f(x) = 1$, or else $\perp$.
- **Verify**(x, m, σ): The verification algorithm takes as input a purported signature σ on an attribute set x and a message m. It outputs 1 (accept) if σ is valid. Otherwise, it outputs 0 (reject).

Correctness Consider all message m, string $x \in \{0,1\}^n$, and depth ℓ circuits f, where $f(x) = 1$. If $sk_f \leftarrow$ **KeyGen**(msk, f) and $\sigma \leftarrow$ **Sign**(sk_f, x, m), then **Verify**(x, m, σ) will output 1, where pp and msk were generated from a call to the setup algorithm.

2.3   Security Models

Generally, an ABS scheme should satisfy two security properties: unforgeability and privacy.

Unforgeability. This notion guarantees that user can sign message on behalf of x if and only if he has a secret key sk_f for f such that $f(x) = 1$. Furthermore, an ABS scheme should can resist a group of colluding users that put their secret keys together. Formally, this notion is defined by the following game which is played by a PPT adversary and a challenger.

1. **Setup**: The challenger runs the setup algorithm to generate the public parameters pp and master key msk, then it gives pp to the adversary and keeps msk to itself.
2. **Key Generation Oracle**: The adversary makes any polynomial number of signing key queries for policy function f of its choice. The challenger returns back $sk_f \leftarrow$ **KeyGen**(msk, f).
3. **Signing Oracle**: The adversary makes any polynomial number of signature queries on inputs a message m and an attribute set x. The challenger returns back $\sigma \leftarrow$ **Sign**(sk_f, x, m), where $f(x) = 1$.

4. **Forgery**: Finally the adversary outputs a forgery (x^*, m^*, σ^*). The adversary succeeds if (1) **Verify**$(x^*, m^*, \sigma^*) = 1$; (2) (m^*, x^*) was never queried to the signing oracle; and (3) $f(x^*) = 0$ for all f queried to the key generation oracle.

The advantage (taken over the random coins of the challenger and adversary) of a PPT adversary in the above game is defined as Pr[the adversary wins].

Definition 1. *An ABS scheme is unforgeable if all PPT adversaries have at most a negligible advantage in the above game.*

We also consider a weaker (selective) variant to the above definition where the (selective) adversary is required to commit to a challenge message/attribute tuple (m^*, x^*) before the setup phase.

Definition 2. *An ABS scheme is selectively unforgeable if all PPT selective adversaries have at most a negligible advantage in the above game.*

Privacy This notion guarantees that from a valid signature which associates with an attribute set x, the verifier learns nothing about the signer more than what is revealed by x known to the verifier. The notion of perfect privacy is defined as follows.

Definition 3. *An attribute-based signature scheme is perfectly private if, for all $(pp, msk) \leftarrow$ **Setup**(1^λ), all policy functions f_0, f_1, all secret keys $sk_{f_0} \leftarrow$ **KeyGen**(msk, f_0), $sk_{f_1} \leftarrow$ **KeyGen**(msk, f_1), all messages m, and all attribute sets x such that $f_0(x) = f_1(x) = 1$, the distributions **Sign**(sk_{f_0}, x, m) and **Sign**(sk_{f_1}, x, m) are equal.*

3   Our Construction Based on Multilinear Maps

In this section we describe our key-policy ABS construction for circuits in the setting of multilinear groups.

3.1   Multilinear Maps

We now give a description of the multilinear maps [9,7]. Let $\vec{\mathbb{G}} = (\mathbb{G}_1, \ldots, \mathbb{G}_k)$ be a sequence of groups each of large prime order p, and g_i be a canonical generator of $\mathbb{G}_i$. We let $g = g_1$. There exists a set of bilinear maps $\{e_{i,j} : \mathbb{G}_i \times \mathbb{G}_j \to \mathbb{G}_{i+j} | i, j \geq 1 \wedge i + j \leq k\}$, which satisfies the following property:

$$e_{i,j}(g_i^a, g_j^b) = g_{i+j}^{ab} : \forall a, b \in \mathbb{Z}_p.$$

When the context is obvious, we drop the subscripts i and j, such as $e(g_i^a, g_j^b) = g_{i+j}^{ab}$. It also will be convenient to abbreviate $e(h_1, h_2, \ldots, h_j) = e(h_1, e(h_2, \ldots, e(h_{j-1}, h_j) \ldots)) \in \mathbb{G}_i$ for $h_j \in \mathbb{G}_{i_j}$ and $i_1 + i_2 + \ldots + i_j \leq k$.

We assume that $\mathcal{G}(1^\lambda, k)$ is a PPT group generator algorithm which takes as input a security parameter λ and a positive integer k to indicate the number of allowed pairing operations, then it outputs the multilinear parameters $mp = (\mathbb{G}_1, \ldots, \mathbb{G}_k, p, g = g_1, g_2, \ldots, g_k, e_{i,j})$ to satisfy the above properties.

We assume that the following assumption holds in the setting described above: Multilinear Computational Diffie-Hellman (MCDH) assumption. This assumption can be viewed as an adaptation of the bilinear computational Diffie-Hellman assumption in the setting of multilinear groups.

Definition 4. *For any PPT algorithm* $\mathcal{B}$, *any polynomial* $p(\cdot)$, *any integer* k, *and all sufficiently large* $\lambda \in \mathbb{N}$,

$$\Pr \left[\begin{array}{l} mp \leftarrow \mathcal{G}(1^\lambda, k); \\ c_1, \ldots, c_k \overset{R}{\leftarrow} \mathbb{Z}_p; \\ T \leftarrow \mathcal{B}(mp, g^{c_1}, \ldots, g^{c_k}) \end{array} \; : \; T = g_{k-1}^{\Pi_{i \in [k]} c_i} \right] < \frac{1}{p(\lambda)},$$

where $c_i \overset{R}{\leftarrow} \mathbb{Z}_p$ *means that* c_i *is randomly and uniformly chosen from the set* $\mathbb{Z}_p$, *and* $[k]$ *is an abbreviation of the set* $\{1, 2, \ldots, k\}$.

3.2   Our Construction

Our construction is of the key-policy form where a secret key is associated with the description of a circuit f and the signatures are associated with some attribute sets x. Our scheme makes use of the structure used in an attribute-based encryption scheme due to Garg, Gentry, Halevi, Sahai, and Waters [10].

We describe our circuit construction for attribute-based signature scheme in terms of four algorithms which include a setup algorithm, a key generation algorithm, a signing algorithm, and a verification algorithm. For readability purposes, we use the same notation for circuits as in [10].

Setup$(1^\lambda, n, \ell, s)$:
The setup algorithm takes as input a security parameter λ, the maximum depth ℓ of a circuit, the number of boolean inputs n, and length s of the messages.

The algorithm then runs $\mathcal{G}(1^\lambda, k = n + \ell + s + 1)$ that produces groups $\overrightarrow{\mathbb{G}} = (\mathbb{G}_1, \ldots, \mathbb{G}_k)$ of prime order p, with canonical generators $g_1, \ldots, g_k$, where we let $g = g_1$. Next it chooses random $\alpha \in \mathbb{Z}_p$ and $(a_{1,0}, a_{1,1}), \ldots, (a_{n,0}, a_{n,1}), (b_{1,0}, b_{1,1}), \ldots, (b_{s,0}, b_{s,1}) \in \mathbb{Z}_p^2$ and computes $A_{i,\beta} = g^{a_{i,\beta}}, B_{j,\beta} = g^{b_{j,\beta}}$ for $i \in [n], j \in [s]$ and $\beta \in \{0, 1\}$.

The public parameters, pp, consist of the group sequence description plus $g_{\ell+1}^\alpha$ and $A_{i,\beta} = g^{a_{i,\beta}}, B_{j,\beta} = g^{b_{j,\beta}}$ for $i \in [n], j \in [s]$ and $\beta \in \{0, 1\}$. The master key, msk, consists of α and $a_{i,\beta}, b_{j,\beta}$ for $i \in [n], j \in [s]$ and $\beta \in \{0, 1\}$.

KeyGen$(msk, f = (n, q, A, B, \texttt{GateType}))$:
The key generation algorithm takes as input the master secret key msk and a description f of a circuit. The circuit has $n + q$ wires with n input wires, q gates and the $(n + q)$-th wire designated as the output wire.

The key generation algorithm chooses random $r_1, \ldots, r_{n+q-1} \in \mathbb{Z}_p$, where we think of the random value r_w as being associated with wire w. It sets $r_{n+q} = \alpha$.

Next, the algorithm generates key components for every wire w. The structure of the key components depends upon whether w is an input wire, an OR gate, or an AND gate. We describe how it generates components for each case.

- *Input wire*
  By our convention if $w \in [n]$ then it corresponds to the w-th input. The key component is:
  $$K_w = g_2^{r_w a_{w,1}}.$$

- *OR gate*
  Suppose that wire $w \in$ Gates and that GateType$(w) =$ OR. In addition, let $j = \text{depth}(w)$ be the depth of the wire. The algorithm will choose random $a_w, b_w \in \mathbb{Z}_p$. Then the algorithm creates key components as:
  $$K_{w,1} = g^{a_w}, K_{w,2} = g^{b_w}, K_{w,3} = g_j^{r_w - a_w \cdot r_{A(w)}}, K_{w,4} = g_j^{r_w - b_w \cdot r_{B(w)}}.$$

- *AND gate*
  Suppose that wire $w \in$ Gates and that GateType$(w) =$ AND. In addition, let $j = \text{depth}(w)$ be the depth of wire w. The algorithm chooses random $a_w, b_w \in \mathbb{Z}_p$ and creates the key components as:
  $$K_{w,1} = g^{a_w}, K_{w,2} = g^{b_w}, K_{w,3} = g_j^{r_w - a_w \cdot r_{A(w)} - b_w \cdot r_{B(w)}}.$$

The secret key sk_f consists of the description of f along with these $n + q$ key components.

Sign$(sk_f, x \in \{0,1\}^n, m \in \{0,1\}^s)$:
Given a message, m, of length s, let $m_1, \ldots, m_s$ be the bits of this message. For exposition we define $H(m) = g_s^{\prod_{i \in [s]} b_{i,m_i}}$ which is computable via the multilinear pairing operation from B_{i,m_i} for $i \in [s]$.

- *Input wire*
  By our convention if $w \in [n]$ then it corresponds to the w-th input. Suppose that $x_w = f_w(x) = 1$. The algorithm computes $E_w = g_{n+1}^{r_w \cdot \prod_{i \neq w} a_{i,x_i}}$. Using the multilinear pairing operation from A_{i,x_i} for $i \in [n] \neq w$. It then computes:
  $$E_w = e(K_w, g_{n-1}^{\prod_{i \neq w} a_{i,x_i}}) = e(g_2^{r_w a_{w,1}}, g_{n-1}^{\prod_{i \neq w} a_{i,x_i}}) = g_{n+1}^{r_w \prod_{i \in [n]} a_{i,x_i}}.$$

- *OR gate*
  Consider a wire $w \in$ Gates and that GateType$(w) =$ OR. In addition, let $j = \text{depth}(w)$ be the depth of the wire. For exposition we define $D(x) = g_n^{\prod_{i \in [n]} a_{i,x_i}}$. This is computable via the multilinear pairing operation from

A_{i,x_i} for $i \in [n]$. The computation is performed if $f_w(x) = 1$. If $f_{A(w)}(x) = 1$ (i.e., the first input evaluated to 1) then it computes:

$$
\begin{aligned}
E_w &= e(E_{A(w)}, K_{w,1}) \cdot e(K_{w,3}, D(x)) \\
&= e(g_{j+n-1}^{r_{A(w)} \prod_{i \in [n]} a_{i,x_i}}, g^{a_w}) \cdot e(g_j^{r_w - a_w \cdot r_{A(w)}}, g_n^{\prod_{i \in [n]} a_{i,x_i}}) \\
&= g_{j+n}^{r_w \cdot \prod_{i \in [n]} a_{i,x_i}}.
\end{aligned}
$$

Otherwise, if $f_{A(w)}(x) = 0$ but $f_{B(w)}(x) = 1$, then it computes:

$$
\begin{aligned}
E_w &= e(E_{B(w)}, K_{w,2}) \cdot e(K_{w,4}, D(x)) \\
&= e(g_{j+n-1}^{r_{B(w)} \prod_{i \in [n]} a_{i,x_i}}, g^{b_w}) \cdot e(g_j^{r_w - b_w \cdot r_{B(w)}}, g_n^{\prod_{i \in [n]} a_{i,x_i}}) \\
&= g_{j+n}^{r_w \cdot \prod_{i \in [n]} a_{i,x_i}}.
\end{aligned}
$$

- *AND gate*
  Consider a wire $w \in \text{Gates}$ and that $\texttt{GateType}(w) = \text{AND}$. In addition, let $j = \texttt{depth}(w)$ be the depth of the wire. The computation is performed if $f_w(x) = 1$ (i.e., $f_{A(w)}(x) = f_{B(w)}(x) = 1$) then it computes:

$$
\begin{aligned}
E_w &= e(E_{A(w)}, K_{w,1}) \cdot e(E_{B(w)}, K_{w,2}) \cdot e(K_{w,3}, D(x)) \\
&= e(g_{j+n-1}^{r_{A(w)} \prod_i a_{i,x_i}}, g^{a_w}) \cdot e(g_{j+n-1}^{r_{B(w)} \prod_i a_{i,x_i}}, g^{b_w}). \\
&\quad e(g_j^{r_w - a_w \cdot r_{A(w)} - b_w \cdot r_{B(w)}}, g_n^{\prod_i a_{i,x_i}}) \\
&= g_{j+n}^{r_w \cdot \prod_i a_{i,x_i}}.
\end{aligned}
$$

The above procedures are evaluated in order for all w for which $f_w(x) = 1$. The final output of these procedures gives a group element

$$
F(sk_f, x) = g_{n+\ell}^{r_{n+q} \prod_{i \in [n]} a_{i,x_i}} = g_{n+\ell}^{\alpha \cdot \prod_{i \in [n]} a_{i,x_i}} \in \mathbb{G}_{n+\ell}.^2
$$

Finally, the signing algorithm computes $e(H(m), F(sk_f, x))$ as the final signature σ. We may note that the final signature is

$$
\sigma = e(H(m), F(sk_f, x)) = g_{n+\ell+s}^{\alpha \cdot \prod_{i \in [n]} a_{i,x_i} \cdot \prod_{j \in [s]} b_{j,m_j}} \in \mathbb{G}_{k-1}.
$$

Verify$(x \in \{0,1\}^n, m \in \{0,1\}^s, \sigma \in \mathbb{G}_{k-1})$:
Given a purported signature σ on a message m and a description of an attribute set x, verify the following equation:

$$
e(\sigma, g) = e(g_{\ell+1}^\alpha, A_{1,x_1}, \ldots, A_{n,x_n}, B_{1,m_1}, \ldots, B_{s,m_s}).
$$

Output 1 (accept) if it holds, else 0 (reject).

---

[2] In fact, $F(sk_f, x)$ is the resulting value of the constrained PRF for circuits in [4].

Correctness The verification of the signature is justified by the following two equations:

$$e(\sigma, g) = e(g_{n+\ell+s}^{\alpha \cdot \prod_{i \in [n]} a_{i,x_i} \cdot \prod_{j \in [s]} b_{j,m_j}}, g)$$
$$= g_{n+\ell+s+1}^{\alpha \cdot \prod_{i \in [n]} a_{i,x_i} \cdot \prod_{j \in [s]} b_{j,m_j}}$$

and

$$e(g_{\ell+1}^{\alpha}, A_{1,x_1}, \ldots, A_{n,x_n}, B_{1,m_1}, \ldots, B_{s,m_s}) = e(g_{\ell+1+n}^{\alpha \cdot \prod_{i \in [n]} a_{i,x_i}}, B_{1,m_1}, \ldots, B_{s,m_s})$$
$$= g_{n+\ell+s+1}^{\alpha \cdot \prod_{i \in [n]} a_{i,x_i} \cdot \prod_{j \in [s]} b_{j,m_j}}.$$

3.3   Proof of Security in the Multilinear Groups

Selective Unforgeability We prove (selective) unforgeability in the security model given in Section 2.3, where the key access structures are monotonic circuits. For length of messages s and a circuit of max depth ℓ and input length n, we prove security under the $k = (n+\ell+s+1)$-Multilinear Computational Diffie-Hellman assumption.

We show that if there exists a PPT adversary $\mathcal{A}$ on our ABS system for messages of length s and circuits of depth ℓ and inputs of length n in the selective security game then we can construct an efficient algorithm $\mathcal{B}$ on the $(n+\ell+s+1)$-MCDH assumption. We describe how $\mathcal{B}$ interacts with $\mathcal{A}$.

Theorem 1. *The ABS construction in the above is selectively unforgeable for arbitrary circuits of depth ℓ and input length n, and messages of length s under the $(n + \ell + s + 1)$-MCDH assumption.*

Proof. The algorithm $\mathcal{B}$ first receives a $k = (n + \ell + s + 1)$-MCDH challenge consisting of the group sequence description $\overrightarrow{\mathbb{G}}$ and $g = g_1, g^{c_1}, \ldots, g^{c_k}$. It also receives challenge attribute set and message pair $(x^* \in \{0,1\}^n, m^* \in \{0,1\}^s)$ from the adversary $\mathcal{A}$.

Setup. Initially, $\mathcal{B}$ chooses random $z_1, \ldots, z_n \in \mathbb{Z}_p$ and sets

$$A_{i,\beta} = \begin{cases} g^{c_i}, & \text{if } x_i^* = \beta \\ g^{z_i}, & \text{if } x_i^* \neq \beta \end{cases}$$

for $i \in [n], \beta \in \{0,1\}$. This corresponds to setting $a_{i,\beta} = c_i$ if $x_i^* = \beta$ and z_i otherwise.

It also chooses random $y_1, \ldots, y_s \in \mathbb{Z}_p$ and sets

$$B_{i,\beta} = \begin{cases} g^{c_{n+\ell+1+i}}, & \text{if } m_i^* = \beta \\ g^{y_i}, & \text{if } m_i^* \neq \beta \end{cases}$$

for $i \in [s], \beta \in \{0,1\}$. This corresponds to setting $b_{i,\beta} = c_{n+\ell+1+i}$ if $m_i^* = \beta$ and y_i otherwise. We observe these are distributed identically to the real scheme. In addition, it will internally view $\alpha = c_{n+1} \cdot c_{n+2} \cdots c_{n+\ell+1}$.

Key Generation Oracle. The adversary $\mathcal{A}$ will query for a secret key for a circuit $f = (n, q, A, B, \texttt{GateType})$, where $f(x^*) = 0$. $\mathcal{B}$ proceeds to make the key. The idea for this oracle is same as in [4,10]. We will think have some invariant properties for each gate. Consider a gate w at depth j and the simulators viewpoint (symbolically) of r_w. If $f_w(x^*) = 0$, then the simulator will view r_w as the term $c_{n+1} \cdot c_{n+2} \cdots c_{n+j+1}$ plus some additional known randomization terms. If $f_w(x^*) = 1$, then the simulator will view r_w as the 0 plus some additional known randomization terms. If we can keep this property intact for simulating the keys up the circuit, the simulator will view r_{n+q} as $c_{n+1} \cdot c_{n+2} \cdots c_{n+\ell}$.

We describe how to create the key components for each wire w. Again, we organize key component creation into input wires, OR gates, and AND gates.

- *Input wire*
  Suppose $w \in [n]$ and is therefore by convention an input wire.
  * If $(x^*)_w = 1$ then we choose random $r_w \leftarrow \mathbb{Z}_p$ (as is done honestly). The key component is:
  $$K_w = g_2^{r_w a_{w,1}}.$$
  * If $(x^*)_w = 0$ then we let $r_w = c_{n+1}c_{n+2} + \eta_w$ where $\eta_w \in \mathbb{Z}_p$ is a randomly chosen value. The key component is:
  $$K_w = (e(g^{c_{n+1}}, g^{c_{n+2}}) \cdot g_2^{\eta_w})^{z_w} = g_2^{r_w a_{w,1}}.$$

- *OR gate*
  Suppose that wire $w \in$ Gates and that $\texttt{GateType}(w) = $ OR. In addition, let $j = \texttt{depth}(w)$ be the depth of the wire.
  * If $f_w(x^*) = 1$, then algorithm will choose random $a_w, b_w, r_w \in \mathbb{Z}_p$. Then the algorithm creates key components as:
  $$K_{w,1} = g^{a_w}, K_{w,2} = g^{b_w}, K_{w,3} = g_j^{r_w - a_w \cdot r_{A(w)}}, K_{w,4} = g_j^{r_w - b_w \cdot r_{B(w)}}.$$
  * If $f_w(x^*) = 0$, then we set $a_w = c_{n+j+1} + \psi_w$, $b_w = c_{n+j+1} + \phi_w$, and $r_w = c_{n+1} \cdot c_{n+2} \cdots c_{n+j+1} + \eta_w$, where ψ_w, ϕ_w, η_w are chosen randomly. Then the algorithm creates key components as:
  $$K_{w,1} = g^{c_{n+j+1}+\psi_w}, K_{w,2} = g^{c_{n+j+1}+\phi_w},$$
  $$K_{w,3} = g_j^{\eta_w - c_{c+j+1}\eta_{A(w)} - \psi_w(c_{n+1}\cdots c_{n+j}+\eta_{A(w)})},$$
  $$K_{w,4} = g_j^{\eta_w - c_{c+j+1}\eta_{B(w)} - \psi_w(c_{n+1}\cdots c_{n+j}+\eta_{B(w)})}.$$
  $\mathcal{B}$ can create the last two key components due to a cancellation. Since both the $A(w)$ and $B(w)$ gates evaluated to 0, we have $r_{A(w)} = c_{n+1} \cdots c_{n+j} + \eta_{A(w)}$ and similarly for $r_{B(w)}$. Note that $g_j^{c_{n+1}\cdots c_{n+j}}$ is always using the multilinear maps.

- *AND gate*
  Suppose that wire $w \in$ Gates and that $\texttt{GateType}(w) = $ AND. In addition, let $j = \texttt{depth}(w)$ be the depth of wire w.

* If $f_w(x^*) = 1$, then the algorithm chooses random $a_w, b_w, r_w \in \mathbb{Z}_p$ and creates the key components as:

$$K_{w,1} = g^{a_w}, K_{w,2} = g^{b_w}, K_{w,3} = g_j^{r_w - a_w \cdot r_{A(w)} - b_w \cdot r_{B(w)}}.$$

* If $f_w(x^*) = 0$ and $f_{A(w)}(x^*) = 0$, then we let $a_w = c_{n+j+1} + \psi_w, b_w = \phi_w$, and $r_w = c_{n+1} \cdot c_{n+2} \cdots c_{n+j+1} + \eta_w$, where ψ_w, ϕ_w, η_w are chosen randomly. Then the algorithm creates key components as:

$$K_{w,1} = g^{c_{n+j+1}+\psi_w}, K_{w,2} = g^{\phi_w},$$
$$K_{w,3} = g_j^{\eta_w - \psi_w c_{n+1} \cdots c_{n+j} - (c_{n+j+1}+\psi_w)\eta_{A(w)} - \phi_w r_{B(w)}}.$$

$\mathcal{B}$ can create the last key component due to a cancellation. Since the $A(w)$ gate evaluated to 0, we have $r_{A(w)} = c_{n+1} \cdots c_{n+j} + \eta_{A(w)}$. Note that $g_j^{r_{B(w)}}$ always computable regardless of whether $f_{A(w)}(x^*)$ evaluated to 0 or 1, since $g_j^{c_{n+1}\cdots c_{n+j}}$ is always using the multilinear maps.

The case where $f_{B(w)}(x^*) = 0$ and $f_{A(w)}(x^*) = 1$ is performed in a symmetric to what is above, with the roles of a_w and b_w reversed.

Signing Oracle. The adversary $\mathcal{A}$ will query for a signature for a message m and an attribute set x, where $(m, x) \neq (m^*, x^*)$. Conceptually, $\mathcal{B}$ will be able to create signature for the adversary, because his query will differ from the challenge message or attribute set in at least one bit. More specifically, $\mathcal{B}$ proceeds to make the signature according to the following two cases.

* *Case 1:* If $x \neq x^*$, we let $x_i \neq x_i^*$ for some $i \in [n]$. $\mathcal{B}$ first computes a temporary value $T = g_{\ell+1}^\alpha = g_{\ell+1}^{c_{n+1} \cdot c_{n+2} \cdots c_{n+\ell+1}}$. This can be computed by using the multilinear pairing operation. Next, it computes $T' = g_{n-1}^{\Pi_{j \in [n]} a_{j,x_j}}$. To do so, $\mathcal{B}$ first uses the pairing operation along with A_{j,x_j} for all $j \neq i \in [n]$ to compute $g_{n-1}^{\Pi_{j \neq i \in [n]} a_{j,x_j}}$. Next, it raises this value to $a_{i,x_i} = z_i$ to get T'. Then, it computes $e(T, T') = g_{n+\ell}^{\alpha \cdot \Pi_{i \in [n]} a_{i,x_i}} = F(sk_f, x)$. Finally, it computes the signature as:

$$\sigma = e(F(sk_f, x), H(m)) = g_{n+\ell+s}^{\alpha \cdot \Pi_{i \in [n]} a_{i,x_i} \cdot \Pi_{j \in [s]} b_{j,m_j}},$$

where $H(m) = g_s^{\Pi_{j \in [s]} b_{j,m_j}}$ can be computed by using the pairing operation from B_{j,m_j} for $j \in [s]$.

* *Case 2:* If $m \neq m^*$, we let $m_i \neq m_i^*$ for some $i \in [s]$. $\mathcal{B}$ first computes a temporary value $V = g_{n+\ell+1}^{\alpha \cdot \Pi_{i \in [n]} a_{i,x_i}} = g_{n+\ell+1}^{c_{n+1} \cdot c_{n+2} \cdots c_{n+\ell+1} \cdot \Pi_{i \in [n]} a_{i,x_i}}$. This can be computed by using the multilinear pairing operation. Next, it computes a temporary $T = g_{s-1}^{\Pi_{j \in [n]} b_{j,x_j}}$. To do so, $\mathcal{B}$ first uses the multilinear pairing operation from B_{j,m_j} for all $j \neq i \in [s]$ to compute $g_{s-1}^{\Pi_{j \neq i \in [s]} b_{j,m_j}}$. Next, it raises this value to $b_{i,m_i} = y_i$ to get T. Finally, it computes the signature as:

$$\sigma = e(V, T) = g_{n+\ell+s}^{\alpha \cdot \Pi_{i \in [n]} a_{i,x_i} \cdot \Pi_{j \in [s]} b_{j,m_j}}.$$

Forgery. Eventually, $\mathcal{A}$ outputs an attribute signature σ^* on message m^* and attribute set x^*. Then $\mathcal{B}$ outputs σ^* as the solution of the given instance of the $k = (n + \ell + s + 1)$-MCDH assumption. According to the public parameters built in the setup phase and the assumption that σ^* is valid, we know that $\sigma^* = g_{k-1}^{\prod_{i \in [k]} c_i}$, implies that σ^* is a solution for the given instance of the k-MCDH problem, and thus $\mathcal{B}$ breaks the k-MCDH assumption.

It is clear that the view of $\mathcal{A}$ simulated by $\mathcal{B}$ in the above game is distributed statistically exponentially closely to that in the real unforgeability game, hence $\mathcal{B}$ succeeds whenever $\mathcal{A}$ does. $\qquad\square$

Perfect Privacy. Given a valid signature σ^* on a message m^* and an attribute set x^*, we show that any secret key sk_f such that $f(x^*) = 1$ could possibly have created it. The proof is straight-forward.

Theorem 2. *The ABS construction in the above is perfectly private.*

Proof. According to the construction of the signing algorithm, for any challenge tuple (f_0, f_1, x^*, m^*) such that $f_0(x^*) = f_1(x^*) = 1$, which was chosen by an unbounded adversary $\mathcal{A}$, both of the signatures created by the signing key sk_{f_0} and sk_{f_1} are $g_{n+\ell+s}^{\alpha \cdot \prod_{i \in [n]} a_{i.x_i^*} \cdot \prod_{j \in [s]} b_{j.m_j^*}}$. Therefore, any secret key sk_f such that $f(x^*) = 1$ can compute a same signature on a given message m^* and an attribute set x^*. The perfect privacy follows easily from this observation. $\qquad\square$

4   Our Construction Based on GGH Graded Algebras

We now describe how to modify our construction to use the GGH graded algebras analogue of multilinear maps. The translation of our scheme above is straightforward to the GGH setting.

4.1   Graded Encoding Systems

We start by providing background on Garg et al.'s [9] graded encoding systems (i.e., GGH framework). For readability purposes, we use the same notation for graded encoding systems as in [9].

In the graded encoding systems, we view a group element g_i^α in a multilinear group family as simply an encoding of α at "level-i". This encoding allows basic functionalities, such as equality testing (i.e., it is easy to check that two level-i encodings encode the same exponent), additive homomorphism (via the group operation in group $\mathbb{G}_i$), and bounded multiplicative homomorphism (via the multilinear map e). They retain the notion of a somewhat homomorphic encoding with equality testing, but they use probabilistic encodings, and replace the multilinear group family with "less structured" sets of encodings.

Abstractly, Garg et al.'s k-graded encoding system for a ring R includes a system of sets $\mathcal{S} = \{S_i^{(\alpha)} \subset \{0,1\}^* : i \in [0, k], \alpha \in R\}$ such that, for every fixed $i \in [0, k]$, the sets $\{S_i^{(\alpha)} : \alpha \in R\}$ are disjoint (and thus form a partition of $S_i = \cup_\alpha S_i^{(\alpha)}$). The set $S_i^{(\alpha)}$ consists of the "level-i encodings of α". More specifically, the graded encoding system consists of the following nine algorithms:

- **Instance Generation.** The randomized $\mathbf{InstGen}(1^\lambda, 1^k)$ takes as input the security parameter λ and integer k. It outputs $(\mathbf{params}, \mathbf{p}_{zt})$ where $\mathbf{params}$ parameterize the graded encoding system and $\mathbf{p}_{zt}$ is a zero-test element.
- **Ring Sampler.** The randomized $\mathbf{samp}(\mathbf{params})$ outputs a "level-0 encoding" $a \in S_0$, such that the induced distribution on α such that $a \in S_0^{(\alpha)}$ is statistically uniform.
- **Encoding.** The (possibly randomized) $\mathbf{enc}(\mathbf{params}, i, a)$ takes as input $i \in [k]$ and a level-0 encoding $a \in S_0^{(\alpha)}$ of $\alpha \in R$ and outputs a level-i encoding $u \in S_i^{(\alpha)}$ of the same α.
- **Re-Randomization.** The randomized $\mathbf{reRand}(\mathbf{params}, i, \alpha)$ re-randomizes a level-i encoding of α leaving it at the same level.
- **Addition and negation.** Given two level-i encodings $u_1 \in S_i^{(\alpha)}$ and $u_2 \in S_i^{(\beta)}$ outputs an encoding $\mathbf{add}(\mathbf{params}, u_1, u_2) \in S_i^{(\alpha+\beta)}$ and $\mathbf{neg}(\mathbf{params}, u_1, u_2) \in S_i^{(\alpha-\beta)}$.
- **Multiplication.** Given two level-i encodings $u_1 \in S_i^{(\alpha)}$ and $u_2 \in S_j^{(\beta)}$ outputs an encoding $\mathbf{mult}(\mathbf{params}, u_1, u_2) \in S_{i+j}^{(\alpha \cdot \beta)}$.
- **Zero-test.** Given $\mathbf{p}_{zt}$ and an encodings u, $\mathbf{isZero}(\mathbf{params}, \mathbf{p}_{zt}, u)$ outputs 1 if $u \in S_k^{(0)}$ and 0 otherwise.
- **Extraction.** This algorithm outputs a canonical and random representation of level-k encodings. Namely, $\mathbf{ext}(\mathbf{params}, \mathbf{p}_{zt}, u)$ outputs $K \in \{0,1\}^\ell$ that, for a particular $\alpha \in R$ is the same for all $u \in S_k^{(\alpha)}$. For a random $\alpha \in R$, and $u \in S_k^{(k)}$, $\mathbf{ext}(\mathbf{params}, \mathbf{p}_{zt}, u)$ outputs a uniform value $K \in \{0,1\}^{\ell(\lambda)}$.

We can extend $\mathbf{add}$ and $\mathbf{mult}$ to handle more than two encodings as inputs, by applying the binary version of $\mathbf{add}$ and $\mathbf{mult}$ iteratively. Also, we use the canonicalizing encoding algorithm $\mathbf{cenc}_\ell(\mathbf{params}, i, a)$ which takes as input encoding of a and generates another encoding according to a "nice" distribution. The assumption of k-Graded Multilinear Computational Diffie-Hellman (GM-CDH) is as follows.

Definition 5. *The k-GMCDH assumption states the following: Given k encodings $\mathbf{cenc}_1(\mathbf{params}, 1, 1), \mathbf{cenc}_1(\mathbf{params}, 1, c_1), \ldots, \mathbf{cenc}_1(\mathbf{params}, 1, c_k)$, it is hard to compute $T = \mathbf{cenc}_1(\mathbf{params}, k-1, \prod_{i \in [k]} c_i)$, with better than negligible advantage (in security parameter λ), where $(\mathbf{params}, \mathbf{p}_{zt}) \leftarrow \mathbf{InstGen}(1^\lambda, 1^k)$ and $c_1, \ldots, c_k \leftarrow \mathbf{samp}(\mathbf{params})$.*

Garg, Gentry, and Halevi [9] realized the graded encoding system based on lattice. Then Coron, Lepoint, and Tibouchi [7] followed the GGH framework and constructed a practical multilinear maps over the Integers.

4.2   Our Construction

We now describe our construction in the GGH $k = (n + \ell + s + 1)$-graded encoding system. For ease of notation on the reader, we suppress repeated **params** arguments that are provided to every algorithm. For example, we will write $\alpha \leftarrow$ **samp**() instead of $\alpha \leftarrow$ **samp**(**params**). Note that in our scheme, there will only ever be a single uniquely chosen value for **params** throughout the scheme, so there is no cause for confusion.

Setup$(1^\lambda, n, \ell, s)$:
The setup algorithm takes as input a security parameter λ, the maximum depth ℓ of a circuit, the number of boolean inputs n, and length s of a the messages.

 The algorithm then runs $\mathbf{p}_{zt} \leftarrow$ **InstGen**$(1^\lambda, 1^{k=n+\ell+s+1})$. Next it samples $\alpha, (a_{1,0}, a_{1,1}), \ldots, (a_{n,0}, a_{n,1}), (b_{1,0}, b_{1,1}), \ldots, (b_{s,0}, b_{s,1}) \leftarrow$ **samp**(). The public parameters, pp, consist of $\mathbf{p}_{zt}$ plus $\mathbf{cenc}_2(\ell + 1, \alpha), \mathbf{cenc}_2(1, a_{i,\beta}), \mathbf{cenc}_2(1, b_{j,\beta})$ for $i \in [n], j \in [s]$ and $\beta \in \{0, 1\}$. The master key, msk, consists of α and $a_{i,\beta}, b_{j,\beta}$ for $i \in [n], j \in [s]$ and $\beta \in \{0, 1\}$.

KeyGen$(msk, f = (n, q, A, B, \mathtt{GateType}))$:
The key generation algorithm takes as input the master secret key msk and a description f of a circuit. The circuit has $n + q$ wires with n input wires, q gates and the $(n + q)$-th wire designated as the output wire.

 The key generation algorithm samples $r_1, \ldots, r_{n+q-1} \leftarrow$ **samp**(), where we think of the random value r_w as being associated with wire w. It sets $r_{n+q} = \alpha$.

 Next, the algorithm generates key components for every wire w. The structure of the key components depends upon whether w is an input wire, an OR gate, or an AND gate. We describe how it generates components for each case.

- *Input wire*
  By our convention if $w \in [n]$ then it corresponds to the w-th input. The key component is:
  $$K_w = \mathbf{cenc}_3(2, r_w a_{w,1}).$$

- *OR gate*
  Suppose that wire $w \in$ Gates and that $\mathtt{GateType}(w) =$ OR. In addition, let $j = \mathtt{depth}(w)$ be the depth of the wire. The algorithm will sample $a_w, b_w \leftarrow$ **samp**(). Then the algorithm creates key components as:
  $$K_{w,1} = \mathbf{cenc}_3(1, a_w), K_{w,2} = \mathbf{cenc}_3(1, b_w),$$
  $$K_{w,3} = \mathbf{cenc}_3(j, r_w - a_w \cdot r_{A(w)}), K_{w,4} = \mathbf{cenc}_3(j, r_w - b_w \cdot r_{B(w)}).$$

- *AND gate*
  Suppose that wire $w \in$ Gates and that $\mathtt{GateType}(w) =$ AND. In addition, let $j = \mathtt{depth}(w)$ be the depth of wire w. The algorithm samples $a_w, b_w \leftarrow$ **samp**() and creates the key components as:
  $$K_{w,1} = \mathbf{cenc}_3(1, a_w), K_{w,2} = \mathbf{cenc}_3(1, b_w),$$
  $$K_{w,3} = \mathbf{cenc}_3(j, r_w - a_w \cdot r_{A(w)} - b_w \cdot r_{B(w)}).$$

The secret key sk_f consists of the description of f along with these $n + q$ key components.

Sign$(sk_f, x \in \{0,1\}^n, m \in \{0,1\}^s)$:

Given a message, m, of length s, let $m_1, \ldots, m_s$ be the bits of this message, we define $H(m) = \mathbf{cenc}_3(s, \prod_{i \in [s]} b_{i,m_i})$.

- *Input wire*

  By our convention if $w \in [n]$ then it corresponds to the w-th input. Suppose that $x_w = f_w(x) = 1$. It then computes:

$$E_w = K_w \cdot \mathbf{cenc}_3(n-1, \prod_{i \neq w} a_{i,x_i}).$$

  Thus, E_w computes a level $n+1$ encoding of $r_w \cdot \prod_{i \in [n]} a_{i,x_i}$.

- *OR gate*

  Consider a wire $w \in$ Gates and that $\mathtt{GateType}(w) = $ OR. In addition, let $j = \mathtt{depth}(w)$ be the depth of the wire. For exposition we define $D(x) = \mathbf{cenc}_3(n, \prod_{i \in [n]} a_{i,x_i})$. Suppose that $f_w(x) = 1$. If $f_{A(w)}(x) = 1$ (i.e., the first input evaluated to 1) then it computes:

$$E_w = E_{A(w)} \cdot K_{w,1} + K_{w,3} \cdot D(x).$$

  Thus, E_w computes a level $n+j$ encoding of $r_w \cdot \prod_{i \in [n]} a_{i,x_i}$. Otherwise, if $f_{A(w)}(x) = 0$ but $f_{B(w)}(x) = 1$, then it computes:

$$E_w = E_{B(w)} \cdot K_{w,2} + K_{w,4} \cdot D(x).$$

  Thus, E_w computes a level $n+j$ encoding of $r_w \cdot \prod_{i \in [n]} a_{i,x_i}$.

- *AND gate*

  Consider a wire $w \in$ Gates and that $\mathtt{GateType}(w) = $ AND. In addition, let $j = \mathtt{depth}(w)$ be the depth of the wire. The computation is performed if $f_w(x) = 1$ (i.e., $f_{A(w)}(x) = f_{B(w)}(x) = 1$) then it computes:

$$E_w = E_{A(w)} \cdot K_{w,1} + E_{B(w)} \cdot K_{w,2} + K_{w,3} \cdot D(x).$$

  Thus, E_w computes a level $n+j$ encoding of $r_w \cdot \prod_{i \in [n]} a_{i,x_i}$.

The above procedures are evaluated in order for all w for which $f_w(x) = 1$. The final output of these procedures gives a level $n + \ell$ encoding of

$$F(sk_f, x) = r_{n+q} \prod_{i \in [n]} a_{i,x_i} = \alpha \cdot \prod_{i \in [n]} a_{i,x_i}.$$

Finally, the signing algorithm runs the algorithm $\mathbf{mult}(H(m), F(sk_f, x))$ and outputs the resultant value as the signature σ. We may note that σ is a level $k - 1 = n + \ell + s$ encoding of

$$\alpha \cdot \prod_{i \in [n]} a_{i,x_i} \cdot \prod_{j \in [s]} b_{j,m_j}.$$

Verify$(x \in \{0,1\}^n, m \in \{0,1\}^s, \sigma)$:

Given a purported signature σ on a message m and description of attribute set x, the algorithm first sets $\tau = \mathbf{cenc}_2(1,1)$, and then tests:

$$\mathbf{isZero}(\mathbf{p}_{zt}, \tau \cdot \sigma - \mathbf{cenc}_2(\ell+1, \alpha \cdot \prod_{i \in [n]} a_{i,x_i} \cdot \prod_{j \in [s]} b_{j,m_j})).$$

Output 1 if and only if the zero testing procedure outputs 1.

Correctness. Correctness follows from the same argument as for the attribute-based signature scheme in the generic multilinear setting.

4.3   Proof of Security in GGH Framework

We now analyze the security of the above scheme in the GGH framework.

Selective Unforgeability. We prove (selective) unforgeability in the security model given in Section 2.3, where the key access structures are monotonic circuits. For length of messages s and a circuit of max depth ℓ and input length n, we prove security under the $k = (n+\ell+s+1)$-Graded Multilinear Computational Diffie-Hellman (GMCDH) assumption.

   We show that if there exist a PPT adversary $\mathcal{A}$ on our ABS system for messages of length s, circuits of depth ℓ and inputs of length n in the selective security game then we can construct an efficient algorithm $\mathcal{B}$ on the $(n+\ell+s+1)$-GMCDH assumption.

Theorem 3. *The ABS construction in the above is selectively unforgeable for arbitrary circuits of depth ℓ and input length n, and messages of length s under the $(n + \ell + s + 1)$-MCDH assumption.*

   Due to the limitation of space, the proof of the above theorem will be given in the full version of this paper.

Perfect Privacy. Given a valid signature σ^* on a message m^* and an attribute set x^*, we show that any secret key sk_f such that $f(x^*) = 1$ could possibly have created it. The proof is straight-forward.

Theorem 4. *The ABS construction in the above is perfectly private.*

Proof. According to the construction of the signing algorithm, for any challenge tuple (f_0, f_1, x^*, m^*), such that $f_0(x^*) = f_1(x^*) = 1$, chosen by an unbounded adversary $\mathcal{A}$, both of the signatures created by the signing key sk_{f_0} and sk_{f_1} are $\mathbf{cenc}_3(n + \ell + s, \alpha \cdot \prod_{i \in [n]} a_{i,x_i^*} \cdot \prod_{j \in [s]} b_{j,m_j^*})$. Therefore, any secret key sk_f such that $f(x^*) = 1$ can compute a same signature on a given message m^* and an attribute set x^*. The perfect privacy follows easily from this observation. $\square$

Acknowledgement. The authors would like to thank anonymous reviewers for their helpful comments and suggestions.

References

1. Bellare, M., Fuchsbauer, G.: Policy-based signatures (2013),
   http://eprint.iacr.org/2013/413
2. Boyle, E., Goldwasser, S., Ivan, I.: Functional signatures and pseudorandom functions. In: Krawczyk, H. (ed.) PKC 2014. LNCS, vol. 8383, pp. 501–519. Springer, Heidelberg (2014)
3. Boneh, D., Silverberg, A.: Applications of multilinear forms to cryptography. IACR Cryptology ePrint Archive, 2002/80 (2002)
4. Boneh, D., Waters, B.: Constrained pseudorandom functions and their applications. In: Sako, K., Sarkar, P. (eds.) ASIACRYPT 2013, Part II. LNCS, vol. 8270, pp. 280–300. Springer, Heidelberg (2013)
5. Boneh, D., Waters, B., Zhandry, M.: Low overhead broadcast encryption from multilinear maps. IACR Cryptology ePrint Archive (2014)
6. Chaum, D., van Heyst, E.: Group signatures. In: Davies, D.W. (ed.) EUROCRYPT 1991. LNCS, vol. 547, pp. 257–265. Springer, Heidelberg (1991)
7. Coron, J.-S., Lepoint, T., Tibouchi, M.: Practical multilinear maps over the integers. In: Canetti, R., Garay, J.A. (eds.) CRYPTO 2013, Part I. LNCS, vol. 8042, pp. 476–493. Springer, Heidelberg (2013)
8. Freire, E.S.V., Hofheinz, D., Paterson, K.G., Striecks, C.: Programmable hash functions in the multilinear setting. In: Canetti, R., Garay, J.A. (eds.) CRYPTO 2013, Part I. LNCS, vol. 8042, pp. 513–530. Springer, Heidelberg (2013)
9. Garg, S., Gentry, C., Halevi, S.: Candidate multilinear maps from ideal lattices. In: Johansson, T., Nguyen, P.Q. (eds.) EUROCRYPT 2013. LNCS, vol. 7881, pp. 1–17. Springer, Heidelberg (2013)
10. Garg, S., Gentry, C., Halevi, S., Sahai, A., Waters, B.: Attribute-based encryption for circuits from multilinear maps. In: Canetti, R., Garay, J.A. (eds.) CRYPTO 2013, Part II. LNCS, vol. 8043, pp. 479–499. Springer, Heidelberg (2013)
11. Goyal, V., Pandey, O., Sahai, A., Waters, B.: Attribute-based encryption for fine-grained access control of encrypted data. In: CCS 2006, pp. 89–98. ACM (2006)
12. Groth, J., Sahai, A.: Efficient non-interactive proof systems for bilinear groups. In: Smart, N.P. (ed.) EUROCRYPT 2008. LNCS, vol. 4965, pp. 415–432. Springer, Heidelberg (2008)
13. Herranz, J., Laguillaumie, F., Libert, B., Ràfols, C.: Short attribute-based signatures for threshold predicates. In: Dunkelman, O. (ed.) CT-RSA 2012. LNCS, vol. 7178, pp. 51–67. Springer, Heidelberg (2012)
14. Hohenberger, S., Sahai, A., Waters, B.: Full domain hash from (Leveled) multilinear maps and identity-based aggregate signatures. In: Canetti, R., Garay, J.A. (eds.) CRYPTO 2013, Part I. LNCS, vol. 8042, pp. 494–512. Springer, Heidelberg (2013)
15. Kumar, S., Agrawal, S., Balaraman, S., Rangan, C.P.: Attribute based signatures for bounded multi-level threshold circuits. In: Camenisch, J., Lambrinoudakis, C. (eds.) EuroPKI 2010. LNCS, vol. 6711, pp. 141–154. Springer, Heidelberg (2011)
16. El Kaafarani, A., Ghadafi, E., Khader, D.: Decentralized traceable attribute-based signatures. In: Benaloh, J. (ed.) CT-RSA 2014. LNCS, vol. 8366, pp. 327–348. Springer, Heidelberg (2014)
17. Li, J., Au, M.H., Susilo, W., Xie, D., Ren, K.: Attribute-based signature and its applications. In: Proceedings of the 5th ACM Symposium on Information, Computer and Communications Security, pp. 60–69. ACM (2010)
18. Li, J., Kim, K.: Attribute-based ring signatures (2008),
    http://eprint.iacr.org/2008/394

19. Lewko, A., Okamoto, T., Sahai, A., Takashima, K., Waters, B.: Fully secure functional encryption: Attribute-based encryption and (hierarchical) inner product encryption. In: Gilbert, H. (ed.) EUROCRYPT 2010. LNCS, vol. 6110, pp. 62–91. Springer, Heidelberg (2010)

20. Maji, H.K., Prabhakaran, M., Rosulek, M.: Attribute-based signatures: Achieving attribute privacy and collusion-resistance (2008),
http://eprint.iacr.org/2008/328

21. Maji, H.K., Prabhakaran, M., Rosulek, M.: Attribute-based signatures. In: Kiayias, A. (ed.) CT-RSA 2011. LNCS, vol. 6558, pp. 376–392. Springer, Heidelberg (2011)

22. Okamoto, T., Takashima, K.: Homomorphic encryption and signatures from vector decomposition. In: Galbraith, S.D., Paterson, K.G. (eds.) Pairing 2008. LNCS, vol. 5209, pp. 57–74. Springer, Heidelberg (2008)

23. Okamoto, T., Takashima, K.: Efficient attribute-based signatures for non-monotone predicates in the standard model. In: Catalano, D., Fazio, N., Gennaro, R., Nicolosi, A. (eds.) PKC 2011. LNCS, vol. 6571, pp. 35–52. Springer, Heidelberg (2011)

24. Okamoto, T., Takashima, K.: Decentralized attribute-based signatures. In: Kurosawa, K., Hanaoka, G. (eds.) PKC 2013. LNCS, vol. 7778, pp. 125–142. Springer, Heidelberg (2013)

25. Rivest, R.L., Shamir, A., Tauman, Y.: How to leak a secret. In: Boyd, C. (ed.) ASIACRYPT 2001. LNCS, vol. 2248, pp. 552–565. Springer, Heidelberg (2001)

26. Shahandashti, S.F., Safavi-Naini, R.: Threshold attribute-based signatures and their application to anonymous credential systems. In: Preneel, B. (ed.) AFRICACRYPT 2009. LNCS, vol. 5580, pp. 198–216. Springer, Heidelberg (2009)

27. Sahai, A., Waters, B.: Fuzzy identity-based encryption. In: Cramer, R. (ed.) EUROCRYPT 2005. LNCS, vol. 3494, pp. 457–473. Springer, Heidelberg (2005)

28. Zhang, L.F., Safavi-Naini, R.: Private outsourcing of polynomial evaluation and matrix multiplication using multilinear maps. In: Abdalla, M., Nita-Rotaru, C., Dahab, R. (eds.) CANS 2013. LNCS, vol. 8257, pp. 329–348. Springer, Heidelberg (2013)

PAEQ: Parallelizable Permutation-Based Authenticated Encryption

Alex Biryukov and Dmitry Khovratovich

University of Luxembourg, Luxembourg
{alex.biryukov,dmitry.khovratovich}@uni.lu

Abstract. We propose a new authenticated encryption scheme PAEQ, which employs a fixed public permutation. In contrast to the recent sponge-based proposals, our scheme is fully parallelizable. It also allows flexible key and nonce length, and is one of the few which achieves 128-bit security for both confidentiality and data authenticity with the same key length.

The permutation within PAEQ is a new design called AESQ, which is based on AES and is 512 bits wide. In contrast to similar constructions used in the SHA-3 competition, our permutation fully benefits from the newest Intel AES instructions and runs at 2.5 cycles per byte if used as the counter-mode PRF.

The full version of the paper is available at [7].

Keywords: Authenticated encryption, SHA-3, permutation-based cryptography, AES.

1 Introduction

It has been known for a while that standard blockcipher modes of operation do not provide any integrity/authenticity protection, and hence additional mechanisms are needed to ensure the receiver that the ciphertext has not been modified of generated by the adversary. Whenever two distinct keys are available, the problem can be solved with a simple combination of encryption and MAC generation [4], and the Encrypt-then-MAC paradigm has become an international standard [24]. In contrast to confidentiality-only modes, the authenticated encryption schemes do not provide any decryption but return "invalid" ($\perp$), if the ciphertext has been created or modified by an adversary.

Since at least the year of 2000, cryptographers have tried to design an authenticated encryption (AE) scheme, which would use a single key and would be at least as efficient as Encrypt-then-MAC. The research went in two directions. The first one deals with new modes of operation which use an arbitrary block cipher. The ISO standards [24] GCM, CCM, and OCB are typical examples. The patented OCB mode runs almost as fast as the counter mode of encryption, which yields the speed below one cycle per byte on modern CPUs if used with AES [18]. The second direction is to design a dedicated AE scheme, like Phelix [25] ALE [9], or AEGIS [27].

S.S.M. Chow et al. (Eds.): ISC 2014, LNCS 8783, pp. 72–89, 2014.

Modern authenticated encryption schemes are also able to authenticate so called *associated data* (AD) without encrypting it [23]. A typical application is Internet packets, whose contents are encrypted, whereas headers are not (for routing purposes), while they still should be bound to the encrypted data.

Blockcipher- vs. Permutation-Based Designs. The most generic AE schemes employ blockciphers. However, they are not the only source of good transformations. If the mode is encryption-only (see above), the transformation does not need to be invertible (cf. the folklore use of hash functions for the CTR mode). Quite recently, the hash function Keccak [6], which employs a 1600-bit permutation, has been selected as the new SHA-3 standard. We expect that it will be widely deployed in the near future, and hence its building block will be readily accessible to other cryptographic applications. On resource-constrained devices, where the space is limited, it would be very tempting to use a single cryptographic primitive, such as the Keccak permutation, for many purposes. Whenever Keccak or AES are considered too expensive for a device, the lightweight hash functions like Spongent [8] and Quark [3] are also based on a single permutation and may offer it for other schemes. A wide permutation also simplifies the security proofs, as additional inputs such as tweaks and counters can be easily accommodated within the permutation input.

This idea also fits the recent paradigm of the *permutation-based cryptography* [10] as opposed to the blockcipher-based cryptography. From the practical point of view, it would allow to have a single permutation for all purposes, whereas it would simplify the analysis as a target for a cryptanalyst would be much simpler. The downside of the permutation-based approach is that the security proof has to be devised in the random oracle/permutation model, and does not rely on the PRP assumption.

Our Contributions. We offer a new mode, called *PPAE*, which employs a public fixed permutation (let us denote its bit width by n). We have tried to make the mode as universal as possible, and to provide the users with almost every capability an authenticated encryption mode might have. Properties of our scheme are summarized as follows.

Key/nonce/tag length. We allow keys, nonces, and tags of arbitrary length, as long as they fit into the permutation, fulfill some minimal requirements, and constitute the integer number of bytes.

Performance. Depending on the key size, the encryption speed is about 6 cycles per byte on modern CPUs with permutation AESQ.

Security level. Depending on the key length and the permutation width, we support a range of security levels from 64 to 256 bits. For the permutation of width n bits we can use a key of about $(n/3 - 6)$ bits and get the same security level for both confidentiality and ciphertext integrity. Hence a permutation of 400 bits width already delivers a security level of 128 bits.

Security proof. Our mode is provably secure in the random permutation model, whereas the security proof is short and verifiable by the third parties.

Parallelism. Our scheme is fully parallelizable: all blocks of plaintext and associated data can be processed in parallel; only the last call of the permutation needs all the operations to finish.

Online processing. Our scheme is fully online, being able to process plaintext blocks or blocks of associated data as soon as they are ready without knowing the final length. Patents. It is not patented, and we are not aware of any patent covering any part of the submission.

Tag update. If the tag is not truncated, then the last permutation call can be inverted given the key, and only two extra permutation calls are needed to encrypt and authenticate a new plaintext with one new block. Inverse. The permutation is used in the forward direction only, with a sole exception of tag update, if this feature is needed.

Nonce misuse. If the nonce is reused, then the integrity is still provided. Additionally, a user may generate a nonce out of the key, the plaintext, and the associated data with a dedicated routine.

1.1   Permutation AESQ

Alongside the mode, we propose an AES-based permutation called $AESQ$, which is 512-bit wide and has been optimized for recent CPU with AES instructions. It can be used in other permutation-based constructions, e.g. the extended Even-Mansour cipher or the sponge construction. It has a large security margin against the most popular attacks on permutation-based schemes: differential-based collision search, rebound attack, and meet-in-the-middle attack. The authenticated encryption scheme $PAEQ$ is the instantiation of PPAE with AESQ.

The wide permutation allows to get a much higher security level compared to the other AES-based designs (AES-GCM, AES-CBC+HMAC-SHA-256, OCB, COPA [1], COBRA, OTR). These schemes can not deliver a security level higher than 64 bits due to the birthday phenomena at the 128-bit AES state. In contrast, PAEQ easily brings the security level of 128 bits and higher. We note that this security level assumes a nonce-respecting adversary, who does not repeat nonces in the encryption requests.

PAEQ allows to encrypt and decrypt the data on the arbitrary number of subprocessors with tiny amount of shared memory. Those processors, threads, or other computation units may perform decryption and encryption of incoming blocks in any order.

PAEQ is based on the AES block cipher, and does not use any operations except those of AES. The mode of operation around the AES-based permutation AESQ uses only XORs and counter increments, and aims to be amongst easiest authenticated encryption modes to implement.

The paper is structured as follows. We provide a formal syntax of authenticated encryption modes and describe PPAE and PAEQ in this context in Section 2. We discuss its performance in Section 3 and give the design rationale in Section 4. The security aspects are discussed in Sections 5 and 6.

2 PAEQ as Authenticated Encryption Scheme

2.1 Notation

The *authenticated encryption scheme* is denoted by Π and is defined as a pair of functions $\mathcal{E}$ and $\mathcal{D}$, which provide encryption and decryption, respectively. The inputs to $\mathcal{E}$ are a plaintext $P \in \mathcal{P}$, associated data $A \in \mathcal{A}$, a nonce $N \in \mathcal{N} = \{0,1\}^r$, a key $K \in \mathcal{K} = \{0,1\}^k$, and a tag length τ. $\mathcal{P}$ is the set of byte strings with length between 1 and 2^{96}, and A is the set of byte strings with length between 0 and 2^{96}. The encryption function outputs ciphertext $C \in \mathcal{C}$ and tag $T \in \mathcal{T} = \{0,1\}^\tau$:

$$\mathcal{E} : \mathcal{K} \times \mathcal{N} \times \mathcal{A} \times \mathcal{P} \to \mathcal{C} \times \mathcal{T}.$$

For fixed (N, A, K) the encryption function is injective and hence defines the decryption function. The decryption function takes a key, a nonce, associated data, the ciphertext, and the tag as input and returns either a plaintext or the "invalid" message $\perp$:

$$\mathcal{D} : \mathcal{K} \times \mathcal{N} \times \mathcal{A} \times \mathcal{C} \times \mathcal{T} \to \mathcal{P} \cup \{\perp\}.$$

For brevity we will write $\mathcal{E}_K^{N,A}(\cdot)$ and $\mathcal{D}_K^{N,A}(\cdot)$.

To accomodate nonce-misuse cases, the user may choose to generate N as a keyed function of plaintext, key, and associated data. It is then called *extra nonce* and denoted by N_e. The user is then supposed to communicate N_e to the receiver, with ordinary nonce transmission rules applied.

Let X be an internal s-bit variable. Then we refer to its bits as $X[1]$, $X[2]$, ..., $X[s]$ and denote the subblock with bits from s_1 to s_2 as $X[s_1 \dots s_2]$. In pictures and formulas the least' significant bits and bytes are at the left, and the concatenation of multi-bit variables is defined as follows:

$$X\|Y = X[1]X[2]\dots X[t]Y[1]Y[2]\dots Y[s].$$

The counter values have their least significant bits as the least significant bits of corresponding variables.

PAEQ is a concrete instantiation of generic mode PPAE, which takes permutation $\mathcal{F}$ of width n.

2.2 PPAE Mode of Operation

For the domain separation the following two-byte constants are used:

$$D_i = (k, (r+i) \pmod{256}), \ i = 1,2,3,4,5,6,$$

where the second value is taken modulo 256.

We refer to Figure 2 for a graphical illustration of the PAEQ functionality.

Encryption.

1. During the first stage the plaintext is split into blocks P_i of $(n - k - 16)$ bits $(n/8 - k/8 - 2$ bytes$)$ $P_1, P_2, \ldots, P_l$. We encrypt block P_i as follows:

$$V_i \leftarrow D_0 || R_i || N || K;$$
$$W_i \leftarrow \mathcal{F}(V_i);$$
$$C_i \leftarrow W_i[17..(n - k)] \oplus P_i.$$

Here V_i, W_i are n-bit intermediate variables, and C_i is a $(n - k - 16)$-bit block. The counter $R_i = i$ occupies the $(n - k - r - 16)$-bit block. If last plaintext block is incomplete and has length t', then D_0 is replaced with D_1.

2. During the second stage we compute intermediate variables for authentication:

$$X_i \leftarrow D_2 || C_i || W_i[(n - k + 1)..n];$$
$$Y_i \leftarrow (\mathcal{F}(X_i))[17..(n - k)].$$

If the last plaintext block P_t is incomplete, then we define $P_t' = P_t || a || a || \ldots || a$, where a is one-byte variable with value equal to t', the byte length of P_t, and define

$$C_t' \leftarrow W_t[17..(n - k)] \oplus P_t';$$
$$X_t \leftarrow D_3 || C_t' || W_i[(n - k + 1)..n].$$

Finally,

$$Y \leftarrow \bigoplus_i Y_i;$$

3. During the third stage we compute an intermediate variable for authenticating associated data. The AD is splitted into blocks $A_1, A_2, \ldots, A_s$ of length $n - 2k - 16$. Then we compute

$$X_i' \leftarrow D_4 || R_i || A_i || K;$$
$$Y_i' \leftarrow (\mathcal{F}(X_i'))[17..(n - k)],$$

where R_i is a k-bit counter starting with 1. If the last AD block is incomplete (has t'' bytes), it is padded with bytes whose value is t'', and the constant D_4 is changed to D_5. Finally,

$$Y' \leftarrow \bigoplus_i Y_i';$$

4. In the final stage we compute the tag. First, we define

$$Z \leftarrow Y \bigoplus Y'.$$

and then,

$$T \leftarrow \mathcal{F}(D_6 || Z || K) \oplus (0^{n-k} || K),$$

where 0^{n-k} stands for $(n - k)$ zero bits.
The tag T is truncated to $T[1 \ldots \tau]$.

The encryption and authentication process is illustrated in Figure 2.

Decryption. The decryption process repeats the encryption process with minor corrections. We decrypt as

$$P_i \leftarrow W_i[17..(n-k)] \oplus C_i$$

with appropriate corrections for the incomplete block if needed. C_i is used as given when composing X_i. Finally, the tag T' is computed and matched with submitted T. If $T \neq T'$ (including length mismatch), the decryption process returns invalid.

The plaintext is returned only if the tags match.

2.3   The AESQ Permutation

The $\mathcal{F}$ permutation in PAEQ is the AESQ permutation, which is defined below. AESQ operates on 512-bit inputs, which are viewed as four 128-bit registers. The state undergoes 20 identical rounds. The rounds use standard AES operations: SubBytes, ShiftRows, and MixColumns, which are applied to the 128-bit registers exactly as in AES.

Input: 128-bit states A, B, C, D, round constants $Q_{i,j,k}$
for $0 \leq i < R = 10$ **do**
    for $0 \leq j < 2$ **do**
        $A \leftarrow$ MixColumns $\circ$ ShiftRows $\circ$ SubBytes(A);
        $A_0 \leftarrow A_0 \oplus Q_{i,j,1}$;
        $B \leftarrow$ MixColumns $\circ$ ShiftRows $\circ$ SubBytes(B);
        $B_0 \leftarrow B_0 \oplus Q_{i,j,2}$;
        $C \leftarrow$ MixColumns $\circ$ ShiftRows $\circ$ SubBytes(C);
        $C_0 \leftarrow C_0 \oplus Q_{i,j,3}$;
        $D \leftarrow$ MixColumns $\circ$ ShiftRows $\circ$ SubBytes(D);
        $D_0 \leftarrow D_0 \oplus Q_{i,j,4}$;
    end
    $(A, B, C, D) \leftarrow$ Shuffle(A, B, C, D)
end

Algorithm 1. Pseudocode for the AESQ permutation with $2R$ rounds

---

The round constants are chosen as follows in the matrix register view:

$$Q_{i,j,k} = \begin{pmatrix} 8i+4j+k & 8i+4j+k & 8i+4j+k & 8i+4j+k \\ 0 & 0 & 0 & 0 \\ 0 & 0 & 0 & 0 \\ 0 & 0 & 0 & 0 \end{pmatrix}$$

Here is the Shuffle mapping that permutes columns of the internal states:

	A				B			
From	$A[0]$	$A[1]$	$A[2]$	$A[3]$	$B[0]$	$B[1]$	$B[2]$	$B[3]$
To	$A[3]$	$D[3]$	$C[2]$	$B[2]$	$A[1]$	$D[1]$	$C[0]$	$B[0]$
	C				D			
From	$C[0]$	$C[1]$	$C[2]$	$C[3]$	$D[0]$	$D[1]$	$D[2]$	$D[3]$
To	$A[2]$	$D[2]$	$C[3]$	$B[3]$	$A[0]$	$D[0]$	$C[1]$	$B[1]$

2.4 Extra Nonce

The extra nonce N_e is a function of the key, the plaintext, and the associated data. It is not online, i.e. it needs to know the length of all inputs and the output. It is the sponge hash function:

$$T_1 = \mathcal{F}(Q_1||0^{2k}), \ T_2 = \mathcal{F}(T_1 \oplus (Q_2||0^{2k})), \ \ldots, \ T_m = \mathcal{F}(T_{m-1} \oplus (Q_m||0^{2k})),$$

and N_e is the truncation of T_m to r bits. The injection blocks Q_i come from the string Q, which is composed as follows:

$$Q = |P| \, || \, |A| \, || \, k_b || \, r_b \, || K || P || A || 10^*1,$$

where $|P|$ is the plaintext length in bytes, $|A|$ is the associated data length in bytes, k_b is the key length in bytes, r_b is the nonce length in bytes, and 10^*1 is the sponge padding: one byte with value 1, then as many zero bytes as needed to fill all but one bytes in the injection block, and then the byte with value 1.

3   Performance

Now we provide speed benchmarks for PAEQ. On the recent Haswell CPU family we obtain the speed of AESQ as 5 cycles per byte (cpb) for the 64-bit key, 6 cpb for the 128-bit key, and 9 cpb for the 256-bit key.

Security level / Key length	PAEQ (20 rounds, cycles per byte)
64	4.9
80	5.1
128	5.8
256	8.9

Our permutation is best suited for the last Intel and AMD processors equipped with special AES instructions. Of the AES-NI instruction set, we use only AES-ENC instruction that performs a single round of encryption. More precisely, AESENC(S, K) applies ShiftRows, SubBytes, and MixColumnsto S and then XORs the subkey K. In our scheme the subkey is a round constant. For two rounds of AESQ, we use the following instructions:

- 8 `aesenc` instructions for AES round calls;
- 8 `vpunpckhdq` instructions to permute the state columns between registers;
- 8 `vpaddq` instructions to update round constants.

This gives a total of 24 instructions per 64 bytes of the state in two rounds. This means that the full $2R$-round AESQ permutation needs $24R$ instructions, and theoretically may run in $3R/8$ cycles per byte if properly pipelined. This speed may be achieved in practice, since the throughput of AES instructions is 1 cycle starting from the Sandy Bridge architecture (2010), and the `vpaddq` instruction may be even faster. The following strategy gives us the best performance on the newest Haswell architecture:

- Process two 512-bit states in parallel in order to mitigate the latency of 8 cycles of the AESENC instruction. Store each state into 4 xmm registers.
- Store round constants in 4 xmm registers and use them in the two parallel computations.
- Use one xmm register to store the constant 1 to update the round constants, and one temporary register for the Shuffle operation.

We note that on the earlier Westmere architecture the AESENC instruction has latency 6 and throughput 2, hence the AES round calls should be interleaved with mixing instructions conducted on several parallel states.

We have made our own experiments and concluded that the 16-round version of AESQ ($R = 8$) runs at 1.5 cycles per byte on a Haswell-family CPU, whereas the 20-round version runs at 1.8 cycles per byte. This can be compared to the speed of the Keccak-1600 permutation. As eBASH reports, on a Haswell CPU the Keccak hash function with rate 1088 runs at 10.6 cpb, which implies that the full 1600-bit permutation runs at approximately $\frac{10.6 \cdot 1088}{1600} = 7.2$ cpb. Therefore, the 20-round AESQ is 4 times as fast as Keccak-1600.

4   Design Rationale

4.1   Design of PPAE

When creating the new scheme, we pursued the following goals:

- Offer high security level, up to 128 bits, ideally equal to the key length,
- Make the mode of operation simple enough to yield compact and reliable security proofs.
- Deliver as many features as possible.

To achieve these goals, we decided to trade performance for clarity and verifiability.

Existing block ciphers were poor choice for these goals. They commonly have a 128-bit block, which almost inevitably results in the loss of security at the level of 2^{64} cipher calls. The 256-bit cipher Threefish could have been used, but the lack of cryptanalysis in the single-key model makes it a risky candidate. Our mode of operation would be also restricted to a single cipher.

Instead, we constructed a permutation-based mode, which takes a permutation of any width if it is at least twice as large as the key. This choice makes the scheme much more flexible, allows for variable key and nonce length, and simplifies the proof. The key update also becomes very easy. The downside of the permutation-based approach is that the security proof has to be devised in the random oracle/permutation model, and does not rely on the PRP assumption. This is inevitable, but the success of the sponge-based constructions tells that it is not necessarily a drawback.

For the encryption stage we have chosen an analogy with the CTR mode, so that we do not have to use the permutation inverse. It also allows us to truncate some parts of the intermediate variables. For the authentication stage we use a parallel permutation-based construction. It takes the yet unused secret input from the encryption stage, which provides pseudo-randomness.

It remains to choose a permutation. Initially we thought of using a family of permutations with different widths. Examples could be Keccak [6], Spongent [8], or Quark [3] permutations. However, the performance loss would be too high given two invocations of a permutation per plaintext block. Instead, we designed our own permutation which shows the best performance on modern CPUs. It can be used in other permutation-based constructions, e.g. the extended Even-Mansour cipher or the sponge construction.

4.2   Design of AESQ

When designing AESQ, we needed a permutation wide enough to accomodate 128-bit keys and nonces. The AES permutation would be too short, while AES-based permutations used in the SHA-3 context would be too large or not well optimized for AES instructions on modern CPUs.

We decided to run 4 AES states in parallel and regularly shuffle the state bytes. Since two AES rounds provide full diffusion, the shuffle should occur every two rounds. The shuffle operation should make each state to affect all four states, resembling the ShiftRows transformation in AES. The recent Intel processors, along with dedicated AES instructions, provide a set of instructions that interleave the 32-bit subwords of 256-bit registers. Those subwords are columns of the AES state, so we shuffled the columns. The shuffle function in this submission is one of the permutations that provide full diffusion and needs the minimal 8 number of processor instructions.

5   Security of PPAE

In this section we provide the security analysis of the PPAE mode of operation, which takes a permutation $\mathcal{F}$ of width n. Though PAEQ fixes the permutation to AESQ and its width to 512 bits, the following proof is useful when defining other scheme within the PPAE mode.

The security of a AE scheme is defined as the inability to distinguish between the two worlds, where an adversary has access to some oracles and a permutation. One world consists of the encryption oracle $\mathcal{E}_K(\cdot, \cdot, \cdot)$ and decryption oracle

$\mathcal{D}_K(\cdot, \cdot, \cdot)$, where the secret key is randomly chosen and shared. The second world consists of the "random-bits" oracle $\$(\cdot, \cdot, \cdot, \cdot)$ and the "always-invalid" oracle $\perp (\cdot, \cdot, \cdot, \cdot)$. In addition, the adversary and all the oracles have an oracle access to the permutation $\mathcal{F}$. The encryption requests must be nonce-respecting, i.e. all the nonces in those requests must be distinct. A decryption request (N, A, C, T) shall not contain the ciphertext previously obtained with (N, A), but it may repeat nonces with other ciphertexts.

We give a security proof for a fixed key length k, nonce length r, and random permutation $\mathcal{F}$ of width n. The variable-length security proof is not given in this paper, but can be mounted thanks to the D_i constants. The proof also does not take into account the calls to $\mathcal{F}$ made in the extra nonce computation.

5.1   Confidentiality

We prove confidentiality as indistinguishability of the pair (ciphertext, tag) from a random string for the tag length n. The result for truncated tags is a trivial corollary. Let $\Pi[\mathcal{F}]$ be the PPAE mode instantiated with permutation $\mathcal{F}$. We do not consider the decryption oracle here, as we later show that with overwhelming probability it always returns $\perp$.

Theorem 1. *Suppose adversary $\mathcal{A}$ has access to $\Pi[\mathcal{F}]$, $\mathcal{F}$, and $\mathcal{F}^{-1}$, and let $K \xleftarrow{\$} \mathcal{K}$. Let σ_Π be the total number of queries to $\mathcal{F}$ made during the calls to the Π oracle, and $\sigma_\mathcal{F}$ be the total number of queries to $\mathcal{F}$ and $\mathcal{F}^{-1}$ oracles together. Then his advantage of distinguishing the oracle $\Pi[\mathcal{F}]$ from the random-bits oracle $\$$ is upper bounded as follows:*

$$\mathbf{Adv}_\Pi^{\mathrm{conf}}(\mathcal{A}) \leq \frac{(\sigma_\mathcal{F} + \sigma_\Pi)^2}{2^n} + \frac{2\sigma_\Pi^2}{2^{n-k-16}} + \frac{2\sigma_\mathcal{F}}{2^k} + \frac{2\sigma_\mathcal{F}\sigma_\Pi}{2^{n-16}}. \tag{1}$$

The proof can be found in the full version of this paper [7].

5.2   Ciphertext Integrity

Theorem 2. *Let an adversary be in the setting of Theorem 1. Suppose he makes q decryption requests to Π. Then his advantage of distinguishing Π^{-1} from the "always-invalid" oracle $\perp$ is upper bounded as follows:*

$$\mathbf{Adv}_\Pi^{\mathrm{int}}(\mathcal{A}) \leq \mathbf{Adv}_\Pi^{\mathrm{conf}}(\mathcal{A}) + \frac{\sigma_\Pi^2}{2^{n-k-16}} + \frac{q}{2^\tau} + \frac{\sigma_\Pi q}{2^{n-k-16}} + \frac{q}{2^k}. \tag{2}$$

The proof can be found in the full version of this paper [7].

5.3   Basic Robustness to Nonce Misuse and the Extra Nonce Feature

The details are available in the full version of this paper [7], where we prove the following theorem:

Theorem 3. *Let an adversary be able to repeat nonces, but not to repeat plaintexts. Suppose he makes q decryption requests to Π. Then his advantage of distinguishing Π^{-1} from the "always-invalid" oracle $\perp$ is upper bounded as follows:*

$$\mathbf{Adv}_{\Pi}^{\text{int}}(\mathcal{A}) \leq 2\frac{(\sigma_{\mathcal{F}} + \sigma_{\Pi})^2}{2^k}. \tag{3}$$

6   Security of AESQ

In this section we discuss the security of the AESQ permutation. We expect that the reader is familiar with properties of the AES internal operations, and refer to [12] in case of questions.

6.1   Structure and Decomposition

First we recall that one round of AES does not provide full diffusion as it mixes together only 32 bits. For instance, consider a column in the AES round that undergoes the MixColumns transformation. Before that its bytes have been permuted by ShiftRows and substituted by SubBytes; afterwards they are xored with a subkey and again go through SubBytes and ShiftRows. These operations can be grouped into a single 32-bit so-called "SuperSBox" parametrized with a 32-bit subkey. As a result, two AES rounds can be viewed as a layer SuperSubBytes of four 32-bit SuperSBoxes followed by ShiftRows and MixColumns, so that we can view AES-128 as a 5-round cipher with larger S-boxes.

The same strategy applies to the AESQ permutation. The two-round groups that process the registers A, B, C, D can be viewed as a single round with $16 = 4 \cdot 4$ parallel SuperSBoxes and a large linear transformation. This yields an R-round SPN permutation with 32-bit S-boxes out of the original $2R$-round one.

We can go further and view as many as four rounds of AESQ as a single round with MegaSBoxes (cf. MegaBoxes in [11]). Indeed, refer to Figure 1 and Algorithm 1. Consider a 128-bit register in round 2, for instance register A. Let us first look at its input. Going into backward direction, the columns spread to all the registers and then each column undergoes MixColumns without any influence from the other parts of the register. Then the values are shuffled by ShiftRows, and finally updated byte-wise by SubBytes. Hence we can recompute the 16 bytes at the beginning of round 1, even though they are located in different registers. Let us now look at the register A at the end of round 3. Its columns again spread to all the registers, and again undergo SubBytes and ShiftRows independently of the other register bytes. As a result, we can view as many as four rounds as a layer MegaSubBytes of four 128-bit MegaSBoxes followed by a MixColumns-based linear transformation, which we call MegaMixColumns (cf. the analysis in [16]). It has branch number 5, as it is exactly a set of MixColumns transformations with reordered inputs and outputs. Note that this decomposition must start with an odd round, and does not work for even rounds.

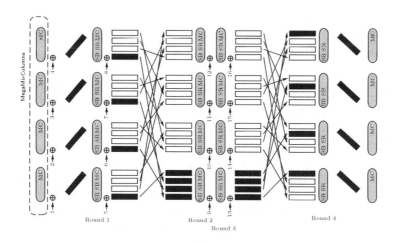

Fig. 1. MegaSBox in AESQ

6.2 Analysis of Permutations in the Attack Context

Only a few permutations as a single and secure object have been designed for the use in practical constructions. The most well-known is the Keccak 1600-bit permutation, which is used in the Keccak/SHA-3 hashing algorithm; the others are used in the SHA-3 competitors: CubeHash [5], Grostl [14], JH [26]. It is worth noticing that a permutation per se can not be formally defined "secure". The best we can make is an informal statement like the 2^l *"flat sponge" claim* [6], which basically states that no attack with complexity below 2^l and specific for the particular permutation exists. The parameter l is used in defining the capacity parameters in sponge functions and in fact measures the designers' confidence.

In our case we claim $l = 256$ or the 256-bit security of AESQ against all attacks. In order to support our claim, we look at the existing attacks on permutation-based designs and check if they apply to AESQ.

Collision attacks. We first consider collision attacks on sponge-based hash functions. The collision attacks on the reduced Keccak [13] strongly rely on high-probability differential trails [21], and only add a couple of rounds over their length with the help of message-modification techniques. The so-called internal-differential attack [22], while exploiting similarities within the internal state, is also limited by the propagation of difference generated by the round constants. Hence to prevent these attacks we have to demonstrate the absence of high-probability differential trails for a high number of rounds.

Let us now consider compression functions based on permutations. For example, Grostl uses functions

$$P(x \oplus y) \oplus Q(y) \oplus x \quad \text{and} \quad x \oplus P(x),$$

where P and Q are AES-based permutations. The main strategy in collision attacks on the AES-based designs [15, 19] is the construction of a truncated differential trail with low input and output Hamming weight. Then the conforming inputs are found with the rebound attack and are tested for a collision.

Preimage attacks. The preimage attacks on sponge-based hash functions have been also based on the differential properties of the permutation. As long as a differential generated by message difference ΔM has high probability in some output bits, it can be used to speed up the preimage search [20]. There are also generic methods that can save a factor of several bits by exploiting incomplete diffusion in the final rounds, but we note that their complexity can not be reduced much. The invariant attacks [2] do not apply because of round constants.

Preimage attacks on Grostl are based on the meet-in-the-middle framework. Whereas it is difficult to formalize the necessary conditions for these attacks to work, we notice that the number of rounds attacked with meet-in-the-middle or bicliques is smaller compared to the rebound attack (even though the attack goals are distinct).

Other attacks. A more generic set of attacks are given by the CICO and multi-CICO problems [6], which require the attacker to find one or more (input, output) pair conforming to certain bit conditions. There is no comprehensive treatment of these attacks, but they seem to be limited by twice the number of rounds needed for full diffusion.

Consider an 8-round version F of AESQ, for instance, that $F(X) = Y$, and we know the last 384 bits of both X and Y. How difficult is it to restore X and Y? The last 384 bits mean three registers B, C, D. Hence, we can compute registers B, C, D through rounds 0,1 and 6,7. Then we know 96 bits of each register in the beginning of round 2 and in the end of round 5. By a simple meet-in-the-middle attack we match between rounds 3 and 4 and recover X and Y with complexity 2^{32}. However, we did not manage to extend this technique to more rounds.

6.3 Possible Attacks

Differential analysis of AESQ. Let us evaluate the differential properties of AESQ. We are backed up with the analysis of AES by Daemen and Rijmen. They prove the following lemma[1].

Lemma 1. *Any differential trail over 4 AES rounds has at least 25 active S-boxes.*

We have demonstrated in Section 6.1 that four rounds of AESQ can be viewed as a layer of 128-bit S-boxes and a large linear transformation, which has branch number 5. The 128-bit MegaSBoxes are exactly four-rounds of AES without the key additions and the last MixColumns. Hence each active MegaSBox has at least 25 active S-boxes.

The branch number of MegaMixColumns implies that eight AESQ rounds starting from an odd round have at least 5 active MegaSBoxes, and hence 125

---

[1] Actually, they prove a more general theorem for SPN ciphers but the AESQ permutation does not satisfy its assumptions.

active S-boxes. Therefore, nine rounds starting from an even round (as required by the permutation) have at least 126 active S-boxes, as the first round must have at least one. More generally, $8R + 1$ rounds must have at least $125R + 1$ active S-boxes. In turn, the last round has at maximum 64 active S-boxes, so the lower bound for $8R$ rounds is $125R - 63$ active S-boxes. The other bounds are as follows:

Rounds	Active S-boxes	Rounds	Active S-boxes
2	5	8R+2	125R +2
4	10	8R+4	125R +7
6	27	8R+6	125R + 27
8R	125R-63	8R+8	125R+68

The table implies that a 20-round trail has at least 257 active S-boxes. Given that the maximal differential probability of each S-box is 2^{-6}, the vast majority of these trails have differential probability close to zero. The effect of clustering these trails into a single differential and the resulting probability is not completely studied, but the bound of maximum expected differential probability for 4 rounds of AES of 2^{-113} [17] indicates that these values should be very low for 16 rounds of AESQ and more unless the round constants accidentally admit a large number of conforming pair for a differential (the so called *height* parameter).

If we consider slide-like attacks, where differences between registers or states shifted by several rounds are considered, we obtain that the round constants generate one-byte difference in every round, so we can not hope for high probability differentials in this case.

Rebound attacks. Rebound attacks aim to construct a conforming input for a truncated differential trail with a little amortized cost. The attack consists of two phases: inbound phase, where a conforming input is constructed for the low-probability part in the meet-in-the-middle manner, and the outbound phase, where the differences are traced through the rest of the primitive without the control of the cryptanalyst.

Again, given the decomposition of AESQ into MegaSBoxes, we can generate a full difference distribution table for MegaSBoxes with the complexity 2^{256}. Consider MegaSBoxes in rounds 5–8 of the permutation. Let us select an active S-box in rounds 4 and 9, and select a one-byte difference in them. Then we apply MegaMixColumns and obtain input/output differences for MegaSBoxes. We obtain actual state values and then recompute the difference in the outer directions. Active S-boxes in round 4 activates one MegaSBox in rounds 1–4, and then activates all the S-boxes of round 0. At the other end, the active S-box in round 9 activates a MegaSBox in rounds 9–12, which activates all S-boxes in round 13. This yields a clear distinguisher for 12 rounds of AESQ with complexity 2^{256}, and possibly a distinguisher for 14 rounds.

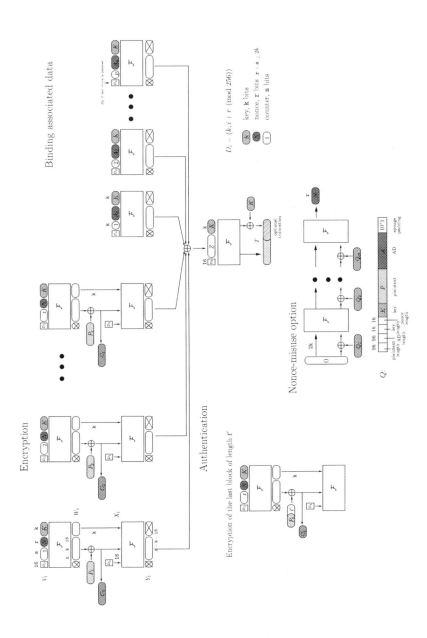

Fig. 2. Encryption and authentication with PAEQ

Final choice. We take the 20-round AESQ permutation within PAEQ, which should provide a 256-bit security against all attacks. After the third-party cryptanalysis, if little improvement over our analysis will be shown, we might consider reducing the number of rounds to 16 for 80- and 128-bit security.

Acknowledgement. We are grateful to Yann Le Corre for help with AESQ implementation.

7   Conclusion

We have proposed a new mode PPAE and its instantiation PAEQ, that delivers virtually all features of a modern authenticated encryption scheme. To the best of our knowledge, no other scheme combines parallelism, 128-bit security, online processing and a quick tag update. The scheme has a compact security proof in the random permutation model.

Unfortunately, these advantages come at some cost. Our scheme employs $(2 + \epsilon)$ calls to a permutation of width n to encrypt n bits of the message. To mitigate this effect, we suggested to use a reduced Keccak permutation or a new AES-like permutation AESQ of width 512 bits. We conduct the basic security analysis of AESQ and concluded that the most relevant attacks leave a significant security margin. The resulting scheme runs at 6-7 cycles per byte on modern processors.

For the future research, we plan to extend our security proof to capture variable-length keys and nonces. We also plan to optimize the permutation AESQ so that it provides the same security with reduced number of operations.

References

1. Andreeva, E., Bogdanov, A., Luykx, A., Mennink, B., Tischhauser, E., Yasuda, K.: Parallelizable and authenticated online ciphers. In: Sako, K., Sarkar, P. (eds.) ASIACRYPT 2013, Part I. LNCS, vol. 8269, pp. 424–443. Springer, Heidelberg (2013)
2. Aumasson, J.-P., Brier, E., Meier, W., Naya-Plasencia, M., Peyrin, T.: Inside the hypercube. In: Boyd, C., González Nieto, J. (eds.) ACISP 2009. LNCS, vol. 5594, pp. 202–213. Springer, Heidelberg (2009)
3. Aumasson, J.-P., Henzen, L., Meier, W., Naya-Plasencia, M.: QUARK: A lightweight hash. In: Mangard, S., Standaert, F.-X. (eds.) CHES 2010. LNCS, vol. 6225, pp. 1–15. Springer, Heidelberg (2010), https://131002.net/quark/quark_full.pdf
4. Bellare, M., Namprempre, C.: Authenticated encryption: Relations among notions and analysis of the generic composition paradigm. In: Okamoto, T. (ed.) ASIACRYPT 2000. LNCS, vol. 1976, pp. 531–545. Springer, Heidelberg (2000)
5. Bernstein, D.J.: CubeHash specification (2.b.1). Submission to NIST (Round 2) (2009)
6. Bertoni, G., Daemen, J., Peeters, M., Van Assche, G.: The Keccak reference, version 3.0 (2011), http://keccak.noekeon.org/Keccak-reference-3.0.pdf

7. Biryukov, A., Khovratovich, D.: Paeq: Parallelizable permutation-based authenticated encryption (full version). Technical report, University of Luxembourg (2014), https://www.cryptolux.org/index.php/PAEQ

8. Bogdanov, A., Knežević, M., Leander, G., Toz, D., Varıcı, K., Verbauwhede, I.: SPONGENT: A lightweight hash function. In: Preneel, B., Takagi, T. (eds.) CHES 2011. LNCS, vol. 6917, pp. 312–325. Springer, Heidelberg (2011)

9. Bogdanov, A., Mendel, F., Regazzoni, F., Rijmen, V., Tischhauser, E.: ALE: AES-based lightweight authenticated encryption. In: Moriai, S. (ed.) FSE 2013. LNCS, vol. 8424, pp. 447–466. Springer, Heidelberg (2014), http://www2.compute.dtu.dk/~anbog/fse13-ale.pdf

10. Daemen, J.: Permutation-based symmetric cryptography and Keccak. Technical report, Ecrypt II, Crypto for 2020 Invited Talk (2013), https://www.cosic.esat.kuleuven.be/ecrypt/cryptofor2020/slides/KeccakEcryptTenerife.pdf

11. Daemen, J., Lamberger, M., Pramstaller, N., Rijmen, V., Vercauteren, F.: Computational aspects of the expected differential probability of 4-round aes and aes-like ciphers. Computing 85(1-2), 85–104 (2009)

12. Daemen, J., Rijmen, V.: The Design of Rijndael. AES— The Advanced Encryption Standard. Springer (2002)

13. Dinur, I., Dunkelman, O., Shamir, A.: New attacks on Keccak-224 and Keccak-256. In: Canteaut, A. (ed.) FSE 2012. LNCS, vol. 7549, pp. 442–461. Springer, Heidelberg (2012)

14. Gauravaram, P., Knudsen, L.R., Matusiewicz, K., Mendel, F., Rechberger, C., Schlffer, M., Thomsen, S.S.: Grøstl – a SHA-3 candidate. Submission to NIST (Round 3) (2011)

15. Ideguchi, K., Tischhauser, E., Preneel, B.: Improved collision attacks on the reduced-round Grøstl hash function. In: Burmester, M., Tsudik, G., Magliveras, S., Ilić, I. (eds.) ISC 2010. LNCS, vol. 6531, pp. 1–16. Springer, Heidelberg (2011)

16. Jean, J., Naya-Plasencia, M., Schläffer, M.: Improved Analysis of ECHO-256. In: Miri, A., Vaudenay, S. (eds.) SAC 2011. LNCS, vol. 7118, pp. 19–36. Springer, Heidelberg (2012)

17. Keliher, L., Sui, J.: Exact maximum expected differential and linear probability for 2-round Advanced Encryption Standard (AES). IACR Cryptology ePrint Archive, 2005:321 (2005)

18. Krovetz, T., Rogaway, P.: The software performance of authenticated-encryption modes. In: Joux, A. (ed.) FSE 2011. LNCS, vol. 6733, pp. 306–327. Springer, Heidelberg (2011)

19. Mendel, F., Rechberger, C., Schläffer, M., Thomsen, S.S.: Rebound attacks on the reduced Grøstl hash function. In: Pieprzyk, J. (ed.) CT-RSA 2010. LNCS, vol. 5985, pp. 350–365. Springer, Heidelberg (2010)

20. Morawiecki, P., Pieprzyk, J., Srebrny, M., Straus, M.: Preimage attacks on the round-reduced Keccak with the aid of differential cryptanalysis. Cryptology ePrint Archive, Report 2013/561 (2013), http://eprint.iacr.org/

21. Naya-Plasencia, M., Röck, A., Meier, W.: Practical analysis of reduced-round KECCAK. In: Bernstein, D.J., Chatterjee, S. (eds.) INDOCRYPT 2011. LNCS, vol. 7107, pp. 236–254. Springer, Heidelberg (2011)

22. Peyrin, T.: Improved differential attacks for ECHO and Grøstl. In: Rabin, T. (ed.) CRYPTO 2010. LNCS, vol. 6223, pp. 370–392. Springer, Heidelberg (2010)

23. Rogaway, P.: Authenticated-encryption with associated-data. In: ACM Conference on Computer and Communications Security 2002, pp. 98–107 (2002)
24. ISO/IEC 19772 JTC 1 SC 27. Information technology – Security techniques – Authenticated encryption (2009)
25. Whiting, D., Schneier, B., Lucks, S., Muller, F.: Phelix: Fast encryption and authentication in a single cryptographic primitive. Technical report
26. Wu, H.: The hash function JH. Submission to NIST (round 3) (2011)
27. Wu, H., Preneel, B.: AEGIS: A fast authenticated encryption algorithm. Cryptology ePrint Archive, Report 2013/695 (2013), http://eprint.iacr.org/2013/695

(Pseudo-) Preimage Attacks on Step-Reduced HAS-160 and RIPEMD-160

Gaoli Wang[1,2] and Yanzhao Shen[3,4]

[1] School of Computer Science and Technology, Donghua University, Shanghai 201620, China
[2] State Key Laboratory of Information Security
Institute of Information Engineering, Chinese Academy of Sciences, Beijing, China
wanggaoli@dhu.edu.cn
[3] Key Laboratory of Cryptologic Technology and Information Security, Ministry of Education, Shandong University, Jinan 250100, China
[4] School of Mathematics, Shandong University, Jinan 250100, China

Abstract. The hash function HAS-160 is standardized by the Korean government and widely used in Korea, and the hash function RIPEMD-160 is a worldwide ISO/IEC standard. In this paper, by careful analysis of the two hash functions, we propose a pseudo-preimage attack on 71-step HAS-160 (no padding) with complexity $2^{158.13}$ and a preimage attack on 34-step RIPEMD-160 (with padding) with complexity $2^{158.91}$. Both of the attacks are from the first step. The improved results are derived from the differential meet-in-the-middle attack and biclique technique etc. As far as we know, they are the best pseudo-preimage and preimage attacks on step-reduced HAS-160 and RIPEMD-160 respectively in terms of the step number.

Keywords: Preimage attack, HAS-160, RIPEMD-160, Differential meet-in-the-middle attack, Biclique, Hash function.

1   Introduction

Hash Functions are the important primitives which are used for numerous security protocols, such as digital signature and message authentications. Hash functions should satisfy several security properties such as collision resistance, preimage resistance and second preimage resistance. With the breakthroughs in the collision attacks on a series of standard hash functions such as MD5, SHA-0 and SHA-1 [22–24], preimage attack has drawn a great amount of attention from many researchers (see [1, 5, 7, 9, 15–17] for example). As an n-bit hash function, it should be required to have preimage resistance up to 2^n computations. Up to now, the meet-in-the-middle technique [1, 3] is the basic tool used in preimage attacks and improved by splice-and-cut technique [1] which consider the first and last steps of the attack target as consecutive steps. The meet-in-the-middle technique is also improved by the other techniques, such as biclique technique [7], initial structure technique [17] and so on, which make the meet-in-the-middle technique skip more steps and attack more steps. Knellwolf and Khovratovich proposed the differential meet-in-the-middle technique [9] at Crypto 2012 which improves the preimage attack on hash functions especially the ones with linear message expansion and weak diffusion properties.

S.S.M. Chow et al. (Eds.): ISC 2014, LNCS 8783, pp. 90–103, 2014.
© Springer International Publishing Switzerland 2014

HAS-160 [19] is the Korean hash standard which was developed by Korean government in 2000 and widely used in Korea. The structure of HAS-160 resembles the structure of SHA-0 and SHA-1. However, the message expansion of HAS-160 is different. At ICISC 2005, Yun et al. [20] presented the first cryptanalysis on HAS-160. By applying techniques introduced by Wang et al. [21], they proposed a collision attack on 45-step HAS-160 with complexity 2^{12}. At ICISC 2006, Cho et al. [2] extended the previous work to 53 steps with complexity about 2^{55}. No significantly new technique was used other than those in [20], but by selecting message differences judiciously and constructing more complicated first round differential path, they extend the result up to 53 steps. Later Mendel and Rijmen [11] improved the attack and reduced the complexity to 2^{35} by using a slightly different message modification technique to fulfill the conditions on the state variables in the first 20 steps. Furthermore, they presented a collision message pair for 53-step HAS-160 and extended the attack to 59 steps. At ICISC 2011, Mendel et al. [12] constructed a long differential characteristic by connecting two short ones by a complex third characteristic and proposed a semi-free-start collision on 65-step HAS-160 with practical complexity. At ICISC 2008, Sasaki and Aoki [16] proposed the first preimage attack on 52-step HAS-160. They obtained a 48-step preimage attack with complexity 2^{145} by using splice-and-cut and partial-matching techniques and a 52-step preimage attack with complexity 2^{153} by using local-collision technique. And Hong et al. [5] improved the attack up to 68 steps at ICISC 2009. This attack is based on the previous preimage attack on 52 steps of HAS-160 [16]. Recently, Kircanski et al. [8] proposed a boomerang attack on the full version of HAS-160 which illustrates the compression function of HAS-160 is non-random.

RIPEMD-160 [4] is the strengthened version of RIPEMD and standardized in ISO/IEC 10118-3:2003 [6] that is yet unbroken and is present in many implementations of security protocols. Its compression function uses a double-branch structure. At ISC 2006, Mendel et al. [10] proposed an analysis on a round-reduced variant of RIPEMD-160 by combining methods from coding theory with the techniques which were successfully used in the attack on SHA-1. At Inscrypt 2010, Ohtahara et al. [15] proposed a second-preimage attack on 30-step RIPEMD-160 with complexity 2^{155} and a preimage attack on 31-step RIPEMD-160 with complexity 2^{155} by using the initial structure and local-collision techniques. At ACNS 2012, Sasaki and Wang [18] proposed 2-dimension sum distinguishers on 42-step RIPEMD-160 and 51-step RIPEMD-160 with complexities of 2^{36} and 2^{158} respectively by using a non-linear differential property of RIPEMD-160 which can avoid the quick propagation of the difference. At ISC 2012, Mendel et al. [13] proposed a semi-free-start near-collision attack for the middle 48-step RIPEMD-160 and a semi-free-start collision attack for 36-step RIPEMD-160 which was the first application of the attacks of Wang et al. on MD5 and SHA-1 [23, 24] to RIPEMD-160. At ASIACRYPT 2013, Mendel et al. [14] presented a semi-free-start collision attack for 42-step RIPEMD-160 and a semi-free-start collision attack for 36-step RIPEMD-160 which starts from the first step by using a carefully designed non-linear path search tool.

In this paper, we propose a pseudo-preimage attack on 71-step HAS-160 (no padding) with complexity $2^{158.13}$ and a preimage attack on 34-step RIPEMD-160 (with padding) with complexity $2^{158.91}$ using the differential meet-in-the-middle and biclique techniques. Both of the attacks are from the first step. Furthermore, the 6-step bicliques (in HAS-160) and 7-step bicliques (in RIPEMD-160) can be constructed in the intermediate step and the biclique search algorithm is given in this paper. How to choose the linear space is a difficult task. The bit positions of the linear space not only affect the bit position of the truncation mask vector and the complexity of the attack but also affect the existence of the truncation mask vector and bicliques. Therefore, we should choose a proper linear space to make sure the truncation mask vector and bicliques are valid. As far as we know, they are the best pseudo-preimage attack on step-reduced HAS-160 and the best preimage attack on step-reduced RIPEMD-160 in term of the step number. The summary of the previous work and our results about the (pseudo-)preimage attacks are given in Table 1.

Paper Outline. The rest of this paper is organized as follows. Section 2 gives a description of differential meet-in-the-middle attack with bicliques. Section 3 gives a brief description of HAS-160 and RIPEMD-160. Section 4 presents the pseudo-preimage attack on step-reduced HAS-160 and preimage attack on step-reduced RIPEMD-160. We conclude the paper in Section 5 and give the discussion in Section 6.

Table 1. Summary of the (pseudo-)preimage attacks on HAS-160 and RIPEMD-160

Hash Function	Steps	Complexity		Reference
		Pseudo-preimage	Preimage	
HAS-160	$48(1\rightarrow48, 21\rightarrow68)$	2^{128}	2^{145}	[16]
	$52(12\rightarrow63)$	2^{144}	2^{153}	[16]
	$65(0\rightarrow64)$	$2^{143.4}$	$2^{152.7}$	[5]
	$67(0\rightarrow66)$	2^{154}	2^{158}	[5]
	$68(12\rightarrow79)$	$2^{150.7}$	$2^{156.3}$	[5]
	$71(0\rightarrow70)$	$2^{158.13}$	-	Section 4.1
RIPEMD-160	$31(50\rightarrow79)$	2^{148}	2^{155}	[15]
	$34(0\rightarrow33)$	$2^{155.81}$	$2^{158.91}$	Section 4.2

2   Differential Meet-in-the-Middle Attack with Bicliques

The differential meet-in-the-middle attack [9] is the extension of the meet-in-the-middle attack by combining the differential framework. It can propose preimage attacks on more steps of hash functions, which mainly benefits from the statistical tool. However, for a fast diffusion hash function, it may not increase the attacked steps significantly compared with the traditional meet-in-the-middle attack. The differential meet-in-the-middle attack is fully compatible with the splice-and-cut technique [1], which connects the last and the first steps via the feed-forward of the Davies-Meyer mode.

2.1   The Meet-in-the-Middle Technique

The meet-in-the-middle preimage attack [1, 3] is a type of birthday attack and makes use of a space-time tradeoff. The meet-in-the-middle preimage attack can be summarized as follows(see also Fig 1).

- Choose neutral words W_b and W_f respectively, and split the compression function into two parts c_b and c_f according to the neutral words, where c_b is independent from neutral word W_f and c_f is independent from neutral word W_b. Here we call the two parts the backward direction and the forward direction respectively.
- Assign random value to the chaining values at the splitting point and fix all other message words except W_b and W_f. For all possible values of W_b, compute backward from the splitting point and obtain the value at the matching point. Store the values in a list L_b.
- For all possible values of W_f, compute forward from the splitting point and obtain the value at the matching point. Check if there exists an entry in L_b that matches the result (all the state bits or only some bits of the state) at the matching point.
- Repeat the above two steps with different initial assignments until find a full match.

The above four steps offer a way to return a pseudo-preimage of the given hash function because the initial value is determined during the attack. Furthermore, we can convert pseudo-preimages to a preimage by using the generic algorithm that converts pseudo-preimages to a preimage, see the detail in [25].

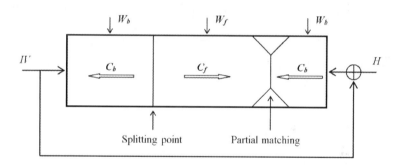

Fig. 1. Schematic view of the meet-in-the-middle attack

2.2   The Differential Meet-in-the-Middle Attack Algorithm

Combined with the biclique technique [7], the differential meet-in-the-middle preimage attack can be summarized as follows (pseudo-preimage attack and see also Fig 2).

- **Split the compression function and construct the linear spaces.** Split the compression function CF into 3 parts, $CF = CF_3 \cdot CF_2 \cdot CF_1$. The bicliques are precomputed and constructed for CF_3. Find the linear spaces LD_1, LD_2 which satisfy $LD_1 \cap LD_2 = \{0\}$. Note that LD_1 is the linear space used in CF_1 and LD_2 is the linear space used in CF_2^{-1}. Furthermore, assume the chaining value P is an input to CF_1 and the chaining value Q is an input to CF_3 (or an output of CF_2).
- **Search for the mask vector T.** Randomly choose M and Q, and compute $P = CF_3(M, Q)$.
  - For each $\delta_1 \in LD_1$, search for the difference Δ_1, such that the equation $\Delta_1 = CF_1(M, P) \oplus CF_1(M \oplus \delta_1, P)$ holds with high probability.

- For each $\delta_2 \in LD_2$, search for the difference Δ_2, such that the equation $\Delta_2 = CF_2^{-1}(M, Q) \oplus CF_2^{-1}(M \oplus \delta_2, Q)$ holds with high probability.
- For each $(\delta_1, \delta_2) \in LD_1 \times LD_2$, compute $\Delta = CF_1(M \oplus \delta_1, P) \oplus \Delta_1 \oplus CF_2^{-1}(M \oplus \delta_2, Q) \oplus \Delta_2$. For each bit i $(0 \leq i \leq n - 1)$, count the number of $\Delta_i = 1$. Set those d bits of T to 1 which have the lowest counters and the other bits of T to 0. Here, $T \in \{0, 1\}^n$ is a truncation mask vector, d is the dimension of LD_1 or LD_2, and n is the length of the hash value.
- **Construct the bicliques in** CF_3. We use the biclique search algorithm (Section 2.3) to construct the bicliques in CF_3 such that for all $(\delta_1, \delta_2) \in LD_1 \times LD_2$, $P[\delta_2] = CF_3(M \oplus \delta_1 \oplus \delta_2, Q[\delta_1])$ hold.
- **Search for the candidate preimage.** Compute the list $L_1 = CF_1(M \oplus \delta_2, P[\delta_2]) \oplus \Delta_2$ and the list $L_2 = CF_2^{-1}(M \oplus \delta_1, Q[\delta_1]) \oplus \Delta_1$. If $L_1 =_T L_2$, then $M \oplus \delta_1 \oplus \delta_2$ is a candidate pseudo-preimage. For a truncation mask vector $T \in \{0, 1\}^n$, $L_1 =_T L_2$ means $L_1 \wedge T = L_2 \wedge T$, where $\wedge$ denotes bitwise AND.

The error is defined as: $M \oplus \delta_1 \oplus \delta_2$ is a preimage, but we reject it falsely. Count the number that $\Delta \neq_T 0^n$ and denote it as c, so the error probability $\bar{\alpha}$ can be estimated as the proportion of c to the total test number. The total complexity of the differential meet-in-the-middle attack is $(2^{n-d}\Gamma + 2^{n-r}\Gamma_{re})/(1 - \bar{\alpha})$, where n is the length of the hash value, d is the dimension of LD_1 or LD_2, Γ is the cost of CF_1 and CF_2, r is the Hamming weight of T, and Γ_{re} is the cost of retesting a candidate preimage.

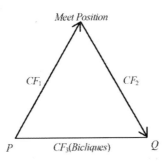

Fig. 2. Schematic view of the differential meet-in-the-middle attack with bicliques

2.3   The Biclique Search Algorithm

The bicliques are constructed for CF_3 which is from step s_1 to step s_2. Assume the chaining value before step s_1 is Q and the chaining value after step s_2 is P. A biclique for CF_3 is a tuple $\{M, LD_1, LD_2, Q, P\}$, where M is a message, LD_1 and LD_2 are linear difference spaces of dimension d respectively, Q is the list of 2^d states $Q[\delta_1]$ for $\delta_1 \in LD_1$ and P is the list of 2^d states $P[\delta_2]$ for $\delta_2 \in LD_2$, such that for all $(\delta_1, \delta_2) \in LD_1 \times LD_2$, $P[\delta_2] = CF_3(M \oplus \delta_1 \oplus \delta_2, Q[\delta_1])$ hold. The dimension of the biclique is equal to the dimension of the linear space. In order to obtain a biclique, we linearize the step function, i.e., replacing $+$ by $\oplus$ and setting P, Q and the constants to 0 to track the differences propagating of the intermediate chaining values. Meanwhile, in each step of the hash function, by adding a set of sufficient conditions, we can make sure most of the differences of the outputs of the non-linear Boolean functions are zero. Then we can get the output differences of each step. The bicliques can be searched from the intermediate step and the search algorithm is described as follows.

- Split the CF_3 into 2 parts, $CF_3 = CF_3^2 \cdot CF_3^1$. Assume CV is the intermediate chaining value which connects CF_3^1 and CF_3^2.
- Assume $FT(\delta_2)$ is the differences of the outputs of CF_3^1 corresponding to δ_2, and $BT(\delta_1)$ is the differences of the outputs of $(CF_3^2)^{-1}$ corresponding to δ_1. Randomly choose M and CV, and make sure a part of the sufficient conditions hold. Search the bicliques as follows.
  - For every $\delta_1 \in LD_1$, search for the chaining value $Q[\delta_1]$, such that $Q[\delta_1] = (CF_3^1)^{-1}(M \oplus \delta_2, CV \oplus FT(\delta_2) \oplus BT(\delta_1))$ hold for every $\delta_2 \in LD_2$.
  - For every $\delta_2 \in LD_2$, search for the chaining value $P[\delta_2]$, such that $P[\delta_2] = CF_3^2(M \oplus \delta_1, CV \oplus FT(\delta_2) \oplus BT(\delta_1))$ hold for every $\delta_1 \in LD_1$.
- If $\{M, LD_1, LD_2, Q, P\}$ constitute a biclique, return $(M, Q[0])$. Otherwise, repeat the above steps until a biclique is constructed.

3 Description of HAS-160 and RIPEMD-160

The HAS-160 and RIPEMD-160 hash functions compress any arbitrary (less than 2^{64}) bit-length message into a 160-bit (five 32-bit words) hash value. Given any message, the two algorithms first pad it into a message with a length that is a multiple of 512 bits. The compression function of HAS-160 and RIPEMD-160 basically consist of two parts: the message expansion and the state update transformation. For the detailed descriptions of the hash functions we refer to [19] and [4].

3.1 The Compression Function of HAS-160

In this section, we mainly describe the compression function of HAS-160.

Message Expansion. The message expansion of HAS-160 splits the 512-bit message block M into 16 words $m_i(i = 0, \ldots, 15)$, and expands them into 80 expanded message words $w_i(0 \le i \le 79)$. The expansion is specified in Table 2.

State Update Transformation. The state update transformation of HAS-160 starts from an initial value $IV = (A_0, B_0, C_0, D_0, E_0)$ of five 32-bit words and updates them in 80 steps. In step $i(0 \le i \le 79)$ the 32-bit word w_i is used to update the state variables A_i, B_i, C_i, D_i, E_i as follows:

$$A_{i+1} = ((A_i \lll s_i^1) + F_i(B_i, C_i, D_i) + E_i + w_i + k_i,$$
$$B_{i+1} = A_i, C_{i+1} = (B_i \lll s_i^2),$$
$$D_{i+1} = C_i, E_{i+1} = D_i.$$

The bitwise Boolean function $F_i(B_i, C_i, D_i)$, constant k_i and rotation s_i^2 used in each step are defined as follows:

Table 2. Message expansion of HAS-60

0	1	2	3	4	5	6	7	8	9	10	11	12	13	14	15	16	17	18	19
$m_8 \oplus m_9 \oplus m_{10} \oplus m_{11}$	m_0	m_1	m_2	m_3	$m_{12} \oplus m_{13} \oplus m_{14} \oplus m_{15}$	m_4	m_5	m_6	m_7	$m_0 \oplus m_1 \oplus m_2 \oplus m_3$	m_8	m_9	m_{10}	m_{11}	$m_4 \oplus m_5 \oplus m_6 \oplus m_7$	m_{12}	m_{13}	m_{14}	m_{15}

20	21	22	23	24	25	26	27	28	29	30	31	32	33	34	35	36	37	38	39
$m_{11} \oplus m_{14} \oplus m_1 \oplus m_4$	m_3	m_6	m_9	m_{12}	$m_7 \oplus m_{10} \oplus m_{13} \oplus m_0$	m_{15}	m_2	m_5	m_8	$m_3 \oplus m_6 \oplus m_9 \oplus m_{12}$	m_{11}	m_{14}	m_1	m_4	$m_{15} \oplus m_2 \oplus m_5 \oplus m_8$	m_7	m_{10}	m_{13}	m_0

40	41	42	43	44	45	46	47	48	49	50	51	52	53	54	55	56	57	58	59
$m_4 \oplus m_{13} \oplus m_6 \oplus m_{15}$	m_{12}	m_5	m_{14}	m_7	$m_8 \oplus m_1 \oplus m_{10} \oplus m_3$	m_0	m_9	m_2	m_{11}	$m_{12} \oplus m_5 \oplus m_{14} \oplus m_7$	m_4	m_{13}	m_6	m_{15}	$m_0 \oplus m_9 \oplus m_2 \oplus m_{11}$	m_8	m_1	m_{10}	m_3

60	61	62	63	64	65	66	67	68	69	70	71	72	73	74	75	76	77	78	79
$m_{15} \oplus m_{10} \oplus m_5 \oplus m_0$	m_7	m_2	m_{13}	m_8	$m_{11} \oplus m_6 \oplus m_1 \oplus m_{12}$	m_3	m_{14}	m_9	m_4	$m_7 \oplus m_2 \oplus m_{13} \oplus m_8$	m_{15}	m_{10}	m_5	m_0	$m_3 \oplus m_{14} \oplus m_9 \oplus m_4$	m_{11}	m_6	m_1	m_{12}

step	$F_i(B_i, C_i, D_i)$	k_i	s_i^2
$0 \le i \le 19$	$(B_i \wedge C_i) \vee (\neg B_i \wedge D_i)$	$0x00000000$	10
$20 \le i \le 39$	$B_i \oplus C_i \oplus D_i$	$0x5a827999$	17
$40 \le i \le 59$	$C_i \oplus (B_i \vee \neg D_i)$	$0x6ed9eba1$	25
$60 \le i \le 79$	$B_i \oplus C_i \oplus D_i$	$0x8f1bbcdc$	30

The rotational constant s_i^1 are specified as follows:

$i \bmod 20$	0	1	2	3	4	5	6	7	8	9	10	11	12	13	14	15	16	17	18	19
s_i^1	5	11	7	15	6	13	8	14	7	12	9	11	8	15	6	12	9	14	5	13

If M is the last block, then $(A_{80} + A_0, B_{80} + B_0, C_{80} + C_0, D_{80} + D_0, E_{80} + E_0)$ is the hash value, otherwise $(A_{80} + A_0, B_{80} + B_0, C_{80} + C_0, D_{80} + D_0, E_{80} + E_0)$ is the inputs of the next message block.

3.2 The Compression Function of RIPEMD-160

In this section, we mainly describe the compression function of RIPEMD-160.

Message Expansion. The message expansion of RIPEMD-160 splits the 512-bit message block M into 16 words $m_i(i = 0, \ldots, 15)$, and expands them into 160 expanded message words w_i^L and $w_i^R(0 \le i \le 79)$. The expansion is specified in Table 3.

State Update Transformation. The state update transformation of RIPEMD-160 starts from an initial value $IV = (A_0, B_0, C_0, D_0, E_0)$ of five 32-bit words and updates them in 80 steps. In step $i(0 \le i \le 79)$ the 32-bit words w_i^L and w_i^R are used to update the state variables $A_i^L, B_i^L, C_i^L, D_i^L, E_i^L, A_i^R, B_i^R, C_i^R, D_i^R, E_i^R$ as follows:

$$B_{i+1}^L = ((A_i^L + F_i^L(B_i^L, C_i^L, D_i^L) + w_i^L + k_i^L) \lll s_i^L) + E_i^L,$$
$$A_{i+1}^L = E_i^L, C_{i+1}^L = B_i^L, D_{i+1}^L = C_i^L \lll 10, E_{i+1}^L = D_i^L.$$

$$B_{i+1}^R = ((A_i^R + F_i^R(B_i^R, C_i^R, D_i^R) + w_i^R + k_i^R) \lll s_i^R) + E_i^R,$$
$$A_{i+1}^R = E_i^R, C_{i+1}^R = B_i^R, D_{i+1}^R = C_i^R \lll 10, E_{i+1}^R = D_i^R.$$

The rotational constant s_i^L and s_i^R are specified in Table 3. And the bitwise Boolean function $F_i^L(B_i^L, C_i^L, D_i^L)$, $F_i^R(B_i^R, C_i^R, D_i^R)$, constant k_i^L and k_i^R used in each step are defined as follows:

step	$F_i^L(B_i^L, C_i^L, D_i^L)$	k_i^L	$F_i^R(B_i^R, C_i^R, D_i^R)$	k_i^R
$0 \leq i \leq 15$	$B_i^L \oplus C_i^L \oplus D_i^L$	$0x00000000$	$B_i^R \oplus (C_i^R \vee (\neg D_i^R))$	$0x50a28be6$
$16 \leq i \leq 31$	$(B_i^L \wedge C_i^L) \vee (\neg B_i^L \wedge D_i^L)$	$0x5a827999$	$(B_i^R \wedge D_i^R) \vee (C_i^R \wedge \neg D_i^R)$	$0x5c4dd124$
$32 \leq i \leq 47$	$(B_i^L \vee \neg C_i^L) \oplus D_i^L$	$0x6ed9eba1$	$(B_i^R \vee \neg C_i^R) \oplus D_i^R$	$0x6d703ef3$
$48 \leq i \leq 63$	$(B_i^L \wedge D_i^L) \vee (C_i^L \wedge \neg D_i^L)$	$0x8f1bbcdc$	$(B_i^R \wedge C_i^R) \vee (\neg B_i^R \wedge D_i^R)$	$0x7a6d76e9$
$64 \leq i \leq 79$	$B_i^L \oplus (C_i^L \vee \neg D_i^L)$	$0xa953fd4e$	$B_i^R \oplus C_i^R \oplus D_i^R$	$0x00000000$

If M is the last block, then $(B_0+C_{80}^L+D_{80}^R, C_0+D_{80}^L+E_{80}^R, D_0+E_{80}^L+A_{80}^R, E_0+A_{80}^L+B_{80}^R, A_0+B_{80}^L+C_{80}^R)$ is the hash value, otherwise $(B_0+C_{80}^L+D_{80}^R, C_0+D_{80}^L+E_{80}^R, D_0+E_{80}^L+A_{80}^R, E_0 + A_{80}^L + B_{80}^R, A_0 + B_{80}^L + C_{80}^R)$ is the inputs of the next message block.

Table 3. Message expansion and rotation of RIPEMD-60

r	$w_r^L, w_{r+1}^L, \ldots, w_{r+15}^L$																$w_r^R, w_{r+1}^R, \ldots, w_{r+15}^R$															
0	0	1	2	3	4	5	6	7	8	9	10	11	12	13	14	15	5	14	7	0	9	2	11	4	13	6	15	8	1	10	3	12
16	7	4	13	1	10	6	15	3	12	0	9	5	2	14	11	8	6	11	3	7	0	13	5	10	14	15	8	12	4	9	1	2
32	3	10	14	4	9	15	8	1	2	7	0	6	13	11	5	12	15	5	1	3	7	14	6	9	11	8	12	2	10	0	4	13
48	1	9	11	10	0	8	12	4	13	3	7	15	14	5	6	2	8	6	4	1	3	11	15	0	5	12	2	13	9	7	10	14
64	4	0	5	9	7	12	2	10	14	1	3	8	11	6	15	13	12	15	10	4	1	5	8	7	6	2	13	14	0	3	9	11

r	$s_r^L, s_{r+1}^L, \ldots, s_{r+15}^L$																$s_r^R, s_{r+1}^R, \ldots, s_{r+15}^R$															
0	11	14	15	12	5	8	7	9	11	13	14	15	6	7	9	8	8	9	9	11	13	15	15	5	7	7	8	11	14	14	12	6
16	7	6	8	13	11	9	7	15	7	12	15	9	11	7	13	12	9	13	15	7	12	8	9	11	7	7	12	7	6	15	13	11
32	11	13	6	7	14	9	13	15	14	8	13	6	5	12	7	5	9	7	15	11	8	6	6	14	12	13	5	14	13	13	7	5
48	11	12	14	15	14	15	9	8	9	14	5	6	8	6	5	12	15	5	8	11	14	14	6	14	6	9	12	9	12	5	15	8
64	9	15	5	11	6	8	13	12	5	12	13	14	11	8	5	6	8	5	12	9	12	5	14	6	8	13	6	5	15	13	11	11

4 Attacks on Step-Reduced HAS-160 and RIPEMD-160

In this section, we present the details of the pseudo-preimage attack on 71-step HAS-160 and the preimage attack on 34-step RIPEMD-160 using the differential meet-in-the-middle technique combined with the biclique technique etc.

4.1 Pseudo-Preimage Attack on 71-Step HAS-160

Split the Compression Function and Choose the Linear Spaces. The forward direction CF_1 of our attack is from step 59 to step 70 connecting with step 0 to step 20 and the linear space of this part is $LD_1 = \{\Delta m_0 \| \ldots \| \Delta m_{15} | \Delta m_{10} = \Delta m_{11} = \Delta m_{12} = \Delta m_{15} = [26 \sim 28], \Delta m_i = 0, 0 \leq i \leq 15, i \neq 10, 11, 12, 15\}$. Here and in the following, Δx denotes the difference corresponding to the variable x, and $\Delta x = [\alpha \sim \beta]$ means the bits from the α-th bit to the β-th bit of Δx take all possible values, and the other bits of Δx are zero. Let CV_i denote the input of the i-th step of HAS-160. In this case, the input difference of the 13-th step of $CF_1(M \oplus \delta_1, CV_{59})$ and $CF_1(M, CV_{59})$ is zero. The backward direction CF_2^{-1} is from step 52 to step 21 and the linear space of this part is $LD_2 = \{\Delta m_0 \| \ldots \| \Delta m_{15} | \Delta m_3 = \Delta m_6 = \Delta m_8 = \Delta m_{15} = [29 \sim 31], \Delta m_i = 0, 0 \leq i \leq 15, i \neq$

$3, 6, 8, 15$}. Obviously, the input difference of the 30-th step of $CF_2^{-1}(M \oplus \delta_2, CV_{53})$ and $CF_2^{-1}(M, CV_{53})$ is zero. The bicliques CF_3 is from step 53 to step 58, and the schematic view of the pseudo-preimage attack on 71-step HAS-160 is shown in Fig 3.

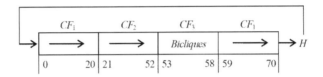

Fig. 3. Schematic view of the pseudo-preimage attack on 71-step HAS-160

Construct 6-Step Bicliques. We construct bicliques CF_3 which covers step 53 to step 58, such that for all $(\delta_1, \delta_2) \in LD_1 \times LD_2$, $CV_{59}[\delta_2] = CF_3(M \oplus \delta_1 \oplus \delta_2, CV_{53}[\delta_1])$ hold. And the dimension of the bicliques is 3. As in the description in Section 2.3, CF_3 is divided into two parts CF_3^1 and CF_3^2. CF_3^1 covers step 53 to step 55 and CF_3^2 covers step 56 to step 58. Then $CV_{56} = (A_{56}, B_{56}, C_{56}, D_{56}, E_{56})$ is the intermediate chaining value. For $\delta_2 \in LD_2$ and $\delta_1 \in LD_1$, we can compute the output difference $FT(\delta_2)$ of CF_3^1 and the output difference $BT(\delta_1)$ of $(CF_3^2)^{-1}$, respectively. Meanwhile, we can deduce a set of sufficient conditions that make sure $FT(\delta_2)$ and $BT(\delta_1)$ hold. Then the bicliques can be constructed by using the biclique search algorithm in Section 2.3. Note that we only set the sufficient conditions in steps 55, 56 and 57 in the biclique search algorithm.

Table 4. Computation of the output difference $FT(\delta_2)$ for HAS-160

step	ΔA	ΔB	ΔC	ΔD	ΔE	Δw	Conditions
53	0	0	0	0	0	(29,30,31)	-
54	(29,30,31)	0	0	0	0	(29,30,31)	-
55	(3,4,5, 29,30,31)	(29,30,31)	0	0	0	0	$D_{55,i} = 0 (i = 29, 30, 31)$, No bit carry
56	(9,10,11, 15,16,17)	(3,4,5, 29,30,31)	(22,23,24)	0	0	(29,30,31)	$D_{56,i} = 0 (i = 3, 4, 5)$, $C_{56,i} \neq B_{56,i} \oplus w_{56,i}$ $(i = 29, 30, 31)$, $D_{56,i} = 1 (i = 29, 30, 31)$, $B_{56,24} \vee (\neg D_{56,24}) \neq C_{56,24} \oplus A_{56,15}$, No bit carry
57	(18,19,20, 22,23,25,26)	(9,10,11, 15,16,17)	(22,23,34, 28,29,30)	(22,23,24)	0	0	No bit carry

We illustrate how to get $FT(\delta_2)$ for $\delta_2 \in LD_2$ and how to deduce the sufficient conditions. $BT(\delta_1)$ and the other conditions can be deduced similarly. When $\delta_2 = 0\|0\|0\|(29, 30, 31)\|0\|0\|0\|(29, 30, 31)\|0\|(29, 30, 31)\| \ldots \|0\|(29, 30, 31)$, the output difference $FT(\delta_2)$ of CF_3^1 can be calculated in Table 4. The XOR difference $\Delta w_{53} = (29,30,31)$ means the i-th $(i = 29, 30, 31)$ bits of Δw_{53} are 1 and the other bits of Δw_{53} are 0. Combined with the algorithm of HAS-160, we can get $\Delta A_{54} = (29,30,31)$. Because $\Delta w_{54} = (29,30,31)$, $\Delta A_{54} = (29,30,31)$, $\Delta B_{54} = 0$, $\Delta C_{54} = 0$, $\Delta D_{54} = 0$ and $\Delta E_{54} = 0$, from $A_{55} = (A_{54} \lll 6) + F_{54}(B_{54}, C_{54}, D_{54}) + E_{54} + w_{54} + k_{54}$, we can get $\Delta A_{55} = (3,4,5,29,30,31)$ if there is no bit carry in ΔA_{55}. Here, no bit carry means the i-th $(0 \le i \le 28, i \neq 3, 4, 5)$

bits of ΔA_{55} equal to 0. Obviously, we can get $\Delta B_{55} = (29,30,31)$ because $\Delta B_{55} = \Delta A_{54}$. From $F_{55}(B_{55}, C_{55}, D_{55}) - C_{55} \oplus (B_{55} \vee \neg D_{55})$, we can get $\Delta B_{55} = (29,30,31)$ results in no difference in CF_{55} if the conditions $D_{55,i} = 0(i = 29, 30, 31)$ are satisfied. Note that $D_{55,i}$ denotes the i-th bit of D_{55}. Then from $A_{56} = (A_{55} \lll 12) + F_{55}(B_{55}, C_{55}, D_{55}) + E_{55} + w_{55} + k_{55}$, we can get $\Delta A_{56} = (9,10,11,15,16,17)$ if there is no bit carry in ΔA_{56}. Similarly, the other conditions in Table 4 can be obtained. By setting the conditions in Table 4, a biclique can be obtained in a few seconds on a PC. So the complexity of constructing a biclique is negligible. An example of the 6-step bicliques is shown in Table 5.

Table 5. An example of the 6-step bicliques in HAS-160

M	0x8d0cae50	0x33de090e	0x39801485	0x1d1e2b3f	0xeb009f21	0x1c3af5a3
	0x092798cd	0xb9a878fa	0x1f5d84c6	0xa88cbad5	0xe053d3e1	0xc0ad415a
	0x21b3c7b3	0xfc7d3660	0x378962c7	0x00ba96c2		
CV_{53}	0x8fc640fe	0xd387c000	0x1dcc8b3b	0xcc96b2a0	0x7cb3634a	

The Process of the Attack. When the bicliques are obtained, we can get $CV_{59}[\delta_2]$ and $CV_{53}[\delta_1]$, and compute $L_1 = CF_1(M \oplus \delta_2, CV_{59}[\delta_2]) \oplus \Delta_2$ and $L_2 = CF_2^{-1}(M \oplus \delta_1, CV_{53}[\delta_1]) \oplus \Delta_1$. Note that $\Delta_1 = \overline{CF_1}(\delta_1, 0)$ and $\Delta_2 = \overline{CF_2^{-1}}(\delta_2, 0)$, where $\overline{CF_1}$ and $\overline{CF_2^{-1}}$ are obtained from CF_1 and CF_2^{-1} respectively by linearizing the step function, i.e., replacing $+$ by $\oplus$ and setting CV_{53}, CV_{59} and the constants to 0. We choose the truncation mask vector as $T_{20} = (0, 0, 0x86, 0, 0)$, then $L_1 =_T L_2$ means that the message $M \oplus \delta_1 \oplus \delta_2$ is a candidate preimage of 71-step compression function, that is, a candidate pseudo-preimage of 71-step HAS-160 hash function. By extensive experiments, the total test number is 2^{26} and $c = 0x1b70c39$, therefore, the error probability is about $0.429 = c/2^{26}$. In order to retest a candidate pseudo-preimage, we need to make 18 steps (step 12 to step 29) calculation, and the complexity of constructing a biclique is negligible. Thus, the complexity of the pseudo-preimage attack is $2^{158.13} \approx (2^{160-3} + 2^{160-3} \times (18/71))/(1 - 0.429)$. Because the dimension of linear space is 3, the storage complexity can be evaluated as follows. Without loss of generality, assume we storage $L_1 : \{CV_{59}[\delta_2], \Delta_2, CF_1(M \oplus \delta_2, CV_{59}[\delta_2]) \oplus \Delta_2\}$. Because the item length of L_1 is 11 $(= 5 + 1 + 5)$ words, then we can get the length of L_1 is $2^3 \times 11$ words. We also storage the random message (16 words) and the biclique ($2^3 \times 2 \times 5$ words), so the storage complexity is at most $2^3 \times 21 + 16$ ($= 2^3 \times 11 + 16 + 2^3 \times 10$) words. Note that the attack uses the message word m_{15}, so it can not satisfy the message padding and can not convert to a preimage attack.

4.2   Preimage Attack on RIPEMD-160

Split the Compression Function and Choose the Linear Spaces. The forward direction CF_1 of our attack is from step 1 to step 33 in left branch connecting with the step of computing the hash value (marked with red dotted box in Fig 4) and the linear space of this part is $LD_1 = \{\Delta m_0 \| \ldots \| \Delta m_{15} | \Delta m_0 = [27 \sim 31], \Delta m_i = 0, 1 \le i \le 15\}$. Let CV_i^L

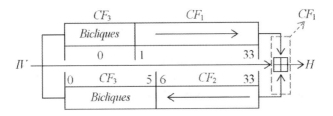

Fig. 4. Schematic view of the preimage attack on RIPEMD-160

and CV_i^R denote the inputs of the i-th step of left branch and right branch respectively. Then the difference corresponding to CV_{25}^L of $CF_1(M \oplus \delta_1, CV_1^L)$ and $CF_1(M, CV_1^L)$ is always zero. The backward direction CF_2^{-1} is from step 6 to step 33 in right branch and the linear space of this part is $LD_2 = \{ \Delta m_0 \| \ldots \| \Delta m_{15} | \Delta m_2 = [4 \sim 8], \Delta m_i = 0, 0 \leq i \leq 15, i \neq 2 \}$. Obviously the difference corresponding to CV_{31}^R of $CF_2^{-1}(M \oplus \delta_2, CV_6^R)$ and $CF_2^{-1}(M, CV_6^R)$ is always zero. The biclique CF_3 covers step 0 in left branch connecting with step 0 to step 5 in right branch. It is noted that the initial value IV is an intermediate value in the bicliques, and IV also affects the last step (marked with red dotted box in Fig 4), so the truncation mask vector can only be chosen in a smaller scope, e.g., the unchanged bits of IV. Therefore, we choose the linear spaces LD_1 and LD_2 as above so that we can get a proper truncation mask vector to make sure the error probability of the attack is as small as possible. The schematic view of the preimage attack on 34-step RIPEMD-160 is shown in Fig 4.

Table 6. An example of 7-step bicliques in RIPEMD-160

M	0x07b30181	0xd82b6a5c	0x5effd80e	0x05b7f446	0x1988e2c7	0xf7962189
	0x6c9bb3bb	0xff1152bb	0xa0d8905f	0xa9bc41d8	0x1407078a	0x7441af76
	0x7d9241a6	0x5ffd5259	0x00000000	0xddf9371d		
CV_6^R	0x8fcfef67	0xc592715e	0x26d8dac7	0x71f4450e	0x6c6740a0	

Construct 7-Step Bicliques. We construct bicliques CF_3 which covers step 0 in left branch connecting with step 0 to step 5 in right branch, such that for all $(\delta_1, \delta_2) \in LD_1 \times LD_2$, $CV_1^L[\delta_2] = CF_3(M \oplus \delta_1 \oplus \delta_2, CV_6^R[\delta_1])$ hold. And the dimension of the bicliques is 5. As in the description in Section 2.3, CF_3 is divided into two parts CF_3^1 and CF_3^2. CF_3^1 covers step 0 in left branch connecting with step 0 to step 2 in right branch and CF_3^2 covers step 3 to step 5 in right branch. Then $CV_3^R = (A_3^R, B_3^R, C_3^R, D_3^R, E_3^R)$ is the intermediate chaining value. For $\delta_2 \in LD_2$ and $\delta_1 \in LD_1$, we can compute the output difference $FT(\delta_2)$ of CF_3^1 and the output difference $BT(\delta_1)$ of $(CF_3^2)^{-1}$, respectively. Meanwhile, we can deduce a set of sufficient conditions that make sure $FT(\delta_2)$ and $BT(\delta_1)$ hold. Then the bicliques can be constructed by using the biclique search algorithm in Section 2.3. Note that in the biclique search algorithm, we only set the sufficient conditions in steps 1, 2 and 3 in right branch. By setting those conditions, a biclique can be obtained in a few seconds on a PC. So the complexity of constructing a biclique is negligible. An example of the 7-step bicliques is shown in Table 6.

The Process of the Attack. When the bicliques are obtained, we can get $CV_1^L[\delta_2]$ and $CV_6^R[\delta_1]$, and compute $L_1 = CF_1(M \oplus \delta_2, CV_1^L[\delta_2]) \oplus \Delta_2$ and $L_2 = CF_2^{-1}(M \oplus \delta_1, CV_6^R[\delta_1]) \oplus \delta_1$. Note that $\Delta_1 = \overline{CF_1}(\delta_1, 0)$ and $\Delta_2 = \overline{CF_2^{-1}}(\delta_2, 0)$, where $\overline{CF_1}$ and $\overline{CF_2^{-1}}$ are obtained from CF_1 and CF_2^{-1} respectively by linearizing the step function, i.e., replacing + by and setting CV_1^L, CV_6^R and the constants to 0. We choose the truncation mask vector as $T_{34} = (0, 0xf80, 0, 0, 0)$, then $L_1 =_T L_2$ means that the message $M \oplus \delta_1 \oplus \delta_2$ is a candidate preimage of 34-step compression function. By extensive experiments, the total test number is 2^{26} and $c = 0x152edbf$, therefore, the error probability is about $0.331 = c/2^{26}$. In order to retest a candidate pseudo-preimage, we need to make 6 steps (step 25 to step 33 in left branch add step 31 to step 33 in right branch, which equals to $6 = (9 + 3)/2$ steps) step calculations, and the complexity of constructing a biclique is negligible. Thus, the complexity of the pseudo-preimage attack is $2^{155.81} \approx (2^{160-5} + 2^{160-5} \times (6/34))/(1 - 0.331)$. The storage complexity of the attack can be evaluated similarly as HAS-160, i.e., the storage complexity is $2^5 \times 21 + 16 (= 2^5 \times 11 + 16 + 2^5 \times 10)$ words. Futhermore, this attack can be converted to a preimage attack with complexity $2^{158.91}$.

5   Conclusions

In this paper, we have shown the application of the pseudo-preimage attack on step-reduced HAS-160 and the preimage attack on step-reduced RIPEMD-160 using the differential meet-in-the-middle attack and biclique technique. Applying these techniques to HAS-160 led to a pseudo-preimage attack for 71 steps of the hash function with complexity $2^{158.13}$. However, our result can not convert to the preimage attack and without padding. As for RIPEMD-160, we obtain a preimage attack for 34 steps of the hash function with complexity $2^{158.91}$ and with padding. Future analysis should be able to explore how to construct bicliques including more steps so that preimage attacks can be implemented on more steps of hash functions.

6   Discussion

We have tried to determine Δ_1, Δ_2 and the truncation mask vector by statistical test using the original hash functions instead of using the linearized hash functions. CF_1, CF_2, CF_3 and the linear spaces LD_1, LD_2 are chosen as above. Then for 71-step HAS-160, the truncation mask vector is chosen as $(0, 0, 0x3, 0, 0x1)$ and the error probability is 0.544 (the total test number is 2^{26} and $c = 0x22ce51b$). Therefore, the complexity of the pseudo-preimage attack on 71-step HAS-160 is $2^{158.46} \approx (2^{160-3} + 2^{160-3} \times (18/71))/(1 - 0.544)$. For 34-step RIPEMD-160, the truncation mask vector is chosen as $(0, 0x1601, 0, 0x10, 0)$ and the error probability is 0.773 (the total test number is 2^{26} and $c = 0x317bf0d$). Then the complexity of the pseudo-preimage attack on 34-step RIPEMD-160 is $2^{157.37} \approx (2^{160-5} + 2^{160-5} \times (6/34))/(1 - 0.773)$. Thus, RIPEMD-160 seems more random than HAS-160 to some extent.

Acknowledgment. The author would like to thank the anonymous reviewers for their valuable suggestions and remarks. This work is supported by the National Natural Science Foundation of China (No. 61103238, 61373142), the MMJJ201301003, the Opening Project of State Key Laboratory of Information Security (Institute of Information Engineering, Chinese Academy of Sciences), and the Fundamental Research Funds for the Central Universities.

References

1. Aoki, K., Sasaki, Y.: Preimage attacks on one-block MD4, 63-step MD5 and more. In: Avanzi, R., Keliher, L., Sica, F. (eds.) SAC 2008. LNCS, vol. 5381, pp. 103–119. Springer, Heidelberg (2009)
2. Cho, H.-S., Park, S., Sung, S.H., Yun, A.: Collision Search Attack for 53-Step HAS-160. In: Rhee, M.S., Lee, B. (eds.) ICISC 2006. LNCS, vol. 4296, pp. 286–295. Springer, Heidelberg (2006)
3. Diffie, W., Hellman, M.E.: Exhaustive cryptanalysis of the NBS data encryption standard. IEEE Computer 10(6), 74–84 (1977)
4. Dobbertin, H., Bosselaers, A., Preneel, B.: RIPEMD-160: A strengthened version of RIPEMD. In: Gollmann, D. (ed.) FSE 1996. LNCS, vol. 1039, pp. 71–82. Springer, Heidelberg (1996)
5. Hong, D., Koo, B., Sasaki, Y.: Improved Preimage Attack for 68-Step HAS-160. In: Lee, D., Hong, S. (eds.) ICISC 2009. LNCS, vol. 5984, pp. 332–348. Springer, Heidelberg (2010)
6. International Organization for Standardization: ISO/IEC 10118-3:2004, Information technology - Security techniques - Hash-functions - Part 3: Dedicated hashfunctions (2004)
7. Khovratovich, D., Rechberger, C., Savelieva, A.: Bicliques for Preimages: Attacks on Skein-512 and the SHA-2 Family. In: Canteaut, A. (ed.) FSE 2012. LNCS, vol. 7549, pp. 244–263. Springer, Heidelberg (2012)
8. Kircanski, A., AlTawy, R., Youssef, A.M.: A heuristic for finding compatible differential paths with application to HAS-160. In: Sako, K., Sarkar, P. (eds.) ASIACRYPT 2013, Part II. LNCS, vol. 8270, pp. 464–483. Springer, Heidelberg (2013)
9. Knellwolf, S., Khovratovich, D.: New Preimage Attacks against Reduced SHA-1. In: Safavi-Naini, R., Canetti, R. (eds.) CRYPTO 2012. LNCS, vol. 7417, pp. 367–383. Springer, Heidelberg (2012)
10. Mendel, F., Pramstaller, N., Rechberger, C., Rijmen, V.: On the Collision Resistance of RIPEMD-160. In: Katsikas, S.K., López, J., Backes, M., Gritzalis, S., Preneel, B. (eds.) ISC 2006. LNCS, vol. 4176, pp. 101–116. Springer, Heidelberg (2006)
11. Mendel, F., Rijmen, V.: Colliding Message Pair for 53-Step HAS-160. In: Nam, K.-H., Rhee, G. (eds.) ICISC 2007. LNCS, vol. 4817, pp. 324–334. Springer, Heidelberg (2007)
12. Mendel, F., Nad, T., Schläffer, M.: Cryptanalysis of Round-Reduced HAS-160. In: Kim, H. (ed.) ICISC 2011. LNCS, vol. 7259, pp. 33–47. Springer, Heidelberg (2012)
13. Mendel, F., Nad, T., Scherz, S., Schläffer, M.: Differential Attacks on Reduced RIPEMD-160. In: Gollmann, D., Freiling, F.C. (eds.) ISC 2012. LNCS, vol. 7483, pp. 23–38. Springer, Heidelberg (2012)
14. Mendel, F., Peyrin, T., Schläffer, M., Wang, L., Wu, S.: Improved Cryptanalysis of Reduced RIPEMD-160. In: Sako, K., Sarkar, P. (eds.) ASIACRYPT 2013, Part II. LNCS, vol. 8270, pp. 484–503. Springer, Heidelberg (2013)
15. Ohtahara, C., Sasaki, Y., Shimoyama, T.: Preimage Attacks on Step-Reduced RIPEMD-128 and RIPEMD-160. In: Lai, X., Yung, M., Lin, D. (eds.) Inscrypt 2010. LNCS, vol. 6584, pp. 169–186. Springer, Heidelberg (2011)

16. Sasaki, Y., Aoki, K.: A Preimage Attack for 52-Step HAS-160. In: Lee, P.J., Cheon, J.H. (eds.) ICISC 2008. LNCS, vol. 5461, pp. 302–317. Springer, Heidelberg (2009)
17. Sasaki, Y., Aoki, K.: Finding Preimages in Full MD5 Faster Than Exhaustive Search. In: Joux, A. (ed.) EUROCRYPT 2009. LNCS, vol. 5479, pp. 134–152. Springer, Heidelberg (2009)
18. Sasaki, Y., Wang, L.: Distinguishers beyond Three Rounds of the RIPEMD-128/-160 Compression Functions. In: Bao, F., Samarati, P., Zhou, J. (eds.) ACNS 2012. LNCS, vol. 7341, pp. 275–292. Springer, Heidelberg (2012)
19. Telecommunications Technology Association. Hash Function Standard Part 2: Hash Function Algorithm Standard, HAS-160 (2000)
20. Yun, A., Sung, S. H., Park, S., Chang, D., Hong, S., Cho, H.-S.: Finding Collision on 45-Step HAS-160. In: Won, D., Kim, S. (eds.) ICISC 2005. LNCS, vol. 3935, pp. 146–155. Springer, Heidelberg (2006)
21. Wang, X., Lai, X., Feng, D., Chen, H., Yu, X.: Cryptanalysis of the Hash Functions MD4 and RIPEMD. In: Cramer, R. (ed.) EUROCRYPT 2005. LNCS, vol. 3494, pp. 1–18. Springer, Heidelberg (2005)
22. Biham, E., Chen, R., Joux, A., Carribault, P., Lemuet, C., Jalby, W.: Collisions of SHA-0 and reduced SHA-1. In: Cramer, R. (ed.) EUROCRYPT 2005. LNCS, vol. 3494, pp. 36–57. Springer, Heidelberg (2005)
23. Wang, X., Yin, Y.L., Yu, H.: Finding Collisions in the Full SHA-1. In: Shoup, V. (ed.) CRYPTO 2005. LNCS, vol. 3621, pp. 17–36. Springer, Heidelberg (2005)
24. Wang, X., Yu, H.: How to Break MD5 and Other Hash Functions. In: Cramer, R. (ed.) EUROCRYPT 2005. LNCS, vol. 3494, pp. 19–35. Springer, Heidelberg (2005)
25. Menezes, A.J., Oorschot, P.C., Vanstone, S.: Handbook of Applied Cryptography. CRC Press (1996)

Revised Algorithms for Computing Algebraic Immunity against Algebraic and Fast Algebraic Attacks

Lin Jiao[1,2], Bin Zhang[1,3], and Mingsheng Wang[4]

[1] TCA, Institute of Software, Chinese Academy of Sciences,
Beijing 100190, China
[2] Graduate University of Chinese Academy of Sciences, Beijing 100049, China
[3] State Key Laboratory of Computer Science, Institute of Software, Chinese
Academy of Sciences, Beijing 100190, China
[4] State Key Laboratory of Information Security, Institute of Information
Engineering, Chinese Academy of Sciences, Beijing 100093, China
jiaolin@tca.iscas.ac.cn

Abstract. Given a Boolean function with n variables, a revised algorithm for computing the algebraic immunity d against conventional algebraic attacks in $O(D^{2\pm\epsilon})$ complexity is described for $D = \sum_{i=0}^{d} \binom{n}{i}$ and a small ϵ, which corrects and clarifies the most efficient algorithm so far at Eurocrypt 2006. An analysis of the success rate of the algorithm for determining the immunity against fast algebraic attacks in the above paper is also provided. Based on the revised algorithm, an algorithm for computing all the overdefined implicit equations on an S-box is given, which is the core of the algebraic attacks on block ciphers.

Keywords: Algebraic immunity, Algebraic attacks, Fast algebraic attacks, Stream ciphers, S-box.

1 Introduction

Algebraic attacks have proved to be a powerful class of attacks which might threaten both block and stream ciphers [2, 4, 6–8]. The idea is to set up an algebraic system of equations verified by the key bits and to try to solve it. The essential point is how to find these equations and how to make them with lower degree, which can be efficiently solved.

For stream ciphers working as LFSR-based filtering and combining generators, the construction itself directly depicts such an algebraic system, and it is crucial for such an attack to work effectively that the filtering or combining functions have low degree multiples or low degree annihilators [7], i.e., the security of such cipher highly depends on the choice of the nonlinear Boolean function deployed. Thus, for an arbitrary Boolean function f, the notion of algebraic immunity (AI) is introduced as a cryptographic criterion to measure its strength against conventional algebraic attacks [13]. But it has been shown that the AI criterion cannot fully assure the immunity against fast algebraic attacks [6], whose emphasis is to

S.S.M. Chow et al. (Eds.): ISC 2014, LNCS 8783, pp. 104–119, 2014.

find a function g of a small degree e such that the multiple gf has degree d not too large. Here came the (e, d) pair for assessing the performance against fast algebraic attacks of a Boolean function. To make a step towards provable security against algebraic and fast algebraic attacks, efficient algorithms for computing the AI against conventional algebraic attacks and the corresponding immunity against fast algebraic attacks are of great importance both in the design and analysis aspects. In [1] presented at Eurocrypt 2006, two paragon algorithms were proposed, which are still regarded as the most efficient so far. We study the algorithms, and find a number of incorrect statements and confusions. In Section 2 and 3, we will highlight these mistakes, refine and further analyze the algorithms on validity, complexity and success rate aspects. While we finished this work, we found [10] at the Cryptology ePrint Archive, which also queried the algorithm in [1], viewed from the equivalence between solving a system of linear equations and finding affine annihilators. However, this paper has not proposed the modification plan, and the new algorithms that they put forward in quadratic-time complexity do not been assessed with the same measure as before. Besides, there are also many different algorithms at the similar aim with quite different methods, such as [11] using the Wiedemann's Algorithm, and [12] a non-existence of an annihilator algorithm using the recursive decomposition of Boolean functions. But, still [1] is the most famous and claimed efficient so far.

For block ciphers with layers of small S-boxes interconnected by linear key-dependent layers, several algebraic attacks were given [4, 8], based on the hypothesis that the S-box can be described by an overdefined system of algebraic equations, so can the whole block cipher in variables of plaintext, ciphertext, round keys and internal states while thinking of the key expansion. Then people can use Gauss elimination, XL, XSL and Gröbner basis algorithms to figure out the key. Thus the main point is to find deterministic low-degree relations between the input and output variables of the S-box. However, previous papers found these equations just by manual analysis. According to our revised algorithm, we will propose a method to compute the graph algebraic immunity [3] of S-boxes meanwhile all the equations with the degree of the immunity in Section 4.

This paper is organized as follows: In Section 2, we further analyze the algorithm for algebraic immunity in [1] and point out several mistakes. In addition, we present a revised algorithm and analyze its complexity. In Section 3, we give an analysis of the success rate of the algorithm for immunity against fast algebraic attacks in [1]. In Section 4, we propose a method for efficiently computing all the equations with the degree of the graph algebraic immunity on an S-box. In Section 5, we conclude the paper.

2   Immunity against Conventional Algebraic Attacks

Let us first specify the notations used hereafter and present some preliminary notations.

- Denote by $\mathbb{F}$ the finite field $GF(2)$ and by $\mathbb{F}^n$ the vector space of dimension n over $\mathbb{F}$.

- Denote by $\alpha = (a_1, \ldots, a_n) \in \mathbb{F}^n$ the multi-index. Let $x = (x_1, \ldots, x_n)$ be a binary variable vector and $z = (z_1, \ldots, z_n) \in \mathbb{F}^n$, then $x^\alpha = x_1^{a_1} \cdots x_n^{a_n}$ is a monomial and $z^\alpha = z_1^{a_1} \cdots z_n^{a_n} \in \mathbb{F}$.
- Denote by $\mathcal{E} = \{\alpha_1, \ldots, \alpha_D\} \subseteq \mathbb{F}^n$ a set of multi-indices. Denote by $x^{\mathcal{E}} = \{x^{\alpha_1}, \ldots, x^{\alpha_D}\}$ the set of associated monomials.
- For $\alpha, \beta \in \mathbb{F}^n$, let $\alpha \subseteq \beta$ be an abbreviation for $supp(\alpha) \subseteq supp(\beta)$, where $supp(\alpha) = \{i \mid a_i = 1\}$. Let $wt(\alpha)$ be the cardinality of $supp(\alpha)$, i.e., the weight of α.
- $\mathcal{E}^d$ is the set of all α of weight equal to d. $\mathcal{E}^{\leq d} = \mathcal{E}^0 \cup \cdots \cup \mathcal{E}^d$, which is ordered by increasing weight. $\mathcal{E}_i = \{\alpha_1, \ldots, \alpha_i\}$, which are the first i elements of $\mathcal{E}^{\leq d}$.
- Denote by $\mathbb{B}_n = \mathbb{F}[x] / < x_i^2 - x_i, i = 1, \ldots, n >$ the ring of Boolean function with n variables.

Definition 1. *The annihilator of a Boolean function f with n input variables is another Boolean function g with n input variables such that $f \cdot g = 0$.*

A necessary and sufficient condition for $f \cdot g = 0$ is that the function g vanishes for all the arguments z for which $f(z) = 1$. Denote $\{z \in \mathbb{F}^n \mid f(z) = 1\}$ by $supp(f)$.

Definition 2. *For a given Boolean function f, the algebraic immunity $AI(f)$ is the minimum value d such that f or $f + 1$ has a nonzero annihilator of degree d.*

It was shown that $AI(f) \leq \min\{\lceil \frac{n}{2} \rceil, \deg f\}$ for any f with n input variables [13].

The previous algorithms almost work as follows: given a guess of $AI = d$ usually not too large, the algebraic normal form (ANF) of a function g with n variables of degree d is

$$g = \bigoplus_{\alpha : wt(\alpha) \leq d} g_\alpha x^\alpha,$$

determined by its coefficients $\{g_\alpha\}$, whose number equals $D = \sum_{i=0}^{d} \binom{n}{i}$. In order to determine the unknown coefficients of an annihilator g, substitute all the arguments x in $g(x)$ with the elements in $supp(f)$. The derived system of linear equations can be solved by Gaussian elimination. This method immediately allows deciding whether there is an annihilator g of degree at most d, and if so, to determine a set of linearly independent annihilators (of degree at most d), if not, to amplify the guess of d. The cost for one such guess is about D^3 [13].

2.1   Analysis of Algorithm 1 in [1]

The idea in [1] is to reduce the minimum annihilator problem to a multivariate Lagrange interpolation problem, as stated below.

Problem 1. Let $\mathcal{E} = \{\alpha_1, \ldots, \alpha_D\} \subseteq \mathbb{F}^n$, $\mathcal{Z} = \{z_1, \ldots, z_D\} \subseteq \mathbb{F}^n$ and $\bar{v} = (v_1, \ldots, v_D) \in \mathbb{F}^D$, find a polynomial $g \in \mathbb{F}[x_1, \ldots, x_n]$ whose monomials are all included in $x^{\mathcal{E}}$ such that $g(z_i) = v_i, \forall i \in \{1, \ldots, D\}$.

Equivalently, the problem can be described as the form below. Let

$$V_{\mathcal{Z},\mathcal{E}} := \begin{pmatrix} z_1^{\alpha_1} & \cdots & z_1^{\alpha_D} \\ \vdots & \ddots & \vdots \\ z_D^{\alpha_1} & \cdots & z_D^{\alpha_D} \end{pmatrix}$$

and $v_{\mathcal{Z},\mathcal{E}} = \det(V_{\mathcal{Z},\mathcal{E}})$. Let $g(x) = \oplus_{j=1}^{D} g_{\alpha_j} x^{\alpha_j}$, where $\bar{g} = (g_{\alpha_1}, ..., g_{\alpha_D})^t$ and $(\cdot)^t$ means the transpose. Then the problem is whether

$$V_{\mathcal{Z},E} \cdot \bar{g} = \bar{v}$$

has a solution. And when $v_{\mathcal{Z},E} \neq 0$, there is a unique solution to the problem.

There are several details incorrect and not clear for Algorithm 1 in [1]. We first give a proof of Remark 2 in [1], which is the basis of the main theory, i.e., Proposition 2 in [1]. The remark is not proved in the original paper and is hard to understand without an explanation.

Definition 3. *Contraposition is a law, which says that a conditional statement is logically equivalent to its contrapositive, i.e., the contrapositive of $P \to Q$ is thus $\neg Q \to \neg P$.*

Proposition 1. *Given the set $\mathcal{Z} = \{z_1, ..., z_D\} \subseteq \mathbb{F}^n$, the existence of a set $\mathcal{E} = \{\alpha_1, ..., \alpha_D\} \subseteq \mathbb{F}^n$ such that $v_{\mathcal{Z},\mathcal{E}} \neq 0$ is ensured.*

Proof. Let I be the ideal generated by the polynomials vanishing at each point of $\mathcal{Z}$, i.e., $< f_i \mid f_i(\mathcal{Z}) = 0, i = 1, 2, ... >$, where $f(\mathcal{Z}) = 0$ means $f(z_i) = 0, i = 1, ..., D$. Then derive the Gröbner basis of I for a graduated order, i.e. $GB = \{g_1, ..., g_s\}$. It is that, for any $f \in I$, there exist $g_i \in GB$ such that the leading monomial of g_i can divide the leading monomial of f, i.e., $LT(g_i)|LT(f)$. Take

$$\mathcal{E} : x^{\mathcal{E}} = \overline{< LT(GB) >},$$

i.e., the complementary of the monomial ideal generated by the leading monomials of the Gröbner basis, then we will test and verify $v_{\mathcal{Z},\mathcal{E}} \neq 0$. For any polynomial g whose monomials are all included in $x^{\mathcal{E}}$, we have $LT(g) \in x^{\mathcal{E}}$. Thus $LT(g) \notin < LT(GB) >$, i.e., for any $g_i \in GB$, $LT(g_i) \nmid LT(g)$, and it deduces $g \notin I$, which is the contrapositive of the equivalent definition of Gröbner basis. That is, there exists $z_i \in \mathcal{Z}$ such that $g(z_i) \neq 0$. For the arbitrariness of g, we derive $V_{\mathcal{Z},\mathcal{E}}$ is of full column rank. Thus $v_{\mathcal{Z},\mathcal{E}} \neq 0$.   □

Based on the proof, we can refine the Proposition 2 in [1] as follows.

Theorem 1. *Let $\mathcal{Z} = supp(f)$, and $\mathcal{E} : x^{\mathcal{E}} = \overline{< LT(GB(< f + 1 >)) >}$ with graduated order. Take one $\beta \notin \mathcal{E}$ with minimum weight, then*

$$R_\beta := \det \begin{pmatrix} x^\beta & x^{\alpha_1} & \cdots & x^{\alpha_D} \\ z_1^\beta & z_1^{a_1} & \cdots & z_1^{a_D} \\ \vdots & \vdots & \ddots & \vdots \\ z_D^\beta & z_D^{a_1} & \cdots & z_D^{a_D} \end{pmatrix}$$

is a minimum-degree annihilator of f, and $R_\beta = x^\beta \oplus g$, where

$$
g(x): \begin{pmatrix} z_1^{\alpha_1} & \cdots & z_1^{\alpha_D} \\ \vdots & \ddots & \vdots \\ z_D^{\alpha_1} & \cdots & z_D^{\alpha_D} \end{pmatrix} \cdot \bar{g} = \begin{pmatrix} z_1^\beta \\ \vdots \\ z_D^\beta \end{pmatrix}.
\tag{1}
$$

Proof. (Sketch.) Obviously, R_β is an annihilator of f and has the minimum degree for the graduated monomial order. According to Proposition 1, the coefficient matrix of equation (1) is nonsingular, then g exists with uniqueness. Comparing every coefficient of R_β and $x^\beta \oplus g$, we find them equal by the Cramer's rule. □

A recursion algorithm is proposed in [1] by applying the above theorem incrementally, claiming a quadratic time complexity for the whole process by the method of LU-decomposition. We have carefully studied the algorithm and found that there are several details unclear and even wrong, which cannot assure the resulting conclusion. Further, the complexity analysis is also incorrect.

Now let us take a closer look at the algorithm proposed in [1]. The notations here are the same as those in [1].

Algorithm 1. ([1]) Computation of an annihilator with the minimum degree

---

Input f, $\mathcal{Z} = supp(f)$, $\mathcal{E}^{\leq \lceil n/2 \rceil}$
Output An annihilator of f of minimum degree.
1. Initialization: $U_1 \leftarrow (z_1^{\alpha_1})$, $v_1 \leftarrow f(z_1) \oplus 1$, $\bar{g} \leftarrow 1$, $P \leftarrow (x_1)$, $i \leftarrow 1$.
2. **while** the polynomial associated to $\bar{g}$ is not an annihilator of f **do**
3. $i \leftarrow i + 1$.
4. $\begin{pmatrix} U_i & P(z_1^{\alpha_i}, \ldots, z_{i-1}^{\alpha_i}) \\ z_i^{\alpha_1}, \ldots, z_i^{\alpha_{i-1}} & z_i^{\alpha_i} \end{pmatrix}$ $row\ op. \mapsto \begin{pmatrix} U_i & P(z_1^{\alpha_i}, \ldots, z_i^{\alpha_i}) \\ & 0 \ldots 0 \end{pmatrix} = U_{i+1}$
5.   Use the same row operations from $(P(z_1^{\alpha_i}, \ldots, z_{i-1}^{\alpha_i}), z_i^{\alpha_i}) \mapsto P(z_1^{\alpha_i}, \ldots, z_i^{\alpha_i})$ to perform the update $(P(v_1, \ldots, v_{i-1}), v_i) \mapsto P(v_1, \ldots, v_i)$
6.   Solve $U_i \bar{g}_i = P(v_1, \ldots, v_i)$ with $\bar{g}_i = (g_1, \ldots, g_i)$.
7. **end while**
8. Output $g(x) = \oplus_{j=1}^i g_j x^{\alpha_j}$.

The algorithm works as applying Theorem 1 for an intermediate set of point $\mathcal{Z}_i$ and an associated set of exponents $\mathcal{E}_i$ with $\beta = \alpha_{i+1}$, resulting in an intermediate annihilator of f on the set $\mathcal{Z}_i$. If it pass the check of the global set $\mathcal{Z}$, then it is a minimum degree annihilator of f, if not, one considers a new point z_{i+1} and associated set $\mathcal{E}_{i+1}$. The main skill of the algorithm is to keep the coefficient matrix U_i upper triangular all along with the increment of i through last row elimination, and here it takes P to record all the row operations, whose idea is based on LU-decomposition and making the most of last recursion.

The advantage of the algorithm is that we no longer need a prior guess of $AI = d$ and repetition to solve the equation systems to calculate the minimum degree annihilator as before. We have the following arguments about the details of Algorithm 1.

(i) The algorithm just assumed that $v_{\mathcal{Z}_i, \mathcal{E}_i} \neq 0$ for all $i \in \{1, ..., |\mathcal{Z}|\}$, but did not specify the order of $\mathcal{Z}$ that keeps the property, which is the condition of the success, for it applies Theorem 1 incrementally. And in the footnote, it is said that the order can be computed incrementally in quadratic time. Actually, we can modify the algorithm without extra operations or pre-computation to satisfy the condition, which will be justified later.

(ii) The algorithm was claimed to base on Theorem 1, however, in the following context of [1], it used $g_i = R_\beta \oplus x^\beta$ as the annihilator, which is a contradiction in terms, and in this way, the result from the algorithm is quite doubtful.

(iii) The algorithm above has not given the updating mode of the vector $\overline{v}$, and in Section 3.3 of [1], it is said that $\overline{v}_{i+1} = (\overline{v}_i, v_{i+1})^t$, which is inconsistent with Theorem 1 where $\beta = \alpha_{i+1}$ for an intermediate set of points $\mathcal{Z}_i$ and an associated set of exponents $\mathcal{E}_i$.

Next, we point out some errors in the complexity analysis of Algorithm 1 and show that it cannot reach the quadratic time complexity as stated in [1].

(i) It is said that the triangulation in Step 4 requires i operations by replacing $z_i^{\alpha_1}, ..., z_i^{\alpha_{i-1}}$ with 0 and $z_i^{\alpha_i}$ with $z_i^{\alpha_i} - \sum_{j=1}^{i-1} z_i^{\alpha_j} \cdot P_{i,j}$, where $(P_{i,1}, ..., P_{i,i-1}) = P(z_1^{\alpha_i}, ..., z_{i-1}^{\alpha_i})$. However, there is no reason to do the row elimination in this way, since we should add all the elements in a row to the last row when we perform one row operation to eliminate the last row. Moreover, we have achieved the matrix by experiments, and it shows no specificity which may allow to do so. Here we give a simple example to illustrate its incorrectness. For matrix

$$\begin{pmatrix} 1 & 0 & 1 & 0 \\ 0 & 1 & 0 & 0 \\ 0 & 0 & 1 & 1 \\ 1 & 1 & 0 & 0 \end{pmatrix},$$

in the normal form of row elimination for Row 4, we first add Row 1 to Row 4 and Row 4 turns $\begin{pmatrix} 0 & 1 & 1 & 0 \end{pmatrix}$, then add Row 2 to Row 4 and Row 4 turns $\begin{pmatrix} 0 & 0 & 1 & 0 \end{pmatrix}$, last add Row 3 to Row 4 and Row 4 turns $\begin{pmatrix} 0 & 0 & 0 & 1 \end{pmatrix}$. If we do the row elimination as stated in [1], then the last row become $\begin{pmatrix} 0 & 0 & 0 & 0 \end{pmatrix}$, for $0 - 1 \times 0 - 1 \times 0 - 0 \times 1 = 0 \pmod 2$, where $P(z_1^{\alpha_i}, ..., z_{i-1}^{\alpha_i}) = (0, 0, 1)$. And by the normal form of row elimination, the computation required is in quadratic, not linear complexity at each step, which leads to the complexity of the algorithm as a cubic behavior.

(ii) It is also said when solving the system in Step 6, it is feasible with linear operations by correcting g_i in order to compute g_{i+1}. However, there are no details illustrating how to do the error-correcting procedure. We suspect its feasibility, since according to Theory 1, $\overline{v_{i+1}} = \{z_1^{\alpha_{i+1}}, ..., z_{i+1}^{\alpha_{i+1}}\}$ is totally changed from $\overline{v}_i = \{z_1^{\alpha_i}, ..., z_i^{\alpha_i}\}$. Moreover, it is known that to solve a linear system with triangular coefficient matrix a number of quadratic operations are requested, which also leads to the complexity of the algorithm as a cubic behavior.

2.2   Our Revised Algorithm

Now, we present our revised algorithm, which has been implemented in C code and verified in experiments.

Algorithm 2. (Revised) Computation of an annihilator with the minimum degree

---

Input $\mathcal{Z} = supp(f)$, $\mathcal{E}^{\leq \lceil n/2 \rceil}$.

Output An annihilator of f of minimum degree.

1. Initialization:

   $U \leftarrow \begin{pmatrix} z_1^{\alpha_1} & z_1^{\alpha_2} \\ z_2^{\alpha_1} & z_2^{\alpha_2} \end{pmatrix}$, where $z_1^{\alpha_2} \neq z_2^{\alpha_2}$. $\bar{v} \leftarrow \begin{pmatrix} z_1^{\alpha_3} \\ z_2^{\alpha_3} \end{pmatrix}$, $P \leftarrow \begin{pmatrix} x_1 \\ x_2 \end{pmatrix}$, and $i \leftarrow 2$. Last

   row eliminate U (i.e., eliminate the fist $i-1$ entries of the last row of U by row operations to make U upper triangular). Record the last row elimination in P. Use the same row operations as the record in P to eliminate $\bar{v}$. Solve the equation system $U\bar{g} = \bar{v} \mapsto \bar{g} \mapsto R = x^{\alpha_{i+1}} \oplus g$.

2. **while** $R(z) \neq 0, z \in \mathcal{Z} \backslash \{z_1, ..., z_i\}$ **do**

3. $i \leftarrow i+1$, exchange $z_i \leftrightarrow z$.

4. $U \leftarrow \begin{pmatrix} U & \bar{v} \\ z_i^{\alpha_1}, ..., z_i^{\alpha_{i-1}} & z_i^{\alpha_i} \end{pmatrix}$, $\bar{v} \leftarrow \begin{pmatrix} z_1^{\alpha_{i+1}} \\ \vdots \\ z_i^{\alpha_{i+1}} \end{pmatrix}$.

5.   Last row eliminate U. Increase one dimension in P and record the last row eliminationin P.

6.   Use all the row operations recorded in P to eliminate $\bar{v}$.

7.   Solve the equation system $U\bar{g} = \bar{v} \mapsto \bar{g} \mapsto R = x^{\alpha_{i+1}} \oplus g$.

8. **end while**

9. Output R.

The row elimination record can be realized by donating P as an array, each line of which describes the associated row operations with U. Here we give a justification of our revised algorithm.

Proposition 2. *The reason for keeping the condition* $v_{\mathcal{Z}_i, \mathcal{E}_i} \neq 0$ *for all* $i \in \{1, 2, ..., |\mathcal{Z}|\}$ *is as follows.*

Proof. First, $z_1^{\alpha_1} = z_2^{\alpha_1} = 1$, since $\alpha_1 = \{0, ..., 0\}$ for $\mathcal{E}^{\leq \lceil n/2 \rceil}$ ordered by increasing weight. Then to ensure $v_{\mathcal{Z}_2, \mathcal{E}_2} \neq 0$, we attach the condition $z_1^{\alpha_2} \neq z_2^{\alpha_2}$, which can be easily satisfied by searching $z : z^{\alpha_2} \neq z_1^{\alpha_2}$ from z_2 in $\mathcal{Z}$ and exchanging z with z_2. Assume $v_{\mathcal{Z}_i, \mathcal{E}_i} \neq 0$, then according to Algorithm 2 and Theorem 1, we derive an unique annihilator

$$R(x) = \det \begin{pmatrix} x^{a_{i+1}} & x^{\alpha_1} & \cdots & x^{\alpha_i} \\ z_1^{a_{i+1}} & z_1^{a_1} & \cdots & z_1^{a_i} \\ \vdots & \vdots & \ddots & \vdots \\ z_i^{a_{i+1}} & z_i^{a_1} & \cdots & z_i^{a_i} \end{pmatrix}$$

If $R(x) = 0, x \in \mathcal{Z} \backslash \{z_1, ..., z_i\}$, then the algorithm outputs R and terminals. If not, there must be a z s.t. $R(z) \neq 0$, i.e.,

$$
\det \begin{pmatrix} z^{a_{i+1}} & z^{\alpha_1} & \cdots & z^{\alpha_i} \\ z_1^{a_{i+1}} & z_1^{a_1} & \cdots & z_1^{a_i} \\ \vdots & \vdots & \ddots & \vdots \\ z_i^{a_{i+1}} & z_i^{a_1} & \cdots & z_i^{a_i} \end{pmatrix} \neq 0
$$

which is just a plus-minus difference from

$$
v_{\mathcal{Z}_{i+1}, \mathcal{E}_{i+1}} = \det \begin{pmatrix} z_1^{a_1} & \cdots & z_1^{a_i} & z_1^{a_{i+1}} \\ \vdots & \ddots & \vdots & \vdots \\ z_i^{a_1} & \cdots & z_i^{a_i} & z_i^{a_{i+1}} \\ z^{\alpha_1} & \cdots & z^{\alpha_i} & z^{a_{i+1}} \end{pmatrix}
$$

after we exchange z with z_i. Thus $v_{\mathcal{Z}_{i+1}, \mathcal{E}_{i+1}} \neq 0$. Moreover row elimination does not change the rank of the matrix, namely, $\det(U_{i+1}) \neq 0$, and the algorithm is sufficient to continue.   □

Our revised algorithm is entirely consistent with Theorem 1, and the correctness of the theorem ensures that of our algorithm. In addition, we also can use Algorithm 2 to find out all the independent minimum degree annihilators by changing 2-9 as follows.

It is based on the fact that all the minimum degree annihilator highest exponent in $\mathcal{E}^d$ and we can execute a further search without α_i if we have already got one annihilator as $R(x) = x^{\alpha_i} \oplus g$. The existence equivalence is due to the fact that there is another annihilator R' contains x^{α_i} if and only if $R \oplus R'$ is an annihilator, which is without x^{α_i} and independent of R.

Now we analyze the maximum complexity of the revised Algorithm 2.

(i) The triangulation in step 4 basically requires $i(i-1)/2$ bit operations.
(ii) The updating process of P requires i bit operations when we treat it as an array.
(iii) The row elimination of $\bar{v}$ costs $i(i-1)/2$ bit operations for all the records in P.
(iv) Solving the linear system with upper triangular coefficient matrix costs $i(i-1)/2$ bit operations basically.
(v) The complexity to check whether R is an annihilator on the global set $\mathcal{Z}$ is not introduced for the same reason as stated in [1].
(vi) Thus, the total complexity is about $1^2 + 2^2 + \cdots + D^2 \approx D^3/3$, where $D = \sum_{j=0}^{d} \binom{n}{j}$

Actually, the algorithm has done a step-by-step row elimination of the whole matrix $V_{\mathcal{Z}_D, \mathcal{E}_D}$. In addition, we further analyze the complexity taking account of the sparsity of the matrix. Here, we also give an order of the support set $\mathcal{Z}$ in the input with increasing weight and when elements with the same weight order them in the lexicographic order (i.e. increasing integer value). We also further order

1. $flag \leftarrow 0$.
2. **for** i from 3 to $\sum_{j=0}^{\lceil n/2 \rceil} \binom{n}{j}$ **do**
3.   **if** $R(z) \neq 0, z \in \mathcal{Z} \backslash \{z_1, ..., z_{i-1}\}$ **do**
4.   Exchange $z_i \leftrightarrow z$.
5. $U \leftarrow \begin{pmatrix} U & \bar{v} \\ z_i^{\alpha_1}, ..., z_i^{\alpha_{i-1}} & z_i^{\alpha_i} \end{pmatrix}, \bar{v} \leftarrow \begin{pmatrix} z_1^{\alpha_{i+1}} \\ \vdots \\ z_i^{\alpha_{i+1}} \end{pmatrix}$.
6.   Last row eliminate U. Increase one dimension in P and record the last row elimination in P.
7.   Use all the row operations recorded in P to eliminate $\bar{v}$.
8.   Solve the equation system $U\bar{g} = \bar{v} \mapsto \bar{g} \mapsto R = x^{\alpha_{i+1}} \oplus g$.
9.   **else do**
10.   Output R.
11.   **if** flag=0 **do**
12. $d \leftarrow wt(\alpha_i), D \leftarrow \sum_{j=0}^{d} \binom{n}{j}$. $flag \leftarrow 1$.
13.   **end if**
14.   **if** $D > i + 1$ **do**
15.   Exchange $\alpha_i \leftrightarrow \alpha_D$. $D - -, i - -$.
16.   **else do**
17.   Output "Termination". **break**.
18.   **end if**
19.   **end if**
20. **end for**

the elements with the same weight in $\mathcal{E}^{\leq \lceil n/2 \rceil}$ in the reverse lexicographic order (i.e. decreasing integer value), which is the fastest order for computing Gröbner basis in general. Since $z^\alpha = 1$ if and only if $\alpha \subseteq z$, there will be many zero elements in the upper triangular since $wt(z) < wt(\alpha)$. Now let us approximately consider the row elimination complexity. The first i-th row can be added to the other rows below at most $D - i$ times according to the row elimination process. Assume $wt(z_i) = k$, then there is $\sum_{j=0}^{d} \binom{k}{j}$ nonzero elements in the i-th row of $V_{\mathcal{Z}_D, \mathcal{E}_D}$. For a randomly chosen balanced Boolean function f with n variables, we expect that one half points of $\mathbb{F}^n$ with weight i is in the support set of f, where $i = 0, 1, ..., n$. Then the complexity can be calculated as follows. Since the complexity is an approximate value, we do not consider the exchanged order in the algorithm.

1. $C \leftarrow 0$.
2. Find minimum l s.t. $\frac{1}{2} \sum_{j=0}^{l} \binom{n}{j} \geq D$.
3. **for** k from 1 to l **do**
4.   **for** i from $\frac{1}{2} \sum_{j=0}^{k-1} \binom{n}{j}$ to $\frac{1}{2} \sum_{j=0}^{k} \binom{n}{j}$ **do**
5. $C \leftarrow C + (D - i) \cdot \sum_{j=0}^{d} \binom{k}{j}$.
6.   **end for**
7. **end for**
8. Output C.

We compute a table of values of the approximate complexity for some n and $d = \lceil n/2 \rceil$. From Table 1, we can derive that $\epsilon \approx 0.5$.

Table 1. Approximate Complexities of Algorithm 2

n	10	11	12	13	14	15	16	17
d	5	6	6	7	7	8	8	9
D	$2^{9.3}$	$2^{10.5}$	$2^{11.3}$	$2^{12.5}$	$2^{13.3}$	$2^{14.5}$	$2^{15.3}$	$2^{16.5}$
C	$2^{22.6}$	$2^{25.7}$	$2^{27.7}$	$2^{30.8}$	$2^{32.8}$	$2^{35.9}$	$2^{37.8}$	$2^{41.0}$

Beyond above analysis, we have done a number of experiments on the inverse functions and the Kasami type functions which were enumerated in [1] to test the actual performance of our revised algorithm. We put a counter to calculate the actual number of computations done and the memory called for in the algorithm (mainly for U and P), and the results verify the approximate complexity $D^{2+\epsilon}$, where $\epsilon \approx 0.5$, as shown in Table 2. Note that there is no explicit results on the actual time complexity in previous works, while we list the experimental counter results here. We list the experimental counter other than the actual time for considering the differential performance of realizing methods and hardware configurations. We also present the number of all the independent minimum degree annihilators. One point to note is that here the algebraic immunity of the vectorial Boolean function $F : \mathbb{F}^n \to \mathbb{F}^n$ is defined as $\min\{AI(v \cdot F) | 0 \neq v \in \mathbb{F}^n\}$. In particular, for x^k, it equals $AI(tr(x^k))$, where $tr(\cdot)$ denotes the trace function over $\mathbb{F}^n$. Similarly, we flip the function value of $(0, ..., 0)$ to convert the Kasami function to a balanced function when $n = 10, 12, 14, 16$ as that in [1]. Here we have not presented more variables, since we have already tested the manifestation of the revised algorithm, and a better implementation needs more programming skills.

About the order that used to rearrange the support set, we have attempted three kinds of orders as follows on the representative ciphers attacked by algebraic method: LILI-128 [7] and WG7 [14]. And the experimental results manifest that the one we finally chosen performs best, which mostly enlarges the sparseness of the equation system. The order of $\mathcal{E}^{\leq \lceil n/2 \rceil}$ is the same as above. Here we just take f for example and omit $f + 1$ for simplicity as shown in Table 3.

(i) Order the support set $\mathcal{Z}$ in the lexicographic order. (lex)
(ii) Order the support set $\mathcal{Z}$ with increasing weight and when elements with the same weight order them in the lexicographic order. (wt-lex)
(iii) Order the support set $\mathcal{Z}$ with increasing weight and when elements with the same weight order them in the reverse lexicographic order. (wt-rev-lex)

From Table 3 we find that for LILI-128, ϵ is negative.

We also use the 0-1 random number generator to build the truth tables of Boolean functions, and compute the Algebraic Immunity of these functions.

Table 2. Complexities of Algorithm 2

Inverse function x^{-1}

n	AI	$d_{[f//f+1]}$	$D_{[f//f+1]}$	counter$_{[f//f+1]}$	mem.$_{[f//f+1]}$	num.$_{[f//f+1]}$
10	5	5/5	$2^{9.3}/2^{9.3}$	$2^{20.6}/2^{20.8}$	$2^{15.7}/2^{15.7}$	126/126
11	5	5/5	$2^{10.0}/2^{10.0}$	$2^{23.5}/2^{23.8}$	$2^{17.5}/2^{17.6}$	33/33
12	5	5/5	$2^{10.6}/2^{10.6}$	$2^{25.4}/2^{25.7}$	$2^{18.7}/2^{18.8}$	7/7
13	6	6/6	$2^{12.0}/2^{12.0}$	$2^{29.9}/2^{29.9}$	$2^{21.7}/2^{21.7}$	3/3
14	6	6/6	$2^{12.7}/2^{12.7}$	$2^{31.6}/2^{31.6}$	$2^{22.8}/2^{22.8}$	91/91
15	6	6/6	$2^{13.3}/2^{13.3}$	$2^{33.4}/2^{33.4}$	$2^{23.9}/2^{23.9}$	23/23
16	6	6/6	$2^{13.9}/2^{13.9}$	$2^{35.1}/2^{35.1}$	$2^{25.0}/2^{25.0}$	4/4
17	7	7/7	$2^{15.3}/2^{15.3}$	$2^{39.2}/2^{39.2}$	$2^{28.0}/2^{28.0}$	323/323

Kasami function $x^{2^{2^k}-2^k+1}, k \le n/2, \gcd(n,k)=1$

n	exp.(k)	AI	$d_{[f//f+1]}$	$D_{[f//f+1]}$	counter$_{[f//f+1]}$	mem.$_{[f//f+1]}$	num.
10	57(3)	4	5/4	$2^{9.3}/2^{8.6}$	$2^{19.8}/2^{17.1}$	$2^{15.2}/2^{13.5}$	159/2
11	993(5)	5	5/5	$2^{10.0}/2^{10.0}$	$2^{24.0}/2^{24.0}$	$2^{17.7}/2^{17.7}$	1/1
12	993(5)	5	6/5	$2^{11.3}/2^{10.6}$	$2^{26.8}/2^{25.5}$	$2^{19.7}/2^{18.7}$	399/2
13	993(5)	6	6/6	$2^{12.0}/2^{12.0}$	$2^{29.9}/2^{29.9}$	$2^{21.7}/2^{21.7}$	3/3
14	993(5)	6	6/6	$2^{12.7}/2^{12.7}$	$2^{31.7}/2^{31.7}$	$2^{22.9}/2^{22.8}$	1/1
15	16257(7)	7	7/7	$2^{14.0}/2^{14.0}$	$2^{35.9}/2^{35.9}$	$2^{25.7}/2^{25.7}$	3/3
16	16257(7)	7	7/7	$2^{14.7}/2^{14.7}$	$2^{37.4}/2^{37.4}$	$2^{26.8}/2^{26.8}$	4/6

Table 3. Comparison of three kinds of order

order	LILI-128: n=10, AI=4			WG7: n=7, AI=3		
	D	counter	memory	D	counter	memory
lex	$2^{8.6}$	$2^{15.6}$	$2^{12.3}$	2^6	$2^{12.6}$	$2^{10.0}$
wt-lex	$2^{8.6}$	$2^{15.2}$	$2^{12.1}$	2^6	$2^{12.3}$	$2^{9.9}$
wt-rev-lex	$2^{8.6}$	$2^{15.6}$	$2^{12.4}$	2^6	$2^{12.6}$	$2^{10.1}$

The complexity is about $D^{2+\epsilon}$, where $\epsilon \approx 0.5$ on average. For the sake of brevity, we would not like to give the tables of these results here.

3 Immunity against Fast Algebraic Attacks

For fast algebraic attacks, we search for the relation as $h = fg$, that h of degree d not too large, and g of degree e small. We represent g and h in the ANF form as $g(x) = \oplus_{\beta:wt(\beta)\leq e} g_\beta x^\beta$ and $h(x) = \oplus_{\gamma:wt(\gamma)\leq d} h_\gamma x^\gamma$. And $E = \sum_{i=0}^{e} \binom{n}{i}$, $D = \sum_{i=0}^{d} \binom{n}{i}$.

Now let us turn to the algorithm for the determination of the immunity against fast algebraic attack proposed in [1]. Firstly, we revisit the theory it is based on. Since $h_\gamma = \oplus_{\alpha \subseteq \gamma} h(\alpha) = \oplus_{\alpha \subseteq \gamma} f(\alpha) \cdot g(\alpha) = \oplus_{\alpha \subseteq \gamma} f(\alpha) \cdot (\oplus_{\beta \subseteq \alpha} g_\beta) = \oplus_{\beta \subseteq \gamma} g_\beta \cdot (\oplus_{\beta \subseteq \alpha \subseteq \gamma} f(\alpha))$, and $h_\gamma = 0, wt(\gamma) > d$, there derives a system of linear equations with variables of $\{g_\beta\}$. There is no g of degree at most e and h of degree at most d such that $fg = h$ if and only if the $\sum_{i=d+1}^{n} \binom{n}{i} \times \sum_{i=0}^{e} \binom{n}{i}$ coefficient matrix of the equation system is of full column rank. The idea of the algorithm is to extract a $\sum_{i=0}^{e} \binom{n}{i} \times \sum_{i=0}^{e} \binom{n}{i}$ matrix from the coefficient matrix and judge if it is of full rank, if so, then obviously the original matrix is of full column rank, if not, it cannot decide whether the original one is of full column rank. Thus, it is a probabilistic algorithm to determine the non-existence of g and h for any f other than the existence as [1] stated, and when the step 12 has a nontrivial solution, the algorithm has no output. But note that [1] has not pointed the probabilistic character explicitly, and given an analysis of the success rate. The algorithm calls for a complexity corresponding to $O(DE^2)$, which is more efficient compared to the complexity $O(D^3)$ of Algorithm 2 in [13]. Let us have a closer look at the algorithm.

Algorithm 3. ([1]) Determine the existence of g and h for any f

Input f with n input variables, two integers $0 \leq e \leq AI(f)$ and $AI(f) \leq d \leq n$.
Output Determine if g of degree at most e and h of degree at most d exist such that $fg = h$.
1. Initialization an $E \times E$ matrix G, and let each entry be zero.
2. Compute an ordered set $\mathcal{I} \leftarrow \{\beta : wt(\beta) \leq e\}$.
3. **for** i from 1 to E **do**
4.   Choose a γ with $wt(\gamma) = d + 1$.
5.   Determine the set $\mathcal{B} \leftarrow \{\beta : \beta \subseteq \gamma, wt(\beta) \leq e\}$.
6.   **for** all β in $\mathcal{B}$ **do**
7.   Determine the set $\mathcal{A} \leftarrow \{\alpha : \beta \subseteq \alpha \subseteq \gamma\}$.
8.   Compute $A \leftarrow \oplus_{\mathcal{A}} f(\alpha)$.
9.   Let the entry of G in row i and column β (in respect to $\mathcal{I}$) be 1 if $A = 1$.
10.   **end for**
11. **end for**
12. Solve the linear system of equations, and output no g and h of the corresponding degree if there is only a trivial solution.

Let us explain the randomness of the extracted matrix. For a fixed column in the matrix, i.e., the same β, consider two different elements A_1 and A_2 for two rows γ_1 and γ_2 with weight $d+1$. If $\beta \not\subseteq \gamma_1$ and $\beta \subseteq \gamma_2$, then $A_1 = 0$ and $A_2 = \oplus_{\beta \subseteq \alpha \subseteq \gamma_2} f(\alpha)$, which is determined by $2^{d+1-wt(\beta)}$ values of f. Since $d \approx n/2$ and $e \ll d$, $2^{d+1-wt(\beta)} \approx 2^{n/2}$. For a randomly chosen f, the value of A_2 is random. If $\beta \subseteq \gamma_1$ and $\beta \subseteq \gamma_2$, then $A_1 = \oplus_{\beta \subseteq \alpha \subseteq \gamma_1 \cap \gamma_2} f(\alpha) \oplus_{\beta \subseteq \alpha \subseteq \gamma_1 \setminus \gamma_1 \cap \gamma_2} f(\alpha)$ and $A_2 = \oplus_{\beta \subseteq \alpha \subseteq \gamma_1 \cap \gamma_2} f(\alpha) \oplus_{\beta \subseteq \alpha \subseteq \gamma_2 \setminus \gamma_1 \cap \gamma_2} f(\alpha)$, where $wt(\beta) \leq wt(\gamma_1 \cap \gamma_2) \leq d$. There are $2 \times 2^{d+1-wt(\gamma_1 \cap \gamma_2)}$ different values of f to make A_1 and A_2 independent for the randomness of f, and obviously for different pairs of γs, the set of different values are not same. Thus the elements in the same column are independent from each other, and the same with the elements in the same row.

Now, let us give an analysis of the success rate.

Proposition 3. *A random $n \times n$ matrix over $\mathbb{F}$ takes the probability of 28.9% to be of full column rank.*

Proof. The n-dimensional vector space contains $2^n - 1$ nonzero vector, and if the matrix is of full column rank, any row of the matrix cannot be a linear combination of other rows. Thus, for the first row, we have $2^n - 1$ choices. Then for the second row, it cannot be an element of the linear span generated by the first row, and there are $2^n - 2$ choices. Similarly, there are $2^n - 2^{i-1}$ choices for the i-th row, and the probability is $(1 - 1/2^n)(1 - 2/2^n)(1 - 2^2/2^n) \cdots (1 - 2^{n-1}/2^n)$, which tends to 28.9%. □

Thus the success rate of Algorithm 3 approximates 28.9%. If the original matrix is of full column rank, then we almost need three times test to find it out.

4   Immunity for S-box against Algebraic Attacks

On the algebraic immunity for vectorial Boolean functions, there are three kinds of definitions. At first, we present a concept of annihilating set [3]. Let U be a subset of $\mathbb{F}^n$, then $\{g \in \mathbb{B}_n \mid g(U) = 0\}$ is the annihilating set of U. Then we give the concept of the algebraic immunity of U as $AI(U) = \min\{\deg g \mid 0 \neq g \in \mathbb{B}_n, g(U) = 0\}$.

Definition 4. *[3, 5] Let F be an (n, m) vectorial Boolean function, and define*

$$AI(F) = \min\{AI(F^{-1}(a)) | a \in \mathbb{F}^m\}$$

as the basic algebraic immunity of F,

$$AI_{gr}(F) = \min\{\deg g \mid 0 \neq g \in \mathbb{B}_n, g(gr(F)) = 0\}$$

as the graph algebraic immunity of F, where $gr(F) = \{(x, F(x)) \mid x \in \mathbb{F}^n\} \subseteq \mathbb{F}^{n+m}$,

$$AI_{comp}(F) = \min\{AI(v \cdot F) | 0 \neq v \in \mathbb{F}^m\}$$

as the component algebraic immunity of F.

Algorithm 4. Computation of all the implicit equations with minimum degree belonging to an S-box

---

Input $\mathcal{Z} = \{x_1, ..., x_n, x_{n+1}, ..., x_{2n}\} = \{(x, S(x)) \mid x \in \mathbb{F}^n\} \subseteq \mathbb{F}^{2n}$, $\mathcal{E}^{\leq d_{gr}}$.

Output All the implicit equations of S with minimum degree.

1. Initialization: $U \leftarrow \begin{pmatrix} z_1^{\alpha_1} & z_1^{\alpha_2} \\ z_2^{\alpha_1} & z_2^{\alpha_2} \end{pmatrix}$, where $z_1^{\alpha_2} \neq z_2^{\alpha_2}$, $\overline{v} \leftarrow \begin{pmatrix} z_1^{\alpha_3} \\ z_2^{\alpha_3} \end{pmatrix}$, $P \leftarrow \begin{pmatrix} x_1 \\ x_2 \end{pmatrix}$,

   and $i \leftarrow 2$. Last row eliminate U. Record the last row elimination in P. Use the same row operations as the record in P to eliminate $\overline{v}$. Solve the equation system $U\overline{g} = \overline{v} \mapsto \overline{g} \mapsto R = x^{\alpha_{i+1}} \oplus g$.

2. $flag \leftarrow 0$.

3. **for** i from 3 to $\sum_{j=0}^{g_{gr}} \binom{2n}{j}$ **do**

4.   **if** $R(z) \neq 0, z \in \mathcal{Z} \backslash \{z_1, ..., z_{i-1}\}$ **do**

5.   Exchange $z_i \leftrightarrow z$.

6. $U \leftarrow \begin{pmatrix} U & \overline{v} \\ z_i^{\alpha_1}, ..., z_i^{\alpha_{i-1}} & z_i^{\alpha_i} \end{pmatrix}$, $\overline{v} \leftarrow \begin{pmatrix} z_1^{\alpha_{i+1}} \\ \vdots \\ z_i^{\alpha_{i+1}} \end{pmatrix}$.

7.   Last row eliminate U. Increase one dimension in P and record the last row elimination in P.

8.   Use all the row operations recorded in P to eliminate $\overline{v}$.

9.   Solve the equation system $U\overline{g} = \overline{v} \mapsto \overline{g} \mapsto R = x^{\alpha_{i+1}} \oplus g$.

10.   **else do**

11.   Output R.

12.   **if** flag=0 **do**

13. $d \leftarrow wt(\alpha_i)$, $D \leftarrow \sum_{j=0}^{d} \binom{2n}{j}$. $flag \leftarrow 1$.

14.   **end if**

15.   **if** $D > i + 1$ **do**

16.   Exchange $\alpha_i \leftrightarrow \alpha_D$. $D--$, $i--$.

17.   **else do**

18.   Output "Termination". **break**.

19.   **end if**

20.   **end if**

21. **end for**

Next, let have a look at the bounds of these three algebraic immunity definitions, which are explained in [3].

Theorem 2. *Let F be an (n, m) vectorial Boolean function, where $n \geq m \geq 1$.*

1. *Let d be the minimum positive integer which satisfies $\sum_{i=0}^{d} \binom{n}{i} > 2^{n-m}$, then $AI(F) \leq d$.*
2. *Let d_{gr} be the minimum positive integer which satisfies $\sum_{i=0}^{d_{gr}} \binom{n+m}{i} > 2^n$, then $AI_{gr}(F) \leq d_{gr}$.*
3. *$AI_{comp}(F) \leq \lceil \frac{n}{2} \rceil$.*

Especially , we just consider the algebraic immunity of S-box S, i.e., $n = m$. Then according to Theorem 2, we derive $AI(S) = 1$. So there is no significance to research the basic algebraic immunity of S-box. And there is no literature that proposes any attack given the basic and component algebraic immunity rather than the graph algebraic immunity [4, 8]. Thus we focus on the graph algebraic immunity of S-box.

To handle an algebraic attack on block ciphers built with layers of small S-boxes interconnected by linear key-independent layers, such as Khazad, Misty1, Kasumi, Camellia, Rijndael and Serpent [4], people want to construct simple algebraic equations that completely describe the cipher. The starting point is the fact that a small set of multivariate polynomials in input and output bits can completely define the only nonlinear element of the cryptosystem, S-box. Here comes the problem, how to find out all the independent equations in input and output bits of the S-box with the minimum degree efficiently unlike the manual way so far. Actually, it can be done by Algorithm 2 just changing a few of parameters, mainly the input of $\mathcal{Z} = \{(x, S(x)) \mid x \in \mathbb{F}^n\} \subseteq \mathbb{F}^{n+m}$ and $\mathcal{E}^{\leq d_{gr}}$, where d_{gr} is the minimum positive integer satisfied $\sum_{i=0}^{d_{gr}} \binom{n+m}{i} > 2^n$, shown as Algorithm 4. We have tested it on the S-box of AES [9], and it takes no more than one second to give out all the 39 quadratic equations.

5   Conclusion

In this paper, the two previous algorithms proposed in the Eurocrypt 2006 paper [1] for computing the immunity of Boolean functions against algebraic attacks and fast algebraic attacks are reviewed. We pointed out several flaws of the first one, and present a revised and corrected version, which has a complexity of $O(D^{2\pm\epsilon})$ for $D = \sum_{i=0}^{d} \binom{n}{i}$ and a small $\epsilon \approx 0.5$. We also analyzed the success rate of the second algorithm, which is approximately 28.9%. At last, we presented an efficient method to compute all the minimum degree equations that describe an S-box.

Acknowledgements. This work was supported by the National Grand Fundamental Research 973 Program of China(Grant No. 2013CB338002, 2013CB834203) and the programs of the National Natural Science Foundation of China (Grant No. 61379142, 11171323, 60833008, 60603018, 61173134, 91118006, 61272476).

References

1. Armknecht, F., Carlet, C., Gaborit, P., Künzli, S., Meier, W., Ruatta, O.: Efficient computation of algebraic immunity for algebraic and fast algebraic attacks. In: Vaudenay, S. (ed.) EUROCRYPT 2006. LNCS, vol. 4004, pp. 147–164. Springer, Heidelberg (2006)
2. Armknecht, F., Krause, M.: Algebraic attacks on combiners with memory. In: Boneh, D. (ed.) CRYPTO 2003. LNCS, vol. 2729, pp. 162–175. Springer, Heidelberg (2003)
3. Armknecht, F., Krause, M.: Constructing single- and multi-output boolean functions with maximal algebraic immunity. In: Bugliesi, M., Preneel, B., Sassone, V., Wegener, I. (eds.) ICALP 2006, Part II. LNCS, vol. 4052, pp. 180–191. Springer, Heidelberg (2006)

4. Biryukov, A., De Cannière, C.: Block ciphers and systems of quadratic equations. In: Johansson, T. (ed.) FSE 2003. LNCS, vol. 2887, pp. 274–289. Springer, Heidelberg (2003)
5. Carlet, C.: Vectorial boolean functions for cryptography. Boolean Models and Methods in Mathematics, Computer Science, and Engineering 134, 398–469 (2010)
6. Courtois, N.T.: Fast algebraic attacks on stream ciphers with linear feedback. In: Boneh, D. (ed.) CRYPTO 2003. LNCS, vol. 2729, pp. 176–194. Springer, Heidelberg (2003)
7. Courtois, N., Meier, W.: Algebraic attacks on stream ciphers with linear feedback. In: Biham, E. (ed.) EUROCRYPT 2003. LNCS, vol. 2656, pp. 345–359. Springer, Heidelberg (2003)
8. Courtois, N.T., Pieprzyk, J.: Cryptanalysis of block ciphers with overdefined systems of equations. In: Zheng, Y. (ed.) ASIACRYPT 2002. LNCS, vol. 2501, pp. 267–287. Springer, Heidelberg (2002)
9. Daemen, J., Rijmen, V.: Aes proposal: Rijndael. In: First Advanced Encryption Standard (AES) Conference (1998)
10. Dalai, D.K.: Computing the rank of incidence matrix and the algebraic immunity of boolean functions
11. Didier, F.: Using wiedemann's algorithm to compute the immunity against algebraic and fast algebraic attacks. In: Barua, R., Lange, T. (eds.) INDOCRYPT 2006. LNCS, vol. 4329, pp. 236–250. Springer, Heidelberg (2006)
12. Didier, F., Tillich, J.-P.: Computing the algebraic immunity efficiently. In: Robshaw, M. (ed.) FSE 2006. LNCS, vol. 4047, pp. 359–374. Springer, Heidelberg (2006)
13. Meier, W., Pasalic, E., Carlet, C.: Algebraic attacks and decomposition of boolean functions. In: Cachin, C., Camenisch, J. (eds.) EUROCRYPT 2004. LNCS, vol. 3027, pp. 474–491. Springer, Heidelberg (2004)
14. Orumiehchiha, M., Pieprzyk, J., Steinfeld, R.: Cryptanalysis of wg-7: A lightweight stream cipher. Cryptography and Communications 4(3-4), 277–285 (2012)

Obfuscation-Based Non-Black-Box Extraction and Constant-Round Zero-Knowledge Arguments of Knowledge

Ning Ding[1,2]

[1] NTT Secure Platform Laboratories, Japan
[2] Shanghai Jiao Tong University, China
ning.ding@lab.ntt.co.jp

Abstract. This paper addresses the issues of constructing zero-knowledge arguments of knowledge (ZKAOK) with properties such as a small number of rounds, public-coin and strict-polynomial-time simulation and extraction, and shows the existence of the following systems for **NP** for the first time under some assumptions.

- There exists a 4-round auxiliary-input ZKAOK with strict-polynomial-time simulation and extraction. Previously even combining the strict-polynomial-time simulation and extraction construction by Barak and Lindell (STOC'02) with the recent 4-round zero-knowledge argument by Pandey *et al.* [ePrint'13] brings such a construction using at least 6 rounds.
- There exists a 3-round bounded-auxiliary-input ZKAOK with strict-polynomial-time simulation and extraction. Previously the extractor of the 3-round construction by Bitansky *et al.* [STOC'14] runs in expected-polynomial-time.
- There exists a 2-round public-coin bounded-auxiliary-input ZKAOK with strict-polynomial-time simulation which extractor works for bounded-size provers and runs in strict-polynomial-time.

We demonstrate a new non-black-box extraction technique based on differing-input obfuscation due to Ananth *et al.* [ePrint'13] to achieve strict-polynomial-time extraction.

Keywords: Cryptographic Protocols, Zero-Knowledge, Proofs of Knowledge, Differing-Input Obfuscation.

1 Introduction

Zero-knowledge (ZK) proof systems, introduced by Goldwasser, Micali and Rackoff [20], are a fundamental notion in cryptography. Goldreich and Oren [19] later refined this notion to *plain* ZK and *auxiliary-input* ZK, where plain ZK requires ZK holds for all uniform PPT verifiers while auxiliary-input ZK requires it holds for all PPT verifiers with polynomial-size auxiliary-inputs (for the two notions distinguishers are always defined to be non-uniform polynomial-time

S.S.M. Chow et al. (Eds.): ISC 2014, LNCS 8783, pp. 120–139, 2014.
© Springer International Publishing Switzerland 2014

algorithms). Since their introduction, there have been a vast number of positive results and some negative results for constructing them. The fundamental positive result is that every language in **NP** has a ZK proof [18]. There are also many works constructing ZK protocols that satisfy some additional properties. One important example of such a property is having a small number of rounds of interaction. Other properties consider many aspects such as whether the protocol satisfies the proof of knowledge property [7] and whether the simulator or extractor can run in strict-polynomial-time (in some literals they run in expected-polynomial-time). We present the positive and negative results on exact round complexity of ZK proofs and arguments ([13]) (of knowledge) possibly with the aforementioned other properties as follows.

Positive Results. For auxiliary-input ZK for **NP**, Goldreich and Kahan [17] presented a 5-round ZK proof. Lindell [24] presented a 5-round ZKPOK. Feige and Shamir [14] gave a 4-round ZKAOK. The simulators (resp. extractors) of these protocols use a verifier's code (resp. a prover's code) in a black-box way and run in expected-polynomial-time. Hada and Tanaka [22] presented a 3-round ZK argument based on two knowledge-exponent assumptions.

Barak [2] presented a constant-round public-coin non-black-box ZK argument, which can be implemented in 6 rounds via round optimization [26] and achieves many properties that black-box ZK is impossible to achieve. For example, it is simultaneously of constant number of rounds and strict-polynomial-time simulation. Pandey et al. [27] presented a 4-round (concurrent) ZK argument with strict-polynomial-time simulation, which uses Barak's paradigm and differing-input obfuscation [1]. Combining the strict-polynomial-time simulation and extraction construction by Barak and Lindell [4] with this result [27], it is possible to construct a (at least) 6-round ZKAOK with strict-polynomial-time simulation and extraction.

For plain ZK for **NP**, Barak et al. [5] presented a 2-round public-coin ZK argument. Binasky et al. [9] presented a 2-round ZK argument and a 3-round ZKAOK, of which the simulators use a verifier's code in a non-black-box way and run in strict-polynomial-time but the extractor (of the 3-round construction) runs in expected-polynomial-time (it can run in strict-polynomial-time only if it knows a prover's noticeable success probability, but in general it does not know). Actually, these protocols can achieve bounded-auxiliary-input ZK by scaling the security parameter, i.e. their simulators can work for verifiers with auxiliary-inputs of bounded-size.

Negative Results. Goldreich and Oren [19] showed there is no 1-round ZK and no 2-round auxiliary-input ZK for any language outside **BPP**. Extending this result, Goldreich and Krawczyk [17] showed that 3-round black-box ZK proofs exist only for languages in **BPP** and Katz [23] showed that 4-round black-box ZK proofs, even with imperfect completeness, exist only for languages whose complements are in **MA**. Barak and Lindell [4] showed black-box simulators and extractors of constant-round protocols cannot run in strict-polynomial-time. Barak et al. [5] presented some trivialities of 2-round ZK proofs from some complexity assumptions.

The current state of the art leaves some problems in constructing ZK for **NP** of small number of rounds. We list some of them on ZKAOK as follows.

1. *Does there exist an auxiliary-input ZKAOK with strict-polynomial-time simulation and extraction which uses fewer rounds than 6?*
2. *Does there exist a 3-round bounded-auxiliary-input ZKAOK with strict polynomial time simulation and extraction?*
3. *Does there exist a 2-round bounded-auxiliary-input ZKAOK?*

This paper will address these problems and attempt to provide an affirmative answer to them.

1.1 Our Results

We present a new non-black-box extraction technique based on the differing-input obfuscator $(di\mathcal{O})$ in [1]. Generally, an obfuscator is a compiler that on input a program outputs a new program of same functionality with some security. Specifically, a $di\mathcal{O}$ is associated with a randomized sampling algorithm Sampler, which samples (M_0, M_1, z) where M_0, M_1 are two machines of same size and run in same time for a same input (we will always consider all machine pairs in this paper satisfy this requirement by padding) and z is an auxiliary input. We say $di\mathcal{O}$ is a differing-input obfuscator for Sampler if when it is hard for any polynomial-size algorithm to find an input x from (M_0, M_1, z) satisfying $M_0(x) \neq M_1(x)$, $di\mathcal{O}(M_0)$ and $di\mathcal{O}(M_1)$ are indistinguishable for any distinguisher that has z.

In this paper we assume the existence of $di\mathcal{O}$ for some natural samplers. Note that Garg *et al.* [16] constructed a contrived sampler that outputs two circuits $C_b, b = 0, 1$ each of which has the verification key vk and outputs b on receiving a message-signature pair (outputs 0 otherwise) and a contrived auxiliary input z that is an obfuscated program which has the signing key sk and on input any circuit C outputs $C(h(C), \sigma)$ where h is a hash function and σ is a signature of $h(C)$. Thus z on input any obfuscation of C_b outputs b and thus can distinguish the obfuscation of C_0, C_1. So if it is hard to find a differing-input for C_0, C_1 by e.g. trying to retrieve sk from z, there is no $di\mathcal{O}$ for this sampler (if it is easy to find a differing-input from (C_0, C_1, z), the non-existence of $di\mathcal{O}$ for this sampler is trivial). However, the auxiliary inputs output by the samplers in this paper are quite straightforward (which are e.g. random strings and commitments), so we take $di\mathcal{O}$ in [1] as a candidate obfuscator for our samplers (a similar assumption on $di\mathcal{O}$ was made also in [27]).

Applying the extraction technique we achieve the following positive results for ZKAOK, in which by $di\mathcal{O}$ we mean the differing-input obfuscator for the samplers specified in our constructions.

Theorem 1. *Assuming the existence of $di\mathcal{O}$ (and other standard assumptions), there is a 4-round auxiliary-input ZKAOK for **NP** with strict-polynomial-time non-black-box simulation and extraction.*

Theorem 2. *Assuming the same primitives as above, there is a 3-round bounded-auxiliary-input ZKAOK for* **NP** *with strict-polynomial-time non-black-box simulation and extraction.*

Theorem 3. *Assuming the existence of diO and 2-round public-coin universal arguments (and other standard assumptions), there is a 2-round public-coin bounded-auxiliary-input ZKAOK for* **NP** *which admits a strict-polynomial-time non-black-box simulator and a strict-polynomial-time non-black-box extractor working (only) for bounded-size provers.*

We note that all simulators and extractors of these results are universal.

Our Techniques. We sketch the technique used in the 4-round protocol. Let us recall the zero-knowledge argument in [27]. Very informally, the protocol in [27] first runs the preamble of Barak's protocol [2] in which V sends a random hash function h_1 and then P computes $Z_1 = \mathsf{Com}(h_1(0^n))$ where Com denotes a commitment scheme and then V responds with a random r_1. Let $\lambda_1 = (h_1, Z_1, r_1)$. Then V sends an obfuscated program $\widetilde{M}_{\lambda_1,s}$ that receives a witness for λ_1 (in a well-defined language) outputs s and outputs 0^n on all other inputs. Lastly P proves to V using a witness w for x that either $x \in L$ or it knows s. (We use notation $\widetilde{M}$ to denote a differing-input obfuscation of M.)

It can be seen the completeness holds. In simulation the simulator having a witness for λ_1 can run $\widetilde{M}_{\lambda_1,s}$ with input the witness to gain s and thus finish the remainder interaction. Since any cheating prover cannot gain s, it cannot finish the remainder interaction when $x \notin L$. So the soundness holds. However, it is unknown if the protocol admits a knowledge extractor.

We present a new non-black-box extraction technique to construct an extractor for it. Basically, we add a more preamble in a reverse direction in the protocol. Let $\lambda_2 = (h_2, Z_2, r_2)$ denote the transcript of this preamble. In the last step, we require P to send two obfuscated programs $\widetilde{W}_{\lambda_2,w}$ and $\widetilde{Q}_{\lambda_2,0^n}$, where $W_{\lambda_2,w}$ is the program that on receiving a witness for λ_2 (in a well-defined language) outputs w and outputs 0^n on all other inputs, and $Q_{\lambda_2,0^n}$ is the program that first verifies if the input is a witness for λ_2 but then outputs 0^n always. Note that when interacting with any prover P', P''s code is a witness for λ_2 (as V's code is a witness for λ_1). So in extraction, the extractor, having P''s code, can run $\widetilde{W}_{\lambda_2,w}$ on input the witness to gain w.

Organizations. The rest of the paper is arranged as follows. For lack of space we omit the preliminaries on zero-knowledge, witness-indistinguishability and commitment etc. and present the definition of differing-input obfuscation for machines in Appendix A. In Section 2, 3 and 4, we present the protocols claimed in the aforementioned three theorems respectively.

2   Four-Round Auxiliary-Input ZKAOK

In this section we present the 4-round ZKAOK. In Section 2.1 we show the construction idea and in Section 2.2 we present the overview of the protocol and in Section 2.3 present the formal description.

Public input: x (statement to be proved is "$x \in L$");
Prover's auxiliary input: w, (a witness for $x \in L$).

---

1. $V \rightarrow P$: Send $h_1 \in_R \mathcal{H}_n$.
2. $P \rightarrow V$: Send Z_1, $\langle y_1, y_2, \alpha \rangle$, $h_2 \in_R \mathcal{H}_n$.
3. $V \rightarrow P$: Send $r_1 \in_R \{0,1\}^n$, β, Z_2, $\langle y_s, \widetilde{M}_{\lambda_1,s}, \widetilde{M'}_{\lambda_1,0^n}, \pi_1 \rangle$.
4. $P \rightarrow V$: Send $r_2 \in_R \{0,1\}^n$, γ, $\langle \widetilde{W}_{\lambda_2,w}, \widetilde{Q}_{\lambda_2,0^n}, \pi \rangle$.

Protocol 1 *The 4-round auxiliary-input ZKAOK.*

2.1 Construction Idea

As sketched previously, our protocol is based on the construction in [27] and a new extraction strategy. Very informally, it can be described as follows. P and V perform the interaction to generate $\lambda_1 = (h_1, Z_1, r_1)$ and $\lambda_2 = (h_2, Z_2, r_2)$, which are the transcripts of the two reverse preambles. Besides, in Step 3 V sends an obfuscated program $\widetilde{M}_{\lambda_1,s}$ and in the last step P sends two obfuscated programs $\widetilde{W}_{\lambda_2,w}$ and $\widetilde{Q}_{\lambda_2,0^n}$ and proves to V that either $\widetilde{W}_{\lambda_2,w}$ is honestly generated or $\widetilde{Q}_{\lambda_2,0^n}$ is an obfuscation of the program $Q_{\lambda_2,s}$ that outputs s if the input is a witness for λ_2 and outputs 0^n else. It can be seen that the completeness holds since P can use a witness w for x to finish the interaction.

In simulation, the simulator commits to the verifier's code in Z_1. Thus the verifier's code is a witness for λ_1. So the simulator can run $\widetilde{M}_{\lambda_1,s}$ on input the witness to recover s. In the last step send $\widetilde{W}_{\lambda_2,0^n}, \widetilde{Q}_{\lambda_2,s}$ and a proof using s as witness, where $W_{\lambda_2,0^n}$ is the program that runs identically to $W_{\lambda_2,w}$ except that it always outputs 0^n. Due to the hardness of finding a differing-input for $W_{\lambda_2,w}$ and $W_{\lambda_2,0^n}$ (resp. $Q_{\lambda_2,s}$ and $Q_{\lambda_2,0^n}$) $\widetilde{W}_{\lambda_2,w}$ and $\widetilde{W}_{\lambda_2,0^n}$ (resp. $\widetilde{Q}_{\lambda_2,s}$ and $\widetilde{Q}_{\lambda_2,0^n}$) are indistinguishable. The zero-knowledge property follows. In extraction, the extractor commits to the prover's code in Z_2. Thus the prover's code is a witness for λ_2. So the extractor can run $\widetilde{W}_{\lambda_2,w}$ on input the witness to gain w.

2.2 Overview

Let $\{\mathcal{H}_n\}$ be a collision-resistant hash function family with $h \in \mathcal{H}_n : \{0,1\}^* \rightarrow \{0,1\}^n$, Com denote a non-interactive perfectly-binding computationally-hiding commitment scheme e.g. one in [10], NIWI denote a non-interactive WI proof for **NP** constructed in [21] or [6]. Let WIPOK denote a 3-round WI proof of knowledge e.g. the one in [11], (α, β, γ) denote the 3 messages of WIPOK. Let $di\mathcal{O}$ denote the differing-input obfuscator in [1] for samplers $\mathsf{Sampler}_i, 1 \leq i \leq 4$ shown below, f denote a one-way function.

Let L denote an arbitrary language in **NP** and our protocol is shown in Protocol 1 which specification is as follows.

1. Through Steps 1 to 4 P and V jointly generate $\lambda_1 = (h_1, Z_1, r_1)$ and $\lambda_2 = (h_2, Z_2, r_2)$, in which h_1, h_2 are represented by a n-bit string each and $Z_1 = \mathsf{Com}(h_1(0^n))$ and $Z_2 = \mathsf{Com}(h_2(0^n))$. Besides, P and V do the following.

2. In Step 2, P computes (y_1, y_2) that are $(f(u_1), f(u_2))$ for random $u_1, u_2 \in \{0,1\}^n$, and then computes α of WIPOK using a random one of u_1, u_2 as witness where WIPOK is to prove that it knows u_1 or u_2. Note that in this scenario WIPOK is witness-hiding [14].

3. In Step 3, V generates β, $y_s = f(s)$ for a random $s \in \{0,1\}^n$ and computes $\widetilde{M}_{\lambda_1, s} \leftarrow di\mathcal{O}(M_{\lambda_1, s})$ and $\widetilde{M}'_{\lambda_1, 0^n} \leftarrow di\mathcal{O}(M'_{\lambda_1, 0^n})$, where $M'_{\lambda_1, 0^n}$ is a program that verifies if an input is a witness for λ_1 but always outputs 0^n.
   Then compute a NIWI proof π_1 using witness s for that either $\widetilde{M}_{\lambda_1, s}$ is the program as specified or $\widetilde{M}'_{\lambda_1, 0^n}$ is an obfuscation of $M'_{\lambda_1, u}$ that on input a witness for λ_1 can output u, a pre-image of y_1 or y_2 w.r.t. f.
   Here by a witness for λ_1, we mean a program Π satisfying $|\Pi| \leq n^{\log \log n}$ and $h_1(\Pi) = \mathsf{Com}^{-1}(Z_1)$ and $\Pi(msg_2)$ outputs r_1 in $n^{\log \log n}$ steps where msg_2 denotes the message of Step 2.

4. In Step 4, P sends γ, $\widetilde{W}_{\lambda_2, w} \leftarrow di\mathcal{O}(W_{\lambda_2, w})$, $\widetilde{Q}_{\lambda_2, 0^n} \leftarrow di\mathcal{O}(Q_{\lambda_2, 0^n})$ and a NIWI proof π using witness w to prove either $\widetilde{W}_{\lambda_2, w}$ is as specified or $\widetilde{Q}_{\lambda_2, 0^n}$ is an obfuscation of $Q_{\lambda_2, s}$ where $f(s) = y_s$.
   Note that in the running of $Q_{\lambda_2, s}$ or $Q_{\lambda_2, 0^n}$, they will verify if the input is a witness for λ_2. By a witness for λ_2, we mean a program Π' satisfying $|\Pi'| \leq n^{\log \log n}$ and $h_2(\Pi') = \mathsf{Com}^{-1}(Z_2)$ and $\Pi'(msg_3)$ outputs r_2 in $n^{\log \log n}$ steps where msg_3 denotes the message of Step 3.

The completeness of the protocol is straightforward. We then sketch the zero-knowledge and argument of knowledge properties from which the soundness follows.

Auxiliary-Input Zero-Knowledge. We show a strict-polynomial-time non-black-box simulator S for any $V^* \in \{0,1\}^{\text{poly}(n)}$ and any $x \in L$. S computes $Z_1 = \mathsf{Com}(h_1(\Pi))$, where Π denotes V^*'s remainder strategy of Step 2 and interacts with V^* of the first three steps. If V^*'s messages are invalid, abort the interaction. Otherwise, let $\widetilde{M}$ and $\widetilde{M}'$ denote the two obfuscated programs in V^*'s message of Step 3. Then either $\widetilde{M}$ is $\widetilde{M}_{\lambda_1, s}$ or $\widetilde{M}'$ is $\widetilde{M}'_{\lambda_1, u}$ where $M'_{\lambda_1, u}$ is the program that outputs u that is a pre-image of y_1 or y_2 if the input is a witness for λ_1 and outputs 0^n otherwise.

If the latter possibility holds with noticeable probability, S can run $\widetilde{M}'$ with input Π (and the coins in generating Z_1) to recover u. This implies even if (y_1, y_2, α) comes from outside i.e. it is not generated by S, S can still recover a pre-image of y_1 or y_2, contradicting the witness-hiding property of WIPOK. So it must be the case that $\widetilde{M}$ is $\widetilde{M}_{\lambda_1, s}$ except for negligible probability. This means S can run $\widetilde{M}$ with input Π to recover s, which provides S an ability to finish the last step.

In the last step, S samples a random r_2, generates $\widetilde{W}_{\lambda_2, 0^n} \leftarrow di\mathcal{O}(W_{\lambda_2, 0^n})$ and $\widetilde{Q}_{\lambda_2, s} \leftarrow di\mathcal{O}(Q_{\lambda_2, s})$ where $W_{\lambda_2, 0^n}$ is a program identical to $W_{\lambda_2, w}$ except that w is replaced by 0^n and $Q_{\lambda_2,s}$ is a program identical to $Q_{\lambda_2,0^n}$ except that 0^n is replaced by s, and generates π using s as witness.

We now consider the indistinguishability of simulation. Let $\mathsf{Sampler}_1$ be a sampler that has (x,w) and adopts P's strategy to interact with V^* except computing $Z_1 = \mathsf{Com}(h_1(\Pi))$ to generate $Q_{\lambda_2,0^n}$ and the auxiliary input $z_1 = (\lambda_1, \lambda_2, y_1, y_2, \alpha, \beta, \gamma)$, and adopts S's strategy to generate $Q_{\lambda_2,s}$ and outputs $(Q_{\lambda_2,s}, Q_{\lambda_2,0^n}, z_1)$. Assume $di\mathcal{O}$ works for $\mathsf{Sampler}_1$. We show any polynomial-size adversary having $(Q_{\lambda_2,s}, Q_{\lambda_2,0^n}, z_1)$ cannot find a witness for λ_2, so $\widetilde{Q}_{\lambda_2,s}, \widetilde{Q}_{\lambda_2,0^n}$ are indistinguishable for any distinguisher knowing z_1.

Let $\mathsf{Sampler}_2$ be a sampler that has (x,w) and adopts S's strategy to interact with V^* to generate $W_{\lambda_2,0^n}$ and the auxiliary input $z_2 = (\lambda_1, \lambda_2, y_1, y_2, \alpha, \beta, \gamma)$ (of this interaction), and adopts P's strategy to generate $W_{\lambda_2,w}$ and outputs $(W_{\lambda_2,w}, W_{\lambda_2,0^n}, z_2)$. Assume $di\mathcal{O}$ works for $\mathsf{Sampler}_2$ too. Similarly, $\widetilde{W}_{\lambda_2,w}, \widetilde{W}_{\lambda_2,0^n}$ are indistinguishable for any distinguisher knowing z_2. Thus with the hybrid argument the WI property of NIWI, the hiding property of Com and the indistinguishability of $di\mathcal{O}$ ensure the indistinguishability of simulation.

Argument of Knowledge. We show there is a strict-polynomial-time non-black-box extractor E such that if P' is a polynomial-size prover that can convince V of $x \in L$ with probability ϵ, $E(P', x)$ outputs a witness for x with probability $\epsilon - \mathsf{neg}(n)$. E works as follows. It adopts V's strategy to finish the interaction except computing $Z_2 = \mathsf{Com}(h_2(\Pi'))$, where Π' denotes P''s remainder strategy of Step 3. Thus Π' (and the coins in generating Z_2) is a witness for λ_2. Let $\widetilde{W}$ and $\widetilde{Q}$ denote the two obfuscated programs in P''s message of the last step. Due to the validity of P''s messages, either $\widetilde{W}$ is $\widetilde{W}_{\lambda_2,w}$ or $\widetilde{Q}$ is $\widetilde{Q}_{\lambda_2,s}$. We claim that the latter possibility holds with negligible probability. Then E runs $\widetilde{W}$ on input Π' to output a witness for $x \in L$ with probability $\epsilon - \mathsf{neg}(n)$. So all that is left is to prove the claim.

Assuming the latter possibility holds with noticeable probability, we can construct an expected polynomial-time algorithm A to reverse f. A first interacts with P' and then runs the extractor of WIPOK to extract u, a pre-image of y_1 or y_2. Then rewind P' to Step 3 in which A independently generates $(r_1, \beta, Z_2, y_s, \widetilde{M}_{\lambda_1,0^n}, \widetilde{M}'_{\lambda_1,u})$ and computes π_1 using witness u.

Let $\mathsf{Sampler}_3$ be a sampler that adopts V's strategy to interact with P' except computing $Z_2 = \mathsf{Com}(h_2(\Pi'))$ to generate $M'_{\lambda_1,0^n}$ and the auxiliary input $z_3 = (\lambda_1, \lambda_2, y_1, y_2, \alpha, \beta, v)$, where v denotes the coins used in computing Z_2, and adopts A's strategy to compute $M'_{\lambda_1,u}$ and outputs $(M'_{\lambda_1,u}, M'_{\lambda_1,0^n}, z_3)$. Let $\mathsf{Sampler}_4$ be a sampler that adopts V's strategy to interact with P' except computing $Z_2 = \mathsf{Com}(h_2(\Pi'))$ and outputs $(M_{\lambda_1,s}, M_{\lambda_1,0^n}, z_4)$, where $z_4 = (\lambda_1, \lambda_2, y_1, y_2, \alpha, \beta, v)$ (of this interaction).

We also show any polynomial-size adversary obtaining $(M'_{\lambda_1,u}, M'_{\lambda_1,0^n}, z_3)$ (resp. $(M_{\lambda_1,s}, M_{\lambda_1,0^n}, z_4)$) cannot find a witness for λ_1. Assume $di\mathcal{O}$ works for $\mathsf{Sampler}_3, \mathsf{Sampler}_4$. Then $\widetilde{M}'_{\lambda_1,u}, \widetilde{M}'_{\lambda_1,0^n}$ (resp. $\widetilde{M}_{\lambda_1,s}, \widetilde{M}_{\lambda_1,0^n}$) are indistinguishable for any distinguisher knowing z_3 (resp. z_4).

With the hybrid argument the WI property of NIWI and the indistinguishability of diO ensure P''s views interacting with V and A are indistinguishable and P''s response of Step 4 is still convincing in which $\widetilde{Q}$ is $Q_{\lambda_2,s}$ with noticeable probability. Thus A can run $\widetilde{Q}$ on input Π' to recover s. Notice that in the rewinding run, A actually uses y_s only, regardless of s. This means if A is given y_{s^*} for an unknown random s^*, A can send $(r_1, \beta, Z_2, y_{s^*}, \widetilde{M}_{\lambda_1,0^n}, \widetilde{M}'_{\lambda_1,u}, \pi_1)$ to P' in the rewinding run and then get P's message to recover s^* (or another pre-image of y_{s^*}). This implies that f can be reversed in polynomial-time with noticeable probability, which is impossible. Thus the claim is proved.

Remark 1. One might think there is possibly a potential attack by a cheating prover P' so that the extractor may not extract a witness in strict-polynomial-time. That is, assume the obfuscation and NIWI are malleable (although we do not know how to perform such attacks to them currently). Then the prover might make use of the program $\widetilde{M}_{\lambda_1,s}$ sent by E to compute a valid $\widetilde{Q}$ even if it does not know s and generate a valid proof π from π_1 when E computes $\widetilde{M}_{\lambda_1,s} = diO(M_{\lambda_1,s}), \widetilde{M}' = diO(M'_{\lambda_1,0^n})$ and uses s as witness to compute π_1 in Step 3. Thus $\widetilde{Q}$ could be $Q_{\lambda_2,s}$ with non-negligible probability.

We argue that such an attack cannot be carried out and show in more detail that in extraction the $\widetilde{Q}$ sent by P' can be $Q_{\lambda_2,s}$ only with negligible probability. Supposing this is not true, we have the following hybrids.

Hybrid 1: This hybrid is identical to the extraction except that E/A first adopts A's strategy above to extract u and then rewinds to Step 3 where E/A generates $\widetilde{M}_{\lambda_1,s} = diO(M_{\lambda_1,s}), \widetilde{M}'_{\lambda_1,u} = diO(M'_{\lambda_1,u})$ and computes π_1 honestly. Then $\widetilde{Q}$ in the rewinding on input Π' (and the coins) outputs s with non-negligible probability. Otherwise, there is a polynomial-size algorithm when having z_3 can distinguish $\widetilde{M}'_{\lambda_1,u}$ and $\widetilde{M}'_{\lambda_1,0^n}$. The algorithm has P''s code hardwired and receives z_3 and $\widetilde{M}'$ which is either $\widetilde{M}'_{\lambda_1,u}$ or $\widetilde{M}'_{\lambda_1,0^n}$. It runs P' internally basically using the messages in z_3 except in Step 3, it generates $y_s, \widetilde{M}_{\lambda_1,s}, \widetilde{M}'$ and computes π_1 using witness s. On gaining P''s message of the last step it generates Π' and outputs $Q(\Pi', v)$ (precisely v is a part of the input). It can be seen that this algorithm can distinguish $\widetilde{M}'_{\lambda_1,u}$ and $\widetilde{M}'_{\lambda_1,0^n}$ with noticeable probability, since the $\widetilde{Q}$ in the extraction and this hybrid with input (Π', v) outputs differently with non-negligible probability. This is impossible.

Hybrid 2: This hybrid is identical to Hybrid 1 except that in Step 3 E/A generates $\widetilde{M}_{\lambda_1,s} = diO(M_{\lambda_1,s}), \widetilde{M}'_{\lambda_1,u} = diO(M'_{\lambda_1,u})$ and computes π_1 using witness u. Then still $\widetilde{Q}$ on input (Π', v) still outputs s with non-negligible probability. Otherwise, there is a polynomial-size algorithm that having z_3 can distinguish the two witnesses, u or s, being used in the NIWI.

Hybrid 3: This hybrid is identical to Hybrid 2 except that in Step 3 E/A generates $\widetilde{M}_{\lambda_1,0^n} = diO(M_{\lambda_1,0^n}), \widetilde{M}'_{\lambda_1,u} = diO(M'_{\lambda_1,u})$ and computes π_1 using witness u. In this hybrid, $\widetilde{Q}$ on input (Π', v) still outputs s with non-negligible probability, due to a similar analysis in Hybrid 1. Otherwise, there is a polynomial-size algorithm when having z_4 can distinguish $\widetilde{M}_{\lambda_1,0^n}$ and $\widetilde{M}_{\lambda_1,s}$.

The above hybrids show that in Hybrid 3 $\widetilde{Q}$ on input (Π', v) outputs s with non-negligible probability. However, if this is true, when receiving y_{s^*} from outside, A can send $y_{s^*}, \widetilde{M}_{\lambda_1,0^n}, \widetilde{M}'_{\lambda_1,u}$ and compute π_1 using witness u in Step 3. When getting $\widetilde{Q}$, run $\widetilde{Q}$ with (Π', v) to gain a pre-image of y_{s^*}, breaking the one-way-ness of f. This is impossible. Thus, in extraction $\widetilde{Q}$ cannot be $\widetilde{Q}_{\lambda_2,s}$ except for an negligible probability. So E runs $\widetilde{W}$ with (Π', v) to gain w.

2.3   Formal Descriptions

Now we formalize Protocol 1. First we present the definitions of the underlying languages and the explanation of the protocol. Then we present the proof of Theorem 1.

Definition 1. *We define language $\mathcal{U}_0$ as follows: $(\lambda_1 = (h_1, Z_1, r_1), msg) \in \mathcal{U}_0$ iff there exist Π, s_1 such that $|\Pi| < n^{\log\log n}$ and $\mathsf{Com}(h_1(\Pi); s_1) = Z_1$ and $\Pi(Z_1, msg)$ outputs r_1 in $n^{\log\log n}$ steps, where h_1 denotes a hash function and s_1 denotes the coins used in computing Com.*

Definition 2. *We define language $\mathcal{U}_1$ as follows: $(\lambda_2 = (h_2, Z_2, r_2), msg') \in \mathcal{U}_1$ iff there exist Π', s_1 such that $|\Pi'| < n^{\log\log n}$ and $\mathsf{Com}(h_2(\Pi'); s_1) = Z_2$ and $\Pi'(Z_2, msg')$ outputs r_2 in $n^{\log\log n}$ steps.*

Here are some programs that are required by the following languages (and we will state them again in specifying the nice machine samplers).

1. $M_{\lambda_1,msg,s}$ is the program that on input x' if x' is a witness for $(\lambda_1, msg) \in \mathcal{U}_0$ where $msg = (y_1, y_2, \alpha, h_2)$ outputs s and outputs 0^n otherwise.
2. $M'_{\lambda_1,msg,u}$ is the program that on input x' if x' is a witness for $(\lambda_1, msg) \in \mathcal{U}_0$ outputs u and outputs 0^n otherwise.
3. $W_{\lambda_2,msg',w}$ is the program that on input x' if x' is a witness for $(\lambda_2, msg') \in \mathcal{U}_1$ where $msg' = (r_1, \beta, y_s, \widetilde{M}_{\lambda_1,s}, \widetilde{M}'_{\lambda_1,0^n}, \pi_1)$ outputs w satisfying $(x, w) \in R_L$ and outputs 0^n otherwise.
4. $Q_{\lambda_2,msg',s}$ is the program that on input x' if x' is a witness for $(\lambda_2, msg') \in \mathcal{U}_1$ outputs s and outputs 0^n otherwise.

Definition 3. *We define language L_1 as follows: $(x, \lambda_2, msg', \widetilde{W}) \in L_1$ iff there exist w, s_1 such that $(x, w) \in R_L$ and $\widetilde{W} = di\mathcal{O}(W_{\lambda_2,msg',w}; s_1)$ where s_1 denotes the coins used in running $di\mathcal{O}$.*

Definition 4. *We define language L_2 as follows: $(y, \lambda_2, msg', \widetilde{Q}) \in L_2$ iff there exist s, s_1 such that $\widetilde{Q} = di\mathcal{O}(Q_{\lambda_2,msg',s}; s_1)$ and $f(s) = y$.*

Definition 5. *We define language L_3 as follows: $(y_1, y_2) \in L_3$ iff there exists u such that $f(u) = y_1$ or $f(u) = y_2$.*

Definition 6. *We define language L_4 as follows: $(msg, y, \lambda_1, \widetilde{M}, \widetilde{M}') \in L_4$ where $msg = (y_1, y_2, \alpha, h_2)$ iff there exist s, s_1 such that $\widetilde{M} = di\mathcal{O}(M_{\lambda_1, msg, s}; s_1)$ and $f(s) = y$ or there exist u, s_2 such that $\widetilde{M}' = di\mathcal{O}(M'_{\lambda_1, msg, u}; s_2)$ in which u satisfies $f(u) = y_1$ or $f(u) = y_2$.*

Definition 7. *We define language L_5 as follows: $(x, \lambda_2, y, msg', \widetilde{W}, \widetilde{Q}) \in L_5$ iff $(x, \lambda_2, msg', \widetilde{W}) \in L_1$ or $(y, \lambda_2, msg', \widetilde{Q}) \in L_2$.*

Note that $\mathcal{U}_0, \mathcal{U}_1 \in \mathbf{Ntime}(n^{O(\log \log n)})$, $L_1, L_2, L_3, L_4 \in \mathbf{NP}$. For simplicity of statement and consistence with the description in Section 2.2, we will omit msg and msg' in the specifications of the involved programs when (λ_1, msg) or (λ_2, msg') is generated in a consecutive running of the protocol. For example, we will simply write $M_{\lambda_1, s}$ instead of $M_{\lambda_1, msg, s}$ and $Q_{\lambda_2, s}$ instead of $Q_{\lambda_2, msg', s}$. Then we can explain Protocol 1 formally.

1. In Step 1, V sends a random $h_1 \in \mathcal{H}_n$.
2. In Step 2, P computes $Z_1 = \mathsf{Com}(h_1(0^n))$ and samples $h_2 \in \mathcal{H}_n$ and $u_i \in \{0,1\}^n$ and computes $y_i = f(u_i)$ for $i = 1, 2$ and α (in which (α, β, γ) is to prove $(y_1, y_2) \in L_3$) and sends them to V.
3. In Step 3, V samples $r_1 \in \{0,1\}^n$, computes β and $Z_2 = \mathsf{Com}(h_2(0^n))$, samples $s \in \{0,1\}^n$ and computes $y_s = f(s)$ and $\widetilde{M}_{\lambda_1, s}, \widetilde{M}'_{\lambda_1, 0^n}$ and a NIWI proof π_1 using s (and the coins) as witness to prove $(msg, y, \lambda_1, \widetilde{M}_{\lambda_1, s}, \widetilde{M}'_{\lambda_1, 0^n}) \in L_4$ where $msg = (y_1, y_2, \alpha, h_2)$. Send them to P.
4. In Step 4, P samples $r_2 \in \{0,1\}^n$, computes $\gamma, \widetilde{W}_{\lambda_2, w}, \widetilde{Q}_{\lambda_2, 0^n}$, a NIWI proof π using w (and the coins) as witness to prove $(x, \lambda_2, y_s, msg', \widetilde{W}_{\lambda_2, w}, \widetilde{Q}_{\lambda_2, 0^n}) \in L_5$, where $msg' = (r_1, \beta, y_s, \widetilde{M}_{\lambda_1, s}, \widetilde{M}'_{\lambda_1, 0^n}, \pi_1)$.

Nice Machine Samplers. Now we formalize the four nice machine samplers (to which we allow arbitrary inputs (x, w, V^*) or (x, P') instead of 1^n and will assume $di\mathcal{O}$ works for the samplers with these arbitrary inputs).

1. $\mathsf{Sampler}_1(x, w, V^*)$: $\mathsf{Sampler}_1$ emulates V^*'s computation to output the verifier's message of each step. In computing the prover's messages, $\mathsf{Sampler}_1$ computes $Z_1 = \mathsf{Com}(h_1(\Pi))$ and other messages honestly in Step 2, where Π denotes V^*'s remainder strategy, and adopts P's strategy to generate the message of the last step. Let $z_1 = (\lambda_1, \lambda_2, y_1, y_2, \alpha, \beta, \gamma)$.
   Let $\widetilde{M}, \widetilde{M}'$ denote the two programs in V^*'s message of Step 3. Then either $\widetilde{M}$ is $\widetilde{M}_{\lambda_1, s}$ or $\widetilde{M}'$ is $\widetilde{M}'_{\lambda_1, u}$. Due to the witness-hiding property of WIPOK, $\widetilde{M}'$ is $\widetilde{M}'_{\lambda_1, u}$ with negligible probability. (Otherwise suppose (y_1, y_2, α) is an input to $\mathsf{Sampler}_1$. Then $\mathsf{Sampler}_1$ can recover a pre-image of y_1 or y_2, contradicting the witness-hiding property of WIPOK.) So $\mathsf{Sampler}_1$ runs $\widetilde{M}$ with input Π (and the coins) to recover s. Let $msg' = (r_1, \beta, y_s, \widetilde{M}_{\lambda_1, s}, \widetilde{M}'_{\lambda_1, 0^n}, \pi_1)$. Finally output $(Q_{\lambda_2, s}, Q_{\lambda_2, 0^n}, z_1)$, in which
   (a) $Q_{\lambda_2, s}$ is the program that on input x' if x' is a witness for $(\lambda_2, msg') \in \mathcal{U}_1$ outputs s and outputs 0^n otherwise.

(b) $Q_{\lambda_2,0^n}$ is the program that runs identically to $Q_{\lambda_2,s}$ except that it always outputs 0^n finally.

2. $\mathsf{Sampler}_2(x, w, V^*)$: $\mathsf{Sampler}_2$ adopts $\mathsf{Sampler}_1$'s strategy to interact with V^* of the protocol and let $z_2 = (\lambda_1, \lambda_2, y_1, y_2, \alpha, \beta, \gamma)$. Generate $W_{\lambda_2,w}, W_{\lambda_2,0^n}$. Finally output $(W_{\lambda_2,w}, W_{\lambda_2,0^n}, z_2)$, in which

   (a) $W_{\lambda_2,w}$ is the program that on input x' if x' is a witness for $(\lambda_2, msg') \in \mathcal{U}_1$ outputs w satisfying $(x, w) \in R_L$ and outputs 0^n otherwise.
   (b) $W_{\lambda_2,0^n}$ is the program that runs identically to $W_{\lambda_2,w}$ except that it always outputs 0^n finally.

3. $\mathsf{Sampler}_3(x, P')$: $\mathsf{Sampler}_3$ emulates P''s computation to generate all prover's messages. For computing the verifier's messages, it adopts V's strategy to send the message of Step 1 and computes $Z_2 = \mathsf{Com}(h_2(\Pi'); v)$ and other messages honestly in Step 3, where Π' denotes P''s remainder strategy. Let $z_3 = (\lambda_1, \lambda_2, y_1, y_2, \alpha, \beta, v)$ and $msg = (y_1, y_2, \alpha, h_2)$. Besides, it runs the extractor of WIPOK to extract u, a pre-image of y_1 or y_2, while preparing other messages honestly. Finally output $(M'_{\lambda_1,u}, M'_{\lambda_1,0^n}, z_3)$, in which

   (a) $M'_{\lambda_1,u}$ is the program that on input x' if x' is a witness for $(\lambda_1, msg) \in \mathcal{U}_0$ outputs u and outputs 0^n otherwise.
   (b) $M'_{\lambda_1,0^n}$ is the program that runs identically to M'_{λ_1,u_1} except that it always outputs 0^n finally.

4. $\mathsf{Sampler}_4(x, P')$: $\mathsf{Sampler}_4$ adopts $\mathsf{Sampler}_3$'s strategy to interact with P' (without the extraction of u). Let $z_4 = (\lambda_1, \lambda_2, y_1, y_2, \alpha, \beta, v)$. Finally, output $(M_{\lambda_1,s}, M_{\lambda_1,0^n}, z_4)$, in which

   (a) $M_{\lambda_1,s}$ is the program that on input x' if x' is a witness for $(\lambda_1, msg) \in \mathcal{U}_0$ outputs s and outputs 0^n otherwise.
   (b) $M_{\lambda_1,0^n}$ is the program that runs identically to $M_{\lambda_1,s}$ except that it always outputs 0^n finally.

Lemma 1. *For* $(M_{\lambda_1,s}, M_{\lambda_1,0^n}, z_4) \leftarrow \mathsf{Sampler}_4(x, P')$, *letting* $\widetilde{M}_{\lambda_1,s} \leftarrow di\mathcal{O}(M_{\lambda_1,s})$, $\widetilde{M}_{\lambda_1,0^n} \leftarrow di\mathcal{O}(M_{\lambda_1,0^n})$, *the following holds:*

- *for any polynomial-size A, $\Pr[M_{\lambda_1,s}(x') \neq M_{\lambda_1,0^n}(x') : A(M_{\lambda_1,s}, M_{\lambda_1,0^n}, z_4) \rightarrow x'] = \mathsf{neg}(n)$.*
- $\widetilde{M}_{\lambda_1,s}$ *and* $\widetilde{M}_{\lambda_1,0^n}$ *are indistinguishable for any distinguisher having z_4.*

Proof. It can be seen that Result 2 follows Result 1 due to the definition of differing-input obfuscation. So we only need to prove 1. Suppose Result 1 is not true. In the following we adopt the argument in [2] to find a collision of h_1.

Suppose there is a polynomial-size A that given $(M_{\lambda_1,s}, M_{\lambda_1,0^n}, z_4)$ can output x' such that $M_{\lambda_1,s}$ and $M_{\lambda_1,0^n}$ disagree with noticeable probability. Then x' is a witness for $(\lambda_1, msg) \in \mathcal{U}_0$, which is of form (Π, s_1). So with the argument in [2], rewind P' to Step 3 with input an independently generated message of Step 3 and finish the interaction. Let λ'_1 denote the new transcript. Then A on receiving $(M_{\lambda'_1,s'}, M_{\lambda'_1,0^n}, z'_4)$ can still output a witness for (λ'_1, msg). Let (Π, s_1)

and (Π^*, s_1^*) denote the two witnesses. Notice that h_1, Z_1 in the two running are identical. So $h_1(\Pi) = h_1(\Pi^*)$ due to the perfectly-binding property of Com. Since $\Pi(Z_1, msg) = r_1$ and $\Pi^*(Z_1, msg) = r_1$ but the two r_1 in the two running are identical only with negligible probability, $\Pi \neq \Pi^*$. This means they are a collision of h_1. That is impossible. $\qquad\square$

Lemma 2. *For* $(M'_{\lambda_1,u}, M'_{\lambda_1,0^n}, z_3) \leftarrow$ Sampler$_3(x, P')$, *letting* $\widetilde{M'}_{\lambda_1,u} \leftarrow di\mathcal{O}$ $(M'_{\lambda_1,u})$, $\widetilde{M'}_{\lambda_1,0^n} \leftarrow di\mathcal{O}(M'_{\lambda_1,0^n})$, *the following holds:*

- *for any polynomial-size* A, $\Pr[M'_{\lambda_1,u}(x') \neq M'_{\lambda_1,0^n}(x') : A(M'_{\lambda_1,u}, M'_{\lambda_1,0^n}, z_3)$ $\to x'] = neg(n)$.
- $\widetilde{M'}_{\lambda_1,u}$ *and* $\widetilde{M'}_{\lambda_1,0^n}$ *are indistinguishable for any distinguisher having* z_3.

Proof. This proof is similar to that of Lemma 1. $\qquad\square$

Lemma 3. *For* $(W_{\lambda_2,w}, W_{\lambda_2,0^n}, z_2) \leftarrow$ Sampler$_2(x, w, V^*)$, *letting* $\widetilde{W}_{\lambda_2,w}$ $\leftarrow di\mathcal{O}(W_{\lambda_2,w})$, $\widetilde{W}_{\lambda_2,0^n} \leftarrow di\mathcal{O}(W_{\lambda_2,0^n})$, *the following holds:*

- *for any polynomial-size* A, $\Pr[W_{\lambda_2,w}(x') \neq W_{\lambda_2,0^n}(x') : A(W_{\lambda_2,w}, W_{\lambda_2,0^n}, z_2)$ $\to x'] = neg(n)$.
- $\widetilde{W}_{\lambda_2,w}$ *and* $\widetilde{W}_{\lambda_2,0^n}$ *are indistinguishable for any distinguisher having* z_2.

Proof. This proof is similar to that of Lemma 1. We only show there is no polynomial-size A that can find a differing-input for $W_{\lambda_2,w}, W_{\lambda_2,0^n}$ with noticeable probability. Suppose this is not true. Then the input output by A is a witness for $(\lambda_2, msg') \in \mathcal{U}_1$ where msg' is of form $(r_1, \beta, y_s, \overline{M}, \overline{M'}, \pi_1)$. So Simpler$_2$ runs the last step independently and let λ'_2 denote the new transcript. Inputting the machines and auxiliary input of the new running to A, A can output a witness for $(\lambda'_2, msg') \in \mathcal{U}_1$, where λ'_2 contains an independent r_2 but (h_2, Z_2) is same. Let Π', Π'^* be the two programs in the two witnesses. With a similar argument shown above, (Π', Π'^*) is a collision of h_2. This is impossible. $\qquad\square$

Lemma 4. *For* $(Q_{\lambda_2,s}, Q_{\lambda_2,0^n}, z_1) \leftarrow$ Sampler$_1(x, w, V^*)$, *letting* $\widetilde{Q}_{\lambda_2,s}$ $\leftarrow di\mathcal{O}(Q_{\lambda_2,s})$, $\widetilde{Q}_{\lambda_2,0^n} \leftarrow di\mathcal{O}(Q_{\lambda_2,0^n})$, *the following holds:*

- *for any polynomial-size* A, $\Pr[Q_{\lambda_2,s}(x') \neq Q_{\lambda_2,0^n}(x') : A(Q_{\lambda_2,s}, Q_{\lambda_2,0^n}, z_1) \to x'] = neg(n)$.
- $\widetilde{Q}_{\lambda_2,s}$ *and* $\widetilde{Q}_{\lambda_2,0^n}$ *are indistinguishable for any distinguisher having* z_1.

Proof. This proof is similar to that of Lemma 3. $\qquad\square$

Theorem 4. *Assuming the existence of* $\{\mathcal{H}_n\}$, Com, NIWI, WIPOK *and* $di\mathcal{O}$ *for* Sampler$_i$, $1 \leq i \leq 4$, *Protocol 1 satisfies all the properties claimed in Theorem 1.*

Proof. We show the completeness, the soundness, zero-knowledge, argument of knowledge properties are satisfied.

Completeness. It can be seen that the honest prover can use the witness w to convince the verifier in polynomial-time.

Computational Soundness. This follows from the argument of knowledge property.

Auxiliary-Input Zero-Knowledge. We now show there is a strict polynomial-time simulator S. For any $x \in L$ and any $V^* \in \{0,1\}^{\mathrm{poly}(n)}$, $S(x, V^*)$ runs as follows. It first emulates V^* to output the first message h_1. Then it computes $Z_1 = \mathsf{Com}(h_1(\Pi))$, where Π denotes V^*'s remainder strategy of Step 2. Sample a random h_2 and compute (y_1, y_2, α) honestly. Send $(Z_1, \langle y_1, y_2, \alpha \rangle, h_2)$ to V^*. Then emulate V^* to output $(r_1, \beta, Z_2, \langle y_s, \widetilde{M}, \widetilde{M}', \pi_1 \rangle)$. If V^*'s message is invalid, abort the interaction. Otherwise, $\widetilde{M}$ is $\widetilde{M}_{\lambda_1,s}$ or $\widetilde{M}'$ is $\widetilde{M}'_{\lambda_1,u}$ where u is a pre-image of y_1 or y_2. If the latter possibility holds with noticeable probability, then S can run $\widetilde{M}'$ with input Π (with some coins) to recover u. This shows even if (y_1, y_2, α) is given from outside i.e. it is not generated by S itself, S can still recover a pre-image of y_1 or y_2, contradicting the witness-hiding property of WIPOK. So it must be the case that $\widetilde{M}$ is $\widetilde{M}_{\lambda_1,s}$ except for an negligible probability. Then S can run $\widetilde{M}$ with input Π to recover s and finish the last step. In the last step, S samples a random r_2, generates $\widetilde{W}_{\lambda_2,0^n} \leftarrow di\mathcal{O}(W_{\lambda_2,0^n})$ and $\widetilde{Q}_{\lambda_2,s} \leftarrow di\mathcal{O}(Q_{\lambda_2,s})$, and generates π using s as witness.

For convenience of statement, let S_0 denote the real interaction of the protocol in which the prover uses w as witness. We now establish the indistinguishability of simulation via the hybrid argument.

Hybrid 1. We construct a hybrid simulator S_1 which has w hardwired and runs identically to S_0 except that S_1 adopts S's strategy in computing Z_1. Then by the hiding property of Com, the outputs by S_0 and S_1 are indistinguishable.

Hybrid 2. We construct a hybrid simulator S_2 which has w hardwired and runs identically to S_1 except that S_2 adopts S's strategy in computing $\widetilde{Q}_{\lambda_2,s}$. Thus by Lemma 4, $\widetilde{Q}_{\lambda_2,s}$ and $\widetilde{Q}_{\lambda_2,0^n}$ are indistinguishable for any distinguisher with $z_1 = (\lambda_1, \lambda_2, y_1, y_2, \alpha, \beta, \gamma)$. Any distinguisher for the outputs of S_2 and S_1 can be transformed to one with z_1 for distinguishing $\widetilde{Q}_{\lambda_2,s}$ and $\widetilde{Q}_{\lambda_2,0^n}$.

Assume D is a distinguisher for the outputs of S_2 and S_1. We construct an algorithm D' that has z_1 and is capable of distinguishing $\widetilde{Q}_{\lambda_2,0^n}$ from $\widetilde{Q}_{\lambda_2,s}$. On given $\widetilde{Q}_{\lambda_2,*}$, D' that has x, w, V^*'s code hardwired invokes an interaction with V^*. First it invokes V^* to send out h_1 and then it responds with $Z_1, y_1, y_2, \alpha, h_2$ which are those in z_1. Then V^* sends $r_1, \beta, Z_2, y_s, \widetilde{M}, \widetilde{M}', \pi_1$, in which r_1, β, Z_2 are also equal to those in z_1. D' computes $\widetilde{W}_{\lambda_2,w}$ and copies $\widetilde{Q}_{\lambda_2,*}$ and computes π using witness w and sends $(r_2, \gamma, \widetilde{W}_{\lambda_2,w}, \widetilde{Q}_{\lambda_2,*}, \pi)$ to V^*, where r_2, γ are those in z_1. Lastly, D' calls D with V^*'s view generated in the interaction and outputs what D outputs. It can be seen that D''s probability of distinguishing $\widetilde{Q}_{\lambda_2,0^n}$ from $\widetilde{Q}_{\lambda_2,s}$ is identical to that of D of distinguishing the outputs of S_2 and S_1. So the two outputs are indistinguishable.

Hybrid 3. We construct a hybrid simulator S_3 which has w hardwired and runs identically to S_2 only except that S_3 adopts S's strategy in computing π. Due to the WI property of NIWI, any distinguisher D for the outputs of S_1 and S_2

can be transformed to a distinguisher D' for the two π's. So the two outputs are indistinguishable.

Hybrid 4. We construct a hybrid simulator S_4 which runs identically to S_3 only except that S_4 adopts S's strategy in computing $\widetilde{W}_{\lambda_2,0^n}$. By Lemma 3, $\widetilde{W}_{\lambda_2,w}$ and $\widetilde{W}_{\lambda_2,0^n}$ are indistinguishable for any distinguisher with $z_2 = (\lambda_1, \lambda_2, y_1, y_2, \alpha, \beta, \gamma)$. With a similar argument in Hybrid 2, any distinguisher for the outputs of S_4 and S_3 can be transformed to one with z_2 for distinguishing $\widetilde{W}_{\lambda_2,w}$ and $\widetilde{W}_{\lambda_2,0^n}$. This is impossible. So the two outputs are indistinguishable.

Actually, $S_4 = S$. Combining all the hybrids, S_0's output is indistinguishable from S's. The zero-knowledge property is satisfied.

Argument of Knowledge. We show there is a strict-polynomial-time non-black-box extractor E such that if P' is a polynomial-size prover that can convince V of $x \in L$ with probability ϵ, $E(P', x)$ outputs a witness for x with probability $\epsilon - \mathsf{neg}(n)$. E works as follows. It adopts V's strategy to finish the interaction but computes $Z_2 = \mathsf{Com}(h_2(\Pi'))$, where Π' denotes P''s remainder strategy of Step 3. Let $\widetilde{W}$ and $\widetilde{Q}$ denote the two obfuscated programs in P''s message of last step.

Due to the validity of P''s messages, either $\widetilde{W}$ is $\widetilde{W}_{\lambda_2,w}$ or $\widetilde{Q}$ is $\widetilde{Q}_{\lambda_2,s}$. We claim that the latter possibility holds with negligible probability. Then E runs $\widetilde{W}$ on input Π' (and the coins in generating Z_2) to output a witness for $x \in L$ with probability $\epsilon - \mathsf{neg}(n)$. So all that is left is to prove the claim.

Assume the latter possibility holds with noticeable probability. Then we can construct an algorithm A that breaks the one-way-ness of f. A is given y_{s^*} for an unknown random s^* and its goal is to recover s^*. It first follows E's strategy to interact with P'. When the interaction finishes, A calls the extractor of WIPOK and in extraction A still generates other messages honestly. This means A can extract a pre-image of y_1 or y_2, denoted u. Then A rewinds P' to Step 3 and at this time it sends messages $y_{s^*}, \widetilde{M}_{\lambda_1,0^n}, \widetilde{M}'_{\lambda_1,u}, \pi_1$ as well as random (r_1, β) to P^* where it generates π_1 using u as witness. When P' sends the message of last step, A runs the second obfuscated program $\widetilde{Q}$ with input Π' to output s^*.

Then A runs in expected polynomial-time. As shown in Remark 1 of which now we employ Lemma 1 and Lemma 2 in the Hybrids 3 and 1 to establish the indistinguishability of obfuscation, A can recover s^* with noticeable probability, contradicting the one-way-ness of f. Thus the latter possibility cannot happen with noticeable probability. The argument of knowledge property holds. □

3   Three-Round Bounded-Auxiliary-Input ZKAOK

Now we present the 3-round ZKAOK. When considering bounded-auxiliary-input ZK, we can modify Protocol 1 by dropping h_1, Z_1 and letting r_1 be chosen from $\{0,1\}^{10n}$ and $\lambda_1 = r_1$ for which a witness now means there is Π of size $5n$ such that Π outputs r_1 in $n^{\log \log n}$ steps. Then Theorem 2 can be proved with a similar argument in the previous section. In this section we present an alternative construction of the 3-round protocol which admits a simpler description.

Public input: x (statement to be proved is "$x \in L$");
Prover's auxiliary input: w, (a witness for $x \in L$).

---

1. $P \to V$: Send $h_2 \in_R \mathcal{H}_n$.
2. $V \to P$: Send $r_1 \in_R \{0,1\}^{10n}$, $Z_2 \leftarrow \mathsf{Com}(h_2(0^n))$, UA_1.
3. $P \to V$: Send $r_2 \in_R \{0,1\}^n$, $\widetilde{W}_{\lambda_2,w}, \widetilde{Q}_{\lambda_2,0^n}$, $Z_3 \leftarrow \mathsf{Com}(0^{\mathrm{poly}(n)})$, π.

Protocol 2 *The 3-round bounded-auxiliary-input ZKAOK.*

3.1 Construction Idea

Let us recall the 2-round public-coin plain zero-knowledge in [5]. The protocol assumes 2-round public-coin universal arguments, which admit a candidate construction i.e. the 2-round variant of Micali's CS proof [25], denoted UA. Let $(\mathsf{UA}_1, \mathsf{UA}_2)$ denote the two messages of UA.

In the protocol V first sends a random $10n$-bit string r and UA_1 and then P computes $Z = \mathsf{Com}(0^{\mathrm{poly}(n)})$ where $\mathrm{poly}(n)$ denotes the bit length of UA_2 and a WI proof using w for that either $x \in L$ or letting $\mathsf{UA}_2 \leftarrow \mathsf{Com}^{-1}(Z)$, $(\mathsf{UA}_1, \mathsf{UA}_2)$ is a valid proof for that there is Π of size $\leq 5n$ and Π outputs r in $n^{\log \log n}$ steps (actually $5n$ can be relaxed to any fixed polynomial). The protocol is zero-knowledge, but it is unknown if it admits an extractor. We will apply the proposed extraction strategy to it.

Similarly, we add a reverse Barak's preamble to the protocol and let P send the two obfuscated programs. Let $\lambda_2 = (h_2, Z_2, r_2)$ and $\lambda_1 = r_1$ denote the transcripts of this preamble and the original preamble respectively. Now we do not let V send the two obfuscated programs, which originally aims at providing the simulator an ability to finish the interaction. Instead, the simulator having any verifier's code can convince the verifier via UA. In extraction, the extractor can use a similar way to extract a witness for x from the two programs.

3.2 Overview

Let $\{\mathcal{H}_n\}, \mathsf{NIWI}, \mathsf{UA}$ be as above and $di\mathcal{O}$ work for two samplers shown below. We follow [5] to assume that Com can be broken totally in time $2^{n^{\epsilon/2}}$ and UA is sound against $2^{n^{\epsilon}}$-sized circuits for some constant $\epsilon > 0$ (which is induced by an appropriate assumption on trapdoor permutation families).

Note that in the 4-round universal argument in [3] when the first message is generated honestly, any P^* cannot cheat the verifier of a false public input adaptively chosen posterior to it. We also assume the soundness of UA holds even if the public input is adaptively chosen after UA_1 is uniformly chosen.

Let L denote an arbitrary language in **NP**. Our protocol is shown in Protocol 2 with the following specification.

1. In Step 1, P sends a hash function $h_2 \in_R \mathcal{H}_n$ to V.

2. In Step 2, V sends $r_1 \in_R \{0,1\}^{10n}$, $Z_2 \leftarrow \mathsf{Com}(h_2(0^n))$ and UA_1.

3. In Step 3, P generates $r_2, Z_3, \widetilde{W}_{\lambda_2, w}, \widetilde{Q}_{\lambda_2, 0^n}$ as specified and the NIWI proof π using witness w for that either $\widetilde{W}_{\lambda_2, w}$ is as specified or letting $\mathsf{UA}_2 = \mathsf{Com}^{-1}(Z_3)$, $(\mathsf{UA}_1, \mathsf{UA}_2)$ is a valid proof for (1) there is Π of size $\leq 5n$ and $\Pi(h_2)$ outputs r_1 in $n^{\log \log n}$ steps and (2) $\widetilde{Q}_{\lambda_2, 0^n}$ is an obfuscation of $Q_{\lambda_2, \Pi}$ where $Q_{\lambda_2, \Pi}$ is the program identically to $Q_{\lambda_2, 0^n}$ except 0^n is replaced by Π (note that the witness for the second statement of NIWI is UA_2).

The completeness of the protocol is straightforward. Now we sketch the zero-knowledge and argument of knowledge properties.

Zero-Knowledge. We show there is a strict polynomial-time non-black-box simulator S for any $x \in L$ and $V^* \in \{0,1\}^{2n}$. $S(x, V^*)$ runs as follows. It samples h_2 and sends it to V^*. Then it samples $u \in \{0,1\}^n$ and runs a pseudorandom generator $\mathsf{PRG}(u)$ to provide V^* enough coins. Let Π denote the program that has V^*'s code and x, PRG, u hardwired and emulates $V^*(x)$'s computing while running $\mathsf{PRG}(u)$ to provide it coins. So Π's size can be less than $5n$. When V^* sends out the message of Step 2, Π is actually a witness for λ_1. In the last step, S samples r_2, generates $\widetilde{W}_{\lambda_2, 0^n}$ and $\widetilde{Q}_{\lambda_2, \Pi}$ and UA_2 using Π as witness and $Z_3 \leftarrow \mathsf{Com}(\mathsf{UA}_2)$ and computes π using witness UA_2.

We consider the indistinguishability of simulation. Let $\mathsf{Sampler}_1$ (we reuse the notations $\mathsf{Sampler}_i, i = 1, 2$ only for simplicity) be a sampler that adopts P's strategy to interact with V^* to generate $Q_{\lambda_2, 0^n}, z_1$ where $z_1 = (\lambda_1, \lambda_2, \mathsf{UA}_1)$, and adopts S's strategy to compute $\widetilde{Q}_{\lambda_2, \Pi}$. Then output $(Q_{\lambda_2, \Pi}, Q_{\lambda_2, 0^n}, z_1)$. Let $\mathsf{Sampler}_2$ be a sampler that adopts S's strategy to interact with V^* to generate $W_{\lambda_2, 0^n}, z_2$ where $z_2 = (\lambda_1, \lambda_2, \mathsf{UA}_1)$ (of this interaction), and adopts P's strategy to compute $W_{\lambda_2, w}$ and outputs $(W_{\lambda_2, w}, W_{\lambda_2, 0^n}, z_2)$.

Assume $di\mathcal{O}$ works for $\mathsf{Sampler}_1$ and $\mathsf{Sampler}_2$. Due to the hardness of finding a witness for λ_2, $\widetilde{Q}_{\lambda_2, \Pi}, \widetilde{Q}_{\lambda_2, 0^n}$ (resp. $\widetilde{W}_{\lambda_2, w}, \widetilde{W}_{\lambda_2, 0^n}$) are indistinguishable for any distinguisher knowing z_1 (resp. z_2). The pseudorandomness of PRG and the WI property of NIWI and the hiding property of Com and the indistinguishability of obfuscation ensure the indistinguishability of simulation. This proves the zero-knowledge property for all $2n$-size V^*. Since the security parameter n can be scaled, this protocol can be easily modified to be bounded-auxiliary-input zero-knowledge for all V^* of size bounded by an a-prior fixed polynomial.

Argument of Knowledge. We show there is a strict-polynomial-time non-black-box extractor E such that if P' is a polynomial-size prover that can convince V of $x \in L$ with probability ϵ', $E(P', x)$ outputs a witness for x with probability $\epsilon' - \mathsf{neg}(n)$. E works as follows. First it emulates P' to send out h_2. In Step 2, E samples r_1, UA_1 honestly and computes $Z_2 \leftarrow \mathsf{Com}(h_2(\Pi'))$, where Π' denotes P''s remainder strategy at this step, and then sends them to P' and lastly finishes the whole interaction. Π' (with some coins) is a witness for λ_2.

Let $\widetilde{W}$ and $\widetilde{Q}$ denote the two obfuscated programs in P''s message of the last step. As shown in [2], for $r_1 \in_R \{0,1\}^{10n}$ there is no Π of size $\leq 5n$ satisfying $\Pi(h_2)$ outputs r_1 except for negligible probability. Due to the soundness of UA

Public input: x (statement to be proved is "$x \in L$");
Prover's auxiliary input: w, (a witness for $x \in L$).

---

1. $V \to P$: Send $r_1 \in_R \{0,1\}^{10n}$, UA_1.
2. $P \to V$: Send $r_2 \in_R \{0,1\}^{10n}$, $\widetilde{W}_{\lambda_2,w}, \widetilde{Q}_{\lambda_2,0^n}$, $Z_3 \leftarrow \mathsf{Com}(0^{\mathrm{poly}(n)})$, π.

Protocol 3 *The 2-round public-coin bounded-auxiliary-input ZKAOK.*

and NIWI, either $\widetilde{W}$ is $\widetilde{W}_{\lambda_2,w}$ or $\widetilde{Q}$ is $\widetilde{Q}_{\lambda_2,\Pi}$ except for negligible probability[1] (note that a part of the public input to UA is $\widetilde{Q}$ that is determined after UA_1 is sampled). Due to the non-existence of such Π, it must be the case $\widetilde{W}$ is $\widetilde{W}_{\lambda_2,w}$. So E can run $\widetilde{W}$ on input Π' to output a witness for $x \in L$ with probability $\epsilon' - \mathrm{neg}(n)$. So we have the following result (with a more assumption UA).

Theorem 5. *Assuming the existence of $\{\mathcal{H}_n\}$, Com, NIWI, $di\mathcal{O}$, UA, Protocol 2 satisfies all the properties claimed in Theorem 2.*

For lack of space we omit the formal descriptions of the assumption on the soundness of UA, Protocol 2, the samplers and the proof of Theorem 5.

4   On Two-Round Bounded-Auxiliary-Input ZKAOK

In this section we present the 2-round protocol claimed in Theorem 3. Let Com, NIWI, $di\mathcal{O}$, UA be as before. The protocol is shown in Protocol 3 that is a simplification of Protocol 2 in which we drop h_2, Z_2 and let $|r_2| = 10n$. Thus $\lambda_2 = r_2$ and a witness for λ_2 changes to be a program Π' of size $\leq 5n$ such that $\Pi'(r_1, \mathsf{UA}_1)$ outputs r_2 in $n^{\log \log n}$ steps. The honest prover P computes the NIWI proof π using w to prove that either $\widetilde{W}_{\lambda_2,w}$ is as specified or letting $\mathsf{UA}_2 = \mathsf{Com}^{-1}(Z_3)$, $(\mathsf{UA}_1, \mathsf{UA}_2)$ is a valid proof for (1) there is Π of size $\leq 5n$ and Π outputs r_1 in $n^{\log \log n}$ steps and (2) $\widetilde{Q}_{\lambda_2,0^n}$ is an obfuscation of $Q_{\lambda_2,\Pi}$.

It can be seen that the simulation strategy of Protocol 2 still works for this protocol. The extractor E runs as follows. For any P' with $|P'| < 2n$ it samples $u \in \{0,1\}^n$ and runs $\mathsf{PRG}(u)$ to provide P' enough coins. Let Π' denote the program that has P''s code and x, PRG, u hardwired and emulates $P'(x)$'s computing while running $\mathsf{PRG}(u)$ to provide it coins. So Π''s size can be less than $5n$. When E sends r_1, UA_1 and P' replies with r_2, Π' is a witness for λ_2. Let $\widetilde{W}, \widetilde{Q}$ denote the two obfuscated programs in P''s message. For the same reason shown in Section 3.1, there is no Π of size $\leq 5n$ satisfying $\widetilde{Q}$ is $\widetilde{Q}_{\lambda_2,\Pi}$ except for

---

[1] Due to the soundness of NIWI, $\widetilde{W}$ is $\widetilde{W}_{\lambda_2,w}$ or the second statement is true. If the second statement is true, we can recover UA_2 in $2^{n^{\epsilon/2}}$-time which ensures $\widetilde{Q}$ is $\widetilde{Q}_{\lambda_2,\Pi}$ except for negligible probability due to the soundness of UA.

negligible probability. Then $\widetilde{W}$ is $\widetilde{W}_{\lambda_2,w}$. So E can run $\widetilde{W}$ on input Π' to output a witness for $x \in L$ except for negligible probability.

Finally let us consider the soundness (since now the extractor works only for all provers of bounded-size, we still need to establish the soundness that holds for all polynomial-size provers). When $x \notin L$, $\widetilde{W}$ cannot be $\widetilde{W}_{\lambda_2,w}$ and for random $r_1 \ \widetilde{Q}$ cannot be $\widetilde{Q}_{\lambda_2,\Pi}$ as specified. Thus the soundness against all polynomial-size provers follows from the soundness of UA and NIWI (refer to footnote 1). Thus we have the following result.

Theorem 6. *Assuming the existence of* Com, NIWI, $di\mathcal{O}$, UA, *Protocol 3 satisfies all the properties claimed in Theorem 3.*

Acknowledgments. The author shows his deep thanks to the reviewers of ISC 2014 for their detailed and useful comments. This work is supported by the National Natural Science Foundation of China (Grant No. 61100209) and Doctoral Fund of Ministry of Education of China (Grant No. 20120073110094).

References

1. Ananth, P., Boneh, D., Garg, S., Sahai, A., Zhandry, M.: Differing-inputs obfuscation and applications. IACR Cryptology ePrint Archive 2013, 689 (2013)
2. Barak, B.: How to go beyond the black-box simulation barrier. In: FOCS, pp. 106–115 (2001)
3. Barak, B., Goldreich, O.: Universal arguments and their applications. In: IEEE Conference on Computational Complexity, pp. 194–203 (2002)
4. Barak, B., Lindell, Y.: Strict polynomial-time in simulation and extraction. In: Reif, J.H. (ed.) STOC, pp. 484–493. ACM (2002)
5. Barak, B., Lindell, Y., Vadhan, S.P.: Lower bounds for non-black-box zero knowledge. J. Comput. Syst. Sci. 72(2), 321–391 (2006)
6. Barak, B., Ong, S.J., Vadhan, S.P.: Derandomization in cryptography. In: Boneh, D. (ed.) CRYPTO 2003. LNCS, vol. 2729, pp. 299–315. Springer, Heidelberg (2003)
7. Bellare, M., Goldreich, O.: On defining proofs of knowledge. In: Brickell, E.F. (ed.) CRYPTO 1992. LNCS, vol. 740, pp. 390–420. Springer, Heidelberg (1993)
8. Bitansky, N., Canetti, R., Chiesa, A., Tromer, E.: Recursive composition and bootstrapping for snarks and proof-carrying data. In: Boneh, D., Roughgarden, T., Feigenbaum, J. (eds.) STOC, pp. 111–120. ACM (2013)
9. Bitansky, N., Canetti, R., Paneth, O., Rosen, A.: On the existence of extractable one-way functions. In: Shmoys, D.B. (ed.) STOC, pp. 505–514. ACM (2014)
10. Blum, M.: Coin flipping by telephone. In: Gersho, A. (ed.) CRYPTO, pp. 11–15. U. C. Santa Barbara, Dept. of Elec. and Computer Eng., ECE Report No 82-04 (1981)
11. Blum, M.: How to prove a theorem so no one else can claim it. In: Proceedings of the International Congress of Mathematicians, pp. 1444–1451 (1987)
12. Boyle, E., Chung, K.-M., Pass, R.: On extractability obfuscation. In: Lindell, Y. (ed.) TCC 2014. LNCS, vol. 8349, pp. 52–73. Springer, Heidelberg (2014)
13. Brassard, G., Chaum, D., Crépeau, C.: Minimum disclosure proofs of knowledge. J. Comput. Syst. Sci. 37(2), 156–189 (1988)

14. Feige, U., Shamir, A.: Witness indistinguishable and witness hiding protocols. In: Ortiz, H. (ed.) STOC, pp. 416–426. ACM (1990)
15. Garg, S., Gentry, C., Halevi, S., Raykova, M., Sahai, A., Waters, B.: Candidate indistinguishability obfuscation and functional encryption for all circuits. In: FOCS, pp. 40–49. IEEE Computer Society (2013)
16. Garg, S., Gentry, C., Halevi, S., Wichs, D.: On the implausibility of differing-inputs obfuscation and extractable witness encryption with auxiliary input. In: Garay, J.A., Gennaro, R. (eds.) CRYPTO 2014, Part I. LNCS, vol. 8616, pp. 518–535. Springer, Heidelberg (2014)
17. Goldreich, O., Kahan, A.: How to construct constant-round zero-knowledge proof systems for np. J. Cryptology 9(3), 167–190 (1996)
18. Goldreich, O., Micali, S., Wigderson, A.: Proofs that yield nothing but their validity and a methodology of cryptographic protocol design (extended abstract). In: FOCS, pp. 174–187. IEEE Computer Society (1986)
19. Goldreich, O., Oren, Y.: Definitions and properties of zero-knowledge proof systems. J. Cryptology 7(1), 1–32 (1994)
20. Goldwasser, S., Micali, S., Rackoff, C.: The knowledge complexity of interactive proof systems. SIAM J. Comput. 18(1), 186–208 (1989)
21. Groth, J., Ostrovsky, R., Sahai, A.: Non-interactive zaps and new techniques for NIZK. In: Dwork, C. (ed.) CRYPTO 2006. LNCS, vol. 4117, pp. 97–111. Springer, Heidelberg (2006)
22. Hada, S., Tanaka, T.: On the existence of 3-round zero-knowledge protocols. In: Krawczyk, H. (ed.) CRYPTO 1998. LNCS, vol. 1462, pp. 408–423. Springer, Heidelberg (1998)
23. Katz, J.: Which languages have 4-round zero-knowledge proofs? In: Canetti, R. (ed.) TCC 2008. LNCS, vol. 4948, pp. 73–88. Springer, Heidelberg (2008)
24. Lindell, Y.: A note on constant-round zero-knowledge proofs of knowledge. J. Cryptology 26(4), 638–654 (2013)
25. Micali, S.: Cs proofs (extended abstracts). In: FOCS, pp. 436–453. IEEE Computer Society (1994)
26. Ostrovsky, R., Visconti, I.: Simultaneous resettability from collision resistance. Electronic Colloquium on Computational Complexity (ECCC) 19, 164 (2012), http://dblp.uni-trier.de/db/journals/eccc/eccc19.html#OstrovskyV12
27. Pandey, O., Prabhakaran, M., Sahai, A.: Obfuscation-based non-black-box simulation and four message concurrent zero knowledge for np. Cryptology ePrint Archive, Report 2013/754 (2013), http://eprint.iacr.org/

A   Preliminaries: Differing-Input Obfuscation for Turing Machines

In this section we follow [27] to present the definition of differing-input obfuscation for Turing machines introduced in [1,12]. Let $\mathsf{Steps}(M; x)$ denote the number of steps taken by a TM M on input x; we use the convention that if M does not halt on x then $\mathsf{Steps}(M; x)$ is defined to be the special symbol ∞. We define the notion of "compatible Turing machines" and "nice sampler".

Definition 8. *(Compatible TMs). A pair of Turing machines (M_0, M_1) is said to be compatible if $|M_0| = |M_1|$ and for every string $x \in \{0,1\}^*$ it holds that $\mathsf{Steps}(M_0, x) = \mathsf{Steps}(M_1, x)$.*

Definition 9. *(Nice TM Sampler). We say that a (possibly non-uniform) PPT (possibly expected-polynomial-time) Turing machine* Sampler *is a nice sampler for Turing machines if the following conditions hold:*
1. the output of Sampler *is a triplet* (M_0, M_1, z) *such that* (M_0, M_1) *is always a pair of compatible TMs, and* $z \in \{0, 1\}^*$ *is a string;*
2. for every polynomial $a : \mathbb{N} \to \mathbb{N}$, *every sufficiently large* $n \in \mathbb{N}$, *and every (possibly non-uniform) TM* A *running in time at most* $a(n)$, *it holds that:* $\Pr[(M_0, M_1, z) \leftarrow \mathsf{Sampler}(1^n), A(M_0, M_1, z) = x; \mathsf{Steps}(M_0, x) \leq a(n); M_0(x) \neq M_1(x).] = \mathsf{neg}(n).$

Definition 10. *(Differing-input Obfuscators for Turing machines) A uniform PPT machine* $di\mathcal{O}$ *is called a Turing machine differing-input Obfuscator defined for a nice sampler* Sampler, *if the following conditions are satisfied:*
1. Polynomial slowdown and functionality: there exists a polynomial a_{dio} *such that for every* $n \in N$, *every* M *in the output of* Sampler, *every input* x *such that* M *halts on* x, *and every* $\widetilde{M} \leftarrow di\mathcal{O}(M)$, *the following conditions hold:*

- $\mathsf{Steps}(\widetilde{M}, x) \leq a_{dio}(n, \mathsf{Steps}(M; x)).$
- $\widetilde{M}(x) = M(x).$

Polynomial a_{dio} *is called the slowdown polynomial of* $di\mathcal{O}$.
2. Indistinguishability: for every polynomial-size D, *for* $(M_0, M_1, z) \leftarrow$ Sampler(1^n) *it holds that:* $|\Pr[D(\widetilde{M_0}, z) = 1] - \Pr[D(\widetilde{M_1}, z) = 1]| = \mathsf{neg}(n).$

A candidate construction for this primitive appears in the work of [1]. Their construction is based on $di\mathcal{O}$ for circuits (constructed in [15]), fully homomorphic encryption and SNARKs [8]. If an a-priori bound on the input is known, then comparatively better constructions are possible [1,12].

Garg *et al.* [16] showed that *general-purpose* $di\mathcal{O}$ for all samplers implies a specific circuit with a specific auxiliary input cannot be obfuscated in a way that hides some specific information. (So such a general $di\mathcal{O}$ and the specific obfuscator aiming at hiding the specific information for the specific circuit cannot exist simultaneously.)

Lightweight Zero-Knowledge Proofs for Crypto-Computing Protocols

Sven Laur[1,*] and Bingsheng Zhang[2,**]

[1] University of Tartu, Estonia
swen@math.ut.ee
[2] National and Kapodestrian University of Athens, Greece
bzhang@di.uoa.gr

Abstract. Crypto-computing is a set of well-known techniques for computing with encrypted data. The security of the corresponding protocols are usually proven in the semi-honest model. In this work, we propose a new class of zero-knowledge proofs, which are tailored for crypto-computing protocols. First, these proofs directly employ properties of the underlying crypto systems and thus many facts have more concise proofs compared to generic solutions. Second, we show how to achieve universal composability in the trusted set-up model where all zero-knowledge proofs share the same system-wide parameters. Third, we derive a new protocol for multiplicative relations and show how to combine it with several crypto-computing frameworks.

Keywords: Universal composability, conditional disclosure of secrets, zero-knowledge, homomorphic encryption, multi-party computation.

1 Introduction

Garbled circuit evaluation and protocols for computing with ciphertexts are two basic approaches for crypto-computing. Garbled circuit evaluation is Turing complete [Yao82,BHR12], while the computational expressibility of ciphertext manipulation depends on the underlying cryptosystem [SYY99,IP07,Gen09]. In this work, we consider only protocols based on ciphertext manipulation. These protocols often rely on the specific properties of the underlying plaintexts. If these conditions are not satisfied, these protocols often break down and even the privacy of inputs is not guaranteed [AIR01,LL07]. Hence, these protocols are often complemented with zero-knowledge proofs to guarantee security.

Conditional disclosure of secrets (CDS) can be used as a lightweight alternative to zero knowledge proofs [GIKM98,AIR01,LL07]. In a nutshell, CDS protocols are used to release secrets only if received ciphertexts satisfy a desired

---

* Partially supported by the European Regional Development Fund through the Estonian Centre of Excellence in Computer Science (EXCS) and by Estonian Research Council through the grant IUT2-1.
** Partially supported by Greek Secretariat of Research and Technology through the project FINER and by ERC through the project CODAMODA.

S.S.M. Chow et al. (Eds.): ISC 2014, LNCS 8783, pp. 140–157, 2014.

relation. If we use these secrets to encrypt replies in the original protocol, these replies become unreadable when the ciphertexts are malformed. The resulting protocol is extremely lightweight, as the transformation adds only few extra ciphertexts and some additional crypto-computing steps. On the flip side, CDS transformation ensures only input-privacy.

In this work, we extend this approach to full-fledged zero-knowledge proofs by adding a few short messages. As relatively efficient CDS protocols exist for proving **NP/poly** relations between plaintexts [AIR01,LL07], our method can be used to get zero-knowledge proofs for any **NP** language.

It is a new and interesting paradigm for constructing zero-knowledge proofs, as resulting CDSZK protocols do not follow standard sigma structure. In particular, parties do not have to agree before the protocol execution whether they aim for input-privacy, honest verifier or full-fledged zero-knowledge. This can be decided dynamically during the protocol with no overhead. For many zero-knowledge techniques, where the full-fledged zero-knowledge is achieved by adding extra messages into the beginning of the protocol, such flexibility is unachievable without additional overhead.

Properties of our zero-knowledge protocols are largely determined by the underlying commitment scheme. Essentially, we must choose between perfect simulation and statistical soundness. For perfect simulatability, the commitment scheme must be equivocal and thus security guarantees hold only against computationally bounded provers. Statistical binding assures unconditional soundness but it also prevents statistical simulatability. In both cases, usage of trusted setup together with dual mode commitments assures universal composability even if the setup is shared between all protocols.

As CDS and CDSZK protocols are mostly applied for protecting crypto-computing protocols against malicious adversaries, we show how the trusted setup can be implemented in the standard or in the common reference string model without losing security guarantees, see Sect. 6.

As the second major contribution, we describe a CDS protocol for a multiplicative relation with the cost of two additional ciphertexts. This is a major advancement, as all previous CDS protocols for multiplicative relation had a quadratic overhead in communication. This result is important as many crypto-computing protocols can be made secure against active attacks by verifying multiplicative relations. These relations naturally occur in the computation of Beaver tuples and shared message authentication codes [DPSZ12].

For clarity, all of our results are formalised in the concrete security framework and statements about polynomial model are obtained by considering the asymptotic behaviour w.r.t. the security parameter.

Due to space constraints, we have omitted some auxiliary results and properties, which are not essential for the exposition. These results and more thorough discussion of various technical details can be found in the full version of this paper achieved in the Cryptology ePrint Archive.

2   Preliminaries

We use boldface letters for vectors and calligraphic letters for sets and algo-rithms. A shorthand $m \leftarrow \mathcal{M}$ denotes that m is chosen uniformly from a set $\mathcal{M}$. For algorithms and distributions, the same notation $x \leftarrow \mathcal{A}$ means that the element is sampled according to the (output) distribution. A shorthand $A \equiv B$ denotes that either distributions are identical. All algorithms are assumed to be specified as inputs (*programs*) to a universal Turing machine $\mathcal{U}$. A *t-time* algo-rithm is an algorithm that is guaranteed to stop in t time steps. In particular note that the program length of a t-time algorithm must be less than t bits.

To make definitions more concise, we use stateful adversaries. Whenever an adversary $\mathcal{A}$ is called, it has read-write access to a state variable $\sigma \in \{0,1\}^*$ through which important information can be sent from one stage of the attack to another. In the beginning of the execution the state variable σ is empty.

Homomorphic Encryption. As all protocols for conditional disclosure of se-crets are based on homomorphic encryption schemes, we have to formalise corre-sponding security notions. A *public key encryption scheme* is specified by a triple of efficient algorithms (gen, enc, dec). The probabilistic key generation algorithm gen generates a public key pk and a secret key sk. The deterministic algorithms $\mathsf{enc}_{\mathsf{pk}} : \mathcal{M}_{\mathsf{pk}} \times \mathcal{R}_{\mathsf{pk}} \rightarrow \mathcal{C}_{\mathsf{pk}}$ and $\mathsf{dec}_{\mathsf{sk}} : \mathcal{C}_{\mathsf{pk}} \rightarrow \mathcal{M}_{\mathsf{pk}}$ are used for encryption and decryption, where the message space $\mathcal{M}_{\mathsf{pk}}$, the randomness space $\mathcal{R}_{\mathsf{pk}}$ and the ciphertext space $\mathcal{C}_{\mathsf{pk}}$ might depend on pk. As usual, we use $\mathsf{enc}_{\mathsf{pk}}(m)$ to denote the distribution of ciphertexts $\mathsf{enc}_{\mathsf{pk}}(m; r)$ for $r \leftarrow \mathcal{R}_{\mathsf{pk}}$.

In this work, we put two important additional restrictions to the encryption scheme. First, encryption scheme must be with *perfect decryption*:

$$\forall(\mathsf{pk}, \mathsf{sk}) \leftarrow \mathsf{gen} \quad \forall m \in \mathcal{M}_{\mathsf{pk}} : \quad \mathsf{dec}_{\mathsf{sk}}(\mathsf{enc}_{\mathsf{pk}}(m)) = m \ .$$

Second, membership for the set $\mathcal{C}_{\mathsf{pk}}$ must be efficiently testable given pk, i.e., everybody should be able to tell whether a message is a valid ciphertext or not.

We say that an encryption scheme is *additively homomorphic* if there exists an efficient binary operation $\cdot$ such that

$$\forall m_1, m_2 \in \mathcal{M}_{\mathsf{pk}} : \quad \mathsf{enc}_{\mathsf{pk}}(m_1) \cdot \mathsf{enc}_{\mathsf{pk}}(m_2) \equiv \mathsf{enc}_{\mathsf{pk}}(m_1 + m_2) \tag{1}$$

where the equivalence means that the corresponding distributions coincide. A cryptosystem is *multiplicatively homomorphic* if there exists an efficient binary operation $\circledast$ such that

$$\forall m_1, m_2 \in \mathcal{M}_{\mathsf{pk}} : \quad \mathsf{enc}_{\mathsf{pk}}(m_1) \circledast \mathsf{enc}_{\mathsf{pk}}(m_2) \equiv \mathsf{enc}_{\mathsf{pk}}(m_1 \cdot m_2) \ . \tag{2}$$

First, these definitions directly imply that if a fixed ciphertext $\mathsf{enc}_{\mathsf{pk}}(m_1)$ is combined with a freshly generated $\mathsf{enc}_{\mathsf{pk}}(m_2)$ then nothing except $m_1 + m_2$ or $m_1 \cdot m_2$ can be deduced from the resulting ciphertext. Second, the message space must be cyclic or a direct product of cyclic subgroups. In the following, we refer to these as *simple* and *vectorised* cryptosystems to make a clear distinction.

An encryption scheme is (t, ε)-*IND-CPA secure*, if for any t-time adversary $\mathcal{A}$, the corresponding distinguishing advantage is bounded:

$$2 \cdot \left| \Pr \left[\begin{array}{l} (\mathsf{sk}, \mathsf{pk}) \leftarrow \mathsf{gen}, (m_0, m_1) \leftarrow \mathcal{A}(\mathsf{pk}), \\ b \leftarrow \{0, 1\}, c \leftarrow \mathsf{enc}_{\mathsf{pk}}(m_b) : \mathcal{A}(c) = b \end{array} \right] - \frac{1}{2} \right| \le \varepsilon \ .$$

The ElGamal [EG85] and Paillier [Pai99] cryptosystems are the most commonly used cryptosystems that satisfy all these requirements under standard number theoretic assumptions. The ElGamal cryptosystem is multiplicatively homomorphic and the ciphertext space has an efficient membership test if it is built on top of elliptic curve with prime number of elements. The Paillier encryption and its extension Damgård-Jurik [DJ01] cryptosystem are additively homomorphic with ciphertext space $\mathbb{Z}_{N^k}^*$, which is also efficiently testable. Both of these cryptosystems have a cyclic message space.

Commitment Schemes. A commitment scheme is specified by triple of efficient probabilistic algorithms $(\mathsf{gen}, \mathsf{com}, \mathsf{open})$. The algorithm gen fixes public parameters ck. The algorithm $\mathsf{com}_{\mathsf{ck}} : \mathcal{M}_{\mathsf{ck}} \to \mathcal{C}_{\mathsf{ck}} \times \mathcal{D}_{\mathsf{ck}}$ maps messages into commitment and decommitment pairs. A commitment is opened by applying an algorithm $\mathsf{open}_{\mathsf{ck}} : \mathcal{C}_{\mathsf{ck}} \times \mathcal{D}_{\mathsf{ck}} \to \mathcal{M}_{\mathsf{ck}} \cup \{\bot\}$ where the symbol $\bot$ indicates that the commitment-decommitment pair is invalid. We assume that

$$\forall \mathsf{ck} \leftarrow \mathsf{gen} \quad \forall m \in \mathcal{M}_{\mathsf{ck}} : \quad \mathsf{open}_{\mathsf{ck}}(\mathsf{com}_{\mathsf{ck}}(m)) = m \ .$$

A commitment scheme is (t, ε)-*hiding*, if for any t-time adversary $\mathcal{A}$, the corresponding distinguishing advantage is bounded by ε:

$$2 \cdot \left| \Pr \left[\begin{array}{l} \mathsf{ck} \leftarrow \mathsf{gen}, (m_0, m_1) \leftarrow \mathcal{A}(\mathsf{ck}), \\ b \leftarrow \{0, 1\}, (c, d) \leftarrow \mathsf{com}_{\mathsf{ck}}(m_b) : \mathcal{A}(c) = b \end{array} \right] - \frac{1}{2} \right| \le \varepsilon \ .$$

A commitment scheme is (t, ε)-*binding*, if for any t-time adversary $\mathcal{A}$, the probability of successful double-openings is bounded by ε:

$$\Pr \left[\begin{array}{l} \mathsf{ck} \leftarrow \mathsf{gen}, (c, d_0, d_1) \leftarrow \mathcal{A}(\mathsf{ck}) : \mathsf{open}_{\mathsf{ck}}(c, d_0) \neq \bot \\ \wedge \ \mathsf{open}_{\mathsf{ck}}(c, d_1) \neq \bot \wedge \mathsf{open}_{\mathsf{ck}}(c, d_0) \neq \mathsf{open}_{\mathsf{ck}}(c, d_1) \end{array} \right] \le \varepsilon \ .$$

A commitment is ε-*binding* if the bound holds for all adversaries.

A commitment scheme is *perfectly equivocal* if there exists a modified setup procedure gen^* that in addition to ck produces an equivocation key ek, which is later used by additional algorithms $\mathsf{com}_{\mathsf{ek}}^*$ and $\mathsf{equiv}_{\mathsf{ek}}$. The algorithm $\mathsf{com}_{\mathsf{ek}}^*$ returns a fake commitment c_* together with a trapdoor information σ such that the invocation of $\mathsf{equiv}_{\mathsf{ek}}(\sigma, m)$ produces a valid decommitment d_* for m, i.e., $\mathsf{open}_{\mathsf{ck}}(c_*, d_*) = m$. More over, for any message $m \in \mathcal{M}_{\mathsf{ck}}$ the distribution (c_*, d_*) must coincide with the distribution generated by $\mathsf{com}_{\mathsf{ck}}(m)$.

In some proofs, we need commitments that are simultaneously equivocal and ε-binding. To achieve such a chameleon-like behaviour, we can use two commitment schemes, which differ only in the key generation. That is, they use the same algorithms com and open for committing and decommitting. Now if commitment parameters ck are (t, ε_1)-indistinguishable and the first commitment scheme is perfectly equivocal and the second ε_2-binding, we can switch the key generation algorithms during security analysis and use both properties. This construction is known as a (t, ε_1)-*equivocal and* ε_2-*binding dual-mode commitment*.

Such commitments can be constructed from additively homomorphic encryption, see [GOS06]. Let $e \leftarrow \mathsf{enc}_{\mathsf{pk}}(1)$ together with pk be the commitment key. Then we can commit $m \in \mathcal{M}_{\mathsf{pk}}$ by computing $c \leftarrow e^m \cdot \mathsf{enc}_{\mathsf{pk}}(0; r)$ for $r \leftarrow \mathcal{R}$. To open c, we have to release m and r. This construction is perfectly binding. To assure equivocality, we can set $e \leftarrow \mathsf{enc}_{\mathsf{pk}}(0; r_*)$ for some $r_* \leftarrow \mathcal{R}$. Then any commitment c is an encryption of zero and can be expressed as $e^m \cdot \mathsf{enc}_{\mathsf{pk}}(0; r)$ provided that r_* is known. It is easy to see that (t, ε)-IND-CPA security is sufficient to guarantee hiding and computational indistinguishability of commitment keys. As you never need to decrypt during the equivocation, the construction can be based on lifted ElGamal or Paillier encryption scheme.

Trusted Setup Model. In this model, a trusted dealer $\mathcal{T}$ computes and privately distributes all public and secret parameters according to some procedure π_{ts}. For instance, $\mathcal{T}$ might set up a public key infrastructure or generate a common reference string. In practical applications, the trusted setup π_{ts} is commonly implemented as a secure two- or multi-party protocol run in *isolation*. Hence, protocols with trusted setup are practical only if many protocols can share the same setup without rapid decrease in security.

3   Conditional Disclosure of Secrets

A *conditional disclosure of secrets* (CDS) is a two-message protocol between a client $\mathcal{P}$ and a server $\mathcal{V}$ where the client learns a secret s specified by the server only if its encrypted inputs $\boldsymbol{x} = (x_1, \ldots, x_n)$ satisfy a public predicate $\phi(\boldsymbol{x})$. The server should learn nothing beyond the vector of encryptions $(q_1, \ldots, q_n)$. We also assume that the client knows the secret key sk, whereas the server knows only the public key pk. These protocols are often used as sub-protocols in more complex crypto-computing protocols, see for example [AIR01,BK04].

The complexity of CDS protocol depends on the predicate. For instance, it is straightforward to construct CDS protocols for all monotone predicates if the input $\boldsymbol{x}$ is a bit vector [AIR01,LL07]. These constructions can be used as a basis for more complex predicates. In particular, note that for any predicate $\phi(\boldsymbol{x})$ there exists a constant depth monotonous predicate $\psi(\boldsymbol{x}, \boldsymbol{w})$ such that

$$\phi(\boldsymbol{x}) = 1 \quad \Leftrightarrow \quad \exists \boldsymbol{w} : \psi(\boldsymbol{x}, \boldsymbol{w}) = 1$$

and $\boldsymbol{w}$ can be efficiently computed from $\boldsymbol{x}$ and ϕ. For the conversion, fix a circuit that computes $\phi(\boldsymbol{x})$. Let $w_1, \ldots, w_k$ denote the output values of all gates in the circuit when the input is $\boldsymbol{x}$. Then you can define a monotonous formula $\psi(\boldsymbol{x}, \boldsymbol{w})$, which states that all gates are correctly evaluated and the output of the circuit is one. Clearly, the circuit complexity of ψ is linear in the circuit complexity of ϕ. As a result, efficient CDS protocols exist for all predicates provided that the client is willing to encrypt $\boldsymbol{w}$ besides $\boldsymbol{x}$ and the server is willing to combine encryptions. See [AIR01,LL07] for more detailed discussions.

Formal Security Definition. A CDS protocol is specified by a triple of algorithms (query, answer, recov). The client first sends a query $\boldsymbol{q} \leftarrow \mathsf{query}_{\mathsf{pk}}(\boldsymbol{x}, \boldsymbol{w})$ for which the server computes a reply $\boldsymbol{a} \leftarrow \mathsf{answer}_{\mathsf{pk}}(\boldsymbol{q}, s; r)$ for $r \leftarrow \mathcal{R}$. The client can recover the secret by computing $\mathsf{recov}_{\mathsf{sk}}(\boldsymbol{a})$. For clarity, let $\mathcal{Q}_{\mathrm{inv}}$ denote the *set of all invalid queries*, i.e., queries for which $\psi(\boldsymbol{x}, \boldsymbol{w}) = 0$ or which contain invalid ciphertext or are otherwise malformed.

All standard implementations of CDS protocols are secure in the relaxed model, where the client can be malicious and the server is assumed to be honest but curious. Hence, a security definition should be expressed in terms of simulator constructions. However, as the protocol structure is so simple, we can be more explicit. Namely, a CDS protocol is (ε, t_a)-*client-private*, if for any t_a-time stateful adversary $\mathcal{A}$, the next inequality holds:

$$2 \cdot \left| \Pr \left[\begin{array}{l} (\mathsf{sk}, \mathsf{pk}) \leftarrow \mathsf{gen}, (\boldsymbol{x}_0, \boldsymbol{w}_0, \boldsymbol{x}_1, \boldsymbol{w}_1) \leftarrow \mathcal{A}(\mathsf{pk}), \\ i \leftarrow \{0, 1\}, \boldsymbol{q} \leftarrow \mathsf{query}_{\mathsf{pk}}(\boldsymbol{x}_i, \boldsymbol{w}_i) : \mathcal{A}(\boldsymbol{q}) = i \end{array} \right] - \frac{1}{2} \right| \le \varepsilon \ .$$

To simulate a malicious client $\mathcal{P}_*$, the simulator can forward its query $\boldsymbol{q}$ to the trusted third party $\mathcal{T}$ who will test it. If $\boldsymbol{q} \notin \mathcal{Q}_{\mathrm{inv}}$, $\mathcal{T}$ releases the secret s and the simulator can compute $\mathsf{answer}_{\mathsf{pk}}(\boldsymbol{q}, s)$. Otherwise, $\mathcal{T}$ will release nothing and we need an efficient algorithm $\mathsf{answer}^*_{\mathsf{pk}}(\boldsymbol{q})$ for faking replies without knowing the secret. A CDS protocol is ε-*server private* if, for all valid public keys pk and secrets $s \in \mathcal{S}$ and for all invalid queries $\boldsymbol{q} \in \mathcal{Q}_{\mathrm{inv}}$, the statistical distance between the distributions $\mathsf{answer}_{\mathsf{pk}}(\boldsymbol{q}, s)$ and $\mathsf{answer}^*_{\mathsf{pk}}(\boldsymbol{q})$ is at most ε. As the statistical distance of replies is at most ε, the joint output distribution is also at most ε apart and thus the protocol is secure against malicious clients [LL07].

Example Protocols. The simplest of CDS protocols is a *disclose-if-one* protocol, where the client $\mathcal{P}$ learns the secret only if the server $\mathcal{V}$ receives encryption of one. A standard way to build such a protocol relies on the fact that

$$\mathsf{enc}_{\mathsf{pk}}(x)^e \circledast \mathsf{enc}_{\mathsf{pk}}(s) \equiv \mathsf{enc}_{\mathsf{pk}}(x^e \cdot s)$$

when the underlying encryption scheme is multiplicatively homomorphic. If the plaintext order is a publicly known prime p then e can be chosen uniformly from $\mathbb{Z}_p$. As a result, $x^e \cdot s$ is uniformly distributed over $\mathcal{M}_{\mathsf{pk}}$ if $x \ne 1$ and $x^e \cdot s = s$ otherwise. The latter forms the core of many CDS constructions [AIR01,BGN05]. For additively homomorphic encryption schemes, we can utilise the equality

$$\mathsf{enc}_{\mathsf{pk}}(x)^e \cdot \mathsf{enc}_{\mathsf{pk}}(s) \equiv \mathsf{enc}_{\mathsf{pk}}(xe + s)$$

to construct a *disclose-if-zero* protocol. As commonly used additively homomorphic encryption schemes have a composite plaintext space $\mathbb{Z}_N$, extra care is required to address cases when x is non-trivial factor of N. In such cases, the noise term xe is uniformly distributed over non-zero additive subgroup $\mathcal{G} \subseteq \mathbb{Z}_N$ and additional randomness is needed to hide the secret. Laur and Lipmaa [LL07] proposed a solution where the secret is first encoded with a randomised substitution cipher encode. Thus, the client learns

$$\mathsf{enc}_{\mathsf{pk}}(x)^e \cdot \mathsf{enc}_{\mathsf{pk}}(\mathsf{encode}(s)) \equiv \mathsf{enc}_{\mathsf{pk}}(xe + \mathsf{encode}(s)) \ .$$

If there exists an efficient function decode such that $\mathsf{decode}(\mathsf{encode}(s)) = s$ for all $s \in \mathcal{S}$, the honest client can still recover the secret. For the security, $\mathsf{encode}(s)$ must contain enough randomness so that an additive noise from a small subgroup can hide the secret when the query is invalid. We say that the encoding is ε-*secure* if for any $s \in \mathcal{S}$ and for any non-zero additive subgroup $\mathcal{G} \subseteq \mathcal{M}_{\mathsf{pk}}$, $\mathsf{encode}(s) + g$ for $g \leftarrow \mathcal{G}$ is statistically ε-close to uniform distribution over $\mathcal{M}_{\mathsf{pk}}$.

Secure Encodings. If the message space $\mathcal{M}_{\mathsf{pk}}$ has a prime order, the only non-zero subgroup is $\mathcal{M}_{\mathsf{pk}}$ and thus the identity function can be used as perfectly secure encoding. For composite message spaces $\mathbb{Z}_N$, we can choose t randomly from $\mathbb{Z}_{\lfloor N/2^\ell \rfloor}$ and set $\mathsf{encode}(s) = s + 2^\ell \cdot t$ to encode ℓ-bit secrets. Laur and Lipmaa showed that this encoding is $2^{\ell-1}/\gamma$-secure where γ is the smallest factor of N and there are no alternative encoding functions with significantly longer secrets [LL07]. Both encodings can be lifted to the vectorised setting provided that all plaintext components have the same public order n as in [SV11,GHS12,DPSZ12]. For that one must use additive secret sharing to split the secret among all components of the plaintext vector. See the full version of our paper for further results about secure encodings.

4   A New CDS Protocol for a Multiplicative Relation

For clarity, we specify the solution for additively homomorphic encryption and then discuss how the same protocol can be modified to work with other types of cryptosystems. Let $\mathsf{enc}_{\mathsf{pk}}(x_1)$, $\mathsf{enc}_{\mathsf{pk}}(x_2)$ and $\mathsf{enc}_{\mathsf{pk}}(x_3)$ be the ciphertexts sent by the client $\mathcal{P}$ and let $s \in \mathcal{S}$ be the secret picked by the server $\mathcal{V}$. Then the client should learn s only if the multiplicative relation $x_1 x_2 = x_3$ holds between plaintexts. Figure 1 depicts the corresponding CDSMUL protocol.

Theorem 1. *If the encryption scheme is (t, ε_1)-IND-CPA secure and* encode *is ε_2-secure, the* CDSMUL *protocol is $(t, 3\varepsilon_1)$-client and ε_2-server private.*

Proof. As the output computed by the client satisfies the following equation

$$d_2 - x_2 d_1 = x_3 e_1 + x_2 e_2 + \mathsf{encode}(s) - x_2(x_1 e_1 + e_2)$$
$$= (x_3 - x_1 x_2)e_1 + \mathsf{encode}(s) \ ,$$

the client is guaranteed to recover the secret s when $x_1 x_2 = x_3$. If $x_1 x_2 \neq x_3$ the term $(x_3 - x_1 x_2)e_1$ adds additive noise to the payload. Clearly, the term

---

GLOBAL PARAMETERS: Both parties know functions encode and decode for secrets. The client $\mathcal{P}$ has a secret key sk and the server $\mathcal{V}$ has the corresponding public key pk. Let n be a publicly known common multiple of all cyclic subgroup sizes in $\mathcal{M}_{\mathsf{pk}}$.

CLIENT'S INPUT: The client $\mathcal{P}$ has inputs x_1, x_2, x_3 such that $x_1 x_2 = x_3$ over $\mathcal{M}_{\mathsf{pk}}$
SERVER'S SECRET: The server $\mathcal{V}$ wants to release a secret $s \in \mathcal{S}$.

QUERY: The client $\mathcal{P}$ sends $\boldsymbol{q} = (q_1, q_2, q_3)$ to the server $\mathcal{V}$ where $q_i = \mathsf{enc}_{\mathsf{pk}}(x_i)$.
ANSWER: The server $\mathcal{V}$ picks $e_1, e_2 \leftarrow \mathbb{Z}_n$ and and sends back

$$u_1 \leftarrow q_1^{e_1} \cdot \mathsf{enc}_{\mathsf{pk}}(e_2), \quad u_2 \leftarrow q_3^{e_1} \cdot q_2^{e_2} \cdot \mathsf{enc}_{\mathsf{pk}}(\mathsf{encode}(s)) \ .$$

RECOVERY: $\mathcal{P}$ computes $d_i \leftarrow \mathsf{dec}_{\mathsf{sk}}(u_i)$ for $i \in \{1, 2\}$ and uses the known plaintext value $x_2 = \mathsf{dec}_{\mathsf{sk}}(q_2)$ to compute the output $s \leftarrow \mathsf{decode}(d_2 - x_2 d_1)$.

Fig. 1. CDS protocol CDSMUL for multiplicative relation

$(x_3 - x_1 x_2)e_1$ belongs to a cyclic additive subgroup $\mathcal{G} = \{(x_3 - x_1 x_2)m : m \in \mathbb{Z}\}$. Since the size of every cyclic subgroup of $\mathcal{M}_{\mathsf{pk}}$ divides n, the term $(x_3 - x_1 x_2)e_1$ must be uniformly distributed over $\mathcal{G}$. Consequently, we have established that $d_2 - x_2 d_1 \equiv \mathsf{encode}(s) + \mathcal{G}$ for $\mathcal{G} \neq \{0\}$. By the assumptions, $\mathcal{G} + \mathsf{encode}(s)$ is ε_2-close to the uniform distribution over $\mathcal{M}_{\mathsf{pk}}$. More importantly, the distribution of $d_2 - x_2 d_1$ is also independent from e_2. As e_2 perfectly masks the term $x_1 e_1$ in d_1, we can simulate the replies u_1 and u_2 by encrypting two random messages. The claim on client-privacy is straightforward.   □

Remarks. First, note that the protocol has perfect server-privacy when the message space is cyclic and has a prime order, since the identity function as encode has perfect privacy. If all non-trivial cyclic subgroups have the same prime order, we can use additive secret sharing to make sure that the multiplicative relation holds for every sub-component of a vectorised cryptosystem.

Second, note that the protocol does not work directly with lifted cryptosystems. In such schemes, the new encryption rule $\overline{\mathsf{enc}}_{\mathsf{pk}}(x) = \mathsf{enc}_{\mathsf{pk}}(g^x)$ is defined in terms of an old multiplicatively homomorphic encryption rule $\mathsf{enc}_{\mathsf{pk}}(\cdot)$ and a generator of the plaintext space g. The resulting scheme is additively homomorphic, but discrete logarithm must be taken to complete the decryption. Hence, we cannot blindly follow the reconstruction phase in CDSMUI protocol.

Still, the client can employ the old decryption algorithm to compute g^{d_1} and g^{d_2} from the lifted ciphertexts u_1 and u_2. Since the client knows x_2, he or she can compute a partial decryption $g^{d_1}(g^{d_2})^{-x_2} = g^{(x_3 - x_1 x_2)e_1} g^{\mathsf{encode}(s)}$. Thus, the honest client learns $g^{\mathsf{encode}(s)}$, which might not be enough to extract s.

For lifted ElGamal with a prime order message space, we can use identity function to encode s. Hence, the secret s can be restored by brute-forcing g^s if the set of secrets is small. Alternatively, both parties can use g^s instead of s to share a randomly distributed value over $\mathcal{M}_{\mathsf{pk}}$. These solutions are not viable for lifted cryptosystems with composite message spaces, such as [BGN05].

TRUSTED SETUP FOR ENCRYPTION. Trusted dealer runs $(\mathsf{pk}, \mathsf{sk}) \leftarrow \mathsf{pkc.gen}$ and sends pk as a public key of $\mathcal{P}$ to everyone. The secret key sk is sent securely to $\mathcal{P}$.

TRUSTED SETUP FOR CDSZK PROTOCOLS. Trusted dealer generates commitment parameters $\mathsf{ck} \leftarrow \mathsf{cs.gen}$ and broadcast ck to everyone.

MESSAGE FORMATION. The CDS protocol is chosen according to the predicate ψ. The prover $\mathcal{P}$ sends $\boldsymbol{q} \leftarrow \mathsf{query}_{\mathsf{pk}}(\boldsymbol{x}, \boldsymbol{w})$ to the verifier $\mathcal{V}$.

PROOF PHASE. A statement to be proved is $\boldsymbol{q} \notin \mathcal{Q}_{\mathrm{inv}}$.

1. $\mathcal{V}$ chooses $s \leftarrow \mathcal{S}$ and $r \leftarrow \mathcal{R}$ and sends $\boldsymbol{a} \leftarrow \mathsf{answer}_{\mathsf{pk}}(\boldsymbol{q}, s; r)$ to $\mathcal{P}$.
2. $\mathcal{P}$ recovers $s \leftarrow \mathsf{recov}_{\mathsf{sk}}(\boldsymbol{a})$, computes $(c, d) \leftarrow \mathsf{cs.com}_{\mathsf{ck}}(s)$ and sends c to $\mathcal{V}$.
3. $\mathcal{V}$ reveals (s, r) to $\mathcal{P}$ who aborts if $\boldsymbol{a} \neq \mathsf{answer}_{\mathsf{pk}}(\boldsymbol{q}, s; r)$.
4. $\mathcal{P}$ reveals decommitment d to $\mathcal{V}$ who accepts the proof only if $s = \mathsf{cs.open}_{\mathsf{ck}}(c, d)$.

Fig. 2. Zero-knowledge proof of correctness CDSZK for encryptions

5 From CDS Protocols to Zero Knowledge

The new zero-knowledge protocol is based on the following observation. The client $\mathcal{P}$ is able to reconstruct the secret s in a CDS protocol only if ciphertexts satisfy a certain relation, i.e., $\boldsymbol{q} \notin \mathcal{Q}_{\mathrm{inv}}$. Hence, if $\mathcal{P}$ sends the secret s back to the honest server $\mathcal{V}$, the server is able to verify whether $\boldsymbol{q} \notin \mathcal{Q}_{\mathrm{inv}}$ or not. This leaks no information to the semihonest server, since $\mathcal{V}$ already knows s. Thus, we can view this simple modification as an honest verifier zero-knowledge proof where the client $\mathcal{P}$ is a prover and the server $\mathcal{V}$ acts as a verifier.

However, if $\mathcal{V}$ acts maliciously, the secret s recovered by the honest $\mathcal{P}$ might leak additional information. To counter this attack, $\mathcal{V}$ should prove knowledge of s and randomness $r \in \mathcal{R}$ needed to compute the reply $\mathsf{answer}_{\mathsf{pk}}(\boldsymbol{q}, s; r)$ before $\mathcal{P}$ sends the secret. Since neither s nor r are among private inputs, $\mathcal{V}$ can release them so that $\mathcal{P}$ can validate the behaviour of $\mathcal{V}$ by recomputing $\mathsf{answer}_{\mathsf{pk}}(\boldsymbol{q}, s; r)$. As the secret becomes public, $\mathcal{P}$ must commit to its reply before the pair (s, r) is released or otherwise we lose soundness. The corresponding idea gives a rise to zero-knowledge protocol, which is depicted in Fig. 2.

5.1 Proper Model for Defining Security

By definition a CDSZK protocol verifies a relation between ciphertexts. As such it is fairly useless without another protocol that uses these ciphertexts in its computations to get an output. As these protocols must share a joint state (at minimum the same public key), standard composability results are not applicable [CR03]. A systematic way to prove security of such compound protocols is to split the proof into phases. In the first phase, we show that instances of CDSZK protocols are concurrently composable with other protocols if the shared state is generated by the trusted setup procedure. In the second phase, we replace the trusted setup with a secure multiparty protocol *run in isolation* and use the standard sequential composability result to prove that the setup protocol followed by the compound protocol remains secure. In this section, we consider only the first phase. The second phase is thoroughly discussed in Sect. 6.

First of all, note that a CDSZK protocol might be used to prove statements that depend on common parameters z shared by all participants after the trusted setup. To handle such cases, the language of all valid statements $\mathcal{L}_z = \{x \mid \exists w : \psi_*(z, x, w) = 1\}$ must be specified by an efficiently decidable ternary predicate ψ_*. As usual x is the *protocol input* (statement) and w is the *witness* and z additional auxiliary input. For a CDSZK protocol, $\boldsymbol{q}$ is the protocol input, sk is the witness, and pk is the common parameter z shared by many CDSZK protocols and crypto-computing protocols using CDSZK. The statement $\boldsymbol{q}$ belongs to $\mathcal{L}_{\mathsf{pk}}$ iff $\boldsymbol{q} \notin \mathcal{Q}_{\mathrm{inv}}(\mathsf{pk})$. Indeed, the knowledge of sk allows us to decrypt $\boldsymbol{q}$ to obtain $\boldsymbol{x}$ and then output $\phi(\boldsymbol{x})$ as $\psi_*(\mathsf{pk}, \boldsymbol{q}, \mathsf{sk})$.

A *proof system* is determined by two algorithms $\mathcal{P}$ and $\mathcal{V}$ which specify the actions of the prover and the verifier, respectively. The prover $\mathcal{P}$ takes (x, w) as inputs and the verifier $\mathcal{V}$ takes x as an input. Both algorithms can additionally access parameters distributed to them by the setup π_{ts}. Let $\langle\!\langle \mathcal{P}, \mathcal{V} \rangle\!\rangle (x)$ denote the verifiers output after the interaction with $\mathcal{P}(x, w)$. Then a proof system is *perfectly complete* if all valid statements are provable, i.e., for all statements $x \in \mathcal{L}_z$ and corresponding witnesses w, $\langle\!\langle \mathcal{P}, \mathcal{V} \rangle\!\rangle (x) \equiv 1$ for all runs of π_{ts}°.

We formalise the security of a zero-knowledge protocol by specifying the ideal functionality. An ideal implementation of a zero knowledge proof is a restricted communication channel $\pi_{\psi_*}^\circ$ from the prover $\mathcal{P}$ to the verifier $\mathcal{V}$. The prover $\mathcal{P}$ can provide an input x to the channel, after which the verifier $\mathcal{V}$ obtains

$$\pi_{\psi_*}^\circ (x) = \begin{cases} x, & \text{if } \exists w : \psi_*(z, x, w) = 1 \ , \\ \bot, & \text{if } \forall w : \psi_*(z, x, w) = 0 \ , \end{cases} \tag{3}$$

where the set of shared parameters z is determined by the trusted setup π_{ts}°.

Let $\mathcal{E}\langle \pi_{\mathsf{ts}}^\circ, \pi_{\psi_1}^\circ, \ldots, \pi_{\psi_\ell}^\circ \rangle$ denote a compound protocol (*computational context*), which internally uses a single instance of π_{ts}° and ideal zero-knowledge protocols $\pi_{\psi_1}^\circ, \ldots, \pi_{\psi_\ell}^\circ$. The compound protocol can model any kind of computational activity that uses the functionality of $\pi_{\psi_1}^\circ, \ldots, \pi_{\psi_\ell}^\circ$. For instance, it can use zero-knowledge to guarantee correctness of an e-voting protocol.

We say that $\mathcal{E}$ is a t_e-*time computational context* if the maximal amount of computation steps done by $\mathcal{E}$ disregarding invocations of π_{ts}° and $\pi_{\psi_i}^\circ$ is bounded by t_e. Now let $\mathcal{E}\langle \pi_{\mathsf{ts}}^*, \pi_{\psi_1}, \ldots, \pi_{\psi_\ell} \rangle$ be a *hybrid implementation*[1] with augmented trusted setup π_{ts}^* and real instantiations of zero knowledge protocols π_{ψ_i}. The augmented setup procedure first runs π_{ts}° and then creates some additional parameters (commitment parameters in our case) shared by $\pi_{\psi_1}, \ldots, \pi_{\psi_\ell}$. Now we would like that both compound protocols have comparable security against plausible attacks. The corresponding formalisation is rather technical. Hence, we only highlight the major aspects in the formal definition and refer to the

---

[1] The name hybrid implementation is used here to distinguish the implementation from the final (real world) setting where π_{ts}^* is replaced with a setup protocol π_{ts}.

manuscripts [Can01,Lin03] for further details. Still, it is important to emphasise that we consider only security against static corruption but semi-adaptively chosen contexts[2] and we assume that the communication is asynchronous.

In a nutshell, let $\boldsymbol{\xi}$ denote inputs of all parties and let $\boldsymbol{\zeta}^\circ$ and $\boldsymbol{\zeta}$ denote joint outputs of all parties in the ideal and the hybrid execution. Let f be a low-degree polynomial. Then a protocol π is $(t_e, t_a, f, t_d, \varepsilon)$-universally composable in the shared setup model if for any t_e-time $\mathcal{E}$ and for any t_a-time adversary $\mathcal{A}$ against π there exists $f(t_a)$-time adversary $\mathcal{A}^\circ$ against π° such that for any input $\boldsymbol{\xi}$ distributions $\boldsymbol{\zeta}^\circ$ and $\boldsymbol{\zeta}$ are (t_d, ε)-indistinguishable. We omit the time bound t_d if $\boldsymbol{\zeta}^\circ$ and $\boldsymbol{\zeta}$ are statistically indistinguishable.

The parameter t_e limits the complexity of protocols where we can safely use zero-knowledge protocols. The parameter t_a shows how much computational power the adversary can have before we lose security guarantees. The polynomial f shows how much extra power the adversary gains by participating in the hybrid protocol. The slower it grows the better. In the standard asymptotic setting, we view all parameters as functions of a security parameter k and we ask whether for any polynomials $t_e(k)$, $t_a(k)$, $t_d(k)$, the distinguishing advantage $\varepsilon(k)$ decreases faster than a reciprocal of any polynomial.

5.2   Soundness Guarantees for CDSZK Protocols

A proof system π is ε-sound if for any prover $\mathcal{P}_*$ and adaptively chosen statement x the deception probability is bounded by ε:

$$\mathsf{Adv}_\pi^{\mathrm{snd}}(\mathcal{P}_*) = \Pr\left[x \leftarrow \mathcal{P}_* : (\!(\mathcal{P}_*, \mathcal{V})\!)(x) = 1 \wedge x \notin \mathcal{L}_z\right] \leq \varepsilon$$

where the probability is taken over all possible runs of the setup procedure π_{ts} and $\mathcal{P}_*$ can choose the statement x based on all public and private parameters received during π_{ts}. An argument system π is (t, ε)-sound if for any t-time $\mathcal{P}_*$ the deception probability is bounded: $\mathsf{Adv}_\pi^{\mathrm{snd}}(\mathcal{P}_*) \leq \varepsilon$.

Theorem 2. *If the commitment scheme is ε_2-binding and CDS protocol is ε_3-server private, CDSZK protocol is $\left(\frac{1}{|\mathcal{S}|} + \varepsilon_2 + \varepsilon_3\right)$-sound for any $\mathcal{P}_*$. If the commitment scheme is perfectly binding, a stronger claim holds:*

$$\forall \mathsf{sk}, \mathsf{ck}, \ (\psi, \boldsymbol{q}) \leftarrow \mathcal{P}_*(\mathsf{sk}, \mathsf{ck}) : \Pr\left[(\!(\mathcal{P}_*, \mathcal{V})\!)(\boldsymbol{q}) = 1 \wedge \boldsymbol{q} \in \mathcal{Q}_{\mathrm{inv}}\right] \leq 1/|\mathcal{S}| + \varepsilon_3 \ .$$

Proof. Consider a modified protocol where $\boldsymbol{a}$ is replaced with $\mathsf{answer}^*_{\mathsf{pk}}(\boldsymbol{q})$ whenever $\boldsymbol{q} \in \mathcal{Q}_{\mathrm{inv}}$. As the CDS protocol is ε_3-server private, the replacement can reduce the deception probability at most by ε_3. Let c denote the the commitment issued in the modified protocol and let $d_* \in \mathcal{D}$ be the first valid decommitment for c according to some fixed ordering over the decommitment space $\mathcal{D}$. Let $\hat{s} = \mathsf{open}_{\mathsf{ck}}(c, d_*)$ be the corresponding opening. Then

$$\Pr\left[(\!(\mathcal{P}_*, \mathcal{V})\!)(\boldsymbol{q}) = 1 \wedge \boldsymbol{q} \in \mathcal{Q}_{\mathrm{inv}}\right] \leq \Pr\left[\hat{s} = s \wedge \boldsymbol{q} \in \mathcal{Q}_{\mathrm{inv}}\right] + \varepsilon_2 + \varepsilon_3 \ .$$

---

[2] In this setting, the adversary must decide before the execution which of two parties it corrupts. However, the adversary can choose the computational context $\mathcal{E}$ based on the outputs received in during the setup π_{ts}^*.

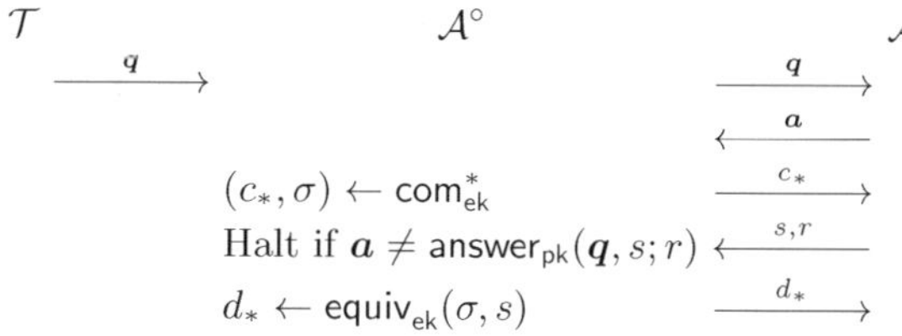

$$\begin{array}{ccc}
\mathcal{T} & \mathcal{A}^{\circ} & \mathcal{A} \\
\xrightarrow{\quad q \quad} & & \\
 & & \xrightarrow{\quad q \quad} \\
 & & \xleftarrow{\quad a \quad} \\
(c_*, \sigma) \leftarrow \mathsf{com}^*_{\mathsf{ek}} & & \xrightarrow{\quad c_* \quad} \\
\text{Halt if } a \neq \mathsf{answer}_{\mathsf{pk}}(q, s; r) & & \xleftarrow{\quad s, r \quad} \\
d_* \leftarrow \mathsf{equiv}_{\mathsf{ek}}(\sigma, s) & & \xrightarrow{\quad d_* \quad}
\end{array}$$

Fig. 3. Simulation construction for corrupted verifier

Indeed, note that $\mathcal{P}_*$ can succeed only if $\mathsf{open}_{\mathsf{ck}}(c, d) = s$ and thus all successful runs with $s \neq \hat{s}$ correspond to valid double openings. As the commitment scheme is statistically binding $\Pr\left[s = \mathsf{open}_{\mathsf{ck}}(c, d) \neq \hat{s}\right] \leq \varepsilon_2$. To complete the proof, note that the fake reply $\mathsf{answer}^*_{\mathsf{pk}}(q)$ is independent of s (which is uniformly chosen in CDSZK) and thus $\Pr\left[\hat{s} = s \wedge q \in \mathcal{Q}_{\mathrm{inv}}\right] \leq \Pr\left[\hat{s} = s | q \in \mathcal{Q}_{\mathrm{inv}}\right] \leq 1/|\mathcal{S}|$.

For the strengthened claim, note that $\mathsf{answer}^*_{\mathsf{pk}}$ and $\mathsf{answer}_{\mathsf{pk}}(q, s)$ are ε_3-close for any valid pk. Now if the commitment is perfectly binding then $s = \hat{s}$ for any ck and we get the desired bound.   □

5.3   Universal Composability of CDSZK Protocols

Usually, one gives a simulation construction for a malicious verifier to prove the zero-knowledge property. Here, we aim for more. Namely, we show that any adversary who attacks the CDSZK protocol π_{ψ_*} can be converted to an efficient adversary against the ideal implementation $\pi^{\circ}_{\psi_*}$ even if some side-computations are done in parallel with π_{ψ_*}. For that we modify the generation of parameters for the dual-mode commitments to get an equivocality key into the simulator.

Lemma 1. *If (t, ε_1)-equivocal and ε_2-binding dual-mode commitment is used in the hiding mode, then for any t_a-time malicious verifier $\mathcal{A}$ there exists $\mathcal{O}(t_a)$-simulator $\mathcal{A}^{\circ}$ that achieves perfect simulation for a single CDSZK protocol.*

Proof. To convert $\mathcal{A}$ to an equivalent ideal world adversary $\mathcal{A}^{\circ}$, we need to simulate π^*_{ts} given access to $\pi^{\circ}_{\mathsf{ts}}$ and π_{ψ_*} given access to $\pi^{\circ}_{\psi_*}$. The simulation of π^*_{ts} is trivial, since $\pi^{\circ}_{\mathsf{ts}}$ delivers all messages as π^*_{ts} except for the commitment key ck. For the latter, the simulator $\mathcal{A}^{\circ}$ computes $(\mathsf{ck}, \mathsf{ek}) \leftarrow \mathsf{gen}^*$ and delivers ck to $\mathcal{A}$ and stores the equivocation key ek for later use.

Figure 3 depicts the simulation construction for π_{ψ_*}. Let (c, d) denote the commitment and decommitment values used in π_{ψ_*} and (c_*, d_*) the corresponding values computed in the simulation. Since the commitment is perfectly equivocal, distributions of c and c_* coincide. Next, note that if $a = \mathsf{answer}_{\mathsf{pk}}(q, s; r)$, then honest prover would have committed s. Consequently, if corrupted verifier $\mathcal{A}$ releases values s, r such that $a = \mathsf{answer}_{\mathsf{pk}}(q, s; r)$, the distribution of (c, d) and (c_*, d_*) coincides, again. Hence, simulation of π_{ψ_*} is perfect.   □

Note that the transformation outlined in Lemma 1 can be applied only once to the malicious verifier, as the resulting adversary $\mathcal{A}^{\circ}$ expects the trusted dealer to engage π_{ts}°. Therefore, we need a separate proof to show that concurrent composition of CDSZK protocols, which share the commitment parameters ck can be replaced with ideal implementations.

Lemma 2. *If (t, ε_1)-equivocal and ε_2-binding dual-mode commitment is used in the hiding mode, then for any t_a-time malicious verifier $\mathcal{A}$ there exists $\mathcal{O}(t_a)$-simulator $\mathcal{A}^\circ$ that achieves perfect simulation even if many CDSZK protocols are concurrently executed.*

Proof. We use the same simulator for the setup phase as in Lemma 1 and share the equivocation key ek for all sub-simulators that simulate π_{ψ_i}. For each protocol, we use the same construction as depicted in Fig. 3. As the commitment is still perfectly equivocal, all claims about perfect simulatability of messages of the sub-protocols π_{ψ_i} hold and the claim follows.   □

Lemma 2 shows how to handle corrupted verifiers but we also need to handle the case where the prover is malicious. The following result follows directly from soundness guarantees of Theorem 2.

Lemma 3. *If (t, ε_1)-equivocal and ε_2-binding dual-mode commitment is used in the binding mode and the underlying CDS protocols is ε_3-server private, then an unbounded prover can succeed with probability $\frac{1}{|S|} + \varepsilon_2 + \varepsilon_3$ when $\boldsymbol{q} \in \mathcal{Q}_{\mathrm{inv}}$.*

Note that a corrupted party $\mathcal{P}_1$ can act simultaneously as a prover and verifier. Let pk_1 and pk_2 be the public key of $\mathcal{P}_1$ and $\mathcal{P}_2$. Then $\mathcal{P}_1$ can prove that encryptions under the public key pk_1 are well formed and let $\mathcal{P}_2$ to convince that encryptions under the key pk_2 satisfy certain relations. Hence, the final claim about the security of CDSZK must provide a simultaneous simulation. For brevity, let us prove the claim for two-party setting as the proof can be easily generalised for multi-party setting. Recall that $t_a + t_e$ is the maximal running time of the adversary and the context where the protocols are executed.

Theorem 3. *If (t, ε_1)-equivocal and ε_2-binding dual-mode commitment is used in the hiding mode and all ℓ instances of CDS protocols are ε_3-server private, then the CDSZK protocol is $(t_e, t_a, \mathcal{O}(t_a), t_d, \ell(\frac{1}{|S|} + \varepsilon_2 + \varepsilon_3) + \varepsilon_1)$-universally composable in the shared setup model provided that $t \geq t_a + t_e + \mathcal{O}(\ell)$.*

Proof. Let $\mathcal{E}\langle \pi_{\mathsf{ts}}^\circ, \pi_{\psi_1}^\circ, \ldots, \pi_{\psi_\ell}^\circ \rangle$ denote a compound protocol with ideal implementations and let $\mathcal{E}\langle \pi_{\mathsf{ts}}, \pi_{\psi_1}, \ldots, \pi_{\psi_\ell} \rangle$ be the corresponding hybrid world implementation. Let $\mathcal{P}_1$ denote the corrupted and $\mathcal{P}_2$ the honest party. Let $\mathcal{A}$ denote a malicious adversary that acts in behalf of $\mathcal{P}_1$.

For the proof, we modify the simulator construction as in Lemma 2 so that it can handle protocol instances where $\mathcal{A}$ is a prover. For that $\mathcal{A}^\circ$ must forward the query $\boldsymbol{q}$ to trusted third party $\mathcal{T}$ and play the role of honest verifier $\mathcal{V}$. The latter is always possible as $\mathcal{V}$ has no inputs. Note that the simulator will accept some proofs where $\boldsymbol{q} \in \mathcal{Q}_{\mathrm{inv}}$ while the $\mathcal{P}_2$ reading the channel will always reject these proofs. Hence, there will be small discrepancy between ideal and hybrid executions. If the commitment scheme is in the binding mode, such events occur with probability $\varepsilon_s = \frac{1}{|\mathcal{S}|} + \varepsilon_2 + \varepsilon_3$ per each protocol instance due to Lemma 3. If the commitment scheme is in the hiding mode, such events can occur at probability $\ell\varepsilon_s + \varepsilon_1$ if $t_a + t_e + \mathcal{O}(\ell) \le t$. Otherwise, we can use the hybrid execution model for distinguishing between commitment keys. Consequently, executions in the ideal and hybrid world can diverge with with probability $\ell\varepsilon_s + \varepsilon_1$ and the claim follows.   □

Theorem 3 assures that CDSZK protocols are concurrently composable zero-knowledge arguments and not proofs, since unbounded prover can always fool verifiers. If the commitment scheme is run in the binding mode, we get zero-knowledge proofs but lose statistical simulatability.

Theorem 4. *If the commitment scheme is used in the binding mode, the CDSZK protocol is* $(t_e, t_a, \mathcal{O}(t_a), t_d, \ell(\frac{1}{|\mathcal{S}|} + \varepsilon_2 + \varepsilon_3) + 2\varepsilon_1)$*-universally composable in the shared setup model provided that* $t \ge t_a + t_e + t_d + \mathcal{O}(\ell)$.

Proof. Note that t_d-time algorithm can distinguish hybrid executions of different commitment modes with probability ε_1 if $t \ge t_a + t_e + t_d + \mathcal{O}(\ell)$. Hence, we can use hybrid argument to show that the simulator from Thm 3 is sufficient.   □

Concluding Remarks. First, note that the trusted setup for public key infrastructure is essential. Although the commitment key ck can be viewed as common reference string, the CDSZK proofs are not guaranteed to be universally composable in the CRS model. The underlying CDS protocol is required to be server-private only if the public key pk is valid. Hence, a prover with an invalid public key may succeed in deception. There are no easy fix for this problem, as the public keys of most additively homomorphic cryptosystems are not efficiently verifiable. Second, note that CDSZK protocols can be proven secure in the standard model. However, we lose universal composability, see the full version of our paper for further details.

5.4   Comparison with Other Zero-Knowledge Protocols

The CDSZK protocol can be viewed as a challenge-response protocol followed by the proof of knowledge to tame cheating verifiers. However, differently from the standard solution [GMR89] where the proof of knowledge is executed before the prover sends the response, the commitment is used to fix the response. As result, the verifier can directly reveal the internal state yielding the challenge, while standard construction needs witness-indistinguishable proofs. Hence, our protocol is clearly superior compared to the standard approach.

More remarkably, our methodology can be used to convert any challenge-response protocol to zero-knowledge proof. By using dual-mode commitments, we can easily get security guarantees analogous to Theorems 3 and 4. If the challenge does not have a nice algebraic form (say we want to establish provers ability to decrypt ciphertexts or invert hash functions) then it is much more

Table 1. The number of elementary operations done in sigma and CDS protocols

Relation	Sigma protocol		CDS protocol	
	Prover	Verifier	Client	Server
$x = 0$	1E + 1M	1E + 1M + 1X	1D	1E + 1M + 1X
$ax = b$	1E + 1M	2E + 2M + 2X	1D	2E + 2M + 2X
$x_1 x_2 = x_3$	2E + 4M	2E + 3M + 4X	2D + 1M	2E + 3M + 3X

efficient to reveal all inputs (to the encryption algorithm or hashing function) directly instead of crafting witness-indistinguishable proofs of knowledge.

There is a strange structural duality between CDS and sigma protocols. Namely, in both cases, protocols for elementary relations can be combined into disjunctive and conjunctive proofs using secret sharing, see [CDS94,LL07]. As there are efficient sigma protocols for affine relations between plaintexts, the overhead of combiner constructions for both types of protocols is comparable.

Table 1 summarises the efficiency of the three most important elementary relations, where E, D, M, X correspond to encryption, decryption, multiplication and exponentiation operations. For the multiplicative relation, we use Damgård-Jurik sigma protocol from [DJ01]. In all these CDS protocols, the client sends only the ciphertexts to be checked, while prover in a sigma protocols must send extra ciphertexts. Thus, CDS protocols have smaller communication complexity, while the computation complexity is comparable[3].

Dual-mode commitments with the trusted setup can be used to convert sigma protocols into universally composable zero-knowledge proofs. For that the prover must commit the challenge ahead of proof. By using extractable commitments during the simulation, we can extract the challenge and thus create a matching sigma protocol transcript without access to the witness.

The resulting zero-knowledge protocol has higher communication complexity than CDSZK proof and cannot be dynamically reconfigured. Namely, assume that $\mathcal{P}_1$ sends some encryptions to $\mathcal{P}_2$. Then $\mathcal{P}_2$ can add CDS transformation to his reply without prior agreement with $\mathcal{P}_1$ to guarantee input privacy or start honest verifier zero knowledge proof (HVZK). If the prover $\mathcal{P}_1$ feels frightened then it can upgrade HVZK proof to a full-fledged zero knowledge proof by committing the reply. Again, there is no need for an agreement before this step. Such dynamic configurability is not attainable for other types of zero-knowledge proofs.

Zero-Knowledge Proofs for Any NP Language. The CDSZK protocol can be directly converted to general purpose zero-knowledge proof for any **NP** statement. Let $\psi(\boldsymbol{x}, \boldsymbol{w})$ be the predicate that checks validity of the witness $\boldsymbol{w}$ and the statement $\boldsymbol{x}$. Now if we encrypt $\boldsymbol{w}$, then we can use CDSZK to prove that we have indeed encrypted a witness to public statement $\boldsymbol{x}$. Since relatively efficient CDS protocols exist for any circuit [AIR01,LL07], the resulting CDSZK

---

[3] Exact comparisons are hard to give, since the cost of decryption operations depends on implementation details, which vary from a cryptosystem to a cryptosystem.

protocol is rather efficient. Since CDSZK protocol can be proved secure in the standard model (see the full version of our paper), the resulting proof system is no worse than any other general purpose zero-knowledge proof.

6   Securing Crypto-Computations against Active Attacks

Assume that we have a crypto-computing protocol $\mathcal{E}_0$ that is secure in the semi-honest setting and our goal is to protect the protocol against malicious client who has the secret key. Moreover, assume that the only way for the client to cheat is to send invalid ciphertexts to the server. For instance, send an incorrectly encrypted vote in the e-voting protocol. The are wide range of client-server protocols that can be formalised such way, see [AIR01,BK04,LL07].

All such crypto-computing protocols can be split into a computational phase and a trusted setup π_{ts}^* that sets up public and private keys. Let π_{ts}° be the augmented trusted setup, which additionally sets up the commitment parameters shared by all CDSZK protocols. Let π_{ts} be a secure two- or multi-party protocol that implements π_{ts}° in the standard model. Clearly, we can modify the protocol $\mathcal{E}_0 \langle \pi_{ts}^\circ \rangle$ by adding ideal zero-knowledge protocols $\pi_{\psi_i}^\circ$ to guarantee that the client cannot cheat. The resulting protocol can be viewed as a context $\mathcal{E} \langle \pi_{ts}^\circ, \pi_{\psi_1}^\circ, \ldots, \pi_{\psi_\ell}^\circ \rangle$ for execution ideal zero knowledge protocols.

By substituting the real implementations of CDSZK, we get a hybrid execution model $\mathcal{E} \langle \pi_{ts}^*, \pi_{\psi_1}, \ldots, \pi_{\psi_\ell} \rangle$. Finally, we can consider the crypto-computing protocol $\mathcal{E} \langle \pi_{ts}, \pi_{\psi_1}, \ldots, \pi_{\psi_\ell} \rangle$ where first π_{ts} is run isolation to get all setup parameters and then the remaining protocols are concurrently scheduled. To prove the security of the resulting implementation, we have to use hybrid argument several times to show indistinguishability.

As π_{ts} is a secure implementation of π_{ts}^*, the standard sequential composability result guarantees that any efficient adversary $\mathcal{A}$ against $\mathcal{E} \langle \pi_{ts}, \pi_{\psi_1}, \ldots, \pi_{\psi_\ell} \rangle$ can be converted to an efficient adversary $\mathcal{A}_1$ against $\mathcal{E} \langle \pi_{ts}^*, \pi_{\psi_1}, \ldots, \pi_{\psi_\ell} \rangle$ so that outputs are computationally indistinguishable. Due to Theorems 3 and 4, we can convert $\mathcal{A}_1$ into an efficient adversary $\mathcal{A}_2$ against $\mathcal{E} \langle \pi_{ts}^\circ, \pi_{\psi_1}^\circ, \ldots, \pi_{\psi_\ell}^\circ \rangle$ such that outputs remain indistinguishable. As the adversary $\mathcal{A}_2$ cannot successfully cheat and we can efficiently mimic the outputs of $\pi_{\psi_i}^\circ$ by knowing the secret key sk, we can convert $\mathcal{A}_2$ to the adversary $\mathcal{A}_3$ against $\mathcal{E}_0 \langle \pi_{ts}^\circ \rangle$. Since $\mathcal{E}_0 \langle \pi_{ts}^\circ \rangle$ is secure protocol, we can convert $\mathcal{A}_3$ into an efficient adversary against the ideal implementation of crypto-computing protocol.

To summarise, we can concurrently schedule CDSZK protocols together with a crypto-computing protocol provided that honest parties do no other computations during the setup phase π_{ts}, which generates global parameters.

7   Conclusions

In brief, we showed how to build universally composable zero-knowledge protocols from dual-mode commitments and homomorphic encryption. As the standard dual-mode commitments are also based on homomorphic encryption, our construction can be based solely on homomorphic encryption.

We acknowledge that the CDSZK protocol is not round-optimal, as there are theoretical constructions for zero-knowledge in three rounds. However, these are not as efficient. Similarly, our protocols are not the best in terms of online communication, as there are zero-knowledge protocols with constant communication in the trusted setup model, e.g. [Gro10]. However, these protocols use stronger security assumptions (bilinear parings combined with unfalsifiable assumptions) and the size of public parameters is really large.

In terms of practical performance our protocols are comparable to the zero-knowledge protocols based on sigma protocols where the security against malicious verifiers is achieved by committing the challenge at the beginning of the protocol. These protocols are more rigid against dynamic change of security levels. The latter is a drawback in the covert model where parties should randomly alter security levels during the computations to discourage cheating.

Protocols CDSMUL and CDSZKMUL have also many direct applications. First, crypto-computing protocols often use oblivious polynomial evaluation, in which a valid query consists of encryptions of $x, \ldots, x^k$. Second, multiplicative relations naturally occur in crypto-computing systems.

Finally, conditional disclosure of secrets (CDS) is often used for going beyond linearity in crypto-computing. For example, it is possible to securely evaluate greater than predicate using CDS protocols [BK04,LL07]. As somewhat homomorphic encryption (SHE) remains additively homomorphic when we reach the multiplication limit, we can combine CDS with SHE to extend the set of functions computable with SHE without extending its multiplicative depth.

References

AIR01.   Aiello, W., Ishai, Y., Reingold, O.: Priced oblivious transfer: How to sell sigital goods. In: Pfitzmann, B. (ed.) EUROCRYPT 2001. LNCS, vol. 2045, pp. 119–135. Springer, Heidelberg (2001)

BGN05.   Boneh, D., Goh, E.-J., Nissim, K.: Evaluating 2-DNF Formulas on Ciphertexts. In: Kilian, J. (ed.) TCC 2005. LNCS, vol. 3378, pp. 325–341. Springer, Heidelberg (2005)

BHR12.   Bellare, M., Hoang, V.T., Rogaway, P.: Foundations of garbled circuits. In: Proc. of ACM CCS, pp. 784–796. ACM (2012)

BK04.   Blake, I.F., Kolesnikov, V.: Strong conditional oblivious transfer and computing on intervals. In: Lee, P.J. (ed.) ASIACRYPT 2004. LNCS, vol. 3329, pp. 515–529. Springer, Heidelberg (2004)

Can01.   Canetti, R.: Universally composable security: A new paradigm for cryptographic protocols. In: Proc. of FOCS 2001, pp. 136–145. IEEE (2001)

CDS94.   Cramer, R., Damgård, I., Schoenmakers, B.: Proof of partial knowledge and simplified design of witness hiding protocols. In: Desmedt, Y.G. (ed.) CRYPTO 1994. LNCS, vol. 839, pp. 174–187. Springer, Heidelberg (1994)

CR03.   Canetti, R., Rabin, T.: Universal composition with joint state. In: Boneh, D. (ed.) CRYPTO 2003. LNCS, vol. 2729, pp. 265–281. Springer, Heidelberg (2003)

DJ01.   Damgård, I., Jurik, M.: A generalisation, a simplification and some applications of Paillier's probabilistic public-key system. In: Kim, K. (ed.) PKC 2001. LNCS, vol. 1992, pp. 119–136. Springer, Heidelberg (2001)

DPSZ12.   Damgård, I., Pastro, V., Smart, N., Zakarias, S.: Multiparty computation
           from somewhat homomorphic encryption. In: Safavi-Naini, R., Canetti, R.
           (eds.) CRYPTO 2012. LNCS, vol. 7417, pp. 643–662. Springer, Heidelberg
           (2012)
EG85.   El Gamal, T.: A public key cryptosystem and a signature scheme based on
           discrete logarithms. In: Blakely, G.R., Chaum, D. (eds.) CRYPTO 1984.
           LNCS, vol. 196, pp. 10–18. Springer, Heidelberg (1985)
Gen09.   Gentry, C.: Fully homomorphic encryption using ideal lattices. In: Proc. of
           STOC 2009, pp. 169–178. ACM (2009)
GHS12.   Gentry, C., Halevi, S., Smart, N.P.: Fully homomorphic encryption with
           polylog overhead. In: Pointcheval, D., Johansson, T. (eds.) EUROCRYPT
           2012. LNCS, vol. 7237, pp. 465–482. Springer, Heidelberg (2012)
GIKM98.   Gertner, Y., Ishai, Y., Kushilevitz, E., Malkin, T.: Protecting data privacy
           in private information retrieval schemes. In: Proc. of STOC 1998, pp. 151–
           160. ACM (1998)
GMR89.   Goldwasser, S., Micali, S., Rackoff, C.: The knowledge complexity of inter-
           active proof systems. SIAM J. Comput. 18(1), 186–208 (1989)
GOS06.   Groth, J., Ostrovsky, R., Sahai, A.: Non-interactive Zaps and New Tech-
           niques for NIZK. In: Dwork, C. (ed.) CRYPTO 2006. LNCS, vol. 4117, pp.
           97–111. Springer, Heidelberg (2006)
Gro10.   Groth, J.: Short pairing-based non-interactive zero-knowledge arguments.
           In: Abe, M. (ed.) ASIACRYPT 2010. LNCS, vol. 6477, pp. 321–340.
           Springer, Heidelberg (2010)
IP07.   Ishai, Y., Paskin, A.: Evaluating branching programs on encrypted data.
           In: Vadhan, S.P. (ed.) TCC 2007. LNCS, vol. 4392, pp. 575–594. Springer,
           Heidelberg (2007)
LL07.   Laur, S., Lipmaa, H.: A new protocol for conditional disclosure of secrets and
           its applications. In: Katz, J., Yung, M. (eds.) ACNS 2007. LNCS, vol. 4521,
           pp. 207–225. Springer, Heidelberg (2007)
Lin03.   Lindell, Y.: General composition and universal composability in secure
           multi-party computation. In: Proc. of FOCS 2003, pp. 394–403 (2003)
Pai99.   Paillier, P.: Public-key cryptosystems based on composite degree residuosity
           classes. In: Stern, J. (ed.) EUROCRYPT 1999. LNCS, vol. 1592, pp. 223–
           238. Springer, Heidelberg (1999)
SV11.   Smart, N.P., Vercauteren, F.: Fully Homomorphic SIMD Operations. IACR
           Cryptology ePrint Archive, 2011:133 (2011)
SYY99.   Sander, T., Young, A.L., Yung, M.: Non-Interactive CryptoComputing For
           NC[1]. In: Proc. of FOCS 1999, pp. 554–567. IEEE Computer Society (1999)
Yao82.   Yao, A.C.-C.: Protocols for secure computations. In: Proc. of FOCS 1982,
           pp. 160–164. IEEE Computer Society (1982)

Efficient Secure and Verifiable Outsourcing of Matrix Multiplications

Yihua Zhang and Marina Blanton

Department of Computer Science and Engineering, University of Notre Dame
Notre Dame, IN, USA
{yzhang16,mblanton}@nd.edu

Abstract. With the emergence of cloud computing services, a resource-constrained client can outsource its computationally-heavy tasks to cloud providers. Because such service providers might not be fully trusted by the client, the need to verify integrity of the returned computation result arises. The ability to do so is called verifiable delegation or verifiable outsourcing. Furthermore, the data used in the computation may be sensitive and it is often desired to protect it from the cloud throughout the computation. In this work, we put forward solutions for verifiable outsourcing of matrix multiplications that favorably compare with the state of the art. Our goal is to minimize the cost of verifying the result without increasing overhead associated with other aspects of the scheme. In our scheme, the cost of verifying the result of computation uses only a single modulo exponentiation and the number of modulo multiplications linear in the size of the output matrix. This cost can be further reduced to avoid all cryptographic operations if the cloud is rational. A rational cloud is neither honest nor arbitrarily malicious, but rather economically motivated with the sole purpose of maximizing its monetary reward. We extend our core constructions with several desired features such as data protection, public verifiability, and computation chaining.

1   Introduction

The emergence of cloud computing technologies enables clients who are unable to procure and maintain their own computing infrastructure to resort to convenient on-demand computing resources. Despite the paradigm being economically sensible for resource-limited clients, it comes with new security and privacy concerns. One of them is the lack of transparency and control over the outsourced computation, which necessitates the need to verify the result to guarantee integrity of the computation. Another is the need to protect confidentiality of the data used in the computation. Addressing these security objectives is the focus of this work.

Computation outsourcing to a cloud computing provider is common today and can take different forms. In particular, in addition to the conventional scenario when a computationally-limited client outsources its computation to the cloud and receives the result, there are many uses of cloud computing that involve multiple entities. For example, a doctor's office might send computation

S.S.M. Chow et al. (Eds.): ISC 2014, LNCS 8783, pp. 158–178, 2014.

associated with a patient's test to a cloud provider, while the patient in question is also entitled to access to the result of the computation and thus should be able to verify integrity of the returned result. This calls for solutions where the integrity of the result can be verified by entities who do not have access to secret keys thus achieving public verifiability. Furthermore, if the task result may be verified by different entities or verified by the same entity multiple times over time (e.g., every time the result is used), then it is desirable to lower the overhead associated with verifying the result of the outsourced computation without increasing other costs.

The specific problem that we treat here is matrix multiplication outsourcing. Because of popularity of matrix multiplication in a number of different domains and relatively high cost of this operation on large inputs, secure matrix multiplication outsourcing has received attention [4, 5, 18, 30]. We continue this line of research and show below how our results compare to the state of the art.

A novel feature that we propose in this work and which is not present in publications on the same topic is as follows: we divide the overall computation in multiple stages and associate a key with each of them. Only if the computation in the current stage is performed correctly and the correct key is recovered, the server can learn the computation used in the next stage. Without the correct key, it is computationally infeasible to proceed to the next stage and pass verification in any of the stages that follow. This feature allows us to achieve several goals:

1. Chaining of computation from one stage to another allows for more efficient verification of the overall task. That is, corrupting a single cell of the product matrix invalidates the values in all other cells that follow, and verifying the result of the final stage is sufficient in ensuring that the entire task was performed correctly. Other publications (such as [9, 18]), on the other hand, require verification of every matrix cell to ensure correctness of the output.

2. If the server misbehaves during the computation and produces incorrect values for one or more cells of the product matrix, in order to proceed with the computation, it has to invest into substantially larger computation. In other words, the effect of the server's misbehavior is enlarged to the maximum extent, where any deviation from the computation substantially increases the computation cost. Thus, this mechanism is designed to deter the server from deviating from the correct computation.

3. When the result is returned to the client and does not pass verification, the client can efficiently identify the first stage during which the server deviated from the prescribed computation and ask the same or different server to rerun the computation starting from that stage.

4. If the server carries out the computation honestly, but gets compromised or infected by malware that corrupts the computation, the server can use the checkpoints between the stages to efficiently determine that corruption took place and quickly recover from it. That is, if the server is unaware of the compromise and continues with the task, the outcome will not pass verification and the computational effort becomes wasted. With the checkpoints, on

the other hand, the server will identify the problem, stop the computation, resolve the problem, and resume the task without wasting its efforts.

Our contributions can be summarized as follows:

- We present the core construction for verifiable outsourcing in presence of malicious adversaries who arbitrarily deviates from the prescribed computation (Section 4). In our scheme, delegating a task requires work linear in the input size, carrying out the computation itself has the same complexity as that of conventional matrix multiplication algorithm (i.e., $O(n^3)$ for matrices of dimension $n \times n$), and verification of the resulting matrix product involves only a single modulo exponentiation and the number of modulo multiplications linear in the output size.
- We present another construction that assumes rational adversaries who are neither honest nor malicious, but rather economically motivated with the sole purpose of maximizing the reward (Section 5.1). Under this adversarial model, verification of the returned result involves only a single comparison.
- We extend the construction in the rational setting to incorporate the chaining feature described above without compromising other properties (Section 5.2). In particular, this has no impact on the complexity of the resulting scheme.
- We also sketch how data privacy and public verifiability, in which any entity with access to public verification key can assess the validity of the returned result, can be added to the constructions in both malicious and rational settings (Section 6). This does not increase asymptotic complexities of the scheme, but the cost of recovering the output or verification time, resp., may increase. Note that when public verifiability is combined with data protection, access to the verification key does not allow for data recovery.

All our schemes achieve public delegatability, which means the entity who runs system setup can be different from the entities who form a task to be outsourced.

In reducing the cost associated with verifiable computation schemes for matrix multiplication outsourcing, our focus was on reducing the cost of verifying the result as this may be a more frequently used operation or an operation performed by weaker clients. The cost of task preparation in our and other schemes requires $O(n^2)$ cryptographic operations for input matrices of size $n \times n$, i.e., linear in the size of input. The server's work for carrying out matrix multiplication is $O(n^3)$ cryptographic operations, i.e., uses the conventional matrix multiplication algorithm.

We note that the cost of $O(n^2)$ cryptographic operations used in task preparation is rather high and will exceed the cost of computing matrix multiplication locally for small matrices. In particular, matrix multiplication algorithms of asymptotic complexity as low as $O(n^{2.373})$ are known, but the huge constants hidden behind the big-O notation prevent most of them from being used in practice (i.e., they require more than $2n^3$ work) [3]. In particular, for matrices of dimension $n < 10^{20}$, only the algorithm by Strassen 1969 and Winograd 1971 of complexity $O(n^{2.807})$ and the technique of trilinear aggregating of complexity $O(n^{2.775})$ result in implementations of matrix multiplications that outperform the conventional $O(n^3)$ algorithm [3, 28]. This means that our and related constructions reduce the cost of matrix multiplication for the client only for large matrices when performing n^2 cryptographic operations is below $O(n^{2.775})$ work.

Before we proceed with the description of our schemes, we discuss related work (Section 2) and provide background information and definitions (Section 3).

2   Related Work

Verifiable Computation. In verifiable computation [2, 13, 15, 16, 18, 20, 21, 32], a client outsources a computationally intensive task to a server and verifies its results in an efficient manner upon its completion. The basic question was proposed in work on interactive proofs (IP) [8, 22], efficient arguments based on probabilistically checkable proofs (PCP) [26, 27], and computationally sound (CS) proofs [29]. Such schemes are not generally suitable for our goal (due to, e.g., vast client's storage requirements or the need for interactive verification). Parno et al. recently introduced Pinocchio [31], which allows execution of a general function represented as a circuit to be delegated to an untrusted worker. The cost of output verification in [31] is linear in the size of the function's input and output, but requires only a constant number of most expensive operations (pairing evaluation), which resembles similarities to our scheme in presence of malicious adversaries. Our verification cost is still lower in practical terms and is further reduced in the schemes with rational adversaries. The asymptotic complexity of the server's computation in [31] is the same as in our scheme, but our solution offers faster performance. Lastly, the setup of [31] uses a different key generation phase from our problem generation phase, which is applied to a function instead of function's input in our solution. The cost of key generation in [31] is higher than the cost of problem generation in our solution, but the key generation algorithm in [31] may be executed less frequently.

Homomorphic MAC and Signature. A homomorphic MAC [1] allows an entity to use a secret key sk to produce a tag σ that authenticates message m with an additional property that, given a set of authenticators $\sigma_1, \sigma_2, \ldots, \sigma_n$ for messages $m_1, m_2, \ldots, m_n$, any entity with possession of public parameters can homomorphically evaluate a function P on $(\sigma_1, \sigma_2, \ldots, \sigma_n)$ to produce a short tag σ' that authenticates correctness of $m' = P(m_1, m_2, \ldots, m_n)$. In the public-key setting, signatures are used to replace MACs and achieve similar functionality [25]. While homomorphic MACs or signatures can be used to realize verifiable computation for problems of certain structure, with such solutions the cost of verification is not smaller than the cost of executing the task. Furthermore, it is not intuitive as to how to protect privacy of the underlying messages using these techniques because neither MACs nor signatures are designed for this purpose.

Matrix Computation. The problem of verifiable matrix computation has been studied in recent literature [4, 9, 18, 30]. In addition to computation verification, existing solutions offer other important security features that are: 1) *data protection*, i.e., protection of both input and output matrices throughout the

Table 1. Comparison with related work

Scheme	Verifiable Computation	Data Privacy	Public Verifiability	Output-Indep. Verification	Deterministic Verification
Atallah et al. [4]	✓	✓	✗	✗	✗
Mohassel [30]	✓	✓	✗	✗	✗
Fiore et al. [18]	✓	✗	✓	✗	✓
Backes et al. [9]	✓	✗	✗	✗	✓
This work	✓	✓	✓	✓	✓

computation; 2) *public verifiability*, i.e., the ability of any entity to verify the result of outsourced computation; and 3) *deterministic verification*, i.e., the ability to detect faulty cells in an output matrix with probability 1 (minus a negligible function of the security parameter due to computational assumptions). Unlike prior work, our scheme achieves all these features. Additionally, we achieve another property called *output-independent efficiency* that allows for a constant number of cryptographic operations (modulo exponentiations or pairing operations) to be used during verification independent of the size of the output matrix. Table 1 summarizes features of our solution and other schemes. Note that [18] has similar security features to ours, and although not specified in that work, it is feasible to incorporate privacy protection into their scheme as realized in [33]. We, however, found the scheme cannot be adjusted to efficiently handle rational adversaries.

Next, we provide a more detailed comparison of our work with closely related schemes. Note that we treat deterministic verification as a property that is nontrivial to achieve and thus consider only the work of Fiore et al. [18] and Backes et al. [9]. The construction of [9], however, is based on the scheme of [18] and would offer the same performance as that of [18] in our setting (the advantage of [9] is that it allows for more flexible function specification and scheduling). Thus, we list only performance of [18] as the representative of both [18] and [9].

The computational overhead for the client and the server is presented in Table 2, where n represents the size of each dimension of input matrices, and c_m, c_e, and c_p denote the time to carry out a modular multiplication, exponentiation, and a pairing operation, respectively. In the table, we use notation VC_m and VC_r to denote our verifiable computation schemes for malicious and rational adversaries, respectively. Note that some of the constructions for rational adversary do not involve any cryptographic operations for verification (and rather perform a single comparison) and that work is listed as $O(1)$. Finally, note that in some of our constructions for the rational adversary, the work for task preparation or server's computation is increased compared to the equivalent constructions for the malicious adversary, but the verification cost is substantially (and asymptotically) reduced.

The timings of elementary cryptographic operations can be inferred from the benchmarks in [14]. For groups that admit an asymmetric bilinear map $e : \mathbb{G}_1 \times \mathbb{G}_2 \to \mathbb{G}_T$, which offer faster performance than groups that admit a symmetric bilinear map $e : \mathbb{G}_1 \times \mathbb{G}_1 \to \mathbb{G}_T$ and are used in this work (as detailed later in the paper) and can be used in the solution of [18], the timings are as follows:

Table 2. Computation in our and mostly closely related schemes

Scheme	Client's Preparation	Server's Computation	Client's Verification
Fiore et al. [18] (priv. ver.)	$(4c_e + 3c_m)n^2$	$c_e n^3$	$(c_e + c_m)n^2$
VC_m (priv. ver.)	$(2c_e + 4c_m)n^2$	$c_p n^3$	$c_e + c_m n^2$
VC_r (priv. ver.)	$(4c_e + 6c_m)n^2$	$c_p n^3 + (c_e + c_m)n^2$	$O(1)$
Fiore et al. [18] (pub. ver.)	$(4c_e + 3c_m)n^2$	$c_e n^3$	$(c_p + c_e + c_m)n^2$
VC_m (pub. ver.)	$(2c_e + 4c_m)n^2$	$c_e n^3$	$c_p n + (c_e + c_m)n^2$
VC_r (pub. ver.)	$(4c_e + 6c_m)n^2$	$c_p n^3 + (c_e + c_m)n^2$	c_e
VC_m (priv. ver. + privacy)	$(2c_e + 6c_m)n^2$	$2c_p n^3$	$c_e + c_m n^2$
VC_r (priv. ver. + privacy)	$(5c_e + 8c_m)n^2$	$2c_p n^3 + c_m n^2$	$O(1)$
VC_m (pub. ver. + privacy)	$(8c_e + 16c_m)n^2$	$4c_e n^3$	$4c_p n + 4(c_e + c_m)n^2$
VC_r (pub. ver. + privacy)	$(5c_e + 8c_m)n^2$	$2c_p n^3 + c_m n^2$	c_e

Table 3. Storage and communication in our and mostly closely related schemes

Scheme	Client's Storage	Server's Storage	Communication
Fiore et al. [18] (priv. ver.)	$n^2\kappa$	$n^2\kappa + 2n^2$	$2n^2\kappa + 3n^2$
VC_m (priv. ver.)	$3n\kappa$	$2n^2\kappa + 2n^2$	$(2n^2 + 1)\kappa + 3n^2$
VC_r (priv. ver.)	κ	$2n^2\kappa + 2n^2$	$(2n^2 + 1)\kappa + 3n^2$
Fiore et al. [18] (pub. ver.)	$n^2\kappa$	$n^2\kappa + 2n^2$	$2n^2\kappa + 3n^2$
VC_m (pub. ver.)	$(n^2 + n)\kappa$	$n^2\kappa + 2n^2$	$(n^2 + n)\kappa + 3n^2$
VC_r (pub. ver.)	κ	$2n^2\kappa + 2n^2$	$(2n^2 + 1)\kappa + 3n^2$
VC_m (priv. ver. + privacy)	$3n\kappa$	$4n^2\kappa$	$(5n^2 + 1)\kappa$
VC_r (priv. ver. + privacy)	$(n^2 + 2n)\kappa$	$4n^2\kappa$	$(5n^2 + 1)\kappa$
VC_m (pub. ver. + privacy)	$4(n^2 + n)\kappa$	$4n^2\kappa + 8n^2$	$4(n^2 + n)\kappa + 12n^2$
VC_r (pub. ver. + privacy)	$(n^2 + 2n)\kappa$	$4n^2\kappa$	$(5n^2 + 1)\kappa$

for 128-bit security, a modulo exponentiation in $\mathbb{G}_1$ or $\mathbb{G}_2$ can be performed in 0.1–0.4ms on a single core of a conventional 2.4GHz desktop machine, a modulo exponentiation in $\mathbb{G}_T$ in 0.4–0.9ms, and the pairing operation in 2.1–2.3ms. The cost of modulo multiplications is at least two orders of magnitude smaller than that of performing modulo exponentiations.

The storage and communication requirements of our constructions and those of [18] are listed in Table 3. In the table, the security parameter κ denotes the bitlength of group elements. The client's storage is computed as the amount of information the client needs to maintain in order to be able to verify the result (i.e., the size of the key). Then the task delegator and each task verifier will require additional storage for their respective input and output matrices. The server's storage corresponds to the amount of storage the server needs to maintain in order to carry out the task. Lastly, communication corresponds to both sending the task to the server and returning the result and the proof of computation to the client.

Rational Computation. In recent years, game theory has been used in cryptographic research [17, 24] to develop a new adversarial model – a rational adversary who is no longer treated as arbitrarily malicious, but who is motivated by some utility function with the sole purpose of maximizing its utility. It is known that under this model protocols can be designed with better efficiency than that of traditional counterparts [19]. As our problem deals with verifiable computation, we are interested in rational proof systems that have been recently studied in [6, 7, 23]. The merits of rational proof systems are that they allow for extremely low communication and verification time [6, 7] and can achieve single-round proofs if the prover is computationally bounded [23]. The basic idea of this line of work is that the prover will send the result of computation to the verifier who will compute the corresponding reward based on the "quality" of prover's result, and the reward will be maximized only if the result is correct. The publications focus on general complexity classes such as uniform TC^0 (polynomial-time, constant-depth threshold circuits) and decision problems in $P^{||NP}$ (polynomial time with access to parallel queries to an NP oracle), while in our case, we target specific matrix computation with the goal of achieving even better efficiency.

3   Background and Definitions

3.1   Basic Definitions

Throughout this work we use notation $x \xleftarrow{R} S$ to denote that x is chosen uniformly at random from the set S. A function $\epsilon(n)$ is said to be negligible if for sufficiently large n its value is smaller than the inverse of any polynomial $poly(n)$. Let F be a family of functions and $\mathsf{Dom}(f)$ denote the domain of function $f \in F$. Also, let κ denote a security parameter. We use $H : \{0,1\}^* \to \{0,1\}^{\ell_1(\kappa)}$ to denote a collision resistant hash function that takes as input a string x and outputs an $\ell_1(\kappa)$-bit hash y. We also use notation $||$ to denote string concatenation. For matrix A, notation A_{ij} refers to the element of A at row i and column j. We use notation PRF to refer to a pseudo-random function family defined as follows.

Definition 1. *Let $F : \{0,1\}^{\kappa} \times \{0,1\}^{\ell_2(\kappa)} \to \{0,1\}^{\ell_2(\kappa)}$ be a family of functions. For $k \in \{0,1\}^{\kappa}$, the function $f_k : \{0,1\}^{\ell_2(\kappa)} \to \{0,1\}^{\ell_2(\kappa)}$ is defined as $F_k(x) = F(k,x)$. F is said to be a family of pseudo-random functions (PRF) if for every probabilistic polynomial time (PPT) adversary $\mathcal{A}$ with oracle access to a function F_k and all sufficiently large κ, $|\Pr[\mathcal{A}^{F_k}(1^{\kappa}) - \Pr[\mathcal{A}^{R}(1^{\kappa})]|$ is negligible in κ, where $k \xleftarrow{R} \{0,1\}^{\kappa}$ and R is a function chosen at random from all possible functions mapping $\ell_2(\kappa)$-bit inputs to $\ell_2(\kappa)$-bit outputs.*

Definition 2 (Bilinear map). *A one-way function $e : \mathbb{G}_1 \times \mathbb{G}_2 \to \mathbb{G}_T$ is a bilinear map if the following conditions hold:*
- *(Efficient) $\mathbb{G}_1$, $\mathbb{G}_2$, and $\mathbb{G}_T$ are groups of the same prime order p and there exists an efficient algorithm for computing e.*
- *(Bilinear) For all $g_1 \in \mathbb{G}_1$, $g_2 \in \mathbb{G}_2$, and $a,b \in \mathbb{Z}_p$, $e(g_1^a, g_2^b) = e(g_1, g_2)^{ab}$.*
- *(Non-degenerate) If g_1 generates $\mathbb{G}_1$ and g_2 generates $\mathbb{G}_2$, then $e(g_1, g_2)$ generates $\mathbb{G}_T$.*

Throughout this work, we assume there exists a trusted setup algorithm Set that, on input a security parameter 1^κ, outputs the setup for groups $\mathbb{G}_1 = \langle g_1 \rangle$ and $\mathbb{G}_2 = \langle g_2 \rangle$ of prime order p that have a bilinear map e, and $e(g_1, g_2)$ generates $\mathbb{G}_T$ of order p. That is, $(p, \mathbb{G}_1, \mathbb{G}_2, \mathbb{G}_T, g_1, g_2, e) \leftarrow Set(1^\kappa)$.

3.2   Computational Assumptions

The first computational assumption used in this work is the Multiple Decisional Diffie-Hellman Assumption (m-M-DDH) [12], which can be stated as follows:

Definition 3 (m-M-DDH assumption). *Let $\mathbb{G}$ be a group of prime order p, $g \in \mathbb{G}$ is its generator, and $m \geq 2$. Also let $D = (g^{x_1}, \ldots, g^{x_m}, \{g^{x_i x_j}\}_{1 \leq i < j \leq m})$ for random $x_1, \ldots, x_m \in \mathbb{Z}_p$, and define random tuple as $D_{rand} = (g_1, \ldots, g_m, \{g_{ij}\}_{1 \leq i < j \leq m})$ in $\mathbb{G}$. Adversary $\mathcal{A}$, whose task is to distinguish an M-DDH tuple from a random tuple, outputs a bit. We define the advantage of adversary $\mathcal{A}$ in solving the M-DDH problem as $\boldsymbol{Adv}_{\mathcal{A}}^{\text{M-DDH}}(\kappa) = |\Pr[\mathcal{A}(g, p, m, D) = 1] - \Pr[\mathcal{A}(g, p, m, D_{rand}) = 1]|$. The M-DDH assumption holds if for every PPT algorithm $\mathcal{A}$, $\boldsymbol{Adv}_{\mathcal{A}}^{\text{M-DDH}}(\kappa)$ is negligible.*

Some of our schemes are built using subgroups of elliptic curves with pairings where the decisional Diffie-Hellman (DDH) problem is hard. The use of DDH-hard pairing groups requires the External Diffie-Hellman (XDH) assumption [10].

Definition 4 (XDH assumption). *Let $(p, \mathbb{G}_1, \mathbb{G}_2, \mathbb{G}_T, g_1, g_2, e) \leftarrow Set(1^\kappa)$. We define the advantage of adversary $\mathcal{A}$ in solving the DDH problem in $\mathbb{G}_1$ as $\boldsymbol{Adv}_{\mathcal{A}}^{\text{XDH}}(\kappa) = |\Pr[\mathcal{A}(p, g_1, g_2, g_1^a, g_1^b, g_1^{ab}) = 1] - \Pr[\mathcal{A}(p, g_1, g_2, g_1^a, g_1^b, g_1^c) = 1]|$, where $a, b, c \overset{R}{\leftarrow} \mathbb{Z}_p$. We say that the XDH assumption holds if for every PPT algorithm $\mathcal{A}$ $\boldsymbol{Adv}_{\mathcal{A}}^{\text{XDH}}(\kappa)$ is negligible.*

The XDH assumption implies that there is no efficiently computable homomorphism from $\mathbb{G}_1$ to $\mathbb{G}_2$. This assumption is also necessary for the M-DDH assumption to hold in groups that admit a bilinear map.

3.3   Verifiable Computation

A verifiable computation scheme VC is a 4-tuple of polynomial-time algorithms (Setup, ProbGen, Compute, Verify) that allows a user to outsource the computation of function $f \in F$ to an untrusted worker. VC is defined as follows:

Setup($1^\kappa, f$) → params: On input a security parameter κ and function f to be outsourced, it produces public parameters params.

ProbGen(x, params) → ($\mathsf{SK}_x, \mathsf{EK}_x, \sigma_x$): Given an input $x \in \mathrm{Dom}(f)$, this algorithm is run by the delegator to produce a secret key SK_x associated with the problem instance for computation outsourcing and output verification, an evaluation key EK_x given to the worker to carry out the outsourced computation, and an encoding σ_x of input x.

Compute(EK_x, σ_x) → σ_y: On an encoded input σ_x and EK_x, the worker runs the algorithm to produce an encoded outcome σ_y, where $y = f(x)$.

Verify$(\mathsf{SK}_x, \sigma_y) \to y \cup \perp$: Given an encoded output σ_y and the secret key SK_x, this algorithm outputs y or an error $\perp$ upon result verification.

The *correctness* requirement is such that the values produced by the algorithms will allow any honest worker who faithfully executes Compute to pass verification of the output it produces. More formally, for any $f \in F$, any params $\leftarrow$ Setup$(1^\kappa, f)$, and any $x \in \mathrm{Dom}(f)$, if $(\mathsf{SK}_x, \mathsf{EK}_x, \sigma_x) \leftarrow$ ProbGen(x, params), $\sigma_y \leftarrow$ Compute$(\mathsf{EK}_x, \sigma_x)$, and $y \leftarrow$ Verify$(\mathsf{SK}_x, \sigma_y)$, then $\Pr[y = f(x)] = 1$.

To formulate *security* of a verifiable computation scheme, we define an interactive security experiment described next. In the experiment, the adversary $\mathcal{A}$ is allowed to query ProbGen algorithms on inputs of its choice x_i and obtain the corresponding evaluation key EK_{x_i} and input encoding σ_{x_i}. In the private key setting, it is also granted oracle access to Verify algorithm, where $\mathcal{O}_{\mathsf{Verify}}(x_i, \sigma_y)$ runs $y \leftarrow$ Verify$(\mathsf{SK}_i, \sigma_y)$ and returns y. Eventually, $\mathcal{A}$ outputs the input x^* on which it would like to be challenged, obtains evaluation key EK_{x^*} and encoding σ_{x^*}, and produces output encoding σ'_y. The adversary succeeds if the output is different from $f(x^*)$ and the verification algorithm does not output an error $\perp$. Note that this definition captures full adaptive security as opposed to weaker selective security where the adversary is required to commit to the challenge input x^* in the beginning of the game.

Experiment $\mathbf{Exp}_{\mathcal{A}}^{\mathsf{Ver}}(\mathsf{VC}, f, \kappa)$
   params $\leftarrow$ Setup$(1^\kappa, f)$
   for $i = 1$ to q do
      $x_i \leftarrow \mathcal{A}^{\mathcal{O}_{\mathsf{Verify}}(\cdot, \cdot)}(\sigma_{x_1}, \mathsf{EK}_1, \ldots, \sigma_{x_{i-1}}, \mathsf{EK}_{i-1})$
      $(\mathsf{SK}_i, \mathsf{EK}_i, \sigma_{x_i}) \leftarrow$ ProbGen(x_i, params)
   $x^* \leftarrow \mathcal{A}(\sigma_{x_1}, \mathsf{EK}_1, \ldots, \sigma_{x_q}, \mathsf{EK}_q)$
   $(\mathsf{SK}_{x^*}, \mathsf{EK}_{x^*}, \sigma_{x^*}) \leftarrow$ ProbGen(x^*, params)
   $\sigma'_y \leftarrow \mathcal{A}(\sigma_{x_1}, \mathsf{EK}_1, \ldots, \sigma_{x_q}, \mathsf{EK}_q, \sigma_{x^*}, \mathsf{EK}_{x^*})$
   $y' \leftarrow$ Verify$(\mathsf{SK}_{x^*}, \sigma'_y)$
   if $y' \neq \perp$ and $y' \neq f(x^*)$ return 1
   else return 0

For any $\kappa \in \mathbb{N}$ and any function $f \in F$, we define the advantage of an adversary $\mathcal{A}$ making at most $q = poly(\kappa)$ queries in the above security game against VC as $\mathbf{Adv}_{\mathcal{A}}^{\mathsf{Ver}}(\mathsf{VC}, f, q, \kappa) = \Pr[\mathbf{Exp}_{\mathcal{A}}^{\mathsf{Ver}}(\mathsf{VC}, f, \kappa) = 1]$.

Definition 5. *A verifiable computation scheme* VC *is secure if for any PPT adversary* $\mathcal{A}$, *any* κ, *and any* $f \in F$, $\mathbf{Adv}_{\mathcal{A}}^{\mathsf{Ver}}(\mathsf{VC}, f, q, \kappa)$ *is negligible in* κ.

In this work we consider two types of adversaries: the first type is the traditional adversary that can arbitrarily deviate from the prescribed computation as defined by Compute functionality. We denote this type of adversary as malicious. While the malicious adversary model leads to strong security guarantees, it has been criticized as overly pessimistic due to neglecting the incentive that could potentially cause computational entities to deviate from the prescribed behavior. We therefore consider the second type of adversary which we denote as rational. A rational adversary is neither honest nor malicious, but only interested in maximizing its reward attained during computation. The rationale behind including this type of adversary is that it allows us to design more efficient solutions if the server can be assumed not to intentionally corrupt the result. Next, we formally define the rational adversary model in verifiable computation, which was initially proposed in [6] and later refined for rational argument systems in [7, 23].

Definition 6. *A function f admits a rational argument with security parameter κ if there exists a protocol (P, V) and a randomized reward function* reward: $\{0,1\}^\star \to \mathbb{R}_{\geq 0}$ *such that for any prover $\hat{P}$ of size $\leq 2^{\kappa(|x|)}$ and input $x \in \{0,1\}^\star$, the following three properties hold:*

- $\Pr[\text{output}(P, V)(x) = f(x)] = 1.$
- *There exists a negligible function $\mu(\cdot)$ such that $E[\text{reward}((P, V)(x))] + \mu(|x|) \geq E[\text{reward}((\hat{P}, V)(x))].$*
- *If there exists a polynomial $p(\cdot)$ such that $\Pr[\text{output}((\hat{P}, V)(x)) \neq f(x)] \geq p(|x|)^{-1}$, then there exists a polynomial $q(\cdot)$ such that $E[\text{reward}((P, V)(x))] \geq E[\text{reward}((\hat{P}, V)(x))] + q(|x|)^{-1}.$*

The first property refers to completeness of the protocol, which says prover P is able to return a correct answer $f(x)$ by following the prescribed protocol. The second property ensures that by deviating from the prescribed protocol in a computationally bounded manner, a dishonest prover $\hat{P}$ will achieve at most negligibly larger gain than a faithful prover P. The last property guarantees if $\hat{P}$ does not report a correct answer $f(x)$ with a noticeable probability, he has to bear a noticeable utility loss. A rational argument system ensures that a rational adversary will maximize the reward if and only if he honestly follows the protocol to report the correct answer. Therefore, to prove security of a protocol, we need to show that it conforms to Definition 6 under reasonable assumptions on cost and utility, which we formulate in the server-client setting as follows:

Assumption 1. – *For each outsourced task, both the monetary reward the client compensates and the computational cost the server bears are polynomial to the size of the input.*
- *As the server aims to make profits by devoting his resources to clients' specific tasks, the monetary reward he gains from a client should be larger than the cost it bears, which also conforms to the business practice for cloud service.*
- *In the event of any inconsistency between the answers the server returns and the answers the client expects, the server will not receive any reward or even undertake utility loss resulted from the violation of Service Level Agreement.*

All additional definitions for data protection and public verifiability are omitted due to space constraints, but can be found in the full version of this work [34].

4   Matrix Multiplication for Malicious Adversary

Problem Formulation. The delegator would like to multiply matrices A and B of dimensions $n_1 \times n_2$ and $n_2 \times n_3$, respectively. It is assumed that the elements of A and B are not sensitive and do not require protection. In our solution, the delegator's work is linear in the size of the input and output, which is optimal.

Our first scheme aims to defend against malicious adversary who tampers with the computation or its results regardless of costs and attempts to pass verification. Also, this basic construction does not achieve public verifiability: only the entity who possesses the secret key is able to attest correctness of returned result. In section 6 we extend the scheme to support public verifiability that allows any entity with access to public key assess the validity of computation results. For notational simplicity, we use VC_m to denote this scheme.

Description of the Scheme. The main idea used in our solution is that the delegator encodes matrix A into matrix X and matrix B into matrix Y. The delegator sends matrices $\{A, B, X, Y\}$ to the server who computes $C = A \times B$ and $D = X \times Y$. The client then verifies correctness of C by checking a secret relationship between the elements of C and D.

In more detail, the secret relationship is formed by generating secret random group elements R_{ij} such that $D_{ij} = R_{ij}^{C_{ij}}$, which are of the form $g_T^{r_i c_j}$, where $\{r_i\}_{i=1}^{n_1}$ and $\{c_j\}_{j=1}^{n_3}$ are two vectors of random elements. Moreover, in order to satisfy the relationship, we need to embed r_i and c_j into each D_{ij}, and this is realized by forming X_{ij} and Y_{ij} to be of the form $g_1^{r_i A_{ij}}$ and $g_2^{c_j B_{ij}}$, respectively. We aim to rely on the M-DDH assumption to show that, given $g_1^{r_i}$ and $g_2^{c_j}$, $g_T^{r_i c_j}$ is indistinguishable from a random value, which is however difficult due to the existence of bilinear pairing operation. To remedy the problem, we completely hide all information about $g_2^{c_j}$'s from the server. As a result, if we rely on the XDH assumption, $(g_T^{r_1}, \ldots, g_T^{r_{n_1}}, \{g_T^{r_i c_j}\}_{1 \leq i \leq n_1, 1 \leq j \leq n_3})$ is a partial $(n_1 + n_3)$-M-DDH tuple and the adversary can have only a negligible advantage in distinguishing the $g_T^{r_i c_j}$'s from random elements of the group. The hiding is achieved by further masking each Y_{ij} by a random value T_{ij}, which should be also in a special form; otherwise, the client will have to compute $X \times T$ to satisfy the relationship, which is the exact workload the client wants to avoid. Therefore, T_{ij}'s are formed as $g_2^{w_j d_i}$ using two random vectors $\{d_i\}_{i=1}^{n_2}$ and $\{w_j\}_{j=1}^{n_3}$, and the client only needs to perform $O(n_1 n_3)$ work to compute $X \times T$.

When the server returns C and $\sum_i \sum_j D_{ij}$, the delegator uses $\{r_i c_j\}_{1 \leq i \leq n_1, 1 \leq j \leq n_3}$ and information about the product $X \times T$ to verify that the sum of the elements in C after proper randomization matches $\sum_i \sum_j D_{ij}$. If the verification succeeds, the delegator uses C as the correct output. The details are given in Figure 1.

The complexity of ProbGen run by the delegator is dominated by computing key vk and matrices X and Y, and is therefore $O(n_1 n_2 + n_2 n_3)$. Compute involves the execution of two matrix multiplications resulting in complexity $O(n_1 n_2 n_3)$. Verify consists of ensuring the validity of C by checking its elements against s that Compute produces and has complexity $O(n_1 n_3)$, i.e., linear in the size of the output. The value of s is computed in such a way that a malicious adversary is unable to construct an incorrect s that passes the verification test.

Security of the scheme can be stated as follows:

Setup($1^\kappa, f$): Given f that indicates matrix multiplication, using the security parameter κ run $(p, \mathbb{G}_1, \mathbb{G}_2, \mathbb{G}_T, g_1, g_2, e) \leftarrow Set(1^\kappa)$ and set Set params $= (p, \mathbb{G}_1, \mathbb{G}_2, \mathbb{G}_T, g_1, g_2, g_T = e(g_1, g_2), f)$. Matrix elements should be representable as values in $\mathbb{Z}_p$.

ProbGen($x = (A, B)$, params): On input two matrices A and B of respective dimensions $n_1 \times n_2$ and $n_2 \times n_3$, perform:

   1. Choose $r_i \xleftarrow{R} \mathbb{Z}_p^*$ for $1 \leq i \leq n_1$, $d_j \xleftarrow{R} \mathbb{Z}_p^*$ for $1 \leq j \leq n_2$, and $c_k, w_k \xleftarrow{R} \mathbb{Z}_p^*$ for $1 \leq k \leq n_3$.

   2. Compute $t_i = \sum_{k=1}^{n_2} A_{ik} d_k$ for $1 \leq i \leq n_1$, $f = \sum_{j=1}^{n_3} w_j$, and $vk_i = r_i t_i f$ for $1 \leq i \leq n_1$.

   3. Compute $X_{ij} = g_1^{r_i A_{ij}}$ for $1 \leq i \leq n_1$ and $1 \leq j \leq n_2$.

   4. Compute $Y_{ij} = g_2^{c_j B_{ij} + w_j d_i}$ for $1 \leq i \leq n_2$ and $1 \leq j \leq n_3$.

   5. Set $\mathsf{SK}_x = (\{c_i\}_{i=1}^{n_3}, \{vk_i\}_{i=1}^{n_1}, \{r_i\}_{i=1}^{n_1})$, $\mathsf{EK}_x = $ params, and $\sigma_x = (A, B, X, Y)$.

Compute($\mathsf{EK}_x, \sigma_x = (A, B, X, Y)$): Given σ_x, execute:

   1. Compute $C = A \times B$.

   2. Compute $s = \prod_{i=1}^{n_1} \prod_{j=1}^{n_3} \prod_{k=1}^{n_2} e(X_{ik}, Y_{kj})$.

   3. Set $\sigma_y = (C, s)$.

Verify($\mathsf{SK}_x = (\{c_i\}_{i=1}^{n_3}, \{vk_i\}_{i=1}^{n_1}, \{r_i\}_{i=1}^{n_1}), \sigma_y = (C, s)$): If $g_T^{\sum_{i=1}^{n_1} r_i \sum_{j=1}^{n_3} c_j C_{ij} + vk_i} = s$, output C; otherwise, output $\perp$.

Fig. 1. Description of the core scheme VC_m in presence of malicious adversaries

Theorem 1. *Assuming that the M-DDH and XDH problems are hard, the verifiable computation scheme VC_m is secure according to Definition 5 in presence of malicious adversaries.*

The proof and correctness analysis of this construction can be found in Appendix A.

5   Matrix Multiplication for Rational Adversary

Our next construction aims at defending against a rational adversary. Because a rational adversary behaves in the most profitable manner by considering both the compensation paid by the client and the cost endured during the computation, it would be to the adversary's advantage to honestly report all computed results to obtain compensation for the work (rather than report a bogus result that could be detected with overwhelming probability and hence yield a lower reward).

In our solution against rational adversaries VC_r, we achieve two features: (i) to force a rational adversary who wishes to maximize its profits to conform to the prescribed protocol and (ii) in case of faulty computation, to pinpoint location of faulty cells by both the server and the client. Realizing both features is achieved by requiring the client to perform only work sublinear in the size of the matrices at the time of computation verification. For the ease of presentation, we describe our solution in a modular manner, where the first scheme support only the first feature and the second presented scheme enhances it with the second feature.

5.1 Description of the Base Scheme

The main idea behind VC_r is similar to that of VC_m: as before, the delegator encodes A into X and B into Y and asks the server to compute products $A \times B$ and $X \times Y$. Similar to VC_m, correctness of $A \times B$ is verified by checking a secret relationship between the two matrix products. However, unlike VC_m, where the delegator performs the verification itself, in VC_r, the delegator further outsources the verification task to the server and only performs one string comparison to confirm correctness of the verification process. The saving in the verification cost comes with slightly increased work during problem generation, but this work is still linear in the size of input and output. This scheme can be suitable in the setting with three entities (besides the server) such as a doctor who delegates problem generation to lab assistants and patients who verifies the result of the computation returned by the server. The entity performing problem generation (i.e., lab assistant in the above example) is willing to put in additional (one-time) work to benefit routine operations (i.e., verification) by end users (patients).

In our solution, the delegator, as before, produces X as a randomized version of A with its elements of the form $g_1^{r_i A_{ij}}$ and Y as a randomized version of B with its elements of the form $g_2^{c_j B_{ij} + w_j d_i}$. The delegator also releases a new matrix Z formed as $g_T^{r_i c_j}$ to aid the server in producing a proof of correct computation. Unlike VC_m, security of which relies on the secrecy of $g_T^{r_i c_j}$, in VC_r, we need to incorporate additional secret information in the solution to guarantee security despite public exposure of matrix Z. If we treat Y as $\hat{B} + T$, where $\hat{B} = g_2^{c_j B_{ij}}$ and $T = g_2^{w_j d_i}$, we want to use $X \times T$ as secret information. In order to do that, the cells of $X \times T$ should be indistinguishable from random group elements, and can be produced by the server only if it follows the prescribed protocol. Towards the goal, we incorporate additional randomization into X and $\hat{B}$ by representing their elements as $g_1^{r_i / m_j A_{ij}}$ and $g_2^{c_j m_i B_{ij}}$, respectively, using another random vector $\{m_i\}_{i=1}^{n_2}$. The verification key is set to be the result of hashing of all cells of $X \times T$. Therefore, to pass verification, the server has to recover all elements of $X \times T$ correctly. Notice that because the server is unable to separate $\hat{B}$ from T in Y and thus compute $X \times T$ by unintended means, the server is forced to compute $X \times Y$ and $A \times B$, determine $X \times \hat{B}$ from $A \times B$ with the help of Z, and remove $X \times \hat{B}$ from $X \times Y$ to recover the key. The scheme is given in Figure 2.

The complexity of ProbGen is $O(n_1 n_2 + n_1 n_3 + n_2 n_3)$, i.e., linear in the size of the input. The complexity of Compute is dominated by two matrix multiplications resulting in $O(n_1 n_2 n_3)$ time. Lastly, the Verify algorithm performs a single string comparison of complexity $O(1)$ and outputs the matrix of size $n_1 \times n_3$.

Security of this solution holds only when each cell A_{ij} of matrix A takes a non-zero value. For that reason, we next describe a mechanism for encoding an arbitrary matrix M into an equivalent matrix M' that contains only non-zero values and decoding the result after M' is used in the computation.

Matrix encoding: Given M of dimensions $n_1 \times n_2$, choose any value ℓ such that $A_{ij} + \ell \neq 0$ for $1 \leq i \leq n_1$ and $1 \leq j \leq n_2$. To form M', add ℓ to each element of M, i.e., $M'_{ij} = M_{ij} + \ell$, and store ℓ for future reference.

$\mathsf{Setup}(1^{\kappa}, f)$: The same as in VC_m.

$\mathsf{ProbGen}(x = (A, B), \mathsf{params})$: On input two matrices A and B of respective dimensions $n_1 \times n_2$ and $n_2 \times n_3$, perform:

1. Choose $r_i \overset{R}{\leftarrow} \mathbb{Z}_p^*$ for $1 \leq i \leq n_1$, $d_j, m_j \overset{R}{\leftarrow} \mathbb{Z}_p^*$ for $1 \leq j \leq n_2$, and $c_k, w_k \overset{R}{\leftarrow} \mathbb{Z}_p^*$ for $1 \leq k \leq n_3$.
2. Compute $X_{ij} = g_1^{r_i/m_j A_{ij}}$ for $1 \leq i \leq n_1$ and $1 \leq j \leq n_2$.
3. Compute $Y_{ij} = g_2^{c_j m_i B_{ij} + w_j d_i}$ for $1 \leq i \leq n_2$ and $1 \leq j \leq n_3$.
4. Compute $Z_{ij} = g_T^{c_j r_i}$ for $1 \leq i \leq n_1$ and $1 \leq j \leq n_3$.
5. Compute $t_i = \sum_{k=1}^{n_2} A_{ik} d_k / m_k$ for $1 \leq i \leq n_1$, and $vk_{ij} = t_i r_i w_j$ for $1 \leq i \leq n_1$ and $1 \leq j \leq n_3$.
6. Compute $sk_{ij} = g_T^{vk_{ij}}$ for $1 \leq i \leq n_1$ and $1 \leq j \leq n_3$, set $sk_j = H(sk_{1j}||sk_{2j}||\ldots||sk_{n_1 j})$ for $1 \leq j \leq n_3$, and $sk = H(sk_1||sk_2||\ldots||sk_{n_3})$.
7. Set $\mathsf{SK}_x = sk$, $\mathsf{EK}_x = (\mathsf{params}, Z)$, and $\sigma_x = (A, B, X, Y)$.

$\mathsf{Compute}(\mathsf{EK}_x = (\mathsf{params}, Z), \sigma_x = (A, B, X, Y))$: Execute the following steps:

1. Compute (i) $V_{ij}^{(1)} = \sum_{k=1}^{n_2} A_{ik} B_{kj}$, (ii) $V_{ij}^{(2)} = \prod_{k=1}^{n_2} e(X_{ik}, Y_{kj}) = g_T^{\sum_{k=1}^{n_2} A_{ik}(c_j r_i B_{kj} + w_j d_k r_i / m_k)}$, and (iii) $\Delta_{ij} = V_{ij}^{(2)} / Z_{ij}^{V_{ij}^{(1)}}$ for $1 \leq j \leq n_3$ and $1 \leq i \leq n_1$.
2. Compute $\hat{sk}_j = H(\Delta_{1j}||\Delta_{2j}||\ldots||\Delta_{n_1 j})$ and $\hat{sk} = H(\hat{sk}_1||\hat{sk}_2||\ldots||\hat{sk}_{n_3})$.
3. Set $\sigma_y = (V^{(1)}, \hat{sk})$.

$\mathsf{Verify}(\mathsf{SK}_x = sk, \sigma_y = (V^{(1)}, \hat{sk}))$: Verify whether $\hat{sk} = sk$. If the check succeeds, output $V^{(1)}$; otherwise, output $\perp$.

Fig. 2. Description of the core scheme VC_r in presence of rational adversaries

Matrix decoding: Let $C' = M' \times N'$, where M' (N') is an encoded version of matrix M (resp., N) using value ℓ_1 (resp., ℓ_2) and has dimensions $n_1 \times n_2$ (resp., $n_2 \times n_3$). Observe that each element $C'_{ij} = \sum_{k=1}^{n_2} (M_{ik} + \ell_1)(N_{kj} + \ell_2)$. To recover $C = M \times N$, compute the offset Δ_{ij} for each element, which equals to $\ell_2 \sum_{k=1}^{n_2} M_{ik} + \ell_1 \sum_{k=1}^{n_2} N_{kj} + n_2 \ell_1 \ell_2$, and set $C_{ij} = C'_{ij} - \Delta_{ij}$. Note that the value $\ell_2 \sum_{k=1}^{n_2} A_{ik}$ ($\ell_1 \sum_{k=1}^{n_2} B_{kj}$) is the same for all elements in a single row of matrix M (resp., single column of matrix N). This means that we only need to compute that value for n_1 rows of matrix M (resp., n_3 columns of matrix N). The overall complexity of computing all offsets is therefore $O(n_1 n_2 + n_2 n_3)$ and the complexity of computing C from C' is $O(n_1 n_3)$.

The decoding computation is simplified when only one of the matrices used in the product was encoded to eliminate zero entries (as in VC_r). In that case, the offset becomes $\Delta_{ij} = \ell_1 \sum_{k=1}^{n_2} N_{kj}$ assuming that M was the encoded matrix, and the overall complexity of decoding is $O(n_1 n_3 + n_2 n_3)$.

Security of our VC_r scheme can be stated as follows:

Theorem 2. *If H is a collision-resistant hash function, the M-DDH and XDH assumptions hold, and all elements of A are non-zero, a server that deviates from the protocol can only pass verification with a probability negligible in κ.*

Assumption 2. *If in VC_r the server returns correct $\hat{sk} = sk$ but incorrect $V^{(1)} \neq C$, the error will be detected by the client with a non-negligible probability.*

This assumption is crucial to our result and can be realized by outsourcing a small (but non-negligible) fraction $0 < \rho < 1$ of all tasks to a second independent server. For example, for each task the client chooses random $v \in [0,1]$, and if $v \leq \rho$, the client uses two (non-colluding) servers for independently outsourcing the task and compares the returned products C afterwards. If the servers are non-colluding and at least one returned C is incorrect, the client will be able to detect misbehavior with non-negligible probability. This implies that if the server performed the work to pass the verification test, it will be incentivized to return correct C.

Theorem 3. *If Assumptions 1 and 2 and assumptions of Theorem 2 hold, VC_r is secure in presence of rational adversaries according to Definition 6.*

Due to space constraints, security proofs and correctness analysis are available only in the full version [34]. The above means that following the protocol and producing correct matrix products $A \times B$ and $X \times Y$ is the most profitable strategy for a rational adversary.

5.2   Description of the Enhanced Scheme

In this section, we propose an enhanced scheme that supports chaining and allows honest parties to pinpoint faulty cells in case of computation corruption or intentional deviation from the prescribed computation using a single mechanism. That is, recall that this feature makes it difficult for a dishonest server to continue with the next stage of the computation if it was not carried out correctly at the current stage and allows the client to efficiently identify the first stage at which a fault occurred. It also allows the cloud itself to detect a problem with the computation (in case of compromise or malware infection). The basic idea behind the solution is that the client divides the entire computation into n_3 sub-computations. The keys sk_i are formed as before, but now the ith key is used to encode the inputs of the $(i + 1)$th sub-computation. The server is able to recover the $(i + 1)$th sub-key only if it executes sub-computations $1, \ldots, i$ correctly. Upon computation completion, the verifier receives the last key from the server and examines its correctness. If the verifier notices a discrepancy between the returned key and its expected value, he will ask the server to return all the keys generated throughout the computation. The verifier then applies a procedure similar to binary search to locate the first incorrect key, which corresponds to the first sub-computation that has been executed incorrectly. This operation can be implemented in $O(\log n)$ steps when the number of sub-computations is n. The server will also be able to examine correctness of the first i sub-computations that have been executed so far by verifying that the inputs of $(i + 1)$th sub-computation could be decoded correctly using sk_i. If this check fails, this serves as a notice of the existence of faulty cells and the server can suspend the computation to identify faulty sells among the computed cells.

The detail of our solution can be described as follows: The client generates five matrices $\{A, B, X, Y, Z\}$ and forms keys sk_i as in the base scheme. Now the computation of the ith column of matrices $A \times B$ and $X \times Y$ is considered to be the ith sub-computation. The client then blinds each element of the $(i+1)$th

---

Setup($1^\kappa, f$): The same as in VC_r.

ProbGen($x = (A, B), \mathsf{params}$): Given matrices A and B, perform:
1. Compute X, Y, Z, and $\{sk_i\}_{i=1}^{n_3}$ as in VC_r.
2. Set $\hat{B}_{i1} = B_{i1}$ and $\hat{Y}_{i1} = Y_{i1}$ for $1 \le i \le n_2$; and also set
   - $\hat{B}_{ij} = \mathsf{PRF}_{sk_{j-1}}(i||j||0) \oplus (B_{ij}||0^\lambda)$
   - $\hat{Y}_{ij} = \mathsf{PRF}_{sk_{j-1}}(i||j||1) \oplus (Y_{ij}||0^\lambda)$

   for $1 \le i \le n_2$ and $2 \le j \le n_3$, where λ is a correctness parameter.
3. Set $\mathsf{SK}_x = \{sk_i\}_{i=1}^{n_3}$, $\mathsf{EK}_x = (\mathsf{params}, Z)$, and $\sigma_x = (A, \hat{B}, X, \hat{Y})$.

Compute($\mathsf{EK}_x = (\mathsf{params}, Z), \sigma_x = (A, \hat{B}, X, \hat{Y})$): Execute steps:
1. Set $U_{i1}^{(1)} = \hat{B}_{i1}$ and $U_{i1}^{(2)} = \hat{Y}_{i1}$ for $1 \le i \le n_2$.
2. For $j = 1, \ldots, n_3$ do:
   (a) For $1 \le i \le n_1$, compute $V_{ij}^{(1)} = \sum_{k=1}^{n_2} A_{ik} U_{kj}^{(1)}$ and $V_{ij}^{(2)} = \prod_{k=1}^{n_2} e(X_{ik}, U_{kj}^{(2)})$.
   (b) Let $\Delta_{ij} = V_{ij}^{(2)}/Z_{ij}^{V_{ij}^{(1)}}$ for $1 \le i \le n_1$. Set $\hat{sk}_j = H(\Delta_{1j}||\Delta_{2j}||\ldots||\Delta_{n_1 j})$.
   (c) If $j \ne n_3$, for $1 \le i \le n_2$ compute
       - $U_{ij+1}^{(1)}||W_1 = \hat{B}_{ij+1} \oplus \mathsf{PRF}_{\hat{sk}_j}(i||j+1||0)$
       - $U_{ij+1}^{(2)}||W_2 = \hat{Y}_{ij+1} \oplus \mathsf{PRF}_{\hat{sk}_j}(i||j+1||1)$

       If W_1 or W_2 is not equal to 0^λ, report an error and abort.
3. Set $\sigma_y = (V^{(1)}, \hat{sk}_{n_3})$.

Verify($\mathsf{SK}_x = \{sk_i\}_{i=1}^{n_3}, \sigma_y = (V^{(1)}, \hat{sk}_{n_3})$): Verify whether $\hat{sk}_{n_3} = sk_{n_3}$. If the check succeeds, output $V^{(1)}$. Otherwise, retrieve all $\{\hat{sk}_i\}_{i=1}^{n_3}$ from the server and find the smallest index i such that $\hat{sk}_i \ne sk_i$ using binary search.

Fig. 3. Description of scheme VC_r' that incorporates chaining and allows for fast location of an error in the computation

column of matrices B and Y by XORing them with a pseudo-random string. This string is produced using sk_i as the secret key to a pseudo-random function PRF, which is evaluated on the cell's row and column indices together with a unique identifier for each matrix to guarantee uniqueness of the input/output.

In order to remove blinding and recover the next sub-computation, the server needs to produce correct sk_i as before (by computing the ith column of two matrix products) and evaluate the PRF on that key to reproduce the random mask. This will allow the server to recover the inputs for the $(i + 1)$th column of matrix B and Y, i.e., the $(i + 1)$th sub-computation, and continue the computation. We make the size of the pseudo-random output of PRF longer than the size of matrix elements they mask so that the remaining bits can be used to verify that the input to the $(i+1)$th sub-computation was decoded correctly. That is, we append a zero string of a predefined size to each input before encoding, and the server can use it to verify that decoding was successful (and if it was not, investigate the reason for the failure). At the end of the computation, the server recovers and returns the last key sk_{n_3}, which the client compares to the expected key and accepts the output if the verification succeeds. The scheme is given in Figure 3.

The complexities of ProbGen and Compute are the same as in VC_r, i.e., $O(n_1 n_2 + n_1 n_3 + n_2 n_3)$ and $O(n_1 n_2 n_3)$, respectively. Verify now performs one string comparison of complexity $O(1)$ and outputs a matrix of size $n_1 \times n_3$ in case of no errors. Otherwise, the client retrieves n_3 keys and additionally performs $O(\log(n_3))$ work. Security is stated below and the proof can be found in [34].

Theorem 4. *If H is a collision-resistant hash function,* PRF *is a pseudo-random function, the M-DDH and XDH assumptions hold, and all elements of A are non-zero, the server that deviates from the correct protocol can only pass verification with a probability negligible in κ.*

As the corollary, we have that if the result does not verify, the client can identify the first faulty sub-computation using $O(\log(n_3))$ string comparisons.

6 Extensions

In this section, we sketch how to extend our schemes to incorporate privacy protection of input/output matrices and public verifiability. The details of the constructions and their analysis are available in [34].

Matrix Privacy. This property must ensure that the server is unable to learn any information about A and B and their product $A \times B$. We thus modify both schemes to achieve this goal, while still preserving computation verifiability.

Malicious Setting. To protect matrices A and B in σ_x, our solution encodes them using a homomorphic encryption scheme (e.g., BGN encryption [11]) that supports one multiplication and an unlimited number of additions on encrypted messages. This allows for matrix multiplication $A \times B$ to be privately carried out on encrypted data. The server also sees X and Y, but no changes to Y are necessary. That is, the term $g_2^{w_j d_i}$ protects each element of matrix B that matrix Y encodes, assuming that the M-DDH assumption holds. Matrix X, however, in its original form can disclose information about the elements of A. To prevent this disclosure, we encode its elements in a form similar to that of matrix Y, by incorporating a random term $g_1^{h_j u_i}$ into each element of X.

Rational Setting. Unlike VC_m, where the computation of product matrix C and verification value s can be carried out independently, in VC_r the server needs to perform these two computations together in order to produce correct key $\hat{sk}$ used for verification purposes. As a direct consequence of the difference, we no longer can apply an arbitrary encryption algorithm to matrices A and B because randomness used in ciphertexts will lead to the delegator's inability to properly compute $\hat{sk}$. To resolve the issue, we encode A and B using a similar mechanism to that of forming matrices X and Y. That is, we use the product of two newly generated random values (such as $w_j d_i$ in Y_{ij}) to protect each individual element in A and B, and update vk_{ij} accordingly to allow for correct verification.

Public Verifiability. Recall that public verifiability allows any entity to verify correctness of the returned result using a public verification key. To incorporate public verifiability into VC_m and VC_r, the client will now need to produce a public verification key PVK_x as a public version of SK_x at ProbGen time, which will roughly be in the form of g^{SK_x}. This key then can be utilized by any auditor at Verify time to assess correctness of the result of outsourced computation.

Rational Setting. Recall that in VC_r, client's secret key SK_x consists of only key sk. Therefore, all that is needed to convert the solution into a publicly verifiable scheme is to make a public version of the key, g^{sk}, publicly available. Then, the verification consists of checking whether $g^{sk} = g^{\hat{sk}}$, where, as before, $\hat{sk}$ is produced by the server during Compute.

Malicious Setting. This time, unlike the rational setting, achieving public verifiability by setting PVK_x to be g^v for every $v \in \mathsf{SK}_x$ does not work. Doing so would reveal crucial key information, which can be easily exploited by the server to compromise integrity of the returned result. In particular, suppose we set PVK_x to consist of $g_T^{r_i c_j}$ for $1 \le i \le n_1$ and $1 \le j \le n_3$ and $g_T^{\sum_{k=1}^{n_1} r_i v k_i}$, and let Verify consists of checking whether $\prod_{i=1}^{n_1} \prod_{j=1}^{n_3} (g_T^{r_i c_j})^{C_{ij}} + g_T^{\sum_{k=1}^{n_1} r_i v k_i} = s$ (where s is returned by Compute).[1] To launch an attack, the adversary first correctly computes C and s, then changes an arbitrary matrix element from C_{ij} to $C_{ij} + \delta$, and multiplies s by $g_T^{r_i c_j \delta}$. This allows the adversary to produce a tuple $(\hat{C}, \hat{s})$ that differs from the correct (C, s), but nevertheless passes verification. Notice that this attack is infeasible in the original VC_m scheme as the adversary has no information about $g_T^{r_i c_j}$ and hence is unable to correctly produce $\hat{s}$. Therefore, further changes are needed to support public verifiability in the malicious setting.

The idea behind the modification is to make $g_T^{r_i c_j}$'s and s belong to two different groups. That is, we make the values of the form $g^{r_i c_j}$ available only in $\mathbb{G}_T$, while s will have to be produced as a number of elements s_i in $\mathbb{G}_2$ (and there is no efficiently computable homomorphism from an element of $\mathbb{G}_T$ to an element of $\mathbb{G}_2$). Then in the construction the delegator produces PVK_x as of $g_1^{r_i}$, $g_T^{r_i c_j}$ and $g_T^{\sum_{i=1}^{n_1} r_i v k_i}$ and provides only three matrices A, B, Y in σ_x to the server. The server computes each $s_i \in \mathbb{G}_2$ by performing a modulo exponentiation using A and Y (instead of the pairing operation in VC_m) and returns them to the client. Because verification of the result now consists of checking whether the product of all $(g_T^{r_i c_j})^{C_{ij}}$'s and $g_T^{\sum_{i=1}^{n_1} r_i v k_i}$ matches the product of $e(g_1^{r_i}, s_i)$'s, it is no longer feasible for the adversary to succeed in the above attack.

The introduced modifications can also be applied to the version of the scheme with data privacy in both adversarial settings to obtain verifiable computation schemes with public delegatability and verifiability and privacy protection. Due to space constraints, detailed constructions of these schemes are provided in [34].

7   Conclusions

This work presents schemes for verifiable outsourcing of matrix multiplications in both malicious and rational adversary models. The complexity and features of our schemes favorably compare to the state of the art, with the solution in the rational setting having a very low verification cost of only a single comparison. Our basic constructions achieve public delegatability and can be extended with

---

[1] Note that because s is returned as an element of $\mathbb{G}_T$ and each C_{ij} is returned as an element in $\mathbb{Z}_p^*$, the values $g_T^{r_i c_j}$ for each i and j and $g_T^{\sum_{k=1}^{n_1} r_i v k_i}$ represent the minimum information the verifier needs to possess to carry out verification.

features of data protection, public verifiability and chaining (supporting all or a subset of the features).

Acknowledgments. This work was supported in part by grants CNS-1319090 and CNS-1223699 from the National Science Foundation and FA9550-13-1-0066 from the Air Force Office of Scientific Research. Any opinions, findings, and conclusions or recommendations expressed in this publication are those of the authors and do not necessarily reflect the views of the funding agencies.

References

1. Agrawal, S., Boneh, D.: Homomorphic MACs: MAC-based integrity for network coding. In: Abdalla, M., Pointcheval, D., Fouque, P.-A., Vergnaud, D. (eds.) ACNS 2009. LNCS, vol. 5536, pp. 292–305. Springer, Heidelberg (2009)
2. Applebaum, B., Ishai, Y., Kushilevitz, E.: From secrecy to soundness: Efficient verification via secure computation. In: Abramsky, S., Gavoille, C., Kirchner, C., Meyer auf der Heide, F., Spirakis, P.G. (eds.) ICALP 2010, Part I. LNCS, vol. 6198, pp. 152–163. Springer, Heidelberg (2010)
3. Atallah, M., Blanton, M. (eds.): Algorithms and Theory of Computation Handbook. General Concepts and Techniques, vol. I, ch. 17. CRC Press (2009)
4. Atallah, M., Frikken, K.: Securely outsourcing linear algebra computations. In: ASIACCS, pp. 48–59 (2010)
5. Atallah, M., Frikken, K., Wang, S.: Private outsourcing of matrix multiplication over closed semi-rings. In: SECRYPT, pp. 136–144 (2012)
6. Azar, P.D., Micali, S.: Rational proofs. In: STOC, pp. 1017–1028 (2012)
7. Azar, P.D., Micali, S.: Super efficient rational proofs. In: EC, pp. 29–30 (2013)
8. Babai, L.: Trading group theory for randomness. In: STOC, pp. 421–429 (1985)
9. Backes, M., Fiore, D., Reischuk, R.M.: Verifiable delegation of computation on outsourced data. In: CCS, pp. 863–874 (2013)
10. Ballard, L., Green, M., Medeiros, B., Monrose, F.: Correlation-resistant storage via keyword-searchable encryption. Cryptology ePrint Archive 2005/417 (2005)
11. Boneh, D., Goh, E.-J., Nissim, K.: Evaluating 2-DNF formulas on ciphertexts. In: Kilian, J. (ed.) TCC 2005. LNCS, vol. 3378, pp. 325–341. Springer, Heidelberg (2005)
12. Bresson, E., Chevassut, O., Pointcheval, D.: Dynamic group diffie-hellman key exchange under standard assumptions. In: Knudsen, L.R. (ed.) EUROCRYPT 2002. LNCS, vol. 2332, pp. 321–336. Springer, Heidelberg (2002)
13. Catalano, D., Fiore, D., Gennaro, R., Vamvourellis, K.: Algebraic (trapdoor) one-way functions and their applications. In: Sahai, A. (ed.) TCC 2013. LNCS, vol. 7785, pp. 680–699. Springer, Heidelberg (2013)
14. CertiVox. Benchmarks and Subs performance for MIRACL library, https://certivox.org/display/EXT/Benchmarks+and+Subs
15. Chung, K.-M., Kalai, Y.T., Liu, F.-H., Raz, R.: Memory delegation. In: Rogaway, P. (ed.) CRYPTO 2011. LNCS, vol. 6841, pp. 151–168. Springer, Heidelberg (2011)
16. Chung, K.-M., Kalai, Y., Vadhan, S.: Improved delegation of computation using fully homomorphic encryption. In: Rabin, T. (ed.) CRYPTO 2010. LNCS, vol. 6223, pp. 483–501. Springer, Heidelberg (2010)
17. Dodis, Y., Halevi, S., Rabin, T.: A cryptographic solution to a game theoretic problem. In: Bellare, M. (ed.) CRYPTO 2000. LNCS, vol. 1880, pp. 112–130. Springer, Heidelberg (2000)

18. Fiore, D., Gennaro, R.: Publicly verifiable delegation of large polynomials and matrix computations, with applications. In: CCS, pp. 501–512 (2012)
19. Garay, J., Katz, J., Maurer, U., Tackmann, B., Zikas, V.: Rational protocol design: Cryptography against incentive-driven adversaries. In: FOCS, pp. 648–657 (2013)
20. Gennaro, R., Gentry, C., Parno, B.: Non-interactive verifiable computing: Outsourcing computation to untrusted workers. In: Rabin, T. (ed.) CRYPTO 2010. LNCS, vol. 6223, pp. 465–482. Springer, Heidelberg (2010)
21. Goldwasser, S., Kalai, Y.T., Rothblum, G.N.: Delegating computation: Interactive proofs for muggles. In: STOC, pp. 113–122 (2008)
22. Goldwasser, S., Micali, S., Rackoff, C.: The knowledge complexity of interactive proof systems. SIAM Journal on Computing 18(1), 186–208 (1989)
23. Guo, S., Hubáček, S., Rosen, A., Vald, M.: Rational arguments: Single round delegation with sublinear verification. In: ITCS, pp. 523–540 (2014)
24. Halpern, J., Teague, V.: Rational secret sharing and multiparty computation: Extended abstract. In: STOC, pp. 623–632 (2004)
25. Johnson, R., Molnar, D., Song, D., Wagner, D.: Homomorphic signature schemes. In: Preneel, B. (ed.) CT-RSA 2002. LNCS, vol. 2271, pp. 244–262. Springer, Heidelberg (2002)
26. Kilian, J.: A note on efficient zero-knowledge proofs and arguments (extended abstract). In: STOC, pp. 723–732 (1992)
27. Kilian, J.: Improved efficient arguments (Preliminary version). In: Coppersmith, D. (ed.) CRYPTO 1995. LNCS, vol. 963, pp. 311–324. Springer, Heidelberg (1995)
28. Landerman, J., Pan, V., Sha, X.-H.: On practical algorithms for accelerated matrix mutiplication. Linear Algebra and Its Applications 162–164, 557–588
29. Micali, S.: CS proofs. In: FOCS, pp. 436–453 (1994)
30. Mohassel, P.: Efficient and secure delegation of linear algebra. Cryptology ePrint Archive 2011/605 (2011)
31. Parno, B., Howell, J., Gentry, C., Raykova, M.: Pinocchio: Nearly practical verifiable computation. In: IEEE Symposium on Security and Privacy (2013)
32. Parno, B., Raykova, M., Vaikuntanathan, V.: How to delegate and verify in public: Verifiable computation from attribute-based encryption. In: Cramer, R. (ed.) TCC 2012. LNCS, vol. 7194, pp. 422–439. Springer, Heidelberg (2012)
33. Zhang, L.F., Safavi-Naini, R.: Private outsourcing of polynomial evaluation and matrix multiplication using multilinear maps. In: Abdalla, M., Nita-Rotaru, C., Dahab, R. (eds.) CANS 2013. LNCS, vol. 8257, pp. 329–348. Springer, Heidelberg (2013)
34. Zhang, Y., Blanton, M.: Efficient secure and verifiable outsourcing of matrix multiplications. Cryptology ePrint Archive 2014/133 (2014)

A   Security Analysis of VC_m Scheme

To demonstrate correctness of VC_m construction, we show that if the computation was performed correctly, Verify outputs product $A \times B$. In Verify, we have:

$$g_T^{\sum_{i=1}^{n_1} r_i \sum_{j=1}^{n_3} c_j C_{ij} + vk_i} = \prod_{i=1}^{n_1} g_T^{r_i \sum_{j=1}^{n_3} c_j C_{ij} + vk_i} = \prod_{i=1}^{n_1} \prod_{j=1}^{n_3} g_T^{r_i c_j C_{ij} + r_i t_i w_j}$$

$$= \prod_{i=1}^{n_1} \prod_{j=1}^{n_3} \prod_{k=1}^{n_2} e(g_1^{r_i A_{ik}}, g_2^{c_j B_{kj} + w_j d_k}) = \prod_{i=1}^{n_1} \prod_{j=1}^{n_3} \prod_{k=1}^{n_2} e(X_{ik}, Y_{kj}) = s$$

Proof (Theorem 1). Our proof follows the hybrid argument. We start with the security experiment $\mathbf{Exp}_{\mathcal{A}}^{\mathsf{Ver}}(\mathsf{VC}_m, f, \kappa)$ and devise a sequence of security games, where the adversary $\mathcal{A}$'s view in one game is indistinguishable from its view in another game. We analyze $\mathcal{A}$'s advantage $\mathbf{Adv}_{\mathcal{A}}^{\mathsf{Ver}}(\mathsf{VC}_m, f, q, \kappa)$ in winning the experiment. Let T_i denote the event that the security experiment returns 1 in game $\mathbf{G_i}$. The security games are defined as follows:

Game $\mathbf{G_0}$. Define $\mathbf{G_0}$ to be the same as $\mathbf{Exp}_{\mathcal{A}}^{\mathsf{Ver}}(\mathsf{VC}_m, f, \kappa)$.

Game $\mathbf{G_1}$. The game is identical to $\mathbf{G_0}$, except that when generating Y_{ij}, the delegator will use random value $r_{ij}^{(1)}$ in $\mathbb{Z}_p$ instead of $w_j d_i$. In other words, each Y_{ij} is formed as $g_2^{c_j B_{ij} + r_{ij}^{(1)}}$ as opposed to $g_2^{c_j B_{ij} + w_j d_i}$ in game $\mathbf{G_0}$. To be able to verify the result of computation, the delegator also changes each vk_i in SK_x from its original value $r_i \sum_{k=1}^{n_2} \sum_{j=1}^{n_3} A_{ik} d_k w_j$ to $r_i \sum_{k=1}^{n_2} \sum_{j=1}^{n_3} A_{ik} r_{kj}^{(1)}$, while keeping the remaining portion of VC_m construction unchanged. This will allow the delegator to verify the result of computation without any changes to Verify.

Comparing $\mathcal{A}$'s view in games $\mathbf{G_0}$ and $\mathbf{G_1}$, we have that $g_1^{w_j d_i}$'s are replaced with random group elements. Now notice that all $g_1^{w_j d_i}$'s collectively form a partial $(n_2 + n_3)$-M-DDH tuple. This gives us that the advantage $\mathcal{A}$ has in game $\mathbf{G_0}$ is at most $\mathbf{Adv}_{\mathcal{A}}^{\mathsf{M\text{-}DDH}}(\kappa)$ larger than in game $\mathbf{G_1}$ and thus any non-negligible difference in the adversary's behavior between the games $\mathbf{G_0}$ and $\mathbf{G_1}$ can be used to break the M-DDH assumption. Therefore, we have that $|\Pr[T_1] - \Pr[T_0]| \leq \mathbf{Adv}_{\mathcal{A}}^{\mathsf{M\text{-}DDH}}(\kappa)$ and based on our assumption that the M-DDH problem is hard the difference in the adversary's view between games $\mathbf{G_0}$ and $\mathbf{G_1}$ is negligible.

Game $\mathbf{G_2}$. The game is identical to $\mathbf{G_1}$ except the delegator removes information about c_j and B_{ij} from each Y_{ij}, i.e., $Y_{ij} = g_2^{r_{ij}^{(1)}}$ instead of $g_2^{c_j B_{ij} + r_{ij}^{(1)}}$. To be able to verify the result of computation, we also update SK_x to compensate for the difference in Y_{ij}'s. We thus add the difference in the vk_i's $\sum_{i=1}^{n_1} \sum_{j=1}^{n_3} r_i c_j (\sum_{k=1}^{n_2} A_{ik} B_{kj})$ to the value of vk_1 in $\mathbf{G_1}$ while keeping the remaining vk_i's the same as in $\mathbf{G_1}$ (note that there are other possibilities because only $\sum_i vk_i$'s is used). Because the $r_{ij}^{(1)}$'s are completely random, the distribution of the Y_{ij}'s in $\mathbf{G_1}$ and in $\mathbf{G_2}$ is identical and thus $\Pr[T_2] = \Pr[T_1]$. Now observe that we removed any information about the c_j's from $\mathcal{A}$'s view.

Let us next analyze $\mathcal{A}$'s success in winning $\mathbf{Exp}_{\mathcal{A}}^{\mathsf{Ver}}(\mathsf{VC}_m, f, \kappa)$ in $\mathbf{G_2}$. Assuming the XDH assumption is true, the M-DDH problem is hard in our setting with bilinear maps, i.e., it is hard in $\mathbb{G}_T$. Thus, while $\mathcal{A}$ can compute $g_T^{r_i}$ for each i, $(\{g_T^{r_i}\}_{1 \leq i \leq n_1}, \{g_T^{r_i c_j}\}_{1 \leq i \leq n_1, 1 \leq j \leq n_3})$ is a partial $(n_1 + n_3)$-M-DDH tuple and $\mathcal{A}$ can have only a negligible advantage in distinguishing the $g_T^{r_i c_j}$'s from random group elements. Now suppose that $\mathcal{A}$ was able to return a tuple $(\hat{C}, \hat{s})$ that differs from correct (C, s), but nevertheless passes verification. Then the returned value satisfies the equation $g_T^{\sum_{i=1}^{n_1} \sum_{j=1}^{n_3} r_i c_j (C_{ij} - \hat{C}_{ij})} = s/\hat{s}$. Because the server has no information about the c_j's and furthermore is unable to distinguish $g_T^{r_i c_j}$'s from random group elements, the only way for $\mathcal{A}$ to create simultaneously valid $C_{ij} - \hat{C}_{ij}$ and $s/\hat{s}$ is to correctly guess the value of $g_T^{r_i c_j}$. The probability of this happening is, however, negligible in κ and thus $\Pr[T_2]$ is negligible as well.

Combined with the previous analysis of the differences in the adversarial success between games $\mathbf{G_0}$ and $\mathbf{G_2}$, we obtain that $\mathcal{A}$'s advantage is negligible in winning the experiment $\mathbf{Exp}_{\mathcal{A}}^{\mathsf{Ver}}(\mathsf{VC}_m, f, \kappa)$ as desired. $\square$

Hybrid Model of Fixed and Floating Point Numbers in Secure Multiparty Computations

Toomas Krips[2,3] and Jan Willemson[1,3]

[1] Cybernetica, Ülikooli 2, Tartu, Estonia
janwil@cyber.ee
[2] Institute of Computer Science, University of Tartu, Liivi 2, Tartu, Estonia
[3] STACC, Ülikooli 2, Tartu, Estonia
toomaskrips@gmail.com

Abstract. This paper develops a new hybrid model of floating point numbers suitable for operations in secure multi-party computations. The basic idea is to consider the significand of the floating point number as a fixed point number and implement elementary function applications separately of the significand. This gives the greatest performance gain for the power functions (e.g. inverse and square root), with computation speeds improving up to 18 times in certain configurations. Also other functions (like exponent and Gaussian error function) allow for the corresponding optimisation.

We have proposed new polynomials for approximation, and implemented and benchmarked all our algorithms on the Sharemind secure multi-party computation framework.

1 Introduction

Our contemporary society is growing more and more dependent on high-speed, high-volume data access. On one hand, such an access allows for developing novel applications providing services that were unimaginable just a decade ago. On the other hand, constant data flow and its automatic processing mechanisms are rising new security concerns every day.

In order to profit from the available data, but at the same time provide privacy protection for citizens, *privacy-preserving data analysis* (PPDA) mechanisms need to be applied. There exist numerous well-established statistical and data mining methods for data analysis. However, adding privacy-preservation features to them is far from being trivial. Many data processing primitives assume access to micro-data records, e.g. for joining different tables or even something as simple as sorting. There exist different methods for partial pre-aggregation and perturbation like k-anonymity [18,20] and ℓ-diversity [17], but they reduce the precision of the dataset, and consequently decrease data utility.

Another approach is to tackle the PPDA problem from the privacy and cryptography point of view. Unfortunately, classical encryption methods (like block and stream ciphers) are meant only to scramble data and do not support meaningful computations on the plaintexts. More advanced methods like homomorphic

S.S.M. Chow et al. (Eds.): ISC 2014, LNCS 8783, pp. 179–197, 2014.
© Springer International Publishing Switzerland 2014

encryption and searchable encryption [4] support some limited set of operations insufficient for the fully-featured statistical data analysis. There also exist methods for fully homomorphic encryption, but they are currently too inefficient to allow for analysis of a dataset of even a remotely useful size [9, 10].

Currently, one of the most promising techniques for cryptography-based PPDA is based on secret sharing and multi-party computations (SMC). There exist several frameworks allowing to work on relatively large amounts of secret-shared micro-data [2, 21]. In order to obtain the homomorphic behavior needed for Turing-completeness, they work over some algebraic structure (typically, a finite ring or field). However, to use the full variety of existing statistical tools, computations over real numbers are needed. Recently, several implementations of real-number arithmetic (both fixed and floating point) have emerged on top of SMC frameworks. While fixed point arithmetic is faster, floating point operations provide greater precision and flexibility. The focus of this paper is to explore the possibility of getting the best of both of the approaches and develop a hybrid fixed-floating point real numbers to be used with SMC applications.

2   Previous Work

Catrina and Saxena developed secure multiparty arithmetic on fixed-point numbers in [7], and their framework was extended with various computational primitives (like inversion and square root) in [7] and [15]. This fixed-point approach has been used to solve linear programming problems with applications in secure supply chain management [6, 12]. However, fixed point numbers provide only a limited amount of flexibility in computations, since they can represent values only in a small interval with a predetermined precision. Dahl $et\ al.$ [8] use an approach that is rather close to fixed-point numbers to perform secure two-party integer division. They also use Taylor series to estimate $\frac{1}{x}$.

In order to access the full power of numerical methods, one needs an implementation of floating point arithmetic. This has been done by three groups of authors, Aliasgari $et\ al.$ [1], Liu $et\ al.$ [16], and Kamm and Willemson [11]. All these approaches follow the same basic pattern – the floating point number x is represented as $x = s \cdot f \cdot 2^e$, where s is the sign, f is the significand, and e is the exponent (possibly adjusted by a bias to keep the exponent positive). Additionally, Aliasgari $et\ al.$ add a term to mark that the value of the floating point number is zero. Then all the authors proceed to build elementary operations of addition and multiplication, followed by some selection of more complicated functions.

Liu $et\ al.$ [16] consider two-party additive secret sharing over a ring $\mathbb{Z}_N$ and only develop addition, subtraction, multiplication and division. Aliasgari $et\ al.$ [1], use a threshold (t, n)-secret-sharing over a finite field and also develop several elementary functions such as logarithm, square root and exponentiation of floating-point numbers. All their elementary function implementations use different methods – square root is computed iteratively, logarithm is computed using a Taylor series and in order to compute the exponent, several $ad\ hoc$ techniques are applied.

The research of Kamm and Willemson is motivated by a specific application scenario – satellite collision analysis [11]. In order to implement it, they need several elementary functions like inversion, square root, exponent and Gaussian error function. The authors develop a generic polynomial evaluation framework and use both Taylor and Chebyshev polynomials to get the respective numerical approximations.

2.1 Our Contribution

When evaluating elementary functions, both [1] and [11] use basic floating point operations as monolithic. However, this is not necessarily optimal, since oblivious floating point addition is a very expensive operation due to the need to align the points of the addends in an oblivious fashion. Fixed-point addition at the same time is a local (i.e. essentially free) operation, if an additively homomorphic secret sharing scheme is used. Hence, we may gain speedup in computation times if we are able to perform parts of the computations in the fixed-point representation. For example, in order to compute power functions (like inversion or square root), we can run the computations separately on the significand and exponent parts, but the significand is essentially a fixed-point number. Proposing, implementing and benchmarking this optimisation is the main contribution of this paper. We also propose new polynomials for various elementary functions to provide better precision-speed trade-offs.

Due to space restrictions, some of the technical details (most notably the particular polynomials) are omitted and are available in the full version of the paper [13].

3 Preliminaries

In the rest of the paper, we will assume a secret sharing scheme involving M parties $P_1, \ldots, P_M$. To share a value x belonging to ring (or field) $\mathbb{Z}_r$, it is split into M values $x_1, \ldots, x_M \in \mathbb{Z}_r$, and the share x_i is given to the party P_i ($i = 1, \ldots, M$). The secret shared vector $(x_1, \ldots, x_M)$ will be denoted as $[\![x]\!]$.

We will also assume that the secret sharing scheme is linear, implying that adding two shared values and multiplying a shared value by a scalar may be implemented component-wise, and hence require no communication between the computing parties. This is essential, since the running times of majority of SMC applications are dominated by the network communication. Note that many of the classical secret sharing schemes (like Shamir or additive scheme) are linear.

We will assume availability of the following elementary operations.

- Addition of two secret-shared values $[\![x]\!]$ and $[\![y]\!]$ denoted as $[\![x]\!] + [\![y]\!]$. Due to linearity, this evaluates to $[\![x + y]\!]$.
- Multiplication of a secret shared value $[\![x]\!]$ by a scalar $c \in \mathbb{Z}_r$ denoted as $c \cdot [\![x]\!]$. Due to linearity, this evaluates to $[\![c \cdot x]\!]$.

- Multiplication of two secret-shared values $[\![x]\!]$ and $[\![y]\!]$ denoted as $[\![x]\!] \cdot [\![y]\!]$. Unlike the two previous protocols, this one requires network communication to evaluate $[\![x \cdot y]\!]$.
- PublicBitShiftRightProtocol($[\![x]\!], k$). Takes a secret shared value $[\![x]\!]$ and a public integer k and outputs $[\![x \gg k]\!]$ where $x \gg k$ is equal to x shifted right by k bits. $x \gg k$ is equal to $\frac{x}{2^k}$ rounded down.
- LTEProtocol($[\![x]\!], [\![y]\!]$). Gets two secret-shared values $[\![x]\!]$ and $[\![y]\!]$ as inputs and outputs a secret-shared bit $[\![b]\!]$. The bit b is set to 1 if $x \leq y$ (interpreted as integers); otherwise, b is set to 0.
- ObliviousChoiceProtocol($[\![b]\!], [\![x]\!], [\![y]\!]$). Gets a secret-shared bit b and two values $[\![x]\!]$ and $[\![y]\!]$ as inputs. If $b = 1$, the output will be set to $[\![x]\!]$, and if $b = 0$, it will be set to $[\![y]\!]$.
- ConvertToBoolean($[\![x]\!]$).Takes in a secret-shared value $[\![x]\!]$ where x is equal to either 0 or 1, and converts it to the corresponding boolean value shared over $\mathbb{Z}_2$.
- ConvertBoolToInt($[\![b]\!]$). Takes in a bit $[\![b]\!]$ secret-shared over $\mathbb{Z}_2$ and outputs a value $[\![x]\!]$ secret-shared over $\mathbb{Z}_r$, where x is equal to b as an integer.
- GeneralizedObliviousChoice($[\![x_1]\!], \ldots, [\![x_k]\!], [\![\ell]\!]$). Takes an array of secret integers $[\![x_1]\!], \ldots, [\![x_k]\!]$ and a secret index $[\![\ell]\!]$ where $\ell \in [1, k]$, and outputs the shared integer $[\![x_\ell]\!]$.
- BitExtraction($[\![x]\!]$). Takes in a secret integer $[\![x]\!]$ and outputs the vector of n secret values $\{[\![u_i]\!]\}_{i=0}^{n-1}$ where each $u_i \in \{0, 1\}$ and $u_{n-1}u_{n-2} \ldots u_0$ is the bitwise representation of $[\![x]\!]$.
- PrivateBitShiftRightProtocol($[\![x]\!], [\![k]\!]$) Takes a secret value $[\![x]\!]$ and a secret integer $[\![k]\!]$ and outputs $[\![x \gg k]\!]$ where $x \gg n$ is equal to x shifted right by k bits. When we apply this protocol to an n-bit secret integer $[\![x]\!]$ and k is not among $0, \ldots, n-1$, the result will be $[\![0]\!]$.
- ConvertToFloat($[\![x]\!]$) Takes a secret integer $[\![x]\!]$ and outputs a floating point number $[\![N]\!] = ([\![s_x]\!], [\![E_x]\!], [\![f_x]\!])$ that is approximately equal to that integer.

Implementation details of these elementary operations depend on the underlying SMC platform. The respective specifications for Sharemind SMC engine and the complexities of some of the protocols can be found in [2,3,14].

4   Fixed-Point Numbers

Our fixed-point arithmetic follows the framework of Catrina and Saxena [7], (for example, our multiplication protocol is based on that paper) but has several simplifications allowing for a more efficient software implementation.

First, instead of a finite field, we will be using a ring $\mathbb{Z}_{2^n}$ for embedding the fixed-point representations. Typically this ring will be $\mathbb{Z}_{2^{32}}$ or $\mathbb{Z}_{2^{64}}$, since arithmetic in these rings is readily available in modern computer architectures. What we will lose is the possibility of using secret sharing over fields (including the popular Shamir's scheme [19]). Since our implementation will be based on Sharemind SMC engine [2], this is fine and we can use additive secret sharing instead.

The second simplification is made possible by our specific application of fixed-point numbers. The essential block we will need to build is polynomial evaluation on non-negative fixed point numbers (e.g. significands of floats). Even though we will occasionally need to cope with negative values, we will only represent non-negative fixed-point numbers. Besides the ring $\mathbb{Z}_{2^n}$ of n-bit integers, we will also fix the number m of bits we will interpret as the fractional part. We will consider the ring elements as unsigned, hence they run over the range $[0, 2^n - 1]$. We let the element $x \in \mathbb{Z}_{2^n}$ represent the fixed point number $x \cdot 2^{-m}$. Hence, the range of fixed point numbers we will be able to represent is $[0, 2^{n-m} - 2^{-m}]$, with granularity 2^{-m}. We will assume that all the fixed-point numbers we work on will be among these numbers. If we have to use some fractional number that cannot be represented in this way, we will automatically use the smallest representable fixed-point number that is greater than the number instead of this number.

We will use the following notation for fixed-point numbers. $\widetilde{x}$ denotes a fixed-point number, while x denotes the integer value we use to store $\widetilde{x}$ — namely, $\widetilde{x} \cdot 2^m$. Thus, when we have introduced some integer x, we have also defined the fixed-point number $\widetilde{x} = x \cdot 2^{-m}$ that it represents and vice versa. Likewise, when we want to denote a secret fixed-point number, we will write $[\![\widetilde{x}]\!]$ — this will be stored as a secret integer $[\![x]\!]$ where $x = \widetilde{x} \cdot 2^m$.

We will also need to denote numbers that are, in essence, public signed real numbers. For that, we will use the notation $s\widetilde{c}$ where $\widetilde{c}$ is the fixed-point number that denotes the absolute value of the real number and $s \in \{-1, 1\}$ is the sign of the real number.

4.1   Basic Operations on Fixed-Point Numbers

We will now introduce the operations of addition and subtraction of two secret fixed-point numbers, multiplication of a secret fixed-point number and a public fixed point number and multiplication of two secret fixed-point numbers.

Addition of two secret fixed-point numbers $[\![\widetilde{x}]\!]$ and $[\![\widetilde{y}]\!]$ is free in terms of network communication, since this addition can be implemented by adding the representing values shared as the ring elements. Indeed, the sum of $\widetilde{x} = x \cdot 2^{-m}$ and $\widetilde{y} = y \cdot 2^{-m}$ is $(x+y) \cdot 2^{-m} = \widetilde{x + y}$. Hence we can compute $[\![\widetilde{x}]\!] + [\![\widetilde{y}]\!] = [\![\widetilde{x + y}]\!]$ just by adding the shares locally. The addition of the representatives takes place modulo 2^n and is unprotected against the overflow since checking whether the sum is too big would either leak information or would be expensive. Likewise, subtraction of two secret fixed-point numbers $[\![\widetilde{x}]\!]$ and $[\![\widetilde{y}]\!]$ is free in terms of network communication and can be implemented by subtracting the representing values shared as the ring elements. $[\![\widetilde{x - y}]\!]$ can be computed as $[\![\widetilde{x}]\!] - [\![\widetilde{y}]\!]$. The subtraction operation is also unprotected against going out of the range of the fixed-point numbers that we can represent and thus must be used only when it is known that $x \geq y$.

However, multiplication of a secret fixed-point number by other fixed point numbers, whether public or secret, is not free. Consider first multiplication of a public fixed-point number $\widetilde{a} = a \cdot 2^{-m}$ by a secret fixed-point number $[\![\widetilde{x}]\!] = [\![x]\!] \cdot 2^{-m}$. We need to calculate $[\![\widetilde{y}]\!]$ as the product of $\widetilde{a} = a \cdot 2^{-m}$ and $[\![\widetilde{x}]\!] =$

$[\![x]\!] \cdot 2^{-m}$ where x is secret. Since we keep data as a and $[\![x]\!]$, we shall perform this computation as $a \cdot [\![x]\!] = \tilde{a}2^m \cdot [\![\tilde{x}]\!]2^m$. However, if we do this multiplication in $\mathbb{Z}_{2^n}$, then we risk losing the most significant bits, since the product $\tilde{a}2^m\tilde{x}2^m$ might be greater than 2^n.

In order to solve this problem, we convert a and $[\![x]\!]$ to $\mathbb{Z}_{2^{2n}}$ and compute the product in $\mathbb{Z}_{2^{2n}}$. Then we shift the product to the right by m bits and convert the number back to $\mathbb{Z}_{2^n}$, since the secret result y should be $\tilde{a}[\![\tilde{x}]\!] \cdot 2^m$, not $\tilde{a}[\![\tilde{x}]\!] \cdot 2^{2m}$. We assume that the product is in the range of the fixed-point numbers we can represent. We do not perform any checks to see that the multiplicands or the product are in the correct range, as this could leak information about the secret data, but instead assume that the user will adequately choose the input. After computing $a \cdot [\![x]\!] = \tilde{a}2^m \cdot [\![\tilde{x}]\!]2^m$ in $\mathbb{Z}_{2^{2n}}$, we note that the result should be $a \cdot [\![\tilde{x}]\!]2^m$ and thus we need to divide the result by 2^m. The cheapest way to do this is shifting the numbers to the right by m bits.

There are two ways for doing that. The first one is using the existing protocol PublicBitShiftRightProtocol($[\![y]\!], m$) for shifting bits to the right. This protocol is not free, but gives the best possible result that can be represented with a given granularity and is guaranteed to give the correct result. The second one is to shift y_i to the right by m bits for every party P_i. This is free, but may be slightly inaccurate. Due to loss of the carry in the lowest bits we risk that the result might be smaller than the real product would be by at most $(M-1) \cdot 2^{-m}$. In most cases, this is an acceptable error. The only case where this error is significant is when our result should be among $[\![\widetilde{0}]\!], [\![\widetilde{2^{-m}}]\!], \ldots, (M-1)[\![\widetilde{2^{-m}}]\!]$ which could then be changed into one of $[\![2^{n-m} - (M-1)2^{-m}]\!], \ldots, [\![2^{n-m} - 2^{-m}]\!]$. To avoid this underflow, we add $[\![\widetilde{(M-1)2^{-m}}]\!]$ to the product after shifting. Note that now a symmetric problem where the shifted result should be among the numbers $[\![2^{n-m} - \widetilde{(M-1)2^{-m}}]\!], \ldots, [\![2^{n-m} - \widetilde{2 \cdot 2^{-m}}]\!]$ or $[\![2^{n-m} - \widetilde{2^{-m}}]\!]$ but would now be changed to one of $[\![\widetilde{0}]\!], [\![\widetilde{2^{-m}}]\!], \ldots, (M-1)[\![\widetilde{2^{-m}}]\!]$ could happen. However, we assume that the user would choose such inputs that these numbers would not arise as the products of any two multiplicands similarly as they would not multiply any two fixed-point numbers so that the product would be greater than $2^{n-m} - \widetilde{2^{-m}}$. Now the user should not multiply any two fixed-point numbers so that the product would be greater than $2^{n-m} - M2^{-m}$. Since M is usually a small number, this additional constraint does not practically affect computation.

The multiplication of two secret fixed-point numbers is similar. More specifically, to multiply two secret fixed-point numbers $[\![\tilde{x}]\!]$ and $[\![\tilde{y}]\!]$, we first convert $[\![x]\!]$ and $[\![y]\!]$ to $\mathbb{Z}_{2^{2n}}$ and then compute the product $[\![x]\!] \cdot [\![y]\!] = [\![\tilde{x}]\!] \cdot [\![\tilde{y}]\!]2^{2m} = [\![\widetilde{xy}]\!] \cdot 2^{2m} = [\![xy]\!] \cdot 2^m$ there. Then we shift $[\![xy]\!] \cdot 2^m$ to the right by m bits and add $[\![\widetilde{(M-1)2^{-m}}]\!]$ so that the result would be correct. After that we convert the result back to $\mathbb{Z}_{2^n}$ so that the product would be in the same ring as the multiplicands. We denote this operation by $[\![\tilde{x}]\!] \cdot [\![\tilde{y}]\!]$.

Data: $[\![\widetilde{x}]\!], m, n, s_i\{\widetilde{c_i}\}_{i=0}^{k}$
Result: Takes in a a secret fixed point number $[\![\widetilde{x}]\!]$, the radix-point m, the number of bits of the fixed-point number n and the coefficients $s_i\{\widetilde{c_i}\}_{i=0}^{k}$ for the approximation polynomial. Outputs a secret fixed-point number $[\![\widehat{y}]\!]$ that is the value of the approximation polynomial at point x.

1 $[\![\widetilde{x^1}]\!] \leftarrow [\![\widetilde{x}]\!]$
2 **for** $j \leftarrow 0$ **to** $\lceil \log_2(k) \rceil$ **do**
3 | **for** $i \leftarrow 1$ **to** 2^j **do** in parallel
4 | | $[\![\widetilde{x^{i+2^j}}]\!] \leftarrow [\![\widetilde{x^{2^j}}]\!] \cdot [\![\widetilde{x^i}]\!]$
5 | **end**
6 **end**
7 $[\![\widetilde{y_0}]\!] \leftarrow \mathsf{Share}(\widetilde{c_0})$
8 **for** $i \leftarrow 1$ **to** k **do** in parallel
9 | $[\![\widetilde{y_i}]\!] \leftarrow \widetilde{c_i} \cdot [\![\widetilde{x^i}]\!]$
10 **end**
11 $[\![\widetilde{y'}]\!], [\![\widetilde{y''}]\!] \leftarrow [\![\widetilde{0}]\!]$
12 **for** $i \leftarrow 0$ **to** k **do** in parallel
13 | **if** $s_i == 1$ **then**
14 | | $[\![\widetilde{y'}]\!] + = [\![\widetilde{y_i}]\!]$
15 | **end**
16 | **if** $s_i == -1$ **then**
17 | | $[\![\widetilde{y''}]\!] + = [\![\widetilde{y_i}]\!]$
18 | **end**
19 **end**
20 $[\![\widehat{y}]\!] \leftarrow [\![\widetilde{y'}]\!] - [\![\widetilde{y''}]\!]$
21 **return** $[\![\widehat{y}]\!]$

Algorithm 1. Computation of a polynomial on fixed-point numbers

4.2 Polynomial Evaluation

We will now present Algorithm 1 for evaluating polynomials with given coefficients. It is based on the respective algorithm described in [11] in the sense that the operations performed are the same but use fixed-point numbers instead of floating-point numbers. It takes in public signed coefficients $\{s_i\widetilde{c_i}\}_{i=0}^{k}$ and a secret fixed-point number $[\![\widetilde{x}]\!]$, and outputs $[\![\widehat{y}]\!] = \sum_{i=0}^{k} s_i\widetilde{c_i} \cdot [\![\widetilde{x^k}]\!]$. Here $s_i \in \{-1, 1\}$. We will now describe the general strategy for that.

First we need to evaluate $[\![\widetilde{x^2}]\!], [\![\widetilde{x^3}]\!], \ldots, [\![\widetilde{x^k}]\!]$. It is trivial to do this with $k-1$ rounds of multiplications, however, we shall do it in $\lceil \log k \rceil$ rounds. Every round we compute the values $[\![\widetilde{x^{2^i+1}}]\!], [\![\widetilde{x^{2^i+2}}]\!], \ldots, [\![\widetilde{x^{2^{i+1}}}]\!]$ by multiplying $[\![\widetilde{x^{2^i}}]\!]$ with $[\![\widetilde{x^1}]\!], [\![\widetilde{x^2}]\!], \ldots, [\![\widetilde{x^{2^i}}]\!]$, respectively. (line 4)

Following that, on line 9 we can multiply the powers of x with the respective coefficients $\widetilde{c_i}$ with one round of multiplication, obtaining the values $[\![\widetilde{c_1 x}]\!], [\![\widetilde{c_2 x^2}]\!]$, $\ldots, [\![\widetilde{c_k x^k}]\!]$. We also set $[\![\widetilde{c_0 x^0}]\!]$ to the tuple of shares $(2^m \cdot c_0, 0, \ldots, 0)$. After that we can compute the sums $[\![\sum_{s_i=1} c_i x^i]\!] = \sum_{s_i=1}[\![\widetilde{c_i x^i}]\!]$ and $[\![\sum_{s_i=-1} c_i x^i]\!] = \sum_{s_i=-1}[\![\widetilde{c_i x^i}]\!]$ locally, respectively, on lines 14 and 17 and find the final result $[\![\widetilde{y}]\!] = [\![\sum_{s_i=1} c_i x^i]\!] - [\![\sum_{s_i=-1} c_i x^i]\!]$, which is also a local operation.

For every function, we face the question of which polynomial to use. Generally we have preferred using Chebyshev polynomials, to avoid the Runge phenomenon. For error function, we used Taylor series. However, sometimes large coefficients of Chebyshev polynomials can cause problems, such as making the result less accurate when the coefficients are very big. The reason for this is that to be able to represent large coefficients, the radix-point must be smaller and thus computing x^i will be more inaccurate for higher powers. We need to find the optimal place for the radix-point for each function. We also note that we will use this algorithm only so that it will output positive fixed-point numbers.

5 Hybrid Versions of Selected Functions

We have used the hybrid techniques to efficiently evaluate the square root, inverse, exponential and the Gaussian error function.

Our floating-point number representation is similar to the one from [11]. A floating-point number N consists of sign s, exponent E and significand f where $N = (-1)^{1-s} \cdot f \cdot 2^{E-q}$. Here q is a fixed number called the bias that is used for making the representation of the exponent non-negative. We require that if $N \neq 0$, the significand f would be normalised — i.e. f is represented by a number in $[2^{n-1}, 2^n - 1]$. If $N = 0$, then $f = 0$ and $E = 0$. If N is secret, then it means that the sign, significand and exponent are all independently secret-shared. We denote it with $[\![N]\!] = ([\![s]\!], [\![E]\!], [\![f]\!])$.

Kamm and Willemson [11] present algorithms for computing the sum and product of two secret floating point numbers, and use these operations to implement polynomial approximations. However, the resulting routines are rather slow. Notably, computing the sum of two floating-point numbers is slower than computing the product. The basic structure of our function implementations is still inspired by [11].

The main improvement of the current paper is converting the significand of the floating-point number to a fixed-point number and then performing polynomial approximation in fixed-point format. The basic algorithm for polynomial evaluation was described in Algorithm 1. However, some extra corrections are needed after converting the result back into floating-point form. The general approach that we use for square root and inverse can easily be generalised for other power functions since for them we can work separately with the significand and the exponent. In order to use this approach for other functions (such as the exponential or the error function) work must be done to tailor specific arrangements for computing these functions in such a way.

5.1   Conversion from Fixed-Point Number to Floating-Point Number and Correction of Fixed-Point Numbers

In three out of our four functions, when we evaluate the polynomial on some fixed-point number $[\![\widetilde{x}]\!]$ where $\widetilde{x} \in [2^v, 2^{v+1})$, and we get $[\![\widetilde{y}]\!]$ as the output, where $\widetilde{y}$ should be in $[2^t, 2^{t+1})$ for some t that depends on the function. For example, for inverse, if the input $[\![\widetilde{x}]\!]$ is in $[0.5, 1)$, then the output should be approximately in $[1, 2)$.

However, due to inaccuracies coming from roundings and the error of the polynomial, the result might be out of that range— it might also be in $[0, 2^t)$ or $[2^{t+1}, 2^{t+2})$. If that should happen we will use the protocol Correction($[\![\widetilde{y}]\!], t, m, n,$ $, b_0)b_1$ to get a result that is in the correct range and is not less accurate. Here b_0 and b_1 are public boolean flags that are set to 1 if the result may be in $[0, 2^t)$ or $[2^{t+1}, 2^{t+2})$, respectively. We omitted this algorithm from this version of the paper due to size constraints. It can be read in the full version of the paper.

This algorithm is necessary in several cases for converting a fixed-point number back to the significand of a floating-point number. If this sort of protocol is not performed and we mistakenly assume that the fixed-point number $\widetilde{x}$ that we got as a result is in some $[2^t, 2^{t+1})$, and thus we set the result to $[\![N]\!] = ([\![s]\!], 2^{n-m-t-1} \cdot [\![x]\!], [\![t+q]\!])$, then it might happen that the floating-point number N is not normalised.

We also omitted an algorithm FixToFloatConversion($[\![\widetilde{x}]\!], t, m, n$) that is used for converting a positive fixed-point number $[\![\widetilde{x}]\!]$ to a floating-point number if we know that $\widetilde{x} \in [2^{t-1}, 2^{t+1})$. If $\widetilde{x} \in [2^{t-1}, 2^t)$, then our result should be $[\![N_1]\!] = ([\![s]\!], [\![E]\!], [\![f]\!]) = ([\![1]\!], [\![t+q]\!], [\![\widetilde{y}]\!] \cdot 2^{n-t})$. If $\widetilde{x} \in [2^t, 2^{t+1})$, then our result should be $[\![N_2]\!] = ([\![s]\!], [\![E]\!], [\![f]\!]) = ([\![1]\!], [\![t+q+1]\!], [\![\widetilde{y}]\!] \cdot 2^{n-t-1})$.

5.2   Inverse

We will describe Algorithm 2 for computing the inverse of a floating-point number $[\![N]\!] = ([\![s]\!], [\![E]\!], [\![f]\!])$ in our setting.

First note that since inverse of zero is not defined, we can assume that the input is not zero and that thus the signicand is always normalised. Second, note that the significand $[\![f]\!]$ can now be considered a fixed-point number where $m = n$ as it represents a number in $[0.5, 1)$ but is stored as a shared value in $[2^{n-1}, 2^n - 1]$. However, if the radix-point is so high, we can not perform most of the operations we need to, so on line 1 we shift the significand to the standard fixed-point format. We lose $n - m$ bits, but since the significand has more bits for its significand than the IEEE standard 754 for both single and double precision, the number of bits we have left is not less than the significand of the IEEE standard has. Let us denote the shifted significand with $[\![f']\!]$. Then, on line 2, we securely compute the number $[\![t]\!]$ so that $\widetilde{t}$ is the inverse of $\widetilde{f}'$ by using polynomial evaluation, as described in Algorithm 1.

The exact polynomials we used for the fixed point inversion can be found in the full version of the paper [13]. We will denote calling the Algorithm 1 on value $[\![\widetilde{x}]\!]$ with the coefficients of that polynomial by FixInverseProtocol($[\![\widetilde{x}]\!], \{s_i \widetilde{c}_i\}_{i=0}^k, m, n$),

---

Data: $\llbracket N \rrbracket = (\llbracket s \rrbracket, \llbracket E \rrbracket, \llbracket f \rrbracket), q, m, \{s_i \widetilde{c_i}\}_{i=0}^{k}, n$

Result: Takes in a a secret floating point number $\llbracket N \rrbracket = (\llbracket s \rrbracket, \llbracket E \rrbracket, \llbracket f \rrbracket)$, the bias of the exponent q and the radix-point of the corresponding fixed-point number m, Chebyshev coefficients $\{\widetilde{c_i}\}_{i=0}^{k}$ for computing the fixed-point polynomial and the number of bits of the fixed-point number n. Outputs a secret floating-point number that is approximately equal to the inverse of N.

1 $\llbracket f' \rrbracket \leftarrow \mathsf{PublicBitShiftRightProtocol}(\llbracket f \rrbracket, n - m)$

2 $\llbracket t \rrbracket \leftarrow \mathsf{FixInverseProtocol}(\llbracket \widetilde{f'} \rrbracket, \{s_i \widetilde{c_i}\}_{i=0}^{k}, m, n)$

3 $\llbracket t' \rrbracket \leftarrow \mathsf{Correction}(\llbracket \widetilde{t} \rrbracket, 0, m, n, 0, 1)$

4 $\llbracket t'' \rrbracket \leftarrow \llbracket t' \rrbracket \cdot 2^{n-m-1}$

5 **return** $\llbracket N' \rrbracket = (\llbracket s \rrbracket, \llbracket 2q - E + 1 \rrbracket, \llbracket t'' \rrbracket)$

Algorithm 2. Inverse of a floating point number

---

where m is the position of the radix point and n is the number of bits in the fixed-point number and where $\{s_i \widetilde{c_i}\}_{i=0}^{k}$ refers to the signed coefficients of the polynomial. Calling $\mathsf{FixInverseProtocol}$ on $\llbracket \widetilde{x} \rrbracket$ gives us $\widetilde{t'}$.

Since $\widetilde{f'} \in [0.5, 1)$, we expect the result $\widetilde{t'}$ to be approximately in $(1, 2]$. However, since the polynomial has a small error, then the result might sometimes be slightly bigger than 2 and thus on line 3 we need to correct the result using the $\mathsf{Correction}$ algorithm with range parameter being 0.

Next we want to divide the result by two and then convert the fixed-point number back into the significand format. We can combine these two operations. The first one would require shifting to the right by one bit and the second one would require shifting to the left by $n - m$ bits. By combining, we just have to shift the result to the left by $n - m - 1$ bits, which is a free operation since it is equivalent with multiplying by 2^{n-m-1} which we do on line 4. The sign of the inverse is the same as the sign of N and the exponent should be the additive inverse of the original exponent, minus one to take into account the division by two that we did in the significand. However, we need to take into account that the bias is added to the exponent and thus the exponent of the result shall be $\llbracket -E + q + 1 \rrbracket$. Thus we obtain Algorithm 2 for computing the inverse of a floating-point number.

5.3 Square Root

We will describe Algorithm 3 for computing the square root of a floating-point number in our setting. Note that since we assume that the sign is positive and thus ignore the sign, we will de facto compute the function $\sqrt{|x|}$. If the input is $-x$ for some non-negative x, then the output will be approximately $\sqrt{x}$.

First we shall describe the case where the input is not zero. We note that the significand $\llbracket f \rrbracket$ can be considered a fixed-point number where $m = n$ as it represents a number in $[0.5, 1)$ but is stored as a shared value in $[2^{n-1}, 2^n - 1]$. However, if the radix-point is so big, we can not perform most of the operations

---

Data: $\llbracket N \rrbracket = (\llbracket s \rrbracket, \llbracket E \rrbracket, \llbracket f \rrbracket), q, m, \{s_i \widetilde{c_i}\}_{i=0}^{k}, n$

Result: Takes in a a secret floating point number $\llbracket N \rrbracket = (\llbracket s \rrbracket, \llbracket E \rrbracket, \llbracket f \rrbracket)$, the bias of the exponent q and the radix-point of the corresponding fixed-point number m, Chebyshev coefficients $\{\widetilde{c_i}\}_{i=0}^{k}$ for computing the fix-point polynomial and the number of bits of the fixed-point number n. Outputs a secret floating-point number that is approximately equal to $\sqrt{N}$.

1 $\llbracket \widetilde{f'} \rrbracket \leftarrow \mathsf{PublicBitShiftRightProtocol}(\llbracket f \rrbracket, n - m)$
2 $\llbracket b \rrbracket \leftarrow \llbracket E \rrbracket \pmod 2$
3 $\llbracket E' \rrbracket \leftarrow \mathsf{PublicBitShiftRightProtocol}(\llbracket E \rrbracket, 1)$
4 $\llbracket \widetilde{t_1} \rrbracket \leftarrow \mathsf{FixSquareRootProtocol}(\llbracket \widetilde{f'} \rrbracket, \{s_i \widetilde{c_i}\}_{i=0}^{k}, m, n)$
5 $\llbracket \widetilde{t_2} \rrbracket \leftarrow \llbracket \widetilde{t_1} \rrbracket \cdot \frac{\widetilde{\sqrt{2}}}{2}$
6 $\llbracket \widetilde{t_2'} \rrbracket \leftarrow \mathsf{Correction}(\llbracket \widetilde{t_2} \rrbracket, -1, m, n, 1, 0)$
7 $\llbracket t' \rrbracket \leftarrow \mathsf{ObliviousChoiceProtocol}(\llbracket b \rrbracket, \llbracket \widetilde{t_1} \rrbracket, \llbracket \widetilde{t_2'} \rrbracket)$
8 $\llbracket t'' \rrbracket \leftarrow \llbracket t' \rrbracket \ll (n - m)$
9 **return** $\llbracket N' \rrbracket = (\llbracket 1 \rrbracket, \llbracket E' + 1 + (q \gg 1) \rrbracket, \llbracket t'' \rrbracket)$

Algorithm 3. Square root of a floating point number

we need to, so on line 1, we shift the significand to the standard fixed-point format. Let us denote the shifted significand by $\llbracket \widetilde{f'} \rrbracket$. While computing the square root, it is natural to halve the exponent by shifting it to the right by one bit on line 3. However, the parity of that last bit may change the result $\frac{\sqrt{2}}{2}$ times and thus we have to remember the last bit before that on line 2 and later use it to perform an oblivious choice on line 7. Like in the case of the inverse, we use a Chebyshev polynomial on line 4 to find such $\llbracket \widetilde{t_1} \rrbracket$ that $\widetilde{t_1}$ is approximately equal to the square root of $\widetilde{f'}$. For that we compute the square root of $\llbracket \widetilde{f'} \rrbracket$ by using polynomial evaluation, as described in Algorithm 1. The exact polynomial for computing the fixed point square root can be found in the full version of the paper [13].

We will denote calling the function 1 on value $\llbracket \widetilde{x} \rrbracket$ with the coefficients of that polynomial with $\mathsf{FixSquareRootProtocol}(\llbracket \widetilde{x'} \rrbracket, \{s_i \widetilde{c_i}\}_{i=0}^{k}, m, n)$ where m is the position of the radix point and n is the number of bits in the fixed-point number and where $\{s_i \widetilde{c_i}\}_{i=0}^{k}$ refers to the signed coefficients of the polynomial. Calling $\mathsf{FixSquareRootProtocol}$ on $\llbracket \widetilde{x} \rrbracket$ gives us $\llbracket \widetilde{t_1} \rrbracket$.

Following that, on line 5 we multiply $\llbracket \widetilde{t_1} \rrbracket$ by $\frac{\widetilde{\sqrt{2}}}{2}$ —we then have the risk of $\llbracket \widetilde{t_1 \cdot \frac{\sqrt{2}}{2}} \rrbracket$ being slightly less than $\widetilde{0.5}$, thus we need to use the Correction with range parameter being -1 on line 6 to correct $\llbracket \widetilde{t_1 \cdot \frac{\sqrt{2}}{2}} \rrbracket$ into the range $[0.5, 1)$. Then, on line 7, we use the saved last bit of the exponent to perform an oblivious choice between $\llbracket \widetilde{t_1} \rrbracket$ and $\llbracket \widetilde{t_1 \cdot \frac{\sqrt{2}}{2}} \rrbracket$ and convert the result back into the significand format by shifting the result left by $n - m$ bits on the line 8. The latter operation may be implemented by multiplying the result by 2^{n-m} which is a local operation.

The sign of a square root is always plus. We correct for the bias and rounding errors by adding $1 + (q \gg 1)$ to $[\![E']\!]$. The added 1 comes from the fact that the bias is odd and we lose 0.5 from the exponent twice when truncating q by a bit. Thus we obtain Algorithm 3 for computing the square root of a floating-point number. The algorithm also gives a correct result if the input is zero but the reasoning for this was omitted due to size constraints and can be read in the full version of the paper.

5.4   Exponent

We will describe Algorithm 4 for computing the exponent of a floating-point number in our setting. Given a secret floating-point number $[\![N]\!] = ([\![s]\!], [\![E]\!], [\![f]\!])$ we wish to compute $[\![e^N]\!] = [\![2^{\log_2 e \cdot N}]\!] = [\![2^y]\!]$ where $y := \log_2 e \cdot N$.

Data: $[\![N]\!] = ([\![s]\!], [\![E]\!], [\![f]\!]), q, m, \{s_i \widetilde{c_i}\}_{i=0}^k, n$

Result: Takes in a a secret floating point number $[\![N]\!] = ([\![s]\!], [\![E]\!], [\![f]\!])$, the bias of the exponent q and the radix-point of the corresponding fixed-point number m, coefficients $\{s_i \widetilde{c_i}\}_{i=0}^k$ for computing the fix-point polynomial and the number of bits of the fixed-point number n. Outputs a secret floating-point number that is approximately equal to e^N.

1   $[\![y]\!] = ([\![s_y]\!], [\![E_y]\!], [\![f_y]\!]) \leftarrow \log_2 e \cdot [\![N]\!]$

2   $[\![z]\!] \leftarrow \mathsf{PrivateBitShiftRightProtocol}([\![f_y]\!], [\![n - (E_y - q)]\!])$

3   $[\![[y]]\!] = ([\![s_{[y]}]\!], [\![E_{[y]}]\!], [\![f_{[y]}]\!]) \leftarrow \mathsf{ConvertToFloat}([\![z]\!])$

4   $[\![\{y\}]\!] = ([\![s_{\{y\}}]\!], [\![E_{\{y\}}]\!], [\![f_{\{y\}}]\!]) \leftarrow [\![y]\!] - [\![[y]]\!]$

5   $[\![\widetilde{w}]\!] \leftarrow \mathsf{PrivateBitShiftRightProtocol}([\![f_{\{y\}}]\!], [\![-E_{\{y\}} + q + n - m]\!])$

6   **begin** in parallel

7      $[\![\widetilde{f'}]\!] \leftarrow \mathsf{FixPowerOfTwoProtocol}([\![\widetilde{w}]\!], \{s_i \widetilde{c_i}\}_{i=0}^k, m, n)$

8      $[\![\widetilde{f''}]\!] \leftarrow \mathsf{FixPowerOfTwoProtocol}([\![\widetilde{1 - w}]\!], \{s_i \widetilde{c_i}\}_{i=0}^k, m, n)$

9   **end**

10 **begin** in parallel

11      $[\![\widetilde{f'}]\!] \leftarrow \mathsf{Correction}([\![\widetilde{f'}]\!], 0, m, n, 1, 1)$

12      $[\![\widetilde{f''}]\!] \leftarrow \mathsf{Correction}([\![\widetilde{f''}]\!], 0, m, n, 1, 1)$

13 **end**

14 $[\![2^{\{y'\}}]\!] = ([\![s_{2^{\{y'\}}}]\!], [\![E_{2^{\{y'\}}}]\!], [\![f_{2^{\{y'\}}}]\!]) \leftarrow ([\![1]\!], [\![q + 1]\!], [\![f' \cdot 2^{n-m-1}]\!]$

15 $[\![2^{\{y''\}}]\!] = ([\![s_{2^{\{y''\}}}]\!], [\![E_{2^{\{y''\}}}]\!], [\![f_{2^{\{y''\}}}]\!]) \leftarrow ([\![1]\!], [\![q + 1]\!], [\![f'' \cdot 2^{n-m-1}]\!]$

16 $[\![2^{[y']}]\!] = ([\![s_{2^{[y']}}]\!], [\![E_{2^{[y']}}]\!], [\![f_{2^{[y']}}]\!]) \leftarrow ([\![1]\!], [\![1 + q + z]\!], [\![100\ldots0]\!])$

17 $[\![2^{[y'']}]\!] = ([\![s_{2^{[y'']}}]\!], [\![E_{2^{[y'']}}]\!], [\![f_{2^{[y'']}}]\!]) \leftarrow ([\![1]\!], [\![q - z]\!], [\![100\ldots0]\!])$

18 $[\![b]\!] \leftarrow \mathsf{ConvertToBoolean}([\![s_y]\!])$

19 **begin** in parallel

20      $[\![2^{\{y\}}]\!] \leftarrow \mathsf{ObliviousChoiceProtocol}([\![b]\!], [\![2^{\{y'\}}]\!], [\![2^{\{y''\}}]\!])$

21      $[\![2^{[y]}]\!] \leftarrow \mathsf{ObliviousChoiceProtocol}([\![b]\!], [\![2^{[y']}]\!], [\![2^{[y'']}]\!])$

22 **end**

23 $[\![2^y]\!] = ([\![s_{2^y}]\!], [\![E_{2^y}]\!], [\![f_{2^y}]\!]) \leftarrow [\![2^{[y]}]\!] \cdot [\![2^{\{y\}}]\!]$

24 **return** $[\![N']\!] = [\![2^y]\!]$

Algorithm 4. Power of e of a floating point number

It is easier to compute a power of 2 in our setting than a power of e so thus on line 1 we first compute the floating-point number $[\![y]\!] = ([\![s_{\{y\}}]\!], [\![E_{\{y\}}]\!], [\![f_{\{y\}}]\!]) = [\![\log_2 e]\!] \cdot [\![N]\!]$. To compute $[\![2^y]\!]$, we split $[\![y]\!]$ into two parts — the integer part $[\![[y]]\!]$ and the fractional part $[\![\{y\}]\!]$. Note that the $n - (E_y - q)$ last bits of the s_y represent the fractional part of y and the rest of the bits represent the integer part, so it is equal to $2^{n-(E_y-q)}([y] + \{y\})$ Thus, on line 2 we privately shift $[\![f_y]\!]$ to the right by $[\![n - (E_y - q)]\!]$ bits to truncate the fractional part and divide by $2^{n-(E_y-q)}$ and thus obtain the integer part of $[\![y]\!]$ that we represent with $[\![z]\!]$. This, however, has integer type and we want to deduce it from a floating-point number so we need to call the ConvertToFloat method on line 3 to cast $[\![z]\!]$ into a floating point-number $[\![[y]]\!] = ([\![s_{[y]}]\!], [\![E_{[y]}]\!], [\![f_{[y]}]\!])$.

We find the fractional part $[\![\{y\}]\!] = ([\![s_{\{y\}}]\!], [\![E_{\{y\}}]\!], [\![f_{\{y\}}]\!])$ on line 4 by $[\![\{y\}]\!] = [\![y]\!] - [\![[y]]\!]$. Note that if $[y]$ is negative, then the fractional part will be in $[-1, 0)$, so we have to subtract 1 from $[\![[y]]\!]$ and add it to $[\![\{y\}]\!]$ in order for $\{y\}$ to be in $[0, 1)$. We shall do the next operations in both the positive and the negative case and use oblivious choice in the end to choose between them. Now we convert $[\![\{y\}]\!]$ to a fixed-point number $[\![\tilde{w}]\!]$ by shifting $[\![f_{\{y\}}]\!]$ to the right by $[\![-E_{\{y\}} + q + n - m]\!]$ bits. The exact polynomial for computing the fixed point exponent can be found in the full version of the paper [13].

We will denote calling the Algorithm 1 on value $[\![\tilde{x}]\!]$ with the coefficients of that polynomial by FixPowerOfTwoProtocol($[\![\tilde{x}]\!], \{s_i \tilde{c}_i\}_{i=0}^{k}, m, n$), where m is the position of the radix point and n is the number of bits in the fixed-point number and where $\{s_i \tilde{c}_i\}_{i=0}^{k}$ refers to the signed coefficients of the polynomial and we call this function in parallel on lines 7 and 8 this polynomial on values $[\![\tilde{w}]\!]$ and $[\![\widetilde{1 - w}]\!]$ and thus obtain $[\![\tilde{f}']\!]$ and $[\![\tilde{f}'']\!]$ respectively and correct them with range parameter 0. We then initialize the possible values for $[\![2^{\{y\}}]\!]$, that is, $[\![2^{\{y'\}}]\!]$ and $[\![2^{\{y''\}}]\!]$ with signs 1, exponents $q + 1$ and significands that are equal to $[\![\tilde{f}']\!]$ and $[\![\tilde{f}'']\!]$ that have been shifted by $n - m - 1$ bits to the left. Likewise, we initialize the possible values for $[\![2^{[y]}]\!]$, that is, $[\![2^{[y']}]\!]$ and $[\![2^{[y'']}]\!]$ with signs 1, exponents that are equal to $1 + q + z$ and $q - z$, respectively and significands $100 \ldots 0 = 2^{n-1}$. $[\![2^{[y']}]\!]$ and $[\![2^{\{y'\}}]\!]$ are the correct values if the input was a positive number, i.e. $s_y = 1$ and $[\![2^{[y'']}]\!]$ and $[\![2^{\{y''\}}]\!]$ are the correct values if the input was a negative number i.e. $s_y = 0$. Thus we convert $[\![s_y]\!]$ to a boolean on line 18 and, in parallel, perform oblivious choice between $[\![2^{\{y'\}}]\!]$ and $[\![2^{\{y''\}}]\!]$ on line 20 and $[\![2^{[y']}]\!]$ and $[\![2^{[y'']}]\!]$ on line 21 to respectively obtain $[\![2^{\{y\}}]\!]$ and $[\![2^{[y]}]\!]$. We then multiply $[\![2^{\{y\}}]\!]$ and $[\![2^{[y]}]\!]$ together on line 23 to obtain the result. Thus we obtain the Algorithm 4 for computing the exponential function of a floating-point number. Note that the algorithm also works on inputs 0 and -0.

5.5   Error Function

Gaussian error function is defined by $\text{erf } x = \frac{2}{\sqrt{\pi}} \int_0^x e^{-t^2} dt$. It is a antisymmetric function — i.e. $\text{erf}(-x) = -\text{erf}(x)$. Thus we can evaluate the function only depending on the exponent and the significand, and in the end, set the sign of the output to be the sign of the input. So, for the sake of simplicity, we will assume that our input is non-negative. However, since $\text{erf}(a \cdot b)$ can not be easily computed from $\text{erf}\,a$ and $\text{erf}\,b$, we can not use the approach we used for inverse and square root. Computing the error function using approximation polynomials on significands does not seem possible, as we would have to be able to represent all numbers with fixed-point numbers of a good precision (conflicting goals) and also use very many approximation polynomials.

However, it turns out that we can bound the range of inputs in which case we have to compute the error function with a fixed-point polynomial. Namely, if x is close to 0 then $\text{erf}\,x$ can be well approximated with $\frac{2}{\sqrt{\pi}}x$—observe that the McLaurin series of the error function is $\text{erf}\,x = \frac{2}{\sqrt{\pi}}\sum_{i=0}^{\infty}\frac{(-1)^n}{n!(2n+1)}x^{2n+1}$ and note that $|\text{erf}\,x - \frac{2}{\sqrt{\pi}}x| = |\frac{2}{\sqrt{\pi}}\sum_{i=1}^{\infty}\frac{(-1)^n x^{2n+1}}{n!(2n+1)}| < \frac{2x}{3\cdot\sqrt{\pi}}\sum_{i=1}^{\infty}x^{2n} = \frac{2}{3\cdot\sqrt{\pi}}\frac{x^3}{1-x^2}$.

If x is small enough, then $\frac{2}{3\cdot\sqrt{\pi}}\frac{x^3}{1-x^2}$ is negligible. On the other hand, $\text{erf}\,x$ is a monotonously growing function that approaches 1 so we can approximate $\text{erf}(x)$ with 1 for large enough x. In our approach, if $x \geq 4$, we set $\text{erf}\,x = 1$. The error we make is at most $2 \cdot 10^{-8}$. Thus, we will only need to compute polynomial approximations for $x \in [2^{-w}, 2^2)$ where w is a previously fixed public parameter that depends on how precise we would like the algorithm to be.

Thus we need approximation polynomials for the range $[0, 4)$ only. We will use four approximation polynomials, p_0, p_1, p_2 and p_3 where $p_i(\tilde{y}) \approx \text{erf}\,y$ if $\tilde{y} \in [i, i+1)$, where $i \in \{0, 1, 2, 3\}$. The exact polynomials p_1, p_2, p_3 and p_4 can be found in the full version of the paper [13].

We shall now describe Algorithm 5 for computing the error function of a floating-point number. First, we shall find the possible corresponding fixed-point numbers on which we compute our polynomial. We will, in parallel, on line 4, compute

$$[\![\tilde{f_i}]\!] = \text{PublicBitShiftRightProtocol}([\![f]\!], n - m + i - 2) \text{ for } i \in [0, w].$$

If $[\![f]\!]$ is the significand of $[\![x]\!]$ then $[\![\tilde{f_i}]\!] \in [2^{-i}, 2^{-i+1})$ if $x \in [2^{-i}, 2^{-i+1})$.

Note that although we need polynomial approximation for values that are in $[2, 4)$, we did not compute any such fixed-point number $\widetilde{f_{-1}} = \text{PublicBitShiftRightProtocol}([\![f]\!], n - m - 3)$ that is equal to x if $x \in [2, 4)$. Instead of computing the polynomial $\sum_{i=0}^{l}s_i a_i \cdot \widetilde{f_{-1}}^i$ we compute the polynomial $\sum_{i=0}^{l}s_i a_i 2^i \cdot \widetilde{f_0}^i$. Note that these two expressions are almost equivalent since $\widetilde{f_0} = \widetilde{f_{-1}} \ll 1$. However, the latter is preferable since we will only be able to represent $\widetilde{f_{-1}}^i$ if i is very small and thus not be able to use good polynomials.

We now wish to compute values $[\![\tilde{g_i}]\!]$ where $\tilde{g_i} \approx \text{erf}\,\tilde{f_i}$ if $x \in [2^{-i}, 2^{-i+1})$ for $i \in [-1, w]$. For $i \in [1, w]$ we compute $\tilde{g_i} = p_0(\widetilde{f_i})$ on line 6. For $i = 0$, we compute $\tilde{g_0} = p_1(\widetilde{f_0})$ on line 7 For $i = -1$, we compute $\widetilde{g_{-1,0}}$ and $\widetilde{g_{-1,1}}$ on lines 8 and 9 by applying the modified versions of the polynomials p_2 and p_3 to $\widetilde{f_0}$, as described before. Note that these values are computed in parallel. Now we need to evaluate $\widetilde{g_{-1}}$ using oblivious choice so that if the result is in $[2, 4)$, $\widetilde{g_{-1}} = \widetilde{g_{-1,0}}$ if $f \leq 2^{n-1}$ and $\widetilde{g_{-1}} = \widetilde{g_{-1,1}}$ if $f > 2^{n-1}$. We note that whether $f \leq 2^{n-1}$ or

Data: $[\![N]\!] = ([\![s]\!], [\![E]\!], [\![f]\!]), q, m,$
$\{s_{i,0}\widetilde{c_{i,0}}\}_{i=0}^{l}, \{s_{i,1}\widetilde{c_{i,1}}\}_{i=0}^{l}, \{s_{i,2}\widetilde{c_{i,2}}\}_{i=0}^{l}, \{s_{i,3}\widetilde{c_{i,3}}\}_{i=0}^{l}, n, w$

Result: Takes in a a secret floating point number $[\![N]\!] = ([\![s]\!], [\![E]\!], [\![f]\!])$, the bias of the exponent q and the radix-point of the corresponding fixed-point number m, coefficients $\{s_{i,j}\widetilde{c_{i,j}}\}_{i=0}^{l}$ for computing the fixed-point values that are accurate in $[j, j+1)$ and an integer w so that we evaluate the function with a polynomial, if $2^w \le N < 4$. Outputs a secret floating-point number that is approximately equal to the error function of N.

1 **for** $k \leftarrow 0$ **to** w **do**
2    $shifts_k \leftarrow n - m + i - 2$
3 **end**
4 $\{[\![f_k]\!]\}_{k=0}^{w} \leftarrow \mathsf{PublicBitShiftRightProtocol}([\![f]\!], \{shifts\}_{k=0}^{w}))$
5 **for** $k \leftarrow 1$ **to** w **do** in parallel
6    $[\![\widetilde{g_k}]\!] \leftarrow \mathsf{FixGaussianErrorFunction}([\![\tilde{f_k}]\!], m, n, \{s_{i,0}\widetilde{c_{i,0}}\}_{i=0}^{l})$
7    $[\![\widetilde{g_0}]\!] \leftarrow \mathsf{FixGaussianErrorFunction}([\![\tilde{f_0}]\!], m, n, \{s_{i,1}\widetilde{c_{i,1}}\}_{i=0}^{l})$
8    $[\![\widetilde{g_{-1,0}}]\!] \leftarrow \mathsf{FixGaussianErrorFunction}([\![\tilde{f_0}]\!], m, n, \{s_{i,2}\widetilde{c_{i,2}}\}_{i=0}^{l})$
9    $[\![\widetilde{g_{-1,1}}]\!] \leftarrow \mathsf{FixGaussianErrorFunction}([\![\tilde{f_0}]\!], m, n, \{s_{i,3}\widetilde{c_{i,3}}\}_{i=0}^{l})$
10 **end**
11 $\{[\![u_i]\!]\}_{i=0}^{n} \leftarrow \mathsf{BitExtraction}([\![f]\!])$
12 $[\![g_{-1}]\!] \leftarrow \mathsf{ObliviousChoiceProtocol}([\![u_m]\!], [\![g_{-1,1}]\!], [\![g_{-1,0}]\!])$
13 $t_{-1} \leftarrow 0$
14 $t_0 \leftarrow 0$
15 **for** $k \leftarrow 1$ **to** w **do**
16    $t_k \leftarrow 2 - k$
17 **end**
18 **for** $k \leftarrow -1$ **to** w **do** in parallel
19    $[\![N_k]\!] = ([\![s_k]\!], [\![E_k]\!], [\![f_k]\!]) \leftarrow \mathsf{FixToFloatConversion}([\![\widetilde{g_k}]\!], t_k, m, n)$
20 **end**
21 $[\![N_{-2}]\!] = ([\![s_{-2}]\!], [\![E_{-2}]\!], [\![f_{-2}]\!]) \leftarrow \frac{2}{\sqrt{\pi}} \cdot [\![N]\!]$
22 $[\![N_{w+1}]\!] = ([\![s_{w+1}]\!], [\![E_{w+1}]\!], [\![f_{w+1}]\!]) \leftarrow 1$
23 **begin** in parallel
24    $b_0 \leftarrow \mathsf{LTEProtocol}([\![E]\!], [\![q - w]\!])$
25    $b_1 \leftarrow \mathsf{LTEProtocol}([\![q + 3]\!], [\![E]\!])$
26 **end**
27 $[\![E]\!] \leftarrow \mathsf{ObliviousChoiceProtocol}([\![b_0]\!], [\![q - w]\!], [\![E]\!])$
28 $[\![E]\!] \leftarrow \mathsf{ObliviousChoiceProtocol}([\![b_1]\!], [\![q + 3]\!], [\![E]\!])$
29 $[\![E']\!] \leftarrow \mathsf{GeneralizedObliviousChoice}([\![E_{-2}]\!], \ldots, [\![E_{w+1}]\!], [\![E - q]\!])$
30 $[\![f']\!] \leftarrow \mathsf{GeneralizedObliviousChoice}([\![f_{-2}]\!], \ldots, [\![f_{w+1}]\!], [\![E - q]\!])$
31 **return** $[\![N']\!] = ([\![s]\!], [\![E']\!], [\![f']\!])$

Algorithm 5. Gaussian error function of a floating point number

not depends only on the last bit of f, thus we use the BitExtract protocol on $[\![f]\!]$ on line 11 to find that bit and use that to perform oblivious choice on line 12 between $\widetilde{g_{-1,0}}$ and $\widetilde{g_{-1,1}}$.

Note that for $i < -1$, if $x \in [2^i, 2^{i+1})$ then erf $x \in [2^i, 2^{i+2})$. If $x \geq 0.5$, then erf $x \in [0.5, 1)$. Thus we can apply the FixToFloat protocol on line 19 to the numbers $[\![\tilde{g_i}]\!]$ to obtain floating point numbers $[\![N_{-1}]\!], \ldots, [\![N_w]\!]$ We additionally compute $[\![N_{-2}]\!] = \frac{2}{\sqrt{\pi}}[\![N]\!]$ and set $[\![N_{w+1}]\!]$ to $[\![1]\!]$ on lines 21 and 22. In order to be able to use only the last $\log_2(2 + w)$ bits for the generalized oblivious choice, we set $[\![E]\!]$ to $q - w$ if it is smaller than $q - w$ and to $q + 3$ if it is larger than $q + 3$ on lines 24, 25, 27 and 28.

Then we use the generalised oblivious choice protocol on both exponents and significands on lines 29 and 30 to choose the final result between $[\![N_{-2}]\!], \ldots, [\![N_{w+1}]\!]$ based on the exponent $[\![E]\!]$. Note that if $x = 0$, then the oblivious choice will choose $[\![N_{-2}]\!] = \frac{2}{\sqrt{\pi}}[\![0]\!] = [\![0]\!]$ and thus the protocol is correct also when the input is zero.

Note that while it would have been possible to shift by a protected number of bits and thus obtain a single fixed-point number on which we could do polynomial evaluation, we still would have to use different polynomials for different value ranges of the input and perform oblivious choices between them and since shifting by a protected number of bits is an expensive operation we decided against it.

6   Results and Comparison

We have implemented four selected functions on the Sharemind 3 computing platform and benchmarked the implementations. To measure the performance of the floating point operations we deployed the developed software on three servers connected with fast network connections.

More specifically, each of the servers used contains two Intel X5670 2.93 GHz CPUs with 6 cores and 48 GB of memory. Since on Sharemind parallel composition of operations is more efficient than sequential composition, all the operations were implemented as vector operations. To see how much the vector size affects the average performance, we ran tests for different input sizes for all our inputs. We did 5 tests for each operation and input size and computed the average.

We compare here our results with previously existing protocols for computing the functions on either fixed-point values or floating-point values. How we reached the error estimates is described in the full version of the paper [13]. The error estimates are relative errors given for the significand, meaning that they will be respectively bigger or smaller when the exponent is bigger or smaller. As for previous work, the protocols of Catrina and Dragulin, Aliasgari *et al*, and Liedel can achieve any precision, given sufficient number of bits. The used numbers of bits (and hence the respective precision) for their respective experiments can be found in the tables. Kamm and Willemson achieve precision that is approximately in the same orders of magnitude as this paper.

Table 1 compares previous results for computing the inverse with our results. Our results are up to 6 times faster than the previously existing implementations. We estimate the error to be no larger than $1.3 \cdot 10^{-4}$ for the 32 bit case and

Table 1. Operations per second for different implementation of the inverse function for different batch sizes

	1	10	100	1000	10000
Catrina, Dragulin, 128 bits, AppDiv2m, LAN(ms) [5]	3.39				
Catrina, Dragulin, 128 bits, Div2m, LAN(ms) [5]	1.26				
Kamm and Willemson, Chebyshev, 32 bits [11]	0.17	1.7	15.3	55.2	66.4
Kamm and Willemson, Chebyshev, 64 bits [11]	0.16	1.5	11.1	29.5	47.2
Current paper, 32 bits	0.99	8.22	89.73	400.51	400.51
Current paper, 64 bits	0.82	8.08	62.17	130.35	130.35

$1.3 \cdot 10^{-8}$ for the 64 bit case. We had $m = 25$ in the 32 bit case and $m = 52$ in the 64 bit case. In the 32 bit case the polynomial has rank 5 and in the 64 bit case it had rank 10.

Table 2. Operations per second for different implementation of the square root function for different input sizes

	1	10	100	1000	10000
Liedel [15], 110 bits	0.204				
Kamm and Willemson, 32 bits [11]	0.09	0.85	7	24	32
Kamm and Willemson, 64 bits [11]	0.08	0.76	4.6	9.7	10.4
Current paper, 32 bits	0.77	7.55	70.7	439.17	580.81
Current paper, 64 bits	0.65	6.32	41.75	78.25	119.99

Table 2 compares previous results for computing the square root with our results. Our results are up to 18 times faster than the best previously existing implementations. We estimate the error to be no larger than $5.1 \cdot 10^{-6}$ for 32 bit case and $4.1 \cdot 10^{-11}$ for the 64 bit case. We had $m = 31$ in the 32 bit case and $m = 52$ in the 64 bit case. In the 32 bit case the polynomial has rank 6 and in the 64 bit case it had rank 16.

Table 3 compares previous results for computing the exponent with our results. Our results are up to 2 times faster than the best previously existing

Table 3. Operations per second for different implementation of the exponential function for different input sizes

	1	10	100	1000	10000
Aliasgari et al. [1], 32 bits		6.3	9.7	10.3	10.3
Kamm and Willemson, (Chebyshev) 32 bits [11]	0.11	1.2	11	71	114
Kamm and Willemson, (Chebyshev) 64 bits [11]	0.11	1.1	9.7	42	50
Current paper, 32 bits	0.24	2.41	24.03	104.55	126.42
Current paper, 64 bits	0.23	2.27	16.66	47.56	44.84

implementations. We estimate the error to be no larger than $6 \cdot 10^{-6}$ for 32 bit case and $1.5 \cdot 10^{-12}$ for the 64 bit case. We had $m = 30$ in the 32 bit case and $m = 62$ in the 64 bit case. In the 32 bit case the polynomial has rank 4 and in the 64 bit case it had rank 8.

Table 4. Operations per second for different implementation of the Gaussian error function for different input sizes

	1	10	100	1000	10000
Kamm and Willemson, 32 bits [11]	0.1	0.97	8.4	30	39
Kamm and Willemson, 64 bits [11]	0.09	0.89	5.8	16	21
Current paper, 32-bit	0.5	4.41	30.65	45.42	49.88
Current paper, 64-bit	0.46	4.13	21.97	19.54	26.11

Table 4 compares previous results for computing the Gaussian error function with our results. Our results are up to 4 times faster than the previously existing implementations. We estimate error for 32 bit case to be no greater than $1.1 \cdot 10^{-6}$ for inputs from $[0, 1)$, no greater than $7 \cdot 10^{-6}$ for inputs from $[1, 2)$, no greater than $1.5 \cdot 10^{-5}$ for inputs from $[2, 3)$ and no greater than $4 \cdot 10^{-6}$ for inputs from $[3, 4)$. We estimate error for 64 bit case to be no greater than $2 \cdot 10^{-8}$ in $[0, 1)$, no greater than $4 \cdot 10^{-9}$ in $[1, 2)$, no greater than 10^{-8} in $[2, 3)$ and no greater than $1 \cdot 10^{-7}$ in $[3, 4)$. We had $m = 26$ in the 32 bit case and $m = 51$ in the 64 bit case and $w = 4$. All the polynomials had rank 12.

7   Conclusion

We developed fixed-point numbers for the Sharemind secure multiparty computation platform. We improved on existing algorithms by [11] for floating-point numbers for the inverse, square-root, exponential and error functions by constructing a hybrid model of fixed-point and floating-point numbers. These new algorithms allow for considerably faster implementations than the previous ones.

References

1. Aliasgari, M., Blanton, M., Zhang, Y., Steele, A.: Secure computation on floating point numbers. In: NDSS (2013)
2. Bogdanov, D., Laur, S., Willemson, J.: Sharemind: A Framework for Fast Privacy-Preserving Computations. In: Jajodia, S., Lopez, J. (eds.) ESORICS 2008. LNCS, vol. 5283, pp. 192–206. Springer, Heidelberg (2008)
3. Bogdanov, D., Niitsoo, M., Toft, T., Willemson, J.: High-performance secure multi-party computation for data mining applications. International Journal of Information Security 11(6), 403–418 (2012)
4. Boneh, D., Di Crescenzo, G., Ostrovsky, R., Persiano, G.: Public key encryption with keyword search. In: Cachin, C., Camenisch, J.L. (eds.) EUROCRYPT 2004. LNCS, vol. 3027, pp. 506–522. Springer, Heidelberg (2004)

5. Catrina, O., Dragulin, C.: Multiparty computation of fixed-point multiplication and reciprocal. In: 20th International Workshop on Database and Expert Systems Application, DEXA 2009, pp. 107–111 (2009)
6. Catrina, O., de Hoogh, S.: Secure multiparty linear programming using fixed-point arithmetic. In: Gritzalis, D., Preneel, B., Theoharidou, M. (eds.) ESORICS 2010. LNCS, vol. 6345, pp. 134–150. Springer, Heidelberg (2010)
7. Catrina, O., Saxena, A.: Secure computation with fixed-point numbers. In: Sion, R. (ed.) FC 2010. LNCS, vol. 6052, pp. 35–50. Springer, Heidelberg (2010)
8. Dahl, M., Ning, C., Toft, T.: On secure two-party integer division. In: Keromytis, A.D. (ed.) FC 2012. LNCS, vol. 7397, pp. 164–178. Springer, Heidelberg (2012)
9. Gentry, C.: Fully homomorphic encryption using ideal lattices. In: STOC 2009, pp. 169–178 (2009)
10. Gentry, C., Halevi, S.: Implementing gentry's fully-homomorphic encryption scheme. In: Paterson, K.G. (ed.) EUROCRYPT 2011. LNCS, vol. 6632, pp. 129–148. Springer, Heidelberg (2011)
11. Kamm, L., Willemson, J.: Secure floating-point arithmetic and private satellite collision analysis. Cryptology ePrint Archive, Report 2013/850 (2013), http://eprint.iacr.org/
12. Kerschbaum, F., Schroepfer, A., Zilli, A., Pibernik, R., Catrina, O., de Hoogh, S., Schoenmakers, B., Cimato, S., Damiani, E.: Secure collaborative supply-chain management. Computer 44(9), 38–43 (2011)
13. Krips, T., Willemson, J.: Hybrid model of fixed and floating point numbers in secure multiparty computations. Cryptology ePrint Archive, Report 2014/221 (2014), http://eprint.iacr.org/
14. Laur, S., Willemson, J., Zhang, B.: Round-Efficient Oblivious Database Manipulation. In: Lai, X., Zhou, J., Li, H. (eds.) ISC 2011. LNCS, vol. 7001, pp. 262–277. Springer, Heidelberg (2011)
15. Liedel, M.: Secure distributed computation of the square root and applications. In: Ryan, M.D., Smyth, B., Wang, G. (eds.) ISPEC 2012. LNCS, vol. 7232, pp. 277–288. Springer, Heidelberg (2012)
16. Liu, Y.-C., Chiang, Y.-T., Hsu, T.S., Liau, C.-J., Wang, D.-W.: Floating point arithmetic protocols for constructing secure data analysis application
17. Machanavajjhala, A., Kifer, D., Gehrke, J., Venkitasubramaniam, M.: L-diversity: Privacy Beyond K-anonymity. ACM Transactions on Knowledge Discovery from Data (TKDD) 1(1) (March 2007)
18. Samarati, P.: Protecting Respondents' Identities in Microdata Release. IEEE Transactions on Knowledge and Data Engineering 13, 1010–1027 (2001)
19. Shamir, A.: How to share a secret. Communications of the ACM 22(11), 612–613 (1979)
20. Sweeney, L.: K-anonymity: A Model for Protecting Privacy. Int. J. Uncertain. Fuzziness Knowl.-Based Syst. 10(5), 557–570 (2002)
21. Zhang, Y., Steele, A., Blanton, M.: Picco: A general-purpose compiler for private distributed computation. In: Proceedings of the 2013 ACM SIGSAC Conference on Computer & Communications Security, CCS 2013, pp. 813–826. ACM, New York (2013)

Exploiting the Floating-Point Computing Power of GPUs for RSA

Fangyu Zheng[1,2,3,*], Wuqiong Pan[1,2,**], Jingqiang Lin[1,2],
Jiwu Jing[1,2], and Yuan Zhao[1,2,3]

[1] State Key Laboratory of Information Security, Institute of Information
Engineering, CAS, China
[2] Data Assurance and Communication Security Research Center, CAS, China
[3] University of Chinese Academy of Sciences, China
{fyzheng,wqpan,linjq,jing,zhaoyuan12}@is.ac.cn

Abstract. Asymmetric cryptographic algorithms (e.g., RSA and ECC) have been implemented on Graphics Processing Units (GPUs) for several years. These implementations mainly exploit the highly parallel GPU architecture and port the integer-based algorithms for common CPUs to GPUs, offering high performance. However, the great potential cryptographic computing power of GPUs, especially by the more powerful floating-point instructions, has not been comprehensively investigated in fact. In this paper, we try to fully exploit the floating-point computing power of GPUs for RSA, by various designs, including the floating-point-based Montgomery multiplication algorithm, the optimization for the fundamental operations and the utilization of the latest thread data sharing instruction `shuffle`. The experimental result on NVIDIA GTX Titan of 2048-bit RSA decryption reaches a throughput of 38,975 operations per second, achieves 2.21 times performance of the existing fastest integer-based work and outperforms the previous floating-point-based implementation by a large margin.

Keywords: GPU, CUDA, Floating-Point, Montgomery Multiplication, RSA.

1 Introduction

With the rapid development of e-commerce and cloud computing, the high-density calculation of digital signature algorithms such as the ECC (Elliptic Curve Cryptography) [15, 19] and RSA [30] algorithms is urgently required. However, without significant development in recent years, CPUs are more and more difficult to keep pace with this rapidly expanding demand. Specialized for compute-intensive and high-parallel computation required by graphics rendering,

* This work was partially supported by the National 973 Program of China under award No. 2014CB340603 and the Strategic Priority Research Program of Chinese Academy of Sciences under Grant XDA06010702.
** Corresponding author.

S.S.M. Chow et al. (Eds.): ISC 2014, LNCS 8783, pp. 198–215, 2014.

GPUs own much more powerful arithmetic capability than CPUs by devoting more transistors to arithmetic processing unit rather than data caching and flow control. With the advent of NVIDIA Compute Unified Device Architecture (CUDA) technology, it is now possible to perform general-purpose computation on GPUs. Due to their powerful arithmetic capability and moderate cost, GPUs are excellent candidates to perform cryptographic acceleration.

Born for high-definition 3D graphics, GPUs require high-speed floating-point processing capability. Therefore, the GPU vendors focus on the development of the floating-point computing power. In NVIDIA GPUs, the floating-point computing power develops rapidly, from 1,300 GFLOPS (Giga Floating-Point Operations Per Second) in 2010 to 5,000 GFLOPS in 2014 [25]. By contrast, the integer multiplication processing power makes slighter progress, achieving only 25% [25, 33] growth.

However, the floating-point instruction set is inconvenient and complicated to realize larger integer modular multiplication which is the core operation of asymmetric cryptography. Furthermore, the floating-point instruction set in the previous GPUs shows no significant performance advantage over the integer one. As far as we know, Bernstein et al. [4] is the first and the only one to utilize the floating-point processing power of CUDA GPUs for asymmetric cryptography. However, compared with their later work [3] based on the integer instruction set, the floating-point-based one only achieves $1/6$ performance. Nevertheless, with the rapid development of GPU floating-point processing power, fully utilizing the floating-point processing resource is a great benefit to asymmetric cryptography implementation in GPUs.

Based on the above observations, in this paper, we propose a new approach to implement high-performance RSA by fully exploiting the floating-point processing power in CUDA GPUs. In particular, we propose a floating-point-based large integer representation and modify the Montgomery multiplication algorithm according to it. Also, we flexibly employ the integer instruction set to supplement the deficiency of the floating-point computing. Besides, we are the first to fully utilize the latest instruction shuffle to share data between threads, which makes the core algorithm Montgomery multiplication a non-memory-access design, decreasing greatly the performance loss in thread communication. With these improvements, the performance of our RSA implementation increases dramatically compared with the previous works. In NVIDIA GeForce GTX Titan, our resulting 2048-bit RSA implementation reaches 38,975 operations per second, which achieves 2.21 times performance of the existing fastest work and performs 13 times faster than the previous floating-point-based implementation.

The rest of our paper is organized as follows. Section 2 introduces the related work. Section 3 presents the overview of GPU, CUDA, floating-point elementary knowledge, RSA and Montgomery multiplication. Section 4 describes our proposed floating-point-based Montgomery multiplication algorithm in detail. Section 5 shows how to implement RSA decryption in GPUs using our proposed Montgomery multiplication. Section 6 analyses performance of proposed algorithm and compares it with previous work. Section 7 concludes the paper.

2   Related Work

Many previous papers report performance results to demonstrate that the GPU architecture can already be used as an asymmetric cryptography workhorse. Large integer modular multiplication is the heart of asymmetric cryptography. Directed at ECC implementation, [1–3, 5, 18, 28, 32] used various methods to implement modular multiplication. Antão et al. [1, 2] and Pu et al. [28] employed Residue Number System (RNS) to parallelize the modular multiplication into several threads. Bernstein et al. [3] and Leboeuf et al. [18] used one thread to handle a modular multiplication with Montgomery multiplication. Bos et al. [5] and Szerwinski et al. [32] used fast reduction to implement modular multiplication over the Mersenne prime fields [31].

Unlike ECC scheme, RSA calculation requires longer and unfixed modulus and depends on modular exponentiation. Many previous work [9–12, 21, 22, 32], made effort to its implementation. Before NVIDIA CUDA was proposed, Moss et al. [21] mapped RNS arithmetic to the GPU to implement a 1024-bit modular exponentiation. Later in CUDA GPUs, Szerwinski et al. [32] and Harrison et al. [9] developed efficient modular exponentiation by both Montgomery multiplication Coarsely Integrated Operand Scanning (CIOS) method and RNS arithmetic. Jang et al. [11] presented a high-performance SSL acceleration using CUDA GPUs. They parallelized Separated Operand Scanning (SOS) method [17] of the Montgomery multiplication by single limb. Jeffrey et al. [12] used the similar technology to implement 256-bit, 1024-bit and 2048-bit Montgomery multiplication. Neves et al. [22] and Henry et al. [10] used one thread to perform single Montgomery multiplication to economize overhead of thread synchronization and communication, however, their implementations resulted in a high latency.

Note that all above works are based on the CUDA integer computing power. Bernstein et al. is the pioneer to utilize CUDA floating-pointing processing power in asymmetric cryptography implementation [4]. They used 28 SPFs (Single Precision Floating-point) to represent 280-bit integer and implemented the field arithmetic. However, the result was barely satisfactory. Their later work [3] in the same platform GTX 295 using integer instruction set performs almost 6.5 times throughput of [4].

3   Background

In this section, we provide a brief introduction to the basic architecture of modern GPUs, the floating-point arithmetic, the basic knowledge of RSA and the Montgomery multiplication.

3.1   GPU and CUDA

CUDA is a parallel computing platform and programming model that enables dramatic increases in computing performance by harnessing the power of GPU. It is created by NVIDIA and implemented by the GPUs that they produce [25].

Our target platform GTX Titan is a CUDA-compatible GPUs with code-name GK-110 [24], which contains 14 streaming multiprocessors (SM). 32 threads (grouped as a *warp*) within one SM can concurrently run in a clock. Following the SIMT (Single Instruction Multiple Threads) architecture, each GPU thread runs one instance of the kernel function. A warp may be preempted when it is stalled due to memory access delay, and the scheduler may switch the runtime context to another available warp. Multiple warps of threads are usually assigned to one SM for better utilization of the pipeline of each SM. These warps are called one *block*. Each SM could access 64KB fast *shared memory*/L1 cache and 64K 32-bit registers. All SMs share 6GB 256-bit wide slow *global memory*, cached read-only *texture memory* and cached read-only *constant memory* [24, 25]. Each SM of GK-110 owns 192 single precision CUDA cores, 64 double precision units, 32 special function units (SFU) and 32 load/store units [24], yielding a throughput of 192 SPF arithmetic, 64 DPF arithmetic, 160 32-bit integer add and 32 32-bit integer multiplication instructions per clock circle [24, 25].

GK-110 also brings a brand new method of data sharing between threads. Previously, shared data between threads requires separated store and load operations to pass data through *shared memory*. First introduced in NVIDIA Kepler architecture, `shuffle` instruction [24, 25] allows threads within a wrap to share data. With shuffle instruction, threads within a wrap can read any value of other thread's in any imaginable permutations [24]. NVIDIA conducted various experiments [26] on the comparison between `shuffle` instruction and *shared memory*, which shows that `shuffle` instruction always gives a better performance than *shared memory*.

3.2   Floating-Point and Integer Arithmetic in GPU

Floating-point arithmetic in CUDA GPUs complies with 754-2008 IEEE Standard for Floating-Point Arithmetic [6]. Among the basic formats which the standard defines, 32-bit binary (single precision floating-point, SPF) and 64-bit binary (double precision floating-point, DPF) are supported in CUDA GPUs.

Table 1. SPF and DPF Basic Formats

	sign	*exp*	biase	*mantissa*
SPF	1 bit	8 bits	0x7F	23 bits
DPF	1 bit	11 bits	0x3FF	52 bits

As demonstrated in Table 1, the real value assumed by a given SPF or DPF data with a sign bit *sign*, a given biased exponent *exp* and a significand precision *mantissa* is $(-1)^{sign} \times 2^{exp-biase} \times 1.mantissa$. Therefore, a SPF and a DPF can respectively represent 24-bit and 53-bit integer.

When executing floating-point multiplication, to avoid precision loss, each SPF or DPF must contain lower than or equal to $\lfloor \frac{24}{2} \rfloor = 12$ or $\lfloor \frac{53}{2} \rfloor = 26$ significant bits. And in NVIDIA Kepler architecture, SPF offers 3 times the multiplication instruction throughput of DPF. Taking all above factors into consideration,

SPF can only provide $(\frac{12}{26})^2 \times 3 = 63.9\%$ multiplication performance of DPF. Therefore, in this paper, we focus on the DPF-based RSA implementation.

3.3   RSA and Montgomery Multiplication

RSA [30] is an asymmetric cryptography algorithm widely used for digital signature, whose core operation is modular exponentiation. In practice, CRT (Chinese Remainder Theorem) [29] technology is widely used to promote the RSA decryption. Instead of calculating k-bit modular exponentiations directly, we can perform two $k/2$-bit modular exponentiations (Equation (1a) & (1b)) and the Mixed-Radix Conversion (MRC) algorithm [16] (Equation (2)) to conduct the RSA decryption.

$$Plain_1 = Cipher^{d \mod (p-1)} \mod p \qquad (1a)$$

$$Plain_2 = Cipher^{d \mod (q-1)} \mod q \qquad (1b)$$

$$Plain = Plain_2 + [(Plain_1 - Plain_2) \times (q^{-1} \mod p) \mod p] \times q \qquad (2)$$

where p and q are $k/2$-bit prime numbers chosen in private key generation ($M = p \times q$). All parameters, p, q, $(d \mod p-1)$, $(d \mod q-1)$ and $(q^{-1} \mod p)$ are part of the RSA private key [13]. Compared with calculating k-bit modular exponentiations directly, the CRT technology gives 3 times performance promotion [29].

Algorithm 1. High-radix Montgomery Multiplication According to [27]

Input:
   $M > 2$ with gcd(M,2)=1, , positive integers n, w such that $2^{wn} > 4M$
   $M' = -M^{-1} \mod 2^w$, $R^{-1} = (2^{wn})^{-1} \mod M$
   Integer multiplicand A, where $0 \le A < 2M$
   Integer multiplier B, where $B = \sum_{i=0}^{n-1} b_i 2^{wi}$, $0 \le b_i < 2^w$ and $0 \le B < 2M$
Output:
   An integer S such that $S = ABR^{-1} \mod 2M$ and $0 \le S < 2M$;
1: $S = 0$
2: **for** $i = 0$ *to* $n - 1$ **do**
3:    $S = S + A \times b_i$
4:    $q_i = ((S \mod 2^w) \times M') \mod 2^w$
5:    $S = (S + M \times q_i)/2^w$
6: **end for**

In RSA, modular multiplication is the bottleneck of its overall performance. In 1985 Peter L. Montgomery proposed an algorithm [20] to remove the costly division operation from the modular reduction. Let $\bar{A} = AR(mod\ M)$, $\bar{B} = BR(mod\ M)$ be the Montgomeritized form of A, B modulo M, where R and M are coprime and $M \le R$. Montgomery multiplication defines the multiplication between 2 Montgomeritized form numbers, $MonMul(\bar{A}, \bar{B}) = \bar{A}\bar{B}R^{-1}$

(mod M). Even though the algorithm works for any R which is relatively prime to M, it is more useful when R is taken to be a power of 2, which lead to a fast division by R.

In 1995, Orup et al. [27] economized the determination of quotients by loosening the restriction for input and output from $[0, M)$ to $[0, 2M)$. Algorithm 1 shows the detailed steps.

4   DPF-Based Montgomery Multiplication

In this section, we propose a DPF-based Montgomery multiplication parallel algorithm in CUDA GPUs. The algorithm includes the large integer representation, the fundamental operations and the parallelism method of Montgomery multiplication.

4.1   Advantage and Challenges of DPF-Based RSA

Large integer modular multiplication is the crucial part of asymmetric cryptographic algorithm and it largely depends on the single-precision multiplication (or multiply-add) especially when using Montgomery multiplication. In GTX Titan, DPF multiplication (or multiply-add) instruction provides double throughput (64/Clock/SM) of the 32-bit integer multiplication instruction (32/Clock/SM) [25]. In CUDA, the product of 32-bit multiplication requires 2 integer multiplication instructions, one for the lower half, the other for the upper half. Thus, DPF can perform about $\frac{64}{32/2} \times (\frac{26}{32})^2 = 2.64$ (DPF supports at most 26-bit multiplication which is discussed in Section 3.2) times multiplication processing power of 32-bit integer in GTX Titan.

However, DPF encounters many problems when used in asymmetric cryptographic algorithm which largely depends on large integer multiplication.

- **Round-off Problem:** Unlike integer, the floating-point add or multiply-add instruction does not support carry flag (CF) bit. When the result of the add instruction is beyond the limit bits of significand (24 for SPF, 53 for DPF), the round-off operation will happen, in which the least significant bits will be left out to keep the significand within the limitation.
- **Inefficient Bitwise Operations:** Floating-point arithmetic does not support bitwise operations which are frequently used in our algorithm. CUDA Math API does support the __fmod function [23], which can be employed to extract the least and most significant bits. But it consumes tens of instructions which is extremely inefficient, while using integer native instructions set, the bitwise operation needs only one instruction.
- **Inefficient Add Instruction:** Unlike multiplication instruction, the DPF add instruction is slower than the integer add instruction. In a single clock circle, each SM can execute respectively 64 DPF and 160 integer add instructions [25]. Even taking the instruction word length (53 bits for DPF, 32 bits for integer) into consideration, DPF add instruction only performs nearly

2/3 processing power of the integer one. In our implementation, we use at most 30-bit add operation, which intensifies this performance disadvantage further.

- **Extra Memory and Register File Cost:** A DPF occupies 64-bit memory space, however, only 26 or lower bits are used. In this way, we have to pay $\frac{64-26}{26} = 138\%$ times extra cost in memory access and utilization of register files. While, in integer-based implementation, this issue is not concerned since every bit of a integer is utilized.

4.2 DPF-Based Representation of Large Integer

In Montgomery multiplication, multiply-accumulation operation $s = a \times b + s$ is frequently used. In CUDA, fused multiply-add (FMA) instruction is provided to perform floating-point multiply-add operation, which can be executed in one step with a single rounding. Note that, when each DPF (a and b) contains w ($w \leq 26$) significant bits, we can execute $2^{53}/2^{2w} = 2^{53-2w}$ times of $s = a \times b + s$ (the initial value of s is zero), free from the round-off problem. For example, if we configure 26 significant bits per DPF, after $2^{53-52} = 2$ times of $s = a \times b + s$ operations, s is 53 bits long. In order to continue executing $s = a \times b + s$, we have to restrict s within 52 bits over and over again. This operations is very expensive, due to the complexity of bit extraction. Therefore, we should carefully configure it to get rid of the expensive bit extraction. In Algorithm 1, there are n loops and each loop contains 2 FMA operations for each limb, where $n = \lceil \frac{1024+2}{w} \rceil$. Thus $2 \times \lceil \frac{1024+2}{w} \rceil$ times of FMA operations are needed. Table 2 shows the number of FMAs supported and needed with varying significant bits in each DPF.

Table 2. Choose the Best w for Proposed 1024-bit Montgomery Multiplication

Bits/DPF w	FMAs Supported 2^{53-2w}	FMAs Needed in Algorithm 1 $2 \times \lceil \frac{1024+2}{w} \rceil$	Radix $R = w \times \lceil \frac{1024+2}{w} \rceil$
26	2	80	1040
25	8	84	1050
24	32	86	1032
23	**128**	**90**	**1035**
22	512	94	1034
21	2048	98	1029
20	8192	104	1040

From Table 2, we find that the supported number of FMAs starts to be larger than which is needed, when w is 23. The lower w means we need more instructions to process the whole algorithm. To achieve the best performance, we choose $w = 23$ and the radix $R = 2^{1035}$.

In this paper, we propose 2 kinds of DPF-based large integer representations, Simplified Format and Redundant Format.

Simplified format formats like $A = \sum_{i=0}^{44} 2^{23i} a[i]$, where each limb $a[i]$ contains at most 23 significant bits. It is applied to represent the input of the DPF FMA instruction.

Redundant format formats like $A = \sum_{i=0}^{44} 2^{23i} a[i]$, where each limb $a[i]$ contains at most 53 significant bits. It is applied to accommodate the output of the DPF FMA instruction.

4.3   Fundamental Operation and Corresponding Optimization

In DPF-based Montgomery multiplication, fundamental operations include multiplication, multiply-add, addition and bit extraction.

- **Multiplication:** In our implementation, native multiplication instruction (`mul.f64`) is used to perform multiplication. We require that both multiplicand and multiplier are in Simplified format to avoid round-off problem.
- **Multiply-Add:** In CUDA GPUs, fused multiply-add (`fma.f64`) instruction is provided to perform floating-point $s = a \times b + c$, which can be executed in one step with a single rounding. In our implementation, when using FMA instruction, we require that multiplicand a and multiplier b are both in Simplified format and addend c is in Redundant format but less than $2^{53} - 2^{46}$.
- **Bit Extraction:** In our implementation, we need to extract the most or least significant bits from a DPF. However, as introduced in Section 4.1, the bitwise operation for DPF is inefficient. To promote the performance, we try 2 improvements. The first one is introduced in [8]. Using round-to-zero, we can perform

$$x = a + 2^{52+53-r}$$
$$u = x - 2^{52+53-r}$$
$$v = a - u$$

  to extract the most significant r bits u and the least significant $(53-r)$ bits v from a DPF a. Note that, in $x = a + 2^{52+53-r}$, the least significant $(53-r)$ bits will be leaved out due to the round-off operation. The second one is converting DPF to integer then using CUDA native 32-bit integer bitwise instruction to handle bit extraction. Through our experiment, we find that the second method always gives a better performance. Therefore, in most of cases, we use the second method to handle bit extraction. There is only one exception when the DPF a is divisible by a 2^r, we can use $a/2^r$ to extract the most significant $53 - r$ bits, which can be executed very fast.
- **Addition:** CUDA GPUs provide the native add instruction to perform addition between two DPFs. But it has two shortcomings: 1. its throughput is low, 64/CLOCK/SM, while, the integer add instruction is 160/CLOCK/SM; 2. it does not provide support for carry flag. Thus, we convert DPF to integer and use CUDA native integer add instruction to handle addition.

4.4 DPF-Based Montgomery Multiplication Algorithm

According to Algorithm 1, in our Montgomery multiplication, $S = ABR^{-1}$ ($mod\ M$), A, B, M and S are all (1024+1)-bit integer (in fact, M is 1024 bits long, we also represent it as a 1025-bit integer for a common format). As choosing $w = 23$, we need $\lceil \frac{1025}{23} \rceil = 45$ DPF limbs to represent A, B, M and S.

In the previous work [11, 12], Montgomery multiplication is parallelized by single limb, that is, each thread deals with one limb (32-bit or 64-bit integer). The one-limb parallelism causes large cost in the thread synchronization and communication, which decreases greatly the overall throughput. To maximize the throughput, Neves et al. [22] performed one entire Montgomery multiplication in one thread to economize overhead of thread synchronization and communication, however, it results in a high latency, about 150ms for 1024-bit RSA, which is about 40 times of [11].

To make a tradeoff between throughput and latency, in our implementation, we try multiple-limb parallelism, namely, using r threads to compute one Montgomery multiplication and each thread dealing with t limbs, where $t = \lceil \frac{45}{r} \rceil$. The degree of parallelism r can be flexibly configured to offer the maximal throughput with acceptable latency. Additionally, we restrict threads of one Montgomery multiplication within a wrap, in which threads are naturally synchronized free from the overhead of thread synchronization and `shuffle` instruction can be used to share data between threads. To fully occupy thread resource, we choose r, where $r \mid 32$.

In our Montgomery multiplication $S = ABR^{-1}(\ mod\ M)$, the input A, B, M are in Simplified format and the initial value of S is 0. We use 2 phases to handle one Montgomery multiplication, *Computing Phase* and *Converting Phase*. In Computing Phase, we use Algorithm 1 to calculate S, whose result is represented in Redundant format. In Converting Phase, we convert S from Redundant format to Simplified format.

Computing Phase. Algorithm 2 is a t-limb parallelized version of Algorithm 1. Using it, we can complete the Computing Phase.

We divide both the input A, B, M and the output S into $r = \lceil \frac{45}{t} \rceil$ groups and distribute the groups into each thread. For Thread k, the variables with index $tk \sim tk + t - 1$ are processed in it. Note that if the index of certain variable is larger than 44, we pad the variable with zero. In this section, the lowercase variables (index varies from 0 to $t - 1$) indicate the private registers stored in the thread and the uppercase ones (index varies from 0 to $tr - 1$) represent the global variables. For example, $a[j]$ in Thread k represents the $A[tk + j]$. And $v_1 = \mathtt{shuffle}(v_2, k)$ means that current thread obtains variable v_2 of Thread k and stores it into variable v_1.

By reference to Algorithm 1, we introduce our parallel Montgomery multiplication algorithm step by step.

(1) $S = S + A \times b_i$: In this step, each Thread k respectively calculates $S[tk : tk + t - 1] = S[tk : tk + t - 1] + A[tk : tk + t - 1] \times B[i]$. In our design, each thread stores a group of B. Thus firstly we need to broadcast the corresponding $B[i]$ from

Algorithm 2. DPF-based parallel 1024-bit Montgomery Multiplication ($S = ABR^{-1} \mod M$) Algorithm: Computing Phase

Input:

Number of processed limbs per thread t ;

Number of threads per Montgomery multiplication r, where $r = \lceil \frac{45}{t} \rceil$;

Thread ID k, where $0 \leq k \leq r - 1$;

Radix $R = 23 \times 45$, Modulus M, $M' = -M^{-1} \mod 2^{23}$;

Multiplicand A, where $A = \sum_{i=0}^{rt-1} A[i]2^{23i}$, $0 \leq A[i] < 2^{23}$ and $0 \leq A < 2M$;

Multiplier B, where $B = \sum_{i=0}^{rt-1} B[i]2^{23i}$, $0 \leq B[i] < 2^{23}$ and $0 \leq B < 2M$;

a, b and m consist of t DPF limb, respectively representing $A[tk : tk + t - 1]$, $B[tk : tk + t - 1]$ and $M[tk : tk + t - 1]$

Output:

Redundant-formatted sub-result $s[0 : t - 1] = S[tk : tk + t - 1]$

1: $S = 0$

2: **for** $i = 0$ *to* 44 **do**

3:     $b_i = \text{shuffle}(b[i \mod t], \lfloor i/t \rfloor)$

       /* Step (1): $S = S + A \times b_i$ */

4:     **for** $j = 0$ to $t - 1$ **do**

5:         $s[j] = s[j] + a[j] \times b_i$

6:     **end for**

       /* Step (2): $q_i = ((S \mod 2^w) \times M')(\mod 2^w)$ */

7:     **if** $k = 0$ **then**

8:         $temp = s[0] \mod 2^{23}$

9:         $q_i = temp \times M'$

10:        $q_i = q_i \mod 2^{23}$

11:    **end if**

12:    $q_i = \text{shuffle}(q_i, 0)$

       /* Step (3): $S = S + M \times q_i$ */

13:    **for** $j = 0$ to $t - 1$ **do**

14:        $s[j] = s[j] + m[j] \times q_i$

15:    **end for**

       /* Step (4): $S = S/2^w$ */

16:    **if** $k \neq r - 1$ **then**

17:        $temp = \text{shuffle}(s[0], k + 1)$

18:    **else**

19:        $temp = 0$

20:    **end if**

21:    **if** $k = 0$ **then**

22:        $s[0] = s[0] >> 23 + s[1]$

23:    **else**

24:        $s[0] = s[1]$

25:    **end if**

26:    **for** $j = 1$ to $t - 2$ **do**

27:        $s[j] = s[j + 1]$

28:    **end for**

29:    $s[t - 1] = temp$

30: **end for**

certain thread to all threads. In the view of single thread, $B[i]$ corresponds to $b[i$ mod $t]$ of Thread $\lfloor i/t \rfloor$. We use shuffle to conduct this broadcast operation. Then each thread execute $s[j] = s[j] + a[j] \times b_i$ where $j \in [0, t-1]$.

(2) $q_i = ((S \mod 2^w) \times M')(\mod 2^w)$: $S \mod 2^w$ is only related to $S[0]$, which is stored in Thread 0 as $s[0]$. Therefore, in this step, we conduct this calculation only in Thread 0. Note that S is in Redundant format, we should firstly extract the least significant 23 bits $temp$ from $S[0]$ then execute $q_i = temp \times M'$. And in next step, q_i will act as a multiplier, hence, we should also extract the least significant 23 bits of q_i.

(3) $S = S + M \times q_i$: In this step, each Thread k respectively calculates $S[tk : tk+t-1] = S[tk : tk+t-1] + M[tk : tk+t-1] \times q_i$. Because q_i is stored only in Thread 0, firstly we should broadcast it to all threads. Similar with Step (1), then each thread executes $s[j] = s[j] + m[j] \times q_i$ where where $j \in [0, t-1]$.

(4) $S = S/2^w$: In this step, each thread conducts a division by 2^w by shifting right operation. In the view of the overall structure, shifting right can done by executing $S[k] = S[k+1]$ $(0 \le k \le rt-2)$ and padding $S[rt-1]$ with zero. In the view of single thread, there are 2 noteworthy points. The first point is that Thread k needs to execute $S[tk+t-1] = S[t(k+1)]$ but $S[t(k+1)]$ is stored in Thread $(k+1)$, thus Thread $(k+1)$ needs to propagate its stored $S[t(k+1)]$ to Thread j. Note that $S[t(k+1)]$ is corresponding to $s[0]$ of Thread $k+1$ and to avoid to override variable Thread k uses $temp$ to store it. The second point is that we represent S in Redundant format, when executing $S = S/2^w$ $(w = 23)$, the upper-30-bit of the least significant limb $S[0]$ needs to be stored. Thus, in Thread 0, we use $s[0] = s[0] >> 23 + s[1]$ instead of $s[0] = s[1]$ in other threads. $S[0]/2^{23}$ is at most $(53-23)=30$ bits long, and due to the right shift operation, the $S[0]$ in each loop is not the same. Thus $S[0]$ can be only accumulated within $45 \times 2 \times 2^{23} \times 2^{23} + 2^{30} < 2^{53}$, which does not cause round-off problem. Note that $S[0]$ is divisible by 2^{23}, as introduced in Section 4.3, we use $S[0]/2^{23}$ to extract the most significant 30 bits.

After Computing Phase, S is in Redundant format. Next, we use Converting Phase to convert S into Simplified format.

Converting Phase. In Converting Phase, we convert S from Redundant format to Simplified format: every $S[k]$ adds the $carry$ ($S[0]$ does not execute this addition) and holds the least significant 23 bits of the sum and propagates the most significant 30 bits as new $carry$ to $S[k+1]$. However, this method is serial, the calculation of every $S[k]$ depends on the $carry$ that $S[k-1]$ produces, which does not comply with the architecture of GPU. In practice, we use parallelized method to accomplish Converting Phase, which is shown in Algorithm 3.

Algorithm 3 uses symbol $split(c) = (h, l) = (\lfloor \frac{c}{2^{23}} \rfloor, c \mod 2^{23})$ to denote that we divide 53-bit number c into 30 most significant bits h and 23 least significant bits l. Firstly, all threads execute a chain addition for its $s[0] \sim s[t-1]$ and store the $carry$ that the last additions produces. Then every Thread $k-1$ (except Thread $(r-1)$) propagates the stored $carry$ to Thread k using shuffle instruction, then repeats chain addition with the propagated $carry$. This step continues until $carry$ of every thread is zero. We use the CUDA __any() voting

Algorithm 3. DPF-based parallel Montgomery Multiplication $(S = ABR^{-1}$ mod $M)$ Algorithm: Converting Phase

---

Input:
  Thread ID k
  Number of processed limbs per thread t ;
  Number of threads per Montgomery multiplication r, where $r = \lceil \frac{45}{t} \rceil$;
  Redundant-formatted sub-result $s[0 : t - 1] = S[tk : tk + t - 1]$
Output:
  Simplified-formatted sub-result $s[0 : t - 1] = S[tk : tk + t - 1]$
 1: $carry = 0$
 2: **for** $j = 0$ to $t - 1$ **do**
 3:   $s[j] = s[j] + carry$
 4:   $(carry, s[j]) = split(s[j])$
 5: **end for**
 6: **while** $carry$ of any thread is non-zero **do**
 7:   **if** $k = 0$ **then**
 8:     $carry = 0$
 9:   **else**
10:     $carry = \texttt{shuffle}(carry, k - 1)$
11:   **end if**
12:   **for** $j = 0$ to $t - 1$ **do**
13:     $s[j] = s[j] + carry$
14:     $(carry, s[j]) = split(s[j])$
15:   **end for**
16: **end while**

---

instruction to check $carry$ of each thread. The number of the iterations is $(r-1)$ in the worst case, but for most cases it takes one or two. Compared with the serialism method, we can save over 75% execution time in Converting Phase using the parallelism method.

After Converting Phase, S is in Simplified format. An entire Montgomery multiplication is completed.

5   RSA Implementation

This section introduces how to implement Montgomery exponentiation using our proposed Montgomery multiplication. We also discuss the CRT computation.

5.1   Montgomery Exponentiation

Using CRT, We need to perform 2 1024-bit Montgomery exponentiation $S = X^Y R \ (mod \ M)$ to accomplish 2048-bit RSA decryption. In Montgomery exponentiation, we represent the exponent Y in integer. Similar with our processing of shared data B in Montgomery multiplication $S = ABR^{-1}$, each thread stores a piece of Y and uses $\texttt{shuffle}$ to broadcast Y from certain thread to all threads.

With the binary square-and-multiply method, the expected number of modular multiplications is $3k/2$ for k-bit modular exponentiation. The number can be reduced with m-ary method given by [14] that scans multiple bits, instead of one bit of the exponent. We have implemented 2^6-ary method and reduced the number of modular multiplications from 1536 to 1259, achieving 17.9% improvement. Using 2^6-ary method, we need to store $(2^6 - 2)$ pre-processed results $(X^2R \sim X^{63}R)$ into memory. The memory space that pre-processed results needed (about 512KB) is far more than the size of *shared memory* (at most 48KB), thus we have to store them into *global memory*. Global memory load and store consume hundreds of clock circles. To improve memory access efficiency, we first convert the Simplified formatted pre-processed results into 32-bit integer format then store the integer-formatted results in global memory. This optimization saves about 50% memory access latency.

5.2   CRT Computation

In our first implementation, GPU only took charge of the Montgomery exponentiation. And the CRT computation (Equation (2)) was offloaded to CPU (Intel E3-1230 v2) using GNU multiple precision (GMP) arithmetic library [7]. But we find the low efficiency of CPU computing power greatly limits the performance of the entire algorithm, which occupies about 15% of the execution time. Thus we integrate the CRT computation into GPU. For CRT computation, we additionally implement a modular subtraction and a multiply-add function. Both functions are parallelized in the threads which take charge of Montgomery exponentiation. The design results in that the CRT computation occupies only about 1% execution time and offers the independence of CPU computing capability. The detailed schematic diagram is shown in Fig. 1.

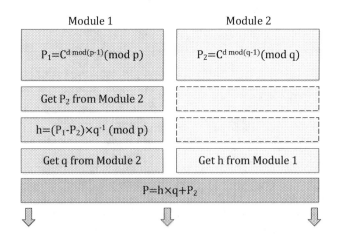

Fig. 1. Intra-GPU CRT implementation

6   Performance Evaluation

In this section we discuss the implementation performance and summarize the results for the proposed algorithm. Relative assessment is also presented by considering related implementation.

6.1   Implementation Result

Using the DPF-based Montgomery multiplication algorithm and RSA implementation respectively described in Section 4 and Section 5, we implement 2048-bit RSA decryption in NVIDIA GTX Titan.

Table 3 summarizes the performance of our DPF-based RSA implementation. $Thrds/RSA=t \times 2$ indicates we use t threads to process a Montgomery multiplication and $t \times 2$ threads to process one RSA decryption. Theoretically, $Thrds/RSA$ can vary from 2×2 to 16×2, however, when using 2 threads to handle a DPF-based Montgomery multiplication, the number of registers per thread surpasses the GPU hardware limitation, 255 registers per thread, thus many variables need to be stored in memory, which decreases greatly the overall performance. Thus we restrict $Thrds/RSA$ from 4×2 to 16×2.

Table 3. Performance of 2048-bit RSA Encryption

Thrds/RSA	Throughput (/s)	Latency (ms)	RSAs/launch	MonMul Throughput $(10^6/s)$
4×2	38,975	22.47	896	110.5
8×2	36,265	18.53	672	102.8
16×2	27,451	16.32	448	78.4

Varying the number of RSA decryptions per launch, we record the performance respectively for 3 configurations of $Thrds/RSA$. We also record its corresponding $Latency$ and $RSAs/launch$, which represent respectively how much time single GPU calculation costs and how many RSA encryptions it contains. For comparison with the work which only implements modular multiplication, we also measure our performance of 1024-bit modular multiplication, which is demonstrated in the $MonMul$ $Throughput$ column.

From Table 3, we can find that the less $Thrds/RSA$ gives higher throughput due to the decreased computational cost for data traffic between threads. When $Thrds/RSA$ is 4×2, our implementation gives the highest throughput of 38,975 RSA decryptions per second.

6.2   Performance Comparison

Proposed vs. Floating-point-based Implementation. Bernstein et al. [4] employed the floating-point arithmetic to implement 280-bit Montgomery multiplication, thus, we scale its performance by floating-point processing power.

Table 4 shows performance of the work [4] and ours. Note that the work [4] implemented the 280-bit modular multiplication, thus we multiply its performance by $(\frac{280}{1024})^2$ as the performance of the 1024-bit one. "1024-bit MulMod (scaled)" indicates the performance scaled by the platform floating-point processing power.

Table 4. Performance Comparison of Bernstein et al. [4] and ours

	GTX 295 Bernstein et al. [4]	GTX Titan Proposed
Floating-point Processing Power (GFLOPS)	1788	4500
Scaling Factor	0.397	1
280-bit MulMod ($\times 10^6$)	41.88	-
1024-bit MulMod ($\times 10^6$)	3.13	110.5
1024-bit MulMod (scaled) ($\times 10^6$)	7.89	110.5

Our implementation achieves 13 times speedup than [4]. Part of our performance promotion results from the advantage DPF achieves over SPF as discussed in Section 3.2. The reason why they did not utilize DPF is that GTX 295 they used does not support DPF instructions. The second reason of our advantage comes from the process of Montgomery multiplication. Bernstein et al. used Separated Operand Scanning (SOS) Montgomery multiplication method which is known inefficient [17]. And they utilized only 28 threads of a wrap (consists of 32 threads), which wastes 1/8 processing power. The third reason is that we used the CUDA latest `shuffle` instruction to share data between threads, while [4] used *shared memory*. As Section 3.1 introduced, the `shuffle` instruction gives a better performance than shared memory. The last reason lies in that Bernstein et al. [4] used floating-point arithmetic to process all operations, some of which are more efficient using integer instruction such as bit extraction. By contrast, we flexibly utilize integer instructions to accomplish these operations.

Proposed vs. Integer-based Implementation. For fair comparison, we firstly evaluate the CUDA platform of each work, and scale their performance into our GPU hardware GTX Titan. [10–12, 22, 32] are all based on integer arithmetic. Thus we scale their CUDA platform performance based on the integer processing power. The parameters in Table 5 origin from [25] and [33], but the integer processing power is not given directly. Taking SM Number, processing power of each SM and shader clock into consideration, we calculate integer multiplication instruction throughput *Int Mul*. Among them, 8800GTS and GTX 260 support only 24-bit multiply instruction, while, the other platforms support 32-bit multiply instruction. Hence, we adjust their integer multiply processing capability by a correction parameter $(\frac{24}{32})^2$ (unadjusted data is in parenthesis). *Scaling Factor* is defined as *Int Mul* ratio between the corresponding CUDA platform and GTX Titan.

Table 5 summarizes the resulting performance of each work. We divide each resulting performance by the corresponding *Scaling Factor* listed in Table 5 as the scaled performance. Note that the RSA key length of Neves et al. [22] is 1024 bits, while ours is 2048 bits, we multiply it by an additional factor $1/4 \times 1/2 = 1/8$ (1/4 for the performance of modular multiplication, 1/2 for the half bits of the exponent). And Szerwinski et al. [32] accomplished non-CRT 1024-bit Montgomery exponentiation with 813/s throughput, thus we divide it by an additional factor 2 as the performance of the CRT 2048-bit RSA decryption.

Table 5. Throughput and Latency of Operations per second

	Szerwinski et al. [32]	Neves et al. [22]	Henry et al. [10]	Jang et al. [11]	Robinson et al. [12]	Ours
CUDA platform	8800GTS	GTX 260	M2050	GTX 580	GTX Titan	
SM Number	16	24	14	16	14	
Shader Clock (GHz)	1.625	1.242	1.150	1.544	0.836	
Int Mul/SM (/Clock)	8 (24-bit)	8 (24-bit)	16	16	32	
Int Mul (G/s)	117 (208)	134 (238)	258	395	375	
Scaling Factor	0.312	0.357	0.688	1.053	1	
1024-bit MulMod (10^6/s)	-	-	11.1	-	49.8	110.5
MulMod (scaled) (10^6/s)	-	-	16.1	-	49.8	110.5
RSA-1024(/s)	813(non-CRT)	41,426	-	-	-	-
RSA-2048(/s)	406.5		-	12,044	-	38,975
RSA (scaled)(/s)	1,303	14,504	-	11,438	-	38,975

From Table 5, we can see that our modular multiplication implementation outperforms the others by a great margin. We achieve nearly 6 times speedup than the work [10], and even at the same CUDA platform, we obtain 221% performance of the work [12]. Our RSA implementation also shows a great performance advantage, 269% performance of the next fasted work [22]. Note that 1024-bit RSA decryption of [22] have latency of about 150ms, while 2048-bit RSA decryption of ours reaches 22.47ms.

The great performance advantage lies mainly in the utilization of floating-point processing power and the superior handling of Montgomery multiplication. Besides, compared with the work using multiple threads to process a Montgomery multiplication [11, 12], another vital reason is that we use more efficient `shuffle` instruction to handle data sharing instead of shared memory and employ more reasonable degree of thread parallelism to economize the overhead of thread communication.

7 Conclusion

In this contribution, we propose a brand new approach to implement high-performance RSA cryptosystems in latest CUDA GPUs by utilizing the powerful floating-point computing resource. Our results demonstrate that the floating-point computing resource is a more competitive candidate for the asymmetric cryptography implementation in CUDA GPUs. In the NVIDIA GTX Titan platform, our 2048-bit RSA decryption achieves 2.21 times performance of the existing fastest integer-based work and performs 13 times faster than the previous floating-point-based implementation. We will try to apply these designs to other asymmetric cryptography, such as the floating-point-based ECC implementation.

References

1. Antão, S., Bajard, J.C., Sousa, L.: Elliptic curve point multiplication on GPUs. In: IEEE International Conference on Application-specific Systems Architectures and Processors (ASAP), pp. 192–199. IEEE (2010)
2. Antão, S., Bajard, J.C., Sousa, L.: RNS-Based elliptic curve point multiplication for massive parallel architectures. The Computer Journal 55(5), 629–647 (2012)
3. Bernstein, D.J., Chen, H.C., Chen, M.S., Cheng, C.M., Hsiao, C.H., Lange, T., Lin, Z.C., Yang, B.Y.: The billion-mulmod-per-second PC. In: Workshop Record of SHARCS, vol. 9, pp. 131–144 (2009)
4. Bernstein, D.J., Chen, T.-R., Cheng, C.-M., Lange, T., Yang, B.-Y.: ECM on graphics cards. In: Joux, A. (ed.) EUROCRYPT 2009. LNCS, vol. 5479, pp. 483–501. Springer, Heidelberg (2009)
5. Bos, J.W.: Low-latency elliptic curve scalar multiplication. International Journal of Parallel Programming 40(5), 532–550 (2012)
6. IEEE Standards Committee, et al.: 754-2008 IEEE standard for floating-point arithmetic. IEEE Computer Society Std. 2008 (2008)
7. Granlund, T.: the gmp development team. gnu mp: The gnu multiple precision arithmetic library, 5.1 (2013)
8. Hankerson, D., Vanstone, S., Menezes, A.J.: Guide to elliptic curve cryptography. Springer (2004)
9. Harrison, O., Waldron, J.: Efficient acceleration of asymmetric cryptography on graphics hardware. In: Preneel, B. (ed.) AFRICACRYPT 2009. LNCS, vol. 5580, pp. 350–367. Springer, Heidelberg (2009)
10. Henry, R., Goldberg, I.: Solving discrete logarithms in smooth-order groups with CUDA. In: Workshop Record of SHARCS, pp. 101–118. Citeseer (2012)
11. Jang, K., Han, S., Han, S., Moon, S., Park, K.: Sslshader: Cheap ssl acceleration with commodity processors. In: Proceedings of the 8th USENIX Conference on Networked Systems Design and Implementation, p. 1. USENIX Association (2011)
12. Jeffrey, A., Robinson, B.D.: Fast GPU Based Modular Multiplication, http://on-demand.gputechconf.com/gtc/2014/poster/pdf/P4156_montgomery_multiplication_CUDA_concurrent.pdf
13. Jonsson, J., Kaliski, B.: Public-key cryptography standards (PKCS)# 1: RSA cryptography specifications version 2.1 (2003)
14. Knuth, D.E.: The Art of Computer Programming: Seminumerical Algorithms, vol. 2, p. 116. Addison-Wesley, Reading (1981)
15. Koblitz, N.: Elliptic curve cryptosystems. Mathematics of Computation 48(177), 203–209 (1987)
16. Koç, C.K.: High-speed RSA implementation. Technical report, RSA Laboratories (1994)

17. Koç, Ç.K., Acar, T., Kaliski Jr., B.S.: Analyzing and comparing Montgomery multiplication algorithms. IEEE Micro 16(3), 26–33 (1996)
18. Leboeuf, K., Muscedere, R., Ahmadi, M.: A GPU implementation of the Montgomery multiplication algorithm for elliptic curve cryptography. In: 2013 IEEE International Symposium on Circuits and Systems (ISCAS), pp. 2593–2596. IEEE (2013)
19. Miller, V.S.: Use of elliptic curves in cryptography. In: Williams, H.C. (ed.) CRYPTO 1985. LNCS, vol. 218, pp. 417–426. Springer, Heidelberg (1986)
20. Montgomery, P.L.: Modular multiplication without trial division. Mathematics of Computation 44(170), 519–521 (1985)
21. Moss, A., Page, D., Smart, N.P.: Toward acceleration of RSA using 3D graphics hardware. In: Galbraith, S.D. (ed.) Cryptography and Coding 2007. LNCS, vol. 4887, pp. 364–383. Springer, Heidelberg (2007)
22. Neves, S., Araujo, F.: On the performance of GPU public-key cryptography. In: 2011 IEEE International Conference on Application-Specific Systems, Architectures and Processors (ASAP), pp. 133–140. IEEE (2011)
23. NVIDIA: NVIDIA CUDA Math API, http://docs.nvidia.com/cuda/cuda-math-api/index.html#axzz308wmibga
24. NVIDIA: NVIDIA GeForce Kepler GK110 Writepaper, http://159.226.251.229/videoplayer/NVIDIA-Kepler-GK110-Architecture-Whitepaper.pdf?ich_u_r_i=e1d64c09bd2771cfc26f9ac8922d9e6d&ich_s_t_a_r_t=0&ich_k_e_y=1445068925750663282471&ich_t_y_p_e=1&ich_d_i_s_k_i_d=1&ich_u_n_i_t=1
25. NVIDIA: CUDA C Programming Guide 5.5 (2013), http://docs.nvidia.com/cuda/cuda-c-programming-guide/index.html
26. NVIDIA: Shuffle: Tips and Tricks (2013), http://on-demand.gputechconf.com/gtc/2013/presentations/S3174-Kepler-Shuffle-Tips-Tricks.pdf
27. Orup, H.: Simplifying quotient determination in high-radix modular multiplication. In: Proceedings of the 12th Symposium on Computer Arithmetic, pp. 193–199. IEEE (1995)
28. Pu, S., Liu, J.-C.: EAGL: An Elliptic Curve Arithmetic GPU-Based Library for Bilinear Pairing. In: Cao, Z., Zhang, F. (eds.) Pairing 2013. LNCS, vol. 8365, pp. 1–19. Springer, Heidelberg (2014)
29. Quisquater, J.J., Couvreur, C.: Fast decipherment algorithm for RSA public-key cryptosystem. Electronics Letters 18(21), 905–907 (1982)
30. Rivest, R.L., Shamir, A., Adleman, L.: A method for obtaining digital signatures and public-key cryptosystems. Communications of the ACM 21(2), 120–126 (1978)
31. Solinas, J.A.: Generalized mersenne numbers. Citeseer (1999)
32. Szerwinski, R., Güneysu, T.: Exploiting the power of GPUs for asymmetric cryptography. In: Oswald, E., Rohatgi, P. (eds.) CHES 2008. LNCS, vol. 5154, pp. 79–99. Springer, Heidelberg (2008)
33. Wikipedia: Wikipedia: List of NVIDIA graphics processing units (2014), http://en.wikipedia.org/wiki/Comparison_of_NVIDIA_Graphics_Processing_Units

On Formally Bounding Information Leakage by Statistical Estimation[*]

Michele Boreale[1],[**] and Michela Paolini[2]

[1] Università di Firenze, Dipartimento di Statistica,
Informatica, Applicazioni (DiSIA), Viale Morgagni 65, 50134 Firenze, Italy
[2] IMT Lucca Institute for Advanced Studies, Piazza S. Ponziano 6,
55100 Lucca (IT), Italy
michele.boreale@unifi.it, michela.paolini@imtluccca.it

Abstract. We study the problem of giving formal bounds on the information leakage of deterministic programs, when only a black-box access to the system is provided, and little is known about the input generation mechanism. After introducing a statistical set-up and defining a formal notion of information leakage *estimator*, we prove that, in the absence of significant a priori information about the output distribution, no such estimator can in fact exist that does significantly better than exhaustive enumeration of the input domain. Moreover, we show that the difficult part is essentially obtaining tight *upper* bounds. This motivates us to consider a relaxed scenario, where the analyst is given some control over the input distribution: an estimator is introduced that, with high probability, gives lower bounds irrespective of the underlying distribution, and tight upper bounds if the input distribution induces a "close to uniform" output distribution. We then define two methods, one based on Metropolis Monte Carlo and one based on Accept-Reject, that can ideally be employed to sample from one such input distribution, and discuss a practical methodology based on them. We finally demonstrate the proposed methodology with a few experiments, including an analysis of cache side-channels in sorting algorithms.

Keywords: Confidentiality, quantitative information flow, statistical estimation, Monte Carlo algorithms.

1 Introduction

Quantitative Information Flow (QIF) [9,10,5,3,4] is a well-established approach to confidentiality analysis: the basic idea is, relying on tools from Information Theory, measuring information leakage, that is, how much information flows from sensitive variables to observables of a program or system. This lays the basis for analyses that are far more flexible than the rigid safe/unsafe classification provided by the classical Noninterference approach [19]. Unfortunately,

---

[*] Work partially supported by the EU project ASCENS under the FET open initiative in FP7 and by Italian PRIN project CINA.
[**] Corresponding author.

S.S.M. Chow et al. (Eds.): ISC 2014, LNCS 8783, pp. 216–236, 2014.

the problem of exactly computing the information leakage of programs, or even proving bounds within a specified threshold, turns out to be computationally intractable, even in the deterministic case [24]. For this reason, there has been recently much interest towards methods for calculation of *approximate* information leakage [8,12,14,13,20].

Köpf and Rybalchenko method [20,21] relies on structural properties of programs: it leverages static analysis techniques to obtain bounds on information leakage. This method has proven quite successful in specific domains, like analysis of cache based side-channels [17]. In general terms, however, its applicability depends on the availability and precision of static analysis techniques for a specific domain of application. In many fields of Computer Science, a viable alternative to static analysis is often represented by simulation. In the case of QIF, this prompts interesting research issues. To what extent can one dispense with structural properties and adopt a black box approach, in conjunction with statistical techniques? What kind of formal guarantees can such an approach provide? And under what circumstances, if any, is it effective? The present paper is meant as a systematic examination of such questions. We are not aware of previous work on QIF that systematically explores such issues. Most related to ours are a series of papers by Tom Chothia and collaborators [8,12,14,13,16], which we will discuss in the concluding section (Section 8).

Our main object of study is maximum information leakage, known as *capacity*, for deterministic programs. For both Shannon- and min-entropy based QIF, capacity is easily proven to equal $\log k$, where k is the number of distinct outputs, or more generally observables, the program can produce, as the input changes. Operationally, this means that, after one observation, an attacker is up to k times more likely to correctly guess the secret input, than it was before the observation (see Section 2). For a more intuitive explanation, high capacity means that, due to unforeseen input configurations, surprising or unexpected outputs can be triggered (think of buffer overflow). In more detail, we give the following contributions.

1. After introducing a black-box, statistical set-up, we give a formal definition of program capacity estimator (Section 3). We then prove a few negative results (Section 4). Depending on the available knowledge about inputs, either no estimator exists (case of unknown input distribution); or no estimator exists that performs significantly better than exhaustive enumeration of the input domain (case of input known to be uniform). Another result indicates that the source of this difficulty lies essentially in obtaining tight *upper* bounds on capacity.

2. Motivated by the above negative results, we introduce a weak estimator, J_t (Section 5): this provides lower bounds with high confidence under *all* distributions, and accurate upper bounds under output distributions that are known to be, in a precise sense, close to uniform. The size of the required samples does not depend on that of the input domain.

3. We propose two methods (one based on Markov Chain Monte Carlo MCMC, and one based on Accept-Reject, AR) which, ideally, can be used to sample

from an *optimal* input distribution, that is an input distribution under which the corresponding output is uniform (Section 6). Under such distribution, J_t can in principle provide both tight lower and upper bounds on capacity. We then discuss how to turn this ideal algorithm into a practical methodology that at least provides good input distributions. This method is demonstrated with a few experiments (Section 7) that give encouraging results.

All in all, these results demonstrate some limits of the black-box approach (1), as well as the existence of positive aspects (2,3) that could be used, we believe, in conjunction with static analysis techniques. For ease of reading, some technical material has been confined to a separate Appendix.

2   Preliminaries on QIF

We assume an input set $\mathcal{X} = \{x_1, x_2, ...\}$ and an output set $\mathcal{Y} = \{y_1, y_2, ...\}$, which are both finite and nonempty. We view a deterministic program P simply as a function $P : \mathcal{X} \to \mathcal{Y}$. The set of "outputs" $\mathcal{Y}$ here should be interpreted in a broad sense: depending on the considered program and adversary, even physical quantities such as execution times, power traces etc. might be included in $\mathcal{Y}$. For simplicity, we restrict our attention to terminating programs, or, equivalently, assume nontermination is a value in $\mathcal{Y}$. We model the program's input as a random variable X taking values in $\mathcal{X}$ according to some probability distribution g, written $X \sim g(x)$. Once fed to a program P, X induces an output $Y = P(X)$: we denote by p the probability distribution of the latter, also written as $Y \sim p(y)$. We denote the set of outputs of nonzero probability as supp(Y) or supp(p). We let $|Y| \triangleq |\text{supp}(p)|$ denote the size of this set. Finally, we let $g(x|y) \triangleq \Pr(X = x|Y = y)$ denote the a posteriori input distribution after observation of an output y (when $p(y) > 0$).

In Quantitative Information Flow (QIF), one compares the average difficulty of guessing a secret for an adversary, before and after the observation of the system, according to the *motto*:

information leakage = prior uncertainty - posterior uncertainty.

Uncertainty is expressed in terms of entropy. Either of two forms of entropy is commonly employed. Shannon entropy of X and conditional Shannon entropy of X given Y are defined thus (in what follows, log denotes base 2 logarithm): $H_{\text{Sh}}(X) \triangleq -\sum_x g(x) \cdot \log g(x)$ and $H_{\text{Sh}}(X|Y) \triangleq \sum_y p(y) H_{\text{Sh}}(X|Y = y)$, where $0 \log 0 = 0$, and $H_{\text{Sh}}(X|Y = y)$ denotes the Shannon entropy of the distribution $g(x|y)$. Min-entropy of X and conditional min-entropy of X given Y are defined thus: $H_\infty(X) \triangleq -\log_2 \max_x g(x)$ and $H_\infty(X|Y) \triangleq -\log \sum_y p(y) \max_x g(x|y)$.

Definition 1 (information leakage, capacity). *Let $i \in \{\text{Sh}, \infty\}$. We let the i-information leakage due to an input distribution g and program P be defined as*

$$L_i(P; g) \stackrel{\triangle}{=} H_i(X) - H_i(X|Y).$$

The i-entropy capacity of P is $C_i(P) \stackrel{\triangle}{=} \sup_g L_i(P; g)$.

There are precise operational guarantees associated with the above definitions. In the case of Shannon entropy, a result by Massey [22] guarantees that, after observation, the effort necessary for a brute force attack against the secret is still greater than $2^{H_{\text{Sh}}(X|Y)}$. In the case of min entropy, leakage is directly related to chances of success of an attacker that can attempt a single guess at the secret: capacity C means that the success probability for an adversary, after one observation, can be up to 2^C times higher than before (see e.g. [23]). The next result is standard and characterizes maximum information leakage of deterministic programs; for the sake of completeness, we report its proof in the Appendix. Some notation is needed. Recall that the image of a program (function) P is the set $\text{Im}(P) \stackrel{\triangle}{=} \{y \in \mathcal{Y} : P^{-1}(y) \neq \emptyset\}$. Let us denote by $[x]$ the inverse image of $P(x)$, that is $[x] \stackrel{\triangle}{=} \{x' \in \mathcal{X} : P(x') = P(x)\}$.

Theorem 1. *Let $i \in \{\text{Sh}, \infty\}$ and $k = |\text{Im}(P)|$. Then $L_i(P; g) \leq C_i(P) = \log k$. In particular: (a) if $g(x) = \frac{1}{k \times |[x]|}$ for each $x \in \mathcal{X}$, then $L_{\text{Sh}}(P; g) = C_{\text{Sh}}(P) = \log k$; (b) if g is uniform then $L_{\infty}(P; g) = C_{\infty}(P) = \log k$.*

In the light of the above result, a crucial security parameter is therefore the image size of P, $k = |\text{Im}(P)|$: with a slight language abuse, we will refer to this value as to *program capacity*.

Example 1 (cache side-channels). Programs can leak sensitive information through timing or power absorption behaviour. Side-channels induced by cache behaviour are considered a particularly serious threat. The basic observation is that the lookup of a variable will take significantly more CPU cycles if the variable is not cached (*miss*), than if it is (*hit*). Consider the following program fragment, borrowed from [1].

```
if ( h>0 )
    z = x;
else
    z = y;
z = x;
```

Assume an adversary can perform accurate time measurements, so that $\mathcal{Y} = \{t_1, ..., t_n\}$, a discrete set of execution times (or buckets thereof). Following the terminology of [17], we refer to this type of adversary as *time*-based. If neither x nor y are cached before execution, there are only two possible execution times for the above program, say t_{short} and t_{long}. Observing t_{short} implies that the if condition is true, while t_{long} implies that the if condition is false. Assuming the value of the variable h is uniformly distributed over signed integers, this behaviour will actually reveal one bit about h.

In a different scenario, related to e.g. power absorption measurements, the adversary might be able to detect if each individual memory lookup instruction causes a miss or a hit. In this case, the set of observables is $\mathcal{Y} = \{H, M\}^*$, where actually only a finite subset of these traces will have positive probability. For the program above, the only two possible traces are $tr_{short} = MH$ and $tr_{long} = MM$ and, again, they can reveal up to one bit about h. We refer to this type of adversary as *trace*-based.

In yet another scenario, the adversary might only be able to observe the final cache state, that is, the addresses of the memory blocks that are cached at the end of the execution - not their content. We refer to this type of adversary as *access*-based. See [17] for additional details on the formalization of cache side-channels.

3   Statistical Set Up

We seek for a method to statistically estimate program capacity k, based on a sample of the outputs. This method should come equipped with formal guarantees about the accuracy and confidence in the obtained results, depending on the size of the sample. We will not assume that the code of the program is available for inspection. Of course, for any such method to be really useful, the size of the sample should be significantly smaller than the size of the input domain. That is, the statistical method should be substantially more efficient than exhaustive input enumeration.

We define the following statistical set up. We postulate that the input and output domains, $\mathcal{X}$ and $\mathcal{Y}$, are known to the analyst, and that P can be independently run a certain number of times with inputs generated according to some probability distribution g. However, we do not assume that the code of P is accessible for analysis. Moreover, we will typically assume the analyst has only some very limited or no information about the input distribution g. These assumptions model a situation where P's code cannot be analyzed, perhaps because it is a "live" application that cannot be inspected; or P and its input generation mechanism are just too complex to fruitfully apply analysis tools[1]. In Section 6, we will relax the assumption on the input distribution, and grant the analyst with some control over g.

Let $\mathcal{D}$ denote the set of all probability distributions on $\mathcal{Y}$. For any $\epsilon \geq 0$, define $\mathcal{D}_\epsilon \stackrel{\triangle}{=} \{p \in \mathcal{D} : \text{ for each } y \in \text{supp}(p), p(y) \geq \epsilon\}$; clearly $\mathcal{D}_\epsilon \subseteq \mathcal{D} = \mathcal{D}_0$. Another useful subset is $\mathcal{D}_u \stackrel{\triangle}{=} \{p \in \mathcal{D} : P(X_u) \sim p \text{ for some program } P\}$, where

---

[1] One must be careful here in distinguishing the position of the *analyst* from that of the *attacker*. The analyst wishes to estimate the capacity of a system he has only a limited, black-box access to. The attacker wants to recover the secret, and the black-box restriction may possibly not apply to him: indeed, it is typically assumed he has got access to the source code of P.

X_u is the uniform distribution over $\mathcal{X}$: this is the set of output distributions generated by considering all programs under the uniform input distributions. It is easy to see that $\mathcal{D}_u \subseteq \mathcal{D}_{\frac{1}{|\mathcal{X}|}}$.

Program outputs are generated a certain fixed number $m \geq 1$ of times, independently (this number may also depend on predefined accuracy and confidence parameter values; see below). In other words, we obtain a random sample of m outputs

$$S \stackrel{\triangle}{=} Y_1, ..., Y_m \quad \text{with } Y_i \text{ i.i.d. } \sim p(y) \tag{1}$$

for some in general *unknown* distribution $p \in \mathcal{D}$.

Our ideal goal is to define a random variable I that estimates the capacity $k = |Y| = |\text{supp}(p)|$. This estimator should be a function solely of the sequence of observed outputs, S: indeed, while the size of the input and output domains, $\mathcal{X}$ and $\mathcal{Y}$, are assumed to be known, both the program and its input distribution are in general unknown. Formally, we have the following definition, where the parameters γ and δ are used to control, respectively, the accuracy of the estimation and the confidence in it: a good estimator would have γ close to 1 and δ close to 0. The parameter $\mathcal{D}' \subseteq \mathcal{D}$ implicitly encodes any a priori partial information that may be available about the output distribution p (with $\mathcal{D}' = \mathcal{D}$ meaning no information).

Definition 2 (estimators). *Let $0 < \delta < 1/2$, $\gamma > 1$ and $\mathcal{D}' \subseteq \mathcal{D}$ be given. We say a function $I : \mathcal{Y}^m \to \mathbb{R}$ ($m \geq 1$) γ-approximates program capacity under $\mathcal{D}'$ with confidence $1 - \delta$ if, for each $p \in \mathcal{D}'$, the random variable $I \stackrel{\triangle}{=} I(S)$, with S like in (1), satisfies the following where $k = |\text{supp}(p)|$:*

$$\Pr\left(k/\gamma \leq I \leq \gamma k\right) \geq 1 - \delta. \tag{2}$$

Note that in the above definition the size of the sample, m, may depend on the given δ, γ and set $\mathcal{D}'$.

4   Limits of Program Capacity Estimation

We will show that, under very mild conditions on γ, the problem of γ-approximation of program capacity cannot be solved, unless nontrivial a priori information about the distribution p is available to the analyst. We first deal with the case where an $\epsilon > 0$ is known such that $p \in \mathcal{D}_\epsilon$. The proof of the next theorem is reported in the Appendix. To get a flavour of it, consider that a capacity estimator under $\mathcal{D}_\epsilon$ can be used to distinguish between two distributions, say p and q: p concentrates all the probability mass on a small set; q is essentially the same, except that a small fraction $\epsilon > 0$ of the mass is spread on a large number of new elements. On average, $1/\epsilon$ extractions from q will be necessary before one of these new elements shows up.

Theorem 2. *Assume $\epsilon > 0$ and $\gamma < \min\{\sqrt{|\mathcal{Y}|}, \sqrt{(1 - \epsilon)/\epsilon}\}$. Then any I that γ-approximates capacity under $\mathcal{D}_\epsilon$ requires a sample size of at least $m \geq \frac{\ln 2}{\gamma^2 \epsilon} - 1$, independently of any fixed confidence level.*

A variation of the above result concerns the case where the analyst knows the output distribution is induced by a uniform *input* distribution; see the Appendix for a proof.

Theorem 3. *Assume* $\gamma < \min\{\sqrt{|\mathcal{Y}|}, \sqrt{|\mathcal{X}| - 1}\}$. *Then any* I *that* γ-*approximates capacity under* $\mathcal{D}_u$ *requires a sample size of at least* $m \geq \frac{\ln 2}{\gamma^2}|\mathcal{X}| - 1$, *independently of any fixed confidence level.*

Hence, for γ close to 1, approximately $|\mathcal{X}|/\gamma^2 \approx |\mathcal{X}|$ i.i.d. extractions of Y are necessary for estimating capacity when it is known that $p \in \mathcal{D}_u$: this shows that no such estimator exists that is substantially more efficient than exhaustive input enumeration. We next give the main result of the section: if no partial information is available about p, then it is just impossible to estimate capacity.

Theorem 4 (non existence of estimators). *Assume that* $\gamma < \sqrt{|\mathcal{Y}|}$. *Then there is no* I *that* γ-*approximates capacity under* $\mathcal{D}$, *independently of any fixed confidence level.*

Proof. Assume by contradiction the existence of one such estimator I, and let m be the size of the sample it uses. Take now $\epsilon > 0$ small enough such that $\frac{1-\epsilon}{\epsilon} > |\mathcal{Y}|$, so that $\gamma < \sqrt{\frac{1-\epsilon}{\epsilon}}$, and such that $m < \frac{\ln 2}{\gamma^2 \epsilon} - 1$. Since I is an estimator under $\mathcal{D}$, it is also an estimator under $\mathcal{D}_\epsilon$. Now applying Theorem 2, we would get $m \geq \frac{\ln 2}{\gamma^2 \epsilon} - 1$, which is a contradiction.

We end the section with a result indicating that the difficulty of estimating k is mostly due to obtaining accurate *upper* bounds on it. Let us say that a function $I : \mathcal{Y}^m \to \mathbb{R}$ is *reasonable* if it achieves its minimum for a string where a single output occurs m times. That is, for some $y_0 \in \mathcal{Y}$, $y_0^m \in \operatorname{argmin}_{\boldsymbol{y} \in \mathcal{Y}^m} I(\boldsymbol{y})$. The meaning of the following theorem is that any reasonable estimator I can nowhere be significantly smaller than the trivial upper bound $|\mathcal{Y}|$.

Theorem 5. *Assume there is a reasonable* I *such that, for some* $\gamma > 1$ *and* $0 < \delta < 1/2$ *and for each* $p \in \mathcal{D}$ *and* $k = |\operatorname{supp}(p)|$, *it holds that* $\Pr(I \geq k/\gamma) \geq 1 - \delta$. *Then it must be* $I \geq |\mathcal{Y}|/\gamma$.

Proof. Assume w.l.o.g. that $|\mathcal{Y}| > 1$. Consider any I satisfying the hypotheses, and let $k^- \triangleq \min\{I(\boldsymbol{y}) : \boldsymbol{y} \in \mathcal{Y}^m\}$, with $I(y_0^m) = k^-$. Choose any $0 < \epsilon < 1$ such that $\ln(1 - \delta)/\ln(1 - \epsilon) > m$. Consider the distribution p over $\mathcal{Y}$ which assigns y_0 probability $1 - \epsilon$ and divides evenly the rest among the remaining elements: $p(y) = \epsilon/(|\mathcal{Y}| - 1)$ for each $y \neq y_0$; clearly, $|\operatorname{supp}(p)| = k = |\mathcal{Y}|$. Now, since $m < \ln(1 - \delta)/\ln(1 - \epsilon)$, it is easy to see that, under p, $\Pr(S = y_0^m) > 1 - \delta > \frac{1}{2}$. Hence, according to the hypotheses on I, it must be $I(y_0^m) = k^- \geq |\mathcal{Y}|/\gamma$.

Note that no similar result can hold for lower bounds of program capacity: the trivial estimator $I \triangleq$ n. of distinct elements occurring in S - which holds even with $\delta = 0$ and $\gamma = 1$ - shows that lower bounds exist that can be arbitrarily larger than the trivial 1. In the next section, we will concentrate our efforts on analysing one such lower bound.

5 A Weak Estimator

The negative results on the existence of general estimators in the preceding section suggest we should relax our goals, and concentrate our efforts on lower bounds, and/or assume that some partial information about the program's output distribution is available. For example, if the source code of P is accessible for inspection, this information could be derived by employing abstract interpretation techniques as suggested in [20]. We seek for an estimator J that, with high confidence: (a) gives a lower-bound of k under *all* output distributions; (b) is close to k under the additional assumption that the underlying output distribution is "close to uniform". The last requirement means that, at least in the ideal situation where all output values are equally likely to be generated, the estimator should approximate k accurately. Finally, we would like the estimator be substantially more efficient than exhaustive input enumeration.

In what follows, we define a weak estimator based on the number of distinct elements occurring in our sample. This is of course not the only statistics one could consider, but it turns out to be easy to analyse and quite effective in practice[2]. Assume the distribution of Y is uniform, or close to uniform, on a subset of k elements. It is well known that, in this case, the *expected* number of distinct elements occurring at least once in a i.i.d. sample of Y is $k(1 - (1 - \frac{1}{k})^m) \approx k(1 - \exp(-m/k))$, where m is the size of the sample (see below). Viewing this expression as a function of k, say $f(k)$, this suggests the estimation $k \approx f^{-1}(D)$, where D is the number of distinct elements that have *actually* been observed in S, rather than its expected value. There are a few points that need to be addressed in order to prove this a weak estimator in the above sense: mainly, the fact that the distribution we are faced with is possibly not uniform, and to what extent one can approximate $E[D]$ via D. We tackle these issues below.

Formally, let S be the sample defined in (1), obtained under a generic output distribution $Y \sim p(y)$. Consider the function defined as $f(x) \overset{\triangle}{=} x(1 - (1 - \frac{1}{x})^m)$, for $x \in [1, +\infty)$. It is easy to see that f is an increasing function of x; hence, its inverse f^{-1} exists and is in turn increasing on the same interval. We consider the following random variables, where the second one depends on a parameter $t \in \mathbb{R}$:

$$D \overset{\triangle}{=} |\{y \in \mathcal{Y} : y \text{ occurs at least once in } S\}|$$
$$J_t \overset{\triangle}{=} f^{-1}(D - t)$$

with the proviso that $J_t \overset{\triangle}{=} 1$ if $D - t < 1$. Let us indicate by $E_u[D]$ the expected value of D under the distribution that is uniform on $\mathrm{supp}(p)$. Moreover, for each $\eta \geq 1$, let us say that p is η-*close to uniform* if for each $y \in \mathrm{supp}(p)$, $p(y) \geq \frac{1}{\eta |\mathrm{supp}(p)|}$. Note that p is uniform on its own support iff $\eta = 1$. The proof of the next lemma is reported in the Appendix.

---

[2] In particular, we have found that statistics based on the number of collisions, such the index of coincidence, although easy to analyse, perform poorly in terms of accuracy.

Lemma 1. *Let $k = |\text{supp}(p)|$. Then: (a) $f(k) = E_u[D] \geq E[D]$; (b) if additionally p is η-close to uniform, then $E[D] \geq f(k)/\eta$.*

The next lemma ensures that the measure of D is quite concentrated around its expected value: roughly, D is unlikely to deviate from $E[D]$ by more than $O(\sqrt{m})$. The proof is a standard application of McDiarmid's theorem, see for example [18, Chapter 5].

Lemma 2. *Let $t > 0$. Then $\Pr(D > E[D] + t) \leq \exp(-2t^2/m)$ and $\Pr(D < E[D] - t) \leq \exp(-2t^2/m)$.*

The next theorem formalizes the fact that J_t is a family of weak estimators.

Theorem 6. *Let $k = |\text{supp}(p)|$. Let $m \geq 1$, $0 < \delta < 1/2$ and t such that $t \geq \sqrt{m \ln(1/\delta)/2}$. Then*

1. $\Pr(J_t \leq k) \geq 1 - \delta$;
2. *assume additionally that p is η-close to uniform. Let $t' \triangleq t + (\eta - 1)D$. Then $\Pr(J_{-t'} \geq k) \geq 1 - \delta$.*

Proof. Concerning part 1, from Lemma 1(a), we have that $f(k) \geq E[D]$. From Lemma 2, $E[D] \geq D - t$ with probability at least $1 - \delta$; hence, with probability at least $1 - \delta$, $f(k) \geq D - t$. Applying $f^{-1}(\cdot)$ on both sides of this inequality, the wanted statement follows.

Concerning part 2, from Lemma 2 we have that, with probability at least $1 - \delta$, $D + t \geq E[D]$. From Lemma 1(b), $E[D] \geq f(k)/\eta$; hence $D + t' = \eta(D + t) \geq f(k)$ with probability at least $1 - \delta$. Applying $f^{-1}(\cdot)$ on both sides of this inequality, the wanted statement follows.

Part 1 of the above theorem says that J_t is in fact an underestimation of k. Moreover, under a p η-close to uniform, part 2 allows us to overapproximate k by $J_{-t'}$. Let us stress that the values of t, t' and m *do not depend on the size of the input* domain, $\mathcal{X}$. Theorem 6 implies that, with high probability, $k \in [J_t, J_{t'}]$. As η approaches 1, this interval gets narrower; for $\eta = 1$, we have the best possible estimation which is $k \in [J_t, J_{-t}]$. Unfortunately, part 2 of Theorem 6 cannot be directly applied when one does not know η. Nevertheless, it implies we can improve the quality of the estimation by tweaking the input distribution so as to make η as close as possible to 1 - that is, the output distribution as close as possible to uniform on $\text{supp}(p)$.

6　Sampling from Good Input Distributions

For a fixed program P, computing k is the same as counting the number of classes of the equivalence relation over $\mathcal{X}$ given by $x \sim x'$ iff $P(x) = P(x')$ – that is, counting the number of nonempty inverse images $P^{-1}(y)$, for $y \in \mathcal{Y}$. The weak estimator J_t implements a Monte-Carlo method to count such classes, based on counting the number of distinct outputs when inputs x are drawn

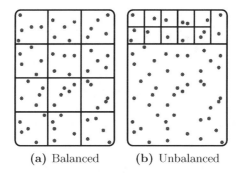

(a) Balanced   **(b)** Unbalanced

Fig. 1. Balanced and unbalanced equivalence relations

according to a distribution $g(x)$. It should be intuitively clear that the accuracy of J_t depends critically on this g. In particular, if g is the uniform distribution on $\mathcal{X}$, as often found in practice, the resulting estimator J implements what we may call a *crude Monte Carlo* (CMC) method. CMC can perform well when the equivalence classes are more or less all of the same size, like in Fig. 1(a). On the other hand, consider a situation like in Fig. 1(b): under a uniform g, most of the sampled x's will fall in the big class, thus giving rise to a low value of D, hence of J_t, that will lead to severely underestimate the true value of k. In terms of Theorem 6(2), the resulting output distribution will have an η much higher than 1. In order to rectify this, one should modify g so as to "squeeze" the probability mass out of the big class and redistribute it on the small classes. Ideally, one should be able to obtain an optimal input distribution g^* such that all outputs (equivalence classes) are equally likely.

These considerations lead us to study a scenario where we grant the analyst with some control over the input distribution. We will define two methods to sample from good input distributions: one based on a Markov Chain Monte Carlo (MCMC) algorithm, and another one based on an Accept-Reject (AR) algorithm. We will then outline a practical methodology based on these algorithms.

6.1   A Metropolis Monte Carlo Method

We first note that, once P is fixed, an optimal input distribution g^* always exists. For example, denoting by $[x]$ the $\sim$-equivalence class of each $x \in \mathcal{X}$ according to P, we can set

$$g^*(x) \triangleq \frac{1}{k \times |[x]|} \,.$$

Of course, this expression requires knowledge of k, hence does not yield *per se* a method for computing g^*. We note, however, that for any two given x and x', the calculation of the ratio $g^*(x')/g^*(x) = |[x]|/|[x']|$ does not require k, which is canceled out. Now, it is reasonable to assume that this ratio can be, at least roughly, approximated. For example, we could count the number of elements so far found in $[x]$ and $[x']$ in a random walk in the state space $\mathcal{X}$ (see below); or we could use abstract interpretation techniques as suggested by [20]. This puts us in a position to apply the *Metropolis* MCMC, as explained below. For a general treatment of MCMC see e.g. [6].

The general idea of MCMC is that sampling from $\mathcal{X}$ according to a given distribution, for instance $g^*(x)$, can be accomplished by defining a Markov Chain over $\mathcal{X}$ whose (unique) stationary distribution is $g^*(x)$. Metropolis MCMC, instantiated to our setting, can be described as follows. We fix a *proposal distribution* $Q(x'|x)$, according to which candidate next-states will be chosen. In the version we consider, this is required to be symmetrical and positive: $Q(x'|x) = Q(x|x') > 0$. Starting from an arbitrary initial state $x_0 \in \mathcal{X}$, a random walk on $\mathcal{X}$ is performed: at each step, starting from the current state x_t, a candidate next state x' is drawn according to $Q(x'|x_t)$. Now, x' can be either *accepted* ($x_{t+1} = x'$) or *rejected* ($x_{t+1} = x_t$): the first event occurs with probability $\alpha(x, x') \stackrel{\triangle}{=} \min\{1, |[x]|/|[x']|\}$, the second one with probability $1 - \alpha(x, x')$.

This defines a random walk $\{X_t\}_{t\geq 0}$ on $\mathcal{X}$, which, so to speak, "keeps off the big classes": once a big class $[x]$ is entered, it is likely to be exited at the next step. This will happen indeed provided the proposal distribution will select a small class $[x']$ at the next step , as $|[x]|/|[x']| > 1$. As a result, after an initial "burn in" period, the random walk will tend to spend the same amount of time on each class. The theorem below is an instance of a far more general result (see e.g. [6, Th.7.2]).

Theorem 7. *The stochastic process $\{X_t\}_{t\geq 0}$ defined above is a time-homogeneous Markov chain over $\mathcal{X}$ whose stationary distribution is g^*. That is, letting $X_t \sim g_t$, one has $\lim_{t\to+\infty} g_t = g^*$.*

A practical consequence of this theorem is that, after a burn in period of T steps, we can use the realizations x_t ($t \geq T$) of the stochastic processes as extractions drawn according to $g^*(x)$. If we insist these extractions to be *independent*, we can run independently say m copies of the Markov chain for T steps each, and only keep the last state of each of the resulting random walks, say $x_T^{(i)}$, for $i = 1, ..., m$. Next, we can compute m independent samples of Y as $y_i \stackrel{\triangle}{=} P(x_T^{(i)})$, for $i = 1, ..., m$.

6.2   An Accept-Reject Method

The Accept-Reject (AR) method can be used to generate samples from an arbitrary probability distribution function $f(x)$, by using an *instrumental* distribution $\hat{f}(x)$, under the only restriction that $f(x) \leq M\hat{f}(x)$, where $M > 1$ is an appropriate bound on $\frac{f(x)}{\hat{f}(x)}$. This sampling method is usually used in cases where the form of the distribution $f(x)$ makes sampling difficult, but computing $\frac{f(x)}{M\hat{f}(x)}$ is easy. We give an outline of the procedure below:

1. sample x according to $\hat{f}(x)$;
2. generate u uniformly at random in $[0,1]$;
3. if $u < \frac{f(x)}{M\hat{f}(x)}$ accept x and return it; else reject x and repeat from 1.

For further details on the correctness of this procedure, see [6, Ch.2]. In our case, we take as an instrumental distribution simply the uniform distribution on $\mathcal{X}$: $\hat{f}(x) \triangleq \frac{1}{|\mathcal{X}|}$. The constant M must satisfies the relation $g^*(x) = \frac{1}{k|[x]|} \leq \frac{M}{|\mathcal{X}|}$ for any $x \in \mathcal{X}$. This is the case if we take $M \triangleq \frac{|\mathcal{X}|}{k \min_x |[x]|}$: this implies that the comparison in step 3 becomes $u < \frac{\min_{x'} |[x']|}{|[x]|}$.

6.3   Methodology

In order to turn the previous algorithms into practical methodologies, we have to cope with two major issues.

1. For both MCMC and AR, the values $|[x]|$ cannot possibly be computed exactly. In practice, we approximate $|[x]|$ during a pre-computation phase, where inputs are sampled uniformly at random and used to compute a histogram of the most frequent outputs. In the actual sampling phase of either methods, the accept rule will make use of this histogram to ensure that such outputs are, in practice, not generated too often.
2. For MCMC, convergence to g^* only holds in the limit: if we fix a maximum number of steps T, the resulting distribution $\hat{g}$ can only approximate g^*. The determination of this "burn in" period T can be nontrivial and often proceeds by trials and errors.

Additional details on the methodology are reported in the Appendix. The above approximations imply that, in general, what one can hope for is to find a "good" input distribution $\hat{g}$, that is, one that does significantly better than the uniform input distribution, in terms of output's J_t. This is *not* to say, however, that we are not able to give *formal* bounds on the program's capacity. Indeed, Theorem 6(1) applies to *any* output distribution, be it optimal or not.

7   Experiments

We have tested the methodology introduced in the previous section on two classes of simple programs. The first class includes programs with artificially highly unbalanced classes; the second class is about cache side channels in sorting algorithms, a problem discussed in [17,1]. The Java code used to run these experiments is available from the authors.

7.1   Unbalanced Classes

We wish to test the behaviour of our estimator and sampling methodologies over programs with increasingly unbalanced classes. To this purpose, we consider a family of programs $P_{n,l,r}(x)$, depending on parameters $n \geq l \geq r \geq 1$, input $x \in \mathcal{X} = \{0,1\}^n$ and output $y \in \mathcal{Y} = \{0,1\}^l$, defined as follows.

```
z=mod(x,2^l);
if mod(z,2^r)==0
    y=z;
else
    y=mod(z,2^r);
return y;
```

The capacity of any such program is easy to compute by inspection. In fact, $P_{n,l,r}(x)$ divides the input domain into $k = 2^{r-1} - 1 + 2^{l-r}$ distinct classes: of these, $2^{r-1} - 1$ are "fat" classes with 2^{n-r} elements each (outputs corresponding to the **else** branch); and 2^{l-r} are "small" classes, with 2^{n-l} elements each (outputs corresponding to the **then** branch). Once fixed n, depending on the value of $l - r$, we will have more or less balanced situations. If $l - r$ is small, we will face a balanced situation, with about 2^l classes of similar size. As $l - r$ gets larger, we will face increasingly unbalanced situations, with relatively few (about 2^r) fat classes and a lot (2^{l-r}) of small classes.

   We have put the estimator J_t at work on several instances of this program, using input distributions sampled according to CMC, Metropolis MCMC and AR. For several values of n, we have tested J_t on both balanced (l close r) and unbalanced (r small) situations. For simplicity, we have just fixed an "affordable" sample size of $m = 5 \times 10^5$ throughout all the experiments and $\delta = 0.001$. Following Theorem 6, we have set $t = \sqrt{m \ln(1/\delta)/2}$. Table 1 displays some outcomes of these experiments.

   A few considerations are in order. AR is quite accurate in all cases, even in terms of absolute error with respect to the true value of k. However, it can be quite time-inefficient in strongly unbalanced situations, where a lot of proposals are rejected (see Appendix). CMC is accurate only in balanced situations and is always time-efficient. Metropolis MCMC never wins in terms of accuracy: however, it can be much more efficient than AR in strongly unbalanced situations, where CMC performs poorly. Therefore, MCMC can still be a sensible choice if time is at stake. Note that the quality of the estimation does not depend at all on the size of the input domain, 2^n. This fact is extremely encouraging from the point of view of the scalability of the methodology. We also notice that in cases with a small k compared to m, not reported in the table, using J_t is in a sense overkill, as the trivial estimator D already performs quite well. Running each experiment (row of the table) required at most a few seconds on a common notebook.

7.2   Cache Side-Channels in Sorting Algorithms

Sorting algorithms are fundamental building blocks of many systems. The observable behaviour of a sorting algorithm is tightly connected with the (sensitive)

Table 1. Capacity lower bounds J_t for $P_{n,l,r}(x)$, with different sampling methods, for $m = 5 \times 10^5$ and $\delta = 0.001$. In each row, the most accurate obtained outcome is highlighted in boldface.

n	l	r	k	CMC	MCMC	AR
	22	22	4.9143×10^6	$\mathbf{4.0320 \times 10^6}$	2.8233×10^6	4.0101×10^6
24	22	20	1.0486×10^6	$\mathbf{1.0317 \times 10^6}$	8.4025×10^5	1.0281×10^6
	22	2	2.0972×10^6	1.1853×10^5	4.0384×10^5	$\mathbf{1.0320 \times 10^6}$
	22	1	2.0972×10^6	2.8133×10^5	7.6395×10^5	$\mathbf{2.0450 \times 10^6}$
	23	23	8.3886×10^6	$\mathbf{7.7328 \times 10^6}$	4.8261×10^6	7.6385×10^6
28	23	20	1.0486×10^6	$\mathbf{1.0339 \times 10^6}$	8.3413×10^5	1.0316×10^6
	23	2	2.0972×10^6	1.2232×10^5	4.8578×10^5	$\mathbf{2.0330 \times 10^6}$
	23	1	4.1943×10^6	2.9472×10^5	9.3437×10^5	$\mathbf{3.9806 \times 10^6}$
	26	26	6.7108×10^7	$\mathbf{3.8772 \times 10^7}$	1.5140×10^7	3.8651×10^7
32	26	23	8.3886×10^6	$\mathbf{7.7432 \times 10^6}$	4.7617×10^6	7.7312×10^6
	26	2	1.6777×10^7	1.2593×10^5	5.9906×10^5	$\mathbf{1.3683 \times 10^7}$
	26	1	3.3554×10^7	3.0841×10^5	1.1975×10^6	$\mathbf{2.4993 \times 10^7}$

data it manipulates. Therefore, designing secure sorting algorithms is challenging and, as discussed in [1], should provide indications on how other algorithms can be implemented securely. An additional reason for analysing sorting algorithms is that this is one of the cases where static analysis alone may fail to provide useful bounds (we discuss this at the end of this subsection).

To give a flavour of the leaks that may arise in this context, consider the following code implementing `InsertionSort`.

```
1.    public static int InsertionSort(int[] v){
2.        int n = v.length;
3.        int i, j, index;
4.        for(i = 1; i<n; i++){
5.            index = v[i];
6.            for(j = i; j>0 && v[j-1]>index; j--){
7.                v[j] = v[j-1];
8.            }
9.            v[j]= index;
10.    } }
```

Suppose all elements of v are distinct, and the secret one wants to protect is the initial ordering of the elements in v, that is, the sequence of indices $(i_1, i_2, ...)$ s.t. $v[i_1] < v[i_2] < \cdots$. One wonders how much information a trace-based adversary, capable of detecting the hit/miss sequence relative to lookups of vector v, can learn about the secret. Assuming for simplicity a cache that is initially empty and large enough that v could fit entirely in it, it is clear that the beginning of each innermost `for` cycle (line 6) will be marked by a miss, corresponding to the lookup of $v[i]$ in line 5. The distance between the last miss and the next will reveal the adversary how many iterations are executed in 6-7 for the current value of i; hence the value of j (if any) s.t., for the current version of the vector,

$v[j-1] > v[i]$. It appears the overall execution will give the adversary significant information about the secret, although it is far from easy to exactly determine how much, manually.

We have put our estimator J_t at work on both InsertionSort and BubbleSort (the code of the latter is reported in the Appendix), for several values of the vector's size n and cache configurations. For simplicity, we have confined ourselves to the AR sampling method. We have considered increasing values of the sample size m, until observing J_t getting essentially stable: this indicates that the maximum leakage has been observed. We have analyzed the case of both access- and trace-based cache side-channel adversaries, introduced in Example 1. In the first case, we have obtained $J_t = 1$ in all configurations, that is a leakage of 0, which is consistent with what known from previous analyses [17]. The outcomes of the experiments for the trace-based adversary is displayed in Table 2. Lower bounds on capacity are expressed in bits, that is, $\log J_t$ is reported; the size in bits of the secret, $\log(n!)$, is also given for reference. The running time of each experiment ranges from few seconds to several minutes on a common notebook.

Table 2. Capacity lower bounds $\log J_t$ (in bits) for several cache configurations and vector lengths. Values obtained with $\delta = 0.001$ and $m = 5 \times 10^3$ (for $n = 8, 16$) or $m = 10^5$ (for $n = 32$). Here, CS and BS are respectively the cache and block size, expressed in 32 bits words. Preload = no means the cache is initially empty. Preload = yes means that the entire vector v is cached before execution of the sorting algorithm. The adopted cache replacement policy is *least recently used* (LRU).

n		8			16			32		
$\log_2(n!)$		15.3			44.25			117.66		
CS BS		Bubble	Insertion		Bubble	Insertion		Bubble	Insertion	
Preload	yes	no	yes	no	yes	no	yes	no	yes	no
64 2	4.75	7.51	4.80	9.76	4.10	11.44	3.90	16.23	6.22	16.01	6.32	18.88
64 4	4.75	4.95	4.80	6.34	3.90	10.05	4.10	14.91	6.35	14.81	6.26	18.89
64 8	4.85	4.80	4.85	4.80	3.80	7.77	3.80	9.65	6.18	12.88	6.22	18.70
128 2	4.75	7.51	4.85	9.68	4.10	11.39	3.80	16.20	6.28	15.99	6.06	18.89
128 4	4.75	4.85	4.80	6.27	4.10	10.04	3.90	14.88	6.18	14.82	6.35	18.92
128 8	4.80	4.80	4.80	4.80	3.90	8.49	3.80	9.67	6.12	12.89	6.18	18.69
256 2	4.80	7.46	4.80	9.79	3.80	11.46	3.80	16.26	6.20	15.99	6.14	18.90
256 4	4.75	4.85	4.85	6.39	3.80	10.06	4.10	14.81	6.20	14.81	6.12	18.89
256 8	4.85	4.80	4.85	4.85	3.70	8.52	3.80	9.67	6.12	12.89	6.33	18.67

As expected, the observed leakage consistently increases as n increases. For fixed n, ignoring small fluctuations, the observed values consistently decrease as the block size BS increases, while they appear to be independent from the cache size CS. As expected, a preload phase drastically reduces the observed leakage. Note however that, even when the vector is completely cached, there is still some variability as to the number of comparison operations, which depends on the initial ordering of v. This explains the positive leakage observed with preload. Finally, we note that BubbleSort is consistently observed to leak less than InsertionSort.

We also note that this is a scenario where current static analysis methods are not precise enough to give useful results. In particular, the upper bounds reported in [17, Fig.9] for the trace-based adversary are vacuous, since larger than the size in bits of the secret[3]. A direct comparison of our results with those of [17] is not possible, because of the different experimental settings - in particular, they analyze compiled C programs, while we simulate the cache behaviour by instrumented Java code. Nevertheless, our results indicate that a black-box analysis of lower bounds is a useful complement to a static analysis of upper bounds, revealing potential confidentiality weaknesses in a program. In the future, we plan to test systematically the examples given in [17], in a comparable experimental setting.

8   Conclusion, Future and Related Work

We have studied the problem of formally bounding information leakage when little is known about the program under examination and the way its input is generated. We have provided both negative results on the existence of accurate estimators and positive results, the latter in the terms of a methodology to obtain good input sampling distributions. This methodology has been demonstrated with a few simulations that have given encouraging results.

The present paper has mostly focused on theoretical issues. In the future, it would be interesting to investigate to what extent the proposed methodology scales to sizeable, real-world applications. Our results indicate that accurate estimators could be obtained if taking advantage of prior or subjective knowledge about the program: it would be interesting to investigate Bayesian versions of our methodology to this purpose.

Besides the already discussed Köpf and Rybalchenko's [20,21], our work is mostly related to some recent papers by Tom Chothia and collaborators. In [8,13], methods are proposed to estimate, given an input-output dataset, confidence intervals for information leakage, and therefore test whether the apparent leakage indicates a statistically significant information leak in the system, or whether it is in fact consistent with zero leakage. The LEAKIEST and LEAK-WATCH tools [14,15,16] are based on these results. Issues related to the existence of estimators in our sense are not tackled, though. Moreover, while handling also probabilistic systems, the resulting analysis methods and tools are essentially based on exhaustive input enumeration. In practice, they can be put into use only when the input space is quite small [11].

Batu et al. [2] consider the problem of estimating Shannon entropy. Their negative results are similar in spirit to ours in Section 4. Technically, they mainly rely on birthday paradox arguments to prove lower bounds of (approximately) $\Omega(\sqrt{|\mathcal{Y}|})$ on the number of required samples, for a class of "high entropy" distributions; these arguments could be easily adapted to our case. On the positive side, they also provide an estimator based on the observed number of collisions

---

[3] The authors there also analyse a third sorting algorithm, which gives results identical to `BubbleSort`.

for the special case of distributions that are uniform on their own support, in which case Shannon and min entropy coincide. In our experience, collision-based estimators perform very poorly in unbalanced situations.

Estimation of program capacity is related to the problem, considered in applied Statistics, of estimating the number of classes in a population, by some form or another of sampling. There exists a large body of literature on this subject, and a variety of estimators have been proposed; in particular, our estimator J_t for $t = 0$ is essentially equivalent to the maximum likelihood estimator for k; see e.g. [7] and references therein. To the best of our understanding, though, the setting and concerns of these works are quite different from ours. Indeed, either a parametric approach is adopted, which requires the class probabilities to fit into certain families of distributions, which cannot be assumed for programs; or the focus is rather on obtaining practical estimations of the number of classes, which at best become exact when the sample is infinite. Confidence intervals in the case of finite samples are often obtained from (empirical) estimations of the variance of the proposed estimators. However, circumstances under which formal guarantees can, or can not, actually be obtained, are scarcely considered. Another crucial difference is that, in our setting, there is an input distribution the analyst can be assumed to exercise some control on, which of course is not the case for population classes.

References

1. Agat, J., Sands, D.: On Confidentiality and Algorithms. In: IEEE Symposium on Security and Privacy, pp. 64–77 (2001)
2. Batu, T., Dasgupta, S., Kumar, R., Rubinfeld, R.: The Complexity of Approximating the Entropy. SIAM J. Comput. 35(1), 132–150 (2005)
3. Boreale, M., Pampaloni, F., Paolini, M.: Asymptotic Information Leakage under One-Try Attacks. In: Hofmann, M. (ed.) FOSSACS 2011. LNCS, vol. 6604, pp. 396–410. Springer, Heidelberg (2011)
4. Boreale, M., Pampaloni, F., Paolini, M.: Quantitative Information Flow, with a View. In: Atluri, V., Diaz, C. (eds.) ESORICS 2011. LNCS, vol. 6879, pp. 588–606. Springer, Heidelberg (2011)
5. Braun, C., Chatzikokolakis, K., Palamidessi, C.: Quantitative Notions of Leakage for One-try Attacks. Electr. Notes Theor. Comput. Sci. 248, 75–91 (2009)
6. Casella, G., Robert, C.: Monte Carlo Statistical Methods, 2nd edn. Springer (2004)
7. Chao, A., Lee, S.-M.: Estimating the number of classes via sample coverage. Journal of the American Statistical Association 87(417), 210–217 (1992)
8. Chatzikokolakis, K., Chothia, T., Guha, A.: Statistical Measurement of Information Leakage. In: Esparza, J., Majumdar, R. (eds.) TACAS 2010. LNCS, vol. 6015, pp. 390–404. Springer, Heidelberg (2010)
9. Chatzikokolakis, K., Palamidessi, C., Panangaden, P.: Anonymity protocols as noisy channels. Inf. Comput. 206(2-4), 378–401 (2008)
10. Chatzikokolakis, K., Palamidessi, C., Panangaden, P.: On the Bayes risk in information-hiding protocols. Journal of Computer Security 16(5), 531–571 (2008)
11. Chothia, T.: Personal communication to the authors (2014)
12. Chothia, T., Guha, A.: A Statistical Test for Information Leaks Using Continuous Mutual Information. In: CSF, pp. 177–190 (2011)

13. Chothia, T., Kawamoto, Y.: Statistical Estimation of Min-entropy Leakage. Manuscript available at http://www.cs.bham.ac.uk/research/projects/infotools/leakiest/
14. Chothia, T., Kawamoto, Y., Novakovic, C.: A Tool for Estimating Information Leakage. In: Sharygina, N., Veith, H. (eds.) CAV 2013. LNCS, vol. 8044, pp. 690–695. Springer, Heidelberg (2013)
15. Chothia, T., Kawamoto, Y., Novakovic, C., Parker, D.: Probabilistic Point-to-Point Information Leakage. In: CSF, pp. 193–205 (2013)
16. Chothia, T., Kawamoto, Y., Novakovic, C.: LeakWatch: Estimating Information Leakage from Java Programs. In: Kutyłowski, M., Vaidya, J. (eds.) ESORICS 2014, Part II. LNCS, vol. 8713, pp. 219–236. Springer, Heidelberg (2014)
17. Doychev, G., Feld, D., Köpf, B., Mauborgne, L., Reineke, J.: CacheAudit: A Tool for the Static Analysis of Cache Side Channels. In: USENIX Security, pp. 431–446 (2013)
18. Dubhashi, D.P., Panconesi, A.: Concentration of Measure for the Analysis of Randomized Algorithms. Cambridge University Press (2009)
19. Goguen, J.A., Meseguer, J.: Security Policies and Security Models. In: IEEE Symposium on Security and Privacy, pp. 11–20 (1998)
20. Köpf, B., Rybalchenko, A.: Automation of Quantitative Information-Flow Analysis. In: Bernardo, M., de Vink, E., Di Pierro, A., Wiklicky, H. (eds.) SFM 2013. LNCS, vol. 7938, pp. 1–28. Springer, Heidelberg (2013)
21. Köpf, B., Rybalchenko, A.: Approximation and Randomization for Quantitative Information-Flow Analysis. In: CSF, pp. 3–14 (2010)
22. Massey, J.L.: Guessing and Entropy. In: Proc. 1994 IEEE Symposium on Information Theory (ISIT 1994), vol. 204 (1994)
23. Smith, G.: On the Foundations of Quantitative Information Flow. In: de Alfaro, L. (ed.) FOSSACS 2009. LNCS, vol. 5504, pp. 288–302. Springer, Heidelberg (2009)
24. Yasuoka, H., Terauchi, T.: On bounding problems of quantitative information flow. Journal of Computer Security 19(6), 1029–1082 (2011)

A   Additional proofs

A.1   Proof of Theorem 1

Theorem A1 (Theorem 1). *Let $i \in \{\mathrm{Sh}, \infty\}$ and $k = |\mathrm{Im}(P)|$. Then $L_i(P; g)$ $\leq C_i(P) = \log k$. In particular: (a) if $g(x) = \frac{1}{k \times |[x]|}$ for each $x \in \mathcal{X}$, then $L_{\mathrm{Sh}}(P; g) = C_{\mathrm{Sh}}(P) = \log k$; (b) if g is uniform then $L_\infty(P; g) = C_\infty(P) = \log k$.*

Proof. The result for $i = \infty$ (min entropy) is in see e.g. [20,23]. For the case $i = \mathrm{Sh}$ (Shannon entropy) first note that, denoting by $I(\cdot; \cdot)$ mutual information and exploiting its symmetry, we have, for each g: $L_{\mathrm{Sh}}(P; g) = I(X; Y) = H_{\mathrm{Sh}}(Y) - H_{\mathrm{Sh}}(Y|X) = H_{\mathrm{Sh}}(Y) \leq \log k$, where in the last but one step we use $H(Y|X) = 0$ (output is entirely determined by input for deterministic programs), and the last step follows from a well known inequality for Shannon entropy. Now, taking the input distribution g specified in the the the theorem, we see that all inverse images of P have the same probability $1/k$, hence Y is uniformly distributed on k outputs. This implies $H_{\mathrm{Sh}}(Y) = \log k$. $\qquad \square$

A.2   Proofs of Section 4

Theorem A2 (Theorem 2). *Assume $\epsilon > 0$ and $\gamma < \min\{\sqrt{|\mathcal{Y}|}, \sqrt{(1 - \epsilon)/\epsilon}\}$. Then any I that γ-approximates capacity under $\mathcal{D}_\epsilon$ requires a sample size of at least $m \geq \frac{\ln 2}{\gamma^2 \epsilon} - 1$, independently of any fixed confidence level.*

Proof. Consider two distributions $p, q \in \mathcal{D}$ defined as follows. The distribution p concentrates all the probability mass in some $y_0 \in \mathcal{Y}$. The distribution q assigns y_0 probability $1 - h\epsilon$, and ϵ to each of h distinct elements $y_1, ..., y_h \in \mathcal{Y}$, where $h \stackrel{\triangle}{=} \lfloor \gamma^2 \rfloor$. Note that $1 \leq h \leq \gamma^2 < h + 1$, hence $h\epsilon \leq \gamma^2 \epsilon < \frac{1-\epsilon}{\epsilon} \cdot \epsilon = 1 - \epsilon$, and $h \leq \gamma^2 < |\mathcal{Y}|$, so that the distribution q is well-defined and in $\mathcal{D}_\epsilon$. Of course, p is in $\mathcal{D}_\epsilon$ as well.

Fix an arbitrary $0 < \delta < 1/2$, and assume by contradiction there is an estimator I for $\mathcal{D}_\epsilon$ with confidence $1 - \delta$ that uses m i.i.d. extractions, and that $m < \frac{\ln(1-\delta)}{\ln(1-\gamma^2\epsilon)}$. Now consider the sequence y_0^m, in which only y_0 occurs m times. Given the above strict upper bound on m, some easy calculations show that under either p or q, y_0^m has probability $> 1 - \delta > 1/2$. Let $a = I(y_0^m)$. By definition of estimator for $\mathcal{D}_\epsilon$, we must have both $1/\gamma \leq a \leq \gamma$ (since I is supposed to estimate $|\mathrm{supp}(p)| = 1$) and $(h + 1)/\gamma \leq a \leq (h + 1)\gamma$ (since I is supposed to estimate $|\mathrm{supp}(q)| = h + 1$). From these inequalities, and using the fact that by construction $\gamma^2 < h + 1$, we have: $a \leq \gamma < (h + 1)/\gamma \leq a$, which implies $a < a$, a contradiction. Hence it must be $m \geq \frac{\ln(1-\delta)}{\ln(1-\gamma^2\epsilon)}$. Given that δ is arbitrary, by letting $\delta \to \frac{1}{2}$ we obtain that $m \geq \frac{\ln(1/2)}{\ln(1-\gamma^2\epsilon)} = -\frac{\ln 2}{\ln(1-\gamma^2\epsilon)}$.

Now by Taylor expansion, one sees that $-1/\ln(1-x) \geq x^{-1} - 1$ for $x \in (0, 1)$, which implies the wanted result. $\qquad \square$

Theorem A3 (Theorem 3). *Assume $\gamma < \min\{\sqrt{|\mathcal{Y}|}, \sqrt{|\mathcal{X}| - 1}\}$. Then any I that γ-approximates capacity under $\mathcal{D}_u$ requires a sample size of at least $m \geq \frac{\ln 2}{\gamma^2}|\mathcal{X}| - 1$, independently of any fixed confidence level.*

Proof. Let $\epsilon = 1/|\mathcal{X}|$, $h = \lfloor \gamma^2 \rfloor$ and consider distinct elements $x_0, x_1, ..., x_h$ in $\mathcal{X}$ and $y_0, y_1, ..., y_h$ in $\mathcal{Y}$. The output distributions p and q over $\mathcal{Y}$ defined in the proof of Theorem 2 are generated, under the uniform distribution over $\mathcal{X}$, by the following two programs (functions) $P_i : \mathcal{X} \rightarrow \mathcal{Y}$, for $i = 1, 2$, respectively:

- $P_1(x) = y_0$ for each $x \in \mathcal{X}$;
- $P_2(x) = y_i$ if $x = x_i$ $(i = 1, ..., h)$, $= y_0$ otherwise.

Note that by virtue of the condition $\gamma < \min\{\sqrt{|\mathcal{Y}|}, \sqrt{|\mathcal{X}| - 1}\}$, P_2 is well defined and $q \in \mathcal{D}_u$. The rest of the proof is identical to that of Theorem 2.

A.3 Proofs of Section 5

Lemma A1 (Lemma 1). *Let $k = |\mathrm{supp}(p)|$. Then: (a) $f(k) = E_u[D] \geq E[D]$; (b) if additionally p is η-close to uniform, then $E[D] \geq \frac{f(k)}{\eta}$.*

Proof. Concerning (a), denoting by $I_{(.)}$ the indicator function, we note that (y below ranges over $\mathrm{supp}(p)$):

$$E_u[D] = E_u[\sum_y I_{\{y \in S\}}] = \sum_y E_u[I_{\{y \in S\}}] = \sum_y (1 - (1 - \frac{1}{k})^m) \qquad (3)$$

$$= k(1 - (1 - \frac{1}{k})^m) = f(k)$$

where in (3) $(1 - \frac{1}{k})^m$ is the probability that y does not occur in the sample S under a uniform on k outputs distribution. Furthermore $E_u[D] \geq E[D]$, indeed

$$E[D] = \sum_y 1 - (1 - p(y))^m = k - \sum_y (1 - p(y))^m = k - k \sum_y \frac{1}{k}(1 - p(y))^m$$

$$\leq k - k(1 - \frac{1}{k})^m = E_u[D] \qquad (4)$$

where in (4) we have applied the Jensen's inequality to the function $x \mapsto (1-x)^m$ which is convex in $[0, 1]$.

Concerning (b), we have

$$E[D] = \sum_y 1 - (1 - p(y))^m = k - \sum_y (1 - p(y))^m$$

$$\geq k(1 - (1 - \frac{1}{k\eta})^m) = \frac{1}{\eta}f(k\eta) \geq \frac{1}{\eta}f(k) \qquad (5)$$

where in (5) we use the fact that p is η-close to uniform and that the function $x \mapsto (1 - x)^m$ is decreasing, while $f(x)$ is non decreasing.

B   Additional Details on the Sampling Methodology

Two aspects of practical concern of the sampling methodology are the following.

(a) For MCMC, the choice of the proposal distribution, $Q(x'|x)$.
(b) The 'waste' involved in rejecting samples.

Aspect (a) is related to the specific input domain $\mathcal{X}$ under consideration. In our experiments, we have considered $\mathcal{X} = \{0,1\}^n$, the set of n-bit vectors. With such a domain, we have found that the random function $Q(x'|x)$ that computes x' by flipping independently each individual bit of the current state x, with a fixed probability $0 < \beta < 1$, has given encouraging results; in particular, the value $\beta = 0.165$ has experimentally been found to give good results.

   Concerning (b) in the case of AR, from step 3 of the algorithm it is easy to deduce that, for x chosen uniformly at random, the probability of acceptance is $\mu \frac{k}{|\mathcal{X}|}$, where $\mu \overset{\triangle}{=} \min_{x'} |[x']|$. Note that, in a perfectly balanced situation ($|[x]| = \frac{|\mathcal{X}|}{k}$ for each x), this probability is 1. In strongly unbalanced situations ($\mu k \ll |\mathcal{X}|$), however, we will observe a high rejection rate and inefficiency. We can mitigate this problem by formally replacing M with $M\theta$, for a parameter $\theta \in (0,1]$ that can be tuned by the analyst. This in practice means that the comparison at step 3 becomes $u < \frac{\mu}{\theta|[x]|}$. This modification will affect the sampled distribution, which will not be g^* anymore, but not the soundness of the method, that is, the formal guarantees provided on lower bounds.

   In the case of MCMC, problem (b) mostly occurs when generating the m i.i.d. extractions, because all but the last state of each of m random walks should be discarded. We have explored the following alternative method: we first run a single Markov chain for a suitably long burn in time T, thus reaching a state x_T and producing a histogram h_T of output frequencies. We then use x_T (as a starting state) and h_T to run again m independent copies of the Markov chain, but each for a much shorter time T'. The results thus obtained have been found experimentally to be equivalent to those obtained by sequentially sampling m times a single same Markov chain, after the burn in period. One will have often to tune the various parameters involved in the sampling algorithm (T, β, m, θ) in order to get satisfactory input distributions and guarantees.

C   Additional Code for Section 7

```
public static int BubbleSort(int[] v){
  int n = v.length; int swap;
  for(int i=0; i<n-1; i++){
    for(int j=0; j<n-i-1; j++){
      if(v[j]>v[j+1]){
        swap = v[j];
        v[j] = v[j+1];
        v[j+1] = swap;
        }
       }
     }
    }
```

Structure Based Data De-Anonymization of Social Networks and Mobility Traces

Shouling Ji[1], Weiqing Li[1], Mudhakar Srivatsa[2],
Jing Selena He[3], and Raheem Beyah[1]

[1] Georgia Institute of Technology, Atlanta, GA, USA
[2] IBM T.J. Watson Research Center, Yorktown Heights, NY, USA
[3] KSU, Kennesaw, GA, USA
{sji,wli64}@gatech.edu, msrivats@us.ibm.com,
jhe4@kennesaw.edu, rbeyah@ece.gatech.edu

Abstract. We present a novel *de-anonymization attack* on mobility trace data and social data. First, we design an Unified Similarity (US) measurement, based on which we present a US based De-Anonymization (DA) framework which iteratively de-anonymizes data with an accuracy guarantee. Then, to de-anonymize data *without the knowledge of the overlap size* between the anonymized data and the auxiliary data, we generalize DA to an Adaptive De-Anonymization (ADA) framework. Finally, we examine DA/ADA on mobility traces and social data sets.

1 Introduction

Social networking is a fast-growing business sector nowadays. To protect users' privacy, social network owners usually anonymize data by removing "Personally Identifiable Information (PII)" before releasing the data to the public. However, this data anonymization is vulnerable to a new *social auxiliary information based data de-anonymization attack* [1][2][3]. For example, just recently, it was reported that poorly anonymized logs revealed New York City cab drivers' detailed whereabouts (June 23, 2014) [5].

A few de-anonymization attacks have been designed for social data [1][2] or mobility trace data [3]. In [1], Backstrom et al. proposed active and passive attacks on social data. Since the proposed active attack relies on sybil nodes to obtain auxiliary information before social data release, it is not practical as analyzed in [2]. For the passive attack designed in [1], it is workable however not scalable [2]. In [2], Narayanan and Shmatikov showed a de-anonymization attack on social network data which can be modeled by directed graphs (where direction information carried by data can be viewed as free auxiliary information for adversaries). In [3], Srivatsa and Hicks presented the first de-anonymization attack to mobility traces by using social networks as a side-channel. Our work improves existing works in some or all of the following aspects. First, we significantly improve the de-anonymizaiton accuracy and decrease the computational complexity by proposing a novel *Core Matching Subgraphs* (CMS) based adaptive de-anonymization strategy. Second, besides utilizing nodes' local property,

S.S.M. Chow et al. (Eds.): ISC 2014, LNCS 8783, pp. 237–254, 2014.
© Springer International Publishing Switzerland 2014

we incorporate nodes' global property into de-anonymization without incurring high computational complexity. Furthermore, we also define and apply two new similarity measurements in the proposed de-anonymization technique. Finally, the de-anonymization algorithm presented in this work is a much more general attack framework. It can be applied to both mobility trace data and social data, directed and undirected data graphs, and weighted and unweighted data sets. We give the detailed analysis and remarks in the related work (due to space limitation, we put the related work in the Technical Report [17] of this paper).

In summary, our main contributions are as follows[1]. (*i*) We define three de-anonymization metrics, namely *structural similarity*, *relative distance similarity*, and *inheritance similarity*, (*ii*) Toward effective de-anonymization, we define a *Unified Similarity* (US) measurement by collectively considering the defined structural similarity, relative distance similarity, and inheritance similarity. Subsequently, we propose a US based **De-Anonymization** (DA) framework, by which we iteratively de-anonymize the anonymized data with an accuracy guarantee provided by a *de-anonymization threshold* and a *mapping control factor*. (*iii*) To de-anonymize data without the knowledge on the overlap size between the anonymized data and the auxiliary data, we generalize DA to an **Adaptive De-Anonymization** (ADA) framework. (*iv*) We apply the presented de-anonymization framework to mobility traces and social data sets. The experimental results demonstrate that the presented de-anonymization attack is very effective and robust. For instance, 93.2% of the users in Infocom06 [13] can be successfully de-anonymized given one seed mapping, and 58% of the users in Google+ can be de-anonymized given five seed mappings.

2   Preliminaries and Model

Anonymized Data Graph. In this paper, we consider anonymized data which can be modeled by an undirected graph[2], denoted by $G^a = (V^a, E^a, W^a)$, where $V^a = \{i | i \text{ is a node}\}$ is the node set (e.g., users in an anonymized Google+ graph [10]), $E^a = \{l_{i,j}^a | i, j \in V^a, \text{ and there is a tie between } i \text{ and } j\}$ is the set of all the links existing between any two nodes in V^a (a link could be a friend relationship such as in Google+ [10]), and $W^a = \{w_{i,j}^a | i, j \in V^a, l_{i,j}^a \in E^a, w_{i,j}^a \text{ is a real number}\}$ is the set of possible weights associated with links in E^a (e.g., in a coauthor graph, the weight of a coauthor relationship could be the number of coauthored papers). If G^a is an unweighted graph, we simply define $w_{i,j}^a = 1$ for each link $l_{i,j}^a \in E^a$.

For $\forall i \in V^a$, we define its neighbor set as $N^a(i) = \{j \in V^a | l_{i,j}^a \in E^a\}$. Then, $\Delta_i^a = |N^a(i)|$ represents the number of neighbors of i in G^a. For $\forall i, j \in V^a$, let

---

[1] Due to the space limitation, we put more discussion and experimental results in the Technical Report [17] of this paper.

[2] Note that, the de-anonymization algorithm designed in this paper can also be applied to directed graphs directly by overlooking the direction information on edges, or by incorporating the edge-direction based de-anonymizatoin heuristic in [2] which could obtain better accuracy.

$p^a(i,j)$ be a shortest path from i to j in G^a and $|p^a(i,j)|$ be the number of links on $p^a(i,j)$ (the number of links passed from i to j through $p^a(i,j)$). Then, we define $\mathbb{P}^a_{i,j} = \{p^a(i,j)\}$ the set of all the shortest pathes between i and j. Furthermore, we define the diameter of G^a as $D^a = \max\{|p^a(i,j)|\forall i,j \in V^a, p^a(i,j) \in \mathbb{P}^a_{i,j}\}$, i.e., the length of the longest shortest path in G^a.

Auxiliary Data Graph. As in [2][3][4], we assume the auxiliary data is the information crawled in current online social networks, e.g., the "follow" relationships on Twitter [2], the "friend" relationships on Facebook [3], etc. Furthermore, similar as the anonymized data, the auxiliary data can also be modeled as an undirected graph $G^u = (V^u, E^u, W^u)$, where V^u is the node set, E^u is set of all the links (relationships) among the nodes in V^u, and W^u is the set of possible weights associated with the links in E^u. As the definitions on the anonymized graph G^a, we can define the neighborhood of $\forall i \in V^u$ as $N^u(i)$, the shortest path set between $i \in V^u$ and $j \in V^u$ as $\mathbb{P}^u(i,j) = \{p^u(i,j)\}$, and the diameter of G^u as $D^u = \max\{|p^u(i,j)|\forall i,j \in V^u, p^u(i,j) \in \mathbb{P}^u(i,j)\}$.

In addition, we assume G^a and G^u are connected. Note that this is not a limitation of our scheme. The designed de-anonymization attack is also applicable to the case where G^a or G^u is not connected. We will discuss this in Section 4.

Attack Model. Our de-anonymization objective is to map the nodes in the anonymized graph G^a to the nodes in the auxiliary graph G^u as accurate as possible. Formally, let $\gamma(v)$ be the *objective reality* of $v \in G^a$ in the physical world. Then, an ideal de-anonymization can be represented by mapping Φ : $G^a \to G^u$, such that for $v \in G^a$, $\Phi(v) = v'$ if $v' = \Phi(v) \in V^u$ and $\Phi(v) = \perp$ if $\Phi(v) \notin V^u$, where $\perp$ is a special *not existing indicator* in the auxiliary data graph. Now, let $\mathcal{M} = \{(v_1, v'_1), (v_2, v'_2), \cdots, (v_n, v'_n)\}$ be the outcome of a de-anonymization attack such that $v_i \in V^a, \cup v_i = V^a, n = |V^a|$ $(i = 1, 2, \cdots, n)$ and $v'_i = \Phi(v_i), v'_i \in V^u \cup \{\perp\}$ $(i = 1, 2, \cdots, n)$. Then, the de-anonymization on v_i is said to be *successful* if $\Phi(v_i) = \gamma(v_i)$ when $\gamma(v_i) \in V^u$ or $\Phi(v_i) = \perp$ when $\gamma(v_i) \notin V^u$; and *failure* if $\Phi(v_i) \in \{u|u \in V^u, u \neq \gamma(v_i)\} \cup \{\perp\}$ when $\gamma(v_i) \in V^u$ or $\Phi(v_i) \neq \perp$ when $\gamma(v_i) \notin V^u$. In this paper, we are aiming to design a de-anonymization attack with a high success rate (accuracy).

3   De-Anonymization

From a macroscopic view, the designed de-anonymization attack framework consists of two phases: *seed selection* and *mapping propagation*. In the seed selection phase, we identify a small number of seed mappings from the anonymized graph G^a to the auxiliary graph G^u serving as landmarks to bootstrap the de-anonymization. In the mapping propagation phase, we de-anonymize G^a through synthetically exploiting multiple similarity measurements.

Seed Selection and Mapping Spanning. Seed selection is possible in our de-anonymization framework because of three reasons. The first reason is the common availability of huge amounts of social data, which is an open and rich source for obtaining a small number of seeds. For instance, the data published for academic and government data mining may also release some auxiliary information [4]. The second reason is the existence of multiple effective channels to obtain a small number of seed mappings (actually, we can obtain much richer auxiliary information), e.g., data leakage [2][4], third party applications [2], etc. The third reason is that a small number of seed mappings is sufficiently helpful (or *enough* depends on the required accuracy) to our de-anonymization framework. As shown in our experiments, a small number of seed mappings (sometimes even one seed mapping) are sufficient to achieve highly accurate de-anoymization. In our de-anonymization framework, we can select a small number of seed mappings by employing multiple seed selection strategies [1][2][3][4] individually or collaboratively, e.g., launching a small scale *sybil attack* [1][2], compromising a small number of nodes [1][2][3], third party applications [6][7][8], etc.

Since seed selection is not our primary contribution in this paper, we assume we have identified κ seed mappings by exploiting the aforementioned strategies individually or collaboratively, denoted by $\mathcal{M}_s = \{(s_1, s_1'), (s_2, s_2'), \cdots, (s_\kappa, s_\kappa')\}$, where $s_i \in V^a$, $s_i' \in V^u$, and $s_i' = \Phi(s_i)$. In the mapping propagation phase, we start with the seed mapping $\mathcal{M}_s$ and propagate the mapping (de-anonymization) to the entire G^a iteratively. Let $\mathcal{M}_0 = \mathcal{M}_s$ be the *initial mapping set* and $\mathcal{M}_k$ ($k = 1, 2, \cdots$) be the mapping set after the k-th iteration. To facilitate our discussion, we first define some terminologies as follows.

Let $M_k^a = \bigcup_{i=1}^{|\mathcal{M}_k|} \{v_i | (v_i, v_i') \in \mathcal{M}_k\}$ and $M_k^u = \bigcup_{i=1}^{|\mathcal{M}_k|} \{v_i' | (v_i, v_i') \in \mathcal{M}_k\} \setminus \{\perp\}$

be the sets of nodes that have been mapped until iteration k in G^a and G^u, respectively. Then, we define the *1-hop mapping spanning set* of M_k^a as $\Lambda^1(M_k^a) = \{v_j \in V^a | v_j \notin M_k^a$ and $\exists v_i \in M_k^a$ s.t. $v_j \in N^a(v_i)\}$, i.e., $\Lambda^1(M_k^a)$ denotes the set of nodes in G^a that have some neighbor been mapped and themselves not been mapped yet. To be general, we can also define the *δ-hop mapping spanning set* of M_k^a as $\Lambda^\delta(M_k^a) = \{v_j \in V^a | v_j \notin M_k^a$ and $\exists v_i \in M_k^a$ s.t. $|p^a(v_i, v_j)| \leq \delta\}$, i.e., $\Lambda^\delta(M_k^a)$ denotes the set of nodes in G^a that are at most δ hops away from some node been mapped and themselves not been mapped yet. Here, $\delta(\delta = 1, 2, \cdots)$ is called the *spanning factor* in the mapping propagation phase of the proposed de-anonymization framework. Similarly, we can define the *1-hop mapping spanning set* and *δ-hop mapping spanning set* for M_k^u as $\Lambda^1(M_k^u) = \{v_j' \in V^u | v_j' \notin M_k^u$ and $\exists v_i' \in M_k^u$ s.t. $v_j' \in N^u(v_i')\}$ and $\Lambda^\delta(M_k^u) = \{v_j' \in V^u | v_j' \notin M_k^u$ and $\exists v_i' \in M_k^u$ s.t. $|p^u(v_i', v_j')| \leq \delta\}$, respectively. Based on the defined δ-hop mapping sets $\Lambda^\delta(M_k^a)$ and $\Lambda^\delta(M_k^u)$, we try to seek a mapping Φ which maps the anonymized nodes in $\Lambda^\delta(M_k^a)$ to some nodes in $\Lambda^\delta(M_k^u) \cup \{\perp\}$ iteratively in the mapping propagation phase of our de-anonymization framework.

Structural Similarity. Since both anonymized data and the auxiliary data can be modeled by graphs, the structural/topological characteristics could be a reference for *coarse granularity* (*high level*) de-anonymization. Here, coarse granularity de-anonymization implies for an anonymized node $v \in V^a$, we de-anonymize it by mapping it to some nodes $\{v' | v' \in V^u \cup \{\perp\}$ and v' in G^u is structurally similar to v in $G^a\}$ even the ideal *one-to-one mapping* cannot be achieved. Structural characteristics based coarse granularity de-anonymization is meaningful since we can employ further techniques to refine the coarse granularity de-anonymization and finally de-anonymize v exactly.

In graph theory, the concept of *centrality* to measure the topological importance and characteristic of a node within a graph is often used. In this paper, we employ three centrality measurements to capture the topological property of a node in G^a or G^u, namely *degree centrality*, *closeness centrality*, and *betweenness centrality*. In the case that the considered data is modeled by a weighted graph, we also define the weighted version of the three centrality measurements. Furthermore, to demonstrate the aforementioned topological properties, we will employ an example data set (St Andrews [11]), which consists of a mobility trace data set of 27 users and a Facebook network of the same 27 users. The mobility trace data set has 18,241 WiFi records. We use the method in [3] to construct an anonymized graph on the 27 users based on the mobility trace data and take the Facebook network as the auxiliary data.

Degree Centrality and Weighted Degree Centrality. The *degree centrality* is defined as the number of ties that a node has in a graph. For instance, in the considered anonymized data graph, the degree centrality of $v \in V^a$ is defined as $d_v = \Delta_v^a = |N^a(v)|$. We calculate the degree centrality of the nodes in St Andrews and their counterparts in Facebook, and the result is shown in Fig.1 (a). From Fig.1 (a), we observe that the degree centrality distributions of the anonymized graph and auxiliary graph are similar, which implies degree centrality can be used for de-anonymization. On the other hand, multiple nodes in both graphs may have similar degree centrality, which suggests that degree centrality can be used for coarse granularity de-anonymization.

When the data being considered is modeled by a weighted graph, the weights on links provide extra information in characterizing the centrality of a node. In this case, the degree centrality defined for an unweighted graph cannot properly reflect a node's structural importance [9]. To consider both the number of links associated with a node and the weights on these links, we define the *weighted degree centrality* for $v \in V^a$ as $wd_v = \Delta_v^a \left(\frac{\sum_{u \in N^a(v)} w_{v,u}^a}{\Delta_v^a} \right)^\alpha$, where α is a positive tuning parameter that can be set according to the research setting and data [9]. Basically, when $0 \leq \alpha \leq 1$, high degree is considered more important, whereas when $\alpha \geq 1$, weight is considered more important. Similarly, we can define the *weighted degree centrality* for $v' \in V^u$ as $wd_{v'} = \Delta_{v'}^u \left(\frac{\sum_{u' \in N^u(v')} w_{v',u'}^u}{\Delta_{v'}^u} \right)^\alpha$.

Closeness Centrality and Weighted Closeness Centrality. From the definition of degree centrality, it indicates the local property of a node since only the adjacent links are considered. To fully characterize a node's topological importance, some centrality measurements defined from a global view are also important and useful. One manner to count a node's global structural importance is by *closeness centrality*, which measures how close a node is to other nodes in a graph and is defined as the ratio between $n - 1$ and the sum of its distances to all other nodes. In the definition, n is the number of nodes and *distance* is the length in terms of hops from a node to another node in a graph. Formally, for

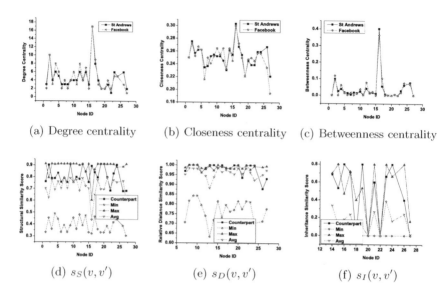

(a) Degree centrality   (b) Closeness centrality   (c) Betweenness centrality

(d) $s_S(v, v')$ (e) $s_D(v, v')$ (f) $s_I(v, v')$

Fig. 1. Structural/relative distance/inheritance similarity

$v \in V^a$, its *closeness centrality* c_v is defined as $c_v = \dfrac{|V^a| - 1}{\sum\limits_{u \in V^a, u \neq v} |p^a(v, u)|}$. Similarly, the *closeness centrality* $c_{v'}$ of $v' \in V^u$ is defined as $c_{v'} = \dfrac{|V^u| - 1}{\sum\limits_{u' \in V^u, u' \neq v'} |p^u(v', u')|}$.

Fig.1 (b) demonstrates the closeness centrality score of the nodes in St Andrews and their counterparts in the corresponding social graph Facebook. From Fig.1 (b), the closeness centrality distribution of nodes in the anonymized graph generally agrees with that in the auxiliary graph, which suggests that closeness centrality can be a measurement for de-anonymization. In the case that the data being considered is modeled by a weighted graph, we generalize the *weighted closeness centrality* for $v \in V^a$ and $v' \in V^u$ as $wc_v = \dfrac{|V^a| - 1}{\sum\limits_{u \in V^a, u \neq v} |p^a_w(v, u)|}$ and $wc_{v'} = \dfrac{|V^u| - 1}{\sum\limits_{u' \in V^u, u' \neq v'} |p^u_w(v', u')|}$, respectively, where $p^a_w(\cdot, \cdot)/p^u_w(\cdot, \cdot)$ is the shortest path between two nodes in a weighted graph.

Betweenness Centrality and Weighted Betweenness Centrality. Besides closeness centrality, *betweenness centrality* is another measure indicating a node's global structural importance within a graph, which quantifies the number of times a node acts as a bridge (intermediate node) along the shortest path between two other nodes. Formally, for $v \in V^a$, its *betweenness centrality* b_v in G^a is defined as $b_v = \dfrac{2}{(|V^a| - 1)(|V^a| - 2)} \cdot \sum\limits_{x' \neq v \neq y'} \dfrac{\sigma^a_{xy}(v)}{\sigma^a_{xy}}$, where $x', y' \in V^a$, $\sigma^a_{xy} = |\mathbb{P}^a(x, y)|$ is the number of all the shortest paths between x and y in G^a, and $\sigma^a_{xy}(v) = |\{p^a(x, y) \in \mathbb{P}^a(x, y) | v$ is an intermediate node on path $p^a(x, y)\}|$ is the number of shortest paths between x and y in G^a that v lies on. Similarly, the *betweenness centrality* $b_{v'}$ of $v' \in V^u$ in G^u is defined as

$$b_{v'} = \frac{2}{(|V^u|-1)(|V^u|-2)} \cdot \sum_{x' \neq v' \neq y'} \frac{\sigma^u_{x'y'}(v')}{\sigma^u_{x'y'}}.$$

According to the definition, we obtain the betweenness centrality of nodes in St Andrews as shown in Fig.1 (c). From Fig.1 (c), the nodes in G^a and their counterparts in G^u agree highly on betweenness centrality. Consequently, betweenness centrality can also be employed in our de-anonymization framework for distinguishing mappings. For the case that the considering data is modeled as a weighted graph, we define the *weighted betweenness centrality* for $v \in V^a$ and $v' \in V^u$ as $wb_v = \frac{2}{(|V^a|-1)(|V^a|-2)} \cdot \sum_{x \neq v \neq y} \frac{\sigma^{wa}_{xy}(v)}{\sigma^{wa}_{xy}}$ and $wb_{v'} =$

$\frac{2}{(|V^u|-1)(|V^u|-2)} \cdot \sum_{x' \neq v' \neq y'} \frac{\sigma^{wu}_{x'y'}(v')}{\sigma^{wu}_{x'y'}}$, respectively, where σ^{wa}_{xy} and $\sigma^{wa(v)}_{xy}$ (respec-

tively, $\sigma^{wu}_{x'y'}$ and $\sigma^{wa(v')}_{x'y'}$) are the number of shortest paths between x and y (respectively, x' and y') and the number of shortest paths between x and y (respectively, x' and y') passing v (respectively, v') in the weighted graph G^a (respectively, G^u).

Structural Similarity. From the analysis on real data sets, the local and global structural characteristics carried by degree, closeness, and betweenness centralities of nodes can guide our de-anonymization framework design. Following this direction, to consider and utilize nodes' structural property integrally, we define a unified structural measurement, namely *structural similarity*, to jointly count two nodes' both local and global topological properties. First, for $v \in V^a$ and $v' \in V^u$, we define two *structural characteristic vectors* $\mathbf{S}^a(v)$ and $\mathbf{S}^u(v')$ respectively in terms of their (weighted) degree, closeness, and betweenness centralities as follows: $\mathbf{S}^a(v) = [d_v, c_v, b_v, wd_v, wc_v, wb_v]$ and $\mathbf{S}^u(v') = [d_{v'}, c_{v'}, b_{v'}, wd_{v'}, wc_{v'}, wb_{v'}]$. In $\mathbf{S}^a(v)$, if G^a is unweighted, we set $wd_v = wc_v = wb_v = 0$; otherwise, we first count d_v, c_v, and b_v by assuming G^a is unweighted, and then count wd_v, wc_v, and wb_v in the weighted G^a. We also apply the same method to obtain $\mathbf{S}^u(v')$ in G^u. Based on $\mathbf{S}^a(v)$ and $\mathbf{S}^u(v')$, we define the *structural similarity* between $v \in V^a$ and $v' \in V^u$, denoted by $s_S(v, v')$, as the *cosine similarity* between $\mathbf{S}^a(v)$ and $\mathbf{S}^u(v')$, i.e., $s_S(v, v') = \frac{\mathbf{S}^a(v) \cdot \mathbf{S}^u(v')}{\|\mathbf{S}^a(v)\|\|\mathbf{S}^u(v')\|}$, where $\cdot$ is the *dot product* and $\| \cdot \|$ is the *magnitude* of a vector.

The structural similarity between the nodes in St Andrews and its auxiliary network Facebook is shown in Fig.1 (d), where *Counterpart* represents $s_S(v, v' = \gamma(v))$ indicating the structural similarity between $v \in V^a$ and its objective reality $\gamma(v)$ in G^u, *Min* represents $\min\{s_S(v, x')|x' \in V^u, x' \neq \gamma(v)\}$, *Max* represents $\max\{s_S(v, x')|x' \in V^u, x' \neq \gamma(v)\}$, and *Avg* represents $\frac{1}{|V^u|-1} \sum_{x' \in V^u, x' \neq \gamma(v)} s_S(v, x')$. From Fig.1 (d), we have the following two basic observations. (i) For some nodes with distinguished structural characteristics, e.g., nodes 2, 16, 24, they agree with their counterparts and disagree with other nodes in the auxiliary graphs significantly. Consequently, this suggests that these nodes can be de-anonymized even just based on their structural characteristics. In addition, this confirms that structural properties can be employed in de-anonymization attacks. (ii) For the nodes with indistinctive structural similarities, e.g., nodes 7, 10, 22, 26, exact node mapping relying on structural property alone is difficult or impossible to achieve from the view of graph theory. Fortunately, even if this is true, structural characteristics can also help us to differentiate these indistinctive nodes from most of the other nodes. Hence, structural similarity based coarse granularity de-anonymization is practical.

Relative Distance Similarity. In the first phase, we select an initial seed mapping $\mathcal{M}_0 = \mathcal{M}_s = \{(s_1, s_1'), (s_2, s_2'), \cdots, (s_\kappa, s_\kappa')\}$. This apriori knowledge can be used to conduct more confident ratiocination in de-anonymization. Therefore, for $v \in V^a \setminus M_0^a$, we define its *relative distance vector*, denoted by $\mathbf{D}^a(v)$ to the seeds in $M_0^a = \{s_1, s_2, \cdots, s_\kappa\}$ as $\mathbf{D}^a(v) = [D_1^a(v), D_2^a(v), \cdots, D_\kappa^a(v)]$, where $D_i^a(v) = \frac{|p^a(v, s_i)|}{D^a}$ is the *normalized relative distance* between v and seed s_i. Similarly, based on the initial seed set $M_0^u = \{s_1', s_2', \cdots, s_\kappa'\}$ in G^u, we can define the *relative distance vector* for $v' \in V^u \setminus M_0^u$ to the seeds in M_0^u as $\mathbf{D}^u(v') = [D_1^u(v'), D_2^u(v'), \cdots, D_\kappa^u(v')]$, where $D_i^u(v') = \frac{|p^u(v', s_i')|}{D^u}$ is the *normalized relative distance* between v' and seed s_i'. Again, we can define the *relative distance similarity* between $v \in V^a \setminus M_0^a$ and $v' \in V^u \setminus M_0^u$, denoted by $s_D(v, v')$, as the *cosine similarity* between $\mathbf{D}^a(v)$ and $\mathbf{D}^u(v')$, i.e., $s_D(v, v') = \frac{\mathbf{D}^a(v) \cdot \mathbf{D}^u(v')}{\|\mathbf{D}^a(v)\| \|\mathbf{D}^u(v')\|}$.

For St Andrews/Facebook, by assuming $\mathcal{M}_s = \{(i, i) | i = 1, 2, \cdots, 6\}$ (which implies $M_0^a = M_0^u = \{1, 2, 3, 4, 5, 6\}$), we can obtain the relative distance similarity scores between the nodes in $V^a \setminus M_0^a$ and the nodes in $V^u \setminus M_0^u$ as shown in Fig.1 (e). From Fig.1 (e), we can observe the following facts. (i) Some anonymized nodes (which may be indistinctive with respect to structural similarity), e.g., nodes 14, 19, 23, highly agree with their counterparts and meanwhile disagree with other nodes in the auxiliary graph, which suggests that they can be de-anonymized successfully with a high probability by employing the relative distance similarity based metric. (ii) For some nodes, e.g., nodes 11, 21, 26, 27, they are indistinctive on the relative distance similarity with respect to the initial seed selection $\{1, 2, 3, 4, 5, 6\}$. To distinguish them, extra effort is expected, e.g., by utilizing structural similarity collaboratively, employing another seed selection, etc. (ii) The nodes that are significantly distinguishable with respect to structural similarity may be indistinctive with respect to relative distance similarity, and vice versa. This inspires us to design a proper and effective multi-measurement based de-anonymization framework.

Inheritance Similarity. Besides the initial seed mapping, the de-anonymized nodes during each iteration, i.e., $\mathcal{M}_k$, could provide further knowledge when de-anonymize $\Lambda^\delta(M_k^a)$. Therefore, for $v \in \Lambda^\delta(M_k^a)$ and $v' \in \Lambda^\delta(M_k^u)$, we define the knowledge provided by the currently mapped results as the *inheritance similarity*, denoted by $s_I(v, v')$. Formally, $s_I(v, v')$ can be quantified as

$$s_I(v, v') = \frac{C}{|N_k(v, v')|} \cdot \left(1 - \frac{|\Delta_v^a - \Delta_{v'}^u|}{\max\{\Delta_v^a, \Delta_{v'}^u\}}\right) \cdot \sum_{(x, x') \in N_k(v, v')} s(x, x') \text{ if } N_k(v, v') \neq \emptyset,$$

and $s_I(v, v') = 0$, otherwise, where $C \in (0, 1)$ is a constant value representing the *similarity loss exponent*, $N_k(v, v') = (N^a(v) \times N^u(v')) \cap \mathcal{M}_k = \{(x, x') | x \in N^a(v), x' \in N^u(v'), (x, x') \in \mathcal{M}_k\}$ is the set of mapped pairs between $N^a(v)$ and

$N^u(v')$ till iteration k, and $s(x, x') \in [0, 1]$ is the overall similarity score between x and x' which is formally defined in the following subsection.

From the definition of $s_I(v, v')$, we can see that (i) if two nodes have more common neighbors which have been mapped, then their inheritance similarity score is high; (ii) we also count the degree similarity in defining $s_I(v, v')$. If the degree difference between v and v' is small, then a large weight is given to the inheritance similarity; otherwise, a small weight is given; and (iii) we involve the similarity loss in counting $s_I(v, v')$, which implies the inheritance similarity is decreasing with the distance increasing (iteration increasing) between (v, v') and the original seed mapping.

Now, for St Andrews/Facebook, if we assume half of the nodes have been mapped (the first half according to the ID increasing order), then the inheritance similarity between the rest of the nodes in the anonymized graph and the auxiliary graph is shown in Fig.1 (f). From the result, we can observe that under the half number of nodes been mapped assumption, some nodes, e.g., nodes 16, 19, 24, agree with their counterparts and meanwhile disagree with all the other nodes significantly in the auxiliary graph, which implies that they are potentially easier to be de-anonymized when inheritance similarity is taken as a metric. Note that, in Fig.1 (f), we just randomly assume that the known mapping nodes are the first half nodes in the anonymized graph and auxiliary graph. Actually, the accuracy performance of the inheritance similarity measurement could be improved. This is because there are no necessary correlations among the randomly chosen mapping nodes in Fig.1 (f). Nevertheless, in our de-anonymization framework, the obtained mappings in one iteration depend on the mappings in the previous iteration. This strong correlation among mapped nodes allows for use of the inheritance similarity in practical de-anonymizaiton.

De-Anonymization Algorithm. From the aforementioned discussion, we find that the differentiability of anonymized nodes is different with respect to different similarity measurements. For instance, some nodes have distinctive topological characteristics, e.g., node 16 in St Andrew, which implies they can be potentially de-anonymized solely based on the structural similarity. On the other hand, for some nodes, due to lacking of distinct topological characteristics, the structural similarity based method can only achieve coarse granularity de-anonymization. Nevertheless and fortunately (from the view of adversary), they may become significantly distinguishable with the knowledge of a small amount of auxiliary information, e.g., nodes 14, 19, and 23 in St Andrews are potentially easy to de-anonymize based on relative distance similarity. In summary, the analysis on real data sets suggests to us to define a unified measurement to properly involve multiple similarity metrics for effective de-anonymization. To this end, we define a *Unified Similarity* (US) measurement by considering the structural similarity, relative distance similarity, and inheritance similarity synthetically for $v \in \Lambda^\delta(M_k^a)$ and $v' \in \Lambda^\delta(M_k^u)$ in the k-th iteration of our de-anonymization framework as $s(v, v') = c_S \cdot s_S(v, v') + c_D \cdot s_D(v, v') + c_I \cdot s_I(v, v')$, where $c_S, c_D, c_I \in [0, 1]$ are constant values indicating the weights of structural similarity, relative distance similarity, and inheritance similarity, respectively, and $c_S + c_D + c_I = 1$.

Algorithm 1. US based **De-Anonymization** (DA)

---

1 $\mathcal{M}_0 = \mathcal{M}_s$, $k = 0$, $\mathit{flag} = \textbf{true}$;
2 **while** $\mathit{flag} = \textbf{true}$ **do**
3   calculate $\Lambda^\delta(M_k^a)$ and $\Lambda^\delta(M_k^u)$;
4   if $\Lambda^\delta(M_k^a) = \emptyset$ or $\Lambda^\delta(M_k^u) = \emptyset$, output $\mathcal{M}_k$, **break**;
5   for $\forall v \in \Lambda^\delta(M_k^a)$ and $\forall v' \in \Lambda^\delta(M_k^u)$, calculate $s(v, v')$;
6   construct a weighted bipartite graph $B_k = (\Lambda^\delta(M_k^a) \cup \Lambda^\delta(M_k^u), E_k^b, W_k^b)$;
7   find a *maximum weighted bipartite matching* $\mathcal{M}'$ of B_k;
8   for every $(x, x') \in \mathcal{M}'$, if $s(x, x') < \theta$, $\mathcal{M}' = \mathcal{M}' \setminus \{(x, x')\}$;
9   let $K = \max\{1, \lceil |\epsilon \cdot \mathcal{M}'| \rceil\}$ and for $\forall (x, x') \in \mathcal{M}'$, **if** $s(x, x')$ *is not the* Top-K *mapping*
    score in $\mathcal{M}'$ **then**
10          $\lfloor$ $\mathcal{M}' = \mathcal{M}' \setminus \{(x, x')\}$;
11   if $\mathcal{M}' = \emptyset$, output $\mathcal{M}_k$ and **break**;
12      $\mathcal{M}_{k+1} = \mathcal{M}_k \cup \mathcal{M}'$, k++;

In addition, we define $s(v, v') = 1$ if $(v, v') \in \mathcal{M}_s$. Now, we are ready to present our US based **De-Anonymization** (DA) framework, which is shown in Algorithm 1.

In Algorithm 1, $B_k = (\Lambda^\delta(M_k^a) \cup \Lambda^\delta(M_k^u), E_k^b, W_k^b)$ is a *weighted bipartite graph* defined on the intended de-anonymizing nodes during the k-th iteration, where $E_k^b = \{l_{v,v'}^b | \forall v \in \Lambda^\delta(M_k^a), \forall v' \in \Lambda^\delta(M_k^u)\}$, and $W_k^b = \{w_{v,v'}^b\}$ is the set of all the possible weights on the links in E_k^b. Here, for $\forall (v, v') \in E_k^b$, the weight on this link is defined as the US score between the associated two nodes, i.e., $w_{v,v'}^b = s(v, v')$. Parameter θ is a constant value named *de-anonymization threshold* to decide whether a node mapping is accepted or not. Parameter $\epsilon \in (0, 1]$ is the *mapping control factor*, which is used to limit the maximum number mappings generated during each iteration. By ϵ, even if there are many mappings with similarity score greater than the de-anonymization threshold, we only keep the $K = \max\{1, \lceil |\epsilon \cdot \mathcal{M}'| \rceil\}$ more confident mappings.

We give further explanation on the idea of Algorithm DA as follows. The de-anonymization is bootstrapped with an initial seed mapping and starts the iteration procedure. During each iteration, the intended de-anonymizing nodes are calculated first based on the mappings obtained in the previous iteration followed by calculating the US scores between nodes in $\Lambda^\delta(M_k^a)$ and nodes in $\Lambda^\delta(M_k^u)$. Subsequently, based on the obtained US scores, a weighted bipartite graph is constructed between nodes in $\Lambda^\delta(M_k^a)$ and nodes in $\Lambda^\delta(M_k^u)$. Then, we compute a *maximum weighted bipartite matching* $\mathcal{M}'$ on the constructed bipartite graph. To improve the de-anonymization accuracy, we apply two important rules to refine $\mathcal{M}'$: (i) by defining a *de-anonymization threshold* θ, we eliminate the mappings with low US scores in $\mathcal{M}'$. This is because we are not confident to take the mappings with low US scores ($< \theta$) as correct de-anonymizaiton, and more improtantly, they may be more accurately de-anonymized in the following iterations by utilizing confident mapping information obtained in this iteration (this can be achieved since we involve inheritance similarity in the US definition); and (ii) we introduce a *mapping control factor* ϵ, or K equivalently, to limit the maximum number of mappings been accepted as correct de-anonymization.

During each iteration, only K mappings with highest US scores will be taken as correct de-anonymization with confidence even if more mappings having US scores greater than the de-anonymizaiton threshold. This strategy has two benefits. On one hand, only highly confident mappings are kept, which could improve the de-anonymization accuracy. On the other hand, for the mappings been rejected, again, they may be better re-de-anonymized in the following iterations by utilizing the more confident knowledge of the Top-K mappings from this iteration.

Time and Space Complexities Analysis. Let $n = \max\{|V^a|, |V^u|\}$ and $m = \max\{|E^a|, |E^u|\}$. Then, according to combinatorial analysis, Algorithm 1's time complexity is $O(n^2 \log n + mn)$ and space complexity is $O(\min\{n^2, m+n\})$.

4   Generalized Scalable De-Anonymization

De-Anonymization on Data Sets without Knowledge of Overlap Size. One predicament in practical de-anonymization, which is omitted in existing de-anonymization attacks, is that we do not actually know how large the overlap between the anonymized data and the auxiliary data even we have a lot of auxiliary information available. Therefore, it is unadvisable to do de-anonymization based on the entire anonymized and auxiliary graphs directly, which might cause low de-anonymization accuracy as well as high computational overhead.

To address the aforementioned predicament, guarantee the accuracy of DA, and simultaneously improve de-anonymization efficiency and scalability, we extend DA to an *Adaptive De-Anonymization* framework, denoted by ADA. ADA adaptively de-anonymizes G^a starting from a *Core Matching Subgraph* (CMS), which is formally defined as follows. Let $\mathcal{M}_s$ be the initial seed mapping between the anonymized graph G^a and the auxiliary graph G^u. Furthermore, define $V_s^a = \bigcup_{x,y \in M_0^a} \{v | v \text{ lies on } p^a(x,y) \in \mathbb{P}^a(x,y)\}$, i.e., V_s^a is the union of all the nodes on the shortest paths among all the seeds in G^a, and $V_c^a = V_s^a \cup \Lambda^\delta(V_s^a)$, i.e., V_c^a is the union of V_s^a and the δ-hop mapping spanning set of V_s^a. Then, we define the initial CMS on G^a as the subgraph of G^a on V_c^a, i.e., $G_c^a = G^a[V_c^a]$. Similarly, we can define $V_s^u = \bigcup_{x',y' \in M_0^u} \{v' | v' \text{ lies on } p^u(x',y') \in \mathbb{P}^u(x',y')\}$ and $V_c^u = V_s^u \cup \Lambda^\delta(V_s^u)$. Then, the initial CMS on G^u is $G_c^u = G^a[V_c^u]$.

The CMS is generally defined for two purposes. First, we can employ a CMS to adaptively and roughly estimate the overlap between G^a and G^u in terms of the seed mapping information. On the other hand, we propose to start the de-anonymization from the CMSs, by which the de-anonymization is smartly limited to start from two small subgraphs with more information confidence, and thus we could improve the de-anonymization accuracy and reduce the computational overhead.

Now, based on CMS, we discuss ADA as shown in Algorithm 2. In Algorithm 2, μ is the *adaptive factor* which controls the spanning size of the CMS during each adaptive iteration. The basic idea of ADA is as follows. We start

---

Algorithm 2. Adaptive De-Anonymization (ADA)

1   generate G_c^a and G_c^u and run DA for G_c^a and G_c^u;

2   **if** *Step 1 is ended on the condition that* $\Lambda^\delta(M_k^a) = \emptyset$ *or* $\Lambda^\delta(M_k^u) = \emptyset$ **then**

3   **if** $\Lambda^\mu(V_c^a) = \emptyset$ or $\Lambda^\mu(V_c^u) = \emptyset$, **return**;

4       $V_c^a = V_c^a \cup \Lambda^\mu(V_c^a)$, $V_c^a = V_c^a \cup \Lambda^\mu(V_c^a)$;

5       $G_c^a = G^a[V_c^a]$, $G_c^a = G^u[V_c^u]$;

6   go to Step 1 to de-anonymize unmapped nodes in updated G_c^a and G_c^u;

---

the de-anonymization from CMSs G_c^a and G_c^u by running DA. If DA is ended with $\Lambda^\delta(M_k^a) = \emptyset$ or $\Lambda^\delta(M_k^u) = \emptyset$, then the actual overlap between G^a and G^u might be larger than G_c^a/G_c^u since more nodes could be mapped. Therefore, we enlarge the previous considering CMS G_c^a/G_c^u by involving more nodes in $\Lambda^\mu(V_c^a)/\Lambda^\mu(V_c^u)$ and repeat the de-anonymization for unmapped nodes. Same as DA, the time and space complexities of ADA are $O(n^2 \log n + mn)$ and $O(\min\{n^2, m+n\})$, respectively.

Disconnected Data Sets. In reality, when we employ a graph G^a/G^u to model the anonymized/auxiliary data, G^a/G^u might be not connected. In this case, G^a and G^u can be represented by the union of connected components as $\bigcup_{i=1}^{m} G_i^a$ and $\bigcup_{j=1}^{n} G_j^u$ respectively, where G_i^a and G_j^u are some connected components. Now, when defining the structural similarity, relative distance similarity, or inheritance similarity, we change the context from G^a/G^u to components G_i^a/G_j^u. Then, we can apply DA/ADA to conduct de-anonymization.

5   Experiments

In this section, we examine the performance of the presented de-anonymization attack on real data sets[3]. In each group of experiments, we specify the employed setup and provide comprehensive analysis. The default settings are: $\alpha = 1.5$, $C = 0.9$, $c_S = 0.2$, $c_D = 0.6$, $c_I = 0.2$, $\theta = 0.6$, $\delta \in \{1, 2\}$, $\mu \in \{1, 2, 3\}$, $\epsilon = 0.5$ and seed number $= 5$.

Data Sets. In this paper, we employ six well known data sets to examine the effectiveness of the designed de-anonymization framework[4]: St Andrews/Facebook [11][3], Infocom06/DBLP [13][3], Smallbule/Facebook [12][3], ArnetMiner [14], Google+ [10], and Facebook [15]. St Andrews, Infocom06, and Smallbule are three mobility trace data sets. An overview of the three mobility traces is shown in Table 1. We employ the same techniques as in [3] to preprocess the three mobility trace data sets to obtain three anonymized data graphs. To de-anonymize

---

[3] Due to the space limitation, we put the detailed experimental settings and more results in the Technical Report [17] of this paper.

[4] Not that it has been shown that the classical mobility traces of the (latitude, longitude, timestamp) form can also be represented by graph models [16].

Table 1. Mobility traces

	St Andrews	Infocom06	Smallblue
Comm. network type	WiFi	Bluetooth	IM
Comm. nodes No.	27	78	125
Contacts No.	18,241	182,951	240,665
Social network type	Facebook	DBLP	Facebook
Social nodes No.	27	616	400

the three anonymized mobility data traces, we employ three auxiliary social network data sets [3] associated with these three mobility traces. For St Andrews, we have a Facebook data set indicating the "friend" relationships among the T-mote users in the trace. For Infocom06, we employ a coauthor data set consisting of 616 authors obtained from DBLP which indicates the "coauthor" relationships among all the attendees of INFOCOM 2005. For Smallblue, we have a Facebook network among 400 employees from the same enterprise as Smallblue. Note that, the social network data sets corresponding to Infocom06 and Smallblue are supersets of them with respect to involved users.

We also apply the presented de-anonymization attack to social data sets: ArnetMiner [14], Google+ [10], and Facebook [15]. ArnetMiner is an online academic social network, which consists of 1,127 authors and 6,690 "coauthor" relationships. For each coauthor relationship, there is a weight associated with it indicating the number of coauthored papers by the two authors. As a new social network, Google+ was launched in early July 2011. We use two Google+ data sets which were created on July 19 and August 6 in 2011 [10], denoted by JUL and AUG respectively. Both JUL and AUG consist of 5,200 users as well as their profiles. In addition, there were 7,062 connections in JUL and 7,813 connections in AUG. By insight analysis [10], some connections appeared in AUG may not appear in JUL and vise versa. This is because a user may add new connections or disable existing connections. Furthermore, the two data sets are preprocessed as undirected graphs. Since we know the hand labeled ground truth of JUL and AUG, we will examine the presented de-anonymization framework by de-anonymizing JUL with AUG as auxiliary data and then de-anonymizing AUG with JUL as auxiliary data. The Facebook data set consists of 63,731 users and 1,269,502 "friend" relationships (links). To use this data set to examine the presented de-anonymization attack, we will preprocess it based on the known hand labeled ground truth.

De-anonymize Mobility Traces. By utilizing the corresponding social networks as auxiliary information, we exploit the presented de-anonymization algorithm DA to de-anonymize the three well known mobility traces St Andrews, Infocom06, and Smallblue. The results are shown in Fig.2 (a)-(c), where DA denotes the presented US-based de-anonymization framework, and DA-SS, DA-RDS, and DA-IS represent the de-anonymization based on structural similarity solely, relative distance similarity solely, and inheritance similarity solely, respectively. From Fig.2 (a)-(c), we can see that (i) the presented de-aonymization framework is very effective even with a small amount of auxiliary information.

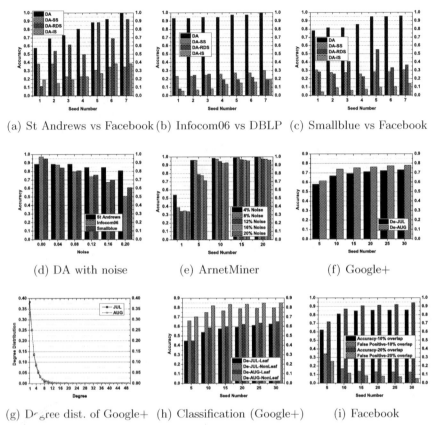

(a) St Andrews vs Facebook (b) Infocom06 vs DBLP (c) Smallblue vs Facebook

(d) DA with noise   (e) ArnetMiner   (f) Google+

(g) D̃egree dist. of Google+ (h) Classification (Google+)   (i) Facebook

Fig. 2. De-anonymize mobility traces

For instance, DA can successfully de-anonymize 93.2% of the Infocom06 data just with the knowledge of one seed mapping. For St Andrews and Smallblue, DA can also achieve accuracy of 57.7% and 78.3% respectively with one seed mapping. Furthermore, DA can successfully de-anonymize all the data in St Andrews and Smallblue and 96% of the data of Smallblue with the knowledge of 7 seed mappings; and (ii) the US-based de-anonymization is much more effective and stable than structural, relative distance, or inheritance similarity solely based de-anonymization. The reason is that US tries to distinguish a node from multiple perspectives, which is more efficient and comprehensive. As the analysis shown in Section 3, the nodes can be easily differentiated with respect to one measurement but might be indistinguishable with respect to another measurement. Consequently, synthetically characterizing a node as in US is more powerful and stable.

We also examine the robustness of the presented de-anonymization attack to noise and the result is shown in Fig.2 (d) (on the knowledge of 5 seed mappings). In the experiment, we only add noise to the anonymized data. According to the same argument in [2], the noise in the auxiliary data can be counted as noise in the anonymized data. To add p percent of noise to the anonymized data, we randomly add $\frac{p}{2} \cdot |E^a|$ spurious connections to and meanwhile delete $\frac{p}{2} \cdot |E^a|$ existing connections from the anonymized graph (a node may become *isolated* after the noise adding process). For instance, in Fig.2 (d), 20% of noise implies we add 10% spurious connections and delete 10% existing connections of $|E^a|$ from the anonymized data. From Fig.2 (d), we can see that the presented de-anonymization framework is robust to noise. Even if we change 20% of the connections in the anonymized data, the achieved accuracies on St Andrews, Infocom06, and Smallblue are still 80.8%, 50.7%, and 60.8%, respectively. Note that, when 20% of the connections have been changed, the structure of the anonymized data is significantly changed. In practical, if the anonymized data release is initially for research purposes, this structural change may make the data useless. However, by considering multiple perspectives to distinguish a node, the anonymized data can still be de-anonymized as shown in Fig.2 (d), which confirms the assertion in [2] that data set structure change may not provide effective privacy protection.

De-Anonymize ArnetMiner. ArnetMiner can be modeled by a weighted graph where the weight on each relationship indicates the number of coauthored papers by the two authors. To examine the de-anonymization framework, we first anaonymize ArnetMiner by adding p percent noise as explained in the previous subsection. Furthermore, for each added spurious coauthor relationship, we also randomly generate a weight in $[1, A_{\max}]$, where $A_{\max}$ is the maximum weight in the original ArnetMiner graph. Then, we de-anonymize the anonymized data using the original ArnetMiner data and the result is shown in Fig.2 (e).

From Fig.2 (e), we can observe that the presented de-anonymization framework is very effective on weighted data. With only knowledge of one seed mapping, more than a half (53.9%) and one-third (34.1%) of the authors can be de-anonymized even with noise levels of 4% and 20%, respectively. Furthermore, when adding 20% of noise to the anonymized data, the presented de-anonymization framework achieves 71.5% accuracy if 5 seed mappings are available and 92.8% accuracy if 10 seed mappings are available; (*ii*) the presented de-anonymization framework is robust to noise on weighted data. When we have 10 or more seed mappings, the accuracy degradation of our de-anonymization algorithm is small even with more noise, e.g., the accuracy is degraded from 99.7% in the 4%-noise case to 96% in the 20%-noise case; and (*iii*) if the available number of seed mappings is 10, the knowledge brought by more seed mappings cannot improve the de-anonymization accuracy significantly. This is because the achieved accuracy on the knowledge of 10 seed mappings is already about 95%. Therefore, to de-anonymize a data set, it is not necessary to spend efforts to obtain a lot of seed mappings. As in this case, to de-anonymize most of the authors, 5 to 10 seed mappings is sufficient.

De-Anonymize Google+. Now, we validate the presented de-anonymization framework on the two Google+ data sets JUL and AUG. We first utilize AUG as auxiliary data to deanonymize JUL denoted by De-JUL, i.e., use future data to de-anonymize historical data, and then utilize JUL to de-anonymize AUG denoted by De-AUG, i.e., use historical data to de-anonymize future data. The results is shown in Fig.2 (f). Again, from Fig.2 (f), we can see that the presented de-anonymization framework is very effective. Just based on the knowledge of 5 seed mappings, 57.9% of the users in JUL and 61.6% of the users in AUG can be successfully deanonymized. When 10 seed mappings are available, the de-anonymization accuracy can be improved to 66.8% on JUL and 73.9% on AUG, respectively.

However, we also have two other interesting observations from Fig.2 (f): (i) when the number of available seed mappings is above 10, the performance improvement is not as significant as on previous data sets (e.g., mobility traces, ArnetMiner) even the de-anonymization accuracy is around 70% for JUL and 75% for AUG; and (ii) De-AUG has a better accuracy than De-JUL, which implies that the AUG data set is easier to de-anonymize than the JUL data set. To explain the two observations, we assert this is because of the structural property of the two data sets. Follow this direction, we investigate the degree distribution of JUL and AUG as shown in Fig.2 (g). From Fig.2 (g), we can see that the degree of both JUL and AUG generally follows a *heavy-tailed distribution*. In particular, 38.4% of the users in JUL and 34.3% of the users in AUG have degree of one, named *leaf users*. This is normal since Google+ was launched in early July 2011, and JUL and AUG are data sets crawled in July and August of 2011, respectively. That is also why JUL has more leaf users than AUG (a user connects more people later). Now, we argue that the leaf users cause the difficulty in improving the de-anonymization accuracy. From the perspective of graph theory, the leaf users limit not only the performance of our de-anonymization framework but also the performance of any de-anonymization algorithm. An explanatory example is as follows. Suppose $v \in V^a$ is successfully de-anonymized to $v' \in V^u$. In addition, the two neighbors x and y of v and the two neighbors x' and y' of v' are all leaf users. Then, even $x' = \gamma(x)$, $y' = \gamma(y)$, and v has been successfully de-anonymized to v', it is still difficult to make a decision to map x (or y) to x' or y' since $s(x, x') \approx s(x, y')$ from the view of graph theory. Consequently, to accurately distinguish x, further knowledge such as semantic information is required.

To support our argument, we take an insightful look on the experimental results. For each successfully de-anonymized user in JUL and AUG, we classify the user in terms of its degree into one of two sets: *leaf user set* if its degree is one or *non-leaf user set* if its degree is greater than one. Then, we re-calculate the de-anonymization accuracy for leaf users and non-leaf users and the results are shown in Fig.2 (h), where De-JUL-Leaf/De-AUG-Leaf represents the ratio of leaf nodes that have been successfully de-anonymized in JUL/AUG while De-JUL-NonLeaf/De-AUG-NonLeaf represents the ratio of non-leaf users that have been successfully de-anonymized in JUL/AUG. From Fig.2 (h), we can see that (i) the successful de-anonymization ratio on non-leaf users is higher than that on leaf users in JUL and AUG. This is because non-leaf users carry more structural information; and (ii) considering the results shown in Fig.2 (f), the de-anonymization accuracy on non-leaf users is higher than the overall accuracy and the de-anonymization accuracy on leaf users is lower than the overall accuracy. The two observations on Fig.2 (h) confirms our argument that leaf users are more difficult than non-leaf users to de-anonymize. Furthermore, this is also why De-AUG has higher accuracy than De-JUL in Fig.2 (f). AUG is easier to de-anonymize since it has less leaf users than JUL.

De-Anonymize Facebook. Finally, we examine ADA on Facebook. Based on the hand labeled ground truth, we partition the data sets into two about-equal parts utilizing the method employed in [2], and then we take one part as auxiliary data to de-anonymize the other part. When the two parts only have 10% and 20% users in common, the achievable accuracy and the induced false positive error of ADA are shown in Fig.2 (i). As a fact, most of the existing de-anonymization attacks are not very effective for the scenario that the overlap between the anaonymized data and the auxiliary data is small or even cannot work totally. Surprisingly, for ADA, we can observe from Fig.2 (i) that (i) based on the proposed CMS, ADA can successfully de-anonymize 62.4% of the common users with false positive error of 34.1% when the overlap is 10% and 71.8% of the common users with false positive error of 25.6% when the overlap is 20% with the knowledge of just 5 seed mappings; (ii) the de-anonymization accuracy is improved to 81.3% (resp., 85.6%) and the false positive error is decreased to 16.8% (resp., 13%) when the overlap is 10% and 10 (resp., 20) seed mappings available, and the de-anonymization accuracy is improved to 87% (resp., 90.8%) and the false positive error is decreased to 11.6% (resp., 8.6%) when the overlap is 20% and 10 (resp., 20) seed mappings available, which demonstrates that ADA is very effective in dealing with the partial data overlap situation; and (iii) ADA has a higher de-anonymization accuracy and lower false positive error in the 20% data overlap scenario than that in the 10% data overlap scenario. This is because a larger overlap size implies a common node will carry much more similar structural information in both graphs, and thus it can be de-anonymized with higher probability and accuracy. From Fig.2 (i), we can also see that 10 seed mappings are sufficient to achieve high de-anonymization accuracy and low false positive error. Therefore, ADA is applicable with efficiency and performance guarantee in practical.

6   Conclusion

In this paper, we present a novel and effective de-anonymization attack based on a Unified Similarity (US) measurement which synthetically incorporates multiple data structural factors. The experimental results demonstrate that the presented de-anonymization framework is very effective and robust to noise.

Acknowledgments. Mudhakar Srivatsa's research was sponsored by US Army Research laboratory and the UK Ministry of Defence and was accomplished under Agreement Number W911NF-06-3-0001. The views and conclusions contained in this document are those of the authors and should not be interpreted as representing the official policies, either expressed or implied, of the US Army

Research Laboratory, the U.S. Government, the UK Ministry of Defense, or the UK Government. The US and UK Governments are authorized to reproduce and distribute reprints for Government purposes notwithstanding any copyright notation hereon. Jing S. He's research is partly supported by the Kennesaw State University College of Science and Mathematics the Interdisciplinary Research Opportunities (IDROP) Program.

References

1. Backstrom, L., Dwork, C., Kleinberg, J.: Wherefore Art Thou R3579X? Anonymized Social Networks, Hidden Patterns, and Structural Steganography. In: WWW 2007 (2007)
2. Narayanan, A., Shmatikov, V.: De-anonymizing Social Networks. In: S&P 2009 (2009)
3. Srivatsa, M., Hicks, M.: Deanonymizing Mobility Traces: Using Social Networks as a Side-Channel. In: CCS 2012 (2012)
4. Narayanan, A., Shmatikov, V.: Robust De-anonymization of Large Sparse Datasets (De-anonymizing the Netflix Prize Dataset). In: S&P 2008 (2008)
5. Goodin, D.: Poorly anonymized logs reveal NYC cab drivers detailed whereabouts, http://arstechnica.com/tech-policy/2014/06/poorly-anonymized-logs-reveal-nyc-cab-drivers-detailed-whereabouts/
6. Singh, K., Bhola, S., Lee, W.: xBook: Redesigning Privacy Control in Social Networking Platforms. In: USENIX 2009 (2009)
7. Hornyack, P., Han, S., Jung, J., Schechter, S., Wetherall, D.: "These Aren't the Droids You're Looking For": Retrofitting Android to Protect Data from Imperious Applications. In: CCS 2011 (2011)
8. Egele, M., Kruegel, C., Kirda, E., Vigna, G.: PiOS: Detecting Privacy Leaks in iOS Applications. In: NDSS 2011 (2011)
9. Opsahl, T., Agneessens, F., Skvoretz, J.: Node Centrality in Weighted Networks: Generalizing Degree and Shortest Paths. Social Networks 32, 245–251 (2010)
10. Gong, N.Z., Talwalkar, A., Mackey, L., Huang, L., Shin, E.C.R., Stefanov, E., Shi, E., Song, D.: Jointly Predicting Links and Inferring Attributes using a Social-Attribute Network (SAN). In: SNA-KDD 2012 (2012)
11. Bigwood, G., Rehunathan, D., Bateman, M., Henderson, T., Bhatti, S.: CRAW-DAD data set st_andrews/sassy (v. 2011-06-03) (June 2011), Downloaded from http://crawdad.cs.dartmouth.edu/~crawdad/st_andrews/sassy/
12. Smallblue, http://domino.research.ibm.com/comm/research_projects.nsf/pages/smallblue.index.html
13. Scott, J., Gass, R., Crowcroft, J., Hui, P., Diot, C., Chaintreau, A.: CRAW-DAD data set cambridge/haggle (v. 2009-05-29) (May 2009), Downloaded from http://crawdad.cs.dartmouth.edu/cambridge/haggle
14. Tang, J., Zhang, J., Yao, L., Li, J., Zhang, L., Su, Z.: ArnetMiner: Extraction and Mining of Academic Social Networks. In: KDD 2008 (2008)
15. Viswanath, B., Mislove, A., Cha, M., Gummadi, K.P.: On the Evolution of User Interaction in Facebook. In: WOSN 2009 (2009)
16. Pham, H., Shahabi, C., Liu, Y.: EBM - An Entropy-Based Model to Infer Social Strength from Spatiotemporal Data. In: Sigmod 2013 (2013)
17. Ji, S., Li, W., Srivatsa, M., He, J., Beyah, R.: Technical Report: Data De-anonymization: From Mobility Traces to On-line Social Networks, http://users.ece.gatech.edu/~sji/Paper/isc14TechReport.pdf

Investigating the Hooking Behavior:
A Page-Level Memory Monitoring Method
for Live Forensics

Yingxin Cheng, Xiao Fu*, Bin Luo, Rui Yang, and Hao Ruan

Software Institute, Nanjing University, China
yingxincheng@gmail.com, {fx,luobin,yr10,mg1332013}@software.nju.edu.cn

Abstract. In intrusion forensics, it is difficult to find the evidences about who placed the hooks and how these hooks were placed simply by analyzing the memory dump. That's because such behavior is transient and the snapshot of memory usually doesn't contain enough information about it. Lack of this information will cause an uncompleted chain of evidence. Although dynamic analysis can trace this behavior by instruction-level analysis, this technique is slow and inconvenient in real forensic cases. And many investigated systems do not run in the virtualization environment that dynamic analysis needed.

So we present a new method to intercept processes and create accurate modification bitmaps to reveal the process's memory behavior based on hardware events. A complete evidence chain is created for forensics by acquiring the process's running snapshots and its file image path during the monitoring. Considering live forensic cases, we apply this method to a novel lightweight hypervisor which can build up a virtualization environment on-the-fly. Without modifying any code in target OS and suspending it, this method can monitor systems running on bare hardware. Its memory usage and performance impact on the investigated system is also proved to be acceptable in our experiment.

Keywords: live forensics, memory behavior, process, virtualization, evidence.

1 Introduction

Traditional computer forensics solves investigation problems in a 'snatch and grab' way [1]. In acquisition phase, researchers have presented several methods to obtain an exact copy of memory [31,15]. And then in the analyzing phase, investigators extract evidences from these snapshots by certain forensic tools such as the Volatility Framework [25]. These two phases are independent, however, researchers [12] have pointed out a huge gap in this process in revealing who committed the crime in the first place. Specifically, it is very difficult to extract software behaviors simply from static memory dumps. So many researchers turn to analyzing live system directly for evidences [11,23].

---

* Corresponding author.

S.S.M. Chow et al. (Eds.): ISC 2014, LNCS 8783, pp. 255–272, 2014.

In intrusion forensics, especially in hook-related cases, above problems are also existent and may be even more serious. That's because hooking behavior is transient and the snapshot of memory usually doesn't contain enough information about it. Lack of this information will cause an uncompleted chain of evidence. In order to solve these problems, we have done some research work. Our work is mainly based on two motivations. Firstly, we believe that a better way to investigate hooking behavior is monitoring the modifications to the key memory areas and identifying related processes immediately. By this way, we can easily know who placed the hooks and how these hooks were placed. More generally, many advanced attack techniques such as DKOM [8] and Return Oriented Programming [22] are stealthy because they can circumvent kernel-level detectors by manipulating underlying memory directly. But they can't avoid making memory modifications. Secondly, we think that the investigation process should not interrupt the normal work, i.e. the system should be investigated without suspension. That's because some evidences in memory will lose if the system is suspended and the suspension may bring about unaffordable cost for system owners, such as e-business companies. Moreover, the criminal may detect the environment changes and hide the traces immediately.

Based on the above two motivations, we present a new method to investigate memory behavior. It can trace memory modifications by controlling page-grained permissions and intercepting hardware events. In addition, our method can create a complete evidence chain which includes not only specific memory modifications, but also the related processes and their EXE file paths. It can also capture corresponding running states of the process for further detailed analysis. These evidences are very important to prove where memory modifications come from. Although the virtualization environment is needed in our method, the Physical to Virtual (P2V) migration can be done with the help of a type-I Virtual Machine Monitor (VMM) which can install itself on-the-fly. The system reconfiguration is not required because only CPUs are virtualized and other hardware environment remains unchanged. Compared with the widely-used instruction-level memory behavior analysis which requires software emulators such as QEMU [7] and complex P2V migration, our method can reach the same accuracy in revealing the processes and the precise memory addresses modified by them. Moreover, the overhead is much lower because our method can select the traced memory range and optimize the performance while target memory is accessed by the same process. We also consider supporting typical Symmetric Multi-Processor (SMP) systems. The implementation of our prototype does not modify any code of target OS and maintains the hardware environment. So it is transparent to target system. Some experiments have proved that the memory usage of our method and its performance impact on the investigated system is acceptable.

2   Related Work

Recent researchers have an interest in collecting evidences from live systems due to the limitations of after-the-fact [1] forensic approaches. User-level information is easy to be modified, system call logger [11] can be circumvented if malware uses DKOM. But malware cannot bypass memory, there are already many works focusing on memory monitoring technique. We classify their works into two categories: asynchronous and synchronous memory monitoring.

Asynchronous memory monitoring methods do not catch up with the system states. Cross-view approach is a best example. This approach detects inconsistent states of the memory and can find out the hidden processes and hooks easily. It is applied by some rootkit detection tools [10] and integrity checkers [9]. Due to the fact that rootkits can stay silent most of time and interfere the system whenever needed, some forensic systems [16] run simultaneously with the target system and monitor the memory continuously. However, if the tool monitors memory asynchronously, transient behaviors could be lost. Moreover, this approach cannot find out the behavior-related processes simply by making comparisons. For example, the detection subsystem of HookScout [30] can periodically check the attacks based on policies generated by dynamic analysis, but it cannot figure out the related processes who perform these attacks.

Synchronous memory monitoring methods monitor every memory activity made by instructions. So it is able to introspect process's state after any instruction. It is convenient to use software VMMs such as QEMU to implement synchronous memory monitoring by interfering the binary translation process. Some works apply this method to record process's memory behavior for dynamic analysis. PoKeR [18] is a kernel rootkit profiler designed for producing multi-aspect profiles including hooking behavior. SigGENE [23] profiles a malware from its memory access patterns related to kernel objects. KernelGuard [17] prevents malicious memory modifications using a similar approach. And HookFinder [29] analyzes impacts introduced by malicious code using instruction-level analysis in a software emulator. Although these methods are powerful, they are slow and not suitable for monitoring systems running on bare hardware.

We classify our method into the third category – para-synchronous method, because this method does not have the ability to analyze every instructions. Instead, it monitors memory according to hardware events triggered by hardware. HookSafe [27] introduces the protection granularity gap between byte-level granularity of memory protection and page-level granularity of hardware protection. HookSafe partially solves the problem by relocating hooks to a dedicated memory space. And we propose a generic way to break the gap. This method is quicker and is not limited by software emulators compared with synchronous methods. And it is able to capture transient behaviors compared with asynchronous methods.

The P2V migration ability of hardware-assisted virtualization [14,3] fills the gap between dynamic analysis and live forensics. The New Blue Pill project [19] implements a novel VMM that can be deployed under a running system. It is noticed by some forensic researchers. HyperSleuth [15] firstly adopts the idea and is a system call logger and memory dumper for forensics. VIS [31] can dump a consistent memory image for memory acquisition. However, there is no solution focusing on monitoring memory behavior in live forensics, which is important in profiling a software and is widely used in the area of dynamic analysis.

3 Design

We proposed two motivations in the introduction section. In order to achieve the goals, we firstly design a method to intercept and record memory modifications and then design a prototype hypervisor to apply this method to live forensic scenario.

3.1 The Tracing Units and Recorders

Our memory modification tracing method is based on page permission control and acquires the accurate modifications each process made in target area.

Most commercial operating systems use paging technique provided by hardware to isolate memory from processes, so the memory is naturally divided by 4KiB pages from both hardware and software perspectives. Based on page permission control, it is natural to use the Second Level Address Translation (SLAT) mechanism to control guest physical memory. We design tracing units as basic components to monitor 4KiB pages, each of them can respond to hardware events, control page permission and record evidences about a page automatically. It simplifies our design because we can focus on the implementation of a single tracing unit. Tracing units are independent from each other, and each of them can be configured to a page.

The recorders are responsible to merge information from tracing units according to the monitored page and process. Multiple writes from a process to the same physical address are merged into a record block. The block size is 512 bytes. Each bit of it corresponds to an address in the monitored page, and indicates whether this address is modified by the process.

3.2 The Memory Modification Tracing Method

The overall algorithm is shown in Fig. 1. Our method leverages the copy-on-write-like operation and makes it transparent to the target OS using SLAT. When a guest process tries to modify the read-only memory, write violation happens. The violation-related tracing unit dumps the original memory page and obtains the process information using Virtual Machine Introspection (VMI). After a series of steps in hypervisor mode, the page resumes to be writable so that the process can modify the memory as it wants. The accurate modifications can be recorded simply by making a quick comparison between the original and the modified image. When the unit detects a different process coming, the former modifications of the page should be submitted to the hypervisor, and the page image should be dumped again for the new process. It is called the submit-and-copy operation, the idea comes from copy-on-write operation, which guarantees the accuracy of the modification records between processes in a core.

The tracing unit should identify the process whenever a different one writes into the same page. It happens only if the page is marked read-only so that the write violation can be triggered. That is to say, the write permissions of the pages should be cancelled when the process is changed in current processor. In uniprocessor systems, the pages are marked read-only whenever context change event

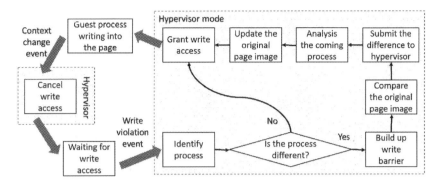

Fig. 1. Trace memory modifications in a single page

happens. The related units can then recheck the process and decide whether to
do the submit-and-copy operation.

In SMP systems, things get complicated because the memory can be accessed
concurrently from different cores. Another mechanism called write-barrier is introduced during submit-and-copy operation to intercept and identify processes
between cores. The purpose is to allow threads from the same process to write
concurrently into the page for efficiency, meanwhile, to make different processes
run sequentially for record accuracy.

According to our design, guest process can run without much interference
whenever it accesses a page it has recently accessed. It is very efficient compared
with instruction-level analysis, because access patterns of typical computer applications have locality of reference [28]. The implementation of important steps
such as write-barrier, copy-and-submit and write access cancelling will be explained in Sect. 4. And our method need supports such as SLAT, P2V migration
and evidence collecting for live forensics. So we also design a forensic hypervisor
to better demonstrate our ideas.

3.3 The Hypervisor Design

In order to test our method in real forensic cases, we must design and implement a proper hypervisor. Since Rutkowska [19] has already presented an open
source hypervisor which supports live P2V migration and can install and uninstall itself on-the-fly. We make use of the similar feature to add supports for our
method. This migration can be applied to laptops, desktops and servers, because
hardware-assisted virtualization is supported by most of latest Intel and AMD
CPUs based on their product lists [14,3].

The overall system design is shown in Fig. 2. Memory tracing module is the
implementation of our method described in previous section. The introspection
module collects information about process and make a snapshot of the process
running state during write violation. Control module receives external commands
to control the tracing module. The commands are used to start, pause, stop
tracing, and to set up monitoring area in memory. SLAT module builds up

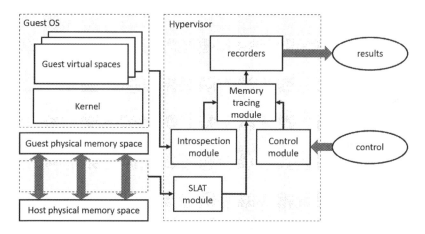

Fig. 2. Hypervisor design

a second level address translation from guest physical memory space to host physical memory space and controls related permissions such as read, write and execute in page granularity. All the tracing records including process information and memory modifications are collected by record module during the submit-and-copy phase. These modification records are classified according to processes and are finally outputted to disk.

4   Implementation

We have implemented the method and applied it to a hypervisor. Current prototype works for Windows 7 64-bit version and supports hardware with intel-VT technology. H2V migration uses intel-VT, and SLAT mechanism is implemented using EPT technology. With the help of EPT technology, any write violation to the guest physical address will cause VM-exit with an EPT violation event. The VM-exits then transfer control from guest to an entry point specified by the VMM to handle the event. The initiation process about P2V migration and the implementation of auxiliary modules will be presented in Sect. 4.4 and 4.5.

It is important to explain our method implementation by showing how to monitor pages in Sect. 4.1, how to adapt to the situations in SMP systems in Sect. 4.2, and how to use our method to collect evidences for forensics in Sect. 4.3.

4.1   Tracing Unit Implementation

Our method is implemented to trace memory modifications in page granularity using tracing units. That is to say, a tracing unit is related to:

- A physical address of the monitored target page.
- A dump area of the page to monitor modifications.
- Guest process information.

Each guest process runs in its own dedicated address space, so the process can be identified by Control Register 3 (CR3) value. When write violation event happens, if the violated-address-related tracing unit detects that the CR3 value is different from the value in guest process information field, it will do the submit-and-copy operation and update the process information field. Otherwise, if the CR3 value matches, the tracing unit simply allows the write attempt and resumes the control to guest.

During submit-and-copy operation, the tracing unit firstly compares the current page with the old dump and generates a modification map. The map labeled with page's base physical address will then be submitted to record module with guest process information. The record module finally merges the records according to the guest process's CR3 and the target page's physical address. If the specific record doesn't exist, the record module will establish one during the submission process. After submission, the tracing unit reinitiates itself by getting information from the new process and dumps the page before any changes are made.

Multiple tracing units can be set up separately, because all the tracing activities are triggered by hardware events, which occur sequentially in a single processor. To optimize the performance, physical addresses of the monitored pages are hashed in order to find the corresponding tracing unit among thousands of them. And the CR3 value and physical address of the records in record module are also hashed, so they can be quickly selected during the submission operation.

4.2   SMP System Support

Implementations are much more elaborated in SMP systems because of two special cases shown in Fig. 3. The first one happens when the tracing unit is recording write operations to page K from one guest process (process A) and another process (process B) tries to write into the same area from a different core. It happens concurrently in SMP systems. And the modification record will be contaminated if there is no write-barrier. The second case is similar, but the attempt is made by another thread of the same process (process A'). Contaminations in the first case will not happen in the second case, so tracing unit should allow the writing attempt for best performance.

Write-barrier requires to control write permission of all the cores separately. When the write-barrier is built, the tracing unit cancels write permissions of the target page in all cores. And the following submit-and-copy operation is protected by the critical zone built by spinlock. So if any processes tries to write into the same page at the same time, the write violation is immediately triggered and they will be trapped in hypervisor mode and wait outside the critical zone until submit-and-copy process is done. After that, the tracing unit checks the coming process's CR3 from the waiting cores. If it matches the current CR3 value, the unit grants write access to that core directly. If not, the write-barrier will be built up. And then the submit-and-copy operation follows again. In current prototype, the barrier cannot be built immediately because the Inter-Processor Interrupt (IPI) hasn't

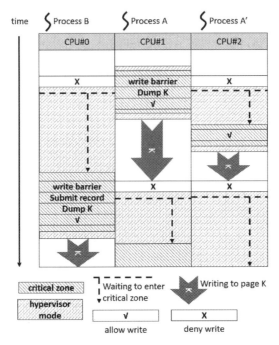

Fig. 3. Timeline of a tracing unit monitoring page K in SMP systems

been implemented yet. The TLB cache will not be synchronized in time until we do it in the next context change event. The false positives caused by lacking of TLB-shootdown will be discussed in case study II.

During context switch events handling, the hypervisor cancels write permissions of all the tracing pages in current processor, so that every unit can check again after the process is changed in the core.

Our tracing method guarantees the accuracy of records by assuring sequential writings among processes. The performance is improved because the writings are concurrent within the same process. Different tracing units are independent to each other. And current prototype supports up to 512 pages covering 2MiB memory space.

4.3   Evidence Collection

The evidence is about WHO, WHAT, WHERE and WHEN. 'WHEN' is the system time when modification happens. 'WHERE' is the places related to modifications. These places can be described by physical addresses, virtual addresses with its CR3 value, or the symbols in the program. The record module provides modification maps about processes. Each map contains a set of modification blocks according to modified pages. Every block records the modified bytes of monitored page into a bitmap. These bitmaps indicate all the modified places of the process in memory. 'WHO' is extracted from the CR3 value of the process when the memory-write behavior is intercepted. This value can be further evaluated by introspection module to indicate the process ID (PID), process name and the image path of the program. 'WHAT' includes detailed information about the behavior, the hypervisor makes a quick snapshot to record the process's running states. The above evidences can be integrated to a chain to link from memory behavior to a process, and finally to an EXE image.

The introspection module extracts process information from guest OS during VM-exit. To obtain the image path of the process, the module firstly gets the address from PsInitialSystemProcess defined in the driver compiler, and traverses the ActiveProcessLinks to find the corresponding EPROCESS structure according to current CR3 value. Then it follows the path from EPROCESS, PEB, to ProcessorParam, until it finds the variable ImageFileName. The memory content starting from PEB resides in user space, so the module translates the virtual address with CR3, and accesses the content by physical address. The data structure may be divided by pages, so an object assembler is implemented to automatically assemble them by analyzing the start address and symbol size. Other information such as PID and process name can be extracted directly from EPROCESS.

The running snapshot of the process contains a stack page dump with a stack pointer, a code page dump with the instruction pointer, and other register contents. The snapshot is captured during VM-exit caused by EPT violation. And the violation-related physical address and virtual address can be referred from VM-exit Information Fields in Virtual-Machine Control Structure (VMCS). The code page dump can be further analyzed to understand process behavior, and the stack information is important in forensics according to Arasteh's work [4].

4.4   H2V Migration

The hypervisor is implemented to run a hardware VMM using virtual-machine extensions (VMX). In order to accomplish H2V migration, the initiation driver firstly builds up a private virtual memory space for the hypervisor, and creates private mappings from host virtual space to guest memory space. Additional mappings are created for hypervisor to access physical memory directly.

Secondly, control structures for hypervisor and paging structures for SLAT are initialized. The driver traces all the memory allocated from kernel, and make them invisible to guest by remapping them to a spare page. Then the initiation thread attaches itself to each processors, and runs the callback functions to migrate the OS to guest VM on each CPU. The migration process on a single CPU includes checking VMX capabilities, allocating VMCS regions, setting VMCS properly, migrating running state of OS to a virtual machine, and finally doing VMLAUNCH to continue target OS running in guest mode.

After all CPUs are subverted, the hypervisor is ready to handle VM-exits according to Exit reason and Exit qualification. The VM-exits caused by EPT violation and Control-register accesses are handled by our method. The hypervisor will dispatch them to memory tracing module. If the hypervisor is commanded to uninstall, it will call unload process on each CPU. It generates trampoline, delivers the hardware control to the OS one core after another, and finally frees all the memories allocated during the initiation process.

4.5 Auxiliary Modules

The auxiliary modules are introspection module, record module, SLAT module and control module providing supports for our method. Introspection module is implemented to extract process information using VMI, and has been explained in Sect. 4.3. Record module is to collect records during monitoring and generate result to create an evidence chain. It is explained in Sect. 4.1 and Sect. 4.3.

The SLAT module provides interfaces to control write permissions of guest physical addresses. After H2V migration, the module enables EPT, sets paging structures to extended-page-table pointer (EPTP). Every core maintains a different set of paging structures, so that the module can control memory permissions separately among cores. The read, write and execute permissions are controlled by setting related bits in EPT Page-Directory Entries. INVEPT operation is needed to invalidate cached mappings after changing these bits. In order to make the tracing method unnoticed by guest, the hypervisor's memory is also hidden by SLAT module.

The control module opens an encrypted tunnel to transfer commands to hypervisor and controls the whole monitor process. This module intercepts normal CPUID operation, decrypts commands and parameters from guest registers, and can respond immediately. The investigators can thus control the underlying VMM inside guest OS without being noticed by guest software. The commands are used to set up the tracing units, start or stop tracing, output results to disk, and to reset the tracing units for a new round.

5 Evaluation

In this section, we will present 2 case studies and evaluate the system. The experiments are done on a Lenovo E47A laptop, with 4GiB RAM and Intel Core i5-2450M CPU 2.50GHz inside. The target OS is 64-bit Window 7 Build 7601.

The hypervisor has a small code base to ensure security. We use SLOC Metrics 3.0.7[1] to measure the Source Lines of Code (SLOC). This system has 9405 SLOC in total, including 2905 lines in initiation, 1480 lines in memory management, 3229 lines in auxiliary modules, 1168 lines in method implementation, and 623 lines in assembler code. Comparing to other lightweight hypervisors, VIS [31] has 5962 SLOC, SecVisor [21] has 4092 SLOC with 97 line-changes in kernel, and Terra [13] has about 13000 lines of code. Heavy VMM such as Xen [6] has about 260000 SLOC, and about 5000 line-changes in kernel.

The current implementation supports up to 512 tracing units to monitor pages, 256 record maps for processes' records, and 16384 record blocks to store modification bitmaps. That is to say, this implementation consumes 512 pages

---

[1] http://microguru.com/products/sloc

for page dumps, 512 pages for code and stack dumps, and 2048 pages for record blocks. The overall initiation process takes 2.5 seconds in average.

Case study I is presented to show our method's ability in live forensics. And case study II will present the performance analysis and the capability in SMP system.

5.1   Case Study I

In order to simplify our experiment and focus on hooking behavior monitoring in real-world cases, we choose kis14.0.0.4651.exe[2] as our sample, because cross-view based tools reported that there are 25 new hooks in shadow SSDT after running kis14.0.0.4651.exe. However, these hooks cannot be proved to have any relevance to their installer, because there are 64 different processes running simultaneously in the system. It could be some other processes deliberately or happened to install hooks when the sample is running. This is the same scenario when an investigator is facing a system filled with unknown software and want to find out the hook process.

In order to investigate these hooks in shadow SSDT, we firstly install our prototype on the target system. We then use CPUID operation to set tracing units to monitor physical pages which contain shadow SSDT structures. After that, we start monitoring and run the sample program. Those tracing units then automatically cancel the write permission of related pages, and they continuously intercept the write attempts in those areas. After the hooks are installed, we stop monitoring and output the records from hypervisor to disk. The abbreviated result in Fig. 4 shows that two tracing units have records of a process. The process information includes CR3 value, process name, PID, and its image path. Each tracing unit has information about the modifications made by the process and a related running snapshot including a stack page dump and a code page dump.

Surprisingly, the process's image path doesn't match kis14.0.0.4651.exe which should be on the desktop. And we do not see any file named 'MsiExec.exe' in folder 'C:/Windows/syswow64/' or any process named 'msiexec.exe' after the experiment. We may speculate that 'MsiExec.exe' is created to install hooks and is deleted during the installation of hooks. That is to say, if a memory dump is made after or before hooks are installed, an investigator cannot get enough information simply from after-the-fact analysis.

A code page dump with an instruction pointer value is recorded by tracing units. This information is collected at the first write violation event of the monitored process. The binary image is disassembled to assembly code shown in Fig. 5, which proves that msiexec.exe is making an atomic write operation at the moment. More static analysis can be made to evaluate the code page to reconstruct the related behavior.

---

[2] MD5 hash: 46cf0b296ab8dd1adc1482f52882f94d

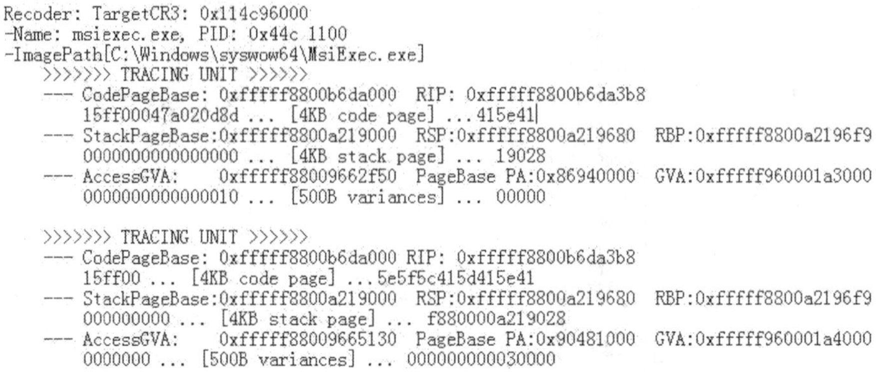

Recoder: TargetCR3: 0x114c96000
-Name: msiexec.exe, PID: 0x44c 1100
-ImagePath[C:\Windows\syswow64\MsiExec.exe]
>>>>>>> TRACING UNIT >>>>>>
--- CodePageBase: 0xfffff8800b6da000 RIP: 0xfffff8800b6da3b8
15ff00047a020d8d ... [4KB code page] ... 415e41
--- StackPageBase:0xfffff8800a219000 RSP:0xfffff8800a219680 RBP:0xfffff8800a2196f9
0000000000000000 ... [4KB stack page] ... 19028
--- AccessGVA: 0xfffff88009662f50 PageBase PA:0x86940000 GVA:0xfffff960001a3000
0000000000000010 ... [500B variances] ... 00000
>>>>>>> TRACING UNIT >>>>>>
--- CodePageBase: 0xfffff8800b6da000 RIP: 0xfffff8800b6da3b8
15ff00 ... [4KB code page] ... 5e5f5c415d415e41
--- StackPageBase:0xfffff8800a219000 RSP:0xfffff8800a219680 RBP:0xfffff8800a2196f9
000000000 ... [4KB stack page] ... f880000a219028
--- AccessGVA: 0xfffff88009665130 PageBase PA:0x90481000 GVA:0xfffff960001a4000
0000000 ... [500B variances] ... 000000000030000

Fig. 4. Abbreviated records made by tracing units

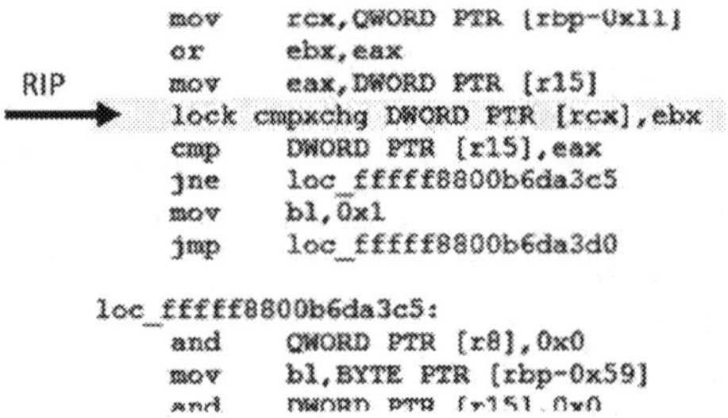

mov rcx,QWORD PTR [rbp-0x11]
or ebx,eax
RIP
mov eax,DWORD PTR [r15]
lock cmpxchg DWORD PTR [rcx],ebx
cmp DWORD PTR [r15],eax
jne loc_fffff8800b6da3c5
mov bl,0x1
jmp loc_fffff8800b6da3d0
loc_fffff8800b6da3c5:
and QWORD PTR [r8],0x0
mov bl,BYTE PTR [rbp-0x59]

Fig. 5. Disassembled code of the running snapshot

Stack information is also available in the records of Fig. 4. The analysis shown in Fig. 6 presents the stack frame at the moment. It proves that this record is accurate because the violated address can be seen in local variables of the stack frame. More detailed analysis can be made according to Arasteh's work [4].

Memory content is compared byte-by-byte in the page. Each byte is marked by a bit to show if it is modified. There are 4096 bits to record 4KiB content of a page, which means the size of a record block is 512B. Analysis in Fig. 7 shows the modifications made by msiexec.exe. The range of shadow SSDT starts from physical address of 0x86940f00. The two pages are not continuous in virtual address space because msiexec.exe remaps them separately. Each digit stands for 4 bits in record and thus 4 bytes in real memory, which is an entry of shadow SSDT. A 4-byte entry stands for an entry point of a handling procedure, and shadow SSDT has 827 those entries in Windows 7. The result shows that 25 hooks are installed by msiexec.exe, and their indexes accurately match the result of an integrity checker PCHunter V1.31[3]. The total number of copy-and-submit operation performed is 2, because target pages are modified by only 1 process. The overhead

[3] `http://www.xuetr.com/`

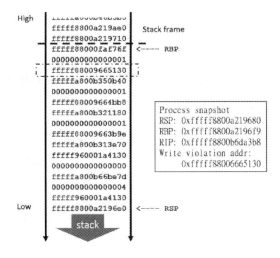

Fig. 6. The stack frame of the running snapshot

is much lower compared with instruction-level analysis, which need to translate every related write instructions.

The information in records is not available in after-the-fact analysis because the hooking behavior is transient and important evidence could be lost in nanoseconds. Our method can catch up with those important events by using hardware memory protection mechanism to intercept processes' write attempts. A complete evidence chain is created for forensics from memory modifications to hooking behavior snapshots, and to the process information including its file image path.

5.2   Case Study II

In case study II, we test the performance of our method in 2 situations. The sequential situation is that the traced page is accessed by only one process. It usually happens when a process installs hooks or accesses local variables, case study I belongs to the first situation. The parallel situation happens when a traced page is accessed by 2 or more processes simultaneously. It happens when the traced page contains data shared by processes. Guest OS will be trapped into the hypervisor very frequently in the second situation, because in order to generate an accurate record, the tracing unit should do the submit-and-copy operation every time when a different process comes.

We implement a test program. It continuously accesses the memory of two predefined pages in a high speed. And we set up two tracing units for these pages

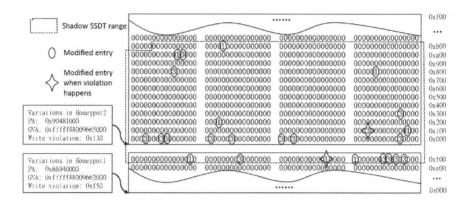

Fig. 7. Memory modifications made by msiexec.exe

in hypervisor. The performance evaluation uses the test program in 2 situations, and the 4 environments in each situation are:

- Env. 1, the OS with the test program(s);
- Env. 2, the OS, the test program(s) and a silent hypervisor;
- Env. 3, the OS, the test program(s) and the hypervisor enables SLAT;
- Env. 4, the OS with the test program(s) traced by hypervisor.

Env. 2 measures the overhead introduced from a simple intel-VT VMM. Env. 3 measures the overhead introduced from EPT mechanism. And Env. 4 measures the overhead coming from our monitor method.

The performance is measured by PCMark 7[4], which is a complete PC benchmark for Windows 7. It can make comprehensive performance tests of a system. The result shown in Fig. 8 indicates the largest impact (24%) comes from the SLAT mechanism, comparing to only 5% impact caused by our tracing method. And the payload mainly comes from computation usage because of the additional VM-exit handling logic. The hypervisor does not interfere external device usage, so the storage score doesn't change so much. Compared with instruction-level analysis, PoKeR [18] introduces 200%-500% performance impact while profiling and KernelGuard [28] introduces 19.4% impact while preventing NIC manipulation and process hiding. And QEMU emulator is slower than the VMM which leverages hardware-assisted virtualization to optimize performance.

To measure the overhead to the traced processes, we add a counter inside test program. The counter prints the memory accessing times in every second. The result is shown in Table 1, and the average counts are calculated when the counts are stable. The result indicates that 56% overhead is caused by SLAT mechanism, and the overhead from our method is strongly based on the access pattern of the program. The result conforms to the fact that the concurrent accesses from different processes are forced sequential in order to guarantee accuracy.

---

[4] http://www.futuremark.com/benchmarks/pcmark

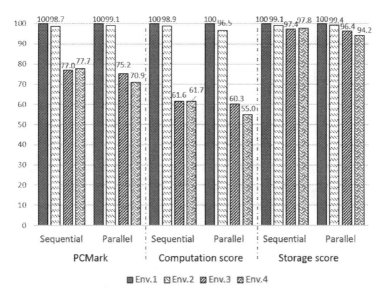

Fig. 8. The performance evaluation

Table 1. Performance of the traced processes

($Accesses per second$)	Sequential	Parallel
Env. 1	240601	200921
Env. 2	240193	197639
Env. 3	106970	86416
Env. 4	105516	17000-4500

Effective write-barrier mentioned in Sect. 4.2 is not built completely in the current prototype because TLB-shootdown has not yet been available. The write operation cannot be blocked immediately because the CPU uses permissions cached in TLB. Thus false positives are introduced in the parallel situation. We measure it by letting two processes write into different addresses in the same page. And the record of two processes shows 208 of 8192 bytes are contaminated in 2 seconds. The contamination speed is 2.54% every 2 seconds. Due to performance impact and false positives in current prototype, our hypervisor is recommended to intercept memory modifications in static memory area similar to what we did in case study I.

During the experiment, the system consumes 13065 4KiB pages, only 1.24% of 4GiB memory. There are 29 pages for hypervisor code, 9824 pages for paging structures, and 3212 pages for dynamic allocations.

6   Limitation

Current method implementation still has many limitations. We do not handle page fault in current prototype. So if the tracing target is paged out to disk, the tracing method will no longer work. Secondly, target traced pages need to be manually set according to its physical address, the lack of automatic scheme is inconvenient for forensics. And we haven't implemented interrupt control of target OS, it will introduce some false positives because the TLB cannot be invalidated immediately in current version. We are in progress to improve our implementation to trace virtual memory page contents and support interrupt control. Our current prototype is meant to prove feasibility of our method to apply in live forensic scenarios.

The hypervisor currently supports specific OS kernel and the hardware with Intel-VT. We do not check the integrity of our hypervisor to prevent malicious modifications, but the hypervisor isolates itself by creating private page tables and by remapping related kernel memory to make it transparent. The effort needed to implement a Trusted Computing Base and to support multi-platform is considerable. We do not want to reinvent the wheel because recent research [5] already shows that the similar P2V migration can be done to Linux and AMD-V. And XMHF [24] provides a powerful lightweight hypervisor platform which can guarantee the hypervisor integrity. Moreover, verifications from System Management Mode [26] are also suitable for the safety of type-I hypervisors.

7   Conclusion

In this paper, we present a para-synchronous method to collect evidences about transient software behaviors, which is unavailable in asynchronous approaches. Moreover, it is a method to monitor processes' modifications according to hardware events. Compared with synchronous approaches, our method introduces less overhead and can be used in live forensic scenario.

We implement the prototype on a type-I lightweight hypervisor. The novel hypervisor leverages hardware-assisted virtualization and can migrate the target OS to a virtual machine without suspending it. The prototype proves that our method can monitor systems running on bare hardware.

In our experiment, the prototype shows its ability to monitor hooking behaviors in shadow SSDT table. It has collected rich information including process's modification map, detailed information about the process, and related running snapshots. A complete evidence chain can be created from the hook to an EXE file image after a quick analysis. The performance overhead and memory usage are also acceptable. The result demonstrates that this method has a promising usage in live forensics.

Acknowledgement. This work is supported by the National Natural Science Foundation of China (61100198/F0207).

References

1. Adelstein, F.: Live forensics: Diagnosing your system without killing it first. Commun. ACM 49(2), 63–66 (2006)
2. Al-Shaer, E., Jha, S., Keromytis, A.D. (eds.): Proceedings of the 2009 ACM Conference on Computer and Communications Security, CCS 2009, Chicago, Illinois, USA, November 9-13. ACM (2009)
3. AMD: Amd virtualization (2014), http://www.amd.com/us/solutions/servers/virtualization/Pages/virtualization.aspx
4. Arasteh, A.R., Debbabi, M.: Forensic memory analysis: From stack and code to execution history. Digital Investigation 4, 114–125 (2007)
5. Athreya, M.B.: Subverting linux on-the-fly using hardware virtualization technology (2010)
6. Barham, P., Dragovic, B., Fraser, K., Hand, S., Harris, T.L., Ho, A., Neugebauer, R., Pratt, I., Warfield, A.: Xen and the art of virtualization. In: Scott, Peterson (eds.) [20], pp. 164–177
7. Bellard, F.: Qemu, a fast and portable dynamic translator. In: USENIX Annual Technical Conference, FREENIX Track, pp. 41–46. USENIX (2005)
8. Butler, J.: Dkom. Black Hat Windows Security (2004)
9. Carbone, M., Cui, W., Lu, L., Lee, W., Peinado, M., Jiang, X.: Mapping kernel objects to enable systematic integrity checking. In: Al-Shaer, et al. (eds.) [2], pp. 555–565
10. Cogswell, B., Russinovich, M.: Rootkitrevealer (2006), http://technet.microsoft.com/en-us/Sysinternals/bb897445.aspx
11. Dinaburg, A., Royal, P., Sharif, M.I., Lee, W.: Ether: malware analysis via hardware virtualization extensions. In: Ning, P., Syverson, P.F., Jha, S. (eds.) ACM Conference on Computer and Communications Security, pp. 51–62. ACM (2008)
12. Garfinkel, S.L.: Digital forensics research: The next 10 years. Digital Investigation 7, S64–S73 (2010)
13. Garfinkel, T., Pfaff, B., Chow, J., Rosenblum, M., Boneh, D.: Terra: A virtual machine-based platform for trusted computing. In: Scott, Peterson, (eds.) [20], pp. 193–206
14. Intel: Hardware-assisted virtualization technology (2014), http://www.intel.com/content/www/us/en/virtualization/virtualization-technology/hardware-assist-virtualization-technology.html
15. Martignoni, L., Fattori, A., Paleari, R., Cavallaro, L.: Live and trustworthy forensic analysis of commodity production systems. In: Jha, S., Sommer, R., Kreibich, C. (eds.) RAID 2010. LNCS, vol. 6307, pp. 297–316. Springer, Heidelberg (2010)
16. Moon, H., Lee, H., Lee, J., Kim, K., Paek, Y., Kang, B.B.: Vigilare: toward snoop-based kernel integrity monitor. In: Yu, T., Danezis, G., Gligor, V.D. (eds.) ACM Conference on Computer and Communications Security, pp. 28–37. ACM (2012)
17. Rhee, J., Riley, R., Xu, D., Jiang, X.: Defeating dynamic data kernel rootkit attacks via vmm-based guest-transparent monitoring. In: ARES, pp. 74–81. IEEE Computer Society (2009)
18. Riley, R., Jiang, X.: Multi-aspect profiling of kernel rootkit behavior. In: Schröder-Preikschat, W., Wilkes, J., Isaacs, R. (eds.) EuroSys, pp. 47–60. ACM (2009)
19. Rutkowska, J., Tereshkin, A.: Isgameover() anyone. Black Hat, USA (2007)
20. Scott, M.L., Peterson, L.L. (eds.): Proceedings of the 19th ACM Symposium on Operating Systems Principles, SOSP 2003, Bolton Landing, NY, USA, October 19-22. ACM (2003)

21. Seshadri, A., Luk, M., Qu, N., Perrig, A.: Secvisor: a tiny hypervisor to provide lifetime kernel code integrity for commodity oses. In: Bressoud, T.C., Kaashoek, M.F. (eds.) SOSP, pp. 335–350. ACM (2007)

22. Shacham, H.: The geometry of innocent flesh on the bone: return-into-libc without function calls (on the x86). In: Ning, P., di Vimercati, S.D.C., Syverson, P.F. (eds.) ACM Conference on Computer and Communications Security, pp. 552–561. ACM (2007)

23. Shosha, A.F., Liu, C.C., Gladyshev, P.: Evasion-resistant malware signature based on profiling kernel data structure objects. In: Martinelli, F., Lanet, J.L., Fitzgerald, W.M., Foley, S.N. (eds.) CRiSIS, pp. 1–8. IEEE Computer Society (2012)

24. Vasudevan, A., Chaki, S., Jia, L., McCune, J.M., Newsome, J., Datta, A.: Design, implementation and verification of an extensible and modular hypervisor framework. In: IEEE Symposium on Security and Privacy, pp. 430–444. IEEE Computer Society (2013)

25. Walters, A.: The volatility framework: Volatile memory artifact extraction utility framework (2007), https://www.volatilesystems.com/default/volatility

26. Wang, Z., Jiang, X.: Hypersafe: A lightweight approach to provide lifetime hypervisor control-flow integrity. In: IEEE Symposium on Security and Privacy, pp. 380–395. IEEE Computer Society (2010)

27. Wang, Z., Jiang, X., Cui, W., Ning, P.: Countering kernel rootkits with lightweight hook protection. In: Al-Shaer, et al. [2], pp. 545–554

28. Wikipedia: Cache (2014), http://en.wikipedia.org/wiki/Cache_(computing)

29. Yin, H., Liang, Z., Song, D.: Hookfinder: Identifying and understanding malware hooking behaviors. In: NDSS. The Internet Society (2008)

30. Yin, H., Poosankam, P., Hanna, S., Song, D.: HookScout: Proactive binary-centric hook detection. In: Kreibich, C., Jahnke, M. (eds.) DIMVA 2010. LNCS, vol. 6201, pp. 1–20. Springer, Heidelberg (2010)

31. Yu, M., Qi, Z., Lin, Q., Zhong, X., Li, B., Guan, H.: Vis: Virtualization enhanced live forensics acquisition for native system. Digital Investigation 9(1), 22–33 (2012)

SystemWall: An Isolated Firewall Using Hardware-Based Memory Introspection

Sebastian Biedermann[1] and Jakub Szefer[2]

[1] Security Engineering Group
Department of Computer Science
Technische Universität Darmstadt, Germany
biedermann@seceng.informatik.tu-darmstadt.de
[2] Computer Architecture and Security Laboratory
Department of Electrical Engineering
Yale University, New Haven, CT, USA
jakub.szefer@yale.edu

Abstract. Memory introspection can be a powerful tool for analyzing contents of a system's memory for any malicious code. Current approaches based on memory introspection have focused on Virtual Machines and using a privileged software entity, such as a hypervisor, to perform the introspection. Such software-based introspection, however, is susceptible to variety of attacks that may compromise the hypervisor and the introspection code. Furthermore, a hypervisor setup is not always wanted. In this work, we present a hardware-based approach to memory introspection. Dedicated hardware is introduced to read and analyze memory of the target system, independent of any hypervisor or OSes running on the system. We apply the new hardware approach to memory introspection to built-up an architecture that uses DMA and fine-grained memory introspection techniques in order to match network connections to the application-layer while being isolated and undetected from the operating system or the hypervisor. We call the proposed architecture SystemWall since it can be a standalone physical device which can be added as an expansion card to the mother board or a dedicated external box. The architecture is transparent and cannot be manipulated or deactivated by potential malware on the target system. We use the SystemWall in the evaluation to analyze the target system for malicious code and prevent unknown (malicious) applications from establishing network connections which can be used to spread viruses, spam or malware and to leak sensitive information.

1 Introduction

Memory introspection is a powerful technique for analyzing code and data contained in memory of a running system. Past approaches have focused on Virtual Machine (VM) based introspection techniques. In this work, we present another type of introspection, based on dedicated hardware components and software that can perform the introspection independent of any software running on the target system.

S.S.M. Chow et al. (Eds.): ISC 2014, LNCS 8783, pp. 273–290, 2014.

We apply the introspection techniques to build the SystemWall, a firewall-like system that can analyze memory of the running target computer, detect malicious or unknown applications and block their connections to the external world. This can prevent spread of viruses, malware, spam or even leakage of sensitive documents by malicious, unknown applications. SystemWall is logically fully external to the target computer system: it can be implemented as a stand-alone box that connects to target system or a dedicated extension card on the motherboard.

1.1   Security through Firewalls

In general, firewalls are either an external device only connected to the network or software-based and installed on a target computer. Firewalls control the incoming and outgoing network traffic depending on network events and predefined rules. Firewalls can be simple packet filters blocking or allowing network packets depending on their header information like the source and the destination. Other firewalls can analyze the content of network packets (deep packet inspection), for example with regular expressions [33], which allows the definition of more sophisticated rules.

External firewalls, however, do not have insight into the contents of physical memory of a target system and cannot make decisions based on what code is accessing or handling the network traffic on that system. Software-based firewalls, on the other hand, monitor the network traffic as well as the application-layer of a target computer system and control the incoming and outgoing traffic depending on rules which refer to protocols and states of the involved applications. Software-based firewalls are widely used as personal firewalls. However, software-based firewalls are installed on a target system and can be the target of attacks themselves. In particular, malware can successfully execute attacks against the operating system and can manipulate deployed rules, disable or change the mode of the installed software-based firewall's operation. This way, the user does not even notice the infiltration of the operating system and deems the system to be secured by trusting the running firewall and assuming its correct operation.

External firewalls with added ability to analyze memory of the target, like software-based firewalls, would combine best of both approaches – this is the motivation for SystemWall design.

1.2   Leveraging DMA for Security

DMA (Direct Memory Access) is a specification that allows hardware devices to bypass the Central Processing Unit (CPU) and access the system memory directly. This brings the advantage that the CPU can perform other useful tasks while DMA operations are in progress and it can also accelerate certain tasks. A lot of hardware devices like graphic cards, disk controllers or network cards use DMA.

DMA has been the focus of security researchers for some years, because it allows to dump the memory of a system through certain external interfaces while

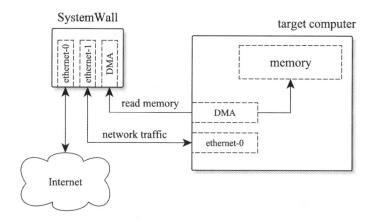

Fig. 1. The SystemWall deployed as an external hardware device which can read the memory of the target and regulate the target's network traffic. Only one ethernet connection is shown for the target in this figure, if target has multiple network interfaces, such as two ethernet ports, SystemWall should regulate each ethernet connection.

bypassing the operating system and any software-based security restrictions. In particular, DMA can be exploited by an attacks on unattended, running systems which provide DMA via an external buses like ExpressCard, FireWire or Thunderbolt to create a dump of the memory. Afterwards, the memory dump can be investigated using forensic techniques in order to retrieve passwords or other sensitive information.

However, DMA can also be used to increase security of a computer system and prevent numerous attacks. In particular, DMA can be used to transparently read the memory contents via the hardware, contents which can later be analyzed for malicious programs or network connections – as we do in SystemWall.

1.3   SystemWall Overview

In this work, we benefit from DMA to setup an isolated firewall-like system which we call SystemWall, shown in Figure 1. The SystemWall can be deployed between the system which it protects and the Internet, and intercept all packets traveling from and to the target system. Placing SystemWall between the Internet and the target system allows for it to, for example, delay network packets going to or from the target while the target's memory is analyzed to validate the packets are related to a legitimate, non-malicious application. To perform the analysis, SystemWall transparently uses DMA and fine-grained memory introspection techniques to match detected initiations of new network connections to applications running on the system. It can use application names, hashes or even scan for shellcode to detect malicious applications and prevent them from making network connections. The SystemWall remains undetected from the operating system and is a combination of a personal software-based firewall and hardware-based memory introspection. This way, the SystemWall cannot be manipulated or disabled by potential malware which could infect the system. Given its access to target's memory, it can monitor the applications and control network connections of these applications to prevent spread of malware, viruses or potentially leakage of sensitive files. Figure 1 shows a block diagram of the SystemWall architecture, fully described in Section 3.

1.4 Paper Organization

The remainder of the paper is organized as follows. Section 2 presents related work. Section 3 explains the architecture and details of the SystemWall implementation. Section 4 evaluates the proposed architecture and Section 5 discusses limitations. Finally, Section 6 concludes.

2 Related Work

This section presents related work in the field of physical memory acquisition with the help of hardware extensions, attacks based on this, countermeasures and methods that use DMA for other purposes. Furthermore, we list some related work in the field of tamper-resistant security architectures that can run isolated from the target system and that are based on hardware extensions or a hypervisor setup.

2.1 Physical Memory Acquisition

DMA has been previously exploited to execute attacks against a running system. Attackers have used buses like USB, FireWire, Thunderbolt or PCMCIA cards to transparently acquire the volatile memory of a running system without being detected and without being the subject to software-based control mechanisms. Afterwards, the memory dump can be analyzed for sensitive data like passwords. However, attackers can also write to the memory pages of the running system and this way modify the system's properties or work-flow on-the-fly. For example, a Windows 7 kernel can be directly manipulated in the memory in order to allow an attacker to log-in using a blank password [2]. Nowadays, standard DMA attacks and further procedures are even implemented in exploitation frameworks [7].

As a consequence of these attacks, several countermeasures have been suggested and are of interest, e.g., [23] or [24]. In particular, hardware-based memory acquisition of specific memory regions can be prevented by modifying the processor's North Bridge's memory map [22]. Also, malware that uses DMA to infiltrate an operating system can be detected using techniques such as those presented by [29] or [28].

Furthermore, memory acquisition can be also used for non-attacking purposes, for example for the transparent acquisition and analysis of volatile memory of a compromised system [8]. Seger et al. [26] presented a memory sampling mechanism based on DMA using a GPU coprocessor as an extension. Schwarz et al. [25] presented an architecture that prevents virtual guest machines accessing memory regions of other virtual guest machines using DMA only by using software and standard hardware. Balogh et al [4] proposed a memory acquisition system which uses DMA based on a custom network interface protocol driver and a network card that can directly send the memory over the network. Chen et al. [9] developed a FPGA based data protection system called sAES which uses DMA to improve throughput and latency of a data protection system.

In this work, we use physical acquisition of specific regions in the volatile memory of a system in order to built an isolated firewall which transparently matches the system's running applications to the network traffic and is able to detect connections from malicious or unknown applications.

2.2 Tamper-Resistant Security Architectures

In order to avoid being manipulated by malware or intruders who successfully execute an attack and gain access, some advanced security architectures use tamper-resistant components. These architectures are somehow isolated from the system which they actually protect and monitor, and they also run transparently and unknown to the target.

Many tamper-resistant security architectures use hardware expansions like a special trusted processor booting the micro kernel of the system [30] or particular co-processors only used for monitoring [20] or cryptography [1]. For example, Yashiro et al. [32] propose to use a tamper-resistant chip which does the sensitive operations in a access control scenario on a file system. The most popular architectures are based on a Trusted Platform Module (TPM) which is a dedicated secure crypto-processor issued by the Trusted Computing Group[1] and used for encryption, attestation and sealing of data on a target system. However, tamper-resistant architectures are also used in embedded devices [21].

In recent years, tamper-resistant security architectures which are based on a hypervisor setup are in the center of interest. These architectures are software-based and use a Virtual Machine Monitor (VMM) as a mechanism to guarantee trustworthy isolation of components. Wang et al. [31] propose a combination of both, a hardware-assisted monitor to verify the integrity of a hypervisor and its isolation mechanisms.

Often, these architectures use Virtual Machine Introspection (VMI) in order to analyze the memory of a user virtual machine from another isolated and privileged virtual machine running co-residently on the same hardware [19]. Baiardi et al. [3] proposed a tamper-resistant intrusion detection architecture that merges target monitoring via VMI and network monitoring. Payne et al. [18] proposed an architecture that allows popular security tools to do active monitoring while being isolated in a trusted virtual machine. Srivastava et al. [27] built an isolated firewall which correlates network packets to applications using VMI in a hypervisor setup.

---

[1] http://www.trustedcomputinggroup.org/

In this work, we secure a target system using an external hardware box or expansion card that regulates the network traffic and additionally uses transparent memory introspection via DMA to match network packets to running applications.

2.3   Memory Introspection

Analysis of memory contents has been explored before in different contexts. Most recently, Virtual Machine Introspection (VMI) was developed as a new technique which uses virtualization and the privileged hypervisor software to analyze memory of guest virtual machine (VM). The ability to analyze memory has been leveraged to detect kernel rootkits [10] or detect malware inside the guest VMs [6]. Such techniques are software-only and do not combine firewall like networking protections with the memory introspection capabilities.

Among hardware-oriented proposals, new architectures have been proposed which leverage extra hardware to perform the memory introspection. Multi-core processors have been extended so that the measurement code can run in a specialized processor local memory [12], and thus not be affected by the malware on the target. Other architectures, e.g. [13,16], have proposed extra hardware to directly monitor memory bus traffic. Outside of the main processor, [20] has proposed a co-processor based solution where a co-processor performs the monitoring. Also, a special piece of hardware that is connected to one of the DRAM sockets has been proposed to be used to transparently analyze the memory contents as the reads and writes happen [14].

Unfortunately, such hardware additions or modifications are not available today, unlike PCI expansion cards such as FireWire or Thunderbolt available today and used in our project. Moreover, most of the previous projects focus on kernel integrity measurement, whereas SystemWall monitors networking applications.

3   Architecture

The SystemWall is a physical device which is connected to the network inbetween the target which it protects and the Internet. Additionally, the SystemWall is connected to an interface of the target computer that allows DMA to the system memory. Figure 1 illustrates the proposed architecture in which the SystemWall is an external standalone device. However, it can be also an internal device added into the target computer like for example a custom PCIe expansion card having two Ethernet interfaces plugged into the target system. When the SystemWall is deployed, it inspects the incoming and outgoing traffic of the target system and matches the traffic to a program running on the system. Currently, the SystemWall focuses on the TCP network protocol.

3.1   Threat Model

We assume a strong threat model where we want protect a computer system even when its applications, OS or hypervisor software may have been compromised.

For example, SystemWall should work even when malware has successfully infected the target system and tries to communicate with a malicious remote party. The potential infection can be based on a drive-by exploit targeting an unpatched browser vulnerability or the malware could be a botnet-client trying to connect to the botnet. It could even be a malicious piece of monitoring software performing a phishing attack and trying to send out sensitive information about a user's banking account.

We assume that the hardware-based memory introspection mechanism, such as through a dedicated PCIe expansion card, cannot be compromised as it is separate from the target system and no software on the target system can manipulate it. It is correctly manufactured and hardware itself is not malicious. The firewall software runs on the dedicated external box (which is connected to the hardware-based memory introspection card in the target), or could be part of a custom expansion card in the target. Since the firewall software runs separate from the target platform it cannot be compromised. We assume the firewall software is correctly written.

If system wall is implemented (as in our prototype) on a separate computer, Trusted Platform Module could be used to attest the SystemWall software on startup. If SystemWall is implemented as a custom PCIe card, the firmware will likely be small enough for formal verification, which would given even more confidence in correctness of the SystemWall system.

In our threat model, we do not address physical attacks which means we do not assume an attacker gaining physical access to the target system and being able to remove the SystemWall, or modify the system's hardware parameters, such as the amount of installed memory (DRAM). SystemWall is not able to protect against denial-of-service attacks.

3.2   Boot-Up and SystemWall Initialization

SystemWall is independent of the target computer, although may share the same power supply if it is implemented as a expansion card on the motherboard. When the target system is offline, the SystemWall does not have to operate, and thus can be off as well. When the target system boots up, there is no special SystemWall operation. We only assume that at boot up time, SystemWall knows the amount of physically installed memory (DRAM), and that the amount will not change after runtime. The SystemWall can block network connections until the memory acquisition is working, thus any potential malware will not be able to use the network before SystemWall has chance to access memory and look for the malware.

3.3   SystemWall Run-Time

Once the SystemWall can detect the initiation of a new outgoing network connection (SYN sent), it starts the following procedures which are illustrated in Figure 2.

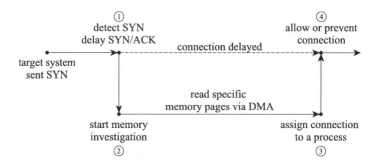

Fig. 2. Once a new outgoing TCP connection initiation is detected (SYN packet), the SystemWall delays the returning SYN/ACK packet while investigating the memory to find the process associated with the connection on the target system

First, it detects new network connection and delays the SYN/ACK packet sent back from a remote system on the Internet. This way it delays the TCP three-way-handshake completion and the establishment of a new connection.

Second, during the time when the reply is being delayed, the SystemWall starts to transparently read specific memory regions on the target system by exploiting DMA. The target memory regions depend on the installed operating system and a corresponding template which is usually used for forensic investigations on a memory dump. By reading this data, the SystemWall retrieves information about the currently running processes and their currently allocated physical memory regions on the target system. With the help of this information, it matches the new initiated connection to the corresponding running process by matching the source port of the SYN network packet to the port used by the process[2].

Once the SystemWall is able to match the new connection to a process, further checks can be initiated and the integrity of the process can be verified. Finally, the SystemWall decides if the establishment of the new connection is allowed because it is made by a process that is allowed to communicate or if the connection establishment should be prevented because it was initiated by an unknown process.

For some applications, such as web servers, the response delay (while SystemWall checks connection) may have impact on usability. For example, a study [17] found that website users can tolerate around 2s delays in waiting for information retrieval. An alternative to design presented in Figure 2 would be to throttle the responses, while the checking is performed. This way remote parties start receiving information, albeit at a slower pace, while the checking is performed. This, however, is less secure as the unknown process is allowed to

---

[2] If the OS is compromised and, for example, there are duplicate data structures that map network ports to applications, SystemWall can scan whole memory of the target system to look for such duplicates and issue a warning.

send back some information, and in the remainder of the paper we focus on the design which delays the whole connection until all checking is done.

As SystemWall is a separate system and has its own network interface, SystemWall is able to use the network connection to notify the administrator or the target system of any security events. Such interface should be limited, so it cannot be exploited by attackers to compromise SystemWall. In a simplest form, SystemWall could send e-mail notifications to a fixed (administrator) e-mail address without exposing complicated web interface that could be attacked.

SystemWall may also need an interface for updating the firewall software or trusted program white lists. Secure software update is a broad research area with many challenges [5], and secure update of SystemWall is outside of scope of this paper.

3.4 Memory Acquisition

The key part of SystemWall is the memory acquisition which is done through DMA and independent of any software running on the target system. The DMA operations go directly from the SystemWall device (such as a FireWire card), through I/O MMU (if present, such as if the system has Intel's VT-d extensions or AMD's IOMMU technology), to DRAM memory. When target system's OS or hypervisor is non-malicious, then the DMA can proceed easily and verify the integrity system.

If, however, the OS or a hypervisor manipulates the memory contents or the I/O MMU configuration, the SystemWall's access to some of the memory pages can be denied, redirected or false memory content can be presented. SystemWall deals with memory acquisition issues in number of ways and can always block network connections to outside world. If any anomalies are encountered, first action is to temporarily delay the network connections until the issue is resolved. SystemWall can also measure the executable code of the OS or a hypervisor in order to detect any malicious modifications or code. This could be tricked, however, through a number of memory manipulation attacks which we discuss in the following sections. To counter any possible memory manipulation attacks by malicious OS or hypervisor, SystemWall can use a number of detection techniques, for example time-based measurements to detect if an OS or hypervisor is doing something malicious with memory mappings.

Memory Swapping Attacks. The software of the target system or a hypervisor may swap memory at some address to disk or another secondary storage and replace it with other memory. For example, when SystemWall is trying to scan the system's memory to determine application binary, the OS or hypervisor may temporarily swap the actual (malicious) running application's memory to swap disk and replace it with some (benign) application's memory. Thus, when SystemWall uses DMA to the memory, it will read the benign application's memory and not detect any problem. This, however, requires precise timing on the part of the malicious OS or hypervisor. Nevertheless, it may be possible. If a malicious OS or hypervisor tries to swap memory, it has to execute three steps:

Table 1. Measured time while reading data from the volatile memory or swapped out data from the hard drive

	16kbyte	32kbyte	64kbyte	128kbyte
memory	$0.003 \pm 0.001s$	$0.006 \pm 0.001s$	$0.007 \pm 0.001s$	$0.007 \pm 0.001s$
hard drive	$0.015 \pm 0.000s$	$0.027 \pm 0.001s$	$0.036 \pm 0.001s$	$0.038 \pm 0.001s$

1. Delay the access request of the SystemWall to a memory region through manipulation of the I/O MMU remapping tables, e.g., deny access to the memory region.
2. Swap current memory contents for (benign) memory content from disk (or another storage device).
3. Allow the SystemWall to proceed and to access the target memory region by re-allowing memory access to the memory region.

SystemWall can detect this memory swapping attack by observing memory access errors and access time delay. If access is denied and re-gained after a period of time, SystemWall can detect the time period when memory was not accessible and issue warning that something malicious is going on. Depending on the design of the underlying hardware of the system, a denied DMA access request can result in returning 0s, which can also be detected as an anomaly by the SystemWall. Furthermore, if access is delayed, SystemWall can simply detect the longer access times and this way reveal the memory swapping attack.

In several test-runs we measured time while the system reads small chunks of data from the volatile memory or swapped out data from the hard drive. Clear and constant differences in the measured time can be seen (Table 1) which can be reliably used for anomaly detection.

A more complicated memory swapping attack could be performed using memory in the GPU or other device that is connected to higher speed bus, such as PCI Express, rather than to disk. Nevertheless, even with PCI Express 2.0, the maximum bandwidth is 8GB/s (16 lanes)[3] whereas main memory can have bandwidth over 16GB/s (DDR3)[4]. Including other overheads, memory accesses are at least over 2x faster than going to devices and time differences can be used for anomaly detection.

In-Memory Redirection Attacks. Malicious software or a hypervisor could also swap two memory regions in the system's memory (e.g. swap address A1 with A2) or redirect DMA requests of the SystemWall from A1 to other memory pages by remapping the adresses in the I/O MMU. Then, SystemWall would think it is accessing machine address A1, whereas it would actually access A2. In Figure 3, the SystemWall tries to read the physical address 0x04, but is redirected by malicious software to read 0x15. This way, the malicious software

---

[3] https://en.wikipedia.org/wiki/PCI_Express, accessed Aug. 8, 2014.
[4] https://en.wikipedia.org/wiki/DDR3_SDRAM, accessed Aug. 8, 2014.

can hide data under the address 0x04. The SystemWall cannot detect the in-memory redirection attack based on deviations in access time, like in the previous section. However, from the perspective of the SystemWall, there are now two regions in the memory which store exactly the same data, namely 0x04 and 0x15. This means the SystemWall can search for duplicates.

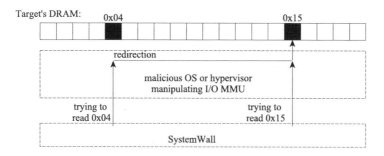

Fig. 3. Example of remapping of requests of the SystemWall to other memory regions by malicious software. The SystemWall reads the same data at 0x04 and 0x15, and can detect duplicates.

In this context, we assume malicious software modified the executable code of a process and stores the original content somewhere else in memory. If the SystemWall wants to read the process's executable code, it is redirected to the original content. In order to detect this attack, the SystemWall uses integrity measurements on the executable code. We assume a SHA1 hash over the original executable code is known. As a potential detection mechanism, the SystemWall executes the following procedures:

- It computes a hash over the process's executable code by reading the target physical addresses via DMA.
- If the hash is valid, the SystemWall subsequently scans the whole memory in chunks of the same size each time computing a hash over the data as well.
- If the same hash occurs in another memory region, a remapping attack was executed which leads to immediate blocking of the process's communication.

At regular time intervals, the SystemWall can execute these procedures and try to detect a potential redirection attack. Scanning for duplicates on a system with 4GB of memory can be performed in our setup in 67.2 ± 2.8 seconds for 512KB of data. This kind of attack detection procedure requires some time, however, it can be halted and continued once the SystemWall needs to perform other operations. The SystemWall continuously tries to detect memory redirection attacks in the background. If other operations need to be performed, the memory redirection attack detection procedure can be halted.

A special case of duplicate code may involve code that is replicated in multiple programs, e.g. statically linked libraries. We assume that in such cases, SystemWall will have sufficient information about the target binaries to be able to count the expected number of code repetitions and detect when there are more than the expected number of code copies when scanning the memory.

Memory Blocking Attacks. Usually, some memory regions are "reserved" and used by devices, thus not accessible. A malicious OS or hypervisor could modify the memory map and pretend that there is some device at address $[addr_1, addr_2]$ range. It could then hide some memory contents there so when SystemWall accessed address A1 it would find some benign code or data, while the actual running code or data would be inside $[addr_1, addr_2]$ range.

However, since SystemWall knows the total amount of memory installed, it can analyze memory to try to find any memory region it cannot access. Next, it can use that information and compare with the correct memory configuration of the target system. If the correct or expected memory configuration differs from the one detected, e.g., there is one too many reserved memory regions, then SystemWall can calculate that some malicious action has been taken to create this fake reserved memory region (which could hide some code).

4   Evaluation

In this section, we evaluate a prototype implementation of the SystemWall and discuss the results.

4.1   Prototype Implementation

In an evaluation, we used a system with a Intel Core 2 Duo 2.33 GHz and 3 Gb of memory as the target system for protection. The SystemWall was a second computer system, running Ubuntu Linux 12.04 (64-bit). Both computers were connected via FireWire PCIe cards (LSI FireStorm FW643e2) [15]. Other interfaces like for example Thunderbolt or PCMCIA could be used as well. The components of the SystemWall are programmed in Python and use forensic1394[5] library and the Volatility framework[6] for forensic investigations. The SystemWall also uses Linux iptables with a netfilter script to dynamically delay (and block if needed) network packets.

4.2   New Connection Detection and Decision

We investigated the times required for the SystemWall to make a decision if the new initiated connection can be allowed or should be blocked. This is actually also the time for how long the establishment of every new outgoing network connection needs to be delayed. The time required to make a decision should be

---

[5] https://github.com/wertarbyte/forensic1394
[6] https://code.google.com/p/volatility/

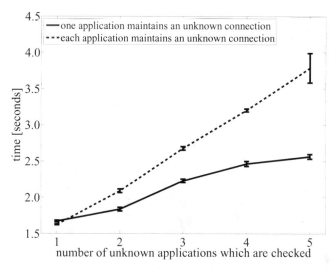

Fig. 4. Time required to decide if a new initiated connection can be allowed or should be blocked based on timely memory investigations (DMA connection overhead, if using FireWire card for example, is not included)

small, since in this case the usability plays a major role. To accelerate the procedures, the SystemWall can maintain a white-list of known processes belonging to the target's OS's standard installation (e.g. Ubuntu Linux in our case) and which are allowed to communicate. The white-list can be deployed from the beginning, or it can be created stepwise by the SystemWall through an enrollment phase, such as on-line evaluation of the programs and building a white list as the target runs. Figure 4 shows the time of different test-runs (10 test-runs in each configuration) and the corresponding standard deviation.

In the first test-runs (continuous line), a new connection was initiated and the SystemWall matched it to 1 of 5 unknown running processes on the target by investigating the memory via DMA. In the seconds test-runs (dashed line), a new connection was initiated and each of the 5 unknown processes already maintains another connection which needs to be matched as well. In the performed test-runs, the maximum required time was around 4 seconds which is feasibly in our scenario and provides an acceptable level of usability.

4.3 Memory Throughput Requirements

In further test-runs, we investigated how much data needs to be read through DMA in order decide if a new connection is allowed (Figure 5). The amount of data that needs to be read strongly influences the performance. For this, we used the same configurations as in the previous test-runs (standard Ubuntu installation and up to five unknown additional processes). Once a single new connection was initiated and needed to be matched to 1 up to 5 unknown processes, up to

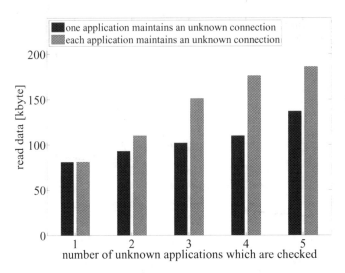

Fig. 5. Amount of memory read by the SystemWall through DMA in order to match a new initiated connection to a running process

140 kilobyte was read (black bars). Once a single new connection was initiated and each of the unknown processes maintained a connection, only up to 190 kilobyte was read (grey bars). This is little data and therefore the throughout of the DMA bus cannot be a bottleneck for the SystemWall.

In general, DMA is executed on the PCI bus and therefore could be influenced by other work-loads on the bus running in parallel. In order to investigate the stability of the technique, we executed benchmarks on the target which focused on CPU, memory, file I/O on the storage device and network utilization. In parallel, we executed the procedures of the SystemWall. In each case, the time remain unaffected and stable, since only small amount of data needs to be read. Only if the SystemWall dumped the whole memory of the target system during the benchmarks, deviations in the ranges of $\pm 1\%$ could be seen.

4.4 Process Identification

In order to verify running processes and to create and maintain a white-list of known processes which are allowed to establish connections, investigating information of a process like the name or other simple properties is not sufficient. Malware could hide itself by presenting information and properties of known processes. In the next test-runs, the SystemWall computed hashes (SHA1) over the processes' executable codes (ELF) stored in memory in order to verify integrity of the code and to detect potential modifications (Figure 6). Computing the hashes over up to 5 different processes could be checked in less than 5 seconds (continuous line). Furthermore, Figure 6 shows the times for computing the hash and additionally scanning the allocated heap and stack memory of each process for

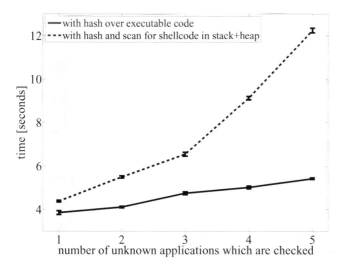

Fig. 6. Time required to decide if a new initiated connection can be allowed or should be blocked using hashes over the executables, and hashes plus scanning for shellcode

a shellcode signature (long sequences of NOPs) indicating an exploited software vulnerability (dashed line). Shellcode which could be found in these memory regions can be the result of a successfully executed attack against the process that could change its work-flow to do unwanted tasks for example revealing sensitive information. Even if the SystemWall executes these additional in-depth checks, the time increase only up to around 6 seconds for 3 processes and go up to 12 seconds for 5 processes (dashed line). However, the time is strongly depend on the processes and their sizes.

5   Limitations

We believe that hardware assisted memory introspection had a very good potential. Currently, a number of limitations exist. We believe these to be very interesting research opportunities which we and other researchers can tackle and make hardware-assisted memory introspection the new standard in memory introspection work. Physical attacks are not in the scope of our work. However, future design could be created which attempt to deal with physical attacks. An attacker could physically disable or disconnect SystemWall from the system being monitored. Integration of SystemWall into vPro [11] or similar solution could tightly integrate it with the target system's hardware. Integration of SystemWall into the networking infrastructure could also be another research direction. The traffic management could be done at switch or router level. If this "distributed" SystemWall does not receive information form a target (such as when there has been a physical attack), then all traffic can be stopped deeper in the network before the malware or sensitive information has ability to spread.

6   Conclusion

In this work we have presented SystemWall which uses hardware memory introspection to transparently analyze the applications running on a target computer. SystemWall is independent of any software running on the target computer, thus cannot be affected by any malware on the target system, even by a compromised hypervisor. The SystemWall can monitor the applications and control network connections of these applications to prevent spread of malware, viruses or potentially leakage of sensitive files. We have built and evaluated a prototype based on a FireWire card and external SystemWall box. The prototype is able to control network connections of the target system and decide if unknown application's connection can be let through or should be blocked. Our ongoing and future work involves addressing the performance and other challenges presented in the paper.

6.1   SystemWall Code

SystemWall code related to this publication and setup instructions are available online at `http://caslab.eng.yale.edu/code`.

Acknowledgements. We would like to thank anonymous reviewers for their feedback, which helped to improve the final version of this paper. We would like to also thank LSI Corporation for the donation of the two LSI FireStorm FireWire PCIe cards.

References

1. Anderson, R., Bond, M., Clulow, J., Skorobogatov, S.: Cryptographic Processors – A Survey. Proceedings of the IEEE 94, 357–369 (2006)
2. Aumaitre, D., Devine, C.: Subverting Windows 7 x64 Kernel with DMA attacks. In: HITB Security Conference Presentation (2010), `http://esec-lab.sogeti.com/dotclear/public/publications/10-hitbamsterdam-dmaattacks.pdf` (accessed August 8, 2014)
3. Baiardi, F., Sgandurra, D.: Building Trustworthy Intrusion Detection through VM Introspection. In: Proceedings of the International Symposium on Information Assurance and Security, pp. 209–214 (August 2007)
4. Balogh, S., Mydlo, M.: New possibilities for memory acquisition by enabling DMA using network card. In: Proceedings of the International Conference on Intelligent Data Acquisition and Advanced Computing Systems (IDAACS), pp. 635–639 (September 2013)
5. Bellissimo, A., Burgess, J., Fu, K.: Secure Software Updates: Disappointments and New Challenges. In: Proceedings of USENIX Hot Topics in Security (HotSec), pp. 37–43 (July 2006)
6. Benninger, C., Neville, S., Yazir, Y., Matthews, C., Coady, Y.: Maitland: Lighter-Weight VM Introspection to Support Cyber-security in the Cloud. In: Proceedings of the International Conference on Cloud Computing (CLOUD), pp. 471–478 (June 2012)

7. Breuk, R., Spruyt, A.: Integrating DMA attacks in exploitation frameworks. Technical Report, System and Network Engineering Research Group, University of Amsterdam (February 2012)
8. Carrier, B.D., Grand, J.: A hardware-based memory acquisition procedure for digital investigations. Digital Investigation 1(1), 50–60 (2004)
9. Chen, Y., Wang, Y., Ha, Y., Felipe, M., Ren, S., Aung, K.M.M.: sAES: A high throughput and low latency secure cloud storage with pipelined DMA based PCIe interface. In: Proceedings of the International Conference on Field-Programmable Technology (FPT), pp. 374–377 (December 2013)
10. Fraser, T., Evenson, M., Arbaugh, W.: VICI – Virtual Machine Introspection for Cognitive Immunity. In: Proceedings of the Annual Computer Security Applications Conference, pp. 87–96 (December 2008)
11. Intel vPro Technology, http://www.intel.com/content/www/us/en/architecture-and-technology/vpro/vpro-technology-general.html (accessed August 8, 2014)
12. Kinebuchi, Y., Butt, S., Ganapathy, V., Iftode, L., Nakajima, T.: Monitoring Integrity Using Limited Local Memory. IEEE Transactions on Information Forensics and Security, 1230–1242 (July 2013)
13. Lee, H., Moon, H., Jang, D., Kim, K., Lee, J., Paek, Y., Kang, B.B.: KI-Mon: A Hardware-assisted Event-triggered Monitoring Platform for Mutable Kernel Object. In: Proceedings of the USENIX Security Symposium, pp. 511–526 (August 2013)
14. Liu, Z., Lee, J., Zeng, J., Wen, Y., Lin, Z., Shi, W.: CPU Transparent Protection of OS Kernel and Hypervisor Integrity with Programmable DRAM. In: Proceedings of the International Symposium on Computer Architecture (ISCA), pp. 392–403 (June 2013)
15. FireStorm FW643/FW533 Evaluation Platform, http://www.lsi.com/downloads/Public/1394%20Products/1394%20Products%20Common%20Files/LSI-FireStorm-PB.pdf (accessed August 8, 2014)
16. Moon, H., Lee, H., Lee, J., Kim, K., Paek, Y., Kang, B.B.: Vigilare: Toward Snoop-based Kernel Integrity Monitor. In: Proceedings of the Conference on Computer and Communications Security, pp. 28–37 (October 2012)
17. Nah, F.F.-H.: A study on tolerable waiting time: how long are web users willing to wait? Behaviour & Information Technology 23(3), 153–163 (2004)
18. Payne, B., Carbone, M., Sharif, M., Lee, W.: Lares: An Architecture for Secure Active Monitoring Using Virtualization. In: Proceedings of the IEEE Symposium on Security and Privacy, pp. 233–247 (May 2008)
19. Payne, B., de Carbone, M., Lee, W.: Secure and Flexible Monitoring of Virtual Machines. In: Proceedings of the Annual Computer Security Applications Conference, pp. 385–397 (May 2007)
20. Petroni Jr., N.L., Fraser, T., Molina, J., Arbaugh, W.A.: Copilot - a Coprocessor-based Kernel Runtime Integrity Monitor. In: Proceedings of the USENIX Security Symposium, pp. 13 (August 2004)
21. Ravi, S., Raghunathan, A., Chakradhar, S.: Tamper resistance mechanisms for secure embedded systems. In: Proceedings of the International Conference on VLSI Design, pp. 605–611 (January 2004)
22. Rutkowska, J.: Beyond The CPU: Defeating Hardware Based RAM Acquisition. Black Hat DC Presentation (2007), http://www.blackhat.com/presentations/bh-dc-07/Rutkowska/Presentation/bh-dc-07-Rutkowska-up.pdf (accessed August 8, 2014)

23. Sang, F., Lacombe, E., Nicomette, V., Deswarte, Y.: Exploiting an I/OMMU vulnerability. In: Proceedings of the International Conference on Malicious and Unwanted Software (MALWARE), pp. 7–14 (October 2010)
24. Sang, F., Nicomette, V., Deswarte, Y.: I/O Attacks in Intel PC-based Architectures and Countermeasures. In: Proceedings of the SysSec Workshop (SysSec), pp. 19–26 (July 2011)
25. Schwarz, O., Gehrmann, C.: Securing DMA through virtualization. In: Proceedings of the Workshop on Complexity in Engineering (COMPENG), pp. 1–6 (June 2012)
26. Seeger, M., Wolthusen, S.: Towards Concurrent Data Sampling Using GPU Coprocessing. In: Proceedings of the International Conference on Availability, Reliability and Security (ARES), pp. 557–563 (August 2012)
27. Srivastava, A., Giffin, J.: Tamper-Resistant, Application-Aware Blocking of Malicious Network Connections. In: Lippmann, R., Kirda, E., Trachtenberg, A. (eds.) RAID 2008. LNCS, vol. 5230, pp. 39–58. Springer, Heidelberg (2008)
28. Stewin, P., Bystrov, I.: Understanding DMA malware. In: Flegel, U., Markatos, E., Robertson, W. (eds.) DIMVA 2012. LNCS, vol. 7591, pp. 21–41. Springer, Heidelberg (2013)
29. Stewin, P., Seifert, J.-P., Mulliner, C.: Poster: Towards detecting DMA malware. In: Conference on Computer and Communications Security, pp. 857–860 (October 2011)
30. Suh, G.E., Clarke, D., Gassend, B., van Dijk, M., Devadas, S.: AEGIS: Architecture for Tamper-evident and Tamper-resistant Processing. In: Proceedings of the International Conference on Supercomputing, pp. 160–171 (June 2003)
31. Wang, J., Stavrou, A., Ghosh, A.: HyperCheck: A Hardware-assisted Integrity Monitor. In: Jha, S., Sommer, R., Kreibich, C. (eds.) RAID 2010. LNCS, vol. 6307, pp. 158–177. Springer, Heidelberg (2010)
32. Yashiro, T., Bessho, M., Kobayashi, S., Koshizuka, N., Sakamura, K.: T-Kernel/SS: A Secure Filesystem with Access Control Protection Using Tamper-Resistant Chip. In: Computer Software and Applications Conference Workshops, pp. 134–139 (July 2010)
33. Yu, F., Chen, Z., Diao, Y., Lakshman, T.V., Katz, R.H.: Fast and Memory-efficient Regular Expression Matching for Deep Packet Inspection. In: Proceedings of the Symposium on Architecture for Networking and Communications Systems, pp. 93–102 (December 2006)

Soundsquatting: Uncovering the Use of Homophones in Domain Squatting

Nick Nikiforakis[1], Marco Balduzzi[2], Lieven Desmet[3],
Frank Piessens[3], and Wouter Joosen[3]

[1] Department of Computer Science, Stony Brook University, Stony Brook, NY, USA
nick.nikiforakis@cs.stonybrook.edu
[2] TrendMicro
marco.balduzzi@iseclab.org
[3] iMinds-DistriNet, KU Leuven, 3001 Leuven, Belgium
firstname.lastname@cs.kuleuven.be

Abstract In this paper we present *soundsquatting*, a previously unreported type of domain squatting which we uncovered during analysis of cybersquatting domains. In soundsquatting, an attacker takes advantage of homophones, i.e., words that sound alike, and registers homophone-including variants of popular domain names. We explain why soundsquatting is different from existing domain-squatting attacks, and describe a tool for the automatic generation of soundsquatting domains. Using our tool, we discover that attackers are already aware of the principles of soundsquatting and are monetizing them in various unethical and illegal ways. In addition, we register our own soundsquatting domains and study the population of users who reach our monitors, recording a monthly average of more than 1,700 non-bot page requests. Lastly, we show how sound-dependent users are particularly vulnerable to soundsquatting through the abuse of text-to-speech software.

1 Introduction

Due to its critical position, DNS has, over the years, attracted many attacks targeting various parts of the protocol and the DNS infrastructure. These attacks can be grouped in the following targeted categories: protocol weaknesses (e.g., DNS cache poisoning [14,25]), vulnerable implementations of DNS servers (e.g., buffer overflows in BIND [20]), and the interaction of users with DNS. Of all the aforementioned categories, we postulate that the attacks targeting the user-DNS interaction are the hardest to eliminate, since they involve the education of the entire current and future Internet population, rather than the technical correction of a protocol shortcoming, or a software vulnerability.

One of the ways users interact with DNS is through the typing of domain names in their browsers' address bar. Attackers realized early on that users make spelling mistakes when typing the domain name of their desired destinations and started registering these "typo-including" domains in order to capitalize on the potential incoming traffic. This practice was named *typosquatting* [19,27],

S.S.M. Chow et al. (Eds.): ISC 2014, LNCS 8783, pp. 291–308, 2014.
© Springer International Publishing Switzerland 2014

and typosquatters use these domains in a wide range of unethical and illegal ways including showing paid ads of competitors [21], and the exfiltration of user credentials through phishing [10]. In addition to typosquatting, other variations of domain squatting have been proposed over time, such as homograph attacks [11,16] where the attacker abuses the visual similarity of two characters from different character sets to construct domains that have the appearance of a popular authoritative domain but lead to different destinations.

In this paper we present *soundsquatting*, a domain squatting technique which we uncovered while researching generic cybersquatting. Soundsquatting takes advantage of the sound similarity of words and the user's confusion of which word represents the desired concept. The attack is based on *homophones*, i.e., sets of words that are pronounced the same but are spelled differently, e.g., {*ate, eight*}. Soundsquatting is different from typosquatting in that it does not rely on typing mistakes and in that not all domains contain homophones and thus, not all domains can be soundsquatted.

To evaluate soundsquatting, we compile an English homophone database and we design *AutoSS* (AutoSoundSquatter), a tool which, given a list of target domains, generates valid soundsquatting domains. For the Alexa top 10,000 websites, AutoSS was able to generate 8,476 soundsquatting domains, 1,823 (21.5%) of which were already registered. Through a series of automatic and manual experiments, we categorize these registered domains and discover that, even though homophone-based domain squatting has not appeared in literature around cybersquatting, its principles are known and practiced by cybersquatters, albeit to a lesser extent than typosquatting. Using data that we obtain through crawling, we show that soundsquatting is being used for displaying ads on parked domains, stealing traffic from target domains, performing affiliate scams, conducting phishing attacks and installing malicious software on unsuspecting visitors.

In addition to studying the use of already registered soundsquatting domains, we register 30 available ones and study the population of users that reached our domains, recording a monthly average of 1,718 requests from real users, originating from 123 countries, showing that users are indeed susceptible to homophone confusion. Finally, we examine six popular software screen readers and show how they can all be abused to perform soundsquatting attacks against sound-dependent users who rely on text-to-speech software.

Overall, our findings show that soundsquatting can be abused in exactly the same way as typosquatting, and thus should be taken into account by owners of large websites when they are trying to protect their brand-names and customers.

Our main contributions are:

- We uncover a, previously unreported, domain-squatting attack, based on homophone-confusion rather than typographic mistakes, which we name *soundsquatting*, and present the architecture of a tool capable of automatically generating soundsquatting domains.
- We perform a systematic, large-scale analysis of the existing soundsquatting domains targeting the Alexa top 10,000 sites and highlight their abuse.

- We actively measure the worldwide population of users who make homophone mistakes, confirming the validity and practicality of the attack.
- We show how soundsquatting can be used against sound-dependent users.

2 Soundsquatting

In this section, we introduce all the necessary terminology for soundsquatting and describe in detail the workings of AutoSS, our tool for the automatic generation of soundsquatting domains. Lastly, we examine the soundsquatting domains that our tool generated, when targeting the Alexa top 10,000 sites.

2.1 Terminology

Homophones are sets of words that have the same pronunciation. Homophones can be spelled differently but have the same meaning, such as {*guarantee, guaranty*} or spelled differently and have a different meaning, such as {*whether, weather*} and {*idle, idol, idyll*}.

Given the aforementioned definition of homophones, we define *soundsquatting* as the practice of registering domain names that contain words that are homophones of authoritative domains and *soundsquatters* the individuals or organizations involved in the activity of soundsquatting. As in generic domain squatting, *authoritative* domains are domains that are targeted by soundsquatters, and usually belong to high-traffic websites with millions of visitors. The more legitimate users a website has, the more users are likely to land on the website connected to the soundsquatting domain. When an authoritative domain is targeted by a soundsquatting attack, this domain is called *soundsquatted*.

For instance, assuming `weatherportal.com`, an authoritative weather site, a soundsquatter can register the domain `whetherportal.com`, in order to capture the traffic of users who mistakenly type the word "whether" instead of "weather". When users type the wrong word and reach the soundsquatting domain, the soundsquatter, like generic domain squatters, can then monetize their visit in a wide range of unethical and illegal ways.

2.2 Differences with Typosquatting

Before moving on to the discovery and study of soundsquatting domains, it is important to differentiate *soundsquatting* from *typosquatting*. As the term reveals, typosquatting involves "typos", i.e., misspellings of domain names, usually associated with typing mistakes. In 2006, Wang et al. categorized the typos involved in typosquatting in five different categories [27]. Assuming the domain `example.com` and the intended URL `www.example.com`, the five proposed categories are the following:

1. **Missing-dot typos**: The dot following "www" is forgotten, e.g., `wwwexample.com`

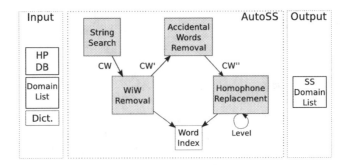

Fig. 1. The architecture of AutoSS. Given a homophone database, a list of target domains and a dictionary, AutoSS outputs a list of possible. soundsquatting domains

2. **Character-omission typos**: One character is omitted, e.g., `www.exmple.com`
3. **Character-permutation typos:** Consecutive characters are swapped, e.g., `www.examlpe.com`
4. **Character-replacement typos:** Characters are replaced by their adjacent ones, given a specific keyboard layout, e.g., `www.ezample.com` where "x" was replaced by the QWERTY-adjacent "z".
5. **Character-insertion typos:** Characters are mistakenly typed twice, e.g., `www.exaample.com`

Later research in typosquatting showed that in addition to the above classes of typos, domain squatters are also registering authoritative domains under different, less-popular top-level domains [4].

In all of the above cases, users *intend* to type a specific URL, but accidentally mistype the URL and initiate a request for the wrong page, before realizing their mistake. Contrastingly, in soundsquatting, users type exactly what they were planning to, but their intended destination is a different one. The mistake happens at a word-level, rather than a character-level, and the substituted words are real dictionary words and not mistypes of other words. The confusion between the intended word and the typed one, is further amplified when a domain contains a homophone belonging to a set of homophones with the same meaning. Consider guarantybanking.com, a domain belonging to a banking website. As mentioned earlier, "guarantee" is a homophone of "guaranty", and guaranteebanking.com is, at the time of this writing, parked and available for sale. In such cases, the typing of the "correct" domain, involves the memorization of one specific spelling, rather than a translation from concepts into words. It is also difficult to predict, if a person hears about "Guarantee Banking" for the first time, which spelling will she choose to use.

2.3   Generating Soundsquatting Domains

Any system that is geared towards the discovery of domain-squatting activity requires at least the following two resources: a set of target, authoritative domains and a list of rules and models for the transformation of authoritative domains and the generation of possible squatting domains. In the case of typosquatting, these rules may be the neighboring characters of every key under a specific keyboard layout and models of character omission, duplication and replacement. For soundsquatting, these resources are the following:

Authoritative Domain List: Under the assumption that popular domains are targeted more than less popular domains, we obtained a list of the top 10,000 Internet websites, according to Alexa. In Section 2.4 we provide the number of unique domains contained in this list.

Dictionary: A dictionary (or wordlist) is required for the extraction of valid words from domain names. For instance, given a sufficiently large dictionary and the domain `youtube.com`, an algorithm can straightforwardly search for the presence of all words in that domain (excluding the top-level domain) and conclude that the domain is comprised out of the words "you" and "tube".

Transformation Rules: Apart from a dictionary, a database of English homophones is also required. We compiled a homophone database, by scraping `homophone.com`, a website dedicated to homophones, and Wikipedia's list of dialect-independent homophones [28]. In addition, we manually added to our homophone database the list of numbers from one to one hundred together with their appropriate word form, e.g., {*9, nine*}, and a few common idioms used regularly in Internet slang, e.g., {*you, u*}.

To automatically generate soundsquatting domains, we created AutoSS (AutoSoundSquatter), a tool which receives as input the aforementioned resources and generates valid soundsquatting domains – see Figure 1. AutoSS operates as follows. After the loading to memory of the homophone database and the dictionary, each entry in the Alexa list of websites is parsed, in order to isolate the main domain, from its domain extension and possible subdomains and paths. If the resulting string contains dashes (-), then we perceive this as a sign, from the domain owner, of separation of words; e.g., `search-results.com` can be separated to the words "search" and "results" without the aid of a dictionary. If the domain does not contain dashes, then we perform a string search for the presence of every word in our dictionary, in the domain. While this is a relatively fast process, the resulting set of candidate words (CW in Figure 1) requires substantial processing, mainly due to candidate words included in other candidate words and the presence of accidental words. Below, we describe these issues and our techniques for automatically detecting and resolving them:

Words-in-Words Removal: Consider the domain `linkedin.com` and the homophone set {*in, inn*}. Even though we would, ideally, want to discover just the words "linked" and "in", a typical dictionary search will discover the words: {*in, ink, inked, ked, link, linked*}. The obvious next step would be to delete all candidate words words that are contained in others. The issue, however, is that, while the words {*in, ink, inked, ked, link*} are all contained in the word "linked", a removal of the word "in" from the candidate words is wrong (since the word exists on its own accord after the word "linked") and by doing so we miss the opportunity of generating the soundsquatting domain linkedinn.com. In our implementation we solve this problem as follows:

Whenever a pair of words $\{a,\ b\}$ is found, where a is included in b, b is replaced, in the domain name, by another string of equal length. After this replacement, the domain name is searched again for the presence of word a. If the word is still found, then a is not deleted from the set of candidate words. Thus, in our earlier example and the pair of words $\{in,\ linked\}$, linkedin.com is transformed to _____in.com. Since the word "in" is still found in the domain name, it is not removed from the candidate words. Before proceeding, our tool also records the index of the word's location in the transformed domain in the "Word Index" component, so that when, at a later step, words are replaced by their homophones, our tool replaces the appropriate "in", avoiding results such as linnkedinn.com, which do not conform to our definition of soundsquatting since "linnked" is neither a valid dictionary word, nor a homophone of any other word. At the end of this process, our candidate words set is reduced to $\{linked, in\}$ (CW' in Figure 1), which is the desirable outcome.

Accidental Words Removal: This module receives the, possibly modified, candidate-words set from the Words-in-Words module and attempts to identify and remove accidental words from the candidate words. The issue of accidental words can be illustrated as follows. Consider the domain leaseweb.com, belonging to a web-hosting provider. The ideal word breakdown would be $\{lease,\ web\}$. Given our dictionary and the previous step of selective removal of words that are included in others, AutoSS would discover the words $\{lease,\ sew,\ web\}$. In this set, the word "sew" is included since it is a dictionary word, which accidentally appears in the domain name, formed by the last two letters of the word "lea<u>se</u>" and the first letter of the word "<u>w</u>eb". We partially solve this problem by attempting to exhaustively create permutations of the candidate words, e.g.,:

```
lease        ✗
...
websew       ✗
...
leasesew     ✗
leaseweb     ✓
```

until either the permutation perfectly matches the target domain name (CW''), or, due to the exponential nature of permutations, the computation reaches a timeout. If the timeout is reached before the module finishes, then AutoSS falls back to the candidate word list that was received by the WiW Removal.

Homophone Replacement: In this last module, AutoSS uses the set of candidate words discovered by the previous modules and generates new domains, through the replacement of one homophone word by another. To this end, the module queries its homophones database for each candidate word, and for each homophone discovered the system generates a new soundsquatting domain, by replacing the candidate word with a homophone. For every candidate word, the module takes into account information found in the Word Index, so as to replace the right word in the aforementioned corner cases.

In addition to single replacements of homophones, AutoSS has a "Level" parameter (as shown in Figure 1), which specifies the number of concurrent homophone replacements for domain names with more than one homophones discovered. Consider the case of `thepiratebay.se`, a popular Torrent tracker. AutoSS will discover the homophones {*the, thee*} and {*bay, bey*}. While these can be used to create the soundsquatting domains `theepiratebay.se` and `thepiratebey.se`, a third domain can be generated by replacing both at the same time, i.e., `theepiratebey.se`. For our experiments we chose a Level of two, in order to limit AutoSS to maximum two homophone replacements at a time, even if a domain contains multiple homophones. While a higher Level would allow significantly more combinations and thus generate many more soundsquatting domains, we reasoned that three or more homophone mistakes in a single domain name are unlikely to occur, and thus decided against it.

AutoSS Limitations: Due to the flexibility of the English language and the freedom of infinite word-plays, our system's techniques for isolating words in domain names are necessarily heuristic-based. In Section 6, we estimate the amount of false positives generated by AutoSS, and briefly discuss possible ways of further lowering false positives that could be pursued in future work.

2.4   Results

From the Alexa list of top 10,000 Internet websites, we extracted 9,926 *Public Suffix + 1* domains. Given these domains and our homophone database which contains 2,913 words belonging to 1,337 sets of homophones, AutoSS extracted a total of 6,418 homophones from the list of domains which, when combined up to a level of two, generated 8,476 soundsquatting domains. Interestingly, in our results we saw that 67.3% of the domains contained no homophones.

The highest ranking domain which contained homophones was `youtube.com`, for which AutoSS generated the domains: `yewtube.com`, `ewetube.com` and `utube.com`. The domain with the most homophones was `wearehairy.com`, with a ranking of 5,663 in the Alexa list of websites, containing 12 different homophones which, when combined, allowed for 32 different soundsquatting domains. From the 1,337 sets of homophones, 568 (42.48%) were used at least once to generate a soundsquatting domain. Table 1 shows the ten homophone sets used the most by AutoSS over the Alexa list of websites.

In Figure 2, we explore the correlation between the ranking of a website and the number of homophones found in its domain name. The scatterplot reveals that there is no significant relationship between the two, meaning that, on average, high ranking websites are as vulnerable to soundsquatting as low-ranking ones. This is an experimental validation of what we intuitively expected: homophones are dependent on the choice of words, which is unrelated to the popularity of a website, at least within the top Alexa sites.

Table 1. Top ten homophone sets for word replacements in the Alexa top 10K sites

Homophone set	#Times Utilized
{2, two, to, too}	735
{1, one, won}	300
{ere, air, aire, are, ayr, ayre, err, eyre, heir}	278
{four, 4, for, fore}	250
{bi, buy, by, bye}	223
{do, dew, due, doe, dough}	208
{whirled, whorled, world}	156
{yew, you, ewe, u}	150
{cite, sight, site}	134
{0, zero, -xero}	134

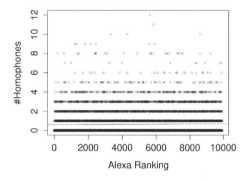

Fig. 2. Scatterplot showing the lack of a significant correlation between a website's popularity and the number of homophones contained in its domain name (r=0.019)

3   Evaluation of Soundsquatting

In this section, through a series of automated and manual experiments, we analyze the existing, i.e., already registered, soundsquatting domains and categorize them according to the purpose that they serve.

3.1   Method of Categorization

In Section 2.3 we described AutoSS, our system for the automatic generation of soundsquatting domains, and we showed that it was able to generate 8,476 soundsquatting domains targeting the Alexa top 10,000 sites. In order to find out whether domain squatters are already aware of homophones and the principles of soundsquatting, we performed a two-step process to identify the domains that were already registered. First, we tried to resolve all domains to their IP addresses. A domain that successfully resolves is obviously registered, however a domain that does not resolve may still be registered but not assigned a valid IP address. Thus, for the set of domains that did not resolve to an IP address, we performed WHOIS lookups, and tried to register them with a popular domain name registrar. At the end of this process, we identified that 1,823 domains (21.5% of the total domains generated) were already registered.

   In order to classify the registered domains, we implemented a crawler based on *PhantomJS* [15] that visited each of the 1,823 domains, waited for ten seconds (to allow the loading of remote content) and then took a screenshot of the page as well as recorded the HTML and final URL for later processing. The final URL was used in order to detect redirections from the visited soundsquatting domain to a different domain. To categorize each site, we followed a semi-automatic approach. We begun by manually skimming over the screenshots of all pages and grouping together images that looked alike. The majority of these were *parked pages*, i.e., pages that show ads, somewhat relevant to the domain name and usually also advertise that the domain may be for sale. Other groups were pages with little content, stating that the site is "under construction", placeholder pages by popular registrars informing their clients how to setup a website on their registered domain, and pages containing generic errors, such as `404 Forbidden`. For each group, we examined the corresponding HTML of a few domains and created generic HTML- and JavaScript-signatures that could automatically categorize the remaining pages within each group. Using this approach, our page-characterizing scripts could eventually automatically classify 77.2% of all the crawled domains. The remaining unclassified domains (417) were classified manually by visiting each website and carefully inspecting the source code, available WHOIS information and any similarities (visual-, content- and audience-based) between the soundsquatting domain and the corresponding authoritative one.

3.2   Categorization Results

By combining the results of the automatic classification and those of our manual investigation, we categorized the registered soundsquatting domains as follows:

Authoritative-Owned Domains: From the 1,823 studied domains, we identified 155 soundsquatting domains which belonged to the companies and organizations behind the corresponding authoritative domains. In the vast majority of cases, the user would be automatically redirected to the correct authoritative domain without warnings or additional dialogues. The redirect almost always happened through a 301/302 HTTP status code and occasionally users were redirected to one or two intermediate hosts, which would in-turn redirect them to the appropriate domain. In these cases, we were able to identify that the intermediate hosts belong to brand-protecting companies which were, most likely, registering the fact that a user did a specific mistake, before redirecting her to the appropriate destination.

In two instances, the owners of the authoritative domains were attempting to educate their users about homophone confusion. For instance, `myfreepaysight.com` is a soundsquatting domain for the adult site `myfreepaysite.com`. When the soundsquatting domain is visited, users greeted with a message that points out the difference between the two domains.

Parked/Ads/For Sale Domains: Parked domains have been identified by prior research as the preferred monetizing way of domain squatters [21,27]. As we mentioned earlier, these domains contain no real content, except ads which are constructed on demand, usually by a domain-parking agency, based on the words included in a domain name and preferences by the owner of the domain. In this category, we also include domains that were found to show ads without being affiliated with a large domain-parking agency, e.g., `net0.net`, a soundsquatting domain targeting `netzero.net`, as well as domains that are listed as "for-sale". In total, ad-driven domains represent the largest chunk of existing soundsquatting domains, with 954 cases (52.3%).

Affiliate-Abusing Domains: While examining the soundsquatting domains that redirected the user to the appropriate authoritative domain, we realized that 32 soundsquatting domains were abusing affiliate programs of the corresponding authoritative domains. In affiliate programs, existing customers of a website are encouraged to bring new customers, through the promise of a small commission for every brought customer.

In affiliate abuse, an attacker takes advantage of an affiliate program of a website by appending his own affiliate identifier to unsuspecting visitors. More specifically, consider the domain `mybrowsercache.com` which is a soundsquatting domain for `mybrowsercash.com`. At the time of this writing, when a user visits the former domain, she will automatically be redirected to `http://www.mybrowsercash.com/index.php?refid=312044` . Notice that a specific referrer identifier is added to the URL. In this way, the attacker who registered `mybrowsercache.com` will gain a commission every time that a user confuses "cash" and "cache", and subsequently registers to the target website. In addition to soundsquatting domains redirecting to sites with affiliate programs, we also recorded some cases where the user was kept on the soundsquatting domain and the authoritative site, together with the attacker's affiliate identifier, was opened in a full-page HTML frame.

Hit-Stealing Domains: From our analysis, we discovered 22 cases where attackers used soundsquatting to capture traffic and feed their own "business-related" domains with hits intended for the authoritative targeted site. In fact, in the majority of cases, both the soundsquatting and corresponding authoritative domain had content belonging to the same category, but they were owned by different organizations and individuals. From our experiments, we found that the most targeted business categories are adult, online shopping and travel. Below, is a list of a few examples:

- `ashemailtube.com` is a soundsquatting domain for the domain `ashemaletube.com`, a transvestite-oriented porn site. When the soundsquatting domain is visited, the user is redirected to `trannydates.com` a dating site specifically for transvestites.
- `video-1.com`, a soundsquatting domain for the adult video portal `video-one.com`, currently hosts an online sex shop
- `todomains.ru` is selling domain-registration services and is a soundsquatting domain for `2domains.ru`, a large Russian domain registrar.
- `gamefive.com` is a soundsquatting domain for `game5.com`, an online gaming site. The soundsquatting domain had been for sale for three years before being turned into an online gaming site.
- `textsail.ru` is a soundsquatting site for `textsale.ru`. Both sites sell articles and stories on a wide range of topics.

In this category, we also included soundsquatting domains that profit from the trustworthiness associated with well-known and popular authoritative domains, in order to advertise non-related websites. In these cases, it is not necessary that the category of the soundsquatting domain matches the category of the targeted authoritative domain. For example, the owner of `freemale.hu` is probably exploiting the popularity of the well-known Hungarian e-mail provider `freemail.hu` to advertise his own web page, in the same way that the sound-squatting domain `tvto.no` is abusing the popularity of the Norwegian TV2 channel `tv2.no` and subsequently redirects the user to an online casino.

Scam-Related Domains: Soundsquatting domains can also be used for scamming purposes. We identified 16 cases of domains used to perform different form of scams, generally by luring visitors into subscribing to fake lotteries and surveys. For instance, `vhone.com`, a soundsquatting domain targeting `vh1.com`, redirects the user to a survey site, where users are promised an opportunity to win high-end electronics in return to their participation in a short survey. The users are then trapped in a series of redirects where they are constantly promised more and more prizes while they divulge more and more private information, such as their names, email addresses and mobile phone numbers.

Promoting-Related Domains: In this category, we included seven domains that were used to promote someone or something related to the authoritative domain. For example, `teambeechbody.com` is a soundsquatting domain for `teambeachbody.com`, an online fitness club where people can subscribe as "fitness coaches" and gain a commission for every user that they successfully coach. At the time of this writing, when the soundsquatting domain is visited, the user is redirected to the page of a specific coach within the authoritative `teambeachbody.com` domain, giving that specific coach higher chances of getting selected to coach a user, over other coaches on the website. In another case, the authoritative `readnovel.com` domain is soundsquatted by `rednovel.com`, which redirects the user to `http://www.lvse.com/site/readnovel-com-3550.html`. The page contains a review of the authoritative website, providing a safety score, user comments and similar websites.

Others Domains: We conclude our analysis with six soundsquatting domains used for malicious purposes, e.g., to install malware and acquire private information. Movreel (`movreel.com`), is a free-of-charge service for streaming movies, and `movreal.com` is one of its soundsquatting domains. At a first glance, `movreal.com` appears to be another streaming provider for movies, but interestingly the user is requested to download a browser plug-in (`AVS_Media_Player.exe`) to watch the video. The offered plug-in, however, is malicious and detected by most antivirus vendors as a variant of the `solimba` malware, i.e., an installer for other malicious software and a provider of adware campaigns. Similarly, `utube.com`, a soundsquatting domain for `youtube.com`, makes use of videos to social-engineer the user into first divulging her personal information and then, depending on the type of browser used, installing a browser extension. The extension installed when using Mozilla Firefox, injects unwanted search results, launches pop-ups, and gathers user statistics.

Among the other cases, we identified two domains that are likely used to acquire private user information, in particular email credentials. One of these is `innbox.lv`, which is the soundsquatting domain of the well-known Latvian service provider *InBox*, where both sites offer free e-mail accounts. Finally, we were able to confirm two soundsquatting domains involved in phishing campaigns against e-commerce and business-related sites.

Summary: Overall, by combining the results of the outlined categories, out of the 1,823 registered soundsquatting domains, 1,037 (56.88%) were categorized as malicious. For the remaining domains, 155 of them belong to the corresponding authoritative domains' owners, 300 domains are registered under different organizations that are using them in a legitimate way, and 331 domains were offline, or showing HTTP errors, or under construction when we visited them.

4   User Characterization

In previous sections, we categorized the registered soundsquatting domains according to their use. We now turn our attention to the users who, due to homophone confusion, land on soundsquatting domains and study their population.

As described in Section 2.4, AutoSS was able to generate 8,476 soundsquatting domains for the Alexa top 10,000 websites. From these, 1,823 domains (21.5%) were already registered, leaving 6,653 unregistered soundsquatting domains. To actively measure the worldwide population of users, and to assess the viability of the soundsquatting attack, we decided to register our own soundsquatting domains and monitor the requests towards them. Since there is no prior research on soundsquatting, there was no objective or historical way of assessing which of the unregistered domains would attract more users than others. For this reason, we manually examined the list of available soundsquatting domains from which we selected a total of 30 domains, trying to cover a wide range of soundsquatting errors.

The first three columns of Table 2 show 20 of the 30 authoritative domains targeted, the pair(s) of used homophones and the registered soundsquatting domains.While three of targeted domains could be associated with typosquatting, e.g., `theefreedictionary.com`, the rest are radically different from domains which researchers have, over the years, associated with typosquatting, e.g., `prophetclicking.com`. Most domains were registered in December 2012 while some additional ones were registered in March 2013. To present a uniform view of traffic, we provide the monthly average number of requests that each domain received, till December 11, 2013.

We resolved all domains, subdomains and requests for specific file paths to a single blank page, while recording each request's details in a set of Apache log files. Users were not automatically redirected to the appropriate authoritative domain, to avoid reinforcing the behavior of typing the wrong domain, but rather to make users aware of their mistake. We discuss our ethical considerations regarding this experiment, in the Appendix of this paper.

The last column of Table 2 shows the monthly average number of human requests received during the monitored period, and the percentage of human requests over all requests. To assess the impact of soundsquatting on humans, we needed to separate visiting bots from visiting human users. To the best of our knowledge, there is no single, generic technique that can perfectly separate bots

Table 2. A list of 20 out of the 30 registered soundsquatting domains, the corresponding authoritative domains, the homophone pair used, the monthly average number of human requests received, as well as the percentage human traffic against total traffic

Auth. Domain	Homophone pair	SS Domain	#Human Req. (per month)	
thefreedictionary.com	{*the,thee*}	theefreedictionary.com	283	(39.86%)
fc2.com	{*2, too*}	fctoo.com	165	(44.84%)
jimdo.com	{*do, doe*}	jimdoe.com	150	(38.27%)
turbobit.net	{*bit, bitt*}	turbobitt.net	132	(36.07%)
leboncoin.fr	{*coin, quoin*}	lebonquoin.fr	110	(74.32%)
adserverplus.com	{*ad, add*}	addserverplus.com	98	(60.49%)
profitclicking.com	{*profit, prophet*}	prophetclicking.com	56	(48.28%)
hostgator.com	{*gator, gaiter*}	hostgaiter.com	45	(45.92%)
sitesell.com	{*sell, cel*}	sitecel.com	44	(40.00%)
discuz.net	{*disc, disk*}	diskuz.net	43	(40.19%)
tube8.com	{*8, ait*}	tubeait.com	42	(43.30%)
clixsense.com	{*sense, scents*}	clixscents.com	40	(44.44%)
a8.net	{*8, eight*}	aeight.net	48	(43.24%)
newegg.com	{*new, gnu*}	gnuegg.com	37	(36.63%)
redtubelive.com	{*red, read*}	readtubelive.com	44	(51.76%)
fiverr.com	{*err, air*}	fivair.com	33	(37.93%)
exoclick.com	{*click, clique*}	exoclique.com	32	(45.71%)
theglobeandmail.com	{*mail, male*}	theglobeandmale.com	35	(38.46%)
pastebin.com	{*bin, been*}	pastebeen.com	35	(39.77%)
ku6.com	{*6, sics*}	kusics.com	28	(33.33%)
...	...	...	...	
Total Requests per Month (30 domains):			1,718	

from humans. If such a technique would exist, attackers would use it to perfectly evade malware researchers, by detecting *all* high-interaction honeypots and never presenting them with malicious code.

For our purposes, we separated the requests as follows: during preliminary manual inspection of the recorded requests, we noted which requests had non-standard user-agents. Using keywords extracted from these requests, we assembled a set of nine generic identifiers, like "spider", "bot" and "crawl", that many bots have in common. In addition to these generic identifiers, we scraped 707 specific bot signatures from useragentstring.com. As a result, if the user agent contained any of the 716 bot signatures in our set, the request was classified as belonging to a bot. To account for bots which do not identify themselves, we also queried the IP address of each request, in the blacklist provided by stopforumspam.com, i.e., a database containing hundreds of thousands of IP addresses, belonging to known forum spamming bots. Lastly, each address was queried in a list of IP addresses used by well-known search engine spiders [1].

Table 2 shows that our set of 30 soundsquatting domains received an average of 1,718 human requests per month. The monthly average of total requests (not shown in the table) was 4,150. The domain that received the most hits, theefreedictionary.com, is a domain that can also be associated with typosquatting and thus naturally attracts more traffic than domains that are just soundsquatting. Apart from requests towards the main page of each website, we recorded many requests towards subdomains within each domain. For instance, `jimdo.com` is a web application where users can create their own websites and host them on subdomains of the main domain. In our `jimdoe.com` logs, we found requests towards 176 subdomains connected to personal sites, such as `awesomegrizzlybears.jimdoe.com`, `karatedojo-oppeln.jimdoe.com` and `armaniwoe.jimdoe.com`. These were all valid subdomains within the main `jimdo.com` domain and thus their visits show that, even though people can accurately type relatively long and obscure subdomains, they can still confuse homophone words.

By geolocating the IP addresses of all requests we discovered that, while there were 42 countries involved in the crawling of our domains, requests from human visitors originated from 123 different countries, demonstrating that users from all countries are prone to homophone confusion and thus vulnerable to soundsquatting attacks.

In general, each soundsquatting domain received between 2 and 283 human requests per month. While these numbers are not incredibly large, and probably smaller than popular typosquatting domains, it is important to remind the reader that soundsquatting and typosquatting are not competing techniques, but rather complementing ones in the arsenal of domain squatters. Moreover, since we are the first ones to study soundsquatting, we registered domains with homophone replacements ranging from more likely, to less likely. Careful attackers could target domains in a better way, and thus acquire more users for a smaller cost.

Finally, it is worthwhile mentioning that we received a significant number of emails sent to our soundsquatting domains. Among others, we received social networking invitations, notifications of the shipment of various products, account-creation emails with credentials, and bills of mobile telephony. In all cases, it was evident that the emails were meant to be sent to accounts belonging to the legitimate domains which we targeted, but were sent to us due to homophone confusion. The receipt of these emails further demonstrates that businesses and users are indeed vulnerable to soundsquatting attacks.

5   Sound-Dependent Users

In previous sections, we described the soundsquatting attack and we investigated the existing soundsquatting domains as well as the users' susceptibility to homophone confusion. In this section, we describe a variation of the attack that is geared towards people that rely on sound.

According to the Word Health Organization, there are currently 285 million people that are visually impaired, of which 39 million are blind [2]. People that are severely visually impaired, cannot properly interact with a computer without the use of assistive technologies. The two most popular assistive technologies for the visually impaired are Braille displays and screen readers [9]. In both cases, the assistive technology converts content that would be consumed by sight, into content that can be consumed by touch or sound. If one thinks of the definition of homophones and their relation to soundsquatting, a new attack becomes clear.

A user that depends on a screen reader in order to consume content in emails, web pages, messages in social networks or instant messaging applications, is vulnerable to links pointing to soundsquatting domains. A soundsquatting domain will be "read" near-identical with the targeted authoritative domain and thus the visually impaired user has no reason not to click on the offered link. While Braille displays are not vulnerable to this attack, the fact that about 90% of the visually impaired people live in developing countries combined with the high cost of Braille displays, suggest that, due to limited resources and possible portability issues, screen readers are used much more than Braille devices. Moreover, apart from visually-impaired users, hundreds of thousands of smartphone users utilize personal assistant software, like Apple's Siri, with text-to-speech capabilities when involved in another activity that makes it hard to operate their smartphones, like driving or running.

To test our theory, we sent to a web-mail account an email with two links, one pointing to `youtube.com` and one pointing to `yewtube.com`. We had our email read to us by five popular free screen readers, i.e., by the built-in screen reader of Windows XP, Windows 7 and Mac OS X, the Linux-based, open-source ORCA [23], and the Thunder screen reader [26]. We also sent a text-message with the same information to an Android smartphone with Skyvi [5], a popular Siri-like application installed by more than 260,000 users.

In all six cases, the two links sounded identical to each other which means that a sound-dependent person would have no means of separating a legitimate link from a malicious one. To further exacerbate the issue, this type of attack can also work with pseudohomophones, i.e., combinations of characters that are not real dictionary words but are purposefully constructed to sound like real words, such as {*joke, joak*} [24]. Thus, pseudo-soundsquatting domains can be crafted even for target domains that contain no homophones, such as `phacebook.com` and `phaceboocc.com`.

Due to the potentially large number of such domain variations and the specificity of this attack, we reason that the responsibility of protecting a sound-dependent user should be on the text-to-speech software. One way of protecting against this threat is for the software to switch to a "spelling mode" whenever a link is encountered, so that the user will realize that the link is not what it sounds like and avoid visiting the malicious website.

6   Limitations and Future Work

In Section 2.3 we described the workings of AutoSS, a tool that automatically generates soundsquatting domains. While we accounted for many corner cases when attempting to identify the words comprising a domain name, there is, unfortunately, still room for false positives, i.e., generated domains that do not conform to our definition of soundsquatting and the intuition behind it. For instance, there are many domains in the Alexa top 10,000 which do not include English words, like `laredoute.fr`, a French e-shop. AutoSS uses an English dictionary and thus will identify the words "lare", "do" and "ute" inside the domain name. The "Accidental Words Removal" module of AutoSS, will successfully combine these words to "laredoute" and thus, the words will then be used as keys in our homophone replacement database resulting to improbable domains, such as `laredewute.fr`. Prior systems proposed to automatically generate typosquatting domains do not suffer from such problems, since they operate at a character level [21,27] whereas soundsquatting operates at a word level.

To estimate the number of false positives, we randomly sampled 5% (424) of the generated soundsquatting domains and manually examined each homophone replacement, classifying each domain as a true or false positive. At the end of this process we identified 80 false positives out of the 424 investigated domains (18.9% with a margin of sampling error ±4.75%). While the number of false positives is not negligible, the main purpose of our work was to investigate a previously unreported domain-squatting technique and to evaluate its practicality and adoption on the web which we believe that we did.

The lack of punctuation in a domain name makes identifying its language, a challenging task. One way around this problem would be to actually inspect the main page of the site, characterize the language of that, and assume that the domain name contains words of the same language. We leave the exploration of this and other techniques of reducing false positives, for future work.

7   Related Work

To the best of our knowledge, this paper is the first one that uncovers the use of homophones as a way of performing domain squatting, and systematically studies the adoption of the attack, as well as the user's susceptibility towards it.

Domain squatting was the first type of cybersquatting involving the registration of domains that were trademarks belonging to other persons and companies, before the latter had a chance to register [6,8,13]. Domain squatting later evolved into *typosquatting* [8,21,27], i.e., the act of registering domains that are mistypes of popular authoritative domains which can be traced back to 1999, through the Anticybersquatting Consumer Protection Act (ACPA) which already mentions URLs that are "sufficiently similar to a trademark of a person or entity." [3]

Apart from typosquatting, there also exist other, less popular, types of domain squatting, such as domains that abuse the visual similarity of characters in different character sets [11,16], and domains that capture the traffic originating from erroneous bit-flips in user devices [7,22].

8   Conclusion

In this paper we uncovered a new type of domain squatting based on the sound similarity of words, rather than typographical mistakes. We named the attack "soundsquatting", described a system for automatically generating soundsquatting domains, and showed that attackers are already familiar with the concepts behind soundsquatting, abusing them in ways similar to known types of domain squatting. By registering our own soundsquatting domains, we showed that it is possible for well-selected soundsquatting domains to attract hundreds of human visitors every month. We also briefly examined the relationship between text-to-speech software and soundsquatting, and showed that attackers could abuse the former to trick sound-dependent users into visiting malicious soundsquatting and pseudo-soundsquatting domains. Overall, our findings verify the practicality of soundsquatting and show that homophone confusion should be accounted for, by website owners, registrars, and cybersquatting countermeasures.

Acknowledgments: We want to thank the anonymous reviewers for the valuable comments. This research was performed with the financial support of the Prevention against Crime Programme of the European Union (B-CCENTRE), the Research Fund KU Leuven, and the EU FP7 projects NESSoS and STREWS.

References

1. IP Addresses of Search Engine Spiders, `http://iplists.com/`
2. WHO — Visual impairment and blindness,
   `http://www.who.int/mediacentre/factsheets/fs282/en/`
3. Anticybersquatting Consumer Protection Act (ACPA) (November 1999),
   `http://www.patents.com/acpa.htm`
4. Banerjee, A., Barman, D., Faloutsos, M., Bhuyan, L.N.: Cyber-fraud is one typo away. In: Proceedings of IEEE INFOCOM (2008)
5. BlueTornado. Skyvi (Siri for Android), `http://www.skyviapp.com`
6. Coull, S.E., White, A.M., Yen, T.-F., Monrose, F., Reiter, M.K.: Understanding domain registration abuses. In: Rannenberg, K., Varadharajan, V., Weber, C. (eds.) SEC 2010. IFIP AICT, vol. 330, pp. 68–79. Springer, Heidelberg (2010)
7. Dinaburg, A.: Bitsquatting: DNS Hijacking without Exploitation. In: Proceedings of BlackHat Security (July 2011)
8. Edelman, B.: Large-scale registration of domains with typographical errors (2003)
9. Even Grounds - How Do Blind People Use The Computer,
   `http://www.evengrounds.com/blog/how-do-blind-people-use-the-computer`
10. Ferguson, R.: Tvviter Typosquatting Phishing Site, `http://countermeasures.trendmicro.eu/tvviter-typosquatting-phishing-site/`
11. Gabrilovich, E., Gontmakher, A.: The homograph attack. Communications of the ACM 45(2), 128 (2002)
12. Gee, G., Kim, P.: Doppelganger Domains, `http://www.wired.com/images_blogs/threatlevel/2011/09/Doppelganger.Domains.pdf`
13. Golinveaux, J.: What's in a domain name: Is cybersquatting trademark dilution? University of San Francisco Law Review 33 U.S.F. L. Rev. (1998-1999)
14. Herzberg, A., Shulman, H.: Fragmentation Considered Poisonous, or: One-domain-to-rule-them-all.org. In: CNS 2013, pp. 224–232. IEEE (2013)
15. Hidayat, A.: PhantomJS: Headless WebKit with JavaScript API
16. Holgers, T., Watson, D.E., Gribble, S.D.: Cutting through the confusion: A measurement study of homograph attacks. In: Proceedings of USENIX ATC (2006)
17. Jakobsson, M., Finn, P., Johnson, N.: Why and How to Perform Fraud Experiments. IEEE Security & Privacy 6(2), 66–68 (2008)

18. Jakobsson, M., Ratkiewicz, J.: Designing ethical phishing experiments: A study of (ROT13) rOnl query features. In: WWW 2006 (2006)
19. Kesmodel, D.: The Domain Game: How People Get Rich from Internet Domain Names. Xlibris Corporation (2008)
20. McMahon, R.: BIND 8.2 NXT Remote Buffer Overflow Exploit (2000)
21. Moore, T., Edelman, B.: Measuring the perpetrators and funders of typosquatting. In: Sion, R. (ed.) FC 2010. LNCS, vol. 6052, pp. 175–191. Springer, Heidelberg (2010)
22. Nikiforakis, N., Acker, S.V., Meert, W., Desmet, L., Piessens, F., Joosen, W.: Bitsquatting: Exploiting bit-flips for fun, or profit? In: WWW 2013, pp. 989–998 (2013)
23. Orca: a free, open source, flexible, and extensible screen reader
24. Seidenberg, M.S., Petersen, A., MacDonald, M.C., Plaut, D.C.: Pseudohomophone Effects and Models of Word Recognition. Journal of Experimental Psychology: Learning, Memory and Cognition 22, 48–62 (1996)
25. Stewart, J.: DNS Cache Poisoning - The Next Generation (2003)
26. ScreenReader.net: freedom for blind and Visually impaired people
27. Wang, Y.-M., Beck, D., Wang, J., Verbowski, C., Daniels, B.: Strider typo-patrol: Discovery and analysis of systematic typo-squatting. In: SRUTI 2006 (2006)
28. List of dialect-independent homophones, http://en.wiktionary.org/wiki/Appendix:List_of_dialect-independent_homophones

A   Appendix

Ethical considerations Registering soundsquatting domains and receiving user traffic to these domains may raise ethical concerns. However, analogous to the real-world experiments conducted by Jakobsson et al. [17,18], we believe that realistic experiments are the only way to reliably estimate success rates of attacks in the real world. Moreover, we believe that our findings will help websites to protect their brands and customers.

The data that was collected for this experiment are the following: For each request to our soundsquatting domains we recorded i) its timestamp ii) the IP address of the host performing the request, iii) domain, path and GET parameters and iv) the user agent, as provided by the Apache web server. This kind of data is collected by every web server on the web in standard server logs and many web developers even share this collected information with third parties, such as Google Analytics, for the purpose of gathering usage statistics. The server logs were only accessible to the authors of this paper. Similarly, the emails were all collected in a single, password-protected email account of one of the authors. We did not attempt to extract any information from these emails, nor trace their senders. Gee and Kim performed a similar experiment in 2011, capturing emails through typosquatting domains and released statistics to the research community, as a demonstration of the dangers of typosquatting [12].

Reducing User Tracking through Automatic Web Site State Isolations

Martin Stopczynski and Michael Zugelder

Technische Universität Darmstadt (CASED), Germany
{martin.stopczynski,michael.zugelder}@cased.de

Abstract. Protecting the privacy of web users against tracking by blocking third-party content has become a cat-and-mouse game. Continuously changing tracking methods make it difficult to block all third-party content. On the other hand, it is necessary to accept some third-party content to ensure web site functionality. In this work we present the concept and an implementation for the automatic isolation of the locally stored web site state into separate containers. This eliminates the ability of trackers to re-identify users across different sites, by isolating HTTP cookies, HTML5 Web Storage, Indexed DB, and the browsing cache. The so-called Site Isolation was implemented for the Chromium browser and in addition secures the browser against CORS, CSRF, and click-jacking attacks, while limiting the impact of cache timing, and rendering engine hijacking. To evaluate the effectiveness of Site Isolation, we visited 1.6 million pages on over 94,000 distinct domains and compared the data saved against usual browsing. We show that top trackers collect enough information to identify billions of users reliably. In contrast, with Site Isolation in place the number of tracked pages can be reduced by 44%.

Keywords: Tracking, Privacy, Browser, Isolation, Security.

1 Introduction

User tracking is extremely widespread in today's web. Third-party trackers exist on most commercial web sites [1] and although the number of different trackers decreases, the reach of large tracking companies increases [2]. The tracking data is used to establish detailed user profiles, e.g. for targeted advertisements, or selling the data in anonymized form [3] [4]. The resulting problems are that advertisers can reconstruct the profiles [5], anonymized data is easily de-anonymized [6], and fraudsters often obtain the unredacted data set by infiltrating the tracking companies [7]. The issue has become a part of growing public privacy concerns, with reports suggesting that 60% of users were concerned about privacy on the web [8]. In contrast to other kinds of surveillance, for example via credit cards or cameras, web tracking is cheap, mostly automated and not limited to purchases done by credit card. It is also a lot faster and can be more reliable, especially when the data can be linked to important and persistent online services, such as social media like Facebook or e-mail accounts, to create detailed user profiles [9].

S.S.M. Chow et al. (Eds.): ISC 2014, LNCS 8783, pp. 309–327, 2014.

Moreover, tracking is often completely invisible or integrated into other features and services, such as advertising, or the Facebook 'Like' button. In effect, most tracking is performed in the background, without the users being able to make an informed choice to opt-in or opt-out [10] [11].

Client Identifiers: As our analysis shows, the most common way for third-party trackers to track users is by storing a unique ID on the client and retrieve it on every page the tracking code is embedded [12]. The most basic method to save this ID is by using an HTTP cookie. Any HTTP request suffices for this, but a tracking pixel, a transparent 1:1 pixel sized image, is the traditional, bandwidth-conserving, and high-performance way of doing it. Once implemented, the visitor's browser will request the tracking cookie each time a page is loaded, because caching this image is forbidden by the server. After the first request where the cookie is set, every request carries the (third-party) cookie and the tracking server can re-identify a user on every page that includes server's tracking pixel.

Even more details can be gathered via JavaScript and browser plugins [13]. With only using JavaScript, it is possible to monitor mouse position and clicks, interaction with forms [14], retrieve the list of installed plugins, and perform canvas fingerprinting [15,16]. Plugins such as Flash or Java provide even more information, for example the list of installed fonts, CPU type or even clock skew [17]. Minimizing information leaks by external plugins is still an important task, despite them being slowly obsoleted by browser vendors and the recent HTML5 and JavaScript improvements, but not within the scope of this work.

Existing Countermeasures: To avoid a customized ID being saved by the trackers, users can disable all mechanisms to save data. This will eliminate tracking based on storing custom identifiers, but cannot protected against using pre-existing data, i.e. browser fingerprinting. More importantly, it affects every site that tries to use the saved data for *any* purpose. Since nearly every interactive web site uses cookies to save session data, this is hardly practical. A less restrictive method is to disable only third-party cookies. Third-party cookies are cookies being set (or retrieved) from a different domain than the one in the browser's address bar. If a cookies is considered first-party or third-party therefore depends not only on the cookie itself, but also the currently opened page. Ambiguities arise when first-party cookies are later used in a third-party context. Consider a user visiting *facebook.com*, which sets first-party cookies for *facebook.com*. On the site *a.com*, which contains an iframe loaded from *facebook.com*, these cookies are now considered third-party. If they are sent back to Facebook depends on the browser: One group of browsers completely blocks third-party cookies, the other group only prohibits third-party web sites from *setting* cookies, but does allow them to send *pre-existing* cookies. To which group a browser belongs can change on short notice, Roesner et al. [1] report Firefox as the only major browser that also blocks the sending of third-party cookies, while Chromium shows the same behavior since version 17. The most serious issue of blocking third-party cookies is the loss of functionally on most web sites that share cookie data with multiple domains [11]. A new approach to provide a similar benefit is to isolate web sites from each other, which will be described as Site Isolation in this paper.

The basic principle of isolation can prohibit some linkage of visited web sites and is already used in different varieties:

- Regularly clearing all site data, for example when exiting the browser. The isolation of web sites visited before cleaning from the web sites visited after cleaning is often useful, but clumsy and cannot protect against data leaks involving longterm identifiers like the linkage of an e-mail address or a Facebook account id.
- Using different browsers to isolate some high profile sites from others, for example to isolate online banking from regular web browsing. Site Isolation aims to scale up this approach to *all* visited sites, because it is not feasible to do it manually.
- The private browsing modes that major browsers implemented over the last years provide an additional isolated profile that is only stored in RAM. This protects the private browsing session from the normal browsing session, but has the big drawback, that trackers can still read all associated cookies in the active private browsing session.

The proposed concept of Site Isolation in this work transfers these principles into the core of the browser to isolate every site by default. It was developed to be similar to having a separate browser for each site, but is integrated into a single browser. Higher privacy than through manual isolation is achieved by combining the strengths of the different isolating varieties and applying them to everyday browsing. Site Isolation principally limits tracking to each individual site and is able to reduce the number of pages tracked by the top 10 most used trackers by over 50%. It cannot provide as much privacy as would be achieved by blocking all third-party cookies, but it breaks less web sites. In addition, the isolated browsing cache also limits various cache-based tracking methods and cache-timing attacks, which are not affected by blocking third-party cookies.

Our Contributions Are: Implementation of automated Site Isolation to reduce tracking. Performing a large scale study on Site Isolation effectiveness, and adding user-defined cookie timeout as well as support for multiple incognito profiles in Chromium.

2   Related Work

Various work has been done in the field of tracking detection [18,10,19,20], showing the difficulty of preventing web tracking with third-party tools. Roesner et al. [1] offer a refreshingly differentiated look into third-party tracking on the web. Blocking third party cookies was found to be effective against most trackers, but it is not easy to avoid tracking by major social networks plugins without breaking their functionality completely. To that end, they developed the *ShareMeNot* browser extension that strips cookies from the initial requests to these resources, but allows them if they are clicked. However, this solution is limited to stop tracking of some services like Twitter or Youtube, while it does not entirely remove the presence of Facebook and Google.

When comparing Site Isolation to tracking prevention tools like Ghostery[1] or AdblockPlus[2], various disadvantages in those tools can be identified. One drawback is that they rely on detecting tracking elements and reactively blocking them. Therefore, elements not categorized as trackers will not be blocked and can still be used for tracking. In addition, if legitimate elements are falsely classified (false positives) as trackers and blocked, web site functionalities may break. The most serious privacy problems exist in the commercial purpose of such tools, i.e., a third-party can avoid that certain tracking elements will be blocked (by making a contract) [21], or by the utilization of the gathered user data [22] to allow building a user web profile. With Site Isolation elements from a web site are isolated from other web sites and will stop cross-domain tracking without breaking the web site functionalities. Furthermore, web site owner can use ads to earn money without enabling trackers to re-identify the user in order to learn the user's web behavior.

The *doublekey* cookie concept proposed on Mozilla Bugtracker [23] is similar to Site Isolation. Its core idea is to replace the cookie domain with a tuple consisting of the cookie domain and the domain of the currently open web page. As a result, each visited first-party site has its own storage location for third-party cookies. The TOR project is currently in the process of implementing this concept [24]. Site Isolation is a superset of this concept, not only isolating cookies, but also other storage methods. Thereby, our solution is limiting many more tracking avenues such as cache based tracking (e.g. using ETags or modified PNG images), Web Storage and others [20]. Site Isolation is also extensible for storage methods not yet isolated, such as data from Flash. Since no other implementation has been done before, we were able to present the first large scale study on this topic and analyze the effectiveness of this concept.

Chen et al. [25] investigated which security benefits are possible using multiple, separate browsers and showed how they can be brought to a single browser using *App Isolation*. Here, the researchers implemented entry-point and state isolation for Chromium, which sets a whitelist of allowed entry-points into a web site, to avoid reflected XSS, session fixation, Cross-Origin Resource Import and CSRF attacks. The disadvantage is that this whitelisting has to be done manually for each web site and every entry point separately. In comparison, our State Isolation brings the same security features and confines each web site automatically into a separate, from other web sites isolated data container, where the cache, cookies and other locally persisted state are saved. On a high level, App Isolation is targeted at protecting users from manipulation of a few select high-profile sites, while Site Isolation is targeted at protecting users against privacy breaches resulting from third-party trackers embedded on all pages with no changes on the web.

---

[1] https://www.ghostery.com
[2] https://adblockplus.org

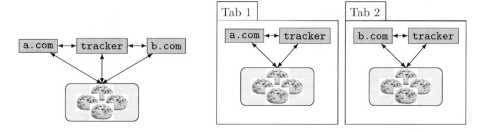

Fig. 1. Regular storage   **Fig. 2.** Isolated storage

Other strategies to gain security by isolating web elements like plugins or JavaScript and better control the execution of those exist [26,27,28,29]. Unfortunately, these strategies do not consider privacy aspects related to tracking, since they allow the communication between storage elements like cookies, which still have access to all tracking related elements. In contrast, our isolation strategy benefits from the same security aspects while protecting the users privacy from being tracked online.

3   Concept and Implementation

Considering the typical methods trackers are employing as well as the privacy impact resulting from their linking data from multiple sites, we developed three methods to reduce tracking and improve the users' privacy. These are first of all the Site Isolation strategy, secondly the user-defined cookie timeout and thirdly the support for multiple incognito profiles.

Site Isolation builds upon the best practices regarding content isolation and pushes the boundaries by applying isolation to every site. Persistent data is stored for each site individually, instead of directly in the browser's profile, together with the data of all other sites visited - see Figure 3. With Site Isolation each site gets its own storage partition, which is essentially a miniature version of the regular browser profile, where the data would be stored usually. A storage partition is specific to a certain web site, other sites have no possibility of accessing it, contributing to it or even checking if it exists. Site Isolation splits up HTTP cookies, HTML5 Web Storage (formerly Local Storage), IndexDB, and caches. Similar mechanisms provided by plugins can (and shall) be isolated in a similar manner, but has not yet been implemented. Figures 1 and 2 illustrate the difference between a global cookie storage and cookies stored in two storage partitions.

The goal of Site Isolation is to avoid automatic data flows between distinct web sites. This is typically used to facilitate user tracking, in the form of third-party cookies using JavaScript or when requesting external resources. But it is also used for many mechanisms users expect to work, such as logins or shopping carts.

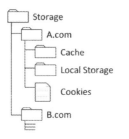

Fig. 3. Folder structure   **Fig. 4.** User Interface

Mechanism: For the implementation we used some of the Chromium infrastructure to support Site Isolation in place. As an entry point to allow storage partitions for any site we extended the *GetStoragePartitionConfigForSite* method of *ChromeContentBrowserClient*. By default, this function checks if the URL has the *extension://* scheme and if isolation is enabled for the extension. It was extended to enable all sites to be isolated. The storage partition class itself was extended to contain a label, which can be shown to the user, to identify the current storage partition. This value corresponds to the host name and is the same as the subfolder used for the storage partition, as seen on disk (illustrated in Figure 3).

User Interface: We implemented the user interface by extending the easily accessible site preference dialog of Chromium. The UI opens by clicking the page icon in the address bar as shown in Figure 4. This dialog is used for example to disable cookies or JavaScript for each individual site. The additional entry allows to set if a web site should be isolated or not. This makes the isolation feature very discoverable, because users already using this dialog to black- or whitelist access to cookies or plugins manually are already familiar with the usage.

User-Defined Cookie Timeout: In order to provide better user privacy, a user-defined cookie timeout was implemented. Instead of removing the cookies after the corresponding tab is closed, cookies will be deleted if they have not been read or changed for a user-defined period of time. Removing cookies automatically from tabs that are still open can result in a better protection, for example when the tabs have not been used for a long time. By giving users the option to limit the lifetime of cookies, previously saved for years, can now being deleted automatically. The per-site timeout can also be set to infinite, in order to white list important sites. A downside to this strategy is that the most popular and therefore the trackers with the biggest privacy impact will be the ones the least affected by this change. If third-party cookies are constantly used on web sites, they will not be deleted unless the user stops browsing for a while.

3.1   Storage Policy

With the isolation mechanism in place, a policy is required to decide which tab should map to which storage partition. Using the *host name* of the visited site as the storage partition name turned out not to be ideal, because it often results in bigger storage partitions then necessary. For example, *a.blogspot.com* and *b.blogspot.com* are both lumped together into *blogspot.com*, although they are *owned* by different persons. This is similar to some registrars allowing or forcing the registration of domains not directly under the Top- Level Domain (TLD), such as *bbc.co.uk*. This could allow dangerously far-reaching cookies to be set.

The solution is Mozilla's public suffix list[3], which is already integrated in Chromium. It can be used to find out the *top private domain* (the name used by Google's Guava library[4]), which is the largest part of the domain, where cookies are allowed to be set to. For example, *a.b.c.co.uk* has the top private domain *c.co.uk*.

In our policy the user can enable/disable the isolation for specific sites. Allowing more customization, such as to switch to arbitrary storage partitions is possible and more powerful, but also difficult to make usable. Another improvement that may have merit would be to allow making the storage partitions finer or coarser, i.e., changing how many subdomains are used to construct the storage partition name, like to allow to isolate *www.google.com* from *maps.google.com* or *play.google.com*. To implement this policy, the mechanism for the existing per-site preferences was reused.

Support of Complex Interaction: Considering realistic browsing sessions where users click links to other pages or where web sites communicate with other domains, we have to make sure not to break features provided by web sites, as soon as there are multiple storage partitions involved. In the next paragraphs we show how Site Isolation reacts to different browsing scenarios to make the site (or feature) usable.

Social Plugins: The variety of social plugins like Facebook *like* or Google *+1* buttons is an important use case for Site Isolation. Considering a *like* button on a web site the user has to perform a login at some point to use the service when clicking the button.

If this login is implemented as an overlay on the main site, which uses further iframe/AJAX techniques to log in without redirecting to a different site, this will work fine with Site Isolation. The address bar and therefore the storage partition was never changed, resulting in the user being logged in, but only for the actual site. If implemented as an iframe, as on most pages, the iframe runs in the same storage partition as the main page, but neither the plugin has access to the page it is embedded on nor has the page access to the plugin iframe's content. When

---

[3] http://publicsuffix.org
[4] https://code.google.com/p/guava-libraries/
   wiki/InternetDomainNameExplained

clicking, a new page is opened, avoiding trivial login phishing. This works also with Site Isolation, if the social plugin is implemented robustly. The following possibilities should cover most existing interaction strategies.

Simple Link: The social plugin always uses a link directly to the intended destination, regardless of any cookies set, for example linking to */like?page=a.com*. This works, because when clicking the link, the storage partition switches and the user is logged in (or will be prompted to login on the new page). Note, that switching or not the storage partition does not have any effect on redirects here, since there are no (cross-domain) redirects involved.

Login Detection: The social plugin detects that there is no login cookie in the currently active storage partition and behaves differently. It might create a link with enough information to redirect back to the original destination, e.g. instead of linking to /like?page=a.com, the link now points to */login?redirect=%2Flike %3Fpage%3Da.com*. The login page will be loaded with a different storage partition and can automatically redirect back to */like?page=a.com*. To avoid redirect problems, navigation events from inside frames could keep the currently active storage partition if the user wants to.

3.2   Storage Strategies

To avoid some of the potential web site breakage, there are many enhancements possible, but it is important to keep their side effects in mind, especially on usability.

Keep Storage Partition on Automatic Redirects: In the current implementation we only change the storage partition in response to user-initiated actions. This seems to be counter-intuitive, as generally more storage partitions lead to more isolation and privacy. However, this solution can isolate an *attack*, which can be used to achieve the effect of third-party cookies, even when they are disabled by the user. Keeping automatic redirects in the same partition will improve the functionality of web sites and is more user friendly, though users will have to login multiple times. Otherwise the user might get confused, about already being logged-in on another site despite using isolation. A disadvantage exists if URL shortener or alternative URLs like fb.com are used instead of the original domain. Here, the user would have to set cookies in both storage partitions.

Keep Storage Partition when Clicking Links Inside Frames: A different strategy is to ensure that pages linked from inside frames have access to the same set of cookies as the frame did. In other words, it means keeping the storage partition, if the redirect originated from inside a frame. It has similar usability problems as all stateful storage partition selection has, namely that visiting exactly the same URL can result in a different storage partition being used. But it avoids the issues of a frame showing a user to be logged in, but after clicking on a link the user is suddenly logged out. Conversely, it also avoids a frame showing a user to be logged out, then switching to logged in after clicking on a link. This might be a usability hindrance, since previously there was no additional login required. But it is also easy to construct an example where it is beneficial. Usually, it takes just a single misplaced click to *like* a page on a dubious web site accidentally. By keeping the storage partition when clicking links inside frames, users are not logged in on the first try to *like* a page. They can now decide to go back or log in, in which case future *likes* from the same page will work directly.

Manual Storage Partition Selection: In the case where automatic selection should not work, users can take control of how storage partitions are assigned. In the advanced user mode, the complete storage partition name can be specified on a per-site basis.

Whitelisting to Allow Communication between Sites: As an alternative the user can define which web sites should share a storage partition by selecting it in a list. The list is built from the set of domains referenced on the current page as resources and the domains it has links to. Here, the user can select sites to be part of the same storage partition.

Additional Link Open Modes: Additional options to open links that explicitly keep or change the current storage partition can help power users to manually work around defects or improve isolation further. These could be available in the context menu of links, as an additional entry, besides opening the link in a new tab, new window, or incognito window.

Keep Storage Partition Mode: Instead of using one of the mentioned modes, a simple switch to disable switching between storage partitions temporarily can help fixing webs site isolation problems. Activating the *storage partition lock* and reloading the page should fix problems caused by Site Isolation.

Multiple Incognito Profiles: The private browsing features that are integrated into most of the browsers, essentially provide a convenient isolation for two profiles. Both profile types are isolated from each other, meaning that web sites opened in the incognito profile cannot read cookies or local storage from the base profile and vice versa. Two sites opened in two incognito tabs, however, can freely share data. Unlike Site Isolation, it requires explicit manual intervention to activate and is less powerful. In Site Isolation we combined both strategies and enabled for each incognito window its own profile. Closing an incognito window would always cause the associated data to be deleted, where previously the data would only be deleted if it was the last incognito window.

Summary: The built-in storage partition mechanism of Chromium 29 was extended with different storages strategies to isolate Cookies, Web Storage, IndexDB and caches of each visited site automatically, based on its top private domain. The benefits from these changes are:

– Improved privacy by limiting cookie-based web tracking to each visited site. The effectiveness of the privacy benefits will be shown in section 5.2.
– Improved security by isolating web sites by default, making it harder for malicious sites to affect security-critical sites opened in the same browser. CORS, CSRF, and click-jacking attacks are no longer possible.
– Improved cookie management, while the locally stored data is split up per site and can be cleaned up selectively.
– Protection against stored or cached tracking mechanisms like Web Storage, IndexDB, ETags or cached images [10] [20].

4   Isolation Classifications

Site Isolation operates on different boundaries than previous isolation strategies. Compared to other isolation strategies Site Isolation has numerous benefits and advantages.

– Compared to *private browsing*, Site Isolation works for every site and is not limited to two partitions. In private browsing, trackers can still retrieve the data from different sites opened in a private browsing session [30]. Site Isolation prohibits this behavior by using a new storage partition (on disk) for each visited web site.
– In contrast to using *multiple browsers*, Site Isolation works automatically, so the user cannot forget to use it. Multiple browsers can be used to have different states for the same site, but this can be achieved more easily by using multiple incognito profiles, presented in Section 3.2.
– Site Isolation works orthogonal in respect to clearing site data (cookies). It partitions in space, while clearing cookies partitions in time. Both can be used complementary to improve privacy further.

Benefits Compared with Third-Party Cookie Blocking: Site Isolation prevents trackers from using a single identifier for users across their entire browsing session. Each time a new site is visited, the cookies will be read from a different, initially empty storage partition. Trackers can no longer follow the user around the web, because the user appears to be a different one on each site. As a result, Site Isolation has largely the same effect as blocking third-party cookies, despite being conceptually different. This results in the following advantages:

– Third-parties can easily check if third-party cookie blocking is enabled and try to use a different tracking method. Site Isolation behaves in this example as if third-party cookie blocking is disabled - an isolated cookie will be sent.
– Blocking third-party cookies can easily break web sites [11]. By not blocking cookies, Site Isolation helps to retain the web site functionalities.
– Site Isolation can interfere with multiple web sites in a similar manner, but only if they actively redirect the top level window. Only when the domain (host) in the address bar changes, there are observable effects. Redirections and links to the same domain and cross-site frames will always keep the

storage partition, meaning that no breakage can occur. Facebook Apps, for example, are always running in a third-party iframe on *facebook.com*. If they are using cookies, blocking third-party cookies will break them; with Site Isolation, they continue to work.

Positive Effects of Cache Isolation: The isolation of the browsing cache severely limits the scope of cache-timing attacks [31,32] that can be used to get information about the browsing history. However, it does not impact other kinds of timing attacks to steal history data, such as the recently discovered rendering time attack [33]. Isolating caches can also limit cache-based tracking methods where trackers try to store a random identifier in the browser cache and retrieve it from the cache later.

While Jackson et al. [34] propose a same-origin restriction for cache access forbidding third-party content to read from the cache, Site Isolation already eliminates most of the privacy issues. Because every site has its own cache, tracking is limited to each individual site. Just as with cookie-based tracking, trackers cannot re-identify a user on a different site by using cached data.

Summary: Site Isolation will provide a similar privacy benefit as blocking third-party cookies, but it is more practical, as it breaks less web sites and enables new privacy features like user-defined cookie timeout and multiple incognito windows. It can limit different tracking strategies, such as cache-based or using additional storage options besides cookies, for example Web Storage, IndexDB or tracking images. Additional security is provided by avoiding CORS, click-jacking, CSRF and limiting history hijacking, cache timing, and rendering engine hijacking, as already analyzed by Chen et al. [25]. With our implementation for Chromium by directly patching the source code Site Isolation is enabled for all web sites by default.

5   Evaluation

Demonstrating the performance of Site Isolation we conduct an analysis of the reliability, the effectiveness, and correct handling of real browsing behavior as well. To test the reliability of usual browsing we use a script to open 1.6 million pages automatically in the patched Chromium. The next aspect is effectiveness. Here, we compare this data in normal browsing mode and with Site Isolation. Lastly, it is also important to ensure that Site Isolation does not break any widely used web techniques that provide features to users, while still hindering web tracking. Automating such tests is hardly possible and depends on a solid definition of *break*, which is why general browsing examples were evaluated manually.

Test Environment: Chromium was compiled and executed on a Linux (Ubuntu 12.04 LTS) system. A Java-based control program started and managed multiple Chromium instances. The test machine, powered by an Intel Core i7, 16 GiB RAM and a 50 MBit/s Internet connection, handled 40 parallel browsers without problems and 75% CPU usage.

Automation Using a Control Program: A Java-based control tool starts and automates the Chromium instances. These are directed to load a custom extension, which sends the following data back and receives new commands:

— The list of all HTTP requests made by the current page, which are used later to identify the biggest privacy leaks, and the content of the address bar are sent whenever it changes. This is done to be able to handle redirects, which could lead to a different site, such as redirects between http:// and https://, subdomains or to forward to the corresponding country TLD.
— The list of all links on the loaded web page. To consider links that are added by other scripts while waiting for the page to be completely loaded, a configurable delay is introduced, before gathering links. We also filtered and validated this list to discard links like '#top' or invalid targets.
— The control tool replies with a random URI from this list. The extension now sends a click event to the associated DOM element, so onclick actions do not get skipped and the correct HTTP Referer is sent with the request.
— Navigation errors and timeouts are directly reported by Chromium to the control tool, and a random previously visited URI is opened.

After each browsing session, Chromium's internal databases in the profile directory are read out and all data is stored by the control tool in an SQLite database.

5.1   Biased Crawling

As entry points for the browsing sessions the Alexa global top 40.000 list is used. Each domain is then selected as the entry point of a browsing session. After opening the home page of each domain, the work flow described above applies, and selected links are clicked until 50 pages have been visited. At this point Chromium is closed, the data of the profile read out and added to the database. When finished, another session is started using an empty profile, until all domains have been visited.

In order to avoid problems with completely random link selected crawling and to produce data that is more useful the following mechanism was used:

1. Only pages that have not yet been visited in any other browsing session are considered for link clicks.
2. Domains that have been visited less often than others are favored, instead of selecting completely at random.
3. To avoid visiting a lot of obscure domains that users are not very likely to ever visit, domains ranking high in the Alexa list are favored.

The resulting list has the links to the least visited sites first, but unpopular sites are slightly disadvantaged. The goal of this mechanism was to visit more of the smaller sites at the expense of larger ones. As pre-tests have shown, completely random link selection resulted in visiting Twitter, Google and Facebook links in 30% of all crawled pages, which is unlikely to produce useful data.

Table 1. Top 10 domains referenced most from other pages

Domain	Present on pages (%)
google-analytics.com	(54.0 %)
doubleclick.net	(26.4 %)
facebook.com	(23.2 %)
google.com	(21.6 %)
gstatic.com	(16.4 %)
fbcdn.net	(14.2 %)
twitter.com	(12.1 %)
googlesyndication.com	(11.2 %)
scorecardresearch.com	(11.1 %)
googleapis.com	(7.8 %)

Table 2. Top 10 domains that had the most cookies set in browsing sessions

Domain	Sessions with Cookies
doubleclick.net	27,844 (97.8 %)
google.com	22,951 (75.6 %)
twitter.com	20,828 (92.2 %)
scorecardresearch.com	19,263 (98.7 %)
youtube.com	15,339 (96.9 %)
adnxs.com	13,417 (98.9 %)
quantserve.com	11,495 (97.3 %)
yahoo.com	10,951 (97.1 %)
addthis.com	10,377 (90.8 %)
yieldmanager.com	9,569 (84.8 %)

We implemented the biased link selection by choosing the index for the list as $i = \lfloor r^b \cdot n \rfloor$, with r being a random floating point number in the interval $[0.0, 1.0[$, b being the bias exponent, and n being the number of elements in the list. The value $b = 1.2$ is used to provide just a slight bias towards the first elements of the list. Note that the sheer number of links to a site still has a big impact on link selection.

The cumulative distribution of the domains visited is plotted as shaded green area in Figure 5. When compared to a real world page view, data by AOL users reformatted from [35], shown as rounded blue circles in Figure 5, has a similar shape, just with a less distinct curvature. For the purposes of this evaluation, this means the data obtained is somewhat realistic, but it is much more evenly spaced than expected from the biased settings. The curvature in the range from 0 to 50 page views can be explained by the 50 link click limit and corresponds to the part of the 40,000 sites, that received very few incoming links from other sites, or that were left early in the session because an external link was followed.

5.2   Crawling Results

To evaluate how Site Isolation performs, a few metrics are needed. Every request to an external domain is considered a tracker, except user initiated clicks, because they direct to a different storage partition. The more domains are referenced by web sites, the bigger their tracking potential. Table 1 shows the results of the 10 domains with the biggest tracking potential (out of 1,608,038 visited pages).

Tracking Data: The occurrence of external content and potential tracking by setting cookies is not a sufficient definition of tracking. To actually perform tracking, the trackers have to be able to re-identify users. They can either use static user data directly, such as the IP address or browser fingerprinting, or they have to store a custom identifier and retrieve it later.

In this section, the latter method is investigated. Table 2 shows the top 10 cookie setting domains. When comparing with Table 1, the Content Delivery

Network (CDN) domains are notably absent, because they do not set any cookies. Also note that *google-analytics.com*, which has by far the biggest tracking potential, in fact does not even set a single tracking cookie for its domain. Instead, it uses first party cookies on each website it is embedded. To find out which of these cookies are used (or are usable) for tracking, we calculate the information content of cookies.

In theory, just 33 bits are enough to uniquely identify 8.5 billion people. To calculate the information content of a web site's cookies, Shannon's information entropy [36] was used. The entropy of the cookie data is equivalent to the number of bits required to encode one symbol (i.e. a byte) of the cookie data and can be calculated using $H = -\sum p_i \log_2 p_i$. Multiplying the entropy H by the total number of symbols results in the size that the data can be compressed to when using a Huffman coding [37]. In order to avoid over-interpreting cookies with a high information content, like web sites storing settings with the similar data for all users, the data gathered from multiple sessions can be used to differentiate between random and highly structured cookies. This can be performed by calculating the information content of the concatenation of a web site's complete set of cookies in all sessions.

The actual string that will be analyzed is a concatenation of the sorted cookie names followed by the sorted cookie values (skipping duplicates). The benefits for this method are:

- All cookies are concatenated because information might be split across multiple cookies.
- Names are included because information may only reside in the cookie names.
- Identical names and values are included only once, to get a smaller size estimation from Huffman coding, which does not check for identical blocks in the uncompressed data.
- Almost identical names and values will not compress well when just using Huffman coding. To avoid this problem a state of the art compression algorithm is used in addition.
- To improve compression, the names and values are sorted to get similar names/values next to each other.

The additional compression algorithm used is LZMA2 [38], because of its great compression ratio [39] and because it is based in a different principle than Huffman-coding. In the calculation LZMA2 tends to win over Huffman-Coding in 39% of the cases, typically when data is large enough to offset the overhead, that is ignored for Huffman-coding. The lower estimate of both algorithms is then used as an estimation of the information content, and if a suspected tracker domain has a very low content, it is analyzed manually. Table 3 summarizes the findings, showing that all of the heavily referenced and cookie setting domains can use their cookies to track users (high information bit ratio).

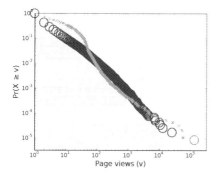

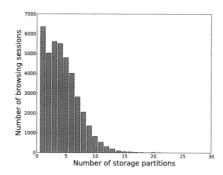

Fig. 5. Pages crawled (crosses) / viewed by AOL users (blue circles)

Fig. 6. Storage partition usage in browsing sessions

Isolation Effectiveness: Site Isolation impacts the previously gathered data in a number of ways, by not using the global profile to store data but instead storing it on a per-site basis. Each of the 40,000 cookie stores (sessions) can now have an unlimited amount of storage partitions. The total number of distinct storage partitions seen across all sessions was 94,701. But since not every storage partition is present in every session, there were only 176,330 non-empty storage partitions in total. The main effect of Site Isolation is that tracking cookies are not global to the browsing session and are instead local to each storage partition. Trackers will assign multiple different identities per session, making the tracking effectively limited to a single site, no more than what would be possible using completely passive server logs or using first-party cookies. The number of storage partitions per browsing session is visualized in Figure 6. The statistics indicate that on average the storage partition is switched every 11.2 pages. As each switch means all trackers up to this point will continue with an empty set of data, this sets the limit for how long users can be tracked. An interesting and unexpected result is the high percentage of empty storage partitions. These numbers, shown in Table 3 were generated by calculating the cookie statistics based not on the whole browsing sessions, but based on the individual storage partitions.

The *Cookies in Sessions* metric (from Table 2) was supplemented by *Cookies in Storage Partitions* and the *Pages tracked* metric was recalculated. While the absolute number of used cookie stores is always higher in the storage partition case, the total number of cookie stores where the domains were referenced is much higher still. This results in a comparatively huge number of storage partitions that do not have any cookies, while many trackers previously had cookies in nearly 100% of sessions. This has a corresponding drastic effect on the pages' tracked metric that fell immensely, indicating that Site Isolation can completely avoid various tracking mechanisms and is not merely reducing it to within-site tracking.

Table 3. Effects of storage partitions on the top trackers. The pages tracked metric dropped by 44 %, compared with not using Site Isolation.

Domain	Cookies in Sessions		Pages tracked	Cookies in S. Partitions		Pages tracked*	Inf. bits
doubleclick.net	27,844	(97.8 %)	414,808	30,104	(51.2 %)	217,320	69
google.com	22,951	(75.6 %)	263,019	24,742	(35.4 %)	123,170	375
twitter.com	20,828	(92.2 %)	178,853	22,624	(49.6 %)	96,178	143
scorecardresearch.com	19,263	(98.7 %)	175,727	20,576	(62.4 %)	111,169	35
facebook.com	8,974	(32.0 %)	119,628	9,390	(15.2 %)	56,743	97
quantserve.com	11,495	(97.3 %)	99,217	12,280	(66.8 %)	68,096	54
adnxs.com	13,417	(98.9 %)	93,785	14,117	(72.9 %)	69,167	357
addthis.com	10,377	(90.8 %)	80,840	10,890	(64.3 %)	57,289	153
youtube.com	15,339	(96.9 %)	56,598	16,012	(66.7 %)	38,929	272
baidu.com	2,152	(95.3 %)	47,923	2,606	(45.6 %)	22,939	104

6   Discussion

Using Site Isolation brings a slightly higher resource usage on disk space, CPU, RAM and bandwidth. As already analyzed by Chen et al. [25], showing that visiting 12 sites in isolated storage partitions instead of a single profile increased the disk usage 4.5-fold, while RAM usage grew by just 1 MB. As our experiments confirm, no noticeable performance loss could be noticed while using Site Isolation on the test computer.

For privacy reasons we advise to use TOR as an addition to Site Isolation, to prevent sending the actual client IP address, since this could reveal the user even when multiple web site containers are used.

Currently the storage partition name is determined by the address bar URI. One option that could improve compatibility and isolation, is to make the selection stateful and consider how the browser arrived at the current page. If a normal link was clicked, the storage partition should change according to the address bar, as usual. If a HTTP, HTML or JavaScript based redirect was used, however, keeping the last storage partition can be beneficial. Reducing the number of storage partition switches cannot make compatibility worse, but tends to reduce isolation, as no switches also mean no isolation. But these redirects are often used to set cookies, giving web sites an easy way of setting identical cookies in multiple storage partitions. Another possibility is to use more information for selecting the storage partition name, for example by using numerical ids and an associated database. Multiple sites could then be mapped into a single storage partition, to allow a user-configurable whitelist. Any changes that affect the behavior of the storage partitions itself, for example allowing sites to read from multiple storage partitions at the same time, are not considered.

Flash usage is widespread in the current web but dropping in recent years. Unfortunately Chromium does not yet confine the Flash plugin to storage partitions yet, which makes data exchange across storage partitions possible using Flash LSOs.

However, instances of Flash LSO based tracking are rare, and in our study only 1,886 domains (0.11%) in total used Flash LSOs. Comparing the top domains using Flash LSO to the tables of sites present on most pages, or the domains having cookies in most sessions, no overlap domains (in the top 10) can be found. We are not aware of any conceptual issues with integrating the PPAPI[5] Flash plugin into the Site Isolation concept, but actually implementing it was left as future work.

7   Conclusion

The data shows that Site Isolation has a huge effect on the abilities of web trackers to follow users across the web. It limits their ability to re-identify users switching to a different site, making the tracking much less invasive. On average, the storage partition was switched every 11.2 pages, providing credibility to that claim. On top of that, the number of pages that can still be tracked, was shown to be reduced by 44%, compared the number of pages tracked by the 10 most prevalent trackers when not using Site Isolation. This was unexpected and suggests that external trackers do not set tracking cookies on all occasions, which helps Site Isolation achieve this huge reduction in the pages' tracked metric. The findings confirm the tracking statistics of Roesner et al. [1], that the biggest trackers are now present on over 20% of web pages. As the data shows, Facebook has stopped indiscriminately tracking even non-users [40]. It set cookies in just 32.0% of sessions it was referenced, dropping further to having cookies in only 15.2% of storage partitions.

The evaluation of countermeasures against Site Isolation in Section 6 demonstrate that the currently known evasion techniques are impractical. Neither using Flash LSOs nor redirects to set first party cookies are well suited for avoiding Site Isolation on a massive scale. Finally, compared with blocking third-party cookies, Site Isolation is more user friendly and does not depend on the definition of what exactly constitutes a third-party cookie. Combined with the security benefits of preventing CORS, CSRF, click-jacking, and limiting the effects of cache-based tracking, cache-timing attacks as well as rendering engine hijacking, Site Isolation is a promising approach for improving user privacy and security.

Acknowledgment. This work was supported by the *European Center for Security and Privacy by Design (EC SPRIDE),* funded by the German Federal Ministry of Education and Research (BMBF), and the *Center for Advanced Security Research Darmstadt (CASED),* funded by the LOEWE program of the Hessian Ministry for Science and the Arts (HMWK).

References

1. Roesner, F., Kohno, T., Wetherall, D.: Detecting and Defending Against Third-Party Tracking on the Web. In: Usenix NSDI (2012)
2. Krishnamurthy, B., Wills, C.: Privacy diffusion on the web: A longitudinal perspective. In: WWW (2009)

---

[5] https://code.google.com/p/ppapi

3. Baviskar, S., Thilagam, P.S.: Protection of Web User's Privacy by Securing Browser from Web Privacy Attacks. IJCTA (2011)
4. nugg.ad AG: Predictive Behavioural Targeting (2014), http://nuggad.net/en/solutions/predictive-behavioural-targeting.html (accessed on August 13, 2014)
5. Castelluccia, C., Ali Kaafar, M., Tran, M.-D.: Betrayed by Your Ads! Reconstructing User Profiles from Targeted Ads. In: Fischer-Hübner, S., Wright, M. (eds.) PETS 2012. LNCS, vol. 7384, pp. 1–17. Springer, Heidelberg (2012)
6. Ohm, P.: Broken Promises of Privacy: Responding to the Surprising Failure of Anonymization. UCLA Law Review (2009)
7. Behar, R.: Never Heard of Acxiom? Chances Are It's Heard of You. How a little-known Little Rock company—the world's largest processor of consumer data—found itself at the center of a very big national security debate (2004), http://money.cnn.com/magazines/fortune/fortune_archive/2004/02/23/362182/index.htm (accessed on October 25, 2013)
8. Communications Consumer Panel: Online Personal Data: the Consumer Perspective. Technical report (2011), http://www.communicationsconsumerpanel.org.uk/Online%20personal%20data%20final%20240511.pdf
9. Steel, E., Fowler, G.A.: Facebook in Privacy Breach (2010), http://online.wsj.com/article/SB10001424052702304772804575558484075236968.html (accessed on October 25, 2013)
10. Mayer, J.R., Mitchell, J.C.: Third-Party Web Tracking: Policy and Technology. In: IEEE Symposium on Security and Privacy (2012)
11. Leon, P.G., Ur, B., Balebako, R., Cranor, L.F., Shay, R., Wang, Y.: Why Johnny Can't Opt Out: A Usability Evaluation of Tools to Limit Online Behavioral Advertising. In: CHI (2012)
12. Scientist, C., Italia, T.: Flash Cookies and Privacy II: Now with HTML5 and ETag Respawning. World Wide Web Internet and Web Information Systems (2009)
13. Eckersley, P.: How unique is your web browser? In: Atallah, M.J., Hopper, N.J. (eds.) PETS 2010. LNCS, vol. 6205, pp. 1–18. Springer, Heidelberg (2010)
14. Anderson, N.: Firm uses typing cadence to finger unauthorized users (2010), http://arstechnica.com/tech-policy/2010/02/firm-uses-typing-cadence-to-finger-unauthorized-users (accessed on October 30, 2013)
15. Mowery, K., Shacham, H.: Pixel Perfect: Fingerprinting Canvas in HTML5. In: W2SP. IEEE Computer Society (2012)
16. Nikiforakis, N., Kapravelos, A., Joosen, W., Kruegel, C., Piessens, F., Vigna, G.: Cookieless Monster: Exploring the Ecosystem of Web-based Device Fingerprinting. In: IEEE Symposium on Security and Privacy (2013)
17. Acar, G., Juarez, M., Nikiforakis, N., Diaz, C., Gürses, S., Piessens, F., Preneel, B.: FPDetective: Dusting the web for fingerprinters. In: CCS (2013)
18. Tran, M., Dong, X., Liang, Z., Jiang, X.: Tracking the trackers: Fast and scalable dynamic analysis of web content for privacy violations. In: Bao, F., Samarati, P., Zhou, J. (eds.) ACNS 2012. LNCS, vol. 7341, pp. 418–435. Springer, Heidelberg (2012)
19. Bau, J., Mayer, J., Paskov, H., Mitchell, J.: A Promising Direction for Web Tracking Countermeasures. In: W2SP (2013)
20. Siddiqui, M.S.: Evercookies: Extremely persistent cookies. IJCSIS (2011)
21. Eyeo GmbH: Allowing acceptable ads in Adblock Plus (2014), https://adblockplus.org/en/acceptable-ads (accessed on August 13, 2014)

22. Bilton, R.: Ghostery: A Web tracking blocker that actually helps the ad industry (2012), http://venturebeat.com/2012/07/31/ghostery-a-web-tracking-blocker-that-actually-helps-the-ad-industry (accessed on August 13, 2014)
23. Witte, D.: (doublekey) Key cookies on setting domain * toplevel load domain (2010), https://bugzilla.mozilla.org/show_bug.cgi?id=565965 (accessed on July 10, 2014)
24. Perry, M.: Apply third party cookie patch (2011), https://trac.torproject.org/projects/tor/ticket/3246 (accessed on July 10, 2014)
25. Chen, E.Y., Bau, J., Reis, C., Barth, A., Jackson, C.: App Isolation: Get the Security of Multiple Browsers with Just One. In: CCS (2011)
26. Reis, C., Gribble, S.D.: Isolating web programs in modern browser architectures. In: EuroSys 2009 (2009)
27. Wang, H., Grier, C., Moshchuk, A.: The Multi-Principal OS Construction of the Gazelle Web Browser. In: Usenix Security Symposium (2009)
28. Jackson, C., Bortz, A., Boneh, D., Mitchell, J.C.: Protecting browser state from web privacy attacks. In: WWW (2006)
29. Grier, C., Tang, S., King, S.T.: Secure Web Browsing with the OP Web Browser. In: IEEE Symposium on Security and Privacy (2008)
30. Aggarwal, G., Bursztein, E., Jackson, C., Boneh, D.: An analysis of private browsing modes in modern browsers. In: Usenix Security Symposium (2010)
31. Felten, E., Schneider, M.: Timing attacks on Web privacy. In: CCS (2000)
32. Weinberg, Z., Chen, E., Jackson, C.: I Still Know What You Visited Last Summer: Leaking Browsing History Via User Interaction and Side Channel Attacks. In: IEEE Symposium on Security and Privacy (2011)
33. Stone, P.: Pixel Perfect Timing Attacks with HTML5. White Paper (2013), http://contextis.co.uk/files/Browser_Timing_Attacks.pdf
34. Jackson, C., Bortz, A., Boneh, D., Mitchell, J.: Protecting Browser State from Web Privacy Attacks. In: WWW (2006)
35. Clauset, A., Shalizi, C.R., Newman, M.E.J.: Power-Law Distributions in Empirical Data. SIAM Rev. (2009)
36. Shannon, C.E.: A Mathematical Theory of Communication. The Bell System Technical Journal (1948)
37. Huffman, D.A.: A Method for the Construction of Minimum-Redundancy Codes. Institute of Radio Engineers (1952)
38. Pavlov, I.: LZMA specification (2013), http://dl.7-zip.org/lzma-specification.zip (accessed on October 30, 2013)
39. Morse Jr., K.G.: Compression Tools Compared. Linux J. (2005)
40. Eferati, A.: 'Like' Button Follows Web Users (2011), http://online.wsj.com/news/articles/SB10001424052748704281504576329441432995616 (accessed on October 30, 2013)

Comprehensive Behavior Profiling
for Proactive Android Malware Detection

Britton Wolfe[1], Karim O. Elish[2], and Danfeng (Daphne) Yao[2]

[1] Information Analytics and Visualization Center,
Indiana Univ.-Purdue Univ. Fort Wayne (IPFW),
2101 E. Coliseum Blvd., Fort Wayne, IN 46825 USA
wolfeb@ipfw.edu
[2] Department of Computer Science,
Virginia Tech., 2202 Kraft Dr., KWII, Blacksburg, VA 24060 USA
danfeng@vt.edu, kelish@vt.edu

Abstract. We present a new method of screening for malicious Android applications that uses two types of information about the application: the permissions that the application requests in its installation manifest and a metric called percentage of valid call sites (PVCS). PVCS measures the riskiness of the application based on a data flow graph. The information is used with machine learning algorithms to classify previously unseen applications as malicious or benign with a high degree of accuracy. Our classifier outperforms the previous state of the art by a significant margin, with particularly low false positive rates. Furthermore, the classifier evaluation is performed on malware families that were not used in the training phase, simulating the accuracy of the classifier on malware yet to be developed. We found that our PVCS metric and the SEND_SMS permission are the specific pieces of information that are most useful to the classifier.

Keywords: android, malware, machine learning, mobile security.

1   Background

The Android operating system continues to gain market share among smart phone users across the world. At the end of 2013, it had reached over 50% market share in the United States and Great Britain and over 70% in Germany and China [22]. In all four countries, Android gained more than 4% market share over the previous year. With an increase in market share also comes an increase in the attention of malware developers. There are hundreds of malicious applications in the official and alternative Android marketplaces [15]. This work presents a new way of detecting malicious Android applications, resulting in higher accuracy than previous methods.

Our technique combines two very different types of information about Android applications. The first one is the set of permissions that the application requests when it is installed (Section 2.2). The second, percentage of valid call

S.S.M. Chow et al. (Eds.): ISC 2014, LNCS 8783, pp. 328–344, 2014.

sites (PVCS) (Section 2.3), is a measure of an application's riskiness, calculated from its data dependence graph. While each kind of information is useful on its own, when combining them, we are able to detect over 83% of malware with only 1% false positive rate, a significant improvement over previous work (Section 3.2).

1.1 Malware Classification

Traditional methods for detecting malware rely upon recognizing a specific signature that has been previously identified as belonging to a specific, known malware. A limitation of this approach is that it cannot recognize previously unknown malware. In contrast, heuristic-based or machine learning methods learn general rules and patterns from examples of malware and clean files, which are then used to automatically recognize previously unseen malware. The capability of identifying unseen new malware is important for realizing proactive defense of the mobile infrastructure.

When using machine learning, each application is represented as a vector of feature values, which can come from dynamic analysis or static analysis. Dynamic analysis involves running the application in a sandbox and recording information about its behavior, such as battery and network usage [3]. Static analysis uses features extracted without running the application, such as the list of permissions that the application requests upon installation or information about the control flow of the program (e.g., 15). Both types of analyses are useful and provide complementary insights about applications' behaviors. Our work uses static analysis.

1.2 Related Work: General Security

Machine learning techniques have been widely adopted in the computer security literature since the work by Lee et al. [12]. Equipped with domain knowledge, the methods extract domain specific features based on empirical observations of malicious programs or traffic patterns.

For example, solutions described by Cova et al. [6] use binary classification techniques to identify malicious Javascript code on the web. The features they extracted from malicious code include the presence of redirection and obfuscation. Xie et al. [24] used a Bayesian network to infer abnormal network traffic patterns. Besides classifying programs and network traffic, learning-based security research also includes database intrusion detection [19] and SMS/social network spam detection [20].

1.3 Related Work: Android Malware

Researchers have applied both static [2, 4, 9] and dynamic [13] approaches to malware detection on Android devices. The approaches differ in the features extracted and the classification algorithms employed, leading to varying degrees of success. The data sets employed by the researchers were also of different qualities, ranging from just a handful of malware that the researchers created themselves up to data sets with hundreds of examples pulled from live marketplaces.

Schmidt et al. [17] used a data set of ELF files. It consisted of approximately 240 malware which targeted Linux systems (i.e., not specifically designed for the mobile ARM architecture), and less than 100 Linux system commands from an Android device. They used static analysis to construct binary features, one for each function called by any file in the data set. That information was extracted using `readelf`. They applied three classifiers (rule inducer, nearest neighbor, and decision tree) to a few subsets of the features. All of their configurations that achieved 80% or higher detection rate (i.e., true positive rate) also suffered a false positive rate over 10%.

Burguera et al. [5] proposed the CrowdDroid system for identifying a specific type of malware: repackaged malware. Repackaged malware is created by taking a benign application and repackaging it with additional malicious code. The CrowdDroid approach uses dynamic features. A central system collects the frequencies of several system calls from several users running the application on different devices. It then uses k-means clustering with $k = 2$ to cluster the results, with the goal of separating the benign instances of the application from the malicious (repackaged) instances. Their experiments used only four author-created malware and two real malware. While CrowdDroid successfully identified all of the author-created malware, it produced a 20% false positive rate on one of the two real malware (the more substantial application of the two).

Shabtai et al. [18] also used a small number of fabricated malware (i.e., four applications) to test their Andromaly system, due to a lack of real malware at that time. They used 88 hand-designed dynamic features, including memory page activity, CPU load, SMS message events, network usage, touch screen pressure, binder information, and battery information, among others. They pared down the features using the information gain and Fisher scores for each individual feature, selecting the features with the best scores. Then they applied several classifiers: decision trees, naïve Bayes, Bayes nets, histograms, k-means, and logistic regression. Even on a synthetic data set, their best configuration—naïve Bayes after using Fisher score to select 10 features—still had over 10% false positive rate, with approximately 88% accuracy.

Later research had the advantage of access to more actual malware. Like Andromaly, Amos et al. [3] used hand-selected dynamic features (e.g., memory, CPU, binder information), but evaluated performance on a larger data set: 1330 malware and 408 benign applications. They compared random forests, naïve Bayes, multilayer perceptrons, Bayes nets, logistic regression, and decision trees. As with the previously mentioned work, their methods suffer from a high false positive rate: over 15% for all of their configurations. Their accuracy was 95% on new traces from applications included in the training set, but no higher than 82% on traces from applications that were not included in the training set.

Sanz et al. [16] used a simple feature set: the permissions and features of the device that the application requests upon installation. They are listed in the downloaded application's manifest, so these features are extracted with static analysis. Their data set consisted of 357 benign and 249 malicious applications. They tried several classifiers: logistic regression, naïve Bayes, Bayes nets, support vector machines with polynomial kernel, k-nearest neighbors, decision trees, random trees, and random forests. As with other work, the false positive rate remains stubbornly high: their false positive rate is never below 11%, and even that classifier only detects 45% of the malware. The best overall accuracy was 86%, using random forests.

Sahs and Khan [15] tried a substantially different approach, training a 1-class support vector machine on benign applications in order to detect malware as anomalies. They used a custom kernel that combines permissions information with control flow graph information, both of which come from static analysis. However, their false positive rate is nearly 50%, making their method untenable.

Wu et al. [23] report much better results—false positive rate below 1% and accuracy of 98%—but they only report on the training set error. Without evaluating on a testing set or using cross-validation, the good results are likely due to overfitting[1] instead of a model that generalizes well to unseen malware.

Peng et al. [14] explored the use of different probabilistic generative models for scoring the risk of different Android applications. They used the permissions requested by the application as the binary features (i.e., static analysis). Each model estimates the probability that an application would request those permissions. Each model is trained on several thousand applications from the marketplace, which the authors assume to all be benign. When a new application requests permissions that have a low probability according to the model, it is flagged as unusual or high risk. The probabilistic models range in complexity from simple naïve Bayes through a hierarchical mixture of naïve Bayes models. They used 378 malware applications mixed with different subsets of the benign set to calculate cross-validation error. The hierarchical mixture of naïve Bayes models performs the best, detecting 78% of malware with a false positive rate of 4%. The simpler models also do well, achieving close to the same results.

1.4   Receiver Operating Characteristics (ROC) Curve

For classifiers that produce probability estimates—e.g., there is a 72% chance this application is malware—instead of just a yes/no decision, the aggressiveness of the overall system can be adjusted without modifying the classifier itself. To do this, one simply adjusts the probability threshold at which an application is declared malware. When the threshold is 0.0, everything is declared malware (i.e., the most aggressive classifier). On the other extreme, when the threshold is 1.0, nothing is declared malware. The default threshold is 0.5, picking the

---

[1] Overfitting is a common problem in machine learning applications where the algorithms memorize characteristics specifically of the training examples instead of general trends. When evaluated on the training data, the algorithm uses those characteristics to re-recognize *the same examples*. This gives a false sense of accuracy, since the real evaluation should be on examples other than the training examples.

Table 1. Summary of related work reporting moderate or low false positive rates. TPR numbers read from a plot are approximate, indicated by ≈.

Citation	AUC	TPR values for FPR $\leq x$					Limitations
		$x = 0.01$	$x = 0.02$	$x = 0.05$	$x = 0.10$	$x = 0.15$	
Schmidt et al. [17]	-	0.77	-	-	0.99	1.00	ELF files only
Shabtai et al. [18]	0.913	-	-	≈ 0.967	-	0.847	author-created malware
Sanz et al. [16]	0.920	-	-	-	-	0.50	
Peng et al. [14]	0.954	< 0.5	≈ 0.59	≈ 0.79	≈ 0.87	≈ 0.90	

most likely category according to the classifier. This ability is important for malware classification because in different situations, different levels of aggressiveness would be appropriate. If one wants very high security, one might pick an aggressive classifier that can detect all of the malware, but also mistakenly flags several benign applications as malware (i.e., high false positives). On the other hand, if the classifier is used as part of a larger security suite, a less aggressive classifier would be preferred, producing fewer false positives.

While there are several ways to measure the quality of a classifier—accuracy, false positive rate, precision, etc.—the receiver operating characteristic curve (ROC curve) illustrates the trade off between false positives and false negatives as one moves from a conservative classifier (i.e., nothing is malware) to an aggressive classifier (i.e., everything is malware). (See Figure 1 for examples of ROC curves.) One can examine the curves in several different ways. The most concise is to calculate the area under the curve (AUC), which summarizes the quality of the classifier at all different levels of aggressiveness. An AUC of 1.0 is optimal, representing a perfect classifier.

One can also examine specific points on the ROC curve to find what fraction of malware can be detected—the true positive rate or TPR—when limiting the false positive rate (FPR) below some threshold. For example, one might want no more than 2% FPR in a particular system, so looking at the TPR value on the ROC curve when FPR=0.02 will estimate the detection rate of such a system.

1.5 Summary of Related Work

Table 1 summarizes the results from previous work. The table lists the TPR for different values of the FPR, along with the AUC. When the ROC is not reported in the given work, the closest FPR column is filled in. The best classifiers from each publication that meet the FPR limit are reported, and the best AUC is reported. Thus, the different columns may represent different classifiers. Publications where all of the FPR values were above 0.15 are omitted. As noted in Section 1.3, there are many factors that influence the results, such as the makeup of the data set and whether or not applications as a whole are classified (static analysis) or execution traces from applications are classified (dynamic analysis). Thus, this table alone is an oversimplification of the results, but it highlights the difficulty in achieving decent detection rates at FPR of 0.02 or less.

2   Methods

Our work utilizes a static analysis feature, called *percentage of valid call sites (PVCS)* [8], described in Section 2.3. We also examine the most common features used in previous static analysis work: the permissions that the Android application requests upon installation (Section 2.2). We compare classifiers' performance when trained using PVCS with classifiers trained using the permissions features, as well as a combination of the two. Adding the PVCS information to the permissions information leads to classifiers that are substantially better than the previous state of the art (Section 3.2).

2.1   Learning Process Overview

For each of the three feature sets—permissions, PVCS, and the combination of both—we use the following learning process. To begin, the vector of feature values for each application in the training set is calculated. Then the set of vectors is given to several classifier learning algorithms. We compared five classifiers: support vector machines (SVMs), random forests, naïve Bayes, k-nearest neighbors (KNN), and boosted decision trees (J48 with Adaboost). We used the Weka implementation of each classifier [10]. These classifiers represent fundamentally different approaches to classification, each of which has its own strengths and weaknesses. Thus, we evaluate all of these classifiers to find the best kind for classifying Android malware.

Three of the classifiers have hyperparameters that the user selects to tune the classifier performance. We used 10-fold cross-validation on the training set to pick these parameters, selecting the values that led to the highest cross-validation AUC. For support vector machines, we explored values of $C \in \{10^{-3}, 10^{-2}, 10^{-1}, ..., 10^3, 10^4\}$ and $\gamma \in \{2^{-1}, 2^{-2}, 2^{-3}, ..., 2^{-10}\}$. For the random forests, we explored numbers of trees in $\{16, 32, 64, 128, 256\}$. For k-nearest neighbors, we picked the best value of k from 1 through 7.

After picking the hyperparameters, there is one trained classifier of each type for each feature set (15 total). These classifiers are then evaluated on the test set, including the malware from families not present in the training set. Section 3.2 presents the results of this testing evaluation, but first we describe the feature sets in more detail.

2.2   Permissions Bits

Each Android application is required to list in its installation manifest the permissions that it will need at any point during its execution. That list of permissions is used by the Android system to restrict or allow access to system-wide resources like network connections, contact list, boot notifications, sending or receiving SMS text messages, etc. Thus, the permissions list indicates what system resources an application is allowed to use. In addition to the standard Android system permissions, users can define their own permissions. For example, there are permissions specific to particular hardware manufacturers.

For each application, we compute a vector of bits, each of which represents a particular permission. A 1 indicates that the application requests that permission and a 0 indicates that it does not. Sanz et al. [16] use the same encoding of permissions information as their feature set. Sahs and Khan [15] use the same encoding for system permissions, combining with their own features from user-defined permissions and control flow graphs. Peng et al. [14] also use the same encoding, but only for the 20 most frequently requested permissions. We include all the standard Android system permissions as well as any user-defined permissions.

2.3   Percentage of Valid Call Sites

In addition to using permissions bits, we utilize a statistic of Android applications called *percentage of valid call sites (PVCS)* [8]. Android applications are characterized by intensive user interaction. Researchers [7, 8] found that the majority of the benign applications require user interaction in order to initiate sensitive operations like network access. On the other hand, malicious applications require little to no user interaction before executing sensitive operations. Hence, the PVCS metric is designed to capture the dependence relations between user triggers and sensitive operations. This metric represents the degree of sensitive operations that are authorized by the user. It is a fine-grained metric which provides more in-depth behavior information about the applications, as opposed to Android permission information which does not capture the applications' behavior.

To define the PVCS metric, we first define some other terms: operation, call site, and valid call site. An *operation* is defined as a function call related to network operations, file operations, and telephony services in an application. For example, operations include APIs related to sending/receiving network traffic, sending text messages, and accessing private information such as location information. These are sensitive API calls that we want to examine in order to detect malicious behavior.

A *call site* is defined as one instance of an operation. Each API operation may have one or more call sites in an application. Each call site is checked to determine if it is triggered by user actions by constructing a data dependence graph. A *data dependence graph (DDG)* is a well-known program analysis technique which represents data flows through a program [11]. The DDG is a directed graph representing data dependence between program statements, where each node represents a program statement, and an edge represents the data dependence between two nodes.

Android has a special mechanism called *Intent* to provide communication between applications or components (Activity, Service, Receiver). Therefore, the DDG needs to be augmented in order to obtain the complete set of operations that depend on user triggers through Intent. The Android Intent-based dependence analysis tracks the control flow between Intent-sending methods in intra- and inter-application communication. This Intent-specific control flow analysis helps to bridge disjoint graph components and captures the data dependence relations across multiple Android components.

The DDG is constructed for each application by utilizing the libraries provided by Soot [1], a static analysis toolkit for Java. Furthermore, the constructed DDG is augmented with the Android Intent-based dependence analysis to get one complete, connected DDG. More implementation details can be found in [8].

After building the DDG, each call site is labeled as valid or not valid. The call site is called *valid* if there is a valid path in the data dependence graph from a user trigger to the call site.

The PVCS metric is defined as follows [8]:

Definition 1. *Percentage of Valid Call Sites* $PVCS \in [0\%, 100\%]$ *of an application is the percentage of valid call sites out of the total number of call sites across all the operations. Let k_i be the number of* valid *call sites for operation i and let l_i be the number of total call sites for operation i. Given the n operations used in an application, PVCS is computed as*

$$PVCS = \frac{\sum_{i=1}^{n} k_i}{\sum_{i=1}^{n} l_i} \qquad (1)$$

For example, assume that there are 10 call sites in an application. If 9 out of 10 call sites are triggered by the user, the PVCS value of the application is 90%. A high PVCS is desirable, as it generally indicates that there are not sensitive operations going on without the user's knowledge.

3   Experiments

3.1   Data Set

We used a collection of 3869 Android applications, which consists of 1433 malicious applications and 2436 benign applications. The malicious Android applications were collected from the VirusShare repository[2] and the Android Malware Genome Project[3] [25]. The benign Android applications are free, real-world applications collected from the Google Play market, covering various application categories. These free applications include different levels of popularity, as determined by the user rating scale. We used two existing malware detection tools [7, 21] to scan the collected free applications. Applications that did not trigger any alerts in those tools are kept in the benign set.

The applications were partitioned into training and test sets. For the clean applications, a random 20% were selected for the test set, with the remainder going into the training set. The malicious applications were split based on the malware family. For each family with just one application, that application was randomly assigned to training or testing. For all the other families, at least one application was assigned to the test set. A few families of varying sizes

---

[2] http://virusshare.com/
[3] http://www.malgenomeproject.org/

were completely held out of the training set (Table 5), so we could evaluate the algorithm's accuracy on completely unseen malware families. For each of the other malware families, 20% of the applications were selected for the test set, with the remainder going into the training set. In the end, there were 1948 benign and 1066 malicious applications in the training set, and there were 488 benign and 367 malicious applications in the test set.

3.2   Classification Results

After picking the best classifier of each type (Section 2.1), we evaluated their accuracies on our test set (i.e., applications that were not used at all in the training or parameter selection). The experiments answer the following questions:

– Which of the three feature sets (permissions, PVCS, or both) is best for malware screening?
– When using the best feature set, which classifier type is best for malware screening?

We ran two evaluations of the classifiers, using two different subsets of the test data. They each use all the benign applications in the test set, but they differ in which malware from the test set is used. The "unfamiliar" comparison only uses the malware from families that were *not* represented in the training data. The "familiar" comparison (Section 3.4) uses the other malware (i.e., their families were represented in the training data).

Of the two, the unfamiliar is more important. In reality, we want to detect new malware families that have been created after training on existing malware families. Table 2 presents the results, including AUC and the true positive rate (TPR) for different levels of false positive rate (FPR). The corresponding ROC curves for the best performers are plotted in Figure 1.

While different classifiers perform the best at different FPR levels, **all of the best performers use both permissions bits and PVCS**. That is, for each FPR level, the model with the best TPR is always one that uses both permissions bits and PVCS. Furthermore, the model with the best AUC also uses the combined feature set.

For practical use, an FPR of even 5% is too high, as it would flag one out of every 20 clean applications as malicious. Thus, the **boosted decision trees classifier trained upon permissions and PVCS is the best option for detecting malware.** It has the highest AUC (0.9850) and the highest TPR for both the FPR=0.01 and FPR=0.02 levels, detecting 83.75% of the malware from unfamiliar families at the FPR=0.01 level. This is in stark contrast to previous work on malware screening of Android APKs, where the TPR at FPR=0.01 is less than 50%, even when testing on malware families used in the training (Table 1, Peng et al. [14]).

Looking at Figure 1 and Table 2, one can see that for FPR values less than 5%, the boosted decision trees and random forest have higher TPR than the other three classifiers by a considerable margin. It is noteworthy that those

Table 2. Evaluation on unfamiliar test data, sorted by the TPR at FPR= 0.01. The best value in each column is highlighted in bold.

Features	Algorithm	AUC	TPR values for FPR= x				
			$x = 0.01$	$x = 0.02$	$x = 0.05$	$x = 0.10$	$x = 0.15$
PVCS	KNN	0.9550	0.1625	0.5875	0.7750	0.9250	0.9750
Permissions	Naïve Bayes	0.9030	0.1875	0.2750	0.5875	0.7125	0.7625
Permissions	SVMs	0.9380	0.2375	0.4750	0.8000	0.8625	0.9000
Permissions	KNN	0.9080	0.4250	0.5125	0.6500	0.8000	0.8125
Permissions	Boosted Dec. Trees	0.9450	0.4750	0.7250	0.8500	0.8750	0.9000
PVCS	Random Forest	0.9580	0.4875	0.6000	0.8125	0.9250	0.9750
PVCS	Boosted Dec. Trees	0.9640	0.5125	0.5125	0.7125	0.8625	0.9875
PVCS	Naïve Bayes	0.9600	0.5375	0.5750	0.6250	0.9000	0.9875
PVCS	SVMs	0.9590	0.5375	0.6250	0.6625	0.7875	0.9875
Permissions	Random Forest	0.9240	0.5500	0.5500	0.8375	0.8750	0.8875
Both	Naïve Bayes	0.9790	0.6125	0.7375	0.8250	0.9875	**1.0000**
Both	KNN	0.9590	0.6250	0.6250	0.8375	0.9000	0.9250
Both	SVMs	0.9840	0.7500	0.7875	0.8250	**1.0000**	**1.0000**
Both	Random Forest	0.9820	0.8125	0.8750	**0.9500**	0.9625	0.9750
Both	Boosted Dec. Trees	**0.9850**	**0.8375**	**0.8875**	0.9250	0.9750	0.9875

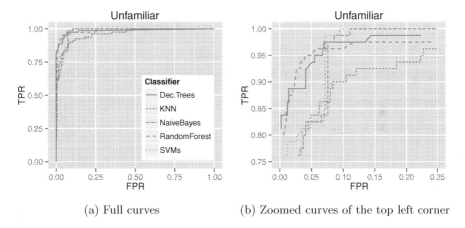

(a) Full curves   (b) Zoomed curves of the top left corner

Fig. 1. The ROC curves for the classifiers trained on both permissions bits and PVCS, evaluated on the unfamiliar data subset: true positive rate (TPR) versus false positive rate (FPR).

two classification algorithms performed the best, since both are based upon decision trees. These decision tree-based learning algorithms are designed to intelligently select *some* of the features to use for classification, in contrast with KNN, SVMs, and naïve Bayes, which each construct a model using *all* of the features to perform classification. This indicates that PVCS and *some* of the permissions bits are useful for detecting malware, but other permissions bits are not useful. The next section examines this issue in more detail.

3.3   Feature Analysis

Since the boosted decision tree model performed the best, we examined its structure to find insights about what contributed toward its good performance. The boosted decision tree model consists of several decision trees; the overall prediction of the model is a weighted vote of the classifications from the decision trees. Each decision tree consists of several decision nodes, where the value of one particular feature (a permission bit or the PVCS value) is examined. Comparing the feature value to a learned threshold decides which branch of the tree is (recursively) used to make a decision. Leaf nodes are the decision: malware or benign.

In order to determine which features are most useful in screening for malware, we assigned a score to each feature (permission bit or PVCS value) that was used in the model. Simply summing the number of times a feature is used in one of the decision trees would be one option, but it does not account for the fact that nodes higher in the tree are deemed more discriminative by the learning algorithm. Thus, we weight each occurrence of the feature according to its depth in the tree. Furthermore, the scores from each tree are weighted according to that tree's contribution to the overall decision of the classifier; those tree weights come from the boosting learning algorithm. In mathematical form, the score for a feature f is

$$\sum_{x \in N(f)} \frac{1}{d(x) + 1} w(x) \qquad (2)$$

where $N(f)$ is the set of decision tree nodes that examine feature f, $d(x)$ is the depth of x, and $w(x)$ is the weight of the tree in which x is contained.

The model used 59 different features out of the 387 features in the training data (15%). The features with the top 20 scores are listed in Table 3. The highest scoring feature by far is PVCS. In fact, it was the root node feature (i.e., most informative feature) in half of the decision trees in the model. System permissions, as opposed to user-defined permissions, dominate the list, filling out the top 10 features. The second highest scoring feature, the SEND_SMS permission, scores very high, about 50% higher than the next permission. This is likely indicative of malware that send out text messages without the user's consent.

In addition to its prominence in the boosted decision trees, the importance of PVCS is also seen when considering the classifiers trained only with PVCS information. When comparing them with the classifiers trained only on permissions (Table 2), the PVCS feature generally did better than the permissions features. Specifically, four out of five of the PVCS classifiers have a higher TPR for FPR= 0.01 than the permissions classifiers, with the notable exception of the permissions random forest. Furthermore, all of the PVCS classifiers have higher AUC than any of the permissions classifiers.

Table 3. Top scoring features from the best model (boosted decision trees trained on permissions and PVCS). System permissions have the "android.permission" prefix removed from their names in the table.

Rank Feature	Score
1 **PVCS**	72.52
2 SEND_SMS	21.07
3 READ_PHONE_STATE	13.79
4 ACCESS_COARSE_LOCATION	11.98
5 RECEIVE_BOOT_COMPLETED	11.45
6 ACCESS_NETWORK_STATE	10.45
7 INTERNET	10.19
8 SET_ORIENTATION	8.91
9 READ_CONTACTS	8.63
10 CAMERA	8.02
11 GET_ACCOUNTS	7.33
12 WAKE_LOCK	7.11
13 com.software.android.install.permission.C2D_MESSAGE	6.84
14 GET_TASKS	6.78
15 READ_SETTINGS	6.65
16 READ_SMS	6.38
17 CHANGE_WIFI_STATE	6.03
18 com.android.browser.permission.READ_HISTORY_BOOKMARKS	5.70
19 INSTALL_PACKAGES	5.59
20 WRITE_EXTERNAL_STORAGE	5.22

3.4 Classification of Known Families

While the ability of the classifier to detect unfamiliar malware families is most important, we also want to verify that the classifier can detect new instances of malware from families upon which it was trained. Table 4 shows the results from evaluating the 15 models on the "familiar" subset of the test data. Figure 2 plots the corresponding ROC curves for the models trained on both permissions and PVCS.

As when testing on the unfamiliar subset, the combination of permissions and PVCS information leads to the best classifiers. Specifically, for each FPR level, the best model uses the combination of feature sets. The best model at screening for unfamiliar malware—the boosted decision trees—also has the best AUC when screening for familiar malware. Furthermore, its TPR for FPR= 0.02 is the best among the 15 models, and the TPR of 0.9617 for FPR= 0.01 is within half of a percent of the best model (0.9652). Thus, that model is not only the best at screening for new malware families, it is also very nearly the best at screening for malware from known families.

3.5 Analyzing the Mistakes

We found 22 applications out of the 855 testing applications (2.6%) are classified incorrectly by the best-performing classifier. This section provides some

Table 4. Evaluation on familiar test data, sorted by the TPR at FPR= 0.01. The best value in each column is highlighted in bold.

Features	Algorithm	AUC	TPR values for FPR= x				
			$x = 0.01$	$x = 0.02$	$x = 0.05$	$x = 0.10$	$x = 0.15$
PVCS	KNN	0.9790	0.1533	0.7875	0.9059	0.9826	0.9930
Permissions	Naïve Bayes	0.9680	0.6411	0.7770	0.8815	0.9373	0.9547
PVCS	Random Forest	0.9860	0.7003	0.7909	0.9129	0.9861	0.9965
PVCS	Naïve Bayes	0.9860	0.7631	0.7909	0.9164	0.9686	0.9965
PVCS	SVMs	0.9820	0.7631	0.7700	0.8223	0.9652	0.9965
PVCS	Boosted Dec. Trees	0.9860	0.7770	0.7770	0.9233	0.9582	0.9965
Both	Naïve Bayes	0.9950	0.8815	0.9268	0.9895	**0.9965**	1.0000
Permissions	Boosted Dec. Trees	0.9830	0.9059	0.9164	0.9617	0.9686	0.9756
Permissions	KNN	0.9880	0.9164	0.9443	0.9547	0.9686	0.9721
Permissions	SVMs	0.9840	0.9199	0.9373	0.9512	0.9617	0.9721
Permissions	Random Forest	0.9900	0.9338	0.9443	0.9652	0.9686	0.9721
Both	KNN	0.9950	0.9547	0.9652	0.9756	0.9756	0.9861
Both	Boosted Dec. Trees	**0.9980**	0.9617	**0.9826**	0.9895	**0.9965**	0.9965
Both	Random Forest	**0.9980**	**0.9652**	0.9756	0.9895	0.9930	0.9930
Both	SVMs	**0.9980**	**0.9652**	0.9686	**0.9930**	0.9930	0.9965

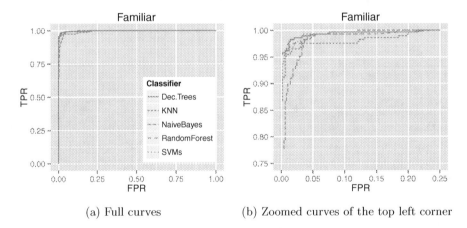

(a) Full curves   (b) Zoomed curves of the top left corner

Fig. 2. The ROC curves for the classifiers trained on both permissions and PVCS, evaluated on the familiar data subset

insights on the reasons why these applications are misclassified. There are 8 free benign applications misclassified as malware. The main reason behind this is that these applications contain ads/analytics libraries in which sensitive operations have no valid user trigger according to our PVCS calculations. Hence, these benign applications have low PVCS values, resulting in benign applications with profiles similar to malware applications. As an example, `com.jaredshack.-androidtimecardfree` is a timesheet application to track the time. It contains the Google ad library in which some sensitive operations, such as `getLatitude()`

and `getLongitude()`, have invalid user triggers. Its PVCS value is 0.3 which is considered low and similar to the values of malware applications.

The reason behind the 14 malware applications misclassified as benign is that these malware applications are repackaged applications. Malware writers bundle malicious code with existing benign applications, producing what is called a repackaged application. Therefore, these malware applications have high PVCS values since most of the sensitive operations inside these applications have valid user triggers according to our calculation of PVCS. As a result, they exhibit similar profiles to benign applications. For example `DroidKungFu` malware is bundled with `com.sniper.awrvvitieetewa` (a game application). Its PVCS value is 0.6 which is considered high, similar to the values of benign applications.

Just as we were able to combine PVCS with permissions information to greatly improve classification accuracy, future work can look for further improvements in accuracy by adding additional information into the feature sets for the classifiers. Our analysis suggests that information about the use of ad libraries or very high similarity with other applications (e.g., repackaged applications) could be important pieces of information to add. In addition, information from dynamic analysis could be added to our static analysis features, providing a more complete picture of the application's behavior. Such work would address limitations of static analysis in general, like code that is dynamically loaded at runtime, the use of native code, or extensive obfuscation. Of course, dynamic analysis has its own weaknesses, such as the difficulty of realistically simulating user behavior in a sandbox environment. Thus, while dynamic analysis could improve detection rates, our work demonstrates that high detection is possible using different kinds of static analysis features.

4   Conclusions and Future Work

We presented a new method for classifying Android applications as malicious or benign that is more accurate than previous work. The method combines two sources of information about the application: the percentage of valid call sites (PVCS) measure and the permissions requested by the application. Both are obtained through static analysis of the application, so there is no need to run the application in order to compute this information. Either set of features alone produced results that were comparable to the previous state of the art.

However, the primary contribution of this work is a demonstration that combining the information from PVCS with the information from permissions results in substantially better performance than previous work. The previous best work detected less than 50% of the malware when limiting false positives to 1% [14], whereas our best classifier detects 83.75% of the malware from unfamiliar families and 96% of the malware from familiar families while maintaining less than 1% false positives.

For future work, we plan to explore other variations of program analysis-based risk features for detecting Android malware, in combination with the permission analysis of applications. We will also perform more extensive evaluation on new applications from the Android Play market.

Appendix: Details of Malware Families in the Data Set

Table 5. Training/testing split for each malware family, listing the number of applications. In addition to the families in the table, the families with only one application were divided as follows: training set included the adrd, andcom, foncy, lovetrap, nickispy, nickyspy, smswatcher, smszombie, youmi, and zsone families; testing set included the airpush, anserverbot, droidcoupon, fakeapp, fakelogo, fakesite, gamblersms, generic, sheridroid, and smsbomber families.

	test	train		test	train
bgserv	1	1	smskey	2	4
crusewin	1	1	droidkungfu2	2	5
gamex	1	1	leadbolt/ropin	2	5
koogame/koomer	1	1	penetho	2	5
walkinwat	2	0	yzhc	2	5
wapsx	1	1	fakedoc	2	6
asroot	1	2	faketimer	2	7
droiddream	3	0	geinimi	2	7
droidkungfusapp	1	2	kmin	9	0
ggtracker	1	2	pjapps	2	7
ksapp	1	2	smssend	2	7
kuguo	1	2	gingermaster	2	8
mania	1	2	fakeplayer	3	8
mobiletx	1	2	jsmshider	3	12
opfake	1	2	zitmo	4	12
gingerbreak	1	3	droiddreamlight	4	13
hipposms	1	3	droidkungfu4	17	0
infostealer	1	3	golddream	4	13
jifake	4	0	droidkungfu1	4	15
imlog	1	4	droidrooter	30	0
tapsnake	1	4	droidkungfu3	9	33
wooboo	5	0	plankton	11	41
adwo	2	4	basebridge	12	45
droidkungfu	2	4	fakeinst	189	752

References

1. Soot: a Java optimization framework (2012), http://www.sable.mcgill.ca/soot/
2. Aafer, Y., Du, W., Yin, H.: DroidAPIMiner: Mining API-level features for robust malware detection in Android. In: Zia, T., Zomaya, A., Varadharajan, V., Mao, M. (eds.) SecureComm 2013. LNICST, vol. 127, pp. 86–103. Springer, Heidelberg (2013)
3. Amos, B., Turner, H., White, J.: Applying machine learning classifiers to dynamic android malware detection at scale. In: 2013 9th Int. Wireless Commun. and Mobile Computing Conf. (IWCMC), pp. 1666–1671 (2013)

4. Arp, D., Spreitzenbarth, M., Hubner, M., Gascon, H., Rieck, K.: Drebin: Efficient and explainable detection of Android malware in your pocket. In: Proc. of 17th Network and Distributed System Security Symposium (NDSS) (2014)
5. Burguera, I., Zurutuza, U., Nadjm-Tehrani, S.: Crowdroid: Behavior-based malware detection system for Android. In: Proc. of the 1st ACM Workshop on Security and Privacy in Smartphones and Mobile Devices, SPSM 2011, pp. 15–26 (2011)
6. Cova, M., Kruegel, C., Vigna, G.: Detection and analysis of drive-by-download attacks and malicious JavaScript code. In: Proc. of 19th Int. World Wide Web Conf. (2010)
7. Elish, K.O., Yao, D., Ryder, B.G.: User-centric dependence analysis for identifying malicious mobile apps. In: Proc. of the IEEE Mobile Security Technologies (MoST) Workshop, in conjunction with the IEEE Symposium on Security and Privacy (2012)
8. Elish, K.O., Yao, D., Ryder, B.G., Jiang, X.: A static assurance analysis of Android applications. Technical Report TR-13-03, Virginia Tech (2013)
9. Grace, M.C., Zhou, Y., Zhang, Q., Zou, S., Jiang, X.: RiskRanker: scalable and accurate zero-day Android malware detection. In: Proc. of the 10th International Conference on Mobile Systems, Applications, and Services (MobiSys), pp. 281–294. ACM (2012)
10. Hall, M., Frank, E., Holmes, G., Pfahringer, B., Reutemann, P., Witten, I.H.: The WEKA data mining software: An update. SIGKDD Explorations 11 (2009)
11. Horwitz, S., Reps, T., Binkley, D.: Interprocedural slicing using dependence graphs. ACM Transactions on Programming Languages and Systems 12, 26–60 (1990)
12. Lee, W., Stolfo, S.J., Mok, K.W.: A data mining framework for building intrusion detection models. In: Proc. of the 1999 IEEE Symposium on Security and Privacy, pp. 120–132. IEEE (1999)
13. Liu, L., Yan, G., Zhang, X., Chen, S.: VirusMeter: Preventing your cellphone from spies. In: Kirda, E., Jha, S., Balzarotti, D. (eds.) RAID 2009. LNCS, vol. 5758, pp. 244–264. Springer, Heidelberg (2009)
14. Peng, H., Gates, C., Sarma, B., Li, N., Qi, Y., Potharaju, R., Nita-Rotaru, C., Molloy, I.: Using probabilistic generative models for ranking risks of android apps. In: Proc. of the 2012 ACM Conf. on Computer and Commun. Security, CCS 2012, pp. 241–252 (2012)
15. Sahs, J., Khan, L.: A machine learning approach to Android malware detection. In: 2012 European Intelligence and Security Informatics Conf. (EISIC), pp. 141–147 (2012)
16. Sanz, B., Santos, I., Laorden, C., Ugarte-Pedrero, X., Bringas, P.G., Álvarez, G.: PUMA: Permission usage to detect malware in Android. In: Herrero, Á., et al. (eds.) Int. Joint Conf. CISIS'12-ICEUTE'12-SOCO'12. AISC, vol. 189, pp. 289–298. Springer, Heidelberg (2013)
17. Schmidt, A.D., Bye, R., Schmidt, H.G., Clausen, J., Kiraz, O., Yuksel, K., Camtepe, S., Albayrak, S.: Static analysis of executables for collaborative malware detection on android. In: IEEE Int. Conf. on Commun., pp. 1–5 (2009)
18. Shabtai, A., Kanonov, U., Elovici, Y., Glezer, C., Weiss, Y.: Andromaly: a behavioral malware detection framework for Android devices. Journal of Intelligent Inform. Syst. 38(1), 161–190 (2012)
19. Srivastava, A., Sural, S., Majumdar, A.K.: Database intrusion detection using weighted sequence mining. Journal of Computers 1(4), 8–17 (2006)
20. Tan, H., Goharian, N., Sherr, M.: $100,000 prize jackpot. Call now!: Identifying the pertinent features of SMS spam. In: Proc. of the 35th Int. ACM SIGIR Conf. on Research and Development in Information Retrieval, pp. 1175–1176. ACM (2012)

21. Virustotal: Virus Total (2013), https://www.virustotal.com/
22. Whitney, L.: iPhone market share shrinks as Android, Windows Phone grow (January 2014), http://news.cnet.com/8301-13579_3-57616679-37/iphone-market-share-shrinks-as-android-windows-phone-grow/
23. Wu, D.J., Mao, C.H., Wei, T.E., Lee, H.M., Wu, K.P.: DroidMat: Android malware detection through manifest and API calls tracing. In: 2012 Seventh Asia Joint Conf. on Inform. Security (Asia JCIS), pp. 62–69 (2012)
24. Xie, P., Li, J.H., Ou, X., Liu, P., Levy, R.: Using Bayesian networks for cyber security analysis. In: 2010 IEEE/IFIP Int. Conf. on Dependable Syst. and Networks (DSN), pp. 211–220. IEEE (2010)
25. Zhou, Y., Jiang, X.: Dissecting Android malware: Characterization and evolution. In: Proc. of the IEEE Symposium on Security and Privacy, pp. 95–109 (2012)

Analyzing Android Browser Apps
for `file://` Vulnerabilities

Daoyuan Wu and Rocky K.C. Chang

Department of Computing, The Hong Kong Polytechnic University,
Hung Hom, Hong Kong
{csdwu,csrchang}@comp.polyu.edu.hk

Abstract. Securing browsers in mobile devices is very challenging, because these browser apps usually provide browsing services to other apps in the same device. A malicious app installed in a device can potentially obtain sensitive information through a browser app. In this paper, we identify four types of attacks in Android, collectively known as File-Cross, that exploits the vulnerable `file://` to obtain users' private files, such as cookies, bookmarks, and browsing histories. We design an automated system to dynamically test 115 browser apps collected from Google Play and find that 64 of them are vulnerable to the attacks. Among them are the popular Firefox, Baidu and Maxthon browsers, and the more application-specific ones, including UC Browser HD for tablet users, Wikipedia Browser, and Kids Safe Browser. A detailed analysis of these browsers further shows that 26 browsers (23%) expose their browsing interfaces unintentionally. In response to our reports, the developers concerned promptly patched their browsers by forbidding `file://` access to private file zones, disabling JavaScript execution in `file://` URLs, or even blocking external `file://` URLs. We employ the same system to validate the ten patches received from the developers and find one still failing to block the vulnerability.

1 Introduction

Using `file://` to browse local files is very common in desktop browsers. However, this file protocol mechanism, when applied to mobile platforms, could cause unexpected security risks. In modern smartphone systems, notably Android and iOS, each app's sensitive files are stored in their own system-provided private file zones, which cannot be accessed by other apps or users. Supporting `file://` without additional access control in mobile browsers, however, will break such security boundaries. This `file://` vulnerability is further aggravated in Android, because Android browsers usually accept external browsing requests which, in the absence of any user interaction, can be issued by another (malicious) app. Unlike Android, these requests in iOS must be invoked by users' clicking.

Supporting external `file://` browsing requests (or termed as external `file://` URLs) is only a necessary condition for realizing actual attacks. In this paper, we show that combining with the capability of accessing private file

S.S.M. Chow et al. (Eds.): ISC 2014, LNCS 8783, pp. 345–363, 2014.

zones through `file://`, JavaScript support, and other browsers' flaws (such as auto-file download), a malicious app in Android can launch four different types of attacks to steal a victim browser's private files (e.g., users' cookies, bookmarks, and browsing histories) or a victim website's private files (e.g., cookie or content). We refer to this class of attacks as *FileCross* , in which all attack vectors are delivered through the `file://` protocol between a browser app and an attack app. The attack app can automatically download a private file to the public SD card for exporting, steal a private file by compromising same-origin policy (SOP [1]) on the "host" level, steal the content of another website by compromising SOP on the protocol level (`file://` and `http(s)://`), and steal a private file by exploiting a SOP flaw in handling symbolic links.

Several isolated incidences on stealing browsers' private files were reported for Chrome and Firefox [2,3,4]. However, as we will show in this paper, these attacks are just instances of the FileCross attacks. To characterize the prevalence and impact of the FileCross attacks, we develop a system based on dynamic analysis to automatically test over 100 browser apps in Android. The main approach is to mimic actual attacks and use them to test the browsers on real smartphones. This system determines whether a browser app is vulnerable to the four FileCross attacks. It also analyzes whether the app, before and after patching, supports `file://`, allows access to private file zones through `file://`, and supports JavaScript.

The main findings obtained from our analysis of 115 browser apps can be summarized below.

1. More than half of the browsers tested are vulnerable to the FileCross attacks. In particular, 50% of the most popular browsers (e.g., Firefox, Baidu, and Maxthon) are also vulnerable. Similarly, many major browsers in different categories could leak out private information through the FileCross attacks. Among the four different attacks, the three attacks that are based on compromising SOP affect 55% of the browsers on Android 4.0, 4.3 or 4.4.

2. The `file://` vulnerabilities are exploitable in all Android versions (including the latest 4.4), and even occur in different web engines. Specifically, our system identifies 46 browsers being vulnerable in 4.4 (across all four FileCross attacks). This result contradicts the general belief that Chrome-based new system engine will no longer contain these flaws by default. We are also contacting Google Android security team to fix one common flaw at the engine level. Moreover, we detect three vulnerable browsers (Firefox, UC Browser HD and Sogou) out of 15 browsers that employ custom engines.

3. A further analysis reveals that 23% of the browsers expose their browsing interfaces unintentionally. Had the developers realized the browser interfaces' exposure, one third of them would not have been vulnerable to the FileCross attacks. Moreover, 65% of the browsers accept external `file://` browsing requests, and 62% even allow `file://` access to the private file zones. The latter is necessary for three FileCross attacks. Moreover, 63% support JavaScript execution in `file://` URLs which makes three FileCross attacks possible.

4. In response to our vulnerability reports, 19 developers followed up with our findings. We have so far received nine patches from them (and will receive more). An analysis of the patches shows that the patching methods include disabling the access to unrenderable private files, blocking external `file://` URLs, or disabling JavaScript execution in `file://` URLs. Most of them could effectively thwart the attacks. However, our system developed for testing browsers finds that one patch failed to block the vulnerability, because the patch missed a second attack entry.

2   The `file://` Vulnerabilities

2.1   The FileCross Attacks

We have discovered from our evaluation, which will be further elaborated in Section 2.2, that 113 out of 115 browsers in Android expose their browsing interfaces, and 75 out of the 113 browsers support external browsing requests from other apps through `file://`. As illustrated in Fig. 1, an attack app can issue a "malicious" browsing request to a victim browser through the `file://` channel. The attack can steal sensitive files directly or indirectly from the victim browser's private file zone by having the URL in the browsing request point to a target sensitive file or a malicious HTML file, respectively.

The direct method exploits the fact that some browsers allow `file://` requests to access their private file zones. The indirect method, on the other hand, exploits the same-origin policy (SOP [1]) flaws in handling `file://` requests, and it also requires the JavaScript support for executing the malicious HTML file. In our evaluation, 71 browser apps (out of the 75 that support `file://`) allow the requests received from `file://` to access their private file zones, and 72 permit JavaScript execution in `file://` URLs. Moreover, the indirect method can be used to steal sensitive files from websites.

Fig. 1 shows examples of four FileCross attack patterns. The first one uses the direct method, whereas the last three use the indirect method by compromising the SOP. The first and fourth attacks are in fact first reported by an individual hacker. We discovered the other two from the Android developer document. We thus do not claim the discovery of these attacks as our main contribution. But we are the first to identify them as a unified attack model (i.e., FileCross) and conduct automated testing to analyze their prevalence in Android browsers. In addition, our system to be presented in Section 3 could be extended to detect new attack patterns.

Attack 1 (A1):   The `file://` URL points to a sensitive file (`Cookies` in the figure) in the victim browser's private file zone. Some browsers automatically download the requested file to the `Download` directory on a SD card. The attack app can use keyword search to find and read the target file from the SD card (see `Cmd 1`). The auto-download feature has been identified as a flaw responsible for a successful FileCross attack against Chrome for Android [2].

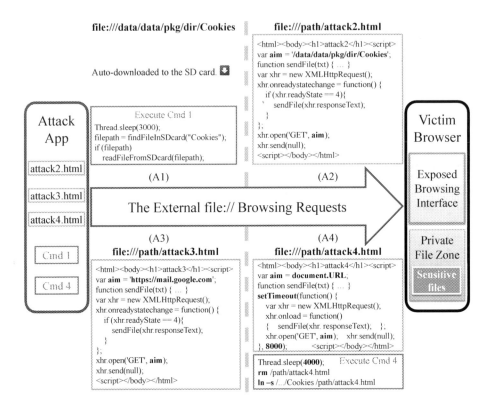

Fig. 1. Examples of four FileCross attacks (A1 to A4)

Attack 2 (A2): The `file://` URL points to a malicious HTML file `attack2.html`. The attacker prepares the HTML file for the browser to retrieve a sensitive file (`Cookies` in the figure) from its private file zone. Once the attack HTML file is loaded, an asynchronous request (e.g., via the XMLHttpRequest API [5]) is issued to retrieve the sensitive file (`xhr.responseText` in the figure). After this, `sendFile(txt)` is invoked to send the file to a remote server that can be accessed by the attacker. The fundamental problem enabling this attack is compromising SOP for `file://` requests (i.e., a local file should not be allowed to read contents of another file). Our evaluation shows that 63 browsers are vulnerable to this attack.

Attack 3 (A3): The `file://` URL points to a malicious HTML file `attack3.html`. The attacker prepares the HTML file for the browser to retrieve sensitive content from a remote website (`mail.google.com` in the figure). Similar to the last attack, the content is retrieved by an asynchronous request and sent to a remote server via `sendFile(txt)`. The fundamental problem is again compromising SOP, but this time on the protocol level (`file://` and `https`). Our evaluation uncovers 56 vulnerable browsers. This attack can also steal cookies of a website, but the details are omitted here.

Attack 4 (A4): The `file://` URL points to a malicious HTML file `attack4.html`. While the objective of this attack is the same as A2, it sets the target (in the `aim` JavaScript variable) as the current URL (i.e., `document.URL` in the figure), thus not violating SOP. However, the codes will not be executed until after 8000 ms. The attack app in the meantime removes `attack4.html` and builds a symbolic link for the removed file using the target sensitive file `Cookies`. Now when the time comes for the browser to execute the codes, it may load `Cookies` according to the link and return its contents to JavaScript. This flaw of loading a symbolic link to a file when the file cannot be found exists in modern browsers, including Chrome [3] and Firefox [4]. Our evaluation reveals 57 vulnerable browsers.

The last three attacks exploit the flaws on enforcing SOP for external `file://` requests. For *webkit*, Android's default web engine, the SDKs prior to 4.1 suffered from flawed SOP enforcement. Although the flaws in attacks A2 and A3 have been fixed by the default setting introduced to Android 4.1, the `file://` vulnerabilities still remain for two reasons. First, we notice that the two new APIs introduced in 4.1 still suffer from the SOP flaws. Therefore, developers may still use these vulnerable APIs, especially when they cannot find the security implications from Google's Developer Document. Second, developers must compile their apps using recent SDKs to block the vulnerabilities. Our evaluation, however, shows that over 30 browsers on Android 4.3 are still vulnerable, because the developers still used the old SDKs to compile their apps.

Starting from the latest Android 4.4, the system web engine is changed to Chrome's Blink engine. A general belief is that Chrome-based engine will no longer contain these flaws by default (we even made this mistake earlier via preliminary manual testing, since file paths are changed in 4.4). But surprisingly, our automated testing finds 46 browsers are still vulnerable in 4.4, across all four FileCross attacks. In particular, we notice Android 4.4 does not provide by-default patches for the SOP flaw (in A4), causing 40 browsers still exploitable in 4.4 by attack A4. We are contacting Google Android security team to fix this common flaw at the engine level. Moreover, similar to the Android 4.3 cases, apps compiled with old SDKs (i.e., below 4.1) cannot be protected by system-level defenses for attacks A2 and A3, even running on Android 4.4. Additionally, the flaw in A1 is application specific. In summary, mitigating the FileCross flaws in all Android versions still require browser developers' careful implementations. Therefore, our evaluation system is designed to test browser implementations but not specific web engines.

2.2   Attack Conditions

Table 1 summarizes the conditions required for launching the four FileCross attacks. Exposing browsing interfaces and supporting `file://` are obviously necessary for all of them. Allowing `file://` access to private file zones is also necessary for major FileCross attacks that aim at stealing browsers' private files. In addition, attacks A2, A3, and A4 require JavaScript execution in `file://` URLs for constructing the corresponding exploits (as shown in Fig. 1). Although it is always possible for some advanced attackers to invent non-JavaScript exploits for these three attacks, we believe this JavaScript condition is currently required and therefore include it into our FileCross threat models.

Table 1. The required conditions for the four FileCross attacks

Attack IDs	Exposed browsing interface	Support file:// URLs	file:// access to private file zones	JavaScript execution in file:// URLs	Major flaws
A1	√	√	√		Auto-download file to SD card
A2	√	√	√	√	SOP bypass for two file:// origins
A3	√	√		√	SOP bypass for file:// and http(s):// origins
A4	√	√	√	√	SOP bypass in handling symbolic links

Before moving to the next section, it is instructive to understand how browsing interfaces are exposed. As mentioned above, 113 of our tested 115 browsers expose their browsing interfaces to other apps. By inspecting their manifest files, we further infer that some browsers expose their browsing interfaces *unintentionally*, although most express *explicit* intentions to accept external browsing requests. We summarize these intentionally and unintentionally exposed patterns in Fig. 2(a), and also give a simple Exposed Browsing Interface (EBI) example in Fig. 2(b). Our inference for intentional exposures is based on the presence of an Intent with the `action` of "VIEW" and the `category` of "BROWSABLE," because this type of Intent is usually delivered to browsers [6].

The unintentionally exposed cases, in our understanding, are mainly caused by the Android's implicit Intent mechanism [7]. Specifically, Android requires each app to register an Intent filter with the `action` of "MAIN" and the `category`

(a) Intentionally or unintentionally exposed browsing interface and their related attributes.

(b) The intentionally exposed browsing interface (`.ViewLink`) in Offline Browser (`it.nikodroid.offline`).

Fig. 2. A summary of EBI patterns and an EBI example

of "LAUNCHER" for the first user interface component, so that the app can be launched by the default launcher. This behavior, however, will implicitly cause the corresponding component to be exposed to other apps. It may happen for some browser developers to register their browsing interfaces with such Intent, thus exposing them as EBIs even without claiming to receive "BROWSABLE" intents. Hence, these EBIs cannot be triggered by normal browsing requests. We thus believe they are unintentionally exposed by developers in terms of serving external browsing requests. Due to the space limitation, we refer readers to Section 5.4 of [8] for a general discussion on such implicit intents.

3   Automated Testing of Android Browsers

We design and implement a system for testing browsers for the `file://` vulnerabilities. In order to test all browser apps available in Android markets, our system can automatically test all of them without human intervention. Using the system, we could test over 100 Android browsers in less than four hours. Since our ultimate goal is to report vulnerable browsers to their developers for patching, it is not enough to just demonstrate that private files can be *accessed* by invoking JavaScript's `alert(content)` function. Instead, our system mimics the actual attacks to *steal* victim browsers' private files and tests the browsers on actual smartphones. Besides detecting the vulnerabilities, the system also helps determine whether the external browsing interfaces are opened intentionally and analyze the patches obtained from the developers.

3.1   The System Design

Fig. 3 shows the architecture and workflow of our testing system. The three main components in this system are *Commander* for controlling the entire testing process, *Attack Executer* for launching the FileCross attacks, and *Web Receiver* for validating whether the attacks are successful. The Commander running in a PC host controls the connected Android devices (which can be emulators or real phones) via Android Debug Bridge (ADB) channels (from ADB host to ADB daemons on devices). We implement Commander in the pure Python language to avoid the unstable issues of MonkeyRunner [9] reported in [10]. Moreover, we implement parts of the failure controlling mechanisms proposed in [10] to improve the stability of ADB over long runtimes and use multiple threads to concurrently control each device for testing multiple Android versions in parallel.

We implement the Attack Executer as an Android app and install it in each tested device. Like a real attack app, it launches the FileCross attacks to steal private files from the target browsers. Moreover, its attack behaviors are fully controlled by the Commander through each incoming attack command (including target browser information and attack parameters). Once receiving the attack commands, it generates the corresponding exploits on-the-fly and loads them into target browsers via the Intent channels. The Web Receiver, on the other hand, is a server-side program responsible for accepting the stolen private files and

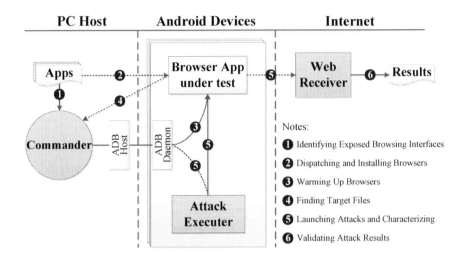

Fig. 3. The architecture and workflow of our testing system

validating the attack results. An attack is considered successful if the stolen file is received.

3.2 The Major Testing Steps

Fig. 3 shows six major testing steps in our system. We discuss them below in three pairs.

Identifying Exposed Browsing Interfaces. We propose a lightweight but effective scoring mechanism to identify EBIs in Android browsers. The basic idea is to score each component based on our summarized EBI patterns in Section 2.2 and select the component with a maximal score as the EBI. That is, a component with the maximal score is most likely to be an EBI. This maximal score also helps us locate the major (or true) browsing interface. For instance, Chrome's `ManageBookmarkActivity` exhibits EBI patterns but is not functional for handling browsing requests. In this case, our scoring mechanism can help identify the right browsing interface `chrome.Main`, which shows more explicit EBI patterns, thus a higher score. When several EBIs have the same score, we handle such case by randomly selecting one EBI for dynamic testing. In addition, if all components score zero, we conclude there is no EBI in the browser. In our experiments, we find that this scoring mechanism can accurately identify the EBIs in 113 browsers out of the tested 116 browsers. For the remaining three cases that have no EBIs, one of them is only a browser add-on, and the other two do not expose their browsing interfaces.

Bit Id	5	4		3	2	1	0
0/1	1	1		1	1	1	1
Score	32	16		8	4	2	1

Bit Id	0	1	2	3	4	5
EBI pattern	MAIN LAUNCHER	file	VIEW BROWSABLE	http	https	MAIN LAUNCHER
Pre condition	Bits $(1-4)$ are all empty	Bit 2 is set	–	Bit 2 is set	Bit 2 is set	One of bits $(1-4)$ is set

Fig. 4. The detailed rules for scoring EBI patterns using six bits

The detailed scoring algorithm works as follows. We use six bits to flag five specific EBI patterns (two bits are set for one pattern under different situations). Fig. 4 illustrates the detailed rules for scoring the EBI patterns under different scenarios. For example, if one component has an Intent filter which defines the `action` of "VIEW" and the `category` of "BROWSABLE," we set bit 2 (i.e., a score of 4). If this Intent filter also registers the `data` scheme of "http," we further set bit 3 (i.e., a score of 8). Now the component has a total score of 12, which can be used also for reversely inferring the EBI patterns using its binary representation.

These scoring rules (with different weights) are summarized according to our manual analysis of a dozen of EBI patterns. First, we treat the basic EBI pattern (i.e., "VIEW" and "BROWSABLE") as a reference pattern. On the basis of this pattern, we further assign weights to three data schemes, if any. Among them, we score the "https" scheme higher than "http," because we find accepting "https" is more likely to represent an EBI. On the other hand, we lower the "file" scheme even below the reference pattern, to remove the potential noises introduced by "file". The noises can occur when "file" is registered for browsing document or video files. So such components are actually document viewers or video players, instead of browser components. Finally, we observe the "LAUNCHER" pattern, if exists, can add more weights when the aforementioned patterns also occur. That is, a component with both "BROWSABLE" and "LAUNCHER" patterns will be always the major EBI, compared with those non-launcher "BROWSABLE" components. In addition, a component with only the "LAUNCHER" pattern should be scored less than other "BROWSABLE" components.

Warming up Browsers and Finding Target Sensitive Files. The goal of warming up browsers is to produce some private files as the target sensitive files. To do so, the system automatically sends several normal browsing requests before launching the attacks. Specifically, the tested browsers are instructed to browse several Alexa top 10 websites using HTTP or HTTPS. This warming-up step can also help validate the EBIs identified by the scoring mechanism. That is, if an EBI is correctly identified, we can effectively warm up the corresponding browser. Otherwise, the browser will not respond according to our external browsing requests.

After warming up the browsers, our system continues to find as many target sensitive files as possible from the newly generated private files. To do so, the system searches browsers' private file zones (i.e., `/data/data/package/`) using a set of prioritized keywords (e.g., "cookie", "password", and "bookmark") and certain file formats (e.g., ".sqlite" and ".db" files). Note that accessing private file zones, which is normally disabled on unrooted phones, is only used for finding target sensitive files in our system (and attackers can also use this method to obtain the same information for their attacks). The actual FileCross exploitation is still conducted by the Attack Executer through the normal Intent channels.

Automatic Attack Validation and Characterization. Another challenge in designing our system is how to *automatically* validate attack results and conduct further characterization. Unlike manual testing, we cannot rely on human intervention, such as naked-eye inspection. To address this issue, we pre-define patterns that describe the attack details given by the Commander and embed them into each attack request sent by exploit scripts, which will be finally received and interpreted by the Web Receiver. In particular, we embed five patterns into the attack requests: an app package's name (for identifying the tested browser), an attack ID (for differentiating different attacks), a device version (for characterizing attacks on different Android versions), contents (for transmitting and validating potential private files), and a key ID (for authentication and differentiating different experiments).

To further characterize the FileCross attacks, we adopt the similar methods as for launching attacks, except that the attack scripts are now replaced by other scripts for characterization purposes. Specifically, we design HTML files to characterize the `file://` support (loaded from SD card or private file zones) and JavaScript execution in `file://` URLs. For example, the following HTML file is for characterizing the `file://` support. The Attack Executer loads this HTML file from both SD card and private file zones (with different attack IDs, such as `atk=5`), and sets the current Android version (e.g., `ver=4.3`).

```
<html><body> <img src='http://ourserver.com/req?pkg=example.package
&atk=5&con=reqflag&ver=4.3&kid=keyid'>  </body></html>
```

Interested readers may refer to Appendix A in the Technical Report [11] version of this paper for the HTML file used to characterize JavaScript execution in `file://` URLs. It is relatively complex.

4   Evaluation

4.1   The Dataset and Experiments

Dataset. Our dataset consists of 115 browser apps collected from Google Play on January 21, 2014. Initially, we searched the keyword "Browser" on Google Play and fetched 139 browsers, after excluding several non-browser apps. We further revisited these 139 browsers on March 21 to characterize their meta information (e.g., the install numbers) using the Selenium scripts [12]. Based on the results, we further excluded 23 browsers in which 14 of them were no longer updated for more than one year, and 9 others had been withdrawn from Google Play. Among the remaining 116 browsers, one more was excluded, because it was only a browser add-on.

Experiments. We run our experiments using three Android phones: Sony Xperia J (with Android 4.0), Google Nexus 4 (with Android 4.3), and Nexus 5 (with Android 4.4). These phones are connected to a Dell Studio XPS desktop machine with Ubuntu 12.04 64-bit system through USB cables. We do not use Android emulators in previous studies [13,14,10,15,16], because they are not stable and a number of apps cannot be correctly installed or run on emulators. However, accessing apps' private file zones via ADB on real phones is disabled by default. We thus root the phones to enable it for our automatic testing.

In this section, we report the results obtained from three independent experiment runs conducted on March 27 and June 18 (when the 4.4 device newly joined). Our system incurs no false positives but may incur some false negatives due to the possible instability of dynamic testing. To mitigate this possibility, our final result is a union of the results from these three runs. Regarding the testing performance, each run takes around four hours (i.e., 3 minutes per app). We use a relatively long timeout (12 seconds) before starting a new browsing request to obtain stable results and duplicate the app testing on three phones for observing possible different results in the three major Android versions.

4.2   Vulnerability Results

Overall Results. Our system identifies 64 vulnerable browsers and a total of 177 FileCross issues, as shown in Fig. 5(a). The results clearly show that the vulnerabilities are prevalent in Android browsers: 55.7% of browsers are affected and on average 2.77 issues per vulnerable browser. Furthermore, according to their distribution by the number of installs, 13 out of 26 popular browsers with over million installs each are found vulnerable. They are from top browser vendors, including Firefox, Baidu, and Maxthon. In other words, the FileCross attacks are not easy to discover and were not known to them before our disclosures.

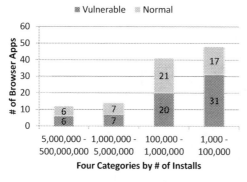

IDs	# of Browsers	
A1	1	
A2	63	62 (4.0)
		35 (4.3)
		25 (4.4)
A3	56	55 (4.0)
		31 (4.3)
		22 (4.4)
A4	57	57 (4.0)
		49 (4.3)
		40 (4.4)
Sum	177	

(a) The distribution of browsers with(out) vulnerabilities

(b) Detailed results for each attack

Fig. 5. Overall detection results in our dataset consisting of 115 Android browsers

Fig. 5(b) shows the detailed results for each FileCross attack. In our dataset, we only discover one auto-file download issue, i.e., attack A1. However, we observe that 71 browsers actually load and display the contents of their private files when challenged by attack A1. Therefore, they will face the potential risk of screen-shot attacks, although we do not consider such risk as a vulnerability in this paper.

For attacks A2, A3 and A4, the number next to (4.0) (or (4.3) and (4.4)) is the number of browsers vulnerable to the attack on Android 4.0 (or 4.3 and 4.4). The number next to these three is the total number of vulnerable browsers for that attack. Some browsers are vulnerable on only one system. These three attacks have a similar number of vulnerable browsers, around 60. Moreover, attack A4 is much less affected by different Android versions than A2 and A3. In the following sections, we thus do not differentiate the results of attack A4 on the three versions. As for attacks A2 and A3, there are over 30 vulnerable browsers for each attack on Android 4.3 and over 20 on Android 4.4, mainly because the developers still use the old SDKs to compile their apps. Thus, their browsers cannot benefit from the webkit patch in Android SDK 4.1.

Representative Vulnerable Browsers. Table 2 summarizes 20 representative vulnerable Android browsers identified by our system. To make it simple, we only use the app package name to refer to each browser, and their full app names can be obtained from Google Play. We also include the number of installs for each browser to underscore the scope of the impact. For each vulnerable browser, we list their detailed assessment results of the four FileCross attacks launched by our system. The red "y" means a successful attack, and the black "n", otherwise. In addition, a blank space represents the case where our attack scripts cannot send response requests to our server, mainly because the target browser is either

Table 2. Representative vulnerable Android browser apps identified by our system

Categories	App Package Names	A1	A2 4.0	4.3	4.4	A3 4.0	4.3	4.4	A4	# of Installs
Popular	org.mozilla.firefox	y				n	n	n		50.000.000 - 100.000.000
	com.baidu.browser.inter	n	y		n	y	n	n	y	5.000.000 - 10.000.000
	com.mx.browser	n	y	y	y	y	y	y	y	5.000.000 - 10.000.000
	com.jiubang.browser	n	y	y	y	y	y	y	y	5.000.000 - 10.000.000
	com.tencent.ibibo.mtt	n	y			n			y	1.000.000 - 5.000.000
	com.boatbrowser.free	n	y	y	y	n	n	y	y	1.000.000 - 5.000.000
	com.ninesky.browser	n	y	y	y	y	y	y	y	1.000.000 - 5.000.000
Tablet	com.uc.browser.hd	n	y	y	y	y	y	y	y	1.000.000 - 5.000.000
	com.baidu.browserhd.inter	n	y		n	y	n	n	y	100.000 - 500.000
	com.boatbrowser.tablet	n	y	y	n	n	n	n	y	100.000 - 500.000
Privacy	com.app.downloadmanager	n	y	n	n	y	n	n	y	10.000.000 - 50.000.000
	nu.tommie.inbrowser	n	y	y	y	y	y		y	500.000 - 1.000.000
	com.kiddoware.kidsafebrowser	n	y	n	n	y	n	n	y	50.000 - 100.000
Fast browsing	com.ww4GSpeedUpInternetBrowser	n	y	y		y	y		y	1.000.000 - 5.000.000
	iron.web.jalepano.browser	n	y	y	y	y	y	y	y	500.000 - 1.000.000
	com.wSuperFast3GBrowser	n	y	y		y	y		y	100.000 - 500.000
Specialized	com.appsverse.photon	n	y	y	y	y	y	y	y	5.000.000 - 10.000.000
	com.isaacwaller.wikipedia	n	y	y	y	n	n	n		1.000.000 - 5.000.000
	galaxy.browser.gb.free	n	y	y		y	y		y	100.000 - 500.000
	com.ilegendsoft.mercury	n	y	n	n	y	n	n	y	100.000 - 500.000

invulnerable or not stable on some Android versions (e.g., 4.3 and 4.4). For such cases, they are assumed invulnerable if no further manual efforts are involved.

We organize these vulnerable browsers into five categories, mainly according to their popularity and unique features. For example, in the "Popular" category, we present several popular browsers with over million installs each. In particular, we identify an auto-file download issue (i.e., attack A1) in Firefox for Android, which is quite popular and has at least 50 million installs. This security issue is ranked by Firefox a high impact one. Moreover, we discover more File-Cross issues in other listed popular browsers. For example, Maxthon Browser (`com.mx.browser`) and Next Browser (`com.jiubang.browser`) suffer from three FileCross attacks in all Android versions we tested, which pose significant security threats to their five million users.

The second category ("Tablet") lists three vulnerable browsers built for Android tablets. Except for UC Browser HD (`com.uc.browser.hd`) that has over million installs, these browsers are not as popular as those in the "Popular" category. However, we notice from Google Play that they are essentially the only choices for users who want to install a dedicated tablet browser. This would entice attackers to launch more targeted attacks at tablet users.

Due to the page limit, the description on the last three categories of vulnerable browsers is available in Appendix B of [11]. Here we only mention two cases. The Kids Safe Browser (`com.kiddoware.kidsafebrowser`) that provides children a safe Internet surfing environment by content filtering jeopardizes children's privacy by the FileCross attacks. Another example is a dedicated browser for browsing Wikipedia, called Wikidroid (`com.isaacwaller.wikipedia`). Attackers can launch the FileCross attacks to infer users' interests and profiles.

4.3 Underlying Engine Analysis

It is useful to determine how many browsers do not use the default engine (which has inherent flaws). Implementing a custom web engine in Android usually requires embedding native codes as shared libraries (`.so` files). For example, Chrome uses `libchromeview.so` as its underlying engine to support browsing functionalities. Determining which `.so` files are web engines is hard and also beyond the scope of this paper. Here, we adopt two strategies to infer which browsers embed their own engines. First, we use regular expression "`native.*loadUrl`" to locate five browsers that implement their own native version of "loadUrl" API, including Chrome, Yandex (`libchromiumkit.so`), Flash Browser (`libxul.so`), and even the vulnerable UC Browser HD (`libWebCore_UC.so`). However, this strategy is not robust enough, because it even misses the Firefox engine. Therefore, we directly inspect each `.so` file name from 24 browsers which have `.so` files. The inspection (combined with existing knowledge) shows that another six browsers embed their own engines, such as Firefox (`libmozglue.so`), Dolphin Browser (`libdolphinwebcore.so`), and three Opera browsers (`libom.so`).

It is also a trend that more Android browsers will use custom engines. Our analysis of five popular Chinese browser apps (which were collected on

May 1) shows that four of them define their own engines. They are QQ (`libmttwebcore.so`), Baidu (`libzeus.so`), Liebao (`libchromeview.z.so`) and Sogou (`libsogouwebcore.so`) browsers. In particular, our system identifies Sogou Browser being vulnerable to FileCross attack A4.

In summary, we have identified 15 (out of the total 120) browsers embedding their custom engines instead of the system default one. In addition, our system identifies three of them being vulnerable: Firefox, UC Browser HD, and Sogou browsers. These findings demonstrate the effectiveness of our system to uncover `file://` vulnerabilities in non-webkit browsers.

5   Further Analysis and Recommendations

5.1   Analyzing the Patches

An Overview. We have devoted considerable efforts on reporting our identified vulnerabilities to the developers (see Appendix C of [11]). Table 3 summarizes the nine patches received so far. Our analysis reveals three kinds of patch methods adopted by the developers. First, similar to the method used by Chrome [3], Firefox's developer disabled the capability of accessing the contents of some unrenderable private files to address the auto-file download issue. However, unlike Chrome, Firefox still allows `file://` access to the private file zone and loading renderable files. We argue that accessing private file zone should be totally banned to mitigate all potential risks. Second, Lightning Browser (`acr.browser.barebones`) and In-Browser (in its beta version, `nu.tommie.inbrowser.beta`) directly blocked the external `file://` URLs from other apps. This fix suggests that supporting external `file://` URLs is not necessary for maintaining some browsers' functionalities. It is interesting to note that the developer of Lightning Browser also applied this method to protect his two other browsers (one is a paid version, and the other an unpublished new browser). Finally, the developers of most patched browsers chose to disable JavaScript execution in `file://` URLs, because it is the easiest way to thwart the three FileCross attacks that require JavaScript support. Although this patch does not eliminate all the possible risks (e.g., screen-shot attacks or origin-crossing attacks without JavaScript), it could be considered effective for the threat models considered in this paper.

Table 3. An overview of the nine patches received from the developers

Package Names	Patched Versions	The Patching Methods
org.mozilla.firefox	28.0.1	Disable accessing unrenderable private files
acr.browser.barebones	3.0.8a	Block external `file://` URLs and alert users
nu.tommie.inbrowser.beta	2.11-55	Block external `file://` URLs
com.baidu.browser.inter	3.1.2.0	Disable JavaScript execution in `file://` URLs
com.jiubang.browser	1.16	Disable JavaScript execution in `file://` URLs
com.baidu.browserhd.inter	1.2.0.1	Disable JavaScript execution in `file://` URLs
easy.browser.classic	1.3.6	Disable JavaScript execution in `file://` URLs
harley.browsers	1.3.2	Disable JavaScript execution in `file://` URLs
com.kiddoware.kidsafebrowser	1.0.4	Disable JavaScript execution in `file://` URLs

An Interesting Patching Process. During the process of analyzing the patches, we identified an interesting case which illustrates the importance of automatic testing even for patches. The developers of Baidu Browser once sent us a version that they thought was patched, because they had disabled the JavaScript execution. However, our system could still successfully exploit this "patched" version. By a careful manual analysis of the patched version, we have found that there were two rendering points in Baidu Browser's browsing interface: one is invoked when users manually input a URL in the browser bar, and the other is for external browsing Intents. Interestingly, the developers disabled the JavaScript support for `file://` URLs only for the first rendering point, thus leaving the real attack point intact. Since the developers did not have an actual attack app, they tested the "patch" manually and mistakenly thought it was patched.

5.2   Exposed Browsing Interfaces

Fig. 6 shows the breakdown of the EBIs in our tested 115 browsers, of which 113 expose their browsing interfaces, meaning that exposing browsing interfaces is a common practice among Android browsers. However, we notice that 26 browsers (23%) expose their browsing interfaces unintentionally. Among them, eight are vulnerable. In other words, these eight browsers could originally avoid the File-Cross issues, if they realized to close their unintentionally exposed interfaces.

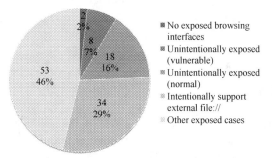

Fig. 6. A breakdown of exposed browsing interfaces in the 115 tested browsers.

We also observe that only 34 browsers (29%) explicitly or intentionally accept external `file://` browsing requests. But our dynamic testing actually finds 75 browsers supporting external `file://` browsing requests. This discrepancy shows that the other 41 browsers may accidentally leak the `file://` channels to other apps. That is, they intend to support `file://` URLs only for internal uses (e.g., when users manually input a `file://` URL).

5.3   `file://` Support in Android Browsers

Based on our analysis, we report three major observations on the `file://` support in Android browsers. First, (at most) 40 of our collected 115 browsers do not support `file://` at all. It is worth noting that 40 is only a upper bound, because our system may not successfully characterize some browsers due to the limitation of dynamic analysis. Among the 40 unsupported ones, Opera Mini and UC Browser Mini are the very popular ones. Opera Mini explicitly mentions *"The protocol "file" is not currently supported"* when a `file://` URL is entered, whereas UC Browser Mini redirects users to a Google search page using the keyword of the entered URL. Other unsupported cases that we manually confirm are dedicated browsers, such as The Pirate Bay Browser for browsing torrents and SkyDrive Browser for accessing Microsoft's SkyDrive service. These cases collectively show that `file://` is generally not supported in lightweight and dedicated browsers, and this practice spares them from the FileCross attacks.

Second, we find that several popular browsers already forbid `file://` access to private file zones. Our system identifies four such cases, including Chrome, Dolphin (`mobi.mgeek.TunnyBrowser`), UC (`com.UCMobile.intl`) and Yandex browsers. All of them allow `file://` access to contents in SD card and permit JavaScript execution in `file://` URLs, but forbid `file://` access to their private file zones. Thanks to this security policy, they are robust to most FileCross attacks (i.e., except A2). We therefore recommend adopting this practice for all Android browsers, because it can better meet the security model of mobile systems.

Finally, we observe three browsers actively disabling the JavaScript execution in `file://` URLs: 3G Browser (`com.mx.browser.free.mx100000004981`) and another two from the same developer (Maxthon Tablet and Maxthon Fast Pioneer browsers). Although the percentage of this practice is currently low (i.e., 3 out of 75), according to our analysis of the patches, we believe that more browsers will follow this practice.

6 Related Work

WebView Security. The closest related works are those on the security of WebView, which uses Android's default web engine (mainly webkit) APIs to help apps display web pages. However, different from our study, most of these studies (e.g., [17,18,19]) mainly concern the insecure invocation between JavaScript and Java levels which may compromise a WebView app by misusing its exposed JavaScript interfaces. In particular, the file-based cross-zone scripting attack reported in [18] is similar to the FileCross attacks, but their attack follows the man-in-the-middle model where malicious JavaScript codes are injected by network adversaries. Without adopting a realistic threat model and proposing detailed attacks, they conclude that file-based cross-zone scripting vulnerabilities are *fortunately* fairly rare. In our study, however, we show that `file://` vulnerabilities are prevalent in Android browsers. Additionally, our study is more general for testing major practices in the Android browser ecosystem (i.e., not limited to WebView flaws), and we also identify non-webkit vulnerable cases (notably Firefox and UC Browser HD).

Android Exposed Component Issues. One important condition for launching FileCross attacks is that browsing interfaces in victim browsers are exposed. Many previous works (e.g., [8,20,21,22,23,24]) have studied the general exposed component problem from the perspective of information flow analysis. They aim at the source-sink problem that other apps can trigger dangerous APIs (i.e., sinks) in an exposed component from its exposed entry points (i.e., sources). Compared to the FileCross attacks, constructing their exploits are less complicated (due to the main focus on the raw Intent fields) and do not require the domain knowledge of browser SOP and file protocol. The exploit for Facebook Next Intent issue in [25] is also launched from `file://`, but it does not aim at stealing Facebook app's private files as the Facebook FileCross attack reported in [26].

Android Dynamic Testing. Besides our system, there are a number of other Android dynamic testing systems proposed for various purposes. Systems from the software engineering community aim at improving the app test rates by covering more code paths (e.g., [15,16,27]) with lower costs [28] and in more flexible ways [29]. In contrast, systems for security testing focus on adding more dedicated components, such as taint tracking in [13], fingerprint generator in [14], and pre-performed static analysis in [10]. In our case, we also embed an EBI scoring module and two dedicated components (i.e., the Attack Executer and Web Receiver) into our system, making it the first system for detecting the `file://` vulnerabilities in Android browsers.

7 Conclusions and Future Works

In this paper, we identified a class of attacks in Android called FileCross that exploits the vulnerable `file://` to obtain user's private files, such as cookies, bookmarks, and browsing histories. We designed and implemented an automatic system to detect the vulnerabilities in 115 browser apps. Our results show that the vulnerabilities are prevalent in Android browsers. More than half of our tested 115 browser apps were found vulnerable. A further detailed analysis yielded more insights into the current browser practices, such as exposed browsing interfaces and allowing `file://` access to private file zones. Our vulnerability reports also helped around ten developers patch their vulnerable browsers promptly. For one browser, our system helped discover that their first patch failed to block the vulnerability.

Our system currently focuses on detecting `file://` vulnerabilities in Android browsers. However, the FileCross attacks may also exist in other kinds of apps that use web engine APIs. For example, Facebook was identified vulnerable to attack A2 [26], although it only suffered with another issue called Next Intent [25]. Detecting `file://` vulnerabilities in these non-browser apps is a future work of our system. We plan to incorporate static analysis techniques to identify "similar" browsing interfaces which may not have clear EBI patterns.

There are another two limitations in our current system and experiments. First, some browsers have the splash or welcome views in the front of their browsing interfaces, which may interfere with our automatic attacks. But we also notice several such cases (e.g., Next and Boat browsers) that actually do not affect the effectiveness of our attacks, because the underlying component is still the browsing interface although it is not visible. Second, our current experiments do not cover the default browsers which are pre-installed in devices, because we do not have enough phones to collect and test them.

Acknowledgements. We thank the three anonymous reviewers for their critical comments. This work is partially supported by a grant (ref. no. ITS/073/12) from the Innovation Technology Fund in Hong Kong.

References

1. Mozilla: Same-origin policy, `https://developer.mozilla.org/en-US/docs/Web/Security/Same-origin_policy`
2. Terada, T.: Chrome for Android download function information disclosure, `https://code.google.com/p/chromium/issues/detail?id=144820`
3. Terada, T.: Chrome for Android bypassing SOP for local files by symlinks, `https://code.google.com/p/chromium/issues/detail?id=144866`
4. Terada, T.: Mfsa 2013-84: Same-origin bypass through symbolic links, `http://www.mozilla.org/security/announce/2013/mfsa2013-84.html`
5. W3C: Xmlhttprequest, `http://www.w3.org/TR/XMLHttpRequest/`
6. Android: Category browsable, `http://developer.android.com/reference/android/content/Intent.html#CATEGORY_BROWSABLE`
7. Android: Intents and Intent Filters, `http://developer.android.com/guide/components/intents-filters.html`
8. Chin, E., Felt, A.P., Greenwood, K., Wagner, D.: Analyzing inter-application communication in Android. In: Proc. ACM MobiSys (2011)
9. Android: MonkeyRunner, `http://developer.android.com/tools/help/monkeyrunner_concepts.html`
10. Sounthiraraj, D., Sahs, J., Greenwood, G., Lin, Z., Khan, L.: SMV-Hunter: Large scale, automated detection of SSL/TLS man-in-the-middle vulnerabilities in Android apps. In: Proc. ISOC NDSS (2014)
11. Wu, D., Chang, R.: Analyzing Android browser apps for file: vulnerabilities (Technical Report) (2014), `http://arxiv.org/abs/1404.4553`
12. Selenium: Selenium - web browser automation, `http://docs.seleniumhq.org/`
13. Rastogi, V., Chen, Y., Enck, W.: AppsPlayground: Automatic security analysis of smartphone applications. In: Proc. ACM CODASPY (2013)
14. Dai, S., Tongaonkar, A., Wang, X., Nucci, A., Song, D.: Networkprofiler: Towards automatic fingerprinting of Android apps. In: Proc. IEEE INFOCOM (2013)
15. Anand, S., Naik, M., Harrold, M., Yang, H.: Automated concolic testing of smartphone apps. In: Proc. ACM FSE (2012)
16. Machiry, A., Tahiliani, R., Naik, M.: Dynodroid: An input generation system for Android apps. In: Proc. ACM FSE (2013)
17. Luo, T., Hao, H., Du, W., Wang, Y., Yin, H.: Attacks on webview in the Android system. In: Proc. ACM ACSAC (2011)
18. Chin, E., Wagner, D.: Bifocals: Analyzing webView vulnerabilities in Android applications. In: Kim, Y., Lee, H., Perrig, A. (eds.) WISA 2013. LNCS, vol. 8267, pp. 129–146. Springer, Heidelberg (2014)

19. Georgiev, M., Jana, S., Shmatikov, V.: Breaking and fixing origin-based access control in hybrid web/mobile application frameworks. In: Proc. ISOC NDSS (2014)
20. Grace, M., Zhou, Y., Wang, Z., Jiang, X.: Systematic detection of capability leaks in stock Android smartphones. In: Proc. ISOC NDSS (2012)
21. Lu, L., Li, Z., Wu, Z., Lee, W., Jiang, G.: CHEX: Statically vetting Android apps for component hijacking vulnerabilities. In: Proc. ACM CCS (2012)
22. Zhou, Y., Jiang, X.: Detecting passive content leaks and pollution in Android applications. In: Proc. ISOC NDSS (2013)
23. Octeau, D., McDaniel, P., Jha, S., Bartel, A., Bodden, E., Klein, J., Traon, Y.: Effective inter-component communication mapping in Android with Epicc: An essential step towards holistic security analysis. In: Proc. Usenix Security (2013)
24. Wu, L., Grace, M., Zhou, Y., Wu, C., Jiang, X.: The impact of vendor customizations on Android security. In: Proc. ACM CCS (2013)
25. Wang, R., Xing, L., Wang, X., Chen, S.: Unauthorized origin crossing on mobile platforms: Threats and mitigation. In: Proc. ACM CCS (2013)
26. Terada, T.: Facebook for Android - information diclosure vulnerability, http://seclists.org/bugtraq/2013/Jan/27
27. Azim, T., Neamtiu, I.: Targeted and depth-first exploration for systematic testing of Android apps. In: Proc. ACM OOPSLA (2013)
28. Choi, W., Necula, G., Sen, K.: Guided GUI testing of Android apps with minimal restart and approximate learning. In: Proc. ACM OOPSLA (2013)
29. Hao, S., Liu, B., Nath, S., Halfond, W., Govindan, R.: PUMA: Programmable UI-automation for large scale dynamic analysis of mobile apps. In: Proc. ACM MobiSys (2014)

Expressive and Secure Searchable Encryption in the Public Key Setting

Zhiquan Lv[1,3], Cheng Hong[1], Min Zhang[1,2], and Dengguo Feng[1]

[1] Trusted Computing and Information Assurance Laboratory, Institute of Software, Chinese Academy of Sciences, Beijing, China
[2] State Key Laboratory of Computer Science, Institute of Software, Chinese Academy of Sciences, Beijing, China
[3] University of Chinese Academy of Sciences, Beijing, China
{lvzhiquan,hongcheng,mzhang,feng}@tca.iscas.ac.cn

Abstract. Searchable encryption allows an untrusted server to search on encrypted data without knowing the underlying data contents. Traditional searchable encryption schemes focus only on single keyword or conjunctive keyword search. Several solutions have been recently proposed to design more expressive search criteria, but most of them are in the setting of symmetric key encryption. In this paper, based on the composite-order groups, we present an expressive and secure asymmetric searchable encryption (ESASE) scheme, which is the first that simultaneously supports conjunctive, disjunctive and negation search operations. We analyze the efficiency of ESASE and prove it is secure in the standard model. In addition, we show that how ESASE could be extended to support the range search and the multi-user setting.

Keywords: Searchable Encryption, Asymmetric Searchable Encryption, Expressive Search.

1 Introduction

In a remote storage system, a user can store his data on the remote server and then access the data using his PC or mobile devices. Since the server cannot always be fully trusted, data containing sensitive information must be encrypted to protect the user's privacy. However, it makes retrieval of such encrypted data difficult. To cope with this problem, searchable encryption schemes have been proposed, which can be divided into two versions: symmetric searchable encryption (SSE) and asymmetric searchable encryption (ASE).

However, most traditional searchable encryption schemes focus only on single keyword search [1–3, 5, 17] or multiple keyword search [4, 8, 9, 15, 16], practical systems desire a more expressive search. In the symmetric key setting, some solutions have been recently designed for general boolean queries on encrypted data [7, 18]. However, in the public key setting, to the best of our knowledge, there are only two related works [6, 12].

Consider a secure searchable email system. To support the keyword search, each email will be defined some keyword fields, such as *"Sender"*, *"Priority"*,

S.S.M. Chow et al. (Eds.): ISC 2014, LNCS 8783, pp. 364–376, 2014.
© Springer International Publishing Switzerland 2014

and "*Month*". Here we use "Z_1", "Z_2" and "Z_3" to denote these fields respectively. Before sending an email, a sender first encrypts the email content using a standard public key encryption algorithm with the receiver's public key, and then appends some additional encrypted keywords of the keyword fields, such as "*Alice*", "*urgent*", and "*October*". Generally speaking, the ASE schemes focus only on the keywords encryption since it is well known that the standard public key encryption is secure. In schemes [6, 12], an email gateway might be given a trapdoor for the boolean formula ($Z_1 = $ "*Alice*" OR $Z_2 = $ "*urgent*"), which indicates that the receiver wants the gateway to return all emails either sent by Alice or having urgent priority. Since the receiver might have read all the emails in September, the trapdoor for boolean formula would actually like (($Z_1 = $ "*Alice*" OR $Z_2 = $ "*urgent*") AND (NOT $Z_3 = $ "*September*")). However, neither of schemes [6, 12] can work in the situations involving negation search operation. Moreover, the former is less secure and the latter is less efficient.

Our Contributions. Our contributions are summarized as follows.

- Based on the composite-order groups, we propose an expressive and secure ASE scheme named ESASE. To the best of our knowledge, ESASE is the first that simultaneously supports conjunctive, disjunctive and negation search operations in the public key setting.

- We give a detail security proof of ESASE in the standard model. Compared with [6], ESASE does not disclose the searching keywords in the trapdoor. Furthermore, the efficiency analysis shows that the overhead on storage, communication, and computation in ESASE is close to [6], but much lower than [12].

- We show how ESASE could be extended to support a class of simple range search, which can help the user choose the search ranges based on different choices of granularity. Furthermore, we extend ESASE to the multi-user setting, which can minimize the overhead on computation and communication.

1.1   Related Work

Song et al. [1] initiate the research on searchable encryption and present the first practical solution in the symmetric key setting. After this work, several works [2, 3, 11, 13, 14, 17] have been proposed to improve the efficiency of the system or provide stronger security. Boneh et al. [5]. first address the concept of asymmetric searchable encryption, where anyone can use the public key to encrypt the data and keywords but only authorized users with secret key can search. However, these early schemes focus only on single keyword search.

Golle et al. [4] propose the first solution for symmetric conjunctive keyword search, where each encrypted file is associated with encrypted keywords that are assigned to separate keyword fields. If a user queries a trapdoor with several keywords, the server can search a file containing those keywords with this trapdoor. Later, Park et al. [16] propose the notion of asymmetric conjunctive keyword search. To improve the efficiency or security in conjunctive keyword search, several solutions are presented in [8, 9, 15].

To design more expressive search criteria, Cash et al. [18] propose the first searchable symmetric encryption protocol that supports conjunctive search and general boolean queries on outsourced data. One of the drawbacks of their scheme is that the trapdoors they generate are deterministic, in the sense that the same trapdoor will always be generated for the same keyword. Thus, it leaks statistical information about the user's search pattern [19]. Based on the orthogonalization of the keyword field according to the Gram-Schmidt process, Moataz et al. [7] present a similar solution which avoids this drawback in [18]. In the public key setting, Katz et al. [12] propose an inner-product predicate encryption scheme, which can enable the disjunctive search over encrypted data using polynomial evaluation. However, as pointed out in [12], the complexity is proportional to d^t, where t is the number of variables and d is the maximum degree (of the resulting polynomial) in each variable. Recently, Lai et al. [6] present a more efficient construction based on the fully secure key-policy attribute-based encryption (KP-ABE) scheme [10]. Compared with [12], the size of a ciphertext (or a trapdoor) in [6] is linear with the number of keyword fields (or the size of the search predicate), not superpolynomial. However, scheme [6] discloses the searching keywords in the trapdoor, which will let the server learn whether the encrypted data contains the keywords in the trapdoor.

2   Preliminaries

2.1   Linear Secret Sharing Schemes

Our construction will make essential use of linear secret-sharing schemes (LSSS). We first describe the definition of access structure [20] that will be used in LSSS. Then, we give the definition of LSSS adapted from [20].

Definition 1 (Access Structure). *Let* $\{P_1, \ldots, P_m\}$ *be a set of parties. A collection* $\mathbb{A} \subseteq 2^{\{P_1,\ldots,P_m\}}$ *is monotone if* $\forall B, C$ *: if* $B \in \mathbb{A}$ *and* $B \subseteq C$ *then* $C \in \mathbb{A}$*. An* access structure *(respectively, monotonic access structure) is a collection (respectively, monotone collection)* $\mathbb{A}$ *of non-empty subsets of* $\{P_1, \ldots, P_m\}$*, i.e.,* $\mathbb{A} \subseteq 2^{\{P_1,\ldots,P_m\}} \backslash \{\emptyset\}$*. The sets in* $\mathbb{A}$ *are called the* authorized sets, *and the sets not in* $\mathbb{A}$ *are called the* unauthorized sets.

Definition 2 (Linear Secret-Sharing Schemes (LSSS)). *Suppose there exists a linear secret sharing structure* $\mathbb{A} = (\mathbf{A}, \rho)$*, where* $\mathbf{A}$ *is a* $\ell \times m$ *matrix and* ρ *is an injective function from* $\{1, \ldots, \ell\}$ *to a party. Let* $S \in \mathbb{A}$ *be any authorized set, and* $I \subset \{1, \ldots, \ell\}$ *be defined as* $I = \{i : \rho(i) \in S\}$*. Therefore, there exist constants* $\{\sigma_i \in \mathbb{Z}_p\}$ *such that* $\sum_{i \in I} \sigma_i A_i = (1, 0, \ldots, 0)$*, where* A_i *is the* i*'th row of matrix* $\mathbf{A}$*. When we consider the column vector* $\upsilon = (s, r_2, \ldots, r_m)$*, where* $s \in \mathbb{Z}_p$ *is the secret to be shared, and* $r_2, \ldots, r_m \in \mathbb{Z}_p$ *are randomly chosen, then* $\mathbf{A}\upsilon$ *can be regarded as the linear secret sharing where each share* $A_i \upsilon$ *belongs to party* $\rho(i)$*. On the other hand, given a party set* S *and its corresponding rows* $I = \{i : \rho(i) \in S\}$ *in the matrix* $\mathbf{A}$*, finding* $\{\sigma_i \in \mathbb{Z}_p\}$ *such that* $\sum_{i \in I} \sigma_i A_i \upsilon = s$ *is called linear secret reconstruction. Note that, for unauthorized sets, no such constants exist.*

In our context, the role of the parties is taken by the keywords. Access structures (i.e., search predicates) might also be described in terms of monotonic boolean formulas. As described in [20], any monotonic boolean formula can be converted into an LSSS representation.

2.2 Composite-Order Bilinear Groups

We will construct our scheme in composite-order bilinear groups [21]. Let $\mathbb{G}$ and $\mathbb{G}_T$ be two multiplicative cyclic groups of order $N = p_1 p_2 p_3 p_4$, where p_1, p_2, p_3 and p_4 are distinct primes. Let g be a generator of $\mathbb{G}$ and $e : \mathbb{G} \times \mathbb{G} \longrightarrow \mathbb{G}_T$ be a bilinear map such that $e(g, g) \neq 1$, and for any $u, v \in \mathbb{G}$, $a, b \in \mathbb{Z}_N$ it holds $e(u^a, v^b) = e(u, v)^{ab}$. We say that $\mathbb{G}$ is a bilinear group if the group operation in $\mathbb{G}$ and the bilinear map e are both efficiently computable.

Let $\mathbb{G}_{p_1}$, $\mathbb{G}_{p_2}$, $\mathbb{G}_{p_3}$ and $\mathbb{G}_{p_4}$ denote the subgroups of order p_1, p_2, p_3 and p_4 in $\mathbb{G}$ respectively. Observe that $\mathbb{G} = \mathbb{G}_{p_1} \times \mathbb{G}_{p_2} \times \mathbb{G}_{p_3} \times \mathbb{G}_{p_4}$. We note that if $h_i \in \mathbb{G}_{p_i}$ and $h_j \in \mathbb{G}_{p_j}$ for $i \neq j$, $e(h_i, h_j) = 1$. To see this, suppose $h_1 \in \mathbb{G}_{p_1}$ and $h_2 \in \mathbb{G}_{p_2}$. We let g denote a generator of $\mathbb{G}$. Then, $g^{p_2 p_3 p_4}$ generates $\mathbb{G}_{p_1}$, and $g^{p_1 p_3 p_4}$ generates $\mathbb{G}_{p_2}$. Hence, we can have:

$$e(h_1, h_2) = e((g^{p_2 p_3 p_4})^{\alpha_1}, (g^{p_1 p_3 p_4})^{\alpha_2}) = e(g^{p_3 p_4 \alpha_1}, g^{\alpha_2})^{p_1 p_2 p_3 p_4} = 1,$$

where $\alpha_1 = \log_{g^{p_2 p_3 p_4}} h_1$ and $\alpha_2 = \log_{g^{p_1 p_3 p_4}} h_2$. This orthogonality property will be a principal tool in our construction and security proof.

2.3 Complexity Assumptions

We now state the complexity assumptions that will be used in our security proof. The first two assumptions are also used in [6] and the third can be easily proved by utilizing the theorems proposed in [12]. We note all of them hold in the generic group model.

Assumption 1. Let $N, \mathbb{G}, \mathbb{G}_T, e$ be defined as in above section. Let $g \in \mathbb{G}_{p_1}$, $X_3 \in \mathbb{G}_{p_3}$, and $X_4 \in \mathbb{G}_{p_4}$ be chosen at random. Given the tuple $(\mathbb{G}, \mathbb{G}_T, N, e, g, X_3, X_4)$, this assumption states that no probabilistic polynomial-time algorithm $\mathcal{B}$ can distinguish the random elements in $\mathbb{G}_{p_1}$ and $\mathbb{G}_{p_1 p_2}$ with non-negligible advantage.

Assumption 2. Let $N, \mathbb{G}, \mathbb{G}_T, e$ be defined as in above section. Let $g, X_1 \in \mathbb{G}_{p_1}$, $X_2, Y_2 \in \mathbb{G}_{p_2}$, $X_3, Y_3 \in \mathbb{G}_{p_3}$, and $X_4 \in \mathbb{G}_{p_4}$ be chosen at random. Given the tuple $(\mathbb{G}, \mathbb{G}_T, N, e, g, X_1 X_2, Y_2 Y_3, X_3, X_4)$, this assumption states that no probabilistic polynomial-time algorithm $\mathcal{B}$ can distinguish the random elements in $\mathbb{G}_{p_1 p_2 p_3}$ and $\mathbb{G}_{p_1 p_3}$ with non-negligible advantage.

Assumption 3. Let $N, \mathbb{G}, \mathbb{G}_T, e$ be defined as in above section. Let $s \in \mathbb{Z}_N$, $g, u \in \mathbb{G}_{p_1}$, $g_2, X_2 \in \mathbb{G}_{p_2}$, $X_3 \in \mathbb{G}_{p_3}$, $X_4, h' \in \mathbb{G}_{p_4}$, and $B_{24}, D_{24} \in \mathbb{G}_{p_2 p_4}$ be chosen at random. Given the tuple $(\mathbb{G}, \mathbb{G}_T, N, e, g, g_2, X_2, uh', g^s B_{24}, X_3, X_4)$, this assumption states that no probabilistic polynomial-time algorithm $\mathcal{B}$ can distinguish $u^s D_{24}$ and a random element in $\mathbb{G}_{p_1 p_2 p_4}$ with non-negligible advantage.

3 Syntax and Security Model

3.1 Syntax

We consider that a user' s encrypted data is outsourced in the storage of an untrusted server, such as an email gateway. To support the keyword search, we will define some keyword fields for the emails, such as "Sender", "Priority", "Month". Suppose each email is associated with a keyword set $W = \{w_1, \ldots, w_n\}$, where w_i is the keyword of the email in the ith keyword field and n is the number of keyword fields. Moreover, we employ the same assumption used in the previous works [4, 6, 9, 16]: every keyword field is defined for every email. An ESASE scheme consists of the following fundamental algorithms:

Setup(1^κ): takes as input a security parameter 1^κ, and generates a public key pk and secret key sk.

Encrypt(pk, W): for the public key pk and a keyword set W, produces a ciphertext C_W.

Trapdoor(pk, sk, $\mathcal{P}$): given the public key pk, the secret key sk, and a predicate $\mathcal{P}$, produces a trapdoor $T_\mathcal{P}$.

Test(pk, C_W, $T_\mathcal{P}$): takes as input the public key pk, the ciphertext C_W, and the trapdoor $T_\mathcal{P}$. It outputs "1" if the keyword set W satisfies the predicate $\mathcal{P}$ and "0" otherwise.

3.2 Security Model

In ESASE, we assume the server who runs the **Test** algorithm is Honest but Curious. That is to say, it will execute correctly the proposed algorithm, but try to find out as much secret information as possible based on its inputs. Moreover, we define the security notion in the sense of indistinguishability security against chosen predicate attacks. Formally, security is defined using the following game between an attacker and a challenger.

Setup. The challenger takes a security parameter 1^κ and runs the Setup algorithm. The public key pk is given to the adversary. The challenger keeps the secret key sk to itself.

Phase 1. The adversary adaptively queries a number of predicates, $\mathcal{P}_1$, ..., $\mathcal{P}_q$, to the challenger. In response, the challenger runs the Trapdoor algorithm and gives the trapdoor $T_{\mathcal{P}_i}$ to the adversary, for $1 \leq i \leq q$.

Challenge. The adversary sends two target keyword sets W_0, W_1 to the challenger. The challenger picks a random bit $\beta \in \{0, 1\}$ and obtains the ciphertext C_{W_β} by running the Encrypt algorithm. Then, he gives C_{W_β} to the adversary. Note that W_0 and W_1 cannot satisfy any of queried predicates in phase 1.

Phase 2. The adversary additionally queries the challenger for trapdoors corresponding to predicates with the restriction that none of them can be satisfied by W_0 and W_1.

Guess. The adversary outputs a guess β' of β and wins the game if $\beta' = \beta$.

The advantage of the adversary in this game is defined as $|\Pr[\beta' = \beta] - \frac{1}{2}|$.

Definition 3. *An ESASE scheme is secure if all polynomial time adversaries have at most a negligible advantage in this security game.*

4   Construction

4.1   Overview

Consider any search predicate with negation operations, we can use the DeMorgan's law to propagate the negation operations. For example, a search predicate $(Z_1 = $ "A" AND (NOT $(Z_3 = $ "B" AND $Z_4 = $ "C"))) equals $(Z_1 = $ "A" AND ((NOT $Z_3 = $ "B") OR (NOT $Z_4 = $ "C"))), where A, B and C denote the keywords in keyword fields Z_1, Z_3 and Z_4 respectively. After the propagation, the negation operations can only be associated with keywords. Thus, the name of the keywords in our scheme may be of two types: either the name is normal (like x) or it is *primed* (like x'). We conceptually associate primed keywords as representing the negation of unprimed keywords. Therefore, the search predicate $(Z_1 = $ "A" AND (NOT $(Z_3 = $ "B" AND $Z_4 = $ "C"))) can be parsed as $(Z_1 = $ "A" AND $(Z_3 = $ "B'" OR $Z_4 = $ "C'")).

As described previously, any monotonic boolean search predicate can be converted into an LSSS representation $\mathbb{A} = (\mathbf{A}, \rho)$, where $\mathbf{A}$ is a $\ell \times m$ matrix and ρ is a map from each row of $\mathbf{A}$ to a keyword field (i.e., ρ is a function from $\{1, \ldots, \ell\}$ to $\{1, \ldots, n\}$). Let $\mathcal{E} = \{t_{\rho(1)}, \ldots, t_{\rho(\ell)}\}$, $\mathcal{F} = \{f_{\rho(1)}, \ldots, f_{\rho(\ell)}\}$, where $t_{\rho(i)}$ is the keyword of keyword field $\rho(i)$ specified by the predicate and $f_{\rho(i)}$ denotes whether the keyword of keyword field $\rho(i)$ is primed or not (We can use " $-$ " and " $+$ " to denote it is primed and not primed respectively).

Thus, in our scheme, any boolean search predicate, which simultaneously supports the operations of conjunctive and disjunctive and negation, can be converted into an LSSS representation $\mathbb{A} = (\mathbf{A}, \rho, \mathcal{E}, \mathcal{F})$. A keyword set $W = \{w_1, \ldots, w_n\}$ satisfies a predicate $\mathbb{A} = (\mathbf{A}, \rho, \mathcal{E}, \mathcal{F})$ if and only if there exist $I \subset \{1, \ldots, \ell\}$ and constants $\{\sigma_i\}_{i \in I}$ such that $\sum_{i \in I} \sigma_i A_i = (1, 0, \ldots, 0)$, where A_i is the i'th row of matrix $\mathbf{A}$. Note that for $\forall i \in I$, if $f_{\rho(i)}$ denotes the keyword of keyword field $\rho(i)$ is not primed, then $t_{\rho(i)} = w_{\rho(i)}$; else, $t_{\rho(i)} \neq w_{\rho(i)}$. In addition, we define $I_{\mathbf{A}, \rho}$ as the set of minimum subsets of $\{1, \ldots, \ell\}$ that satisfies $(\mathbf{A}, \rho, \mathcal{E}, \mathcal{F})$. Figure 1 shows an example of converting a search predicate into an LSSS representation, where $\ell = 3$ and $I_{\mathbf{A}, \rho} = \{\{1, 2\}, \{1, 3\}\}$.

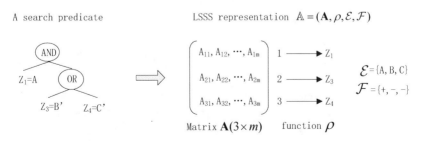

Fig. 1. An example of the conversion

4.2 Main Construction

Setup(1^κ): Take as input a security parameter 1^κ, it returns `params` $= (p_1, p_2, p_3, p_4, \mathbb{G}, \mathbb{G}_T, e)$ with $\mathbb{G} = \mathbb{G}_{p_1} \times \mathbb{G}_{p_2} \times \mathbb{G}_{p_3} \times \mathbb{G}_{p_4}$, where $\mathbb{G}$ and $\mathbb{G}_T$ are cyclic groups of order $N = p_1 p_2 p_3 p_4$. Next it chooses $g, u_1, \ldots, u_n \in \mathbb{G}_{p_1}, X_3 \in \mathbb{G}_{p_3}$, $X_4, h_1, \ldots, h_n \in \mathbb{G}_{p_4}$ and $\alpha \in \mathbb{Z}_N$ uniformly at random. The public key pk is published as:

$$N, g, g^\alpha, \{H_i = u_i \cdot h_i\}_{1 \le i \le n}, X_4.$$

The secret key $sk = \{u_1, \ldots, u_n, \alpha, X_3\}$.

Encrypt(pk, W): The encryption algorithm chooses $s \in \mathbb{Z}_N$ and $h, Z_0, \{Z_{1,i}\}_{1 \le i \le n}$ $\in \mathbb{G}_{p_4}$ uniformly at random. Let the keyword set be $W = (w_1, \ldots, w_n) \in \mathbb{Z}_N^n$, then the corresponding ciphertext $C_W = (\widetilde{C}, C_0, \{C_i\}_{1 \le i \le n})$ is computed as:

$$\widetilde{C} = e(g, g^\alpha)^s, C_0 = (gh)^s \cdot Z_0, C_i = (H_i^{w_i})^s \cdot Z_{1,i}.$$

Trapdoor$(pk, sk, \mathcal{P})$: Suppose that the predicate $\mathcal{P}$ is corresponding to $(\mathbf{A}, \rho, \mathcal{E}, \mathcal{F})$, where $\mathbf{A}$ is the $\ell \times m$ matrix, $\rho(i)$ is a map from each row A_i of $\mathbf{A}$ to $\{1, \ldots, n\}$, $\mathcal{E} = \{t_{\rho(1)}, \ldots, t_{\rho(\ell)}\} \in \mathbb{Z}_N^\ell$, and $\mathcal{F} = \{f_{\rho(1)}, \ldots, f_{\rho(\ell)}\}$. The algorithm first chooses a random vector $\upsilon \in \mathbb{Z}_N^m$ such that $\mathbf{1} \cdot \upsilon = \alpha$, where $\mathbf{1} = (1, 0, \ldots, 0)$. Then, for each row A_i of $\mathbf{A}$, it chooses a random $r_i \in \mathbb{Z}_N$ and random elements $V_{1,i}, V_{2,i}, V_{3,i} \in \mathbb{G}_{p_3}$. The trapdoor key $T_{\mathcal{P}} = ((\mathbf{A}, \rho, \mathcal{F}), \{D_{1,i}, D_{2,i}, D_{3,i}\}_{1 \le i \le \ell})$ is computed as

$$D_{1,i} = g^{A_i \cdot \upsilon} (u_{\rho(i)})^{t_{\rho(i)} \cdot r_i} \cdot V_{1,i}, D_{2,i} = g^{r_i} \cdot V_{2,i}, D_{3,i} = (u_{\rho(i)})^{t_{\rho(i)} \cdot r_i} \cdot V_{3,i}.$$

Test$(pk, C_W, T_{\mathcal{P}})$: Given a ciphertext C_W and a trapdoor $T_{\mathcal{P}}$, the algorithm first calculates $I_{\mathbf{A},\rho}$. Then, it checks if there exists an set $I \in I_{\mathbf{A},\rho}$ that satisfies

$$\widetilde{C} = \prod_{i \in I} (U_i)^{\sigma_i},$$

where $\sum_{i \in I} \sigma_i A_i = (1, 0, \ldots, 0)$. For $i \in I$: if $f_{\rho(i)}$ denotes the keyword of keyword field $\rho(i)$ is not primed, the above $U_i = e(D_{1,i}, C_0)/e(D_{2,i}, C_{\rho(i)})$; else, $U_i = e(D_{1,i}, C_0)/e(D_{3,i}, C_0)$ and the equality $e(C_0, D_{3,i}) = e(C_{\rho(i)}, D_{2,i})$ does not hold.

If no element in $I_{\mathbf{A},\rho}$ satisfies it, it outputs "0". Otherwise, it outputs "1".

4.3 Discussion

In Lai et al. [6], the searching keywords are disclosed in the trapdoor, which will cause a Test algorithm executor (i.e., an email gateway) to learn whether the encrypted data contains the keywords in the trapdoor. However, ESASE just discloses whether the keywords in the trapdoor are primed or not. We note that it might be inevitable in an ASE scheme that can support negation search operation, but it is acceptable if the scheme is proved secure in the security model defined in Section 3.2.

5   Security

Theorem 1. *If Assumptions 1, 2 and 3 hold, then ESASE is secure.*
Proof. The proof consists of the following three steps:

1. **Constructing the semi-functional ciphertexts and keys**
   We define two additional structures: semi-functional ciphertexts and keys,
   which will not be used in the real system, but will be needed in our proof.
   Semi-functional Ciphertext. A semi-functional ciphertext is formed as fol-
   lows. Firstly, a normal ciphertext $C'_W = (\widetilde{C}', C'_0, \{C'_i\}_{1\le i\le n})$ is generated
   by the Encrypt algorithm. Then, for each component C'_i, a random value
   $\gamma_i \in \mathbb{Z}_N$ is chosen. Let g_2 be a generator of $\mathbb{G}_{p_2}$ and $c \in \mathbb{Z}_N$ be a random
   exponent, the semi-functional ciphertext $C_W = (\widetilde{C}, C_0, \{C_i\}_{1\le i\le n})$ is set to
   be:

   $$\widetilde{C} = \widetilde{C}', C_0 = C'_0 \cdot g_2^c, C_i = C'_i \cdot g_2^{c\gamma_i}.$$

   Note that the values $\{\gamma_i\}_{1\le i\le n}$ are chosen randomly once and then fixed in
   the following semi-functional keys.
   Semi-functional Key. To create a semi-functional key, a normal trapdoor
   key $T'_{\mathcal{P}} = ((\mathbf{A}, \rho, \mathcal{F}), \{D'_{1,i}, D'_{2,i}, D'_{3,i}\}_{1\le i\le \ell})$ should be first generated by
   the Trapdoor algorithm. Then, it chooses a random vector $z \in \mathbb{Z}_N^m$ and
   sets $\delta_i = A_i \cdot z$. For each row i of the $\ell \times m$ matrix $\mathbf{A}$, a random value
   $\eta_i \in \mathbb{Z}_N$ is chosen. Finally, a semi-functional key $T_{\mathcal{P}} = ((\mathbf{A}, \rho, \mathcal{F}), \{D_{1,i}, D_{2,i}, D_{3,i}\}_{1\le i\le \ell})$ is divided into two types of forms. The type 1 is set to be:

   $$D_{1,i} = D'_{1,i} \cdot g_2^{\delta_i + \eta_i \gamma_{\rho(i)}}, D_{2,i} = D'_{2,i} \cdot g_2^{\eta_i}, D_{3,i} = D'_{3,i} \cdot g_2^{\eta_i \gamma_{\rho(i)}}.$$

   The type 2 is similarly formed as:

   $$D_{1,i} = D'_{1,i} \cdot g_2^{\delta_i}, D_{2,i} = D'_{2,i}, D_{3,i} = D'_{3,i}.$$

2. **Defining a sequence of attack games**
   We organize our proof as a sequence of games. The first game, $Game_{real}$, is
   the real security game defined in Section 3.2. In the next game, $Game_0$, all of
   the trapdoor keys are normal, but the challenge ciphertext is semi-functional.
   Let q be the number of trapdoor key queries made by the attacker. For k
   from 1 to q and ξ from 1 to n, we define:
   - $Game_{k,1}$: the first $k-1$ trapdoor keys are semi-functional of type 2, the
     k^{th} trapdoor key is semi-functional of type 1, and the remaining trapdoor
     keys are normal. In addition, the challenge ciphertext is semi-functional.

   - $Game_{k,2}$: the first k trapdoor keys are semi-functional of type 2, and the
     remaining trapdoor keys are normal. In addition, the challenge ciphertext
     is semi-functional.

   - $Game_{final,\xi}$: all the trapdoor keys are semi-functional of type 2, and the
     challenge ciphertext $C_{W_\beta} = (\widetilde{C}, C_0, \{C_1, \ldots, C_n\})$ is a semi-functional
     encryption of W_β with $C_1, \ldots, C_\xi$, each of which is randomly chosen
     from $\mathbb{G}_{p_1 p_2 p_4}$.

Note that $Game_{0,2}$ and $Game_{final,0}$ can be considered as another way of denoting $Game_0$ and $Game_{q,2}$ respectively.

3. **Proving these attack games are indistinguishable with the real game**
   We prove that these games are indistinguishable in four lemmas, which formal descriptions and proofs are given in the full version of this paper [24]. It is clear that in $Game_{final,n}$, the attacker's advantage is negligible since the challenge ciphertext C_{W_β} is independent of the keyword sets W_0 and W_1. Therefore, we conclude that the advantage of the adversary in $Game_{real}$ is negligible. This completes the proof of Theorem 1.

6   Efficiency

We compare our scheme with Katz et al. [12] and Lai et al. [6] in Table 1.

Storage and Communication Overhead. The storage and communication overhead is mainly determined by the sizes of the ciphertext and trapdoor respectively. Table 1 shows the size of a ciphertext (resp. a trapdoor), both in ESASE and [6], is *linear* with the number of keyword fields n (resp. the number of rows ℓ) rather than *superpolynomial* in [12].

Computation Overhead. The computation overhead is mainly determined by three algorithms, including Encryption, Trapdoor, and Test. Table 1 shows that the computation overhead of Encryption (resp. Trapdoor), both in ESASE and [6], is *linear* with n (resp. ℓ), not *superpolynomial*. In addition, since ESASE supports the negation search, the Test cost of it is a little bigger than [6].

It can be concluded that the efficiency of ESASE is close to Lai et al. [6], but much better than Katz et al. [12]. Furthermore, compared with these two schemes, ESASE can simultaneously support conjunctive, disjunctive and negation search operations.

7   Extensions

7.1   Range Search

Besides the boolean search, range search is also an important requirement for the searchable encryption. However, there is no expressive ASE scheme for the range search. Here, we show how ESASE could be extended to support a class of simple range search. Although supporting an arbitrary range search can be achieved by using the technique in scheme [22], high complexity will be introduced when there are lots of disjunctive operations in a predicate. Therefore, we are not aiming at it.

We first introduce the concept of the keyword hierarchy. Let Z_i be a numerical keyword field, a keyword hierarchy over Z_i is a balanced tree $\Gamma(Z_i)$, where each internal node represents a range that is the union of the ranges of its children nodes and each leaf node is the keyword. In addition, we define some notations as follows.

Table 1. Comparisons of efficiency of existing expressive ASE schemes

Schemes	Katz et al. [12]	Lai et al. [6]	ESASE
Size of ciphertext	$(1+2m)L_1$	$(1+n)L_1 + L_2$	$(1+n)L_1 + L_2$
Size of trapdoor	$(1+2m)L_1$	$2\ell \cdot L_1$	$3\ell \cdot L_1$
Encryption	$(1+4m) \cdot$ e	$2(1+n) \cdot$ e	$(2+n) \cdot$ e+p
Trapdoor	$6m \cdot$ e	$4\ell \cdot$ e	$4\ell \cdot$ e
Test	$(1+2m) \cdot$ p	$\leq \psi_1 \cdot$ e $+ 2\psi_2 \cdot$ p	$\leq \psi_1 \cdot$ e $+ 2(\psi_2 + \psi_3) \cdot$ p
Security	secure in the standard model	less secure	secure in the standard model
Expressiveness	AND, OR	AND, OR	AND, OR, NOT

[1] n: the number of keyword fields; ℓ: the number of rows in the matrix $\mathbf{A}$; L_1 (or L_2): a bit length of a group element in $\mathbb{G}$ (or $\mathbb{G}_T$); m: the length of the vector corresponding to the ciphertext in Katz et al. [12]. Note that m is proportional to d^ℓ, where d is the maximum degree (of the resulting polynomial) in each variable in [12].

[2] e: an exponentiation operation in $\mathbb{G}$ or $\mathbb{G}_T$; p: a pairing operation; ψ_1: the number of elements in $I_{\mathbf{A},\rho} = \{I_1, \dots, I_{\psi_1}\}$; ψ_2: $|I_1| + \dots + |I_{\psi_1}|$; ψ_3: the number of primed keywords in a search predicate.

- $\mathbb{R}(id)$: the range of a internal node id.
- $\mathbb{P}(z)$: the path from a leaf node z to the root. For every node id in $\mathbb{P}(z)$, it has $z \in \mathbb{R}(id)$.
- $\Gamma_l(Z_i)$: the node set in the l-th level of the tree, it is also called a "*level-l field*". Any node in $\Gamma_l(Z_i)$ is called a "*level-l keyword*" and denotes a "*level-l simple range*".

Suppose there is a "*Hour*" field Z_4 with keywords from number 0 to 23, Figure 2 depicts the keyword hierarchy. The path of the leaf node "12" is ("0-23", "12-17", "12-13", "12"), and "12-17", "12-13" are level-2 and level-3 simple range respectively.

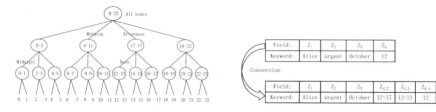

Fig. 2. The keyword hierarchy of the "*Hour*" field

Fig. 3. An example of field structure conversion for the "*Hour*" field

To support the simple range search based on the keyword hierarchy in ESASE, the original field structure should be converted in the following way: for each hierarchical field Z_i, let φ be its maximum level and z_i be a leaf node, we expand Z_i into $(\varphi - 1)$ subfields: $Z_{i,2}, \cdots, Z_{i,\varphi}$, where the value set of $Z_{i,l}$ is $\Gamma_l(Z_i)$ and the value of $Z_{i,l}$ is the l-th element in $\mathbb{P}(z_i)$. Note that since there is only one keyword in $Z_{i,1}$, we ignore this subfield. Figure 3 shows that the *Hour* field is expanded into 3 subfields. As a result, the parameter n in the Setup algorithm will be set to 6 and the keyword set to be encrypted will be $\{$"*Alice*", "*urgent*", "*October*","12$-$17","12$-$13","12"$\}$.

Moreover, before we use a predicate with range search to generate a trapdoor, we should convert it based on the system-defined simple ranges. For example, an original search predicate $(Z_1 = $ "*Alice*" AND ("6" $\le Z_4 \le$ "13")) can be converted into $(Z_1 = $ "*Alice*" AND $(Z_{4,2} = $ "6-11" OR $Z_{4,3} = $ "12-13")). Since ESASE supports the negation search, a search predicate ($Z_1 = $ "*Alice*" OR (NOT "6" $\le Z_4 \le$ "13")) can be converted into $(Z_1 = $ "*Alice*" OR $(Z_{4,2} = $ "(6-11)'" AND $Z_{4,3} = $ "(12-13)'")).

We note this class of simple range search will bring two benefits. On the one hand, the hierarchy on a numerical field can be well-designed depending on different usages. Take Figure 2 for example, the node "6-11" stands for the morning and "12-17" for the afternoon. On the other hand, a user can choose the search ranges based on different choices of granularity.

7.2   Multi-user Setting

We briefly outline how we extend ESASE scheme to the multi-user setting [9]. In a Single-user ESASE scheme, if a sender wants to send an encrypted email to multiple receivers in the example of a secure email system, he needs to encrypt the same email and the same keywords with each receiver's public key respectively. A multi-user ESASE scheme can eliminate these repeated operations and minimize the size of the public key material that needs to be obtained. We use the *randomness re-use* technique [23] as follows.

We observe that in the Trapdoor algorithm, the parameter α is the secret in the LSSS representation $\mathbb{A} = (\mathbf{A}, \rho, \mathcal{E}, \mathcal{F})$. In addition, only one element, g^α in the public key depends upon α. Therefore, we can use different α to distinguish between different users. Specifically, in the Setup algorithm of a multi-user ESASE scheme, the public parameter shared by all the users is published as: $PP = \{N, g, \{H_i\}_{1 \le i \le n}, X_4\}$. while the public/secret key pair for user x is:

$$\{pk_x, sk_x\} = \{g^{\alpha_x}, (u_1, \ldots, u_n, \alpha_x, X_3)\}.$$

Note that all the users also share the same parameters $u_1, \ldots, u_n$, and X_3.

When a sender is to encrypt a keyword set W for k users, he first uses the public parameter PP to encrypt W to obtain the components C_0 and $\{C_i\}_{1 \le i \le n}$, and then uses each user's public key pk_x to generate the components $\widetilde{C}_1, \ldots, \widetilde{C}_x, \ldots, \widetilde{C}_k$ respectively. Finally, for the user x $(1 \le x \le k)$, the corresponding ciphertext will be: $C_W = (\widetilde{C}_x, C_0, \{C_i\}_{1 \le i \le n})$.

The rest two algorithms of the multi-user ESASE scheme are the same with ESASE respectively. We compare the efficiency of the Single-user and Multi-user ESASE in Table 2. It shows that the multi-user ESASE is more efficient when there are multiple receivers.

Table 2. Comparisons of efficiency for the Single-user and Multi-user ESASE

Schemes	Single-user ESASE	Multi-user ESASE
OE	$(2+n) \cdot k\mathsf{e} + k \cdot \mathsf{p}$	$(1+n+k) \cdot \mathsf{e} + k \cdot \mathsf{p}$
SPK	$(3+n)k \cdot L_1$	$(2+n+k) \cdot L_1$

k: the number of the receivers; n, e, p, L_1: defined as in Section 6; OE: the encryption overhead; SPK: the size of the public key material that a sender needs to obtain.

8   Conclusion and Future Work

In this paper, we present a new scheme ESASE, to solve the problem of expressive search in the public key setting. To the best of our knowledge, ESASE is the first that can simultaneously support conjunctive, disjunctive and negation search operations. Then we prove ESASE is secure in the standard model. By analyzing, the efficiency of ESASE compares favorably to that of existing, less-expressive ASE schemes. Finally, we make two useful extensions: one is to support the range search, and another is to the multi-user setting.

Acknowledgment. This work was supported by National Natural Science Foundation of China under Grant No.61232005, No.61100237 and No.91118006.

References

1. Song, D.X., Wagner, D., Perrig, A.: Practical techniques for searcheson encrypted data. In: IEEE Symposium on Security and Privacy, S&P 2000, pp. 44–55 (2000)
2. Goh, E.J.: Secure indexes. IACR Cryptology ePrint Archive 2003, 216 (2003)
3. Curtmola, R., Garay, J., Kamara, S., Ostrovsky, R.: Searchable Symmetric Encryption: Improved Definitions and Efficient Constructions. In: ACM CCS 2006, pp. 79–88 (2006)
4. Golle, P., Staddon, J., Waters, B.: Secure conjunctive keyword search over encrypted data. In: Jakobsson, M., Yung, M., Zhou, J. (eds.) ACNS 2004. LNCS, vol. 3089, pp. 31–45. Springer, Heidelberg (2004)
5. Boneh, D., Di Crescenzo, G., Ostrovsky, R., Persiano, G.: Public key encryption with keyword search. In: Cachin, C., Camenisch, J. (eds.) EUROCRYPT 2004. LNCS, vol. 3027, pp. 506–522. Springer, Heidelberg (2004)
6. Lai, J., Zhou, X., Deng, R.H., Li, Y., Chen, K.: Expressive search on encrypted data. In: ACM ASIACCS 2013, pp. 243–252 (2013)
7. Moataz, T., Shikfa, A.: Boolean symmetric searchable encryption. In: ACM ASIACCS 2013, pp. 265–276 (2013)
8. Cai, K., Hong, C., Zhang, M., Feng, D., Lv, Z.: A Secure Conjunctive Keywords Search over Encrypted Cloud Data Against Inclusion-Relation Attack. In: IEEE CloudCom 2013, vol. 1, pp. 339–346 (2013)
9. Hwang, Y.-H., Lee, P.J.: Public key encryption with conjunctive keyword search and its extension to a multi-user system. In: Takagi, T., Okamoto, T., Okamoto, E., Okamoto, T. (eds.) Pairing 2007. LNCS, vol. 4575, pp. 2–22. Springer, Heidelberg (2007)

10. Lewko, A., Okamoto, T., Sahai, A., Takashima, K., Waters, B.: Fully secure functional encryption: Attribute-based encryption and (hierarchical) inner product encryption. In: Gilbert, H. (ed.) EUROCRYPT 2010. LNCS, vol. 6110, pp. 62–91. Springer, Heidelberg (2010)
11. Lu, Y.: Privacy-preserving logarithmic-time search on encrypted data in cloud. In: Network and Distributed System Security Symposium (NDSS) (2012)
12. Katz, J., Sahai, A., Waters, B.: Predicate encryption supporting disjunctions, polynomial equations, and inner products. In: Smart, N. (ed.) EUROCRYPT 2008. LNCS, vol. 4965, pp. 146–162. Springer, Heidelberg (2008)
13. Kamara, S., Papamanthou, C., Roeder, T.: Dynamic searchable symmetric encryption. In: ACM CCS 2012, pp. 965–976 (2012)
14. Kamara, S., Papamanthou, C.: Parallel and dynamic searchable symmetric encryption. In: Sadeghi, A.-R. (ed.) FC 2013. LNCS, vol. 7859, pp. 258–274. Springer, Heidelberg (2013)
15. Ballard, L., Kamara, S., Monrose, F.: Achieving efficient conjunctive keyword searches over encrypted data. In: Qing, S., Mao, W., López, J., Wang, G. (eds.) ICICS 2005. LNCS, vol. 3783, pp. 414–426. Springer, Heidelberg (2005)
16. Park, D.J., Kim, K., Lee, P.J.: Public key encryption with conjunctive field keyword search. In: Lim, C.H., Yung, M. (eds.) WISA 2004. LNCS, vol. 3325, pp. 73–86. Springer, Heidelberg (2005)
17. Chang, Y.-C., Mitzenmacher, M.: Privacy Preserving Keyword Searches on Remote Encrypted Data. In: Ioannidis, J., Keromytis, A.D., Yung, M. (eds.) ACNS 2005. LNCS, vol. 3531, pp. 442–455. Springer, Heidelberg (2005)
18. Cash, D., Jarecki, S., Jutla, C., Krawczyk, H., Roşu, M.-C., Steiner, M.: Highly-Scalable Searchable Symmetric Encryption with Support for Boolean Queries. In: Canetti, R., Garay, J.A. (eds.) CRYPTO 2013, Part I. LNCS, vol. 8042, pp. 353–373. Springer, Heidelberg (2013)
19. Islam, M., Kuzu, M., Kantarcioglu, M.: Access pattern disclosure on searchable encryption: Ramification, attack and mitigation. In: Network and Distributed System Security Symposium (NDSS) (2012)
20. Beimel, A.: Secure schemes for secret sharing and key distribution, Doctoral dissertation. PhD thesis, Israel Institute of Technology (1996)
21. Boneh, D., Goh, E.-J., Nissim, K.: Evaluating 2-DNF formulas on ciphertexts. In: Kilian, J. (ed.) TCC 2005. LNCS, vol. 3378, pp. 325–341. Springer, Heidelberg (2005)
22. Lewko, A., Waters, B.: New proof methods for attribute-based encryption: Achieving full security through selective techniques. In: Safavi-Naini, R., Canetti, R. (eds.) CRYPTO 2012. LNCS, vol. 7417, pp. 180–198. Springer, Heidelberg (2012)
23. Bellare, M., Boldyreva, A., Staddon, J.: Randomness re-use in multi-recipient encryption schemeas. In: Desmedt, Y.G. (ed.) PKC 2003. LNCS, vol. 2567, pp. 85–99. Springer, Heidelberg (2002)
24. Lv, Z., Hong, C., Zhang, M., Feng, D.: Expressive and Secure Searchable Encryption in the Public Key Setting (Full Version). Cryptology ePrint Archive, Report 2014/614 (2014), http://eprint.iacr.org/

Graded Encryption, or How to Play "Who Wants To Be A Millionaire?" Distributively

Aggelos Kiayias[2], Murat Osmanoglu[1], and Qiang Tang[1]

[1] University of Connecticut, USA
[2] National and Kapodistrian University of Athens, Greece
{aggelos,murat,qiang}@cse.uconn.edu

Abstract. We propose a new identity-based cryptographic primitive which we call graded encryption. In a graded encryption scheme, there is one central (mostly offline) authority and a number of sub-authorities holding master keys that correspond to different levels. As in identity-based encryption, a sender can encrypt a message using only the identity of the receiver (plus public parameters) but it may also specify a numerical grade i. Users may decrypt messages directed to their identity at grade i as long as they have executed a *key-upgrade* protocol with sub-authorities $1, \ldots, i$. We require a grade i ciphertext to be secure in a strong sense: as long as there is one sub-authority with index $j \leq i$ that is not corrupted, the plaintext should be hidden from any recipient that has not properly upgraded her identity.

Graded encryption is motivated by multi-stage games (e.g., "who wants to be a millionaire") played in a distributed fashion. Players unlock ciphertexts that belong to a certain stage i only if they have met all challenges posed by sub-authorities $\{1, \ldots, i\}$. This holds true even if players collaborate with some of the sub-authorities. We give an efficient construction that has secret key and ciphertext size of a constant number of group elements. We also demonstrate further applications of graded encryption such as proving that a certain path was followed in a graph.

1 Introduction

In the game show "who wants to be a millionaire", participants earn money as rewards by answering a series of consecutive multiple-choice questions. The questions are distributed in several rounds, and the difficulty and the amount of money of the questions increase with the rounds. After correctly answering the question in each round, the contestants can either choose "walk away" with the current rewards or choose "continue" to go after a higher reward with the understanding that if they fail in the next round they will loose everything. The above description is a simplified version of the game, but it captures the essential idea of it.

In a live game that operates in stages with an increasing sequence of rewards, one can implement a scheme that the audience can audit and a trusted host can ensure the distribution of the questions and the reward. In a more general

S.S.M. Chow et al. (Eds.): ISC 2014, LNCS 8783, pp. 377–387, 2014.

setting, in multi-stage computer games a player has to advance the game stage-by-stage, and receive a corresponding reward for advancing. The rewards could be monetary, but also virtual devices that make the player more competitive in the next stages so that the player expects to "unlock" such device from some existing elements when they complete some tasks at a certain level. In a distributed version of all those games, managing multi-stage games for a large amount of players may be quite complicated, and especially in some computational intensive 3-D video games, it may be required that multiple servers are needed to handle the game management for different stages. Moreover, the servers can be targeted by attackers who want to pass the game or directly "steal" the rewards. In our work, we raise the problem whether we can design the reward distribution mechanism for such online games distributively in a way that the damage from the corruption of the servers can be reduced to as little as possible and the task of managing the games and rewards becomes as simple as possible.

1.1   Our Results

Definition and Security Model: We formalize a new cryptographic primitive as an extension of IBE that we call graded encryption (GE). In a graded encryption scheme, there is a central authority and a number of sub-authorities corresponding to fixed indices. GE enables a sender to specify a grade for a ciphertext depending on the importance of the plaintext. The receiver of the ciphertext can decrypt it only if he holds a secret key with the same grade (or higher grade depending on the implementation). Here the level of a secret key determines the number of the authorities the receiver should contact (we will use both terms, grade and level interchangeably). More importantly, every user can start with a secret key generated by one authority and then can upgrade the level of his secret key by sequentially contacting other authorities depending on their indices. The authority can provide some auxiliary information to the user so that the user can upgrade the level of his key.

We highlight some requirements of a GE scheme here that makes the primitive challenging to implement. The security requirement we emphasize is that a secret key with grade i can not be useful in decrypting a ciphertext with grade j when $j > i$; in our formalization we even consider the stronger requirement that, any authority except the central authority can be corrupted as long as one authority in the chain of $1, 2, \ldots, i$ remains honest. We require that a ciphertext with grade i would remain secure in such conditions. This strong requirement is necessary in applications where we want to protect against an adversary who might directly corrupt the authority at level i while ignoring authorities corresponding to stage-1 to stage-$(i-1)$ in the hope of unlocking ciphertexts of level i.

Constructions of Efficient GE: We present an efficient scheme of graded encryption that has both secret key and ciphertext of constant size. We further extend it to a two-mode version that there are two types of partial keys that each authority can provide, and depending on that, there are two types of secret keys of each level. A Type-1 secret key is used to decrypt a ciphertext with the same level, but it can not be upgraded anymore; while a Type-2 secret key can be upgraded to higher level depending on the type of partial secret key, but it is not useful for decrypting ciphertexts. The two-mode GE is suitable for the application of online games like "who wants to be a millionaire", i.e., the Type-1 secret keys correspond to the choice of "walk away"; and a Type-2 partial key is used as long as the player chooses "continue" and advances levels (without receiving a reward).

Applications of GE: Our primitive is motivated by several applications.

Graded Rewarding System: In a graded reward system, there are multiple stages of tasks, and there are rewards corresponding to the task of each stage. Once the user completes a task, he should be given access to the corresponding reward. In many such systems, the tasks have hierarchy, i.e., a user has to finish task $1, \ldots, i$, respectively, in order to start task $(i + 1)$. Our graded encryption will enable us to utilize a simple model that there are n different authorities who manages the tasks for each stage individually, and provides partial secret keys to the users. Rewards are provide distributively by third parties at each level using the graded encryption. A user starts his journey of conquering the challenges the game presents by registering the first level authority and getting a bunch of rewards locked in some ciphertexts. Once he finishes the task of stage i, the authority i acknowledges the achievement and provides a partial key which can be used together with the secret key that the user currently holds, to unlock the reward associated to level i.

Who wants to be a millionaire: In this application, we need a mechanism that allows players to get rewards only when they choose "walk away with the current reward". Besides, there should be another mechanism such that it not only allows players to choose "continue", but also ensures that if players choose to continue the game instead of getting the current reward, they have to pass the next level challenge in order to get a chance to be able to choose again. Our two-mode GE can be used for this application. Similar to the graded rewarding system, players register with the first level authority and gets all rewards associated to different levels as encrypted with Type-1 public keys. After providing true answers to the corresponding questions, they make a choice: either "walk away" or "continue." If the players decide to "continue", the authority prepares the partial secret key of type-2, which can only be used to upgrade the secret key to an upper level, and is not useful for decrypting any current level ciphertexts. If they want to take the current rewards and walk away, the authority prepares the partial secret key of type-1 which together with the current key of the user, can be used to decrypt the reward ciphertexts at that level; note that this key can not be upgraded thus the rewards of higher levels cannot be acquired.

Path proof: Our graded encryption schemes can also be used to do a sequential authentication or proof of a certain path, like a passport control, in which there are multiple gates that travelers have to pass the checks. In order to pass the check, the traveler needs to provide a valid proof that he already passed all previous gates. Note that this problem can be solved by the multi-use proxy re-signature scheme in [1]. As the multi-use proxy re-signature provides a neat solution for the problem, we just observe that our new primitive can be used as an alternative candidate. A simple observation is that the user can use his secret key as the proof. The i-th gate controller can simply encrypt a random message r under the user identity, using the master public key with level $(i-1)$, and give the encryption to the user. If the user's secret key is upgraded from level 1 to level $i-1$ already, he will be able to decrypt the encryption as r. The gate controller then provides a partial secret key which will be used to upgrade the secret key of the user for the next gate.

The semantic security of graded encryption guarantees that the traveler needs all partial secret keys psk_j up to level i to answer the challenge of the verifier. Therefore, holding a valid secret key of level i is enough to prove that the owner of the secret key passed all checkpoints from 1 to i. So the traveler only keeps one secret key at a time. Moreover, each checkpoint keeps only two keys; one to check the validity of the secret key the traveler has, another to produce the partial secret key.

1.2   Related Work

Identity based encryption [4,15,10] could be possibly used to build a graded encryption generically, but a black-box construction incurs an overhead which is linear in the number of levels (more detailed discussion is in section 3). In an hierarchical IBE [11], a user with identity a_1 can derive secret key for user with identity $\langle a_1, a_2 \rangle$, but not the other way around. One may think of using an HIBE to construct a GE, starting from an identity with a largest level $\langle a_1, \ldots, a_n \rangle$ and gradually improving his grade by obtaining keys for $\langle a_1, \ldots, a_{n-i} \rangle$. However, it is not obvious how to achieve this efficiently. We want to emphasize that in a GE scheme, there is no hierarchy among sub-authorities in the sense of HIBE, i.e., sub-authority of level i cannot decrypt ciphertexts of level $i-1$.

In threshold public key encryption systems [7,5,2,3], at least t parties jointly decrypt a ciphertext. The threshold of the system should be fixed during setup. In [6], an extension of threshold public key encryption to the dynamic setting was proposed so that any user can dynamically join the system, and the sender can dynamically set the threshold in each encryption. Once the adversary corrupts more than the threshold shareholders, the security collapses. In a graded encryption, on the other hand, even if all except one authority is corrupted, we can guarantee the security of the ciphertext which has a level greater than the index of the uncorrupted authority. Furthermore, the authorities in a GE do not communicate or can not decrypt messages, only provide assistance to the user to upgrade his decrypting ability. The threshold encryption schemes do not provide an "ordered" mechanism suitable for the type of sequential upgrading that we consider here is needed.

In multiple encryption [13,12,9], a message is encrypted several times with independent keys or even using different encryption algorithms to provide better security. As suggested by [8], the message m can be split into shares $m_1, ..., m_i$ and all shares can be encrypted using a public key pk_i associated to the authority i. A user is only able to decrypt the ciphertexts when they have all corresponding secret keys $sk_1, ..., sk_i$ taken from the associated authorities. This trivial solution has the secret key size and the ciphertext size linear in the grade of the ciphertext.

Ateniese and Hohenberg [1] showed how multi-use proxy re-signatures, which allow a proxy to convert a valid signature of a message from Alice into a valid signature of same message from Bob, can be elegantly used to prove that a certain path was taken in a graph (our third application). Our graded encryption scheme provides an alternative solution for this application. Our construction also applies to the applications that require encryption and as such they are out of the scope of the scenarios that proxy re-signatures are applicable.

2   Definition and Security Model

In this section, we will formally define our new primitive and the security requirements of it.

2.1   Definition of Graded Encryption Schemes

As elaborated in the introduction, we are considering a new cryptographic primitive we call graded encryption, which can be seen as an extension of the canonical identity based encryption [14,4,15]. In a graded encryption scheme, each ciphertext encrypted for an identity is associated with an integer j, called the grade of the ciphertext, according to the importance of the content. Every secret key of an identity is also associated with an integer i, which we call the grade of the secret key. We require that a ciphertext encrypted under identity id with grade j can only be decrypted by secret key of id with the same grade. Furthermore, there is a mechanism that the user with identity id can upgrade his secret key with grade i to grade $(i+1)$ if the secret key with grade i was updated in same manner all the way from a key with grade 1 .

We present the formal definition of a graded encryption scheme as the following five algorithms: Setup, KeyGen, Upgrade, Enc, and Dec.

- **Setup**(λ, n) This algorithm takes a security parameter λ and a maximum grade n as inputs and outputs the system parameters P, the master public keys $\{mpk_i\}_{i=1,...,n}$ and the master secret keys denoted by $\{msk_i\}_{i=1,...,n}$ where each (mpk_i, msk_i) correspond to a level i.
- **KeyGen**(id, i, msk_i, P) This algorithm takes user id and an integer i, the corresponding master secret key msk_i and the system parameter P as input, and outputs a partial secret key psk_i for grade i.
- **Upgrade**(sk_{i-1}, psk_i) This algorithm takes a secret key sk_{i-1} with grade $(i-1)$ and a partial secret key psk_i, and outputs the secret key sk_i for level i.
- **Enc**(m, mpk_i, id, P) This randomized algorithm takes the level i public key mpk_i, a message $m \in M$, where M is the message space, the user's identity id and the system parameter P as inputs. It outputs a ciphertext c_i with grade i.

- **Dec**(c_i, sk_i, P) This deterministic algorithm takes a ciphertext c_i with grade i encrypted under the identity id, the secret key sk_i with grade i for id and the system parameter P as inputs. It outputs message m.

We first define the correctness that using a secret key with grade i for identity id, one should be able to decrypt a grade i ciphertext encrypted under id, i.e.,:
$\forall m \in M, \Pr[\mathbf{Dec}(\mathbf{Enc}(m, mpk_i, id, P), sk_i, P) = m] = 1$.

2.2   Security Model

Next, we present the security model of a graded encryption scheme. From the exemplary applications in the introduction, we can see that the only way a user can decrypt a graded ciphertext with a grade i is if she executes a sequential update for her secret key from level 1 to level i. In our security model we allow the adversary to corrupt a set of authorities, and get partial secret key of an identity for any level. But we require that as long as there exists an $i_0 \leq i$ such that neither the adversary has the master secret key msk_{i_0} nor she has the partial secret key for the challenge identity in this level, she would not be able to have significant advantage decrypting a ciphertext with grade i. Thus, on top of the basic security requirement for any identity based encryption scheme, we have to capture this new security requirement.

We adapt the standard ID-IND-CPA security model for our purpose. There are two modified queries allowed for the adversary. *Corrupt queries* allow the adversary to get a master secret key corresponding to a level i of her choice; and *KeyGen queries* allow the adversary not only to get secret keys for other identities, but also allow her to learn some of the secret keys corresponding to certain grades for the challenge identity. Consider the following game between an adversary $\mathcal{A}$ and a challenger $\mathcal{C}$,

- Setup. The challenger $\mathcal{C}$ runs the **Setup** algorithm and returns all the master public keys $\{mpk_i\}_{i=1,\dots,n}$ and the maximum level n.
- The adversary is allowed to interleave the following queries.
  - Corrupt query. The adversary $\mathcal{A}$ ask a subset S of level master secret keys, where $S \subset [1, n]$ is of her choice and with size q_1, and the challenger $\mathcal{C}$ returns $\{msk_i\}_{i \in S}$.
  - KeyGen query. The adversary $\mathcal{A}$ asks a number of KeyGen queries for identities $id_1, \dots, id_{q'}$ w.r.t. level $i_1, \dots, i_{q'}$ of her choice, and the challenger returns $psk_{i_1}(id_1), \dots, psk_{i_{q'}}(id_{q'})$, where $psk_i(id_j)$ is the KeyGen query on identity id_j for level i.
- Challenge. The adversary $\mathcal{A}$ sends an identity id^*, an integer i^*, and two messages m_0, m_1. The challenger flips a coin b and sends back $c_b = \mathbf{Enc}(m_b, mpk_{i^*}, id^*, P)$.
- The adversary continues asking Corrupt and KeyGen queries.
- Guess. The adversary outputs a guess b'.

We do allow the adversary to ask *KeyGen* queries for id^* w.r.t some levels, but we need to rule out attacks that trivialize the adversarial goal. Suppose S' is the subset of levels that $\mathcal{A}$ asks *KeyGen* queries for id^*. We require that $\{1, \ldots, i^*\} \setminus (\{1, \ldots, n\} \setminus S \cup S') \neq \emptyset$, i.e., there exists an $i_0 \leq i^*$, the adversary does not have msk_{i_0}, and also she does not have a partial secret key for id^* at level i_0.

Definition 1. *If $b = b'$, the adversary wins. A graded encryption scheme is (t, q, q', ϵ)-fully ID-IND-CPA secure if all t-time adversaries making at most q corrupt queries and q' KeyGen queries have advantage at most ϵ in winning the above game.*

In the game defined above, the adversary does not need to declare the target identity or the target grade at the beginning of the game, we call the above game a fully ID-IND-CPA game. Other possible variants can also be considered, i.e., if the adversary needs to declare the target identity at the beginning of the game, we call it selective-ID IND-CPA game; if the adversary needs to declare the target grade at the beginning of the game, we call it selective-grade ID-IND-CPA game; if both of them need to be declared at the beginning, we call it selective IND-CPA game.

Remark. that our security model explicitly states that there is no hierarchy among the master secret keys, having a master secret key with a higher grade does not grant the ability to decrypt a lower grade ciphertext.

3 Constructions of Graded Encryption

In this section, we set forth to construct efficient graded encryption schemes and briefly describe how we could deploy such schemes. First, we can have a simple generic construction of a graded encryption scheme from any IBE scheme by doing a message splitting. Briefly, in **Setup**, n independent master key pairs $\{(mpk_i, msk_i)\}_{i=1,\ldots,n}$ for each authority are generated by calling the Setup algorithm of the underlying IBE. To construct a ciphertext c_i with the grade i, the message is simply split into $m_1, \ldots, m_i$, and each part m_j is encrypted under the receiver's identity by using corresponding master public key mpk_j. A secret key with grade i consists of i partial secret key pk_j such that each of them is generated by calling the KeyGen algorithm of the underlying IBE. Furthermore, the decryption algorithm can be easily done by decrypting each piece of ciphertext using the corresponding key and combine the plaintexts to return the message.

This solution is easy and generic, however, it incurs large overhead (linear in the number of levels) of the secret key and the ciphertext for each level. A much more preferable solution would achieve a constant size for the secret key, ciphertext and each master key pair.

Intuitively, if an IBE scheme has some kind of key homomorphism so that one can aggregate secret keys to reduce the key size; also, if the master public keys can be aggregated, then the ciphertext size could also be brought down to constant.

Actually, most of the existing IBE schemes satisfy those requirements, thus we pick the Waters IBE [15] as an example to present the idea. We leave as an interesting open problem whether it is possible to have a black-box construction of a constant overhead graded encryption scheme from *any* IBE scheme. The detailed construction (we call S-I) based on Waters IBE is as follows:

- **Setup**(λ, n). The algorithm first chooses a random generator $g \in \mathbb{G}$ and a random element g_2 of $\mathbb{G}$. It also chooses a set of random numbers $a_1, \ldots, a_n \in Z_p$. For $i = 1, \ldots, n$, the algorithm computes $mpk_i = g^{\sum_{j=1}^{i} a_j}$, and $msk_i = g_2^{a_i}$. The algorithm also chooses a random value $u' \in G$ and a random l-length vector $U = (u_i)_{i=1,\ldots,\ell}$ where $u_i \in \mathbb{G}$ and ℓ is the length of the identities. It outputs the master key pairs $\{mpk_i, msk_i\}_{i=1,\ldots,n}$ and the system parameter $P = (g, g_2, u', U)$.
- **KeyGen**(id, i, msk_i, P). Suppose $V \subseteq [n]$ is the set of all indices j for which $id_j = 1$. The algorithm then chooses a random $r \in Z_p$, and computes psk_i as $(psk_i^1, psk_i^2) = (msk_i \cdot H(id)^r, g^r)$, where $H(id) = u' \prod_{i \in V} u_i$.
- **Upgrade**(sk_{i-1}, psk_i): If $i = 1$, $sk_1 = psk_1$; If $i > 1$, the algorithm computes $sk_i = (sk_i^1, sk_i^2)$ where $(sk_{i-1}^1 \cdot psk_i^1, sk_{i-1}^2 \cdot psk_i^2)$.
- **Enc**(id, m, mpk_i, P) To encrypt a message m under identity id for grade i, the algorithm computes c_i as: (C_1, C_2, C_3) equal to:

$$(e(mpk_i, g_2)^s \cdot m, g^s, H(id)^s) = (e(g^{\sum_{j=1}^{i} a_j}, g_2)^s \cdot m, g^s, H(id)^s).$$

  It returns (C_1, C_2, C_3) as the grade i ciphertext c_i.
- **Dec**(c_i, sk_i, P). Taking the level i ciphertext $c_i = (C_1, C_2, C_3)$ and the level i secret key $sk_i = (sk_i^1, sk_i^2)$ as input, this algorithm computes $C_1 \cdot e(sk_i^2, C_3)/e(sk_i^1, C_2) = m$. It returns m as the plaintext.

Deployment: Based on our applications, our graded encryption scheme can be implemented as follows: a central authority initializes and sets up the system at the beginning, and distributes each level master keys to the individual authorities. Then, the central authority goes offline. The **KeyGen** algorithm is executed by each individual authority which only has a master secret corresponding to one level. When a user has a secret key with grade i, together with the partial secret key with grade $(i + 1)$ generated by authority $(i + 1)$, he can upgrade his key to grade $(i + 1)$ locally. Observe that in the above construction, msk_i does not correspond to mpk, thus if the adversary does not update a secret key from grade 1, corrupting an intermediate authority does not give much advantage. Furthermore, the master secret key with grade i is independent of the master public keys with smaller grade, thus the online authorities can not decrypt the user's ciphertext as in the normal IBE setting unless all of them collude. This reveals an interesting fact that our graded encryption scheme provides some defense to the key escrow problem in IBE systems.

Theorem 1. *S-I is fully ID-IND-CPA secure if Waters IBE [15] is ID-IND-CPA secure.*

Proof. (Sketch:) Due to space limit, we only provide a sketch here showing how to reduce the security of S-1 to that of Waters IBE.

The simulator $\mathcal{S}$ first makes a random guess for the index j of the sub-authority which the adversary will not corrupt, and $\mathcal{S}$ will abort if the guess is incorrect. For other indices, $\mathcal{S}$ selects a random a_i and generates the master key pair using (a_i, g^{a_i}) multiplying all g^{a_i} with the Waters IBE master public key, $\mathcal{S}$ gives the system mpk. Note that $\mathcal{S}$ knows msk for all $i \neq j$ thus can answer all Corrupt and KeyGen queries trivially, and KeyGen query for j can be answered by asking Waters IBE challenger secret key query.

In the challenge phase; the adversary submits two messages m_0, m_1 and an identity id^* with a grade i^*. if $j > i^*$, the simulator aborts. Otherwise, the simulator uses the Waters IBE challenge ciphertext (c_1, c_2, c_3) by asking m_0, m_1 on id^* to derive $c_i^{(b)} = (c_1.e(c_2, g_2)^{\sum_{k=1, k \neq j}^{i} a_k}, c_2, c_3)$ to the adversary as the challenge ciphertext where all a_k are integers generated at the beginning of the game for the master secret keys. Furthermore, it is easy to see that the probability of which the simulator aborts is bounded by $\frac{1}{n}$.

3.1 Two-Mode Graded Encryption

In the "who wants to be a millionaire" show, the contestants have two options: either they can leave the game with the rewards of the current level, or they can continue the game by requesting a higher level question from the next server and thus pursuit higher level rewards. The construction given above does not allow to the contestants to have these opportunities. So we extend it in a way that the new one enables servers to provide two types of partial secret keys to the contestants according to the their choices, either "leave", or "continue." Our two-mode construction, which will have two **KeyGen** algorithms so that each of them provides different type of partial secret key, is as follows:

- **Setup**(λ, n). The algorithm first chooses a random generator $g \in \mathbb{G}$ and a random element g_2 of $\mathbb{G}$. It also chooses a set of random numbers $a_1, b_1 \ldots, a_n, b_n \in Z_p$. For $i = 1, \ldots, n$, the algorithm computes $mpk_i = g^{\sum_{j=1}^{i-1} b_j + a_i}$, and $msk_i^{(1)} = g_2^{a_i}, msk_i^{(2)} = g_2^{b_i}$. The algorithm also chooses a random value $u' \in G$ and a random l-length vector $U = (u_i)_{i=1,\ldots,\ell}$ where $u_i \in \mathbb{G}$ and ℓ is the length of the identities. It outputs the master key pairs $\{mpk_i, msk_i^{(1)}, msk_i^{(2)}\}_{i=1,\ldots,n}$ and the system parameter $P = (g, g_2, u', U)$.
- **KeyGen1**$(id, i, msk_i^{(1)}, P)$. Suppose $V \subseteq [n]$ is the set of all indices j for which $id_j = 1$. The algorithm then chooses a random $r \in Z_p$, and computes the type-1 partial secret key $psk_i^{(1)}$ as $(psk_{i,1}^{(1)}, psk_{i,2}^{(1)}) = (msk_i^{(1)} \cdot H(id)^r, g^r)$, where $H(id) = u' \prod_{i \in V} u_i$.
- **KeyGen2**$(id, i, msk_i^{(2)}, P)$. Suppose $V \subseteq [n]$ is the set of all indices j for which $id_j = 1$. The algorithm then chooses a random $r \in Z_p$, and computes the type-2 partial secret key $psk_i^{(2)}$ as $(psk_{i,1}^{(2)}, psk_{i,2}^{(2)}) = (msk_i^{(2)} \cdot H(id)^r, g^r)$, where $H(id) = u' \prod_{i \in V} u_i$.

- **Upgrade**$(sk_{i-1}, psk_i^{(b)})$: If $i = 1$, $sk_1 = psk_1$; If $i > 1$, for both $b = 1, 2$, the algorithm computes $sk_i = (sk_i^{(1)}, sk_i^{(2)})$ where $(sk_{i-1}^{(1)} \cdot psk_{i,1}^{(b)}, sk_{i-1}^{(2)} \cdot psk_{i,2}^{(b)})$.
- **Enc**(id, m, mpk_i, P) To encrypt a message m under identity id for grade i, the algorithm computes c_i as (C_1, C_2, C_3) to be equal to:

$$(e(mpk_i, g_2)^s \cdot m, g^s, H(id)^s) = (e(g^{\sum_{j=0}^{i-1} b_j + a_i}, g_2)^s \cdot m, g^s, H(id)^s).$$

It returns (C_1, C_2, C_3) as the grade i ciphertext c_i.
- **Dec**(c_i, sk_i, P). Taking the level i ciphertext $c_i = (C_1, C_2, C_3)$ and the level i secret key $sk_i = (sk_{i,1}, sk_{i,2})$ as input, this algorithm computes $C_1 \cdot e(sk_{i,2}, C_3)/e(sk_{i,1}, C_2) = m$. It returns m as the plaintext.

Note. that the type-1 secret key $sk_i^{(1)}$ is accumulated from $(i-1)$ type-2 partial secret keys and one type-1 partial secret key as the last component. Besides, it corresponds to the exponent in the corresponding master public key, thus can be used for decrypting the ciphertext with level i. We can also see that the type-1 secret key cannot be upgraded anymore for the next level since there is no way to eliminate the type-1 partial secret key in the middle. Furthermore, since the type-2 secret keys are accumulated from only type-2 partial secret keys, it can not be used to decrypt ciphertexts prepared by type-1 master public keys unless they are upgraded at the end using some type-1 partial secret key. The security analysis of this construction is similar to the basic construction, and we ignore the details here.

Acknowledgements. Research partly supported by NSF award 0831306, and ERC Project CODAMODA.

References

1. Ateniese, G., Hohenberger, S.: Proxy re-signatures: new definitions, algorithms, and applications. In: ACM Conference on Computer and Communications Security, pp. 310–319 (2005)
2. Baek, J., Zheng, Y.: Identity-based threshold decryption. In: Bao, F., Deng, R., Zhou, J. (eds.) PKC 2004. LNCS, vol. 2947, pp. 262–276. Springer, Heidelberg (2004)
3. Boneh, D., Boyen, X., Halevi, S.: Chosen ciphertext secure public key threshold encryption without random oracles. In: Pointcheval, D. (ed.) CT-RSA 2006. LNCS, vol. 3860, pp. 226–243. Springer, Heidelberg (2006)
4. Boneh, D., Franklin, M.: Identity-based encryption from the weil pairing. In: Kilian, J. (ed.) CRYPTO 2001. LNCS, vol. 2139, pp. 213–229. Springer, Heidelberg (2001)
5. Canetti, R., Goldwasser, S.: An efficient *threshold* public key cryptosystem secure against adaptive chosen ciphertext attack. In: Stern, J. (ed.) EUROCRYPT 1999. LNCS, vol. 1592, pp. 90–106. Springer, Heidelberg (1999)
6. Delerablée, C., Pointcheval, D.: Dynamic threshold public-key encryption. In: Wagner, D. (ed.) CRYPTO 2008. LNCS, vol. 5157, pp. 317–334. Springer, Heidelberg (2008)

7. Desmedt, Y., Frankel, Y.: Threshold cryptosystems. In: Brassard, G. (ed.) CRYPTO 1989. LNCS, vol. 435, pp. 307–315. Springer, Heidelberg (1990)

8. Dodis, Y., Katz, J.: Chosen-ciphertext security of multiple encryption. In: Kilian, J. (ed.) TCC 2005. LNCS, vol. 3378, pp. 188–209. Springer, Heidelberg (2005)

9. Dodis, Y., Katz, J., Xu, S., Yung, M.: Key-insulated public key cryptosystems. In: Knudsen, L.R. (ed.) EUROCRYPT 2002. LNCS, vol. 2332, pp. 65–82. Springer, Heidelberg (2002)

10. Gentry, C.: Practical identity-based encryption without random oracles. In: Vaudenay, S. (ed.) EUROCRYPT 2006. LNCS, vol. 4004, pp. 445–464. Springer, Heidelberg (2006)

11. Horwitz, J., Lynn, B.: Toward hierarchical identity-based encryption. In: Knudsen, L.R. (ed.) EUROCRYPT 2002. LNCS, vol. 2332, pp. 466–481. Springer, Heidelberg (2002)

12. Maurer, U.M., Massey, J.L.: Cascade ciphers: The importance of being first. J. Cryptology 6(1), 55–61 (1993)

13. Merkle, R.C., Hellman, M.E.: On the security of multiple encryption. Commun. ACM 24(7), 465–467 (1981)

14. Shamir, A.: Identity-based cryptosystems and signature schemes. In: Blakely, G.R., Chaum, D. (eds.) CRYPTO 1984. LNCS, vol. 196, pp. 47–53. Springer, Heidelberg (1985)

15. Waters, B.: Efficient identity-based encryption without random oracles. In: Cramer, R. (ed.) EUROCRYPT 2005. LNCS, vol. 3494, pp. 114–127. Springer, Heidelberg (2005)

Adding Controllable Linkability
to Pairing-Based Group Signatures for Free*

Daniel Slamanig, Raphael Spreitzer, and Thomas Unterluggauer

Graz University of Technology, IAIK, Austria
{daniel.slamanig,raphael.spreitzer,thomas.unterluggauer}@iaik.tugraz.at

Abstract. Group signatures, which allow users of a group to anonymously produce signatures on behalf of the group, are an important cryptographic primitive for privacy-enhancing applications. Over the years, various approaches to enhanced anonymity management mechanisms, which extend the standard feature of opening of group signatures, have been proposed.

In this paper we show how pairing-based group signature schemes (PB-GSSs) based on the sign-and-encrypt-and-prove (SEP) paradigm can be generically transformed in order to support one particular enhanced anonymity management mechanism, *i.e.*, we propose a transformation that turns every such PB-GSS into a PB-GSS with *controllable linkability*. Basically, this transformation replaces the public key encryption scheme used for identity escrow within a group signature scheme with a modified all-or-nothing public key encryption with equality tests scheme (denoted AoN-PKEET*) instantiated from the respective public key encryption scheme. Thereby, the respective trapdoor is given to the linking authority as a linking key. The appealing benefit of this approach in contrast to other anonymity management mechanisms (such as those provided by traceable signatures) is that controllable linkability can be added to PB-GSSs based on the SEP paradigm *for free*, *i.e.*, it neither influences the signature size nor the computational costs for signers and verifiers in comparison to the scheme without this feature.

1 Introduction

The concept of group signature schemes (GSSs) has been introduced by Chaum and van Heyst [12] in 1991. Members within a predefined group are able to sign messages on behalf of the group anonymously. Verifiers can determine whether a signature indeed has been produced by a group member, but are not able to determine the actual identity of the signer. However, the so-called group manager (GM) is able to open a given signature in order to determine the identity of the actual signer in case of dispute. Early GSSs were static in terms of having the group members fixed at setup time [4], whereas more recent constructions

---

* The full version of this extended abstract including more details as well as all formal models, definitions and proofs is available as Cryptology ePrint Archive Report 2014/607 [24].

S.S.M. Chow et al. (Eds.): ISC 2014, LNCS 8783, pp. 388–400, 2014.
© Springer International Publishing Switzerland 2014

consider dynamic groups [5], *i.e.*, members may be added to and deleted from the group over time. Moreover, in some cases it is also desirable to have distributed authorities, *i.e.*, one party only receives the opening key and a distinct party receives the issuing key required to add new members or to revoke existing members. For both, the static as well as the dynamic setting, there are constructions under generic assumptions [4, 5] based on the *sign-and-encrypt-and-prove* (SEP) paradigm [11]. In schemes following this paradigm, a user on joining a group receives a signature from the GM, which is known as the membership certificate. A group signature produced by a user is then an encryption of the membership certificate of the user under the GM's public key and a non-interactive zero-knowledge proof of knowledge of well-formedness of the respective values.

Today, pairing-based group signature schemes (PB-GSSs) [6, 8, 9, 14, 17] are prevalent due to performance benefits compared to earlier constructions. Essentially, most of the PB-GSSs are variations of the BBS [8] or the BBS$^+$ scheme [3], secure under the SDH (or a slightly modified SDH$^+$) assumption. The core schemes essentially differ in the use of various IND-CPA or IND-CCA secure public key encryption schemes that support efficient zero-knowledge proofs of knowledge about encrypted plaintexts and are secure in the random oracle model (ROM). With the exception of [6] all existing schemes follow the SEP paradigm.

Over the years, various approaches to enhanced anonymity management mechanisms for group signatures extending the standard opening feature have been proposed. For instance, it may be desirable to either support linkability across different groups in the context of multi group signatures [2] or to publicly link signatures of users without identifying them [21] or even to allow public tracing for signers that have produced a number of signatures above a certain threshold $k \geq 2$ [27]. Besides these authority-free linking approaches, there also exist schemes, and in particular [17, 18], supporting so-called *controllable linkability*. This means that there are designated linking authorities (LAs), *i.e.*, parties in possession of a so-called *linking key*, who are able to link two signatures by means of this key, but no verifier is able to do so. A LA thereby can *only* decide whether two given signatures have been issued by the same *unknown* signer, *i.e.*, signers stay anonymous. Hwang *et al.* [17, 18] argued that the concept of controllable linkability can be useful in vehicular adhoc networks (VANETs), for instance, to prevent Sybil attacks. It can further be beneficial in the context of data mining, e.g., service providers might want to establish statistics regarding buying patterns, while still preserving the customer's privacy, and "smart cities", e.g., public transport systems may support anonymous traveling where a valid group signature represents a valid ticket. However, service providers might also be interested in analyzing traveling patterns, *i.e.*, some kind of flow control analysis. Thereby, the concept of controllable linkability allows the service provider to efficiently link signatures (ticket showings), while the customer of the public transport system still remains anonymous. Traceable signatures [19], on the other hand, pursue to add selective traceability features to group signatures, *i.e.*, they allow an authority to compute a tracing trapdoor for every user such that only signatures produced by this user can be linked using the tracing trapdoor, but all other signatures from remaining users stay unlinkable. A somewhat orthogonal approach to restrict the power of the opener are group signatures with message-dependent opening [22]. They distribute the power of the opener by introducing another authority (the admitter). Thereby, the admitter can issue a token which corresponds to a particular message and by using this token, the opener can extract the identity of the signer from a signature for this message.

Contribution: In this paper we show how PB-GSSs based on the SEP paradigm can be generically transformed in order to support controllable linkability. Basically, this transformation replaces the public key encryption scheme used for the identity escrow within a group signature scheme with a modified *all-or-nothing public key encryption with equality tests* scheme (denoted AoN-PKEET*) instantiated from the respective public key encryption scheme. Thereby, the LA receives a trapdoor and can perform trapdoor-equality tests on the ciphertexts without learning the respective plaintexts and thus is able to link signatures without being able to identify the signers.

The techniques which are the basis for our generic approach to controllable linkability have been inspired by the works of Hwang *et al.* [17, 18]. The appealing benefit of our approach is that controllable linkability can be added to PB-GSSs based on the SEP paradigm *for free*, *i.e.*, it neither influences the signature size nor the computational costs for signers and verifiers. Therefore, we present and formalize this generic transformation based on AoN-PKEET* that can be used to turn any PB-GSS following the SEP paradigm into a GSS with controllable linkability. The used mechanisms, *i.e.*, the AoN-PKEET*, may also be of independent interest for other applications.

In comparison to other approaches such as traceable signatures, which have a different goal but may be "casted" to be used to achieve controllable linkability, our generic approach is more efficient and simplistic when only requiring controllable linkability. Finally, we note that our transformation has no influence whatsoever on the used revocation mechanism of the group signature scheme.

2   Background

2.1   GSSs with Controllable Linkability

For *GSSs with controllable linkability* we use the model of Hwang *et al.* [17, 18] that extends [5]. The model requires an additional authority, namely a *linking authority* (LA), who is capable of linking signatures. The corresponding private key of this linking authority is denoted as *master linking key* (mlk). A GSS with controllable linkability is specified as a tuple $\mathcal{GS} = (\mathsf{GkGen}, \mathsf{UkGen}, \mathsf{Join}, \mathsf{Issue}, \mathsf{GSig}, \mathsf{GVf}, \mathsf{Open}, \mathsf{Judge}, \mathsf{Link})$. Subsequently, we present the algorithms that change relative to the original GSS as well as the additional Link algorithm.

$\mathsf{GkGen}(\lambda)$: On input of a security parameter $\lambda \in \mathbb{N}$, the algorithm generates the public parameters and outputs a tuple (gpk, mok, mik, mlk), representing the group public key, the master opening key, the master issuing key, and the master linking key.

Link(gpk, M, σ, M', σ', mlk): On input of the group public key gpk, a pair of tuples (M, σ) and (M', σ'), as well as the master linking key mlk, the algorithm first verifies both signatures by calling GVf(gpk, M, σ) and GVf(gpk, M', σ'). If both signatures are valid for messages M and M' under the group public key gpk, the algorithm uses mlk to determine whether σ and σ' have been produced by the same *unknown* signer. If both signatures are valid and can be linked to the same *unknown* signer, the algorithm returns true and false otherwise.

Modified and Additional Properties for GSSs with Controllable Linkability.

- **Correctness**: Signatures generated by honest group members should be valid, the Open algorithm should correctly identify the signer, and the proof returned by the Open algorithm should be accepted by the Judge algorithm. Furthermore, the Link algorithm should correctly link two signatures from the same *unknown* signer.
- **Linkability**: The authority in possession of the master linking key mlk should neither be able to gain any useful information for opening a signature nor for generating a Judge proof τ. Furthermore, colluding parties—including users, the linking authority, and/or the opening authority—should not be able to generate pairs of messages and signatures (M, σ) and (M', σ') that violate the correctness property mentioned above.

2.2 Sign-and-Encrypt-and-Proof Paradigm in the ROM

We use the representation of [20] for illustration. GSSs based on the SEP paradigm consist of a secure signature scheme $\mathcal{DS} = (\mathsf{KeyGen}_s, \mathsf{Sign}, \mathsf{Vrfy})$, and an (at least IND-CPA secure) public key encryption scheme $\mathcal{AE} = (\mathsf{KeyGen}_e, \mathsf{Enc}, \mathsf{Dec})$. Additionally, we require zero-knowledge proofs of knowledge (PKs) which are converted to signatures of knowledge (SPKs) using the Fiat-Shamir transform in the ROM. Furthermore, let $f(\cdot)$ be a one-way function.

The group public key gpk includes a public encryption key pk_e, and a signature verification key pk_s. The master opening key mok is the decryption key sk_e, and the master issuing key mik is the signing key sk_s. During the execution of the Join protocol a user i generates a secret x_i and sends $f(x_i)$ to the issuer. The issuer in turn returns a signature cert $\leftarrow \mathsf{Sign}(\mathsf{sk}_s, f(x_i))$ by signing $f(x_i)$ with the signing key sk_s.

A group signature $\sigma = (T, \pi)$ for a message M is computed as follows: Compute a ciphertext $T \leftarrow \mathsf{Enc}(\mathsf{pk}_e, X_i)$ and

$$\pi \leftarrow \mathsf{SPK}\{(x_i, \mathsf{cert}) : \mathsf{cert} = \mathsf{Sign}(\mathsf{sk}_s, f(x_i)) \quad \wedge \quad T = \mathsf{Enc}(\mathsf{pk}_e, X_i)\}(M)$$

where X_i can either be $g(x_i)$ for some one-way function $g(\cdot)$ or cert. We note that the membership certificate typically refers to the issuer's signature (cert) but might also be a commitment to a user's secret that has been signed by the issuer. Nevertheless, we always refer to T as an encryption of the membership certificate.

3   Adding Controllable Linkability to GSSs Generically

In all PB-GSSs following the SEP paradigm, the used encryption scheme depends on the respective bilinear map setting, *i.e.*, types of used pairings, as well as whether the construction targets to achieve weak or full anonymity. If one assumes the decisional Diffie-Hellman (DDH) problem to be easy in $\mathbb{G}_1$ and $\mathbb{G}_2$, then one usually relies on linear encryption variants of ElGamal encryption [8, 18] which are IND-CPA secure under the decision linear (DLIN) or some related assumption. However, if one assumes the external Diffie-Hellman (XDH) assumption to hold for the respective group used for encryption, *i.e.*, DDH is assumed to be hard in this group, then one can use standard IND-CPA secure El-Gamal encryption. Weak anonymity (CPA-full-anonymity) is basically achieved by IND-CPA secure encryption and full anonymity (CCA-full-anonymity) when using IND-CCA secure encryption schemes.

3.1   Trapdoor Equality Test for Public-Key Encryption

At the heart of our controllable linkability is a means to extend IND-CPA/IND-CCA secure public key encryption schemes with a feature that allows a designated party (holding a trapdoor) to check whether two ciphertexts under the same public key contain the same message, but without being able to decrypt the ciphertexts, *i.e.*, still providing one-wayness (OW). Our idea is related to the concept of probabilistic public key encryption with equality tests (PKEET) [28], but differs in that their equality tests are *public* (not only feasible for one holding a trapdoor) and need to work for ciphertexts under *different* public keys. Consequently, their constructions can no longer satisfy any meaningful notion of indistinguishability. However, in our approach indistinguishability supported by the respective encryption scheme still needs to hold for all parties except the one holding the trapdoor, who clearly can test against any potential message. However, if the messages are not guessable (the message space has large enough entropy) they can still be hidden from the party holding the trapdoor, *i.e.*, provide OW-CPA security. Our idea is also related to all-or-nothing PKEET (AoN-PKEET) [25]. Their focus is on allowing a semi-trusted proxy to compare ciphertexts for two distinct parties by obtaining a trapdoor from each party (but also works for a single user). However, firstly they target applications for searchable encryption and consequently their Type-I adversary (the party holding the trapdoor(s)) is very powerful, *i.e.*, they require OW-CCA security, and also for an outsider who does not know the trapdoor(s) (Type-II adversary), they always require IND-CCA security. Secondly, due to their focus on comparison of ciphertexts from distinct users, their construction is quite involved and costly. Besides inefficiency, the most important difference between the AoN-PKEET construction and our approach is that we need compatibility with efficient proofs of knowledge of encrypted messages, which are not possible in [25] as their approach involves encrypting hashes of the messages. This is not applicable in our setting as it cannot be efficiently proven that the correct hash of the message (membership certificate) has been included by the user.

3.2   Modified All-Or-Nothing PKEET (AoN-PKEET*)

In brief, our approach may be considered as a restricted version of AoN-PKEET, where we only allow comparison of ciphertexts of the same user, *i.e.*, under the same public key. Furthermore, against a Type-I adversary (the trapdoor holder) we do not require OW-CCA but OW-CPA security (that the opener cannot be used as a decryption oracle by the LA is reasonable, but one may extend the approach to OW-CCA). Against Type-II adversaries (outsiders) we require either IND-CPA or IND-CCA security (depending on the underlying public key encryption scheme). Additionally, we require that there are efficient (honest-verifier) zero-knowledge proofs of knowledge about messages hidden in ciphertexts, which is clearly true for all encryption schemes used for the SEP approach with PB-GSSs. We denote such a modified scheme as an AoN-PKEET* scheme.

Formal Model: An AoN-PKEET* scheme (KeyGen, Enc, Dec, Aut, Com) is a conventional public key encryption scheme (KeyGen, Enc, Dec) augmented by two polynomial time algorithms (Aut, Com), which are defined as follows, and where (sk, pk) have been generated by KeyGen:

Aut(sk): The authorization algorithm Aut takes a private key sk and outputs a trapdoor tk.
Com(c,c',tk): The comparison algorithm takes two ciphertexts c and c' (of two messages m and m') produced under pk, and a trapdoor tk produced with the corresponding sk, and outputs true if $m = m'$ and false otherwise.

Security Definition: For our modified setting, the *soundness* definition of [25] reduces to the fact that besides correctness of the public key encryption scheme, we have that for all (pk, sk) ← KeyGen(λ) we require that Com(Enc(pk, m), Enc(pk, m'), Aut(sk)) = true if and only if $m = m'$. Against a Type-I adversary, we require OW-CPA security, which is defined as follows.

Definition 1 (AoN-PKEET* OW-CPA). *For all PPT adversaries $\mathcal{A}$ and security parameters λ there is a negligible function ϵ such that:*

$$\Pr\left[\begin{array}{c} (\mathsf{pk}, \mathsf{sk}) \leftarrow \mathsf{KeyGen}(\lambda), \mathsf{tk} \leftarrow \mathsf{Aut}(\mathsf{pk}), m \xleftarrow{R} \{0,1\}^\lambda, \\ m^* \leftarrow \mathcal{A}(\mathsf{pk}, \mathsf{tk}, \mathsf{Enc}(\mathsf{pk}, m)) : m^* = m \end{array}\right] \leq \epsilon(\lambda).$$

Against a Type-II adversary, we require IND-CPA security or IND-CCA security (depending on the used public key encryption scheme) as it is defined for conventional public key encryption schemes.

Definition 2. *An AoN-PKEET* scheme is called secure if it is sound, provides OW-CPA security against Type-I adversaries and provides the respective IND-CPA/IND-CCA security provided by (KeyGen, Enc, Dec) against Type-II adversaries.*

We present an AoN-PKEET* scheme based on ElGamal encryption and refer the reader to the full version [24] for security proofs and other constructions.

ElGamal Encryption: We consider ElGamal encryption in $\mathbb{G}_1$ in a bilinear map setting $e : \mathbb{G}_1 \times \mathbb{G}_2 \to \mathbb{G}_T$ such that the XDH assumption holds for $\mathbb{G}_1$. Let the private key be $\xi \in \mathbb{Z}_p^*$ and the public key be $h \leftarrow g^\xi \in \mathbb{G}_1$ and a ciphertext for message $m \in \mathbb{G}_1$ be $(T_1, T_2) \leftarrow (g^\alpha, mh^\alpha) \in \mathbb{G}_1^2$ for random $\alpha \in \mathbb{Z}_p^*$. Decryption works in $\mathbb{G}_1$ by computing $m \leftarrow T_2/(T_1^\xi)$. Now, we describe Aut and Com.

Aut : Given ξ, the trapdoor is computed as $\mathsf{tk} \leftarrow (r, t = r^\xi)$ for a random $r \in \mathbb{G}_2$.

Com : Given two ciphertexts $(T_1, T_2) = (g^\alpha, mh^\alpha)$ and $(T_1', T_2') = (g^{\alpha'}, m'h^{\alpha'})$ and trapdoor $\mathsf{tk} = (r, t = r^\xi)$ check:

$$\frac{e(T_2, r)}{e(T_1, t)} \overset{?}{=} \frac{e(T_2', r)}{e(T_1', t)}.$$

If the check holds return `true` and `false` otherwise.

3.3   Transformation to PB-GSSs with Controllable Linkability

To generically add controllable linkability to PB-GSSs following the SEP paradigm, we replace the used public key encryption scheme with its AoN-PKEET* version. The required additional linking key mlk is the trapdoor tk computed by Aut and is given to the linking authority (LA). Consequently, for linking the linking authority is given two message-signature pairs (M, σ) and (M', σ') where the signatures $\sigma = (T, \pi)$ and $\sigma' = (T', \pi')$ contain ciphertexts T and T' as well as the non-interactive proofs π and π'. Then, by running $\mathsf{Com}(T, T', \mathsf{mlk})$, the LA can decide whether the two signatures have been produced by the same *unknown* signer.

In order to convert a PB-GSS $\mathcal{GS} =$ (GkGen, UkGen, Join, Issue, GSig, GVf, Open, Judge) following the SEP paradigm into a PB-GSS with controllable linkability, we have to add a linking authority LA as well as an additional algorithm Link (cf. Section 2.1). Below we present the required modifications:

GkGen(λ): On input of a security parameter $\lambda \in \mathbb{N}$, the algorithm generates the public parameters and outputs a tuple (gpk, mok, mik, mlk), where mok is generated by running KeyGen(λ) of the AoN-PKEET* scheme (where pk is integrated into gpk) and the master linking key is computed as mlk $\leftarrow$ Aut(mok). All the remaining steps remain unchanged.

Link(gpk, M, σ, M', σ', mlk): On input of the group public key gpk, a pair of tuples (M, σ) and (M', σ'), as well as the master linking key mlk, the algorithm first verifies both signatures via the GVf algorithm. If at least one of these verifications fails, the algorithm outputs `false`. Otherwise, the algorithm extracts the ciphertexts T and T' from σ and σ' and runs $\mathsf{Com}(T, T', \mathsf{mlk})$ and outputs whatever Com outputs.

We state the following abstract assumption and discuss it below:

Assumption 1. *Honestly computed membership certificates (used for identity escrow) of users are uniformly distributed over the respective group and are unknown to the linking authority[1].*

Note that for all BBS variations the membership certificates are of the form $A = g_1^{\frac{1}{x_i+\text{mik}}}$ where g_1 may be the product of other group elements representing a BB signature [7]. In [15] it is shown that this represents a verifiable random function (VRF) (for a superlogarithmic sized input). However, in our setting we do not want to realize a pseudo-random function, but we only require that A is uniformly distributed over the group for uniformly at random sampled values of messages (that are not known to the linking authority). In case of BB signatures A will be a random element of the group, if the unknown value of x_i is chosen uniformly at random from the integers in the order of the group.

We note that the CL group signature scheme in [10] uses Cramer-Shoup encryption in $\mathbb{G}_T$ as they defined it over Type 1 pairings, which would not work with our approach. However, it can easily be adapted to asymmetric pairings such that the encryption scheme no longer needs to work on elements in $\mathbb{G}_T$ and then our transformation applies. Anyways, in a CL signature the membership certificate to be encrypted is not the signature (as in BBS variants), but a commitment to a user's secret. As this secret is chosen uniformly at random, our assumption clearly holds. The above stated assumption also seems to be reasonable for other constructions that one may envision, where the membership certificate is a signature for a uniformly at random chosen message.

Relation to Hwang et al.: The controllable linkability feature of the group signature schemes in [17, 18] uses two ciphertexts (sharing the same randomness), one for opening and one for linking, and consequently an additional ciphertext in the actual group signature. Recall, that in the BBS scheme a membership certificate is of the form $A_i = g_1^{\frac{1}{x_i+\text{mik}}}$ for x_i randomly chosen by the issuer and when requiring strong exculpability $A = (g_1 h^{-z_i})^{\frac{1}{x_i+\text{mik}}}$ where z_i is randomly chosen by the user and given in form of a commitment h^{-z_i} to the issuer. In the construction in [17] Hwang et al. add an additional issuer-chosen element y_i and their membership certificates are of the form $A = (g_1 h^{-z_i} h'^{-y_i})^{\frac{1}{x_i+\text{mik}}}$. For realizing controllable linkability, a user encrypts g^{y_i} and provides an additional non-interactive zero-knowledge proof of knowledge that the unrevealed value y_i in the second ciphertext is included in the membership certificate and linking basically represents a plaintext equality check based on ElGamal encryption (the construction in [18] works analogously but uses another encryption scheme).

Consequently, they apply their equality testing for linkability not directly on the membership certificate but on the second ciphertext to the message g^{y_i} and

---

[1] The latter case is just to rule out trivial cases where the membership certificate that is encrypted is a well known public key. Note that in such cases the user could always commit to his secret using a base different from that in the public key and proof the equality of the respective discrete logarithms during joining.

thus do not require Assumption 1 for their LO-linkability. However, this is not generic and makes the group signature more expensive, whereas our construction comes at no additional cost at all.

3.4 Security Analysis

We note that all proofs for security properties of the respective PB-GSS that are not related to the property of linkability remain valid as well as untouched by the generic transformation. In GSSs with controllable linkability [17, 18], the additional security properties related to *linkability* are formally captured by three security notions, called LO-linkability (Link-Only linkability), JP-Unforgeability (Judge-Proof Unforgeability), and E-linkability (Enforced linkability).

LO-linkability: A linking key should be used only for linking signatures, not for gaining useful information for opening.
JP-Unforgeability: A linking key cannot be used for generating a Judge proof.
E-linkability: Colluding users should not be able to generate two message-signature pairs satisfying any of the following two conditions (even with the help of authorities such as the linking authority or the opening authority): (1) Open yields identical identities which are successfully judged, while Link outputs `false` or (2) identities output by Open are different and both are successfully judged, while Link outputs `true`.

In the full version [24] we prove the following:

Theorem 1. *If AoN-PKEET* is secure, PB-GSS is secure and Assumption 1 holds, then the generic transformation yields a secure PB-GSS with controllable linkability.*

4   Comparison with Other Approaches and Future Work

Linkable Group Signatures: *Tracing-by-linking* (TbL) schemes allow for public tracing of members if they sign $k \geq 2$ times [26]. However, the signer's identity cannot be revealed by any authority. Similarly, *link-but-not-trace* group signatures [21] allow the detection of double signing but require an authority to reveal the signer's identity. Both approaches allow for public linking to detect double signing and, thus, cannot be employed to achieve controllable linkability.

Traceable Signatures [19]: Besides opening signatures, these schemes support two additional functionalities: 1) *user tracing*, where a tracing trapdoor published by the GM allows public tracing of signatures generated by a specific user, and 2) *signature claiming*, meaning that signatures can be provably claimed by their corresponding signers. By giving such a tracing trapdoor for every user to the linking authority, computational costs linear in the number of group members are required to implement controllable linkability. *Real traceable signatures* [13] employ a tracing trapdoor to deterministically recompute every tag of a member's signature, thereby allowing to trace users. However,

Table 1. Related concepts and their applicability for controllable linkability (CL)

Mechanism	Suitable for CL	Overhead for LA		
		Update on JOIN	Link	Memory
Tracing-by-linking group signatures [26]	✗	-	-	-
Link-but-not-trace group signatures [21]	✗	-	-	-
Traceable signatures [19]	✓	✓	$\mathcal{O}(N)$	$\mathcal{O}(N)$
Real traceable signatures [13]	✓	✓	$\mathcal{O}(N)$	$\mathcal{O}(\ell \cdot N)$
Public key anonymous tag systems [1]	✓	✓	$\mathcal{O}(N)$	$\mathcal{O}(N)$
Verifier local revocation [9]	✓	✓	$\mathcal{O}(N)$	$\mathcal{O}(N)$
Controllable linkability	✓	✗	$\mathcal{O}(1)$	$\mathcal{O}(1)$

the drawback of this approach is that only ℓ unlinkable signatures can be issued by every user in the system and every signature requires an additional zero-knowledge range proof per signature, *i.e.*, that the used value for the tag is less than ℓ, where ℓ is a fixed parameter in the system. The linking authority could hold lists of ℓ tags for every user and controllable linkability can be achieved by checking whether two tags of different signatures can be found on the same list.

Public Key Anonymous Tag Systems: Abe *et al.* [1] suggested a scheme where users as well as an authority in possession of a dedicated link key can compute link tokens for specific members. Signatures of specific members can be traced publicly with the help of these link tokens. Again, controllable linkability can be implemented with public key anonymous tag systems, but the linking requires costs linear in the size of the number of group members.

Verifier Local Revocation: In verifier local revocation (VLR) [9], verifiers test signatures against entries on a revocation list (RL). In order to achieve controllable linkability the linking authority could check two signatures against all entries on the RL (consisting of all members within the group). If both signatures can be linked to the same entry on the RL then these two signatures have been generated by the same signer. The drawback of this mechanism is that the computational effort for the LA grows linearly with the size of the RL.

Comparison: Table 1 compares the above outlined concepts regarding their applicability for the implementation of *controllable linkability* (CL). The concepts of *tracing-by-linking* as well as *link-but-not-trace* aim at preventing signers from signing more than k times and cannot be employed to achieve *controllable linkability*. The other concepts can be employed to achieve controllable linkability, but are rather inefficient in terms of overhead of communication required every time a new user joins the group (communication of trapdoors to the LA), in terms of additional costs for the actual linking of signatures as well as in terms of memory requirements to achieve the desired functionality of linking. We do not mention the trivial approach of opening group signatures for the purpose of linking due to the serious privacy concerns.

In conclusion, the proposed construction to convert PB-GSSs based on the SEP paradigm into group signature schemes with controllable linkability is superior to "emulating" controllable linkability by means of features provided by existing constructions. In particular, the approach proposed in this paper comes *for free*, meaning that it does not add additional costs to signature generation and verification, does not enlarge the signature size, and the linking authority can perform linking efficiently in constant time and without additional memory requirements. Furthermore, as the linking authority holds a single trapdoor in form of a linking key, there is no necessity to communicate user specific trapdoors from the group manager to the linking authority when a new user joins the group.

Future Work: It would be interesting to investigate group signatures with controllable linkability in a model such as [23], which prevents signature hijacking, being stronger than the BSZ model. Furthermore, it could be valuable to extend the model for controllable linkability in a way that it is required by the linking authority to provide a publicly verifiable proof that two signatures have indeed been produced by the same signer (similar to the proof τ required by the Open algorithm). On a theoretical side, as group signature schemes secure in the dynamic group setting are known to imply public key encryption with noninteractive opening (PKENO) [16], it would also be interesting to study whether group signatures with controllable linkability imply AoN-PKEET*.

Acknowledgments. The authors would like to thank the anonymous reviewers for their valuable comments and suggestions. Daniel Slamanig has been supported by the European Commission through project FP7-FutureID, grant agreement number 318424. Raphael Spreitzer and Thomas Unterluggauer have been supported by the Austrian Government through the research program FIT-IT under the grant number 835917 (NewP@ss).

References

[1] Abe, M., Chow, S.S.M., Haralambiev, K., Ohkubo, M.: Double-Trapdoor Anonymous Tags for Traceable Signatures. In: Lopez, J., Tsudik, G. (eds.) ACNS 2011. LNCS, vol. 6715, pp. 183–200. Springer, Heidelberg (2011)

[2] Ateniese, G., Tsudik, G.: Some Open Issues and New Directions in Group Signatures. In: Franklin, M.K. (ed.) FC 1999. LNCS, vol. 1648, pp. 196–211. Springer, Heidelberg (1999)

[3] Au, M.H., Susilo, W., Mu, Y.: Constant-Size Dynamic k-TAA. In: De Prisco, R., Yung, M. (eds.) SCN 2006. LNCS, vol. 4116, pp. 111–125. Springer, Heidelberg (2006)

[4] Bellare, M., Micciancio, D., Warinschi, B.: Foundations of Group Signatures: Formal Definitions, Simplified Requirements, and a Construction Based on General Assumptions. In: Biham, E. (ed.) EUROCRYPT 2003. LNCS, vol. 2656, pp. 614–629. Springer, Heidelberg (2003)

[5] Bellare, M., Shi, H., Zhang, C.: Foundations of Group Signatures: The Case of Dynamic Groups. In: Menezes, A. (ed.) CT-RSA 2005. LNCS, vol. 3376, pp. 136–153. Springer, Heidelberg (2005)

[6] Bichsel, P., Camenisch, J., Neven, G., Smart, N.P., Warinschi, B.: Get Shorty via Group Signatures without Encryption. In: Garay, J.A., De Prisco, R. (eds.) SCN 2010. LNCS, vol. 6280, pp. 381–398. Springer, Heidelberg (2010)

[7] Boneh, D., Boyen, X.: Short Signatures Without Random Oracles. In: Cachin, C., Camenisch, J.L. (eds.) EUROCRYPT 2004. LNCS, vol. 3027, pp. 56–73. Springer, Heidelberg (2004)

[8] Boneh, D., Boyen, X., Shacham, H.: Short Group Signatures. In: Franklin, M. (ed.) CRYPTO 2004. LNCS, vol. 3152, pp. 41–55. Springer, Heidelberg (2004)

[9] Boneh, D., Shacham, H.: Group Signatures with Verifier-Local Revocation. In: ACM CCS, pp. 168–177 (2004)

[10] Camenisch, J., Lysyanskaya, A.: Signature Schemes and Anonymous Credentials from Bilinear Maps. In: Franklin, M. (ed.) CRYPTO 2004. LNCS, vol. 3152, pp. 56–72. Springer, Heidelberg (2004)

[11] Camenisch, J., Stadler, M.A.: Efficient Group Signature Schemes for Large Groups. In: Kaliski Jr., B.S. (ed.) CRYPTO 1997. LNCS, vol. 1294, pp. 410–424. Springer, Heidelberg (1997)

[12] Chaum, D., van Heyst, E.: Group Signatures. In: Davies, D.W. (ed.) EURO-CRYPT 1991. LNCS, vol. 547, pp. 257–265. Springer, Heidelberg (1991)

[13] Chow, S.S.M.: Real Traceable Signatures. In: Jacobson Jr., M.J., Rijmen, V., Safavi-Naini, R. (eds.) SAC 2009. LNCS, vol. 5867, pp. 92–107. Springer, Heidelberg (2009)

[14] Delerablée, C., Pointcheval, D.: Dynamic Fully Anonymous Short Group Signatures. In: Nguyên, P.Q. (ed.) VIETCRYPT 2006. LNCS, vol. 4341, pp. 193–210. Springer, Heidelberg (2006)

[15] Dodis, Y., Yampolskiy, A.: A Verifiable Random Function with Short Proofs and Keys. In: Vaudenay, S. (ed.) PKC 2005. LNCS, vol. 3386, pp. 416–431. Springer, Heidelberg (2005)

[16] Emura, K., Hanaoka, G., Sakai, Y., Schuldt, J.C.N.: Group signature implies public-key encryption with non-interactive opening. Int. J. Inf. Sec. 13(1), 51–62 (2014)

[17] Hwang, J.Y., Lee, S., Chung, B.-H., Cho, H.S., Nyang, D.: Short Group Signatures with Controllable Linkability. In: LightSec, pp. 44–52 (2011)

[18] Hwang, J.Y., Lee, S., Chung, B.-H., Cho, H.S., Nyang, D.: Group signatures with controllable linkability for dynamic membership. Inf. Sci. 222, 761–778 (2013)

[19] Kiayias, A., Tsiounis, Y., Yung, M.: Traceable Signatures. In: Cachin, C., Camenisch, J.L. (eds.) EUROCRYPT 2004. LNCS, vol. 3027, pp. 571–589. Springer, Heidelberg (2004)

[20] Nakanishi, T., Fujii, H., Hira, Y., Funabiki, N.: Revocable Group Signature Schemes with Constant Costs for Signing and Verifying. In: Jarecki, S., Tsudik, G. (eds.) PKC 2009. LNCS, vol. 5443, pp. 463–480. Springer, Heidelberg (2009)

[21] Nakanishi, T., Fujiwara, T., Watanabe, H.: A Linkable Group Signature and Its Application to Secret Voting. Trans. of Information Processing Society of Japan 40(7) (1999)

[22] Sakai, Y., Emura, K., Hanaoka, G., Kawai, Y., Matsuda, T., Omote, K.: Group Signatures with Message-Dependent Opening. In: Abdalla, M., Lange, T. (eds.) Pairing 2012. LNCS, vol. 7708, pp. 270–294. Springer, Heidelberg (2013)

[23] Sakai, Y., Schuldt, J.C.N., Emura, K., Hanaoka, G., Ohta, K.: On the Security of Dynamic Group Signatures: Preventing Signature Hijacking. In: Fischlin, M., Buchmann, J., Manulis, M. (eds.) PKC 2012. LNCS, vol. 7293, pp. 715–732. Springer, Heidelberg (2012)

[24] Slamanig, D., Spreitzer, R., Unterluggauer, T.: Adding Controllable Linkability to Pairing-Based Group Signatures for Free. IACR Cryptology ePrint Archive, 2014:607 (2014)

[25] Tang, Q.: Public key encryption supporting plaintext equality test and user-specified authorization. Security and Communication Networks 5(12), 1351–1362 (2012)

[26] Teranishi, I., Furukawa, J., Sako, K.: k-Times Anonymous Authentication (Extended abstract). In: Lee, P.J. (ed.) ASIACRYPT 2004. LNCS, vol. 3329, pp. 308–322. Springer, Heidelberg (2004)

[27] Wei, V.K.: Tracing-by-Linking Group Signatures. In: Zhou, J., López, J., Deng, R.H., Bao, F. (eds.) ISC 2005. LNCS, vol. 3650, pp. 149–163. Springer, Heidelberg (2005)

[28] Yang, G., Tan, C.H., Huang, Q., Wong, D.S.: Probabilistic Public Key Encryption with Equality Test. In: Pieprzyk, J. (ed.) CT-RSA 2010. LNCS, vol. 5985, pp. 119–131. Springer, Heidelberg (2010)

"To Share or not to Share" in Client-Side Encrypted Clouds

Duane C. Wilson and Giuseppe Ateniese

Johns Hopkins University,
Computer Science Department Information Security Institute,
Baltimore, Maryland, USA

Abstract. With the advent of cloud computing, a number of cloud providers have arisen to provide Storage-as-a-Service (SaaS) offerings to both regular consumers and business organizations. SaaS (different than Software-as-a-Service in this context) refers to an architectural model in which a cloud provider provides digital storage on their own infrastructure. Three models exist amongst SaaS providers for protecting the confidentiality of data stored in the cloud: 1) no encryption (data is stored in plain text), 2) server-side encryption (data is encrypted once uploaded), and 3) client-side encryption (data is encrypted prior to upload). Through a combination of a Network and Source Code Analysis, this paper seeks to identify weaknesses in the third model, as it claims to offer 100% user data confidentiality throughout all data transactions. The weaknesses we uncovered primarily center around the fact that the cloud providers we evaluated (Wuala, Tresorit, and Spider Oak) were each operating in a Certificate Authority capacity to facilitate data sharing. In this capacity, they assume the role of both certificate issuer and certificate authorizer as denoted in a Public-Key Infrastructure (PKI) scheme - which gives them the ability to view user data contradicting their claims of 100% data confidentiality. We have collated our analysis and findings in this paper and explore some potential solutions to address these weaknesses in these sharing methods. The solutions proposed are a combination of best practices associated with the use of PKI and other cryptographic primitives generally accepted for protecting the confidentiality of shared information.

1 Introduction

Cloud computing is a model for enabling convenient, on-demand network access to a shared pool of configurable computing resources (e.g., networks, servers, storage, applications, and services) that can be rapidly provisioned and released with minimal management effort or service provider interaction. This cloud model promotes availability and is composed of five essential characteristics (On-demand self-service, Broad network access, Resource pooling, Rapid elasticity, Measured Service); three service models (Software as a Service, Platform as a Service, Cloud Infrastructure as a Service); and four deployment models (Private cloud, Community cloud, Public cloud, Hybrid cloud) [1].

S.S.M. Chow et al. (Eds.): ISC 2014, LNCS 8783, pp. 401–412, 2014.

With the advent of cloud computing, a number of Cloud Storage Providers (CSPs) have arisen to provide Storage-as-a-Service (SaaS) offerings to both regular consumers and business organizations. SaaS refers to an architectural model in which a cloud provider provides digital storage on their own infrastructure [2]. Three models exist amongst SaaS providers for protecting the confidentiality of data stored in the cloud: 1) no encryption (data is stored in plain text), 2) server-side encryption (data is encrypted once uploaded), and 3) client-side encryption (data is encrypted prior to upload). Dropbox is the most popular version of a cloud storage provider that adheres to the first confidentiality model. In this paper we examine secure alternatives to Dropbox that provide client-side encryption. Our primary motivation is based on consistent claims made by CSPs that guarantee the confidentiality of data stored in the cloud. The major claims are as follows:

- "No one unauthorized not even the cloud storage provider can access the files." [3]
- "Our 'zero-knowledge' privacy environment ensures we can never see your data. Not our staff. Not government. Not anyone." [4]
- "Contrary to other solutions, no storage provider or network administrator, no unauthorized hacker, not even we can read your files." [5]

These principles of confidentiality hold true in use cases where data is not shared with other cloud users or with entities outside of the cloud storage environment (e.g., non-members). In the evaluated SaaS environments, data sharing is accomplished in three ways: Web Link, Folder, or Group, the latter two being the focus of this research. Through our analysis we discovered that for each data sharing mechanism, there are inherent weaknesses that can expose user data to the cloud provider which directly contradicts the aforementioned CSP claims. What would prompt a Cloud Provider to compromise the trust of its users in this way?

1. **National Security**: In the interest of National Security, governments will often collect citizen data for the purposes of confirming a threat. This collection typically consists of activities such as wire tapping, data harvesting, and other forms of information collection. Data Harvesting (usually associated with Cloud-Storage attacks) refers to the collection of disparate data from a homogeneuous location. This approach is advantageous for a Government seeking data points from multiple entities such as is the case in a co-resident Cloud Storage environment.
2. **Oppressive Government**: Under certain government regimes, there may exist situations in which a Government might "force" a Cloud Provider to comply with new or existing disclosure regulations. Reasons for this may include: 1) Company Sanctions, 2) Periodic Evaluations, or 3) to Limit Monopoly business practices.
3. **Data Leakage Confirmation** As is common in the "Networked" world, data that is accessible via the Internet is subject to data breaches. The Data

Breach Notification Laws require states to report the occurrence of company and state-related data breaches. In order to confirm and subsequently comply with these laws, Cloud Providers could potentially need to access user data in plain text form.

Our primary contributions are as follows:

- We present an overview of the various sharing scenarios employed by each evaluated CSP
- We highlight the weaknesses found in each CSP sharing scenario. Our research focus is on exposing the weaknesses in private Group and Folder sharing scenarios.
- We describe how an attack against private Group or Folder sharing functions could work in practice.
- We provide evidence of Certificate Authority functionality via network traffic and source code analysis.
- We reverse engineer various CSPs code to reveal evidence that substantiates our claims that user data is not 100% safe from being read and/or manipulated by the CSP.
- We provide suggestions for addressing the inherent weaknesses discovered in the design of the CSP sharing functions.

The rest of this paper is organized as follows: Section 2 discusses the work related to our analysis area, Section 3 provides an overview of the CSPs we evaluated and the sharing scenarios they support, Section 4 discusses the attack and associated threat model, Section 5 provides a detailed review of our analysis methodology and associated results, and Sections 6 concludes the paper and provides areas for future work.

2   Related Work

This section puts our analysis in perspective by examining preceding work that relates to the analysis of Cloud Storage Providers (CSPs). Prior work falls into three specific categories: 1) General Analysis of CSP security capabilities, 2) Analysis of the Design an Implementation of CSPs, and 3) Analysis of the Weaknesses associated with CSPs.

In [6], Borgmann and Waidner of the Fraunhofer Institute for Secure Information Technology, studied the security mechanisms of seven Cloud Storage Services: CloudMe, CrashPlan, Dropbox, Mozy, TeamDrive, Ubuntu One, and Wuala. The study focused on the following security requirements: Registration and Login, Transport Security, Encryption, Secure File Sharing, and Secure Deduplication. Similar to our approach, they examine several aspects of the CSP's file sharing mechanisms for security flaws. Specifically, they highlight the fact that if client-side encryption is used, sharing should not weaken the security level. In particular, the CSP should not be able to read the shared files. The scope of their work covered sharing with other subscribers of the same service, sharing files with a closed group of non-subscribers, and sharing files with everybody. As we also discovered in our research, sharing via secret web link (e.g., closed group of non-subscribers) or making data public reveals shared data to the CSP. Their analysis, however, did not yield the fact that sharing files with subscribers could also result in a breach of confidentiality with the CSP due to the CSP acting in a Certificate Authority (CA) capacity. This is proven in our analysis and can serve as a viable extension to their work.

Mager et al. in [7] examine the design and implementation of an online backup and file sharing system called Wuala. The goals of their work consist of four primary items: Characterization of the Infrastructure, Understanding the Data Placement Methodology, Identifying the Coding Techniques relating to Data Availability (accessibility of files at any time) and Durability (ensuring files are never lost), and the Determination of the Data Transport Protocol used. The scope of their evaluation does not necessarily include security considerations, although the data structure employed by Wuala for sharing data is mentioned as well as the type of encryption used for performing client-side encryption. Overall, this research provided useful insight into methods employed by the CSP to facilitate client usage of their infrastructure, however, it does not examine the infrastructure or design in detail for security weaknesses. As it pertains to their goals, our work differs in that it looks at the CSP infrastructure security features for consistency with user expectations. We focus on the data sharing (not placement) methodology employed by the CSPs. The coding techniques we identified relate to the discovery of evidence that the CSP is operating in a CA capacity during sharing transactions.

Finally in [8], Kholia and Wegrzyn analyze the Dropbox cloud-based file storage service from a security perspective. Their research presents novel techniques to reverse engineer frozen Python applications (to include Dropbox). They specifically describe a method to bypass Dropbox's two factor authentication and hijack Dropbox accounts. Additionally, they introduce generic approaches to intercept SSL data using code injection techniques and monkey patching. This work is consistent with our analysis of CSPs for major security flaws as it results in the exposure of private user data and enables the CSP to have access to shared user data.

3   Cloud Storage Provider Overview

This section provides an overview of the Cloud Storage Providers (CSPs) we evaluated and the 'confidentiality model' they each employ to protect the confidentiality of user data. We define a confidentiality model as the method employed by the CSP to protect the confidentiality of users' stored data. According to [9,10], there are several CSPs that offer client-side encryption. For our analysis, we focused on Wuala, Spider Oak, and Tresorit. Wuala encryption is performed with AES256 prior to files being uploaded. Encryption comprehensively includes not only file content, but also: file names, preview images, folders and metadata. An RSA2048 key is used for signatures and key exchange when folders are shared while SHA-256 is used for data integrity. Spider Oak can be used to share and back up files. Data is encrypted on a user computer with AES256 in CFB mode and HMAC-SHA256. As mentioned previously, the company claims to have no knowledge of what data is stored in their servers or user passwords. Their software works in smart phones, the Linux operating system, and Windows [9]. Lastly, Tresorit is a Hungarian-based company that uses AES256 to encrypt data before uploading it to the cloud. The company is offering $10,000.00 (US) to anyone who can break their security software. Similar to Spider Oak, data can be accessed via smart phone or desktop computer. Each CSP's confidentiality model is discussed in the following subsections. All information in this section has been adapted from the CSP's website and/or documentation unless otherwise specified.

Table 1 summarizes the confidentiality model employed by each CSP based on each data sharing scenario provided. It also presents an overview of each sharing scenario according to the level of effort required to 'exploit' the scenario (Trivial or Non-Trivial). Our research focuses on the sharing scenarios that the CSPs highlight as offering 100% data confidentiality (i.e., Private Group and Private Folder sharing). "N/A" represents a feature that is not implemented or has no documentation to support the scenario.

Table 1. Trivial vs. Non-Trivial Sharing Scenarios

	Sharing Scenario					
CSP	Public Web Link	Private Web Link	Public Folder	Private Folder	Public Group	Private Group
Wuala	Public URL	Private URL	Public Folder	Public Key Encryption	Public Group	Public Key Encryption
Spider Oak	Public URL	Public URL with Passwd	N/A	N/A	N/A	N/A
Tresorit	N/A	Encrypted Link	N/A	RSA or TGDH	N/A	ICE Protocol
Difficulty	Trivial	Trivial	Trivial	**Non-Trivial**	Trivial	**Non-Trivial**

As noted above for all CSPs, when data is shared via Private web link, the CSP requires the transmission of some sensitive data in order to process the user's request. Similarly, each of the Public Sharing Scenarios (i.e., Web Link, Group, Folder) require implicit trust of the CSP in order to make the data public. As a result, these scenarios were discussed in this section for completeness but fall outside of the scope of our research due to this requirement of trust.

4   Threat Model

In this section we provide a threat model which describes how an attack against a Cloud Storage Provider (CSP) offering client-side encryption could potentially be executed. Rating the threats we identify in this section are outside of the scope of our research. We make the following assumptions with regards to the threat model:

1. The CSP client is trusted and has not been previously modified by a malicious insider or outsider. In the case of a modified client, users could easily be redirected to a rogue server upon client startup and be forced to perform all sharing transactions through this server.
2. The CSP server is trusted and has not been compromised. Similar to the case of a rogue client, a rogue server would allow an adversary access to user certificates and gain access to decryption keys.
3. Other CSP members can be trusted. When searching for an individual user in the CSP database, it is assumed that the users identified are legitimately who they appear to be.
4. The User certificates issued cannot be trusted because they are issued by the CSP.
5. Public or Private Web Link Sharing scenarios (as mentioned above) require users to reveal some form of sensitive information to the CSPs to enable decryption in a browser environment. Similarly, Public Groups and Folders are accessible by all members and pose no threats to users.

Our analysis consists primarily of network traffic analysis and reverse-engineering the source code of each CSP client through decompilation or disassembly. We selected this course of analysis due to fact that most of the CSP client code used some form of code obfuscation — making the replication of a rogue client challenging. Additionally, we could not produce counterfeit certificates 'on behalf of' other users, because we do not possess the signing key of the CSPs we analyzed. This signing key would be required to produce certificates that CSP client applications would trust. We discuss below a scenario that would enable a Cloud Provider to maliciously access a user's data.

1. User A signs up for Cloud Account
2. User A initiates sharing request with User B
3. Cloud client returns User B's contact information to User A
4. Cloud Provider substitutes its Public Key for User B
5. Without User A's knowledge his/her data is encrypted with the Public Key of the Cloud Provider
6. Cloud Provider is able to decrypt User A's data and view its contents
7. Cloud Provider then re-encrypts data with User B's Public Key and Cloud Client sends sharing request to User B
8. User B decrypts the data sent by User A w/o knowledge of the above-stated attack.

This scenario simply show that because the CSPs are operating as Certificate Authorities, certificate manipulation, spoofing, or substitution of any kind can be accomplished. It is also important to emphasize that a malicious CSP cannot be detected since it can perform a standard *man-in-the-middle* attack in which the encrypted information flow from the sender is first decrypted and then re-encrypted on the fly under the recipient's public key (and vice-versa). In essence, the CSP is able to spoof the identity of any user through the use of certificates. In the next section, we show evidence to support our claim that each evaluated CSP is serving in a CA capacity -- making all users susceptible to the aforementioned attacks.

5   Analysis Methodology and Results

In this section we provide a detailed review of our analysis results. It is important to note that although our results do show that each Cloud Storage Provider is in some capacity operating as a Certificate Authority (i.e., could issue fake certificates to its users) - we did not find any evidence to support any claims of misuse by the Providers themselves (i.e., they are not actively exploiting users via this attack). Our analysis consisted of three phases examining the sharing functions employed by each CSP: 1) Network Traffic Analysis (using Wireshark [12] and Fiddler Web Proxy Debugger [13]), 2) Source Code Decompilation (using AndroChef Java decompiler [14] and Boomerang C++ decompiler), and 3) Source Code Disassembly (using HopperApp [15] and Synalyze [16]). Evidence of CA functionality validates our assertion that when data is shared, it is possible for the CSP to view user data which contradicts their 100% data confidentially claims. Evidence of CA functionality includes (but is not limited to the following): Certificates (i.e., Root or Intermediary), Certificate Issuance, Certificate Validation, Certificate Revocation, Certificate Authentication, Certificate Renewal, Certificate Registration, Obtaining Certificates, Encryption/Decryption, and Digital Signature (signing or verification). The processes for performing the network traffic, source code decompilation, and source code disassembly analyses are discussed below.

A CA is a Trusted Third Party (TTP) organization or company that issues digital certificates used to create digital signatures and public-private key pairs. The role of the CA in this process is to guarantee that the individual granted the unique certificate is, in fact, who he or she claims to be. CAs are a critical component in data security and electronic commerce because they guarantee that the two parties exchanging information are really who they claim to be [17]. Serving as a CA, the CSPs would potentially be able to issue counterfeit certificates to users – pretending to be a legitimate entity that data is shared with. The code we identified falls into five categories: 1) Code Libraries, Certificate Files, Certificate Operations, Keystore Files/Definitions, and Peripheral Cryptographic Functions. Lastly, we focused on disassembling code that could not be reproduced in its original form through Decompilation. The code samples we extracted were only those that could potentially represent CA functionality. We primarily analyzed the main binary files (.app, .exe files) associated with each CSP.

5.1   Wuala Analysis

We started our analysis of Wuala by examining the network traffic that was generated between the client and server during sharing transactions and extracting any evidence of CA functionality. Through our analysis, we discovered evidence of a Wuala CA certificate which was used to confirm the validity of the other certificates below it in the certificate chain. A root certificate is either an unsigned public key certificate or a self-signed certificate that identifies the Root Certificate Authority (CA). It is the top-most certificate of the tree, the private key of which is used to "sign" other certificates. All certificates immediately below the root certificate inherit the trustworthiness of the root certificate [18].

Secondly, during disassembly of the Wuala binaries, we identified multiple references to a 'WualaCACerts' file which contains several certificates that the Wuala client uses during its execution to include the Wuala CA file — mentioned previously. The reference below is to the bouncy castle (Java Crypto Library) implementation of a Keystore which also contains individual certificate entries. The contents of this particular file is discussed in the decompilation section below. Lastly, during the decompilation phase, we discovered that Wuala uses a combination of Name Obfuscation (e.g., changing method and variable names while maintaining the functionality). To overcome this, we performed searches across the entire codebase for files that fit into the aforementioned categories. The primary results of the Wuala decompilation process yielded the WualaCACerts file. This file contains 4 certificate entries: *wualaadminca*, *wualaserverca*, *wualaca*, and *wualaclientca*. Each certificate is named to denote the function it serves within the operation of the Wuala application (i.e., administrative, client, server, and CA). Each certificate entry has an 'Extension' section with an ObjectID and a specification of what the key (within the certificate file is used for). In each of these, the key is used for Certificate Signing. According to IBM's *Key usage extensions and extended key Usage* page, the Certificate Signing key usage extension is to be used when the subject public key is used to verify a signature on certificates. This extension can be used only in CA certificates [19]. We summarize the contents of the Wuala CA certificate below:

```
Alias name: wualaca, Creation date: Jan 6, 2012
Entry type: trustedCertEntry, Owner: CN=Wuala CA, OU=Wuala,
EMAILADDRESS=cert@wuala.com, Issuer: CN=Wuala CA, OU=Wuala, O=LaCie
EMAILADDRESS=cert@wuala.com, Signature algorithm name: SHA1withRSA
```

As indicated by the algorithms identified in the certificates as well as the key usage extension, this certificate is used for creating and verifying digital signatures. However, the same certificates could be issued by Wuala for performing other PKI-related functions to include both encryption and decryption of user data or the creation of private groups.

5.2   Spider Oak Analysis

In the case of Spider Oak, the disassembly of its binary did not yield results to substantiate our claims, thus we focused primarily on the results of decompilation and network traffic analysis in this section. During the code analysis of Spider Oak, we uncovered a 'Public Key' folder containing DSA files, RSA files, a public key file, and other crypto-related files. These files were each disassembled and evaluated for evidentiary purposes. The 'RSA.pyc' file is presented below as a representative sample. It shows a call to (or definition of) a 'generate' function used to generate an RSA public/private key pair programmatically. There is a similar file for generating a DSA key pair which we excluded. The generation of certificates in a programmatic function confirms that users are not allowed to use their own certificates for cryptographic purposes which can allow Spider Oak to generate certificates 'on behalf' of legitimate users. As a result, a user could be tricked into sharing data with Spider Oak instead of the intended user.

```
db "generate(bits:int, randfunc :callable, progress_func:callable)
Generate an RSA key of length 'bits',  using 'randfunc' to get
random data and 'progress_func', if present, to display the
progress of the key generation. l\x02"
```

Unlike Wuala, Spider Oak does not appear to use any form of code obfuscation. The files we examined contained the .pyc extension which is the equivalent of a Java Byte Code file for the Python programming language. To analyze these files, we had to decompile them to the original Python Source code. Similar to Wuala, Spider Oak uses a file called 'certs.pyc' to store its certificates. This file contained 5 certificate entries: Equifax Secure CA, GeoTrust Global CA (2), RapidSSL CA, and the Spider Oak Root CA. Contents of the Spider Oak Root CA certificate are as follows:

```
Data:  Version: 1 (0x0), Serial Number: ea:14:d7:ad:6a:cf:dc:35
Signature Algorithm: sha1WithRSAEncryption
Issuer: emailAddress=ssl@spideroak.com, organizationName=SpiderOak
```

This analysis phase revealed a self-signed root certificate from Spider Oak. A self-signed certificate is an identity certificate that is signed by the same identity it certifies. As mentioned above, a typical PKI scheme requires certificates to be signed by a Trusted Third Party (TTP) for verification purposes. Self-signed certificates cannot (by nature) be revoked, which may allow an attacker who has already gained access to monitor and inject data into a connection to spoof an identity if a private key has been compromised [20]. In this case, the attacker could potentially be Spider Oak (e.g., spoofing the identity of another user and gaining access to user data).

Lastly, examining the network traffic between the Spider Oak client and server produced the same root certificate identified during Decompilation. As a root certifier, Spider Oak is also responsible for validating any certificate in its trusted certificate chain – to include Spider Oak client certificates issued to users. Finding this type of certificate also demonstrates the fact that as a certificate issuer, Spider Oak can issue user certificates that spoof the identity of another user resulting in a data confidentiality breach when data is shared with another Spider Oak user. Similar to the Wuala example (discussed above), the Spider Oak Client certificate is also validated by its respective Spider Oak Root CA.

In a recent article, Eric Snowden urges consumers to adopt more secure file storage systems which are less susceptible to government surveillance - mentioning Spider Oak specicially [21]. However, he fails to disclose (or realize) the fact that data confidentiality could be breached if data is shared (as shown above) within these same environments. Additionally, Spider Oak was interviewed by Senior Writer Brian Butler of Network World concerning some of our findings [22]. The company claims that the features we evaluated fall outside of the scope of features that they have implemented [23], however, we assert that in order to share encrypted data - some form of Public Key Encryption must be used. As a result, Spider Oak users are still subject to this type of flaw.

5.3   Tresorit Analysis

During the Decompilation of Tresorit, we did not identify any independent certificate files as was the case with Wuala and Spider Oak. According to Tresorit's documentation, a roaming profile file contains both the SSL/TLS client certificate and the associated private key required to communicate with Tresorit's servers, and agreement certificates and private keys to securely access and decrypt the contents of a Tresor. Hence, there were not any specific certificate entries to examine during this phase.

Disassembly of the Tresorit binary contained a number of references to certificate operations, files, and uses with application. The entry below shows the loading of a certificate in the traditional X.509 format. This confirms that X.509 certifications are utilized in the Tresorit client application for cryptographic functions, however, does not fully substantiate the fact that Tresorit is operating in a Certificate Authority (CA) capacity.

```
db "y_OBJ", 0, db  "X509_get_pubkey_parameters", 0
db "X509_load_cert_crl_file", 0, db "X509_load_cert_file", 0
db "X509_load_crl_file", 0
```

In the network analysis of Tresorit, we identified a certificate issued by Tresorit — the Tresorit User CA certificate. This certificate differs from the Wuala and Spider Oak certificates in that it is validated by the StartCom CA which would imply that they are using a TTP to ensure 100% data confidentiality during sharing transactions. However, the fact that this certificate was issued by Tresorit verifies that they also can view user data whenever the keys are used for encryption/decryption purposes. Additionally, the fact that users are neither allowed to use their own certificates or enter credentials to generate their own certificates upon signing up with any of the three CSPs gives credence to our claims that counterfeit certificates could be issued. This would lead users to implicitly trust that other users are who they say they are (i.e., the CSP is not acting on their behalf). In the case of Tresorit, StartCom CA is serving as the Root CA with Tresorit serving as an Intermediate CA. This differs from both Wuala and Tresorit, however, is still indicative of Tresorit operating as a CA in some capacity.

6   Conclusions and Future Work

In this paper we examined three major Cloud Storage Providers (CSPs) for weaknesses associated with their sharing functions. Each CSP claims to offer 100% user data confidentiality through all data transactions executed within their respective cloud instances. We discovered that these principles of confidentiality hold true in use cases where data is not shared with other cloud users or with entities outside of the cloud storage environment (e.g., non-members). We presented several scenarios in which data confidentiality could be breached to include: when data is shared via Public Web Link, when data is shared via Private Web Link, when CSP is accessed via Web Browser (i.e., web access), and when data is shared via Group or Private folder (due to the CSP being the issuer of the certificates used for cryptographic operations – hence could allocate counterfeit certificates to users). We provide evidence in the form of Network Traffic and Source Code analysis for each CSP to substantiate these claims.

To address the weaknesses discussed in the previous sections, users should be issued standard certificates via a Trusted Third Party which is typical in a traditional Public/Private Key implementation or should be allowed to use their own certificates (e.g., PGP). Alternatively, out-of-band verification mechanisms should be used such as those provided by several implementations of off-the-record (OTR) messaging [21]. Similarly, Tresorit makes use of ICE — a proprietary protocol that is designed to work in a semi-trusted environment. It does not rely solely on the trustworthiness of the certificate authority issuing the users' certificates, but also an invitation secret and optional pairing password. Lastly, to enhance the security associated with sharing data via Private Web Link, it is possible to require password authentication prior to listing shared files. Spider Oak employs such a strategy, although the CSP would still be required to verify that the password was entered correctly. In the future, we plan to extend this work by examining alternative methods of secure file sharing that do not rely on a CA for identity-verification - which will consequently overcome a number of weaknesses we have identified in this research.

References

1. Bohn, R.: NIST Cloud Computing Program. Cloud Computing. National Institute of Standards and Technology (December 02, 2011), Web (February 06, 2014)
2. SearchStorage. Storage as a Service (SaaS). What Is Storage as a Service. SearchStorage (February 2009), Web (February 06, 2014)
3. Lacie. Wuala. Lacie (January 01, 2014), Web (February 06, 2014)
4. Spider Oak. 100% Private Online Backup, Sync & Sharing. SpiderOak (2014), Web (February 06, 2014)
5. Tresorit. Secure File Sync and Share. Tresorit (2014), Web (February 06, 2014)
6. Borgmann, M., Waidner, M.: On the Security of Cloud Storage Services. Fraunhofer-Verl., Stuttgart (2012) (print)
7. Mager, T., Biersack, E., Michiardi, P.: A Measurement Study of the Wuala On-line Storage Service. In: Peer to Peer IEEE International Conference Proceedings, pp. 237–248 (2012) (print)

8. Kholia, D., Wegrzyn, P.: Looking inside the (Drop) Box. In: 7th USENIX Workshop on Offensive Technologies (2013)
9. Hacker10. List of USA Cloud Storage Services with Client Side Encryption. Hacker 10 Security Hacker (September 12, 2013), Web (February 18, 2014)
10. Hacker10. List of Non USA Cloud Storage Services with Client Side Encryption. Hacker 10 Security Hacker (September 12, 2013), Web (February 18, 2014)
11. Tresorit. Tresorit: White Paper. Tresorit (2012), Web (February 18, 2014)
12. Wireshark Foundation. WireShark. Wireshark Foundation (1998), Web (February 18, 2014)
13. Telerik. Fiddler. The Free Web Debugging Proxy by Telerik. Telerik (2002), Web (February 18, 2014)
14. AndroChef Java Decompiler. AndroChef Java Decompiler, n.d. Web (February 18, 2014)
15. Bénony, V.: Hopper. Vincent Bénony, n.d. Web (February 18, 2014)
16. Synalysis. Synalyze It! Reverse Engineering and Binary File Analysis Made Easy. Synalysis (2010), Web (February 18, 2014)
17. Froomkin, A.M.: 1996 A. Michael Froomkin: The Essential Role of Trusted Third Parties in Electronic Commerce. N.p. (October 14, 1994), Web (February 18, 2014)
18. Microsoft. What Are CA Certificates? Technet Library. Microsoft Technet (March 3, 2003), Web (February 18, 2014)
19. IBM Lotus Domino and Notes Information Center. IBM Lotus Domino and Notes Information Center. N.p. (August 14, 2008), Web (February 18, 2014)
20. The IEEE P1363 Home Page. IEEE P1363 – Standard Specifications for Public Key Cryptography. N.p. (October 10, 2008), Web (February 18, 2014)
21. Kiss, J.: Snowden: Dropbox Is Hostile to Privacy, unlike 'zero Knowledge' Spideroak. Theguardian.com. Guardian News and Media (July 17, 2014), Web (August 13, 2014)
22. Butler, B.: Even the Most Secure Cloud Storage May Not Be so Secure, Study Finds. Network World. Network World Inc. (April 21, 2014), Web (August 13, 2014)
23. Fairless, A.: Comments on Study Citing Design Flaw That Puts Your Privacy at Risk - SpiderOak Blog. SpiderOak Blog. Spider Oak (April 22, 2014), Web (August 13, 2014)
24. Goldberg, I.: Off-the-Record Messaging. Off-the-Record Messaging. OTR Development Team (2012), Web (February 25, 2014)
25. Grolimund, D., Meisser, L., Schmid, S., Wattenhofer, R.: Cryptree: A Folder Tree Structure for Cryptographic File Systems. Reliable Database Systems. Computer Engineering and Networks Laboratory (October 4, 2006), Web (February 25, 2014)

eavesROP: Listening for ROP Payloads in Data Streams

Christopher Jämthagen, Linus Karlsson, Paul Stankovski, and Martin Hell

Dept. of Electrical and Information Technology, Lund University,
P.O. Box 118, 221 00 Lund, Sweden
{christopher.jamthagen,linus.karlsson,paul.stankovski,
martin.hell}@eit.lth.se

Abstract. We consider the problem of detecting exploits based on return-oriented programming. In contrast to previous works we investigate to which extent we can detect ROP payloads by only analysing streaming data, i.e., we do not assume any modifications to the target machine, its kernel or its libraries. Neither do we attempt to execute any potentially malicious code in order to determine if it is an attack. While such a scenario has its limitations, we show that using a layered approach with a filtering mechanism together with the Fast Fourier Transform, it is possible to detect ROP payloads even in the presence of noise and assuming that the target system employs ASLR. Our approach, denoted eavesROP, thus provides a very lightweight and easily deployable mitigation against certain ROP attacks. It also provides the added merit of detecting the presence of a brute-force attack on ASLR since library base addresses are not assumed to be known by eavesROP.

Keywords: Return-Oriented Programming, ROP, Pattern Matching, ASLR.

1 Introduction

Buffer overruns [1] have for a long time been a common source of software vulnerabilities. The buffer overrun vulnerability may be exploited to perform a code injection attack, where the goal is to inject arbitrary data and replacing the return address with the address of the injected data. There are several well-known and widely used mitigations against this approach. Since the injected code should not be executable, but rather considered as data, the memory pages corresponding to this data can be marked as non-executable. Data Execution Prevention (DEP) [14] and the $W \oplus X$ security feature [22] are approaches implementing this idea. While this will prevent the classic code injection attacks, it will not prevent code reuse attacks. In these attacks, the adversary will not inject the code to be executed, but will instead direct the program flow to code that is already loaded by the process, typically a shared library. One example is the return-to-libc attack [4] in which the attack points to existing code in the libc library.

S.S.M. Chow et al. (Eds.): ISC 2014, LNCS 8783, pp. 413–424, 2014.

A more advanced code reuse attack is Return-Oriented Programming (ROP), in which the attacker identifies small pieces of usable code segments, called *gadgets*, and chains them together using a `ret` instruction. A `ret` instruction will pop an address from the stack and continue execution at that address.

One available countermeasure against code injection, which can also be applied to prevent code reuse attacks, is Address Space Layout Randomization (ASLR) [22]. ASLR will randomize the base address of the program's text, stack, and heap segments and the adversary will not know where the gadgets will be located. However, as described in Section 2.2, ASLR can sometimes be bypassed.

There have been several proposed defenses against ROP attacks, all taking slightly different approaches and using different assumptions. A typical mitigation is to identify some specific features in the attack that distinguishes it from benign code execution and then build a mitigation technique based on those distinguishing features [7,8,10,12,21,23]. Another approach is to rewrite libraries or targeted code such that it is not usable in an attack [17,19] or to randomize addresses which are needed by an attacker [13,15,20,32].

Instead of detecting the attacks on the target systems, another goal may be to detect ROP attacks in data. In [23] data was scanned and possible exploits were speculatively executed in order to determine if they were exploits. This requires a snapshot of the virtual memory of the process that is protected. In [29] the authors consider a detection approach where documents are analyzed to find ROP attacks. Documents are collected and sent to a separate virtual machine, where they are opened in their native application, and the memory is then analyzed for ROP payloads.

In this paper we present eavesROP, which is a more lightweight approach where no execution takes place. We try to identify ROP payloads by looking at network traffic only, i.e., we do not make any modifications to machines, programs, libraries or operating systems, nor do we try to execute any of the received data. We do not even require any kind of access to the machines. Scenarios could be an implementation in a gateway to a corporate network, ROP payload detection in switches or at an ISP before data is forwarded to the end user. The question that we try to answer is: How much information can we deduce by just looking at the data? We target ROP exploits where gadget addresses can be explicitly found in the data sent to the application. We assume that ASLR is enforced by the operating system, and that the attacker has somehow acquired information about the location of libraries. Of course, our detection mechanism has no such information. We show how to filter out possible ROP payloads and how to determine if the candidate payload is a ROP attack or not. Even with just a moderate number of gadgets, we can detect the payload efficiently. This is true even if there is a large amount of noise present.

We have tested eavesROP using available exploits and it is able to detect exploits with no false negatives.

2 Background

In this section we provide the necessary background on return-oriented programming and address space layout randomization.

2.1 Return-Oriented Programming

Return-oriented programming [26] is an exploitation technique that allows for arbitrary code execution without having to inject code into the vulnerable process. To achieve this, an attacker constructs a payload of addresses, each pointing to a small sequence of instructions reachable and executable by the affected process. These instruction sequences are called gadgets and typically consist of very few instructions, ending with a return instruction (`ret`). This return instruction will pop the next dword from the stack, put it into the instruction pointer register (EIP) and continue execution at the next gadget. Gadgets do not have to be aligned with the intended instructions. Any byte that represents the opcode of a `ret` can potentially be used as a gadget.

Not only `ret` instructions can be used in these types of attacks. It is also possible to use jump-based instructions as in [2,6]. We will not consider these type of attacks in this paper, but note that it would be possible to extend our algorithms to detect these gadgets as well.

2.2 Address Space Layout Randomization (ASLR)

ASLR protects from buffer overflow attacks by randomizing the location of the stack, the heap and the location of all dynamically loaded libraries. The term was coined by the PaX project [22] which also has a well-known implementation.

Following the introduction of ASLR in Windows XP SP2 (2004) and in the Linux Kernel (since version 2.6.12, 2005), writing exploits has become much more difficult.

However, the efficiency of ASLR is limited. First, some small amount of code is not randomized, leaving the possibility to still use gadgets in the code where the location is predictable. Even though this code is rather small, it has been shown that it is possible to find usable gadgets in it [24]. Randomizing the application code is one kind of protection against these attacks [32]. Another aspect of ASLR, as was shown in [27], is the limited entropy in the address space, which makes it possible to brute-force absolute locations.

In addition to brute-forcing the ASLR, it has been shown that information leakage can occur through e.g., buffer and heap overrun bugs [11,31] and other types of vulnerabilites [25,28]. This could give an attacker at least partial information about the location of ASLR-affected code.

The exact means by which an attacker bypasses ASLR, may it be through brute force or information leakage, are independent of our payload detection algorithm.

3   Our Approach

In this section we give a description of the different parts of eavesROP. The idea is based on the observation that many gadgets are typically taken from the same library which results in gadget addresses being located relatively close to each other. Note though, that our approach only require a few gadgets to be located in the same library, other gadgets can be taken from other libraries. A more detailed description of eavesROP can be found in the full version of this paper [16].

3.1   Optional Data Pre-filter

Certain input data can be expected to exhibit properties that make them look like addresses close to each other in the memory space—thus looking like ROP payloads—even though the data is actually non-malicious. Our goal is to filter out this data before it reaches later steps in the algorithm, to reduce the total computational overhead of our system.

Of special interest are printable ASCII characters, not only because much data is readable text, but also because large portions of adjacent ASCII data may—when combined into 32-bit words—look like adjacent addresses. Filtering is however a trade-off between performance and false negatives. There are ways to make ROP payloads printable [18]. Such a payload would be removed if a filter for printable characters is enabled. This is why the filtering step should be considered optional.

If the pre-filter is enabled, it removes blocks of UTF-8 strings. In our implementation, we define a block as a sequence of five or more adjacent, printable UTF-8 characters. When a matching block is found, the complete block is removed from the input. This leads to potential noise as non-adjacent bytes become adjacent after the data between them is removed. This does, however, only affect a few addresses, which does not cause any problems in practice since our ROP pattern matching is very precise and noise tolerant.

3.2   Cluster Detection

An actual exploit payload will contain several gadget addresses that lie close together with respect to the entire addressable memory space. The purpose of the address cluster detection is to find and isolate the congested parts of the memory space for further processing by generating a binary address vector, P_{obs}, of size L, i.e., the size of the largest targeted library. These vectors indicate addresses found in the data.

We let M denote the maximum size of a ROP payload in 4-byte words that our detection is guaranteed to support. A naïve approach to detect the gadget addresses is to pick M words of data, map them to P_{obs} and match this vector with a known library pattern P_{lib}. Doing this byte by byte in the data would produce the correct maximum matching, but it is a very slow approach. Moreover, all words but one will repeat every 4 bytes. Another problem is that the addresses contained within the data window can be spread out over the entire ASLR address space (N bytes), making P_{obs} very large. We propose to use an algorithm that is much more efficient, and will still always find the correct maximum matching.

Instead of considering M addresses, we pick a data window of size $D = 2M$. Thus, we consider twice as much data as the maximum payload, but in return we consider $M + 1$ possible payloads simultaneously. Doubling the data window size introduces more noise (more data in one window), so a few more data windows will pass the cluster detection stage, but this effect is marginal compared to the significant gain in processing efficiency.

When the data window slides over the next data chunk of size M, we begin by extracting potentially viable addresses. As the offset of a ROP payload in the data buffer is unknown, but we know that each address is four bytes and addresses are aligned inside a payload, we create a list for each offset.

We need to keep track of eight such address lists, four for each of the two M-word data chunks covered by the D-word data window. Separating the lists per data chunk allows for incrementing the data window in steps of M words, while reusing the four lists corresponding to the previous M words.

The four new address lists are sorted using an efficient linear-time sorting algorithm such as bucket sort [9]. Such efficient sorting is possible since all addresses are of the same size. Once sorted, we slide an address window of size L (same size as executable part of the largest library) down the combination of the two lists for each offset. Since each of the two lists is individually sorted, we can easily traverse the sorted combination of the lists with time complexity $\mathcal{O}(D)$.

Let T be a threshold value that determines the minimum number of gadgets in an exploit that we want to be able to detect. A small T leads to detection of more exploits, but it also results in more pattern matching, slowing down the detection algorithm. In practice, the lowest value that our algorithm can handle is $T \approx 6$, depending on the instruction size of each gadget (see Section 3.3) and the error probabilities (see Section 3.4).

If we find an address window that contains at least T unique addresses, the binary vector P_{obs} is constructed by entering a '1' in each position corresponding to an address in the address window. To minimize redundant checks, P_{obs} is normalized to always start with a '1'. Then we proceed to perform pattern matching via FFT, as described in Section 3.3.

3.3   Pattern Matching

The vector P_{obs} from the previous step is matched with binary library vectors. The relative distances of the memory addresses in an address window form a very distinct pattern. This pattern is matched with the gadget address patterns of libraries P_{lib}.

Since we allow ASLR, the precise memory location of the library is unknown. This is handled by using the Fast Fourier Transform to compute the maximum matching between P_{obs} and P_{lib}.

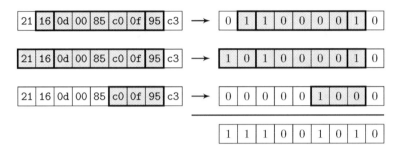

Fig. 1. Translation of maximal length gadget sequences to binary pattern

Identifying Gadgets in a Library. In order to find all possible gadgets in a library, the executable part of it is scanned for the opcode of different types of return instructions, namely 0xC2 (`retn imm16`), 0xC3 (`retn`), 0xCA (`retf imm16`) and 0xCB (`retf`). For each position of these bytes in the library we search backwards one byte at a time and try to assemble a legal instruction flow ending with the return. We define the *entry zone*, z, as the number of instructions we allow for each gadget, not including the return instruction. This means that we can find many gadgets ending at the same return instruction due to the possibility of instruction overlapping in the x86 architecture.

The starting byte of every possible gadget is used to construct the binary vector P_{lib}. This is the vector that is used for pattern matching with P_{obs}, which is the output of the address cluster detection algorithm.

To understand how the gadget structure in a library is translated into a binary pattern, consider the following sequence of nine bytes (hex):

$$\texttt{21 16 0d 00 85 c0 0f 95 c3}$$

Using an entry zone of size $z = 3$ (at most three instructions), we construct maximal gadget chains by interpreting the bytes preceding the return instruction `c3` as consecutive instructions. There are three possible maximal gadget chains in the above byte sequence, as illustrated in Figure 1.

The top two gadget chains are both of length three. While the top chain begins with a single-byte instruction `16`, the second chain extends this to a two-byte instruction `21 16`. The third chain is of length 1, but it is maximal since it cannot be further extended.

A sequence of bytes belonging to a library is translated into a binary pattern. A '1' in the array represents a gadget and a '0' is used for the other positions.

Pattern Matching via FFT. Perfect pattern matching can be performed efficiently using a Fast Fourier Transform (FFT). Pattern matching, here, means that we want to find the maximum weight of the overlap between two patterns that are overlaid. We also want this matching to be perfect, which is to say that all actual gadget addresses that are used in an exploit will be counted. All actual gadget addresses in an exploit will, in the general case, contribute positively to the weight of the maximal pattern match.

Focusing on one library, P_{lib} and P_{obs} are binary vectors of length L. If both patterns are aligned, the maximum matching can be calculated as the dot product between P_{lib} and P_{obs} according to

$$P_{\text{lib}} \cdot P_{\text{obs}} = \sum_{i=0}^{L-1} P_{\text{lib}}[i] P_{\text{obs}}[i].$$

However, we have no way of knowing if the alignment is correct, so we rather need to try all alignments to see which one produces the highest fit. That is, we need to calculate the dot products for all possible shifts of the two patterns. This can be accomplished by using the Fast Fourier Transform (FFT). The FFT computes the circular discrete convolution c of two vectors a and b of length L,

$$c[t] = (a * b_L)[t] = \sum_{i=0}^{L-1} a[i] b[(t-i) \bmod L]. \tag{1}$$

For this to be applicable to our situation, we need to adjust two things. First of all, we need to reverse one of the vectors, say P_{lib}. Secondly, since indices in Eq. (1) are taken modulo L, we need to pad both P_{lib} and P_{obs} with zeros to double length. Without this zero padding, the tails and fronts of the two vectors will contribute to the maximum matching in an undesirable way, effectively bringing more noise into our result.

The FFT approach (see [3]) has time complexity $O(L \lg L)$, compared to $O(L^2)$ for the naïve approach. Letting $\mathcal{F}$ denote the FFT version of the Discrete Fourier Transform (DFT), we may compute c as

$$c = \mathcal{F}^{-1}\left(\mathcal{F}(a) \odot \mathcal{F}(b)\right),$$

where $\odot$ denotes componentwise multiplication.

We let a and b be the vectors P_{lib} and P_{obs} respectively after the zero padding as described above. The weight of the maximum matching is given as the maximum component of c,

$$c_{\text{max}} = \max_i c[i]. \tag{2}$$

Note that P_{lib} is known beforehand, so we can precompute $\mathcal{F}(a)$ for efficiency.

3.4   Statistical Test

The maximum overlap is given by the maximum value of the inverse Fourier transform as given in Eq. (2). In order to find an expression for the number of overlaps we make the following approximations.

- Locations corresponding to gadgets in P_{lib} are uniformly distributed.
- The entries in the convolution vector c are approximated as independent events, all with the same probability.

Using these approximations, the number of overlaps between P_{lib} and P_{obs} is binomially distributed, $X^{(w)} \sim \text{Bin}(w, \frac{G}{L})$, where w and G denote the Hamming weights of P_{obs} and P_{lib}, respectively. Note that G should here be understood as the number of gadgets in a library for a given entry zone, and w is the number of addresses in an address window. Thus, the probability that there are s overlaps is given by

$$\Pr(X^{(w)} = s) = \binom{w}{s}\left(\frac{G}{L}\right)^{s}\left(1 - \frac{G}{L}\right)^{w-s}. \tag{3}$$

Since P_{lib} and P_{obs} are convolved, each convolution consists of L such binomially distributed samples. In order to find the probability distribution for the maximum value of the convolution array, we write the probability that any single value is at most s as

$$\Pr(X^{(w)} \leq s) = \sum_{t=0}^{s} \binom{w}{t}\left(\frac{G}{L}\right)^{t}\left(1 - \frac{G}{L}\right)^{w-t}.$$

The probability that all values are at most s is then, using the second approximation above, $\Pr(X^{(w)} \leq s)^{L}$. From this it follows that the probability distribution for the maximum value of the convolution vector $c_{\text{max}}^{(w)}$ is given by

$$f_{c_{\text{max}}^{(w)}}(s) = \Pr(c_{\text{max}}^{(w)} = s) = \Pr(X^{(w)} \leq s)^{L} - \Pr(X^{(w)} \leq s - 1)^{L} \tag{4}$$

with cumulative distribution function

$$F_{c_{\text{max}}^{(w)}}(s) = \Pr(c_{\text{max}}^{(w)} \leq s) = \Pr(X^{(w)} \leq s)^{L}.$$

A threshold value for $c_{\text{max}}^{(w)}$ is chosen, denoted $\hat{c}_{\text{max}}$. If $c_{\text{max}}^{(w)} \geq \hat{c}_{\text{max}}$ the payload is considered a ROP. Associated with this decision are false positives and false negatives. The false positive rate, denoted α, is defined as the probability that non-malicious data is considered malicious (i.e., a ROP payload) while the false negative rate, denoted β, is the probability that a malicious payload is mistaken for non-malicious data. To write expressions for α and β, let the Hamming weight w of P_{obs} be written as $w = w_G + w_N$, where w_G is the number of ROP gadgets and w_N is the number of noise addresses. The distribution of $c_{\text{max}}^{(w)}$ for non-malicious data is given by Eq. (4). The value of $c_{\text{max}}^{(w)}$ for a ROP payload is given by $c_{\text{max}}^{(w)} = w_G + X^{(w_N)}$, where $X^{(w_N)}$ is distributed according to Eq. (3). Now, we can write the two error probabilities as

$$\alpha = \Pr(c_{\text{max}}^{(w)} \geq \hat{c}_{\text{max}}) = 1 - \Pr(c_{\text{max}}^{(w)} \leq \hat{c}_{\text{max}} - 1)$$
$$= 1 - \Pr(X^{(w)} \leq \hat{c}_{\text{max}} - 1)^{L} \tag{5}$$
$$\beta = \Pr(X^{(w_N)} < \hat{c}_{\text{max}} - w_G) = \Pr(X^{(w_N)} \leq \hat{c}_{\text{max}} - w_G - 1)$$

The false positives rate α is only for one library. If we want to test the payload against a set of ℓ libraries, the total false positive rate α_ℓ is given by $\alpha_\ell = 1 - (1 - \alpha)^{\ell}$.

By choosing $\alpha = 0.0001$ we allow $\ell = 100$ libraries to be supported, still keeping the total false positive rate α_ℓ below 0.01. (We assume here that all libraries are of approximately equal size.) We let the false negative rate $\beta = 0.01$ since this is not affected by multiple libraries (the payload will only match one library). Using these values for α and β allows us to compute the threshold $\hat{c}_{\max}$ and the minimum number of gadgets w_G that are required for successful detection. Table 1 gives these numbers for $z = 3$ and some different choices of w. Note that for all values w such that $\hat{c}_{\max} = w_G$, we have $\beta = 0$. Thus we will only obtain false negatives for very large noise values. For $z = 1$ we will get $\hat{c}_{\max} = w_G = 6$, which is the lowest possible threshold T for eavesROP.

Table 1. Threshold $\hat{c}_{\max}$ and minimum number w_G of gadgets needed for ROP payload detection in an address window of weight w. Error rates $\alpha \leq 0.0001$ and $\beta \leq 0.01$. The example library used is libc 2.18 of size $L = 1224144$.

	entity		values								
weight of P_{obs}	w	7	10	15	20	25	30	50	100	200
threshold	$\hat{c}_{\max}$	7	8	9	10	11	12	15	20	27
min num gadgets	w_G	7	8	9	10	11	12	15	20	26

The standard deviation of Eq. (4) turns out to be very small, with almost all probability mass concentrated to only a few values for s. This makes the detection algorithm efficient, allowing us to choose small error rates while still requiring few gadgets to succeed, even in the presence of a large amount of noise.

It can be noted that the required number of gadgets w_G is very close to the threshold $\hat{c}_{\max}$. Thus, as a rule of thumb, the number of gadgets required for successful detection is approximately equal to the threshold $w_G \approx \hat{c}_{\max}$.

The false positive rate has been simulated using the data from Table 2. The simulations indicate that the actual false positive rate is slightly larger than that given by Eq. (5). This is not surprising, since the theoretical model assumes that gadget addresses are uniformly distributed. Due to data redundancy and the proximity coupling between gadgets and return instructions, a slightly larger $c_{\max}^{(w)}$ is expected. Still, according to our simulations, increasing the threshold value by 2 will remove virtually all false positives. This shows that the theoretical model is adequate.

3.5 Performance

The performance of eavesROP depends on the parameters used in the various stages of the system. All simulations have been performed on an Intel Core i7 4770 @ 3.4 GHz with 16 GB of RAM.

A more aggressive filtering in each step will reduce the amount of data sent to the next stage, which will increase the overall performance. Simulations have been performed on various types of data. The data pre-filter has a throughput

Table 2. Number of matching address windows per GiB of input data, for a data window of size D, and with at least T addresses within distance $L = 1224144$, for different types of data. L is here the size of libc 2.18

type of data	$D = 50$			$D = 200$			$D = 1000$		
	$T = 6$	8	10	$T = 6$	8	10	$T = 6$	8	10
random	0	0	0	12	0	0	24749	53	0
web (HTML, JPG,...)	1795	689	590	5589	1878	1208	40795	8007	3292
mp3	42	8	2	631	106	10	162014	8472	1023
pdf	4068	248	61	34718	5266	1316	1011850	176992	45289
mkv (H.264/MPEG-4) ·	354	2	0	513	81	66	35545	841	125

of approximately 35 MiB/s for compressed or random data, and approximately 50 MiB/s for web data (HTML, JPEG, ...). The input/output size ratio varies between 0.965 for random or compressed data, to 0.068 for web data.

After the optional data pre-filter—which may have reduced the total amount of data—the data is passed to the cluster detection step. This step has a throughput of around 10 MiB/s. The output of the cluster detection step is multiple matched windows, i.e. multiple P_{obs}. Table 2 shows how many P_{obs} vectors that are passed to the pattern matching layer, for some different parameters.

Each P_{obs} outputted from the cluster detection stage will be passed to the pattern matching step. Each pattern matching sequence takes roughly 1 second using FFT implemented in software. If necessary, this step could be accelerated using a hardware FFT implementation.

All parts of eavesROP have been implemented and tested using real-world exploits. We are able to detect all exploits of at least 6 gadgets, using a threshold value $\hat{c}_{max} = 6$, for example [5] and [30]. More simulation results can be found in the full version of the paper [16].

4 Conclusions

We have investigated to which extent it is possible to detect a ROP payload by only analysing data, and assuming that ASLR is used on the target system. If we have the set of libraries and binaries that can be used to find gadgets, we show that it is possible to detect a ROP payload even in the presence of noise and by applying suitable data filters and the Fast Fourier Transform the detection has acceptable performance. Naturally, encrypted traffic, obfuscated ROP payloads and locally generated exploits are out of scope for our detection approach. The exact performance will depend on the type of data and the number of gadgets that are required for an exploit to be detected depends on the maximum allowed size for the payload and the amount of noise. Furthermore performance may be optimized with greater pre-filtering and dedicated hardware for FFT calculations.

References

1. One, A.: Smashing the stack for fun and profit, phrack, 49 (1996)
2. Bletsch, T., Jiang, X., Freeh, V.W., Liang, Z.: Jump-oriented programming: A new class of code-reuse attack. In: Proceedings of the 6th ACM Symposium on Information, Computer and Communications Security, ASIACCS 2011, pp. 30–40. ACM, New York (2011)
3. Bracewell, R.: The Fourier Transform and its Applications, 3rd edn. McGraw-Hill Series in Electrical and Computer Engineering. McGraw-Hill Science/Engineering/Math. (June 1999)
4. cOntex: Bypassing non-executable-stack during exploitation using return-to-libc, http://www.infosecwriters.com/text_resources/pdf/return-to-libc.pdf
5. Cantoni, L.: BigAnt Server 2.52 SP5 - SEH Stack Overflow ROP-based exploit (ASLR + DEP bypass), http://www.exploit-db.com/exploits/22466/
6. Checkoway, S., Davi, L., Dmitrienko, A., Sadeghi, A.R., Shacham, H., Winandy, M.: Return-oriented programming without returns. In: Proceedings of the 17th ACM Conference on Computer and Communications Security, CCS 2010, pp. 559–572. ACM, New York (2010)
7. Chen, P., Xiao, H., Shen, X., Yin, X., Mao, B., Xie, L.: DROP: Detecting return-oriented programming malicious code. In: Prakash, A., Sen Gupta, I. (eds.) ICISS 2009. LNCS, vol. 5905, pp. 163–177. Springer, Heidelberg (2009)
8. Cheng, Y., Zhou, Z., Miao, Y., Ding, X., Deng, R.: ROPecker: A generic and practical approach for defending against ROP attack. In: NDSS. Research Collection School of Information Systems (2014)
9. Cormen, T., Leiserson, C., Rivest, R., Stein, C.: Introduction to Algorithms, 3rd edn. MIT Press (2009)
10. Davi, L., Sadeghi, A., Winandy, M.: ROPdefender: A detection tool to defend against return-oriented programming attacks. In: Proceedings of the 6th ACM Symposium on Information, Computer and Communications Security, ASIACCS 2011 (2011)
11. Durden, T.: Bypassing PaX ASLR protection, phrack, 59 (2002)
12. Fratric, I.: Ropguard: Runtime prevention of return-oriented programming attacks (2012)
13. Gupta, A., Kerr, S., Kirkpatrick, M., Bertino, E.: Marlin: Making it harder to fish for gadgets. In: Proceedings of the 2012 ACM Conference on Computer and Communications Security, CCS 2012. ACM (2012)
14. Hensing, R.: Understanding DEP as a mitigation technology (2009), http://blogs.technet.com/b/srd/archive/2009/06/12/understanding-dep-as-amitigation-technology-part-1.aspx
15. Hiser, J., Nguyen-Tuong, A., Co, M., Hall, M., Davidson, J.: Ilr: Where'd my gadgets go? In: 2012 IEEE Symposium on Security and Privacy (SP) (2012)
16. Jämthagen, C., Karlsson, L., Stankovski, P., Hell, M.: eavesROP: Listening for ROP payloads in data streams (full version) (2014), http://lup.lub.lu.se/record/4586662
17. Li, J., Wang, Z., Jiang, X., Grace, M., Bahram, S.: Defeating return-oriented rootkits with "return-less" kernels. In: Proceedings of the 5th European Conference on Computer Systems, EuroSys 2010. ACM (2010)
18. Lu, K., Zou, D., Wen, W., Gao, D.: Packed, printable, and polymorphic return-oriented programming. In: Sommer, R., Balzarotti, D., Maier, G. (eds.) RAID 2011. LNCS, vol. 6961, pp. 101–120. Springer, Heidelberg (2011)

19. Onarlioglu, K., Bilge, L., Lanzi, A., Balzarotti, D., Kirda, E.: G-free: Defeating return-oriented programming through gadget-less binaries. In: Proceedings of the 26th Annual Computer Security Applications Conference, ACSAC 2010, pp. 49–58. ACM (2010)

20. Pappas, V., Polychronakis, M., Keromytis, A.: Smashing the gadgets: Hindering return-oriented programming using in-place code randomization. In: IEEE Symposium on Security and Privacy. IEEE Computer Society (2012)

21. Pappas, V., Polychronakis, M., Keromytis, A.: Transparent ROP exploit mitigation using indirect branch tracing. Presented as part of the 22nd USENIX Security Symposium (USENIX Security 2013). USENIX (2013)

22. PaX Team: Address space layout randomization (2003), http://pax.grsecurity.net/docs/aslr.txt

23. Polychronakis, M., Keromytis, A.: ROP payload detection using speculative code execution. In: Proceedings of the 2011 6th International Conference on Malicious and Unwanted Software, MALWARE 2011. IEEE Computer Society (2011)

24. Schwartz, E., Avgerinos, T., Brumley, D.: Q: Exploit hardening made easy. In: Proceedings of USENIX Security 2011 (2011)

25. Serna, F.J.: CVE-2012-0769, the case of the perfect info leak (2009), http://zhodiac.hispahack.com/my-stuff/security/Flash_ASLR_bypass.pdf

26. Shacham, H.: The geometry of innocent flesh on the bone: Return-into-libc without function calls (on the x86). In: Proceedings of the 14th ACM Conference on Computer and Communications Security, CCS 2007, pp. 552–561. ACM (2007)

27. Shacham, H., Page, M., Pfaff, N., Goh, E., Modadugu, N., Boneh, D.: On the effectiveness of address-space randomization. In: Proceedings of the 11th ACM Conference on Computer and Communications Security, CCS 2004, pp. 298–307. ACM (2004)

28. Snow, K., Monrose, F., Davi, L., Dmitrienko, A., Liebchen, C., Sadeghi, A.: Just-in-time code reuse: On the effectiveness of fine-grained address space layout randomization. In: 2013 IEEE Symposium on Security and Privacy (SP), pp. 574–588 (May 2013)

29. Stancill, B., Snow, K.Z., Otterness, N., Monrose, F., Davi, L., Sadeghi, A.-R.: Check my profile: Leveraging static analysis for fast and accurate detection of ROP gadgets. In: Stolfo, S.J., Stavrou, A., Wright, C.V. (eds.) RAID 2013. LNCS, vol. 8145, pp. 62–81. Springer, Heidelberg (2013)

30. Sud0: Audio converter 8.1 0day stack buffer overflow PoC exploit ROP/WPM, http://www.exploit-db.com/exploits/13763/

31. Vreugdenhil, P.: Pwn2Own 2010 Windows 7 Internet Explorer 8 exploit (2010), http://vreugdenhilresearch.nl/Pwn2Own-2010-Windows7-InternetExplorer8.pdf

32. Wartell, R., Mohan, V., Hamlen, K., Lin, Z.: Binary stirring: Self-randomizing instruction addresses of legacy x86 binary code. In: Proceedings of the 2012 ACM Conference on Computer and Communications Security, CCS 2012 (2012)

Defining Injection Attacks

Donald Ray and Jay Ligatti

University of South Florida
Department of Computer Science and Engineering
Tampa, FL, USA
{dray3,ligatti}@cse.usf.edu

Abstract. This paper defines and analyzes injection attacks. The definition is based on the *NIE property*, which states that an application's untrusted inputs must only produce Noncode Insertions or Expansions (i.e., NIEs) in output programs. That is, when applications generate output programs (such as SQL queries) based on untrusted inputs, the NIE property requires that inputs only affect output programs by inserting or expanding noncode tokens (such as string and float literals, lambda values, pointers, etc). This paper calls attacks based on violating the NIE property *BroNIEs* (i.e., Broken NIEs) and shows that all code-injection attacks are BroNIEs. In addition, BroNIEs contain many malicious injections that do not involve injections of code; we call such attacks *noncode*-injection attacks. In order to mitigate both code- and noncode-injection attacks, this paper presents an algorithm for detecting and preventing BroNIEs.

Keywords: Code-injection attacks, noncode-injection attacks, language-based security.

1   Introduction

According to multiple sources, the most commonly reported software attacks are injection attacks, including SQL injections, cross-site scripting, and OS-command injections [1–3]. MITRE-SANS considers these attacks to result from three of their top four "most dangerous software errors" [1].

Applications vulnerable to injection attacks generate output programs based on untrusted inputs. By providing a malicious input, an attacker can cause the application to output a malicious program. A classic example involves a simple web application for a bank; the application inputs a password and returns the balance of accounts with the given password. On a typical, benign input such as 123456, the banking application outputs the following program (throughout this paper, input symbols injected into the output program are underlined).

SELECT balance FROM accts WHERE pw='<u>123456</u>'

Unfortunately, if this application does not validate its input, it can be manipulated into creating malicious programs. For example, on input ' OR 1=1 --, the application outputs the following program.

S.S.M. Chow et al. (Eds.): ISC 2014, LNCS 8783, pp. 425–441, 2014.

```
SELECT balance FROM accts WHERE pw='' OR 1=1 --'
```

This output program circumvents the password check and returns the balances of all accounts because (1) the 1=1 subexpression is a tautology, making the entire WHERE clause true, and (2) in SQL, the -- sequence begins a comment, which removes the final apostrophe and makes the program syntactically valid. Because some of the injected symbols in this output program are code symbols (i.e., disjunction and equality operators), this is an example of a *code*-injection attack.

Due to their prevalence, much research has been performed to define code-injection attacks formally (e.g., [4–9]). Many additional papers have described tools for detecting and preventing code-injection attacks (e.g., [10–18]).

Interestingly, a related class of attacks exists but has not yet, as far as we are aware, been explored. These related attacks have many of the symptoms of code-injection attacks but do not involve injecting *code*. Because these attacks are performed by injecting noncode symbols, we call them *noncode-injection attacks*. Although noncode-injection attacks do not involve injecting malicious code, they may nonetheless cause other parts of the output program to execute maliciously.

For example, the following web application[1] is vulnerable to noncode-injection attacks.

```
$attackerControlledString = input();
$code = ''\\$data = '$attackerControlledString'; ''
       . ''securityCheck(); \\$data .= '&f=exit#';\n f();'';
eval($code);
```

On a benign input such as Hello!, this application outputs the following program.

```
$data = 'Hello!'; securityCheck(); $data .= '&f=exit#';\n f();
```

This output program sets $data to a value, calls the securityCheck() function, appends a string to $data, and then invokes the f function. However, if an attacker enters the string \, the application outputs the following program.

```
$data = '\'; securityCheck(); $data .= '&f=exit#';\n f();
```

No code has been injected in this alternative output program; the injected \ is part of a (noncode) string literal. However, because the injected symbol escapes the apostrophe that would have terminated the first string literal, the string literal continues until the next (non-escaped) apostrophe. As a result, the call to securityCheck is bypassed, the function f is updated to be the exit function, and finally, because # begins a comment that continues until the line break, the exit function (i.e., f) is invoked, shutting down the web server and causing a denial of service.

---

[1] We are grateful to Mike Samuel of Google for creating and sharing the first version of this application.

1.1 Related Work

Although much effort has been made to understand and prevent injection attacks, the previous work in this area has focused on *code*-injection attacks.

For example, Halfond et al. and Nguyen-Tuong et al. detect code-injection attacks based on whether keywords or operator symbols have been injected into output programs [7, 8]. False positives and negatives may arise with these techniques [4]; for example, these works consider injections of string literals to be attacks (because the beginning and ending apostrophes in string literals are classified as operators) and consider injections of function names to be non-attacks.

Other works detect code-injection attacks based on whether input symbols span tokens, or parse trees, in output programs. For example, Xu et al. detect code injections based on whether untrusted input spans tokens in the output program [9], while SqlCheck detects code injections based on whether untrusted input, injected into an output program P, spans subtrees in P's parse tree [5]. False positives and negatives again may arise with these techniques [4].

Candid detects code-injection attacks based on whether a second copy of the application, which is only given strings of a's (or 1s) as input but is forced to follow the same control-flow path as the original application, outputs a syntactically different program [6]. Again, false positives and negatives may arise with this technique [4].

To summarize the related work discussed to this point, all exhibit false positives and negatives when detecting code-injection attacks. Furthermore, all of these related works exhibit false positives and negatives when detecting general (i.e., code and noncode) injection attacks; every example false positive/negative given in [4] serves as an example false positive/negative with respect to general injection attacks.

Our previous work detects code-injection attacks based on whether untrusted inputs get used as non-values (i.e., non-normal-form terms) in output programs [4]. Although we believe that this technique detects code-injection attacks precisely (i.e., lacks false positives and negatives), it is tailored to code-injection attacks and cannot detect noncode-injection attacks, because it considers any symbol injected into a noncode value as safe. Hence, false negatives arise when using the techniques of [4] to detect noncode-injection attacks. In contrast, the present paper presents definitions and techniques for detecting injection attacks in general, including noncode-injection attacks.

In practice, a commonly recommended technique for preventing injection attacks is to use *parameterized queries* [19], with which applications create output programs that contain placeholders, or "holes", for untrusted inputs. For example, an application might create an output program that has a single placeholder for a string literal; to output a program, the application provides a string to fill that placeholder. In this way, parameterized queries can limit injections to filling in holes for string, numeric, or other kinds of literals, thus preventing both code- and noncode-injection attacks.

While effective at preventing injection attacks, parameterized queries have significant disadvantages:

- Although parameterized queries are a standard feature of many SQL dialects [20–23], they are not supported by other common output-program languages such as HTML or bash. An output-program language must provide support for parameterized queries before an application can make use of them.
- It is the responsibility of application programmers to make use of parameterized queries. Unfortunately, parameterized queries are of little use to programmers that are either ignorant of, or apathetic to, their benefits. Whatever the reason, many application programmers are not using parameterized queries, as evidenced by the prevalence of injection vulnerabilities [1–3].
- Once the decision to use parameterized queries is made, modifying existing applications to use them is a manual, time-consuming process. Application programmers must find all possible ways for programs to be output, and replace those programs with new versions containing the appropriate placeholders. If even one output program is not replaced, the application will remain vulnerable to injection attacks.

1.2   Summary of Contributions and Roadmap

This paper defines and proves properties of a broad class of injection attacks, one that includes both code- and noncode-injection attacks. The definition of injection attacks is based on whether untrusted inputs affect output programs in any way besides inserting or expanding noncode tokens (such as string, integer, or float literals). When untrusted inputs only insert or expand noncode tokens in an output program, we say that the output program satisfies the *NIE property* (Noncode Insertion or Expansion). Intuitively, the NIE property restricts untrusted inputs to filling in "holes" in output programs that are reserved for noncode tokens.

The NIE property thus restricts untrusted inputs in ways that are similar to parameterized queries; both techniques require untrusted inputs to fill noncode-token "holes" in output programs. With parameterized queries, the holes for untrusted-input tokens are manually specified by application programmers; with this paper's techniques, all holes for untrusted-input tokens are automatically confined to being noncode (e.g., a string or numeric literal). Because this paper's techniques are widely applicable and require no modifications to existing applications, the techniques avoid the disadvantages of parameterized queries. However, the tradeoff here is that this paper's techniques rely on runtime monitoring for injection-attack detection, implying higher runtime overhead than parameterized queries.

This paper considers inputs that violate the NIE property to have "broken" the NIE property and calls such attacks BroNIEs. BroNIEs are therefore considered a general class of injection attacks, one that includes both code- and noncode-injection attacks.

After defining BroNIEs in Section 2, this paper presents several examples in Section 3. Section 4 explores the implications of the definitions in Section 2 and shows that (1) BroNIEs are indeed a strict superset of code-injection attacks, (2) all applications that blindly copy-and-paste an untrusted input into a SQL output program are vulnerable to BroNIEs, and (3) a linear-time algorithm exists for precisely (i.e., soundly and completely) detecting BroNIEs. Section 5 concludes with a brief discussion.

2 Definitions

This section formalizes criteria for determining when a BroNIE has occurred. Section 2.1 presents preliminary notation and assumptions. Then, because BroNIEs occur when injected symbols affect output programs beyond inserting or expanding noncode tokens, it is critical to know which symbols are *injected* and which tokens are *noncode*; Sections 2.2 and 2.3 present these subdefinitions. Finally, Sections 2.4 and 2.5 use the subdefinitions of injected symbols and noncode tokens to define code-injection attacks and BroNIEs.

2.1 Notation and Assumptions

An application vulnerable to BroNIEs outputs programs in some language L (e.g., SQL) that has a finite concrete-syntax alphabet Σ_L (e.g., the set of printable Unicode characters). Following standard convention, these output programs, which we also call L-programs, are finite sequences of Σ_L symbols that each form an element of L. For L-program $p = \sigma_1\sigma_2..\sigma_n$, $|p| = n$ and $p[i] = \sigma_i$. That is, $|p|$ denotes the length of p, and $p[i]$ denotes the ith symbol of p. For a sequence S, the replacement of item t with item t' (i.e., the substitution of t' for t in S) is denoted $[t'/t]S$.

This paper makes a few assumptions about output-program languages. All output-program languages under consideration have well-defined functions for:

- Computing the free variables of program terms. Terms are called *open* if they contain free variables, and are otherwise *closed* (e.g., the term 1+2 is closed, while 1+x is open).
- Testing whether program terms are *values*. Values are the "fully evaluated" terms of a programming language, such as literals, pointers, objects, lists and tuples of other values, lambda terms, etc.
- Tokenizing output programs. Function $tokenize_L(\sigma_1..\sigma_n)$ returns the sequence of tokens within the string $\sigma_1..\sigma_n$ (assuming $\sigma_1..\sigma_n$ is lexically valid; otherwise, $tokenize_L(\sigma_1..\sigma_n)$ returns the empty sequence). A token of *kind* τ composed of symbols $\sigma_i..\sigma_j$ is represented as $\tau_i(\sigma_i..\sigma_j)_j$. For example, for program $q = $ SELECT * FROM orders WHERE s='' OR i<3, $tokenize_{SQL}(q)$ returns tokens $SELECT_1(\texttt{SELECT})_6$, $STAR_8(\texttt{*})_8$, $FROM_{10}(\texttt{FROM})_{13}$, $ID_{15}(\texttt{orders})_{20}$, $WHERE_{22}(\texttt{WHERE})_{26}$, $ID_{28}(\texttt{s})_{28}$, $EQUALS_{29}(\texttt{=})_{29}$, $STRING_{30}(\texttt{''})_{31}$, $OR_{33}(\texttt{OR})_{34}$, $ID_{36}(\texttt{i})_{36}$, $LESS_{37}(\texttt{<})_{37}$, and $INT_{38}(\texttt{3})_{38}$.

Predicate $TR_L(p, i)$ (tokenizer-removed) holds iff i is not within the bounds of any token in $tokenize_L(p)$. For example, $TR_{SQL}(q, i)$ holds for all $i \in \{7, 9, 14, 21, 27, 32, 35\}$.

This paper omits the L subscript from the $tokenize_L$ and TR_L functions when the output-program language is clear from context.

2.2   Defining Injection

Applications vulnerable to BroNIEs output programs based on inputs from trusted and untrusted sources. *Injected* symbols are those that originate from untrusted sources and propagate unmodified through an application into its output program. A BroNIE occurs when injected symbols affect output programs in any way besides inserting or expanding noncode tokens.

We rely on the well-studied concept of taint tracking [7–9, 17, 18, 24, 25] to determine which output-program symbols originate from untrusted sources and are therefore injected. At a high level, a *taint-tracking application* works by replacing all symbols input from untrusted sources with their tainted versions and preserving these taint metadata during all copy and output operations. Provided that the application only taints symbols when they are being input from an untrusted source, and never untaints symbols, the injected symbols in the output program are exactly those that are tainted.

As we have been underlining injected symbols, we use the same notation to mark tainted symbols.

Definition 1 ([4]). *For all alphabets Σ, the* tainted-symbol alphabet $\underline{\Sigma}$ *is* $\{\sigma \mid \sigma \in \Sigma \vee (\exists \sigma' \in \Sigma : \sigma = \underline{\sigma'})\}$.

Next, the language L must also be augmented to allow programs to contain tainted symbols.

Definition 2 ([4]). *For all languages L with alphabet Σ, the* tainted output language $\underline{L}$ *with alphabet* $\underline{\Sigma}$ *is* $\{\sigma_1..\sigma_n \mid \exists \sigma'_1..\sigma'_n \in L: \forall i \in \{1..n\}: (\sigma_i = \sigma'_i \vee \sigma_i = \underline{\sigma'_i})\}$.

Finally, an output-program symbol is injected if and only if it is tainted.

Definition 3 ([4]). *For all alphabets Σ and symbols $\sigma \in \underline{\Sigma}$, the predicate* $injected(\sigma)$ *is true iff $\sigma \notin \Sigma$.*

For example, taint-tracking application `output('Hello' + input() + '!')` on untrusted input `World` will output the program `Hello World!` because the initial input is replaced by its tainted version `World`, and the taint metadata are preserved by the string-concatenation and output operations.

Because taint tracking is a well-studied technique, the remainder of this paper assumes that applications can track taints and thus output programs in which the injected (i.e., tainted) symbols are underlined.

2.3   Defining Noncode

Intuitively, noncode symbols in output programs are those that are dynamically passive; they do not specify dynamic computation to be performed when the program is executed. Code symbols, on the other hand, are dynamically active; they specify dynamic computation that could be performed when the program is executed.

This paper considers a program's noncode symbols to be exactly those that are either (1) removed by the tokenizer or (2) within a closed value:

1. Although previous work sometimes allowed tokenizer-removed symbols such as whitespace or comments to be code [4], we believe that, because tokenizer-removed symbols are dynamically passive and cannot specify computation, it is more intuitive and accurate to consider such symbols noncode.
2. Closed values are operationally irreducible and thus specify no dynamic computation during program execution. Typical values include string and numerical literals, pointers to other values, lists and tuples of other values, and objects. Open values are excluded because they specify the dynamic computation of substituting a term for a free variable during execution; closed values specify no such substitution operations.

When $p[i]$ is noncode (where p is an output program), we write $Noncode(p, i)$. Otherwise, we write $Code(p, i)$.

Definition 4. *For all L-programs $p = \sigma_1..\sigma_n$ and position numbers $i \in \{1..|p|\}$, predicate $Noncode(p, i)$ is true iff either $TR_L(p, i)$ or there exist low and high symbol-position numbers $l \in \{1..i\}$ and $h \in \{i..|p|\}$ such that $\sigma_l..\sigma_h$ is a closed value in p.*

Tokens composed entirely of noncode symbols are called *noncode tokens*. The set of all noncode tokens in a program p is *noncodeToks(p)*.

2.4   An Aside: Defining *Code*-injection Attacks

Using the predicates for determining which symbols are injected (Definition 3), and which are noncode (Definition 4), we can define Code-Injection Attacks on Output programs (CIAOs), as occurring exactly when an output program contains an injected code symbol.

Definition 5 ([4]). *A CIAO occurs exactly when a taint-tracking application outputs $\underline{L}$-program $p = \sigma_1..\sigma_n$ such that $\exists i \in \{1..n\} : (injected(\sigma_i) \wedge Code(p, i))$.*

2.5   Defining BroNIEs

Because BroNIEs occur when injected symbols affect output programs beyond inserting or expanding noncode tokens, they can be detected by observing how a program's sequence of tokens is affected by the removal of its injected symbols. Intuitively, removing all injected symbols from the output program should only affect the sequence of tokens in the following ways:

1. Some noncode tokens may no longer be present.
2. Some noncode tokens may become smaller but should not change kind (e.g., string literals should not become integer literals).

To formalize this intuition, we need to consider the sequence of tokens obtained by removing all injected symbols from an output program. The injected symbols cannot simply be deleted; doing so would affect the indices of tokens that follow the injected symbols. Instead, each injected symbol is replaced with an ε. The sole purpose of an ε symbol is to hold the place of an injected symbol; ε's are otherwise ignored. Because the resulting string contains only uninjected symbols, it can be thought of as a *template* of the original program.

Definition 6. *The* template *of a program p is obtained by replacing each injected symbol in p with an ε. In other words, the template of p refers to p with its injected symbols removed, but without affecting the indices of uninjected symbols. Abusing notation, we denote the template of p as $[\varepsilon/\underline{\sigma}]p$.*

The definition of BroNIEs also relies on notions of token insertion and expansion. Token insertion is straightforward, but we need to be clear about the meaning of token expansion. Injected symbols may expand noncode tokens by increasing their ranges of indices and corresponding strings of program symbols.

Definition 7. *A token $t = \tau_i(v)_j$ can be expanded into token $t' = \tau'_{i'}(v')_{j'}$, denoted $t \preceq t'$, iff:*

- $\tau = \tau'$
- $i' \leq i \leq j \leq j'$ *and*
- v *is a subsequence of v'.*

We can now formally specify when an output program exhibits only noncode insertion or expansion (NIE). Given a program p and its template $[\varepsilon/\underline{\sigma}]p$, it should be possible to get to the sequence of tokens in p from the sequence of tokens in $[\varepsilon/\underline{\sigma}]p$ by only inserting or expanding noncode tokens. If the sequence of tokens in p can be reached from the sequence of tokens $[\varepsilon/\underline{\sigma}]p$ in this way, we say that p satisfies the NIE property.

Definition 8. *An L-program p satisfies the NIE property iff there exist:*

- $I \subseteq noncodeToks(p)$ *(i.e., a set of p's inserted noncode tokens),*
- $n \in \mathbb{N}$ *(i.e., a number of p's expanded noncode tokens),*
- $\{t_1..t_n\} \subseteq tokenize([\varepsilon/\underline{\sigma}]p)$ *(i.e., a set of template tokens to be expanded), and*
- $\{t'_1..t'_n\} \subseteq noncodeToks(p)$ *(i.e., a set of p's expanded noncode tokens)*

such that:

- $t_1 \preceq t'_1, \ldots, t_n \preceq t'_n$, *and*
- $tokenize(p) = ([t'_1/t_1]..[t'_n/t_n]tokenize([\varepsilon/\underline{\sigma}]p)) \cup I.$

At last, BroNIEs can be defined as occurring whenever an application outputs a program that violates the NIE property.

Definition 9. *A BroNIE (Broken NIE) occurs exactly when a taint-tracking application outputs a program that violates the NIE property.*

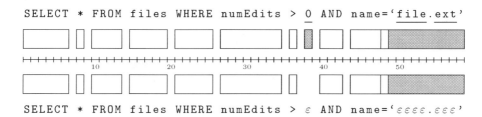

```
SELECT * FROM files WHERE numEdits > 0 AND name='file.ext'
```

Fig. 1. A depiction of how the Example-1 program (top) and its template (bottom) are tokenized. Noncode tokens are shaded.

3 Examples

Let us consider several examples of how Definition 9 classifies programs as either attacks or non-attacks. Although all but one of this section's examples are presented in SQL, the underlying concepts apply to other languages as well.

Example 1. We begin with the following simple output program.

```
SELECT * FROM files WHERE numEdits > 0 AND name='file.ext'
```

This query returns files named `file.ext` that have been edited at least once. Figure 1 shows how this program and its template are tokenized. This program does not exhibit a CIAO; all injected symbols are part of integer or string values. Neither is it a BroNIE; the injected symbols only cause noncode insertion and expansion, as depicted in Figure 1. More formally, the sequence of tokens in the template can be made into the sequence of tokens in the program by inserting the noncode token $INT_{38}(0)_{38}$ and expanding the token $STRING_{49}(`.`)_{58}$ into the (noncode) token $STRING_{49}(`\underline{file}.\underline{ext}`)_{58}$. Hence, the definition of BroNIEs matches our intuition that this program does not exhibit an attack.

Example 2. Turning our attention to programs that do exhibit injection attacks, let us reconsider the first malicious output program presented in Section 1.

```
SELECT balance FROM accts WHERE pw='' OR 1=1 --'
```

This query returns all balances from the `accts` table because the `WHERE` clause is a tautology. The application that output this program tries to access only balances for which a password is known, but the malicious input circumvents the password check.

Figure 2 shows how this program and its template are tokenized. The program exhibits a CIAO and a BroNIE: a CIAO because the injected 0, R, and = symbols are code, and a BroNIE because the injected symbols insert code tokens and *contract* the $STRING_{36}(`')_{48}$ token into $STRING_{36}(`')_{37}$.

Example 3. Because injections that cause CIAOs must also cause BroNIEs (as will be shown in Theorem 1), the remainder of this section focuses on non-CIAO examples of BroNIEs, such as the following.

```
SELECT  balance  FROM  accts  WHERE  pw=' '  OR  1=1  --'
```

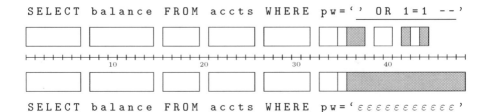

```
SELECT  balance  FROM  accts  WHERE  pw='εεεεεεεεεε'
```

Fig. 2. A depiction of how the Example-2 program (top) and its template (bottom) are tokenized. Noncode tokens are shaded.

```
SELECT * FROM t WHERE c=''' AND now()<exp --this code is smokin'
```

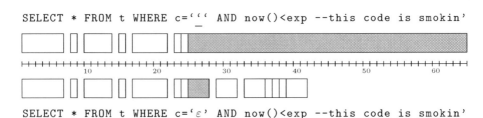

```
SELECT * FROM t WHERE c='ε' AND now()<exp --this code is smokin'
```

Fig. 3. A depiction of how the Example-3 program (top) and its template (bottom) are tokenized. Noncode tokens are shaded.

```
SELECT * FROM t WHERE c=''' AND now()<exp --this code is smokin'
```

The application that outputs this program attempts to find all unexpunged records with a given column value. For instance, the application could be querying a table of juvenile crimes (such as truancy) that can legally only be displayed when the offenders are not yet adults. However, the application appends a seemingly harmless, boastful comment to all output queries that allows an injected apostrophe to circumvent the expungement check by escaping the string literal's terminator (in SQL, apostrophes within string literals are escaped by a second apostrophe). If the attacker has control over the column being queried (e.g., it could be a comment column), then he or she can illegally access records after they have been expunged.

Figure 3 presents the tokenizations of this example program and its template. This program does not exhibit a CIAO; the injected symbol is part of a noncode (string) token. On the other hand, this program does exhibit a BroNIE; the injected symbol violates the NIE property by deleting code tokens.

Example 4. The following output program also contains a non-CIAO BroNIE.

```
INSERT INTO users VALUES ('evilDoer', TRUE)--', FALSE)
```

This program creates a new user by inserting a record into the users table. Each record contains a field for the username and a boolean flag indicating whether the user has administrator privileges. By hardcoding a FALSE value for the second element of new-user records, the application that output this

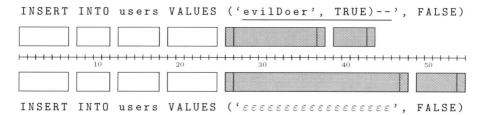

Fig. 4. A depiction of how the Example-4 program (top) and its template (bottom) are tokenized. Noncode tokens are shaded. The tokens for commas and parentheses are considered noncode because they are within a tuple value; tuple terms are values when all subterms are values.

program attempts to ensure that all accounts it creates do not have administrator privileges. However, in this case an attacker has supplied a malicious username that causes the application to create an administrator-privileged account.

Figure 4 shows how this program and its template are tokenized. None of the injected symbols in the program are code symbols, so it does not exhibit a CIAO. However, it does exhibit a BroNIE; the injected symbols cause noncode contraction and deletion. For example, the last three noncode tokens in the template are not present in the output program. Thus, Definition 9 correctly considers this output program to be an attack.

Example 5. The following output program will serve as a final example of a non-CIAO BroNIE in SQL.

```
INSERT INTO trans VALUES (1,- 5E-10);
INSERT INTO trans VALUES (2, 5E+5)
```

Here, an application outputs programs to handle money transfers from one account (e.g., account number 1) to another (e.g., account number 2). The application applies a service charge of $10 to the paying account, but is currently running a special that transfers an extra $5 into the receiving account. However, by appending an 'E' to the money transfer amount, a malicious user can drastically affect the transfer process—in the above program, the paying account is only charged $0.0000000005 while the receiving account is credited $500,000.

Figure 5 shows that this output program exhibits a non-CIAO BroNIE. The output program does not exhibit a CIAO because no code has been injected; only components of float literals have been injected. On the other hand, the output program does exhibit a BroNIE because the injections delete the $MINUS_{33}(\text{-})_{33}$ and $PLUS_{70}(\text{+})_{70}$ tokens and transform two INT tokens into $FLOAT$ tokens (recall that Definition 7 does not allow tokens of one kind to expand into tokens of another kind).

Example 6. To demonstrate the effectiveness of these techniques in other languages, we return to the following output program from Section 1.

```
INSERT INTO trans VALUES (1,- 5E-10); INSERT INTO trans VALUES (2, 5E+5)
```

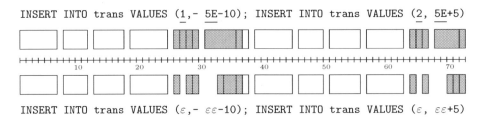

```
INSERT INTO trans VALUES (ε,- εε-10); INSERT INTO trans VALUES (ε, εε+5)
```

Fig. 5. A depiction of how the Example-5 program (top) and its template (bottom) are tokenized. Noncode tokens are shaded. The tokens for commas and parentheses are considered noncode because they are within a tuple value; tuple terms are values when all subterms are values.

```
$data = '\'; securityCheck(); $data .= '&f=exit#';\n f();
```

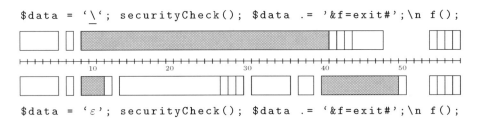

```
$data = 'ε'; securityCheck(); $data .= '&f=exit#';\n f();
```

Fig. 6. A depiction of how the Example-6 program (top) and its template (bottom) are tokenized. Noncode tokens are shaded.

```
$data = '\'; securityCheck(); $data .= '&f=exit#';\n f();
```

Because apostrophes within string literals are escaped by backslashes in this language, the injected symbol bypasses the call to the `securityCheck` function, garbles the contents of the `data` variable, and causes the `exit` function to be invoked instead of the `f` function.

The tokenizations of this program and its template are depicted in Figure 6. The program does not exhibit a CIAO; the only injected symbol is part of a string literal (i.e., a noncode value). However, the program does exhibit a BroNIE; the injected symbol deletes code and noncode tokens, and inserts code tokens, thus violating the NIE property.

4   Analysis of the BroNIE Definition

This section explores implications of the definitions in Section 2.

The first theorem formalizes our claim that all code-injection attacks are also BroNIEs. In other words, every application vulnerable to CIAOs is also vulnerable to BroNIEs

Theorem 1. *If a program exhibits a CIAO, then it exhibits a BroNIE.*

Proof. Let p be an output program that exhibits a CIAO. Then at least one of p's code symbols is injected; let σ_c be one of these injected code symbols and t be the token containing σ_c. Program p satisfies the NIE property iff the sequence of tokens in $\lfloor\varepsilon/\underline{\sigma}\rfloor p$ can be made equal to the sequence of tokens in p, subject to the constraints of Definition 8. Observe that t cannot be in $\lfloor\varepsilon/\underline{\sigma}\rfloor p$; no token in $\lfloor\varepsilon/\underline{\sigma}\rfloor p$ can have the same text as t unless its begin and/or end indices are different because, in $\lfloor\varepsilon/\underline{\sigma}\rfloor p$, σ_c will be replaced by an ε. Because t is a code token (it contains the code symbol σ_c), Definition 8 does not allow t to be inserted into the token sequence of $\lfloor\varepsilon/\underline{\sigma}\rfloor p$, nor for a token in $\lfloor\varepsilon/\underline{\sigma}\rfloor p$ to be expanded into t. Hence, the sequence of tokens in $\lfloor\varepsilon/\underline{\sigma}\rfloor p$ cannot be made equal to the sequence of tokens in p by only inserting and/or expanding noncode tokens, so p exhibits a BroNIE. ◻

The second theorem provides a formal basis for the widely-accepted rule of thumb that it is unsafe to use unvalidated input during query construction [19, 26]. Theorem 2 states that if an application always includes an untrusted input (i_m) verbatim in its output (without even inspecting the input), and the same application has some input ($v_1, .., v_n$) for which it outputs a valid SQL program, then there exists a way to construct an attack input (a_m) such that the application's output will exhibit a CIAO and therefore a BroNIE. This theorem is a generalization of Theorem 9 in [4], which was limited to CIAOs in an idealized subset of SQL called SQL Diminished; in contrast, Theorem 2 below applies to full SQL.

Theorem 2. *For all n-ary functions A and (n-1)-ary functions A' and A'', if $\forall i_1, .., i_n\colon A(i_1, .., i_n) = A'(i_1, .., i_{m-1}, i_{m+1}, .., i_n)\underline{i_m}A''(i_1, .., i_{m-1}, i_{m+1}, .., i_n)$, where $1 \leq m \leq n$, and $\exists v_1, .., v_n\colon (v_m \in \Sigma_{SQL}^+ \wedge A(v_1, .., v_n) \in SQL)$, then $\exists a_1, .., a_n\colon A(a_1, .., a_n) \in SQL$ and $A(a_1, .., a_n)$ exhibits a CIAO and a BroNIE.*

Proof. Observe that changing v_m to any a_m, without changing any of the other inputs to A, will cause A to output the same program but with a_m instead of v_m, because A' and A'' are independent of i_m. Now construct $a_m = v_m A''(v_1, .., v_{m-1}, v_{m+1}, .., v_n)\backslash n;\ drop\ table\ t;\ A'(v_1, .., v_{m-1}, v_{m+1}, .., v_n)v_m$. This construction causes $A(v_1, .., v_{m-1}, a_m, v_{m+1}, .., v_n)$ to output the string $p = A(v_1, .., v_n)\backslash n;\ drop\ table\ t;\ A(v_1, .., v_n)$. Because $A(v_1, .., v_n) \in SQL$, so too is output-program p. By Definition 5, p exhibits a CIAO due to the injected DROP statement, as well as any other code symbols in a_m. By Theorem 1, p also exhibits a BroNIE. ◻

It is also straightforward to prove, using the same techniques from [4], that neither static nor black-box mechanisms can precisely prevent BroNIEs. That is, precise detection of BroNIEs requires dynamic, white-box mechanisms.

4.1 An Algorithm for Precisely Detecting BroNIEs

Given that applications commonly fail to validate untrusted inputs [26], it would be beneficial to have mechanisms for automatically preventing injection attacks.

Algorithm 1 : A BroNIE-preventing mechanism.

Input: Taint-tracking application A and inputs T, U (trusted, untrusted)
Ensure: A's output is executed iff it doesn't exhibit a BroNIE
1: $Output \leftarrow A\,(T,\,\mathrm{Taint}(U))$
2: $pgmTokens \leftarrow \mathrm{tokenize}(Output)$
3: $temTokens \leftarrow \mathrm{tokenize}([\varepsilon/\underline{\sigma}]Output)$
4: $\mathrm{MarkNoncodeToks}(pgmTokens)$
5: $i \leftarrow j \leftarrow 1$
6: **while** $i \leq pgmTokens.\mathrm{length}$ **and** $j \leq temTokens.\mathrm{length}$ **do**
7:   **if** $pgmTokens[i] = temTokens[i]$ **then**
8:     $i \leftarrow i + 1$
9:     $j \leftarrow j + 1$
10:   **else if** $pgmTokens[i].\mathrm{isNoncode}$ **and** $temTokens[i] \preceq pgmTokens[i]$ **then**
11:     $i \leftarrow i + 1$
12:     $j \leftarrow j + 1$
13:   **else if** $pgmTokens[i].\mathrm{isNoncode}$ **then**
14:     $i \leftarrow i + 1$
15:   **else**
16:     throw $BronieException$
17:   **end if**
18: **end while**
19: // Handle any trailing noncode tokens in the program
20: **while** $i \leq pgmTokens.\mathrm{length}$ **and** $pgmTokens[i].\mathrm{isNoncode}$ **do**
21:   $i \leftarrow i + 1$
22: **end while**
23: **if** $i > pgmTokens.\mathrm{length}$ **and** $j > temTokens.\mathrm{length}$ **then**
24:   $\mathrm{Execute}(Output)$
25: **else**
26:   throw $BronieException$
27: **end if**

At a high level, BroNIEs can be precisely and automatically prevented by:

- instrumenting the target application with a taint-tracking mechanism,
- interposing between the target application and the environment that evaluates the application's output programs,
- detecting whether output programs satisfy the NIE property, and
- only executing programs that do not exhibit BroNIEs.

Algorithm 1 is a psuedocode implementation of this mechanism. The algorithm detects BroNIEs by iterating through the sequences of tokens in the output program and its template while ensuring that the two token streams can be made equal subject to the constraints of Definition 8 (i.e., obtaining the sequence of tokens in the output program by only inserting noncode tokens into, or expanding noncode tokens within, the sequence of tokens in the program's template). Algorithm 1 relies on auxiliary functions for tainting untrusted inputs, tokenizing output programs, and marking which tokens are noncode.

4.2 Analysis of the BroNIE-detection Algorithm

The following theorems relate to the correctness and execution time of Algorithm 1.

Theorem 3. *Algorithm 1 executes output-program p iff p does not exhibit a BroNIE.*

Proof. We prove the if direction; the only-if direction is similar. By assumption, p satisfies the NIE property. Hence, the sequence of tokens in $[\varepsilon/\underline{\sigma}]p$ can be made equal to the sequence of tokens in p by expanding tokens in $[\varepsilon/\underline{\sigma}]p$ into noncode tokens in p, and/or inserting noncode tokens in p into $[\varepsilon/\underline{\sigma}]p$. Variables i and j are initialized to 1. For every token present in p and $[\varepsilon/\underline{\sigma}]p$, both i and j will be incremented by the first branch of the `if` statement (Lines 7–9). For every token in $[\varepsilon/\underline{\sigma}]p$ that needs to be expanded into a noncode token in p, i and j will be incremented by the second branch of the `if` statement (Lines 10–12). For every noncode token in p that needs to be inserted into $[\varepsilon/\underline{\sigma}]p$, i will be incremented by either the third branch of the `if` statement (Lines 13–14) or the second `while` loop (Lines 20–22). After both `while` loops have completed, both i and j will be greater than the lengths of their token sequences, so p will be executed on Line 24. □

To examine the running time of Algorithm 1, we exclude the time taken to execute the target application (i.e., Line 1), as this operation is independent from the process of detecting BroNIEs. The parts of Algorithm 1 that are directly related to BroNIE detection are the calls to *tokenize* (Lines 2 and 3), the invocation of *MarkNoncodeToks* (Line 4), the initialization of the i and j variables (Line 5), the two `while` loops (Lines 6–22), and the final `if` statement (Lines 23–27). To the best of our knowledge, every commonly-used programming language requires no more than linear time to partition tokens as code or noncode, so the proof of Theorem 4 assumes that the language-dependent *MarkNoncodeToks* function runs in time linear in the number of tokens.

Theorem 4. *The BroNIE-detection part of Algorithm 1 (i.e., Lines 2–27) executes in $O(n)$ time, where n is the length of the output program.*

Proof. Each call to *tokenize* executes in $O(n)$ time and produces $O(n)$ tokens. After the program tokens are marked as either code or noncode (taking $O(n)$ time), the two `while` loops execute; each iteration takes constant time, and each loop executes $O(n)$ times. Afterward, if a BroNIE has not yet been detected, deciding whether a BroNIE has occurred takes constant time. Thus, Algorithm 1 runs in $O(n)$ time. □

5 Discussion

This paper has presented BroNIEs, a general class of injection attacks in which injected symbols affect output programs beyond inserting or expanding noncode tokens. BroNIEs include not only all code-injection attacks on output programs (CIAOs), but also noncode-injection attacks on output programs.

Precise detection of BroNIEs can be accomplished in practice by using a taint-tracking mechanism (e.g., [7–9, 17, 18, 24, 25]) in conjunction with Algorithm 1. DIGLOSSIA [18] appears to be a good candidate framework for detecting and preventing BroNIEs in practice—DIGLOSSIA (which has not yet been publicly released) already detects and prevents CIAOs, so we expect that it could detect and prevent BroNIEs with relatively minor modifications. Because DIGLOSSIA performs taint tracking through modified system libraries, existing applications need not be manually rewritten.

Fundamentally, we believe that the definition of BroNIEs more closely matches our intuition of malicious injections than just *code* injections, due to the capability of attackers to cause malicious behavior by injecting noncode symbols (as demonstrated in Examples 3–6).

Furthermore, it appears that the definition of BroNIEs imparts similar advantages as parameterized queries, without the significant disadvantages of requiring application-programmer compliance (which has historically been unreliable). Parameterized queries and BroNIE-preventing mechanisms both limit injections to inserting or expanding noncode tokens. However, whereas applications must be manually modified to use parameterized queries, BroNIEs can be automatically prevented by instrumenting applications and/or system libraries with a taint-tracking mechanism and running a lightweight BroNIE-detection algorithm prior to executing output programs.

References

1. The MITRE Corporation: CWE/SANS Top 25 Most Dangerous Software Errors (2011), http://cwe.mitre.org/top25/archive/2011/2011_cwe_sans_top25.pdf
2. Open Sourced Vulnerability Database: OSVDB: Open Sourced Vulnerability Database (2014), http://osvdb.org/
3. The OWASP Foundation: OWASP Top 10 - 2013 (2013), http://owasptop10.googlecode.com/files/OWASPTop10-2013.pdf
4. Ray, D., Ligatti, J.: Defining code-injection attacks. In: Proceedings of the Symposium on Principles of Programming Languages (POPL), pp. 179–190 (2012)
5. Su, Z., Wassermann, G.: The essence of command injection attacks in web applications. In: Proceedings of the Symposium on Principles of Programming Languages (POPL), pp. 372–382 (2006)
6. Bisht, P., Madhusudan, P., Venkatakrishnan, V.N.: CANDID: Dynamic candidate evaluations for automatic prevention of SQL injection attacks. Transactions on Information and System Security (TISSEC) 13(2), 1–39 (2010)
7. Halfond, W., Orso, A., Manolios, P.: Wasp: Protecting web applications using positive tainting and syntax-aware evaluation. Transactions on Software Engineering (TSE) 34(1), 65–81 (2008)
8. Nguyen-Tuong, A., Guarnieri, S., Greene, D., Shirley, J., Evans, D.: Automatically hardening web applications using precise tainting. In: Sasaki, R., Qing, S., Okamoto, E., Yoshiura, H. (eds.) Security and Privacy in the Age of Ubiquitous Computing. IFIP AICT, vol. 181, pp. 295–307. Springer, Boston (2005)

9. Xu, W., Bhatkar, S., Sekar, R.: Taint-enhanced policy enforcement: A practical approach to defeat a wide range of attacks. In: Proceedings of the USENIX Security Symposium, pp. 121–136 (2006)

10. Bravenboer, M., Dolstra, E., Visser, E.: Preventing injection attacks with syntax embeddings. Science of Computer Programming (SCP) 75(7), 473–495 (2010)

11. Jovanovic, N., Kruegel, C., Kirda, E.: Precise alias analysis for static detection of web application vulnerabilities. In: Proceedings of the Workshop on Programming Languages and Analysis for Security, pp. 27–36 (2006)

12. Buehrer, G., Weide, B.W., Sivilotti, P.A.G.: Using parse tree validation to prevent SQL injection attacks. In: Proceedings of the Workshop on Software Engineering and Middleware (SEM), pp. 106–113 (2005)

13. Hansen, R., Patterson, M.: Stopping injection attacks with computational theory. In: Black Hat Briefings Conference (2005)

14. Kieżun, A., Guo, P.J., Jayaraman, K., Ernst, M.D.: Automatic creation of SQL injection and cross-site scripting attacks. In: Proceedings of the International Conference on Software Engineering (ICSE), pp. 199–209 (2009)

15. Luo, Z., Rezk, T., Serrano, M.: Automated code injection prevention for web applications. In: Mödersheim, S., Palamidessi, C. (eds.) TOSCA 2011. LNCS, vol. 6993, pp. 186–204. Springer, Heidelberg (2012)

16. Oracle: How to write injection-proof PL/SQL. An Oracle White Paper (2008)

17. Pietraszek, T., Berghe, C.V.: Defending against injection attacks through context-sensitive string evaluation. In: Valdes, A., Zamboni, D. (eds.) RAID 2005. LNCS, vol. 3858, pp. 124–145. Springer, Heidelberg (2006)

18. Son, S., McKinley, K.S., Shmatikov, V.: Diglossia: detecting code injection attacks with precision and efficiency. In: Proceedings of the Conference on Computer and Communications Security (CCS), pp. 1181–1192 (2013)

19. The Open Web Application Security Project (OWASP): SQL injection prevention cheat sheet (2012),
https://www.owasp.org/index.php/SQL_Injection_Prevention_Cheat_Sheet

20. The PostgreSQL Global Development Group: PostgreSQL 9.3.4 Documentation (2014), http://www.postgresql.org/docs/9.3/static/index.html

21. Microsoft: Transact-SQL Reference, Database Engine (2014),
http://msdn.microsoft.com/en-us/library/bb510741(v=sql.120).aspx

22. Oracle: MySQL 5.7 Reference Manual (2014),
https://dev.mysql.com/doc/refman/5.7/en/index.html

23. Oracle: Oracle Database SQL Language Reference, 12c Release 1 (12.1) (2013),
http://docs.oracle.com/cd/E16655_01/server.121/e17209.pdf

24. Dalton, M., Kannan, H., Kozyrakis, C.: Raksha: A flexible information flow architecture for software security. In: Proceedings of the International Symposium on Computer Architecture (ISCA), 482–493 (2007)

25. Clause, J., Li, W., Orso, A.: Dytan: A generic dynamic taint analysis framework. In: Proceedings of the International Symposium on Software Testing and Analysis (ISSTA), pp. 196–206 (2007)

26. Scholte, T., Robertson, W., Balzarotti, D., Kirda, E.: An empirical analysis of input validation mechanisms in web applications and languages. In: Proceedings of the Symposium on Applied Computing (SAC), pp. 1419–1426 (2012)

Efficient Attack Forest Construction for Automotive On-board Networks

Martin Salfer[1], Hendrik Schweppe[1], and Claudia Eckert[2]

[1] BMW Forschung und Technik GmbH, Munich, Germany
{martin.salfer,hendrik.schweppe}@bmw.de
[2] Technische Universität München, Germany
claudia.eckert@in.tum.de

Abstract. Software-intensive, modern vehicles comprise about 100 computers, which allow a plethora of attack combinations. This paper proposes an efficient attack forest construction method for a vehicle's on-board network security evaluation, based on our system model, and predictions about attractiveness, exploitability, and attackers. We compiled various vehicle development databases and documents to a homogeneous system model. Our algorithm implementation can construct attack forests with typically sized system models usually within a few minutes and with an asymptotic, computational complexity of $O(n * \log(n))$. Attack forests are a foundation for further security analysis and evaluation.

Keywords: Security, Embedded Systems, Attack Tree, Attack Graph.

1 Introduction

The automotive industry drives ECU (Electronic Control Unit) consolidation, extension and connectivity for economic, functional and environmental reasons. Yet, the complexity of deep integration and internetworking also brings hardly foreseeable security implications. Some were revealed by practical attack studies [9,3]. Security researchers demand more objective security engineering instead of mere expert intuition [17]. The German National Road Map for Embedded Systems explicitly demands reliable, quantified security statements for embedded systems [4].

A vehicle's attack surface increases due to the internetworking of control units with the environment and due to the integration of abundant services and functionality, e.g., for highly automated driving. The attack motivation rises due to asset accumulation: A common car will bear payment credentials for tolling, parking, and electric charging and have access to cloud services including communications and sensitive and personal data. Upcoming infotainment functionality shares software with mobile phones, so the potential consequences of mobile system malware become relevant for vehicle security, too. Recent studies reveal a mobile system malware growth of about 614% from 2012 to 2013 and state that exploitation will continue to grow and that black markets are in certain respects "more profitable than the illegal drug trade"[8,1].

S.S.M. Chow et al. (Eds.): ISC 2014, LNCS 8783, pp. 442–453, 2014.

Security fixes are very expensive to create and to deploy for the automotive industry, even compared to enterprise systems: Every change requires heavy testing for guaranteed side-effect free safety quality. As a vehicle's life cycle spans over more than a decade, security support can become costly. Hence, the automotive industry invests in comprehensive engineering and security assessments with a long-term foresight. One method is the software-supported creation and analysis of attack trees. A high grade of automation and efficiency is necessary for handling the large automotive system models.

Our contributions address the complexity problem of a comprehensive security assessment with attack trees for automotive on-board networks.

- We formalized **an attacker and a system model** for an efficient and comprehensive analysis of the automotive on-board network security.
- We created an **algorithm** that searches for the most reasonable attack paths and constructs attack forests efficiently. Our algorithm's computational, asymptotic complexity is only $O(n * \log(n))$, with n as the number of *Software + Communication Media + Asset Nodes*.
- We designed and implemented **a proof of concept** that is able to construct attack forests for large system models with several thousand nodes within a few minutes.

Section 2 defines our system model, Section 3 defines our attacker profile, and Section 4 defines our exploitability anticipation. Section 5 describes, defines and benchmarks our attack forest generation algorithm. In Section 6, we discuss related work. And Section 7 concludes our attack forest generation method.

2   System Model

Our *System Model* represents an automotive on-board network. The model comprises *Software, Asset, ECU,* and *Communication Medium Nodes* and is shown in Figure 1. Our *System Model* is a boiled-down generalization of various engineering models into a single *System Model*, which is required for a homogeneous attack forest. Our *System Model* takes mainly inherent, energy-efficient, protective means into account: memory protection, ECU separation, network segmentation and controlled gateways. A well designed separation concept contributes significantly to the overall security; One analogy is a castle and its wall layout giving protection inherently. Each *Software Node* is a piece of code that runs independently and within a certain memory protection, so a code injection exploit's effect will be contained primarily within this *Software.* Apart from over-the-air software updates, a vehicle's software installation is rather static; It usually stays the same for the entire model life cycle, which spans more than ten years. An *Asset* can not only model financially marketable *Assets,* but also safety-relevant *Assets,* i.e., how much would an adversary pay for endangering (specific) passengers or pedestrians. Some inspiration for *Assets* and related threats can be found in the BSI IT-Grundschutz Catalogues [2].

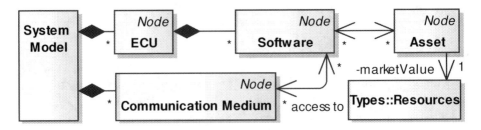

Fig. 1. Our **System Model** represents an automotive on-board network: its *Software*, *Communication Media* and *Assets*. The *Software Nodes* are part of *ECUs* and exchange any information over *Communication Media*, e.g., CAN, MOST, Ethernet, gateway connections, or sockets. *Assets* are valuables and are associated with a market value for a reference worth on how much money would be paid on the (black) market for the access or control of the *Asset*.

3   Attacker Profile

Our *Attacker Profile* represents human attackers with economic reasoning. See Figure 2 for a model of our *Attacker Profile*.

Every *Motivation* points to one or several *Assets* of our *System Model*. A *Motivation* for an *Asset* is optionally annotated with a specific price that the attacker is willing to pay for it. An attacker's valuation of an *Asset* can greatly deviate from the (black) market. For example, if an attacker is unable to sell or utilize a stolen car, his or her *Motivation* price for the car theft *Asset* is far below market value, possibly zero.

An *Access* models entry points, at which an attacker can initially attack the system. Each *Access* a_a points to at least one *System Model Node* and is annotated with an application-wise cost that is necessary for being able to apply an exploit onto one vehicle. This cost is expressed by our function $\mathcal{C}_A(a_a)$.

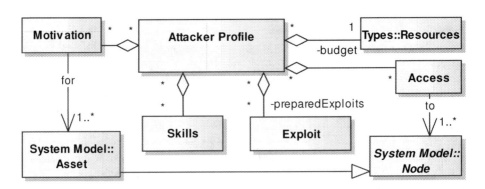

Fig. 2. Our **Attacker Profile** characterizes a human attacker by a set of *Motivation*, *Accesses* to *System Model Nodes*, *Budget*, weighted *Skills*, and their already *Prepared Exploits*

Skills represent an attacker's knowledge and techniques relevant for exploiting *Software Nodes*. Each *Skill* is proportionally weighted with a floating point number $x \in \mathcal{R}^+$ as a grade. It rates an *Attacker Profile's Skill* in comparison to the corresponding *Skill* of the reference *Attacker Profile* a_1. The meaning of the reference attacker skill, having always the qualifier 1, can be arbitrarily defined, yet this definition must be precisely known to the ones who define *Attacker Profiles* and the ones who assess the exploitability of *Software*.

4   Exploitability Anticipation

The *Exploitability Anticipation* represents the set of anticipated *Exploits* that our *System Model* could be confronted with, see Figure 3 for a visual overview. The individual *Resources* estimation for an *Exploit* is given according to the reference *Attacker Profile* a_1, i.e., every *Skill* $e_s \in e_S$ is priced with regards to a *Skill* grade of 1. The individual *Resources* estimation for a given *Exploit* e and a given *Skill* $e_s \in e_S$ is expressed as $C_E(e, e_s)$ and maps to an appropriate *Resource* $r \in R$, that represents a monetary value and is derived from a set of input parameters. Some of the input parameters for r are a target's code size and complexity, known software and hardware defects and vulnerabilities in programming frameworks, processors, memory architecture, operating system and security elements. The cost for an *Exploit* e is specific for every *Attacker Profile* a due to the individual set of *Skills* a_S and is expressed as $\mathcal{C}_E(a, e)$. Every *Exploit* $e \in E$ requires a specific *Skill Set* e_S and each *Attacker Profile* a bears a specific *Skill Set* a_S. Each required *Skill* $e_s \in e_S$ is matched against the corresponding *Skill* $a_s \in a_S$. An *Attacker Profile's Skill* always bears a certain degree and influences the cost $\mathcal{C}_E(a, e)$: If an *Attacker Profile's Skill* is rated higher than the required *Skill* of an *Exploit*, an *Attacker Profile* is awarded with a lower exploit cost $\mathcal{C}_E(a, e)$ and vice versa. The cost for an *Exploit* e for an *Attacker Profile* a is expressed as

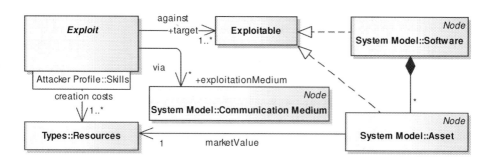

Fig. 3. Our **Exploitability Anticipation** represents anticipated *Exploits* an attacker could create and use for gaining control. Each *Exploit* can only be launched against specific *Software* or *Asset Nodes*, and via specific *Communication Media*. An *Exploit's* creation cost is defined by a map of required *Skills* and *Resource* estimations. The *Exploit* expectations are derived by analysing software and ECU code and meta data.

$\mathcal{C}_E(a, e)$ and is defined as the sum of skill-specific basic costs $\mathcal{C}_E(e, e_s)$, divided by the corresponding *Skills* of the *Attacker Profile* as in

$$\mathcal{C}_E(a, e) := \sum_i \left(\frac{\mathcal{C}_E(e, e_{si})}{a_{si}} \right) \text{ with } a_{si}, e_{si} \in e_S. \tag{1}$$

5   Attack Forest Construction Algorithm

Our *Attack Forest Construction Algorithm* generates attack trees from compiling our *System Model* with a given *Exploitability Anticipation*, and a specific *Attacker Profile*. Our algorithm traverses the *System Model* in two phases; see Figure 4 for an activity diagram and an overview of our attack forest construction algorithm.

Our *Attack Trees* model the root as the initial cause of an attack, i.e., the attack through an *Attacker Profile's Access* $a_a \in a_A$, and span edges towards the set of reachable *Assets* $s_{A1} \in s_A$ for an *Attacker Profile* a and *System Model* s. Each attack path in our *Attack Forest* represents the most reasonable combination of *Exploits* for reaching the *Assets*.

5.1   Phase I: Heuristics Preparation

Our algorithm initially runs over all *System Model Software Nodes* to enrich those with distance information to *Assets*. The traversal starts at the *Asset Nodes* and runs over all connected *Software Nodes*. The traversal over the *System Model* is exhaustive, i.e., all nodes will be visited. The computational complexity of *Phase I* is moderate as no constraint checking is necessary, i.e., node expansions are cheap. At each *Software Node*, the algorithm records the number of steps of the shortest path to all given *Asset Nodes*. The step counters are later fed into the heuristics of *Phase II* and help it avoid costly expansions in remote and *Asset*-free parts of the *System Model*.

5.2   Phase II: Attack Forest Construction

Our algorithm in *Phase II* combines our models: a *System Model* s with the *Phase I* heuristic enrichment, a set of *Exploit Anticipations* E, and a specific *Attacker Profile* a. Our algorithm in *Phase II* traverses the *System Model* expansion-optimally with an admissible and consistent heuristic [7]. *Phase II* starts traversing at the *Attacker Profile's Access Nodes* a_A, which refer to a subset of *Software Nodes* $s_{Si} \in s_S$, runs down towards the *System Model's Asset Nodes* s_A and collects affordable attack paths.

Node Expansion and Cost Calculation. At each node expansion in *Phase II*, the algorithm has to combine cost estimations for a given *Attacker Profile* a, *System Model* s, and set of *Anticipated Exploits* E. The first node expansion are

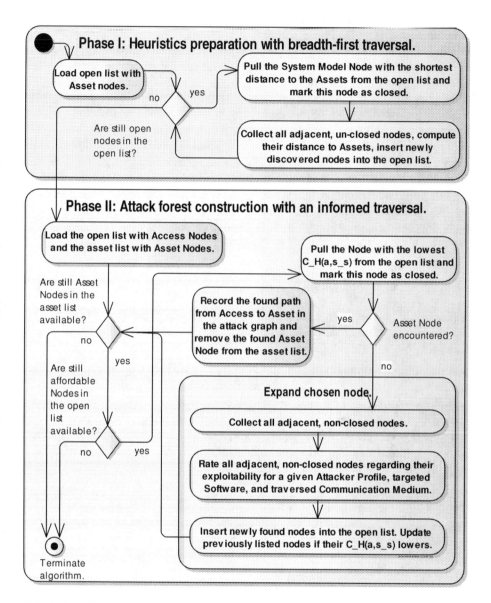

Fig. 4. Our **Attack Forest Construction Algorithm** works in two phases: *Phase I* prepares heuristic information to be used in *Phase II* with a breadth-first and lightweight behaviour, and *Phase II* traverses the *System Model* in order to transform it into attack trees under the proposition of specialized *Attacker Profiles* and given constraints. Each *Attacker Profile* results in specific attack trees and demands an additional run of our *Phase II*. As the constraint checking is computationally intensive for large *System Models*, the heuristic information from *Phase I* helps to reduce the complexity by traversing nodes in *Phase II* expansion-optimally.

on *Attacker Profile's Accesses* a_A. The cost for expanding those is defined above by our function $C_A(a_a)$. The expansion of an *Access* a_a reveals attached *System Model Software Nodes* s_{Si}. At each subsequent node expansion in *Phase II*, our algorithm collects, filters and assesses the exploitation costs of adjacent *Software Nodes*. The filtering depends on the set of *Anticipated Exploits E*, which defines which *Exploitable Node* is targetable and what *Communication Media* is required for. The filtered *Software* and *Asset Nodes* are being priced by the estimated effort $C_E(a, e)$ an *Attacker Profile* a would have to invest. The prices are defined in the set of *Anticipated Exploits E*. The application cost for a software exploit on subsequent nodes is neglected intentionally as the application cost is usually insignificant compared to an exploit's creation cost. If several, different *Anticipated Exploits* $E_1 \in E$ are admissible for a target node, the most reasonable *Anticipated Exploit* $e \in E_1$ is chosen. We define $C_E(a, s_s)$ as the minimum cost for exploiting a *Software Node* s_s with one *Anticipated Exploit* e for an *Attacker Profile* a. We define

$$C_E(a, s_{si}, s_{sj}) := \sum_i C_E(a, e_i) \text{ with } e_i \in E_1 \tag{2}$$

as the minimum cost for an *Attacker Profile* a for a successful attack path with a set of *Anticipated Exploits* E_1 from *Software Node* s_{si} to *Software Node* s_{sj}. We define

$$C(a, s_s) := C_A(a_a) + \sum_i C_E(a, e_i) \text{ with } e_i \in E_1 \tag{3}$$

as the minimum cost for a successful attack from an *Attacker Profile* a against a given *System Model Software Node* s_s. The $C(a, s_s)$ is the sum of the access cost $C_A(a_a)$ plus the cost for unique, previously created and applied *Exploits* $E_1 = \cup_{i=0}^n e_i$ against all intermediate nodes of our *System Model Software Nodes* s_{Si} from an *Access node* a_a to the targeted *Software node* s_s.

Heuristics and Optimality. The *Phase II* algorithm is (result-)optimal and expansion-optimal due to the admissible and consistent heuristic as proven in [7]. The algorithm collects distance information in *Phase I*, i.e., every *Software Node* s_s is annotated with the distance j to *Asset Nodes*. The heuristic estimates the cost for exploiting the remaining *Software Nodes* and *Asset Node* and is expressed for an *Attacker Profile* a at *Software Node* s_s as

$$C_h(a, s_s) := j * r_i \text{ with } 0 < r_i \le C_E(a, e) \forall e \in E. \tag{4}$$

The estimation for the overall cost is expressed as

$$C_H(a, s_s) := C(a, s_s) + C_h(a, s_s). \tag{5}$$

The bigger r_i the more directed the expansions will traverse towards the *Asset Nodes*. Yet, r_i must be small enough to never overestimate the remaining exploitation cost, for the algorithm constructs optimal paths. Ideally, r_i is set as the lower bound of $\mathcal{C}_E(a, e) \forall e \in E$ as an attacker must minimally invest the cost for the most feasible exploit $e_i \in E$ for his targets among the remaining, reachable *System Model Nodes*, i.e., $\mathcal{C}_E(a, e_i) \leq \mathcal{C}_E(a, e) \forall e \in E$. The algorithm in *Phase II* is optimal, i.e., the found attack paths are most feasible paths for a given *Attacker Profile* a, as the heuristic is admissible, i.e., it never overestimates the exploit cost $\mathcal{C}_E(a, e)$ for the remaining nodes in the *System Model*, as shown in [7]. The algorithm in *Phase II* is also expansion-optimal as the *System Model* has only positively weighted edges and as the cost heuristic is consistent by considering the *Phase I* distance information, i.e., $\mathcal{C}_h(a, s_{si}) \leq \mathcal{C}_E(a, s_{si}, s_{sj}) + \mathcal{C}_h(a, s_{sj})$, as shown in [7]. The algorithm in *Phase II* needs the expansion optimality to save on computational power as each *Phase II* expansion requires constraint checks.

Termination. The algorithm terminates in three situations: i) all *Asset Nodes* s_A have been expanded, ii) all reachable *Software Nodes* $s_{Si} \in s_S$ have been expanded, or iii) the *Attacker Profile's budget* a_b is insufficient for further node expansions. At each expansion, a node is being marked as *"closed"* in order to avoid loops and guarantee algorithm termination. Our attack forest generation algorithm compares the *Attacker Profile's* budget a_b with the total cost of all *Software Nodes* in the algorithm's *"open list"* and terminates if necessary. The algorithm can also run without regards to the *Attacker Profile's budget* a_b for estimating the cost for a successful attack on each *Asset Node* $s_a \in s_A$.

5.3 Asymptotic Complexity

The asymptotic, computational complexity of our whole algorithm is:

$$O\left(n * \left(\log\left(n\right) + m\right)\right) \text{ and } \Omega(n + m)$$

with n being the power of the set of all *Software Nodes* s_S plus the power of the set of all *Communication Medium Nodes* s_C plus the power of the set of all *Asset Nodes* s_A and with m being the power of the set of *Exploitability Anticipations* E. *Phase I* alone is similar with $O(n * \log(n))$ and $\Omega(n)$.

5.4 Proof of Concept Benchmark

We implemented a proof of concept in Java and created four benchmarks. All measurements were conducted on a CPU from 2011 with 2.5 GHz and with a single thread. The benchmark in Figure 5 show clearly an efficiency gain because of the heuristics gathered from *Phase I*. Figure 6 shows the results in a logarithmic overview. All benchmarks were conducted in ascending order; first a *System Model* with 500 nodes is measured and subsequently one with 1 000 nodes. The runtime measurements satisfied our expectation: The attack forest construction with a *System Model* with several thousand *Software Nodes* is drastically sped up by our heuristics and is executed within a few minutes. The runtime depends as expected on the number of *Software Nodes* and their grade of interconnection. The *"mesh"* and *"chain"* *System Model* are edge cases. They mark the maximum run time of the algorithm for any *System Model*.

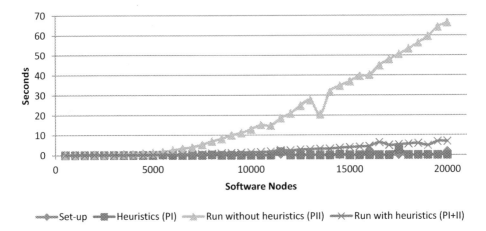

Fig. 5. Benchmark results for the attack forest construction algorithm split up in several phases for a lightly connected *System Model*: **"slim star"**, a conjunction of ten chains on one end and with an *Asset Node* in only one of their ends. The blue line ("Set-up") denotes how much time was spent for loading the corresponding *System Model*. The red line ("Heuristics (PI)") denotes how much time was spent for *Phase I*, which prepares the heuristics. The green line ("Run without heuristics (PII)") denotes how much time was spent for graph construction with *Phase II* not having heuristics. And the violet line ("Run with heuristics (PI+II)") denotes how much time was spent for graph construction with *Phase II* having heuristics active and prepared before, i.e., it denotes the time for a combined run of *Phase I* and *Phase II*. The runtime was as expected: *Phase I* and the set-up were fast and insignificant in comparison to *Phase II*, and the heuristics significantly sped up the graph construction.

6   Related Work

Attack trees originate from fault tree analysis techniques from as early as 1981 and became popular at last by publications in the 1990s [10]. Attack trees are seen as a good security engineering method [5,11,15,16,6].

Oleg Mikhail Sheyner et al. worked on the evaluation of stationary enterprise computer networks with known exploits for assessing the overall network security [18] [19]. Yet, the computational complexity rises exponentially with the amount of hosts and makes a large scale evaluation impossible: 5 minutes for a network of 2 hosts and 30 minutes for a network of 4 hosts. Ou et al. published an approach with logic programming and Prolog with only $O(n^2)$ [12].

Stuart Schechter worked on applying economic approaches onto IT security and defined a lean attacker profile with only four floating point number parameters: ranks, incentives, attack risks, and attack costs [14]. Yet, our approach uses a more versatile attacker profile as we link and annotate motivation and attractors more individually for *Assets*.

Attack graph-based security evaluation is a current focus of research in the setting of enterprise networks for which [18,13] has been exemplified. Such work

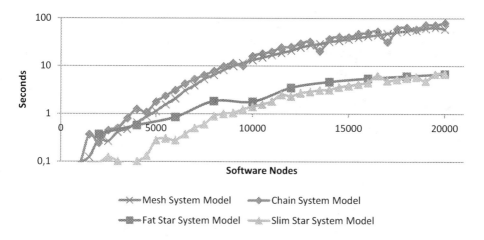

Fig. 6. Benchmark results overview for the attack forest construction algorithm for our four implemented, variably sized *System Model* benchmarks: a minimally connected one *("chain")*, a lightly connected one *("slim star")*, a densely connected one *("fat star")*, and a fully connected one *("mesh")*. *"Chain"* is a concatenation of *Software Nodes*, each with a separate *Communication Medium* in between, an *Access Node* on the one end, and an *Asset Node* on the other end. *"Slim star"* is a conjunction of ten chains on one end and with an *Asset Node* in only one of the arms. *"Fat star"* is a conjunction of 2 000 chains on one end and with an *Asset Node* in only one of the arms. *"Mesh"* is a fully connected *System Model* benchmark, i.e., every *Software Node* is connected to all other *Software Nodes*. The obvious performance deviations were repeatable, which we believe stem from Java's dynamic performance optimisation, e.g., garbage collection, collections reorganisation, and just-in-time compilation.

aims more at supporting IDS (Intrusion Detection Systems) and on how to handle the dynamic change of those systems and inclusion of public vulnerability databases. Yet, automotive systems are too heterogeneous to be reflected in standard scanning tools or public vulnerability databases and need other means for deriving attack graphs.

7   Conclusion

Vehicles become rolling computer networks and their architectural heterogeneity makes security hard to assess. Attack trees are a popular security engineering method, but the plethora of attack combinations makes manual construction cumbersome and likely to miss relevant attack paths. A missing attack vector might enable extra, unseen attack paths that drastically facilitate compromising a security goal. We have proposed a method for automatically constructing attack forests for automotive on-board networks for software attacks that utilize multiple exploits. Our attack forest generation algorithm finds the optimal paths for an attacker through our *System Model* to the *Assets*. All *Software* of a vehicle is going to be set in relation to each other *Software* and boiled down into our homogenized *System Model* for a security assessment. The more detailed input data is, the more detailed and valid attack forests will be. We designed and implemented a proof of concept that evaluates large, synthetic *System Models* even in the worst case within a few minutes. An evaluation supports vehicle manufacturers in investing security hardening efforts more effectively: Engineers and architects can design more securely, and penetration testers can identify potentially heavily attacked targets more precisely. Comprehensive attack forests that are automatically generated from development databases and documents are a foundation for further vehicle security assessment automation.

References

1. Ablon, L., Libicki, M.C., Golay, A.A.: Markets for cybercrime tools and stolen data. Technical Report RR-610-JNI, RAND National Security Research Divison (2014)
2. BSI. IT-Grundschutz-Kataloge. 13. Ergänzungslieferung (September 2013)
3. Checkoway, S., McCoy, D., Kantor, B., Anderson, D., Shacham, H., Savage, S., Koscher, K., Czeskis, A., Roesner, F., Kohno, T.: Comprehensive experimental analyses of automotive attack surfaces. In: Proceedings of the 2011 Usenix Security (2011)
4. Damm, W., Achatz, R., Beetz, K., Broy, M., Daembkes, H., Grimm, K., Liggesmeyer, P.: Nationale roadmap embedded systems. In: Broy, M. (ed.) Cyber-Physical Systems. acatech DISKUTIERT, pp. 67–136. Springer, Heidelberg (2010)
5. Evans, S., Wallner, J.: Risk-based security engineering through the eyes of the adversary. In: Proceedings from the Sixth Annual IEEE SMC Information Assurance Workshop, IAW 2005, pp. 158–165 (June 2005)
6. Grunske, L., Joyce, D.: Quantitative risk-based security prediction for component-based systems with explicitly modeled attack profiles. Journal of Systems and Software 81(8), 1327–1345 (2008)
7. Hart, P.E., Nilsson, N.J., Raphael, B.: A formal basis for the heuristic determination of minimum cost paths. IEEE Transactions on Systems Science and Cybernetics 4(2), 100–107 (1968)
8. Juniper Networks, Inc. Juniper networks third annual mobile threats report - March 2012 through March 2013 (June 2013)
9. Koscher, K., Czeskis, A., Roesner, F., Patel, S., Kohno, T., Checkoway, S., McCoy, D., Kantor, B., Anderson, D., Shacham, H., Savage, S.: Experimental security analysis of a modern automobile. In: 2010 IEEE Symposium on Security and Privacy (SP), pp. 447–462 (May 2010)
10. Mauw, S., Oostdijk, M.: Foundations of attack trees. In: Won, D.H., Kim, S. (eds.) ICISC 2005. LNCS, vol. 3935, pp. 186–198. Springer, Heidelberg (2006)
11. Nicol, D.M., Sanders, W.H., Trivedi, K.S.: Model-based evaluation: from dependability to security. IEEE Transactions on Dependable and Secure Computing 1(1), 48–65 (2004)
12. Ou, X., Boyer, W.F., McQueen, M.A.: A scalable approach to attack graph generation. In: Proceedings of the 13th ACM Conference on Computer and Communications Security, pp. 336–345 (2006)
13. Roschke, S., Cheng, F., Schuppenies, R., Meinel, C.: Towards unifying vulnerability information for attack graph construction. In: Samarati, P., Yung, M., Martinelli, F., Ardagna, C.A. (eds.) ISC 2009. LNCS, vol. 5735, pp. 218–233. Springer, Heidelberg (2009)

14. Schechter, S.E.: Toward econometric models of the security risk from remote attacks. IEEE Security Privacy 3(1), 40–44 (2005)
15. Schneier, B.: Attack trees. Dr. Dobb's Journal of Software Tools, 21–22, 24, 26, 28–29 (1999)
16. Schneier, B.: Secrets and Lies: Digital Security in a Networked World. John Wiley & Sons (March 2004)
17. Schneier, B.: The importance of security engineering. IEEE Security Privacy 10(5), 88–88 (2012)
18. Sheyner, O., Haines, J., Jha, S., Lippmann, R., Wing, J.M.: Automated generation and analysis of attack graphs. In: Proceedings of the 2002 IEEE Symposium on Security and Privacy, SP 2002, pp. 273–284. IEEE Computer Society, Washington, DC (2002)
19. Sheyner, O.M.: Scenario graphs and attack graphs. PhD thesis, Carnegie Mellon University, Pittsburgh, PA, USA, AAI3126929 (2004)

Winnowing Double Structure
for Wildcard Query in Payload Attribution[*]

Yichen Wei, Fei Xu, Xiaojun Chen, Yiguo Pu, Jinqiao Shi, and Sihan Qing

Institute of Information Engineering, Chinese Academy of Sciences, Beijing, China
weiyichen@nelmail.iie.ac.cn,
{xufei,chenxiaojun,puyiguo,shijinqiao,qsihan}@iie.ac.cn

Abstract. Payload attribution is a kind of traceback mechanism which has the ability to find the sources and destinations of packages containing certain excerpts and times of their occurrence. Existing payload attribution methods either fail to support wildcard query or suffers from low response speed. We propose Winnowing Double Structure with Wildcard Query (WDWQ), which support wildcard query efficiently and achieve higher wildcard query speed as well as lower false positive rate under an acceptable data reduction ratio. Our experiment shows that WDWQ is 20 times faster than the state-of-the-art payload attribution techniques.

Keywords: Payload Attribution, Bloom Filter, Winnowing, Wildcard Query, False Positive.

1 Introduction

With the quick development of network, cybercrime affects our lives severely. Due to the limitation of technology, it is impossible to forbid cybercrime absolutely, therefore, it is necessary to perform forensics analysis and punish the cybercriminals afterwards. Payload attribution system is an emerging method for performing network forensics. Payload refers to the actual cargo of the network traffic, carrying the transmitted data which serves the fundamental purpose of the transmission, and excludes metadata (sometimes referred to as overhead data) which is solely for facilitating delivery [1]. Payload attribution is a traceback process for finding the sources and destinations of packets that contain specific excerpts.

It is infeasible to store all the data for forensics analysis due to the cost of too much storage space. Existing methods, such as Hierarchical Bloom Filter, Fixed Block Shingling, Winnowing Multihashing (WMH) [2] etc., can compress the traffic to some degree under an acceptable false positive rate, however, they are not able to support wildcard query well. It is common that we want to query an excerpt in which some characters are not clear by using wildcards to denote these unknown characters in the query procedure.

---

[*] This work is supported by National Key Technology R&D Program (Grant No. 2012BAH37B04), National Natural Science Foundation of China (Grant No.61303260, No. 61170282), and Strategic Priority Research Program of the Chinese Academy of Sciences (Grant No. XDA06030200).

S.S.M. Chow et al. (Eds.): ISC 2014, LNCS 8783, pp. 454–464, 2014.

Our contribution is the invention of data structure called Winnowing Double with Wildcard Query (WDWQ) which supports wildcard query better than the state-of-the-art method. Also, our methods have higher query speed and lower false positive rate under an acceptable data reduction ratio. In addition, it solves the first block offset problem, the alignment problem, and the consecutiveness problem (to be explained in Section 3). We believe the proposed methods can be used for network forensics traffic processing in large scale networks and can improve the efficiency of network forensics processing and analysis.

This paper is organized as follows. Section 2 gives the preliminaries and reviews some related work. In Section 3, we describe our data structure Winnowing Double with Wildcard Query in details. Subsequently, we report our implementation and the performance evaluation in Section 4. Finally, the conclusion is made in Section 5.

2   Preliminaries

In this section, we introduce the basic techniques involved in this paper.

2.1   Winnowing

In order to support query as well as data reduction, we divide the payload into blocks according to the fingerprint algorithm called Winnowing, then insert them into a data structure called Bloom Filter [3]. Many applications show that winnowing [2] has the best block result among fingerprint algorithms. Initially, it is used for generating the fingerprint of the documents, its algorithm is briefly recalled below:

Slide a window [6] of size k on the original text, compute a hash value for k characters each time, then store these generated hash values into an array in turn. Slide a window of size w on the array of hash values, and choose the minimum value in each window. If there are more than one minimum values, choose the rightmost one, and all these values generates the fingerprint of the document.

This strategy ensures taking enough fingerprint information as well as keeping the fingerprint not too large. When partitioning payloads, winnowing is better than other fingerprint algorithms especially for low-entropy payloads, such as long blocks of zero [2]. Winnowing ensures that there is at least one boundary being selected, the distance between each two approaching selected boundaries is at least one window's length. Therefore, the block structures using this algorithm will have proper block size which will not be too large. On the contrary, short excerpts cannot be processed if the payload block is too long. The size of the blocks generated by winnowing is upper bounded by the length of hash window, Fingerprints selected by winnowing are better for document fingerprinting than the subset of Rabin fingerprints, which contains hashes equal to 0 mod p, for some fixed p, because winnowing guarantees that in any window of size w there is at least one hash selected [2]. Rabin fingerprint may produce blocks which are too long to deal with small excerpts.

2.2   Character Dependent Multi-Bloom Filters

Haghighat et al. proposed a method called Character Dependent Multi-Bloom Filters (CMBF) which was seemed as the state-of-the-art technique. Its algorithm is briefly recalled below:

Slide a window on the payload and compute one fingerprint in each window. Then aggregate each g fingerprints to generate aggregated fingerprints. Subsequently, insert these aggregated fingerprints into 1 of the 256 Bloom filters according to the first byte in the aggregated fingerprints [4]. An illustration is shown in Figure 1.

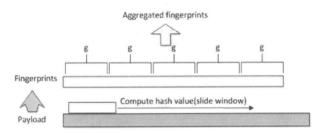

Fig. 1. The illustration of CMBF

3   Winnowing Double Structure Supporting Wildcard Query

3.1   The Existing Problems

The first block offset problem emerges in the methods which use offset number to connect the adjacent blocks. As an illustration, the payload blocks divided by block-based Bloom filter (BBF) method are inserted into the Bloom filters with continuous offset numbers. In the excerpt phase, we do not know the first block of the excerpt match with which block in the payload, so we need to connect every possible offset with the first block then verify them one by one. Our method does not use offset, so it is free from this problem and obtain higher speed.

To support wildcard query, note that CMBF has fixed block size, its excerpt query phase has to try all possible positions, because the beginning position of the first matching block in the excerpt and the generating combinations of aggregated finger-prints are unknown. This is known as the alignment problem, which requires large computational resource. To avoid this problem, our method divides the payload into variable blocks based on winnowing algorithm.

Similar to other techniques, we use Bloom filters to store the blocks divided by the algorithm. So, the time complexity of querying is constant in the number of inserted data elements. We have proposed Winnowing Multihashing Structure with Wildcard Query (WMWQ) in our former paper [5]; however, its query speed is not fast enough because it needs to run multiple instances.

To see, suppose there are two payload blocks with the same overlaps or offsets in different payload packets, however, the connection of these two blocks is not in the payload actually. If the excerpt is just the connection one by coincidence, the system will answer "yes" and make false positive in the fact that the excerpt is not in the payload. This is so-called consecutiveness problem. Although WMWQ has dealt with the consecutiveness problem to some extent via variable blocks divided by winnowing algorithm, because of modulo operation, the probability of meeting consecutiveness problem is a bit high. As a solution, multiple instances of WMWQ will be executed, but it slows down the query speed.

3.2   Our Design

For the purpose of accelerating the wildcard query speed as well as maintaining the high accuracy, we propose a new method called Winnowing Double with Wildcard Query (WDWQ) which employs 256 Bloom filters and double blocks based on WMWQ. WDWQ is an improvement of WMWQ, we bring in another layer which is made up of double blocks.

Suppose that there is a payload $"c_1c_2...c_n"$, in which n refers to the length, the steps of our method are as follows. At first, we slide a window of size k over the payload and compute the hash value $h_i = F(c_i, c_{i+1}, ..., c_{i+k-1})$ over each window, and then we record the hashes into an array. Secondly, we slide another window of size w over the hash array, find the value h_j within each window and insert a boundary after the corresponding character c_j. If there is more than one minimum value in the window, we choose the rightmost one. Consequently, the contents between each two boundaries consist of the original blocks. Thirdly, we concatenate each two nearby blocks to generate double blocks. In the fourth step, we run a hash function modulo 256 over each original and double block, the results refer to the corresponding Bloom filters which are numbered from 0 to 255, and then we insert these blocks. An illustration is given in Figure 2, and the pseudo-code is shown in Table 1.

When we get an excerpt, the query steps are similar. The former three steps are the same and the difference is in Step 4. Instead of inserting the blocks, we check whether every mapping position of the relating Bloom filters is set to 1. We will give out a positive answer only if all the corresponding positions are filled with 1. The pseudo-code is shown in Table 2.

The function of supporting wildcard query mainly depends on modulo operation in hash function $F(c_i, c_{i+1}, ..., c_{i+k-1})$. In detail,

$$F(c_i, c_{i+1}, ..., c_{i+k-1}) = (c_i \bmod q) \times p^{k-1} + (c_{i+1} \bmod q) \times p^{k-2} +$$
$$... + (c_{i+k-1} \bmod q) \times p^0$$

where p is a fixed prime and q is a constant, $q \le p < 256$. For the purpose of saving computations, the hash value of the latter payload window can be computed from the former one leveraging the property of the polynomial, that is:

$$F(c_{i+1}, c_{i+2}, ..., c_{i+k}) = pF(c_i, c_{i+1}, ..., c_{i+k-1}) + (c_{i+k} \bmod q) - (c_i \bmod q) \times p^k$$

If an excerpt contain an unknown character "?", we replace the corresponding position with 0 to q-1 and check all these q substrings. We consider the excerpt to be in the payload if any substring is found.

In addition, WDWQ can deal with more complex query in a simple manner. For example, "xy[n-p]cba" is the excerpt. The real process is to query 3 substrings "xyncba", "xyocba" and "xypcba".

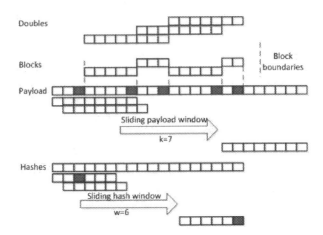

Fig. 2. Illustration of WDWQ

Table 1. The pseudo-code of WDWQ algorithm(In process phase)

Algorithm 1. Gen_Fingerprint_and_Insert(Bloomfilter bloomfilters[1,...,256], Packet packet)

```
1:   Variable global overlap_length,module;
2:   Generate a list L of positions partitioning the packet by winnowing;
3:   start_pos=0, start_pos2=0, start=1;
4:   if (L[next]==0)
             start=2;
5:   end if
6:   next=start;
7:   while (next < L.length) do
8:           if (L[next]+overlap_length>packet.payload.length)
9:                   break;
10:          end if
11:      segment = packet.payload[start_pos: L[next]+overlap_length];
12:          for each i=0,1,...,segment.length do
13:                  segment[i]=segment[i]%module;
14:          end for
15:          hash_value=generate_hash(segment);
16:          Bloomfilter bf = blommfilters[hash_value%256+1];
17:          Insert_Bloomfilter(bf,hash);
18:          start_pos=L[next];
19:          if (next>start)
20:                  segment=packet.payload[start_pos2:L[next]+overlap_length];
21:                  for each i=0,1,...,segment.length do
22:                          segment[i]=segment[i]%module;
23:                  end for
24:                  hash_value=generate_hash(segment);
25:                  Bloomfilter bf=blommfilters[hash_value%256+1];
26:                  Insert_Bloomfilter(bf,hash);
27:                  start_pos2=L[next-1]
28:          end if
29:          next=next+1;
30:  end while
```

Table 2. The pseudo-code of WDWQ algorithm(In query phase)

Algorithm 2. Search_Fingerprint(Bloomfilter bloomfilters[1,...,256], QueryString str)

```
 1:    global overlap_length,module;
 2:    Generate a string list SL for each possible string according to wildcards in str;
 3:    for each string in SL do
 4:            Generate a list L of positions segmenting the string by winnowing;
 5:            if (L.length<2)
 6:                    return false;
 7:            end if
 8:            start_pos=0, start_pos2=0, start=1;
 9:            if (L[next]==0)
10:                    start=2;
11:            end if
12:            next=start;
13:            while (next < L.length) do
14:                    if (L[next]+overlap_length>string.length)
15:                            break;
16:                    end if
17:                    segment = string[start_pos: L[next]+overlap_length];
18:                    for each i=0,1,...,segment.length do
19:                            segment[i]=segment[i]%module;
20:            end for
21:                    hash_value=generate_hash(segment);
22:                    Bloomfilter bf = blommfilters[hash_value%256+1];
23:                    if (!Check_Exist (bf,hash) && start_pos !=0)
24:                            return false;
25:                    end if
26:                    start_pos=L[next];
27:                    if (next>start)
28:                            segment=string[start_pos2:L[next]+overlap_length];
29:                            for each i=0,1,...,segment.length do
30:                                    segment[i]=segment[i]%module
31:                            end for
32:                            hash_value=generate_hash(segment);
33:                            Bloomfilter bf=blommfilters[hash_value%256+1];
34:                            if (!Check_Exist(bf,hash) && start_pos2 != 0)
35:                                    return false;
36:                            end if
37:                            start_pos2=L[next-1];
38:                    end if
39:                    next=next+1;
40:            end while
41:            if (next==L.length)
42:                    return true;
43:            endif
44:    end for
45:    return false;
```

3.3   Performance Evaluation

As higher bandwidth links are being deployed more often, fast response has become a vital factor. Even if some methods have high accuracy and large data reduction ratio, they will be useless when they could not respond the query within proper time. WDWQ has solved the alignment problem due to their variable blocks determined by winnowing algorithm according to the content, its response speed is faster than CMBF which is claimed as the state-of-the-art method. In addition, WDWQ generates another double blocks layer instead of running several instances in WMWQ, the query speed is faster than WMWQ as well. Table 3 shows the order analysis of process and query costs on WDWQ, WMWQ, CMBF and WMH [2].

Table 3. Processing and query cost

	WDWQ	WMWQ	CMBF	WMH
processing	$2n$	$2tn$	$(n,2n)$	$2tn$
query	$2n$	$2tn$	$2gn$	$2tn$

For the notation in the table, n refers to the number of blocks processed by BBF [10], t refers to the number of Winnowing instances deployed, g refers to the CMBF aggregation factor. Since CMBF has the alignment problem, it needs to try all the possibilities which is g times slower than our methods.

4   Experimental Results

The theoretical analysis of our method's advantage follows a similar analysis as our earlier work [5]. In this section, we report our experimental evaluation of our method. For the purpose of demonstrating the advantages of our proposed data structure, we implemented WMH [2] and CMBF [4] which are considered as the state-of-art techniques, in addition to WMWQ and WDWQ.

4.1   Distribution of Block Size

The graph in Figure 3 shows the distribution of WDWQ's block sizes. We set the length of sliding hash window to 20 bytes, the length of payload window is 10 bytes and the modulus is 8.

It shows that the winnowing algorithm leads to a uniform distribution where the block sizes are bounded by the winnowing window size plus the overlap [2], which confirms our usage of the average value in the deduction in our former paper [5] and it is suitable to choose winnowing to partition the payload.

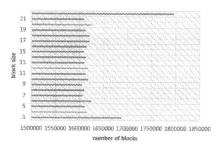

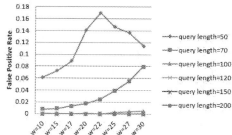

Fig. 3. Block Size Distribution

Fig. 4. Impact of hash window size and excerpt length in WDWQ

4.2   Proper Parameters

Our intention here is to find out experimentally an appropriate hash window size resulting in a minimum false positive rate, to be used in the following experiments.

Firstly, we collected 45MB traffic from our laboratory by WireShark and stored them according to Algorithm 1 in Section 3. Then we constructed 10000 different random strings as excerpts to query the existing Bloom filters for each possible lengths of 50, 70, 100, 120, 150, 200 bytes. As shown in [4], the probability of random strings actually inserted into the Bloom filters is less than 10^{-46} for a 25B excerpt, and it would be smaller for larger excerpts.

Figure 4 shows how the false positive rate changes with different hash window sizes. In general, larger query length results in lower false positive rate. We eventually used the parameter $w = 10$ in the following experiments.

4.3   Response Time

For the purpose of evaluating the wildcard query response time, we implemented WDWQ, WMWQ, and CMBF methods separately, then we recorded different query time with different number of unknown characters for comparison as shown in Figure 5. The query space will increase exponentially with the raising modulus; therefore, we set the modulus to 8 as illustrated in [4].

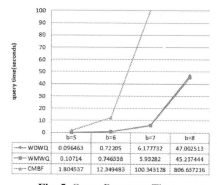

	b=5	b=6	b=7	b=8
WDWQ	0.096463	0.72205	6.177732	47.002513
WMWQ	0.10714	0.746338	5.93282	45.237444
CMBF	1.804537	12.349483	100.343128	806.637216

Fig. 5. Query Response Time

It is noticeable that the platform and hardware have great influences on the response speed of these methods. Our main purpose here is to show the relative performance of different algorithms. The above figure shows that WDWQ runs faster than WMWQ, and is about 20 times faster than the state-of-the-art CMBF method when the numbers of wildcards are 7, 8 or more.

4.4   Accuracy

Larger data reduction ratio is available through changing the number of Bloom filters in WDWQ and WMWQ. In this paper, we use 256 Bloom filters. We can achieve larger data reduction ratio if we decrease the number of the Bloom filters. The program produced 10000 random excerpts for each of the 8 possible lengths, ranging from 50 to 300 bytes. In order to evaluate the accuracy of WDWQ, we carried the same procedures for WMWQ, CMBF and WMH. The result is summarized in Table 4. As discussed earlier in this section, the probability that random strings have been inserted into the Bloom filters is less than 10^{-46}, thus we can deem that the system make false positive (answering "Yes") if all the corresponding bit positions in the Bloom filters were set to 1 after the process of related methods. On the contrary, the answer is exactly correct (answering "No") if the system cannot find the random excerpts, and the statistically figure of accuracy is shown in Figure 6. Note that we set the parameter t in WMWQ to be 2 here. The result shows that WDWQ has lower false positive rate.

Table 4. False Positive Rate

Query Length	WDWQ		WMWQ		CMBF		WMH	
	Yes	No	Yes	No	Yes	No	Yes	No
50	696	8801	3532	6468	4368	5632	4292	5708
70	184	9816	1808	8192	1925	8075	1847	8153
100	14	9986	562	9438	496	9504	472	9528
120	2	9998	265	9735	174	9826	157	9843
150	0	10000	88	9912	43	9957	31	9969
200	0	10000	12	9988	4	9996	1	9999
250	0	10000	4	9994	0	10000	0	10000
300	0	10000	0	10000	0	10000	0	10000

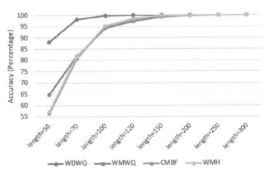

Fig. 6. Accuracy Comparison

4.5   Scope of Application

In the environment which response time is a crucial factor, we recommend to choose WDWQ which achieves faster speed and maintains excellent correctness as well. On the other hand, if the requirement is high accuracy, WMWQ may be better, but its speed is slower than WDWQ since t invocations are needed for the high accuracy. In general, WDWQ achieve faster query speed as well as lower false positive rate than WMWQ when the parameter t in WMWQ is not too large.

Overall, WDWQ and WMWQ accelerate the wildcard query speed 20 times than the state-of-the-art payload attribution technique CMBF, WDWQ is faster than WMWQ, meanwhile, WDWQ has lower false positive rate under an acceptable data reduction ratio.

5   Conclusion

In this paper, we propose WDWQ for the application of payload attribution. We use the winnowing partition algorithm, so the block sizes vary with the payload content, moreover, the generating blocks will not be too long due to the mathematical property of winnowing. In addition, because we use Bloom filters to store the blocks, the query complexity is constant and does not depend on the number of elements inserted into the Bloom filters. Our experimental evaluation shows the high performance of WDWQ. Compared with existing approaches, our method supports wildcard query better and achieves higher query speed as well as lower false positive rate under an acceptable data reduction ratio. Furthermore, it is possible to tune down the number of Bloom filters for larger data reduction ratio while still maintaining acceptable false positive rate.

References

1. Wiki. Payload (computing) (March 15, 2014), http://en.wikipedia.org/wiki/Payload_computing
2. Ponec, M., Giura, P., Wein, J., Brönnimann, H.: New Payload Attribution Methods for Network Forensic Investigations. ACM Transactions on Information and System Security 13(2), Article 15 (2010)
3. Broder, A., Mitzenmacher, M.: Network Applications of Bloom Filters: A Survey. Internet Mathematics 1(4), 485–509 (2004)
4. Haghighat, M.H., Tavakoli, M., Kharrazi, M.: Payload Attribution via Character Dependent Multi-Bloom Filters. IEEE Transactions on Information Forensics and Security 8(5), 705–716 (2013)
5. Wei, Y., Fei, X., Chen, X., Shi, J., Qing, S.: Winnowing Multihashing Structure with Wildcard Query. In: Han, W., Huang, Z., Hu, C., Zhang, H., Guo, L. (eds.) APWeb 2014 Workshops. LNCS, vol. 8710, pp. 265–281. Springer, Heidelberg (2014)
6. Schleimer, S., Wilkerson, D.S., Aiken, A.: Winnowing: Local algorithms for document fingerprinting. In: Proc. of the ACM SIGMOD International Conference on Management of Data (SIGMOD 2003), pp. 76–85. ACM (2003)

7. Bloom, B.: Space/time tradeoffs in hash coding with allowable errors. CACM, 422–426 (1970)
8. Broder, A., Mitzenmacher, M.: Network Applications of Bloom Filters: A Survey. Internet Mathematics 1(4), 485–509 (2004)
9. Jiao, M.: The concept and theory of Bloom Filter, http://blog.csdn.net/jiaomeng/article/details/1495500.2007-01-27
10. Shanmugasundaram, K., Brönnimann, H., Memon, N.: Payload attribution via hierarchical bloom filters. In: Proc. of the 11th ACM Conf. Computer and Communications Security, pp. 31–41. ACM (2004)

An Evaluation of Single Character Frequency-Based Exclusive Signature Matching in Distinct IDS Environments

Weizhi Meng[1,2,*], Wenjuan Li[1], and Lam-For Kwok[1]

[1] Department of Computer Science, City University of Hong Kong, Hong Kong, China
[2] Infocomm Security Department, Institute for Infocomm Research, Singapore
yuxin.meng@my.cityu.edu.hk

Abstract. The signature-based intrusion detection systems are one of the most commonly used software to protect computer networks by comparing incoming traffic with stored signatures. However, the process of signature matching is a key challenge, in which the workload is generally at least linear to the size of a target string. To solve this problem, exclusive signature matching (ESM) has been proposed based on the observation that most network packets would not match any IDS signatures. But this kind of schemes like the single character frequency-based ESM has not been extensively evaluated. In this paper, our interests are to verify the observation above and evaluate the single character frequency-based ESM in regular networks and hostile environments respectively. In the hostile experiment, we specifically design two malicious situations to test the scheme performance. The experimental results show that the single character frequency-based ESM works fine in a regular network, but its performance would be greatly decreased in a hostile environment.

Keywords: Intrusion Detection, Exclusive Signature Matching, Performance Evaluation, Single Character Frequency, Network Security.

1 Introduction

Intrusion detection has been developed for over thirty years, which aims to detect a variety of computer and network attacks such as worms, Trojans, DoS attacks, etc. Generally, an intrusion detection system (IDS) can be classified into two categories: *signature-based IDS* and *anomaly-based IDS*. A signature-based IDS like Snort [20] detects an attack mainly through comparing current events or packet payloads with its stored and predefined rules, i.e., attack signatures. Thus, it is also known as misuse detection and rule-based detection. By contrast, an anomaly-based IDS like Bro [15] attempts to identify an intrusion by detecting great deviations between current events and predefined normal profiles.

Intrusion detection systems have become an important and indispensable component of the modern security infrastructure. Due to relatively less false alarms

---

* Corresponding author and is previously known as Yuxin Meng.

S.S.M. Chow et al. (Eds.): ISC 2014, LNCS 8783, pp. 465–476, 2014.
© Springer International Publishing Switzerland 2014

produced, the signature-based IDSs are much more popular than the anomaly-based approach in practical applications [18]. However, the process of signature matching is a big bottleneck for the signature-based IDSs, in which the workload is at least linear to the size of the target string [17]. Taking Snort as an example, which is a lightweight signature-based network intrusion detection system [20]. It has the ability to perform real-time traffic analysis, content matching and packet logging on Internet Protocol (IP) networks. In general, it spends abut 30 percent of its total processing power in conducting signature matching, whereas the CPU burden would be greatly increased to over 80 percent when deployed in an intensive web-traffic environment [5,6]. In such cases, Snort has to drop a large number of packets, which would cause many security issues and severely degrade the whole security level of a network.

To address this issue, one direct approach is to design more efficient signature matching algorithms (or called *string matching*). The most commonly used way is to find an exact match between signatures and packet strings like [2,6,7,23]. In contrast, another way for constructing a signature matching scheme is to identify a mismatch. For instance, Markatos et al. [10] first proposed an exclusion-based signature matching algorithm called *ExB*, in which the basic idea is to check whether an input string contains all fixed-size bit-strings of a signature. The *ExB* is mainly based on an observation that most network packets would not match any IDS signatures. This work we denote it as packet-signature (*PackSig*) observation. The *ExB* was only implemented in hardware which is not easy for adapting to a specific network.

Motivation. We previously coined this concept as *exclusive signature matching* (ESM) and first developed a software-based scheme, called single character frequency-based ESM [12]. This scheme consists of four statistical tables and uses a decision algorithm to output an appropriate single character in finding a mismatch. Later, we developed this scheme to be adaptive with single and two-length consecutive characters, and tested it in a distributed environment [13]. However, the ESM like the single character frequency-based scheme has still not extensively evaluated. In addition, few work in literature paid attention to the validation of the observation above, while many articles like [19,21] conduct research relying on this observation. Thus in this paper, our interests mainly focus on the single character frequency-based ESM and we have three targets.

- *T1.* To verify the *PackSig* observation that most packets in a regular network would not match any IDS signatures.
- *T2.* To evaluate the single character frequency-based ESM in a regular and a heavy traffic network environment respectively.
- *T3.* To evaluate the performance of single character frequency-based ESM in a hostile network environment.

Contributions. In this paper, we limit our discussions to the single character frequency-based ESM. We do not consider other exclusive signature matching schemes at this stage. The main goal is to explore the effect of single character frequency on ESM more comprehensively. Our contributions of this work can be summarized as follows.

- This work we implement a software-based prototype of the single character frequency-based ESM, which can run in any java-enabled machines.
- Our targets *T1* and *T2* are related, so that we evaluate the single character frequency-based ESM in two regular networks to investigate its performance and verify the *PackSig* observation.
- To achieve *T3*, we design two experiments in different hostile environments: one environment contains much more malicious packets than usual, while in the other environment, we launch an attack called *character padding attack* on the single character frequency-based ESM.

The rest of this paper is organized as follows. Section 2 introduces the background of single character frequency-based ESM and reviews related work. Section 3 describes our implementation prototype of the single character frequency-based ESM. Section 4 evaluates the single character frequency-based ESM in a regular and a heavy web traffic environment, while Section 5 evaluates the single character frequency-based scheme in hostile environments. Finally, we conclude our work and present an outlook to future work in Section 6.

2   Background and Related Work

2.1   Single Character Frequency-Based ESM

The main idea of exclusive signature matching is to identify a mismatch rather than confirm an accurate match in the process of signature matching. As shown in Fig. 1, this scheme mainly contains four statistical tables: *the table of stored NIDS signatures (denoted SNS), two tables of character frequency (denoted SCQ1, SCQ2), the table of matched NIDS signatures (denoted MNS)* and one functional component called *decision component*. These four tables are responsible for calculating character frequency and recording relevant statistical data. Particularly, *SCQ1* aims to count the single character frequency according to *SNS*, while *SCQ2* is used to compute the single character frequency based on *MNS*. In a network, the character frequency would be different in these two tables, and can be arranged in descending order from the most frequent to the least frequent.

The *decision component* is responsible for deciding the most appropriate single character from two *SCQ* tables to compare with incoming packet payloads. Generally, *SCQ2* would be checked first since this table can be adaptive to a specific environment (i.e., the matched IDS signatures would keep updating). In real settings, IDS signatures are usually organized into a decision tree divided by packet type and packet port number. Therefore, signature matching is mainly conducted between a packet payload and one/several IDS signature(s), rather than all signatures. The specific *decision algorithm* can be referred to [12].

It is found that if the scheme identifies at least one character is not included in a packet payload, then the matching process can be completed. However, if no mismatch is detected, a traditional matching algorithm should be used to confirm whether it is an exact match. To conclude, in the best scenario, the scheme can identify a mismatch using only one round, but in the worst scenario, a traditional signature matching algorithm should be applied.

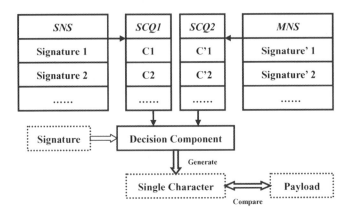

Fig. 1. The architecture of single character frequency-based ESM

2.2   Related Work

In this work, we divide the signature matching schemes into two categories: one is to identify an exact match, while the other is to identify a mismatch.

For the first category, a lot of algorithms have been proposed. For example, Boyer and Moore algorithm [2] is the most widely used single pattern matching algorithm to reduce the number of searches in the matching process by performing the search beginning from the rightmost character of a target string. Then, Horspool [7] improved the Boyer-Moore algorithm by using only the bad character heuristic to further improve its speed. For multi-pattern matching, Aho and Corasick [1] designed a multi-pattern matching algorithm to search all strings at the same time by means of a deterministic finite automaton (DFA) without the need of an additional search structure. Later, Wu and Manber [23] improved the Aho-Corasick algorithm using hash tables and bad character heuristic to search patterns in the matching process. With respect to intrusion detection, Fisk and Varghese [6] first proposed a NIDS-specific string matching algorithm called *Set-wise Boyer-Moore-Horspool*. The experimental results showed that their designed algorithm was faster than both Boyer-Moor and Aho-Corasick algorithms when dealing with medium size pattern sets. Chen et al. [3] then proposed a system by combining fingerprinting with pattern matching techniques aiming to accelerate pattern matching for an IDS. Several related work including regular expression can be referred to [4,8,9,16,21].

Another category aims to identify a mismatch during the matching process. For example, Markatos et al. [10] first proposed a multiple-string matching algorithm designed for NIDSs, called *ExB*. For each incoming packet, *ExB* first creates an *occurrence bitmap* to guarantee that each fixed-size bit-string exists in the packet. Then, the bit-strings for each signature are matched against the occurrence bitmap. Later, they presented an algorithm called E^2xB [11] as an improvement for the *ExB* above in better using cells and supporting case-sensitive matching. The two schemes were implemented on hardware and a time reduction

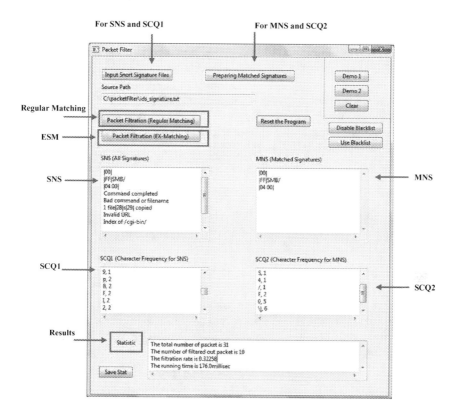

Fig. 2. The interface of the implemented single character frequency-based ESM

rate between 10% and 36% could be achieved. However, hardware-based techniques are expensive and hard to reconfigure in real-world applications. To address the issues, we previously proposed an exclusive signature matching scheme (ESM) based on single character frequency to improve the process of signature matching, which was implemented on software. To further investigate this scheme, we test the scheme using single and two-length consecutive characters in a distributed network [13]. The experimental results indicated that the achieved time reduction rate was ranged from 11% to 37%. But in literature, the ESM schemes like the the single character frequency-based ESM has still not been extensively studied. Several issues are still remained such as how it performs in a hostile environment, which are our motivations in this paper.

3   Scheme Implementation

In Fig. 2, we present the prototype of our implemented single character frequency-based ESM and highlight the major components in the interface. This prototype is designed based on Snort and its functions are described as below:

- *Statistical tables.* There are four tables: *SCQ1*, *SCQ2*, *MNS* and *SNS*. To establish *SNS* and *SCQ1*, we can input signatures by clicking the button of *Input Snort Signature Files*, while for *MSN* and *SCQ2*, we can prepare them using the button of *Preparing Matched Signatures*. After clicking on buttons, the tables' contents can be shown in the interface.
- *Selection of signature matching schemes.* Our prototype can provide two schemes for signature matching. One is the regular matching based on Boyer-Moor algorithm[1] (with the button of *Regular Matching*) while the other is the single character frequency-based ESM (with the button of *EX-Matching*).
- *Result details.* To evaluate the performance of different matching schemes, our prototype can show the detailed results at the bottom of the interface such as filtration rate and running time.

The prototype system was developed in Java and can be compressed to a *jar* file, thus, it is scalable to different java-enabled operating systems. In the interface as shown above, we give a simple example to illustrate the performance of single character frequency-based ESM by means of a small packet dataset (including only 31 packets). There are totally 8 signatures while we assume 3 signatures have been matched. The information of statistical tables is shown in the interface. After choosing *EX-Matching*, the results indicate that the single character frequency-based ESM can complete the matching within 176 millisecond and a reduction rate of about 32.5% can be achieved.

4   Evaluation in Regular Network Environments

Our targets *T1* and *T2* are correlated, thus, this section we attempt to verify the *PackSig* observation, and evaluate the performance of single character frequency-based ESM in regular network environments including a heavy web-traffic environment.

4.1   Regular Networks

In this section, we deploy our prototype into two network environments. One is a college network while the other is a company network (including about 100 personnel). The network structures of these environments are summarized in Fig. 3. We mainly evaluate the single character frequency-based ESM in three levels: *router level*, *server level* and *terminal level*.

In Table 1, we present the packet rate in each level for the two different network environments. It is easily visible that the packet rate in the router level is the highest, since all traffic would go through it. Besides, the packet rate in the server level is generally higher than that in the terminal level. For the different networks, it is found that the packet rate in the company network is often larger than that in the college network. One of the major reasons is that many transactions would be finished online for the company.

---

[1] Snort usually adopts Boyer-Moor algorithm by default.

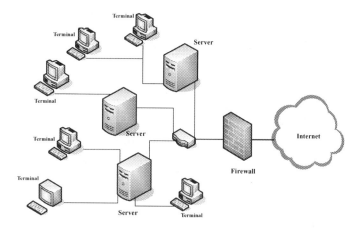

Fig. 3. The structure of network environments

Table 1. The packet rate for different levels and networks

Levels/Packet Rate (packets/sec)	College Network	Company Network
Router Level	6500-8000	5000-12000
Server Level	4000-6800	4600-8000
Terminal Level	1500-2400	1700-3030

To evaluate the performance, we randomly deploy the prototype into several end-points of each level in different networks. Full Snort signatures are applied (version 2.9.3.1). The results regarding packet filtration rate and time-saving rate are shown in Table 2. In general, it is expected to achieve a high packet filtration rate, which means reducing more target packets for an IDS. In addition, a higher filtration rate usually achieves a higher time-saving rate.

Table 2 shows that the single character frequency-based ESM can achieve a filtration rate in the range from 32.3% to 47.9% and from 23.2% to 39.4% for the college network and the company network respectively. We find that the specific filtration rate is depending on the incoming packet payloads. Usually, a longer payload may contain more characters which can decrease the performance of the single character frequency-based ESM, while some special characters in a signature like '/' may make the matching process finish more quickly.

For the time-saving rate, the scheme can achieve a rate ranged from 21.5% to 35.5% and from 18.8% to 29.8% for the college and the company network respectively. On the whole, the performance achieved here is similar, but a bit better than the results in [12]. By considering both filtration rate and saving time in each environment, we consider that the single character frequency-based ESM performs well in a regular network.

PackSig Observation Verification. This observation can be described in a more formal way as follows. Suppose a packet payload pl and a set of IDS signatures $S = \{s1, ..., sn\}$. The observation is that the possibility of pl containing

Table 2. The packet filtration rate and time-saving rate for different levels and networks for a whole day

College Network	Filtration Rate (%)	Time-Saving Rate (%)
Router Level	42.42	33.5
Server Level: Point A	34.23	24.3
Server Level: Point B	42.92	34.6
Terminal Level A	47.82	35.5
Terminal Level B	41.35	32.3
Terminal Level C	32.42	21.5
Company Network	Filtration Rate (%)	Time-Saving Rate (%)
Router Level	35.76	28.1
Server Level: Point A	24.10	19.4
Server Level: Point B	33.14	22.5
Terminal Level A	23.20	18.8
Terminal Level B	35.72	24.4
Terminal Level C	39.32	29.8

Table 3. The results of average packet filtration rate for different levels and networks

Average Filtration Rate (%)	College Network	Company Network
Router Level	73.42	78.57
Server Level (3 points included)	68.3	73.18
Deviation	5.3	4.8
Terminal Level (50 points included)	72.3	68.32
Deviation	3.3	6.8

sn $(n = 1, 2, ...)$ is very small. That is, most network packets will not match any IDS signatures in real deployment.

Based on the same environments above, we use regular matching to verify this observation. In Table 3, we present the results of filtration rate in the network environments. In particular, a higher filtration rate can indicate that more network packets do not match any signatures. Full Snort signature database is also applied. As shown in Table 3, it is visible that the average filtration rate is generally over 70%, indicating that most network packets would not match any signatures. In addition, we notice that the filtration rate in the company network is very close to 80%. To conclude, we consider that the *PackSig* observation is applicable in real network environments.

4.2 Heavy Traffic Networks

In this section, we aim to evaluate the single character frequency-based ESM in a heavy traffic network environment. In this work, we define 'heavy traffic' for a matching scheme as the situation that the packet rate of the monitored network exceeds the handling capability of the matching scheme. As a study, we conduct this evaluation based on a small data center, where the packet rate is considered as larger than usual network environments. The data center does not belong to

Table 4. The packet rate in the data center.

Levels/Packet Rate (packets/sec)	Data Center
Main Router Level	16500-18000
Server Level	6000-14800

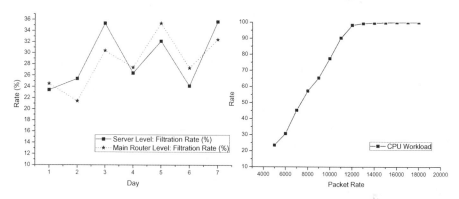

Fig. 4. This figure shows the average filtration rate for a week

Fig. 5. This figure shows the CPU workload versus packet rate

any company or college mentioned above. For privacy reasons, in Table 4 we only provide some information about the packet rate within this center. To evaluate the performance, we deploy the prototype in the server and main router level for a week and the results are presented in Fig. 4 and Fig. 5.

Fig. 4 shows the result of filtration rate for a week, which is in a range from 23% to 36% and from 21% to 35% for server level and main router level respectively. In Fig. 5, we present the CPU workload versus the packet rate and it is found that the workload becomes heavier with the increase of packet rate. When the packet rate is over 11,000 packets/sec, the scheme cannot provide any advantages. During the evaluation, positively, it is found that the prototype does work during this heavy traffic environment, but the performance would be greatly downgrade when the packet rate increases. The major reason we identified in the experiment is that the probability of containing all single characters of a signature is largely increased for a packet payload, when in such a huge traffic network. To conclude, we consider that the single character frequency-based ESM can improve the performance of traditional signature matching, but it still needs to be further improved in a heavy traffic environment.

5   Evaluation in Hostile Environments

The experiments above indicate that the single character frequency-based ESM can perform fine in regular networks. In this section, we attempt to evaluate this scheme in hostile environments. Specifically, we consider two situations: one is the network under attacks and the other is the scheme itself under attacks.

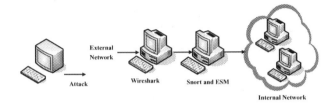

Fig. 6. The network environment under attacks

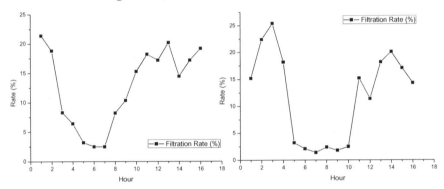

Fig. 7. Network under attacks: the results of filtration rate under attacks

Fig. 8. ESM under attacks: the results of filtration rate under attacks

5.1 Network under Attacks

The network environment is shown in Fig. 6. Snort and our prototype are implemented in a computer while Wireshark [22] is deployed in front aiming to provide statistical information. In this experiment, we simulate a number of attacks like scanning attacks to this environment. Our purpose is that by conducting a series of attacks, packet payloads would contain most single characters with regard to a signature. In real-world applications, this is the worst scenario for the single character frequency-based ESM.

The results of filtration rate are described in Fig. 7. The attacks are launched from the 3rd hour to the 8th hour and it is visible that the filtration rate of the scheme is significantly decreased. For example, the filtration rate is reduced from 18.8% in the 2nd hour to 8.3% in the 3rd hour. The experimental results show that the single character frequency-based ESM would be greatly affected under a malicious network environment. The main reason is that most packet payloads contain all single characters of a signature. To further improve the performance, two or more consecutive-length characters could be used.

5.2 ESM under Attacks

In this experiment, we design an attack called *character padding attack* on the single character frequency-based ESM. The attackers can padding required characters to manipulate their packets, so that the ESM cannot filter out them. The purpose of this attack is to degrade the effectiveness of the ESM. The results of filtration rate are described in Fig. 8.

This attack is launched from the 5th hour to the 10th hour, it is seen that the filtration rate is decreased much more quickly than the results of *network under attacks*. The main reason is that the *character padding attack* is more efficient to take advantage of the worst scenario of the single character frequency-based ESM. In this case, we consider that this attack is a big challenge to impede the performance of this scheme in a hostile environment. To address this problem, similarly, two or more consecutive-length characters could be used as a matching pattern to increase the effectiveness of finding a mismatch.

6   Conclusion and Future Work

In this paper, we focus on the single character frequency-based ESM and propose three targets: 1) to verify the *PackSig* observation that most packets in a regular network would not match any IDS signatures; 2) to evaluate the single character frequency-based scheme in a regular and a heavy traffic environment respectively; and 3) to evaluate this scheme in hostile network environments. After several experiments, we find that *PackSig* observation is applicable in most cases, and that the single character frequency-based ESM could perform well in a regular network environment, but would be weak in a hostile network where the scheme should be performed in the worst scenario. To address these challenges, we recommend that two or more consecutive-length characters should be used as a matching pattern to increase the probability of identifying a mismatch.

There are a lot of potential future work, which could include formalizing the designed *character padding attack* and testing its effect on other exclusive signature matching schemes. In addition, we also plan to identify effective solutions in strengthening the performance of ESM schemes like the single character frequency-based ESM under hostile environments.

Acknowledgments. The authors were fully funded by the Innovation to Realization Funding Scheme of the City University of Hong Kong (under the project number 6351018). We thank all anonymous reviewers for their valuable comments in improving this paper.

References

1. Aho, A.V., Corasick, M.J.: Efficient string matching: An aid to bibliographic search. Communications of the ACM 18(6), 333–340 (1975)
2. Boyer, R.S., Moore, J.S.: A fast string searching algorithm. Communications of the ACM 20(10), 762–772 (1977)
3. Chen, Z., Zhang, Y., Chen, Z., Delis, A.: A digest and pattern matching-based intrusion detection engine. Computer Journal 52(6), 699–723 (2009)

4. Commentz-Walter, B.: String Matching Algorithm Fast on the Average. In: Maurer, H.A. (ed.) ICALP 1979. LNCS, vol. 71, pp. 118–132. Springer, Heidelberg (1979)
5. Dreger, H., Feldmann, A., Paxson, V., Sommer, R.: Operational experiences with high-volume network intrusion detection. In: Proceedings of ACM CCS, pp. 2–11 (2004)
6. Fisk, M., Varghese, G.: An analysis of fast string matching applied to content-based forwarding and intrusion detection. Technical Report CS2001-0670, University of California, San Diego (2002)
7. Horspool, R.: Practical fast searching in strings. Software Practice and Experience 10, 501–506 (1980)
8. Kim, K., Kim, Y.: A fast multiple string pattern matching algorithm. In: Proceedings of AoM/IAoM Conference on Computer Science (1999)
9. Liu, X., Liu, X., Sun, N.: Fast and compact regular expression matching using character substitution. In: Proceedings of ANCS, pp. 85–86 (2012)
10. Markatos, E.P., Antonatos, S., Polychronakis, M., Anagnostakis, K.G.: Exclusion-based signature matching for intrusion detection. In: Proceedings of International Conference on Communications and Computer Networks, pp. 146–152 (2002)
11. Anagnostakis, K.G., Antonatos, S., Markatos, E.P., Polychronakis, M.: E^2xB: A Domain-Specific String Matching Algorithm for Intrusion Detection. In: Gritzalis, D., De Capitani di Vimercati, S., Samarati, P., Katsikas, S. (eds.) Security and Privacy in the Age of Uncertainty. IFIP, vol. 122, pp. 217–228. Springer, Boston (2003)
12. Meng, Y., Li, W., Kwok, L.-F.: Single Character Frequency-based Exclusive Signature Matching Scheme. In: Lee, R. (ed.) Computer and Information Science 2012. SCI, vol. 429, pp. 67–80. Springer, Heidelberg (2012)
13. Meng, Y., Li, W.: Adaptive Character Frequency-based Exclusive Signature Matching Scheme in Distributed Intrusion Detection Environment. In: Proceedings of TrustCom, pp. 223–230. IEEE (2012)
14. Meng, Y., Li, W., Kwok, L.F.: Towards Adaptive Character Frequency-based Exclusive Signature Matching Scheme and its Applications in Distributed Intrusion Detection. Computer Networks 57(17), 3630–3640 (2013)
15. Paxson, V.: Bro: A System for Detecting Network Intruders in Real-Time. Computer Networks 31(23-24), 2435–2463 (1999)
16. Ramakrishnan, K., Nikhil, T., Jignesh, M.: SigMatch: fast and scalable multi-pattern matching. VLDB Endowment 3(1-2), 1173–1184 (2010)
17. Rivest, R.L.: On the worst-case behavior of string-searching algorithms. SIAM Journal on Computing 6, 669–674 (1977)
18. Sommer, R., Paxson, V.: Outside the closed world: On using machine learning for network intrusion detection. In: Proceedings of IEEE Symposium on Security and Privacy, pp. 305–316 (2010)
19. Sourdis, I., Dimopoulos, V., Pnevmatikatos, D., Vassiliadis, S.: Packet pre-filtering for network intrusion detection. In: Proceedings of ANCS, pp. 183–192 (2006)
20. Snort, The Open Source Network Intrusion Detection System, http://www.snort.org/
21. Stakhanova, N., Ren, H., Ghorbani, A.A.: Selective Regular Expression Matching. In: Burmester, M., Tsudik, G., Magliveras, S., Ilić, I. (eds.) ISC 2010. LNCS, vol. 6531, pp. 226–240. Springer, Heidelberg (2011)
22. Wireshark, Network Protocol Analyzer, http://www.wireshark.org
23. Wu, S., Manber, U.: A Fast Algorithm for Multi-Pattern Seaching. Technical Report TR-94-17, Department of Computer Science. University of Arizona (1994)

transAD: An Anomaly Detection Network Intrusion Sensor for the Web

Sharath Hiremagalore, Daniel Barbará, Dan Fleck,
Walter Powell, and Angelos Stavrou

Department of Computer Science, George Mason University,
Fairfax, Virginia, USA
{shiremag,dbarbara,dfleck,wpowell,astavrou}@gmu.edu

Abstract. Content-based Anomaly Detection (AD) techniques are regarded as a promising mechanism to detect 'zero-day' attacks. AD sensors have also been shown to perform better than signature-based systems in detecting polymorphic attacks. However, the False Positive Rates (FPRs) produced by current AD sensors have been a cause of concern. In this paper, we introduce and evaluate transAD, a system of network traffic inspection AD sensors that are based on Transductive Confidence Machines (TCM). Existing TCM-based implementations have very high FPRs when used as NIDS.

Our approach leverages an unsupervised machine-learning algorithm to identify anomalous packets and thus, unlike most AD sensors, transAD does not require manually labeled data. Moreover, transAD uses an ensemble of TCM sensors to achieve better detection rates and lower FPRs than single sensor implementations. Therefore, transAD presents a hardened defense against poisoning attacks.

We evaluated our prototype implementation using two real-world data sets collected from a public university's network. TransAD processed approximately 1.1 million packets containing real attacks. To compute the ground truth, we manually labeled 18,500 alerts. In the course of scanning millions of packets, our sensor's low FPR would significantly reduce the number of false alerts that need to be inspected by an operator, while maintaining a high detection rate.

1 Introduction

Web applications have become an integral part of our lives providing us a platform to accomplish various essential tasks. For example Internet banking, e-commerce, and e-prescription services exchange personally identifiable information that need to be safeguarded. Many widely deployed out-of-the box web product solutions have become easy and profitable targets for attackers. Attacks on these web based applications have been on the rise [12, 19]. Although existing Network Intrusion Detection Systems (NIDS) play an important role in preventing attacks on web services, the rising number and new types of attacks highlight the need for improved web defenses.

S.S.M. Chow et al. (Eds.): ISC 2014, LNCS 8783, pp. 477–489, 2014.

Industry relies primarily on signature based NIDS [4, 9, 13, 15, 17, 18] to secure their networks. These systems rely on previously known signatures to identify attacks. Without prior known signatures, signature based NIDS are unable to detect new attacks, i.e., "zero-day" attacks. TransAD is one of a class of NIDS that use algorithms based on Anomaly Detection (AD) to combat zero-day attacks. AD sensors analyze traffic and identify deviations from "normal" patterns as potential attacks. However, designing these promising systems presents several challenges. First, implementations generally rely on highly labor intensive labeled training data as a basis for establishing the normal traffic pattern [16]. Second, high False Positive Rates (FPRs) are seen as a major drawback due to the amount of effort required to analyze false positives [14]. Third, AD sensors can potentially be subject to poisoning attacks that intentionally modify the learned normal model to allow various types of attacks [23]. Current AD sensors have failed to completely address all of these issues, and therefore, have not been widely deployed in industry [14, 16].

In this paper, we present transAD – a new self-learning, ensemble based AD system for the web that solves all three problems. Unlike many other AD systems, our system features a completely unsupervised learning algorithm that does not require any labeled training data. Given the volume of data processed by a NIDS, labeling training data for NIDS requires a prohibitive amount of manual effort and is a time consuming process. Additionally, most of the supervised learning algorithms learn a static model of traffic and are unable to adapt over time to changes in the system. Adapting to changing traffic patterns requires additional sets of manually labeled training data at regular intervals for re-training the supervised AD sensor. This additional time and effort required to keep supervised AD sensors updated makes their use impractical in the real world. Consequently, these methods have been heavily criticized [16]. Our method does not use labeled training data and works with network traffic collected without any manual intervention. Therefore, our method does not require any manual effort to label a training data set; and transAD is capable of continuously self-adapting to the changes in the system without additional manual intervention or effort.

In addition to obviating the manual effort and time required for labeling training data sets, our system is designed to generate low FPRs. TransAD does this by using a combination of ensemble of sensors called micro-models, an unsupervised AD algorithm, and our hash based distance metric. Each micro-model uses the unsupervised AD algorithm to identify potential attacks. To identify potential attacks, each packet is evaluated by all of the micro-models independently. Decisions made by individual micro-models in the ensemble are combined using a voting mechanism. The use of an ensemble gives robustness to the determination of attacks (those that are considered anomalous by a majority of micro-models), and serves to reduce the errors, FPR in particular.

It is important to note that the direct application of the comparable AD algorithm for NIDS produces unacceptably high FPRs of approximately 3% [10]. Our AD sensor, on the other hand, yields an average FPR of 0.28% when evaluated using two real world data sets. It is the judicious combination of the AD technique, with ensemble methods, and a proper distance metric that allows us to obtain demonstratively outstanding low FPRs and high detection rates in the evaluation section of this paper. The novelty of our approach comes from the successful combination of the above techniques.

In order to evaluate our system, we conducted experiments using two large data sets collected from a public university's web servers. These data sets contain 1.1 million packets of which thousands were attack packets. To statistically analyze the data from our experimental results, we manually labeled over 18,500 alert packets. Our results demonstrate that we are able to achieve high detection rates, while keeping the FPR very low. More importantly, the reduction in the average number of false alarms, reduces the amount of effort wasted by operators.

The following is a summary of the primary contributions of this work:

- Developed a novel Anomaly Detection based Network Intrusion Detection system for the Web that is based on unsupervised learning and does not require labeled training data.
- Applied a novel combination of our hash based distance metric, unsupervised AD system, and ensemble based techniques to produce outstandingly low false positive rates.

2   Related Work

There are many approaches used for network intrusion detection systems (NIDS). In general, systems are categorized into signature-based and anomaly-based systems. Signature-based systems rely on a database of attack signatures and compare new data against these existing signatures. Many signature-based detectors are available commercially (e.g. CISCO [4], Juniper [9], McAfee [13], Snort [17], Suricata [18], Bro [15]). These approaches are widely used, provide high detection rates, and low FPRs for previously known attacks.

Signature-based systems contrast with anomaly-based sensors which typically approach the problem by creating a normal model of the system, and then detect variations from normal [6]. The advantage of anomaly based systems is the ability to detect previously unseen (i.e. "zero-day") attacks. However, the resulting algorithms frequently suffer from high FPRs making them burdensome for human operators [14, 16].

Anomaly detectors designed as NIDS include the Anagram [23] and PAYL [22, 24] anomaly detection systems. Anagram stores the n-grams, a contiguous sequence of n characters of a given string, in a Bloom filter [2]. Anagram scores new packets by the percentage of the n-grams not seen in the training data. PAYL models the frequency of 1-grams in the "normal" model and compares the frequency distribution of an incoming packet to the training data. Both these algorithms requires clean training data which places a large burden on the operator forcing them to label a large number of packets.

The AD sensor we introduce has its roots in existing machine learning principles for transduction [1]. In transduction, potential attacks are determined by comparing the strangeness of incoming packets to the baseline model. The strangeness represents how different an incoming packet is from the normal baseline model. This process is further described in Section 3.3. The authors in [10] present DNIDS a TCM based IDS that uses a strangeness definition taken from the work of TCM classifiers [21]. In [1] an improved strangeness function is introduced which results in performance improvements of the AD sensor. The KDD 1999 benchmark data set used to evaluate DNIDS is outdated and known to contain attacks that are easily separable [20]. Therefore the evaluation presented in [21] is unreliable. Additionally, DNIDS yields a 3% FPR which is an unacceptably high value for a NIDS.

3   System Architecture

Our AD sensor consists of the following main components: A filtering and a pre-processing stage, an ensemble of micro-models, a voting mechanism, and a drift scheme. The system architecture is shown in Figure 1. At the filtering stage, only packets of interest are allowed to pass through the filter and the rest of the packets are discarded. The pre-processing stage extracts the content of the packet and prepares it for use by the AD algorithm. The pre-processed packets are now used to build the ensemble of micro-models. Each micro-model consists of packets collected for a fixed period of time (epoch) called the micro-model duration. Multiple micro-models are created from sequential epochs to form an ensemble. Each micro-model in the ensemble evaluates each test packet, and their decisions are combined using a voting mechanism.

Initially, our AD sensor starts by collecting a baseline of packets that represents the normal incoming pattern of traffic. TransAD builds ensembles in real time by breaking the baseline data into timed "epochs" called a *micro-model*. Many micro-models are built by collecting packets for multiple epochs to form an ensemble. Once the initial ensemble is ready, new packets are evaluated by the ensemble of micro-models. Every micro-model individually acts as an AD sensor that uses packets contained in the respective micro-model as the normal sample. In order to evaluate if a packet is a potential attack, it is individually evaluated by each micro-model in the ensemble using the *transduction* technique. Individual decisions of each micro-model in the ensemble are combined by a weighted voting scheme to arrive at a final decision. If a packet is voted as an attack, an alert is generated by our sensor.

The final component of our sensor helps keeps our sensor up-to-date with respect to the changes in the monitored environment over time. The normal traffic patterns may change due to changes in the types of services offered and/or changes in the behavior of users. If the system does not adapt to these changes, the detection rate and FPR of the system may be adversely affected. Our sensor adapts to these changes by building new micro-models using tested packets.

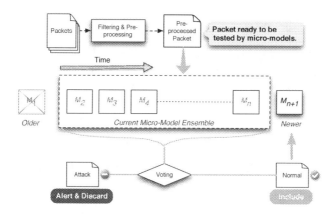

Fig. 1. transAD Architecture

3.1   Bootstrapping transAD

Bootstrapping is the process of building the initial ensemble of micro-models of our AD sensor. Our sensor targets web traffic containing GET requests because they are a common attack vector used by perpetrators. All the GET request packets received by transAD are pre-processed using a filtering and normalization process to enable the AD algorithm to easily identify potential attack packets. Next, these packets are used to build the initial ensemble of micro-models. This is followed by a sanitization process where the micro-model ensemble self-cleanses and removes potential attacks. This sub-section describes the bootstrapping process in greater detail.

Although, our system uses GET requests for evaluation, it can be extended to work with other protocols. POST requests, another popular HTTP method, consist of approximately 0.025% of the total requests received by our web server. Since this makes up a very small portion of the requests seen by our AD sensor, we leave consideration of POST requests for the future.

Filtering and Normalization of GET Requests: As the packets arrive, the filter removes packets other than those containing GET requests with user supplied arguments. Only GET requests with user supplied arguments are allowed through the filter because user arguments is the only part of the request that contains attacks.

Once the packets have been filtered they are normalized. The normalization process involves the following steps: replacing escaped hex characters in GET requests with their respective ASCII characters; discarding numbers in the GET request parameters because strings form the core of a potential attack; and converting the characters into lower case characters to simplify comparison. Normalizing the GET request parameters helps our AD sensor to better discriminate between normal and abnormal packets [3]. Therefore, the filtering and normalization processes are applied to all the packets in the bootstrap phase and also to the packets that are tested during normal operation.

Building Micro-models: Once the packets are normalized, they are ready to be included in a micro-model. Each micro-model is built using the packets collected for a fixed time duration (epoch) called the micro-model duration. Several micro-models are built using packets collected from consecutive, but disjoint, epochs to form an ensemble of micro-models. Since real network traffic is used to build micro-models, they may potentially contain attack packets. In order to ensure we have clean data in our micro-models, we use a sanitization process.

Sanitizing the Micro-models: The sanitization process self-cleanses the initial micro-models without any manual intervention. In the sanitization process, each packet in a micro-model is evaluated by all the other micro-models in the ensemble by using the *transduction* method. The individual decision of the other micro-models in the ensemble are combined using a weighted voting scheme. The weights for each micro-model are in proportion to the number of packets used to build the micro-model. The packets that are voted as potential attacks by the rest of the ensemble are discarded from the micro-model. Only those packets voted as normal are retained in the respective micro-models. This process is repeated for every packet in every micro-model in the initial ensemble.

The sanitization process results in relatively clean data because the majority of the attacks are short-lived, and a given attack is usually limited to a small subset of micro-models. Thus, the sanitization process self-cleans the micro-model by removing packets that are considered anomalous by a majority of the ensemble. Sanitization has also shown to improve the performance in comparable systems [5].

At the end of this bootstrap process, we have an ensemble of filtered, preprocessed packets that are included in **sanitized** micro-models that is ready to diagnose future packets.

3.2 Model Drift

Once the bootstrapping process is complete, we have a sanitized micro-model ensemble. However, the initial ensemble of micro-models may become stale over time and may no longer represent the normal traffic pattern. Micro-models become stale for two reasons: first, changes are made to the services offered within a network; second, the behavior of users interacting with the network changes. As a result, if the ensemble is not updated, it may not conform to the normal traffic pattern and AD sensors could produce more FPs. To adapt to this model drift, we used a scheme where older micro-models are discarded as new micro-models become available.

3.3 Distance Metric and Strangeness

To decide if a packet is a potential attack, our sensor utilizes a technique called *transduction* [8]. The *transduction* method computes the fitness of a test packet with respect to a micro-model. The fitness is computed utilizing a function called **strangeness** that measures the uniqueness (or isolation) of the packet. With respect to the normal packets in each micro-model, the fitness test takes the form of a hypothesis test [25], in which the null hypothesis is that 'the test packet fits the sample distribution.' When the result of the test is statistically significant, we can reject the null hypothesis and therefore consider the test packet a potential attack.

The *strangeness* function can take many forms. In the experiments documented in this paper we use the sum of the distances to a packet's k-nearest neighbors (k-NN). This strangeness function has been shown to work efficiently in [1]. Our sensor uses an improved hash-based distance to measure the distance between two packets. Greater the distance between two packets, the less similar they are. This hash-distance is used to identify a test packet's k-NNs in each micro-model; the strangeness of a test packet is the sum of the hash-distance to the k-nearest neighbors.

In order to compute the nearest neighbors to test packets, we need a suitable distance measure. Since the objective of our implementation is to find abnormal GET request parameter strings, we must use a suitable distance measure that is designed to works with strings. Initially, we tried using Levenshtein string edit distance [11]. However, our experiments using the Levenshtein distance resulted in unacceptably high FPRs. We solved this problem by introducing a hash-based distance that produced a 43% reduction in FPRs when compared to Levenshtein distance.

The hash-based distance works on n-grams, sub-sequences of 'n' characters, of the normalized GET request parameters. A sliding window is used to extract all the n-grams from a GET request. A hash table is created with n-grams of the GET request parameters as the key and packet identifier as the value. FNV1a 32-bit [7] hashing algorithm is used to generate the hash table. We chose FNV1a 32-bit hashing algorithm for its efficiency and low collision rate.

To compute the distance between two normalized GET requests, each of the n-grams of the test request is looked up in the hash table. If the hash bucket that contains that n-gram has the identifier of a micro-model request, then an n-gram match is recorded. The distance is computed by subtracting the total number of n-gram matches normalized by the number of n-grams in the larger GET request from 1 as shown in Equation 1. For example, in Figure 2, the two strings "abcdefg" and "ahbcdz" have one common 3-gram "bcd" that is hashed to the same bucket and is recorded as a match. The number of 3-grams in the larger string is 5. Therefore, the distance between the two strings is 0.8 $(1 - 1/5)$.

$$\text{Distance} = 1 - \frac{\text{total number n-gram matches}}{\text{number of n-grams in the larger string}} \qquad (1)$$

In the example shown in Figure 2, the hash distance performs a simple n-gram match without considering the position of the n-grams in both the requests. Simple n-gram matching is a very relaxed measure of similarity and does not take into account the context of the n-grams present in the two requests; and it does not produce a very accurate representation of distance between the two strings. In order to address this issue, we considered the positions of the n-grams in the request while counting the number of matches.

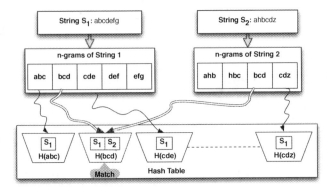

Fig. 2. Hash Distance: A simple n-gram match is shown between two strings

To improve on simple n-gram matching, a relative-distance delta $(r\Delta)$ parameter is introduced to consider a range of possible n-gram positions that should be counted as a match. The simple n-gram matching $(r\Delta = \infty)$ does not consider the n-gram positions. The most restrictive version would require the position of n-grams in the two requests to be the same $(r\Delta = 0)$. If n-gram at position x in the test request has at least one occurrence of the same n-gram in the range of positions $[x - r\Delta, x + r\Delta]$ in the micro-model request, the match count is incremented by one. Matches are considered for all the occurrences of the each n-gram in the test request. In order to eliminate the duplicate n-gram matches, the number of matches cannot exceed the occurrence count of the same n-gram in either of the two requests. Therefore, the number of n-gram matches is defined as the minimum of the following values for the n-gram in question:

- the number of matches counted for each occurrence
- the occurrence count in the test request
- the occurrence count in the micro-model request

The hash distance using the relative position parameter is computed as shown in Equation 1. The sum of the hash distances of the k-nearest neighbors is used to compute the strangeness of the test packet, and the fitness test uses of the strangeness of the test packet to determine if the packet is a potential attack using a hypothesis test.

4   Evaluation

In this section we evaluated the performance of our AD sensor using two real world data sets. Our results show that using transAD yields high detection rates and low FPRs. Additionally, we compared the performance of the transAD sensor with another AD sensor, Sanitization Tool for ANomaly Detection (STAND) [5].

4.1   Data Sets

We used two data sets [1] consisting of 461 million packets to calibrate and evaluate our AD sensor. The details of the two data sets containing network traffic arriving at George Mason University's main web server are shown in Tab. 1.

Table 1. Data sets

	Description	Total Packets	GET Requests	GET Requests w/ User Arguments
Data set 1	13 days in Oct, 2010	223 million	25 million	445,000
Data set 2	14 days in Sep, 2012	238 million	19 million	717,000

In order to evaluate our AD sensor, unique alerts generated by transAD and STAND for the both data sets were manually labeled as attacks or benign packets. To assist in the expert manual inspection of the alert packets, each alert packet's content was compared to attacks seen at honey pot sites and it's source IP was checked against black lists and Offensive IP databases. Manually inspecting the contents of all these alert packets was a time consuming process.

Of the **18,500+** unique alerts generated by both transAD and STAND, **13,443** were unique True Positive (TP) alerts. It was relatively easy to identify TPs, but identifying all of the False Negatives (FNs) was prohibitive as it required analyzing every packet in each data set. In order to estimate the FNs, we identified transAD's FNs as the TPs identified by STAND but missed by transAD. Similarly, STAND's FNs were identified as the TPs identified by transAD but missed by STAND. While this method does not give us a perfect count, it was feasible and represents a lower bound on FNs.

4.2   Parameter Evaluation

In order to study the performance of our AD sensor with respect to the parameters described in Table 2, we conducted experiments to explore the parameter space. The first data set with labeled alerts was used to explore the parameter space. Initially, the parameter values were set to the initial values based on existing literature for comparable AD sensors as shown in the Table 2. Experiments were conducted by varying one parameter at a time while rest of the parameters were set to initial values. When a better parameter was observed, it was noted as the default value (see Table 2). After the default values for all parameters were defined, the experiments were re-run using the default parameters and varying the parameter of interest. The results form these experiments were used to plot Receiver Operating Characteristics (ROC) curves. Each point in a curve is plotted by computing the True Positive Rate (TPR) and FPR at the following confidence levels used for the hypothesis test: 100%, 98%, 95%, 90%, 85%, 80%, 75%, 70%, and 65%. We analyze the ROC curve and the Area Under the Curve (AUC) values to select the default parameters for our experiments. As an example we show how the micro-model duration parameter was selected.

---

[1] An IRB approval was received to collect, store, and conduct experiments.

Table 2. Parameters of transAD

Parameter	Description	Initial Value	Default Value
Number of Nearest Neighbors (k)	'k' is the number of nearest neighbors used by the k-NN algorithm. Based on the results from [1]. we chose 3 for all of our experiments.	3	3
Micro-Model Duration (Δ)	Micro-models used in our sensor are built using packets received in a fixed epoch. The length of this epoch is the duration of each micro-model.	3 h	4 h
n-gram Size (g)	The hash based distance scheme presented earlier uses n-grams to compute the distance between two GET request parameters. This specifies the number of characters in the n-gram.	6	6
Relative n-Gram Position Matching ($r\Delta$)	For n-grams with matching content. $r\Delta$ specifies the range of positions between the n-grams in the request for them to be considered a match.	∞	10
Confidence Level (c)	Confidence level is used by the hypothesis to evaluate if a given packet is normal. Confidence level estimates the reliability of the decisions made by the hypothesis test.	-	80%
Voting Threshold (T)	The voting threshold is the percentage of micro-models in the ensemble that must agree for a packet to be labeled abnormal.	0.66	0.66
Ensemble Size (e)	The number of micro-models used in the ensemble.	25	25
Drift Parameter (r)	The drift parameter denotes the number of old micro-models in the ensemble discarded and the number of staged micro-models inducted into the ensemble at a time.	1	1

Micro-Model duration(Δ): A good micro-model duration needs to be chosen to accurately and fully characterize the normal traffic pattern. Figure 3 shows a magnification of the left hand portion of the Receiver Operating Characteristics (ROC) curve for Δ's between 1 and 5 hours. The FPR considerably improves as micro-model duration is increased and reaches a minimum at 4 and 5 hours, and the AUC reaches a maximum for 4 and 5 hours with the value 0.9989. Further analysis of the graph indicates that our AD sensor has similar performance for Δs of 4 and 5 hours. Since there is no additional improvement in the FPR with a 5 hour micro-model duration, we set the default value of Δ to 4 hours.

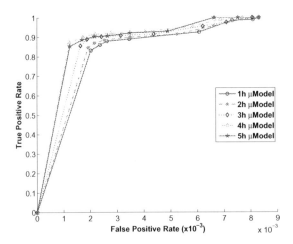

Fig. 3. Magnified section of the ROC Curve for different micro-model duration (Δ)

N-gram Size: We tested the performance of our system with n-gram sizes of 5 through 9. The maximum AUC value of 0.9989 is observed for n-gram size 6. After analyzing the ROC curve, n-gram size 6 is set as the default value.

Relative Position Matching ($r\Delta$): Three relative position matching values were used in the Hash Distance metric: exact position matching ($r\Delta = 0$); $r\Delta$ = 10; and simple n-gram matching ($r\Delta = \infty$). Simple n-gram matching has the maximum AUC value of 0.999 and $r\Delta = 10$ has the next best AUC of 0.9989. Although, $r\Delta = 10$ has a slightly lower AUC compared to simple n-gram matching, it improves our system's defense against poisoning attacks. Therefore, we set $r\Delta$ to 10 as the default value.

Voting Threshold: The performance of transAD was tested with the following voting thresholds: 0.66 (absolute majority), 0.5 (simple majority), 0.4, and 0.3. The AUC is maximum for the absolute majority voting threshold at 0.9989. After analyzing the ROC curves, we use 0.66 as the default voting threshold.

Ensemble Size: We tested our sensor using ensemble sizes 15, 25, and 35. Ensemble size 25 has the maximum AUC value 0.9989. After analyzing the ROC curve, we chose ensemble size 25 as the default parameter.

Model Drift: We examined two different drift settings of 1 and 5. For drift setting 'x', as soon as the 'x' new micro-model(s) are ready, they are included in the ensemble and the 'x' oldest micro-models are discarded. The AUC for both the drift polices have the same value of 0.9989. After analyzing the ROC curve, 1 is selected as the default drift parameter value.

While it is possible to tune the parameters for an individual network as we did, in the future we plan to evaluate our sensor across different networks with the goal of automatically tuning parameters to observed traffic.

4.3 Comparison to Sanitization Tool for ANomaly Detection (STAND)

In this subsection, we compare the detection and false positive rates of transAD with STAND. The TP, FP, TN, and FN counts and rates generated by transAD and STAND for the two data sets are computed. This data is shown Figures 4(a) and 4(b) for transAD and Figures 4(c) and 4(d) for STAND. The number of packets tested by the two algorithms are different. Because STAND requires a smaller baseline than transAD, STAND analyzed more test packets.

For the first data set the TPR (detection rate) and False Positive Rate (FPR) for transAD are **92.78%** and **0.40%** respectively. As presented in the above tables, the TPR and FPR for STAND are **92.52%** and **0.57%** respectively. Using transAD reduces the FPR by 29.82%; the TPR for transAD and STAND were similar.

The TPR and FPR for the second data set for transAD are **94.77%** and **0.15%** respectively; for STAND they are **77.97%** and **0.13%** respectively. Using transAD increases the detection rate by 18% when compared to STAND. The FPR for transAD and STAND were similar. It is interesting to note that for this data set, STAND has a 14.5% lower detection rate when compared to the first data set results. However, transAD performs consistently with over 90% detection rate for both the data sets.

	Packet Count	Rate
True Positives	12,056	92.78%
False Positives	1,125	0.40%
True Negatives	258,758	99.56%
False Negatives	938	7.21%

(a) Results for transAD - data set one

	Packet Count	Rate
True Positives	41,519	94.77%
False Positives	722	0.15%
True Negatives	251,252	99.71%
False Negatives	2,288	5.22%

(b) Results for transAD - data set two

	Packet Count	Rate
True Positives	11,870	92.52%
False Positives	1,475	0.57%
True Negatives	244,278	99.39%
False Negatives	959	7.47%

(c) Results for STAND - data set one

	Packet Count	Rate
True Positives	38,261	77.97%
False Positives	756	0.13%
True Negatives	563,551	99.86%
False Negatives	10,806	22.02%

(d) Results for STAND - data set two

Fig. 4. Results for transAD and STAND

From these two data sets, transAD performed better than STAND. In the first data set where transAD and STAND had comparable detection rates, transAD had a lower FPR. For the second data set where transAD and STAND had comparable FPRs, transAD performed better with a much higher detection rate.

5   Conclusions

We introduced a new AD sensor, transAD, that combines proven technology with new methods to achieve high detection rates and low FPRs. Our sensor consistently yielded good results for two real world data sets. For the two data sets, our AD sensor detected actual attacks; none of which have were previously known to transAD and therefore demonstrated that transAD is able to detect zero-day attacks. Additionally, transAD was shown to be more effective than STAND a leading AD sensor.

TransAD is a new content-based Anomaly Detection sensor for network intrusion detection based on *transduction*. Our AD sensor uses an unsupervised learning algorithm which obviates the need for labeled training data. Therefore, our sensor does not require manual labeling of training data and adapts to changes in the system without additional manual effort.

We conducted a thorough evaluation of our sensor using two real-world data sets. To validate it's performance we manually label over 18,500 alerts. After tuning the parameters used by transAD, we found our sensor achieves consistently high detection rates for both the data sets while having generally lower FPRs than STAND. This unequivocally shows that our AD sensor which is a novel combination of transduction, hash based distance, and ensemble based techniques, produces very low FPRs while achieving high detection rates is suitable for deployment.

References

1. Barbará, D., Domeniconi, C., Rogers, J.P.: Detecting outliers using transduction and statistical testing. In: ACM SIGKDD, KDD 2006, pp. 55–64 (2006)

2. Bloom, B.H.: Space/time trade-offs in hash coding with allowable errors. Commun. ACM 13(7), 422–426 (1970), http://doi.acm.org/10.1145/362686.362692
3. Boggs, N., Hiremagalore, S., Stavrou, A., Stolfo, S.J.: Cross-domain collaborative anomaly detection: So far yet so close. In: Sommer, R., Balzarotti, D., Maier, G. (eds.) RAID 2011. LNCS, vol. 6961, pp. 142–160. Springer, Heidelberg (2011)
4. Cisco: Cisco security products (July 2012), http://www.cisco.com/en/US/products/hw/vpndevc/products.html
5. Cretu, G., Stavrou, A., Locasto, M., Stolfo, S., Keromytis, A.: Casting out demons: Sanitizing training data for anomaly sensors. In: IEEE S&P, pp. 81–95 (May 2008)
6. Denning, D.: An intrusion-detection model. IEEE Transactions on Software Engineering SE-13(2), 222–232 (1987)
7. Fowler, G., Noll, L.C., Vo, P.: Fowler/Noll/Vo (FNV) Hash (1991)
8. Gammerman, A., Vovk, V.: Prediction algorithms and confidence measures based onalgorithmic randomness theory. Theoretical Computer Science 287 (2002)
9. Juniper: Juniper network security products (July 2012), http://www.juniper.net/us/en/products-services/security/
10. Kuang, L.: Dnids: A dependable network intrusion detection system using the csi-knn algorithm. Queen's University (2007)
11. Levenshtein, V.I.: Binary codes capable of correcting deletions, insertions, and reversals. Tech. Rep. 8 (1966)
12. McAfee: McAfee Threats Report: First Quarter 2012 (2012), http://www.mcafee.com/us/resources/reports/rp-quarterly-threat-q1-2012.pdf
13. McAfee: Network intrusion prevention (July 2012), http://www.mcafee.com/us/products/network-security-platform.aspx
14. Patcha, A., Park, J.M.: An overview of anomaly detection techniques: Existing solutions and latest technological trends. Computer Networks 51(12) (2007)
15. Paxson, V.: Bro: A System for Detecting Network Intruders in Real-Time. Computer Networks 31(23-24), 2435–2463 (1999)
16. Sommer, R., Paxson, V.: Outside the closed world: On using machine learning for network intrusion detection. In: IEEE S&P (May 2010)
17. Sourcefire: Snort intrusion detection system (July 2012), http://www.snort.org/
18. Suricata: Suricata intrusion detection (July 2012), http://www.openinfosecfoundation.org/
19. Symantec: Internet Security Threat Report, vol. 17 (2012), http://www.symantec.com/threatreport/
20. Tavallaee, M., Bagheri, E., Lu, W., Ghorbani, A.: A detailed analysis of the kdd cup 99 data set. In: CISDA 2009, pp. 1–6 (July 2009)
21. Vovk, V., Gammerman, A., Saunders, C.: Machine-learning applications of algorithmic randomness. In: ICML 1999, pp. 444–453 (1999)
22. Wang, K., Cretu, G., Stolfo, S.J.: Anomalous payload-based worm detection and signature generation. In: Valdes, A., Zamboni, D. (eds.) RAID 2005. LNCS, vol. 3858, pp. 227–246. Springer, Heidelberg (2006)
23. Wang, K., Parekh, J.J., Stolfo, S.J.: Anagram: A content anomaly detector resistant to mimicry attack. In: Zamboni, D., Kruegel, C. (eds.) RAID 2006. LNCS, vol. 4219, pp. 226–248. Springer, Heidelberg (2006)
24. Wang, K., Stolfo, S.J.: Anomalous payload-based network intrusion detection. In: Jonsson, E., Valdes, A., Almgren, M. (eds.) RAID 2004. LNCS, vol. 3224, pp. 203–222. Springer, Heidelberg (2004)
25. Wasserman, L.: All of Statistics: A Concise Course in Statistical Inference. Springer

Using Machine Language Model
for Mimimorphic Malware Detection

Pinghai Yuan, Qingkai Zeng, and Yao Liu

State Key Laboratory for Novel Software Technology, Nanjing University,
Nanjing 210023, China
Department of Computer Science and Technology, Nanjing University,
Nanjing 210023, China
pinghaiyuan@gmail.com, zqk@nju.edu.cn, storm1945@hotmail.com

Abstract. Binary obfuscation is widely employed by malware distributors against malware analysis and detection. Mimimorphism is a novel technique arising from this arms race and technically opens a new line of obfuscation. It aims to defeat feature-based detection by generating malicious codes sharing features of benign programs. In order to forecast whether mimimorphic malware will burst out in the future, we evaluate current mimimorphic engines by measuring their mimicry outputs, as well as propose an approach to detection. In practice, we build a machine language model by doing a statistical study on code patterns of normal binary and use this model as a foundation for applying machine learning to detect mimimorphic malware. Experimental results show current mimimorphic engines are not powerful enough and mimicry executables can be effectively detected by our approach.

Keywords: Binary obfuscation, Mimimorphism, Machine language model.

1 Introduction

In the continuing arms race between malware distributors and defenders, binary obfuscation and its detection has always been the core problem. As [2] shows, polymorphism is a dominant mechanism to drive up the number of malware variants in common circulation.

On the co-evolution of malware versus anti-malware, encryption, oligomorphism, polymorphism and metamorphism [29] are invented and employed by malware writers. Although these obfuscation techniques make malware instances look different from each other, they can't make these variants look like normal. Byte frequency based detection methods [31] and byte entropy based detection methods [20] have been proposed to detect statistical anomaly code. Moreover, encrypted or compressed code are no longer executable and can be easily classified as obfuscated code by advanced disassembler [19]. Meanwhile, each metamorphic malware variant contains functionally equivalent code. This feature is exploited by semantic analysis techniques [9,35].

S.S.M. Chow et al. (Eds.): ISC 2014, LNCS 8783, pp. 490–501, 2014.
© Springer International Publishing Switzerland 2014

With the advancements in malware detection, Wu et al proposed mimimorphism to binary obfuscation [32]. With instruction-syntax-aware mimic function, mimimorphism can transform a malware executable into mimic-binaries that resembles benign programs in terms of byte frequency statistical properties and control flow fingerprints. Therefore, this novel obfuscation technique has the potential of frustrating both statistical- and semantic-based detections. Moreover, taking into account the complexity of machine instructions and program structures, there is a large space for optimization and mimimorphism enhancement. Technically, Wu's work opens a new line of obfuscation.

This work tries to give an answer to the prospect of mimimorphism. From the view of "code pattern", we evaluate current mimimorphic engines by measuring the distance between their output and normal programs. Here, "normal" means codes match patterns we observed in common binaries. In practice, we do a statistical study on codes of open source files and use n-gram language modeling algorithm to digest all human-observable and unobservable code patterns. This means the model contains instruction sequence level profiles of normal binaries. Consequently, we obtain a machine language model representing the patterns. With the help of this model, we further employ machine learning to detect mimimorphic malware.

We also note that many literatures propose detection techniques based on statistical study of n-gram byte sequences or n-gram opcodes [29,26,27,24]. However, our approach makes good use of semantic information while those solutions make "limited" use or even ignore. Meanwhile, there are many works employing machine learning techniques for malware detection [28,14]. But their scoring schemes are very different from ours.

Our experimental results show that more than 10% malicious codes is not successfully camouflaged by current mimimorphic engines and mimicry executables can be effectively detected with the help of a machine language model.

The contributions of this paper are summarized as follows:

- A machine language model for normal binaries. A model represents instruction sequence features of normal binary. Hence, it is a combination of statistical study and semantic analysis on binary.
- A method to evaluate the strength of mimimorphic engine. Both theories of mimimorphism and experimental results show that the strength of current mimimorphic engine can be evaluated with few mimicry samples.
- An approach to detect mimimorphic malware. With the machine language model as a foundation for scoring systems, we employ machine learning to detect mimimorphic malware.

The remainder of this paper is organized as follows. Section 2 discusses the background and challenge of our solution. In section 3, we present our machine language model and propose the methodologies for evaluating mimimorphic engine and detecting mimicry executables. We show our experimental results in section 4 and discuss them in section 5. Section 6 overviews the related work. Section 7 concludes the paper.

2   Background and Challenge

In this section, we introduce the concept of Machine Language Feature, summarize mimimorphism and highlight the challenges of this work.

2.1   Code Pattern

We use this term to represent all code patterns in common binaries, including all human-observable and unobservable ones. To illustrate this concept, let us fill the 3-gram instruction combination *"push 10h / ? / call sub_foo"*. As we all know, there are many choices. However, experience tells us that a "push eax" instruction makes the combination looking like normal while either a "ret" or a privileged instruction makes it abnormal. Here, "normal" means the combination satisfies some patterns we observed in practice. In this work, we depict code patterns with a model. In the following, we use *Machine Language Feature* instead of *code pattern*.

2.2   Mimimorphism

Mimimorphism aims to circumvent existing statistical and semantic detection techniques. It tries to generate variants that share statistical properties and semantic features of normal binaries.

According to current implementation [32], mimimorphism works like a cryptography system with a code book representing instruction sequence level features of normal programs. In fact, it has a digesting phase in addition to the encoding and decoding phases. In the digesting phase, a high-order mimic function constructs a collection of Huffman trees for instruction sequence level features of the benign programs. In the encoding phase, the mimic function applies the Huffman decoding operation on the randomized input data, and produces the mimicry executables by referring to the Huffman tree. In the decoding phase, the mimic function applies the Huffman encoding operation, referring to the same Huffman tree, and uncovers the original input data from the mimicry executables.

2.3   Challenge

As mimimorphism studies instruction sequence level features of benign programs and applies them when coding and decoding mimicry executables, it is necessary to study these features in our approach if we want to evaluate the effect of mimimorphism. Hence, we build a machine language model for normal binaries. It uses weight value (probability) to denote whether the occurrence of an instruction sequence is frequent or rare. However, it is a challenge to build the model as many factors should be considered, including which parts of an instruction are extracted, a representative file set for studying and an efficient modeling algorithm.

3   Our Approach

3.1   Overview

Our approach is based on statistical machine language processing and supervised machine learning as shown in figure 1. Our solution consists of three major components: i) input processor disassembles binary programs or suspicious data; ii) feature extractor reveals distinguishable features from the instruction sequence and outputs its corresponding data representation named *infocode* sequence; iii) trainer utilizes n-gram language model to build a model of normal programs, and utilizes support vector machines to train detection and classification models based on training set and classifiers that use trained models to detect and classify suspicious data.

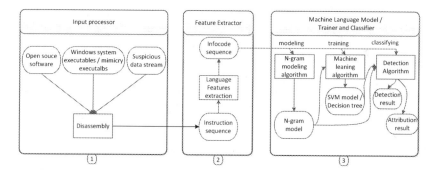

Fig. 1. Approach overview

3.2   Disassembly

The first step of our approach is to disassemble binary, a method that extracts semantics out of machine code and outputs human-readable disassembly. Two basic disassembly algorithms are widely adopted, i.e. linear sweep disassembly and recursive descent disassembly. However, both of them are not enough to disassemble mimicry executables. Mimimorphism blurs the boundaries of functions thus hinder linear sweep disassembly. Meanwhile, because of mimimorphic engine's poor generation algorithm for immediate operands, recursive descent disassembly may lose clues for next passes of disassembling. We realize that heuristic analysis methods are very useful in such situations, and develop a plugin based on IDA [1].

3.3   Machine Language Feature

Generally, machine language features represent statistical and semantic properties of normal binaries, and we use infocode standing for semantic information extracted from instruction.

Typically, infocode consists of opcode, register index and operand type, size, and access mode if they exist; immediate values are excluded out. Such a definition is consistency with code generation techniques in compiler theory. After all, instruction set and registers severely affect the compiling result. Moreover, in logical, immediate values of instructions can be symbolized, because they always represent symbols/labels/constants in programs. Consequently, immediate values are excluded out as we don't distinguish different symbols. We also ignore register indexes to simplify the situation introduced by register substitution. Table 1 gives an example to illustrate the definition.

Table 1. A demo for extracting infocode

machine code	assembly code	info code
E8 50 FE FF FF	call sub_4AD05F31	call_I:d4
6A 10	push 10h	push_I:d1
68 D0 61 D0 4A	push offset word_4AD061D0	push_I:d4
E8 23 C1 FF FF	call sub_4AD02210	call_I:d4
33 DB	xor ebx, ebx	xor_RR:r4;r4

3.4   Machine Language Model

For simplicity, we use n-gram machine language modeling algorithm. Such an algorithm scans infocode sequence with a sliding window of size n. Hence, it picks up a series of n-gram combinations. In order to build a model, we feed the algorithm with a big data set. In order to evaluate the quality of the set, we further check whether the frequencies of infocode combinations is consistent with Zipf's Law [21]. This law states that given some corpus of language utterances, the frequency of any word is inversely proportional to its rank in the frequency table. As show in figure 2. Note that, Mandelbrot's formula is an extension of Zipf's law.

Statistically, the modeling algorithm accumulates the occurrences of all combinations and translates them into a series of probability values. Thanks to the smoothing techniques, all possible combinations are taken into account and assigned a non-zero probability [8].

3.5   Attributing Code Snippet and Classifying

Each code snippet $C = w_1 \cdots w_n$ is assigned a pair of attributes $(mean, var)$, which are defined as follows:

$$mean(C) = \frac{1}{n} \sum_{i=1}^{n} ln \frac{1}{p(w_i|w_{i-2}w_{i-1})} \tag{1}$$

$$var(C) = \frac{1}{n} \sum_{i=1}^{n} (ln \frac{1}{p(w_i|w_{i-2}w_{i-1})} - mean)^2 . \tag{2}$$

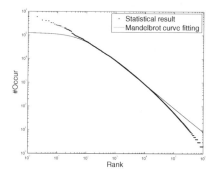

Fig. 2. The rank of occurrences of 3-gram combinations. Also, a fitting curve of Mandelbrot's formula is provided for comparison.

In order to label a snippet, we train a simple classifier Φ:

$$\Phi(\delta(mean, var), \theta) \mapsto \{NORMAL, ODD\}.$$

Here, δ is the distance function of Φ and takes the properties of snippet as parameters. Meanwhile, θ is the threshold for decision. Nominal properties "NORMAL" and "ODD" labels all possible snippets into two categories.

Naturally, function (routine) is code snippet. In fact, all code snippets used in our experiments are functions identified with IDA. We mix the use of code snippet and function as it does not introduce confusion.

3.6 Evaluating the Strength of Mimimorphic Engine

We carry out this work by evaluating the similarity between mimicry codes and normal binaries from the view of machine language features. Due to the power of mimimorphism, most mimicry codes look like normal. But when evaluating the strength of engine, we focus on the percentage of abnormal codes. In practice, we can transform it to evaluate the percentage ODD functions of mimicry samples. That is:

$$p_{ODD}(f) = \frac{\sum_{f \in Fun} |\Phi(f) \; labeled \; as \; ODD|}{|Fun|} \; .$$

The smaller the percentage, the more powerful the engine.

According to the theory and implementation of mimimorphism [32], the input of mimimorphic engine is randomized data, so the percentage of "ODD" functions subject to some statistical distribution. From the view of statistics, malware variants over 3Mb in size own a relatively stable percentage of abnormal code. Our experimental results support this assumption.

3.7 Detecting Mimicry Executables

Codes of a mimicry executable can be decomposed into many functions. We can also label these functions into two categories $\{NORMAL, ODD\}$ with our classifier. We also know that very few functions of normal binary are labeled as "ODD" mainly caused by disassembler fail to distinguish data mixed in code. Therefore, we detect mimicry executables by setting up a classifier to check p_{ODD} against a threshold value.

4   Experimental Evaluation

We first discuss the data collection and implementation details of our proof-of-concept system. Then, we present the mimimorphic malware detection and attribution results.

4.1   Data Collection and Implementation

We have three kinds of file sets: open source software for building Machine Language Model; Windows system programs and mimicry executables for training and classifying.

The generality of the machine language model is the primary element we considered. The model should be as irrelevant as possible with the origin of binary and compiler which generates the binary. To get a set of representative binaries, we collect 4552 PE files (1293 EXEs, 3256 DLLs and 3 SYSs) from open source software downloaded from Sourceforge [5]. Moreover, we need extra data for training a code snippet classifier, evaluating the strength of mimimorphic engine and training mimicry malware classifier. These data are extracted from two file sets. One is Windows system executables and the other is mimimorphic executables. Table 2 lists information about the file sets. As a side note, mimicry executables consist of two subsets downloaded from Wu's homepage. And we encapsulate them as executables. The last column gives the number of functions (unique in MD5). To make each function unique is very important for avoiding over-fitting code patterns of library functions. As referred above, all these binaries are disassembled for extracting infocode sequences.

Table 2. Information of file sets

file set	#file	#function (MD5 unique)
Open-source software	4552	589120
Windows executable	339	971079
Mimicry executable	5 + 5	29053 + 18508

In our experiments, we use IDA [1] for disassembling, MySQL [4] for data management, CMU-SLM [10] for building language model, Matlab [3] for data visualization and data processing, and Weka [15] (includes LIBSVM [7]) for training code snippet classifier.

4.2 Machine Language Model

We build a 3-gram machine language model with help of CMU-SLM. As the modeling algorithm is well recognized, we do not detail it. In fact, we place more interest in the properties of the data set (called corpus in this context) fed the modeling algorithm. In summary, this corpus has 2710 words and 589120 unique functions. And its 1802656 3-gram infocode combinations have 74376117 tokens. As show in figure 2, this corpus satisfies Zipf's law, thus can reflect the statistical properties of benign programs.

This model scores a given code sequence. For instance, the probability of $p(call\_I : d4; push\_I : d1; push\_I : d4) = 0.0701315 \times 0.0149673 \times 0.0822951$.

4.3 Attributing Code Snippet and Classifying

We illustrate the attribution analysis technique from several different aspects including visualization, data comparison. We use data set "sysexe" and "mimi8" in this experiment, each of them has 6000 and 29053 functions respectively. "mimi8" is functions of 8_{th} order mimicry executables. Notes that the 6000 functions of "sysexe" are randomly selected from the file set "Windows executables". We do this for three reasons: i) the original data set is too big (971079 functions); ii) make a difference between those used in detecting mimicry executable; iii) 6000 is almost the number of functions extracted from each $8_{th} order$ mimicry executable.

Data Visualization
The visualization of data samples could tell us the differences of code snippet between normal files and mimicry executables in an intuitive way. As the left subfigure 3 shows, mimimorphic engine fails to camouflage part of the data.

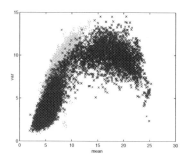 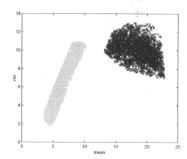

Fig. 3. An overview of data for attributing and classifying. Each circle and cross represent an element from *sysexe* and *mimi8* respectively.

Function Classifier
To train a function classifier, we select 5357 and 2250 samples which are labeled as NORMAL and ODD respectively. The right subfigure 3 illustrates it. Then we calculate the percentage of odd function of each category. The respective number

of ODD functions are 49 and 2979, which are 0.82% and 10.2% in percentage.
Note that the percentage values have a dramatic difference.

4.4   Evaluating the Strength of Mimimorphic Engine

We also evaluate functions of 7_{th} order mimicry executables. The result of classifying code snippets of mimicry executables is listed in table 3. This table shows that 8_{th} order engine is stronger than 7_{th} order engine, thus has low percentage of ODD function. The percentage of ODD function in each sample of a specific engine is relatively stable.

Our conclusion to this result is that: current mimimorphic engines are not powerful enough; their strength for mimicking can be evaluated with some samples.

Table 3. Evaluate the percentage of abnormal code snippets in mimicry executables and normal binaries. The first 5 and last 5 files are 8_{th} and 7_{th} order mimicry executables respectively.

mimicry executable	#function	#oddfun	percentage
ETest1004161217	5780	636	10.9862 %
ETest1005249448	5905	570	9.6528 %
ETest1041411991	6060	636	10.495 %
ETest1056019325	5632	551	9.7834 %
ETest1074220180	5676	578	10.1832 %
ETest18013734	3885	466	11.9949 %
ETest52991750	3438	446	12.9727 %
ETest57530501	4154	474	11.4107 %
ETest57658370	3600	481	13.3611 %
ETest85382868	3431	438	12.766 %
Detection results of normal binaries			
calc.exe	1492	1	0.067 %
cmd.exe	421	0	0
explorer.exe	3234	2	0.0618 %
notepad.exe	122	0	0
regedit.exe	388	0	0
taskmgr.exe	370	0	0

4.5   Detecting Mimicry Executables

Our decision tree to classify mimicry executables works as follows:

if *percentage of odd function is less than* γ:

　f is normal program;

else:

　f is mimicry executable;

Here, f is a binary file we need to classify and γ is a threshold value. According to percentages of ODD functions shown in table 3, we simply set γ to 5%.

We apply this decision tree to distinguish mimicry executables and some Windows applications. As show in table 3, we can correctly identify them.

5   Discussion

In order to find the reason why some functions are labeled as *ODD*, we analysis dozens of *ODD* functions. Combining the results of our experiments, we make the following explanations.

- Odd functions of normal programs. It mainly due to the disassembler incorrectly identify the boundary of a function. For example, there is an instance ended with a "pusha" instruction. As this instruction save registers on the stack, we hardly believe it is a normal function. However, the ratio of ODD functions is no more than 1%.
- Odd functions of mimicry executables. These functions usually contain privileged instructions, which are illegal in user-mode applications. Therefore, we deduce that code bytes of these functions do not look like normal and mislead the disassembler.
- The percentage of odd functions of mimicry executable seems stable. Because mimimorphic engine is fed with randomized byte streams to generate mimicry payload, thus the payload may exhibit some statistical features.

In summary, we can detect almost all mimicry executables generated by current engines with the help of machine language model.

6   Related Work

On the co-evolution of malware versus anti-malware, binary obfuscation techniques are extensively employed by malware writers to evade signature-based detection systems [29,33,25]. Moreover, many polymorphic and metamorphic engines, such as ADMmutate [12] and Mistfall [34], are available for novices to generate mutable malicious variants. Also, new framework of metamorphic engine was proposed [22]. Virtualization-based obfuscators, such as VMProtect [6], are also employed by malware distributors.

Since obfuscation techniques focus more on making variants look different from each other than making them look like normal, anomaly-based approaches to malware detection can provide good defense [28,16,30,31,20,9]. Recently, semantic analysis methods extracting higher-level information about executable receive more and more attentions [18,35,13,9,17].

However, Wu et al introduced a novel binary obfuscation mimimorphism [32]. Mimimorphic malware employs instruction-syntax-aware high-order mimic functions to encode its binaries into mimicry executable. Thus, it shows a potential of evading both statistical anomaly detection and semantic analysis detection techniques under static circumstance.

7   Conclusion

To answer the prospect of mimimorphism, we evaluate the power of mimimorphic engine and propose an approach to detect mimicry executables. Our solution is based on the concept of machine language model, which does statistical study on normal binary to learn all human-observable and unobservable code patterns. Thus, it can be seen as a combination of statistical study and semantic analysis. The model provides a foundation for applying machine learning techniques to detect mimimorphic malware. Experiment results demonstrate current implementation of mimimorphism is not powerful enough to camouflage malicious code.

Acknowledgments. This work has been partly supported by National NSF of China under Grant No. 61170070, 61321491; National Key Technology R&D Program of China under Grant No. 2012BAK26B01 and National High Technology Research and Development Program of China under Grant No.2011AA1A202.

References

1. IDA, http://www.hex-rays.com/products/ida/index.shtml
2. Malicious code trends 2011 (2011), http://www.symantec.com/threatreport
3. Matlab, http://www.mathworks.com/products/matlab
4. MySQL, http://www.mysql.com/
5. Sourceforge, http://sourceforge.net
6. VMProtect, http://vmpsoft.com/
7. Chang, C.C., Lin, C.J.: LIBSVM: A Library for Support Vector Machines
8. Chen, S.F.: An empirical study of smoothing techniques for language modeling
9. Christodorescu, M., Jha, S., Seshia, S.A., Song, D.X., Bryant, R.E.: Semantics-aware malware detection. In: IEEE Symposium on Security and Privacy, p. 32–46
10. Clarkson, P., Rosenfeld, R.: Statistical language modeling using the CMU-cambridge toolkit. In: Proceedings of EUROSPEECH, vol. 97, pp. 2707–2710
11. Intel Corportation: IA-32 intel architecture software developer's manual
12. CREW, A.: Admmutate: A polymorphic shellcode engine, can evade NIDS, http://www.ktwo.ca/readme.html
13. Feng, H.H., Kolesnikov, O.M., Fogla, P., Lee, W., Gong, W.: Anomaly detection using call stack information. In: Proceedings of the 2003 IEEE Symposium on Security and Privacy (2003)
14. Gavrilut, D., Cimpoesu, M., Anton, D., Ciortuz, L.: Malware detection using perceptrons and support vector machines. In: Future Computing, Service Computation, Cognitive, Adaptive, Content, Patterns (COMPUTATIONWORLD 2009), pp. 283–288 (2009)
15. Hall, M., Frank, E., Holmes, G., Pfahringer, B., Reutemann, P., Witten, I.H.: The WEKA data mining software: An update 11(1), 10–18
16. Idika, N., Mathur, A.P.: A survey of malware detection techniques, p. 48
17. Kinder, J., Katzenbeisser, S., Schallhart, C., Veith, H.: Detecting Malicious Code by Model Checking

18. Kruegel, C., Kirda, E., Mutz, D., Robertson, W., Vigna, G.: Polymorphic worm detection using structural information of executables. In: Valdes, A., Zamboni, D. (eds.) RAID 2005. LNCS, vol. 3858, pp. 207–226. Springer, Heidelberg (2006)
19. Kruegel, C., Robertson, W., Valeur, F., Vigna, G.: Static disassembly of obfuscated binaries. In: Proceedings of USENIX Security, pp. 255–270
20. Lyda, R., Hamrock, J.: Using entropy analysis to find encrypted and packed malware 5(2), 40–45
21. Manning, C.D., Schütze, H.: Foundations of statistical natural language processing. MIT Press
22. Mohan, V., Hamlen, K.W.: Frankenstein: Stitching malware from benign binaries. In: WOOT, pp. 77–84
23. Moser, A., Kruegel, C., Kirda, E.: Limits of Static Analysis for Malware Detection
24. Moskovitch, R., Feher, C., Tzachar, N., Berger, E., Gitelman, M., Dolev, S., Elovici, Y.: Unknown malcode detection using OPCODE representation. In: Ortiz-Arroyo, D., Larsen, H.L., Zeng, D.D., Hicks, D., Wagner, G. (eds.) EuroISI 2008. LNCS, vol. 5376, pp. 204–215. Springer, Heidelberg (2008)
25. Nachenberg, C.: Computer virus-coevolution 50(1), 46–51
26. Santos, I., Penya, Y., Devesa, J., Bringas, P.: N-grams-based file signatures for malware detection. In: Proceedings of the 11th International Conference on Enterprise Information Systems (ICEIS), vol. AIDSS, pp. 317–320
27. Santos, I., Sanz, B., Laorden, C., Brezo, F., Bringas, P.G.: Opcode-sequence-based semi-supervised unknown malware detection. In: Herrero, Á., Corchado, E. (eds.) CISIS 2011. LNCS, vol. 6694, pp. 50–57. Springer, Heidelberg (2011)
28. Siddiqui, M., Wang, M.C., Lee, J.: A survey of data mining techniques for malware detection using file features. In: Proceedings of the 46th Annual Southeast Regional Conference on XX, ACM-SE 46, pp. 509–510. ACM
29. Szor, P.: The art of computer virus research and defense. Addison-Wesley Professional
30. Teufl, P., Payer, U., Lackner, G.: From NLP (natural language processing) to MLP (Machine language processing). In: Kotenko, I., Skormin, V. (eds.) MMM-ACNS 2010. LNCS, vol. 6258, pp. 256–269. Springer, Heidelberg (2010)
31. Wang, K., Parekh, J.J., Stolfo, S.J.: Anagram: A content anomaly detector resistant to mimicry attack. In: Zamboni, D., Kruegel, C. (eds.) RAID 2006. LNCS, vol. 4219, pp. 226–248. Springer, Heidelberg (2006)
32. Wu, Z., Gianvecchio, S., Xie, M., Wang, H.: Mimimorphism: A new approach to binary code obfuscation. In: Proceedings of the 17th ACM Conference on Computer and Communications Security, CCS 2010, pp. 536–546. ACM (2010)
33. You, I., Yim, K.: Malware obfuscation techniques: A brief survey. In: Int. Conf. on Broadband, Wireless Computing, Communication and Applications, pp. 297–300
34. Z0mbie: Automated reverse engineering: Mistfall engine,
    http://vxheavens.com/lib/vzo21.html
35. Zhang, Q., Reeves, D.: MetaAware: Identifying metamorphic malware. In: Twenty-Third Annual Computer Security Applications Conference, ACSAC 2007, pp. 411–420 (2007)

CodeXt: Automatic Extraction of Obfuscated Attack Code from Memory Dump

Ryan Farley and Xinyuan Wang

Department of Computer Science
George Mason University
Fairfax, VA 22030, USA
{rfarley3,xwangc}@gmu.edu

Abstract. In this paper, we present *CodeXt*—a novel malware code extraction framework built upon selective symbolic execution (S2E). Upon real-time detection of the attack, CodeXt is able to automatically and accurately pinpoint the exact start and boundaries of the attack code even if it is mingled with random bytes in the memory dump. CodeXt has a generic way of handling self-modifying code and multiple layers of encoding, and it can automatically extract the complete hidden and transient code protected by multiple layers of sophisticated encoders without using any signature or pattern of the decoder. To the best of our knowledge, CodeXt is the first tool that can automatically extract code protected by Metasploit's polymorphic xor additive feedback encoder Shikata-Ga-Nai, as well as transient code protected by multi-layer incremental encoding.

Keywords: Malware Forensics, Binary Analysis, Symbolic Execution.

1 Introduction

Automatically recovering malware attack code is critical to improving effective malware analysis, forensics, and reverse engineering. Existing methods involve substantial manual effort. Given the sheer number of new malwares that arrive every year, it would be invaluable to be able to automate recovery (from the run-time memory) upon detection of an attack.

First, we must automatically pinpoint the exact start and boundaries of the attack code, possibly spread across disjoint segments and intermingled in random surrounding data and/or code bytes within memory, per Figure 1. Second, the attack code can be easily obfuscated with self-modification such as encoding and packing which renders static analysis ineffective. Third, the attack code may never reveal its complete unpacked version in run-time memory at any singular moment, such as multiple layers of decoding or unpacking where each layer only extracts a portion of the final attack code, such as in Figure 2. Such an incremental decoding makes it very difficult to automatically recover the complete attack code even if one can dump run-time memory at any time.

A number of approaches [26,9,31] detect attack code in network traffic. These methods can detect some snippets of code from the packet data, but neither determine the exact start, boundaries, nor recover the complete attack code.

S.S.M. Chow et al. (Eds.): ISC 2014, LNCS 8783, pp. 502–514, 2014.

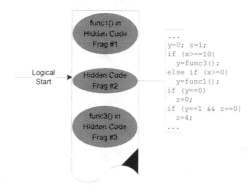

```
        ...
        y=0; z=1;
        if (x>==10)
            y=func3();
        else if (x>=0)
            y=func1();
        if (y==0)
            z=0;
        if (y==1 && z==0)
            z=4;
        ...
```

Fig. 1. Multiple Disjoint and Misaligned Code Fragments Mingled with Random Bytes

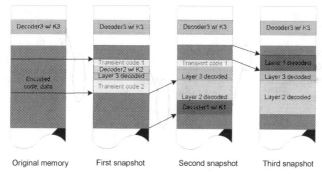

Fig. 2. Transient Code with Multiple Layer of Self-Modifying Code

Existing dynamic analysis based unpacking approaches (e.g., PolyUnpack [21], Renovo [14], OmniUnpack [19]) recover hidden code from packed executables, and thus assume knowledge of the exact starting point and are not effective when the packed byte code is mingled with random data or code. Traditional dynamic analysis approaches normally cover only one execution path and may miss hidden code and data in other (unexplored) paths. To the best best of our knowledge, no existing unpacking method has been shown to be able to recover the complete hidden code protected by the incremental encoding we present is this paper. Since existing software emulators (e.g., QEMU, S2E [10]) do not have full support of FPU instructions, they can neither execute nor recover the code protected by encoders (e.g., Shikata-Ga-Nai [2]) that use FPU instructions. This is the reason why SHELLOS [24] abandoned emulation for hardware virtualization.

In this paper, we seek to advance the current capability of automatic attack code extraction from run-time memory. We present *CodeXt*—a novel malware forensics framework based on selective symbolic execution (S2E) [10]. CodeXt uses two key techniques to achieve unprecedented capability in automatic attack code recovery: 1) combination of concrete and symbolic execution to recover potentially disjoint, misaligned, self-modifying code from all execution paths within a given memory range; 2) intelligent memory update clustering and multi-layer snapshots to recover all the code fragments of incremental decoding.

We have empirically validated the effectiveness of CodeXt with real world attack code and 9 well-known third party encoders (e.g., Shikata-Ga-Nai [2]) as well as 3 encoders (e.g., multi-layer incremental encoding) developed by ourselves. CodeXt is able to accurately locate the attack code that is mingled with random bytes and extract the complete (including transient) code hidden by all of the 12 encoders we have tested. To the best of our knowledge, CodeXt is the first tool that can automatically extract the code protected by Metasploit's polymorphic xor additive feedback encoder Shikata-Ga-Nai and the transient code protected by multi-layer incremental encoding.

2   Overview

Our approach does not seek to determine if a given piece of code is malicious or not, but rather extract hidden attack code from run-time memory upon real-time detection of malware or attack. CodeX thas the following advantageous features. 1) It can automatically identify the exact start and boundaries of all hidden code fragments, even if they are mingled with random data in the run-time memory dump. 2) It can automatically recover the complete attack code, including transient code, protected by sophisticated self-modifying code such as multi-layer incremental encoding and/or packing with overlapped ranges and different keys. 3) It can automatically collect relevant intermediate results during multi-layered decoding, revealing obfuscations used at each layer. 4) It can merge all hidden code fragments into logically related collections. 5) It can recover the complete attack code protected by advanced polymorphic encoders that typically evade emulation, such as those that use FPU instructions or self-modify the current basic block of the run-time decoder. 6) It can validate the extracted hidden code via symbolic execution to verify that execution of extracted hidden code will lead to any detection conditions reported by the intrusion or malware detection system. 7) It is generic and does not rely on any signature or pattern of any particular decoder.

We assume there is some intrusion or malware detection system that can detect the execution of attack code in real-time (e.g., [18,30]) and it will dump the memory around the instruction (e.g., system call) where the attack has been detected and other attack context information. Since many intrusion detection systems (e.g., [13,32,22,28,12,18,30]) use system calls to detect an attack, we assume the attack context information includes some system call triggered by the attack code and corresponding register values. We further assume the dumped memory is large enough to contain all hidden attack code present in the run-time memory when the attack was detected. To avoid the undecidability problem in unpacking determination [21], we assume there is no infinite loop in the the attack code and our system will terminate after a configurable maximum number of instructions have been executed.

2.1   Overall CodeXt Architecture

CodeXt uses a combination of symbolic and concrete execution during analysis. Symbolic execution allows CodeXt to pinpoint the exact code start and boundaries

Fig. 3. Overall CodeXt Architecture

by exploring all the legitimate execution start points and paths. On the other hand, concrete execution enables CodeXt to handle potential dynamic binary transformation and self-modifying code. We choose to build CodeXt upon Selective Symbolic Execution (S2E) [10] which supports in-vivo multi-path analysis and allows us to execute any basic block either concretely with QEMU or symbolically with KLEE [8].

Figure 3 shows the overall architecture of CodeXt with an online and offline component. The online component consists of a number of S2E plugins which can monitor, track, and direct the selective symbolic execution of any given byte stream by exploring all execution paths from all offsets. It filters out impossible code snippets (e.g., invalid instruction, invalid memory access) and records those that are feasible and satisfy the attack context information given. To handle self-modifying code, CodeXt detects and records all instructions dynamically generated before execution and takes snapshots of each layer of self-modification outputting intermediate results. The offline component further analyzes the online results to derive the hidden code's start and boundaries.

3   Design and Implementation

How to Locate Hidden Code? We need to determine the existence of, exact start, and the boundaries of any hidden code from a given memory dump. The hidden code is usually mingled with random data/code. These constraints are different from that of traditional unpacking tools (e.g., PolyUnpack [21], Omni-Unpack [19] and Renovo [14]) which assume the execution start point is already known. We need to treat every offset in the memory dump as a possible logical start, or entry point, of the hidden code we are looking for. To leverage the system call information from the IDS, we have developed a S2E plugin to catch all the system calls triggered from within a given memory dump.

To reliably locate the logical start of the attack code in the memory dump, we use S2E and make the offset of the memory dump symbolic. This also employs S2E's efficient built-in state forking. To avoid unnecessary symbolic execution, we have the following online kill conditions that immediately terminate an offset's execution: exception due to an invalid instruction; invalid memory access such as a segmentation fault; any instruction does not align to the system call we know; detected system call number or address does not match given context from the IDS; execution of end of path system calls (e.g., exit, exec); and, jumps out of bounds of the memory buffer (we assume it contains the complete attack code).

Because any application level attack code must execute one or more segments of privileged code (i.e., system calls) to cause any real harm, we record the

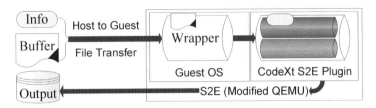

Fig. 4. Wrapper to Run Arbitrary Byte Stream in CodeXt

symbolically executed instructions that end with a system call as a *code fragment* for each starting offset. We leverage research from [5] and assert that valid fragments should have at least 6 instructions and contain at least 15 bytes. To model code with multiple system calls, we define a *code chunk* as a sequence of code fragments in a control flow. To extract code with multiple system calls, we merge adjacent code fragments into a *code chunk* if the following conditions are met: each code fragment itself is a code chunk; code chunks X and Y are adjacent if the start of Y is immediately after the logical end of X; and, if code chunks X and Y are adjacent and the end of X is not an end of path system call (e.g., exit, exec). This process may generate a number of code chunks, subsets of others are eliminated.

How to Handle Self-modifying Code? In order to recover transient code involved in multiple layers of self-modification, we need to take snapshots for each layer of decoding. Since a defining characteristic of self-modifying code is executing dynamically generated instructions, we can reliably identify it if any instruction consists of bytes written by the code under observation. This can be achieved by tracking all the memory updates within the memory buffer range at run-time. However, we do not want to take a snapshot for each dynamically generated instruction as one layer of decoding normally consists of multiple correlated instruction blocks (e.g., strcpy). Instead we developed a clustering based approach for obtaining appropriate snapshots of self-modifying code.

We maintain a global counter of all the instructions executed, and we assign the current global counter to each to be executed instruction as its unique sequence number, which reflects the temporal order of the execution of all instructions. The memory updates within one layer of decoding tend to be clustered to each other in both time and space. Our heuristic clusters writes from memory update instructions whose execution sequence numbers are no more than Δ (e.g., $\Delta = 10$) apart and combines clusters with adjacent memory update ranges.

We treat one cluster of writes as one snapshot. We mark those snapshots from which we executed any instructions after the snapshot was created. These marked snapshots correspond to each layer of self-modifying code executed.

By stringing the snapshots together, we can generate a memory map to show the changes over time. Specifically, we can see all the values of all memory bytes translated, executed, or written, even if the same memory location has been overwritten multiple times during the execution.

Table 1. Accuracy and speed for searching for the start of hidden code within a buffer

Surrounding Type	Run-time Hints	Code Found?	Sec. per Offset
Nulls	EIP, EAX	Yes	0.92
	EIP	Yes	0.94
	EAX	Yes	0.98
	Neither	Yes	0.98
Random	EIP, EAX	Yes	1.08
	EIP	Yes	1.09
	EAX	Yes	1.13
	Neither	Yes	1.11
Captured	EIP, EAX	Yes	1.04
	EIP	Yes	1.08
	EAX	Yes	1.00
	Neither	Yes	1.09

Implementation. We have implemented the prototype of CodeXt upon the
S2E engine with 4,006 lines of C/C++ code. We have also extended the functionalities of S2E, KLEE and QEMU with 444 lines of C/C++ code. We have
chosen to use the Strictly Consistent System-level Execution (SC-SE) consistency model that is "both strict and complete" [10]. Currently our prototype
only monitors Linux system calls but it can be extended to monitor Windows
system calls. Our prototype consists of a wrapper, shown in Figure 4, for loading
an arbitrary byte stream for execution, a S2E plugin for online analysis, and a
number of offline analysis modules. Our S2E plugin hooks into S2E's "CorePlu-
gin" signals to define custom instructions and their handlers. It conducts deep
analysis of the execution of any given byte stream from all offsets until a kill
condition is reached, forwarding each offset's result to further offline analysis.

4 Empirical Evaluation

4.1 Locating the Hidden Code from Memory Dump

In this section, we evaluate CodeXt's capability in pinpointing the start and
boundaries of potentially disjoint, misaligned code hidden in multiple execution
paths from a given memory buffer. We vary two primary factors: the unrelated
bytes surrounding the hidden code; and, amount of attack code context information (i.e., run-time hint). Specifically, we put our sample malware code into
three types of buffers of different fill values for the bytes surrounding the code:
1) all nulls (0x00); 2) uniformly random; 3) capture (memory dump) from a real
world code injection attack. Then we combined the three surrounding data types
with varying amount of run-time hints: 1) with the address of a known system
call (EIP) plus the system call number (EAX); 2) with the address of a known
system call only; 3) with the system call number only; 4) neither.

Because it is easier to locate longer attack code, we deliberately used short
attack code in our experiment: a 41-byte (per section 4.2) and 81-byte in-the-
wild (ghttpd exploit) shell. We used a buffer of size 1024 bytes. We made the
offset variable symbolic which directed CodeXt to explore 1024 potential paths.

Table 2. Encoding Techniques Tested. MSF is Metasploit Framework. Multi-layered varied nesting combinations and depths of junk code insertion and ranged xor.

Technique	Extracted?
Junk code insertion	Yes
Ranged XOR	Yes
Multi-layered	Yes
Incremental	Yes
ADMmutate	Yes
Clet	Yes
Alpha2	Yes
MSF Call+4 Dword XOR	Yes
MSF Single-byte XOR Countdown	Yes
MSF Variable-length Fnstenv/mov Dword XOR	Yes
MSF JMP/CALL XOR Additive Feedback Encoder	Yes
MSF BloXor	Yes
MSF Shikata-Ga-Nai	Yes

For each of the four context information scenarios we experimented twenty runs with the random surrounding bytes, one run with fixed null surrounding bytes and one run with fixed captured bytes. CodeXt has successfully located the hidden code without any false positive in all runs for all the combinations.

Table 1 shows the average time needed to search each offset for all the run combinations. It shows that the run-time hints do not have significant impact on the performance in any combination. It took about the same time (1 second per offset) to search through null, random and fixed captured surrounding bytes.

To validate CodeXt recovering multiple execution paths, we embedded the algorithm shown in Figure 1 into a 1KB buffer, and marked x to be symbolic. CodeXt successfully explored all the three feasible execution paths (it detected a fourth, if (y==1 && z==0), as infeasible) and recovered their byte code.

We also investigated the probability of false positives. The probability for two consecutive random bytes to be the system call instruction int 0x80 (0xcd80) is $2^{-16} = 1.52 \times 10^{-5}$. It is highly likely that a long string of random bytes contains some executable instructions, but far less likely to contain a false positive (i.e., set EAX to a specific value or range, and end in a system call before EAX is clobbered). In addition, previous research details that 90% of random strings should fail execution within 6 instructions [5]. We tested CodeXt with buffers of pure, uniformly random bytes of 1KB, 10KB and 100KB respectively. Specifically, we input 20 different 1KB, one 10KB and one 100KB random bytes. CodeXt did not reported any hidden code detected from these random bytes.

4.2 Extracting Encoded Bytes

To evaluate CodeXt extracting encoded bytes, we have used 12 different encoders to encode a running example of byte code, called hw.shell, that prints "Hello, world!" to the standard output via the write system call.

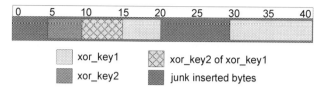

Fig. 5. Multiple layers of ranged xor encoding that overlap each other and use different keys, all layered on top of a junk code inserted encoding

Besides using 9 well-known third party encoders (e.g., ADMmutate [1], Clet [11], Shikata-Ga-Nai [2]), we developed 3 encoders ourselves: the *junk code insertion* encoder and the *ranged xor* encoder based on [3], and a novel *incremental* encoder. Table 2 lists all 12 encoders plus an entry to represent multi-layer combinations (nested) of our in-house encoders. CodeXt was able to automatically recover the original shellcode in all tested cases. In the following subsections, we elaborate the experiments with multi-layer combinations of junk code insertion and ranged xor, the incremental encoder, and Shikata-Ga-Nai.

Multi-layer Combinations. To evaluate CodeXt's capability in extracting code protected with multiple layers of encoding, we tested combinations of our in-house encoders: junk code insertion and ranged xor. Junk code insertion interjects a random length of random value bytes between each input byte, such that $junk(i)$ means to generate encoded output from input i. Junk code insertion, while very rudimentary, effectively interferes with common disassemblers. Ranged xor uses a 1B key to iteratively encrypt a specified range of input bytes, such that $xor(k_n, o, b, i)$, means to encode input i with key k_n at offset o for b bytes; also we will use $xor(k_n, i)$ to mean encoding all bytes in i.

In our trials we tested combinations such as 1) xor of junk: $xor(k_1, junk(i))$, 2) junk of xor: $junk(xor(k_1, i))$, 3) xor of xor: $xor(k_2, xor(k_1, i))$, and, as illustrated in Figure 5, 4) xor of xof of xor of junk: $xor(k_2, 5, 10, xor(k_1, 30, 10, xor(k_1, 10, 10, junk(i))))$. CodeXt was able to recover the original shellcode from all tested multiple layer combinations of encodings.

Incremental Encoder. We have developed a sophisticated incremental encoder, such that during decoding it will incrementally de-obfuscate one portion (or segment) of the original code at a time. After executing the decoded segment, it will decode another code segment into the same buffer and so on. Except the final decoded code segment, all other decoded segments are transient in that they will be overwritten right after execution. Therefore, a memory dump or snapshot at any moment will never reveal the entire decoded form. In order to extract the complete code protected by the incremental encoder, we need to take multiple snapshots at the right moments and places during run-time.

We have used the incremental encoder to encode a popular TCP based reverse connect shellcode with 5 system calls into 4 code segments, roughly representing its basic blocks. CodeXt generated 8 snapshots when executing the incrementally encoded shellcode. This accurately represents the algorithm copying each encoded segment into a buffer and then decoding it. CodeXt has successfully extracted the complete code protected by the incremental encoding.

Shikata-Ga-Nai. Shikata-Ga-Nai is a polymorphic xor additive feedback encoder within the Metasploit Framework. This encoder offers three features that provide advanced protection when combined. First, the decoder stub generator uses metamorphic techniques, through code reordering and substitution, to produce different output each time it is used, in an effort to avoid signature recognition. Second, it uses a chained self modifying key through additive feedback. This means that if the decoding input or keys are incorrect at any iteration then all subsequent output will be incorrect. Third, the decoder stub is itself partially obfuscated via self-modifying of the current basic block as well as armored against emulation using FPU instructions.

```
Offset Bytecode    Mnemonic            ; Comment
0000   DAD4         fcmovbe st4         ; any fpu insn
0002   B892BA1E5C   mov eax,0x5c1eba92  ; key = 92ba1e5c
0007   D97424F4     fnstenv [esp-0xc]   ; write fpu records to
                                        ; put EIP on top of stack
000B   5B           pop ebx             ; ebx = EIP
000C   29C9         sub ecx,ecx         ; clear ecx
000E   B10B         mov cl,0xb          ; loop 11 times
0010   83C304       add ebx,byte +0x4   ; PC += 4
0013   314314       xor [ebx+0x14],eax  ; [0x0018] = [0x0018]^key
0016   034386       add eax,[ebx-0x7a]  ; key += [ebx + Encoded Byte]
0019   58           pop eax             ; False Instruction, Encoded Byte
001A   EBB7         jmp short 0xfffffffd3 ; False Instruction, Encoded Bytes
```

The above code snippet shows the instructions from an instance of Metasploit's polymorphic xor additive feedback encoder Shikata-Ga-Nai [2]. It uses FPU instructions as a way to get EIP, via fnstenv/pop. Since most emulators, including S2E/QEMU, do not fully support FPU instructions, they have been used to detect emulated environments. For such FPU instruction combinations (e.g., fpstenv/pop), we implemented special handlers in our S2E plugin to emulate their semantics (e.g., update FPU internal values).

The instruction at address 0x0013 changes 4 bytes starting from address 0x0018, which changes the subsequent 3 instructions to be executed from address 0x0016. Therefore, Shikata-Ga-Nai dynamically modifies the instructions in the current basic block. S2E/QEMU was not able to handle such intra-basic block modification. To fix this problem, we extended the S2E translation mechanism so that our S2E plugin can force a re-translation should we detect any write within the memory range of upcoming instructions in the current basic block. With the added support of FPU instructions and intra-basic block modification, CodeXt has successfully extracted the hw.shell protected by Shikata-Ga-Nai.

5   Related Work

Attack Code Detection. A number of methods [26,9,31,20] have been proposed to detect attack code from network packet payloads based on various heuristics. Work [9] used static analysis based approach to look for the NOP sledge in the packet payload. SigFree [31] detects the presence of code from packets by checking the push-call instruction pattern and data flow anomaly from the static disassembled instruction sequences, and it depends on static disassembling. SHELLOS [24] utilizes hardware virtualization and KVM to detect the existence of injected code. Because it directly executes the instructions on the CPU, it can execute Shikata-Ga-Nai encodings that most CPU emulators (e.g., QEMU, S2E) can not. While existing attack code detection methods are able to detect the existence of attack code even if the attack code is mingled with random data, they are not able to determine the exact location and boundary of the attack code. Therefore, existing attack code detection methods can not automatically extract the attack code.

Binary Code Extraction. BCR [7] is not designed to extract the complete hidden code from a given memory dump, but rather to extract certain reusable code fragments from a given binary program. It requires the knowledge of the entry point to be effective and it can not handle self-modifying code.

Automated Unpacking of Hidden Code. Automated unpacking hidden code [6,34] has been an active research area and many methods [15,21,14,19,29,4,33,16] have been proposed to address the unpacking issue. Earlier methods (e.g., [15]) have used static analysis, and later approaches have used combination of static and dynamic analysis. Notably, PolyUnpack [21] detects the self-modifying code by checking if the to-be-executed instruction sequence is part of the static code model generated before execution. Because of the need of static modeling, it is not easy to apply PolyUnpack to code packed with multiple layers. OmniUnpack [19] detects unpacking by looking for written-then-execute pattern. It ignores intermediate layers of unpacking and only takes actions upon the invocation of some dangerous system call, which is assumed to be after the innermost layer of unpacking. This is a similar with Justin [16] that assumes a known code entry point and takes action upon a singular confluence of events; which does not account for incremental unpacking. OmniUnpack operates at the granularity of memory page, and it does not give any information about the intermediate layers of unpacking. As a result, it is faster. Renovo [14] also uses the written-then-execute pattern to detect the unpacking. It checks at the granularity of basic block. Specifically, it dumps the memory pages that contain the current basic block and has been written recently. Eureka [25] is a coarse-grained unpacking approach that uses Windows specific heuristics and x86 code statistical pattern.

To the best of our knowledge, all existing automatic unpacking mechanisms require the knowledge of the exact start of the code, and they are not effective when the hidden code is mingled with other bytes (i.e., the exact start of the hidden code is unknown). In addition, most existing unpacking methods only recover the hidden code and data on one execution path. No existing generic unpacking approach has been shown to be able to handle Metasploit's polymorphic xor additive feedback encoder Shikata-Ga-Nai and the incremental encoder that encodes only a segment of the hidden code in each layer of encoding. In fact, SHELLOS [24] abandoned attempted QEMU-based approach because of QEMU's incapability in handling FPU instructions and Shikata-Ga-Nai.

In contrast, CodeXt is able to explore multiple execution paths via combination of symbolic execution and concrete execution and recover the hidden code and data on multiple execution paths. To the best of our knowledge, it is the first system that can automatically recover hidden code protected by Shikata-Ga-Nai and the incremental encoder that encodes only a segment of the hidden code in each layer of encoding.

6   Conclusion

Extracting attack code is indispensable for effective malware analysis, forensics and reverse engineering. No existing approach has been shown to be able to automatically recover 1) disjoint, misaligned attack code mingled with random bytes; and 2) those transient code protected by multi-layer incremental encoding.

In this paper, we have presented CodeXt to address the above mentioned challenges. Based on selective symbolic execution and its unique multi-layer snapshot, CodeXt is designed to accurately pinpoint the exact start and the boundaries of the attack code and recover the hidden and transient code protected by various multiple layers of self-modification. Our experiments with real world shellcode and shellcode encoders have demonstrated that CodeXt is able to accurately extract the hidden code mingled with random bytes even if the code is protected by sophisticated encoders such as polymorphic xor additive feedback mechanisms like Shikata-Ga-Nai. In addition, CodeXt is able to automatically recover the transient code protected by multi-later incremental encoding.

Acknowledgments. This work was supported by NSF grant CNS-0845042.

References

1. ADMmutate Polymorphic Shellcode Engine, http://ktwo.ca/security.html
2. Polymorphic XOR Additive Feedback Encoder. In the Metasploit Framework, http://metasploit.com/modules/encoder/x86/shikata_ga_nai
3. Simple Obfuscation, http://funoverip.net/2011/09/simple-shellcode-obfuscation
4. Bania, P.: Generic Unpacking of Self-modifying, Aggressive, Packed Binary Programs, http://piotrbania.com/all/articles/pbania-dbi-unpacking2009.pdf
5. Barrantes, G., Ackley, D., Forrest, S., Stefanovic, D.: Randomized Instruction Set Emulation. ACM Trans. on Information Systems Security 8(1), 3–40 (2005)
6. Broch, T., Morgenstern, M.: Runtime Packers: The Hidden Problem? http://blackhat.com/presentations/bh-usa-06/BH-US-06-Morgenstern.pdf
7. Caballero, J., Johnson, N., McCamant, S., Song, D.: Binary Code Extraction and Interface Identification for Security Applications. In: Proc. of the 17th Netw. and Dist. System Security Symp. (February 2010)
8. Cadar, C., Dunbar, D., Engler, D.: KLEE: Unassisted and Automatic Generation of High-Coverage Tests for Complex Systems Programs. In: Proc. of the 8th Symp. on Operating Systems Design and Implementation, pp. 209–224 (December 2008)

9. Chinchani, R., van den Berg, E.: A Fast Static Analysis Approach to Detect Exploit Code Inside Network Flows. In: Valdes, A., Zamboni, D. (eds.) RAID 2005. LNCS, vol. 3858, pp. 284–308. Springer, Heidelberg (2006)

10. Chipounov, V., Kuznetsov, V., Candea, G.: S2E: A Platform for In-Vivo Multi-Path Analysis of Software Systems. In: Proc. of the 16th Int. Conf. on Architectural Support for Programming Languages and Operating Systems, pp. 265–278 (2011)

11. Detristan, T., Ulenspiegel, T., Malcom, Y., von Underduk, M.: Polymorphic Shell-code Engine Using Spectrum Analysis. Phrack (61), id 9 (August 2003)

12. Feng, H., Kolesnikov, O., Fogla, P., Lee, W., Gong, W.: Anomaly Detection Using Call Stack Information. In: Proc. of the IEEE Symp. on Security and Privacy (2003)

13. Forrest, S., Hofmeyr, S.A., Somayaji, A., Longstaff, T.A.: A Sense of Self for Unix Processes. In: Proc. of the IEEE Symp. on Security and Privacy (1996)

14. Kang, M., Yin, P.: Renovo: A Hidden Code Extractor for Packed Executables. In: Proc. of the 2007 ACM Workshop on Recurring Malcode, pp. 46–53 (2007)

15. Kruegel, C., Robertson, W., Valeur, F., Vigna, G.: Static Disassembly of Obfuscated Binaries. In: Proc. of the 13th USENIX Security Symp. (August 2004)

16. Guo, F., Ferrie, P., Chiueh, T.-C.: A Study of the Packer Problem and Its Solutions. In: Lippmann, R., Kirda, E., Trachtenberg, A. (eds.) RAID 2008. LNCS, vol. 5230, pp. 98–115. Springer, Heidelberg (2008)

17. Linn, C., Debray, S.: Obfuscation of Executable Code to Improve Resistance to Static Disassembly. In: Proc. of the 10th ACM Conf. on Computer and Commun. Security, pp. 272–280 (October 2003)

18. Linn, C., Rajagopalan, M., Baker, S., Collberg, C., Debray, S., Hartman, J.: Protecting against Unexpected System Calls. In: Proc. of the 14th USENIX Security Symp. (August 2005)

19. Martignoni, L., Christodorescu, M., Jha, S.: Omniunpack: Fast, Generic, and Safe Unpacking of Malware. In: Proc. of the 23rd Annu. Computer Security Applications Conf., pp. 431–441 (2007)

20. Polychronakis, M., Anagnostakis, K., Markatos, E.: Network-level Polymorphic Shellcode Detection Using Emulation. In: Proc. of the IEEE Conf. on Detection of Intrusions and Malware and Vulnerability Assessment, pp. 54–73 (July 2006)

21. Royal, P., Halpin, M., Dagon, D., Edmonds, R., Lee, W.: PolyUnpack: Automating the Hidden-Code Extraction of Unpack-Executing Malware. In: Proc. of the 22nd Annu. Computer Security Applications Conf. (2006)

22. Sekar, R., Bendre, M., Bollineni, P.: A Fast Automaton-Based Method for Detecting Anomalous Program Behaviors. In: Proc. of the 2001 IEEE Symp. on Security and Privacy (May 2001)

23. Sharif, M., Lanzi, A., Giffin, J., Lee, W.: Impeding Malware Analysis Using Conditional Code Obfuscation. In: Proc. of the 15th Network and Distributed System Security Symp. (2008)

24. Snow, K., Krishnan, S., Monrose, F., Provos, N.: SHELLOS: Enabling Fast Detection and Forensic Analysis of Code Injection Attacks. In: Proc. of the the 20th USENIX Security Symp. (August 2011)

25. Sharif, M., Yegneswaran, V., Saidi, H., Porras, P., Lee, W.: Eureka: A Framework for Enabling Static Malware Analysis. In: Jajodia, S., Lopez, J. (eds.) ESORICS 2008. LNCS, vol. 5283, pp. 481–500. Springer, Heidelberg (2008)

26. Tóth, T., Kruegel, C.: Accurate buffer overflow detection via abstract payload execution. In: Wespi, A., Vigna, G., Deri, L. (eds.) RAID 2002. LNCS, vol. 2516, pp. 274–291. Springer, Heidelberg (2002)

27. Udupa, S., Debray, S., Madou, M.: Deobfuscation: Reverse Engineering Obfuscated Code. In: Proc. of the 12th Working Conf. on Reverse Engineering (2005)
28. Wagner, D., Dean, D.: Intrusion Detection via Static Analysis. In: Proc. of the 2001 IEEE Symp. on Security and Privacy (May 2001)
29. Wang, X., Feng, D., Su, P.: Reconstructing a Packed DLL Binary for Static Analysis. In: Bao, F., Li, H., Wang, G. (eds.) ISPEC 2009. LNCS, vol. 5451, pp. 71–82. Springer, Heidelberg (2009)
30. Wang, X., Jiang, X.: Artificial Malware Immunization Based on Dynamically Assigned Sense of Self. In: Burmester, M., Tsudik, G., Magliveras, S., Ilić, I. (eds.) ISC 2010. LNCS, vol. 6531, pp. 166–180. Springer, Heidelberg (2011)
31. Wang, X., Pan, C., Liu, P., Zhu, S.: SigFree: A Signature-free Buffer Overflow Attack Blocker. In: Proc. of the 15th USENIX Security Symp. (August 2006)
32. Warrender, C., Forrest, S., Pearlmutter, B.: Detecting Intrusions Using System Calls: Alternative Data Models. In: Proc. of IEEE Symp. Security Privacy (1999)
33. Wu, Y., Chiueh, T.-C., Zhao, C.: Efficient and Automatic Instrumentation for Packed Binaries. In: Park, J.H., Chen, H.-H., Atiquzzaman, M., Lee, C., Kim, T.-H., Yeo, S.-S. (eds.) ISA 2009. LNCS, vol. 5576, pp. 307–316. Springer, Heidelberg (2009)
34. Yan, W., Zhang, Z., Ansari, N.: Revealing Packed Malware. IEEE Security Privacy 6(5), 65–69 (2008)

SIACHEN: A Fine-Grained Policy Language for the Mitigation of Cross-Site Scripting Attacks

Ashar Javed, Jens Riemer, and Jörg Schwenk

Chair for Network and Data Security Horst Görtz Institute for IT-Security,
Ruhr-University Bochum, Germany
{ashar.javed,jens.riemer,joerg.schwenk}@rub.de

Abstract. Cross-Site Scripting (XSS) attacks are at number three in the OWASP Top 10 2013 list [1] and according to a recent report by WhiteHat, 53% of the web applications are vulnerable to XSS attacks [2]. In this paper, we propose SIACHEN, a fine-grained, white-list and browser-enforced security policy language for the mitigation of XSS attacks. SIACHEN's syntax is similar to Cascading Style Sheets (CSS) and its semantics is based on Content Security Policy (CSP) directives. CSP is a coarse-grained policy language and gives web site administrators a page-level control. Our policy language operates on *per-id* or *per-class* of web page's HTML elements. SIACHEN also supports input validation and output encoding, which is missing in case of CSP. At the same time, SIACHEN leverages ECMAScript's object freezing feature from the earlier work done by Heiderich *et al.* in [3]. SIACHEN glues together a number of disparate technologies into a single framework.

We implemented our proposal in the form of a client-side JavaScript library. Web site administrators can deliver the SIACHEN policy to the browser via a new header named "X-Siachen-Policy". To show the applicability of our solution, we have added support of SIACHEN policy language in three open source web applications (i.e., PHPBB, PHPList & Damn Vulnerable Web App). Our evaluation shows reasonably low overhead is incurred by web applications and requires less amount of effort from developers' side. We have tested our prototype against a large number of state-of-the-art, obfuscated and unobfuscated XSS attack vectors and found no bypass. To assist web site administrators, we present SIACHEN AiDer, an online service for the automated recommendation of policies. Further, this paper also presents results of a short survey of fifty popular desktop web applications and their mobile versions (100 in total). We have found an XSS in all surveyed sites but the main purpose of the survey is to find suitable venues for our prototype.

1 Introduction

Cross-Site Scripting (XSS) — a common security issue web applications are facing since their discovery in 2000 [4]. WhiteHat Security's[1] statistics report about dominant web applications vulnerabilities, published in May 2013, shows that 53% of the web applications are vulnerable to XSS attacks [2]. According to OWASP Top 10 2013, XSS

---

[1] https://www.whitehatsec.com/

S.S.M. Chow et al. (Eds.): ISC 2014, LNCS 8783, pp. 515–528, 2014.

is at number three in the list [1]. FireHost[2] report says that there is a 160% rise in XSS attacks, if we compare Quarter (Q3) and (Q4) of year 2012 [5]. In [20], Javed *et al.* have found XSS vulnerabilities in 81 out of 100 surveyed mobile web applications. Recently Apple developers' website[3] and Ubuntu forums[4] have been hacked by exploiting XSS vulnerabilities [32,33]. The motivation for this work is based on many observations.

Developers' Knowledge Web application developers are often *short of* security knowledge. This has already been noted in academic researchers' community. According to Engin Kirda *et al.* [6]:

"... *Developers of web-based applications often have little or no security background. moreover, business pressure forces these developers to focus on the functionality for the end user and to work under strict time constraints, without the resources (or the knowledge) necessary to perform a thorough security analysis of the applications being developed. ...*"

Lack of Fine-grained Access Control A web browser's default security policy, the Same Origin Policy (SOP) [7], does not provide fine-grained access control over the web page's elements. The SOP does not differentiate the boundary of trustworthiness in web page's elements [8]. This may allow untrusted script or attacker-supplied code to execute in the context of web page. The attacker-supplied injected script can steal users' session cookies, spread worms[5] and deface pages[6].

Security in the form of layers XSS mitigation in the form of "*layered solution*" will increase the cost and raise the bar for the attacker. The CSP authors state in [12]:

"...*CSP is intended to be one protection layer and **not** the only protection mechanism.*"

We have observed a similar type of statement in case of ModSecurity i.e., the worlds number one firewall engine with more than one million installations [17]:

"*ModSecurity is a web application firewall engine that provides very little protection on its own. ...*"

1.1 Related Work

There has been a great amount of academic work [15,16,18,19,21,6] on the mitigation of XSS attacks. In this section, in favor of space restriction, we compare our approach with the closely related proposals.

Security Style Sheets The idea of using CSS as a security policy language syntax is not new. Terri Oda *et al.* proposed Security Style Sheets (SSS) in her PhD thesis [10] and a technical report available at [11]. Our proposal is consistent with SSS as far as leveraging CSS like syntax for security policies. The main reason to have CSS like syntax is the clearly defined and structured policy language syntax. Terri Oda *et al.* have proposed three directives: `domain-channels` (deals with whitelisting of allowed domains related to images, JavaScripts etc), `page-channels` (deals with whitelisting

---

[2] http://www.firehost.com

[3] https://developer.apple.com/

[4] http://ubuntuforums.org/

[5] https://www.whitehatsec.com/resource/whitepapers/ XSS_cross_site_scripting.html

[6] https://www.zone-h.org/archive/special=1

of HTML elements on a web page) and `execution` (flag that controls execution of JavaScript) in [10]. We have extended the work and have proposed CSP style directives for our policy language semantics. This gives better and more fine-grained control to web adminstrators over page's elements. Further, SIACHEN supports input validation, output encoding and ECMAScript's object freezing properties.

Content Security Policy Sterne *et al.* have proposed Content Security Policy (CSP) in [12]. CSP is now a W3C standard [13] and Chrome, Firefox and Blink-based (webkit-based browser engine developed as a part of chromium project) Opera now fully support W3C CSP specifications. Internet Explorer (IE) has limited CSP support. CSP is a white-list, coarse-grained and page-wise policy language and gives web adminstrators better control over their resources like scripts, iframes, images, objects etc. The CSP provides directives (i.e., *directive name: directive value*) to control different types of resources on the web page. SIACHEN's policy language semantics in the form of directives is based on CSP's directives with more adjustable fine-grained control.

The CSP level 2.0 supports inline-scripting via "`nonce`" attribute in a safe manner [14]. A nonce is a randomly generated string which should be difficult for an attacker to predict. In CSP 2.0 developer may include inline scripts, but only if they specify the script with a valid nonce in the CSP policy. In this way, the browser will only execute those scripts whose nonce matches with the one that is in the defined policy. In SIACHEN, web applications developers can also write an inline script code. SIACHEN may be treated as an extension of CSP 2.0 in terms of access control. SIACHEN operates on "`id`" or "`class`" of HTML elements and we leverage CSP for page level access control. At the same time, SIACHEN supports input validation and output encoding, which is missing in case of CSP.

The CSP is vulnerable to HTML tag injection attacks[7]. The following attack vector has `<plaintext>` tag and by using this tag the attacker may break the entire DOM. In [25], Heiderich *et al.* have shown that in order to ex-filtrate users' sensitive information, script injection is not necessary. The attacker may ex-filtrate private information by using an HTML, CSS and web fonts.

```
'"><marquee>CSP is vulnerable to an HTML Tag Injection</marquee><plaintext><b↩
     >HTMLi</b>
```

The attacker can also execute script with the help of "`data URI`" e.g.,

```
<a href=data:text/html;base64,PHN2Zy9vbmxvYWQ9YWxlcnQoMik+click</a>
```

The phishing attack via "`data URI`" has been already known[8]. SIACHEN supports input validation and output encoding and is not vulnerable to an HTML tag injection or scriptless attacks. Further, SIACHEN can work under the umbrella of CSP and may be used as an additional protection layer.

ICESHIELD Heiderich *et al.* have proposed IceShield for the detection and mitigation of malicious sites. It is available in the form of JavaScript library that is making use of ECMAScript's object freezing feature [3]. Authors have successfully shown that ECMAScript's [23] object freezing features enable dynamic JavaScript code anaylsis inside the DOM and may help in mitigation of common web vulnerabilities. The purpose

---

[7] https://www.owasp.org/index.php/HTML_Injection
[8] http://klevjers.com/papers/phishing.pdf

of ECMAScript's freezing properties is to stop runtime modifications of objects and modern browsers supports this feature. Our policy language leverages ECMAScript's object freezing properties from IceShield. This provides an additional layer of security.

1.2 Design Goals

By keeping in mind the above observations and discussions about related proposals, we outline the following design goals for the policy language.

- The solution has to be designed by keeping in mind "*security unaware developers*". Web applications developers are good designers and can make eye-candy websites using HTML, CSS and JavaScript but they often lack security knowledge.
- The solution's access control model has to be fine-grained. It will give better control to the web adminstrators over their contents on the web page and may provide clear privilege separation [8]. At the same time, solution should be flexible enough so that developers can adjust granularity according to their needs. The granularity of the access control mechanism should be per "id" (single element), per "class" (group of elements of same type), or per "page"[9].
- The solution should consists of different layers of protections for the mitigations of XSS attacks. In case, an attacker is able to bypass one layer by leveraging browsers' quirks, the other layer may help in mitigating XSS.
- The performance impact of the solution should be minimal.
- The solution should not break web applications and it should be compatible with the popular and existing protection mechanisms in the wild.

1.3 Our Approach

In this paper, we propose SIACHEN, a fine-grained (per-id or per-class of HTML elements), white-list and browser-enforced policy language whose syntax is similar to CSS and whose language semantics is based on CSP's directives. SIACHEN also leverages object freezing properties from earlier work done by Heiderich *et al.* in [3] (see Section 4.1). SIACHEN glues together a number of disparate technologies into a single proposal. One of the design goal is not to break existing web applications and by keeping this in mind, SIACHEN is a server *opt-in* proposal. In SIACHEN, web administrators define the policy and browser enforced it.

By keeping security unaware developers of web applications in mind, SIACHEN's syntax is similar to CSS. We believe that developers are already familiar with CSS and HTML and it would be helpful for them to enforce a XSS mitigation solution, if approach closely resembles already known technologies. Our language semantics is also based on CSP's directives. The underlying reason to choose CSP style directive is same (see Section 3). Mozilla proposed CSP three years ago and since browser vendors and developers are investing time to learn and implement the CSP. SIACHEN can work under the umbrella of CSP and does not break existing CSP policies (if any). In-fact, SIACHEN enables additional capabilities like flexible access control (developers can either use "id" or "class" attribute or simply "do not specify granularity at all", all according to their needs), object freezing properties and input/output (I/O) validation.

---

[9] In our solution, we leverage CSP for *page-wise* control and as an extra protection layer.

Listing 1.1. Inclusion of siachen.js file

```
1  <html><head><title>Running Example</title></head><body>
2  // HTML code of running example goes here
3  // SIACHEN protection script at the end of the page along with script-nonce
4  <script type="text/javascript" src="siachen.js" class="script-nonce:795490"↵
       ></script></body></html>
```

We implemented SIACHEN in the form of client-side JavaScript library (see Section 4). Web adminstrators can define the SIACHEN policy and deliver the policy to the browser with the help of new security header named "X-Siachen-Policy". The policy may also be delivered as a separate policy file having an extension of "siachen" i.e., *policyfilename.siachen*. The recommended way is to use header for policy delivery. In order to protect our client-side JavaScript code and to overcome implementation's limitation, we have also modified the Firefox browser and added support of CSP 2.0 attribute i.e., "nonce" [14] in it (see Section 4.3). At the time of writing, only Chrome supports "nonce" attribute [26] while Firefox does not support this directive [37]. Web administrators can specify a valid nonce (which is a use-once and randomly generated number) for inline-script(s) on the page and browser will take care of the rest. The listing 1. shows how web application developers may include SIACHEN JavaScript library in code.

XSS filter is also part of client-side JavaScript library. The main purpose of XSS filter is to validate user-supplied input. The filter immediately stops the user-supplied input, if found malicious (see Section 4.2). We borrow the XSS filter from our earlier work in [20]. Our filter is based on black-listing of XSS attack vectors' categories and regular expressions. The filter implemented in SIACHEN policy langauge is part of OWASP ModSecurity (web application firewall engine) Core Rule Set (CRS)[10]. Our JavaScript implemention also contains output escaping routine which is based on Yahoo's UI library [30,31] (see Section 4.2). The input filtering and output encoding functions fulfill one of our design goal i.e., security in the form of layers.

To show the deployment of SIACHEN, we have added support of our policy language in three popular and open source web applications i.e., PHPBB [27], PHPList [28] & Damn Vulnerable Web App (DVWA) [29]. Our evaluation on real world applications show reasonably low overhead incurred by these applications (see Section 5). At the same time, SIACHEN is compatible with the existing defence mechanisms in PHPBB and PHPLIST. The Damn Vulnerable Web App (DVWA) has no existing XSS protection mechanism. We have also tested our solution against large number of XSS attack vectors and found no bypass. It includes state-of-the-art, obfuscated and unobfuscated XSS attack vectors (see Section 6).

To ease adoption of SIACHEN policy language and to assist web administrators, this paper also presents an *online service* named SIACHEN AiDer. The SIACHEN AiDer takes URL as input and recommends respective SIACHEN policy along with a report. It also points out areas in web page that needs attention from the developers' side, if they wish to implement SIACHEN (see Section 7). Though SIACHEN AiDer has some practical implementation's limitations but it may well assist web administrators in order to define the site's policy. We also present a short survey of fifty popular desktop web ap-

---

[10] https://github.com/SpiderLabs/owasp-modsecurity-crs/blob/
master/base_rules/modsecurity_crs_41_xss_attacks.conf

plications and their mobile versions. We found XSS in all surveyed sites and it includes sites like Nokia Maps, SoundCloud, StatCounter, The New York Times, Yellow Pages, Homes and Dictionary etc. The complete list of surveyed sites is available at `http://pastebin.com/hTZRMtwy`. The main purpose of survey is to find suitable venues for our prototype (see Section 8).

1.4 Contributions

This paper makes the following contributions.

- We propose a new fine-grained, white-list and browser-enforced policy language named SIACHEN. SIACHEN has been built on different existing approaches. It includes CSS as syntax, CSP as semantics, ECMAScript's DOM freezing properties, input filter and output encoding as an additional security layers.
- As a part of empirical evaluation and to show the applicability of our approach, we have added support of policy language in three open source, real world web applications (i.e., PHPBB, PHPLIST and DVWA). Our implementation is available in the form of JavaScript library. In order to overcome practical limitation i.e., to protect client-side JavaScript implementation, we have also modified Firefox browser and have added support of "`nonce`" in it.
- We have tested SIACHEN against 10K XSS attack vectors and found no bypass.
- We present SIACHEN AiDer — *an online service*. It assists web administrators to spell-out SIACHEN policies.
- We also present results of short survey of fifty popular desktop web sites and their mobile variants (100 in total).

2   Threat Model and Running Example

2.1 Threat Model

This section describes the capabilities of an attacker that we assume for the rest of this paper. In XSS, an attacker can inject arbitrary JavaScript code on the client-side. Hence, our focus within this paper is to stop client-side JavaScript injection in particular and HTML tag injection or scriptless attacks in general since we also consider this as useful attack vector [25].

2.2 Running Example

In this section, we show a running example of a popular news website available in Figure. at `http://i.imgur.com/zwPdZSA.png`. Later we will use this toy example to demonstrate our policy language based protection (see Figure. available at `http://i.imgur.com/e0VnKia.jpg`). In the toy example available at `http://policy.bplaced.net/policytest/`, we try to simulate the real web application having different types of resources like images, videos, flash animations, scripts and so on. The toy example has a search form that is explicitly left vulnerable to XSS for demonstration purpose. The output of search query also reflects back on the page. Our toy example also contains a log-in form.

3   SIACHEN Policy Language

SIACHEN is a fine-grained and structured policy language for the mitigation of XSS attacks.

3.1   SIACHEN's Directives

Our policy language is available in the form of directives. Each directive has a name and value. SIACHEN's directives are inspired from CSP's directives. Each directive controls particular type of resource available on HTML page. The directive are:

- **image-source-allow-from:** controls requests that will be loaded as images via an HTML's <img> tag. The listing 2. excert from our toy example shows the use of this directive along with corresponding HTML code. The values of the directive should be an absolute file path to the desired resource. This is a requirement because SIACHEN operates on *per-id* or *per-class* of HTML elements.

  SIACHEN policy syntax is similar to CSS. Cascading Style Sheets (CSS), invented in the late 1990s (contemporaneous with JavaScript). CSS provides control over every aspect of the appearance of a page. In CSS # sign is used to identify single element on the page i.e., #identifiername {property: value}. The id selector uses the id attribute of the HTML element.

Listing 1.2. Image-source-allow-from (*per-id* example)

```
1   // HTML Code:
2   <img src="http://policy.bplaced.net/policytest/images/logo.png" id="↩
        logoImage" width="296" height="82"/>
3   // Corresponding SIACHEN Policy:
4   #logoImage{image-source-allow-from:'http://policy.bplaced.net/policytest/↩
        images/logo.png';}
```

In order to apply ECMAScript's object freezing protection, SIACHEN expects that developers will specify a unique identifier on HTML element. If similar identifier exists for more than one HTML elements, only the first one found will be enforced by our implementation, discarding all following (i.e., protection will not apply). In CSS, the class selector is used to specify a style for a group of elements. The class selector uses the HTML class attribute, and is defined with a "." (dot) in CSS. Our policy language also supports class-based identifers (i.e., the value of "class" attribute of an HTML element, if specified) but with the help of keyword "siachenClass:".

Many sites use transparent 1x1 pixel spacer images (e.g., *<img src="spacer.gif" width="1" height="1">*) for arranging contents on their web page e.g., a very popular Pakistani news website i.e., http://www.geo.tv/ is using "spacer.gif"[11] images on their home page 36 times. Further reasons, we found could be custom bullet types, menu icons, line separators and many other items used to structure a web page. Normally, each occurrence of a spacer element would demand a correspondent SIACHEN rule. This will blow-up the policy language and will be hard to enforce. To make things easier and to keep SIACHEN policies shorter, such

---

[11] http://www.geo.tv/images/spacer.gif

elements can use *class-based* identifiers. The listing 3. excert from our toy example shows the use of *class-based* identifier along with corresponding policy language code.

Listing 1.3. Image-source-allow-from (*per-class* example)

```
1  // HTML Code:
2  <img src="http://policy.bplaced.net/policytest/images/spacer.gif" class="↵
       siachenClass:spacer" width="1" height="1" border="0"/>
3  // Corresponding SIACHEN Policy:
4  #siachenClass:spacer{image-source-allow-from:'http://policy.bplaced.net/↵
       policytest/images/spacer.gif';}
```

SIACHEN expects that developers will specify the keyword "`siachenClass:`" along with their own defined class-name in a manner as described in listing 3. In order to facilitate developers for defining SIACHEN policies, this paper also presents SIACHEN AiDer — *an online service* for the recommendation of SIACHEN policies (see Section 7).

– **script-source-allow-from:** controls requests that will be loaded as an external scripts via <`script`> tag. The listing 4 shows the excert from running example. As mentioned above that SIACHEN expects developers of web applications will specify the unique "`id`" of the script. The "`id`" attribute is not a standard HTML5 script attribute [34]. SIACHEN uses this attribute in order to achieve *fine-grained* access control over elements. The SIACHEN does not have a default restriction of "`no inline scripting`" as W3C CSP 1.0 does [13]. It has been noted that *not* allowing inline scripting is the biggest hurdle in CSP's adoption [36]. Recently, a Mozilla security researcher Frederik Braun has surveyed Alexa top 25K sites for the presence of inline JavaScripts and found 96.2% sites are using inline JavaScripts [9].

Listing 1.4. Script-source-allow-from

```
1  // HTML Code:
2  <script src="http://policy.tipido.net/tipido.js" id="tipidoScript"></↵
       script>
3  // Corresponding SIACHEN Policy:
4  #tipidoScript{script-source-allow-from:'http://policy.tipido.net/tipido.js↵
       ';}
```

In this paper, we have modified the Firefox browser and added support of "`nonce`" in it. The underlying reason is: It fullfills one of our design goal i.e., protection in the form of different layers and with the help of "`nonce`", we were able to add extra layer of protection on our client-side SIACHEN script (i.e., our client-side JavaScript implementation script is itself protected via "`nonce`") (see listing 6.). The listing 5. from our running example shows how developers may use "`nonce`" in SIACHEN. We will discuss more about Firefox modification and the attribute "`nonce`" in Section 4.3.

Listing 1.5. script-source-allow-from along with script-nonce

```
1  // HTML Code:
2  <script src="http://policy.square7.ch/square7.js" id="square7Script" class="↵
       script-nonce:318822"></script>
3  // Corresponding SIACHEN Policy:
4  #square7Script{script-source-allow-from:'http://policy.square7.ch/square7.js↵
       ';}
```

Listing 1.6. CSP's script-nonce protection on siachen.js file

```
1  <script type="text/javascript" src="siachen.js" class="script-nonce:318822"↩
      ></script>
```

- **style-source-allow-from:** controls requests that will be loaded as an external style sheets via <link> tag.
- **object-source-allow-from:** controls requests that will be loaded as plug-ins (e.g., Flash) via <object> tag.
- **embed-source-allow-from:** controls requests that will be loaded as plug-ins (e.g., Flash) via <embed> tag. Sites often used <embed> and <object> interchangeably.
- **frame-source-allow-from:** protects requests that will be loaded as an iframe via <iframe> tag.
- **media-source-allow-from:** protects requests that will be loaded as media contents via <audio> and <video> tags of HTML. Web administrators may define the policy for <audio> and <video> tags in a similar manner as described in above cases.
- **form-source-allow-from:** protects <form> tag's "action" and "onsubmit" attributes. Internally "form-source-allow-from" is treated as "form-sou rce-to" but for consistency regarding directives' names, we have chosen the "form-source-allow-from". The listing 7. from our running example shows how to define the SIACHEN policy on <form> tag. This directive, if specified in the policy, supports input validation and output encoding automatically.

Listing 1.7. Form-source-allow-from

```
1  // HTML Code:
2  <form id="loginForm" method="POST" action="http://policy.bplaced.net/↩
      example1/index.php" onsubmit="return i_validate()">
3  // Corresponding SIACHEN Policy:
4  #loginForm{form-source-allow-from:'http://policy.bplaced.net/example1/index.↩
      php';
5  form-source-allow-from:'i_validate()'; }
```

By closely looking at the defined SIACHEN policy in listing 7. it can be seen that "form-source-allow-from" has been declared twice along with respective values i.e., one corresponds to <form> tag's action attribute and one for <form> tag's onsubmit eventhandler. The "i_validate()" is the function that does the job of input validation and output encoding automatically.

Once developers define *per-id* or *per-class* based policy language on page's elements manually or with the help of SIACHEN AiDer, the ECMAScript's object freezing properties will freeze the current state of the objects so that they will not be modified at runtime. Heiderich *et al.* in [3] have already shown that by using these features we can create tamper resistant DOM layer. We will discuss Object freezing properties in Section 4.1. We have choosen self-explanatory names for SIACHEN's directives. The underlying reason is to avoid confusion with CSP's directives' names. Though it is possible to change the names of directives as CSP have. This could be done by changing some string values in our JavaScript implementation.

4   Implementation

In this section, we discuss SIACHEN's implementaion. Our core implementaion consists of 1041 lines[12] of client-side JavaScript code including comments. It includes implementation of SIACHEN's directives parser, ECMAScript's object freezing features, XSS filter and output encoding function. In order to facilitate web administrators, XSS filter and output encoding functions are also available in PHP. The underlying reason to make them available in PHP is: in case web administrators only wish to use input/output validation without object freezing properties.

4.1   ECMAScript's Object Freezing Functions

SIACHEN leverages ECMAScript's object freezing features from the earlier work done by Heiderich *et al.* in [3]. Authors have shown that by using these features one can create tamper resistant DOM layer in modern browsers. All modern browsers i.e, Firefox, Chrome, and Internet Explorer (IE) support object freezing properties[13].

4.2   XSS Filter and Output Encoding

If "`form-source-allow-from`" is part of specified policy, SIACHEN invokes input validation function i.e., "`i_validate()`" (input validation plus output encoding) along with freezing properties. XSS filter is based on computationally fast regular expressions and black-list approach. We borrow the XSS filter from our earlier work in [20]. The complete code of XSS filter is available at [38]. Black-list and regular expressions based approaches are always subject to bypasses e.g., see NoScript's changelog[14]. In case of XSS filter bypass, our implementation automatically passes control to the output encoding module before the attacker-supplied data hit the target. In SIACHEN, output encoding module is based on Yahoo's UI library [30,31]. Output encoding function converts potentially dangerous characters e.g., $<$, $>$ and & etc into their respective HTML entities.

4.3   The Nonce and Firefox Browser's Modification

In this paper, we have modified Firefox's scripting logic and added support of "`nonce`" in it. In Firefox, we have implemented "`nonce`" as a value of "`class`" attribute (see Listing 1.).

5   Evaluation

In this section, we evaluate SIACHEN by adding support of policy language in three popular, open source web applications i.e., PHPBB [27], PHPList [28] & Damn Vulnerable Web App (DVWA) [29]. The goal is to measure the affect on performance, to see

---

[12] We used Chrome browser's developer tool to count the number of lines of JavaScript code.

[13] http://kangax.github.io/es5-compat-table/

[14] http://noscript.net/changelog

amount of effort needed from developers' side (see Figure. available at `http://i.imgur.com/rj04NKZ.jpg`) and to check compatibility with built-in XSS protections.

6   Testing

In this section, we present SIACHEN's effectiveness as a defence mechanism against a very large number of obfuscated, unobfuscated and state-of-the-art XSS attack vectors. For testing, we have used the following three resources for XSS attack vectors.

1. XSS Filter Evasion Cheatsheet by OWASP at `https://www.owasp.org/index.php/XSS_Filter_Evasion_Cheat_Sheet`
2. HTML5 Security Cheatsheet available at `http://html5sec.org/`
3. @XSSVector Twitter Account `https://twitter.com/XSSVector`. It has 140 plus latest XSS vectors that work across browsers.

None of the vectors from these above mentioned resources were able to bypass our layered defence mechanism. The Figure available at `http://i.imgur.com/e0vJkbf.jpg` shows SIACHEN correctly captures the XSS attack vector and displays warning message: "`XSS Vector Detected`".

7   SIACHEN AiDer

In order to assist web site administrators and to ease adoption, we also present an automated approach for the construction of security policies i.e., SIACHEN in web applications. The architecture diagram of SIACHEN AiDer is available at `http://i.imgur.com/wk0nL6K.jpg`. The SIACHEN AiDer is available in the form of an online service (see Figure. available at `http://i.imgur.com/vapOCJQ.jpg`). The SIACHEN AiDer uses the PHP cURL library[15] to fetch the page for analysis (includes all types of resources). This has the advantage that the contents which are added dynamically like iframes or code generated from scripts, are "seen" by the browser. Though SIACHEN AiDer has some limitations (e.g., browser specific code) but we still believe that it may assist web site administartors in order to find out SIACHEN policy for a given page.

8   Survey

In this section, we discuss the results of survey of fifty popular desktop sites and their mobile versions. The mobile version means site's URL starts with a letter "m" or ends in a word "`mobi`". The purpose of our survey is two fold:

1. To see prevalence of XSS in desktop sites and their mobile variants.
2. To identify sites that may easily adopt SIACHEN security policies.

The survey includes sites like MailChimp, Yellow Pages, Vodafone, MTV, StatCounter, Answers, SoundCloud and New York Times etc. The complete list of surveyed sites is available at `http://pastebin.com/hTZRMtwy`. We found XSS in all fifty surveyed desktop sites and their mobile variants (in total 100).

---

[15] `http://php.net/manual/en/book.curl.php`

8.1   Identification of Potential SIACHEN Venues

During source code analysis (exemplified at common elements i.e., images, scripts, iframes and forms etc) of fifty popular desktop web applications and their mobile versions, we found that mobile applications and sites that use absolute resources and static script blocks may adopt SIACHEN policies easily as compare to the sites that rely on dynamic resources.

9   Limitations of SIACHEN and Future Work

The fine granularity of SIACHEN comes with cost. In case of big sites, maintaining a SIACHEN policy is a tedious task and it requires effort in case of policy updation. At the same time, SIACHEN's output encoding module is not context sensitive. It only supports output encoding in standard HTML context. We leave integration of context sensitive output encoding module as a part of future work. We do not provide formal validation of SIACHEN policy language but we do plan as a part of future work also. We also plan to implement "SIACHEN's reporting module" like "CSP Report-Only Mode" as a part of future work. The XSS filter SIACHEN used has also some false positive issues.

10   Conclusion

The Cross-Site Scripting (XSS) vulnerabilities are ubiquitous in desktop and mobile web applications. The bad guys are and will continue to exploit XSS issues in the wild. In this paper, we have presented SIACHEN policy language and have shown that our prototype is compatible with CSP, adds extra layers of security and provides CSP's missing features like I/O validation to the developers of desktop and mobile applications. We believe that layered XSS mitigation solution will raise the bar for the attacker.

References

1. OWASP Top 10 2013, https://www.owasp.org/index.php/Top_10_2013-Top_10
2. WhiteHat Security's Website Security Statistics Report (May 2013), https://www.whitehatsec.com/assets/WPstatsReport_052013.pdf
3. Heiderich, M., Frosch, T., Holz, T.: ICESHIELD: Detection and Mitigation of Malicious Websites with a Frozen DOM. In: Sommer, R., Balzarotti, D., Maier, G. (eds.) RAID 2011. LNCS, vol. 6961, pp. 281–300. Springer, Heidelberg (2011)
4. Cross-site Scripting Overview, http://ha.ckers.org/cross-site-scripting.html
5. Cross-site scripting attacks up 160%, http://www.net-security.org/secworld.php?id=14320
6. Kirda, E., Kruegel, C., Vigna, G., Jovanovic, N.: Noxes: A client-side solution for mitigating cross-site scripting attacks. In: ACM SAC 2006 (2006)
7. Same Origin Policy, http://www.w3.org/Security/wiki/Same_Origin_Policy

8. Jayaraman, K., Du, W., Rajagopalan, B., Chapin, S.J.: ESCUDO: A Fine-grained Protection Model for Web Browsers. In: ICDCS 2010 (2010)
9. Inline JavaScript on Alexa top 25K sites, https://twitter.com/freddyb/status/304878658345107456
10. Oda, T.: Simple Security Policy for the Web, PhD Thesis (October 24, 2011), http://terri.zone12.com/doc/academic/TerriOda-PhDThesis-WebSecurity.pdf
11. Oda, T., Somayaji, A.: Enhancing Web Page Security with Security Style Sheets: SCS Technical Report TR-11-04, http://terri.zone12.com/doc/academic/TR-11-04-Oda.pdf
12. Stamm, S., Sterne, B., Markham, G.: Reining in the Web with Content Security Policy. In: WWW 2010 (2010)
13. Content Security Policy 1.0, http://www.w3.org/TR/CSP/
14. Content Security Policy Level 2.0, http://www.w3.org/TR/CSP11/
15. Louw, M.T., Venkatakrishnan, V.N.: BLUEPRINT: Robust Prevention of Cross-site Scripting Attacks for Existing Browsers. In: IEEE S&P 2009 (2009)
16. Jim, T., Swamy, N., Hicks, M.: BEEP: Browser-Enforced Embedded Policies. In: WWW 2007 (2007)
17. ModSecurity Core Rules, http://www.modsecurity.org/documentation/modsecurity-apache/2.1.3/html-multipage/ar01s02.html
18. Oda, T., Wurster, G., Van Oorschot, P., Somayaji, A.: SOMA: Mutual Approval for Included Content in Web Pages. In: CCS 2008 (2008)
19. Nadji, Y., Saxena, P., Song, D.: Document Structure Integrity: A Robust Basis for Cross-site Scripting Defense. In: NDSS 2009 (2009)
20. Javed, A., Schwenk, J.: Towards Elimination of Cross-Site Scripting on Mobile Versions of Web Applications. In: Kim, Y., Lee, H., Perrig, A. (eds.) WISA 2013. LNCS, vol. 8267, pp. 95–114. Springer, Heidelberg (2014)
21. Van Gundy, M., Chen, H.: Noncespaces: Using randomization to defeat cross-site scripting attacks. In: NDSS 2009 (2009)
22. Kc, G.S., Keromytis, A.D., Prevelakis, V.: Countering code-injection attacks with instruction-set randomization. In: CCS 2003 (2003)
23. ECMAScript Programming Language, http://www.ecmascript.org/
24. Kirda, E., Kruegel, C., Vigna, G., Jovanovic, N.: Noxes: A client-side solution for mitigating cross-site scripting attacks. In: ACM SAC 2006 (2006)
25. Heiderich, M., Niemietz, M., Schuster, F., Holz, T., Schwenk, J.: Scriptless Attacks— Stealing the Pie Without Touching the Sill. In: ACM CCS 2012 (2012)
26. Blink now has CSP 1.1 script nonce support, https://src.chromium.org/viewvc/blink?view=revision&revision=150541
27. PHPBB – Free and Open Source Forum Software, https://www.phpbb.com/
28. PHPLIST – The world's most popular open source email campaign manager, http://www.phplist.com/
29. Damn Vulnerable Web App (DVWA), http://www.dvwa.co.uk/
30. Yahoo! UI Library, http://yuiblog.com/sandbox/yui/3.3.0pr3/api/Escape.html
31. There's more to HTML escaping than &, <, >, and ", http://wonko.com/post/html-escaping
32. Apple Developer Website Hacked - What Happened? http://mytechblog.com/2013/07/apple-developer-website-hacked-what-happened/
33. Ubuntu Forums are back up and a post mortem, http://blog.canonical.com/2013/07/30/ubuntu-forums-are-back-up-and-a-post-mortem/

34. Scripts, http://www.w3.org/TR/REC-html40/interact/scripts.html
35. Tip of the tree Blink now has CSP 1.1 script nonce support! And what the hell does that mean? http://joelweinberger.us/blog/archives/35
36. Weinberger, J., Barth, A., Song, D.: Towards Client-side HTML Security Policies. In: HotSec 2011 (2011)
37. CSP 1.1: Nonce-source (experimental), https://bugzilla.mozilla.org/show_bug.cgi?id=855326
38. XSS Filter Code, https://github.com/SpiderLabs/owasp-modsecurity-crs/blob/master/base_rules/modsecurity_crs_41_xss_attacks.conf#L11

Security Issues in OAuth 2.0 SSO Implementations

Wanpeng Li and Chris J. Mitchell

Information Security Group, Royal Holloway, University of London, UK
Wanpeng.Li.2013@live.rhul.ac.uk, C.Mitchell@rhul.ac.uk

Abstract. Many Chinese websites (relying parties) use OAuth 2.0 as the basis of a single sign-on service to ease password management for users. Many sites support five or more different OAuth 2.0 identity providers, giving users choice in their trust point. However, although OAuth 2.0 has been widely implemented (particularly in China), little attention has been paid to security in practice. In this paper we report on a detailed study of OAuth 2.0 implementation security for ten major identity providers and 60 relying parties, all based in China. This study reveals two critical vulnerabilities present in many implementations, both allowing an attacker to control a victim user's accounts at a relying party without knowing the user's account name or password. We provide simple, practical recommendations for identity providers and relying parties to enable them to mitigate these vulnerabilities. The vulnerabilities have been reported to the parties concerned.

1 Introduction

Since OAuth 2.0 was published in 2012 [1], it has been used by many websites worldwide to provide single sign-on (SSO) services. By using OAuth 2.0, websites can ease password management for their users, as well as saving them the inconvenience of re-typing attributes that are instead stored by identity providers and provided to relying parties as required.

OAuth 2.0 is very widely used on Chinese websites, and there is a correspondingly rich infrastructure of identity providers (IdPs) providing identity services using OAuth 2.0. For example, some relying parties (RPs), such as the travel site Ctrip, support as many as eight different IdPs. At least ten major IdPs offer OAuth 2.0-based identity management services. RPs wishing to offer users identity management services from multiple IdPs must support the peculiarities of a range of different IdP implementations of OAuth 2.0.

Use of OAuth 2.0 by Facebook, Google and Microsoft has previously been studied, and issues have been identified [2–5]. However, despite the wide use of OAuth 2.0 for SSO in China, the authors are not aware of any published research on the properties of Chinese implementations. The very large and essentially self-contained OAuth 2.0 infrastructure in China is an important area for study, motivating the work described here. Also, as an early adopter of OAuth 2.0, lessons learnt from studying the Chinese infrastructure may apply globally.

S.S.M. Chow et al. (Eds.): ISC 2014, LNCS 8783, pp. 529–541, 2014.

OAuth 2.0 is used to protect access to hundreds of millions of user accounts in China alone, and so its security in practice is very important. Assessing practical security is non-trivial, especially as system operation relies on closed code and proprietary specifications and implementation guidance. In the absence of detailed specifications, security assessments require exhaustive experimental evaluation and analysis. In this paper we report on such investigations, including a detailed discussion of serious vulnerabilities found. We also provide recommendations for system improvements that address the identified vulnerabilities.

The paper is structured as follows. §2 introduces OAuth 2.0 and describes related work. In §3 we give two general classes of vulnerability in OAuth 2.0 SSO systems, both of which have been observed in practice. §4 covers our study of real-world OAuth 2.0 systems in China, including details of instances of the classes of vulnerability described in §3. Possible reasons for these vulnerabilities are considered in §5, together with proposed mitigatations.

2   Background and Related Work

OAuth 2.0. OAuth 2.0 [1] allows an application to access resources protected by a resource server on behalf of the resource owner, by consuming an access token issued by the authorisation server. OAuth 2.0 involves four roles. The *Resource Owner* is a host acting on behalf of an end user who can grant access to protected resources. The *Resource Server* is a server which stores the protected resources and consumes access tokens provided by an authorisation server. The *Client* is an application running on a server, which makes requests on behalf of the resource owner (the *Client* is the RP when OAuth 2.0 is used for SSO). The *Authorisation Server* generates access tokens for the client, after authenticating the resource owner and obtaining its authorisation (the *Resource Server* and *Authorisation Server* together constitute the IdP when OAuth 2.0 is used for SSO).

In order to use OAuth 2.0 for SSO, the resource server and authorisation server together play the IdP role, the client plays the role of the RP, and the resource owner corresponds to the user. OAuth 2.0 SSO systems build on user agent (UA) redirections, where a user (U) wishes to access services protected by the relying party (RP) which consumes the access token generated by the identity provider (IdP). The UA is typically a web browser. The IdP provides ways to authenticate the user, asks the user to allow the RP to access the user's attributes, and generates an access token. The RP uses the access token to access the user's attributes using an API provided by the IdP.

OAuth 2.0 supports four ways for RPs to obtain access tokens, namely Authorisation Code Grant, Implicit Grant, Resource Owner Password, and Client Credentials Grant. In this paper we are only concerned with the Authorisation Code Grant procedure, outlined below.

1. U → RP: The user clicks a login button on the RP website, as displayed by the UA, causing the UA to send a HTTP request to the RP.

2. RP → UA: The RP produces an OAuth 2.0 authorisation request and sends it back to the UA. The authorisation request includes *client_id*, the identifier for the client, which the RP registered with the IdP previously; *response_type= code*, indicating the Authorisation Code Grant method; *redirect_uri*, the URI to which the IdP redirects the UA after access is granted; *state*, an opaque value used by the RP to maintain state between request and callback (step 6 below); and the *scope* of the requested permission.

3. UA → IdP: The UA redirects the request received in step 2 to the IdP.

4. IdP → UA: If the user has already been authenticated by the IdP, then steps 4/5 are skipped. If not, the IdP returns a login form used to collect user authentication data.

5. U → UA → IdP: The user completes the login form and grants permission for the RP to access the attributes stored by the IdP.

6. IdP → UA: After using the login form data to authenticate the user, the IdP generates an authorisation response and sends it to the UA. This contains *code*, the IdP-generated authorisation code, and *state*, sent in step 2.

7. UA → RP: The UA redirects the response received in step 6 to the RP.

8. RP → IdP: The RP produces an access token request and sends it to the IdP token endpoint directly (i.e. not via the UA). The request includes the *client_id*, the *code* generated in step 6, the *redirect_uri* and also a *client_secret* shared between the IdP and the RP.

9. IdP → RP: The IdP checks the *client_id*, *client_secret*, *code* and *redirect_uri* and responds to the RP with *access_token*, an access token.

10. RP → IdP: The RP passes *access_token* to the IdP via a defined API to request the user attributes.

11. IdP → RP: The IdP checks *access_token* and, if satisfied, sends the requested user attributes to the RP.

Identity Federation for OAuth 2.0. Like OpenID [6], OAuth 2.0 does not support identity federation as defined in Shibboleth [7] or SAML [8]. A commonly used means of achieving identity federation involves the RP locally binding the user's RP-managed account with the user's IdP-managed account, using the unique identifier for the user generated by the IdP. After binding, a user can log in to the RP-managed account using his or her IdP-managed account.

Such a federation scheme operates as follows. After receiving the access token, the RP retrieves the user's IdP-managed account identifier and binds the user's RP-managed account identifier to the IdP-managed account identifier. When the user next tries to use his or her IdP-managed account to log in to the RP, the RP looks in its account database for a mapping between the supplied IdP-managed identifier and an RP-issued identifier. If such a mapping exists, then the RP simply logs the user in to the corresponding RP-managed user account.

In real-world OAuth 2.0 SSO systems supporting federation, RPs typically use one of two ways to perform the binding. Firstly, suppose a user chooses to log using SSO. After finishing the authorisation process with the IdP, the user is asked either to bind the IdP-managed account to his or her RP-managed account or to log in to the RP directly. The user will need to provide his/her RP-managed account information (e.g. account name and password) to complete the binding. Alternatively, after a user has already logged into an RP, he or she can initiate a binding operation. After being authenticated by the IdP and granting permission to the RP, the user can bind his or her RP-managed account to the IdP-managed account. After binding, many RPs allow users to log in to their websites using an IdP-managed account.

Related Work. The OAuth 2.0 specification [1] and threat model [9] describe possible threats and countermeasures. Pai et al. [10] confirm a security issue described in the OAuth 2.0 Threat Model ([9] §4.1.1) using the Alloy framework [11]. Chari et al. [12] analyse OAuth 2.0 in the Universal Composability Security framework [13], and show that OAuth 2.0 is secure if all communications links are SSL-protected. Frostig and Slack [14] discovered a cross site request forgery attack in the Implicit Grant flow of OAuth 2.0, using the Murphi framework [15]. However, all this work is based on abstract models of OAuth 2.0, and so delicate implementation details are ignored.

To understand the real-world security of OAuth 2.0, Wang et al. [5] examined a number of deployed SSO systems, focussing on a logic flaw present in many such systems, including OpenID. In parallel, Sun & Beznosov [4] also studied deployed systems. Both these studies restricted their attention to systems using English. Indeed, very little research has been conducted on the security of OAuth 2.0 systems using other languages, some of which, like those in Chinese, have very large numbers of users. In this paper, we redress this imbalance by reporting on an analysis of Chinese-language OAuth 2.0 systems.

Like Sun & Beznosov [4], this paper considers the security of deployed OAuth 2.0 systems; however, there are two major differences in approach. First, we do not exploit specific web browser and application vulnerabilities. Second, Sun & Beznosov focus on attacks involving stealing the user's access token from an RP; specific web browser and application vulnerabilities are used to allow the attacks. By contrast, this paper focuses on the security of OAuth 2.0 SSO systems supporting identity federation, and the security flaws identified do not exploit browser or application vulnerabilities.

3   Threats to OAuth 2.0 Identity Federation

OAuth 2.0 is intended to let an RP gain limited access to a service either on behalf of the user or for the RP's own purposes. Hence identity federation, as in Shibboleth [7] or SAML [8], is not supported. As discussed in §2, in order to provide identity federation for OAuth 2.0, RPs typically employ ad hoc means to bind an RP-managed account to an IdP-managed account.

Cross Site Request Forgery Attacks. A cross site request forgery (CSRF) attack [16–22] operates in the context of an ongoing interaction between a target web browser (running on behalf of a target user) and a target website. In the attack, a malicious website somehow causes the browser to initiate a request of the attacker's choice to the target site. This can cause the target site to execute actions without the involvement of the user. In particular, if the target user is currently logged in to the target site, the browser will send cookies containing the target user's authentication tokens, along with the attacker-supplied request, to the target site. The target site will process the malicious request as if it was initiated by the target user. The target browser could be made to send the spurious request in various ways; e.g., a malicious site visited by the browser could use the HTML <img> tag's src attribute to specify the URL of a malicious request, causing the browser to silently use a GET method to send the request.

The OAuth 2.0 specification ([1], §10.12) describes a possible CSRF attack in which the target website corresponds to *redirect_uri*, i.e. the URI to which the target browser is directed by OAuth 2.0. The attack involves an attacker causing the target browser to send the target site a request containing the attacker's own authorisation code or access token. As a result, the target site might associate the attacker's protected resources with the target user's current session; possible undesirable effects could include saving user credit card details or other sensitive user data to an attacker-controlled location.

In this paper we show that a CSRF attack could also be used to attack the federation process of an OAuth 2.0 SSO system, with potentially very serious effects. Suppose a target UA is logged in to a target RP. The UA visits the malicious site, perhaps by following a link on the target RP's site. The malicious site now forces the UA (unbeknownst to the user) to send a request to the target site containing a binding request for the attacker's IdP account. If not appropriately secured, the target website might now bind the attacker's IdP-managed account to the target user's RP-managed account. The attacker can now log in to the target user's RP-managed account at will. If the vulnerability is present, this simple attack could be launched on a very large scale to take control of multiple RP-managed accounts. Note that the attacker would need to use a distinct IdP-managed account for each instance of the attack, although this should not be an issue in practice.

The OAuth 2.0 specification recommends inclusion of a *state* parameter in the authorisation request to protect against CSRF attacks. This allows the RP to verify the source of a request by matching the *state* value to the user-agent's authenticated state (as recorded in a session cookie). However, for this to work the *state* value must not be guessable; otherwise the attacker could include the guessed value in its fraudulent request. However, despite this advice, we have found that many real-world RPs either omit the *state* parameter from the autho-risation request or fail to use *state* correctly (e.g., some RPs allocate a fixed value to *state*). We have also observed that some RPs do not check the correctness of the *state* value even if it is non-guessable. As a result, many RPs supporting identity federation are vulnerable to a CSRF attack against the RP's redirect URL, allowing an attacker to gain full access to the victim's RP-managed ac-count without knowing the user's account name or password.

Logic Flaws. To achieve identity federation, the RP must support a way to bind the user's RP-managed and IdP-managed accounts. The binding operation is clearly security-critical since, after binding, the owner of the IdP-managed account has full control over the RP-managed account. Design flaws in binding could allow an attacker to bind the victim user's RP-managed account to the attacker's IdP-managed account, without the knowledge of the user.

Binding security largely depends on the RP, since binding is done by the RP and the IdP simply provides an access token. The RP chooses how binding works, and decides whether or not to perform it. Since there is no standard for binding, different RPs use different ways of completing it. As a result the security of binding largely depends on the security awareness of the implementers. This is clearly dangerous, and the almost inevitable result is that some RP implementations of OAuth 2.0 SSO contain serious logic flaws, potentially enabling an attacker to bind its IdP- managed account to any RP-managed account. The consequences of such an attack could be very serious indeed.

Adversary Model. We assume all RPs and IdPs are benign, i.e. we only consider attacks involving third parties. However, we suppose an attacker can share malicious links and/or post comments which could contain malicious content on a benign RP website, and send malicious links to the victim, e.g. via email. The malicious content constructed by the attacker could cause the browser to initiate an HTTP request to either the RP or the IdP (or both).

4   Case Studies

We report on an investigation of the security of real-world implementations of SSO systems using OAuth 2.0, including both RPs and IdPs. In particular we looked for vulnerabilities of the types described in §3 above. We focussed our study on RPs using OAuth 2.0 for identity federation, especially those supporting the second method of binding specified in §2. This is because the first method requires a user to provide account information to complete binding, which seems to make using a CSRF attack to achieve a false binding much more difficult.

Conducting a security analysis of commercially deployed OAuth 2.0 SSO systems requires a number of challenges to be addressed. These include lack of access to detailed specifications for the SSO systems, undocumented RP and IdP source code, and the complexity of APIs and/or SDK libraries in deployed SSO systems. The methodology we used is similar to that employed by Wang et al. [5] and Sun & Beznosov [4], i.e. we analysed the browser relayed messages (BRMs). We treated the RPs and IdPs as black boxes, and analysed the BRMs produced during binding to look for possible exploit points.

We used Fiddler (`http://www.telerik.com/fiddler`) to capture the BRMs sent between RPs and IdPs; we also developed a Java program to parse the BRMs to simplify analysis and to avoid mistakes resulting from manual inspection. After confirming an exploit point, we used widely deployed browsers, including IE, Safari, Firefox, and Chrome, to replay or relay the browser request. At no time during our experiments did we access any user accounts without the explicit permission of the user concerned.

Renren Network. Renren Network (http://www.renren.com) is a Chinese social networking service which has been described as the 'Facebook of China'. It claims to have about 320 million active users. Renren Network supports several SSO IdPs, including Baidu [23] and China Mobile [24]. A user can thus sign in to Renren Network using a Baidu or China Mobile account.

A Renren-Baidu account binding attack. In order to use an IdP-managed account to log in to Renren via OAuth 2.0, a user's Renren-managed account must first be bound to an IdP-managed account. Suppose a user already logged in to Renren wants to bind his or her Renren-managed and Baidu (IdP) accounts (step 1 in §2.2). Renren generates an OAuth 2.0 authorisation request (step 2) and redirects the user browser to Baidu (step 3). The authorisation request generated by Renren does not contain a *state value*. After authenticating the user (steps 4 and 5), Baidu generates the authorisation response (step 6), which only contains the *redirect_uri* and *code*. The user agent will send the authorisation response to Renren (step 7) with cookies containing the user's session identifier. Renren uses the *code* to exchange an access token with Baidu (steps 8 and 9). Renren then uses the access token to retrieve the user's Baidu account's identifier (steps 10 and 11), and employs the user's session identifier to retrieve the user's Renren account identifier. Finally, Renren binds the user's Renren-managed and Baidu-managed accounts, based on the identifiers it received earlier.

The RP needs to know the identifiers of the user's RP-managed and IdP-managed accounts in order to complete binding. Renren does not implement any measures to protect against a CSRF attack on the *redirect_uri*. Thus if an attacker can replace the *code* in the authorisation response with its own IdP-generated *code*, then the identifier that the RP retrieves from the IdP will correspond to the attacker's IdP-managed account. This will cause the victim user's RP-managed account to be bound to the attacker's IdP-managed account.

We tested the viability of such an attack by initiating the Renren-Baidu authorisation process. We used a Baidu account to perform authentication to Baidu (acting as the IdP). Baidu then generated and sent a response (as in step 6 of §2.2) containing a *redirect_uri* and *code*. We intercepted this response and posted it as a link on a web forum. If a victim user who has previously logged in to Renren clicks on the link, the victim's browser will submit the request with the cookie containing the victim's session identifier to the *redirect_uri* of Renren. When we tested this, Renren successfully bound the victim's account to our IdP-account. We could thus access the victim's account via our IdP-managed account, without knowing the victim user's account name or password.

A Renren-China Mobile account binding attack. We analysed the data flow for Oauth 2.0 SSO performed between Renren Network (RP) and China Mobile (IdP). Unlike Renren-Baidu, both the authorisation request (step 1) and the authorisation response (step 6) contain a *clientState* value, which we assume is used by Renren to try to prevent CSRF attacks.

However, we observed that the *clientState* value is the same for multiple requests and responses (in fact *clientState*=9 in all requests and responses we observed). That is, the *clientState* is guessable. Thus, and as we observed in practical tests, Renren-China Mobile federation is also susceptible to a CSRF attack that enables an attacker to bind his or her own China Mobile-managed account to a victim user's Renren-Managed account.

For both the above scenarios, the response generated in step 6 begins with the Renren host name. Thus, if posted on a website it will resemble a benign sharing link, so a victim user will have no reason not to click on it.

Ctrip. Ctrip (www.ctrip.com) is a China-focused travel agency with around 60 million members and 2.5 million user reviews. Its services cover around 9,000 flight routes and 200,000 hotels across the world. In order to access Ctrip services, a user must have a membership with either Ctrip itself or with one of the SSO systems it supports. Ctrip supports eight OAuth 2.0 SSO IdPs, including Renren [25], Wangyi [26], Taobao [27], MSN [28] and Sina [29].

A logic flaw in Ctrip. To study the security of the Ctrip-supported SSO systems, we analysed BRMs exchanged between Ctrip (the RP) and Renren (the IdP) while the user is binding his or her Ctrip-managed and Renren-managed accounts using the second method described in §2. As for the Renren-Baidu binding, the OAuth authorisation request in step 2 and the authorisation response in step 6 do not contain the *state* value. This immediately suggested that Ctrip-Renren binding might be vulnerable to a CSRF attack. To test this, we relayed an intercepted IdP-generated authorisation response to a victim user agent which had already logged in to Ctrip. The user agent sent the authorisation response to Ctrip, along with the cookies containing the victim user's session identifier. However, instead of binding the attacker's Renren account to the victim user's Ctrip account, Ctrip just responded with a web page asking the user to input his or her account name and password. We also tried to perform the attack on other IdPs supported by Ctrip. In each case, Ctrip responded with a web page requesting the user to input his or her account name and password. Hence Ctrip, by some means, resists the attack described above.

However, we observed that the request generated in step 1 contains a *Uid*, the Ctrip-generated user identifier. Observing that Ctrip account identifiers are guessable, we conjectured that if we could replace the *Uid* value in the request generated in step 1 with the *Uid* corresponding to the victim user, then it might be possible to force Ctrip to bind the attacker's IdP-managed account to the victim user's Ctrip-managed account. We therefore tested this approach. In order not to cause damage to a real user of the Ctrip website, we modified the *Uid* value to correspond to an account created for the purposes of the experiment. We relayed the request to Ctrip and completed the authorisation procedure with the IdP. Ctrip responded with a blank web page with the URL *http://RP@Recp=0*, indicating that Ctrip had successfully bound the IdP-managed and Ctrip-managed accounts.

http://accounts.ctrip.com/member/RenrenLogin/Authorize.aspx?Action=B
&Uid=E60444782&BackUrl=http://my.ctrip.com/Home/Third/ThirdTransfer.aspx

Fig. 1. The request generated in step 1

To understand why Ctrip is vulnerable to this attack, we analysed all BRMs exchanged in both a normal binding operation (where a logged-in user initiates a binding operation) and an attack binding operation (where an attacker initiates the request in Fig. 1 without logging in to the *Uid* account). We observed that, in a normal binding operation the browser sent Ctrip the request in step 1 with cookies containing the user's session identifier. However, in the attack binding operation, as no cookies had previously been set for the *Uid* account, the user agent just sent the request (step 1). Ctrip generated the authorisation request and set a session identifier cookie for the *Uid* account (step 2). After receiving the IdP-generated authorisation response (step 6), the browser sent both the authorisation response and the cookie containing the session identifier to Ctrip. Ctrip treated the combination of session identifier and authorisation response as a legal binding operation, and so it bound the IdP-managed account to the victim user's Ctrip account. From this we deduced that Ctrip fails to verify the validity of the request in step 1 before generating the authorisation request, i.e. Ctrip does not check the request is initiated by the real owner of *Uid*. An attacker can thus successfully forge a request to bind his or her IdP-managed account to the *Uid* account, i.e. an attacker can circumvent Ctrip's user authentication.

A generic Ctrip binding attack. We used our observations regarding the operation of the Ctrip website to devise the following attack on federation. When a user initiates a binding operation to a different IdP, only the *IdPLogin* value (the *RenrenLogin/Authorize.aspx* in Fig. 3) changes in the request. An attacker can use this to control the binding between RP and IdP. That is, an attacker can bind any RP-managed account to any IdP just by replacing the *IdPLogin* value and the *Uid* value in the request sent in step 1. We further observed that Ctrip provides a user forum to share information and initiate events. An attacker can readily find user *Uid* values by examining the forum, since Ctrip does not effectively conceal them. Using a simple guessing attack, many *Uid* values can be recovered from the poorly-protected forum entries.

We reported these flaws to the Ctrip Security Response Centre and helped Ctrip fix them. Ctrip has listed this report on its acknowledgement page.

5 Discussion and Recommendations

Scope of Study. We studied a total of 60 Chinese RPs supporting SSO via identity federation to an IdP using OAuth 2.0. Of these, 14 only support the first method of binding described in §2.3, and so are not vulnerable to the CSRF attack in §4. Of the remaining 46, a total of 21, i.e. almost half, are vulnerable to the CSRF attack. Many millions of users were potentially affected by this vulnerability, since Renren alone has around 320 million active users.

We further analysed the BRMs to find out why the 21 RPs are vulnerable. Since these RPs support an average of at least three IdPs, we had to analyse 68 distinct sets of RP-IdP browser relayed messages. Of these 68 OAuth 2.0 authorisation processes, 48 do not involve the use of any countermeasures to a CSRF attack. However even in the 20 cases where countermeasures were employed, poor implementation means that the attack remains possible.

One possible reason why some implementers use a constant value for *state* is that the IdP-provided documentation [23, 25–27, 29, 30] does not describe how to generate it. In the absence of guidance on the use of *state*, implementers may reasonably, but falsely, believe they have implemented effective protection against CSRF attacks by using a constant value. Secondly, some RPs which use the same *redirect_uri* for multiple IdPs use the *state* value to distinguish between IdPs, i.e. so they can determine to which IdP the RP-managed account should be bound. That is, they do not appear to understand the intended purpose of the *state* variable, and the need for such values to be non-guessable; as a result they may use guessable *state* values, which again represents a possible vulnerability. Thirdly, even if the *state* value is 'opaque' (i.e. non-guessable), problems can still arise if the RP does not perform the necessary checks. In particular, we discovered that some RPs fail to check that the *state* value in the request used to trigger binding correctly maps to the user's session identifier.

In summary, there are a variety of ways in which the binding vulnerability can arise. The common element is the lack of clear and detailed guidance for the use of CSRF countermeasures in the context of identifier binding for federation. This is hardly surprising since identity binding is not standardised within the OAuth specifications. This lack of clear standards for identity federation is the main underlying source of all the vulnerabilities we have observed.

Recommendations. OAuth 2.0 SSO systems have been widely deployed by Chinese RPs and IdPs, and it appears likely that increasing numbers of Chinese RPs and IdPs will implement OAuth 2.0 for SSO. However, our study has revealed serious vulnerabilities in existing systems, and there is a significant danger that these vulnerabilities will be replicated in future systems. Below we make a number of recommendations, directed at both RPs and IdPs, designed to address the identified vulnerabilities. These recommendations should help to address problems in current systems was well as assist in ensuring that future systems are built in a more robust way. Ideally, a standardised federation system for OAuth 2.0 would be developed, and these recommendations are also intended as input to such work.

In OAuth 2.0 SSO systems supporting identity federation, RPs design the binding process. We have the following recommendations for RPs.

- **Deploy Countermeasures against CSRF Attacks.** One reason the OAuth 2.0 systems we investigated are vulnerable to CSRF attacks is that the RPs do not implement countermeasures. Many IdPs [23, 25, 26, 29] recommend RPs to include the *state* parameter in the OAuth 2.0 authorisation request, and RPs should follow such recommendations.

- **Do not Use a Constant or Predictable *state* Value.** Some RPs include a fixed *state* value in the OAuth 2.0 authorisation request. In this case an attacker can forge a response, since the RP cannot distinguish a legitimate response produced by a valid user from a forged response. Thus the inclusion of the *state* value does not mitigate CSRF attacks. Thus RPs must generate a non-guessable *state* value bound to the user's session identifier, so that the *state* value can be used to verify the validity of the response.
- **Check the *state* Value.** RPs that include an opaque *state* value in their OAuth 2.0 request should check the *state* value in the response before completing binding. We recommend that RPs use a session-dependent *state* value, although such a procedure slightly enlarges the state table which the RP must maintain in order to validate the *state* value.
- **Require the User to Input Account Information.** Perhaps the simplest way to prevent the CSRF attack is to require users to input their account names and passwords before completing binding. However, the user will then be required to 'log in' twice during a single session, damaging the user experience; this also goes against the OAuth 2.0 design goals.

In an OAuth 2.0 SSO system, the IdP designs the OAuth 2.0 protocol process and provides the API for RPs. An RP wishing to support a particular IdP must therefore comply with the requirements of that IdP, and so the IdPs play a critical role in the system. We have the following recommendations for IdPs.

- **Include the *state* in Sample Code.** IdPs typically provide sample code to help RP developers correctly code interactions with the IdP. However, many [23–26, 28–30] fail to include the *state* value in their sample code. This may be the main reason why more than half of the RP-IdP interactions we analysed are vulnerable to CSRF attacks. Including the *state* value in IdP sample code should help encourage RPs to reduce the risk of CSRF attacks.
- **Emphasise the Consequences of CSRF Attacks.** Since IdPs are responsible for designing the way in which OAuth 2.0 is used, RP developers must use the IdP-provided documentation to enable interoperation. In the examples of IdP documentation we examined, many simply mention the possibility of CSRF attacks without emphasising the potentially very serious consequences. This may help explain why some RPs do not appear to take the CSRF threat as seriously as they should.

Concluding Remarks. We studied the security of 60 implementations of OAuth 2.0 for federation-based SSO, as deployed by leading Chinese websites. We discovered that nearly half are vulnerable to CSRF attacks against the federation process, allowing serious compromises of user accounts. These attacks allow a malicious third party to bind its IdP-managed account to a user's IdP-managed account, without knowing the user's account name or password. As a result of the lack of a standardised federation process, we have further discovered logic flaws in real-world implementations of federation, which again allow binding of an attacker's IdP-managed account to a user's RP-managed account.

We reported our findings to all RPs and IdPs affected by the attacks; we also provided them with possible mitigations. We hope our study will be of broader value in warning IdPs and RPs of the dangers of CSRF attacks on OAuth 2.0 identity federation process. Ideally, a robust federation process for OAuth 2.0 will be standardised, helping to reduce the likelihood of future problems.

References

1. Hardt, D.: The OAuth 2.0 authorization framework (2012), http://tools.ietf.org/html/rfc6819
2. Hanna, S., Shin, R., Akhawe, D., Boehm, A., Saxena, P., Song, D.: The emperor's new APIs: On the (in)secure usage of new client-side primitives. In: Proc. W2SP 2010 (2010)
3. Miculan, M., Urban, C.: Formal analysis of Facebook Connect Single Sign-On authentication protocol. In: Proc. SofSem 2011, OKAT, pp. 99–116 (2011)
4. Sun, S.T., Beznosov, K.: The devil is in the (implementation) details: An empirical analysis of OAuth SSO systems. In: Yu, T., Danezis, G., Gligor, V.D. (eds.) Proc. CCS 2012, pp. 378–390. ACM (2012)
5. Wang, R., Chen, S., Wang, X.: Signing me onto your accounts through facebook and google: A traffic-guided security study of commercially deployed single-sign-on web services. In: Proc. IEEE Symp. on Security and Privacy 2012. IEEE (2012)
6. Recordon, D., Fitzpatrick, B.: Open ID Authentication 2.0 — Final (2007), http://openid.net/specs/openid-authentication-2_0.html
7. Morgan, R., Cantor, S., Carmody, S., Hoehn, W., Klingenstein, K.: Federated security: The Shibboleth approach. Educause Quarterly 27, 12–17 (2004)
8. Scott, C., Kemp, J., Philpott, R., Maler, E.: Assertions and Protocols for the OASIS Security Assertion Markup Language (SAML) V2.0 (2005), http://docs.oasis-open.org/security/saml/v2.0/saml-core-2.0-os.pdf
9. Lodderstedt, T., McGloin, M., Hunt, P.: OAuth 2.0 Threat Model and Security Considerations (2013), http://tools.ietf.org/html/rfc6749
10. Pai, S., Sharma, Y., Kumar, S., Pai, R.M., Singh, S.: Formal vericication of OAuth 2.0 using alloy framework. In: Proc. CSNT 2011, pp. 655–659. IEEE (2011)
11. Jackson, D.: Alloy 4.1 (2010), http://alloy.mit.edu/community/
12. Chari, S., Jutla, C.S., Roy, A.: Universally composable security analysis of OAuth v2.0. IACR Cryptology ePrint Archive 2011, 526 (2011)
13. Canetti, R.: Universally composable security: A new paradigm for cryptographic protocols. In: Proc. FOCS 2001, pp. 136–145. IEEE Computer Society (2001)
14. Slack, Q., Frostig, R.: Murphi Analysis of OAuth 2.0 Implicit Grant Flow (2011), http://www.stanford.edu/class/cs259/WWW11/
15. Dill, D.L.: The murϕ verification system. In: Alur, R., Henzinger, T.A. (eds.) CAV 1996. LNCS, vol. 1102, pp. 390–393. Springer, Heidelberg (1996)
16. Burns, J.: Cross site reference forgery: An introduction to a common web application weakness. Security Partners, LLC (2005), http://dl.packetstormsecurity.net/papers/web/XSRF_Paper.pdf
17. Jovanovic, N., Kirda, E., Kruegel, C.: Preventing cross site request forgery attacks. In: Proc. SecureComm 2006, pp. 1–10. IEEE (2006)
18. Barth, A., Jackson, C., Mitchell, J.C.: Robust defenses for cross-site request forgery. In: Ning, P., Syverson, P.F., Jha, S. (eds.) Proc. CCS 2008, pp. 75–88. ACM (2008)

19. Zeller, W., Felten, E.W.: Cross-site request forgeries: Exploitation and prevention. Bericht, Princeton University (2008)
20. Mao, Z., Li, N., Molloy, I.: Defeating cross-site request forgery attacks with browser-enforced authenticity protection. In: Dingledine, R., Golle, P. (eds.) FC 2009. LNCS, vol. 5628, pp. 238–255. Springer, Heidelberg (2009)
21. Shahriar, H., Zulkernine, M.: Client-side detection of cross-site request forgery attacks. In: Proc. ISSRE 2010, pp. 358–367. IEEE Computer Society (2010)
22. De Ryck, P., Desmet, L., Joosen, W., Piessens, F.: Automatic and precise client-side protection against CSRF attacks. In: Atluri, V., Diaz, C. (eds.) ESORICS 2011. LNCS, vol. 6879, pp. 100–116. Springer, Heidelberg (2011)
23. Baidu Inc.: Baidu Open Connect (2014), `http://developer.baidu.com/wiki/index.php?title=docs/oauth/authorization`
24. China Mobile Communications Corporation: ChinaMobile Open Connect (2014), `http://dev.10086.cn/wiki/?p5_01_02`
25. Renren Network: Renren Open Connect (2014), `http://wiki.dev.renren.com/wiki/Authentication`
26. Wangyi Inc.: Wangyi Open Connect (2014), `http://reg.163.com/help/help_oauth2.html`
27. Taobao Marketplace: Taobao Open Connect (2014), `http://open.taobao.com/doc/detail.htm?id=118`
28. Microsoft: Microsoft Live Connect (2014), `http://msdn.microsoft.com/en-us/library/live/hh243647.aspx`
29. Sina Corp.: Sina Open Connect (2014), `http://open.weibo.com/wiki/Oauth2/authorize`
30. Douban.com: Douban Open Connect (2014), `http://developers.douban.com/wiki/?title=oauth2`

A Practical Hardware-Assisted Approach
to Customize Trusted Boot for Mobile Devices

Javier González[1], Michael Hölzl[2], Peter Riedl[2],
Philippe Bonnet[1], and René Mayrhofer[2]

[1] IT University of Copenhagen, Denmark
{jgon,phbo}@itu.dk
[2] University of Applied Sciences Upper Austria, Campus Hagenberg, Austria
{michael.hoelzl,peter.riedl,rene.mayrhofer}@fh-hagenberg.at

Abstract. Current efforts to increase the security of the boot sequence
for mobile devices fall into two main categories: (i) secure boot: where
each stage in the boot sequence is evaluated, aborting the boot process if
a non expected component attempts to be loaded; and (ii) trusted boot:
where a log is maintained with the components that have been loaded
in the boot process for later audit. The first approach is often criticized
for locking down devices, thus reducing users' freedom to choose software. The second lacks the mechanisms to enforce any form of run-time
verification. In this paper, we present the architecture for a two-phase
boot verification that addresses these shortcomings. In the first phase,
at boot-time the integrity of the bootloader and OS images are verified and logged; in the second phase, at run-time applications can check
the boot traces and verify that the running software satisfies their security requirements. This is a first step towards supporting usage control
primitives for running applications. Our approach relies on off-the-shelf
secure hardware that is available in a multitude of mobile devices: ARM
TrustZone as a Trusted Execution Environment, and Secure Element as
a tamper-resistant unit.

Keywords: Secure Boot, Trusted Boot, Secure Element, TrustZone.

1   Introduction

Today, mobile devices are designed to run a single Operating System (OS).
Typically, Original Equipment Manufacturers (OEMs) lock their devices to a
bootloader and OS that cannot be substituted without invalidating the device's
warranty. This practice is supported by a wide range of service providers, such as
telecommunication companies, on the grounds that untested software interacting
with their systems represents a security threat[1]. In the few cases where the OEM
allows users to modify the bootloader, the process is time consuming, requires a
computer, and involves all user data being erased[2]. This leads to OEMs indirectly

---

[1] http://news.cnet.com/8301-17938_105-57388555-1/
verizon-officially-supports-locked-bootloaders/
[2] http://source.android.com/devices/tech/security/

S.S.M. Chow et al. (Eds.): ISC 2014, LNCS 8783, pp. 542–554, 2014.
© Springer International Publishing Switzerland 2014

deciding on the functionalities reaching the mainstream. As a consequence, users seeking the freedom of running the software that satisfies their needs, tend to root their devices. However, bypassing the security of a device means that these users lose the capacity to certify the software running on it. This represents a risk for all parties and services interacting with such a device [17, 21].

This topic has been widely discussed by Cory Doctorow in his talk *Lockdown: The Coming Civil War over General Purpose Computing* [6]. Here, he argues that hardware security platforms such as Trusted Platform Module (TPM) have been misused to implement what he calls the lock-down mode: *"Your TPM comes with a set of signing keys it trusts, and unless your bootloader is signed by a TPM-trusted party, you can't run it. Moreover, since the bootloader determines which OS launches, you don't get to control the software in your machine."*. Far from being taken from one of his science fiction dystopias, the lock-down mode accurately describes the current situation in mobile devices: users cannot always modify the software that handles their sensitive information (e.g. pictures, mails, passwords), control the hardware peripherals embedded in their smart-phones (e.g. GPS, microphone, camera), or connect to their home networked devices (e.g. set-top boxes, appliances, smart meters). This raises obvious privacy concerns.

In the same talk, Doctorow discusses an alternative implementation for hardware security platforms - the certainty mode -, where users have the freedom to choose the software running in their devices, and the *certainty* that this software comes from a source they trust. What is more, he envisions the use of context-specific OSs, all based on the user's trust. The issue is that the trust that a user might have in a given OS, does not necessarily extended to the third parties interacting with it (e.g., private LANs, cloud-based services, etc.).

In this paper we present an approach to allow users choosing the OS they want to run in their mobile devices while (i) giving them the certainty that the OS of their choice is effectively the one being booted, and (ii) allowing running applications to verify the OS in run-time. Put differently, we extend Doctorow's certainty mode idea to all the parties interacting with a mobile device. While modern Unified Extensible Firmware Interface (UEFI) platform support the installation of new OSs and their corresponding signing key by means of the BIOS, this is not the case for mobile platforms, where TPM is not present. The same applies for user space applications doing integrity checks on the running system. In order to address this issue in mobile devices we propose a two-phase verification of the boot process: in the first phase, boot components are verified and logged in the same fashion as trusted boot; in the second phase, the boot trace can be checked by running applications in order to verify the running OS. We base the security of our design in hardware security extensions present on a wide range of mobile devices: ARM TrustZone as a Trusted Execution Environment (TEE), and Secure Element (SE) as a tamper-resistant unit.

2   Related Work

The architecture of a secure boot process for desktop PCs was first proposed by Arbaugh et al. in [3]. Their architecture, called AEGIS, describes a way to verify

the integrity of a system by constructing a chain of integrity checks. Every stage
in the boot process has to verify the integrity of the next stage. After this first
description of a secure bootstrap process, multiple specifications and implementations of such a system have been created. One of the first specifications that
support this feature was proposed by the Trusted Computing Group (TCG) in
conjunction with the Trusted Platform Module (TPM) standard [22]. A TPM is a
secure cryptographic-coprocessor embedded in the PC architecture and provides
a set of functionalities, such as generation of cryptographic key-pairs, a random
number generator and protected storage. **Trusted boot** [7] is the implementation which makes use of this hardware module to verify the boot sequence.
A machine running with trusted boot sends the hash of each following stage in
the boot sequence to the TPM where it will be appended to the previous hash.
This creates a hash chain, called Platform Configuration Register (PCR), that
can be used for various purposes. For example, it can be used to decrypt data
only when the machine reached a specific stage in the boot sequence (*sealing*)
or to verify that the system is in a state that is trusted (*Remote Attestation*).
The TCG Mobile Phone Working Group proposed a concept on how to implement such a trusted boot also on mobile devices using a TPM-like hardware
component called Mobile Trusted Module (MTM) [19]. Another boot verification protocol is **secure boot**, described in the UEFI specifications since version
2.2 [23]. UEFI secure boot verifies the integrity of each stage by computing a
hash and comparing the result with a cryptographic signature. A key database
of trustworthy public keys needs to be accessible during boot time with which
the signature can be verified. If verification fails, the system will abort the boot
process. Due to this reason, and the fact that only a limited amount of keys
are pre-installed on the platform, the implementation of this system has been
criticized of preventing users to install an OS of their choice.

Although both systems, TPM trusted boot and UEFI secure boot, are widely
spread on desktop computers, they still did not reach the mobile platform: the
efforts to port UEFI to ARM devices - mainly driven by Linaro - have been
publicly restricted to ARMv8 servers, not considering mobile or embedded platforms[3]. Also the MTM, though especially designed for mobile devices, has not
been integrated into current device hardware (except for the Nokia N900 in
2009). This leads to the necessity for different solutions in current off-the-shelf
mobile devices.

3   Background

3.1   Secure Hardware Support

The goal of the two-phase boot verification is to have a bootstrap architecture
that can be trusted and easily customized by the owner of the mobile device.
For our approach to be deployable only by means of a software update, it is
necessary that it is based on off-the-shelf secure hardware components already
deployed in mainstream mobile devices.

---

[3] http://www.linaro.org/blog/when-will-uefi-and-acpi-be-ready-on-arm/

Secure Element (SE). The SE is a special variant of a smart card, which is usually shipped as an embedded integrated circuit in mobile devices together with Near field Communication (NFC) [18] and is already integrated in a multitude of mobile devices (e.g., Samsung Galaxy S3, S4, Galaxy Nexus, HTC One X, etc.). Furthermore, a secure element can also be added to the device with a microSD or an Universal Integrated Circuit Card (UICC). The main features of a SE are: **data protection** against unauthorized access and tampering, **execution of program code** in form of small applications (applets) directly on the chip and the **hardware supported execution of cryptographic-operations** (e.g., RSA, AES, SHA, etc.) for encryption, decryption and hashing of data without significant run-time overhead [12].

Trusted Execution Environment (TEE). A TEE is a secure environment inside a computing device that ensures that sensitive data is only stored, processed and protected by authorized software. The secure environment is separated by hardware from the device's area running the main OS and user applications, which is denoted rich environment.

An example of TEE is ARM TrustZone [4]. TrustZone relies on the so-called NS bit, an extension of the AMBA3 AXI Advanced Peripheral Bus (APB), to separate rich and secure environments. The NS bit distinguishes those instructions stemming from the secure environment and those stemming from the rich environment. Access to the NS bit is protected by a gatekeeper mechanism referred to as the secure monitor, which is triggered by the System Monitor Call (SMC). The OS thus distinguishes between user space, kernel space and secure space, where only authorized software runs in secure space, without interference from user or kernel space. Also, any peripheral connected to the APB (e.g., interrupt controllers, timers, and user I/O devices) can be configured by means of the TrustZone Protection Controller (TZPC) virtual peripheral to have prioritized - or even exclusive - access from the secure environment. Since TrustZone enabled processors boot always in secure mode, secure code executes before the general purpose bootloader booting the rich OS (e.g., u-boot) is even loaded in memory. This allows to define a security perimeter formed by code, memory and peripherals, making TrustZone a good candidate to support trusted boot solutions as the one presented in this paper.

While TrustZone was introduced 10 years ago; it is first recently that Trustonic, Xilinx and others have proposed hardware platforms and programming frameworks that makes it possible for the research community [8] as well as the industry to experiment and develop innovative solutions with TrustZone.

3.2   Threat Model

The main motivating goal of our approach for a two-phase boot verification is to give the user the certainty that the OS of their choice is indeed the one being booted and that a malicious entity is not able to get access to sensitive data. Hence, our threat model on mobile devices considers several kind of software and hardware attacks:

Software Attacks. Our threat model for software attacks only concerns attacks within the software stack of the mobile device. This includes attacks on application level, OS level and down to kernel/driver level.

With the increase of mobile devices and amount of security sensitive applications running on them, we can expect a growing number of mobile malware **exploiting errors of the OS** [15]. Attacks might be carried out to read sensitive data within standard application permissions or - in the worst case -, exploit privilege escalation [11,20]. Certain types of mobile malware are able to directly infect the OS to achieve their malicious goal (i.e. *rootkits*). **Communication** between applications, libraries, kernel drivers, etc. is also subject to different kinds of attacks. Mobile malware such as trojans or viruses running on the device might have the ability to interfere with the data path and gain full control over exchanged messages (e.g., service hijacking [5], library call interception [16]). This is specially relevant when the communication affects hardware components handling sensitive information (i.e., secure data path). An adversary seeking to compromise the secure data path can be assumed to be able to perform various types of attacks: eavesdropping, data injection, denial-of-service, man-in-the-middle, etc.

Hardware Attacks. As mobile devices become ubiquitous, the chances of them being lost or stolen have increased substantially. This results in physical attacks being increasingly relevant. Malicious hardware possession enables threats through both physical tampering of the hardware and software modification of the device. An attacker can take advantage of having physical access to the device to either load a malicious OS or bootloader in order to bypass the hardware protection, or directly attempt to access sensitive information (i.e., keys and secrets) from secondary storage (e.g., flash memory). Even if the file system is encrypted using tools provided by the OS, it is possible for an attacker with physical access to the device to circumvent them [10]. Protection against these attacks using hardware based solutions assuring tamper resistance have been proposed in order to increase the protection of sensitive data on mobile devices [13].

4   Example Scenario

The solution we propose in this paper enables providers of services or applications with security concerns to adapt an OS to meet their particular security requirements. This customized OS is offered by a trusted issuer; therefore we call it a **certified OS**. Creating a certified OS could mean to restrict the installation of applications, restrict network access, restrict access to hardware, etc. By allowing users to exchange OSs that can be certified, we enable services and organizations to establish restrictions concerning the software that interacts with their systems, while preserving the user's right to choose - and certify - the software handling their personal information.

Bring Your Own Device (BYOD). One practical application for a certified OS could be the BYOD problem. This refers to users wanting to interact from their personal mobile devices with their company IT services. Recent studies show that the BYOD problem is growing [1]. Given the heterogeneity of the current mobile device OS landscape, this can create overhead for system administrators and impede productivity. The multitude of different versions of each OS combined with adaptions to the OS by hardware manufacturers create this fragmented OS landscape. Concrete challenges include porting services to different platforms, having to deal with platform-specific security threats, or increasing the complexity of the enterprise LAN since the devices connecting to it cannot be trusted. By using a customized OS that is preconfigured to interact with all company services, enterprises could save time and money while increasing the security of their services. The effort to adapt OSs can be reduced by supporting only one or few versions of the most common OS (Android and iOS combined currently have a market share of 96%[4]). When employees are not using company services, they can switch to an OS they trust to handle their personal information. In this way one single device can be used both privately and professionally with the minor trade-off of having to switch OS.

The aforementioned BYOD applications serve as good examples for sensitive data. All credentials needed to authenticate to enterprise services like email and VPN private keys could be stored in the tamper resistant SE (see Section 3.1). Access to that data would only be granted if the request comes from a certified OS and all contextual requirements are met (e.g., being inside of the enterprise's VPN).

5   Architecture

Figure 1a depicts how hardware and software components interact with each other in the two-phase boot verification. In normal operation, both, the rich and the secure environment are booted. This is, the signatures of the mobile OS and the OS bootloader (OSBL) have been validated by the SE. Note that the verification process is carried out by the trusted execution environment (TEE), consisting of its bootloader (TEEBL) and an OS (TEE OS), and the SE as root of trust. Additionally, we assume the manufacturer bootloaders, first (FSBL) and second stage (SSBL), also as root of trust components. Applications running in the rich area can be installed and executed without restrictions just as we are used to see in current mobile OSs. We do not make any assumptions regarding the trustworthiness of these applications. We assume however that the secure tasks running in the TEE can be trusted. We describe the communication between the rich and the trusted execution environment in more detail in Section 3.1.

The SE is configured as a trusted peripheral and therefore only accessible from secure tasks in the TEE. Note that while the SE is not a peripheral in itself, it is connected via the APB (e.g., I2C, etc.), and therefore treated as such. In this way, rich applications make use of secure tasks to access the SE. As a consequence, if the TEE becomes unavailable, rich applications would be unable to communicate with the SE. The TEE becoming unavailable could be a

---

[4] http://www.idc.com/prodserv/smartphone-os-market-share.jsp

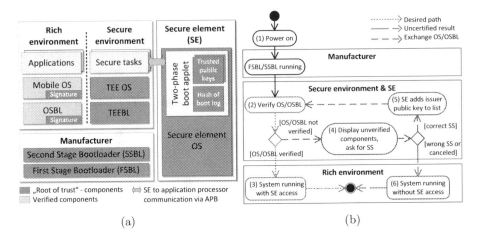

(a) (b)

Fig. 1. Components of our proposed customizable two-phase boot verification for off-the-shelf mobile devices in (a). The activity diagram in (b) describes the first phase. Transitions show the desired path (dotted), an uncertified result (long-dashed) and steps used for exchanging OS (dashed). "SS" stands for shared secret.

product of a failure, but also a defense mechanism against unverified software. If the OS image cannot be verified, the OS would still be booted, however the TEE will not respond to any attempt of communication. In this case we give up on availability in order to guarantee the confidentiality and integrity of the sensitive data stored in the SE (e.g., corporate VPN keys stored in the tamper-resistant hardware). Finally, the TEE also serves as a medium between certificates of the OS issuers in an untrusted storage (e.g., SD card, flash memory) and the SE. This gives users the option to add additional trusted public keys to the SE when the first-phase verification failed (see Section 5.3 for details).

5.1 First Phase Verification

The first phase verification is responsible for verifying and starting the TEE and the OS, and for enabling the user to install a customized OS and flag it as trusted. Figure 1b depicts the process (numbers in the diagram correlate with the enumeration, borders indicate which component is responsible for each step).

1. The user presses the power button. FSBL, SSBL, and TEE are started.
2. The TEE attempts to verify OS/OSBL by sending their signatures and the hash of the images to the SE. The SE attempts to verify the signatures with all trusted public keys of OS/OSBL issuers and compares it to the received hash. The result of this operation is reported to the TEE. In addition, the hash of the booted OS for the second-phase is stored in the SE. Several hashes using different algorithms are calculated to better support phase two (Section 5.2). Since the images can be cached, the transmission cost is only paid once.

3. Once the OS/OSBL are verified, the OS is booted with full access to the SE through the TEE. This is the end of the expected course of actions.
4. If the verification of OS/OSBL (step 2) fails, a message, explaining which component is unverified, is displayed. Now the user can choose to either certify the OS/OSBL by entering the shared secret and provide an issuer certificate, or continue to boot the uncertified OS/OSBL. For further details see Section 5.3.
5. If a legitimate user (authenticated by shared secret) flags an OS/OSBL as trusted, the SE adds the public key of the issuers' certificate to the list of trusted keys and continues to verify the newly added OS.
6. If the user enters a wrong shared secret or cancels the operation, the uncertified OS/OSBL is booted without access to the SE and a message informs the user about the uncertified state. This is the end of the uncertified course of actions, where the user can still use the device (uncertified OS/OSBL is booted), but due to missing verification, access to the secured data is denied by the TEE.

This architecture ensures that only a combination of verified OS and OSBL are granted access to the SE. If one of the components is not verifiable, the device can still be used (uncertified OS is booted), but access to the SE is denied. In the first phase all executed commands are logged in the SE by maintaining a hash chain. This approach is similar to trusted boot, and enables the second phase verification. The process of installing custom OS is explained in Section 5.3.

5.2   Second Phase Verification

In the second phase verification, rich applications can verify the system in which they are running before executing. To do this, rich applications make use of a secure task in the TEE to validate the running OS by checking the boot traces in the SE. This secure task is not application-specific, but a *secure* system primitive that any rich application running in user space can request to use.

In ARMv7, the SMC call needs to be issued from a privilege mode; this means that either the user application executes a Supervisor Call instruction (SCV) to enter in supervisor mode, or the SMC call stems from a place already executing in privilege mode. Relying on user applications to call the monitor and handle the communication with a secure task that is shared among different user applications can compromise the verification coming from the secure environment (e.g., by compromising application binaries). Another alternative would be letting applications handling the boot trace verification by having each of them implementing their own secure task. However, this solution would not only result in code replication, but in an unnecessary increase of the Trusted Computing Base (TCB), and therefore in the secure environment having a larger attack surface. Also, any update concerning the verification of the boot trace would need to propagate to many different secure tasks, therefore slowing down the dissemination of critical security patches[5]. By defining the system verification

---

[5] Cases like the Heartbleed Bug (http://heartbleed.com) are good examples of how a rapid dissemination of a security patch is necessary.

as a system primitive, the kernel makes the process transparent to user applications, addressing all these issues. This approach does not assume the kernel to be trusted; on the other hand, it limits the attack surface to the communication channel (the kernel), avoiding sensitive code to be delegated to user space. In Section 6 we address this is detail as part of the security analysis.

In this way, rich applications can communicate with the kernel, who forwards the call to the secure task that verifies the OS image. Rich applications pass the list of hashes they trust and a parameter defining the hash algorithm used to calculate them (e.g., SHA, DJB2, MD5). This allows for several hashing algorithms being supported and new ones being easily introduced. When the secure task executes, it request the OS image hash to the SE specifying the algorithm, and compares it with the hashes trusted by the rich application calling it. If the hash using the required algorithm is not precomputed, the TEE requests a reboot and adds the algorithm to the SE. Applications can also verify the boot traces to check that all the components they require (e.g., biometric sensor) have been correctly initialized. As a result, rich applications are able to make decisions in run-time depending on both the OS and the available peripherals. This can be seen as a first step towards supporting usage control.

Since the SE is configured as a trusted peripheral, rich applications cannot directly communicate with it; they need to do it through the TEE. Additionally, the SE signs the retrieved hashes using its own private key in a similar manner as the TPM using the Attestation Identity Key (AIK). To distribute the corresponding public keys of the SE, an infrastructure similar to other public-key infrastructures is required (e.g., openPGP).

5.3   Exchange OS/OSBL

We share the concern that current secure boot implementations necessarily lock devices to a specific OS chosen by the OEM. In order to avoid this in our architecture, we propose the **configuration mode**. TEEs are suitable candidates to implement this mode, since they support features for secure user interactions (see Section 3.1). The sequence of actions for the configuration mode starts by the user flashing a new mobile OS or OS bootloader (e.g., using uboot) with a signature that is not trusted by the platform (public key of the OS issuer is not in the list of trusted keys). As depicted in the transition from step 2 to 4 in Figure 1b, the OS will not be booted in that case. The user will now be given the possibility to either manually verify the signature of the new mobile OS or cancel the process within a secure task of the TEE. In case of a requested manual verification, the user will be asked to point to the certificate of the OS issuer on the untrusted memory (e.g., SD card) and enter a shared secret within the secure UI of the TEE (step 4 in Figure 1b). This shared secret could be a PIN or password that has been shipped together with the SE. With an appropriate secure channel protocol (e.g., SRP [24]), the user will be authenticated to the SE and a secure communication between TEE and the applet will be established. If the user does not want to verify the OS, the system would still be booted without access to the sensitive data in the SE (step 6 in Figure 1b). After successful authorization, the secure task sends the public key of the OS issuer to the applet, where it will then be added to the list of trusted keys (step 5 in Figure 1b). If users do not have access to the certificate, or do not want to completely trust the issuer, they can also exclusively sign the specific OS instance with the private/public key pair of the SE. Attacks attempting to flag an OS as trusted will fail as long as the shared secret remains unknown to the attacker. An adversary could also try to manipulate the certificate which is stored in the untrusted memory. However, as the TEE has full network capabilities, it can verify the certificate validity with a correspondent public-key infrastructure, such as the web of trust from PGP.

6 Security Analysis

There are four kinds of attacks that can be perpetrated against the two-phase verification architecture design: (i) attacks against the manufacturer bootloaders to prevent verifying and logging the loaded components during the boot process; attacks against the TEE rich - secure interface through (ii) attacks against the secure monitor, and (iii) attacks against the secure data path; and (iv) physical attacks against the SE to steal the secrets stored in it. Most of the OS specific services we describe either as exploits or defenses assume a Linux based OS. While not all of these services are available in all OSs, they are conceptually independent to a specific implementation.

Bootloaders. Since the root of trust begins with the FSBL and SSBL, a sophisticated software attack that supplants the SSBL could prevent the boot of a legitimate TEE, and therefore prevent the verification and logging of booted components depicted in Figure 1b. While the attacker would gain control of the device and the communication to the SE, the secrets stored in the SE would remain inaccessible at first. Indeed, the SE applet is configured to wait for a trusted system state, thus it will not reveal sensitive information if the correspondent boot hashes are sent. The attacker would need to modify the SSBL so that it is sending the hashes of a normal trusted boot. As the SE is only a passive component, it does not have methods to verify the trustworthiness of the source of a received message. Signing the messages would also not prevent these attacks due to the inability to securely store the private key on the mobile device. While intricate, the attack is theoretically possible. However, we assume that the SSBL is locked by the OEM and additionally verified by the FSBL, as it is the case in current devices. Still, the lack of source verification capability of the SE applet remains an open challenge for the research community. If an attacker only substitutes the OS bootloader, an untrusted OS would be booted without access to the SE. This is already one of the scenarios contemplated as normal operation (i.e. step 6 in Figure 1b).

Secure Monitor. The secure monitor is in charge of switching between the rich and secure environments, thus representing TrustZone's most vulnerable component. If compromised, illegitimate applications could run while the processor is executing in secure mode. This involves prioritized access to all peripherals and memory. The fact that the SMC call for ARMVv7 architectures is implemented in software together with the lack of a standard implementation has led to bad designs like the one reported in October 2013, affecting Motorola devices running Android 4.1.2. (The attack reached the National Vulnerability Database and scored an impact of 10.0 (out of 10.0)[6], and it can be found in the attacker's blog[7]). For ARMv8, however, ARM is providing an open source reference implementation of secure world software called ARM Trusted Firmware. This effort includes the use of Exception Level 3 (EL3) software [9] and a SMC Calling Convention. Organizations such as Linaro are already pushing for its adoption.

Secure Data Path. Attacks to the data paths between rich applications and secure tasks fall into three main categories: channel hijacking, man-in-the-middle (MITM), and denial of service (DoS). Since the SMC call has to stem from a privilege mode, the kernel will always lead the rich-secure communication. This represents a threat since an attacker with root access to the system could modify the kernel at run-time to intercept a data path. Examples of possible attacks include: superseding the return of a secure task, manipulating a rich application's internal state and/or memory (e.g., by using *ptrace*), or attempting a return-oriented programming (ROP) attack. These attacks are only viable if root can inject code to the kernel and access the system's main memory. However, if loadable kernel models such as Linux's LKMs, and access to kernel and physical memory (*/etc/kmem* and */dev/mem* respectively) are disabled, these attacks are not possible. We consider that in the context of mobile devices it is not common to dynamically load and unload kernel modules, and therefore enforcing a static kernel at run-time is a fair compromise. Indeed, popular mobile OSs such as Android are beginning to take similar steps towards limiting kernel functionality; from Android 4.3 debugging tools such as *ptrace* are restricted by SELinux[8], making application code more deterministic and resilient to manipulation and hijacking. If we assume that the OS image that is verified at boot time can be trusted, the fact that the kernel cannot be modified at run-time allows us to guarantee that once a system call is being processed in kernel space, an attacker would not be able to tamper with the secure data paths that the kernel establishes with the secure area. This technique has been used before to protect the kernel from a malicious root [25] [14]. Under this scenario, if rich applications make use of secure system primitives to evaluate the running system and access peripherals securely, and secure tasks to access sensitive data and communicate with their IT infrastructure, channel hijacking and MITM attacks can be prevented. Finally, we consider that preventing DoS attacks when using TEEs is a challenge, and an interesting topic for future research.

---

[6] http://web.nvd.nist.gov/view/vuln/detail?vulnId=CVE-2013-3051
[7] http://blog.azimuthsecurity.com/2013/04/
unlocking-motorola-bootloader.html
[8] http://lwn.net/Articles/491440/

Hardware Attacks. TrustZone is not tamper-resistant, and while SEs are tamper-resistant, they are not tamper proof. The assumption here should be that with enough time, money and expertise and attacker could steal the secrets in the SE by means of a sophisticated physical attack (e.g., a laboratory attack). Having physical access, such a sophisticated attack to reveal the secrets in the SE would mean to either break the security methods (e.g., lock screen) of the running and trusted OS or to imitate a trusted system state to the SE. In the second case, the adversary would need to remove the SE in order to be able to bypass the TZPC, and directly send the boot hashes of the original trusted system to the applet. The SE applet does not have the capability to detect the malicious source of the boot hashes and would therefore grant access to the sensitive data.

7 Conclusion

In this paper, we introduce a two-phase boot verification for mobile devices. The goal is to give users the freedom to choose the OS they want to run in their mobile devices, while giving them the certainty that the OS comes from a source they trust. We extend this certainty to running applications, which can verify the environment where they are executing. This is a first step towards usage control. By not locking devices to specific software, users can switch OSs depending on their social context (e.g., work, home, public network). This protects user's privacy, as well as service providers from untrusted devices. We contemplate this as a necessary change in the way we use mobile devices today, and a natural complement to multi-boot virtualization architectures like *Cells* [2]. One device might fit all sizes, but one OS does definitely not. Finally, since our approach is based on off-the-self hardware, it can be implemented in currently deployed mobile devices.

Acknowledgements. This work has been carried out within the scope of u'smile, the Josef Ressel Center for User-Friendly Secure Mobile Environments. We gratefully acknowledge funding and support by the Christian Doppler Gesellschaft, A1 Telekom Austria AG, Drei-Banken-EDV GmbH, LG Nexera Business Solutions AG, and NXP Semiconductors Austria GmbH.

References

1. The Privacy Engineer's Manifesto, pp. 242–243. Apress (2014)
2. Andrus, J., Dall, C., Hof, A.V., Laadan, O., Nieh, J.: Cells: A virtual mobile smartphone architecture. In: Proceedings of the Twenty-Third ACM Symposium on Operating Systems Principles, pp. 173–187. ACM (2011)
3. Arbaugh, W., Farber, D., Smith, J.: A secure and reliable bootstrap architecture. In: Symposium on Security and Privacy, pp. 65–71 (May 1997)
4. ARM Security Technology. Building a secure system using trustzone technology. Technical report, ARM (2009)
5. Chin, E., Felt, A.P., Greenwood, K., Wagner, D.: Analyzing inter-application communication in android. In: Proceedings of the 9th International Conference on Mobile Systems, Applications, and Services, MobiSys 2011, pp. 239–252. ACM, New York (2011)

6. Doctorow, C.: Lockdown, the coming war on general-purpose computing
7. Gasser, M., Goldstein, A., Kaufman, C., Lampson, B.: The digital distributed system security architecture. In: Proceedings of the 12th National Computer Security Conference, pp. 305–319 (1989)
8. González, J., Bonnet, P.: Towards an open framework leveraging a trusted execution environment. In: Wang, G., Ray, I., Feng, D., Rajarajan, M. (eds.) CSS 2013. LNCS, vol. 8300, pp. 458–467. Springer, Heidelberg (2013)
9. Goodacre, J.: Technology preview: The armv8 architecture. White paper. Technical report, ARM (2011)
10. Halderman, J.A., Schoen, S.D., Heninger, N., Clarkson, W., Paul, W., Calandrino, J.A., Feldman, A.J., Appelbaum, J., Felten, E.W.: Lest we remember: Cold-boot attacks on encryption keys. Commun. ACM 52(5), 91–98 (2009)
11. Höbarth, S., Mayrhofer, R.: A framework for on-device privilege escalation exploit execution on android. In: Proceedings of IWSSI/SPMU (June 2011)
12. Hölzl, M., Mayrhofer, R., Roland, M.: Requirements for an open ecosystem for embedded tamper resistant hardware on mobile devices. In: Proc. MoMM 2013: International Conference on Advances in Mobile Computing Multimedia, pp. 249–252. ACM, New York (2013)
13. Khan, S., Nauman, M., Othman, A., Musa, S.: How secure is your smartphone: An analysis of smartphone security mechanisms. In: Intl. Conference on Cyber Security, Cyber Warfare and Digital Forensic (CyberSec 2012), pp. 76–81 (2012)
14. King, S.T., Chen, P.M.: Backtracking intrusions. ACM SIGOPS Operating Systems Review 37, 223–236 (2003)
15. La Polla, M., Martinelli, F., Sgandurra, D.: A survey on security for mobile devices. IEEE Communications Surveys Tutorials 15(1), 446–471 (2013)
16. Lee, H.-C., Kim, C.H., Yi, J.H.: Experimenting with system and libc call interception attacks on arm-based linux kernel. In: Proceedings of the 2011 ACM Symposium on Applied Computing, pp. 631–632. ACM (2011)
17. Liebergeld, S., Lange, M.: Android security, pitfalls and lessons learned. In: Information Sciences and Systems (2013)
18. Madlmayr, G., Langer, J., Kantner, C., Scharinger, J.: NFC Devices: Security and Privacy, pp. 642–647 (2008)
19. Mobile Phone Work Group. TCG mobile trusted module sepecification version 1 rev 7.02. Technical report (April 2010)
20. Poeplau, S., Fratantonio, Y., Bianchi, A., Kruegel, C., Vigna, G.: Execute this! analyzing unsafe and malicious dynamic code loading in android applications. In: Proceedings of the ISOC Network and Distributed System Security Symposium (NDSS), San Diego, CA (February 2014)
21. Rouse, J.: Mobile devices - the most hostile environment for security? Network Security 2012(3), 11–13 (2012)
22. Trusted Computing Group. TPM main specification version 1.2 rev. 116. Technical report (March 2011)
23. Unified EFI. UEFI specification version 2.2. Technical report(November 2010)
24. Wu, T.: The secure remote password protocol. In: Proc. of the 1998 Internet Society Network and Distributed System Security Symposium, pp. 97–111 (November 1998)
25. Wurster, G., Van Oorschot, P.C.: A control point for reducing root abuse of file-system privileges. In: Proceedings of the 17th ACM Conference on Computer and Communications Security, pp. 224–236. ACM (2010)

MobiHydra: Pragmatic and Multi-level Plausibly Deniable Encryption Storage for Mobile Devices

Xingjie Yu[1,2,3], Bo Chen[4], Zhan Wang[1,2,*],
Bing Chang[1,2,3], Wen Tao Zhu[1,2], and Jiwu Jing[1,2]

[1] State Key Laboratory of Information Security,
Institute of Information Engineering, Chinese Academy of Sciences, China
[2] Data Assurance and Communication Security Research Center,
Chinese Academy of Sciences, China
[3] University of Chinese Academy of Sciences, China
[4] Department of Computer Science, Stony Brook University, USA
{xjyu,zwang,changbing12,jing}@is.ac.cn, chen@chenirvine.org,
wtzhu@ieee.org

Abstract. Nowadays, smartphones have started being used as a tool to collect and spread politically sensitive or activism information. The exposure of the possession of such sensitive data shall pose a risk in severely threatening the life safety of the device owner. Particularly, the data owner may be caught and coerced to give away the encryption keys. Under this circumstances, applying the encryption to data still fails to mitigate such risk.

Plausibly deniable encryption (PDE) promisingly helps to circumvent the coercive attack by allowing the data owner to deny the existence of certain data. In this work, we present MobiHydra, a more pragmatic PDE scheme featuring multi-level deniability on mobile devices. MobiHydra is pragmatic in that it remarkably supports hiding opportunistic data without necessarily rebooting the device. In addition, MobiHydra favorably mitigates the so-called booting-time defect, which is a whistleblower to expose the usage of PDE in previous solutions. We implement a prototype for MobiHydra on Google Nexus S. The evaluation results demonstrate that MobiHydra introduces very low overhead compared with other PDE solutions for mobile devices.

Keywords: Mobile security, plausibly deniable encryption (PDE), data secrecy, coercive attack, countermeasures.

1 Introduction

Many people today perform a majority of their daily communications, web browsing, and financial transactions via their mobile devices, and leave a large amount of sensitive data stored in those devices. Particularly, as mobile phones

---

* Corresponding author.

S.S.M. Chow et al. (Eds.): ISC 2014, LNCS 8783, pp. 555–567, 2014.

spread across the globe, human rights activists are increasingly turning to mobile technology for the help of documenting, visualizing, and prosecuting human rights abuses. Unfortunately, human rights violators often harness similar technologies to silence activists [1]. In order to protect sensitive data, major mobile operating systems now provide different levels of encryption [2–4]. This simple encryption-based solution may work well when a mobile device gets lost or stolen. However, when a mobile device owner is caught and coerced into disclosing his/her decryption keys (which is known as a coercive attack), encryption alone becomes inadequate for protecting the owner's sensitive data. An example of coercive attack against mobile devices is that a human rights worker uses his/her mobile device to collect evidences of atrocities in a region of oppression, but unfortunately he/she is captured and forced to hand over the evidences.

Plausibly deniable encryption (PDE) [5] is a promising tool that helps to circumvent coercive attack and allows the data owner to deny the existence of certain data. In the literature [6–8], PDE has been investigated extensively for desktop operating systems. However, there are very few developments of PDE on mobile devices where coercive attacks are more likely to occur. Mobiflage [9] was the first PDE solution designed for mobile devices by customizing Android full disk encryption (FDE) [4] to offer plausible deniability. A mobile device which is equipped with Mobiflage works in two operation modes, the standard mode and the PDE mode. The standard mode is used to manage the regular data and accessed by entering a public password, which can be disclosed in emergency. The PDE mode, however, is used to manage sensitive data that can be denied of their existence. The PDE mode can only be activated by entering a hidden password. When facing a coercive attacker, the smartphone owner can simply disclose the public password, such that the attacker is only able to access the standard mode. Since the PDE-enabled device does not have any indication of the existence of hidden deniable files, the attacker will likely be convinced that the device owner has not kept any sensitive data and release the owner.

Although Mobiflage [9] initiates the research of PDE on mobile devices, it has some limitations. First of all, hiding data requires booting the device into the PDE mode, which may bring inconvenience to users and even worse, a user may not have enough time to reboot his/her device when an attacker suddenly appears. Second, Mobiflage can only support one deniability level, that is, a Mobiflage user can either keep or disclose all the sensitive data in the presence of a coercive attack. This will be problematic if an attacker insists the existence of the sensitive data. Third, Mobiflage is vulnerable to a new attack due to its design flaw during booting, which may compromise plausible deniability.

In this work, we take a holistic view over the existing PDE solutions, and propose MobiHydra, a novel PDE system for mobile devices, in which we alleviate the limitations of Mobiflage [9]. Our MobiHydra can support secure data hiding in the standard mode for emergency, offering a convenient feature to MobiHydra users. In this way, device owners can work at the standard mode and transfer the sensitive data to the PDE mode without rebooting. In addition, MobiHydra introduces another feature, multiple deniability levels, with which users can choose to store sensitive data at different deniability levels. To the best of our knowledge, we are the first to design a PDE storage system for mobile devices offering both features of hiding data without rebooting and multi-level deniability. Moreover, we integrate certain countermeasures into MobiHydra to mitigate the booting-time defect.

In this paper, we make the following technical contributions:

- We identify a previously unreported booting-time defect, which can be exploited as a whistle-blower to compromise plausible deniability offered by previous PDE solutions for mobile devices. We also develop countermeasures to mitigate this attack.

- We propose MobiHydra, in which we design novel techniques to support hiding data without rebooting and multiple deniability levels. Particularly, we utilize an additional *shelter* volume to support hiding data in the standard mode, and we implement multiple hidden volumes to support multiple deniability levels, such that multi-level deniability can be achieved.

- We theoretically analyze MobiHydra's security guarantee. We also implement MobiHydra on a Google Nexus S smartphone powered by Android 4.0 and experimentally evaluate its performance.

Due to the space limit, a full technical description and a complete security analysis can be found in our technical report [10].

2   Background and Related Work

To support Android full disk encryption (FDE), the Android encryption layer is implemented with dm-crypt [4]. First of all, a randomly chosen master volume key is encrypted with another key derived from 2000 iterations of PBKDF2 (a password-based key derivation function) [11] digest of the user's screen-unlock password and a salt value. Both the encrypted master key and the salt value are stored in the footer, which is located in the last 16KB of the userdata partition. When an Android device is booted and cannot find a valid file system on the userdata partition, it will require the user to enter the password. If a valid file system is then found in the dm-crypt target, it will be mounted and the system will boot regularly.

Deniable encryption was firstly introduced by Canetti et al. [5]. Deniable encryption allows an encrypted message to be decrypted to different meaningful plaintexts, some of which can be used as decoy messages depending on the key or passphrase being used. Anderson et al. [8] designed the first file encryption scheme with PDE support, which is termed as steganographic file system. The steganographic file system can achieve the same objective as a PDE cipher by hiding files in random data. Rubberhose [12] for Linux is the first known instance of a PDE-enabled storage system. Other steganographic file systems [7,13,14] focused on improving the efficiency and reliability. Moreover, a few desktop disk encryption tools [6] can support PDE by allocating hidden volumes on the storage devices.

3   Threat Model and Assumptions

Threat Model. We consider an adversary who is able to fully control a mobile device after capturing it, including a root-level access to the device and a full physical control over the device's internal and external storage. In addition, the adversary can coerce the device's owner for the secrets to access the device.

Assumptions. We rely on several assumptions:

- The adversary cannot capture a mobile device working in the PDE mode and has no knowledge of the PDE key and password which can allow accessing to the PDE mode. Moreover, the adversary is not able to snapshot a mobile device's encrypted physical storage before having captured the device.
- The adversary is rational: it has motivations to coerce the mobile device's owner to reveal the encryption keys and passwords (e.g., secrets for unlocking the screen of the mobile device), but will stop forcing the owner once it is convinced that the secrets have been revealed; it will not hold the user indefinitely or simply punish the user without any evidence of the existence of hidden data.
- MobiHydra must be merged with the default Android code stream, such that its capability is widespread, i.e., an attacker cannot simply distinguish a device with PDE capability. The mobile device has a physical or an emulated FAT32 external (SD or eMMC) storage partition. In Section 5.5, we will discuss more on this limitation. In addition, the device is malware-free in both the standard mode and the PDE mode.

4   Booting-Time Attack

Attack Design. For a mobile device enabling the regular full disk encryption (FDE) without PDE support, when the device's user enters a password, the device will try to mount a valid file system (Section 2). It will test the correctness of the password by decrypting the master volume key to mount a valid file system. The system will boot successfully if the password is correct, otherwise, it will require the user to enter the password again. Thus, in FDE, the pre-boot authentication for both the valid password and the wrong password will not differentiate too much. For a mobile device enabling PDE, when a valid public password is provided, it will boot into the standard mode by successfully mounting the outer volume. However, when a wrong password is provided, the system will first try to boot into the standard mode, and will then try to boot into the PDE mode by trying to mount the hidden volume. Only after the password has failed to activate the PDE mode, the user will be required to enter the password again. A wrong password in Mobiflage is tested for both the public password and the hidden password, which requires much more pre-boot authentication, and hence takes much longer time than when a valid public password is provided.

Based on the above observations, we identify a novel booting-time attack, which can compromise the plausibly deniability offered by Mobiflage. Suppose a mobile device is equipped with Mobiflage, and the device's owner (i.e., the Mobiflage user) has been captured by a coercive attacker $\mathcal{A}$.

1. $\mathcal{A}$ coerces the owner to disclose the password. $\mathcal{A}$ then boots the system with the public password disclosed by the owner several times and records each time interval between the moment when the password is entered and the moment when the operating system starts to boot up. We denote this time interval as t_succ.

2. $\mathcal{A}$ then uses wrong passwords to boot the system multiple times and records each time interval between the moment when the wrong password is entered and the moment when the retry screen appears. We denote this time interval as t_retry.

3. $\mathcal{A}$ makes a decision on whether PDE is present based on the statistical analysis of t_succ and t_retry. If significant statistical deference is observed between instances of t_retry and those of t_succ, then $\mathcal{A}$ can conclude that the device is protected by PDE.

Experimental Validation. To evaluate the effectiveness of the booting-time attack on Mobiflage, we targeted a Google Nexus S Android phone powered by Mobiflage. We depict the instances of t_succ and t_retry in Figure 1(a). As a comparison, we perform the same experiment on the same device with FDE enabled only and depict the instances of t_succ and t_retry in Figure 1(b). We observe that for Mobiflage, t_retry is at least 50% longer than t_succ. However, for a regular FDE-enabled system, t_retry is approximately 30% shorter than t_succ. The difference between Mobiflage and FDE, unfortunately, will offer the adversary a clear indication of the deployment of PDE. Note that for different devices, the specific time characteristics may be a little different, but the statistical time characteristics will be similar.

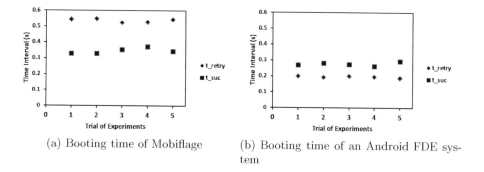

(a) Booting time of Mobiflage   (b) Booting time of an Android FDE system

Fig. 1. A booting-time attack on Nexus S

5   MobiHydra Design

5.1   Overview

MobiHydra works in two modes: the standard mode for daily operations, providing encrypted storage without deniability; and the PDE mode for storing sensitive data with deniable encryption. The standard mode and the PDE mode are activated by a public password and one of the multiple hidden passwords, respectively. The public password and the hidden passwords are specified by the user at the initialization. When booting the device, a password is entered by the user. The system first makes an attempt to decrypt the master public volume key with a password-derived key for mounting the public volume. If it fails, MobiHydra will calculate an offset with the supplied password, and try to mount a hidden volume onto the file system mount point where physical storage would be normally mounted.

MobiHydra enables encryption storage with plausible deniability by hiding sensitive data in hidden volumes. Such volumes are located in the empty space on the mobile device's external storage, and each of those is coordinated to one deniability level. If the user get caught with the device working in the standard mode or powered-off, he/she can feign compliance by relinquishing the public password. In some circumstances, the adversary may not be convinced by the public password, and thus continue intimidating the user to reveal the hidden passwords. The user can give away one or two hidden passwords which are associated with relatively lower deniability levels to avoid safety threat and keep more sensitive data in secret.

To cover up such hidden volumes, the external storage is first filled with randomly generated bits, ensuring that the encrypted hidden volumes are indistinguishable from empty space. The number of hidden volumes is specified by the user at the initialization. Then MobiHydra will format and encrypt the public and hidden volumes as specified by the user. Moreover, a special partition called *shelter* volume on the external storage is also allocated and used as a temporary storage for data transfer from the standard mode to the PDE mode.

MobiHydra provides a safe mechanism for hiding opportunistic files when emergency events take place. In the standard mode, files which are needed to be denied of their existence are stored in the *shelter* volume. Such files are encrypted and cannot be decrypted in the standard mode. When the device is booting into the PDE mode in a safe environment, the opportunistic files will be retrieved and decrypted in the PDE storage. After a successful retrieval, the *shelter* volume will be wiped out securely, to ensure data secrecy and leave enough storage space for future opportunistic files.

5.2   Pragmatic Hiding Support

Principle. The basic idea for hiding data without rebooting the device into the PDE mode is that a sensitive file can be saved in a *shelter* volume temporarily in the standard mode, and will be transferred to the hidden volume automatically when the user boots the device into the PDE mode. The opportunistic file is protected by the same symmetric encryption function used for hidden volumes to keep it undistinguishable from the noise-like random numbers.

In the standard mode, if the user has to hide opportunistic data for emergency, MobiHydra will encrypt such data with a randomly generated symmetric key and save the ciphertext in a *shelter* volume. In order to keep such files unreadable in the standard mode. The random symmetric key will be encrypted with a public key which is generated at the initialization of MobiHydra, and saved in the *shelter* volume. The asymmetric private key is kept in hidden volumes which cannot be accessed from the standard mode. When the PDE mode is activated, MobiHydra will decrypt the symmetric key of the *shelter* volume with the private key saved in the mounted hidden volume to decrypt the opportunistic data saved in the *shelter* volume. After that, such decrypted data will be transferred to the hidden volume and then securely deleted from the *shelter* volume to release storage space. Therefore, the opportunistic data will be transferred to any hidden volume mounted at the first time of activating the PDE mode after storing these data in the standard mode.

Shelter Volume. A small part of the external storage is allocated specifically for the *shelter* volume while other remaining storage is used for the public volume and hidden volumes. To avoid overwriting the hidden volume by writing opportunistic data in the *shelter* volume across the volume border, the offset of this *shelter* volume is close to the end of the external storage. The *shelter* volume is mounted as a block device in the standard mode.

However, since *shelter* volume is mounted as an independent block device in the standard mode, the available space of the external storage is less than its real physical storage in the file system. It may make the adversary suspect the existence of a *shelter* volume. A deniable explanation for the unavailable storage could be given by attributing it to system influence. In this way, the *shelter* volume should be limited to a size that small enough to be reasonable for system influence, which may reduce the data transfer capability. However, since the *shelter* volume is used for emergency data transfer, there is no need to assign a massive storage space for such volume. We suggest that the user need to transfer the opportunistic data to hidden volume as soon as convenient to ensure enough space for next emergency.

5.3   Multi-level Deniability

MobiHydra offers multiple hidden volumes which are corresponding to different deniability levels, and this is how the name MobiHydra comes where "Hydra" is a many-headed serpent in Greek mythology [15].

Storage Layout. The storage space can be regarded as the concatenation of independent encrypted volumes, including one public volume, one shelter volume and multiple hidden volumes. We formalize the storage layout as follows:

$$E_{K_p}(Vol_{pub})||E_{K_1}(Vol_{h1})||E_{K_2}(Vol_{h2})||...||E_{K_n}(Vol_{hn})||E_{K_s}(Vol_{shel})$$

Here, $E_K(\cdot)$ represents a symmetric encryption function with key K and $\parallel$ represents concatenation. Vol_{pub}, Vol_{shel} and Vol_{hi} denote the public volume, *shelter* volume and the i-th hidden volume, respectively. K_p and K_s represent a master volume key of the public volume and *shelter* volume. n represents the required number of hidden volumes specified by the user, and K_1 to K_n represent the master keys of hidden volumes.

The deniability of Vol_{hi} enhances along with the increment of i. When a right password (either a public password or a hidden password) is supplied for system boot, to avoid a visible limit on the mounted volume, the volume decrypted by a given key will appear to consume all remaining space (except the space allocated for the shelter volume), and other volumes will appear to random noise. Thus, the existence of $Vol_{hi,i>1}$ could be denied by relinquishing the password associated with $Vol_{hx,x \in [1,i-1]}$, and the existence of Vol_{h1} may be denied if the user gives away the public password. In addition, the exposure of $Vol_{hi,i>1}$ may expose the existence of $Vol_{hx,x \in [1,i-1]}$ which is detailed in Section *Offset Calculation*. Therefore, if the adversary forces the user to reveal hidden password besides the public password, the user can fake compliance by giving away one or two hidden passwords associated with relatively lower deniability levels, and the data saved in the hidden volumes relatively associated with higher deniability levels are still secret.

Note that the hidden volumes may be overwritten by writing to the currently mounted volume past the volume boundary. This issue is inherent to PDE storage solutions, which is usually addressed by keeping data replicas on desktop [7,12]. However, considering the limited storage space on mobile devices, MobiHydra only protects data secrecy rather than data integrality.

Offset Calculation. The offset for shelter volume is is derived from the password provided by the user at the PDE installation time. The offset calculation for the i-th hidden volume is formalized as follows:

$$offset(i) = m \times vlen + (i-1)/N \times (1-m) \times vlen + H(pwd_i \| salt) \bmod \lfloor 1/kN \times (1-m) \times vlen \rfloor$$

Where $vlen$ denotes the number of 512-byte sectors on the logical block storage device; m represents the proportion of the storage space that is only occupied by the public volume, which is decided by the system; H is a PBKDF2 iterated hash function; pwd_i is the password of the hidden volume with the deniability level of i; N is the largest number of hidden volumes supported by the system; $salt$ (which is also used for the public volume key derivation) is a random value for PBKDF2 and k is a real number larger than 1 and helps to randomize the offset point. We further explain each variable as follows:

– N is a constant decided by the system, and each hidden volume's size is approximately $(1-m)/N$ of external storage size except for the n-th one (consuming the storage space between its offset and the *shelter* volume's offset) regardless of the number of needed hidden volumes specified by the user (represented by n in Section *Storage Layout*). Note that in our design, if the user gives away the passwords associated with Vol_{hi}, the adversary can calculate the space between the offset of this volume and the end of the public volume which may indicate the presence of other hidden volumes $(Vol_{hx,x\in[1,i-1]})$.

- i is not supplied by the user for system boot for the sake of keeping consistence with Android FDE, since any different user interaction will give an indication of the usage of PDE tool. Instead, i will be traversed from 1 to N for offset generation until a valid volume is mounted. Moreover, this design ensures that the required time for testing a wrong password gives no indication of the value of n.

- k should be specific based on a trade-off between security requirement and storage utilization. The offset of a certain hidden volume randomizes in the range of 0 to $\lfloor 1/kN \times (1 - m) \times vlen \rfloor$, therefore, to avoid overwriting the next volume, we suggest that the data stored in a hidden volume should be within $(1 - m)/N \times vlen \times (1 - 1/k) \times 512$ bytes except for the n-th hidden volume. However, it results in a waste of storage space. For the purpose of reducing the waste of storage, k should be set to a relatively large integer which will result in a corresponding small offset random range.

5.4 Mitigating the Booting-Time Attack

In order to make the time characteristic of password verification identical between MobiHydra and a regular Android FDE system, we try to mitigate the booting-time defect by manipulating the verification time for correct password. According to our experiment, if $3\times N$ extra invocations of PBKDF2 are performed as a dump operation after a password is authenticated as the right password (where N is the supported level of deniability in MobiHydra), the characteristics of t_retry and t_succ will be identical in MobiHydra and an Android FDE system as shown in Section 6.

5.5 Known Issues

MobiHydra shares parts of limitations of Mobiflage scheme, including the need of a separate physical FAT32 storage partition and the choice between allowing the user to configure the volume size and avoiding the need to store the offset. Besides, our current MobiHydra design has the following unsolved issues:

1. The size of the *shelter* volume cannot be changed once it has been set at the initialization of the PDE system. Dynamically scaling the size of the *shelter* volume may overwrite the data stored in the hidden volumes.
2. For the sake of simplicity, the opportunistic data stored in the *shelter* volume are visible to all the hidden volumes, so that the user cannot set the deniability level of them. A possible solution is to generate a unique pair of RSA keys for each deniability level, save the private key in each corresponding hidden volume and save the public keys in the *shelter* volume. When a file is stored in the *shelter* volume, MobiHydra will encrypt the file with a random master key and wrap the random master key with its corresponding public key. In this fashion, only the hidden volume with correct private key can retrieve the opportunistic data appropriately.

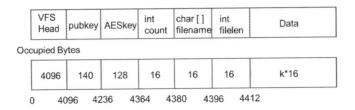

Fig. 2. Content of the *shelter* volume

6   Implementation and Evaluation

We implemented a prototype of MobiHydra on a Google Nexus S smartphone based on Android 4.0 (ICS). Our Nexus S has 1 GB internal and 15GB eMMC external storage (i.e., eMMC partition).

6.1   Prototype Implementation

Pragmatic Hiding Support. We mount the *shelter* volume as a block device, which is linked to */dev/shelter_vol* through a hard link. This volume consumes the last 128MB of the external storage. The content of the *shelter* volume is shown in Figure 2. For experimental purpose, we develop a camera app which can save images to this *shelter* volume when the device works in the standard mode.

MobiHydra generates a pair of 1024-bit RSA keys during its initialization, and saves the public key in the *pubkey* field. Meanwhile, the private key is saved in a file /SDcard/prkey, which will be stored in each hidden volume. In the standard mode, once the camera app has captured an opportunistic image, MobiHydra will encrypt the image by a randomly chosen AES-XTS key, and save the encrypted image in the *Data* field. This AES-XTS key will be encrypted by the public key and saved in the *AESkey* field. The number of files will be kept in the *int count* field. For simplicity, our current prototype only supports 16 files. In addition, the *AESkey* field in the current prototype can only store one AES-XTS key. These limitations can simply be addressed by extending such fields to be able to support more files and AES-XTS keys.

When the PDE mode is activated, MobiHydra will decrypt the *AESkey* field using the private key stored in */SDcard/prkey* and decrypt the *int count* field with the acquired AES-XTS key. If the number of files falls within the range of 1 to 16, those files saved in the *shelter* volume will be decrypted and saved to the corresponding hidden volume. After that, the data stored in the *shelter* volume, starting from the *AESkey* field, will be wiped and replaced by random data. If the number of files does not fall within the range of 1 to 16, no further process for data transfer will be performed.

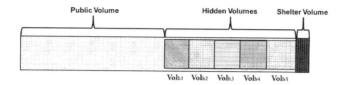

Fig. 3. Storage layout of our MobiHydra prototype

Table 1. Performance comparison between Android FDE, Mobiflage and MobiHydra

system	Key length (bits)	Initialization (s)	IO speed (MB/s)	Boot time (wrong password) (s)	Boot time (public password) (s)
Default FDE	128	85.4	8.5	0.19	0.27
Mobiflage	512	7399	8.2	0.54	0.35
MobiHydra	512	7421	8.1	1.49	1.98

Multi-level Deniability. For simplicity, our prototype only supports up to 5 deniability levels, and it is easy to be extended to support more deniability levels. The storage layout for our prototype is shown in Figure 3. In our prototype, 50% of the external storage is only occupied by the public volume on the account of frequent usage of it for daily operations. If the user initializes 5 hidden volumes at the initialization, the size of each hidden volume is around 1GB. If the user needs more space for one hidden volume, he/she can choose less hidden volumes during initialization. For this case, the last hidden volume can consume the remaining external storage starting from its own offset (except the space allocated for the *shelter* volume). The master volume key for each hidden volume is stored at the storage sector located at the corresponding hidden volume's offset. The master volume key for the public volume is encrypted and stored in a footer located in the last 16KB of the userdata partition. The random salt is also stored in this area. We fix k at 8 to achieve a good tradeoff between the security requirement and the storage utilization (only resulting in a waste of 1GB storage space).

6.2   Experimental Evaluation

Initialization Performance. We average the required time for initialization by initializing each system three times, and the results are recorded in Table 1. The initialization time significantly increases in both MobiHydra and Mobiflage compared to Android FDE, as a result of an expensive two-pass random wipe of the external storage and encryptions over the hidden volumes.

I/O Performance. We use the adb-shell tool to test the I/O performance for each aforementioned system. We run 10 trials, during each of which we write a 100MB file to the external storage (i.e., eMMC partition). The results in Table 1 show that MobiHydra has a similar I/O performance compared to Mobiflage.

Table 2. Booting time for the PDE mode

PDE system	Booting time with hidden password(s)	
Mobiflage	10.7	
MobiHydra	pde_1	0.62
	pde_2	0.86
	pde_3	1.09
	pde_4	1.31
	pde_5	1.62

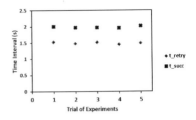

Fig. 4. Booting time of MobiHydra

System Boot Performance. We average the boot time in Table 1 and 2, in which pde_n represents the time for booting the device into the deniability level n. MobiHydra needs to try the wrong password for the public volume and all the hidden volumes during pre-boot authentication, which results in a longer time for testing a wrong password compared to either Mobiflage or Android FDE. Meanwhile, the countermeasures to the booting-time attack used in MobiHydra result in a longer boot time for the standard mode. In addition, the time needed to boot the PDE mode in MobiHydra increases with the deniability levels, because MobiHydra tests a given password starting from the lowest deniability level, and keeps testing until it can find a target deniability level. Moreover, the time needed to boot the PDE mode of MobiHydra ranges around 1 second in MobiHydra, which is significantly shorter than 10.7 seconds as it is in Mobiflage. This is because Mobiflage always tries to un-mount a persistent log partition when booting a system. We optimize this process by judging whether or not the device has such a partition before trying to un-mount it.

Mitigating the Booting-Time Attack. To mitigate the booting-time attack, MobiHydra performs 15 additional invocations of PBKDF2 as dump operations when a correct public password is provided. Figure 4 shows the statistical characteristics of t_retry and t_succ for MobiHydra. t_retry is approximately 30% shorter than t_succ, i.e., the deviation between t_retry and t_succ is similar to that of a default FDE system (Figure 1(b)). Note that the slow boot could be attributed to the poor device performance if it arouses suspicion.

7   Conclusion

We propose MobiHydra to help the device owner circumvent the coercive attack. MobiHydra provides a secure mechanism for pragmatic data hiding, allowing a user to save deniable data in the standard mode without rebooting the device. In addition, MobiHydra supports multi-level deniability, which reduces the possibility of losing all the sensitive data at a time. Moreover, MobiHydra can mitigate the booting-time attack, which is a common vulnerability for all the previous PDE solutions relying on hidden volumes to offer plausible deniability.

Acknowledgments. The authors were partially supported by the Independent Research Project of State Key Laboratory of Information Security under Grant Y4D0021203, the Strategic Priority Research Program of Chinese Academy of Sciences under Grant XDA06010702 and the National Natural Science Foundation of China under Grant 61272479. Xingjie Yu, Zhan Wang and Jiwu Jing were partially supported by the National 973 Program of China under Grant 2014CB340603. Wen Tao Zhu was partially supported by the National 973 Program of China under Grant 2013CB338001.

References

1. SXSW Schedule. Caught in the Act: Mobile Tech & Human Rights (2014), www Document: http://schedule.sxsw.com/2014/events/event_IAP21063

2. Windows Inc. BitLocker Drive Encryption (2014), www Document: http://windows.microsoft.com/en-us/windows7/products/features/bitlocker

3. Google Inc. Linux Unified Key Setup (2014), www Document: https://code.google.com/p/cryptsetup/

4. Google Inc. dm-crypt: Linux kernel device-mapper crypto target (2014), www Document: https://code.google.com/p/cryptsetup/wiki/DMCrypt

5. Canetti, R., Dwork, C., Naor, M., Ostrovsky, R.: Deniable Encryption. In: Kaliski Jr., B.S. (ed.) CRYPTO 1997. LNCS, vol. 1294, pp. 90–104. Springer, Heidelberg (1997)

6. FreeOTFE. FreeOTFE - Free disk encryption software for PCs and PDAs. version 5.21 (2012), Project website: http://www.freeotfe.org/

7. Pang, H., Lee Tan, K., Zhou, X.: StegFS: A Steganographic File System. In: 19th International Conference on Data Engineering, ICDE 2002 (2002)

8. Anderson, R., Needham, R., Shamir, A.: The Steganographic File System. In: Aucsmith, D. (ed.) IH 1998. LNCS, vol. 1525, pp. 73–82. Springer, Heidelberg (1998)

9. Skillen, A., Mannan, M.: On Implementing Deniable Storage Encryption for Mobile Devices. In: 20th Annual Symposium on Network and Distributed System Security (NDSS 2013) (2013)

10. Yu, X., Chen, B., Wang, Z., Chang, B., Zhu, W.T., Jing, J.: MobiHydra Technical Report (2014), www Document: http://www.chenirvine.org/publications/MobiHydra_technical_report.pdf

11. Kaliski, B.: PKCS 5: Password-based cryptography specification, version 2.0. RFC 2898 (informational) (2000)

12. Assange, J., Weinmann, R.-P., Dreyfus, S.: Rubberhose: Cryptographically Deniable Transparent Disk Encryption System (1997), Project website: http://marutukku.org

13. Han, J., Pan, M., Gao, D., Pang, H.: A Multi-user Steganographic File System on Untrusted Shared Storage. In: 26th Annual Computer Security Applications Conference, ACSAC 2010 (2010)

14. Zhou, X., Pang, H., Tan, K.-L.: Hiding Data Accesses in Steganographic File System. In: 20th International Conference on Data Engineering, ICDE 2003 (2003)

15. Wikipedia. Hydra (2014), www Document: http://en.wikipedia.org/wiki/Hydra

Author Index

Physics of Semiconductors

B. Sapoval   C. Hermann

Physics of
Semiconductors

With 157 illustrations

Springer-Verlag
New York Berlin Heidelberg London Paris
Tokyo Hong Kong Barcelona Budapest

B. Sapoval
C. Hermann
Laboratoire de Physique de la
Matière Condensée
Unité de Recherche Associée du
Centre de la Recherche Scientifique
Ecole Polytechnique
91128 Palaiseau Cedex
France

Translator
A.R. King
Department of Astronomy
University of Leicester
Leicester LE1 7RH
United Kingdom

French edition published by Ellipses, Paris, 1988.

Library of Congress Cataloging-in-Publication Data
Sapoval, B.
   [Physique des semi-conducteurs. English]
   Physics of semiconductors B. Sapoval, C. Hermann.
      p. cm.
   Translation of: Physique des semi-conducteurs.
   Includes bibliographical references and index.
   ISBN-13:978-0-387-40630-5   e-ISBN-13:978-1-4612-4168-3
   DOI: 10.1007/978-1-4612-4168-3
   1. Semiconductors I. Hermann, C. II. Title.
QC611.S2613 1993
537.6'22 − dc20   93-286

Printed on acid-free paper.

Production managed by Henry Krell; manufacturing supervised by Jacqui Ashri.
Photocomposed copy prepared from the authors' LaTeX files.

9 8 7 6 5 4 3 2 1

ISBN 0-387-94024-3 Springer-Verlag New York Berlin Heidelberg
ISBN 3-540-94024-3 Springer-Verlag Berlin Heidelberg New York

Preface

The discovery of semiconductors is one of the great scientific and technological breakthroughs of the 20th century. It has caused major economic change, and has perhaps changed civilization itself. Silicon, for example, now plays as important a role in our lives as carbon did in the 19th century. Most of the information technology depends on the properties of semiconductors. One can only be struck by the contrast between our world and one without transistors, computers, rockets, medical image processing and heart pacemakers.

We can see that this development is built on the combination of new and old concepts: miniaturization and printing. If we wish to handle information by a machine, it is clear that the machine's "moving parts" must be as small as possible. Here the parts are the electrons. The invention of the transistor at the end of the 1940s, that made use for the first time of the physics of semiconductors, was the key to miniaturization.

Producing such a machine was difficult because of its small size, and it was not feasible to produce on a mass scale. The introduction of planar technology at the beginning of the 1960s changed that situation. It allowed for the use of photogravure techniques that resemble the printing process. Instead of having to link components one by one, like the individual letters were before Gutenberg, we can now make an entire machine such as a microprocessor through a limited number of processes. The very cheap mass production of these machines has begun to cause industrial and cultural changes that stem from, and are limited by, the physics of semiconductors.

Why should we teach the physics of semiconductors in a course at the Ecole Polytechnique? The main reason is that it applies the most fundamental concepts of quantum and statistical mechanics. We hope to show that it is possible to use these concepts easily so as to meet the needs of the engineer. For this reason, several devices that make use of this physics are described. We give a simple explanation of the principle of the most common systems based on semiconductors.

The appendices that we have included serve two distinct functions: they may give detailed justifications for results given in the main text or illustrate various applications. This book can therefore be used in either an elementary or a more advanced manner. In the first form it is at the level of the second cycle of the Ecole, while the more advanced form is at the third cycle in French universities.

The contents of this book are more in depth than what is currently taught at the Ecole Polytechnique. This follows the tradition of the courses at the Ecole that provide the engineer of tomorrow with a scientific basis for much of his career.

This work owes much to the remarks and criticisms of Yves Quéré, Henri Alloul, Hervé Arribart, Henri-Jean Drouhin, Guy Fishman, Georges Lampel, Gilles de Rosny, Jacques Schmidt and Claude Weisbuch. Some of the problems are based on work by Hervé Arribart, Maurice Bernard, Jean-Noël Chazalviel and Georges Lampel. Very special thanks are due to Jean-Noël Chazalviel and François Ozanam for their numerous comments.

The French version of this book was produced by the Ecole Polytechnique Press and published by Edition Marketing Ellipses (Paris).

Contents

1.

Simple Ideas about Semiconductors

1.1 Definition and Importance of Semiconductors

The solids known as semiconductors have been the subject of very extensive research over recent decades, not simply because of their intrinsic interest but also because of ever more numerous and powerful applications: rectifiers, transistors, photoelectric cells, magnetometers, solar cells, reprography, lasers, and so forth.

A main feature of many of these applications is the possibility of miniaturization of the devices. Miniaturization is more than a convenience: if we are faced with the problem of coding and transmitting messages from a satellite, the complete system of computer and transmitter must be made small. It must work properly for long periods without maintainance. The power available on board the satellite must come from radiation, the only source possible in space. Semiconductor devices, transistors, and solar cells provide solutions to these problems. Similarly the electronic components of a heart pacemaker have to consume little power and be very small. But the most spectacular and most important application of semiconductors is the development of information technology. These developments have only been possible because of the miniaturization of the logic elements allowing the construction of compact systems with great computing power or memory.

Miniaturization has become possible through the perfection of "planar" fabrication techniques. These allow "integration" of circuits and thereby the production of devices containing thousands of elements on a few mm^2. All this industrial development has come into existence only because physics allows us to understand the specific properties of semiconductors, and then use this understanding to create "electron machines" in the form of semiconductor devices.

Semiconductors, as we shall see, are insulators whose "forbidden bands" or "gaps" are sufficiently narrow that thermal excitation allows a small number of electrons to populate the "conduction band." The working element in a semiconductor is this small number of electrons. It is clear that this small number of electrons can be influenced by a small number of chemical impurities or even by the surface of the crystal.

This sensitivity hindered the understanding of the properties of these materials for a long time. For several decades the crystals studied were not pure enough, so that a number of their properties appeared to be impossible to reproduce with other apparently "identical" crystals. Thus the development of semiconductor physics has had to await progress in semiconductor chemistry, and indeed the chemistry of solids in general. Semiconductors are now the purest and most reproducible solids we can make. The techniques perfected in their manufacture are frequently applied in other branches of chemistry and solid-state physics. There is a very close connection in this subject between industrial requirements, control of materials, and the understanding of the phenomena.

The definition of semiconductors as "insulators with narrow forbidden bands" should be supplemented by a description of the essential physical properties of these materials, namely:

1. their resistivity decreases as the temperature rises, at least for a certain temperature range, unlike metals;

2. semiconductors are sensitive to visible light but transparent in the infrared. When irradiated their resistivity decreases. If they are inhomogeneous, an induced electric field may appear;

3. they often give rise to rectifying or non-ohmic contacts;

4. they exhibit a strong thermoelectric effect, i.e., an electric field induced by a temperature gradient;

5. their resistivity lies between 10^{-5} and 10^6 ohm·cm.

The materials possessing such properties are the elements of column IV of the Periodic Table, silicon and germanium; III – V compounds of the type GaAs, GaSb, InSb, InP, and so forth; IV – VI compounds such as PbS, PbSe, PbTe, and so forth; II – VI compounds such as CdSe, CdTe, and Cu_2S; ternary compounds such as $Al_xGa_{1-x}As$; and quaternary compounds.

There are several important dates in the history of semiconductor physics.

1897: Discovery of the Hall effect: When a magnetic field is applied to a conductor carrying a current perpendicular to the field, an electric field appears in the direction perpendicular to the current and the magnetic field. The strength of the electric field allows one to measure the number of mobile charge carriers carrying the electric current. Measurements made at the beginning of the century show the existence of a small or very small number of mobile charges varying from sample to sample in an apparently incoherent way. This number is of the order of 10^{-3} to 10^{-7} per atom. The sign of the Hall electric field also allows one to determine the sign of the charge carriers. Surprisingly, in certain crystals this sign was observed to be positive, suggesting that these charges were cations. However, the observed mobilities were very large, much greater than the mobilities of ions in liquid electrolytes. Indeed, they were comparable to those measured in apparently identical crystals but in which the negative sign of the mobile charges indicated that the carriers were electrons. This was an unexplained paradox.

1926: Bloch's theorem, with its fundamental consequence: a Bloch wave packet can traverse a perfect crystal without colliding with the crystal ions. Collisions result only from crystal defects or vibrations in a perfect crystal. This idea allowed an understanding of the large mobilities observed for electrons.

1931: Wilson lays the foundations of the modern theory of semiconductors as insulators with narrow forbidden bands and introduces the idea of a hole.

1948: Discovery of the transistor effect by John Bardeen and William H. Brattain.

1960: Appearance of planar technology.

1982: World production of 3×10^{13} binary units of active memory in the form of 64 kilobyte (512 kilobit) units alone.

1990: Manufacture of Dynamical Random Access Memories ("DRAM") of 4 megabits per chip.

1991: High Definition Television camera with 2 Megapixel Charge Coupled Device Sensor.

1992: Semiconductor component sales, worldwide: $60 billion.

1993: World production of transistors: 2.10^{17}

1995: Manufacture of 64-megabit DRAM (estimated).

1.2 A Chemical Approach to Semiconductors

Even though some properties of semiconductors were discovered experimentally in the course of the 19th century, an understanding of the origin of this behavior had to await the advent of quantum mechanics.

The first, classical, theory of electrical conductivity in solids, proposed by Drude in 1900, assumed the current to be transported by a fixed number of electrons that behave like classical particles obeying Maxwell–Boltzmann statistics. In the presence of an applied electric field the electrons attain a velocity proportional to the field (Ohm's law) as they constantly undergo collisions that brake their motion (see Chap. 5). While a number of properties of metals could be understood, nothing in this model predicted the increase in the number of charge carriers with temperature in semiconductors observed via the Hall effect. One might appeal to thermal ionization of the electrons from individual atoms of the solid, but since the ionization energies are of order 10 eV, this effect is too weak at room temperature to account for the observed concentrations.

In any case, the Drude model had no explanation at all for the fact that in some samples the mobile charges were positive.

1.2a The Contribution of Quantum Mechanics

We know that in a physical system, here a solid, electrons occupy stationary energy levels which are the solutions of the Schrödinger equation

$$\mathcal{H}\psi = E\psi, \tag{1.1}$$

where the Hamiltonian

$$\mathcal{H} = \frac{p^2}{2m} + V(\mathbf{r}) \tag{1.2}$$

takes account of both the kinetic energy $p^2/2m$ of the electrons, and the potential $V(\mathbf{r})$ of their interaction with the ions of the solid. Here ψ is the wave function and E the energy associated with ψ. The mass of a free electron is m.

The electrons have half-integer spin and are thus fermions: at most two electrons of opposite spins can occupy each orbital state ψ satisfying Schrödinger's Eq. (1.1). In a solid in thermal equilibrium at temperature T, the energy levels must be populated according to Fermi–Dirac statistics. To understand the properties of semiconductors (or solids in general) we therefore have to first find the energy levels satisfying Eq. (1.1). The second step is to find the state of the system at temperature T by populating these levels according to Fermi–Dirac statistics. We can then examine the properties of this system, which is the aim of this book. This procedure is simple in principle, but its implementation encounters considerable difficulties, and we have to resort to approximations.

1.2b Qualitative Description

One can imagine two main ways of understanding at least qualitatively the properties of quantum electron states in solids. We can think of the crystal constructed from atoms, which we bring together by introducing a coupling between them: this is the "chemical" approach we shall follow initially. Or instead we can start with a solid viewed as a "box of electrons," initially empty of ions, and progressively impose the attraction of the ions. This is the "nearly free electron" method, which we shall develop in Sect. 1.3.

1.2c The Chemical Approach

In the chemical approach we first consider two distant atoms each having one electron. As the atoms are brought closer together, the electrons around each of the nuclei will begin to feel the potential caused by the other nuclei. This potential is a perturbation which lifts the degeneracy more and more effectively as the distance between the atoms is decreased (Fig. 1.1). This holds both for the ground and excited states.

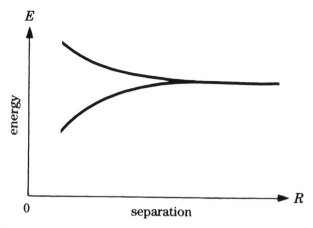

Fig. 1.1. Splitting of an energy level through coupling between two atoms.

In the ground state, the descending branch of Fig. 1.1 is populated by two electrons of opposite spins. This causes a decrease of the system energy when the two atoms are brought closer together, and thus gives rise to a chemical bond, the **covalent bond**. If there are three atoms we start with triply degenerate levels; and in a crystal, with N-fold degeneracy. When the atoms are close, the degeneracy is lifted but several energy levels of the same atom are mixed. This is shown schematically in Fig. 1.2 where the vertical bars show the allowed energies as a function of the separation R of the atoms. The permitted energy bands are separated by forbidden bands. This picture corresponds to **purely covalent** semiconductors such as silicon and germanium. When the atoms constituting the crystal are different, for example in gallium arsenide GaAs, we start with non-degenerate atomic levels. The bond is then **partially covalent**. We shall return in Chap. 2 to the covalent bond. If the relative positions of the bands and their filling by the electrons are such that at zero temperature a band is just full and the band immediately above is empty, we have an insulator, as in such a system a weak electric field cannot increase the electron energies by accelerating them. In fact the only available states are very far away in energy because of the existence of a forbidden energy band: energies near to the initial value are not allowed. Acceleration is thus impossible. The electrons cannot respond to a small electric field, and the system is an insulator. The lower, full, band is called the valence band, while the upper empty band is the conduction band. The forbidden energy region between them is often called the band gap.

The exact determination of the energy bands is difficult; worse, the physical behavior of the solid will be determined by the nature and properties of the levels in small fringes ΔE_c and ΔE_v around the forbidden band. In fact in thermal equilibrium at finite temperature a number of electrons

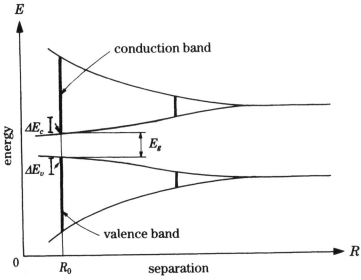

Fig. 1.2. Energy levels in a system of N atoms as a function of the separation R between the atoms. The equilibrium atomic separation is R_0.

populate the conduction band and leave empty states in the valence band. It is clear that the most populated states in the conduction band will be those at the lowest energies, while the empty places in the valence band will correspond to the highest energies allowed in that band. We thus see that observable electron properties will depend on the details of the band structure around the energy extrema on each side of the gap. We thus need to know the band structure very well in the neighborhood of the band gap, something which is very difficult in view of the approximations we have to adopt.

The other approach to understanding the properties of quantum states of the crystal starts from solution of Eq. (1.1) in the case where the crystal potential $V(\mathbf{r})$ is zero, and introduces this potential as a perturbation. We thus get solutions which are linear combinations of plane waves. But the same difficulty appears: the energy scale of the plane waves we have to consider is of the order of atomic energies, i.e., 10 eV. Correct states are linear combinations of a large number of states distributed over a wide energy range, while we are interested in the detailed behavior of only one part of the spectrum. Here too we are faced with approximations which are in conflict with the desired accuracy.

1.3 Quantum States of a Perfect One-Dimensional Crystalline Solid

We consider the case of a linear chain of period a, whose total length L contains N periods, where N is an integer. We seek the eigenstates of the Hamiltonian

$$\mathcal{H} = \frac{p_x^2}{2m} + V(x) \tag{1.3}$$

with a periodic potential

$$V(x + a) = V(x). \tag{1.4}$$

1.3a Nearly Free Electron Model

One way to proceed is to assume that $V(x)$ is small, and regard this term as a perturbation by comparison with the kinetic energy term. For the free electron Hamiltonian

$$\mathcal{H}_0 = \frac{p_x^2}{2m}, \tag{1.5}$$

the eigenfunction of the state $|k>$ is a plane wave:

$$\psi_k(x) = \frac{1}{\sqrt{L}} \exp ikx \tag{1.6}$$

and the corresponding energy is

$$E = \frac{\hbar^2 k^2}{2m}. \tag{1.7}$$

The dispersion curve $E(k)$ is a parabola. To express the fact that the chain of atoms has a finite length $L = Na$, we use the periodic boundary conditions of Born and von Kármán, which consist of joining the chain to itself. We then have

$$\psi(x + L) = \psi(x). \tag{1.8}$$

From Eq. (1.6) we find that the allowed values of k are quantized:

$$k = l_x \frac{2\pi}{Na} \tag{1.9}$$

with l_x an integer.

In a qualitative approach to bringing out the characteristic properties of a crystalline solid, we may confine ourselves to the simplest approximation of a periodic potential: a sine potential of period a:

$$V(x) = 2V_1 \cos \frac{2\pi x}{a}, \qquad (1.10)$$

$$= V_1 \left[\exp\left(i\frac{2\pi x}{a} \right) + \exp\left(-i\frac{2\pi x}{a} \right) \right]. \qquad (1.11)$$

For a macroscopic crystal N is very large, of the order of $(10^{23})^{1/3}$, and the allowed values of k are thus extremely close: the characteristic distance in k space, defined starting from the period (1.10) in x, is of the order of $2\pi/a$, so that the allowed values of k given by Eq. (1.9) are spaced by an amount $2\pi/L$ and are indeed very close.

The origin of the abscissa is chosen as one of the ions, and the sign of the Coulomb interaction between the ions and the electrons (attractive) requires $V_1 < 0$. We assume that V_1 is small compared with the relevant kinetic energies and use perturbation theory to find its effect on a state $|k>$. Let us look for states $|k'>$ coupled to $|k>$ by V: the matrix element of V between the two states $|k>$ and $|k'>$ is

$$< k'|V|k > = \frac{1}{L} \int_0^L \exp\left(-ik'x\right) V(x) \exp\left(ikx\right) dx$$

$$= \frac{V_1}{L} \int_0^L \left[\exp i\left(k - k' + \frac{2\pi}{a} \right) x + \exp i\left(k - k' - \frac{2\pi}{a} \right) x \right] dx.$$

$$(1.12)$$

The two exponentials are periodic functions, and the integral is non-zero only for

$$k - k' = \pm \frac{2\pi}{a}, \qquad (1.13)$$

for which its value is V_1. The potential V thus only couples states differing in k by $\pm 2\pi/a$, values of the wave vectors which appear in the Fourier series for V and which express the periodicity of the crystal. In particular, the first-order correction $< k|V|k >$ to the state $|k>$ vanishes.

If the energy of the state $|k'>$ before the introduction of the crystalline potential differs from that of $|k>$, the introduction of the periodic potential has essentially no effect. If by contrast the two independent states coupled by V have the same unperturbed energy, this simultaneously imposes

$$\frac{\hbar^2 k^2}{2m} = \frac{\hbar^2 k'^2}{2m} \qquad (1.14)$$

and

$$k' = -k.$$

Then Eq. (1.13) requires

$$k = \pm \frac{\pi}{a} \tag{1.15}$$

and the presence of V lifts the degeneracy between the states $|\pi/a>$ and $|-\pi/a>$. Projected onto this basis the Hamiltonian becomes

$$\mathcal{H} = \begin{pmatrix} \frac{\hbar^2}{2m}\left(\frac{\pi}{a}\right)^2 & V_1 \\ V_1 & \frac{\hbar^2}{2m}\left(\frac{\pi}{a}\right)^2 \end{pmatrix}. \tag{1.16}$$

The eigenstates of $\mathcal{H}$ are

$$\psi_- = \frac{1}{\sqrt{2}}\left(\psi_{\frac{\pi}{a}} + \psi_{-\frac{\pi}{a}}\right) = \sqrt{\frac{2}{L}} \cos\frac{\pi}{a}x,$$

$$\text{with } E_- = \frac{\hbar^2}{2m}\left(\frac{\pi}{a}\right)^2 - |V_1|, \tag{1.17}$$

$$\psi_+ = \frac{1}{\sqrt{2}}\left(\psi_{\frac{\pi}{a}} - \psi_{-\frac{\pi}{a}}\right) = i\sqrt{\frac{2}{L}} \sin\frac{\pi}{a}x,$$

$$\text{with } E_+ = \frac{\hbar^2}{2m}\left(\frac{\pi}{a}\right)^2 + |V_1|. \tag{1.18}$$

The lowest energy state has an enhanced probability density at the ions (Fig. 1.3).

We see that in the energy interval

$$\frac{\hbar^2}{2m}\left(\frac{\pi}{a}\right)^2 - |V_1| < E < \frac{\hbar^2}{2m}\left(\frac{\pi}{a}\right)^2 + |V_1| \tag{1.19}$$

there is no stationary energy eigenstate. The presence of the periodic electrostatic potential $V = 2V_1\cos(2\pi x/a)$ leads to the existence of a forbidden energy region, or band gap, near $k = \pm\pi/a$. Figure 1.4 shows the resulting lifting of the degeneracy.

We note that applying the same reasoning to the periodic potential

$$V = 2V_l \cos\frac{2\pi}{a}lx, \text{ with } l \text{ an integer} \tag{1.20}$$

we obtain coupling between the states

$$k = \pm l\frac{\pi}{a} \tag{1.21}$$

producing a forbidden band for energies:

$$\frac{\hbar^2}{2m}\left(\frac{l\pi}{a}\right) - |V_l| < E < \frac{\hbar^2}{2m}\left(\frac{l\pi}{a}\right)^2 + |V_l|. \tag{1.22}$$

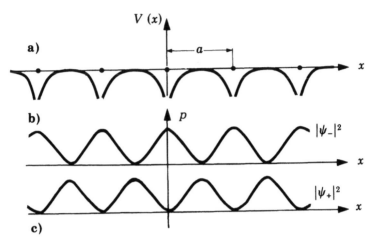

Fig. 1.3. (a) Variation of the electrostatic potential energy of an electron in the Coulomb field of the ions of a linear chain of atoms; (b) and (c) probability densities for the electron for the states ψ_- and ψ_+ (Eqs. (1.17) and (1.18)) after the introduction of a small periodic potential of the form $V(x) = 2V_1 \cos 2\pi x/a$. In the lowest-energy ψ_- state the probability density is maximal at the ions.

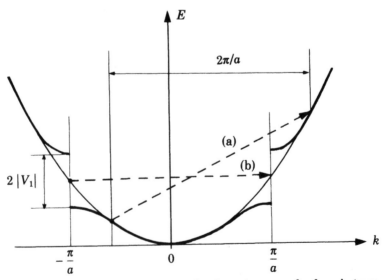

Fig. 1.4. The lightly drawn parabola is the dispersion curve for free electrons. The heavy curve takes account of the effect of the crystalline potential $V = 2V_1 \cos 2\pi x/a$. States coupled by this potential (relation (1.13)) are linked by a dashed arrow: (a) states which are non-degenerate in the absence of the periodic potential; (b) states which are degenerate in the absence of $V(x)$.

More generally, any periodic potential with period a can be written as

$$V = \sum_{l'} 2V_{l'} \cos \frac{2\pi l' x}{a}, \text{ with } l' \text{ an integer.} \tag{1.23}$$

The wave vectors $2\pi l'/a$ appearing in the Fourier decomposition (1.23) of V define a periodic set of period $2\pi/a$ in the space of wave vectors (reciprocal space). The set of points $2\pi l'/a$ constitutes the reciprocal lattice whose generalization to three dimensions we will see in Sect. 2.1. The vectors k defined by Eq. (1.21), around which a band gap opens, are the basic wave vectors of the decomposition (1.23) of V.

By an approximate method, completely different from the chemical approach described in Sect. 1.2b, we have just shown the existence of band gaps in crystalline solids. We will now show that band gaps are expected on general grounds on the basis of Bloch's theorem.

1.3b Bloch's Theorem

In the Bloch theory the electrons, assumed independent, feel a **periodic** crystalline potential $V(x)$. According to Bloch's theorem, the eigenfunctions of

$$\left[\frac{p^2}{2m} + V(x) \right] \psi(x) = E\psi(x), \tag{1.24a}$$

where $V(x + a) = V(x)$ have the form

$$\psi_k(x) = \exp(ikx)\, u_k(x). \tag{1.24b}$$

The term $\exp(ikx)$ describes the variation at large scales, while $u_k(x)$, which has the periodicity of the crystal,

$$u_k(x + a) = u_k(x) \tag{1.25}$$

expresses the variation of the wave function within an elementary cell. The wave vector k, which is real, no longer plays the same role as for the plane wave (1.6). We note that the plane wave is a solution of the type (1.24) for which $u_k(x) = \text{constant}$.

Applying the Hamiltonian to the Bloch function yields the equation satisfied by $u_k(x)$:

$$\left[\left(\hbar k - i\hbar \frac{\partial}{\partial x} \right)^2 + V(x) \right] u_k(x) = E_k\, u_k(x). \tag{1.26}$$

In this equation k plays the role of a parameter, and we can limit x to the interval $[0, a[$ since $u_k(x)$ is periodic. The equation has a form similar to the Schrödinger equation for an atom, and we know that this has discrete

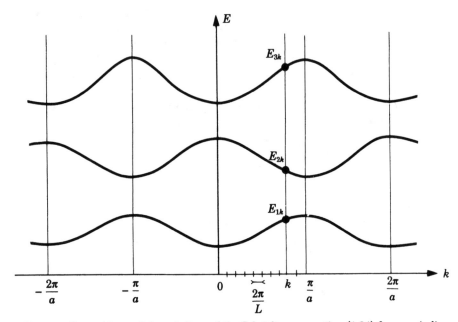

Fig. 1.5. General form of the solutions of the Schrödinger equation (1.24) for a periodic crystalline potential. The solutions $E(k)$ are periodic, with period $2\pi/a$.

eigensolutions, denoted $u_{n,k}(x)$, of energy $E_{n,k}(x)$ $(n = 1, 2, ...)$ (Fig. 1.5). As k is varied, the energy eigenvalues describe an "energy band" for each integer value of n. Whenever there is no overlap in energy between bands with different indices n there appear energy ranges with no stationary values $E_{n,k}$. These are the band gaps.

Their definitions show that the two quantum numbers n and k have quite different meanings. The band index n is an integer. The allowed values of the wave vector k are fixed by the surface boundary conditions of the solid: we could take once again the periodic Born–von Kármán conditions (1.8), which impose the same values (1.9) on k as before; in that case, k varies in a quasi-continuous fashion, with jumps of $2\pi/L$.

The energies $E_{n,k}$ of the Bloch states are periodic functions of k. The Bloch function $\psi_{n,k}$ can also be written as

$$\psi_{n,k}(x) = \exp i\left(k + l\frac{2\pi}{a}\right)x \ u_{n,k}(x) \ \exp\left(-il\frac{2\pi}{a}x\right), \qquad (1.27)$$

where l is an integer, or equivalently,

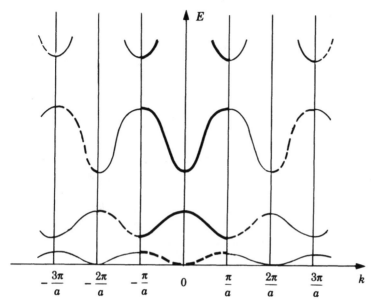

Fig. 1.6. Dispersion curves $E(k)$ in a crystal. The periodicity of the solutions $E(k)$ appears in the lighter curves. The heavy curves show the reduced zone scheme, while the dashed curves represent the extended zone scheme.

$$\psi_{n,k}(x) = \exp i\left(k + l\frac{2\pi}{a}\right)x \cdot u_{n,k+l2\pi/a}(x). \tag{1.28}$$

As $u_{n,k+l2\pi/a}$ is periodic, $\psi_{n,k}$ is also a Bloch function for $k' = k + l2\pi/a$. The associated energy $E_{n,k}$ is thus also an eigenvalue for the state $|k + 2\pi l/a >$, and the energies are periodic in the space of wave vectors k, with period $2\pi/a$.

There are thus two equivalent ways of classifying the eigenstates of the crystal Hamiltonian. Using the periodicity of the solutions of its eigenvalue equation we can arrange to confine ourselves simply to the variation of k over the interval $[-\pi/a, \pi/a[$: for each value of k in this interval there exist discrete solutions labelled by the integer index n. This situation is described by the heavy curves in Fig. 1.6; it is called the **reduced zone scheme**. By contrast we may retain just one branch of the successive curves $E(k)$ for each value of k: the lowest for $-\pi/a \leq |k| < \pi/a$, the second branch for $\pi/a < |k| \leq 2\pi/a$, and so on. We thus obtain the dashed curves in Fig. 1.6; this is called the **extended zone scheme**.

These two descriptions are strictly equivalent. The extended zone description is reminiscent of the parabola $E(k)$ of free electrons in a box. The reduced zone description carries the discrete index n and is reminiscent of the quantization of atomic levels: for $k = 0$, Eq. (1.26) actually gives solutions of atomic type.

We have used quantum mechanics to show that the energy levels of a solid group themselves into permitted and forbidden bands. Let us now see how Fermi–Dirac statistics dictate the filling of these levels.

1.3c Level Filling

The boundary conditions (1.8) on the wave function at the sample surface confine k to the values $k = l_x 2\pi/Na$, where l_x is an integer. Moreover the discussion of crystalline solids in Sect. 1.3b allows us to characterize each electron eigenstate by its band index n and its wave vector k. As we have seen, restricting k to the interval $[-\pi/a, \pi/a[$ gives a full description of the properties of the solid. l_x takes exactly N values over this range. For each value of l_x and thus of k there are two independent allowed spin states, leading to $2N$ possible states per band.

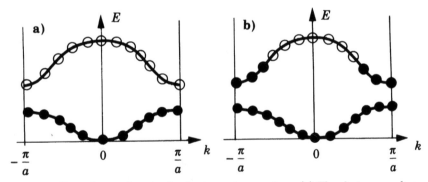

Fig. 1.7. Filling of crystal energy levels at zero temperature. (a) The electron number z per cell is even, and the solid is an insulator; (b) z is odd, and the solid is a metal. Filled circles represent occupied states and hollow circles empty states.

For a linear crystal composed of a medium with z electrons per elementary cell, the total number of electrons in the line is Nz. At zero temperature, the Pauli principle dictates the filling of the levels, starting with the states of lowest energy. If the Nz electrons completely fill one or more bands but leave others empty, i.e., if z is even and the bands do not overlap (Fig. 1.7(a)), it would require an electric field E of order E_g/ea, where E_g is the width of the band gap and e the electron charge, to give an electron enough energy to lift it into the first available energy state, at the bottom of the first empty band. For a normal medium, where E_g is of the order of 1 eV, this requires an electric field $\mathcal{E} \simeq 3 \times 10^9$ V/m, more than the breakdown field! In such a medium there is no acceleration in an external field, thus no conduction, and the medium is an **insulator**.

We denote as a semiconductor an insulator whose band gap between the last filled band, called the valence band, and the first empty band, called the conduction band, is narrow enough to allow thermal excitation of charge carriers at room temperature, i.e., a significant concentration of electrons in the conduction band.

If by contrast z is an odd number (Fig. 1.7(b)), the filling leaves one or more bands only partly full at zero temperature. The first excited state is very close to the last filled state, and the system can respond to even a weak electric field. The medium is therefore metallic, as the electrons can be accelerated by an electric field.

In summary, this discussion of a one-dimensional crystal introduces the following basic ideas:

– an electron state is characterized by two quantum numbers (n, k);

– the range of variation of k can be restricted to the reduced zone $[-\pi/a, \pi/a[$;

– because of the crystalline potential, not all energies are accessible to the electrons. Permitted bands are separated by band gaps;

– the filling of the electron energy levels according to the Pauli principle shows that, depending on the medium and the form of the energy bands, the solid behaves as an insulator or a metal. If the number of electrons per cell is even and the bands do not overlap, the solid is an insulator. It is a metal in the opposite case. A semiconductor is an insulator with a "narrow" band gap.

The existence of a band gap thus explains three characteristics of semiconductors:

– The conductivity rises with temperature since thermal excitation gives a conduction band population which increases with T. In this band there are many empty electron states able to accomodate electrons which can be accelerated by even a weak electric field.

– The existence of a forbidden energy band explains the transparency of semiconductors to infrared radiation; photons with energy $h\nu$ smaller than E_g cannot be absorbed, as the electron cannot reach a final state within the forbidden band.

– In contrast, if $h\nu$ exceeds E_g, electrons can be excited into the conduction band by absorbing photons. This explains the existence of "photoconductivity" in semiconductors.

2.

Quantum States of a Perfect Semiconductor

We have seen in the case of a one-dimensional periodic solid that quantum mechanics provides a basis for understanding the properties of semiconductors. Here we discuss the physics of electrons in perfect semiconductors, i.e., those without defects or impurities. We will see in later chapters that certain properties of these materials depend quite directly on the presence of such impurities or defects in the crystal lattice.

The characteristic properties of a solid are determined by the distribution of energy levels called the band structure. From this band structure we can say whether the medium is an insulator, a conductor, or a semiconductor.

We have to know the band structure for other reasons. In the presence of the periodic crystal potential the response of an electron to an external force is no longer determined by its mass, but by an "effective mass" imposed by the band structure. The effective mass of an electron in a crystal may be very different from the electron mass itself: it can even be negative.

2.1 Quantum States of a Three-Dimensional Crystal

A crystal is made up of the periodic repetition of an elementary basis. More precisely, if a particular point of the structure in equilibrium has coordinates $\mathbf{r}$, all the points in equilibrium having the same physical and chemical "environment" as $\mathbf{r}$ are expressible as

$$\mathbf{r}' = \mathbf{r} + m_1\mathbf{a}_1 + m_2\mathbf{a}_2 + m_3\mathbf{a}_3, \tag{2.1}$$

where the m_i $(i = 1, 2, 3)$ are whole numbers and the $\mathbf{a}_i$ are three non-coplanar vectors specific for the particular crystal. The rhombohedron constructed from $\mathbf{a}_1, \mathbf{a}_2, \mathbf{a}_3$ makes up the cell. The "smallest" cell, with the smallest volume, is called the primitive or unit cell. The vectors $\mathbf{a}_i$ which generate this primitive cell are the periods of the Bravais lattice, or direct lattice, of the crystal.

Semiconductors formed from elements of column IV of the Periodic Table (Si, Ge), and most semiconductors formed from elements of column III and elements from column V (GaAs, InP) crystallize in the face-centered cubic system shown in Fig. 2.1. Some semiconductors of type II–VI like CdS crystallize in a hexagonal form.

In the face-centered cubic system the primitive cell is not a cube but the rhombohedron shown in Fig. 2.1 (or any equivalent cell, e.g., that formed from the basis vectors $-\mathbf{a}_1, \mathbf{a}_2, -\mathbf{a}_3$). For the most usual semiconductors the cell consists of two atoms, one at the position $(0,0,0)$, and the other one-quarter of the way along the diagonal of the cube (Fig. 2.2). The primitive cell thus contains a basis of two atoms. For simplicity, and to show the crystal symmetries, we generate the crystal from a cubic cell which is not the primitive cell. **The lattice of points generated by the translations (2.1) of the primitive cell is called the Bravais lattice.**

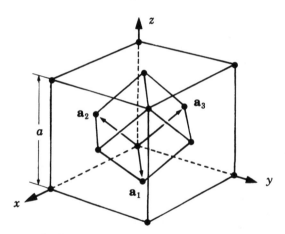

Fig. 2.1. Face-centered cubic lattice. The rhombohedron constructed from the vectors $\mathbf{a}_1 = (a/2)(\mathbf{i}+\mathbf{j})$, $\mathbf{a}_2 = (a/2)(\mathbf{j}+\mathbf{k})$, $\mathbf{a}_3 = (a/2)(\mathbf{i}+\mathbf{k})$ forms the primitive cell.

A macroscopic solid, of characteristic size L contains a very large number of elementary cells, of the order of $(L/a)^3$, where a is the characteristic size of the cell.

In an infinite crystal, electrons at displacements of

$$\mathbf{T} = m_1\mathbf{a}_1 + m_2\mathbf{a}_2 + m_3\mathbf{a}_3 \tag{2.2}$$

feel the same crystal potential, implying that the Hamiltonian is invariant under translations $\mathbf{T}$ of the lattice.

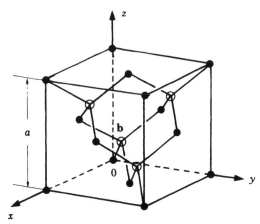

Fig. 2.2. Crystalline structure of silicon. The basis consists of two atoms, that at the origin and atom **b** with coordinates $(a/4)(1, 1, 1)$. The Bravais lattice is face-centered cubic.

2.1a Bloch's Theorem

Solutions of the Schrödinger equation

$$\left[\frac{p^2}{2m} + V(\mathbf{r}) \right] \psi = E\psi, \tag{2.3}$$

where the crystal potential $V(\mathbf{r})$ has the periodicity of the crystal, are called "Bloch functions," and have the form

$$\psi_{n,\mathbf{k}}(\mathbf{r}) = \exp(i\mathbf{k}.\mathbf{r})\, u_{n,\mathbf{k}}(\mathbf{r}), \tag{2.4}$$

where the function $u_{n,\mathbf{k}}$ is periodic in the direct lattice:

$$u_{n,\mathbf{k}}(\mathbf{r} + \mathbf{T}) = u_{n,\mathbf{k}}(\mathbf{r}). \tag{2.5}$$

A state is thus specified by four quantum numbers, n and the three components k_x, k_y, k_z of the vector $\mathbf{k}$. From Eq. (2.5),

$$\psi_{n,\mathbf{k}}(\mathbf{r} + \mathbf{T}) = \exp(i\mathbf{k}.\mathbf{T})\, \psi_{n,\mathbf{k}}(\mathbf{r}). \tag{2.6}$$

The points $\mathbf{r}$ and $\mathbf{r}+\mathbf{T}$ thus have the same physical properties, the functions differing only by a phase factor independent of $\mathbf{r}$. We note further that $|\psi_{n,\mathbf{k}}(\mathbf{r})|^2$ is periodic in space.

As in Chap. 1, we choose as boundary conditions at the edges of the macroscopic solid the periodic boundary conditions of Born and von Kármán:

$$\psi(\mathbf{r} + \mathbf{L}) = \psi(\mathbf{r}). \tag{2.7}$$

We assume that the vector $\mathbf{L}$ is a vector of the Bravais lattice, corresponding to a very large number of periods. For simplicity we choose an elementary cell and a solid in the shape of a rectangular parallelepiped. Imposing condition (2.7) on the form (2.4) of the Bloch function requires the values of $\mathbf{k}$ to be quantized:

$$k_x = l_x \frac{2\pi}{L_x}, \quad k_y = l_y \frac{2\pi}{L_y}, \quad k_z = l_z \frac{2\pi}{L_z}, \tag{2.8}$$

where (l_x, l_y, l_z) are whole numbers or zero. For a macroscopic solid these values of $\mathbf{k}$ are very close on the scale of $1/a$, since $L >>> a$. The vectors $\mathbf{k}$ play an essential role in the description of the properties of solids. We can regard them as belonging to a space "reciprocal" to the crystal itself: **the reciprocal space is generated by the basis vectors $\mathbf{a}_i^*$ derived from the basis vectors $\mathbf{a}_j$ of the Bravais lattice by**

$$\mathbf{a}_i^* \cdot \mathbf{a}_j = 2\pi \delta_{ij}. \tag{2.9}$$

Any vector $\mathbf{G}$ of the reciprocal lattice has the form

$$\mathbf{G} = h\mathbf{a}_1^* + k\mathbf{a}_2^* + l\mathbf{a}_3^* \tag{2.10}$$

(h, k, l) being whole numbers. In one dimension we would have $\mathbf{G} = (2\pi/a)\mathbf{i}$.

The reciprocal lattice of a face-centered cubic lattice with side a is a body-centered cubic lattice with side $4\pi/a$ (as in Si, GaAs, ...) (Fig. 2.3). The reciprocal lattice of a hexagonal lattice is a hexagonal lattice (as in CdS, ...).

From the definition (2.9) it follows that for any translation $\mathbf{T}$ of the direct lattice defined by Eq. (2.2) and for any vector $\mathbf{G}$ of the reciprocal lattice given by Eq. (2.10), we have

$$\exp i\mathbf{G} \cdot \mathbf{T} = 1. \tag{2.11}$$

The relation (2.11) shows the symmetrical roles of the direct and reciprocal lattices (the reciprocal of the reciprocal lattice is the direct lattice). We recall that the real crystal is a lattice of atoms or molecules, or more generally of bases. The reciprocal lattice is a network of points independent of the basis in the real crystal.

Let us consider a Bloch function relative to a wave vector $\mathbf{k}$, such that $\mathbf{k} = \mathbf{k}_0 + \mathbf{G}$, where $\mathbf{G}$ is a vector of the reciprocal lattice, and show that this Bloch function $\psi_{n,\mathbf{k}}(\mathbf{r})$ is also a Bloch function for the wave vector $\mathbf{k}_0$:

$$\begin{aligned} \psi_{n,\mathbf{k}} &= \exp(i\mathbf{k} \cdot \mathbf{r})\, u_{n,\mathbf{k}}(\mathbf{r}) \\ &= \exp(i\mathbf{k}_0 \cdot \mathbf{r})\, \exp(i\mathbf{G} \cdot \mathbf{r})\, u_{n,\mathbf{k}}(\mathbf{r}). \end{aligned} \tag{2.12}$$

The function $\exp i\mathbf{G} \cdot \mathbf{r}\, u_{n,\mathbf{k}}(\mathbf{r})$ has the property (2.5). In fact by using Eqs. (2.5) and (2.11) we deduce

$$\exp i\mathbf{G} \cdot (\mathbf{r} + \mathbf{T})\, u_{n,\mathbf{k}}(\mathbf{r} + \mathbf{T}) = \exp i(\mathbf{G} \cdot \mathbf{r})\, u_{n,\mathbf{k}}(\mathbf{r}) \tag{2.13}$$

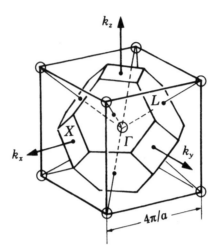

Fig. 2.3. Reciprocal lattice of the face-centered cubic lattice of silicon. This is a body-centered cubic lattice of side $4\pi/a$. We also show the first Brillouin zone. This is bounded by the perpendicular bisector planes of the vectors $(2\pi/a)(1,1,1)$ and their equivalents, producing the hexagonal faces with center L and also by the perpendicular bisector planes of the vectors $(4\pi/a, 0, 0)$ and their equivalents, corresponding to the square faces with center X. The origin Γ is chosen at the atom of the cubic lattice situated at the center of the cube.

which shows that the latter function is periodic. In consequence the function (2.12) $\exp(i\mathbf{k}_0 \cdot \mathbf{r})\ \exp(i\mathbf{G} \cdot \mathbf{r})\ u_{n,\mathbf{k}}\ (\mathbf{r})$ is a solution of the Schrödinger equation simultaneously for $\mathbf{k}$ and for $\mathbf{k}_0 = \mathbf{k} - \mathbf{G}$. The reciprocal space is thus inconveniently "too large" for classifying the Bloch functions, since it contains all the points $\mathbf{k}$ and $\mathbf{k} - \mathbf{G}$, for any $\mathbf{G}$, whose quantum states are identical.

2.1b The Brillouin Zone

We can thus reduce the area of study of Bloch states to those values of $\mathbf{k}$ belonging to the "**first Brillouin zone.**" **The first Brillouin zone is the volume of the reciprocal space closer to the original node $k = 0$ than to any other point of the reciprocal lattice.** This first Brillouin zone is bounded by the perpendicular bisector planes of the shortest vectors $\mathbf{G}$ of the reciprocal lattice. It has the same volume as the elementary cell of the reciprocal lattice. In one dimension it corresponds to the interval $[-\pi/a, \pi/a[$. In three dimensions one of the edges is part of the first zone, but not the one opposite to it.

The first Brillouin zone is shown for a face-centered cubic Bravais lattice (Si, Ge, GaAs, ...) in Fig. 2.3. If the side of the cube is a, the coordinates of the points of high symmetry of the reciprocal lattice shown on the figure are $(\pi/a)(1,1,1)$ for L and $(2\pi/a,0,0)$ for X. The coordinates of points equivalent under the symmetries of the cube are easily deduced.

If we restrict the range of $\mathbf{k}$ to the first Brillouin zone we say we are in the "restricted zone scheme," in contrast to the "extended zone scheme." Similarly, as in one dimension, for a fixed value of $\mathbf{k}$ the Schrödinger equation has a discrete set of solutions, which define, as $\mathbf{k}$ varies, the permitted energy bands, separated by forbidden bands. The index n of the Bloch function (2.4) is thus the band index; when $\mathbf{k}$ varies the eigenvalue $E_{n,\mathbf{k}}$ of Eq. (2.3) spans the nth energy band. The accessible values (2.8) of $\mathbf{k}$ are very close on the scale of the Brillouin zone, so we can regard $E_{n,\mathbf{k}}$ as a quasi-continuous function $E_n(\mathbf{k})$.

We assume without proof that **the functions $\psi_{n,\mathbf{k}}$ (r) form a complete orthonormal basis:**

$$< n,\mathbf{k}|n',\ \mathbf{k}' > = \delta_{n,n'}\delta_{\mathbf{k},\mathbf{k}'}$$

$$\sum_{n\mathbf{k}} |n\mathbf{k}> < n\mathbf{k}| = 1. \tag{2.14}$$

The matrix elements of a periodic operator between states $|n,\mathbf{k}>$ and $|n',\mathbf{k}'>$ of Bloch function form vanish unless $\mathbf{k} = \mathbf{k}'$. This is shown in Appendix 2.1.

2.1c Inversion Symmetry of Constant Energy Surfaces in k-space

If $\psi_{n,\mathbf{k}}$ (r) satisfies

$$\left[\frac{p^2}{2m} + V\ (\mathbf{r})\right] \psi_{n,\mathbf{k}}\ (\mathbf{r}) = E_{n,\mathbf{k}}\ \psi_{n,\mathbf{k}}\ (\mathbf{r}), \tag{2.15}$$

we see on taking the complex conjugate of this equation

$$\left[\frac{p^2}{2m} + V(\mathbf{r})\right] \psi_{n,\mathbf{k}}^*\ (\mathbf{r}) = E_{n,\mathbf{k}}\ \psi_{n,\mathbf{k}}^*\ (\mathbf{r}) \tag{2.16}$$

that $\psi^*(\mathbf{r})$ is an eigenstate with the same eigenvalue. But $\psi_{n,\mathbf{k}}^*$ (r) $= u_{n,\mathbf{k}}^*$ (r) $\exp(-i\mathbf{k}\cdot\mathbf{r})$ is a Bloch function for the point $-\mathbf{k}$ of the Brillouin zone, and thus orthogonal to $\psi_{n,\mathbf{k}}$ (r), with the corresponding eigenvalue $E_{n,-\mathbf{k}}$. We deduce that

$$E_{n,\mathbf{k}} = E_{n,-\mathbf{k}}. \tag{2.17}$$

We can thus confine ourselves to studying the dispersion relations $E_n(\mathbf{k})$ in one-half of the zone. This property is a consequence of the reality of the Hamiltonian, which arises from the invariance under time reversal of the laws of microscopic mechanics. We deduce that at the point $\mathbf{k} = 0$ the relations $E_n(\mathbf{k})$ have the forms 1, 2, or 3, the set of curves $E_n(\mathbf{k})$ being even (Fig. 2.4).

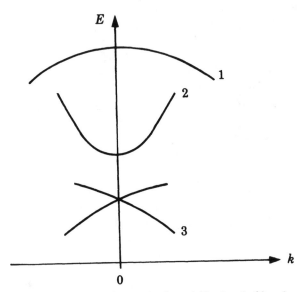

Fig. 2.4. Behavior of dispersion relations in the neighborhood of $\mathbf{k} = 0$.

2.1d Symmetries of Constant Energy Surfaces

Besides translation symmetry, crystals possess point symmetries, i.e., they are invariant under certain symmetry transformations: rotations through $\pi/2, 2\pi/3, ...$, symmetries with respect to a point, an axis, or a plane. Use of the symmetry properties of the Hamiltonian is fundamental in simpifying the study of the system eigenstates, but it requires the use of group theory, which is beyond the scope of this book. Appendix 2.2 gives a proof of the following fundamental result: **the surfaces of constant energy have the symmetries of the crystal.**

Because of these crystal symmetries and the inversion symmetry of the constant energy surfaces we need only study a volume of $\mathbf{k}$ space equal to the volume of the Brillouin zone divided by twice (because of the inversion symmetry) the number of symmetries of the crystal.

2.2 Dynamics of a Bloch Electron. The Crystal Momentum

We begin by noting that the wave vector $\mathbf{k}$ introduced by the Bloch function (2.4) is a quantum number with three components which is not related in a simple way to the momentum operator $\mathbf{p} = i\hbar\nabla$, unlike in the case of plane waves. In fact, if the potential $V(\mathbf{r})$ were constant, the solution of the Schrödinger equation would be a pure plane wave:

$$\psi_{\mathbf{k}} = C\exp(i\mathbf{k}\cdot\mathbf{r}), \tag{2.18}$$

the constant C ensuring the normalization of the wave function over the macroscopic solid. In this case the mean value of the operator $\mathbf{p}$ would be $\hbar\mathbf{k}$:

$$\mathbf{p}\,\psi_{\mathbf{k}} = -i\hbar\nabla(C\exp i\mathbf{k}\cdot\mathbf{r}) = \hbar\mathbf{k}\,\psi_{\mathbf{k}}. \tag{2.19}$$

By contrast, for a Bloch function,

$$<\mathbf{p}>_{n,\mathbf{k}} = <n,\mathbf{k}\,|\mathbf{p}|\,n,\mathbf{k}>, \tag{2.20}$$

$$= \int \psi_{n,\mathbf{k}}^*(\mathbf{r})\,(-i\hbar\nabla)\psi_{n,\mathbf{k}}(\mathbf{r})\,d^3r,$$

$$= \int u_{n,\mathbf{k}}^*(\mathbf{r})[\hbar\mathbf{k} - i\hbar\nabla]u_{n,\mathbf{k}}(\mathbf{r})\,d^3r$$

$$= \hbar\mathbf{k} - i\hbar\int u_{n,\mathbf{k}}^*(\mathbf{r})\nabla u_{n,\mathbf{k}}(\mathbf{r})d^3r$$

$$\neq \hbar\mathbf{k}. \tag{2.21}$$

In a crystal there is no simple relation between the mean value of the momentum in a Bloch state and the wave vector $\mathbf{k}$ which defines the state. However the vector $\hbar\mathbf{k}$ plays a very special role, as we shall see. We call this vector the **crystal momentum**.

2.2a True Momentum, Group Velocity

We show here that the mean value of the momentum is related in a simple manner to the behavior of the dispersion relations $E_n(\mathbf{k})$. Using a first-order Taylor series we write the energy at the point $\mathbf{k} + \mathbf{q}$:

$$E_n(\mathbf{k}+\mathbf{q}) = E_n(\mathbf{k}) + \mathbf{q}\cdot\nabla_{\mathbf{k}}\,E_n(\mathbf{k}) + \cdots. \tag{2.22}$$

Now the second term can be easily calculated by perturbation theory. Trivially generalizing Eq. (1.26) to three dimensions we write the Hamiltonian $\mathcal{H}_{\mathbf{k}}$ satisfied by the periodic part $u_{n,\mathbf{k}}(\mathbf{r})$ of the Bloch function:

$$\mathcal{H}_{\mathbf{k}} = \frac{1}{2m}(\hbar\mathbf{k} - i\hbar\nabla)^2 + V(\mathbf{r}). \tag{2.23}$$

Similarly, at the point $(\mathbf{k} + \mathbf{q})$ we can write

$$\mathcal{H}_{\mathbf{k}+\mathbf{q}} = \mathcal{H}_{\mathbf{k}} + \frac{\hbar^2}{m}\mathbf{q} \cdot (\mathbf{k} - i\nabla) + \frac{\hbar^2 q^2}{2m}. \tag{2.24}$$

We see that for small $|\mathbf{q}|$ the second term is small and the third is negligible. We can thus obtain $E_n(\mathbf{k} + \mathbf{q})$ from $E_n(\mathbf{k})$ by using first-order perturbation theory, which gives

$$E_n(\mathbf{k} + \mathbf{q}) = E_n(\mathbf{k}) + \int u_{n,\mathbf{k}}^* \frac{\hbar^2}{m}\mathbf{q} \cdot (-i\nabla + \mathbf{k}) \, u_{n,\mathbf{k}} \, d^3\mathbf{r} + \cdots \tag{2.25}$$

or equivalently:

$$E_n(\mathbf{k} + \mathbf{q}) = E_n(\mathbf{k}) + \int \psi_{n,\mathbf{k}}^* \mathbf{q} \cdot \left(-\frac{\hbar^2}{m}i\nabla\right) \psi_{n,\mathbf{k}} \, d^3\mathbf{r} + \cdots. \tag{2.26}$$

Equating the coefficients of $\mathbf{q}$ in Eqs. (2.22) and (2.26) we obtain the relation

$$\nabla_{\mathbf{k}} E_n(\mathbf{k}) = \frac{\hbar}{m} <\mathbf{p}>, \tag{2.27}$$

where the mean is taken over $\psi_{n,\mathbf{k}}$. Ehrenfest's theorem gives the group velocity as

$$\mathbf{v} = \frac{d <\mathbf{r}>_{n,\mathbf{k}}}{dt} = \frac{1}{m} <\mathbf{p}>_{n,\mathbf{k}} \tag{2.28}$$

from which we deduce the velocity of an electron in the Bloch state $\psi_{n,\mathbf{k}}$:

$$\mathbf{v} = \frac{1}{\hbar}\nabla_{\mathbf{k}} E_n(\mathbf{k}). \tag{2.29}$$

In fact the motion of an electron regarded as a particle should be described by a wave packet. For an electron in a crystal this is a packet of Bloch waves centered on $\mathbf{k} = \mathbf{k}_0$, which is constructed by introducing other neighboring states $\mathbf{k}$ belonging to the same band n.

We note that, as the Bloch states are eigenstates of $\mathcal{H}$, the velocity of an electron in a Bloch state is constant: an electron in such a state suffers no collisions in the crystalline potential included in $\mathcal{H}$. This is a fundamental discovery: **a periodic potential does not scatter Bloch electrons**; it determines their constant velocity through Eq. (2.29).

In a perfectly periodic crystal electrons suffering no collisions would thus have infinite conductivity. The deviations from periodicity determine the finite value of the conductivity. The defects that are most effective in producing scattering are the presence of impurities and the fact that at a finite temperature the crystal undergoes thermal vibrations which deform the perfect crystalline lattice. The latter excitations, called phonons, are not studied in this book, but play an important role in limiting the mobility of electrons and holes in semiconductors.

2.2b Acceleration Theorem in the Reciprocal Space

Under the effect of an external electric field $\mathcal{E}$ the energy of an electron is modified. Let us assume that $\mathcal{E}$ varies little over the scale of the cell, and only slowly with time at the scale of the transition frequencies between permitted energy bands. The work dW done on an electron of speed $\mathbf{v}$ and charge $-e$ over the time interval dt changes its energy $E_n(\mathbf{k})$ by modifying the value of $\mathbf{k}$ and thus the crystal momentum. Hence we have the relation

$$dW = -e\mathcal{E} \cdot \mathbf{v}dt = \frac{dE_n(\mathbf{k})}{dt} \, dt. \tag{2.30}$$

Using expression (2.29) for $\mathbf{v}$ we deduce

$$-e\mathcal{E} \cdot \frac{1}{\hbar} \nabla_{\mathbf{k}} \, E_n(\mathbf{k}) = \nabla_{\mathbf{k}} \, E_n(\mathbf{k}) \cdot \frac{d\mathbf{k}}{dt} \tag{2.31}$$

so that

$$\hbar \frac{d\mathbf{k}}{dt} = -e\mathcal{E} = \mathbf{F}, \tag{2.32}$$

where $\mathbf{F}$ is the applied force. This is the **acceleration theorem in the reciprocal space**. The essential result is that **the response to an external force varying slowly in space and time is equal to the derivative of the crystal momentum and not the derivative of the electron momentum**. In the presence of an electric field and a magnetic field $\mathbf{B}$ it is possible to generalize Eq. (2.32) to

$$\frac{\hbar d\mathbf{k}}{dt} = -e \left(\mathcal{E} + \mathbf{v} \times \mathbf{B} \right) = \mathbf{F}. \tag{2.33}$$

In fact, Eqs. (2.32) or (2.33) describe the behavior of a packet of Bloch waves, localized to about Δr in real space, and to about $\Delta k \sim 1/(\hbar\Delta r)$ in the reciprocal space. If the force $\mathbf{F}$ varies in time at the scale of the transition frequencies between bands the evolution of the system can no longer be described by the motion of the point $\mathbf{k}$ within a given band, but by transitions between bands. This is true of the effect of light on a semiconductor, studied in Chaps. 6 and 7: optical frequencies are of the order of 10^{14} Hz.

2.2c Effective Mass and Acceleration in Real Space

Differentiating the velocity $\mathbf{v}$ given by Eq. (2.29) with respect to time, and using the acceleration theorem (2.33), we obtain

$$\frac{d\mathbf{v}}{dt} = (\nabla_{\mathbf{k}} \, \mathbf{v}) \cdot \frac{d\mathbf{k}}{dt} = \frac{1}{\hbar^2} \nabla_{\mathbf{k}} \left[\nabla_{\mathbf{k}} \, E_n(\mathbf{k}) \right] \cdot \mathbf{F} \tag{2.34}$$

or

$$\frac{dv_\alpha}{dt} = \sum_\beta \left(\frac{1}{m^*}\right)_{\alpha\beta} F_\beta \tag{2.35}$$

with

$$\left(\frac{1}{m^*}\right)_{\alpha\beta} = \frac{1}{\hbar^2}\frac{\partial^2 E_n(\mathbf{k})}{\partial k_\alpha \partial k_\beta} \tag{2.36}$$

which defines the **effective mass tensor** at a point $\mathbf{k}$ of a given band n. Expressions (2.35) and (2.36) constitute the **acceleration theorem in real space**, which is subject to the same restrictions of slow variations in space and time on the force $\mathbf{F}$ as Eq. (2.33) from which it results. From its definition we see that the effective mass tensor is symmetric.

The notion of effective mass is of most interest in the vicinity of an extremum of the band, where, to lowest order,

$$E_n(\mathbf{k}) - E_n(\mathbf{k}_0) \simeq \sum_{\alpha,\beta} \frac{\hbar^2}{m_{\alpha\beta}} \Delta k_\alpha \Delta k_\beta. \tag{2.37}$$

Here $\mathbf{k}_0$ is the wave vector of the extremum, and $\mathbf{k} = \mathbf{k}_0 + \Delta\mathbf{k}$.

At the zone center of a cubic crystal, if the energy is not degenerate, the constant energy surfaces are spheres; the effective mass is thus isotropic and has a value m^*. It is positive near a band minimum and negative near a band maximum. A negative effective mass implies, from Eq. (2.35), that the velocity resulting from the action of F_β is in the opposite direction from that acquired by an electron in vacuum acted on by F_β (cf. the definition of a hole in Chap. 3). This apparently paradoxical behavior should be compared with Hall effect experiments which in certain materials imply the existence of positive charge carriers. It shows that in a solid the response of an electron to an applied force is strongly influenced by the reaction to the crystal potential. Even when the effective mass has the same sign as m, the mass of a free electron, we can have values of m^*/m very different from unity. While in metals $m^*/m \sim 1$, this is not always true of semiconductors: the effective mass of the conduction band is $+0.067m$ in GaAs, and $+0.014m$ in InSb.

We note that the momentum formula for the simplest case can be written

$$m^* \frac{d\mathbf{v}}{dt} = \mathbf{F} \tag{2.38}$$

so that it is not the derivative of the ordinary momentum $md\mathbf{v}/dt$ which is equal to the external force. In this special case, the velocity (2.29) and crystal momentum are related by

$$\mathbf{v} = \frac{\hbar\mathbf{k}}{m^*}. \tag{2.39}$$

2.3 Metal, Insulator, Semiconductor

The results we have obtained for the Bloch states have the following fundamental consequence: **a full band cannot carry a current**. In fact from the definition of the current density,

$$\mathbf{j} = -e \sum_{\mathbf{k} \in \text{full band}} \mathbf{v}(\mathbf{k}), \tag{2.40}$$

$$= -\frac{e}{\hbar} \sum_{\mathbf{k} \in \text{full band}} \nabla_{\mathbf{k}} E_n(\mathbf{k}) = 0, \tag{2.41}$$

as the functions $E_n(\mathbf{k})$ are even in $\mathbf{k}$ (2.17): from Eq. (2.29),

$$\mathbf{v}(-\mathbf{k}) = -\mathbf{v}(\mathbf{k}) \tag{2.42}$$

and the total current is zero. Let us now apply an electric field $\mathcal{E}$. Using the acceleration theorem in the reciprocal space, each vector $\mathbf{k}$ is modified by $d\mathbf{k} = -e\mathcal{E}\,dt$ in time dt, the shift being the same for all the vectors. From the definition of the first Brillouin zone, states leaving this zone are equivalent up to a vector $\mathbf{G}$ of the reciprocal lattice to those which become empty, and the band remains full. A full band does not react to an applied electric field and does not participate in the current. If the occupied bands of a solid are completely full, the solid is an insulator.

To find the filling factor of the band states of a three-dimensional solid we will use reasoning similar to that used for a one-dimensional system in Sect. 1.3c; the situation is slightly complicated by the fact that we have to consider all directions of the wave vector $\mathbf{k}$.

2.3a Density of States in the Reciprocal Space

The Born–von Kármán boundary conditions (2.8) result in a uniform density of states in the space of wave vectors $\mathbf{k}$: we consider again the example of a crystal of macroscopic dimensions L_x, L_y, L_z. The accessible wave vectors have components:

$$k_i = l_i \frac{2\pi}{L_i} \quad \text{with } i = x,\, y,\, z. \tag{2.43}$$

Two spin states correspond to each wave vector. The number of accessible states in a volume $d^3\mathbf{k}$ of the reciprocal space, assumed large compared to the volume $(2\pi)^3/L_x L_y L_z$ per orbital state, is thus:

$$\underset{\text{spin}}{2} \quad \times \quad \underset{\text{number of orbital states}}{\frac{L_x L_y L_z}{(2\pi)^3} d^3\mathbf{k}} \quad = \quad n\,(\mathbf{k})\,d^3\mathbf{k}. \tag{2.44}$$

The number of orbital states contained in a band is equal to the number of elementary cells contained by the crystal, independent of the number of atoms per elementary cell.

Let us show this for the face-centered cubic lattice, corresponding to the most common semiconductors. If a is the side of the cube, the volume of the primitive cell of Fig. 2.1 is $a^3/4$, since in each cube we have $(8 \times 1/8) + (6 \times 1/2) = 4$ nodes of the Bravais lattice. A macroscopic parallelepiped L_x, L_y, L_z of the crystal thus contains $N = 4L_xL_yL_z/a^3$ primitive cells. The reciprocal lattice is a body-centered cubic of side $4\pi/a$, with $(8\times1/8)+1 = 2$ nodes, so the volume of the primitive cell, which is also that of the first Brillouin zone, is $(1/2)(4\pi/a)^3 = 32\pi^3/a^3$. The density of orbital states in the reciprocal space is, by Eq. (2.44), $L_xL_yL_z/8\pi^3$. The number of states in a primitive cell of the reciprocal lattice is thus $(32\pi^3/a^3) \times (L_xL_yL_z/8\pi^3) = 4L_xL_yL_z/a^3 = N$, the number of primitive cells in the crystal. Hence each band of index n contains N orbital states, or $2N$ states taking account of spin, where N is the number of primitive cells of the crystal.

To find the state of the whole crystal we fill up the states at zero temperature in accordance with the Pauli principle, starting with the states of lowest energy. The number of electrons per elementary cell, and also the position of the bands in the various directions of the reciprocal space and any overlaps, will fix the behavior, conducting or insulating, of the material. Diamond, silicon, germanium, and grey tin have four valence electrons, and crystallize in a face-centered cubic lattice, with two atoms per elementary cell. We thus have eight electrons for each elementary cell. Among the bands constructed from the valence states for C, Si, Ge, the four lowest bands have no energy overlap with the higher bands. We can thus fill them with the $8N$ electrons with two electrons per orbital. These materials are insulators at zero temperature.

The band gap E_g of diamond is 5.4 eV, that of silicon 1.1 eV, and that of germanium 0.67 eV. As we shall see in Chap. 4 the thermal excitation probability of conduction electrons at temperature T is proportional to $\exp(-E_g/2kT)$ where k is the Boltzmann constant. At room temperature (for $T = 300$ K, i.e., $kT = 25$ meV) this probability is 1×10^{-47} for diamond, 3×10^{-10} for silicon, 1.5×10^{-6} for germanium, while the number of electrons per m^3 in a solid is of order 10^{28}. Diamond is thus a very good insulator, silicon and germanium are semiconductors. By contrast in grey tin the valence band overlaps a higher partially filled band. This makes it a metallic conductor. The conduction properties of semiconductors are determined by the states with energies close to the extrema of the bands since these are the states most easily populated by thermal excitation. We therefore have to count these states.

2.3b Density of States in Energy

Starting from the dispersion relation for a given band $E_n(\mathbf{k})$, we deduce the number dN of states in the band n whose energy is between E and $E+dE$:

$$dN = n(E)\,dE = \int_{\delta v(E)} n(\mathbf{k})\,d^3\mathbf{k} = \frac{L_xL_yL_z}{4\pi^3}\int_{\delta v(E)} d^3\mathbf{k}. \qquad (2.45)$$

This relation defines the density of states in energy $n(E)$; $\delta v(E)$ is the volume of the reciprocal space contained between the constant energy surfaces $S(E)$ and $S(E+dE)$. We can easily find the density of states in energy $n(E)$ for a free electron: from the dispersion relation

$$E\,(\mathbf{k}) = \frac{\hbar^2 k^2}{2m} \qquad (2.46)$$

we deduce

$$dE = \frac{\hbar^2}{m}k\,dk. \qquad (2.47)$$

If the electron moves in three dimensions, the volume $\delta v(E)$ between two constant energy spheres is $4\pi k^2 dk$. Relation (2.44) is then

$$2\left(\frac{L}{2\pi}\right)^3 4\pi k^2 dk = n(E)\,dE \qquad (2.48)$$

and we get

$$n(E) = 4\pi\left(\frac{L}{h}\right)^3 (2m)^{3/2}\sqrt{E}. \qquad (2.49)$$

For an electron in a periodic solid the dispersion relation does not in general have a simple analytic form. The density of states of silicon is shown in Fig. 2.5. The volume element $\delta v(E)$ of Eq. (2.45) can be decomposed into d^2S, the elementary area on $S(E)$, multiplied by the distance along the normal to the surface

$$d^3\mathbf{k} = d^2S \cdot \frac{d\mathbf{k}}{dE}dE$$
$$= d^2S \cdot \frac{1}{|\nabla_{\mathbf{k}}E|}dE \qquad (2.50)$$

so that, taking account of the spin,

$$n(E) = \frac{L_xL_yL_z}{4\pi^3}\int_{S(E)} \frac{d^2S}{|\nabla_{\mathbf{k}}E|}. \qquad (2.51)$$

If we can define an isotropic effective mass m_e in the vicinity of the minimum E_c of the conduction band, the density of states there can be found directly from expression (2.49) applying for the free electron, by substituting m_e for m:

$$n(E) = 4\pi \left(\frac{L}{h}\right)^3 (2m_e)^{3/2} \sqrt{E - E_c}. \tag{2.52}$$

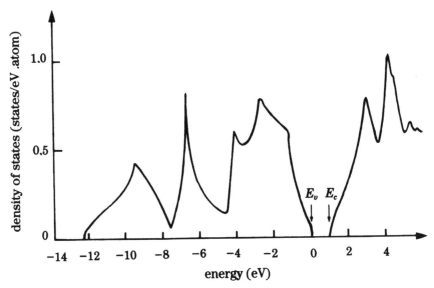

Fig. 2.5. Density of states of the valence and conduction bands of silicon, calculated by J.R. Chelikowsky and M.L. Cohen, Physical Review B **14**, 556 (1976). The energy origin is at the maximum E_v of the valence band. In the neighborhood of E_v, the maximum of the valence band, and E_c, the minimum of the conduction band, the density of states varies parabolically with energy.

2.4 Theoretical Determination of Band Structure

We have seen that one can deduce the dynamical properties of the electrons and the density of states from the relations $E_n(\mathbf{k})$. To understand the properties of semiconductors, we therefore have to determine their band structure. This is a complex task, simultaneously involving first principles, semi-empirical calculations, and experimental data.

We sketch below several methods used to calculate the band structure of semiconductors. These can be skipped in a first reading.

2.4a The Tight Binding Approximation

This is also called the LCAO ("linear combination of atomic orbitals") method. It corresponds to a "chemical" point of view, i.e., to deducing the crystal properties from the already known eigenstates of the constituent atoms and their chemical bonds, which are calculated now.

Simple Cubic Crystal with One Atom per Cell

We consider first the hypothetical case of a simple cubic crystal having one monovalent atom per cell, and assume that the Hamiltonian $\mathcal{H}_{at}$ of an isolated atom only has one eigenvalue E_0, associated with a non-degenerate eigenstate $\phi(\mathbf{r})$:

$$\mathcal{H}_{at} \ \phi(\mathbf{r}) = \left[\frac{p^2}{2m} + V(\mathbf{r}) \right] \phi(\mathbf{r}) = E_0 \ \phi(\mathbf{r}). \tag{2.53}$$

The crystal potential is the sum of the atomic potentials centered at the various sites j of the lattice:

$$\mathcal{H} = \frac{p^2}{2m} + \sum_j V(\mathbf{r} - \mathbf{R}_j) \tag{2.54}$$

and we seek a solution of the Schrödinger equation for the crystal of the form

$$\psi_{\mathbf{k}}(\mathbf{r}) = \sum_j A_{\mathbf{k},j} \ \phi(\mathbf{r} - \mathbf{R}_j). \tag{2.55}$$

For simplicity, we will sometimes write

$$\phi(\mathbf{r} - \mathbf{R}_j) = \phi_j. \tag{2.56}$$

Bloch's theorem fixes the form of the $A_{\mathbf{k},j}$: $\psi_{\mathbf{k}}$ must be a Bloch function, so that for each translation $\mathbf{T}$ of the direct lattice we have

$$\begin{aligned} \psi_{\mathbf{k}}(\mathbf{r} + \mathbf{T}) &= \exp\left(i\mathbf{k} \cdot \mathbf{T}\right) \ \psi_{\mathbf{k}}(\mathbf{r}) \\ &= \sum_j A_{\mathbf{k},j} \ \phi(\mathbf{r} - \mathbf{R}_j + \mathbf{T}). \end{aligned} \tag{2.57}$$

Replacing $\psi_{\mathbf{k}}(\mathbf{r})$ by expression (2.55) and setting $\mathbf{R}_{j'} = \mathbf{R}_j - \mathbf{T}$ we obtain

$$A_{\mathbf{k},j+\mathbf{T}} = \exp\left(i\mathbf{k} \cdot \mathbf{T}\right) \ A_{\mathbf{k},j}. \tag{2.58}$$

For this relation to hold for all $\mathbf{T}$ we require

$$A_{\mathbf{k},j} = C \exp\left(i\mathbf{k} \cdot \mathbf{R}_j\right), \tag{2.59}$$

where C is a constant. The normalization of the function $\psi_{\mathbf{k}}$ can thus be written as

$$< \psi_{\mathbf{k}} | \psi_{\mathbf{k}} > = C^2 \sum_{j,j'} \exp(-i\mathbf{k} \cdot \mathbf{R}_{j'}) \exp(i\mathbf{k} \cdot \mathbf{R}_j) \times$$

$$\int \phi^*(\mathbf{r} - \mathbf{R}_{j'}) \, \phi(r - \mathbf{R}_j) \, d^3r, \tag{2.60}$$

$$= C^2 \sum_{j,j'} \exp i\mathbf{k} \cdot (\mathbf{R}_j - \mathbf{R}_{j'}) < \phi_{j'} | \phi_j > . \tag{2.61}$$

If we make the hypothesis that states centered on different atoms overlap very little, then

$$< \phi_j | \phi_{j'} > = \delta_{jj'} \tag{2.62}$$

and

$$< \psi_{\mathbf{k}} | \psi_{\mathbf{k}} > = C^2 \sum_j < \phi_j | \phi_j > = NC^2 = 1, \tag{2.63}$$

hence

$$\psi_{\mathbf{k}}(\mathbf{r}) = N^{-1/2} \sum_j \exp(i\mathbf{k} \cdot \mathbf{R}_j) \ \phi(\mathbf{r} - \mathbf{R}_j). \tag{2.64}$$

The matrix representing the Hamiltonian (2.54) on the basis of Bloch states is diagonal since the Hamiltonian is a periodic operator (cf. Appendix 2.1). Its eigenvalues are therefore directly given by

$$E(\mathbf{k}) = < \mathbf{k} | \mathcal{H} | \mathbf{k} > = N^{-1} \sum_{j,\, j'} \exp i\mathbf{k} \cdot (\mathbf{R}_j - \mathbf{R}_{j'}) < \phi_{j'} | \mathcal{H} | \phi_j > .$$

$$\tag{2.65}$$

To evaluate the terms of Eq. (2.65) we decompose the crystal Hamiltonian:

$$\mathcal{H} = \frac{p^2}{2m} + V(\mathbf{r} - \mathbf{R}_j) + \sum_{j'' \neq j} V(\mathbf{r} - \mathbf{R}_{j''}) \tag{2.66}$$

so that

$$E(\mathbf{k}) = E_0 + N^{-1} \sum_{j,j',j'' \neq j} \exp i\mathbf{k} \cdot (\mathbf{R}_j - \mathbf{R}_{j'}) < \phi_{j'} | V(\mathbf{r} - \mathbf{R}_{j''}) | \phi_j > .$$

$$\tag{2.67}$$

The summation (2.67) contains terms of two types: integrals involving three different centers $j \neq j' \neq j''$, and terms where $j'' = j'$. The terms with three different centers are small as the functions ϕ and the potential V decrease with distance, so we neglect them. There remain either diagonal terms:

$$< \phi_j | \sum_{j' \neq j} V(\mathbf{r} - \mathbf{R}_{j'}) | \phi_j > = -\alpha < 0 \tag{2.68}$$

or non-diagonal terms

$$<\phi_{j'}|V\ (\mathbf{r} - \mathbf{R}_{j'})|\phi_j > =$$

$$\int \phi^*(\mathbf{r} - \mathbf{R}_{j'})\ V(\mathbf{r} - \mathbf{R}_{j'})\ \phi(\mathbf{r} - \mathbf{R}_j)d^3\mathbf{r} = I_{jj'}. \qquad (2.69)$$

Here again, as the atomic functions decrease rapidly with distance, the only non-negligible terms come from the nearest neighbors, for which we set $I_{jj'} = -\gamma$. We thus write $E(\mathbf{k})$ as

$$E(\mathbf{k}) = E_0 - \alpha + N^{-1} \sum_{j' \text{ next to } j} \exp i\mathbf{k}\cdot(\mathbf{R}_j - \mathbf{R}_{j'})\ (-\gamma). \qquad (2.70)$$

The quantities α and γ are the fundamental parameters of the chemical bond.

We can calculate γ for the 1s states of hydrogen. We find

$$\gamma = 2\ E_1\left(1 + \frac{a}{a_1}\right)\exp\left(-\frac{a}{a_1}\right), \qquad (2.71)$$

where E_1 is the binding energy of the 1s level, a_1 the Bohr radius for this level ($E_1 = 13.6$ eV, $a_1 = 0.53$ angstrom) and a the distance between nearest neighbors.

The order of magnitude of the energy spread of the band $E(\mathbf{k})$ is about 2γ. For deep electron states this spread is extremely small as $a \gg a_1$. Let us take the case of silicon: $a_1 = (0.53/Z)$ angstrom with $Z = 14$, $a = 2.53$ angstrom (distance between nearest neighbors). We find $\gamma = 6 \times 10^{-24}$ eV, showing that there is only an infinitesimal overlap between orbitals: the 1s level remains essentially an atomic level within the solid, without energy dispersion. The valence band of silicon is formed by 3s and 3p states, but because of the screening of the nuclear charge by electrons in levels $n = 1$ and $n = 2$, we can regard the effective charge as $Z' = 4$. For a 3s state, $a_3 = (3 \times 0.53/Z')$ Å $\simeq 0.4$ Å. We thus obtain a broadening or bandwidth of the order of one electron volt.

For the case of a simple cubic lattice where $\mathbf{R}_{j'} - \mathbf{R}_j$ takes the six values $\pm a$ along the three axes

$$E(\mathbf{k}) = E_0' - 2\gamma\left[\cos k_x a + \cos k_y a + \cos k_z a\right] \qquad (2.72)$$

with

$$E_0' = E_0 - \alpha.$$

Exercise: Show that the constant energy surfaces given by Eq. (2.72) are orthogonal to the faces of the Brillouin zone.
The width of the band is 12γ. For small $|k|$,

$$E(\mathbf{k}) \simeq E_0' - 6\gamma + \gamma a^2(k_x^2 + k_y^2 + k_z^2). \qquad (2.73)$$

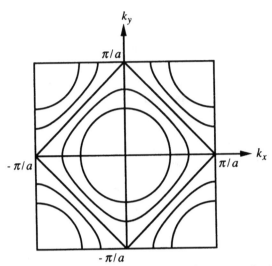

Fig. 2.6. Representation of the constant energy curves for a square lattice in two dimensions. We see that for small k these are circles (spheres in three dimensions).

The constant energy surfaces are therefore spheres. The effective mass (2.36) is isotropic and takes the simple form

$$m^* = \frac{\hbar^2}{2\gamma a^2}.$$

(2.74)

For larger values of the energy the form of the constant energy surfaces is shown schematically in Figs. 2.6 and 2.7.

Exercise: Using the above formulas and appropriate formulas from quantum mechanics, find the effective mass at the bottom of the lowest band formed from the lowest hydrogen-like orbital (principal quantum number $n = 1$) for an atom of atomic number Z:

Result: $m^*/m = 2^{-1}(a_1/Za)^2[1 + (a_1/Za)]^{-1} \exp(aZ/a_1).$ (2.75)

Applying the result to silicon, assuming a simple cubic lattice as above, gives for the $1s$ level an enormous value $m^*/m \sim 1.9 \times 10^{21}$! This expresses the fact that the energy is effectively independent of **k**.

In this calculation, each atomic level of energy E_0 has a corresponding permitted energy band centered around $E_0 - \alpha$. **Permitted bands and forbidden gaps appear only if we couple systems with several atomic levels** (cf. Fig. 1.1). This is the case when several orbitals are taken into account. It is also true for crystals built from dissimilar atoms, or for crystals built from the same kind of atom, but with an asymmetric cell.

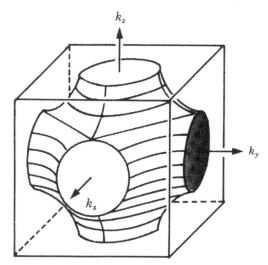

Fig. 2.7. Constant energy surface in the tight binding approximation for a cubic lattice, for the energy value $E(\mathbf{k}) = E_0'$ (Eq. (2.72)).

One-Dimensional AB Crystal

We will bring out some basic ideas for the example of a one-dimensional AB crystal by following step-by-step the above reasoning for the cubic crystal (Eqs. (2.53) to (2.72)).

The crystal is shown schematically in Fig. 2.8. The elementary cell has period $2d$. We take as the quantum state of the valence electron of an isolated atom A (respectively B) the non-degenerate orbital ϕ_A of energy E_A (respectively ϕ_B of energy E_B):

$$\mathcal{H}_A \, \phi_A = \left[\frac{p^2}{2m} + V_A \, (\mathbf{r} - \mathbf{R}_A)\right] \phi_A(\mathbf{r} - \mathbf{R}_A) = E_A \, \phi_A(\mathbf{r} - \mathbf{R}_A),$$

$$\mathcal{H}_B \, \phi_B = \left[\frac{p^2}{2m} + V_B \, (\mathbf{r} - \mathbf{R}_B)\right] \phi_B(\mathbf{r} - \mathbf{R}_B) = E_B \, \phi_B(\mathbf{r} - \mathbf{R}_B).$$

$$(2.76)$$

The Hamiltonian $\mathcal{H}$ of the crystal is

$$\mathcal{H} = \frac{p^2}{2m} + \sum_{j=1}^{N} V_{Aj} + \sum_{j=1}^{N} V_{Bj}.$$

$$(2.77)$$

We seek an eigenstate $|\phi_k >$ as a linear combination of the atomic orbitals ϕ_{Aj} and ϕ_{Bj}, centered on the atoms A_j and B_j:

$$|\psi_k> = \sum_{j=1}^{N} \xi_{kj}|\phi_{Aj}> + \eta_{kj}|\phi_{Bj}> .$$

$$(2.78)$$

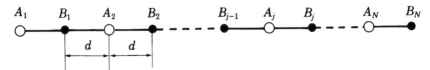

Fig. 2.8. Schematic view of a linear crystal formed by the repetition of atoms A and B at distances d.

The use of Bloch's theorem allows us to write

$$\xi_{kj} = a_k \exp(ik2jd) \quad \text{and} \quad \eta_{kj} = b_k \exp(ik2jd). \tag{2.79}$$

We assume again that the wave functions of neighboring sites overlap only weakly, i.e.,

$$< \phi_{Ai}|\phi_{Aj} > = \delta_{ij}, \quad < \phi_{Bi}|\phi_{Bj} > = \delta_{ij}, \quad < \phi_{Ai}|\phi_{Bj} > = 0. \tag{2.80}$$

The normalized wave function is therefore

$$|\phi_k > = N^{-1/2} \sum_{j=1}^{N} \exp(ik2jd) \, (a_k|\phi_{Aj} > + b_k|\phi_{Bj} >) \tag{2.81}$$

with

$$|a_k|^2 + |b_k|^2 = 1. \tag{2.82}$$

The coefficients a_k and b_k express the weights of the atomic states A and B in ψ_k, i.e., the degree of hybridization. We seek solutions of the eigenvalue equation

$$[\mathcal{H} - E(k)]|\psi_k > = 0. \tag{2.83}$$

It is convenient, as in Eq. (2.66), to write the Hamiltonian $\mathcal{H}$ by separating out one atomic Hamiltonian $\mathcal{H}_{Aj}$:

$$\mathcal{H} = \mathcal{H}_{Aj} + \sum_{j'' \neq j} V_{Aj''} + \sum_{j} V_{Bj}. \tag{2.84}$$

Acting with $< \phi_{Aj}|$ on the left of Eq. (2.83) and using Eqs. (2.78) and (2.84) we write

$$[E_A - E(k)]a_k \exp(2ikjd) + < \phi_{Aj}| \sum_{j'' \neq j} V_{Aj''} + \sum_{j} V_{Bj} |\psi_k > = 0.$$
$$\tag{2.85}$$

The most important terms in the summations involve only the nearest neighbors if we neglect the integrals with three centers and the overlap of distant states. The remaining matrix elements are of four types:

$$< \phi_A | V_B | \phi_A > = -\alpha_A , \quad < \phi_B | V_A | \phi_B > = -\alpha_B,$$
$$< \phi_A | V_A | \phi_B > = -\gamma_A , \quad < \phi_B | V_B | \phi_A > = -\gamma_B, \tag{2.86}$$

where the four quantities $\alpha_A, \alpha_B, \gamma_A$, and γ_B are positive. Moreover we note that $\gamma_A = \gamma_B = \gamma$ since

$$< \phi_A | \frac{p^2}{2m} + V_A + V_B | \phi_B > = E_A < \phi_A | \phi_B > + < \phi_A | V_B | \phi_B >$$
$$= -\gamma_A$$
$$= E_B < \phi_A | \phi_B > + < \phi_A | V_A | \phi_B >$$
$$= -\gamma_B. \tag{2.87}$$

Replacing the wave function $|\psi_k >$ by its expression (2.81) in Eq. (2.85), and limiting ourselves to coupling between nearest neighbors, we get

$$a_k \exp(ik2jd) \; [E_A - E(k) > + < \phi_{Aj} | V_{Bj-1} + V_{Bj+1} | \phi_{Aj} >] +$$
$$b_k \; [\exp(ik2jd) < \phi_{Aj} | V_{Aj} | \phi_{Bj} > +$$
$$\exp[ik2(j-1)d] < \phi_{Aj} | V_{Aj} | \phi_{Bj-1} >] = 0 \tag{2.88}$$

or

$$[E_A - 2\alpha_A - E(k)] \, a_k - \gamma \, [1 + \exp (-2ikd)]b_k = 0. \tag{2.89}$$

Similarly the projection of Eq. (2.83) on to $< \phi_{Bj}|$ gives

$$-\gamma \, [1 + \exp(2ikd)]a_k + [E_B - 2\alpha_B - E(k)]b_k = 0. \tag{2.90}$$

Equations (2.89) and (2.90) are two homogeneous equations for two unknowns a_k and b_k. For them to be compatible requires their eigenvalues $E(k)$ to be solutions of the quadratic equation:

$$[E_A - 2\alpha_A - E(k)][E_B - 2\alpha_B - E(k)]-$$
$$\gamma^2[1 + \exp(-2ikd)][1 + \exp(2ikd)] = 0. \tag{2.91}$$

This equation has two solutions; there are therefore two energy bands for k in the first Brillouin zone, namely,

$$E(k) = \frac{E_A - 2\alpha_A + E_B - 2\alpha_B}{2} \pm$$
$$\frac{1}{2}\sqrt{(E_A - 2\alpha_A - E_B + 2\alpha_B)^2 + 16\gamma^2 \cos^2 kd}. \tag{2.92}$$

At $k = 0$ the state of lowest energy E_- is a binding orbital, the state E_+ an antibonding orbital. At the edge of the Brillouin zone, the energies

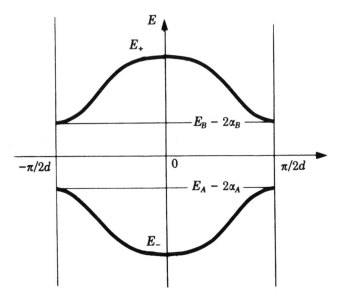

Fig. 2.9. Band structure of the one-dimensional compound AB. We have assumed $E_A < E_B$.

$E_A - 2\alpha_A$ and $E_B - 2\alpha_B$ differ little from those of isolated atoms. The band gap, situated at the edge of the zone, has a width of $|E_B - E_A - 2(\alpha_B - \alpha_A)|$.

By this simple example we have shown that if the elementary cell of a crystal has two different atoms, a band gap appears. Depending on how the bands are filled, i.e., on the number z of electrons per elementary cell, we have a metal (z odd) or an insulator (z even). These two bands are symmetrical, as shown in Fig. 2.9.

Distorted Linear Chain

Let us now take the example of the crystal shown in Fig. 2.10. It consists of just one type of atom A but is distorted.

The period is still $2d$, and we now have $E_A = E_{A'}$, but the values of the matrix elements must be reconsidered: we set

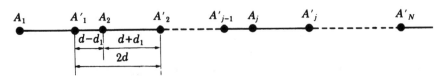

Fig. 2.10. Linear chain made up of one type of atom, but distorted.

$$<\phi_{Aj}|V_{A'j} + V_{A'j-1}|\phi_{Aj}> = -2\alpha_A,$$
$$<\phi_{A'j}|V_{Aj} + V_{Aj+1}|\phi_{A'j}> = -2\alpha_{A'},$$
(2.93)
$$<\phi_{Aj}|V_{Aj}|\phi_{A'j}> = -\gamma(1-\epsilon) = <\phi_{Aj}|V_{A'j}|\phi_{A'j}>,$$
$$<\phi_{Aj}|V_{Aj}|\phi_{A'j-1}> = -\gamma(1+\epsilon) = <\phi_{Aj}|V_{A'j-1}|\phi_{A'j-1}>.$$
(2.94)

The equations analogous to Eqs. (2.89) and (2.90) that yield a_k and a'_k, the coefficients of ψ_k, have one non-vanishing solution: we obtain from it the equation giving the eigenvalues $E(k)$:

$$E(k) = E_A - (\alpha_A + \alpha_{A'}) \pm$$
$$\sqrt{(\alpha_A - \alpha_{A'})^2 + 4\gamma^2[\cos^2 kd\,(1 - \epsilon^2) + \epsilon^2]}.$$
(2.95)

A band gap opens between the two solutions at $k = \pi/2d$ which exists only if α_A differs from $\alpha_{A'}$ and/or ϵ is non-zero (Fig. 2.11).

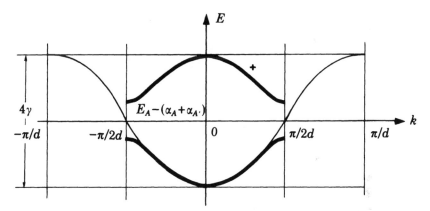

Fig. 2.11. Band structure of the distorted compound of Fig. 2.9. The first Brillouin zone is the segment $[-\pi/2d, \pi/2d[$. $|\alpha_A - \alpha'_A|$ is assumed to be small with respect to $2|\gamma|$.

In the absence of distortion the period of the lattice is d and the first Brillouin zone consists of the segment $[-\pi/d, \pi/d[$. The distortion doubles the spatial period, which becomes $2d$, and reduces the first zone to the segment $[-\pi/2d, \pi/2d[$. The band structure of the compound can be regarded as a folding of the band: the upper branch denoted $+$ then arises from the regions $|k| > \pi/2d$ of the dispersion curve $E(k)$ of the undistorted compound.

If the periodic chain has one electron per cell, the electron states are filled up to $k = \pi/2d$. In the absence of distortion the chain is a conductor. The distortion opens a band gap at precisely this value of k, and the system becomes an insulator. This is the "Peierls transition."

2.4b Plane Wave Expansion

We are interested in states close to the band gap and seek a solution of the Schrödinger equation of the form

$$\psi = \sum_{\mathbf{k}} a_{\mathbf{k}} |\mathbf{k}>, \tag{2.96}$$

where $|\mathbf{k}>$ is any plane wave of the form $\Omega^{-1/2} \exp i\mathbf{k} \cdot \mathbf{r}$, where Ω the volume of the crystal. Substituting this into the crystal Hamiltonian and multiplying by $<\mathbf{k}'|$ we get a set of homogeneous linear equations in $a_{\mathbf{k}}$:

$$\frac{\hbar^2}{2m} k'^2 a_{\mathbf{k}'} + \sum_{\mathbf{k}} <\mathbf{k}'|V|\mathbf{k}> a_{\mathbf{k}} = E\, a_{\mathbf{k}'}. \tag{2.97}$$

Expanding V in a Fourier series,

$$V = \sum_{\mathbf{G}} V_{\mathbf{G}} \exp i\,\mathbf{G} \cdot \mathbf{r}, \tag{2.98}$$

where $\mathbf{G}$ is a vector of the reciprocal lattice (cf. Appendix 2.1), and regarding the crystal as a parallelepiped of sides L_x, L_y, L_z, we can write

$$
\begin{aligned}
<\mathbf{k}'|V|\mathbf{k}> &= \frac{1}{\Omega} \sum_{\mathbf{G}} V_{\mathbf{G}} \int \exp\left[i(\mathbf{k} - \mathbf{k}' + \mathbf{G}) \cdot \mathbf{r}\right]\, d^3\mathbf{r} \\
&= \sum_{\mathbf{G}} V_{\mathbf{G}} \prod_{x,y,z} \frac{2\sin\,(k_x - k'_x + G_x)L_x/2}{(k_x - k'_x + G_x)L_x} \\
&= \sum_{\mathbf{G}} V_{\mathbf{G}}\, \delta_{\,\mathbf{k}-\mathbf{k}'+\mathbf{G}} \cdot
\end{aligned}
\tag{2.99}
$$

The matrix element $<\mathbf{k}'|V|\mathbf{k}>$ is then zero unless $\mathbf{k}' - \mathbf{k}$ is a vector of the reciprocal lattice. In Eq. (2.97) we can limit ourselves to summing over those $\mathbf{k}$ which differ from $\mathbf{k}'$ by a vector of the reciprocal lattice. This divides the number of terms by the number of cells in the crystal and is an enormous simplification. But the remaining sum is still in principle infinite. In practice, however, the Fourier components $V_{\mathbf{G}}$ of the potential decrease for large vectors $\mathbf{G}$, and we can truncate the summation at a few hundred terms. There thus remain for each zone point several hundred equations giving several hundred energy eigenvalues and the first few hundred bands. Even with modern computers this method remains relatively inaccurate.

We can greatly improve the convergence by looking for states as linear combinations of **orthogonalized plane waves** $\Phi_{\mathbf{k}}$ defined as

$$\Phi_{\mathbf{k}} = \exp(i\,\mathbf{k}\cdot\mathbf{r}) + \sum_{c} b_c\, \psi^c_{\mathbf{k}}\,(\mathbf{r}), \tag{2.100}$$

where $\psi^c_{\mathbf{k}}\,(\mathbf{r})$ is a Bloch state corresponding to the core of the ions (hence with very negative energy). We require $\Phi_{\mathbf{k}}$ to be orthogonal to the states $\psi^c_{\mathbf{k}}$, i.e.,

$$\int d^3\mathbf{r} \ \psi_\mathbf{k}^{c*}(\mathbf{r}) \ \Phi_\mathbf{k} = 0$$

or

$$b_c = -\int d^3\mathbf{r} \ \psi_\mathbf{k}^{c*}(\mathbf{r}) \ \exp(i \ \mathbf{k} \cdot \mathbf{r}). \tag{2.101}$$

The $\Phi_\mathbf{k}$ functions are known precisely if we know the $\psi^c{}_\mathbf{k}$ accurately. This is true if we use the tight binding method for calculating the deep Bloch states, as the overlap of atomic functions of very negative energies is small. The tight binding approximation is excellent in this case.

The $\Phi_\mathbf{k}$ functions have, by construction, the properties we seek: They are orthogonal to the deep states. They have a very localized part which oscillates rapidly, like the wave functions of the atomic core, and between atoms they appear like plane waves. The method then consists of seeking a solution to the Schrödinger equation for the crystal of the form

$$\psi = \sum_\mathbf{k} c_\mathbf{k} \ \Phi_\mathbf{k}. \tag{2.102}$$

Again, the $\Phi_\mathbf{k}$ are only coupled by the periodic potential to functions $\Phi_{\mathbf{k}+\mathbf{G}}$ so that this method is analogous to the plane wave method described above. But we obtain excellent results using expansions of ψ in a few tens of orthogonalized plane waves. This is at present the most powerful method of calculating the band structure.

Appendix 2.4 describes another semiempirical method of determining the band structure, the so-called $\mathbf{k} \cdot \mathbf{p}$ method.

2.5 The True Band Structure

Before describing real band structure for silicon we note that Appendix 2.3 shows, using the tight binding method, that crystals of the face-centered cubic form like silicon have triply degenerate bands at $\mathbf{k} = 0$ (levels x_2 and x_6). This occurs at the top of the valence band. We also know from the $\mathbf{k} \cdot \mathbf{p}$ method (Appendix 2.4) that for $\mathbf{k} = 0$ the states resemble atomic states, and are thus possibly degenerate. In fact, we have up to now neglected the role of spin in the electron Hamiltonian by taking account of its influence only via the Pauli principle. It can be shown that in its motion in the electric potential, the spin sees a magnetic field which results in a Hamiltonian of the form

$$\mathcal{H}_{\text{s.o.}} = \frac{1}{4 \ m^2 \ c^2} \ \boldsymbol{\sigma} \times \mathbf{grad} \ V(\mathbf{r}) \cdot \mathbf{p}, \tag{2.103}$$

where $\boldsymbol{\sigma} = (\sigma_x, \sigma_y, \sigma_z)$ are the Pauli matrices.

This so-called "spin-orbit" interaction has the effect of partially lifting the degeneracy mentioned above. This is seen in Fig. 2.12 which shows the band structure of silicon around the band gap of width E_g. We note that the maximum of the valence band is doubly degenerate (the point $\Gamma^{25'}$). The spin-orbit interaction has split one of the valence bands by an amount $\Delta E_{s.o.}$.

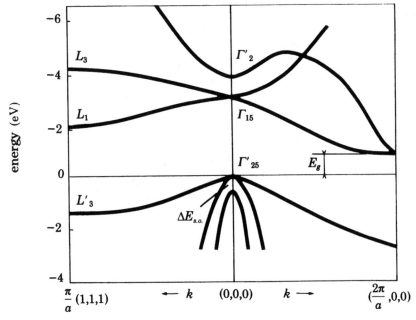

Fig. 2.12. Band structure of silicon: the letters L, Γ denote particular points of high symmetry in the zone. Γ is the center of the zone. L is the point at the edge of the zone in the direction (111). The distance $L_1 L_3'$ is experimentally determined, while the absolute energies of the L_3' and L_1 levels are found by a calculation by the method of orthogonalized plane waves (D. Brust, J.C. Phillips, and F. Bassani, Physical Review Letters **9**, 94, 1962).

Some of the energy values given in the figure are found from experiment and others from calculations based on the method of orthogonalized plane waves. The form of the constant energy surfaces near the top of the valence band is complex (to describe it we need a $\mathbf{k} \cdot \mathbf{p}$ theory for a degenerate level). We assume that it consists of two spheres, one called "heavy holes" and the other "light holes" (the concept of a hole will be introduced in Chap. 3).

We note that the minimum of the conduction band is not at $\mathbf{k} = 0$, in contrast to the maximum of the valence band. In such a case the gap is called **indirect**. The value of E_g is 1.12 eV for silicon. The mass of the heavy holes is $m_{hh} = 0.49m$ and the mass of the light holes is $m_{lh} = 0.16m$.

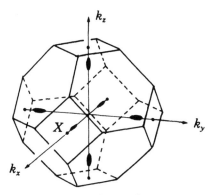

Fig. 2.13. Constant energy surfaces near the bottom of the conduction band of Si.

The minimum of the conduction band is in the direction [100], and by symmetry on the equivalent directions $< 100 >$. There are thus six minima of the conduction band around $|k| \sim 0.8 \times (2\pi/a)$.

By symmetry each ellipsoid of constant energy in the conduction band must have two equal axes. These are prolate ellipses as shown in Fig. 2.13. The dispersion relation near a minimum of the conduction band has the form

$$E\left(\mathbf{k}\right) = \frac{\hbar^2}{2} \left(\frac{k_x^2 + k_y^2}{m_T} + \frac{\Delta k_z^2}{m_L} \right) \tag{2.104}$$

for the ellipsoids [001] and [00$\bar{1}$]. Here we have set $\Delta k_z = k_z - k_{z0}$, with $k_{z0} = (0, 0, 1.6\pi/a)$. There appears a longitudinal effective mass $m_L = 0.98m$ and a transverse mass $m_T = 0.19m$.

This shows the complexity of the real situation. However to understand many properties it is often enough to consider a band structure with a direct gap as shown below, using appropriate effective masses. We will call this representation "standard band structure" in the remainder of this book (Fig. 2.14).

2.6 Experimental Study of Band Structure

We confine ourselves to indicating two important methods.

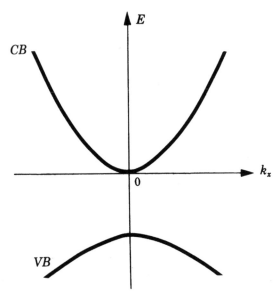

Fig. 2.14. Standard band structure: the maximum of the valence band (VB) and the minimum of the conduction band (CB) are at $k = 0$.

2.6a Optical Methods: Energy Level Determinations

In the same way that atomic spectroscopy helps to determine atomic energy levels, spectroscopy of semiconductors allows us to fix the energy levels of the crystal. Consider first the effect of a light beam of wavelength λ incident on the surface of a semiconductor. If the frequency of the light is such that $h\nu < E_g$ the light beam will traverse the crystal without attenuation. If by contrast $h\nu > E_g$ the photons can be absorbed by excitation of valence-band electrons into the conduction band. In addition a part of the incident beam is reflected.

The intensity of the light beam I varies with distance as

$$I = I_0 \exp(-\alpha x), \tag{2.105}$$

where α is the absorption coefficient (Fig. 2.15). One can show (e.g., Wooten: Optical Properties of Solids, Academic Press, New York, 1972) that the absorption coefficient can be expressed as

$$\alpha \sim (h\,\nu - E_g)^\gamma, \tag{2.106}$$

where γ is a constant which depends on the nature of the transitions. A calculation of the absorption coefficient is presented in Sect. 6.1.

Figure 2.16 shows the various possible transitions: (a) vertical permitted transitions between extrema, called direct transitions (see Chap. 4): $\gamma =$

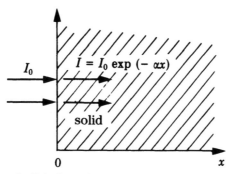

Fig. 2.15. Absorption of a light beam by a solid.

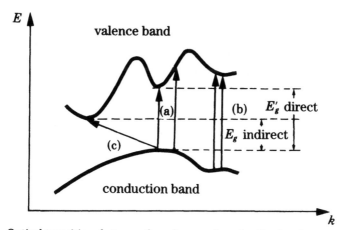

Fig. 2.16. Optical transitions between the valence and conduction bands.

1/2; (b) vertical forbidden transitions: $\gamma = 3/2$; (c) indirect transitions between extrema situated at different points in the zone: $\gamma = 2$.

The latter transitions can only occur if accompanied by the emission or absorption within the crystal of sound waves known as phonons.

When $h\nu$ is much larger than the fundamental absorption these effects overlap for the various gaps E_{g1}, E_{g2} which occur in the band structure.

Absorption curves for various semiconductors are shown in Fig. 2.17. We see that absorption rises very rapidly with energy. We reach absorption coefficients of order 10^5 to 10^6 cm^{-1}, i.e., the beam decreases in intensity by a factor $1/e$ over a distance of 100 or 1000 angstroms. It is then difficult to measure the absorption, which requires a very thin sample, and it is preferable to study the reflectivity of the crystal surface. Each time the photon energy reaches a critical value for the band structure (the distance between band extrema) a structure is observed in the reflectivity. Figure

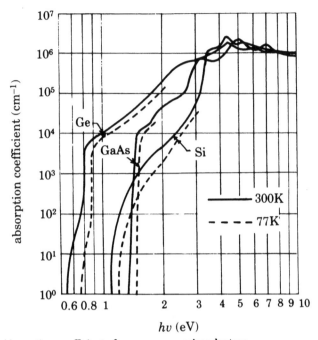

Fig. 2.17. Absorption coefficients for common semiconductors.

2.18 shows a schematic view of a reflectivity experiment, which may use a light source with a non-constant wavelength intensity.

An elliptical mirror with foci A and B is mounted on a rotating axis. The monochromator supplies at A a beam of wavelength λ but whose intensity may vary with λ. As the elliptical mirror rotates, the beam from A alternately falls on the sample under study, when the detector only receives light reflected from the sample, or directly on the detector after a rotation

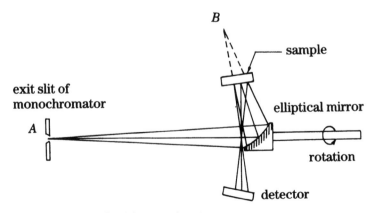

Fig. 2.18. Principle of a reflectivity experiment.

of the mirror through π. The rotation thus produces a sequence of reflected and incident signals at the detector, allowing one to measure their ratio as a function of wavelength.

2.6b The Cyclotron Resonance: Measuring the Effective Mass

Let us consider for example a carrier of charge q placed in a constant magnetic field $\boldsymbol{B}$ and an alternating electric field orthogonal to $\boldsymbol{B}$ with frequency ω. Let us assume that its effective mass m^* is isotropic. The equation of motion (2.38) is

$$m^* \frac{d\mathbf{v}}{dt} = q\,(\boldsymbol{\mathcal{E}} + \mathbf{v} \times \boldsymbol{B}). \tag{2.107}$$

In the absence of the electric field the motion is circular in the plane perpendicular to $\boldsymbol{B}$ or helical around $\boldsymbol{B}$ with frequency ω_c:

$$\omega_c = \frac{q\,B}{m^*}. \tag{2.108}$$

We call these orbits "cyclotron orbits." The orbits are helices around the magnetic field. In the presence of an electric field we have to take account of collisions, which we can do by introducing an average frictional force; in the Drude model (Sect. 5.2) this is $-m^*\mathbf{v}/\tau$.

The equation of motion is then

$$m^* \left(\frac{d\,\mathbf{v}}{dt} + \frac{\mathbf{v}}{\tau} \right) = q\,(\boldsymbol{\mathcal{E}} + \mathbf{v} \times \boldsymbol{B}). \tag{2.109}$$

Setting $B_x = B_y = 0$; $B_z = B$; $\mathcal{E}_y = \mathcal{E}_z = 0$; $\mathcal{E}_x = \mathcal{E}_0 \exp i\omega t$; $v_x = A_1 \exp i\omega t$; $v_y = A_2 \exp i\omega t$, substitution in Eq. (2.109) gives

$$A_1 \left(i\,\omega + \frac{1}{\tau} \right) = \frac{q\,\mathcal{E}_0}{m^*} + A_2\,\omega_c,$$

$$A_2 \left(i\,\omega + \frac{1}{\tau} \right) = -A_1\,\omega_c,$$

and the conductivity for this carrier

$$\sigma_{xx} = \frac{j_x}{\mathcal{E}_x} = \frac{q\,v_x}{\mathcal{E}_x} = \frac{qA_1}{\mathcal{E}_0} = \frac{q^2\,\tau}{m^*} \frac{1 + i\,\omega\,\tau}{1 + \tau^2(\omega_c^2 - \omega^2) + 2\,i\,\omega\,\tau}. \tag{2.110}$$

A resonance appears for $\omega = \omega_c$, in the form of a conductivity increase. This corresponds to a resonant absorption of energy by the carriers, which occurs when the charge rotates at exactly the same frequency as the electric field. Measurement of the resonant frequency gives a determination of the effective mass (in practice one works at fixed frequency and scans the magnetic field). For this phenomenon to be important requires $\omega_c \tau > 1$, i.e., we have to work with strong magnetic fields and at low temperatures where collisions are rarer. Appendix 3.1 describes the results of a cyclotron resonance experiment for silicon.

Appendix 2.1

Matrix Element of a Periodic Operator between Two Bloch States

Periodicity in Three Dimensions and Fourier Expansion

We show first that the Fourier expansion of a function $f(\mathbf{r})$ which is periodic on a crystal picks out the vectors $\mathbf{G}$ of the reciprocal lattice and no others.

The Fourier expansion of $f(\mathbf{r})$, for any $\mathbf{r}$, can be written

$$f(\mathbf{r}) = \sum n_{\mathbf{K}} \exp i\mathbf{K} \cdot \mathbf{r}. \tag{2.111}$$

The periodicity of $f(\mathbf{r})$, for any translation $\mathbf{T}$ of the direct lattice given by Eq. (2.2), can be written as

$$f(\mathbf{r} + \mathbf{T}) = f(\mathbf{r}) \tag{2.112}$$

or

$$\sum_{\mathbf{K}} n_{\mathbf{K}} \exp(i\mathbf{K} \cdot \mathbf{r}) \exp(i\mathbf{K} \cdot \mathbf{T}) = \sum_{\mathbf{K}} n_{\mathbf{K}} \exp(i\mathbf{K} \cdot \mathbf{r}). \tag{2.113}$$

The relation (2.113) will hold for all $\mathbf{r}$ and $\mathbf{T}$ only for those vectors $\mathbf{K}$ such that

$$\exp i\mathbf{K} \cdot \mathbf{T} = 1. \tag{2.114}$$

This is precisely the property (2.11) of the reciprocal lattice, and $\mathbf{K}$ can be identified with the vector $\mathbf{G}$ defined by Eq. (2.10). Hence the crystal potential has a Fourier expansion involving only the reciprocal lattice vectors. Similarly the periodic part of a Bloch function has the expansion

$$u_{n,\mathbf{k}}(\mathbf{r}) = \sum_{\mathbf{G}} U_{n,\mathbf{k},\mathbf{G}} \exp i\mathbf{G} \cdot \mathbf{r}. \tag{2.115}$$

Action of a Periodic Operator on a Bloch Function

A periodic operator $A(\mathbf{p}, \mathbf{r})$ is an operator for which

$$A(\mathbf{p},\ \mathbf{r} + \mathbf{T}) = A(\mathbf{p}, \mathbf{r}). \tag{2.116}$$

If we write it in the form

$$A(\mathbf{p}, \mathbf{r}) = \sum_i f_i(\mathbf{r})\, A_i(\mathbf{p}), \tag{2.117}$$

the definition (2.116) requires that the $f_i(\mathbf{r})$ should be periodic functions. Expanding these functions in a Fourier series, we can always write a periodic operator in the form

$$A(\mathbf{p}, \mathbf{r}) = \sum_G \exp(i\mathbf{G}\cdot\mathbf{r})\, A_G(\mathbf{p}). \tag{2.118}$$

The action of such an operator on a Bloch function $\exp(i\mathbf{k} \cdot \mathbf{r})\, u_{n,\mathbf{k}}(\mathbf{r})$ will produce a periodic function multiplied by $\exp(i\mathbf{k} \cdot \mathbf{r})$. In fact $A_G\,(\mathbf{p})$ is a combination of derivative operators which transform a periodic function into a periodic function, while retaining a factor $\exp(i\mathbf{k} \cdot \mathbf{r})$. In general, we will have

$$A(\mathbf{p}, \mathbf{r}) \exp(i\mathbf{k} \cdot \mathbf{r})\, u_{n,\mathbf{k}}\,(\mathbf{r}) = \exp(i\mathbf{k} \cdot \mathbf{r})\, \mathcal{U}_{n,\mathbf{k}}(\mathbf{r}), \tag{2.119}$$

where $\mathcal{U}_{n,\mathbf{k}}(\mathbf{r})$ is a periodic function which can be expanded in a Fourier series:

$$\mathcal{U}_{n,\mathbf{k}}(\mathbf{r}) = \sum_G \mathcal{U}_{n,\mathbf{k},G}\, \exp(i\mathbf{G} \cdot \mathbf{r}). \tag{2.120}$$

Matrix Elements of a Periodic Operator

We wish to calculate

$$\begin{aligned}
< n'\mathbf{k}'|A|n\ \mathbf{k} > &= \int_\Omega u'^*_{n',\mathbf{k}'}(\mathbf{r})\, \mathcal{U}_{n,\mathbf{k}}(\mathbf{r})\, \exp[-i(\mathbf{k}' - \mathbf{k}) \cdot \mathbf{r}]\, d^3\mathbf{r} \\
&= \sum_{G,G'} U'^*_{n',\mathbf{k}',G},\ \mathcal{U}_{n,\mathbf{k},G} \times \\
&\qquad \int_\Omega d^3\mathbf{r}\ \exp i(\mathbf{k} - \mathbf{k}' + \mathbf{G} - \mathbf{G}') \cdot \mathbf{r}.
\end{aligned} \tag{2.121}$$

We assume a cubic lattice with lattice constant a, and let $\Omega = L_x L_y L_z$ be the volume of the crystal, taken to be a parallelepiped of edges L_x, L_y, L_z. We calculate the integral

$$\frac{1}{\Omega} \int \exp i(\mathbf{k} - \mathbf{k}' + \mathbf{G} - \mathbf{G}') \cdot \mathbf{r} \, d^3\mathbf{r}$$

$$= \frac{1}{\Omega} \prod_{x,y,z} \int_{-Lx/2}^{+Lx/2} dx \exp i(k_x - k_x' + G_x - G_x')x$$

$$= \prod_{x,y,z} \frac{2 \sin \frac{1}{2}(k_x - k_x' + G_x - G_x')L_x}{(k_x - k_x' + G_x - G_x')L_x}. \qquad (2.122)$$

Using the fact that $(\mathbf{G} - \mathbf{G}')$ is a reciprocal lattice vector, $G_x - G_x' = m_x \times 2\pi/a$. Further, $L_x = N_x a$ and the components k_x and k_x' have the forms $n_x 2\pi/L_x$ and $n_x' 2\pi/L_x$ respectively, with N_x, m_x, n_x, n_x' whole numbers; the expression (2.122) then becomes

$$\prod_{x,y,z} \left[\frac{2 \sin \pi(n_x - n_x' + m_x N_x)}{2\pi(n_x - n_x' + m_x N_x)} \right], \qquad (2.123)$$

so that the argument of the sine is always an integer multiple of π. We have factors of the form $\sin X / X$, which are 1 for $X = 0$, and identically zero for $X \neq 0$. We calculate the argument of the sine. If k and k' are in the first Brillouin zone, either they are equal and the above expression becomes $\Pi_{x,y,z} \sin(m_x \pi N_x)/(m_x \pi N_x) = \Pi_{x,y,z} \delta_{m_x}$, or they differ but are within the zone so that $|n_x - n_x'| < N_x$ and the expression vanishes. It follows that

$$\frac{1}{\Omega} \int \exp i(\mathbf{k} - \mathbf{k}' + \mathbf{G} - \mathbf{G}')\mathbf{r} \, d^3\mathbf{r} = \delta_{\mathbf{k},\mathbf{k}'} \, \delta_{\mathbf{G},\mathbf{G}'}, \qquad (2.124)$$

and the expression (2.121), after replacement of the integral by Eq. (2.124) and carrying out the summation over $\mathbf{G}'$, becomes

$$< n' \, \mathbf{k}'|A|n \, \mathbf{k} > = \Omega \sum_{\mathbf{G}} U_{n',\mathbf{k}',\mathbf{G}}'^* \, \mathcal{U}_{n,\mathbf{k},\mathbf{G}} \, \delta_{\mathbf{k},\mathbf{k}'} \qquad (2.125)$$

which shows that **a periodic operator has zero matrix elements between Bloch functions of different k.**

Appendix 2.2

Symmetries of the Band Structure

Transformation of the Wave Function under Symmetry Operations of the Direct Space

Consider a symmetry operation $\mathcal{R}$ and a state $\psi_{n,\mathbf{k}}$ corresponding to a point $\mathbf{k}$ in the first Brillouin zone. We will show that the application of $\mathcal{R}$ to the wave function rotates the wave vector in the reciprocal space in the same way that the point $\mathbf{r}$ is rotated by the operator R in the real space:

$$\mathcal{R}\,\psi_{n,\mathbf{k}}\,(\mathbf{r}) = \psi_{n,R\mathbf{k}}\,(\mathbf{r}). \qquad (2.126)$$

To demonstrate this result we have to define the effect of a symmetry operation on a quantum state in general. Consider for example a rotation in ordinary space (here in two dimensions) taking the point A to the point B (Fig. 2.19(a)). Let R be the geometrical transformation $B = R(A)$ and a physical system, for example an electron, be in the state $\psi_1(\mathbf{r})$ localized around A. In Fig. 2.19(b) the hatched region represents the space where the probability density $\psi_1^*(\mathbf{r})\psi_1(\mathbf{r})$ is not negligible. Rotating the system amounts to bringing it into the state ψ_2 that we seek, and which is represented in Fig. 2.19(c). It is clear from the figure that

$$\psi_2(B) = \psi_1(A) = \psi_1(R^{-1}B). \qquad (2.127)$$

This holds for any point $\mathbf{r}$, whence

$$\mathcal{R}\,\psi(\mathbf{r}) = \psi(R^{-1}\mathbf{r}). \qquad (2.128)$$

This relation allows us to associate with a symmetry operation R in the ordinary space a (unitary) operation $\mathcal{R}$ in the state space. Applying this relation to $\psi_{n,\mathbf{k}}\,(\mathbf{r})$ we obtain

$$\begin{aligned} \mathcal{R}\,\psi_{n,\mathbf{k}}\,(\mathbf{r}) &= \exp i\mathbf{k}\cdot(R^{-1}\mathbf{r})\ u_{n,\mathbf{k}}\,(R^{-1}\mathbf{r}) \\ &= \exp i(R\,\mathbf{k})\cdot\mathbf{r}\ u_{n,\mathbf{k}}\,(R^{-1}\mathbf{r}) \end{aligned} \qquad (2.129)$$

which demonstrates Eq. (2.126). The symmetry operation has given us a new state which is to be associated with the vector $R\mathbf{k}$.

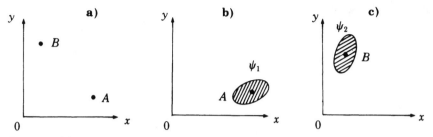

Fig. 2.19. (a) B is the transformation of A under the rotation R. The hatched region (c) is deduced from that of (b) by the same operation.

Symmetry of Constant Energy Surfaces

We show that $\mathcal{R}\psi_{n,\mathbf{k}}\,(\mathbf{r})$ is an eigenstate of the crystal Hamiltonian $\mathcal{H}$: let us apply the operator $\mathcal{R}$ to each side of the Schrödinger equation:

$$\mathcal{H}\,\psi_{n,\mathbf{k}}\,(\mathbf{r}) = E_{n\mathbf{k}}\,\psi_{n,\mathbf{k}}\,(\mathbf{r}), \tag{2.130}$$

$$\mathcal{R}\,\mathcal{H}\,\psi_{n,\mathbf{k}}\,(\mathbf{r}) = E_{n\mathbf{k}}\,\mathcal{R}\,\psi_{n,\mathbf{k}}\,(\mathbf{r}). \tag{2.131}$$

As $\mathcal{H}$ is invariant under the symmetry operation $\mathcal{R}$, it commutes with $\mathcal{R}$, so that

$$\mathcal{H}\,\mathcal{R}\,\psi_{n,\mathbf{k}}\,(\mathbf{r}) = E_{n\mathbf{k}}\,\mathcal{R}\,\psi_{n,\mathbf{k}}\,(\mathbf{r}) = E_{n\mathbf{k}}\,\psi_{n,R\mathbf{k}}\,(\mathbf{r}). \tag{2.132}$$

The state $\mathcal{R}\psi_{n,\mathbf{k}}$ deduced from $\psi_{n\mathbf{k}}$ by the symmetry operation corresponds to the same energy and is orthogonal to $\psi_{n,\mathbf{k}}$ since it corresponds to a different point of the zone. We can thus find for each symmetry operation a state degenerate with the original state and situated at $R\mathbf{k}$. This shows that **the constant energy surfaces have the symmetries of the crystal, making it sufficient to study only a fraction of the Brillouin zone.**

Appendix 2.3

Band Structure of Column IV Elements Calculated by the LCAO Method

We give here a first idea of the band structure of column IV elements, calculated by the tight binding method, for energies close to the band gap, and thus obtain a theory of the covalent bond in semiconductors. This calculation formally holds for the sequence diamond, silicon, germanium, and grey tin. (See G. Leman, Annales de Physique, Paris, **7** 1962 p. 505.)

The crystal is face-centered cubic with two atoms per cell, one at the origin, the other translated in the direction [1 1 1] of the quarter **b** of the principal diagonal of the cube (see Figs. 2.1 and 2.2). Each atom is at the center of a regular tetrahedron. The primitive cell, shown in Fig. 2.1, is rhombohedral with basis vectors:

$$\mathbf{a}_1 = (a/2)(\mathbf{i} + \mathbf{j}),$$
$$\mathbf{a}_2 = (a/2)(\mathbf{i} + \mathbf{k}),$$
$$\mathbf{a}_3 = (a/2)(\mathbf{j} + \mathbf{k}). \tag{2.133}$$

The electron configuration of column IV atoms is ns^2np^2 with the atomic Hamiltonian

$$\mathcal{H}_{\text{at}} = \frac{p^2}{2m} + V(\mathbf{r}), \quad \text{or} \quad \mathcal{H}_{\text{at}} = \frac{p^2}{2m} + V(\mathbf{r} - \mathbf{b}), \tag{2.134}$$

and eigenstates ϕ_s, ϕ_p of energies E_s, E_p. We consider

$$\phi_0 = (\phi_s + \phi_x + \phi_y + \phi_z)/2,$$
$$\phi_1 = (\phi_s - \phi_x - \phi_y + \phi_z)/2,$$
$$\phi_2 = (\phi_s - \phi_x + \phi_y - \phi_z)/2,$$
$$\phi_3 = (\phi_s + \phi_x - \phi_y - \phi_z)/2. \tag{2.135}$$

These functions, called hybrid orbitals sp_3, are not atomic eigenfunctions but form a basis for the tensor product $s \otimes p$. They have the essential property of "pointing," respectively, in the directions [1 1 1], [-1,-1,1],

[-1,1,-1], [1,-1,-1], i.e., toward the vertices of a tetrahedron in the observed direction of the covalent bond. The matrix elements of $\mathcal{H}_{at}$ in this basis are

$$< \phi_i|\mathcal{H}_{at}|\phi_i > = \frac{E_s + 3E_p}{4},$$

$$< \phi_i|\mathcal{H}_{at}|\phi_l > = \frac{E_s - E_p}{4}, \qquad (2.136)$$

where l denotes an orbital other than i localized at the same atom.

Similarly we consider hybrid orbitals pointing in the opposite directions:

$$\begin{aligned}
\phi'_0 &= (\phi_s - \phi_x - \phi_y - \phi_z)/2 && \text{pointing toward } [-1, -1, -1], \\
\phi'_1 &= (\phi_s + \phi_x + \phi_y - \phi_z)/2 && \text{pointing toward } [+1, +1, -1], \\
\phi'_2 &= (\phi_s + \phi_x - \phi_y + \phi_z)/2 && \text{pointing toward } [+1, -1, +1], \\
\phi'_3 &= (\phi_s - \phi_x + \phi_y + \phi_z)/2 && \text{pointing toward } [-1, +1, +1]. \qquad (2.137)
\end{aligned}$$

We seek a Bloch function solution for the crystal of the form

$$\psi_{n,k} = C\sum_j \exp(i\, \mathbf{k} \cdot \mathbf{R}_j) \sum_{i=0,1,2,3} [A_i\, \phi_i(\mathbf{r} - \mathbf{R}_j) + A'_i\, \phi'_i(\mathbf{r} - \mathbf{R}_j - \mathbf{b})], \qquad (2.138)$$

where the index j denotes the particular site in the lattice, the index i one of the four orbitals Eq. (2.135), and C is a normalization coefficient.

The crystal Hamiltonian is

$$\mathcal{H} = -\frac{\hbar^2}{2m}\Delta + \sum_j V(\mathbf{r} - \mathbf{R}_j) + V(\mathbf{r} - \mathbf{R}_j - \mathbf{b}), \qquad (2.139)$$

whence the eigenstate problem:

$$\mathcal{H}\,\psi_{n,k}\,(\mathbf{r}) = E_{n,k}\,\psi_{n,k}\,(\mathbf{r}). \qquad (2.140)$$

Consider the functions ϕ_i and ϕ'_i centered, respectively, at the origin and at $(1/4, 1/4, 1/4)$, and let us take their product with Eq. (2.140). We get

$$< \phi_i|\mathcal{H}|n, \mathbf{k} > = E_{n,k} < \phi_i|n, \mathbf{k} >, \qquad (2.141)$$

$$< \phi'_i|\mathcal{H}|n, \mathbf{k} > = E_{n,k} < \phi'_i|n, \mathbf{k} > . \qquad (2.142)$$

We can rewrite Eq. (2.139) in the form

$$\mathcal{H} = \mathcal{H}_{at} + \sum_{j\neq0} V(\mathbf{r} - \mathbf{R}_j) + \sum_j V(\mathbf{r} - \mathbf{R}_j - \mathbf{b}), \qquad (2.143)$$

where $\mathcal{H}_{at}$ is the electron Hamiltonian for the atom at the origin. One can thus see that the only contributions on the left in Eqs. (2.141) and (2.142) come either from the atomic terms or from the interaction term coupling ϕ_i or ϕ'_i to the only orbital from the nearest neighbor which points toward it (neglecting terms of the form $< \phi_i|V_{i\neq j}|\phi_i >$ and interactions between orbitals centered on two neighboring atoms but not pointing to each other). The interaction or "transfer" integrals are all equal and negative since the potential is attractive and the functions ϕ and ϕ' have the same sign in the region where they overlap. We set

$$< \phi|V|\phi' > = -\lambda. \tag{2.144}$$

Let us consider the site at the origin: ϕ_0 points toward the site $\mathbf{b} = (a/4)(1,1,1)$ and is thus only coupled to the orbital ϕ_0' centered at $\mathbf{b}$, with the coupling coefficient $\exp i\mathbf{k} \cdot \mathbf{a}_0 = 1$. ϕ_1, being directed along [-1,-1,1] is coupled to the orbital centered at the site $(a/4)(-1,-1,1) = \mathbf{b} - \mathbf{a}_1$, or $\phi_1'[\mathbf{r} - (-\mathbf{a}_1)]$ with the coefficient $\exp i\mathbf{k} \cdot \mathbf{R}_j = \exp(-i\mathbf{k} \cdot \mathbf{a}_1)$, etc. Finally Eqs. (2.141) and (2.142) can be written

$$\left[\frac{E_s + 3 E_p}{4} - E_{n,\mathbf{k}} \right] A_i + \frac{E_s - E_p}{4} \sum_{l \neq i} A_l -$$
$$\lambda \exp(-i \cdot \mathbf{k} \cdot \mathbf{a}_i)A_i' = 0, \tag{2.145}$$

$$-\lambda \exp i\, \mathbf{k} \cdot \mathbf{a}_i\, A_i + \left[\frac{E_s + 3 E_p}{4} - E_{n,\mathbf{k}} \right] A_i' +$$
$$\frac{E_s - E_p}{4} \sum_{l \neq i} A_l' = 0. \tag{2.146}$$

This system of homogeneous equations gives the eight coefficients A_i and A_i' which specify the wave function Eq. (2.138), provided the determinant of the coefficients vanishes. One thus obtains the following secular equation, with

$$x = E_{n,\mathbf{k}} - E_p, \qquad \delta = \frac{E_p - E_s}{4}, \qquad \text{and} \quad \alpha_n = \exp(-i\, \mathbf{k} \cdot \mathbf{a}_n), \tag{2.147}$$

$$\begin{vmatrix} x+\delta & \delta & \delta & \delta & \lambda\,\alpha_0 & 0 & 0 & 0 \\ \delta & x+\delta & \delta & \delta & 0 & \lambda\,\alpha_1 & 0 & 0 \\ \delta & \delta & x+\delta & \delta & 0 & 0 & \lambda\,\alpha_2 & 0 \\ \delta & \delta & \delta & x+\delta & 0 & 0 & 0 & \lambda\,\alpha_3 \\ \lambda\,\alpha_0^* & 0 & 0 & 0 & x+\delta & \delta & \delta & \delta \\ 0 & \lambda\,\alpha_1^* & 0 & 0 & \delta & x+\delta & \delta & \delta \\ 0 & 0 & \lambda\,\alpha_2^* & 0 & \delta & \delta & x+\delta & \delta \\ 0 & 0 & 0 & \lambda\,\alpha_3^* & \delta & \delta & \delta & x+\delta \end{vmatrix} = 0. \tag{2.148}$$

Setting $\Phi = |\frac{1}{4}\Sigma_n \alpha_n|$, a careful calculation shows that Eq. (2.148) can be written as

$$(x^2 - \lambda^2)^2(x^2 + 4\,\delta x - \lambda^2 + 4\,\delta\,\lambda\,\Phi),$$
$$(x^2 + 4\,\delta x - \lambda^2 - 4\,\delta\,\lambda\,\Phi) = 0, \tag{2.149}$$

with

$$\Phi(\mathbf{k}) = \frac{1}{2} \left[1 + \cos \frac{k_x\, a}{2} \cos \frac{k_y\, a}{2} \right.$$

$$\left. + \cos \frac{k_x\, a}{2} \times \cos \frac{k_z\, a}{2} + \cos \frac{k_y\, a}{2} \cos \frac{k_z\, a}{2} \right]^{1/2} \tag{2.150}$$

Equation (2.149) gives $x(\Phi)$, i.e., the dispersion relation $E_n(\mathbf{k})$.

Resulting Band Structure

We obtain four flat bands (for which E_n does not depend on $\mathbf{k}$) which correspond to the doubly degenerate solutions $x_1 = \lambda$ and to the similarly doubly degenerate solutions $x_2 = -\lambda$.

The broad bands are associated with the other solutions of Eq. (2.149), i.e.,

$$x_3 = -2\,\delta + \sqrt{4\,\delta^2 + \lambda^2 + 4\,\delta\,\lambda\,\Phi},$$

$$x_4 = -2\,\delta - \sqrt{4\,\delta^2 + \lambda^2 + 4\,\delta\,\lambda\,\Phi},$$

$$x_5 = -2\,\delta + \sqrt{4\,\delta^2 + \lambda^2 - 4\,\delta\,\lambda\,\Phi},$$

$$x_6 = -2\,\delta - \sqrt{4\,\delta^2 + \lambda^2 - 4\,\delta\,\lambda\,\Phi}. \tag{2.151}$$

The band structure has a different shape for $\lambda > 2\delta$ or $\lambda < 2\delta$. It is shown in Fig. 2.20(a) for two directions of the vector $\mathbf{k}$ in the zone and in the special case $\lambda = \delta$. There are $8N$ electrons to be placed in these levels since the primitive cell contains two atoms each possessing four valence electrons. Taking account of the spin degeneracy of each atomic level we see that at zero temperature the bands x_4, x_6, and x_2 are filled. We are thus dealing here with a metal, as the band x_5 is empty and very near in energy to the filled levels.

By contrast if $\lambda > 2\delta$ we have a semiconductor. This is shown in Fig. 2.20(b) for the special case $\lambda = 3\delta$. The $8N$ electrons fill the x_4, x_6, and x_2 bands, and the material is an insulator at zero temperature. The width of the band gap is

$$E_g = x_{5,k=0} - x_2 = 2\,\lambda - 4\,\delta. \tag{2.152}$$

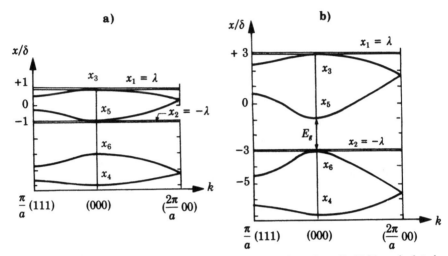

Fig. 2.20. Band structure of an element of column IV of the Periodic Table, calculated by the LCAO method. The transfer integral is $-\lambda$, and the distance between the atomic s and p levels is 4δ. The bands x_4, x_6, and x_2 are full. (a) $\lambda = \delta$; the material is a conductor; (b) $\lambda = 3\delta$; the material is an insulator.

Application: Binding Energy of Semiconductors of Column IV

This very simplified picture of band structure already gives an approximate understanding of several properties of these solids, for example the relation between binding energy and band gap width. The following values are measured for the binding energy E_C and the width of the band gap E_g, expressed in eV (P. Manca, Journal of Physics and Chemistry of Solids **20**, 268, 1961).

	C	Si	Ge	Sn
E_C	14.7	7.55	6.52	5.5
E_g	5.2	1.12	0.66	~ 0

These values are plotted in Fig. 2.21. The experimental points fall approximately on a line with the equation

$$E_C(\text{eV}) = 1.85\, E_g + 5.36. \tag{2.153}$$

In our model the binding energy is the difference between the energy $2E_s + 2E_p$ of an atom (2 s and 2 p electrons) and the energy per atom of the crystal. This energy E_X is, taking account of Eqs. (2.147) and (2.151),

$$E_X = 4\, E_p + 2\, x_2 + (\text{zone average of } x_6 + x_4). \tag{2.154}$$

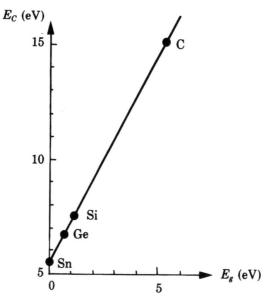

Fig. 2.21. The relation between cohesive energy and band gap width obtained by the LCAO method for the sequence of crystals of elements of column IV of the Periodic Table.

As a first approximation for the average we can take the value $(x_{6,k=0} + x_{4,k=0})$. We then find

$$E_X = 4\,E_p - 4\,\delta - 4\,\lambda, \tag{2.155}$$

and using Eqs. (2.152) and (2.155) the binding energy can be written as

$$E_C = 2(E_s + E_p) - E_X,$$
$$E_C = 2\,E_g + 4\,\delta. \tag{2.156}$$

This relation should be compared with the experimental relation Eq. (2.153). The energy $4\delta = E_p - E_s$ is the excitation energy for an atomic electron from level s to level p. This quantity, provided by atomic physics, is almost constant over column IV and is about 5 eV. The agreement with the experiment is very good. This description of the binding energy neglects the effect of the repulsive term that must exist if the crystal structure is not to collapse. If the repulsive interactions vary very rapidly with distance then the above description holds; this is for example true of the hard-sphere repulsive potential.

We can understand also, at least qualitatively, the influence of the temperature or pressure on E_g. When the temperature decreases or the pressure increases, the mean distance between atoms decreases, raising the value of the integral λ, and simultaneously the band gap.

However, the above calculation does not allow us to get more than the order of magnitude of these variations. The fundamental reason is that by limiting ourselves to combinations of a small number of atomic orbitals and only considering nearest neighbors we are working in a Hilbert space of too small a dimension. Nonetheless, this method gives an excellent approximation for the deep levels in solids.

Remark: for tin (the allotropic form called "grey tin") the gap is actually negative. This means that there is an overlap in energy between the top of the valence bands and the bottom of the conduction band. In this situation the highest energy electrons in the valence band populate the lowest energy states of the conduction band. There can thus exist a metallic-type conductivity due simultaneously to conduction-band electrons and to "holes" or empty places in the valence band, with equal numbers (see Chap. 3). We say that we are dealing with a "semimetal."

Appendix 2.4

The **k·p** Method

The $\mathbf{k} \cdot \mathbf{p}$ method is a semi-empirical method which uses quantities found from experiment in the theoretical calculation of the band structure. We start from the equation

$$\left[\frac{p^2}{2m} + V(\mathbf{r})\right] \psi_{n,\mathbf{k}}(\mathbf{r}) = E_{n,\mathbf{k}}\, \psi_{n,\mathbf{k}}(\mathbf{r}). \tag{2.157}$$

Replacing $\psi_{n,\mathbf{k}}$ by $u_{n,\mathbf{k}}(\mathbf{r}) \exp(i\mathbf{k} \cdot \mathbf{r})$ we note that

$$\mathbf{p}\, \psi_{n,\mathbf{k}} = \exp(i\mathbf{k}\cdot\mathbf{r})\ (\mathbf{p} + \hbar\, \mathbf{k})\ u_{n,\mathbf{k}}(\mathbf{r}),$$
$$p^2\, \psi_{n,\mathbf{k}} = \exp(i\mathbf{k}\cdot\mathbf{r})\ (\mathbf{p} + \hbar\, \mathbf{k})^2\, u_{n,\mathbf{k}}(\mathbf{r}), \tag{2.158}$$

and thus

$$\left[\frac{(\mathbf{p} + \hbar\, \mathbf{k})^2}{2m} + V(\mathbf{r})\right] u_{n,\mathbf{k}}(\mathbf{r}) = E_{n,\mathbf{k}}\ u_{n,\mathbf{k}}(\mathbf{r}). \tag{2.159}$$

The periodic part of the Bloch function obeys an equation resembling the original equation apart from the vector $\hbar\mathbf{k}$. We can rewrite Eq. (2.159) in the form

$$\left[\frac{p^2}{2m} + \frac{\hbar\, \mathbf{k} \cdot \mathbf{p}}{m} + \frac{\hbar^2\, k^2}{2m} + V(\mathbf{r})\right] u_{n,\mathbf{k}}(\mathbf{r}) = E_{n,\mathbf{k}}\, u_{n,\mathbf{k}}(\mathbf{r}). \tag{2.160}$$

For a free electron in a box $V(\mathbf{r}) = 0$ an obvious solution is $u = $ constant or $E_k = \hbar^2 k^2/2m$ and $\psi = \Omega^{-1/2} \exp i\mathbf{k} \cdot \mathbf{r}$. Equation (2.160) takes a particularly simple form for $\mathbf{k} = 0$:

$$\left[\frac{p^2}{2m} + V(\mathbf{r})\right] u_{n,0}(\mathbf{r}) = E_{n,0}\, u_{n,0}(\mathbf{r}). \tag{2.161}$$

We note that when the atoms are very far apart, the $E_{n,0}$ are the atomic levels and the $u_{n,0}(\mathbf{r})$ atomic eigenfunctions. We note also that the equation giving $u_{n,0}(\mathbf{r})$ has the symmetries of the crystal potential $V(\mathbf{r})$.

The $\mathbf{k} \cdot \mathbf{p}$ method assumes that we know the values $E_{n,0}$ either from theory or experiment. We then consider small values of $\mathbf{k}$ close to $\mathbf{k} = 0$ and treat the operator $(\hbar/m)\mathbf{k} \cdot \mathbf{p}$ as a perturbation in the Hamiltonian.

Although the $\mathbf{k} \cdot \mathbf{p}$ method is more general, we shall assume for simplicity that the crystal has a center of symmetry (in the diamond structure this center is halfway between two atoms) and we confine ourselves to the study of a given non-degenerate level n.

We first use the symmetry of Eq. (2.161). The Hamiltonian is invariant under inversion, the symmetry operation sending $\mathbf{r}$ to $-\mathbf{r}$. Then if $u_{n,0}(\mathbf{r})$ is an eigenfunction of energy $E_{n,0}$, $u_{n,0}(-\mathbf{r})$ is also an eigenfunction of the same energy, hence also $[u_{n,0}(\mathbf{r}) + u_{n,0}(-\mathbf{r})]$ and $[u_{n,0}(\mathbf{r}) - u_{n,0}(-\mathbf{r})]$. The eigenfunctions can thus be classified into even and odd functions.

In perturbation theory one starts by considering the first-order diagonal matrix elements of the perturbation Hamiltonian $\mathcal{H}_p$. If we consider a non-degenerate level $|n, 0 >$ the first-order term $< n, 0|\mathbf{k} \cdot \mathbf{p}|n, 0 >$ vanishes:

$$\int u_{n,0}^*(\mathbf{r}) \frac{\partial}{\partial x} u_{n,0}(\mathbf{r}) \, d^3\mathbf{r} = 0 \tag{2.162}$$

because $u_{n,0}$ is either odd or even. We note that $u_{n,0}(\mathbf{r})$ is an eigenfunction of Eq. (2.160) with the eigenvalue $E_{n,0} + \hbar^2 k^2/2m$. There remains only the second-order correction. To this order the energies are given by

$$E_{n,} = E_{n,0} + \frac{\hbar^2 k^2}{2m} +$$
$$\hbar^2 \sum_{n' \neq n} \frac{< n', 0|\mathbf{k} \cdot \mathbf{p}/m|n, 0 >< n, 0|\mathbf{k} \cdot \mathbf{p}/m|n', 0 >}{E_{n,0} - E_{n',0}}, \tag{2.163}$$

and thus

$$E_{n,\mathbf{k}} = E_{n,0} + \sum_{\alpha,\beta} \frac{\hbar^2}{2m} \left(\frac{m}{m^*}\right)_{\alpha\beta} k_\alpha \, k_\beta, \tag{2.164}$$

with

$$\left(\frac{m}{m^*}\right)_{\alpha\beta} = \delta_{\alpha\beta} + \frac{2}{m} \sum_{n'} \frac{< n', 0|p_\alpha|n, 0 >< n, 0|p_\beta|n', 0 >}{E_{n,0} - E_{n',0}}. \tag{2.165}$$

This so-called $\mathbf{k} \cdot \mathbf{p}$ method allows us to find the effective masses directly either from the energy spectrum at $k = 0$ (the values of $E_{n,0}$) and from the parameters $| < n', 0|p_\alpha|n, 0 > |^2$. This is the most useful theoretical procedure for predicting and analyzing details of the band structure of semiconductors near the band extrema, and thus in the region of interest.

We note that if we know the $u_{n,0}(\mathbf{r})$ we can calculate the parameters directly. Generally these matrix elements are deduced from experiment. The energy differences which appear in the denominator are most usually found from optical absorption or reflectivity.

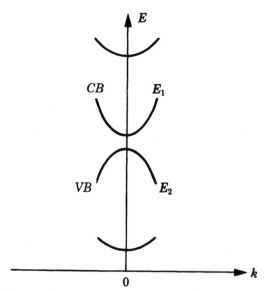

Fig. 2.22. The curvatures at $k = 0$ of the bands E_1 and E_2 are principally determined by their mutual interaction: the other bands are clearly further away, and the energy denominators make their interactions negligible (cf. Eq. (2.166)). In consequence the curvatures, and thus the effective masses of the bands E_1 (conduction) and E_2 (light holes) are practically opposite.

The order of magnitude of m^* is given by

$$\frac{m}{m^*} \sim 1 + 2 < \frac{p_x^2}{m} > \frac{1}{E_g}. \qquad (2.166)$$

Here p_x^2/m is of the order of the ionization energy of an atom, for example 5 eV, thus for $E_g = 0.5$ eV, m/m^* will be of order 20.

If we are interested in two bands 1 and 2 close in energy with their extrema at $\mathbf{k} = 0$, we can confine ourselves to including just their interaction in the expression (2.165), the corresponding term being much bigger than all the others (Fig. 2.22).

Example: for a cubic crystal the constant energy surfaces are spheres around $k = 0$, and we have

$$\frac{1}{m_1^*} \simeq \frac{1}{m} + \frac{2}{m^2} \frac{| < 1|p_x|2 > |^2}{E_1 - E_2},$$

$$\frac{1}{m_2^*} \simeq \frac{1}{m} + \frac{2}{m^2} \frac{| < 1|p_x|2 > |^2}{E_2 - E_1}, \qquad (2.167)$$

or

$$\frac{1}{m_1^*} = -\frac{1}{m_2^*} + \frac{2}{m}. \tag{2.168}$$

For germanium at $\mathbf{k} = 0$ the "light hole" band (see Chap. 3) has an effective mass $m_2^* = -0.042m$, while the conduction band has an effective mass m_1^* (at $k = 0$) of $0.036m$, in good agreement with the preceding expression: we sometimes say that in the region of small $\mathbf{k}$, these two bands are "mirror images" of each other, since $m_1^* \simeq -m_2^*$.

For the perturbation theory to provide a good approximation the matrix elements of the perturbation $(\hbar/m)\mathbf{k} \cdot \mathbf{p}$ between the functions $u_{n,0}$ and $u_{n',0}$ must be such that

$$\left| \frac{< n,0 \ |\hbar \ \mathbf{k} \cdot \mathbf{p}/m| \ n',0 >}{E_n - E_{n'}} \right| < 1, \tag{2.169}$$

or, for two close bands, from Eqs. (2.167) and (2.169):

$$\frac{1}{m_1^*} \sim \frac{2}{m^2} \frac{| < 1|p_x|2 > |^2}{E_1 - E_2},$$

$$\frac{\hbar \ k_x}{m} < \frac{E_1 - E_2}{| < 1|p_x|2 > |}.$$

Combining these two relations we get

$$\frac{\hbar^2 \ k^2}{2m_1^*} < E_1 - E_2. \tag{2.170}$$

The quantity $\hbar^2 k^2/2m_1^*$ is the kinetic energy in the E_1 band. For perturbation theory to give a good approximation, we require the energy in the band to be small compared with the band gap.

Example: $E_1 - E_2 = E_g = 1$ eV. For typical values of the electron energy in the band, of order the thermal energy $kT = 25$ meV at room temperature, the approximation easily holds.

3.

Excited States of a Pure Semiconductor and Quantum States of Impure Semiconductors

A semiconductor at zero temperature is an insulator. At room temperature we know that the system is not in its ground state. We thus have to consider the first excited states of a semiconductor. In these excited states a few electrons occupy the conduction band rather than the valence band, where they leave empty states. These empty states, called "holes," play a fundamental role in the conduction process.

Experiments show that the purity of semiconductors most often determines their behavior. The understanding of quantum states resulting from the presence of impurities is thus essential. Control of the concentration of selected impurities, "doping," is the main engineering tool for practical applications of semiconductors.

3.1 The Hole Concept

Up to now we have considered the ground state of a semiconductor at zero temperature. In this state the valence band is full and the conduction band empty. We are now interested in the first states accessible at non-zero temperature. The simplest excited state has one electron in the conduction band and one empty place in the valence band. Such a state can be obtained at low temperature by illuminating the crystal with electromagnetic radiation of energy greater than the width of the band gap. A photon can then be absorbed and excite an electron from the valence band into the conduction band.

Consider a solid in which we have created such a pair (electron in the conduction band + empty place in the valence band). The electron unbound from a covalent bond and placed in the conduction band (Fig. 3.1(a)) can then carry a current in the presence of an electric field because there exist nearby empty energy levels. If we put this electron into a (non-stationary) state formed by a wave packet we can consider that there is a Si$^-$ ion in the crystal and that this charge can move in the crystal by displacing the extra electron from one atom to another (Fig. 3.1(b)).

Now consider the valence band. There are normally four electrons per atom, the bond being shown schematically in Fig. 3.1(a). If we have formed a Si$^-$ ion somewhere in the crystal, there remains the equivalent of a Si$^+$ ion at the place where the bond was broken. This represents the lack of an electron in a valence bond. It is clear that in the presence of an electric field the empty place can be filled by an electron from another bond which moves under the effect of the field. We then have the equivalent of the displacement of Si$^+$ in the direction of the electric field, hence the motion of a positive charge (Fig. 3.1(b)). **We see that we can speak of the motion of the lack of an electron as the displacement of a positive charge which we call a hole.**

We now describe the notion of a positive hole more precisely. We have seen in Sect. 2.3, Eqs. (2.42) and (2.43), that the total current of the electron states of a given band is zero. Let us consider a valence band with a single empty place at the state $\mathbf{k} = \mathbf{k}_e$. The current $\mathbf{j}_{total}$ corresponding to the sum of the states of this band can be decomposed as

$$\mathbf{j}_{total} = -e \sum_{occupied} \mathbf{v}(\mathbf{k}) - e\, \mathbf{v}(\mathbf{k}_e) = 0. \tag{3.1}$$

In the present case where $\mathbf{k}_{empty} = \mathbf{k}_e$, the current $\mathbf{j}$ transported by the occupied valence states is then (using the definition of $\mathbf{v}(\mathbf{k}_e)$)

$$\mathbf{j} = e\, \mathbf{v}(\mathbf{k}_e)$$
$$= \frac{e}{\hbar} \nabla_{\mathbf{k}} E_e(\mathbf{k}_e). \tag{3.2}$$

$E_e(\mathbf{k}_e)$ is the dispersion relation of the valence electrons at the point $\mathbf{k}_e$.

We define a quasi-particle, the hole, by

hole = [valence band full except for one empty state].

The current carried by the hole is given by Eq. (3.2) above. If E is the total energy of the full valence band, the energy of the system [full band except for one empty state at the point $\mathbf{k}_e$] is $E - E_e(\mathbf{k}_e)$. The deeper the empty state lies in the valence band the larger the energy of the system (as $E_e(\mathbf{k}_e)$ is smaller). **The energy E of the quasi-particle we call the hole can thus be defined as**

$$E_h = -E \text{ (electron missing in the state } \mathbf{k}_e) + \text{constant.} \tag{3.3}$$

We wish to attribute a positive charge to the hole. For the hole current to have the form (3.2), the wave vector $\mathbf{k}_h$ must be equal to $-\mathbf{k}_e$, i.e., opposite to that of the missing electron. Then we can write

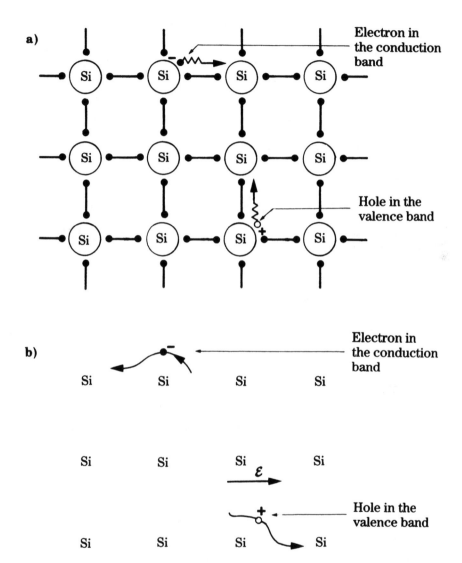

Fig. 3.1. (a) Formation of an electron-hole pair in silicon by absorption of a photon. A thick bar symbolizes a homopolar chemical bond, in which an electron pair with antiparallel spins is shared between two atoms; (b) displacement of these charges under the action of an applied electric field $\mathcal{E}$.

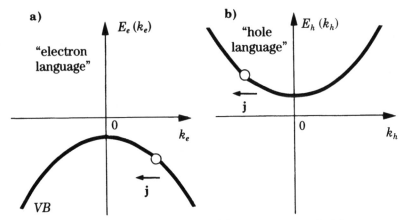

Fig. 3.2. Definition (a) in "electron language"; (b) in "hole language" of the wave vector, energy, and current of a hole in the valence band.

$$j = \frac{e}{\hbar} \nabla_{\mathbf{k}e} \, E_e(\mathbf{k}_e) = \frac{e}{\hbar} \nabla_{\mathbf{k}h} \, E_h(\mathbf{k}_h). \tag{3.4}$$

Figure 3.2 illustrates these definitions. The hole velocity is

$$\mathbf{v}_h = \frac{1}{\hbar} \nabla_{\mathbf{k}h} \, E_h(\mathbf{k}_h). \tag{3.5}$$

This is the velocity of the electron missing at $\mathbf{k}_e$. The dynamical equation, i.e., the evolution of $\mathbf{k}_h$ and $\mathbf{v}_h$ in the presence of electric and magnetic fields, can be found from Eq. (2.33) with a change of sign, since $\mathbf{k}_h = -\mathbf{k}_e$:

$$\hbar \frac{d\,\mathbf{k}_h}{dt} = -\hbar \frac{d\,\mathbf{k}_e}{dt} = e \, (\boldsymbol{\mathcal{E}} + \mathbf{v}_h \times \boldsymbol{B}). \tag{3.6}$$

This is the equation of motion of a positive charge of wave vector $\mathbf{k}_h$, moving with velocity $\mathbf{v}_h$, under fields $\boldsymbol{\mathcal{E}}$ and $\boldsymbol{B}$.

The dynamical equation in the real space results from Eq. (3.5) and the preceding equation. This allows us to define the effective mass of the hole:

$$\frac{dv_{h\alpha}}{dt} = \sum_{\beta} \left(\frac{1}{m_h^*} \right)_{\alpha\beta} F_\beta \quad \text{with} \quad \left(\frac{1}{m_h^*} \right)_{\alpha\beta} = \frac{1}{\hbar^2} \frac{\partial^2 E_h}{\partial k_{h\alpha} \, \partial k_{h\beta}}. \tag{3.7}$$

The effective mass of the hole is positive in the region of the Brillouin zone where the function $E_e(\mathbf{k})$ has a negative second derivative, i.e., near maxima. This holds in particular for the top of the valence band of semiconductors.

In summary, the electron-hole correspondence is as follows:

$E_h = -E$ of the **missing electron**,
$k_h = -k$ of the **missing electron**,
hole velocity = velocity of missing electron,
hole charge = +e,
effective hole mass = − effective mass of missing electron.

The effective mass of the hole is thus positive for a negative curvature of $E(\mathbf{k})$, which holds near a maximum of $E(\mathbf{k})$, i.e., at a band maximum. We then have

$$\mathbf{v}_h = \frac{1}{\hbar} \nabla_{\mathbf{k}_h} E_h(\mathbf{k}_h), \tag{3.8}$$

$$\mathbf{j} = e\, \mathbf{v}_h, \tag{3.9}$$

$$\hbar \frac{d\,\mathbf{k}_h}{dt} = e(\boldsymbol{\mathcal{E}} + \mathbf{v}_h \times \boldsymbol{\mathcal{B}}), \tag{3.10}$$

$$\frac{d\,\mathbf{v}_{h,\alpha}}{dt} = \sum_{\beta} \left(\frac{1}{m_h}\right)_{\alpha\beta} e(\boldsymbol{\mathcal{E}} + \mathbf{v}_h \times \boldsymbol{\mathcal{B}})_{\beta}. \tag{3.11}$$

An experimental determination of the effective masses of electrons and holes in silicon by means of cyclotron resonance is described in Appendix 3.1 (cf Sect. 2.6d).

We can qualitatively illustrate these notions by the example of photo-conductivity: optical excitation by direct transition of an electron in a state $\mathbf{k}_e$ of the valence band into a state of the same wave vector of the conduction band leaves a hole in the valence band of wave vector $\mathbf{k}_h = -\mathbf{k}_e$. Let us apply an electric field $\boldsymbol{\mathcal{E}}$ to this system; we show that the currents from the electron and hole add together. For a standard band scheme like that of Fig. 3.3 the effective masses m_e of the electrons and m_h of the holes are positive and isotropic.

Using the acceleration theorem in the real space [in the present case Eq. (3.38)], the electron motion in the presence of $\boldsymbol{\mathcal{E}}$ is given by

$$m_e \frac{d\mathbf{v}_e}{dt} = -e\, \boldsymbol{\mathcal{E}}, \tag{3.12}$$

giving the velocity and current after time Δt:

$$\Delta\, \mathbf{v}_e = -\frac{e\,\Delta t}{m_e} \boldsymbol{\mathcal{E}}, \tag{3.13}$$

$$\Delta\, \mathbf{j}_e = -e\, \Delta\, \mathbf{v}_e = \frac{e^2 \Delta t}{m_e} \boldsymbol{\mathcal{E}}. \tag{3.14}$$

For the hole,

$$m_h \frac{d\mathbf{v}_h}{dt} = e\, \boldsymbol{\mathcal{E}}, \tag{3.15}$$

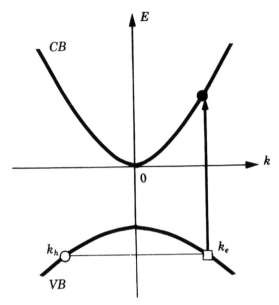

Fig. 3.3. Excitation of an electron-hole pair by direct optical transition.

$$\Delta \, \mathbf{v}_h = \frac{e \, \Delta \, t}{m_h} \mathcal{E}, \tag{3.16}$$

$$\Delta \, \mathbf{j}_h = \frac{e^2 \Delta t}{m_h} \mathcal{E}. \tag{3.17}$$

The electric drift currents for the electron and the hole have the same sign and add together. Figure 3.4 shows this result schematically. Transport problems, such as the calculation of conductivity, will be treated in detail in Chap. 5.

We return now to the real band structure of semiconductors (Sect. 2.4d). The valence band of Si or GaAs is actually degenerate in the neighborhood of its extremum at $k = 0$ (see Fig. 2.12). There are then two hole systems, and hence two effective hole masses. We distinguish heavy holes of mass m_{hh} and light holes of mass m_{lh}. In silicon, $m_{hh} = 0.49m$ and $m_{lh} = 0.16m$.

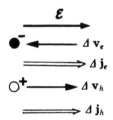

Fig. 3.4. Electric drift currents deriving for the electron ($\Delta \, \mathbf{j}_e$) and hole ($\Delta \, \mathbf{j}_h$) in the presence of an applied electric field $\mathcal{E}$.

3.2 Impurities in Semiconductors

Until now we have only considered perfect crystalline solids, without defects or impurities. What can we say about imperfect crystals? There is no general answer to this question: it all depends on the nature of the defects or impurities. First, they act as scattering centers, scattering electrons because they break the periodicity of the crystal potential. Second, and even more important, impurities play the essential role of modifying the electron content in semiconductors: a pure silicon crystal at room temperature would "naturally" (i.e., in thermal equilibrium) contain of 10^{-12} free electrons per atom. In real crystals the majority of the free electrons will in fact come from impurities.

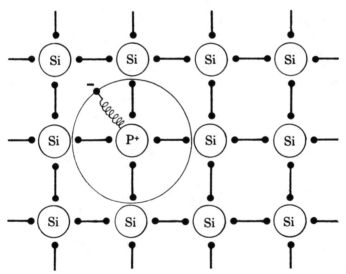

Fig. 3.5. Phosphorus atom substituted in a silicon lattice. Each black dot denotes an electron. At low temperature the extra electron is bound to the phosphorus nucleus. The radius of the orbit is in reality much larger than the interatomic distance. At high temperature the electron is released into the conduction band: phosphorus is a donor.

Consider for example a phosphorus atom replacing a silicon atom in a crystal (Fig. 3.5). This replacement can occur easily since the atoms have approximately the same size. We call this a substitutional impurity. The phosphorus atom has five valence electrons. We can to first approximation assume that four of these electrons, which fill states fairly similar to those of silicon, will participate in four covalent bonds with the four neighboring atoms. The P–Si bonds differ little from the Si–Si bond. These four electrons thus form part of the valence band by replacing the four silicon electrons which have disappeared. But there remains an electron, normally bound to the phosphorus atom which has an additional nuclear charge $+e$ (Fig. 3.5).

The "internal" ionization of this system consists of the removal of the electron from its state bound to the phosphorus nucleus into the conduction band (Fig. 3.5). **Atoms which can give an extra electron to the crystal on ionization are called donors.** There is a "pseudo-atom" of hydrogen in the medium. Its internal "ionization" energy will at most be equal to the width of the band gap.

This result is important as it shows that impurities in a semiconductor will self-ionize at temperatures lower than those required for intrinsic ionization between the valence and conduction bands. The very reason for this is that the energy necessary for this process is not the ionization energy of a phosphorus atom in vacuum (about 10 eV) but the energy to ionize (i.e., unbind an electron from the positive charge) within the semiconductor. This requires at most an energy of order E_g, the width of the band gap, about one electron volt. The ionization energy for a hydrogen-like system is of the order of $e^2/8\pi\epsilon_0 a$, where a is the orbital radius. If the energy is reduced by a factor of 10, the size of the bound orbit should be of the order of $10a_1$, where a_1 is the radius of the first Bohr orbit of the hydrogen atom. As $a_1 = 0.53$ angstrom the size of the orbit is about 5 angstroms. But the reasoning above is no longer valid as the attracting force seen by an electron is no longer $-e^2/4\pi\epsilon_0 r^2$ but $-e^2/4\pi\epsilon_0\epsilon_r r^2$, where ϵ_r is the relative dielectric constant of the medium. Here ϵ_r is of the order of 10, and the potential is reduced by a factor of 10. Because of this the orbit is still bigger. Moreover the electron is then bound by a potential varying slowly over interatomic distances. We know that the response to a force of this type involves not the free electron mass but the effective electron mass in the crystal.

We may then regard this system as a pseudo-hydrogen atom, with the Hamiltonian

$$\mathcal{H} = \frac{p^2}{2\,m^*} - \frac{e^2}{4\,\pi\,\epsilon_0\,\epsilon_r\,r}. \tag{3.18}$$

The energy eigenvalues are given by quantum mechanics as

$$E_n = -\frac{e^4}{2\,(4\,\pi\,\epsilon_0)^2\,\hbar^2\,n^2} \times \frac{m^*}{\epsilon_r^2}. \tag{3.19}$$

The Bohr radius of the orbit of quantum number $n = 1$ is

$$a_1^* = \frac{4\,\pi\,\epsilon_0\,\epsilon_r\,\hbar^2}{e^2\,m^*} \tag{3.20}$$

or

$$a_1^* = \epsilon_r \left(\frac{m}{m^*}\right) a_1, \quad \text{with } a_1 = 0.53\text{Å}. \tag{3.21}$$

For a state with $n > 1$

$$E_n = \frac{m^*}{m} \frac{1}{\epsilon_r^2} \frac{E_{1H}}{n^2}, \quad \text{where } E_{1H} = -13.6 \text{ eV}. \tag{3.22}$$

and the wave function extension is of order $n^2 a_1^*$.

We see that the binding energy E_1 of the ground state is greatly reduced since $\epsilon_r = 16$ for germanium and 11.7 for silicon; $m^*/m \sim 0.2$ for germanium and 0.4 for silicon. We thus predict from this simple theory an ionization energy independent of the nature of the donor, of 0.01 eV for elements of column V (P, As, Pb, Bi) in germanium and an energy of 0.04 eV in silicon (see the excerpt from the Periodic Table). This is the energy that must be supplied to the electron to ionize the atom in the crystal. The donor occupies a level at a distance from the conduction band small compared with the gap; we call this a "shallow donor."

Excerpt from the Periodic Table

columns:	IIb	III	IV	V	VI
		B	C	N	O
		Al	Si	P	S
	Zn	Ga	Ge	As	Se
	Cd	In	Sn	Sb	Te

The following table shows the very good agreement of the experimental values of E_1 in germanium and silicon, given in meV, with this theory:

	P	As	Sb	Bi
Ge	12	12.7	9.6	
Si	44	49	39	69

The radius of the ground state orbit is increased by the factor $\epsilon_r m/m^*$, which is about 50. This justifies the above approximations.

We can argue in a similar fashion for the substitutional impurities of group III of the Periodic Table (boron, aluminium, gallium, thallium). Group III atoms have only three valence electrons. Thus to fill all the valence states and realize for example four B–Si bonds, an electron has to be taken from a nearby Si–Si bond. For this reason **a substitutional element of column III in a semiconductor of column IV is an acceptor**. The electron taken from the Si–Si bond belongs to the valence band, and leaves a hole in its place. The binding energy of the captured electron is large, since this is the energy of the chemical valence bond. After the capture of this electron the ensemble (boron atom + captured electron) is negatively charged and attracts the free hole. This is shown in Fig. 3.6. The resulting bound state is the lowest energy state of the system. The acceptor is therefore neutral, since the electron captured to satisfy the bond and the hole localized near

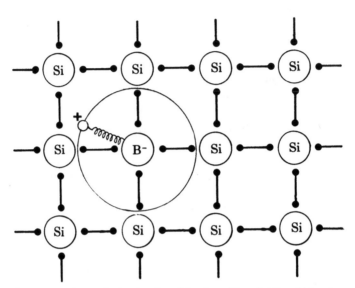

Fig. 3.6. Boron atom in a substitutional position in a silicon lattice. At low temperature the hole is bound to the boron nucleus in an orbit which covers many atomic sites. At high temperature the hole is released into the valence band; boron is an acceptor.

the boron atom form a system of zero net charge. To dissociate this neutral acceptor requires energy, but for the same reasons as for the donors the binding energy of the hole to the charged center of the impurity is weak. It is given by expression (3.22), where we have to replace the effective mass of the electron by the effective mass of the hole.

At very low temperature the hole will remain fixed in a hydrogen-like orbit and the crystal will not be a conductor, but **at room temperature the system will be ionized and the crystal will have a free hole** in the valence band for each element of column III present in the crystal. Then the center B will be negatively charged because of the captured electron.

The table below gives the experimental values in meV of the ionization energies of acceptors in germanium and silicon.

	B	Al	Ga	In	Tl
Ge	10.4	10.2	10.8	11.2	10
Si	46	57	65	160	260

The energy levels due to the presence of donors or acceptors are illustrated in Fig. 3.7.

The level E_d is below the minimum E_c of the conduction band (Fig. 3.7(a)). The fundamental level E_a of the hole should have a negative energy relative to the top of the valence band (in hole language), thus a positive

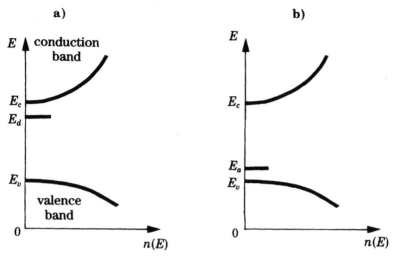

Fig. 3.7. Density of states curve in the vicinity of the band gap in the presence of (a) donors, (b) acceptors.

energy in electron language. This corresponds to the fact that we have to supply energy $E_a - E_v$ to an electron of the valence band to compensate the hole bound to the acceptor, thus ionizing the acceptor and creating a free hole (Fig. 3.7(b)).

The following remark is important for the statistics of semiconductors which we shall study in the next chapter. In the presence of N crystal sites of which N_a are acceptors, the N_a acceptors each capture an electron and create a hole, which at zero temperature remains trapped by the charged nucleus of the acceptor. The crystal is an insulator and its valence band is therefore full. Now, the number of electrons in the valence band is $4N - N_a$. Then in the presence of N_a acceptor sites out of a total of N atomic sites the number of places in the valence band is $4N - N_a$.

The above discussion of donor and acceptor ionization energies does not amount to a proof. In particular we have not shown under what conditions we can pass from the exact Schrödinger equation of the problem

$$\left[\frac{p^2}{2m} + V_p(\mathbf{r}) + V_I(\mathbf{r}) \right] \psi = E\,\psi, \tag{3.23}$$

where m is the mass of the free electron, V_p the potential of the crystal, and $V_I(\mathbf{r})$ the potential of the impurity, to the equation

$$\left[\frac{p^2}{2m^*} + V_I(\mathbf{r}) \right] \phi = (E - E_c)\phi, \tag{3.24}$$

where m^* is the effective mass, E_c the energy of the bottom of the conduction band, and ϕ a pseudo-wave function. It can be shown that the real wave function ψ can be written in the form $\psi(\mathbf{r}) = \phi(\mathbf{r})u_0(\mathbf{r})$, where ϕ plays the role of an envelope function ($u_0(\mathbf{r})$ is the periodic part of the Bloch function for $\mathbf{k} = 0$ in the case of a standard band). The theory taking us from Eq. (3.23) to Eq. (3.24) is called **effective mass theory**. A recent application of this theory to the study of "quantum wells" and "superlattices" is given in Appendix 3.2.

The fact that the experimental values of the ionization energies of the donors or acceptors vary from one impurity to another shows that some of our assumptions do not rigorously hold:

1 – The donor–Si bond differs from the Si–Si bond.

2 – The size of the atoms is not the same.

3 – The relative dielectric constant is not constant over all space. It varies from one very close to the impurity to ϵ_r over a few atomic layers.

Further, we should take into account the fact that the conduction band of Si or Ge has several minima (see W. Kohn, in Solid State Physics, Volume 5, Academic Press, New York, 1957).

If we consider impurities of column II or VI the simple picture of shallow levels no longer holds for several reasons. The internal levels are very different from those of column IV elements. The nuclear charge is stronger, the radius of the orbit is smaller, and the dielectric constant effect is reduced. This leads to "deep level" states, i.e., levels far from the band edges. These impurities, which exist in all semiconductors, are very important in the recombination of electron-hole pairs but they are relatively difficult to ionize thermally, as their energy distance from a band is much greater than the thermal energy kT. The case of amorphous semiconductors is more subtle, as geometrical defects play the role of chemical impurities. A qualitative description is given in Appendix 3.3.

3.3 Impurity Bands

Up to now we have discussed the effect of substitutional impurities, donors or acceptors, diluted within the semiconductor lattice. They give rise to discrete energy levels within the band gap. We can imagine that if the concentration, e.g., of donors becomes large enough in real space for the orbits to interact, the electrons from the donors will be delocalized within the crystal, even at low temperature. This happens for a donor concentration N_{do} such that $(4\pi/3)a_1^{*3} \cdot N_{do} \simeq 1$. The crystal is then a conductor at any temperature. An insulator–metal transition is thus observed as a function of concentration: for example, for $N_d < N_{do} = 3.7 \times 10^{24}$ m^{-3} silicon is an

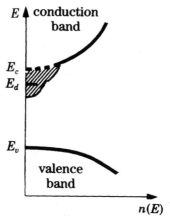

Fig. 3.8. Density of states in the presence of a strong donor concentration. An impurity band forms, shown hatched in the figure.

insulator at low temperature, while above it is a conductor. The presence of many donors leads to a broadening of the level E_d, which is then no longer separated from the conduction band (Fig. 3.8). An "impurity band" has formed.

Appendix 3.1

Problems on Cyclotron Resonance in Silicon

To study cyclotron resonance in a semiconductor one irradiates a single crystal with electromagnetic waves of frequency $\nu = \omega/2\pi$ from a waveguide. The crystal is placed between the poles of an electromagnet (Fig. 3.9). When the magnetic field is such that $\omega = \omega_c = eB/m_e$, where m_e is an effective mass, the crystal conductivity increases rapidly, leading to partial absorption of the waves, which can be observed by the power meter.

The experiment is performed at very low temperatures using a very pure single crystal, and the crystal is illuminated with light of energy greater than the band gap width. With a silicon crystal oriented with the **B** field in the bisecting plane of the cube $(1\bar{1}1)$ at an angle of $30°$ with the direction $[001]$ (see Fig. 3.10), one observes the signal indicated in Fig. 3.11 as a function of magnetic field.

1. Why do we illuminate the sample to observe a signal?

2. Assume that in silicon there exists a minimum of the conduction band inside the first Brillouin zone at $k_z = k_{z,0}; k_x = k_y = 0$. Are there any others? If so, why and how many? (x, y, z are the axes of the cubic cell of diamond.)

Show that the constant energy ellipsoids are ellipsoids of revolution and specify their axis of symmetry.

3. Consider first the ellipsoid of constant energy E centered at $k_{z,0} > 0$ described by the equation

$$E = \frac{\hbar^2}{2}\left[\frac{k_x^2}{m_T} + \frac{k_y^2}{m_T} + \frac{(k_z - k_{z,0})^2}{m_L} \right], \tag{3.25}$$

where m_T and m_L are the transverse and longitudinal masses. We call θ the angle between the magnetic field **B** and the axis of symmetry. We study the electron cyclotron resonance of electrons from this ellipsoid.

Show that the equation of motion of the electrons

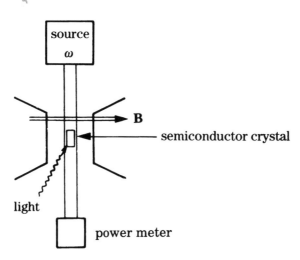

Fig. 3.9. Schematic view of a cyclotron resonance experiment.

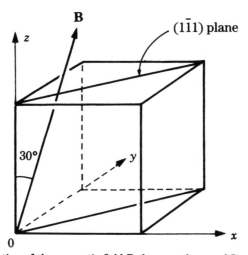

Fig. 3.10. Orientation of the magnetic field **B** the crystal axes of Si.

$$\frac{d\mathbf{v}}{dt} = -\left\|\frac{1}{M}\right\| (e\mathbf{v} \wedge \mathbf{B}) \qquad (3.26)$$

has an oscillating solution:

$$\mathbf{v} = \mathbf{v}_0 \exp i\omega t \qquad (3.27)$$

with $\omega = \omega_c = eB/m_e$ and

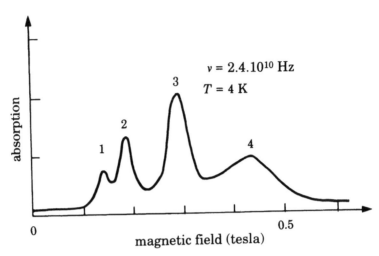

Fig. 3.11. Experimentally measured absorption.

$$m_e = \sqrt{\frac{m_T^2 \, m_L}{m_T \sin^2 \theta + m_L \cos^2 \theta}}. \tag{3.28}$$

(The calculation is simplest if, for this question, the ellipsoid of revolution is referred to axes $k_{x'}$ and $k_{y'}$ such that $\mathbf{B}$ is in the plane $k_{x'}, 0, k_z$.)

4. Taking account of the conduction band structure found in the second question, how many electron cyclotron resonance lines should one observe for an arbitrary orientation of the magnetic field? Answer the same question for the particular direction described for the experiment.

5. When the magnetic field is rotated from the direction described for the experiment we observe that peaks 1 and 4 do not change while peaks 2 and 3 do move, the first sometimes splitting into two. What is the cause of resonances 2 and 3? What is the cause of resonances 1 and 4 which are independent of the field orientation?

6. Deduce from the experimental figure the effective masses of the electrons and holes in silicon. We recall that the field of the cyclotron resonance for a free electron is about 0.86 T for the frequency $\nu = 2.4 \times 10^{10}$ Hz.

7. What is the effect of collisions on this experiment? Why is it necessary to work at low temperature?

Solutions

1. At low temperature the electrons and holes are trapped and there are no free carriers. To observe the resonance requires free electrons or holes, which are created by photoexcitation.

2. The crystal is cubic and the constant energy surfaces must have the same symmetries as the cube. By symmetry with respect to the center of the zone we pass from $+k_{z,0}$ to $-k_{z,0}$. By rotating through $\pi/2$ we pass to $\pm k_{x,0}$ and $\pm k_{y,0}$, with $|k_{x,0}| = |k_{y,0}| = |k_{z,0}|$. Each ellipsoid must be rotationally symmetric around the axis of the cube on which it is centered (this is a result of the invariance under rotations through $\pi/2$ around this axis). The set of all ellipsoids has indeed the symmetry of the crystal (see Sect. 2.4 and Fig. 2.13).

3. Using the expression for the effective mass tensor deduced from Eq. (3.25) and replacing Eq. (3.27) in Eq. (3.26) we get

$$
\begin{bmatrix} i\omega v_{0,x'} \\[4pt] i\omega v_{0,y'} \\[4pt] i\omega v_{0,z} \end{bmatrix} = \begin{Vmatrix} \dfrac{1}{m_T} & 0 & 0 \\[6pt] 0 & \dfrac{1}{m_T} & 0 \\[6pt] 0 & 0 & \dfrac{1}{m_L} \end{Vmatrix} \begin{bmatrix} -ev_{o,y'}B\cos\theta \\[4pt] ev_{o,x'}B\cos\theta - ev_{0,z}B\sin\theta \\[4pt] ev_{o,y'}B\sin\theta \end{bmatrix}
$$

We have three homogeneous equations whose determinant vanishes if

$$
\omega^2 = e^2\,B^2\left(\frac{\sin^2\theta}{m_L m_T} + \frac{\cos^2\theta}{m_T^2}\right) \tag{3.29}
$$

or

$$
\omega = \frac{eB}{m_e} \tag{3.30}
$$

with

$$
m_e = \sqrt{\frac{m_T^2\, m_L}{m_T\sin^2\theta + m_L\cos^2\theta}}. \tag{3.31}
$$

4. As the only relevant angle is between the magnetic field and the principal axis, and there are three principal axes, each common to two ellipsoids, we ought in general to observe three electron peaks. For the particular orientation of the experiment with B in a bisecting plane, the angle of B to the ellipsoids centered on $\pm k_{x0}$ and $\pm k_{y0}$ is the same and we expect two electron peaks.

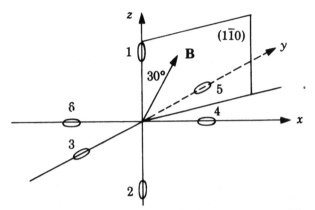

Fig. 3.12. Orientation of **B** relative to the ellipsoids of constant energy of the conduction band.

5. Peaks 1 and 4 are unmodified as they arise from holes, whose two bands are degenerate at $k = 0$ and are spherical. Thus $m_L = m_T$ and the resonance does not move when the magnetic field is rotated. Peaks 2 and 3 which move with the orientation are therefore the electron peaks.

6. For the ellipsoids 1 and 2 centered on Oz: $\theta = 30°$ and

$$m_{1e} = \sqrt{\frac{m_T^2\, m_L}{1/4m_T + 3/4m_L}} \tag{3.32}$$

for the electron resonance of ellipsoids 1 and 2.

The vector **B** has components $B(\sqrt{2}/4, \sqrt{2}/4, \sqrt{3}/2)$. The cosine of the angle between **B** and Ox is $\sqrt{2}/4$; similarly for the four ellipsoids in the plane and

$$m_{3e} = \sqrt{\frac{m_T^2\, m_L}{7/8m_T + 1/8m_L}} \tag{3.33}$$

for the electrons populating the ellipsoids 3, 4, 5, and 6.

We observe peaks 2 and 3 at about 0.18 T and 0.29 T. The strongest peak 3 must correspond to the four ellipsoids in the plane xOy. Comparing with the resonance field for a free electron of mass m at the same frequency we then have

$$\frac{8\, m_T^2\, m_L}{(7\, m_T + m_L)m^2} = \left(\frac{0.29}{0.86}\right)^2 \quad \text{and} \quad \frac{4\, m_T^2\, m_L}{(m_T + 3\, m_L)m^2} = \left(\frac{0.18}{0.86}\right)^2 \tag{3.34}$$

from which we obtain $m_L/m_T = 4.75$ and $m_T \simeq 0.19\, m; m_L = 0.9\, m$.

Peak 1 appears at 0.13 T. It corresponds to an effective mass of $(0.13/0.86)m$, or about $0.15m$, and thus to light holes. Peak 4 at 0.44 T corresponds to heavy holes of mass $(0.44/0.86)m \simeq 0.5m$.

7. To observe the resonance, collisions must be rare enough that a circular or elliptical orbit can be completed between collisions. The effect of collisions is to broaden the resonance. One works at low temperature so that the time τ between collisions is long enough that $\omega_c \tau > 1$.

Appendix 3.2

Quantum Wells and Semiconducting Superlattices

The effective mass theory mentioned in Sect. 3.2 allows us to start discussion of the physics of quantum wells and superlattices. These refer to stacks of alternating crystalline layers, possibly as small as a few atomic monolayers, of semiconductors of differing chemical compositions. In a superlattice the stacking is periodic, as shown in Fig. 3.13, which represents a GaAs–$Al_x Ga_{1-x}$As superlattice. We can only obtain such systems if epitaxy, the crystalline growth of one semiconductor on another, is possible. This requires semiconductors of different chemical composition that are very close geometrically: the same type of crystalline lattice, and the same size of elementary cell. The thickness of the layers may be of the order of 50 angstroms, but this can be varied.

Such structures have a paradoxical physical property: we would expect the optical absorption threshold to be that of the material with the narrower band gap, or in other words that a quantum well would be less transparent than any of its constituents. Experiment shows that this is not the case, and we shall see below that effective mass theory allows us to clear this paradox.

As the two material constituents are different they do not have the same value of band gap. For example in GaAs, E_g is 1.42 eV while in $Al_{0.4}Ga_{0.6}$As the gap E_g is 2 eV. These objects are made by a technique called molecular beam epitaxy (MBE), which gave its first promising results in the 1970s. In this process the crystal grows in an ultra-high vacuum chamber in which are placed crucibles containing gallium, aluminium, and arsenic (Fig. 3.14). The temperature of each crucible is controlled independently, so that one can control the speed of evaporation of each atomic species and thus the flux of atoms of each species. In the ultra-high vacuum chamber the atoms strike a GaAs substrate and one observes crystallization occurring atomic layer by atomic layer. It is possible to control the deposition to rates for example of one atomic layer per second. One can thus create tailored semiconducting structures on demand, in what is called "band gap engineering."

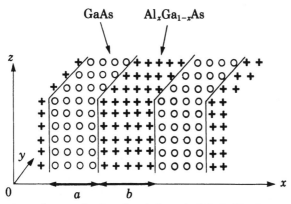

Fig. 3.13. Structure of a superlattice of period $a + b$ of GaAs/Al$_x$Ga$_{1-x}$As.

Why do we have to work in ultra-high vacuum? We can determine from the kinetic theory of gases the rate at which a surface becomes polluted by an atomic monolayer by assuming that every atom which hits the surface sticks to it. For a pressure of 10^{-13} bar the pollution time of the order of an hour. To avoid reaching such low pressures, we could think of increasing the flux of the atoms making up the semiconductor layers, but we have to allow time for the atoms to reach their equilibrium positions on the surface.

The crystalline potentials acting on the electrons are different in the two materials, so we expect that the band structure itself, and particularly the

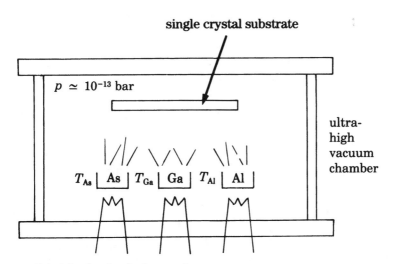

Fig. 3.14. Principle of molecular beam epitaxy.

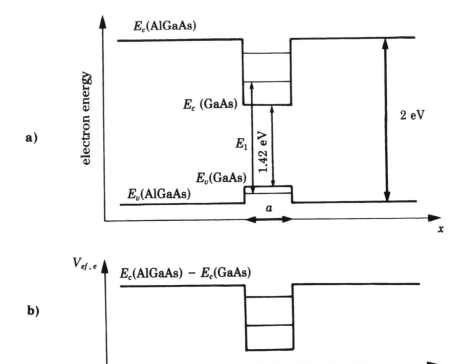

Fig. 3.15. Individual quantum well of GaAs in $Al_xGa_{1-x}As$. We show (a) the electron energy, and (b) the effective potential seen by the electrons, as functions of the distance perpendicular to the layers.

bottom of the conduction band and the top of the valence band would be at different energies. This is shown in Fig. 3.15(a) for the case of an isolated quantum well. Effective mass theory allows us to calculate the energy levels in this structure. We replace the real Schrödinger equation by an effective equation involving the effective Hamiltonian:

$$\left[\frac{p^2}{2\,m_e} + V_{ef,e}\,(x) \right] \phi_e(\mathbf{r}) = (E_e - E_c)\,\phi_e(\mathbf{r}). \qquad (3.35)$$

Here m_e is the electron effective mass and $V_{ef,e}$ the effective potential seen by the electrons, which takes account of the different nature of the two materials. This potential is represented in Fig. 3.15(b). The functions ϕ are the "envelope" wave functions, which take the form

$$\phi_e(\mathbf{r}) = \phi_e(x)\exp(ik_y y)\,\cdot\exp(ik_z z), \qquad (3.36)$$

where $\phi_e(x)$ is the solution of a one-dimensional Schrödinger equation in the potential of Fig. 3.15(b). We are led back to a problem in elementary quantum mechanics, that of a rectangular potential well in which one finds both delocalized levels extending above the barrier and localized levels E_e^α, E_e^β with $\alpha, \beta = 1, 2, \ldots$ inside the well.

In a structure with a periodic stacking of layers, like that of Fig. 3.13, two situations can occur. If the material with the large gap is much thicker than the material with the small gap ($b >> a$), then the eigenstates are bound states in each individual quantum well (1), (2), ..., of energy $E_e^\alpha, E_e^\beta, \ldots$. (See also Appendix 6.3.) On the other hand if b is small enough we have to take account of the overlap between functions localized in neighboring wells and thus make a theory of "superbands" in the "superperiodic" potential of Fig. 3.16. This is what we call a "superlattice." The levels E_e^α, E_e^β widen into bands $E_e^{\alpha'}, E_e^{\beta'}, \ldots$ called "minibands."

Similarly we can find the hole states by using the hole effective potential and the effective mass of the hole in an equation analogous to Eq. (3.35):

$$\left[\frac{p^2}{2\,m_h} + V_{ef,h}(x) \right] \phi_h(\mathbf{r}) = (E_h - E_v)\phi_h(\mathbf{r}). \qquad (3.37)$$

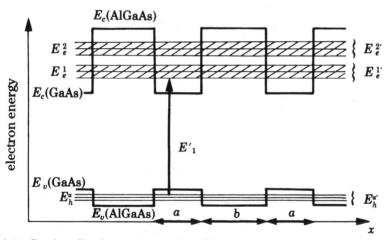

Fig. 3.16. Band profile of a superlattice GaAs/Al$_x$Ga$_{1-x}$As, of period $a + b$.

We obtain in this way hole states $E_h^\alpha \ldots$ or bands $E_h^{\alpha'} \ldots$. Placing all these electron energy levels in Fig. 3.15(a) or 3.16, we see that the new absorption threshold is E_1 or E_1' above the band gap of GaAs. For energies less than E_1 the structure is transparent and our paradox is resolved. The physical effect at the basis of this phenomenon is the increase of energy of a quantum state when it is confined, a general consequence of the Heisenberg principle.

Note that the price of an MBE machine is about one million dollars.

Appendix 3.3

Amorphous Semiconductors

An amorphous solid is one without the property of periodicity in space. Most of the properties of solids mentioned up to now were based on invariance under translations, which gives rise to Bloch's theorem. Should we conclude that a solid of given composition would change all its properties depending on the existence or not of long-range order? To answer this question it is useful to return to the chemical approach (cf. Sect. 1.2). In an amorphous as well as in a crystalline solid, there will be a "chemical bond" formed by the hybridization of the orbitals. This bond is a short-range property, while periodicity is a long-range one.

Let us take the example of amorphous silicon. The nearest neighbors of a silicon atom will still be at the vertices of a tetrahedron, but there will be distortions of the bond angles at the second, third neighbors and so on. This is shown in Fig. 3.17. We will arrive at a distribution of the atoms where some of the chemical bonds cannot be established for geometrical reasons. We say that we have "dangling bonds."

The energy levels corresponding to the dangling bonds are in the band gap of the crystal, as it is precisely the hybridization which creates the band gap. This is shown schematically in Figs. 3.18(a) and 18(b), where we compare the density of states in two semiconductors of the same chemical composition, one crystalline and the other amorphous.

An amorphous semiconductor, even though chemically pure, is thus "electronically" impure, since it has many states in the band gap. However, if we can replace the broken bonds with a strong chemical bond, each dangling bond we suppress will remove a state from the gap, the corresponding energy state now lying inside the valence band. This occurs for example when amorphous silicon is exposed to hydrogen at high temperatures. The hydrogen reacts and gives Si–H bonds, the corresponding states disappearing from the band gap (Fig. 3.19). Paradoxically, hydrogenated amorphous silicon is then more "electronically" pure than pure amorphous silicon. The density of states approaches that of crystalline silicon.

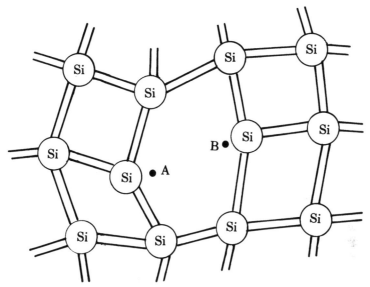

Fig. 3.17. Schematic representation of a disordered lattice of amorphous silicon. Most of the atoms establish four bonds, but because of the deformations situations exist in which atoms A and B are too far apart to establish a bond.

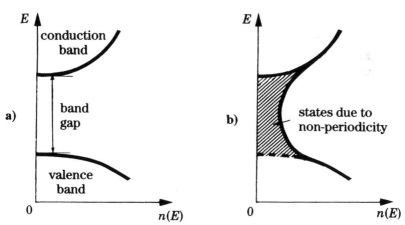

Fig. 3.18. Density of states in semiconductors (a) crystalline, and (b) amorphous, of the same chemical composition.

In practice hydrogenated amorphous silicon is made directly by decomposing silane (SiH_4) in a radiofrequency plasma. This is a very cheap semiconductor which is used particularly in the manufacture of pocket calculator photocells.

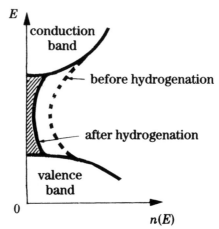

Fig. 3.19. Reduction of the density of states in the band gap by hydrogenation of amorphous silicon. It is reduced from about 10^{20} eV$^{-1}\cdot$cm^{-3} to about 10^{15} eV$^{-1}\cdot$cm^{-3}.

The present price for amorphous silicon cells is between one-tenth and one-hundredth of the current crystalline panels.

4.

Statistics of Homogeneous Semiconductors

4.1 Occupation of the Electron Levels

Knowing the band structure, at least in the region close to the band gap, as well as the localized quantum states caused by the presence of shallow impurities, calculation of the electrical conductivity of the semiconductor now requires to find the number of mobile charges and their nature, electrons or holes, at thermal equilibrium. For this we must calculate at temperature T the occupation probabilities of the accessible energy levels.

The electrons have half-integer spin and are fermions, so the system state only contains one electron per single-particle quantum state. For a given value of $\mathbf{k}$ there are two quantum states with different spins which can be simultaneously occupied. We recall that in such non-interacting Fermi–Dirac gases we define the Fermi factor f as the mean value of the operator measuring the occupation number of an electron state of energy E. If the chemical potential is E_F, hereafter called the Fermi level, we can show (see Appendix 4.1) that

$$f = \frac{1}{1 + \exp(E - E_F)/kT}, \tag{4.1}$$

where k is the Boltzmann constant.

We note that if $(E - E_F)$ is large compared with kT, i.e., if E exceeds E_F by several times the thermal energy, the Fermi factor of the electron can be approximated by a Maxwell–Boltzmann expression:

$$f \simeq \exp\left(\frac{E_F}{kT}\right) \exp\left(-\frac{E}{kT}\right). \tag{4.2}$$

This means that if f is small compared with 1 the effect of exclusion is negligible: then from a statistical point of view the electron behaves like a classical particle.

In semiconductors if we consider the mean occupation rate of donor levels which are **localized** states, Eq. (4.1) does not apply. In fact there are two degenerate spin states corresponding to each donor, but the two states cannot be occupied simultaneously because of the Coulomb repulsion between localized electrons (the corresponding state has a very high energy). In this case we can show (Appendix 4.1) that the Fermi function must be replaced, if we limit ourselves to the ground state of the donor, by

$$f' = \frac{1}{1 + (1/2)\exp(E - E_F/kT)}. \tag{4.3}$$

4.2 Hole Occupation

If the occupation probability of a level is given by Eq. (4.1), the probability that this level will not be occupied by an electron, i.e., the probability of occupation by a hole, is

$$f_h(E) = 1 - f(E) = \frac{1}{\exp(E_F - E/kT) + 1}. \tag{4.4}$$

If we measure the energy of a hole in the opposite sense from the electron energy (cf. Sect. 3.1) we have

$$f_h(E_h) = \frac{1}{1 + \exp(E_h - E_{Fh}/kT)} \tag{4.5}$$

which behaves similarly to the Fermi function Eq. (4.1).

As for an electron, these expressions can be approximated once the Fermi level is at few kT from the hole energy ($E_h - E_{Fh} \gg kT$):

$$f_h \simeq \exp\left(-\frac{E_h - E_{Fh}}{kT}\right) = \exp\left(\frac{E_{Fh}}{kT}\right)\exp\left(-\frac{E_h}{kT}\right). \tag{4.6}$$

We recover the exponential law of classical particles: the exclusion effect is negligible, as the average occupancy is much smaller than 1.

4.3 Determination of the Chemical Potential

Because a semiconductor is by definition an insulator at zero temperature we know that **the Fermi level must be within the band gap at T = 0**. However if we consider a real, and therefore impure, crystal, its properties, including the electron chemical potential, will be determined by its nature and impurity content. We will regard the electron gas occupying various energy states (valence band, impurity levels, conduction band) as a statistical canonical ensemble and determine its chemical potential. A given crystal will in general simultaneously contain "donor" and "acceptor" impurities.

Consider a crystal of silicon. Let N be the number of sites of the crystal, N_d the number of substitutional sites occupied by donors, e.g., phosphorus P, and N_a the number of substitutional sites occupied by acceptors, e.g., boron B. Leaving aside the electrons in deep atomic levels, the total number N_T of electrons present in the crystal is

$$N_T = 4N - 4N_d + 5N_d - 4N_a + 3N_a$$
$$= 4N + N_d - N_a. \tag{4.7}$$

At finite temperature the electrons can be in states of four types: in the valence band forming Si–Si bonds, in acceptor levels to form the four bonds of B–Si type (the acceptor is assumed ionized), in donor levels which are then neutral, or in the conduction band (cf. Sect. 3.2). We call p the concentration of holes, n_a^0 the concentration of neutral acceptors, n_a^- the concentration of ionized acceptors, n_d^0 the concentration of neutral donors, n_d^+ the concentration of ionized donors, and n the concentration of electrons.

Let us think in "electron language." We know (see Sect. 3.2) that when the valence band is full it contains $4N - N_a$ electrons. Writing N_T as the sum of the number of electrons in the valence band $4N - N_a - p$, the number of ionized acceptors n_a^-, the number of neutral donors n_d^0, and the n conduction electrons, we have

$$N_T = 4N - N_a - p + n_a^- + n_d^0 + n. \tag{4.8}$$

On the other hand,

$$N_a = n_a^0 + n_a^-,$$
$$N_d = n_d^0 + n_d^+. \tag{4.9}$$

From Eqs. (4.7), (4.8), and (4.9) we deduce

$$p + n_d^+ = n + n_a^-. \tag{4.10}$$

Writing Eq. (4.8) amounts to writing

$$\int f(E)n(E)\, dE = N_T, \tag{4.11}$$

where $n(E)$ is the density of states of energy E. The states are either delocalized in the valence and conduction bands, or localized for the discrete acceptor and donor states.

This equation expresses the conservation of electron number in the system at a given temperature (canonical ensemble) and therefore must give the chemical potential or Fermi level through the use of Eq. (4.10). We note that since the crystal remains neutral when we substitute donors or acceptors, the appearance of positive charges in the form of free holes and ionized donors must be compensated by the appearance of negative charges: conduction electrons and ionized acceptors.

The conservation of electron number in a homogeneous semiconductor is thus equivalent to electrical neutrality. We shall see later that in inhomogeneous semiconductors local electrical neutrality does not hold.

4.4 Statistics of Pure or Intrinsic Semiconductors

We describe as intrinsic a (hypothetical) semiconductor without any impurity or defect. We shall see that some of the properties demonstrated below hold for real semiconductors in certain temperature ranges.

We call $n_v(E)$ and $n_c(E)$ the densities of states for valence and conduction electrons (Fig. 4.1). We have seen in Sect. 2.3, Eq. (2.52), that near the band edges, at E_c and E_v, the curve for the density of states is parabolic.

The total number n of electrons in the conduction band is

$$n = \int_{CB} n_c(E) \, f(E) \, dE. \tag{4.12}$$

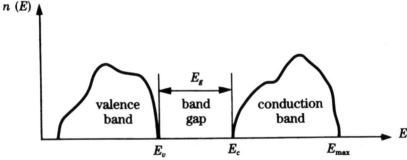

Fig. 4.1. Density of states of the bands near the gap.

If the constant energy surfaces are spheres (that is, if the effective mass is isotropic) and if there is only one minimum, at $k = 0$, the value of $n_c(E)$ for the two spin orientations and unit volume is given by Eq. (2.52) as

$$n_c(E) = 4\pi (2m_e)^{3/2} \frac{1}{h^3} (E - E_c)^{1/2}. \tag{4.13}$$

Let us assume that the Fermi level is several times kT below E_c. Then we can use Eqs. (4.2) and (4.12):

$$n = \exp \frac{E_F - E_c}{kT} \int_{E_c}^{\infty} dE \, n_c(E) \exp[-(E - E_c)/kT]. \tag{4.14}$$

Because of the exponential factor in the integral we were able to extend the upper limit to infinity and use Eq. (4.13) which holds in principle only at the bottom of the band. Setting $x = (E - E_c)/kT$ and using the result

$$\int_0^\infty x^{1/2} e^{-x} dx = \frac{\sqrt{\pi}}{2}$$

we obtain

$$n = N_c \exp \frac{E_F - E_c}{kT}, \tag{4.15}$$

where

$$N_c = 2 \left(\frac{2\pi m_e kT}{h^2} \right)^{3/2}. \tag{4.16}$$

We see that N_c plays the role of the degeneracy in a single level of energy E_c. We call it the effective or equivalent density of states of the conduction band. We note however that N_c is a function of temperature.

We can make an exactly similar calculation for the holes in the valence band. The number of empty electron places is given by Eq. (4.4), which tends to $\exp(E - E_F)/kT$ for $(E_F - E) \gg kT$ (electron population close to 1, hole population low). We find the hole number p by an analogous equation:

$$p = N_v \exp \frac{E_v - E_F}{kT}, \tag{4.17}$$

where N_v is the equivalent density of states of the valence band

$$N_v = 2 \left(\frac{2\pi m_h kT}{h^2} \right)^{3/2} \tag{4.18}$$

with m_h the hole mass.

In the case where, as in silicon or germanium, the bottom of the conduction band comprises several ellipsoids centered at various points of the Brillouin zone, we must take account on the one hand of the number of minima, by multiplying N_c by this number, and on the other hand of the anisotropy of the effective mass. The constant energy surfaces around these minima are ellipsoids with equations:

$$E = \frac{\hbar^2}{2} \left(\frac{k_x^2}{m_x} + \frac{k_y^2}{m_y} + \frac{k_z^2}{m_z} \right), \tag{4.19}$$

where the origin of the wave vector is taken at the energy minimum considered. The volume of an ellipsoid of semiaxes a, b, c is $(4/3)\pi abc$. The volume inside a surface of energy E is thus

$$\frac{4}{3}\pi \left(\frac{2m_x E}{\hbar^2}\right)^{1/2} \left(\frac{2m_y E}{\hbar^2}\right)^{1/2} \left(\frac{2m_z E}{\hbar^2}\right)^{1/2} \tag{4.20}$$

and the density of states per unit volume for the two spin orientations per ellipsoid is

$$n(E) = \pi(2)^{3/2} \frac{(m_x m_y m_z)^{1/2}}{h^3} E^{1/2}. \tag{4.21}$$

For silicon we have to replace the factor $m_e^{3/2}$ which appears in expression (4.16) by $(m_L m_T^2)^{1/2} \times 6$ to take account of the six ellipsoids. We get finally

$$N_c = 2.8 \ 10^{25} \ \text{m}^{-3} \ \text{at 300 K.}$$

If the valence band is degenerate at $k = 0$ as is the case in Si, Ge, GaAs, we have in first approximation two spherical bands each with an isotropic mass $m_{h,h}$ and $m_{h,l}$. We simply have to sum over the two corresponding densities of states. The factor $m_h^{3/2}$ in Eq. (4.18) must be replaced by $m_{h,h}^{3/2} + m_{h,l}^{3/2}$. With this change N_v is $10^{25} \ \text{m}^{-3}$ in silicon at 300 K.

The relations (4.15) and (4.17) hold **even if the crystal is not pure** on the condition that the Fermi level is several times kT from the band edges. Taking the product we thus always have

$$np = N_c N_v \exp\left(-\frac{E_g}{kT}\right) = n_i^2. \tag{4.22}$$

The quantity n_i^2 plays the role of the constant in the law of mass action for the reaction

$$\text{electron} + \text{hole} \rightleftarrows \emptyset + \text{energy.}$$

If the crystal is pure (intrinsic) Eq. (4.10) is

$$n = p = n_i = p_i \tag{4.23}$$

or, taking account of Eqs. (4.15) and (4.17),

$$N_c \exp\frac{E_F - E_c}{kT} = N_v \exp\frac{E_v - E_F}{kT}.$$

We thus find the position of the Fermi level

$$E_F - E_c = -\frac{1}{2}E_g + \frac{kT}{2}\text{Ln}\frac{N_v}{N_c}. \tag{4.24}$$

For "standard" bands (Sect. 2.5)

$$E_F - E_c = -\frac{1}{2}E_g + \frac{3}{4}kT\,\text{Ln}\frac{m_h}{m_e}. \tag{4.25}$$

We see that for an intrinsic semiconductor the Fermi level lies close to the middle of the band gap whatever the temperature. For a band structure exactly symmetrical with respect to the band gap the Fermi level is independent of the temperature. Substituting Eq. (4.25) into Eq. (4.15) or Eq. (4.17) we obtain the intrinsic concentrations

$$n_i = p_i = 2 \left(\frac{2\pi}{h^2} \right)^{3/2} m_e^{3/4} m_h^{3/4} (kT)^{3/2} \exp \left(-\frac{E_g}{2kT} \right). \tag{4.26}$$

The case of multiple minima, or of degeneracy at $k = 0$, have to be taken into account in the expression for N_c and N_v but in all cases the relation (4.22) will hold.

As we have seen in Sect. 3.1b, the electrical conductivity of a crystal is the sum of the conductivities due to the electrons and the holes, σ_e and σ_h:

$$\sigma = \sigma_e + \sigma_h \tag{4.27}$$

or

$$\sigma = n\mu_e e + p\mu_h e. \tag{4.28}$$

In this expression μ_e is the electron mobility and μ_h that of the holes.

We see from this formula that if the mobilities do not vary too rapidly with temperature (Sect. 5.4) the variation of σ will be very rapid, as $\exp(-E_g/2kT)$. This explains the very strong increase in conductivity with temperature and provides a method of measuring E_g. In contrast to metals, where the number of carriers is constant, the conductivity of semiconductors increases with temperature mainly through the increased number of carriers. Figure 4.2 shows the variation of intrinsic concentration with temperature in several semiconductors. We note the small concentration at room temperature: for silicon

$$n_i \simeq 1.6 \times 10^{16} \text{ m}^{-3} \text{ at } 300 \text{ K}.$$

The concentration of silicon atoms in silicon is of the order of 10^{28} m^{-3}. At room temperature the intrinsic band-to-band ionization is thus very low, of the order of 10^{-12} in relative concentration. We then understand how shallow impurity levels may dominate the properties of silicon at room temperature, even at very great purity, as their internal ionization energy is very small.

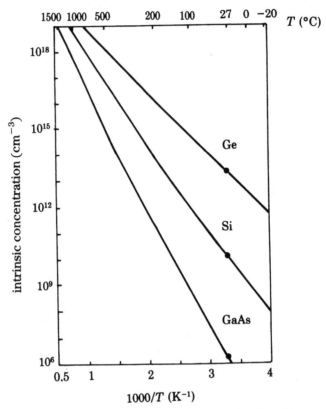

Fig. 4.2. Concentration of intrinsic charge carriers as a function of temperature in Si, Ge, GaAs. In this Arrhenius plot, the slopes of the straight lines are proportional to $E_g/2$, as long as the influence of the $T^{3/2}$ term is not significant.

4.5 Statistics of a Semiconductor Containing Impurities: The Notion of Majority and Minority Carriers

The relations (4.15), (4.17), and (4.22) actually only use the property of the Fermi level of being several times kT distant from the band edges, so that we can neglect the 1 in the denominator of the Fermi function. In this case we say that the semiconductor is **non-degenerate**, a **degenerate** semiconductor corresponding to the case where the Fermi level falls close to or even within a band. The latter situation occurs in heavily doped semiconductors, for which the donor (or acceptor) level is broadened up to merging into the conduction (or valence) band. We shall consider here only **impure** but **non-degenerate** semiconductors where the relations (4.15), (4.17), and (4.22) remain valid.

We note that the word degenerate recurs several times in semiconductor physics in referring to different things. This is the normal usage, but it is important that this restricted vocabulary should not cause confusion. For this reason we recall below the various meanings of the word that may be encountered:

1 – In any crystal the states $\mathbf{k}$ and $-\mathbf{k}$ are degenerate; this is a consequence of the fact that the Hamiltonian is real.

2 – If there are symmetries of the crystal potential (rotation, symmetry with respect to a point...) the surfaces of constant energy reflect this symmetry, and a state deduced from $\psi_{n,\mathbf{k}}(\mathbf{r})$ by one of the symmetry transformations is degenerate with $\psi_{n,\mathbf{k}}(\mathbf{r})$. There is thus a degeneracy associated with the crystal symmetry: for example the minima of the six conduction bands of silicon. At the center of the zone this shows itself through the possible existence of degenerate levels.

3 – Finally there is the meaning "degenerate electron gas" in a solid when the population of the states is not small compared with 1.

Sentences 1 and 2 mean that the states have the same energy; sentence 3 means that Fermi–Dirac statistics cannot be approximated by Maxwell–Boltzmann statistics.

n-type Semiconductor

Consider a semiconductor containing a concentration N_d of donors whose ionization energy is E_d but which is completely free of acceptors ($N_a = 0$). At very low temperature the electrons are in the lowest energy states and thus bound to the donor centers and in the valence band.

At higher temperatures the donors will progressively ionize. Let us assume that the Fermi level is below E_d so that all the donors are ionized. Then the number of conduction electrons is

$$n = N_d \tag{4.29}$$

and the relation (4.15) determines the Fermi level:

$$E_F - E_c = kT \log \frac{N_d}{N_c}. \tag{4.30}$$

We see that if N_d is of order 10^{20} m^{-3} the quantity $(E_F - E_c)$ is of order 25 meV $\cdot$ (Ln 10^{-5}) ~ -0.25 eV at room temperature. The Fermi level is well below E_d for a wide range of temperatures. When the temperature is increased the Fermi level moves away from the conduction band and the assumption of complete ionization of the donors remains valid.

However beyond a certain temperature, which depends only on the donor concentration, intrinsic ionization is no longer negligible. Then we have to write Eq. (4.10) in the form

$$n = n_d^+ + p \quad \text{with} \quad n_d^+ = N_d. \tag{4.31}$$

This relation expresses the fact that the electrons populating the conduction band originate from the donors and the valence band. Using Eqs. (4.22) and (4.31) we obtain

$$n = \frac{1}{2}N_d + \left(\frac{N_d^2}{4} + n_i^2\right)^{1/2}. \tag{4.32}$$

In the preceding limit of intermediate temperature ($n_i << N_d$),

$$n \approx N_d + \frac{n_i^2}{N_d} \approx N_d, \tag{4.33}$$

$$p \approx \frac{n_i^2}{N_d}. \tag{4.34}$$

The temperature range for which the electron number remains equal to N_d, independent of the temperature, is called the **saturation regime**. The number of holes p is $n_i \cdot (n_i/N_d) << n_i$. The ratio of the number of electrons to the number of holes is

$$\frac{n}{p} = \frac{N_d^2}{n_i^2}. \tag{4.35}$$

This ratio is very large: for $N_d = 10^{23}$ m^{-3}, $n/p \sim 10^{14}$. **For this reason the electrons are called majority carriers and the holes minority carriers. Then a silicon crystal containing donors is called n-type silicon, as the electric current is carried by electrons, which greatly outweigh the holes. It is also called n-doped.** Doping of a crystal may be the result of precise manufacturing techniques or occur by accident.

We obtain the remarkable result that **the conductivity of such a crystal only depends on its concentration of impurities in this temperature range.** We say that **the conductivity is n-type extrinsic.**

At **high temperature** we can have $n_i >> N_d$; then Eqs. (4.32) and (4.22) can be approximated as

$$n \approx n_i + \frac{1}{2}N_d,$$

$$p \approx n_i - \frac{1}{2}N_d, \tag{4.36}$$

so that we recover the intrinsic regime.

p-doped Semiconductor

Let us consider a crystal where instead of "donor" impurities we have only a concentration N_a of "acceptors," for example boron in silicon. We can discuss the situation either in terms of the electrons in the electron energy diagram or in terms of the holes in the hole energy diagram. In the electron energy diagram let us assume that at room temperature the Fermi level is above the acceptor level, and thus a few times kT from the valence band. The acceptor levels will thus be populated by electrons from the valence band which leave holes. We then have

$$p = N_a \qquad (4.37)$$

and from Eq. (4.17),

$$E_F - E_v = kT \operatorname{Ln} \frac{N_v}{N_a}. \qquad (4.38)$$

At higher temperature we have to take account of intrinsic ionization by writing Eq. (4.10) in the form

$$n + N_a = p, \qquad (4.39)$$

which takes into account the fact that free holes can be created in the valence band, either by trapping of an electron in an acceptor level or by excitation of a free electron into the conduction band. Taking account of Eq. (4.22) we then have

$$p = \frac{1}{2} N_a + \left(\frac{N_a^2}{4} + n_i^2 \right)^{1/2} \qquad (4.40)$$

and if $N_a \gg n_i$,

$$p \approx N_a + \frac{n_i^2}{N_a} \sim N_a. \qquad (4.41)$$

In this case $n \sim (n_i^2/N_a) \sim p \cdot (n_i^2/N_a^2)$. The holes have concentration N_a and the electrons have the very low concentration $n_i \cdot (n_i/N_a)$. In this case the electrons are **minority carriers**. The holes are the **majority carriers**. In this p-doped crystal electric current is carried by the **majority hole carriers**. We say that the conductivity is **p-type extrinsic**.

This explains why certain semiconducting crystals behave under the Hall effect for example as if the current were transported by "positive" electrons. There is a deep symmetry between the properties of n-doped and p-doped crystals.

We could have reasoned entirely in terms of the statistics of holes. Then Fig. 3.7(b) becomes Fig. 4.3 (inverting energies): the holes are trapped by the acceptors at very low temperatures, but if E_{Fh} is situated as in Fig. 4.3 the acceptors are ionized and

$$p = N_a = N_v \exp \frac{E_{Fh} - E_h}{kT}. \tag{4.42}$$

For p-type semiconductors we also have a saturation regime, with the number of charge carriers equal to N_a and independent of temperature.

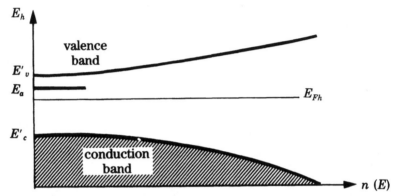

Fig. 4.3. Density of states $n(E)$ for the different hole energies E_h and Fermi level position E_{Fh} near room temperature.

At **high temperature** where $n_i > N_a$ we recover an intrinsic regime in which

$$p = n_i + \frac{1}{2}N_a,$$

$$n = n_i - \frac{1}{2}N_a. \tag{4.43}$$

Figure 4.4 summarizes the results for a semiconductor at ambient temperature, depending on the type of doping.

4.6 Compensated Semiconductor at Intermediate Temperature

Of course in any real crystal, even the purest one can make, there are various kinds of impurities. We must now consider the statistics of electrons in such a crystal.

Let us suppose we are at an intermediate temperature in the saturation regime of an n-type semiconductor containing N_d donors per unit volume. The Fermi level is several kT below the conduction band. Let us introduce some acceptors into the crystal at concentration N_a. For each acceptor there is an energy level situated close to the valence band which will be populated

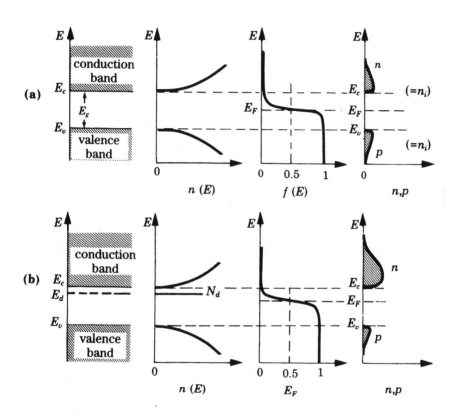

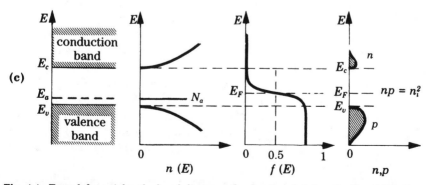

Fig. 4.4. From left to right, the band diagram, the density of states, the Fermi function, and the carrier concentration as functions of energy. (a) Intrinsic case; (b) n-type semiconductor; (c) p-type semiconductor. We note that $np = n_i^2$ in all three cases. (After Sze, Semiconductor devices, J. Wiley, 1985.)

by an electron, since this level is well below the Fermi level. The acceptor has "trapped" an electron. All the acceptors ($<< N_d$) will thus be ionized and the number of free electrons will be decreased accordingly. Thus

$$n = N_d - N_a \tag{4.44}$$

and from Eq. (4.13),

$$E_c - E_F = kT \operatorname{Ln} \frac{N_c}{N_d - N_a}. \tag{4.45}$$

We say that the material is **compensated**. Here we have a partially compensated n-type material. The majority carriers are still the electrons; the number of holes is still very small as the Fermi level is high in the band gap. We note however that as the compensation is increased by raising N_a, the Fermi level decreases. If $N_a = N_d$ the Fermi level will take up the intrinsic position given by Eq. (4.24).

If the number of acceptors becomes larger than the number of donors all the donors will be ionized by trapping an electron in an acceptor level. We will then have $N_a - N_d$ effective acceptors and the concentration of free holes will be

$$p = N_a - N_d, \tag{4.46}$$

where the Fermi level is given by Eq. (4.17) as

$$E_F - E_v = kT \operatorname{Ln} \frac{N_v}{N_a - N_d}. \tag{4.47}$$

E_F will be quite low in the band gap, and the electron number, given by Eq. (4.22) with $p = N_a - N_d$, will remain very small. We will then have a compensated p-type semiconductor.

These ideas allow us to understand the role of purity in the properties of semiconductors at intermediate temperatures. The band structure of the material determines three characteristic concentrations N_c, N_v, and $n_i = (N_c N_v)^{1/2} \exp(-E_g/kT)$.

If a material contains N_d shallow donors, they will ionize and release $n = N_d$ electrons into the conduction band. Let us consider the simultaneous presence of deep donors. If their concentration is not too high ($\lesssim N_c$) they will not play any role because their energy is such that they will not be ionized. Consider further the presence of deep acceptors; these will capture electrons coming from the donors and partially compensate the semiconductor (Appendix 4.2). As long as the concentration of acceptors remains small compared with that of the donors we will still have an n-type semiconductor.

The arguments above show why pure crystals are important in semiconductor applications: until it was possible to manufacture crystals with impurity concentrations less than N_c the phenomena described above were not understood. Section 4.9 deals with the fabrication of pure semiconductors.

4.7 Semiconductor at Low Temperatures

At **low temperatures** we have to consider **partial ionization** of the donors or acceptors (the Fermi level is close to the donor or acceptor levels). We shall confine ourselves for simplicity to uncompensated n-type semiconductors. Then we always have $n_a^- = N_a = 0$. Electrical neutrality (4.10) becomes

$$p + n_d^+ = n. \tag{4.48}$$

Using Eqs. (4.3) and (4.9) we get

$$n_d^+ = N_d - n_d^0, \tag{4.49}$$

$$n_d^0 = N_d \frac{1}{1 + \frac{1}{2}\exp(E_d - E_F)/kT}, \tag{4.50}$$

$$(n - p) = N_d \frac{\frac{1}{2}\exp(E_d - E_F)/kT}{1 + \frac{1}{2}\exp(E_d - E_F)/kT}. \tag{4.51}$$

Substituting E_F from Eq. (4.15) into Eq. (4.51) we get

$$(n - p)n = \frac{N_c}{2}(N_d - n + p)\exp\left(\frac{E_d - E_c}{kT}\right). \tag{4.52}$$

At **zero temperature** $n = 0; p = 0$; the semiconductor is an insulator. The Fermi level must lie above the donor levels.

At **very low temperature** the ionization of the donors is weak and the hole concentration is negligible, as the Fermi level is very high in the band. Neglecting p and n compared with N_d, Eq. (4.52) becomes

$$n = \left(\frac{N_c N_d}{2}\right)^{1/2}\exp\left(\frac{E_d - E_c}{2kT}\right). \tag{4.53}$$

The electron number increases, with an activation energy equal to half the binding energy of the donor. The Fermi level may be obtained by replacing n by its expression (4.15):

$$E_F - E_c = \frac{E_d - E_c}{2} + \frac{kT}{2}\operatorname{Ln}\frac{N_d}{2N_c}. \tag{4.54}$$

At **intermediate temperature** the exponential is of order 1 and the hole number is still negligible; we can write Eq. (4.52) in the form

$$(N_d - n) \sim \frac{2n^2}{N_c \exp(E_d - E_c/kT)}. \tag{4.55}$$

For $N_d \ll N_c$ the solution $n = N_d$ is a very good approximation, and we recover the saturation regime.

At **high temperature** we recover the intrinsic regime (4.36): we can replace $\exp[(E_d - E_c)/kT]$ by 1 and must now retain p in Eq. (4.52). Further, we use Eqs. (4.52) and (4.22) in the case where n_i and N_d are both small compared with N_c.

For a partially compensated crystal such that

$$(N_c/2)\exp[-(E_c - E_d)/kT]N_a \ll N_d$$

one obtains at very low temperature a regime where the electron number varies as $\exp[(E_d - E_c)/kT]$.

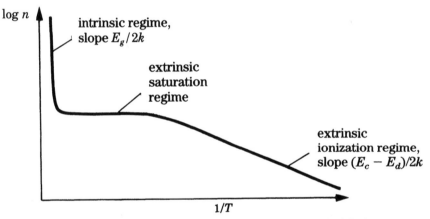

Fig. 4.5. Variation of the logarithm of the concentration as a function of the inverse temperature for an n-type semiconductor. In practice room temperature is in the saturation regime. (From Smith "Semiconductors," Cambridge University Press, 1968.)

Figure 4.5 shows the behavior of the concentration as a function of the inverse temperature. At zero temperature the concentration tends to zero. Similar results hold for a p-type semiconductor.

Exercise: using the data of Chaps. 3 and 4 (and a pocket calculator), show that in silicon doped with 10^{21} atoms of phosphorus per m^3, the saturation regime where the number of carriers is constant extends from about 90 K, where $n \sim 0.9N_d$, to 500 K, where $n \sim 1.1N_d$.

Figure 4.6 shows the variation of Fermi level with temperature for an n-type semiconductor (upper figure) and a p-type semiconductor (lower figure). At zero temperature in the n-type semiconductor the Fermi level is between the donor level and the conduction band.

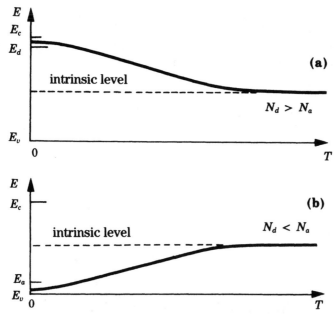

Fig. 4.6. Position of the Fermi level as a function of temperature in the cases of (a) n doping, and (b) p doping. Note the different units of the abscissa compared with Fig. 4.5. In the saturation regime the number of carriers is fixed but the Fermi level changes. (From Smith, "Semiconductors," Cambridge University Press, 1968.)

The problem in Appendix 4.2 shows how a semiconductor of given resistivity is manufactured in practice by choosing the type of dopants (deep or shallow) and their concentration.

4.8 Application: The Semiconducting Thermometer

These results obviously lead to the principle of low-temperature thermometers which measure the resistance of an extrinsic semiconductor. At and above room temperature one generally uses metal resistors as thermometers (such as the etalon thermometer using a platinum resistor), but this is not possible at low temperature because metal resistivity does not depend strongly on the temperature in this range.

The reason for this is that in a metal the number of carriers is constant and the collision time appearing in the resistivity is $\tau(E_F)$, where E_F is the Fermi energy of the crystal. What appears in the resistivity, then, is the collision time of electrons with the Fermi velocity v_F, independent of the temperature. This time varies with temperature at high temperatures, as the thermal vibrations of the lattice are the main source of collisions, and their amplitude increases with temperature. By contrast at low temperature this collision process becomes negligible, and the mean free path is correspondingly infinite. The mean free path, and thus $\tau(E_F)$ is then determined by collisions of electrons of velocity v_F with impurities, so that the metal resistivity is then independent of the temperature.

On the other hand, for a semiconductor in the extrinsic regime the conductivity, which varies with the number of free carriers, remains a function of the temperature at low temperature as we have seen in Fig. 4.5. Arsenic-doped germanium is used currently as a thermometer between 0.1 and 100 K (Fig. 4.7).

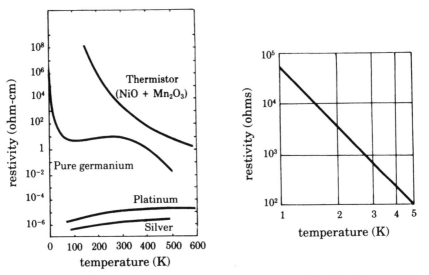

Fig. 4.7. (a) Variation of the resistivity of pure germanium as a function of temperature. Note the rapid rise of the resistivity at very low temperature. (b) Calibration curve for a germanium thermometer. Note the log scale.

4.9 Growth of Pure Crystals

We have seen how important it is to have pure single crystals so as to obtain properly controlled resistivities. Silicon is the basic material for almost all products involving semiconductor components throughout the world. The use of these materials implies a degree of purity out of all comparison with what is normally required in other domains. In many applications the presence of deep recombination centers must be avoided and purities of the order of 10^{-9} have to be reached for some elements.

These single crystals are almost always obtained using a molten bath with a composition close to the desired one. This removes the complications associated with non-uniform materials. To reach sufficient purity one has to start with fairly pure polycrystalline material, itself obtained by hydrogen reduction of $SiCl_4$, which is liquid and can be distilled several times for purification.

One of the main problems for the growth of single crystals of large size is the control of the temperature gradients which must exist for solidification to occur. When the crystal forms from the liquid, the enthalpy of fusion ΔH_f is released and must be removed from the system while preserving its homogeneity. Usually one uses the extraction or Czochralsky method illustrated in Fig. 4.8. A small seed crystal is mounted on a shaft and brought into contact with the surface of the bath. The temperature gradients are controlled so that the crystal grows at the surface of the seed.

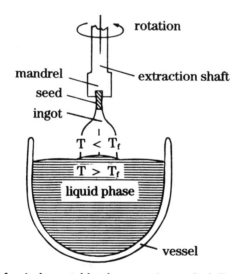

Fig. 4.8. Growth of a single crystal by the extraction method. From Leturcq and Rey, "Physique des Composants Actifs à Semi-Conducteurs," Dunod, 1978.

The crystal is then slowly pulled while being rotated. Crystals up to 50 cm in diameter have been obtained by this method; present standard diameters for microelectronic applications are from 4 to 6 inches (10 to 15 cm).

However in this method the vessel remains as a source of contamination. There is almost always pollution by the vessel which may be slightly

attacked at the bath temperature (1418°C for silicon). It is therefore preferable to eliminate the vessel. This is possible using the ingenious melting
zone technique shown in Fig. 4.9. A polycrystalline ingot is heated locally
by induction until a narrow melting zone forms. The slow displacement of
the heating coil moves the liquid region as the polycrystalline material fuses
and solidifies to a single crystal.

Once the crystal is obtained it is possible to purify it further by zone
refining, a technique developed by William Pfann at the Bell Laboratories.
This procedure has improved the attainable purity limits by several orders
of magnitude and one can now obtain silicon crystals with a purity of 10^{-10}.
It is true to say that without this technique a large part of present electronics and information technology would not exist. Like many discoveries,
zone refining seems obvious once seen, but required a stroke of genius to
think of it.

Let us consider a solid in equilibrium with the liquid of the melting
zone. We define a coefficient of liquid–solid segregation:

$$K = \frac{C_L}{C_S},\qquad(4.56)$$

where C_L and C_S are the concentrations in the liquid and the solid. In general K is much greater than 1, as the impurities are much more soluble in
the liquid than the solid. One starts at $z = 0$ with a crystal having a uniform
impurity concentration C_0. The first melting zone has concentration C_0. By
contrast the first solidified layer obtained as the zone rises has a concentration $C_0' = C_0/K$ and is therefore purer. As the melting zone moves, it gains

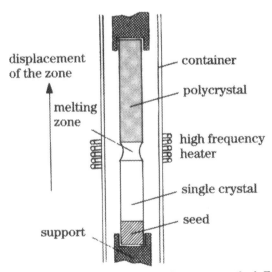

Fig. 4.9. Purification of a single crystal by the melting zone method. From Leturcq and
Rey, "Physique des Composants Actifs à Semi-Conducteurs," Dunod, 1978.

in impurities and leaves behind a purified crystal until the liquid reaches a concentration KC_0. Subsequent motions of the zone do not purify but make the crystal uniform. After a passage along the crystal, it is allowed to cool and one starts again at $z = 0$, heating again from the concentration $C_0' = C_0/K$. The first crystallized zone after the second passage thus has a concentration $C_0'' = C_0/K^2$. One thus makes several passages, beginning at the origin, obtaining the remarkable purities we have mentioned.

The cost of electronic grade single crystals of silicon is about \$100/kg, a remarkably low price for such a technological achievement.

Appendix 4.1

Occupation Number of a Donor Level

We first recall the definition of the Fermi factor. We consider the grand canonical ensemble where the density matrix D is written

$$D = \frac{\exp[-\beta(\mathcal{H} - E_F N)]}{\text{Tr}\exp[-\beta(\mathcal{H} - E_F N)]}, \qquad (4.57)$$

where $\beta = 1/kT$, $\mathcal{H}$ is the Hamiltonian and N the number of particle operators. The operators $\mathcal{H}$ and N have the eigenvalues $\Sigma_{\text{states}} n_i E_i$ and $\Sigma_{\text{states}} n_i$, respectively.

Here the states i are completely defined (energy, orbital quantum number, and spin determined), and the eigenvalues of n_i are 0 and 1 for each state i. Then for each state

$$< n_i > = \text{Tr } n_i D, \qquad (4.58)$$

$$= \frac{\text{Tr } n_i \exp[-\beta(n_i E_i - E_F n_i)] \times \text{Tr}_{j \neq i} \exp[\Sigma_j - \beta(n_j E_j - E_F n_j)]}{\text{Tr}\exp[-\beta(n_i E_i - E_F n_i)] \times \text{Tr}_{j \neq i} \exp[\Sigma_j - \beta(n_j E_j - E_F n_j)]}$$

$$= \frac{\text{Tr } n_i \exp[-\beta(n_i E_i - E_F n_i)]}{\text{Tr}\exp[-\beta(n_i E_i - E_F n_i)]} = \frac{1}{1 + \exp\beta(E_i - E_F)}. \qquad (4.59)$$

For a donor level we wish to know if an electron can occupy any of the localized levels (Fig. 4.10), whatever its spin.

Consider first the donor ground state, a hydrogen-like $1s$ state. In fact, taking account of the spin, we have two states, $1s \uparrow$ and $1s \downarrow$. Because of electron repulsion, there is only one electron to place, and there are thus three possible states: empty, one electron in $1s \uparrow$, and one electron in $1s \downarrow$.

The occupation number $< n_{1s} >$ of the $1s$ level can be expressed by using Eq. (4.59) for the levels $1s \uparrow$ and $1s \downarrow$:

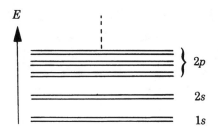

Fig. 4.10. Schematic donor levels including possible orbital and spin degeneracies.

$$< n_{1s} > \; = \; < n_{1s\uparrow} > + < n_{1s\downarrow} >, \tag{4.60}$$

$$= \sum_{i=0,s\uparrow,s\downarrow} \frac{\mathrm{Tr}\, n_i \exp[-\beta(n_i E_i - E_F n_i)]}{\mathrm{Tr} \exp[-\beta(n_i E_i - E_F n_i)]}, \tag{4.61}$$

$$= \frac{2 \exp[-\beta(E_{1s} - E_F)]}{1 + 2 \exp[-\beta(E_{1s} - E_F)]},$$

$$< n_{1s} > \; = \; \frac{1}{1 + (1/2) \exp[\beta(E_{1s} - E_F)]}. \tag{4.62}$$

This is expression (4.3), which we use to calculate the occupation number of the donor levels.

A fully rigorous treatment takes account of all the localized donor states. For the $2p$ level for example, there are six sublevels of the same energy ($l = 1, m = -1, 0, 1$, and $s = \uparrow$ or $\downarrow$). Each of these sublevels may be occupied by zero or one electron. When there is one electron the $2p_{-1\uparrow}$ term in the occupation number is $\exp[-\beta(E_{2p} - E_F)]$, and the six sub-levels contribute in the same way. For a level α of degeneracy $g_\alpha = 2(2l + 1)$, where l is the orbital quantum number, the term $\exp[-\beta(E_\alpha - E_F)]$ occurs g_α times. Further, as the localized donor levels can be occupied by one electron at most, there is in all only a single empty level. Hence the mean occupation number $< n >$ of the ensemble of localized donor levels can be written as

$$< n > \; = \; \Sigma_\alpha < n_\alpha >$$

$$= \frac{2 \exp[-\beta(E_{1s} - E_F)] + \ldots + g_\alpha \exp[-\beta(E_\alpha - E_F)]}{1 + 2 \exp[-\beta(E_{1s} - E_F) + \ldots + g_\alpha \exp[-\beta(E_\alpha - E_F)]}, \tag{4.63}$$

$$< n > \; = \; \frac{1}{1 + \{\Sigma_\alpha g_\alpha \exp[-\beta(E_\alpha - E_F)]\}^{-1}}. \tag{4.64}$$

In fact the number and degeneracy of the localized excited states is such that the sum $\Sigma_\alpha g_\alpha \exp[-\beta(E_\alpha - E_F)]$ diverges. This means that if we consider **all** the localized excited states we will always find $< n > = 1$, i.e., the electron will never be in the conduction band. This paradox, well known for the hydrogen atom, can be resolved. The Bohr radius grows with α, and we should count in the sum only **those states which do not overlap**; if this condition fails the electrostatic potential will no longer go as $1/r$ and the calculation of the donor states is no longer correct.

In all practical cases the numerical values deduced from formulas (4.62) and (4.64) differ very little, and we shall use Eq. (4.62) in this book.

Appendix 4.2

Problem: Substrates for Microelectronics

The microelectronics industry makes great use of epitaxy, the crystalline growth of semiconductors, or semiconductor devices, on a substrate of similar type. The substrate acts as the mechanical and thermal support, but must not short-circuit the components. Thus it must have a very high resistance.

The objective of this problem is showing how to use appropriate doping by deep impurities to render insulating a single crystal of impure gallium arsenide (containing residual impurities). We recall that the electrical resistivity ρ is related to the conduction electron and hole number densities n, p by the relation

$$\rho = (n e \mu_e + p e \mu_h)^{-1},$$

where μ_e and μ_h denote the mobilities of these carriers, and e the electron charge.

The questions in this problem aim at numerical estimates of the resistivity. We simplify the problem as much as possible by adopting adequate approximations (for the Fermi factors in particular).

We give the following characteristics of gallium arsenide: relative dielectric constant $\epsilon_r = 13$; band gap $E_c - E_v = 1.4$ eV; effective electron mass $m_e = 7 \times 10^{-2} m$; electron mobility $\mu_e = 8500$ cm$^2 \cdot$ V$^{-1} \cdot$ s^{-1}; effective hole mass $m_h = 0.5 m$; hole mobility $\mu_h = 400$ cm$^2 \cdot$ V$^{-1} \cdot$ s^{-1}; here m is the electron mass in vacuo. It is convenient to use the following numerical values:

$$N_0 = \frac{2}{h^3} (2\pi m k T)^{3/2} = 2.10^{25} \text{ m}^{-3} \text{ for } T = 300 \text{ K},$$

$$E_H = \frac{e^4 m}{2(4\pi\epsilon_0 \hbar)^2} = 13.6 \text{ eV}.$$

E_H is the binding energy of the hydrogen ground state.

First Question

Assume the semiconductor is **completely** pure. What is its resistivity at $T = 300$ K?

Second Question

Because of the different affinities of Ga and As atoms, it is impossible *in practice* to produce $GaAs$ crystals with less than 10^{20} impurities per m^3. We assume that these impurities are shallow donors. They then introduce energy levels close to the conduction band which we can calculate by likening the donor to a modified hydrogen atom. For this atom we take account of the screening effect through the dielectric constant, and use the fact that the conduction electrons are quasi-particles of effective mass m_e, as shown in the book. To calculate the statistical distribution of the electrons we assume that each impurity has the effect of introducing into the band gap *exactly one state* (without spin degeneracy) of energy E_d whose position with respect to the bottom of the conduction band is that of the ground level of this pseudo-hydrogen atom.

What is the position of this ground level E_d and how does the binding energy of the electron in this level compare with kT for $T = 300$ K?

What is the position of the chemical potential at $T = 300$ K in the purest material obtainable in practice? What is the corresponding resistivity?

Third Question

We can *deliberately* introduce some chromium atoms into the semiconductor. We assume that their number density N_{Cr} is 10^{23} m^{-3}. Each chromium atom introduced into the crystal brings an electron with it. The earlier approximation of an equivalent hydrogen atom does not work for these atoms. We assume that to each chromium atom there corresponds in the band gap an energy level E_{Cr}, this time doubly degenerate, situated 0.7 eV below the bottom of the conduction band.

Write down the conservation of the electron number in this system. Deduce that the Fermi level is very close to E_{Cr}, and that the introduction of chromium increases the resistivity. What is the maximum resistivity one can reach at $T = 300$ K?

Solutions

First Question

We assume that the chemical potential is in the band gap, i.e., we set $\beta = 1/kT, \beta(E_c - E_F) \gg 1$, and $\beta(E_F - E_v) \gg 1$ and thus approximate by neglecting the $+1$ term in the denominator of the Fermi factor

$$f(E) = \frac{1}{\exp \beta(E - E_F) + 1} \simeq \exp[-\beta(E - E_F)]. \tag{4.65}$$

Under these conditions the quasi-particles obey ideal gas statistics, and the electron and hole number densities are given by

$$n = \frac{2}{h^3}(2m_e \pi kT)^{3/2} \exp \beta(E_F - E_c), \tag{4.66}$$

$$p = \frac{2}{h^3}(2m_h \pi kT)^{3/2} \exp \beta(E_v - E_F), \tag{4.67}$$

so that $n = (m_e/m)^{3/2} N_0 \exp \beta(E_F - E_c)$ and $p = (m_h/m)^{3/2} N_0 \times \exp \beta(E_v - E_F)$.

We eliminate the chemical potential by taking the product of these two expressions:

$$np = N_0^2 \left(\frac{m_e}{m}\right)^{3/2} \left(\frac{m_h}{m}\right)^{3/2} \exp[-\beta(E_c - E_v)] = n_i^2. \tag{4.68}$$

For a completely pure semiconductor the condition of electrical neutrality, expressing the conservation of the total number of electrons, can be written

$$n = p.$$

We get

$$n = p = N_0 \left(\frac{m_e m_h}{m^2}\right)^{3/4} \exp[-\beta(E_c - E_v)/2]. \tag{4.69}$$

Numerically we have $n = p = n_i \simeq 1.1 \cdot 10^{12}$ m^{-3} at $T = 300$ K. The chemical potential E_F is given by

$$\beta(E_F - E_c) = \text{Ln}\left[(n/N_0)(m/m_e)^{3/2}\right]$$
$$= \text{Ln}\left[(p/N_0)(m/m_h)^{3/2}\right]. \tag{4.70}$$

Given $n = p \ll N_0$, we justify a posteriori the original assumption. The resistivity ρ is $\rho = [ne(\mu_e + \mu_h)]^{-1}$. We find $\rho \sim 6.10^6 \Omega \cdot$ m.

Second Question

The binding energy of the ground level of the pseudo-hydrogen atom is

$$E_c - E_d = E_H \times \frac{m_e}{m} \times \frac{1}{\epsilon_r^2}. \tag{4.71}$$

We have now taken account of the effective mass m_e of the quasi-particles; the screening effect caused by all the electrons of the solid leads to an attractive potential of the form $e^2/4\pi\epsilon_0\epsilon_r r$ instead of $e^2/4\pi\epsilon_0 r$.

Numerically we have $E_c - E_d \sim 8.10^{-3}$ eV. This is less than the energy kT for $T = 300$ K, which is 25×10^{-3} eV. The donor is neutral when the level E_d is occupied; it is positively charged when it is empty. The neutrality condition for a material containing N_d donors per m^3 is

$$n = p + [1 - f(E_d)]N_d. \tag{4.72}$$

The chemical potential is now closer to the bottom of the conduction band (E_c). Assuming that p is negligible, we can then see that:

• if $1 - f(E_d) \ll 1$, then the chemical potential lies above E_d, and is thus close to the conduction band and p is indeed negligible;

• if $1 - f(E_d) \sim 1$ then $n \gtrsim N_d$ from Eq. (4.72). If $N_d > n_i$, then from Eq. (4.68) $p < n$, i.e., $p < 1.1 \times 10^{12}$ m^3, while $N_d \geq 10^{20}$ m^{-3}, so that $p \ll N_d$.

Now, given that at 300 K, $\beta(E_v - E_d)$ is less than 1, the Fermi factor $f(E)$ has the same order of magnitude for $E = E_c$ and $E = E_d$:

$$f(E_c) \sim f(E_d).$$

For $N_d = 10^{20}$ m^{-3} and $N_0 = 2 \times 10^{25}$ m^{-3}:

$$f(E_d)N_d \ll f(E_c) \left(\frac{m_e}{m}\right)^{3/2} N_0 = n.$$

Expression (4.72) thus reduces to

$$n \simeq N_d$$

and we can clearly neglect p by comparison with N_d (the second assumption above).

The chemical potential is given by

$$\beta(E_c - E_F) = \text{Ln}[(N_0/N_d)(m_e/m)^{3/2}], \tag{4.73}$$

and we find $E_c - E_F \simeq 0.2$ eV and thus $E_F - E_v \simeq 1.2$ eV.

The chemical potential is below E_d, closer to the conduction band than the valence band, justifying the approximation $n \gg p$, but it is far enough from E_c for the approximation of the Fermi factor made in the first question to hold.

The corresponding resistivity is then

$$\rho = \frac{1}{N_d e \mu_e},$$

where $N_d = 10^{20} \text{m}^{-3}$ so that $\rho \simeq 0.07\,\Omega\cdot\text{m}$.

Third Question

The chromium levels are clearly deeper in the band gap than the donor levels; thus the donor electrons are trapped in these levels and E_F is decreased. The energy E_{Cr} is lower than the value of the chemical potential calculated in the preceding question. The electron number n will thus decrease and the resistivity increase. The electrons come either from the valence band (p in number) or the chromium levels (N_{Cr} electrons) or the donors (N_d). At equilibrium they are either delocalized in the conduction band (n) or localized in the chromium sites $[2f(E_{Cr})N_{Cr}]$ or donor sites $[f(E_d)N_d]$. Hence the balance:

$$p + N_{Cr} + N_d = n + 2f(E_{Cr})N_{Cr} + f(E_d)N_d. \tag{4.74}$$

Since $N_{Cr} \sim 10^{23} \text{ m}^{-3}$ is much larger than all the concentrations N_d, n, p under consideration, this equation requires

$$N_{Cr} \simeq 2f(E_{Cr})N_{Cr}, \tag{4.75}$$

that is, $f(E_{Cr}) \simeq 1/2$, and thus that E_F coincides with E_{Cr} to within a few kT. We say that the chemical potential is **pinned** at the level E_{Cr} in the center of the band. We deduce

$$n = \left(\frac{m_e}{m_0}\right)^{3/2} N_0 \exp\beta(E_{Cr} - E_c), \tag{4.76}$$

$$p = \left(\frac{m_h}{m_0}\right)^{3/2} N_0 \exp\beta(E_v - E_{Cr}). \tag{4.77}$$

With $E_c - E_{Cr} = E_{Cr} - E_v = 0.7$ eV we obtain $n = 2.5 \times 10^{11} \text{ m}^{-3}$ and $p \sim 5 \times 10^{12} \text{ m}^{-3}$. Thus $\rho \simeq 1.5 \cdot 10^7\,\Omega\cdot\text{m}$.

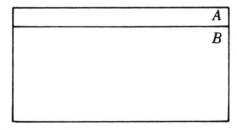

Fig. 4.11. Epitaxial structure: A: active layer; B: semi-insulating substrate.

The introduction of chromium allowed us to obtain a larger resistivity than that of the perfectly intrinsic material. There are a few less very mobile electrons and a few more fairly immobile holes than in the pure material. It is thus a method for obtaining a material of resistivity comparable to that of the pure material, starting from a material which necessarily contains impurities. This kind of material is called a **semi-insulating** semiconductor. We thus have a method allowing us to transform an impure substrate B into an insulator by strongly doping it with deep impurities. We can then grow by epitaxy a high quality crystalline layer A, whose electrical properties can be controlled by selective doping with shallow impurities (Fig. 4.11).

5.

Transport Phenomena in Semiconductors

5.1 Introduction

Whenever an electric field, a concentration gradient, or temperature gradient is present in a semiconductor we observe charge transport (electric current) mass transport (diffusion of carriers), or energy transport (heat conduction). Further, when a semiconductor is subject to photoexcitation there occurs what we may call an internal photochemical reaction: the creation of electron-hole pairs through the excitation of an electron from the valence band into the conduction band.

For each of these different types of excitation there exist mechanisms restoring the system to equilibrium, with characteristic rates which are, however, different. We shall see that the average time between collisions, measureable from the mobility $\mu = e\tau/m^*$ or from the width of the cyclotron resonance, is of the order of 10^{-13} s. This is very short compared with the lifetime of an electron-hole pair created by light. The latter time can for example be found from the decay of "photoconductivity," the variation of conductivity connected with the presence of additional electrons and holes, following a very short flash of light (cf. Chap. 6). In this way we measure lifetimes between 10^{-3} and 10^{-9} seconds.

It is important to understand that even in the absence of any applied excitation, there is ceaseless creation and destruction of electron-hole pairs by thermal motion within the semiconductor. We shall see that the characteristic time of this spontaneous process is exactly the lifetime, and thus very long compared with the time between collisions. **Because these times are very different we can study these processes separately, i.e., determine the transport properties such as conductivity, by regarding the electrons and holes as non-interacting gases.** This is the approach of the present chapter. As a second stage we can study the much slower effects of creations and annihilations (recombinations) of electron-hole pairs by light or through other causes: this is the subject of Chap. 6.

We first recall the elementary transport model constructed by Drude in 1900, which relates the finite value of the electrical conductivity to microscopic electron collisions (Sect. 5.2). The predictions of this model will be more rigorously justified in Sect. 5.3 by use of the Boltzmann transport equation. The main results obtained for semiconductors, and the orders of magnitude of the characteristic times and lengths, are given in Sec. (5.4). On a first reading the latter section can be studied after reading only Sects. 5.2 and 5.3d.

5.2 Drude's Model of Conductivity and Diffusion

We have seen in the preceding chapter that the occupation probability of conduction band states of energy E can be approximated by a function proportional to a Boltzmann factor $\exp[-E/kT]$. Moreover we saw in Chap. 2 that electrons respond to an external force in accord with classical dynamics (2.38), provided we replace their mass by the effective mass m_e. This is also true of the holes. We may therefore try to use a classical treatment to understand transport properties. This is the basic idea in the application of Drude's model to semiconductors.

We thus consider a perfect electron gas, obeying classical mechanics, and confined within a solid by the Coulomb interaction with the ions, which appear only as a potential well of the size of the crystal. Like classical gas particles, the electrons are subject to random collisions. We know (Sect. 2.2) that these collisions only occur with imperfections in the crystal and the boundaries of the potential well.

Let us consider the example of Coulomb scattering of an electron by ionized impurities that deflect its path. If the impurities are randomly distributed in the solid, the scattering probability for an electron by impurities during the time interval dt is independent of the observation time t, and proportional to dt. We write it as dt/τ, where the characteristic time τ depends on the impurity concentration. The electron will be more sensitive to an impurity if it "feels" it for longer, and thus if its kinetic energy E is lower: this must be reflected in the dependence of τ on E.

In Drude's model, whatever the collision mechanism, we take the scattering probability for an electron in the interval dt to be dt/τ, where the collision time is assumed to be independent of E. We call $p(t)$ the probability that an electron suffers no collision between $t = 0$ and the time t. The probability of surviving until $t + dt$ without collisions is the product of the probability of reaching t without collision, and that of suffering no collision between t and $t + dt$, so that

$$p(t + dt) = p(t)(1 - dt/\tau). \tag{5.1}$$

We get

$$dp = -p \, dt/\tau, \tag{5.2}$$

$$p(t) = \exp(-t/\tau). \tag{5.3}$$

The probability density $P(t)dt$ that an electron has suffered no collisions between 0 and t but then collides between t and $t + dt$ is

$$P(t) \, dt = \exp(-t/\tau) \, dt/\tau. \tag{5.4}$$

The probability density $P(t)$, an exponential distribution, gives the mean time $<t>$ between collisions:

$$<t> = \int_0^\infty tP(t) \, dt = \tau. \tag{5.5}$$

The average distance λ_c between collisions, called the mean free path, is the product of τ and the average velocity. We will see in Sect. 5.4 that τ is of the order of 10^{-13} s. For an effective mass $m_e \sim 0.1m$, where m is the free electron mass, and a thermal speed v at 300 K of several 10^5 m/s, $\lambda_c = v\tau$ is about 20 nm.

Using the exponential distribution we can calculate the electrical conductivity and diffusivity.

5.2a Electrical Conductivity

In the absence of an electric field, collisions make all directions of motion equally probable, each one on average occurring without memory of the preceding electron velocity. The electron motion thus produces no global electric current.

In the presence of an electric field $\mathcal{E}$ the electrons feel a constant force:

$$\mathbf{F} = -e\mathcal{E} = m_e \frac{d\mathbf{v}}{dt}. \tag{5.6}$$

From the argument of Sect. 2.2 the mass involved here is the effective mass m_e of the electron in the solid. The electrons thus acquire an additional velocity component v in the direction of $\mathcal{E}$ whose magnitude grows linearly with time until the following collision (Fig. 5.1): this component grows proportionally with the time for an interval Δt_1, is abruptly annihilated, and then begins to grow again during the interval Δt_2, and so on.

The mean value of the velocity in this process is, from Eq. (5.5),

$$<\mathbf{v}> = -\frac{e<t>}{m_e}\mathcal{E}, \tag{5.7}$$

$$<\mathbf{v}> = \mathbf{v}_e = -\frac{e\tau}{m_e}\mathcal{E} = -\mu_e\mathcal{E}, \tag{5.8}$$

which is the drift velocity $\mathbf{v}_e$. This quantity is proportional to the electric field $\mathcal{E}$. Here we have introduced the electron mobility μ_e:

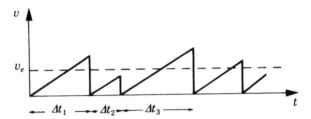

Fig. 5.1. Increase of the velocity component of a free electron along the direction of the applied electric field as a function of time in the Drude model (the resulting velocity is antiparallel to the field). The drift velocity is v_e.

$$\mu_e = \frac{e\tau}{m_e}. \tag{5.9}$$

A convenient way of deriving the results of the Drude model is to introduce into the dynamical equation (5.6) a viscous friction force $-m_e\mathbf{v}/\tau$, where τ is the collision time:

$$m_e \frac{d\mathbf{v}}{dt} = -e\boldsymbol{\mathcal{E}} - m_e \frac{\mathbf{v}}{\tau}. \tag{5.10}$$

The stationary solution ($d\mathbf{v}/dt = 0$) gives again the result (5.8). Equation (5.10) can be extended to more complex cases: sinusoidal driving terms and applied electric and magnetic fields (cf. Appendices 2.6b and 5.1).

The current density $\mathbf{J}_e$ is

$$\mathbf{J}_e = -ne\mathbf{v}_e, \tag{5.11}$$

where n is the carrier density; $\mathbf{J}_e$ is thus proportional to $\boldsymbol{\mathcal{E}}$, allowing us to define the conductivity σ_e:

$$\mathbf{J}_e = \sigma_e \boldsymbol{\mathcal{E}} \tag{5.12}$$

with

$$\sigma_e = ne^2 \frac{\tau}{m_e} = ne\mu_e. \tag{5.13}$$

A solid is in the most general case anisotropic, so that the conductivity is a tensor (cf. Appendix 5.1).

Similarly for a semiconductor possessing free holes of effective mass m_h and concentration p, with collision time τ_h, a discussion analogous to that for the electrons gives

$$\mathbf{v}_h = \frac{e\tau_h}{m_h} \boldsymbol{\mathcal{E}} = \mu_h \boldsymbol{\mathcal{E}}, \tag{5.14}$$

where $\mathbf{v}_h$ is the hole drift velocity in the electric field $\boldsymbol{\mathcal{E}}$ and μ_h their mobility, defined by

$$\mu_h = \frac{e\tau_h}{m_h}. \tag{5.15}$$

The hole current $\mathbf{J}_h$ is proportional to $\boldsymbol{\mathcal{E}}$, and we define the hole conductivity σ_h by

$$\mathbf{J}_h = \sigma_h \, \boldsymbol{\mathcal{E}} \tag{5.16}$$

with

$$\sigma_h = \frac{p \, e^2 \tau_h}{m_h}. \tag{5.17}$$

We can verify (cf. Sect. 3.1) that **the drift velocities of the electrons and holes are in opposite directions, but the current densities $\mathbf{J}_h$ and $\mathbf{J}_e$ are both in the same direction as $\boldsymbol{\mathcal{E}}$ and therefore add**. The total conductivity is thus

$$\sigma = \sigma_e + \sigma_h,$$
$$\sigma = ne\mu_e + pe\mu_h. \tag{5.18}$$

5.2b Diffusion

Let us consider a set of identical particles (e.g., gas molecules) with total concentration n. Assume that some of these particles differ from the others, for example that they are radioactive. Let n_1 be the concentration of these particles. In equilibrium the particles are uniformly distributed in the allowed volume so that the concentrations n and n_1 are independent of position. Let us assume now that the distribution of n_1 is not uniform, but depends on position, e.g., $n_1 = n_1(x)$, while n remains constant. This is not an equilibrium situation. There must therefore be particle motions tending to increase the entropy, i.e., trying to make the concentration n_1 uniform, although there is no net motion of matter. Let us call the particle current $\mathbf{J}_N$. If n_1 is not constant we expect as a first approximation that the current will be proportional to the concentration gradient, here chosen parallel to the x axis:

$$J_{Nx} = -D\frac{\partial n_1}{\partial x}. \tag{5.19}$$

The coefficient of proportionality D is called the diffusion coefficient. For a positive diffusion coefficient, a positive gradient $\partial n_1/\partial x$ gives a negative flux tending to equalize the concentration. This equation, called Fick's law, describes diffusion in very many cases.

If we combine Eq. (5.19) and the equation of particle number conservation:

$$\frac{\partial n_1}{\partial t} = -\nabla \cdot \mathbf{J}_N, \tag{5.20}$$

we obtain the **diffusion equation**

$$\frac{\partial n_1}{\partial t} = D\frac{\partial^2 n_1}{\partial x^2}. \tag{5.21}$$

We immediately see from this equation that if the total particle number remains constant, diffusion moves particles from dense regions, where $\partial^2 n_1/\partial x^2$ is negative, to dilute ones, where it is positive.

5.2c Diffusion in the Drude Model

We consider a classical gas of non-uniform concentration in the Drude model, and attempt to demonstrate Fick's law (5.19) in this case. We assume that the concentration gradient is along x and seek the particle flux across unit surface perpendicular to the x axis at time t. This flux consists of all the particles directed towards this surface from the left since their last collision, from which we must subtract all the particles similarly coming from the right. As the current is proportional to the particle velocity, the total current can be written as

$$J_{Nx} = \int_{-\infty}^{t} n_1[x - v_x(t - t')]v_x \exp[-(t - t')/\tau]\frac{dt'}{\tau}$$
$$- \int_{-\infty}^{t} n_1[x + v_x(t - t')]v_x \exp[-(t - t')/\tau]\frac{dt'}{\tau}. \tag{5.22}$$

This simply states that for a particle to cross the surface at time t with speed v_x requires that it should not suffer a further collision after its last collision at time t' [using the exponential distribution, cf. Eq. (5.4)]. At this time the particle had abscissa $x - v_x(t - t')$, and the number of such particles is $n_1[x - v_x(t - t')]$. Expanding to first order, which is valid only if the gradient is approximately constant over a mean free path, we have

$$J_{Nx} = -\frac{\partial n_1}{\partial x} \int_{-\infty}^{t} 2v_x^2(t - t') \exp[-(t - t')/\tau]\frac{dt'}{\tau}, \tag{5.23}$$

$$J_{Nx} = -\frac{\partial n_1}{\partial x} 2v_x^2 \tau. \tag{5.24}$$

We now have to average over velocities, but counting only positive v_x, since the particles on the left with velocities to the left do not contribute to the current (Fig. 5.2). For an ideal gas the mean value is

$$< v_x^2 > = < v_y^2 > = < v_z^2 > = kT/m \tag{5.25}$$

and here we have to take one-half of this value. We thus have

$$J_{Nx} = -\frac{\partial n_1}{\partial x}\frac{kT}{m}\tau. \tag{5.26}$$

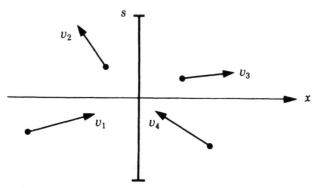

Fig. 5.2. Drude model for diffusion. Electrons with velocities like v_1 cross the surface s at time t and contribute positively to the diffusion current J_x. We have to subtract the contribution of electrons with velocities such as v_4.

We have thus established Fick's equation (5.19) within the framework of the Drude model, and obtained a theoretical value for the diffusion coefficient

$$D = \frac{kT}{m}\tau. \tag{5.27}$$

If we now consider the conduction-band electrons we can give a completely analogous discussion to the extent that the electrons obey classical statistics (Chap. 4). Here, the mean free path is determined by collisions with lattice defects, e.g., impurities or deformations caused by thermal vibrations. The only difference is that we have to replace the mass by the effective mass on taking the mean of v_x^2. We thus get for electrons

$$D_e = \frac{kT}{m_e}\tau_e, \tag{5.28}$$

and for holes

$$D_h = \frac{kT}{m_h}\tau_h. \tag{5.29}$$

If we compare the relations (5.9), (5.28) and (5.15), (5.29) we see that we can write a relation of the form

$$D = \frac{kT}{e}\mu \tag{5.30}$$

for each type of carrier.

This relation has a more general applicability than the restricted cases for which we have derived it, and is called the **Einstein relation**.

Altogether, if there is simultaneously an electric field and a concentration gradient, the two currents, drift and diffusion, add together and we have for the charge current the expression

$$\mathbf{J}_e = en\mu_e\boldsymbol{\mathcal{E}} + eD_e\nabla n. \tag{5.31}$$

For positively charged holes the total current (drift + diffusion) is

$$\mathbf{J}_h = pe\mu_h\boldsymbol{\mathcal{E}} - eD_h\nabla p. \tag{5.32}$$

Note the sign difference between Eqs. (5.31) and (5.32) which reflects the fact that the diffusion of holes or electrons is always opposite to the concentration gradient, whereas this is not true of the associated electric current, because of the sign of the charge carried by each particle.

5.2d Limitations of the Drude Model

This simple model is based on an exponential distribution in which a unique collision time τ appears, and allows one to describe a conductivity, i.e., Ohm's law, and the diffusion constant appearing in Fick's law. It assumes that the electrons obey classical statistics, although this is not always true in semiconductors. We can check its validity by analyzing the behavior of the mobility, as measured, e.g., by the Hall effect: one observes that the mobility varies with temperature (see Sect. 5.4), which implies that τ is a function of electron energy. This is true of collisions with ionized impurities, among others. We thus need a more rigorous formalism based on the Boltzmann transport equation, which we shall give in Sect. 5.3a. We shall recover an expression for the conductivity (Sect. 5.3b) similar to that given by the Drude model, with an interpretation of τ as a certain average of the collision times $\tau(E)$ over the energy E. We shall also be interested in the phenomenon of particle diffusion linked to the presence of a concentration gradient (Sect. 5.3c) and we will show that the Einstein relation is very general.

5.3 Semiclassical Treatment of Transport Processes

The theory of transport processes deals with the relation between currents and the forces which produce them. The formulations used for the calculation of transport properties are based essentially on classical mechanics in the sense that one regards the electrons and holes as particles with well-defined positions and crystal momenta, except for the duration of a collision, which is assumed negligible compared to the time between collisions: this system is completely analogous to an ideal classical gas. However, a part of the discussion requires quantum mechanics, namely the treatment of the collision process itself. It is in this sense that the treatment is called semiclassical.

The physical conditions for a semiclassical treatment are first, the distribution of energy levels must be continuous, or at least the distance between energy levels much smaller than kT. Second, interactions or correlations between particles must be weak. Next, the time that a particle spends in a state of kinetic energy E, i.e., the mean time between collisions, must be large compared with $\hbar/E$, so that the energy of the state is well-defined. Also, the spatial variation of the applied fields must be small over a mean free path. Finally, the time variation of these fields must be small during a collision event.

These conditions are satisfied in semiconductors for a very wide range of temperatures and fields. We can thus apply the classical treatment to the electron and hole gas in semiconductors. We recall here the main lines of this treatment.

In (a) we give the evolution equation for the probability that an electron has position $\mathbf{r}$ and momentum $\mathbf{k}$ at time t. Conductivity is introduced in (b), and diffusion in (c). The main results are summarized in (d), and may be assumed if it is desired to skip Sects. 5.3a,b,c in a first reading.

Let us first make the terminology precise. The treatment below was conceived by Boltzmann to describe the transport properties of gases. In solid state physics the term "Boltzmann equation" is always used to describe the semiclassical transport equation, even for metals, where the electrons obey Fermi–Dirac statistics.

5.3a The Boltzmann Equation for a Semiconductor

Let $f(\mathbf{k}, \mathbf{r}, t)$ be the phase density (or distribution function), such that at time t the number of particles with momentum within $d^3\mathbf{k}$ of $\mathbf{k}$ and position within $d^3\mathbf{r}$ of $\mathbf{r}$ is $f(\mathbf{k}, \mathbf{r}, t)d^3\mathbf{k}\, d^3\mathbf{r}$: this represents the number of particles in the volume element of the one particle phase-space at time t. The distribution function may be either Maxwell–Boltzmann or Fermi–Dirac, according to the system under consideration. To clarify these ideas, Fig. 5.3 shows a section of phase-space along the (x, k_x) plane. The Boltzmann transport equation which governs the evolution of $f(\mathbf{k}, \mathbf{r}, t)$ can be obtained as follows. Between the times t and $t + dt$ the points representing the particles move smoothly towards the volume element $d^3\mathbf{k}'d^3\mathbf{r}'$, equal to $d^3\mathbf{k}d^3\mathbf{r}$ to second order, under the action of external forces and diffusion.

However because of collisions some particles (a) are "lost," while others (b) are "gained." The conservation of particle number imposes only the time independent integral relation

$$\int f(\mathbf{k}, \mathbf{r}, t)\, d^3\mathbf{k}\, d^3\mathbf{r} = \text{constant}. \tag{5.33}$$

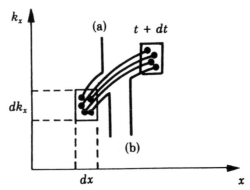

Fig. 5.3. Phase-space with two dimensions x and k_x. Time evolution of particles in the range $dx\,dk_x$.

Comparison of the number of particles within $d^3k\,d^3r$ of $(\mathbf{k},\mathbf{r})$ at time t, and the number within $d^3k\,d^3r$ of $(\mathbf{k}+d\mathbf{k},\mathbf{r}+d\mathbf{r})$ at $t+dt$ requires that

$$f(\mathbf{r}+d\mathbf{r},\mathbf{k}+d\mathbf{k},t+dt) = f(\mathbf{r},\mathbf{k},t) + \left(\frac{\partial f}{\partial t}\right)_{\mathrm{coll}} dt, \qquad (5.34)$$

where $(\partial f/\partial t)_{\mathrm{coll}}$ is the variation of particle number caused by collisions. This gives to first order

$$\frac{\partial f}{\partial \mathbf{r}}\frac{d\mathbf{r}}{dt} + \frac{\partial f}{\partial \mathbf{k}}\frac{d\mathbf{k}}{dt} + \frac{\partial f}{\partial t} = \left(\frac{\partial f}{\partial t}\right)_{\mathrm{coll}} \qquad (5.35)$$

which is the **Boltzmann equation**. Using Eq. (2.33), the equation of motion in the reciprocal space, this can be rewritten as

$$\mathbf{v}\cdot\nabla_r f + \frac{1}{\hbar}q(\boldsymbol{\mathcal{E}}+\mathbf{v}\times\boldsymbol{B})\cdot\nabla_\mathbf{k} f + \frac{\partial f}{\partial t} = \left(\frac{\partial f}{\partial t}\right)_{\mathrm{coll}}. \qquad (5.36)$$

The first term represents the density variation in phase-space when the point under consideration moves in a spatially inhomogeneous system. This term is related to the diffusion. The second term shows the variation of f under the effect of an electromagnetic force applied to a charge q. The third term gives the explicit time dependence, for example in the case of sinusoidal driving force. This term vanishes in a steady state. The collision term is difficult to calculate exactly. One often uses the so-called **relaxation time approximation**, in which

$$\left(\frac{\partial f}{\partial t}\right)_{\mathrm{coll}} = -\frac{f-f_0}{\tau(\mathbf{k})}, \qquad (5.37)$$

that is, if we impose a distribution function f differing from the equilibrium distribution f_0, the system will return to equilibrium in a characteristic time

$\tau(\mathbf{k})$. This hypothesis gives a good description of most collision processes in semiconductors. Thus the steady-state distribution function f is the solution of

$$\mathbf{v} \cdot \nabla_{\mathbf{r}} f + \frac{q}{\hbar}(\boldsymbol{\mathcal{E}} + \mathbf{v} \times \boldsymbol{\mathcal{B}}) \cdot \nabla_{\mathbf{k}} f = -\frac{f - f_0}{\tau(\mathbf{k})}. \tag{5.38}$$

5.3b Conductivity

This is the effect produced by an electric field in a homogeneous semiconductor at constant temperature in the absence of a magnetic field. If there is no temperature gradient we have $\nabla_{\mathbf{r}} f = 0$ and Eq. (5.38) gives for the stationary state

$$f = f_0 - \tau \hbar^{-1} q \, \boldsymbol{\mathcal{E}} \cdot \nabla_{\mathbf{k}} f(\mathbf{k}). \tag{5.39}$$

We assume that the deviation of f from the equilibrium distribution function f_0 is small, so that in the last term of Eq. (5.39) we can replace f by f_0, which depends on k only through the energy. Then

$$f = f_0 - \tau \hbar^{-1} \frac{\partial f_0}{\partial E} \, q \boldsymbol{\mathcal{E}} \cdot \nabla_{\mathbf{k}} E_{\mathbf{k}} \tag{5.40}$$

so that for an electron, with the field taken along the x direction,

$$f = f_0 + \tau e \mathcal{E} v_x \frac{\partial f_0}{\partial E}. \tag{5.41}$$

The electric current density J_x corresponding to a state $\mathbf{k}$ is $-2ev_x$ (the factor 2 comes from the spin). Since the density in $\mathbf{k}$-space is $(1/2\pi)^3$ the total current is

$$J_x = \frac{-e^2 \mathcal{E}}{4\pi^3} \int d^3\mathbf{k} \, v_x^2 \, \frac{\partial f_0}{\partial E} \tau. \tag{5.42}$$

Noting that the total electron number N_T in the volume Ω is

$$N_T = n\Omega = \frac{1}{4\pi^3} \int f d^3 k \int_\Omega d^3 \mathbf{r} = \frac{\Omega}{4\pi^3} \int f_0 d^3 k, \tag{5.43}$$

we get

$$J_x = -e^2 \mathcal{E} n \frac{\int d^3\mathbf{k} \, v_x^2 \, (\partial f_0/\partial E) \, \tau}{\int d^3\mathbf{k} \, f_0}. \tag{5.44}$$

The current J_x is proportional to the electric field $\boldsymbol{\mathcal{E}}$, and we can deduce the electric conductivity (cf. Eq. (5.12)) or the mobility μ. The expression for μ can be found from Eq. (5.44), and is positive by definition.

As we have seen in Chap. 4, the Fermi level is usually sufficiently far from the the conduction and valence bands that we can approximate the distribution function f_0 by a Maxwell–Boltzmann law:

$$f_0 = \text{const. } \exp(-E/kT), \tag{5.45}$$

$$\frac{\partial f_0}{\partial E} = -\frac{1}{kT} f_0. \tag{5.46}$$

The **semiconductor** is thus **non-degenerate**. The mobility μ can then be written, using Eq. (5.44), as

$$\mu = +J_x/ne\mathcal{E} = \frac{e}{kT} \frac{\int d^3k \, v_x^2 \, \tau(E) f_0}{\int d^3k f_0}. \tag{5.47}$$

If the effective mass is a scalar, i.e.,

$$E = \frac{\hbar^2 k^2}{2m_e} = \frac{1}{2} m_e v^2, \tag{5.48}$$

we have the relations

$$\int v_x^2 \phi(E) d^3k = \int v_y^2 \phi(E) d^3k = \int v_z^2 \phi(E) d^3k = \frac{1}{3} \int v^2 \phi(E) d^3k$$

$$= \frac{2}{3m_e} \int E\phi(E) d^3k. \tag{5.49}$$

Further $d^3k = \text{const. } E^{1/2} dE$, so that Eq. (5.47) can be written

$$\mu = \frac{2}{3m_e} \frac{1}{kT} \frac{\int_0^\infty E^{3/2} \tau(E) f_0(E) \, dE}{\int_0^\infty E^{1/2} f_0(E) \, dE}. \tag{5.50}$$

Noting that

$$\int_0^\infty E^{3/2} e^{-E/kT} dE = \frac{3kT}{2} \int_0^\infty E^{1/2} \exp(-E/kT) \, dE \tag{5.51}$$

we finally get

$$J_x = ne^2 \mathcal{E} \frac{<\tau>}{m_e} = \sigma \mathcal{E} \tag{5.52}$$

with

$$<\tau> = \frac{\int_0^\infty \tau(E) E^{3/2} \exp(-E/kT) \, dE}{\int_0^\infty E^{3/2} \exp(-E/kT) \, dE}. \tag{5.53}$$

We see that the conductivity depends on the effective mass and the mean relaxation time defined by Eq. (5.53).

For electrons the mobility is given by

$$\mu_e = e \frac{<\tau_e>}{m_e} \tag{5.54}$$

and similarly for the holes by

$$\mu_h = e\frac{<\tau_h>}{m_h}.\tag{5.55}$$

Expressions (5.54) and (5.55) for the mobilities resemble those of the Drude model (definitions (5.9) and (5.15)). However the collision times are averages which use the non-trivial weighting function $E^{3/2}\exp(-E/kT)dE$. Starting from the observed temperature dependence of these averages, we can deduce $\tau(E)$, and thus identify the microscopic collision mechanisms. This will be done in Sect. 5.4, where we shall also give the orders of magnitude of the collision times.

The Boltzmann equation method can be extended to the calculation of magnetic effects like the Hall effect, magnetoresistance (variation of the conductivity because of the bending of the current lines by the Lorentz force), or thermoelectric effects. One finds that the important quantities involve not just $<\tau>$ but also $<\tau^{-1}>, <\tau^2>/<\tau>^2, ...$, where the mean is defined as in Eq. (5.53), i.e., using the weighting function $E^{3/2}\exp(-E/kT)dE$. Some phenomena such as magnetoresistance for a semiconductor with a single carrier disappear if the relaxation time is independent of the energy.

For the case of a **degenerate semiconductor**, for which f_0 is a Fermi–Dirac distribution function, which cannot be approximated as Eq. (5.45), we have, to the same order in the field as before

$$\int_0^\infty dE\ \tau(E)\frac{\partial f_0}{\partial E}E^{3/2} = -\tau(E_F)E_F^{3/2},$$

$$\int_0^\infty dE\ E^{1/2}f_0 = \frac{2}{3}E_F^{3/2}\tag{5.56}$$

so that

$$J_x = ne^2\mathcal{E}\frac{\tau(E_F)}{m_e}.\tag{5.57}$$

The conductivity and the mobility thus involve the collision time of electrons with the Fermi energy, which itself lies in an allowed band in this case. These results are intuitively understandable: from the Pauli principle, motion of carriers requires accessible quantum states, which can only be found near E_F. (See Sect. 4.8.)

5.3c Diffusion

Equation (5.38) gives the form of the distribution function f for the case of a small concentration gradient: f thus differs little from f_0, and depends on $\mathbf{r}$. In the absence of external forces

$$f = f_0 - \tau \hbar^{-1} \nabla_r f \cdot \nabla_k E_k, \tag{5.58}$$

where f_0 is constant in space, and the velocity has been replaced by its expression (2.29) as a function of E_k.

We will calculate the particle current density $\mathbf{J}_N$. The current corresponding to a state $\mathbf{k}$ is $2\mathbf{v}(\mathbf{k})$ (2 for the spin); taking into account the density of states $(1/2\pi)^3$ the total current density is

$$\mathbf{J}_N = \frac{1}{4\pi^3} \int f\mathbf{v}(\mathbf{k}) d^3\mathbf{k}, \tag{5.59}$$

$$\mathbf{J}_N = \frac{1}{4\pi^3} \int f_0 \mathbf{v}(\mathbf{k}) d^3\mathbf{k} - \frac{1}{4\pi^3} \int \mathbf{v}(\mathbf{k}) \tau v_x \frac{\partial f}{\partial x} d^3\mathbf{k}. \tag{5.60}$$

The first term vanishes as it involves a symmetrical integral of an odd function: f_0 depends on $\mathbf{k}$ only through the energy, which is an even function of $\mathbf{k}$. In any case, $\mathbf{J}$ must vanish in equilibrium. Similarly, in the second term only the velocity component v_x enters, and the (diffusion) current is in the x direction. It thus has the form

$$J_{N,x} = -\frac{1}{4\pi^3} \int v_x^2 \tau \frac{\partial f}{\partial x} d^3\mathbf{k}. \tag{5.61}$$

In the case considered, where there is a concentration gradient along x allowing a **local** equilibrium, the phase-space density can be written as

$$f = A(x) \exp(-E/kT) \tag{5.62}$$

so that the density at the point $\mathbf{r}$ is

$$n(x) = \frac{A(x)}{4\pi^3} \int \exp(-E/kT) d^3\mathbf{k}. \tag{5.63}$$

Comparing Eqs. (5.58) and (5.62), we see that $\tau v_x (\partial A/\partial x)/A(x)$ must be small compared with unity: the assumption of a small concentration gradient is equivalent to taking $A(x)$ as slowly varying over a mean free path τv_x.

The current becomes

$$J_{Nx} = -\frac{1}{4\pi^3} \frac{\partial A}{\partial x} \int v_x^2 \tau \exp(-E/kT) d^3\mathbf{k}. \tag{5.64}$$

From Eq. (5.63),

$$\frac{1}{4\pi^3} \frac{\partial A(x)}{\partial x} = \frac{\partial n/\partial x}{\int \exp(-E/kT) d^3\mathbf{k}}. \tag{5.65}$$

We thus have

$$J_{Nx} = -\frac{\partial n}{\partial x} \frac{\int v_x^2 \tau \exp(-E/kT)d^3\mathbf{k}}{\int \exp(-E/kT)d^3\mathbf{k}} \tag{5.66}$$

which defines the diffusion coefficient in Fick's law:

$$D = \frac{\int v_x^2 \tau(E) \exp(-E/kT)d^3\mathbf{k}}{\int \exp(-E/kT)d^3\mathbf{k}}. \tag{5.67}$$

Comparing with expression (5.45) for the mobility we deduce the Einstein relation for electrons:

$$D = \frac{kT}{e}\mu. \tag{5.68}$$

We have thus generalized Fick's law and the Einstein relation, which we had previously only demonstrated in the context of the ultra-simplified Drude model. Obviously we get a similar result for holes. All that enters in the relation for D is the hole mobility.

For a **degenerate** system we find

$$D = \frac{\tau(E_F)v_F^2}{3} \tag{5.69}$$

so that

$$D = \frac{2}{3}\frac{E_F}{e}\mu. \tag{5.70}$$

We note that the existence of a relation between the mobility and the diffusion coefficient is extremely general: it shows that in fact there exists only a single transport coefficient relating the current of charged particles, the concentration gradient and the electrostatic potential gradient. This is a general property of transport of independent particles, whose proof is beyond the scope of the present treatment. The two gradients are the two components of the gradient of the electrochemical potential.

5.3d Summary

The semiclassical Boltzmann model allows a precise interpretation of the relaxation time appearing in the expressions for the mobility and diffusion coefficient in the Drude model (expressions (5.9) and (5.28)). What appears is a mean $< \tau >$ of the collision times $\tau(E)$, defined by

$$< \tau > = \frac{\int_0^\infty \tau(E)E^{3/2} \exp(-E/kT)\, dE}{\int_0^\infty E^{3/2} \exp(-E/kT)\, dE}. \tag{5.53}$$

We should remember that other averages over $\tau(E)$ appear for other properties such as the Hall effect and magnetoresistance, and the Drude model cannot be straightforwardly applied.

We shall now discuss several of the collision mechanisms determining $<\tau>$, and hence μ, in semiconductors.

5.4 The Mobility of Semiconductors

5.4a Collision Mechanisms

Any disruption of the crystal periodic structure is a source of collisions. In a given crystal there may be various types of collisions with characteristic times τ_1, τ_2.... We thus have to calculate an equivalent time as

$$\frac{1}{\tau} = \frac{1}{\tau_1} + \frac{1}{\tau_2} + \cdots, \tag{5.71}$$

since the probabilities of collisions caused by different mechanisms add if the events have small probability. For metals this is called Matthiesen's rule.

The main collisions mechanisms are:

(α) Scattering by **crystal vibrations (phonons)**. This is the main source of collisions at intermediate and high temperatures. The amplitude of the vibrations increases with temperature and we expect the collision probability $1/\tau$ to increase with T. In fact one can show that

$$\tau = aE^{-1/2}T^{-1} \tag{5.72}$$

implying mobilities which decrease as the temperature rises. For a non-degenerate semiconductor we get

$$\mu = \mu_0 \left(\frac{T_0}{T}\right)^{3/2}. \tag{5.73}$$

Exercise: Show that Eqs. (5.53), (5.54), and (5.72) give a law of the form (5.73).

(β) Collisions with **ionized impurities** — the effect of the Coulomb field of the impurity. As mentioned above, a rapid particle feels less the potential of an impurity, so $1/\tau$ decreases as E grows. One can show that

$$\tau = aE^{3/2} \tag{5.74}$$

which, via Eqs. (5.53) and (5.54), leads to mobilities varying as $T^{3/2}$. The mobility increases with temperature because electrons with larger velocities are less sensitive to the Coulomb fields of the ionized impurities.

Exercise: Show that Eq. (5.74) implies mobilities varying as $T^{3/2}$.

(γ) Neutral impurities and dislocations can also contribute to collisions.

Figure 5.4 shows the dependence of mobility on temperature for variously doped samples of silicon. The law $T^{-3/2}$ (Eq. (5.73)) does not hold

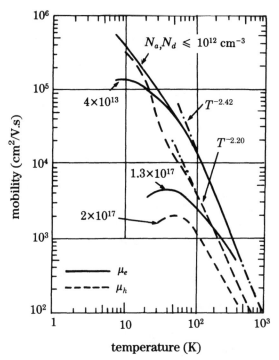

Fig. 5.4. Temperature dependence of the electron and hole mobilities μ_e, μ_h for silicon samples with different doping levels. The electron mobility is shown as continuous curves and the hole mobility as dashed. The dash–dot curves are the best fits to the experimental results.

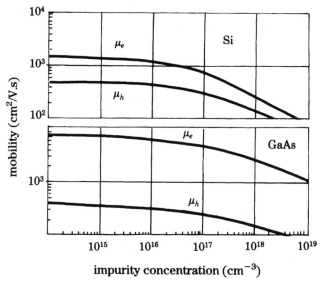

Fig. 5.5. Electron and hole mobility in silicon and gallium arsenide at room temperature, as a function of the impurity concentration.

very well. At low temperature crystal vibrations have very small amplitudes ($\tau_{\text{vibration}}$ increases) and collisions on impurities become dominant (compare Eqs. (5.72) and (5.74)). The larger the impurity concentration, the higher the temperature at which they control the mobility, the weaker the maximum mobility.

At high temperatures, on the other hand, mobilities depend little on the concentration, as shown by the measurements at 300 K given in Fig. 5.5. Using mobility data and effective masses, found, e.g., by cyclotron resonance, we can deduce collision times. We note that the variation of mobility at high temperature is much smaller than the concentration variation caused by intrinsic ionization. This is why the conductivity increases very strongly at high temperature despite the decrease of mobility.

Exercise: Use the data of Fig. 5.5 and the effective masses of Sect. 2.4 to show that the collision time is of order 10^{-13} s in Si and GaAs, both for electrons and holes. Assume that the mobility of the conduction electrons of silicon is essentially determined by the transverse effective mass.

The Einstein relation gives the orders of magnitude of the diffusion coefficients. We have $D = \mu(kT/e)$. At room temperature kT/e is 25 meV; if $\mu = 0.1$ m^2/V.s, the diffusion coefficient is $25 \cdot 10^{-4}$ m^2/s.

5.4b Selective Doping of Superlattices

We shall see at the end of this book that the operation speed of certain semiconductor devices improves with mobility. One might think to operate at low temperature, noting the data of Sect. 5.4a, but we are limited by collisions with ionized impurities. One might try to purify the semiconductors, but the shallow impurities cannot be removed without also removing the conduction electrons. The ideal would be to have electrons without impurities. What appeared to be fantasy has been realized following an idea by Störmer (1980).

We saw in Sect. 3.3 that present modern growth techniques by molecular beam epitaxy allow us to construct superlattices in composition. Let us assume that some donor atoms are added during the controlled growth, but only in the semiconductor with the largest gap, $Al_x Ga_{1-x}$ As (Fig. 5.6). Then the ground state for the electrons is not localized at the donors, in the $Al_x Ga_{1-x}$ As region. The electrons fill the free states in the GaAs quantum wells. But these are states of conduction parallel to the GaAs layers, with very little penetration into the large-gap material containing the impurities. We can say that under these conditions the electrons are localized in zones where the impurities are absent and so their mobility should be increased. This is indeed observed. The present record (1993) is $\mu_e = 1.5 \times 10^2$ m$^2 \cdot$ V$^{-1} \cdot$ s^{-1} at low temperature, while the best mobility measured in a single crystal is of the order of 10 m^2.V^{-1}. s^{-1}. These are huge values. For comparison we recall that the mobility of the best room temperature conductor, metallic silver, is of the order of 10^{-1} m$^2 \cdot$ V$^{-1} \cdot$ s^{-1}.

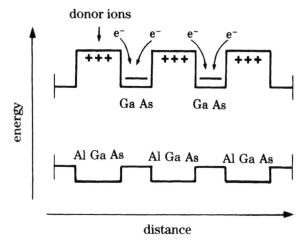

Fig. 5.6. Selective doping of a superlattice.

In the GaAs layer the collision time is increased by the same factor as the mobility, and the mean free path reaches a fraction of a micrometer. If one can manufacture sufficiently small structures, one has electrons that suffer no collisions in the plane of the layer. Such electrons are called "ballistic." The mobility is not increased in the direction perpendicular to the layers, so the mean free path remains of order 10 nm, precisely of the order of the thickness of the GaAs layer constituting the quantum well. Devices are in development (ballistic electron transistors) which use this property.

Appendix 5.1

Problems on the Hall Effect and Magnetoresistance of Semiconductors in the Drude Model

In these problems we study the conductivity originating from carriers which may have different charges (negative electrons or positive holes), in the presence of a magnetic field. We consider the transport equation determining the *mean* velocity $\mathbf{v}$ of a given type of carrier in the simplest form of the Drude model:

$$\frac{d\mathbf{v}}{dt} + \frac{\mathbf{v}}{\tau} = \frac{\mathbf{F}}{m^*}. \tag{5.75}$$

Thus in a steady state (which we assume throughout these problems)

$$\mathbf{v} = \frac{\tau}{m^*}\mathbf{F}, \tag{5.76}$$

where τ and m^* are positive scalars (independent of the magnitude and direction of $\mathbf{v}$) and $\mathbf{F}$ is the external force applied to the carriers. This formulation, although extremely simple, does not alter the essential physics of the problems.

First Part

(1) We consider a crystal with only one kind of carrier, of charge q ($= \pm 1.6 \times 10^{-19}$ C) and concentration n. We apply a fixed electric field $\mathcal{E}$. Derive the relations between $\mathbf{v}$ (mean velocity), $\mathbf{J}$ (current density), and $\mathcal{E}$ in terms of the mobility μ and the conductivity σ. Pay particular attention to signs (by convention the mobility μ is a positive quantity).

(2) We now apply a fixed magnetic field along Oz ($B_x = 0, B_y = 0, B_z = B$). Show that Ohm's law generalizes to $\mathcal{E} = \bar{\bar{\rho}}\mathbf{J}$, or equivalently $\mathbf{J} = \bar{\bar{\sigma}}\mathcal{E}$, and show explicitly the resistivity and conductivity tensors $\bar{\bar{\rho}}, \bar{\bar{\sigma}}$ referred to axes $Oxyz$. In the literature the angle θ, with $|\tan\theta| = \mu B$, is called the "Hall angle." Does this suggest to you a geometrical interpretation?

(3) Slender-bar geometry: We consider a bar of length L along Ox with small cross sections Δy, Δz, such that the current density $\mathbf{J}$ lies along Ox. (a) Give the relation between J_x and $\mathcal{E}_x$ (longitudinal conductivity). (b) Although the current is purely longitudinal, we note that a transverse electric field (the "Hall field") appears; give an expression for it.

(4) Practical application: in a silicon bar with dimensions $L = 2$ cm, $\Delta z = 0.2$ cm, $\Delta y = 0.2$ cm, with magnetic field $B = 0.1$ tesla, we pass a total current of $I_x = 10$ mA. We measure a voltage $V_Q - V_P = 4.15$ V between the ends of the bar, and a voltage $V_N - V_M = 0.21 \times 10^{-3}$ V across the faces MN of the bar. Calculate the characteristics of the silicon sample.

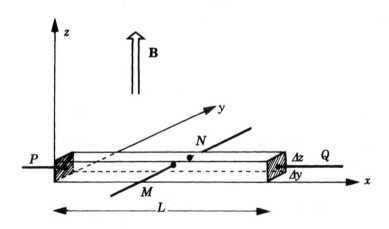

Remark: for a voltmeter to perturb as little as possible the potential difference it measures, its internal resistance must be large. In practice every voltmeter will have various sensitivities; by construction, its internal resistance will be smaller for greater sensitivity. Thus a voltmeter with an internal resistance of 1000 Ω in the range 0–1 V will only have internal resistance 10 Ω in the range 0–1 mV. But the cost of a voltmeter increases as its internal resistance per volt. Thus a cheap ($\simeq \$50$) modern transportable voltmeter has an internal resistance of 10 kΩ/V, a good instrument with 10 MΩ/V costs around \$180, while for greater internal resistances one has to use an electronic voltmeter ($\simeq \$1000$).

What types of voltmeters are required in practice for these measurements? (The contacts M and N are welds of area about 0.1 cm $\times$ 0.1 cm.)

Second Part

(1) We assume now that there are two types of carriers (1) and (2), with charges, densities, and mobilities q_1, n_1, μ_1 and q_2, n_2, μ_2, respectively (for example, two types of electrons or electrons and holes). In all cases $|q_1| = |q_2| = e =$ electron charge. Write down the conductivity tensor.

(2) Consider again the slender bar. Calculate the longitudinal conductivity $\sigma_{long} = J_x/\mathcal{E}_x$. Calculate the Hall constant $R_H = \mathcal{E}_y/J_x B$.

   Note: the exact calculation is laborious. Use the fact that in general $\mu_1 B$ and $\mu_2 B$ are smaller than 1 and calculate to second order in these quantities. Do this explicitly for the case of a semimetal or intrinsic semiconductor ($q_1 = -q_2$ and $n_1 = -n_2$).

(3) Discuss what happens at the surface of the sample in the case of a single carrier type, and in the case of equal electron and hole densities.

Solutions

First Part

(1) Under an applied field $\mathcal{E}$ the force is $\mathbf{F} = q\mathcal{E}$ and the mean velocity is

$$\mathbf{v} = \frac{q\tau}{m^*}\mathcal{E}. \tag{5.77}$$

The mobility μ is defined in Eqs. (5.9) and (5.15) as the ratio of $\mathbf{v}$ and $\mathcal{E}$. As indicated in the problem this quantity is conventionally taken as a positive quantity. We thus have

$$\mu = \frac{|q|\tau}{m^*} \quad \text{and} \quad \mathbf{v} = \frac{|q|}{q}\mu\mathcal{E}. \tag{5.78}$$

The current density $\mathbf{J}$ is given by $\mathbf{J} = nq\mathbf{v}$, therefore $\mathbf{J} = n|q|\mu\mathcal{E}$. We recover Ohm's law

$$\mathbf{J} = \sigma\mathcal{E}, \tag{5.79}$$

where the conductivity $\sigma = n|q|\mu = nq^2\tau/m^*$ is positive whatever the sign of the charge and whatever the convention adopted.

(2) When both an electric and magnetic field are applied, the force $\mathbf{F}$ becomes $\mathbf{F} = q(\mathcal{E} + \mathbf{v} \times \mathbf{B})$, and the velocity $\mathbf{v}$ is given by the equation

$$\mathbf{v} = \frac{q\tau}{m^*}(\mathcal{E} + \mathbf{v} \times \mathbf{B}). \tag{5.80}$$

Using the relations $\mathbf{J} = nq\mathbf{v}$, $\mu = |q|\tau/m^*$, and $\sigma = n|q|\mu$, we find

$$\mathcal{E} = \frac{1}{\sigma}\left(\mathbf{J} + \frac{|q|}{q}\mu\mathbf{B} \times \mathbf{J}\right). \tag{5.81}$$

This formula can be written as $\boldsymbol{\mathcal{E}} = \bar{\bar{\rho}}\mathbf{J}$, where the resistivity tensor referred to axes xyz is

$$
\bar{\bar{\rho}} = \frac{1}{\sigma}
\begin{vmatrix}
1 & -\dfrac{|q|}{q}\mu B & 0 \\[2mm]
\dfrac{|q|}{q}\mu B & 1 & 0 \\[2mm]
0 & 0 & 1
\end{vmatrix}.
\tag{5.82}
$$

We note that in the plane xOy the vectors $\boldsymbol{\mathcal{E}}$ and $\mathbf{J}$ are not collinear: as shown in the figure below, illustrating the vector relation (5.81), the angle θ between $\mathbf{J}$ and $\boldsymbol{\mathcal{E}}$ is given in sign and magnitude by

$\tan\theta = (q/|q|)\mu B$

The conductivity tensor $\bar{\bar{\sigma}}$ is given by simply inverting the tensor $\bar{\bar{\rho}}$:

$$
\bar{\bar{\sigma}} = \sigma
\begin{vmatrix}
\dfrac{1}{1+\mu^2 B^2} & \dfrac{q}{|q|}\dfrac{\mu B}{1+\mu^2 B^2} & 0 \\[3mm]
-\dfrac{q}{|q|}\dfrac{\mu B}{1+\mu^2 B^2} & \dfrac{1}{1+\mu^2 B^2} & 0 \\[3mm]
0 & 0 & 1
\end{vmatrix}.
\tag{5.83}
$$

(3) In the slender-bar geometry we assume $J_y = J_z = 0$. Using the vector relation (5.81) or the matrix (5.82) we find

$$
J_x = \sigma\mathcal{E}_x.
\tag{5.84}
$$

This is the same relation as in the absence of a magnetic field. Conduction is not changed by the presence of a magnetic field for a single carrier type (in the simple transport model we have adopted): there is no "magnetoresistance" effect. However we see that a transverse electric field appears, which from the definitions of σ and μ takes the form

$$
\mathcal{E}_y = \frac{J_x B}{nq}.
\tag{5.85}
$$

Measurement of this "Hall field" $\mathcal{E}_y$ thus allows us to measure the carrier density n and sign. Knowing n, a measurement of $\sigma = n|q|\mu$ gives the mobility of the carriers.

(4) Application. From the sign of the measured voltages we see that the carriers are positively charged holes. The bar is therefore made of p-type

silicon. The numerical application of the above formulas gives a Hall angle $\mu B \sim 5 \times 10^{-4}$ rad and $\mu = 0.05$ m^2 V^{-1}. s^{-1}, $\sigma = 12$ siemens. m^{-1}, $n = 1.5 \times 10^{21}$ m^{-3}. This is a typical material for the manufacture of transistors.

Remark: the voltmeters used must not disturb the measurement. For the longitudinal voltage this is relatively straightforward. We have to measure a voltage of order 5 V with a total current of 10 mA: an instrument with about 20,000 Ω/V will divert about 0.05 mA and thus perturb the experiment by less than 0.5%.

By contrast the measurement of the transverse voltage is much more delicate: the internal resistance between the contacts M and N, of size approximately 0.1 cm $\times$ 0.1 cm, is

$$R_i = \frac{1}{\sigma} \times \frac{L}{S} = \frac{2 \times 10^{-3}}{12 \times 10^{-6}} \simeq 160 \ \Omega.$$

For an error less than 2%, the internal impedance of the voltmeter must be at least $50 R_i$, i.e., 8 kΩ for 0.2 mV. This corresponds to a sensitivity of about 40 MΩ/V. We therefore need an electronic voltmeter for this measurement.

Second Part

(1) The **currents** of the two carrier types add:

$$\mathbf{J} = \mathbf{J}_1 + \mathbf{J}_2 = (\bar{\bar{\sigma}}_1 + \bar{\bar{\sigma}}_2)\mathcal{E}. \tag{5.86}$$

We have

$$\bar{\bar{\sigma}} = \bar{\bar{\sigma}}_1 + \bar{\bar{\sigma}}_2.$$

We must first add the conductivity tensors (matrices of type (5.83)) to obtain $\bar{\bar{\sigma}}$, then invert this matrix to find ρ_{xx} and ρ_{xy}.

Setting $q_1\mu_1 B/|q_1|\mu_1 B = \theta_1$ and $q_2\mu_2 B/|q_2|B = \theta_2$, both assumed small, and expanding to second order in θ_1 and θ_2 we get

$$\bar{\bar{\sigma}} = \begin{vmatrix} \sigma_1(1-\theta_1^2) + \sigma_2(1-\theta_2^2) & \sigma_1\theta_1 + \sigma_2\theta_2 & 0 \\ -(\sigma_1\theta_1 + \sigma_2\theta_2) & \sigma_1(1-\theta_1^2) + \sigma_2(1-\theta_2^2) & 0 \\ 0 & 0 & \sigma_1 + \sigma_2 \end{vmatrix}. \tag{5.87}$$

(2) As $J_y = J_z = 0$, the required quantities follow immediately from the tensor $\bar{\bar{\rho}}$:

$$\mathcal{E}_x = \rho_{xx} J_x \quad \text{and} \quad \mathcal{E}_y = \rho_{yx} J_x.$$

Using the fact that $\sigma_{xx} = \sigma_{yy}$ and $\sigma_{xy} = -\sigma_{yx}$, we find

$$\rho_{xx} = \frac{\sigma_{xx}}{(\sigma_{xx})^2 + (\sigma_{xy})^2} \quad \text{and} \quad \rho_{yx} = \frac{\sigma_{yx}}{(\sigma_{xx})^2 + (\sigma_{xy})^2}. \tag{5.88}$$

Expanding to second order this gives

$$\sigma_{\text{long}} = \frac{J_x}{\mathcal{E}_x} = \sigma_1 + \sigma_2 - (\sigma_1 \theta_1^2 + \sigma_2 \theta_2^2) + \frac{(\sigma_1 \theta_1 + \sigma_2 \theta_2)^2}{\sigma_1 + \sigma_2}.$$

After a little algebra this can be written

$$\sigma_{\text{long}} = \sigma_1 + \sigma_2 - \frac{\sigma_1 \sigma_2}{\sigma_1 + \sigma_2}(\theta_1 - \theta_2)^2$$

or with $\sigma_1 = |q|n_1\mu_1$ and $\sigma_2 = |q|n_2\mu_2$,

$$\sigma_{\text{long}} = |q|(n_1\mu_1 + n_2\mu_2)\left[1 - \frac{n_1\mu_1 n_2\mu_2(q_1\mu_1 - q_2\mu_2)^2 B^2}{e^2(n_1\mu_1 + n_2\mu_2)^2}\right]. \tag{5.89}$$

We see that in this case the longitudinal conductivity depends to second order on the magnetic field unless the carriers have the same sign $(q_1 = q_2)$ and the same mobility $(\mu_1 = \mu_2)$. This result, the variation of resistance with magnetic field, or magnetoresistance, is generic as soon as we are not dealing with a single carrier type with a single relaxation time.

In the same way, the Hall constant is given (still to second order) by

$$\frac{\mathcal{E}_y}{J_x} = B \cdot R_H = \rho_{yx} = \frac{\sigma_1 \theta_1 + \sigma_2 \theta_2}{(\sigma_1 + \sigma_2)^2}, \tag{5.90}$$

$$R_H = \frac{q_1 n_1 \mu_1^2 + q_2 n_2 \mu_2^2}{q^2(n_1\mu_1 + n_2\mu_2)^2}. \tag{5.91}$$

In the domain of validity of these formulas we see that:

• The magnetoresistance is always positive (resistance **increases** with magnetic field).

• We can measure the resistance for $B = 0$, the magnetoresistance (which under our assumptions is a very small effect), and the Hall constant R_H. From these three independent measurements it is in general impossible to determine the four unknowns (n_1, μ_1, n_2, μ_2) even if it is possible to guess the sign of the charges.

• If we have reason to believe that $n_1 = n_2 = n$, for opposite charge carriers (semimetal or intrinsic semiconductor), we can write

$$\sigma_{\text{long}} = |q|n(\mu_1 + \mu_2)(1 - \mu_1\mu_2 B^2), \tag{5.92}$$

$$R_H = \frac{\mu_1^2 - \mu_2^2}{(\mu_1 + \mu_2)^2}\frac{1}{nq}. \tag{5.93}$$

We find that for the Hall effect it is the carriers with the greater mobility that dominate. If $\mu_1 = \mu_2$ the Hall effect disappears (but conductivity and magnetoresistance remain).

For the more realistic case of a distribution of relaxation times we must generalize the above treatment by using the Boltzmann equation. We expect the Hall effect to be slightly modified: there appears in Eq. (5.85) a correction factor of the form $< \tau^2 > / < \tau >^2$. Even for a single carrier type, a distribution of relaxation times creates magnetoresistance. This magnetoresistance must be calculated using the Boltzmann equation.

(3) With a single type of carriers we have $J_y = 0$; the electric field $\mathcal{E}_y$ which compensates the Lorentz force is caused by the charges which accumulate on the faces M and N once the magnetic field is switched on.

When there are two carrier types we have $J_y = J_{1y} + J_{2y} = 0$, but J_{1y} and J_{2y} are not separately zero: there are transverse currents of both carrier types which cancel. It is clear that charges cannot accumulate indefinitely on the faces, and we have a stationary regime where the two types of charges reaching the surface (e.g., electrons and holes) recombine in pairs. This recombination may give rise to measureable effects, such as light emission at the energy of the band gap.

6.

Effects of Light

In Chap. 5 we studied the transport properties of electron and hole gases regarded as independent of each other. We will now study the dynamical equilibrium between these two gases, particularly in the presence of light. This problem is important in practice, since many devices such as detecting and emitting diodes use the electro-optical properties of semiconductors.

Two reactions can occur within a semiconductor subject to intrinsic luminous irradiation (with energy greater than the band gap): the first,

$$\text{light} \rightarrow \text{electron} + \text{hole} \tag{6.1}$$

is that of **optical absorption**, which we shall discuss first. The second reaction,

$$\text{electron} + \text{hole} \rightarrow \text{energy} \tag{6.2}$$

is the process called **recombination**. If it results in the emission of light, it is called radiative recombination. The annihilation of an electron–hole pair can also occur via non-radiative processes. We shall study recombination in Sect. 6.2.

In Sect. 6.3 we will explain the working of several common optoelectronic devices that use homogeneous semiconductors: photoelectric cells, photocopiers, and television screens.

6.1 Light Absorption by Semiconductors

When a semiconductor is irradiated by light, electrons can be excited from the valence band into the conduction band by the absorption of photons, provided the photon energy is greater than E_g. Here we shall calculate this absorption in the simplest case, that of so-called "direct" absorption.

We know that a transverse electromagnetic wave propagating in a medium has a corresponding vector potential:

$$\mathbf{A}(t) = A \, \mathbf{a}_0 \cos(\omega t - \mathbf{K} \cdot \mathbf{r}), \tag{6.3}$$

where $\mathbf{a}_0$ is a unit vector orthogonal to $\mathbf{K}$, the wave vector of the light. In the following we shall be interested in the absorption of light by a small sample of area s and small thickness dx normal to the luminous flux. Under these conditions the wave is weakly attenuated and its constant amplitude A can be chosen as real by suitable choice of the origins of space and time. The wave's electric and magnetic fields have the forms

$$\mathcal{E} = -\frac{\partial \mathbf{A}}{\partial t} = A\omega\, a_0 \sin(\mathbf{K} \cdot \mathbf{r} - \omega t), \tag{6.4}$$

$$\mathcal{B} = \nabla \times \mathbf{A} = -A\,(\mathbf{K} \times \mathbf{a}_0) \sin(\mathbf{K} \cdot \mathbf{r} - \omega t). \tag{6.5}$$

The energy flux is given by the Poynting vector

$$\Pi = \mathcal{E} \times \frac{\mathcal{B}}{\mu_0}. \tag{6.6}$$

Averaging over a period and using the relation $\epsilon_0 \mu_0 c^2 = 1$ we get the mean energy flux

$$\overline{\Pi} = \frac{1}{2}\epsilon_0 \omega^2 A^2\, cn, \tag{6.7}$$

where n is the refractive index of the medium.

We also know that in the presence of the vector potential (6.3) the Hamiltonian of an electron is

$$\mathcal{H} = \frac{1}{2m}(\mathbf{p} + e\mathbf{A})^2 + V(\mathbf{r}), \tag{6.8}$$

$$\mathcal{H} = \mathcal{H}_0 + \frac{e}{2m}(\mathbf{p} \cdot \mathbf{A} + \mathbf{A} \cdot \mathbf{p}) + \frac{e^2}{2m}\mathbf{A}^2. \tag{6.9}$$

The operator $\mathcal{H}_0$ is the crystal Hamiltonian, whose eigenfunctions are the Bloch states. For light of sufficiently low intensity, the last term in Eq. (6.9) can be neglected and the second term regarded as a small time-dependent perturbation of $\mathcal{H}_0$. We easily verify that

$$[\mathbf{p}, \mathbf{A}(t)] = -i\hbar\nabla \cdot \mathbf{A}(t). \tag{6.10}$$

Now with the choice (6.3), the vector potential is such that $\nabla \cdot \mathbf{A} = 0$, and $\mathbf{p}$ and $\mathbf{A}$ commute, so Eq. (6.9) can be rewritten as

$$\mathcal{H} = \mathcal{H}_0 + \frac{e}{m}\mathbf{A} \cdot \mathbf{p}. \tag{6.11}$$

We thus have a quantum mechanical system subject to a sinusoidal perturbation, because of expression (6.3) for $\mathbf{A}$. Indeed, the Hamiltonian describing the interaction of an electron with a travelling electromagnetic wave has the form

$$\mathcal{H}_p(t) = \mathcal{H}_p \cos(\omega t - \mathbf{K} \cdot \mathbf{r}) \tag{6.12}$$

or

$$\mathcal{H}_p(t) = C \exp(i\omega t) + C^+ \exp(-i\omega t) \tag{6.13}$$

with

$$C^+ = \frac{eA}{2m} \, (\mathbf{a}_0 \cdot \mathbf{p}) \, (\exp i\mathbf{K} \cdot \mathbf{r}). \tag{6.14}$$

As we show in Appendix 6.1, the transition probability from an initial state $|i>$ to a final state $|f>$ of higher energy can be written

$$P_{i \to f}(t) = \frac{|<f|\mathcal{H}_p|i>|^2}{\hbar^2} \frac{\sin^2(1/2)(\omega_{if} - \omega)t}{(\omega_{if} - \omega)^2} \tag{6.15}$$

with

$$\hbar\omega_{if} = (E_f - E_i). \tag{6.16}$$

We note from this formula that **the transition probability between two states depends sinusoidally on the time.** This probability is proportional to the square of the matrix element of the perturbation. Replacing $\mathcal{H}_p$ by its expression we obtain the following expression for the absorption transition probability, which we shall use below:

$$P_{i \to f}(t) = \frac{4 \, |<f|C^+|i>|^2}{\hbar^2} \frac{\sin^2(1/2)(\omega_{if} - \omega)t}{(\omega_{if} - \omega)^2}. \tag{6.17}$$

For the case of optical excitation of a semiconductor, the initial state $|\mathbf{k}_v>$ is in the valence band, and the final state $|\mathbf{k}_c>$ in the conduction band.

6.1a The Fermi Golden Rule

We consider the probability, for a **given** initial state $|i>$, of finding an electron in **one or another** of a set of final states of energy close to E_f. If $n(E_f)$ is the density of these states the **total** probability of finding the electron in any one of these states is

$$P(t) = \frac{4}{\hbar} \int |<f|C^+|i>|^2 \, n \, (E_f) \frac{\sin^2(1/2)(\omega_{if} - \omega)t}{(\omega_{if} - \omega)^2} \, d\omega_{if}. \tag{6.18}$$

In this integral the only non-negligible contribution comes from the region where ω_{if} is very close to ω, because of the denominator $(\omega_{if} - \omega)^2$. The variation of the function $(\Delta\omega)^{-2} \sin^2 \Delta\omega t/2$ versus $\Delta\omega = \omega_{if} - \omega$ is shown in Fig. 6.1. We see that only a narrow frequency bandwidth of order $1/t$ contributes to the integral. We can therefore extend the limits of the integral to $\pm\infty$ and use the fact that both $n(E_f) = n(E_i + \hbar\omega_{if})$ and the matrix element are approximately constant over the interval $\hbar/t$ to write

$$P(t) = \frac{4}{\hbar} | < f|C^+|i > |^2 n\,(E_f) \int_{-\infty}^{+\infty} \frac{\sin^2(1/2)(\omega_{if} - \omega)t}{(\omega_{if} - \omega)^2} \, d\omega_{if} \quad (6.19)$$

or, setting

$$x = \frac{1}{2}(\omega_{if} - \omega)t,$$

$$P(t) = \frac{1}{\hbar} | < f|C^+|i > |^2 n(E_f)\, 2\pi\, t, \quad (6.20)$$

where we have used the fact that

$$\int_{-\infty}^{+\infty} \frac{\sin^2 x}{x^2} = \pi.$$

The result (6.20) is remarkable in that while the probability of transition from a given state to another given state is a sinusoidal function of time (6.17), **the total probability of transition from a given initial state to one or another of a set of final states which are very close to each other is proportional to the time. We may therefore define a transition probability per unit time W:**

$$W = \frac{P(t)}{t} = \frac{2\pi}{\hbar} | < f|C^+|i > |^2 n\,(E_f). \quad (6.21)$$

This equation is called the **Fermi Golden Rule**. In the proof leading to this formula we have not used the nature of the system subject to the radiation. This formula applies in several areas of physics, whenever a system possessing nearby states is illuminated by monochromatic radiation.

The fact that the only significant contributions to the integral (6.18) come from energies such that $\omega_{if} \simeq \omega$ is an expression of conservation of energy for states which were eigenstates of the unperturbed system. This conservation is exact in the limit of weak perturbations and long time scales. The central peak of Fig. 6.1 has height $t^2/4$ and becomes sharper and sharper as its width decreases as t^{-1}. The area under the peak increases linearly in time. In this limit we can write Eq. (6.21) as

$$W = \frac{2\pi}{\hbar} | < f|C^+|i > |^2 \delta\,(E_f - E_i - \hbar\omega) \quad (6.22)$$

and the transition probability is obtained by integration over the final states.

6.1b Selection Rules

Using expression (6.14) for C^+ and replacing $|i>$ and $|f>$ by the corresponding Bloch states, the perturbation matrix element becomes

$$\frac{eA}{2m} \int \exp(-i\mathbf{k}_f \cdot \mathbf{r}) u_{c,\mathbf{k}_f}^*(\mathbf{r}) \exp(i\mathbf{K} \cdot \mathbf{r}) \times$$

$$(\mathbf{a}_0 \cdot \mathbf{p}) \exp(i\mathbf{k}_i \cdot \mathbf{r}) u_{v,\mathbf{k}_i}(\mathbf{r}) d^3\mathbf{r}, \quad (6.23)$$

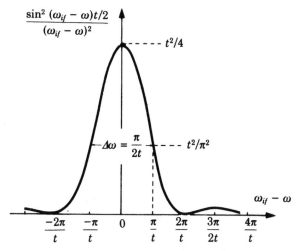

Fig. 6.1. Variation of the function $(\omega_{if} - \omega)^{-2} \sin^2(\omega_{if} - \omega)t/2$.

where $u(\mathbf{r})$ are the periodic parts of the Bloch functions, indexed v for the valence band and c for the conduction band. This integral can be regarded as the matrix element of the periodic operator $\mathbf{a}_0 \cdot \mathbf{p}$ between two functions of Bloch form, one characterized by $\mathbf{k}_f - \mathbf{K}$, the other by $\mathbf{k}_i$. This matrix element vanishes unless, from Appendix 2.1,

$$\mathbf{k}_f = \mathbf{k}_i + \mathbf{K}. \tag{6.24}$$

Now, the light wave vector is very small on the scale of the Brillouin zone, as the wavelength of light with $\hbar\omega \sim E_g$ is around 10^4 times larger than the lattice constant. We may therefore neglect $\mathbf{K}$ in Eq. (6.24) and write

$$\mathbf{k}_i = \mathbf{k}_f \tag{6.25}$$

for allowed transitions.

Transitions of this kind can therefore only occur when the electron wave vector after excitation is approximately equal to the wave vector before excitation. For this reason the transitions are called vertical or direct. They are represented in Fig. 6.2 by the arrow (a). An "oblique" transition such as (b) is forbidden for the above process.

In comparing matrix elements or estimating their order of magnitude, it is useful to remember (see Chap. 2) that the periodic parts of Bloch functions resemble atomic wave functions, and write

$$\psi_{v,\mathbf{k}_i}(\mathbf{r}) = N^{-1/2} u_{v,\mathbf{k}_i}(\mathbf{r}) \exp(i\mathbf{k}_i \cdot \mathbf{r}), \tag{6.26}$$

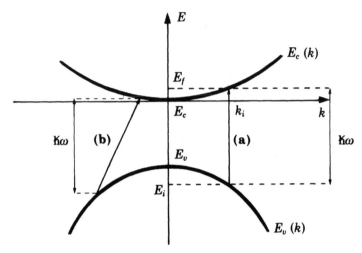

Fig. 6.2. Direct (a) and indirect (b) absorption in a semiconductor.

where N is the number of unit cells of the crystal. With this convention the periodic parts are normalized over the unit cell and have an amplitude comparable to the atomic functions. Only matrix elements with $\mathbf{k}_i = \mathbf{k}_f$ appear, i.e., matrix elements of $\mathbf{p}$ between valence and conduction functions with the form

$$\frac{1}{N} \int_{crystal\ volume} u_{v,\mathbf{k}_i}^*(\mathbf{r})\ \mathbf{p}\ u_{c,\mathbf{k}_i}(\mathbf{r})d^3\mathbf{r}$$

$$= \int_{unit\ cell} u_{v,\mathbf{k}_i}^*(\mathbf{r})\ \mathbf{p}\ u_{c,\mathbf{k}_i}(\mathbf{r})\ d^3\mathbf{r}$$

$$= <\mathbf{p}>_{\mathbf{k}_i}. \tag{6.27}$$

The components of $<\mathbf{p}>_{\mathbf{k}_i}$ will have the same order of magnitude as the matrix elements of $\mathbf{p}$ between two atomic states. Finally the matrix element appearing in the transition probability will be

$$<\mathbf{k}_{f,c}|C^+|\mathbf{k}_{i,v}> = \frac{eA}{2m}\mathbf{a}_0 \cdot <\mathbf{p}>_{\mathbf{k}_i} \delta(\mathbf{k}_i - \mathbf{k}_f). \tag{6.28}$$

To simplify the notation we shall write

$$\mathbf{a}_0 \cdot <\mathbf{p}>_{\mathbf{k}_i} = p_{\mathbf{k}_i}. \tag{6.29}$$

6.1c Calculation of the Absorption Coefficient

The total probability W_T of photon absorption is calculated by summing Eq. (6.22) over all the initial states of the solid (valence band states) and all possible final states (conduction band states):

$$W_T = \sum_{\substack{\mathbf{k}_{i,v} \\ \mathbf{k}_{f,c}}} \frac{2\pi}{\hbar} \frac{e^2 A^2}{4m^2} \, p_{\mathbf{k}_i}^2 \, \delta(\mathbf{k}_i - \mathbf{k}_f) \, \delta(E_f - E_i - \hbar\omega) \tag{6.30}$$

with

$$E_f - E_c = \frac{\hbar^2 k_{f,c}^2}{2m_e}, \tag{6.31}$$

$$E_i - E_v = -\frac{\hbar^2 k_{i,v}^2}{2m_h} \tag{6.32}$$

for a "standard" band structure (Fig. 6.2).

Summing over $\mathbf{k}_i$ we get

$$W_T = \frac{2\pi}{\hbar} \frac{e^2 A^2}{4m^2} \sum_{\mathbf{k}_f} p_{\mathbf{k}_f}^2 \, \delta \left[\frac{\hbar^2 k_f^2}{2} \left(\frac{1}{m_e} + \frac{1}{m_h} \right) + E_g - \hbar\omega \right]. \tag{6.33}$$

We can transform the discrete summation into an integral by using the density of states in $\mathbf{k}$ space. For a sample of volume sdx this density is $sdx/4\pi^3$ so that

$$W_T = \frac{2\pi}{\hbar} \frac{e^2 A^2}{4m^2} \frac{s\,dx}{4\pi^3} \int d^3\mathbf{k}_f \, p_{\mathbf{k}_f}^2 \, \delta \left[\frac{\hbar^2 k_f^2}{2} \left(\frac{1}{m_e} + \frac{1}{m_h} \right) + E_g - \hbar\omega \right]. \tag{6.34}$$

The periodic functions u_v and u_c vary slowly with $\mathbf{k}$ (see for example the $\mathbf{k} \cdot \mathbf{p}$ method, Appendix 2.4), so we can regard $p_{\mathbf{k}_f}^2$ as independent of k and write

$$p_{\mathbf{k}_f}^2 = p^2. \tag{6.35}$$

We define the reduced mass m_r by

$$m_r^{-1} = m_e^{-1} + m_h^{-1}. \tag{6.36}$$

We set $x = (\hbar^2/2m_r)k_f^2 + E_g - \hbar\omega$. It remains to integrate

$$\frac{e^2 A^2 p^2}{8\pi^2 m^2 \hbar} s \, dx \cdot 2\pi \left(\frac{2m_r}{\hbar^2} \right)^{3/2} \int (\hbar\omega - E_g + x)^{1/2} \, \delta\,(x) \, dx. \tag{6.37}$$

We obtain

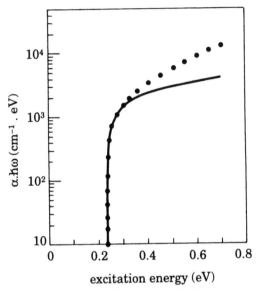

Fig. 6.3. Absorption threshold of InSb at $T = 5$ K. The measurements (points) are compared with the predictions of Eq. (6.40). The deviation at high energy is suppressed by taking account of the correct density of states, which is no longer parabolic at large $\mathbf{k}$, and of the variation of the matrix element $p_{\mathbf{k}_f}$ with $\mathbf{k}_f$.

$$W_T = \frac{s\,dx}{4\pi} \frac{e^2 A^2}{\hbar^4 m^2} (2\,m_r)^{3/2} p^2 (\hbar\omega - E_g)^{1/2}. \tag{6.38}$$

The average incident photon flux is $\overline{\varPi}/\hbar\omega$ per unit area, and by definition the fraction of incident photons absorbed over a depth dx is

$$W_T = \alpha \frac{\overline{\varPi}}{\hbar\omega} s\,dx. \tag{6.39}$$

Substituting Eqs. (6.7) and (6.38) in Eq. (6.39) we get

$$\alpha = \frac{(2m_r)^{3/2}}{2\pi\epsilon_0\,cn} \frac{e^2 p^2}{m^2} \frac{1}{\hbar^3\omega} (\hbar\omega - E_g)^{1/2}. \tag{6.40}$$

This establishes Eq. (2.106) for the case of direct transitions. This result was obtained by J. Bardeen, F.J. Blatt, and L.H. Hall (Atlantic City Photoconductivity Conference, 1954).

Absorption varies from one semiconductor to another through the values m_r, p and the refractive index n. An average order of magnitude is

$$\alpha(\mathrm{cm}^{-1}) = 4.10^4 [(h\nu - E_g)(\mathrm{eV})]^{1/2}. \tag{6.41}$$

For $h\nu = 1.1$ eV and $E_g = 1$ eV we find $\alpha \sim 10^4$ cm^{-1}, showing that a semiconductor is an efficient absorber. It is sometimes used as a filter transparent to long wavelengths.

The experimental dependence of α $(h\nu - E_g)^{1/2}$ can be seen as proof of the existence of direct permitted transitions between band extrema. Figure 6.3 shows the results found for indium antimonide, InSb. We see that near the band gap, which is 0.22 eV for InSb, Eq. (6.40) is well satisfied. But when the initial or final states have energies in the valence or conduction bands comparable with the band gap, the above approximations are no longer sufficient.

6.1d Excitons

Figure 6.4 shows the absorption coefficient of GaAs for photon energies close to E_g ($E_g = 1.52$ eV at low temperature). We note the presence of a peak near E_g, which is not predicted by the calculation of Sect. 6.1c. This results from the creation through photon absorption of **excitons**, whose definition we now give.

The ground state of the semiconductor corresponds to a full valence band with an empty conduction band. According to what we have said, the first excited state should correspond to an electron at the bottom of the conduction band with a hole at the top of the valence band. In this state these two charges are assumed to be far away in real space. We have created a non-interacting electron-hole pair by absorbing a photon of energy $h\nu = E_g$.

In fact the electron, because of its negative charge, feels the Coulomb attraction of the positively charged hole. It is feasible that a bound electron-hole state can exist whose energy will be less than that of the dissociated electron-hole pair. To find the binding energy of this **exciton** pair we solve a "hydrogen-like" problem of two charged particles of effective masses m_e and m_h, which also involves the dielectric constant of the medium (the problem is analogous to that of the binding energy of an electron and a donor, treated in Sect. 3.2). To obtain the motion about the center of mass, we take the reduced mass (6.36), so that the binding energy is

$$E_{\text{exc}} = -\frac{m_r}{m_0} \frac{1}{\epsilon_r^2} \frac{E_1}{n^2}, \qquad (6.42)$$

where $E_1 = -13.6$ eV. As $m_h > m_e$ in general we have $m_r \lesssim m_e$, and E_{exc} is of the order of the donor binding energy (a few meV for the III–V compounds of Sect. 3.5). We thus obtain a hydrogenic series of levels. To energy E_{exc} we must add the kinetic energy of the center of mass, with a mass $m_e + m_h$. The lowest optical excitation energy of the solid will thus not be $h\nu = E_g$, but $h\nu = E_g + E_{\text{exc}}$, and in some cases we get a series of peaks corresponding to different values of n.

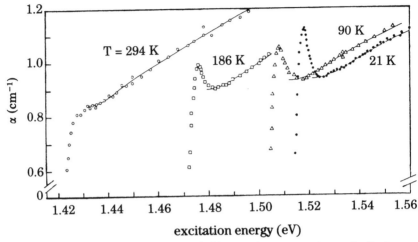

Fig. 6.4. Absorption in GaAs near threshold, at various temperatures. As the temperature increases the band gap decreases, basically because of the thermal expansion of the lattice (Appendix 2.3). The peak is caused by absorption leading to the creation of excitons and is more prominent at low temperature: if kT is of the order of the exciton binding energy, broadening of the peak will become significant.

Figure 6.5 indeed shows the exciton levels $n = 1, 2, 3$ as well as an even lower excitation energy, corresponding to the creation of an exciton bound to a neutral donor $(D^0 - X)$, a system consisting of the donor nucleus, two electrons, and a hole.

It can be shown that the formation of an exciton is possible only for vertical transitions (cf. Fig. 6.2), which occur in most III–V semiconductors. From Eq. (6.42) the binding energy of the exciton is greater, and the exciton thus more readily observable, the larger m_r (and thus m_e) is, and the smaller ϵ_r is. As m_r and E_g are small in the same materials (cf. Sect. 2.4), excitons are more easily observed in semiconductors with large band gaps.

6.2 Recombination

When a semiconductor is illuminated, its properties are modified by the creation of electron-hole pairs, and we study these changes here. We expect qualitatively that as the number of electrons and holes increases the conductivity should rise: this is called photoconductivity. It is easily observable, and is currently used to measure light intensity in photoelectric cells. If a crystal is submitted to continuous irradiation the number of electron-hole pairs should steadily increase in time, and the conductivity would tend to infinity, contrary to experiment. The limiting factor is **the destruction of electron-hole pairs by recombination processes.**

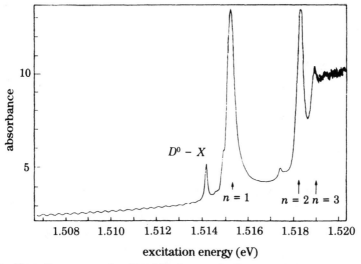

Fig. 6.5. Absorption measured at $T = 6$ K of a sample of pure GaAs containing residual donors. The absorbance is proportional to the absorption coefficient α. Note the peaks corresponding to the levels $n = 1, 2$, and 3 of the exciton, as well as a peak associated with an exciton bound to a neutral donor ($D^0 - X$).

There are other means of establishing electron and hole concentrations departing from thermal equilibrium: injecting excess carriers through a metallic contact, or bombarding the semiconductor with charged particles, which can create excess electron-hole pairs as they decelerate. The latter effect is used to detect particles. In all these cases the return to thermodynamic equilibrium once the external excitation stops must involve the disappearance of the excess electrons and holes. These recombination processes play an important role in semiconductor devices: diodes, transistors, and light-sensitive devices. When light is emitted by the semiconductor we speak of photoluminescence, electroluminescence, thermoluminescence, etc., depending on the type of excitation.

Consider a semiconductor in which we have created deviations Δn and Δp from the equilibrium concentrations. We call $G_0(T)$ the rate of creation of electron-hole pairs through thermal agitation. If we leave the system to evolve freely, the concentrations will return to the equilibrium values n_0, p_0. We may thus express the variation of n and p by the equations

$$n = n_0 + \Delta n, \tag{6.43}$$

$$\frac{dn}{dt} = \frac{d\,\Delta n}{dt} = G_0\,(T) - \frac{n}{\tau_n}, \tag{6.44}$$

where τ_n is a quantity which can vary with p. In equilibrium $d(\Delta n)/dt = 0$, $n = n_0$ and

$$G_0 = \frac{n_0}{\tau_{n_0}}, \tag{6.45}$$

where τ_{n_0} is the value of τ_n for $n = n_0$. Equation (6.45) can thus be written as

$$\frac{d\,\Delta n}{dt} = \frac{n_0}{\tau_{n_0}} - \frac{n}{\tau_n}. \tag{6.46}$$

Similarly we can write

$$\frac{d\,\Delta p}{dt} = \frac{p_0}{\tau_{p_0}} - \frac{p}{\tau_p}. \tag{6.47}$$

Equations (6.46) and (6.47) do no more than introduce the *ad hoc* quantities τ_n, τ_p. These quantities may well depend on n and p and in the general case we can say no more without explicit analysis of the microscopic recombination process. The kinematics of recombination can indeed be complex.

Things are simpler if we consider a **doped** semiconductor subject to moderate light excitation. We consider for example an n-type semiconductor, i.e., one where $n_0 \gg p_0$. If the deviations Δn and Δp are much smaller than n_0, n will change little during the return to equilibrium. We may thus regard τ_p as independent of Δn and write $\tau_p = \tau_{p_0}$. Then

$$\frac{d\,\Delta p}{dt} = -\frac{\Delta p}{\tau_p} \quad \text{with} \quad \Delta p = p - p_0. \tag{6.48}$$

For this reason τ_p is called the lifetime of the minority carriers, here the holes.

The solution of Eq. (6.48) has the form

$$\Delta p = \Delta p_0 \exp(-t/\tau_p) + \text{const} \tag{6.49}$$

and if electrical neutrality holds at all times, $\Delta n = \Delta p$.

We can make this description more precise by considering a particular recombination process: the direct recombination of electron-hole pairs where the recombination involves the simultaneous disappearance of an electron and a hole. The evolution of the electron number is given by the equation

$$\frac{dn}{dt} = \frac{dp}{dt} = -A\,np + G, \tag{6.50}$$

where G is the total generation rate. Indeed the number of electrons recombining per unit time is proportional to the number of electrons present, as each electron has the same probability of recombining. Further this probability is, for an electron, proportional to the number of free holes it may encounter, hence the form of the first term. The factor A depends on the semiconductor but not on n or p. We have added the quantity G which represents the generation rate for pairs.

In thermodynamic equilibrium, i.e., in the absence of excitation, $dn/dt = 0$; $n = n_0$; $p = p_0$. Equation (6.50) gives the thermal pair generation rate as a function of the annihilation rate:

$$G_0(T) = A n_0 p_0 = A n_i^2. \tag{6.51}$$

If there is an additional generation rate g, e.g., by light, Eq. (6.50) allows us to write Eq. (6.51) as

$$\frac{dn}{dt} = \frac{d\,\Delta n}{dt} = -A np + G_0 + g,$$

$$\frac{d\,\Delta n}{dt} = \frac{d\,\Delta p}{dt} = -A(n_0 + \Delta n)(p_0 + \Delta p) + G_0 + p,$$

$$\frac{d\,\Delta n}{dt} = \frac{d\,\Delta p}{dt} = -A p_0\,\Delta n - A(n_0 + \Delta n)\,\Delta p + g. \tag{6.52}$$

In an n-type sample, where $n_0 >> p_0$, and if the injection of carriers is small (Δn and $\Delta p << n_0$) we have

$$\frac{d\,\Delta n}{dt} = \frac{d\,\Delta p}{dt} = -A n_0\,\Delta p + g. \tag{6.53}$$

One defines

$$\frac{1}{\tau_p} = A n_0. \tag{6.54}$$

In the absence of excitation

$$\frac{d\,\Delta n}{dt} = -\frac{\Delta n}{\tau_p}, \tag{6.55}$$

$$\frac{d\,\Delta p}{dt} = -\frac{\Delta p}{\tau_p}. \tag{6.56}$$

The time constant for the disappearance of **electrons** and **holes** is the **same in an n-type semiconductor**, but it is the **lifetime of the minority carriers**. This lifetime decreases as the concentration of majority carriers increases.

Remarks

If g is constant in time, then in a steady state,

$$p = p_0 + g\,\tau_p. \tag{6.57}$$

If at some arbitrary time t_1 the light is switched off then, setting $t_1 = 0$,

$$p(t) = p_0 + g\,\tau_p \exp\left(-\frac{t}{\tau_p}\right). \tag{6.58}$$

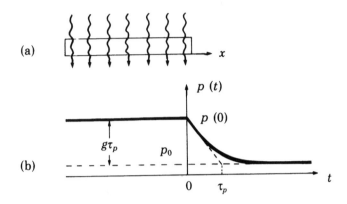

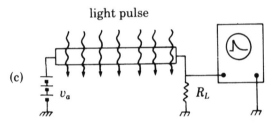

Fig. 6.6. (a) n-type sample subject to continuous irradiation; (b) behavior of the concentration of minority carriers when the irradiation is switched off; (c) measurement of the lifetime. One injects a light pulse of duration short compared with the lifetime. The variation of the resistance of the sample follows that of the number of carriers. (After Sze: "Semiconductor Devices," J. Wiley, 1985.)

This is the principle of the measurement of the lifetime of the minority carriers illustrated in Fig. 6.6.

The result we have obtained may seem paradoxical. **The lifetime of the minority carriers governs the dynamics of recombination.** We can understand this by referring to the recombination probabilities of Eq. (6.53). The recombination probability of an electron, no matter what its origin, is $An_0\Delta p/n_0 = A\Delta p$, and thus depends on Δp. It is therefore not uniquely determined. The recombination probability of an additional electron is $An_0\Delta p/\Delta n = An_0 = \tau_p^{-1}$. This is the inverse lifetime of the minority carriers. The recombination probability of a hole, which must have been created by photoexcitation since we neglected $p_0 \ll \Delta p$, is $An_0\Delta p/\Delta p$, i.e., the same. We should remark that the recombination probability of a hole is large because it sees many electrons.

Of course in a p-type semiconductor the reverse occurs; recombination is governed by the lifetime of the minority carriers which are then the electrons.

Recombination Processes

Recombination processes can be classified into three main categories.

direct recombination

Direct recombination is the reverse of the creation of an electron-hole pair by a photon. We consider the ensemble (photon + crystal). If we neglect the interaction between radiation and matter resulting from the force exerted by the electric field of the light on the electrons, the state (photon + valence-band electron) is degenerate with the state where the photon is absorbed and the electron is in the conduction band. If we leave the system in one of these states it will stay there indefinitely.

This is no longer true if we take account of the interaction between radiation and matter. As we saw in Sect. 6.1 and Appendix 6.1, in the presence of this interaction the state (photon + electron in the valence band) is no longer an eigenstate of the system. If we leave the system in this state at time $t = 0$, then after time t the quantum state will have a non-zero projection on the state (absorbed photon + conduction-band electron). There is thus a finite probability of absorption of the photon and the creation of a pair. Conversely an electron in the conduction band can relax to the valence band by the same effect, with light emission. We call this **radiative recombination**. Appendix 6.2 gives the calculation of the radiative recombination probability. We can also have direct non-radiative recombination via processes involving several electrons. An electron falls back into the valence band, the corresponding energy being given to another conduction electron whose energy in the band increases. This is called Auger recombination.

recombination through traps or deep impurities

In this process recombination occurs in two stages well separated in time. For example a conduction electron may first be captured by an impurity whose level lies deep in the band gap. At a later time this occupied center may capture a hole from the valence band (or, equivalently, emit an electron into the valence band), finally ensuring the recombination of an electron-hole pair. As shown in Appendix 6.2 the probability of a radiative transition decreases very rapidly as the photon energy increases. Performing

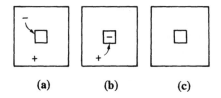

(a)   (b)   (c)

Fig. 6.7. Recombination by trapping: (a) electron capture; (b) hole capture; (c) recombination.

the transition in two stages, thanks to the trapping, can make this process much more probable than the emission of a single photon.

The process is shown schematically in Fig. 6.7: a recombination center alternately captures an electron and a hole and thus "catalyzes" their recombination.

In fact, in the intermediate stage it is possible for an electron to be re-excited into the conduction band before capturing a hole. In this case we speak of **a slow trap**, reserving the term **recombination center** for the case where the capture probability of a hole is larger than the probability of re-emission of the electron towards the conduction band. It appears that the kinematics of recombination by deep centers can be complex.

We give here some orders of magnitude to emphasize the importance of recombination via deep centers: we shall see later that the direct recombination time in germanium is of the order of 1 s. In the purest known crystals the measured time is several milliseconds. It has been shown that the presence of a relative concentration of 10^{-7} copper atoms in germanium reduces this time to 10^{-6} s. The study of recombination processes via deep centers is a very active field of the physics of semiconductors.

Given that most crystalline imperfections (impurities, gaps, dislocations, grain boundaries) can produce states in the band gap, we can see here once again why the manufacture of **pure single crystals** (cf. Sect. 4.9) is so important (both terms are necessary: chemically pure single crystals without crystallographic defects) if one wishes to obtain long lifetimes.

surface recombination

Even a perfectly pure and perfectly regular crystal has an external surface which breaks the periodicity. This break means that not all the chemical bonds of the surface atoms can be satisfied. There are therefore quantum states, localized near the surface, with an energy within the band gap. The presence of surface impurities also contributes to the presence of "surface states" which can be very effective recombination states.

Figure 6.8 illustrates the various recombination processes.

6.3 Photoconductivity and its Applications

6.3a Detection of Electromagnetic Radiation

A direct consequence of the creation of electron-hole pairs is photoconductivity. If there is an excess Δn and Δp of carriers, the photoconductivity of the semiconductor varies as $\Delta\sigma$:

$$\Delta\sigma = \Delta n \, \mu_e \, e + \Delta p \, \mu_h \, e. \qquad (6.59)$$

A commonly used detector of visible light is cadmium sulphide CdS. In CdS the band gap E_g is direct (extrema of the valence and conduction

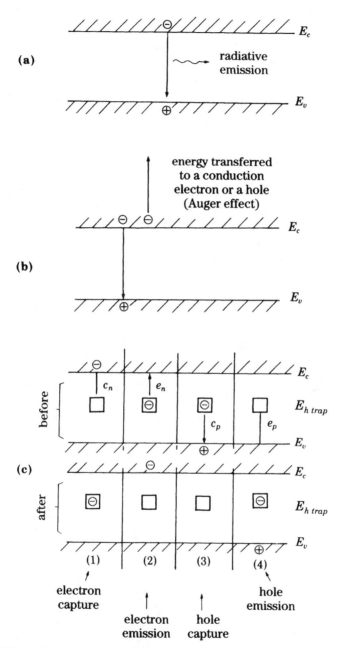

Fig. 6.8. The various recombination processes: (a) direct radiative recombination; (b) direct Auger recombination; (c) recombination via trapping on a deep center. In the latter case the figure shows the state of the system before and after each of the stages (1), (2), (3), (4). (After S.M. Sze, Physics of Semiconductor Devices.)

bands are at $k = 0$) and the width of the gap is 2.4 eV at 300 K. The absorption threshold is therefore 2.4 eV, corresponding to a wavelength of 520 nm, in the green region of the visible spectrum: green and blue are absorbed while longer wavelengths (yellow, red) are transmitted. A CdS crystal thus has a yellow–orange color when seen in transmission.

There are many semiconductors with band gaps less than 1 eV, whose absorption thresholds are in the infrared. For example the three lead salts PbS, PbSe, and PbTe have band gaps of order 0.2 eV. Some alloys $Hg_{1-x}Cd_xTe$ have an E_g of 0.04 eV, which corresponds to $\lambda = 30$ μm, in the far infrared. These materials are used to build infrared detectors.

In a material containing impurities with quantum states in the band gap, light can excite electrons from these localized states into the conduction band. In this case there is light absorption at energies less than E_g and creation of conduction electrons without the creation of free holes. Recombination occurs when the electron falls back to the localized level. In this case one speaks of **extrinsic** absorption and photoconductivity. Germanium doped with gold is used as a sensitive infrared detector between 10 and 20 μm.

6.3b Electrophotography

The xerographic photocopying process is an important application of photoconductivity. The photoconducting material is amorphous selenium, a semiconductor whose band gap is around 2.1 eV at 300 K. This material is deposited by vacuum evaporation in the form of a thin film on a metallic substrate. Selenium has very high resistivity of around 10^{14} Ω.m. The

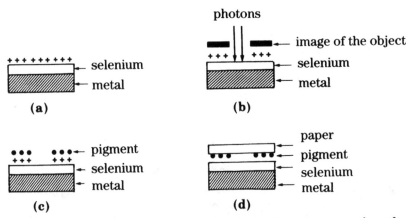

Fig. 6.9. Principle of electrophotography: (a) deposit of positive charges over the surface; (b) recording the image; (c) deposit of pigment; (d) transfer of the pigment to paper. (After Dalven, "Introduction to Applied Solid State Physics," Plenum Press, 1980.)

first stage of electrophotography consists of depositing a positive charge on the semiconductor surface using an electric discharge (corona effect). The charge stays on the surface because of selenium's high resistance. This is shown in Fig. 6.9(a). The second stage consists of sending the luminous image one wishes to record on to the selenium (Fig. 6.9(b)). The illuminated regions become conducting and the charge moves away while the dark parts of the image remain charged. The resistivity of the semiconductor is so high that the charge does not leak laterally. The result of this operation is a distribution of positive charges that replicates the dark regions of the image being copied. We have obtained an **electric image**. The third stage (Fig. 6.9(c)) consists of coating the electric image with black pigment which preferentially fixes itself electrostatically to the charged regions that have not been illuminated in the second stage. The last stage is the transfer of the black pigment to ordinary paper (Fig. 6.9(d)). The pigment is finally fixed to the paper by heating and provides a permanent positive image.

6.3c The Vidicon Tube

The Vidicon tube is the light-sensitive part of the early TV camera. The tube, shown schematically in Fig. 6.10, contains a thin layer of a photoconductor sensitive in the visible — for example As_2S_3 or PbO, for which E_g is, respectively, 2.5 and 2.3 eV. The material must have a very high resistivity $\sim 10^{13}$ $\Omega.m$ in the dark. Photons reach the photoconducting material after crossing a transparent conducting electrode. The photoconductor is continuously scanned from below by an electron beam, which scans its surface sequentially. If the electron beam arrives at a point that is illuminated on the front, the material conducts, and a current proportional to the light intensity at this point will be produced.

We thus obtain an electrical signal containing information about the light received at a given point of the screen. This signal is then processed,

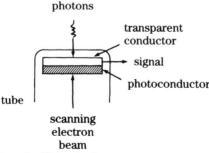

Fig. 6.10. Schematic view of a Vidicon tube.

and then may for example be reproduced on the cathode-ray tube of a TV set.

We note that the production of the image on the cathode-ray tube also uses semiconductors. The screen is covered with powder, usually alloys such as ZnS–ZnSe, which emit visible light when excited by a high-energy beam of electrons. This phenomenon is called cathodoluminescence. ZnS and ZnSe are semiconductors whose band gap is in the ultraviolet or the blue, and which are doped by various elements, generally copper or silver associated with a halogen. These impurities create deep levels which cannot be described using effective mass theory. Luminescence occurs between a delocalized band state and a deep level, or between two deep levels. By judicious choice of the doping agents one can cover the visible spectrum. The three color bands of color television are obtained by using three different powders with adapted doping chemicals.

Image converters, which allow night vision by transforming an infrared image into a visible one, consist of a Vidicon tube followed by a cathode-ray tube.

Appendix 6.1

Quantum System Submitted to a Sinusoidally Varying Perturbation

We are interested in a system (here a semiconductor) submitted to a travelling electromagnetic wave. The system Hamiltonian is

$$\mathcal{H} = \mathcal{H}_0 + \mathcal{H}_1. \tag{6.60}$$

Here $\mathcal{H}_0$ describes the semiconductor in the absence of the electromagnetic field and $\mathcal{H}_1$ its interaction with the field. We assume $\mathcal{H}_1$ small compared with $\mathcal{H}_0$; it has time dependence:

$$\mathcal{H}_1(t) = \mathcal{H}_1 \cos(\omega t - \mathbf{K} \cdot \mathbf{r}), \tag{6.61}$$

where ω is the wave frequency and $\mathbf{K}$ its wave vector.

We assume that at the initial time t_0 the system is in one of the known eigenstates $|n>$ of $\mathcal{H}_0$:

$$\mathcal{H}_0 |n> = E_n |n> . \tag{6.62}$$

As this state is not an eigenstate of $\mathcal{H}$ the system must evolve according to the Schrödinger equation:

$$i\hbar \frac{d}{dt} |\psi(t)> = \mathcal{H} |\psi(t)> . \tag{6.63}$$

We seek the probability of finding the system in the eigenstate $|f>$ at time t(with $t > t_0$). Hence we project the state $|\psi(t)>$ along the basis $|n>$ of eigenstates of $\mathcal{H}_0$:

$$|\psi(t)> = \sum_n \gamma_n(t) \exp(-iE_n t/\hbar)|n> . \tag{6.64}$$

If the Hamiltonian $\mathcal{H}_1$ were time-independent, the γ_n would be constant. Applying the Schrödinger equation to $|\psi(t)>$ we find the relation between the coefficients $\gamma_n(t)$ and their time derivatives $d\gamma_n(t)/dt$:

$$i\hbar \sum_n \frac{d}{dt}\gamma_n(t)\exp(-iE_nt/\hbar)|n>$$

$$+ i\hbar \sum_n \gamma_n(t)\left(-\frac{iE_n}{\hbar}\right)\exp(-iE_nt/\hbar)|n>$$

$$= \sum_n (\mathcal{H}_0 + \mathcal{H}_1)\gamma_n(t)\exp(-iE_nt/\hbar)|n>,$$

which, on using Eq. (6.62), give

$$\sum_n i\hbar\frac{d}{dt}\gamma_n(t)\exp(-iE_nt/\hbar)|n>$$

$$= \sum_n \mathcal{H}_1\gamma_n(t)\exp(-iE_nt/\hbar)|n> . \tag{6.65}$$

Multiplying by $< k|$ we get

$$i\hbar\frac{d}{dt}\gamma_n(t) = \sum_n \gamma_n \exp i(E_k - E_n)t/\hbar < k|\mathcal{H}_1|n> . \tag{6.66}$$

We thus have a set of differential equations to solve, with the initial condition that at $t = t_0 = 0$ the system is in the state $|i>$, i.e., $\gamma_i(t_0) = 1, \gamma_k(t_0) = 0$ for $k \neq i$. The probability $P_{i\to f}$ of finding the system in the final state f at t when it was in the initial state i at $t = 0$ is

$$P_{i\to f}(t) = |< f|\psi(t)>|^2 = |\gamma_f(t)|^2. \tag{6.67}$$

We have assumed $\mathcal{H}_1$ small compared with $\mathcal{H}_0$, so we set

$$\mathcal{H}_1 = \lambda\mathcal{H}_1', \tag{6.68}$$

where λ is a small parameter. We can expand in powers of λ, and solve Eq. (6.66) by equating coefficients of powers of λ:

$$i\hbar\left(\frac{d}{dt}\gamma_k^0(t) + \lambda\frac{d}{dt}\gamma_k^1(t) + \lambda^2\frac{d}{dt}\gamma_k^2(t) + \ldots\right)$$

$$= \sum_n [\gamma_n^0(t) + \lambda\gamma_n^1(t) + \lambda^2\gamma_n^2(t) + \ldots] \times$$

$$\exp[i(E_k - E_n)t/\hbar] < k|\lambda\mathcal{H}_1|n> . \tag{6.69}$$

To order zero:

$$i\hbar\frac{d}{dt}\gamma_k^0(t) = 0. \tag{6.70}$$

To first order:

$$ i\hbar\lambda\frac{d}{dt}\gamma_k^1(t) = \sum_n \gamma_n^0(t)\exp[i(E_k - E_n)t/\hbar] < k|\lambda\mathcal{H}_1'|n >, \tag{6.71}$$

and so on.

Retaining just the first order in λ and using the initial conditions we get

$$\gamma_k^0(t) = \delta_{ik},$$

$$i\hbar\frac{d}{dt}\gamma_k^1(t) = \gamma_i\exp[i(E_k - E_i)t/\hbar] < k|\mathcal{H}_1|i > . \tag{6.72}$$

For the final state $|f>$, such that $\gamma_f(t_0) = 0$,

$$\gamma_f^1(t) = \frac{1}{i\hbar}\int_0^t \exp[i(E_f - E_i)t/\hbar] < f|\mathcal{H}_1|i > dt. \tag{6.73}$$

In this approximation, called the Born approximation, the transition probability from $|i>$ to $|f>$ is

$$P_{i\to f}(t) = |\gamma_f^1(t)|^2. \tag{6.74}$$

We assumed $\gamma_f^1(t)$ small, so $P_{i\to f} << 1$.

We now give the results more explicitly for the interaction of a solid with the electromagnetic field. The interaction Hamiltonian is (cf. Eqs. (6.13) and (6.14)):

$$\mathcal{H}_1 = \frac{eA}{2m}\mathbf{a}_0[\exp i(\omega t - \mathbf{K}\cdot\mathbf{r}) + \exp[-i(\omega t - \mathbf{K}\cdot\mathbf{r})]]\cdot\mathbf{p}, \tag{6.75}$$

so that

$$\mathcal{H}_1 = C\exp i\omega t + C^+\exp(-i\omega t) \tag{6.76}$$

with

$$C^+ = \frac{eA\exp i\mathbf{K}\cdot\mathbf{r}}{2m}\mathbf{a}_0\cdot\mathbf{p}. \tag{6.77}$$

We recall that with the gauge Eq. (6.3) $\mathbf{A}$ and $\mathbf{p}$ commute. We set

$$E_f - E_i = \hbar\omega_{if} > 0 \tag{6.78}$$

as we are interested in **photon absorption** from the initial state. Expression (6.73) for $\gamma_f^1(t)$ is then

$$\gamma_f^1(t) = \frac{1}{i\hbar}\int_0^t \exp i\omega_{if}t[< f|C\exp i\omega t|i > + < f|C^+\exp(-i\omega t)|i >]\, dt$$

$$= \frac{1}{i\hbar}\Bigg[< f|C|i > \frac{\exp i(\omega_{if} + \omega)t - 1}{i(\omega_{if} + \omega)} +$$

$$< f|C^+|i > \frac{\exp i(\omega_{if} - \omega)t - 1}{i(\omega_{if} - \omega)}\Bigg]. \tag{6.79}$$

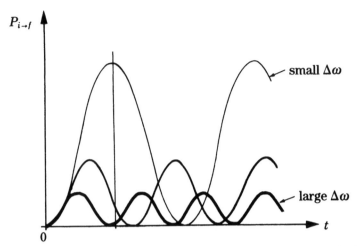

Fig. 6.11. Variation of the absorption transition probability $P_{i\to f}$ versus time for different values of $\Delta\omega = \omega_{if} - \omega$.

We consider the neighborhood of $\omega = \omega_{if}$: The first term is of the order of ω^{-1}, the period of the electromagnetic wave, while the second is of the order of t, much longer than the period of the wave. The first term is therefore negligible compared with the second, so that

$$P_{i\to f}(t) = \frac{4}{\hbar^2} |<f|C^+|i>|^2 \frac{\sin^2[(\omega_{if} - \omega)t/2]}{(\omega_{if} - \omega)^2} \tag{6.80}$$

and

$$P_{i\to f}(t) =$$
$$\left(\frac{eA}{2\hbar m}\right)^2 |<f|\exp(i\mathbf{K}\cdot\mathbf{r})\mathbf{a_0}\cdot\mathbf{p}|i>|^2 \frac{\sin^2[(\omega_{if} - \omega)t/2]}{[(\omega_{if} - \omega)/2]^2}. \tag{6.81}$$

The variation of $P_{i\to f}(t)$ as a function of time for several values of $\Delta\omega = \omega_{if} - \omega$ is shown in Fig. 6.11. $P_{i\to f}$ is a sinusoidal function of time whose peak value is proportional to $1/(\Delta\omega)^2$. For very small t ($t\Delta\omega << 1$ for all $\Delta\omega$) all the curves coincide, and grow as t^2, getting out of phase for larger t. Conversely $P_{i\to f}$ can be regarded as a function of ω with a resonance of width $\Delta\omega = \pi/2t$ around $\omega = \omega_{if}$ (cf. Fig. 6.1).

To calculate the **emission** probability from an excited state we have to distinguish between induced and spontaneous emission. The calculation of induced emission is symmetrical with the one we have just performed for absorption: one simply reverses the roles of $|i>$ and $|f>$. The calculation of spontaneous emission is complex but involves the same matrix element as for absorption, so that the same selection rules do apply.

Appendix 6.2

Calculation of the Radiative Recombination Probability

As we have seen in Sects. 6.1 and 6.2, there is a close connection between light absorption by a semiconductor and radiative recombination, as both involve the same quantum states and the same electromagnetic interaction.

We show here that if the absorption coefficient is known one can deduce the radiative lifetime. The basis of the proof consists of noting that in thermal equilibrium the rate of creation of electron-hole pairs is equal to the recombination rate. Using the **detailed balance** of these processes for each frequency interval $\nu, \nu+d\nu$ we can write that the emission rate $R(\nu)d\nu$ is equal to the pair generation rate

$$R(\nu)\,d\nu = P(\nu)\rho(\nu)d\nu, \tag{6.82}$$

where $P(\nu)$ is the probability per unit time of absorbing a photon of energy $h\nu$ and creating a pair, and $\rho(\nu)d\nu$ is the photon density between frequencies ν and $\nu+d\nu$ for a unit volume. In thermal equilibrium this is given by the Planck blackbody law modified to account for the refractive index n of the medium:

$$\rho(\nu)d\nu = \frac{8\pi\,\nu^2\,n^2}{c^3}\,\frac{1}{\exp\left(h\nu/kT\right)-1}\,\frac{d(n\nu)}{d\nu}d\nu. \tag{6.83}$$

This formula is easily proved by noting that the density of photon states in the volume Ω, $(\Omega/8\pi^3)4\pi k^2 dk$, multiplied by 2 (for the 2 polarization states of light), should be multiplied by the Bose factor $[\exp(h\nu/kT)-1]^{-1}$. As the wave vector k is $2\pi n\nu/c$, the number of occupied states between ν and $\nu+d\nu$ is

$$2\times\frac{\Omega}{8\pi^3}4\pi\left(2\pi n\frac{\nu}{c}\right)^2 d\left(2\pi n\frac{\nu}{c}\right)\times\frac{1}{\exp\left(h\nu/kT\right)-1} \tag{6.84}$$

giving Eq. (6.83).

We also have to relate the probability $P(\nu)$ per unit time of absorbing a photon of energy $h\nu$ to the light-absorption coefficient $\alpha(\nu)$. The electric field of a wave propagating a distance x into a medium of refractive index n and absorption coefficient α is

$$\mathcal{E} = \mathcal{E}_0 \exp i\,(2\pi\,\nu t - kx)\,\exp(-\alpha\,x/2). \tag{6.85}$$

Here α and n depend on ν. The travelling energy is proportional to

$$I(x) = \overline{(\mathrm{Re}\,\mathcal{E})^2} \simeq \frac{\mathcal{E}_0^2}{2}\exp(-\alpha x). \tag{6.86}$$

The quantity α^{-1}, the inverse of the absorption coefficient, can be regarded as the mean free path of a photon of energy $h\nu$ in the medium before it produces a band-to-band transition at this energy. In fact $\alpha\,dx$ is the probability that a photon is absorbed between x and $x + dx$. The change of intensity over dx is

$$dI = I(x + dx) - I(x) = -I(x)\alpha\,dx \tag{6.87}$$

and integration of this equation gives just the form (6.86). The probability that a photon **will not be absorbed** over the distance x is thus $\exp(-\alpha x)$ and the probability that it is absorbed **exactly** between x and $x+dx$ is then

$$\alpha\,dx\exp(-\alpha\,x). \tag{6.88}$$

The photon mean free path is then

$$\int_0^\infty dx\;\alpha\,x\;\exp(-\alpha\,x) = \alpha^{-1}. \tag{6.89}$$

The mean lifetime $\tau(\nu)$ of such a photon is thus $\alpha^{-1}v_g^{-1}$ where v_g is the wave group velocity in the medium

$$v_g = \frac{d\omega}{dk} = c\frac{d\nu}{d\,n\nu}, \tag{6.90}$$

finally giving the inverse photon lifetime

$$\alpha\,v_g = \frac{1}{\tau(\nu)} = \alpha\,c\frac{d\nu}{d\,n\nu}. \tag{6.91}$$

The probability $P(\nu)$ per unit time of absorbing a photon is the inverse of the lifetime:

$$P(\nu) = \frac{1}{\tau(\nu)}, \tag{6.92}$$

so, substituting Eqs. (6.83), (6.91), and (6.92) into Eq. (6.82),

$$R(\nu)\,d\nu = \frac{8\pi}{c^2}\frac{\nu^2 n^2\alpha}{\exp\,(h\nu/kT) - 1}d\nu. \tag{6.93}$$

This is called the Van Roosbroeck–Shockley relation [Physical Review **94** 1558, (1954)]. It is the fundamental relation between the expected emission spectrum and the absorption spectrum. The total number of recombinations per unit volume and per second is obtained by integrating over frequency. Setting $u = h\nu/kT$:

$$R = \int_0^\infty R_\nu \, d\nu = \frac{8\pi\, n^2}{c^2} \left(\frac{kT}{h}\right)^3 \int_0^\infty \frac{n^2(\nu)\alpha(\nu)u^2}{e^u - 1} du. \tag{6.94}$$

We note in Eq. (6.93) that if $\alpha(\nu)$ vanishes, corresponding to $h\nu < E_g$, the emission will also vanish at frequency ν. Formula (6.93) allows us to transform the absorption spectrum of a semiconductor of known refractive index into its emission spectrum. The integral in Eq. (6.94) is negligible except over a very narrow frequency band of width kT/h near the fundamental absorption because of the factor e^u in the denominator and because $\alpha = 0$ for $h\nu < E_g$. We see that the light emission by a semiconductor is relatively monochromatic. This is the reason why the electroluminescent diodes used in control panel lights or in fiber optic telecommunications emit well defined colors.

Although the calculation above was for band-to-band transitions, it also holds for transitions between any pairs of states whatsoever. We see from Eq. (6.93) that if we consider a transition from the conduction band to a deep level in the band gap, the first stage of recombination through impurity centers, the factor $\exp(h\nu/kT)$ is much smaller, as $h\nu \sim E_g$ for the band–band transition is replaced by $h\nu \sim E_g/2$ for recombination by centers; hence the probability of the latter process is greatly enhanced.

In thermal equilibrium (Eq. (6.51))

$$R = G_0(T) = A\, n_i^2. \tag{6.95}$$

The radiative lifetime defined by Eq. (6.54) is then given for n-type material by

$$\frac{1}{\tau_p} = \frac{R}{n_i^2} n_0 = \frac{R}{p_0}. \tag{6.96}$$

In an intrinsic material with $\Delta p = \Delta n << n_i$ we have from Eq. (6.53),

$$\frac{d\,\Delta n}{dt} = -2A\, n_i\, \Delta n + g \tag{6.97}$$

and the intrinsic lifetime is defined by

$$\frac{1}{\tau_i} = \frac{2R}{n_i}. \tag{6.98}$$

The following table gives the values of the lifetimes calculated from Eqs. (6.94), (6.96), and (6.98).

Radiative Recombination at 300 K

Material	E_g (eV)	n_i ($\times 10^{20}\,\mathrm{m}^{-3}$)	τ (intrinsic)	τ for $10^{23}\,\mathrm{m}^{-3}$ majority carriers (μs)
Si	1.08	0.00015	4.6 h	2500
Ge	0.66	0.24	0.61 s	150
GaSb	0.71	0.043	0.009 s	0.37
InAs	0.31	16	15 μs	0.24
InSb	0.18	200	0.62 μs	0.12
PbS	0.41	7.1	15 μs	0.21
PbTe	0.32	40	2.4 μs	0.19
PbSe	0.29	62	2.0 μs	0.25
GaP	2.25			3000

R. N. Hall, Proc. Institution of Electrical Engineering 106B Suppl. No. *17, 923* (1959).

We observe the wide range of recombination times for different semi-conductors.

Appendix 6.3

Semiconducting Clusters for Non-Linear Optics

In recent years much interest has been focussed on "optronics," a new technology in which electronics would be replaced by optics (or electron current by photon flux). This needs non-linear optical devices using non-linear optical materials. In such a material the absorbed radiation power is not a linear function of the incident power. This can be realized in particular with small clusters of semiconducting materials.

For usual semiconducting samples or devices of macroscopic size, the size effects are negligible. Indeed the quantum confinement effect (as measured by the energy of the lowest state in a finite cubic box) of a microelectronic transistor of typical size of $L = 3$ μm corresponds to an energy shift $\delta E = (\hbar^2/2m)(k_x^2 + k_y^2 + k_z^2) = 3\hbar^2/2mL^2 = 10^{-8}$ eV. This energy is very small compared to the thermal energy and plays no role. Even if m is replaced by the effective mass m^* of order $m/10$ this confinement energy is small. However if one considers very small clusters of size ranging from 1 to 10 nm the confinement effect becomes large and can be observed. Using the effective mass approximation one can write the energy of the lowest state corresponding to the conduction band as

$$E_{c,\text{cluster}} = E_{c,\text{bulk}} + 3\hbar^2/2m_e L^2. \tag{6.99}$$

In the same manner one can write the highest energy corresponding to the valence band as

$$E_{v,\text{cluster}} = E_{v,\text{bulk}} - 3\hbar^2/2m_h L^2. \tag{6.100}$$

In consequence the new energy for the gap is

$$E_{g,\text{cluster}} = E_{g,\text{bulk}} + (3\hbar^2/2L^2)(m_e^{-1} + m_h^{-1}). \tag{6.101}$$

This increase in energy may be partially compensated by excitonic effects that we neglect here for the sake of simplicity (see Chap. 6.1d for excitonic effects).

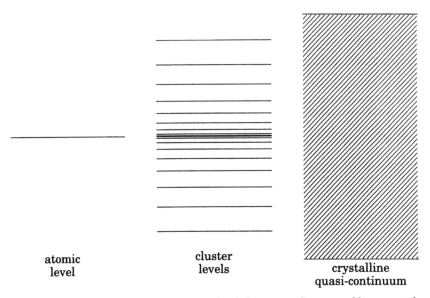

atomic
level

cluster
levels

crystalline
quasi-continuum

Fig. 6.12. Comparative scheme of the energy levels for atoms, clusters, and large crystals. In large crystals, the distance between energy levels is so small that one speaks of a quasi-continuum.

The next effect is that the splitting between conduction (or valence) levels is increased drastically for the same reason. There no longer exists a quasi-continuum of states like in a large crystal but a finite number of discrete levels. This is schematized in Fig. 6.12. Finally the energy level scheme near the new gap is shown in Fig. 6.13.

The non-linear absorption of these small clusters is related to the fact that their optical absorption is very large because the "oscillator strength" is spread over a limited number of states. This means that the absorption probability for photons is larger than in crystals of macroscopic size. This can be understood from Eq. (6.21). The absorption probability for photons corresponding to the excitation from a single level to a single final level of width ΔE is proportional to the density of final states $n(E_f) = 1/\Delta E$. In a bulk semiconductor the term which corresponds to ΔE is of the order of the bandwidth (a few eV) whereas in the cluster the width of the discrete levels is typically 10 to 100 hundred times smaller. Correspondingly the absorption probability is very large.

The absorption spectra of glasses containing clusters of $CdS_{1-x}Se_x$ is shown in Fig. 6.14. The smaller the size of the cluster, the higher the absorption energy. Because of the enhanced absorption, it is possible to "saturate" the optical transition, that is to equalize the populations of the fundamental and excited states, using moderate radiation power. Here the fundamental

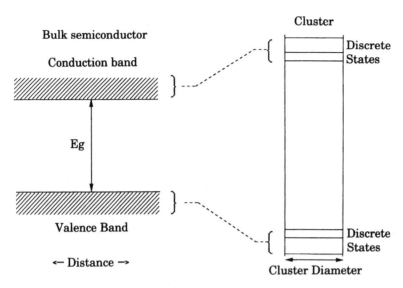

Fig. 6.13. Energy levels near the forbidden band gap for bulk and small clusters of semiconductors. The increase in the distance between valence states and excited levels is a consequence of quantum confinement.

state corresponds to a full valence band and the excited state corresponds to the occupation of the lower state in the conduction band. Of course saturation can be achieved if the recombination is slow enough or if the intensity of the laser irradiation (pump laser) is large enough. Once saturation exists it is no longer possible to absorb photons of the same wavelength because the number of electrons excited by light from the valence to the conduction band is equal to the number of electrons which transit from a state of the conduction band to an empty state of the valence band under the light illumination. This phenomenon, which is called "stimulated emission," is due to the fact that the transition probability from i to f, $P_{i,f}$ that one can compute from Eqs. (6.79)–(6.81) is equal to the transition probability from f to i: $P_{f,i}$. In this saturation situation, the system has become non-linear. There exists a "self-induced transparency" because no more light can be absorbed at that wavelength.

Another illustration of quantum confinement is the recent observation of light emission in the visible range by porous silicon. The band gap of ordinary crystalline silicon is situated in the infrared range and for this reason silicon cannot be used as a light-emitting material. Porous silicon is usually obtained by electropolishing of silicon in aqueous hydrofluoric acid and has been shown recently (1990) to emit light in the visible range. This fact (emission of light at an energy larger than the gap energy of bulk silicon) is explained by the fact that this material which is highly porous (50% porosity) is made of an irregular columnar structure with a typical

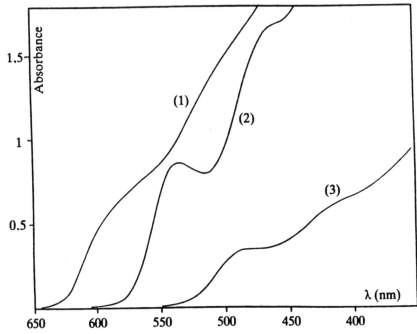

Fig. 6.14. Absorption spectra of three glasses containing small clusters of $CdS_{1-x}Se_x$. The cluster diameter is 12 nm for (1), 5 nm for (2), and 2.5 nm for (3). The cluster diameter can be measured by electron microscopy. The quantum confinement effect due to the size reduction from (1) to (3) creates a "blue shift" of the absorption threshold. The apparition of structure is a consequence of the fact that the distance between energy levels is also increased.

size of a few nanometers. Also, in this material the quantum confinement is the essential mechanism for the increase of the effective band gap, as schematized in Fig. 6.13.

7.

Carrier Injection by Light

When a semiconductor is subject to a flux of radiation of energy exceeding that of the band gap the rate of creation of electron-hole pairs varies spatially as light is progressively absorbed. Under these conditions there appear diffusion currents due to the non-equilibrium injection of carriers by light.

We shall see that the system rearranges itself in order to maintain electrical neutrality. The excess carriers created near the surface redistribute themselves over a finite distance, the diffusion length, determined by the diffusion coefficient and the recombination time of the minority carriers. The concepts introduced in this chapter will be useful for the study of more complex systems, such as inhomogeneous semiconductors, which are the building blocks of solid state electronic devices: junctions and transistors. These devices are studied in Chaps. 8–10.

7.1 Basic Equations for Semiconductor Devices

We recall here the current Eqs. (5.31) and (5.32) involving the drift and diffusion currents:

$$\mathbf{J}_e = ne\mu_e \, \boldsymbol{\mathcal{E}} + eD_e\nabla n, \tag{7.1}$$

$$\mathbf{J}_h = pe\mu_h\boldsymbol{\mathcal{E}} - eD_h\nabla p, \tag{7.2}$$

$$\mathbf{J} = \mathbf{J}_e + \mathbf{J}_h. \tag{7.3}$$

The continuity equations, which express the particle number conservation, have to include recombination terms. These have to be considered for each type of particle. For the minority carriers, n_p electrons in p-type material, p_n holes in n-type material:

$$\frac{\partial n_p}{\partial t} = g_n - \frac{n_p - n_p^0}{\tau_n} + \frac{\nabla \cdot \mathbf{J}_e}{e}, \tag{7.4}$$

$$\frac{\partial p_n}{\partial t} = g_p - \frac{p_n - p_n^0}{\tau_p} - \frac{\nabla \cdot \mathbf{J}_h}{e}, \tag{7.5}$$

where g_n and g_p are the creation rates arising from external excitations, and n_p^0, p_n^0 the equilibrium concentrations of minority carriers. If the system is electrically neutral (a question discussed in Sect. 7.2 and Appendix 7.1), the concentration of majority carriers is obtained from the neutrality relation

$$n_n - n_n^0 = p_n - p_n^0 \tag{7.6}$$

and an analogous relation in p-type material:

$$n_p - n_p^0 = p_p - p_p^0. \tag{7.7}$$

7.2 Charge Neutrality

Consider a homogeneously doped semiconductor in which one has created carrier concentrations $n_0 + \Delta n$ and $p_0 + \Delta p$. If Δn differs from Δp there will be a space charge of density

$$\rho = e(\Delta p - \Delta n). \tag{7.8}$$

We wish to show that in fact $\Delta n = \Delta p$ except in the presence of a very strong field. The charge and electric field are related by Poisson's equation

$$\nabla \cdot \boldsymbol{\mathcal{E}} = \frac{\rho}{\epsilon} = \frac{e\,(\Delta p - \Delta n)}{\epsilon_0\,\epsilon_r}. \tag{7.9}$$

We take $p = 10^{18}\mathrm{m}^{-3}$, $\Delta p - \Delta n = 10^{-2}p$, and $\epsilon_r \sim 10$, $\epsilon_0 \simeq 10^{-11}$ F.m^{-1}. Then $\nabla \cdot \boldsymbol{\mathcal{E}}$ is of the order of 1.6×10^7 V $\cdot$ m^{-2} and the electric field produced by the charges in a slab of thickness d is $1.6 \times 10^7 d$ (V $\cdot$ m^{-1}). Over a 1 cm slab, $\mathcal{E} = 1.6 \times 10^5$ V $\cdot$ m^{-1}. For ordinary fields we will thus have charge neutrality, or more precisely **charge quasi-neutrality**, even in the presence of fields and currents.

It is also interesting to consider the time variation of the charge density ρ in the simple case where we neglect recombination and diffusion. The usual charge conservation law then applies:

$$\nabla \cdot \mathbf{J} = \nabla \cdot \sigma \boldsymbol{\mathcal{E}} = -\frac{d\rho}{dt} \tag{7.10}$$

or using Eq. (7.8),

$$\frac{d\,(\Delta n - \Delta p)}{dt} = -\frac{\sigma}{\epsilon_0\,\epsilon_r}(\Delta n - \Delta p). \tag{7.11}$$

If at time $t = 0$ there is a charge imbalance $(\Delta n - \Delta p)_0$, this imbalance decreases according to

$$(\Delta n - \Delta p) = (\Delta n - \Delta p)_0 \ \exp(-t/\tau_0), \tag{7.12}$$

where

$$\tau_0 = \frac{\epsilon_0 \ \epsilon_r}{\sigma} \tag{7.13}$$

is a characteristic time for the medium called the **dielectric relaxation time**. This is the time required for the charges to screen the perturbation. This time is very short, even in a semiconductor where the conductivity is lower than in a metal. In a silicon crystal with $\sigma = 100 \ \Omega \cdot m$ and $\epsilon_0 \epsilon_r \sim 10^{-10}$ $F \cdot m^{-1}$ we get $\tau_0 \sim 10^{-12}$ s. If a space charge is created it will therefore disappear very quickly.

We may therefore assume that in homogeneously doped materials under normal conditions **charge neutrality holds even in the presence of external excitation. This is nevertheless only an approximation, and we should really refer to it as charge quasi-neutrality.** We give a more complete treatment including diffusion in Appendix 7.1, which shows that the approximation is excellent.

We shall see in Chap. 8 that in materials with strongly inhomogeneous doping there can exist large electric fields and large net charge densities.

7.3 Injection or Extraction of Minority Carriers

We can now discuss the electron and hole currents in a semiconductor where n or p depart from their equilibrium values. We shall consider one-dimensional motion in the x direction resulting from concentration gradients along x. We consider only extrinsic materials, e.g., n type, and as an example we study the effect of irradiation by light at the end of a semiconducting bar (Fig. 7.1). If the photon energy is high enough (3 or 4 eV) the absorption coefficient is about 10^6 cm^{-1}. This means that the light intensity is multiplied by $1/e$ over 10 nm, and that electron-hole pairs are created at the surface only. Appendix 7.2 gives a more detailed treatment of this phenomenon. In steady state there is a concentration gradient near the surface.

The equations governing the phenomenon are the current equation (7.2) and the charge conservation equation (7.5). In the bulk of the material the generation rate g_p is zero and if there is no electric field in the system, substitution of Eq. (7.2) in Eq. (7.5) gives

$$\frac{\partial p_n}{\partial t} = -\frac{p_n - p_n^0}{\tau_p} + D_h \frac{\partial^2 p_n}{\partial x^2}. \tag{7.14}$$

In steady state we have

$$0 = -\frac{p_n - p_n^0}{\tau_p} + D_h \frac{\partial^2 p_n}{\partial x^2}. \tag{7.15}$$

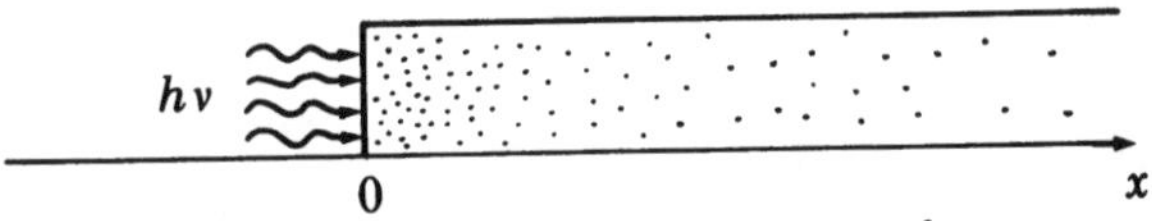

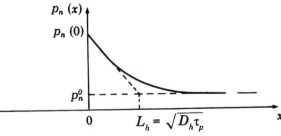

Fig. 7.1. n-type semiconductor crystal illuminated at one end.

The boundary conditions are $p_n(x = 0) = p_n^0 + \Delta p_x(0)$ and $p_n(x \rightarrow \infty) = p_n^0$. The solution of Eq. (7.15) is

$$p_n(x) = p_n^0 + \Delta p_n(0)\exp(-x/L_h) \tag{7.16}$$

with

$$L_h = \sqrt{D_h\,\tau_p}. \tag{7.17}$$

The length L_h is called the **minority carrier diffusion length**.

For the irradiation of Fig. 7.1 the diffusion profile is shown in Fig. 7.2. The length L_h characterizing the injection effects is relatively large. With $D_h = 25 \times 10^{-4}$ m^2/s and $\tau_p = 10^{-6}$ s we get $L_h = 5 \times 10^{-5}$ m. As there exists a hole concentration gradient there will be a diffusion current depending on x:

$$J_h(x) = -D_h\,e\frac{\partial\,\Delta p}{\partial x}. \tag{7.18}$$

At the surface

$$J_h(x = 0) = e\frac{D_h}{L_h}\Delta p_n(0). \tag{7.19}$$

Expression (7.19) is needed for an understanding of the p–n junction and the junction transistor. In the present example the "diffusion velocity" D_h/L_h is 50 m·s^{-1}. The electric field $\mathcal{E}$ which would give the same value of drift velocity is given by $\mathcal{E} \sim D_h/(L_h\mu_h) \sim 500$ V·m^{-1} for silicon. This large value shows that the transport of minority carriers may in some cases be essentially a diffusion current. This discussion is taken up more thoroughly in Appendix 7.1. The electron concentration is found from charge quasi-neutrality (7.6). In short: light creates a packet of excess holes, minority carriers, at $x = 0$. The majority carriers (electrons) surround them and screen them, so as to maintain charge quasi-neutrality in the sample.

Fig. 7.2. Hole concentration profile.

Of course we get an equivalent situation for the injection of electrons in p-type material. Conversely we shall show in the next chapter that in an inhomogeneously doped semiconductor, where this screening cannot occur, charge neutrality cannot be enforced everywhere.

Appendix 7.1

Charge Quasi-Neutrality

We give here a simple example showing that the surface injection of carriers leads not only to a diffusion current of the minority carriers, but also to the existence of an electric field which appears spontaneously in the system. However this field is very small, and we can show that the deviation from charge neutrality remains very small.

We take the geometry of Fig. 7.1. We assume that irradiation creates a departure from equilibrium at the surface of n-type material. For simplicity in the following we suppress the index n on the concentrations.

We start from the charge conservation equation and the expression for the current, which are (cf. Sect. 7.1)

$$\frac{\partial p}{\partial t} = -\frac{p - p_0}{\tau_p} - \frac{\nabla \cdot \mathbf{J}}{e}, \tag{7.20}$$

$$\mathbf{J}_h = pe\mu_h \, \boldsymbol{\mathcal{E}} - eD_h\nabla p, \tag{7.21}$$

which in one dimension gives

$$\frac{\partial p}{\partial t} = -\frac{p - p_0}{\tau_p} - \mu_h \, \mathcal{E}\frac{\partial p}{\partial x} + \mu_h \, p\frac{\partial \mathcal{E}}{\partial x} + D_h\frac{\partial^2 p}{\partial x^2}. \tag{7.22}$$

Up to now we have omitted terms containing the electric field under the assumption of strict charge neutrality. We thus found the stationary solution (7.16):

$$p(x) = p + \Delta p(0)\exp(-x/L_h) \tag{7.23}$$

and

$$J_h(x) = -e \, D_h\frac{\partial p}{\partial x} = e\frac{D_h}{L_h}\Delta p(0)\exp(-x/L_h). \tag{7.24}$$

We consider the total current $\mathbf{J} = \mathbf{J}_e + \mathbf{J}_h$. From Eqs. (7.1) and (7.2):

$$\mathbf{J} = \mathbf{J}_e + \mathbf{J}_h$$

$$= n\,e\,\mu_e\,\boldsymbol{\mathcal{E}} + p\,e\,\mu_h\,\boldsymbol{\mathcal{E}} + e\,D_e\,\nabla n - e\,D_h\,\nabla p. \tag{7.25}$$

In steady state

$$\nabla \cdot \mathbf{J} = -\frac{d\rho}{dt} = 0 \tag{7.26}$$

and $\mathbf{J}$ is constant along the bar. But the current vanishes far from the surface, and therefore must vanish everywhere, so that from Eq. (7.25):

$$(n\,e\,\mu_e + p\,e\,\mu_h)\,\boldsymbol{\mathcal{E}} = e\,D_h\,\nabla p - e\,D_e\,\nabla n. \tag{7.27}$$

Equation (7.27), which is rigorous, shows that we can have strict charge neutrality—i.e., $\nabla n = \nabla p$ at each point, and thus $\boldsymbol{\mathcal{E}} = 0$ since in this case there is no charge in the system—only if the electron and hole diffusion coefficients are equal. The physical reason is that electrons and holes are created in equal numbers at the surface, but as the diffusion coefficients are different the electrons and holes diffuse differently, leading to the appearance of a space charge. This charge creates the field $\boldsymbol{\mathcal{E}}$ of Eq. (7.27) and thus allows the drift current on the left-hand side to compensate the differences between the diffusion currents.

To evaluate the order of magnitude of the electric field $\boldsymbol{\mathcal{E}}$ we note that we can neglect the hole drift current in Eq. (7.27), as there are very few holes in n-type material. We then have

$$\boldsymbol{\mathcal{E}} = \frac{1}{n\,\mu_e}(D_h\,\nabla p - D_e\,\nabla n). \tag{7.28}$$

We assume that charge neutrality holds to a first approximation, so that $\nabla n = \nabla p$ and

$$\boldsymbol{\mathcal{E}} \sim \frac{D_h}{n\,\mu_e}\left(1 - \frac{D_e}{D_h}\right)\nabla p. \tag{7.29}$$

We set

$$\frac{D_e}{D_h} = \frac{\mu_e}{\mu_h} = b \tag{7.30}$$

with b a parameter of order unity. The charge required to create the field $\boldsymbol{\mathcal{E}}$ is obtained from Poisson's equation:

$$\rho(x) = \epsilon_0\,\epsilon_r \nabla \cdot \boldsymbol{\mathcal{E}} = \frac{\epsilon_0\,\epsilon_r}{n\,\mu_e}D_h(1-b)\frac{\Delta p\,(0)}{L_h^2}\exp(-x/L_h). \tag{7.31}$$

To obtain Eq. (7.31) we used Eq. (7.29) and the approximate solution (7.23) of the systems (7.20) and (7.21). At $x = 0$, using $L_h^2 = D_h\tau_p$:

$$\rho(0) = \frac{\epsilon_0\,\epsilon_r}{n\,\mu_e}(1-b)\frac{\Delta p\,(0)}{\tau_p}. \tag{7.32}$$

The charge density ρ is the deviation from strict electrical neutrality:

$$\rho = (\Delta p - \Delta n)e \tag{7.33}$$

or

$$(\Delta p - \Delta n)_{x=0} = \frac{\epsilon_0\,\epsilon_r}{e\,n\,\mu_e}(1-b)\frac{\Delta p(0)}{\tau_p}, \tag{7.34}$$

which using Eq. (7.13) becomes

$$\frac{(\Delta p - \Delta n)_{x=0}}{\Delta p(0)} = \frac{\tau_0}{\tau_p}(1-b). \tag{7.35}$$

This value is very small, as the dielectric relaxation time is always very short compared with the lifetime τ_p: for example $\tau_0 \sim 10^{-12}$ s while $\tau_p \sim 10^{-6}$ s. The deviation from strict equality of the electron and hole concentrations is in this case only 10^{-6} times the deviation of the minority carriers from the equilibrium concentration. It is of course exactly zero if $b = 0$, i.e., if $D_h = D_e$ because then the electrons and holes diffuse at precisely the same speed.

Appendix 7.2

Problems on Photoexcitation, Recombination, and Photoconductivity

We consider a semi-infinite sample of n-type silicon with electron concentration n_0, steadily illuminated by light of wavelength λ or energy $h\nu$ and intensity I_0 (Fig. 7.3). The wavelength considered here is 500 nm (blue–green), with a corresponding absorption coefficient $\alpha = 10^4$ cm^{-1}.

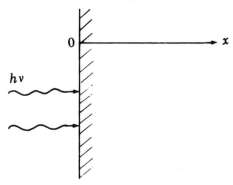

Fig. 7.3. A semiconducting sample is illuminated by photons of energy $h\nu$.

Problems

1. Why is the electron–hole pair creation rate g non-uniform in the sample? Find its variation as a function of x. Assume that each photon absorption produces an electron-hole pair.

    2. Write down the differential equation for the excess hole concentration $\Delta p(x)$ (with respect to the equilibrium concentration). Show that the solution of this equation is the sum of two decreasing exponentials in x, and compare their characteristic lengths. Take the hole recombination time as $\tau = 1 \ \mu\text{s}$ and their diffusion coefficient as $D_h = 25 \times 10^{-4} \ \text{m}^2 \ \text{s}^{-1}$.

As boundary condition assume that there is no recombination at the surface of the sample, i.e., the hole current vanishes at $x = 0$. Give an expression for $\Delta p(x)$ and sketch its variation.

3. Calculate the surface density of excess holes, and show from the results of **2** that Eq. (7.16) describes the variation of $\Delta p(x)$ correctly. For $I_0 = 1 \ \text{W/m}^2$ what is the surface hole concentration? What is the surface electron concentration?

4. An external bias is used to set up an electric field $\mathcal{E}$ between the electrodes M and N, in the geometry of Fig. 7.4. The thickness l of the sample is assumed large compared with L_h, and its width is d.

Calculate the current i through the cross section of the illuminated semiconductor, and show under the above assumptions that it exceeds the dark current i_0 by

$$\Delta i = i_0 \left(1 + \frac{\mu_h}{\mu_e}\right) \frac{I_0 \ \tau}{h \ \nu \ n_0 \ l}, \tag{7.36}$$

where μ_h and μ_e are the hole and electron mobilities, respectively. This expression does not involve the absorption coefficient α. Also give the form of Δi when $h\nu$ is close to the band gap energy.

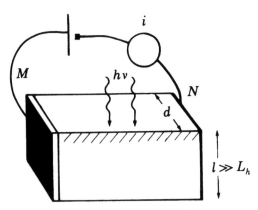

Fig. 7.4. Principle of the measurement of photoconductivity.

5. Keeping the illumination and the sample as in **4** we now extract the excess carriers from the back face via an electrode P, (Fig. 7.5). Assuming again that $l >> L_h$, and following **3**, find the excess hole concentration $\Delta p(x)$, using the facts that the hole current vanishes at $x = 0$ and the excess holes are extracted at $x = l$. What is the current $J_p(l)$?

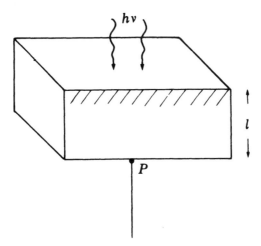

Fig. 7.5. Extraction of the excess carriers at the back of the sample. The front surface is covered with a transparent conductor such as I.T.O. (indium–tin–oxide).

Solutions

1. The intensity reaching depth x is

$$I = I_0 \exp(-\alpha x). \tag{7.37}$$

Between x and $x + dx$ an intensity

$$dI = \alpha\, I_0\, dx\; \exp(-\alpha x) \tag{7.38}$$

is absorbed. We verify that $\int_0^\infty dI = I_0$. The creation energy for an electron-hole pair is $h\nu = hc/\lambda$, so that the number of pairs $g(x)dx$ created per unit area between x and $x + dx$ is

$$g(x)\ dx = \alpha\, I_0\ dx \exp(-\alpha x)/h\nu \tag{7.39}$$

with $g(x)$ expressed in particles/m^3 s.

2. The conservation equation for the minority carriers (holes) involves the creation rate $g(x)$:

$$\frac{\partial \Delta p\,(x)}{\partial t} = D_h \frac{\partial^2\, \Delta p(x)}{\partial x^2} - \frac{\Delta p}{\tau} + \frac{\alpha\, I_0}{h\nu} \exp(-\alpha x). \tag{7.40}$$

We seek the steady-state solution. This is the solution of a second order differential equation with constant coefficients and a function of x on the right-hand side. The solution of the homogeneous equation is

$$\Delta p\,(x) = A \exp(-x/L_h) \tag{7.41}$$

with $A = $ constant and $L_h = (D_h \tau)^{1/2}$. In fact, when x is large compared with L_h and α^{-1} there are no more excess holes and $\Delta p(x)$ vanishes. A particular integral of Eq. (7.40) is

$$\Delta p\,(x) = C\exp(-\alpha x), \tag{7.42}$$

where C is fixed by the condition

$$C\left(\alpha^2\,D_h - \frac{1}{\tau}\right) + \frac{\alpha\,I_0}{h\nu} = 0, \tag{7.43}$$

$$C = \frac{\alpha\,\tau\,I_0}{h\nu\,(1 - \alpha^2\,L_h^2)}. \tag{7.44}$$

Numerically, with $D_h = 25 \times 10^{-4}$ m^2/s, $\tau = 1$ μs, we get $L_h = 50$ μm, the characteristic length for diffusion. Optical absorption occurs over a distance of order $\alpha^{-1} = 10^{-4}$ cm $= 1$ μm, much smaller than L_h. The product αL_h is much larger than 1, and C is negative:

$$C \simeq \frac{-\tau\,I_0}{\alpha\,L_h^2\,h\nu}. \tag{7.45}$$

The general solution of Eq. (7.40) is thus the sum of two terms, the first varying rapidly and the second slowly:

$$\Delta p\,(x) = C\exp(-\alpha x) + A\exp(-x/L_h), \tag{7.46}$$

where C is given by Eq. (7.44). The constant A is determined by the condition that the current should vanish at $x = 0$, with

$$J_h\,(x) = -D_h\frac{\partial \Delta p}{\partial x}, \tag{7.47}$$

$$= D_h[\alpha\,C\exp(-\alpha x) + (A/L_h)\exp(-x/L_h)]. \tag{7.48}$$

The first component of the current pulls holes out of the solid, the second pulls them into it. From $J_h(0) = 0$ we deduce

$$A = -\alpha\,C\,L_h, \tag{7.49}$$

$$\Delta p(x) = \frac{\alpha\tau\,I_0}{h\nu\,(\alpha^2\,L_h^2 - 1)}[\alpha L_h\exp(-x/L_h) - \exp(-\alpha x)]. \tag{7.50}$$

3. The excess hole density at $x = 0$ is

$$\Delta p(0) = \frac{\alpha\tau\,I_0}{h\nu\,(\alpha\,L_h + 1)}. \tag{7.51}$$

Since $\alpha L_h \gg 1$, we have

$$\Delta p(0) \simeq \frac{\tau\,I_0}{h\nu\,L_h}, \tag{7.52}$$

$$\Delta p(x) \simeq \Delta p(0)\exp(-x/L_h) \tag{7.53}$$

which is Eq. (7.16). This expression correctly describes the variation of $\Delta p(x)$ for $x > \alpha^{-1}$ (cf. Fig. 7.6).

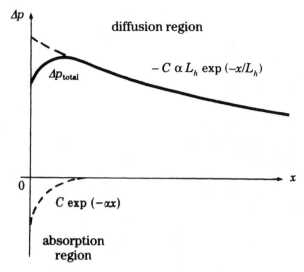

Fig. 7.6. The heavy curve shows $\Delta p(x)$, and the two dashed curves the two terms of Eq. (7.50) which build $\Delta p(x)$.

The energy of a blue–green photon of 500 nm is given by $hc/\lambda = 2.48$ eV. For $I_0 = 1$ W/m² we have 2.5×10^{18} photons/m².s. The excess surface hole concentration given by Eq. (7.52) is about $5 \times 10^{16}/\text{m}^3$. It equals the excess electron concentration.

4. The current through the cross section of the semiconductor and the electric field $\mathcal{E}$ are related by

$$i = \mathcal{E}d \int_0^l \sigma(x)\, dx. \tag{7.54}$$

For the dark current

$$\sigma_0 = e(n_0\, \mu_n + p_0\, \mu_p), \tag{7.55}$$

$$\sigma_0 \simeq e\, n_0\, \mu_n \tag{7.56}$$

since the semiconductor is n type. When illuminated,

$$\Delta\sigma(x) = e\, \mu_n\, \Delta n(x) + e\, \mu_p\, \Delta p(x), \tag{7.57}$$

$$\Delta\sigma(x) = e\, \Delta p(x)\, (\mu_n + \mu_p), \tag{7.58}$$

$$\frac{\Delta i}{i_0} = \frac{i - i_0}{i_0} = \frac{\int_0^l \Delta\sigma(x)\, dx}{l\, \sigma_0}, \tag{7.59}$$

$$\frac{\Delta i_0}{i_0} = \frac{\mu_e + \mu_h}{\mu_e} \frac{\int_0^l \Delta p(x)\, dx}{l\, n_0}. \tag{7.60}$$

The integral in the numerator can be found from Eq. (7.50). Because $\alpha^{-1} \ll L_h \ll l$,

$$\int_0^l \Delta p\,(x)\,dx \simeq \int_0^\infty \Delta p\,(x)\,dx = -C\left(\alpha\,L_h^2 - \frac{1}{\alpha}\right). \tag{7.61}$$

Since $\alpha L_h \gg 1$, we get

$$\frac{\Delta i_0}{i_0} \simeq \left(1 + \frac{\mu_h}{\mu_e}\right)\frac{\tau I_0}{h\nu}\frac{1}{n_0 l}. \tag{7.62}$$

This expression involves only the photon number $I_0/h\nu$, and not the variation of α with $h\nu$. This comes from the approximation $\alpha L_h \gg 1$. We have a detector of light that is fairly insensitive to photon energies.

In contradistinction, when we use photons with energy close to the band gap, α is small, and we can have $\alpha^{-1} > l$, where L_h is not modified. Then $L_h \ll l \ll \alpha^{-1}$, $\alpha L_h \ll 1$,

$$\int_0^l \Delta p\,(x)\,dx = -C\,\alpha\,L_h^2 + C\frac{1 - \exp(-\alpha l)}{\alpha}, \tag{7.63}$$

$$\simeq -C\,\alpha\,L_h^2 + Cl \simeq Cl. \tag{7.64}$$

Moreover C is now positive:

$$C \simeq \frac{\alpha\,\tau\,I_0}{h\nu} \tag{7.65}$$

so that the photoconductivity signal

$$\Delta i_0 \simeq \frac{i_0}{n_0}\left(1 + \frac{\mu_h}{\mu_e}\right)\frac{\alpha\,\tau\,I_0}{h\nu} \tag{7.66}$$

is proportional to αI_0 in the energy range where α is small. The photoconductivity spectrum, i.e., the curve $\Delta i_0(h\nu)$, for a source with photon number varying weakly with $h\nu$, will exhibit the variation of α close to the absorption threshold, and in particular the exciton peaks (see Sect. 6.1d).

5. The general steady solution of the differential Eq. (7.40) now has the form

$$\Delta p\,(x) = A\exp(-x/L_h) + B\exp(x/L_h) + C\exp(-\alpha x), \tag{7.67}$$

where C is given by Eq. (7.44), and A and B are determined by the boundary conditions

$$J_h = 0, \quad \text{i.e.,} \quad D_h[(A-B)/L_h + C\alpha] = 0 \quad \text{at} \quad x = 0; \tag{7.68}$$

$$\Delta p = 0, \quad \text{i.e.,} \quad A\exp(-l/L_h) + B\exp(l/L_h) + C\exp(-\alpha l) = 0$$
$$\text{at} \quad x = l. \tag{7.69}$$

We still have $\alpha^{-1} \ll L_h \ll l$, hence $\exp(-\alpha l) \ll \exp(-l/L_h) \ll 1$, so that

$$A \simeq -C\alpha\,L_h\exp(l/L_h)\,/\,[\exp(l/L_h) + \exp(-l/L_h)], \tag{7.70}$$

$$B \simeq C\,\alpha\,L_h\exp(-l/L_h)\,/\,[\exp(l/L_h) + \exp(-l/L_h)], \tag{7.71}$$

hence

$$J_h\,(l) = \frac{D_h}{L_h}[A\exp(-l/L_h) - B\exp(l/L_h) + \alpha\,L_h\,C\exp(-\alpha l)], \quad (7.72)$$

$$J_h\,(l) \simeq -2\,D_h\,C\,\alpha/[\exp(l/L_h) + \exp(-l/L_h)]. \qquad\qquad (7.73)$$

Replacing C by its expression in the same approximation we get

$$J_h\,(l) \simeq 2\frac{D_h\,\alpha^2\,\tau\,I_0}{\alpha^2\,L_h^2\,h\nu[\exp\,(l/L_h) + \exp(-l/L_h)]},$$

$$J_h\,(l) \simeq 2\frac{I_0}{h\nu}\exp(-l/L_h). \qquad\qquad (7.74)$$

The current collected at the back face of the illuminated sample is a diffusion current. The situation is comparable with that of a p–n–p transistor (see Chap. 10) in which holes are injected from the emitter to the base and extracted at the collector.

8.

The p–n Junction

8.1 Introduction: Inhomogeneous Semiconductors

In the remaining chapters we shall discuss the physics of various "electron machines" such as diodes, transistors, etc. These devices are now vital for everyday life and for the industrial development of the second half of the 20th century. They mostly consist of semiconducting structures which are inhomogeneous either through their chemistry or because of doping. While classical mechanical machines are assembled from variously shaped components, the "electron machines" we will study are manufactured essentially by spatially varying the concentration of impurities.

The p–n junction is a vital component of all these devices, both because of its direct applications and because an understanding of its physics explains the junction transistor and many other devices. A p–n junction consists of a semiconducting crystal whose concentration of shallow impurities varies with x so that a p-type region, where $N_a - N_d > 0$, is next to an n-type region, i.e., to one where $N_a - N_d < 0$. This is therefore an inhomogeneous semiconductor, where the electron and hole concentrations vary with position, even in thermodynamic equilibrium, i.e., in the absence of a current. Now, if there is a concentration gradient there must be a diffusion current. For the electrons, for example, this is

$$\mathbf{J}_{e,d} = e \, D_e \, \nabla n.$$

For the electron current to vanish, the drift current $n e \, \mu_e \, \boldsymbol{\mathcal{E}}$ must exactly cancel it:

$$n \, \mu_e \, \boldsymbol{\mathcal{E}} = -D_e \, \nabla n. \tag{8.1}$$

There must therefore be an electric field, which in the absence of external charges can only result from a charge density ρ within the system. This shows that:

In thermodynamic equilibrium, in an inhomogeneously doped system, charge neutrality cannot be preserved everywhere.

Once charge neutrality is violated an electrostatic potential $V(x)$ appears, given by Poisson's equation

$$\Delta V(x) = -\frac{\rho(x)}{\epsilon_r \, \epsilon_0} \tag{8.2}$$

with

$$\rho(x) = e[\, p\,(x) - n\,(x) + n_d^+\,(x) - n_a^-\,(x)\,], \tag{8.3}$$

where the concentrations of free carriers n, p and ionized impurities n_d^+, n_a^- now depend on position.

This macroscopic potential is added to the usual crystalline potential and shifts the energy levels by an amount $-eV(x)$. The general shape of the bands is shown in Fig. 8.1.

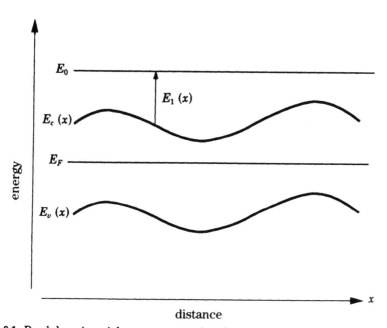

Fig. 8.1. Band shape in an inhomogeneous semiconductor at thermodynamic equilibrium.

For a system in thermodynamic equilibrium the chemical potential, or Fermi level, which is the partial derivative of the thermodynamic potential with respect to the particle number, is spatially constant.

We have to consider an energy origin E_0 which is independent of the charge distribution, for example, the energy of a static electron at infinity (this is the zero of energy in the hydrogen atom, compared with which the ground state has energy -13.6 eV). Then if the energy of a given conduction-band electron is E and $E_1(x)$ is its energy with respect to the bottom of the conduction band, we have

$$E = E_c(x) + E_1(x)$$
$$= E_{co} - eV(x) + E_1(x) \tag{8.4}$$

and the Fermi function is

$$f = \frac{1}{1 + \exp[(E - E_F)/kT]} = \frac{1}{1 + \exp[(E_c + E_1 - E_F)/kT]}. \tag{8.5}$$

The number of conduction electrons at abscissa x is then given by

$$n(x) = \int_{E_c(x)}^{\infty} n[E - E_c(x)] f \, dE$$
$$= \int_0^{\infty} \frac{n(E_1) \, dE_1}{1 + \exp[(E_1 + E_c(x) - E_F)/kT]}. \tag{8.6}$$

There will be analogous Eqs. (8.6'), (8.6'')... giving $p(x), n_d^0(x)$, $n_a^-(x)$ as we have seen in Chap. 4. The total charge density is given by Eq. (8.3). Solving the equilibrium junction problem then reduces to solving Eqs. (8.2), (8.3), (8.6), (8.6'),... so as to find the potential distribution and the carrier densities as functions of the coordinates.

When the Fermi level is far from the band levels (see Sect. 4.4), Eq. (8.6) gives for each n:

$$n(x) = N_c \exp[-(E_c(x) - E_F)/kT], \tag{8.7}$$

$$n(x) = \text{const.} \exp[e \, V(x)/kT], \tag{8.8}$$

and

$$\frac{1}{n}\nabla n = \frac{e}{kT}\nabla V = -\frac{e}{kT}\boldsymbol{\mathcal{E}}. \tag{8.9}$$

Substituting in Eq. (8.1) we recover the Einstein relation (5.30). There is of course an equivalent result for the holes. Formula (8.7) shows that the electrons are more numerous in regions where the Fermi level is closer to the bottom of the conduction band.

8.2 The Equilibrium p–n Junction

In the following we shall limit ourselves to what is called an abrupt junction. It consists of a semiconducting crystal in which the impurity concentration changes discontinuously from a majority concentration of donors to a majority concentration of acceptors at $x = 0$. Of course in reality it is impossible to set up such a system and the transition from the n to the p region is fairly gradual, as shown schematically in Fig. 8.2(a), but we shall see that we can often approximate this type of profile by a discontinuous one (Fig. 8.2(b)).

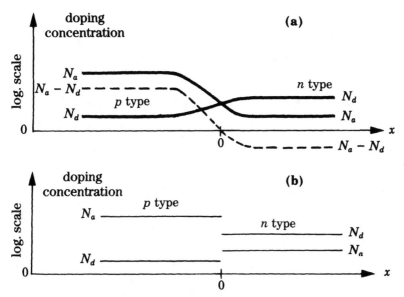

Fig. 8.2. Doping profile of a p–n junction: (a) real profile; (b) approximation as a discontinuous profile.

8.2a Qualitative Discussion

Let us assume that we have made an abrupt junction by bringing together two crystals of different types. The band diagram of the two separate systems is shown in Fig. 8.3. The Fermi levels in the two crystals are different. We assume we are in the saturation regime where $n = N_d$ and $p = N_a$, with the Fermi levels in the band gap; the materials are assumed to be non-compensated (this simplifies the notation without loss of generality in the following).

We denote by n_p^0 and p_p^0 the electron and hole concentrations in the p-type material and n_n^0, p_n^0 the corresponding concentrations in the n-type material, before they are brought into contact. The carrier concentrations are given by Eqs. (4.14), (4.16), (4.31), and (4.37):

$$p_p^0 = N_a = N_v \exp[-(E_{F,p} - E_{v,p})/kT], \tag{8.10}$$

$$n_p^0 = n_i^2/p_p^0 = N_c \exp[-(E_{c,p} - E_{F,p})/kT], \tag{8.11}$$

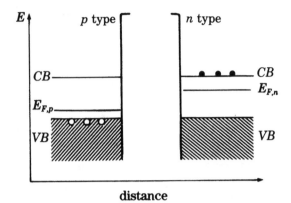

Fig. 8.3. Energy levels in two separated n- and p-type crystals. The hatched zones represent filled states, the heavy dots conduction electrons, and the open dots holes. (After Dalven, "Introduction to Applied Solid State Physics," Plenum Press, 1980.)

$$p_n^0 = n_i^2/n_n^0 = N_v \exp[-(E_{F,n} - E_{v,n})/kT], \tag{8.12}$$

$$n_n^0 = N_d = N_c \exp[-(E_{c,n} - E_{F,n})/kT], \tag{8.13}$$

$$n_n^0\, p_n^0 = n_p^0\, p_p^0 = N_c\, N_v \exp(-E_g/kT) = n_i^2. \tag{8.14}$$

As the chemical potentials differ from left to right, a reaction will occur if the two systems are brought into contact to make a junction; electrons will pass from the system where the chemical potential is higher (on the right) towards the p region on the left, and the holes will move from left to right. This movement of free carriers may also be described in an exactly equivalent fashion by saying that the concentration gradient creates a diffusion current of electrons from right to left and of holes from left to right. The additional electrons in the p material are the excess minority carriers: they recombine with the majority holes, with a similar effect for the additional holes in the n region. The electrons leave ionized donors of concentration N_d in the n material, and the holes leave ionized acceptors of concentration N_a in the p material. A region of non-zero charge is then created near the junction. This is called the space-charge region and is shown in Fig. 8.4.

However, this process cannot continue indefinitely as the space charge creates an electric field that opposes the diffusion of majority carriers. This is how equilibrium is achieved. The space-charge zone is a region where the number of free carriers is very low; for this reason it is also called the depletion region. The space-charge zone bears a charge, namely that of the fixed donor or acceptor sites that are no longer neutralized.

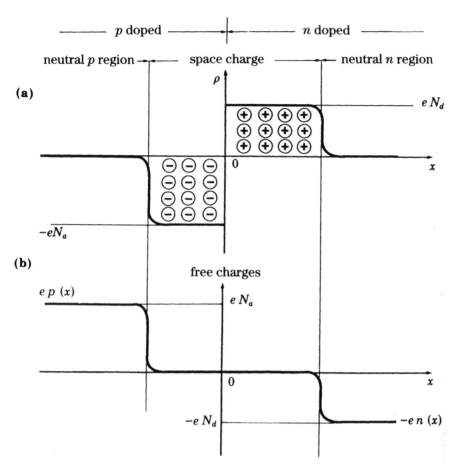

Fig. 8.4. (a) Net charge density $\rho(x)$, and (b) free charge density, in a p–n junction.

8.2b Potential Difference between the p and n Regions

Equilibrium thermodynamics gives the internal electrostatic potential difference ϕ between the p and n regions. The overall shift of the bands in the junction region must have the shape shown in Fig. 8.5 as in equilibrium the chemical potential must be constant throughout the system. Far from the junction the n and p materials must retain their original properties, i.e., constant free carrier concentrations, equal to the initial donor and acceptor concentrations $(p_p^0 = N_a, n_n^0 = N_d)$, and constant electrostatic potential as the current must vanish.

If the electrostatic potentials in the n and p regions far from the junction are V_n and V_p, then setting

$$\phi = V_n - V_p, \tag{8.15}$$

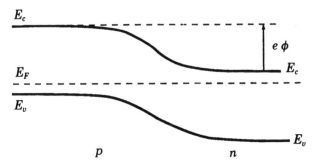

Fig. 8.5. Band profile across an equilibrium p–n junction.

we have

$$E_{c,p} - E_{c,n} = -e\,(V_p - V_n) = e\,\phi \qquad (8.16)$$

and using Eqs. (8.10), (8.11), (8.13), and (8.14),

$$\phi = \frac{E_g}{e} + \frac{kT}{e}\,\log\frac{N_a\,N_d}{N_c\,N_v}. \qquad (8.17)$$

The internal potential ϕ is thus slightly smaller than the band gap. In silicon at room temperature $E_g = 1.12$ eV; $N_c \sim 3 \times 10^{25}$ m^{-3}; $N_v \sim 10^{25}$ m^{-3}; $kT/e = 0.025$ V; if $N_a = 3 \times 10^{23}$ m^{-3}, $N_d = 10^{21}$ m^{-3} we have $\phi = 0.775$ V. We note that we also have from Eqs. (8.10)–(8.13),

$$\frac{n_p^0}{n_n^0} = \frac{p_n^0}{p_p^0} = \exp(-e\,\phi/kT). \qquad (8.18)$$

8.2c Calculation of the Space Charge

Far from the junction electrical neutrality holds between mobile and fixed charges. In contrast, close to the junction, the Fermi level is far from the bands. This is why there are few mobile charges here (see Eqs. (8.10)–(8.13)). Further, the density of carriers given by Eq. (8.8) varies exponentially, i.e., very rapidly with the potential. This means that if the potential has the shape shown in Fig. 8.6(a) the space-charge density given by Eq. (8.3) will have the shape shown by the thick curves in Fig. 8.6(b).

In Sect. 8.1 we set ourselves the aim of integrating the Poisson equation for the potential with a local charge depending on this potential through the laws of statistical mechanics. We see that to a good approximation we can regard ρ as constant and equal to $-eN_a$ in the p region over the range between $-d_p$ and 0, and equal to $+eN_d$ between 0 and d_n; it remains to determine the distances d_p, d_n. We replace the real charge density curve of

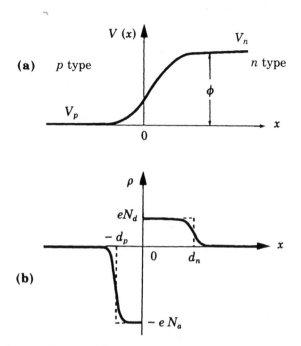

Fig. 8.6. (a) Electrostatic potential across an equilibrium p–n junction; (b) corresponding charge density. The dashed curve is the approximate form of ρ used in the calculation in the text.

Fig. 8.6(b) by the dashed curve. We then have to integrate the very simple equations

$$-\frac{\partial^2 V}{\partial x^2} = \frac{\partial \mathcal{E}}{\partial x} = -\frac{e\,N_a}{\epsilon_0\,\epsilon_r} \qquad \text{for} \quad -d_p \leq x \leq 0, \tag{8.19}$$

$$-\frac{\partial^2 V}{\partial x^2} = \frac{\partial \mathcal{E}}{\partial x} = +\frac{e\,N_d}{\epsilon_0\,\epsilon_r} \qquad \text{for} \quad 0 \leq x \leq d_n. \tag{8.20}$$

Noting that the electric field vanishes in equilibrium for $x = -d_p$ and d_n, since the current is zero at these points, we get

$$\mathcal{E} = -\frac{e\,N_a}{\epsilon_0\,\epsilon_r}(x + d_p) \quad \text{for} \quad -d_p \leq x \leq 0, \tag{8.21}$$

$$\mathcal{E} = \frac{e\,N_d}{\epsilon_0\,\epsilon_r}(x - d_n) \quad \text{for} \quad 0 \leq x \leq d_n. \tag{8.22}$$

We note that at $x = 0$:

$$N_a\,d_p = N_d\,d_n. \tag{8.23}$$

This equation states that the total space charge of the p region exactly compensates that of the n region. It reflects the global neutrality of the junction, since the "manufacture" began with two electrically neutral regions. A second integration gives the potentials, with $V = V_p$ for $x = -d_p$:

$$V(x) = \frac{e}{2\,\epsilon_0\,\epsilon_r} N_a (x + d_p)^2 + V_p \quad \text{for} \quad -d_p \leq x < 0, \tag{8.24}$$

$$V(x) = \frac{e\,N_a}{2\,\epsilon_0\,\epsilon_r} d_p^2 - \frac{e\,N_d}{\epsilon_0\,\epsilon_r}\left(\frac{x^2}{2} - d_n\,x\right) + V_p \quad \text{for} \quad 0 \leq x \leq d_n. \tag{8.25}$$

For $x = d_n$ we have $V = V_n$ and

$$\phi = (V_n - V_p) = \frac{e}{2\,\epsilon_0\,\epsilon_r}[N_a\,d_a^2 + N_d\,d_b^2]. \tag{8.26}$$

The quantities d_n, d_p are found from Eqs. (8.17) and (8.23). The form of the electric field is shown in Fig. 8.7. It is negative, repelling the electrons coming from the n-type region on the right and the holes coming from the p-type region on the left, and thus separates the free carriers from the space-charge region.

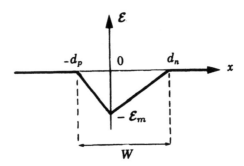

Fig. 8.7. Electric field across an equilibrium p–n junction.

The width of the space-charge region can be found from Eqs. (8.23) and (8.26):

$$W = d_p + d_n = \left(\frac{2\,\epsilon_0\,\epsilon_r\,\phi}{e}\right)^{1/2}\left(\frac{N_a + N_d}{N_a\,N_d}\right)^{1/2}. \tag{8.27}$$

With $\phi \sim 1$ V; $N_d, N_a \sim 10^{21}$ m^{-3}; $\epsilon_0\epsilon_r \sim 10^{-10}$ F·m^{-1} we get $W \sim 2 \times 10^{-6}$ m. If $|\mathcal{E}|$ is the modulus of the maximum field at the junction,

$$|\mathcal{E}_m| = \frac{e\,N_d\,d_n}{\epsilon_0\,\epsilon_r} = \frac{e\,N_a\,d_p}{\epsilon_0\,\epsilon_r}. \tag{8.28}$$

We note that the triangle in Fig. 8.7 has area ϕ, or

$$\phi = \frac{1}{2}|\mathcal{E}_m|(d_n + d_p) = \frac{1}{2}|\mathcal{E}_m|W, \tag{8.29}$$

where $W = (d_n + d_p)$ is the width of the space-charge region.

Let us consider the example of a junction where $N_a >> N_d$, called a p^+–n junction; then $d_p << d_n$, and the space-charge region mainly extends into the less doped region, with $W \sim (2\epsilon_0\epsilon_r\phi/eN_d)^{1/2}$. The maximum electric field in the less doped region is of the order of ϕ/W or about 10^6 V/m for a junction where the space charge extends over 1 μm.

For the "abrupt-junction" approximation used in this section to hold we require that in a real junction the region over which the doping varies from p type to n type should be much smaller than the size W we have just calculated. Suppose that the junction was made by diffusion of phosphorus into silicon originally of p type. If the diffusion operates for time t, the width of the profile is of the order of $L = (D_{\text{phosph}}t)^{1/2}$, where D_{phosph} is the diffusion coefficient of phosphorus. For phosphorus in silicon at 1000° C we have $D_{\text{phosph}} \sim 10^{-17}\text{m}^2 \cdot \text{s}^{-1}$. If diffusion lasts about 10^3 s, we get $L = 10^{-7}$ m: one can therefore manufacture junctions where the width of the space-charge region is much larger than that of the region where the doping changes. Such junctions can reasonably be regarded as abrupt.

A real junction is never made by assembling two different crystals (n- and p-type) because their surface properties (surface states to be discussed in Sect. 9.4) would totally perturb the above situation.

8.2d Currents in the Equilibrium Junction

As we have seen above, the total current through the junction vanishes because the chemical potential is constant. This is true for the electron and hole currents separately. It is interesting to understand the mechanism by which each of these currents cancels out. In the uniform regions where the charge density is zero, left and right of the deserted region, the current of each type of carrier vanishes because the electric field is zero. By contrast there is a very strong electric field in the deserted zone. Although the number of carriers is small, the drift current for each type is large, and the total current vanishes because there is an equally large opposing diffusion current. The diffusion current is large because the concentration gradient is large in this region (Eq. (8.9)). It is interesting to find the order of magnitude of these currents. For simplicity let us consider a symmetrical junction. Then at $x = 0$ the Fermi level is exactly in the middle of the band gap and the concentrations are the intrinsic ones. The drift current $J = n_i e \mu \mathcal{E}$ is then about 0.1 A/cm^2 for silicon where $n_i \sim 10^{16}$ m^{-3}; $\mu \sim 0.1$ m$^2 \cdot$ V^{-1} $\cdot$s^{-1} and for an electric field of about 10^6 V$\cdot$m^{-1} as we have just seen. This is the typical value for the drift current at the center of the junction. This reasoning holds for electrons and holes.

In summary, in equilibrium, far from the junction, the current vanishes because the carriers have a constant concentration and there is no electric field. In the center of the space-charge region two very large currents, the drift and diffusion currents, cancel exactly.

8.3 The Non-Equilibrium Junction

Let us now assume that we apply an external voltage by means of a dc generator connected across the junction. We say that the junction is biased. The external voltage V_e is counted positive if it tries to make the p-doped side positive with respect to the n-doped side, and thus tends to decrease the height of the potential barrier between the two regions. The chemical potential cannot be constant in the whole system and a current will circulate. However the space-charge region has very high resistivity since it has few free carriers. We can thus say that the potential drop will occur in the space-charge region, the width of this zone possibly being modified by the presence of the external potential. Outside this region of very high resistivity the p and n regions can be regarded as approximately equipotential as their resistance is low.

The effect of applying an external potential V_e is shown schematically in Fig. 8.8. The distances d_n and d_p are modified by the existence of V_e and become d'_n and d'_p, which are both functions of V_e.

A calculation analogous to that of Sect. 8.2c gives

$$\phi - V_e = \frac{e}{2\,\epsilon_0\,\epsilon_r}[N_a\,d'^2_p + N_d\,d'^2_n] \tag{8.30}$$

and

$$N_a\,d'_p = N_d\,d'_n. \tag{8.31}$$

We note that **the electric fields and size of the space-charge region are modified without changing their order of magnitude.**

8.3a Energy Level Diagram

To understand the operation of a biased junction, it is useful to consider the energy diagram in the three cases of zero, forward, and reverse bias, as shown in Figs. 8.9(a), 8.9(b), and 8.9(c). Applying a potential difference across the junction means imposing a difference in chemical potential between left and right. We can do this in practice by making two contacts of the same metal at the extremities of the junction and holding them at different potentials.

The band profiles are arcs of parabolas as shown by Eqs. (8.24) and (8.25).

Outside the space-charge region, the p and n conductors remain equipotential (apart from the small variation corresponding to the ohmic drop in the two materials when a current flows). We then have

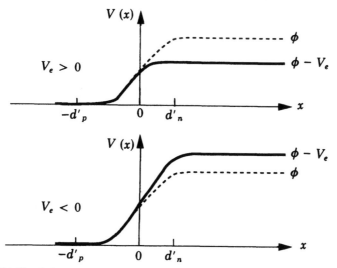

Fig. 8.8. Profile of the electrostatic potential of the junction for an applied positive or negative voltage V_e. The dashed curves show the potential for $V_e = 0$.

$$E_{F,p} - E_{F,n} = -e\, V_e. \tag{8.32}$$

8.3b Calculation of the Current

To calculate the current we make use of our description of the equilibrium currents in the unbiased diode. We saw that in the space-charge region **two very large currents** (drift and diffusion) exactly cancel. To calculate the current in the presence of an external voltage we assume first that the resultant current is small compared with the opposing components of the equilibrium current. Then in the current equation for (for example) the electrons

$$\mathbf{J}_e = n\, e\, \mu_e\, \boldsymbol{\mathcal{E}} + e\, D_e \nabla\, n, \tag{8.33}$$

we can neglect $\mathbf{J}_e$ compared with the two terms on the right-hand side and write in the presence of currents

$$n\, e\, \mu_e\, \boldsymbol{\mathcal{E}} \#-e\, D_e \nabla\, n \tag{8.34}$$

or, using the Einstein relations,

$$n\,(x) \#\text{ const. } \exp \frac{e\, V\,(x)}{kT}. \tag{8.35}$$

The latter equation shows that for currents that are not too large (we will specify this later) the electrons are in thermal equilibrium in the space-charge region. Then using Eq. (8.35) at the points $-d'_p$ and d'_n, we have

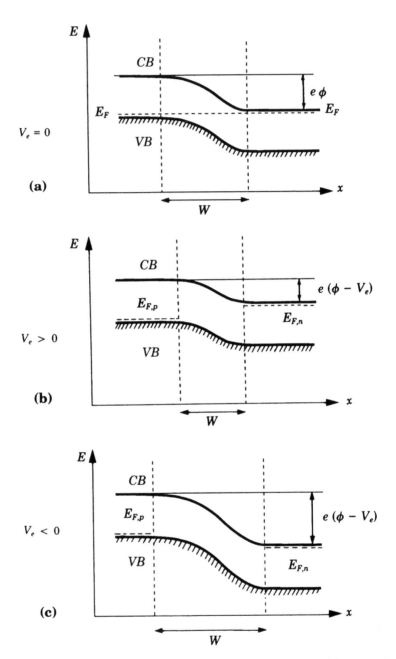

Fig. 8.9. Comparison of band profiles for a *p–n* junction which is (a) unbiased; (b) forward biased; and (c) reverse biased.

$$\frac{n_{(x=-d'_p)}}{n_{(x=+d'_n)}} = \exp\left[-e(\phi - V_e)/kT\right].$$ (8.36)

But the electron concentration in the n-type material, for $x \geq d'_n$, has the fixed value N_d, so that

$$n_{(x=-d'_p)} = N_d \exp\left[-e(\phi - V_e)/kT\right].$$ (8.37)

Using Eq. (8.18),

$$n_{(x=-d'_p)} = n_p^0 \exp\frac{e\,V_e}{kT}.$$ (8.38)

This shows that in the presence of the external voltage the concentration of minority carriers is modified at the edge of the space-charge region. For $x = -d'_p$ we have a deviation from the equilibrium population of amount

$$\Delta\, n_{(x=-d'_p)} = n_{(x=-d'_p)} - n_p^0 = n_p^0\left[\exp\left(\frac{e\,V_e}{kT}\right) - 1\right].$$ (8.39)

Depending on whether V_e is positive or negative, the concentration of minority carriers is increased or decreased. We thus say that there is **injection** or **extraction of minority carriers**. This injection or extraction takes place from the majority carrier region. We can say that the concentration deviation is created by the change in the height of the potential barrier, which changes the number of electrons able to diffuse from the n region towards the semiconducting p region. We expect this number to depend on V_e through a Boltzmann factor. The p region behaves as a semiconducting strip whose surface population of minority carriers is maintained out of equilibrium. There is therefore diffusion and an associated diffusion current. This situation was considered in the preceding chapter (Sect. 7.3 and Appendix 7.2). The current is given by Eq. (7.19), or here, for $x = -d'_p$:

$$J_e = e\,\Delta\, n_{(x=-d'_p)} \cdot \frac{D_e}{L_e},$$ (8.40)

$$J_e = e\, n_p^0 \frac{D_e}{L_e}\left[\exp\left(\frac{e\,V_e}{kT}\right) - 1\right].$$ (8.41)

Similarly for the holes, analogous reasoning gives the diffusion current at the edge of the n region, so that for $x = +d'_n$:

$$J_h = e\, p_n^0 \frac{D_h}{L_h}\left[\exp\left(\frac{e\,V_e}{kT}\right) - 1\right].$$ (8.42)

If there is no recombination in the space-charge region these two currents are constant in this zone and add. The total current is therefore

$$J = J_e + J_h$$

$$= e \left(n_p^0 \frac{D_e}{L_e} + p_n^0 \frac{D_h}{L_h} \right) \left[\exp\left(\frac{e V_e}{kT} \right) - 1 \right], \tag{8.43}$$

$$= J_s \left[\exp\left(\frac{e V_e}{kT} \right) - 1 \right]. \tag{8.44}$$

The quantity J_s is called the **saturation current**:

$$J_s = e \left(n_p^0 \frac{D_e}{L_e} + p_n^0 \frac{D_h}{L_h} \right) = e \, n_i^2 \left(\frac{D_e}{L_e N_a} + \frac{D_h}{L_h N_d} \right). \tag{8.45}$$

Because of the factor n_i^2, J_s varies very rapidly with temperature. The saturation current is small compared with the current components at the center of the barrier. Taking $N_a = N_d = 10^{21}$ m^{-3} and $D/L \sim 5$ m· s^{-1} we get $J_s \sim 10^{-11}$ A/cm^2 for a silicon junction, or six orders of magnitude below the drift current at the center of the barrier. This order of magnitude justifies the hypotheses made in writing Eq. (8.37). Even for forward bias $(V_e > 0)$ the law (8.44) will be correct as long as $J_s \exp(eV_e/kT)$ remains small enough compared with the drift current at the center of the barrier, and thus remains applicable over a very wide range of current (from 10^{-11} A/cm^2 up to about 10^{-3} A/cm^2 in the example considered).

The law (8.44) is called Shockley's law, and dates from 1949. Its form is shown in Fig. 8.10. When V_e is very large and negative we say that the voltage is reversed and the current is $-J_s$. For $V_e > 0$ we say that the voltage is direct; the current is then positive and increases exponentially with the applied voltage. This behavior is shown in Figs. 8.10(a) and 8.10(b) at large and small scales. Figure 8.10(b) is drawn for $J_s = 2.19 \times 10^{-11}$ A/cm^2. Notice the difference in the scales of the ordinates for direct and reverse voltages.

It is interesting to note that for a strong reverse bias the electrons, the minority carriers in the p region, that reach the barrier are accelerated by the electric field. They cross the space-charge region from right to left, while the electrons in the n region, where they are the majority carriers, cannot diffuse into the p region because they cannot cross the very high potential barrier. Then for $x = -d_p'$ the electron concentration falls to zero and there is a concentration gradient for $x < -d_p'$. We thus have a semi-conducting strip in which we have created a deviation $\Delta n_p = -n_p^0$ from the equilibrium populations at $x = -d_p'$. In this situation electron-hole pairs are generated to compensate this deviation and diffusion results. The electron contribution to the saturation current can then be interpreted as the maximum electron current one can draw from p-type material by extracting all the minority electrons at its surface. This current can only flow continuously if there is continuous reappearance of the minority electrons and thus generation of electron-hole pairs by the material. For this reason this current is often called the "electron generation current." Similarly the other component of J_s is often called the "hole generation current."

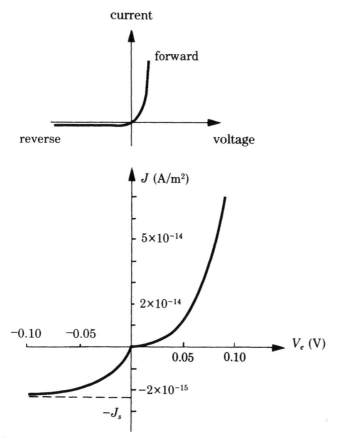

Fig. 8.10. Current–voltage characteristic of a p–n junction. (a) Linear scale; (b) enlarged view of the region near the origin. The current density scale is expanded by a factor 10 for the reverse region, causing the change of slope at the origin.

We have assumed that recombination is negligible in the space-charge region. To justify this we note that if the mean electric field in this zone is of the order of 10^6 V/m an electron of mobility 0.1 $\mathrm{m^2 \cdot V^{-1} \cdot s^{-1}}$ has a drift velocity of 10^5 m/s and the region of 10^{-6} m thickness is crossed by the minority carriers in 10^{-11} s, much shorter than the recombination time. By contrast the diffusion of the majority carriers takes a time of order W^2/D or about 10^{-9} s. The recombination time may not always be extremely long compared with this time and the above theory must be modified.

Shockley's equation shows that the p–n junction is an efficient rectifier, as the current depends very strongly on the sign of V_e; for this reason it is often called a p–n diode. In real diodes there are several limitations which we have neglected:

– we assumed generation and recombination of pairs to be negligible in the space-charge region;

– we limited ourselves to moderate currents;
– we neglected the sides of the junction, where surface recombinations occur.

These effects, and the resistance of the p and n regions, may modify the current–voltage characteristics of the diode.

8.3c Current and Concentration Distributions

We note from Eq. (8.39) that if V_e is positive Δn is > 0: We say that minority carriers are injected. If the potential is strictly constant (by charge neutrality) for $x < -d'_p$ and $x > d'_n$ the current at the edge of the space charge is purely a diffusion current. This is shown in Fig. 8.11.

In a steady state we have

$$\nabla \cdot \mathbf{J} = -\frac{d\rho}{dt} = 0, \tag{8.46}$$

and the total current (electrons + holes) must be constant along x.

If there is injection of excess minority carriers, the carriers injected at the edge of the space-charge region diffuse and give rise to the diffusion current shown. As the total current must be constant, the majority carrier current must also vary with x, as shown in Fig. 8.11. An excess of majority carriers, extremely small in relative value, must therefore appear to compensate the charge of the injected minority carriers. Far from the junction the diffusion currents are small and the current is thus entirely a drift current of majority carriers, with uniform concentrations again. Hence there are five successive regions in a non-equilibrium junction: homogenous p type, p diffusion, space charge, n diffusion, and homogenous n type; by contrast there are no diffusion regions in an equilibrium junction, and thus only three regions.

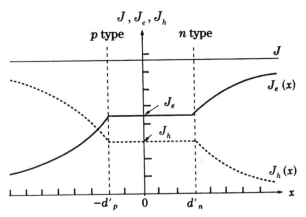

Fig. 8.11. Total current density (J), electron current density (J_e), and hole current density (J_h) across a p–n junction. (After Dalven, "Introduction to Applied Solid State Physics," Plenum Press, 1980.)

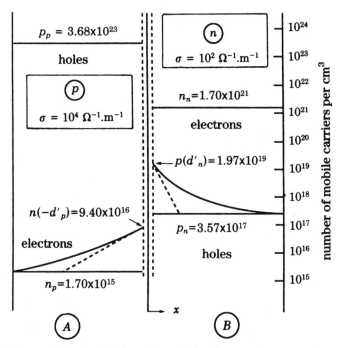

Fig. 8.12. Concentration profile in a forward biased germanium p–n junction ($V_e = 0.1$ V). (After Middlebrook, "An Introduction to Junction Transistor Theory," J. Wiley, 1957, 1965.)

Figure 8.12 gives the carrier densities in a germanium p–n junction as a function of x for a forward bias of 0.1 V. Note the logarithmic scale of the ordinate. The excess of majority carriers is negligible. The excess of minority carriers resulting from injection is very small compared with the majority carrier concentrations but very large compared with the equilibrium concentration of minority carriers. We see the five spatial regions: at the left and right of the figure the unperturbed n and p zones (not shown); then closer to the space charge, regions A and B where the current is mainly a diffusion current; and finally at the center the very narrow space-charge region.

Remark: For the non-equilibrium p–n junction it is useful to introduce the notion of "quasi-Fermi level": as this is not an equilibrium, the Fermi level is not constant in space. However from the concentration $n(x)$, by analogy with expression (8.7) we can define the electron quasi-Fermi level E_{Fe} by

$$n(x) = N_c \exp[-(E_c(x) - E_{Fe})/kT] \tag{8.47}$$

or

$$E_{Fe} = E_{c_0} - eV(x) + kT \log [n(x)/N_c]. \tag{8.48}$$

Similarly we define the quasi-Fermi level E_{Fh} for the holes:

$$E_{Fh} = E_{v_0} - eV(x) - kT \log [p(x)/N_v]. \tag{8.49}$$

For an equilibrium junction E_{Fe} and E_{Fh} are equal, so that

$$E_{c_0} - E_{v_0} + kT \log [n(x)\, p(x)/N_c\, N_v] = 0 \tag{8.50}$$

which is the law of mass action (Eq. (4.22)). In this case the quantities E_{Fe}, E_{Fh} do not depend on position, so that we recover the relation (8.7) or its equivalent for holes. In a non-equilibrium junction the quasi-Fermi levels vary with position. We now show that the spatial variation of E_{Fe} is related to the corresponding current J_e:

$$J_e = e\, \mu_e\, n\, \mathcal{E} + e\, D_e \frac{\partial n}{\partial x}, \tag{8.51}$$

$$J_e = -e\, \mu_e\, n\, \frac{\partial V}{\partial x} + e\, \mu_e \frac{kT}{e} \frac{\partial n}{\partial x}, \tag{8.52}$$

$$= n(x)\, \mu_e \frac{\partial}{\partial x}[-e\, V(x) + kT \operatorname{Ln} n(x)]. \tag{8.53}$$

Using the definition (8.48) of E_{Fe}, we obtain

$$J_e = n(x)\, \mu_e \frac{\partial E_{Fe}}{\partial x}. \tag{8.54}$$

Similarly we have

$$J_p = p(x)\, \mu_h \frac{\partial E_{Fh}}{\partial x}. \tag{8.55}$$

The quasi-Fermi levels are actually the electrochemical potentials of transport theory.

The notion of quasi-Fermi levels provides a description of the carrier density even within the space-charge region. We show that J_e varies little in this region: from the conservation equation for the minority carriers in steady state:

$$\frac{\partial n}{\partial t} = \frac{1}{e} \frac{\partial J_e}{\partial x} - \frac{n - n_0}{\tau_n} = 0 \tag{8.56}$$

or

$$\frac{\partial J_e}{\partial x} = e \left(\frac{n - n_0}{\tau_n} \right),$$

we deduce

$$\frac{\partial J_e}{\partial x} < \frac{e}{\tau_n} \delta n_{\max}. \tag{8.57}$$

The variation of J_e occurs over a distance of order of the diffusion length L_e, or about 0.1 mm, while the width of the space-charge region $d_p + d_n$ is about one micron. Then using the Einstein relation we have

$$J_e \leq \frac{eL_e}{\tau_n} \delta n_{\max}, \tag{8.58}$$

$$\frac{\partial E_{Fe}}{\partial x} \leq \frac{eL_e}{\tau_n \, \mu_e} \frac{\delta n_{\max}}{n} = \frac{\delta n_{\max}}{n} \frac{kT}{L_e} < \frac{kT}{L_e}. \tag{8.59}$$

The variation of E_{Fe} over the distance $(d_p + d_n)$ is certainly less than $kT(d_p + d_n)/L_e$, and thus much smaller than kT. We can say nothing about the variation of E_{Fe} in diffusion regions except that on the n side $\delta n << N_d$ (and $\delta p << N_a$ on the p side), so that E_{Fe} coincides with the Fermi level on the n side, and the quasi-Fermi levels have the variation shown in Fig. 8.13 across a non-equilibrium junction.

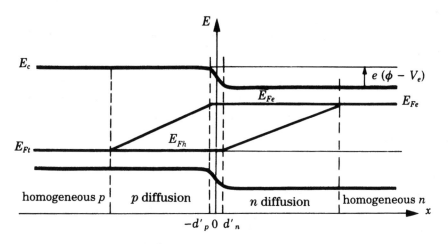

Fig. 8.13. Variation of the quasi-Fermi levels across a forward biased p–n junction. The slopes are not to scale. The width of the p (or n) diffusion zone is several times L_h (or L_e).

At the limit of the space charge we recover Eq. (8.35). The quasi-Fermi level of each species is approximately constant across the space-charge region because, if we neglect recombination, there is no other process allowing the electrons and holes to interact. The quasi-Fermi levels, which are the electrochemical potentials for the electrons and holes, are then decoupled.

8.3d Capacitance of a Junction

In the presence of external bias, the size of the space-charge region is altered. In the case $N_a \gg N_d$ we get from Eq. (8.33),

$$d_n'^2 = \frac{2\,\epsilon_0\,\epsilon_r\,(\phi - V_e)}{e\,N_d}. \tag{8.60}$$

There is an accumulation of charge between the two conducting materials, and hence a capacitance. The charge per unit area is $eN_d d_n'$ or

$$Q_c = [2\epsilon_0\,\epsilon_r(\phi - V_e)e\,N_d]^{1/2}. \tag{8.61}$$

If we change the bias V_e by a small amount dV_e we have a differential capacitance per unit area:

$$C_d = \frac{d\,Q_c}{dV_e} = 2^{-1/2}[\epsilon_r\,\epsilon_0\,e\,N_d]^{1/2}[\phi - V_e]^{-1/2}. \tag{8.62}$$

For $\epsilon_r\epsilon_0 \sim 10^{-10}$ F·m^{-1}; $N_d \sim 10^{21}$ m^{-3}; $(\phi - V_e) \sim 1$ V, the capacitance is $C_d \sim 0.01\ \mu$F/cm^2. This shows that in the equivalent electrical circuit of a p–n junction there is a capacitor in parallel across the junction. The value of the capacitance depends on the bias voltage V_e.

A device of this type is called a "varactor," a contraction of "variable reactor." The variation of capacitance with applied voltage is shown in Fig. 8.14. The ac behavior of p–n junctions is studied in Appendix 8.1. The control of a capacitance by an applied voltage has many applications; for example, varactors are used in the automatic frequency control of radio receivers.

8.3e Breakdown of a p–n Junction

If we apply a reverse voltage exceeding a certain value to a diode, the inverse current increases very rapidly, as shown in Fig. 8.15(a). This sudden increase is called "breakdown," and can occur as a result of two different mechanisms.

The first of these is the Zener effect, the direct tunneling of charge carriers between bands under large reverse bias. Figure 8.15(c) shows the band diagram in this case. The inverse polarization lowers the energies on the n side with respect to those on the p side. For a sufficient voltage there will be occupied states in the valence band on the p side at the same energy as empty states in the conduction band on the n side. There is therefore the possibility of tunnelling between the p and n sides. The width d of the energy barrier decreases as the reverse bias increases, and the probability of tunnelling depends exponentially on the width of the barrier. Thus the inverse current increases rapidly above the threshold V_Z.

Another mechanism which can cause breakdown is the multiplication of the number of carriers through an "avalanche." Within the junction there

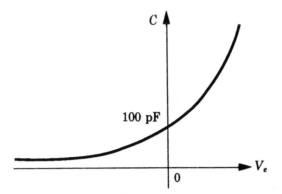

Fig. 8.14. Capacitance of a p–n junction as a function of applied voltage.

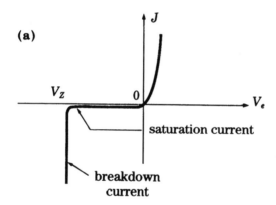

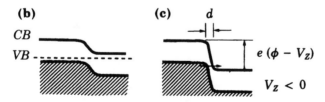

Fig. 8.15. (a) Current–voltage characteristics of a Zener diode; (b) band profile for $V_e = 0$; (c) band profile for $V_e = V_Z \ll 0$. (After Dalven, "Introduction to Applied Solid State Physics," Plenum Press, 1980.)

is a very strong electric field. The electrons can be accelerated by this field until they acquire very large kinetic energies, at which point the electron gas is no longer in thermodynamic equilibrium with the lattice. If an electron has a kinetic energy greater than the band gap it can create an electron-hole pair by giving up its energy. These new carriers are themselves accelerated, and can create new pairs, hence the name "avalanche" for this process, which clearly leads to a rapid increase of the reverse current.

8.3f Transient Response of a p–n Junction

An important characteristic of the behavior of a p–n diode in the large signal regime is its recovery time. Consider a junction forward biased at voltage $V_1(>0)$, in steady state, and assume that at time $t = 0$ the applied voltage is suddenly changed to a new reverse value $V_2(<0)$. The non-equilibrium carrier distributions will reach a new steady state only after a characteristic time τ_r.

Due to the rapid transit time, we can assume that the free carrier concentrations at the edges of the space region instantaneously take their new equilibrium values. We can describe the evolution of the non-equilibrium carrier densities from the initial steady state $(n_1(x), p_1(x))$ to the final steady state $(n_2(x), p_2(x))$ by introducing the distributions $\Delta n = n - n_1$ and $\Delta p = p - p_1$. These distributions obey the equations

$$\frac{\partial \Delta n}{\partial t} = D_e \frac{\partial^2 \Delta n}{\partial x^2} - \frac{\Delta n}{\tau_n} \quad \text{(p region)}, \tag{8.63}$$

$$\frac{\partial \Delta p}{\partial t} = D_h \frac{\partial^2 \Delta p}{\partial x^2} - \frac{\Delta p}{\tau_p} \quad \text{(n region)}, \tag{8.64}$$

with the boundary conditions

$$t = 0 \rightarrow \Delta n = \Delta p = 0, \quad \text{everywhere except at } x = 0, \tag{8.65}$$

$$x = 0 \rightarrow \left|
\begin{array}{l}
\Delta n = n_p[\exp(eV_2/kT) - \exp(eV_1/kT)], \\[4pt]
\Delta p = p_n[\exp(eV_2/kT) - \exp(eV_1/kT)].
\end{array}
\right. \tag{8.66} \tag{8.67}$$

The solution of these equations is shown schematically in Fig. 8.16. The distribution Δn (or Δp), initially confined near $x = 0$, diffuses into the p (or n) region, while $\Delta n(x = 0)$ (or $\Delta p(x = 0)$) remains constant. At times $t \ll \tau_n, \tau_p$, Δn and Δp have a spatial extent of the order of $(D_e t)^{1/2}$ and $(D_h t)^{1/2}$, respectively.

This gives a current density

$$\Delta J \sim \frac{eD_e \, \Delta n \, (0)}{\sqrt{D_e t}} + \frac{eD_h \, \Delta p \, (0)}{\sqrt{D_h \, t}} \propto \frac{1}{\sqrt{t}}. \tag{8.68}$$

For $t \gtrsim \tau_n, \tau_p$ recombination is important, bringing Δn and Δp to the new steady state

$$\Delta n \, (x) = \Delta n \, (x = 0) \times \exp(x/L_e),$$
$$\Delta p \, (x) = \Delta p \, (x = 0) \times \exp(-x/L_h). \tag{8.69}$$

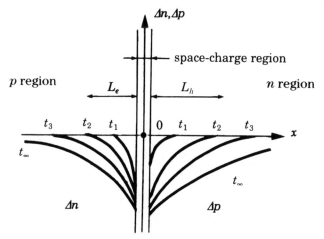

Fig. 8.16. Non-equilibrium densities of electrons (Δn) and holes (Δp) as functions of position for various times, when the applied voltage changes from $V_1 > 0$ to $V_2 < 0$. For clarity the curves are not drawn to $x = 0$ where they take the same value.

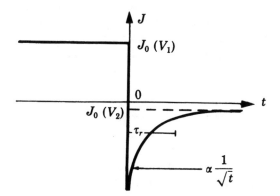

Fig. 8.17. Response of the current J when the applied voltage changes from $V_1 > 0$ to $V_2 < 0$ at time $t = 0$.

For $t \gg \tau_n, \tau_p$ we have $J \simeq J(V_2)$. The recovery time is the larger of the two times τ_n, τ_p. For short times the carriers injected earlier maintain the junction as apparently conducting, even though V_2 is negative, and a significant reverse current flows. This is shown in Fig. 8.17.

This effect is obviously a hindrance to using a $p-n$ junction as a rectifier, since it will not work above a frequency $f \sim 1/\tau_r$. High-frequency $p-n$ rectifiers have to be made with short recombination times, for example, by introducing impurities creating recombination centers in the semiconductor.

Appendix 8.1

Problem: Non-Stationary p–n Junctions and their High-Frequency Applications

Problems **3** and **4** are independent of **1** and **2**. Solution of problem **1** is enough for attempting the second part of this study.

Aim of the problem: p–n junctions are currently used as ac rectifiers. However, when the ac frequency is raised, new phenomena appear that make p–n junctions difficult to use as rectifiers, but make them valuable for other applications. Here we present an introduction to these phenomena.

In all of these problems we assume that the junction is subject to a time-dependent voltage of the form $V = V_p - V_n = V_0 + \mathrm{Re}(\delta V e^{i\omega t})$, where δV is infinitesimally small (the "small signal" domain) and we consider the linear response of the system, i.e., terms up to the first order in δV. (The use of the diode as a rectifier clearly departs from these assumptions.) As in the steady case, we must distinguish on the one hand phenomena associated with charges stored in the space-charge region, and on the other hand phenomena involving transfer of carriers across the junction. We assume that the voltage applied to the device appears fully and instantaneously at the edges of the space-charge region.

We shall adopt the following notation: ϵ_r is the relative dielectric constant of the semiconductor, N_a and N_d the doping of the p and n regions, respectively, N_c and N_v the equivalent densities of states of the conduction and valence bands, $\mu_e, D_e, \tau_n, L_e, n_p$ (and $\mu_h, D_h, \tau_p, L_h, p_n$) the mobility, diffusion coefficient, recombination time, diffusion length, and equilibrium concentration of electrons in the p region (and of holes in the n region), d_p and d_n the respective widths of the space-charge regions on the p and n sides.

First Part: Abrupt p–n Junction

The energy level scheme of an abrupt planar p–n junction in equilibrium is given in Fig. 8.5. The plane of the junction is taken as $x = 0$ and we call $\phi(> 0)$ the internal potential of the equilibrium junction.

1. Using the standard approximation of the space charge leading to bending of the bands in parabolic arcs, give an expression for the charge Q per unit area stored as fixed charges in the band-bending regions on either side of the junction plane as a function of V and the dopings N_a, N_d.

2. Deduce from the foregoing an expression for the current density $\delta J_1(t)$ associated with the variation of Q over time, and show how for small signals this effect gives the junction the behavior of a capacitor whose capacitance depends on V_0 (a "varactor"). Give the values of the "differential capacitance" C (defined in Sect. 8.3) per unit area and of the associated complex impedance Z_1. We recall that the complex impedance per unit area is the ratio $Z = \Delta V / \Delta J$, where ΔV and ΔJ are δV and δJ expressed in complex notation.

3. First neglecting δJ_1, determine the current density $J = J_0 + \delta J_2(t)$ corresponding to the passage of mobile carriers across the junction. One makes the usual (Shockley) approximations, which imply that the electrons and holes are each separately in thermal equilibrium in the space-charge region (Eq. (8.35)) and neglects recombination in this region. Write down the equations governing the evolution of the minority carrier densities (electrons in the p region and holes in the n region) in neutral regions, and the boundary conditions. Using the fact that δV is infinitesimal, write these equations and boundary conditions to zeroth and first orders in δV.

4. Solve these equations and find expressions for J_0 and δJ_2. Deduce as a function of V_0 the contribution Z_2 to the complex impedance per unit junction area associated with the transfer of mobile carriers. Explain physically the ω dependence of this impedance.

5. Combining the results of **2** and **4**, give an expression for the impedance per unit area of the junction. In a real diode the impedance does not tend to zero for infinite ω. Why?

6. We have so far taken δV as "infinitesimal." Up to what order of magnitude of δV_L (in volts) is this an acceptable approximation?

Second Part: Extension to Large Signal Regime: The p–i–n Diode

For high frequencies the allowable range of δV for a linear response can be extended beyond δV_L because of an effect related to the width of the space charge. This effect is particularly important in the pin diode studied here. The pin (p–i–n) diode, generally made of silicon, is derived from the p–n junction by adding an undoped silicon zone (i = "intrinsic") of width Δ $\sim 100 \ \mu$m between the p and n regions (Fig. 8.18).

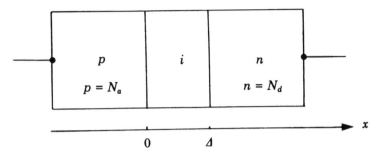

Fig. 8.18. Schematic view of a *pin* diode.

The three following questions deal with the energy level scheme of the *pin* diode in equilibrium. Lacking an analytic solution of the band bending in an *i* region, we attempt to understand it intuitively in the next two questions.

7. A simple approximation to a *p–i* junction is a *p–n* junction with $N_d \ll N_a$ (a $p-n^-$ junction). Using the results of **1**, what can you say about the width of the space-charge zone in the n^- region compared with the width of the corresponding zone in the *p* region? In which spatial region does most of the variation of the electrostatic potential occur?

8. Consider next the behavior of the space charge zone in an intrinsic semiconductor. Write down in one dimension the Poisson equation and the equation expressing the charge density as a function of the electrostatic potential, and deduce the differential equation satisfied by the potential. This equation has no analytic solution. Nonetheless, one can estimate the order of magnitude of the characteristic distance over which the bands curve. To this end, rewrite the equation in an approximate form in the limit of very small potential variation and give a solution in terms of a characteristic distance λ. Express the characteristic distance λ (the "screening length") as a function of the intrinsic equilibrium carrier density $2n_i$. Calculate λ for silicon at room temperature ($\epsilon_r = 11.7$; $n_i = 1.4 \times 10^{16}$ m^{-3}).

9. For typical concentrations N_a, N_d (see the main text), what is the order of magnitude of the width of the space-charge regions for the *n* and *p* parts of the junction? Assuming that the results of **8** remain valid for $V > kT/e$, give the form of the energy level scheme for an equilibrium *pin* diode.

10. When the diode is reverse biased ($-V_0 \gg kT/e$), the *i* region is almost completely depleted of carriers. Deduce a very simple approximation for the form of the electric field in the three regions p, i, and n. The differential capacitance of the diode then reduces to that of a plane capacitor. Give without calculation the expression for this capacitance per unit area.

11. Consider a forward biased diode ($V_0 \gg kT/e$), and assume for the moment $\delta V = 0$. Because of the bias of the *p–i* and *i–n* junctions, a significant number of carriers is injected into the i region: holes from the p region and electrons from the n region. In which sense does the screening length λ (found in **8**) vary? Deduce that the i region is neutral except for two layers of width λ, and that the electric field is either uniform or zero. We wish to show now that it is zero. For this we assume as in the main text that the electrons and holes are each separately in equilibrium in the space-charge regions and in the intrinsic region, and also that the recombination time is infinite. Applying Eq. (8.35) to the electrons and its equivalent to the holes, show that the potential difference between $x = 0_+$ and $x = \Delta_-$ is zero.

Sketch the shape of the energy bands as a function of x.

12. Assuming complete symmetry between electrons and holes ($N_c = N_v, N_a = N_d$), calculate the carrier density in the various regions. Show that the current is still given by the expression found for J_0 in **4** (Shockley's law).

13. The response of a *pin* diode to an ac excitation δV is essentially dominated by the i region. (You may assume this without proof.) This region is electrically equivalent to a resistance R_i in series with a capacitance C_i. We calculate C_i in this question and R_i in the next one.

If the diode is biased forward with a voltage V_0, the i region can store free carriers, electrons, or holes, which flow to the n side for electrons (and to the p side for holes) when the voltage V_0 is reduced to zero. We thus have a rather peculiar capacitor, which stores a hole charge Q_i and an electron charge $-Q_i$ in the same region of space. Calculate the value of Q_i per unit area, and the differential capacitance $C_i = dQ_i/dV_0$. Compare C_i with the capacitance found in **10** for a reverse bias.

14. For forward bias the i region does not behave simply as a capacitance C_i but also as a resistance R_i whose value is determined by the stored charge density. Give the value of R_i per unit surface as a function of V_0. Above what frequency ω is the impedance of a *pin* diode dominated by R_i rather than by the impedance of C_i? Show that the corresponding time can be regarded as a diffusion time for the i region.

Practical Importance of the p-i-n Diode

For a sufficiently high forward bias the impedance of the diode is small. This is therefore a short circuit controllable by the applied voltage V_0. The *pin* diode is thus an electronic switch controlled by the voltage. On the other hand if we apply a large ac voltage, since the impedance varies very rapidly with the voltage we have a non-linear device suitable for harmonic generation.

Solutions

1. In the presence of an applied voltage V, the potential difference between the n and p sides becomes $\phi - V$. The charge density is $-eN_a$ on the p side and $+eN_d$ on the n side. The respective widths d_p and d_n of these regions obey $(N_a d_p^2 + N_d d_n^2)e/2\epsilon_r\epsilon_0 = \phi - V$ (from continuity of the potential) and $N_a d_p = N_d d_n$ (field continuity). We deduce

$$d_p N_a = d_n N_d = \left[\frac{2\,\epsilon_r\,\epsilon_0\,(\phi - V)}{e} \frac{N_a\,N_d}{N_a + N_d} \right]^{1/2}. \tag{8.70}$$

The charge stored in the junction is $Q_p = -eN_a d_p$ on the p side, and the opposite on the n side, with

$$|Q| = Q_n = -Q_p = [2\,e\,\epsilon_r\,\epsilon_0\,(\phi - V)\,N_a\,N_d/(N_a + N_d)]^{1/2}. \tag{8.71}$$

2. Counting the current as positive in the direction $p \to n$ we have

$$\delta J_1(t) = \frac{d}{dt} Q_p = \mathrm{Re}\,(i\omega\,\delta V\,e^{i\omega t}) \left. \frac{dQ_p}{dV} \right|_{V=V_0}, \tag{8.72}$$

$$= \mathrm{Re}\,(i\omega\,\delta V\,e^{i\omega t}) \times \left[\frac{e\,\epsilon_r\,\epsilon_0}{2(\phi - V_0)} \frac{N_a\,N_d}{N_a + N_d} \right]^{1/2}, \tag{8.73}$$

which is of the form

$$\mathrm{Re}\,(iC\omega\,\delta V\,e^{i\omega t}) = C\,dV/dt.$$

This gives a differential capacitance per unit surface:

$$C = \left[\frac{e\,\epsilon_r\,\epsilon_0}{2\,(\phi - V_0)} \frac{N_a\,N_d}{N_a + N_d} \right]^{1/2}, \tag{8.74}$$

associated with a complex impedance $Z_1 = [i\omega C]^{-1}$.

3. For the p region the transport equation for the carriers is

$$\frac{\partial n}{\partial t} = D_e \frac{\partial^2 n}{\partial x^2} - \frac{n - n_p}{\tau_n}. \tag{8.75}$$

The boundary conditions are $\begin{cases} x \to -\infty & n \to n_p \\ x \to 0_- & n \to n_p \exp(eV/kT). \end{cases}$ (8.76)

With $V = V_0 + \mathrm{Re}\,(\delta V e^{i\omega t})$, and setting $n = n_0 + \mathrm{Re}(\delta n e^{i\omega t})$ we have to zero order in δV:

$$0 = D_e \frac{\partial^2 n_0}{\partial x^2} - \frac{n_0 - n_p}{\tau_n}. \tag{8.77}$$

The boundary conditions are $\begin{cases} x \to -\infty & n_0 \to n_p \\ x \to 0_- & n_0 \to n_p \exp\dfrac{eV_0}{kT}. \end{cases}$ (8.78)

To first order in δV:

$$i\omega \delta n = D_e \frac{\partial^2 \delta n}{\partial x^2} - \frac{\delta n}{\tau_n} \tag{8.79}$$

with boundary conditions $\begin{cases} x \to -\infty & \delta n_0 \to 0 \\ x \to 0_- & \delta n \to \dfrac{e\delta V}{kT} n_p \exp \dfrac{eV_0}{kT}. \end{cases}$ (8.80)

Similarly, setting $p = p_0 + \mathrm{Re}(\delta p e^{i\omega t})$ in the n region, we have, to zero order in δV:

$$0 = D_h \frac{\partial^2 p_0}{\partial x^2} - \frac{p_0 - p_n}{\tau_p} \tag{8.81}$$

with boundary conditions $\begin{cases} x \to +\infty & p_0 \to p_n \\ x \to 0_+ & p_0 \to p_n \exp \dfrac{eV_0}{kT} \end{cases}$ (8.82)

and to first order in δV:

$$i\omega \delta p = D_h \frac{\partial^2 \delta p}{\partial x^2} - \frac{\delta p}{\tau_p} \tag{8.83}$$

with boundary conditions $\begin{cases} x \to +\infty & \delta p \to 0 \\ x \to 0_+ & \delta p \to \dfrac{e\delta V}{kT} p_n \exp \dfrac{eV_0}{kT}. \end{cases}$ (8.84)

4. The solutions of the differential equations written down in **3** are: to zero order in δV:

$$p \text{ side: } n_0\,(x) = n_p \left\{ 1 + \left[\exp\left(\frac{eV_0}{kT}\right) - 1 \right] \exp\left(\frac{x}{L_e}\right) \right\}, \tag{8.85}$$

$$n \text{ side: } p_0\,(x) = p_n \left\{ 1 + \left[\exp\left(\frac{eV_0}{kT}\right) - 1 \right] \exp\left(-\frac{x}{L_h}\right) \right\} \tag{8.86}$$

with $L_e = \sqrt{D_e\,\tau_n}$ and $L_h = \sqrt{D_h\,\tau_p}$

to first order in δV:

$$p \text{ side: } \delta n\,(x) = n_p \frac{e\delta V}{kT} \exp(eV_0/kT) \exp(x\,\sqrt{1+i\omega\tau_n}/L_e), \tag{8.87}$$

$$n \text{ side: } \delta p\,(x) = p_n \frac{e\delta V}{kT} \exp(eV_0/kT) \exp(-x\,\sqrt{1+i\omega\tau_p}/L_h). \tag{8.88}$$

This gives the following currents at $x = 0$:

$$\begin{aligned} J_0 &= e\left(D_e \frac{\partial n_0}{\partial x} - D_h \frac{\partial p_0}{\partial x} \right)\bigg|_{x=0} \\ &= e\left(\frac{D_e\,n_p}{L_e} + \frac{D_h\,p_n}{L_h} \right) \left[\exp\left(\frac{eV_0}{kT}\right) - 1 \right], \end{aligned} \tag{8.89}$$

$$\delta J_2 = e \left(D_e \frac{\partial \delta n}{\partial x} - D_h \frac{\partial \delta p}{\partial x} \right)\Big|_{x=0}$$

$$= \frac{e^2 \delta V}{kT} \exp \frac{eV_0}{kT} \times$$

$$\mathrm{Re} \left[\left(\frac{D_e\, n_p\, \sqrt{1 + i\omega\tau_n}}{L_e} + \frac{D_h\, p_n\, \sqrt{1 + i\omega\tau_p}}{L_h} \right) e^{i\omega\tau} \right]. \qquad (8.90)$$

The complex impedance Z_2 is the ratio of δV to δJ_2, both taken as complex:

$$Z_2 = \frac{\delta V}{\delta J_2} = \frac{kT}{e^2} \exp\left(-e\frac{V_0}{kT} \right) \left(\frac{D_e n_p \sqrt{1 + i\omega\tau_n}}{L_e} + \frac{D_h p_n \sqrt{1 + i\omega\tau_p}}{L_h} \right)^{-1}$$

$$(8.91)$$

If $\omega \ll 1/\tau_n, 1/\tau_p$, the junction has the time to reach a steady state over a time of order half a period; Z_2 is thus real and close to dV_0/dJ_0.

If $\omega \gg 1/\tau_n, 1/\tau_p$, a steady state is never reached: diffusion only affects a thickness of order $(D/\omega)^{1/2}$ (the distance over which carriers diffuse in a time of order $1/\omega$), giving the dependence of δJ_2 on $\omega^{1/2}$.

Remark: an impedance with $(i\omega)^{-1/2}$ dependence is typical of diffusion processes. Here it simultaneously describes the resistance and capacitance associated with the injected charge.

5. $\delta J_{\mathrm{total}} = \delta J_1 + \delta J_2$. The impedances Z_1, Z_2 are thus combined in parallel:

$$Z = \left[\frac{1}{Z_1} + \frac{1}{Z_2} \right]^{-1}$$

$$= \left\{ i\omega \left[\frac{e\, \epsilon_r\, \epsilon_0}{2(\phi - V_0)} \frac{N_a\, N_d}{N_a + N_d} \right]^{1/2} + \right.$$

$$\left. \frac{e^2}{kT} \exp \frac{eV_0}{kT} \left(\frac{D_e\, n_p\, \sqrt{1 + i\omega\tau_n}}{L_e} + \frac{D_h\, p_n\, \sqrt{1 + i\omega\tau_p}}{L_h} \right) \right\}^{-1}$$

$$(8.92)$$

In practice the impedance of a real diode does not tend to zero for infinite ω as the resistance R_S of the semiconductor bulk is in series with Z:

$$Z_{\mathrm{total}} = Z + R_S. \qquad (8.93)$$

6. In the region in which Z is dominated by the capacitance of the space-charge region (that is, typically for $eV_0/kT \le 0$ for reverse bias), the linear approximation is acceptable for $\delta V \ll \phi - V_0 \sim 1$ V, typically. For a forward bias of the junction, Z_2 becomes dominant. The linear approximation means to replace the factor $\exp(e\delta V/kT)$ of the boundary conditions (8.76) by $1 + (e\delta V/kT)$ in the boundary conditions (8.80). This is acceptable only for $\delta V \ll kT/e \sim 25$ mV at room temperature.

7. From **1** we have $d_n/d_p = N_a/N_d \gg 1$, and the space-charge region is much larger on the weakly doped side (n^-). The variations of electrostatic potential similarly compare as

$$\frac{(e/2\epsilon_r\epsilon_0)\, N_d\, d_n^2}{(e/2\epsilon_r\epsilon_0)\, N_a\, d_p^2} = \frac{d_n}{d_p} \gg 1. \tag{8.94}$$

Most of the potential variation occurs on the less-doped side (n^-).

8. The Poisson equation is

$$\frac{d^2V(x)}{dx^2} = -\frac{\rho}{\epsilon_r\,\epsilon_0}. \tag{8.95}$$

In the presence of a potential difference V with respect to the intrinsic region, the densities of electrons and holes are given by

$$n\,(x) = n_i \exp[eV(x)/kT], \tag{8.96}$$
$$p\,(x) = n_i \exp[-eV(x)/kT]. \tag{8.97}$$

Hence

$$\rho\,(x) = e(p-n) = -2\, e\, n_i\, \mathrm{sh}[eV(x)/kT]. \tag{8.98}$$

The differential equation satisfied by the electrostatic potential is thus

$$\frac{d^2V(x)}{dx^2} = \frac{2\, e\, n_i}{\epsilon_r\,\epsilon_0}\mathrm{sh}[eV(x)/kT]. \tag{8.99}$$

If $eV(x) \ll kT$ we get the approximate equation

$$\frac{d^2V}{dx^2} = \frac{2\, n_i\, e^2}{\epsilon_r\,\epsilon_0\, kT}V, \tag{8.100}$$

whose general solution is

$$V = \alpha \exp(x/\lambda) + \beta \exp(-x/\lambda) \tag{8.101}$$

with

$$\lambda = (\epsilon_r\,\epsilon_0\, kT/2\, n_i\, e^2)^{1/2}. \tag{8.102}$$

The characteristic length λ is the screening or Debye length. This length characterizes the screening effect, i.e., the rapid spatial attenuation of the effect of an applied electrostatic perturbation on a conductor. This length is shorter the higher the free carrier density (whatever the sign of their charge). Here the total free carrier density is $2n_i$. Equation (8.102) gives $\lambda \simeq 24\ \mu$m for silicon at room temperature. This length is of the same order as Δ.

9. For typical dopings N_a, N_d the width of the space-charge regions on the n and p sides (see the text) is about $0.1 - 1\ \mu$m, and thus much smaller

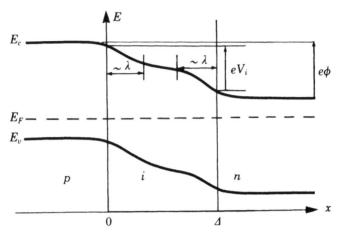

Fig. 8.19. Band profile of an equilibrium *pin* diode.

than λ and Δ. Similarly as in **7**, we deduce that the largest part V_i of the variation of the potential ϕ occurs in the intrinsic region (Fig. 8.19).

10. Under reverse bias, since the i region is almost completely depleted of carriers, we have $\rho \simeq 0$ there, giving $d^2V/dx^2 = 0$ from Poisson's equation. The electric field is constant in the i region. The only space charges are the narrow zones (cf. **7**) at the edges of the n and p regions. The potential variation is negligible in these zones, and we can approximate the electric field as

$$\mathcal{E} = 0 \ (n \text{ and } p \text{ regions}), \tag{8.103}$$

$$\mathcal{E} = \frac{\phi - V_0}{\Delta} \ (i \text{ region}). \tag{8.104}$$

The diode's capacitance is identical to that of a plane capacitor whose plates are the p–i and i–n boundaries, hence

$$C = \frac{\epsilon_r \, \epsilon_0}{\Delta}. \tag{8.105}$$

11. If there are many injected carriers, the carrier concentration clearly exceeds $2n_i$ (strong injection regime), and from **8** the screening length becomes much smaller than Δ. The space charge in the i region concentrates at the edges, and most of the i region becomes neutral $(n = p)$.

Since the i region is neutral, $\mathcal{E}$ is constant. We assume that $\mathcal{E}$ is non-zero, and show that this is contrary to the condition $n = p$.

Under the hypothesis that λ is small and $\mathcal{E}$ is constant, the band profile should have the form shown in Fig. 8.20 for forward bias V_0.

In the approximation where the electrons and holes are separately in equilibrium on either side of the space charge around $x = 0$, we have

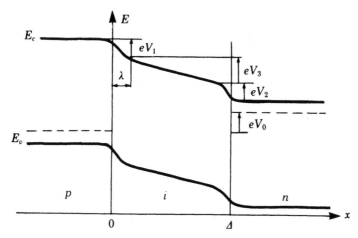

Fig. 8.20. Band profile in a forward biased *pin* diode, assuming non-zero electric field in the *i* region.

$$p\left(\lambda\right) = N_a \exp\left(-\frac{e\,V_1}{kT}\right). \tag{8.106}$$

Similarly

$$n\left(\Delta - \lambda\right) = N_d \exp\left(-\frac{eV_2}{kT}\right). \tag{8.107}$$

Further we also have equilibrium across the *i* region, so that

$$n\left(\lambda\right) = N_d \exp\left[-\frac{e\left(V_2 + V_3\right)}{kT}\right], \tag{8.108}$$

$$p\left(\Delta - \lambda\right) = N_a \exp\left[-\frac{e\left(V_1 + V_3\right)}{kT}\right]. \tag{8.109}$$

Charge neutrality implies the equality of Eqs. (8.106) and (8.108), and of Eqs. (8.107) and (8.109). We deduce

$$-\frac{eV_1}{kT} = -\frac{e(V_2 + V_3)}{kT} + \log\frac{N_d}{N_a}, \tag{8.110}$$

$$-\frac{eV_2}{kT} = -\frac{e(V_1 + V_3)}{kT} + \log\frac{N_a}{N_d}. \tag{8.111}$$

Adding Eqs. (8.111) and (8.110) gives $V_3 = 0$, showing that the bands are flat in the *i* region. The electron and hole concentrations are constant and equal in this region, and the electric field vanishes (Fig. 8.21).

12. Setting $V_3 = 0$ and equating (8.106) and (8.108) we have

$$N_a \exp\left(-\frac{eV_1}{kT}\right) = N_d \exp\left(-\frac{eV_2}{kT}\right). \tag{8.112}$$

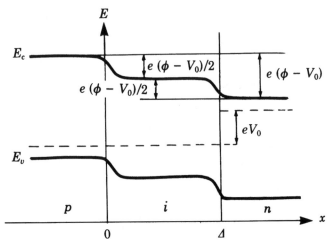

Fig. 8.21. Real band profile in a forward biased *pin* diode.

Further

$$e(V_1 + V_2) = e(\phi - V_0). \tag{8.113}$$

This fixes V_1 and V_2. If we assume for simplicity that $N_a = N_d$, then $V_1 = V_2 = (\phi - V - 0)/2$. In this case the electron and hole concentrations in the i region are equal, and

$$n = p = N_d \exp[-e(\phi - V_0)/2\, kT], \tag{8.114}$$
$$= n_i \exp(eV_0/2\, kT). \tag{8.115}$$

Charge neutrality holds again, but with charge densities higher than in equilibrium: the product np differs from n_i^2.

The minority carriers in the n and p regions are governed by the same equations and the same boundary conditions as in **3**. Then we have, as in **4**:

$$p \text{ region: } n = n_p[1 + (\exp(eV_0/kT) - 1)\exp(x/L_e)]$$
$$\text{and } p \approx N_a \tag{8.116}$$

$$n \text{ region: } p = p_n[1 + (\exp(eV_0/kT) - 1)\exp(-x/L_h)]$$
$$\text{and } n \approx N_d. \tag{8.117}$$

As there is no recombination in the i region, we can find the current as in **4** and in the main text:

$$J_0 = e\left(\frac{D_e\, n_p}{L_e} + \frac{D_h\, p_n}{L_h}\right) [\exp(eV_0/kT) - 1] \quad \text{(Shockley's law)}. \tag{8.118}$$

13. The total hole charge stored per unit area in the i region is

$$Q_i \simeq e\, n_i\, \Delta\, \exp(eV_0/2\, kT). \tag{8.119}$$

The total electron charge stored in this region is $-Q_i$. The variation of Q_i with V gives a capacitance per unit area:

$$C_i = \frac{dQ_i}{dV}\bigg|_{V=V_0} = \frac{e^2\, n_i\, \Delta}{2\, kT} \exp(eV_0/2kT). \qquad (8.120)$$

This capacitance can be much larger than that found in **10**. The transition between the two regimes occurs at

$$V_0 = \frac{2kT}{e} \text{Ln} \frac{2kT\, \epsilon_r\, \epsilon_0}{e^2\, n_i\, \Delta^2} = \frac{4kT}{e} \text{Ln} \frac{2\lambda}{\Delta} \qquad (8.121)$$

which is of the order of kT/e. In practice the capacitance of the junction will be dominated by this effect once $V_0 > 0$.

The existence of a capacitance when the i region is essentially neutral everywhere ($n = p$) may seem surprising at first sight. In contrast to a normal capacitor, electrons and holes are stored in the same spatial region, but with different chemical potentials (the quasi-Fermi levels E_{Fe}, E_{Fh}). They do not react because recombination is considered to be infinitely slow.

14. The series resistance of the i region per unit area is

$$R_i = \frac{\Delta}{n_i \exp(eV_0/2kT)\, e\, (\mu_e + \mu_h)}. \qquad (8.122)$$

This resistance dominates the impedance of the diode if $R_i \gg 1/C_i\omega$, i.e., for frequencies such that

$$\omega \gg 1/R_i C_i = \frac{2kT\, (\mu_e + \mu_h)}{e\Delta^2} = \frac{2\,(D_e + D_h)}{\Delta^2}. \qquad (8.123)$$

The associated response time $\Delta^2/2(D_e + D_h)$ represents the mean time an electron and a hole initially placed at opposite ends of the i region take to meet if they travel by diffusion. It is also the characteristic time taken by the i region to reach a steady state by diffusion. By imagining the inverse process (dissociation of an electron-hole pair into an n-region electron and a p-region hole) we can also regard this time as an effective lifetime for the i region.

9.

Applications of the p–n Junction and Asymmetrical Devices

The p–n junction has many applications which we sketch below. There are of course other asymmetrical devices such as the metal–semiconductor contact. We encounter this latter structure as soon as we try to use the electrical properties of semiconductors in circuits, and we must thus understand its properties. The metal–semiconductor contact properties are also at the basis of the Schottky diode operation.

Many of the properties of these devices depend directly on the surface properties of the semiconductors, which we shall summarize. These properties also allow the use of semiconductors as light detectors. Finally we shall describe junctions between chemically distinct semiconductors, or "heterojunctions."

9.1 Application of p–n Junctions

There are many applications of the p–n junction. The most obvious is their use as ac rectifiers, which produce the dc current required for motors, electrolysis, and all types of electronic devices. These applications make direct use of Shockley's law

$$J = J_s \left[\exp\left(\frac{e V_e}{kT} \right) - 1 \right]. \tag{9.1}$$

Other applications involve the physics we discussed in Chap. 8.

9.1a Photovoltaic Cells and Solar Cells

Consider an unbiased p–n junction. The junction is in equilibrium and there is an energy barrier $e\phi$ between the n side and the p side. The electrical field of the barrier is $\mathcal{E}$. If we illuminate the semiconductor with light of sufficient photon energy $\hbar\omega$ we create electron-hole pairs. Excess minority carriers are then accelerated by the internal field $\mathcal{E}$, which separates them, repelling the electrons to the n region and the holes to the p region. This is shown in Fig. 9.1(a). The separated carriers create an electric field $\mathcal{E}'$ which opposes the field $\mathcal{E}$. The resultant field is then $\mathcal{E} - \mathcal{E}'$, meaning that the potential drop between the p and n sides is reduced from ϕ to $\phi - V_f$ as shown in Fig. 9.1(b).

The effect is the same as a forward bias V_f applied to the junction, the net result being the creation of a potential difference V_f at the extremities of the junction. The appearance of this potential difference at the edges of an illuminated junction is called the "photovoltaic effect." The maximum value of V_f is ϕ, which is in turn less than E_g/e (Eq. (8.17)).

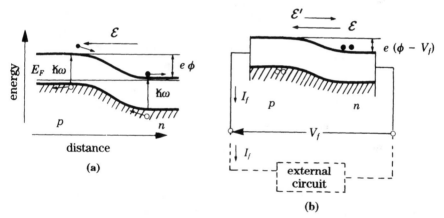

Fig. 9.1. Band profile of an illuminated p–n junction. (a) The electric field $\mathcal{E}$ of the junction separates the carriers created by the illumination. (b) The photocreated carriers induce an electric field $\mathcal{E}'$, and thus a voltage V_f usable in an external circuit.

When the junction is illuminated we may then say that a photocurrent $-I_f$ appears, in the opposite sense from the direct current. Connecting an external circuit to the illuminated diode allows us to measure this current. If the I–V characteristic of the junction is of the form (9.1), and if P is the photon flux and η the quantum efficiency, i.e., the number of pairs collected for one photon, then the photocurrent I_f is

$$I_f = 2\eta e P, \tag{9.2}$$

the total current under illumination is

$$I = I_s \left[\exp \left(\frac{eV_f}{kT} \right) - 1 \right] - I_f \tag{9.3}$$

and the voltage V_f is given by

$$V_f = \frac{kT}{e} \, \log \left[\frac{I + I_s + I_f}{I_s} \right]. \tag{9.4}$$

The current–voltage characteristic of the illuminated junction is shown in Fig. 9.2. The junction behaves as a generator in the region $V_f > 0, I < 0$.

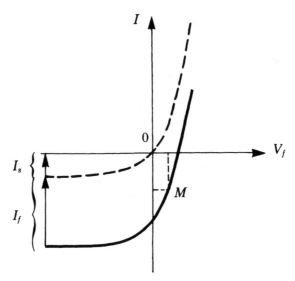

Fig. 9.2. Current–voltage characteristic of a photovoltaic cell. The full curve is that obtained under illumination, and the dashed curve without illumination. The distance between the two curves increases with I_f, and thus with illumination.

A junction diode can be used as a photoelectric cell and as a light detector, but its most important application is the direct conversion of light energy into electrical energy in a solar cell. The power of such a cell, i.e., the product $V_f \times I$, is maximal when the area of the rectangle OM is maximal.

The global yield of a solar cell depends on the fraction of radiation absorbed by the semiconductor and on the quantum efficiency. The solar spectrum has its maximum intensity at $\lambda = 0.5$ μm, and the spectral energy is still about half of this maximum for $\lambda = 1$ μm. This means that a diode with band gap E_g of about 2 eV is efficient in transforming the high-energy part of the solar spectrum but does not collect the infrared energy. By contrast a semiconductor with a band gap of 0.5 eV absorbs most of the photons arriving from the sun but the quantum efficiency decreases: solar photons whose energy greatly exceeds E_g create very energetic electron-hole pairs which can lose their energy by exciting crystal vibrations, i.e., by heating the crystal, and this energy is lost. Crystalline silicon is a good compromise; the theoretical quantum efficiency is 20% and cells have been constructed with efficiencies of 15%. Of course if the recombination time is too short the current I_f is smaller, the pairs recombining on the spot before the junction field separates them.

The power supplied by a solar cell is proportional to its surface: the flux of solar energy at the Earth is about 1 kW/m^2, so a 1 cm^2 cell will yield about 10 mW. It is expensive to produce large areas of crystalline silicon. This is why solar cells in everyday use (e.g., in calculators) use amorphous silicon (see Sect. 3.3), which is cheaper to produce. One can make junctions out of it, and the efficiency of such solar cells can reach 10%. These cells are currently used in pocket calculators.

9.1b Electroluminescent Diodes and Lasers

Electroluminescent diodes are devices that directly convert electrical energy into luminous energy via radiative recombination. This is the inverse transformation to that occurring in a photovoltaic cell. We have seen in Chap. 6 that in the presence of excess electron-hole pairs recombinations will occur, and some of these will be radiative. In a forward biased junction one can cause injection of excess carriers and thus light emission. The emission will be intense if the radiative efficiency is relatively large, i.e., if the efficiency of the non-radiative processes is not too large. This requires pure semiconductors, but also, from the formula of Appendix 6.1, semiconductors whose absorption coefficient is large at energies near the band gap. Semiconductors with direct gaps (see Sect. 2.6a), i.e., where the minimum of the conduction band and the maximum of the valence band are at the same point of the Brillouin zone, have large radiative efficiencies. This is the case for GaAs which emits in the very near infrared. The main materials now used for electroluminescent diodes are, besides GaAs, GaP and alloys $\text{GaAs}_{1-x}\text{P}_x$ which have band gaps between 1.4 and 2.2 eV. We thus get emissions from the near infrared to the green according to the amount of arsenic in the alloy.

To use electroluminescent diodes for electronic displays, we seek emission in a range where the human eye is most sensitive, i.e., in the green region of the spectrum.

These diodes are also used in optical telecommunications for converting an electrical signal into an optical one. The signal is then conveyed by optical fibers and transformed into an electrical signal by a photovoltaic cell. One wants to use the wavelength where the attenuation in the fiber is a minimum, in general 1.5 μm. For this one uses the compound $\text{Ga}_{0.47}\text{In}_{0.53}\text{As}$. The quantum efficiency in the best cases lies between 10^{-3} and 10^{-2}. All of this shows the importance of the study of recombination processes for practical optimization of such devices.

The radiative recombinations referred to above are spontaneous recombinations, i.e., incoherent. However we have seen in Appendix 6.1 that they are fairly monochromatic, the relative emission width being of order kT/E_g.

Besides these processes there exists a completely different emission process, stimulated emission which occurs when a photon of energy E_g meets an electron-hole pair of equal energy. There is then emission of a second photon in phase with the first one. If the number of pairs is large enough this process can be cumulative and lead to the emission of coherent monochromatic radiation, i.e., to a laser (Light Amplification by Stimulated Emission of Radiation). A threshold current must be reached to start the phenomenon in a diode. Semiconducting lasers are small and can work from a very simple energy source, an ordinary battery. Quantum well lasers made of GaAs between barriers of $Al_x Ga_{1-x}$ As, or of $Ga_{0.47} In_{0.53}$ As between InP barriers (cf. Appendix 3.2; Sect. 5.4 and Sect. 9.6) have particularly low threshold currents and are used in fiber optic telecommunications.

9.2 The Metal–Semiconductor Contact in Equilibrium

An understanding of the properties of a junction between a metal and a semiconductor, also named "Schottky diode," becomes necessary as soon as we weld metallic contacts on a semiconductor in order to use it in an electrical circuit. Will the contact be a rectifying one or an ohmic one, i.e., behaving as a pure resistance?

First we study the equilibrium junction and then the effect of applying a voltage. We must use the same energy scale for the quantum states of the metal and the semiconductor. We choose as energy reference the energy of a static electron at infinity (this is the energy zero in the hydrogen atom). We call this the "vacuum level."

Let us begin by some definitions. In a metal we call the work function ϕ_m the energy required to remove an electron that is close to the surface and at the metal Fermi energy E_{Fm}, and leave it in vacuum with zero velocity (see Fig. 9.3). In truth the energy of an electron close to the surface is not the same as that of a static electron at infinity as there is an interaction between the moving electrons of the metal and this electron. This "image charge effect" is small and we shall neglect it in the following.

An analogous definition applies to the semiconductor, and allows us to define its work function ϕ_s. However in a non-degenerate semiconductor the Fermi level E_{Fs} is in the band gap, and there are therefore no electrons at this level. Also the position of E_{Fs} relative to the maximum E_v of the valence band and the minimum E_c of the conduction band depends on the

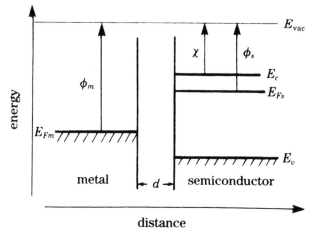

Fig. 9.3. Band profile of separated metal (left) and semiconductor (right). We have assumed that the work function ϕ_m of the metal exceeds that of the semiconductor (ϕ_s). The electron states occupied at $T = 0$ K are hatched.

doping. The work function ϕ_s is therefore not an intrinsic quantity and we prefer another quantity, which is intrinsic, the affinity χ which is the difference between E_c and the vacuum level E_{vac}. We have the following relations:

$$\phi_m = E_{\mathrm{vac}} - E_{Fm}, \tag{9.5}$$

$$\phi_s = E_{\mathrm{vac}} - E_{Fs}, \tag{9.6}$$

$$\chi = E_{\mathrm{vac}} - E_c = \phi_s - (E_c - E_{Fs}). \tag{9.7}$$

In the usual metals and semiconductors ϕ_m, ϕ_s, and χ are a few eV. To form a metal–semiconductor junction we imagine bringing the two solids closer, i.e., we reduce the distance d of Fig. 9.3. We assume in the following that the semiconductor is of type n, and in Sect. 9.2a we take $\phi_m > \phi_s$ as in Fig. 9.3. The other case is treated in Sect. 9.2b.

9.2a $\phi_m > \phi_s$

As d becomes very small the electrons from the semiconductor can pass into the metal by tunnelling until the Fermi levels line up (Fig. 9.4). There is then in the semiconductor a space-charge region of width W, empty of electrons, where there are only the fixed positive charges of the uncompensated donors. As the number of available electron states in the metal is very large, the screening is very strong and only the metal surface is affected. The electron current forms a negative surface charge layer, which, with the positive space charge of the semiconductor, creates an electric field confined almost entirely within the space-charge region.

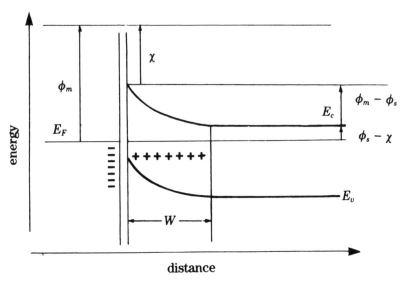

Fig. 9.4. Band profile for a metal and a semiconductor almost in contact.

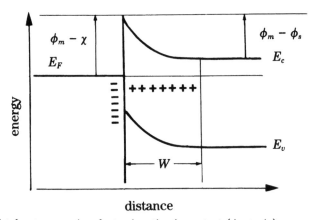

Fig. 9.5. Metal–n-type semiconductor junction in contact ($\phi_m > \phi_s$).

An electric potential and thus a bending of the bands results. In the semiconductor the energy of an electron at the bottom of the conduction band is greater at the surface, where this electron is an amount χ below E_{vac}, than in the interior, where it is $\phi_s - \chi$ above the Fermi level, and thus at $\phi_m - (\phi_s - \chi) = \chi + (\phi_m - \phi_s)$ below E_{vac}. The energy of the band bending is thus $(\phi_m - \phi_s)$. When the contact is made the band profile has the form shown in Fig. 9.5.

As the figure shows, the barrier $\phi_m - \chi$ for crossing from the metal to the semiconductor is larger than the one separating the semiconductor from the metal, which is $\phi_m - \phi_s$. The width of the band bending is found by solving the Poisson equation: in the space-charge region the charge density is eN_d, where N_d is the donor concentration.

From

$$\frac{d^2V}{dx^2} = -\frac{e\,N_d}{\epsilon_0\epsilon_r} \tag{9.8}$$

and the boundary conditions for $x = W$:

$$\frac{dV}{dx} = 0, \tag{9.9}$$

$$-e[V(W) - V(0)] = -(\phi_m - \phi_s), \tag{9.10}$$

we deduce that the bending is parabolic and that

$$\frac{e^2 N_d W^2}{2\,\epsilon_0\,\epsilon_r} = \phi_m - \phi_s \tag{9.11}$$

or

$$W = [2\,\epsilon_0\,\epsilon_r(\phi_m - \phi_s)/e^2\,N_d]^{1/2}. \tag{9.12}$$

This expression is very similar to that for the half-width of a p–n junction (Eq. (8.27)). The accumulated charge per unit area in this region is

$$Q_{sc} = e\,W\,N_d = [2\epsilon_0\epsilon_r N_d(\phi_m - \phi_s)]^{1/2}. \tag{9.13}$$

9.2b $\phi_m < \phi_s$

We now turn to the case where the metal work function is less than that of the semiconductor. When the two solids are far apart the energy level diagram is that of Fig. 9.6, which we should compare with Fig. 9.3. We imagine the two solids brought close together.

When the distance d becomes very small, electrons pass from the metal to the semiconductor: this creates a positive charge layer at the metal surface and a negative one near the semiconductor surface, until there is only a single common Fermi level. At the surface the conduction band remains a distance χ from E_{vac}, as shown in Fig. 9.7.

Because of the existence of these charges there is an electric field and hence a bending of the bands. The transferred electrons occupy the region where E_c is below the Fermi level, and constitute an accumulation layer whose thickness W' is much less than the width W of the band-bending region. Because of the large density of states in the conduction band, a small W' is enough to accommodate the electrons transferred from the metal. Once contact is made, the energy profile becomes as in Fig. 9.8.

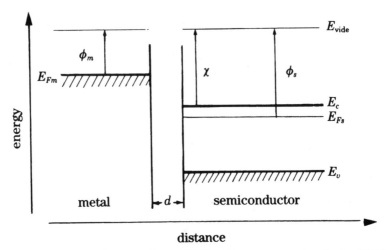

Fig. 9.6. Band profiles of separated metal (left) and an n-type semiconductor (right), in the case $\phi_m < \phi_s$.

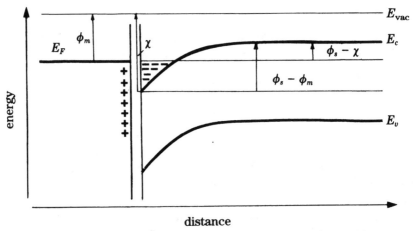

Fig. 9.7. Metal–n-type semiconductor with contact almost made ($\phi_m < \phi_s$).

9.3 Non-Equilibrium Metal–Semiconductor Diode

9.3a $\phi_m > \phi_s$

In this case there is a very resistive space-charge region, and the applied external potential V_s appears exclusively across this region; in the rest of the system the bands remain horizontal on the energy diagram, and are simply shifted. The structure of Fig. 9.5 becomes that of Fig. 9.9(a) for forward bias (potential more negative on the semiconductor side, $V_e > 0$) and that of Fig. 9.9(b) for reverse bias ($V_e < 0$).

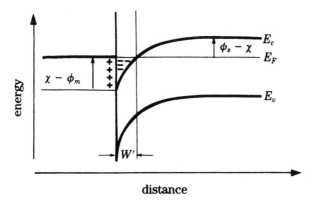

Fig. 9.8. Metal–n-type semiconductor junction in contact ($\phi_m < \phi_s$).

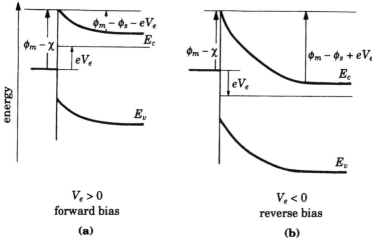

Fig. 9.9. Polarized metal–semiconductor junction ($\phi_m > \phi_s$): (a) $V_e > 0$, semiconductor negatively biased with respect to the metal, forward sense; (b) $V_e < 0$, reverse bias.

In the absence of V_e the current vanishes, so the current J_1 from the metal to the semiconductor exactly balances the current J_2 from the semiconductor to the metal. Applying V_e changes the barrier between the semiconductor and the metal by an amount $-eV_e$; the corresponding current becomes J_2'. However the current J_1 stays the same, since the barrier in the metal–semiconductor direction is unchanged. As the current J_2' corresponds to an activation process, the number of electrons able to cross the barrier is given by a Boltzmann factor, such that

$$J_2' = J_1 \exp \frac{e V_e}{kT}. \tag{9.14}$$

The resulting current $J = J_1 + J_2'$ is

$$J = J_1 \left[\exp\left(\frac{e V_e}{kT} \right) - 1 \right] \tag{9.15}$$

The current–voltage characteristic $J(V_e)$ is analogous to that of a p–n junction and gives rise to rectifying behavior. The saturation current is to first approximation the thermionic emission current of the metal below the $\phi_m - \chi$ barrier.

9.3b $\phi_m < \phi_s$

The width of the space-charge region is small, so this region has very low resistance. Thus the voltage V_e appears over the bulk semiconductor and produces a "sloping" band profile. The structure of Fig. 9.8 becomes Fig. 9.10(a) or that of Fig. 9.10(b) according to the sign of V_e.

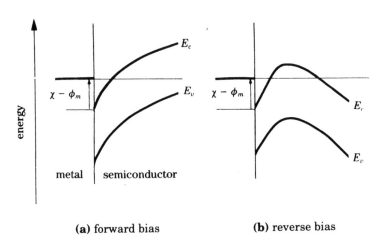

(a) forward bias   (b) reverse bias

Fig. 9.10. Metal–semiconductor junction ($\phi_m < \phi_s$): (a) forward biased, $V_e > 0$; (b) reverse biased, $V_e < 0$. A way of remembering the shape of the figure is to imagine that the band structure is made of rubber bars attached to the metal side and one seeks to bend them up or down to the right.

For forward bias, there is no barrier to cross from the semiconductor to the metal since near to the contact all of the electron states are full.

For reverse bias, there is a barrier to cross of order $\phi_s - \chi$, the equilibrium distance of the Fermi level from the conduction band. In a semiconductor of typical n doping, this barrier is no more than one hundred meV; it is thus easily crossed by the bias electrons. In consequence the current passes in a similar way for both: one observes an approximately ohmic behavior (non-rectifying).

In summary, a metal–n-type semiconductor junction behaves very differently, depending on whether the work function ϕ_m of the metal is less than or greater than that of the semiconductor (ϕ_s). If we want to make a simple ohmic contact on a given semiconductor we must choose a metal whose work function ϕ_m is less than ϕ_s. By contrast if we wish to make a diode we take a metal with ϕ_m greater than ϕ_s.

We can easily show that these conclusions reverse if we consider metal contact with a p-type semiconductor. Note that the above discussion neglected the presence of surface states in the semiconductor. We will consider their effect in Sect. 9.4.

We note finally that in contrast to the p–n junction, the currents crossing metal–semiconductor junctions are always majority carrier currents.

9.4 The Semiconductor Surface

A free surface constitutes a rupture of the periodicity of the solid. The crystal potential vanishes outside the solid, thus, _a priori_, we would expect to represent it as the dashed curve in Fig. 9.11 near the surface. In reality quantum mechanics predicts a non-zero probability of finding the electron in the vacuum near the surface. There is then a lack of negative charge, thus a positive charge, in the first atomic layers of the solid. We thus have a surface dipole, whose effect is to create a potential jump. The shape of the potential is therefore that of the full curve in Fig. 9.11(b), the surface potential jump being directly related to the work function of the solid.

Moreover the presence at the surface of dangling (unsaturated) bonds or atomic rearrangements induces surface electron states. These states can be acceptors or donors. In usual semiconductors, e.g., Si or GaAs, the density N_S of these states is very high, of the order of a fraction of a state per surface atomic site. For example, at the surface (110) of GaAs where the total atom density is 8.8×10^{14} cm^{-2}, the density N_S of quantum surface states is of order 10^{13} cm^{-2}.

Let us consider the effect on the band profile of the introduction of surface states. We start with a semiconductor with no surface states: its bands are flat (Fig. 9.12(a)). We then consider the introduction of a few surface states, e.g., surface acceptors whose energy level is Δ below the conduction band, in an n-doped semiconductor. Electrons from the bulk will occupy these states since their energy is below the Fermi level.

There will remain an uncompensated charge on the semiconductor, and thus a depletion zone of height ϕ_b as in Fig. 9.12(b). Charge neutrality requires these surface charges Q_S to be compensated by the total volume charge:

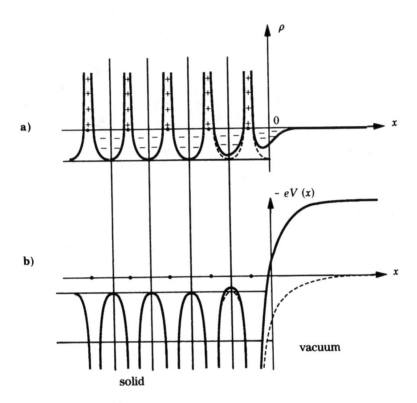

Fig. 9.11. (a) Charge density ρ and (b) electrostatic energy near a solid–vacuum interface. The dashed curves take no account of the probability of finding an electron just outside the solid, while the full curves correspond to the real situation. The dashed and solid curves differ only within two interatomic distances of the surface.

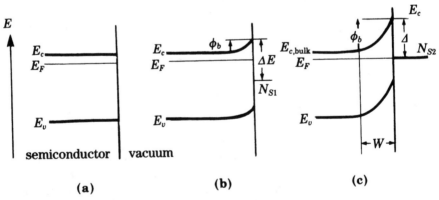

Fig. 9.12. Band profile at a semiconductor–vacuum interface: (a) semiconductor without surface states; (b) low density N_{S1} of surface states; (c) high density N_{S2} of surface states. The Fermi level is then said to be pinned by these states.

$$Q_{\mathrm{vol}} = (2\epsilon_0\epsilon_r N_d \phi_b)^{1/2} = -Q_S. \tag{9.16}$$

For average doping, $N_d \sim 10^{22}$ m^{-3} and $\phi_b \sim 1$ eV, the charge accumulated in the space-charge region is of the order of 3×10^{11} electrons per cm^2. **We obtain a remarkable result: the charge which can be accumulated in the volume, for fixed doping, is limited, as ϕ_b is always smaller than E_g.** Consequently the surface charge Q_S is limited, independently of the number of surface states N_S. These surface states are thus only partially filled if their number exceeds $(\epsilon_0\epsilon_r N_d \Delta)^{1/2}/e$: the Fermi level is thus a few kT from the surface quantum levels. This is the situation depicted in Fig. 9.12(c). We say that the Fermi level is **pinned** by the surface states. In fact the Fermi level of the solid is still determined by the concentrations of dopants and not by the surface. As the surface states must be within a few kT from the Fermi level, it is the band bending ϕ_b that adjusts its value to fulfill this condition. The Fermi level is at a distance ΔE (to within a few kT) of the surface conduction band. We have

$$\Delta E = \phi_b + E_{c,\mathrm{bulk}} - E_F, \tag{9.17}$$

where $E_{c,\mathrm{bulk}}$ denotes the position of the bulk Fermi level. If we make a junction between a metal and this semiconductor the height of the barrier from the semiconductor side will be ϕ_b, independent of the work function of the metal.

The possible existence of surface states may have a considerable influence on properties of semiconductor interfaces, junctions. In particular the discussion of Sect. 9.3 on the metal–semiconductor diode has to be modified to account for their influence.

9.5 Photoemission from Semiconductors

In the photoemission process, or photoelectric effect, a photon gives energy to a bound electron and releases it from the solid. Let us consider a p-type semiconductor. If the electron is in the valence band (Fig. 9.13) the required energy is $E_g + \chi$, where χ is the affinity defined in Sect. 9.2. For p doping the bands bending caused by the pinning of the Fermi level is downwards, so that the surface affinity is $\chi_{\mathrm{surf}} = \chi_{\mathrm{volume}} + \phi_b$. For a clean surface χ_{1bulk} and χ_{1surface} are of order 5 eV, and only an ultraviolet photon has enough energy $h\nu_1$ to liberate electrons.

We have seen in Sect. 9.4 that the value of the work function or the affinity is determined by the size of the surface charge dipole. This dipole can be reduced by the surface deposition of a monolayer of cesium, an easily ionized alkaline metal. The vacuum level is shifted relative to the semiconductor bands, and we can even reach a situation in which the vacuum level E_{2vac} is lower than the level E_c of the bottom of the conduction band in

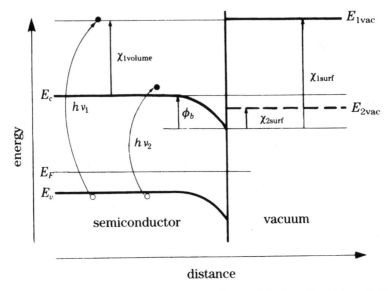

Fig. 9.13. Vacuum–*p*-type semiconductor interface and the lowering of the affinity. For an excitation energy $h\nu_2$ near the forbidden band, photoemission does not occur unless the affinity has been reduced so that the vacuum level $E_{2\text{vac}}$ is lower than E_c.

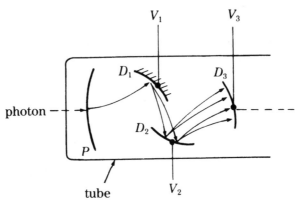

Fig. 9.14. Schematic view of a photomultiplier tube. The photoemissive surface P is followed by dynodes $D_1, D_2, D_3, \ldots$ held at potentials $0 < V_1 < V_2 < V_3, \ldots$ which multiply the electrons. (From Dalven, "Introduction to Applied Solid State Physics," Plenum Press, 1980.)

the bulk of the solid. The affinity is then $\chi_{2\text{surface}}$. To remove an electron it therefore suffices to excite it with a photon of energy $h\nu_2$ just exceeding E_g, and thus in the visible range. If the electron is excited into the conduction band close enough to the surface it may be emitted into vacuum: this is possible if the electron reaches the surface after its random walk in the bulk, before recombination.

This is the principle of the photocathodes in GaAs photomultipliers. These are very sensitive light detectors, whose quantum efficiency can reach 30%, or 0.3 electrons per photon. After being extracted from the cathode by photoemission, the electron passes through several multiplier stages, called dynodes, held at increasing positive potentials: at each stage an electron gives rise to several secondary electrons. This is shown in Fig. 9.14. These detectors, which can be used in the visible or the ultraviolet range (for any photons with energy larger than the band gap of the semiconductor) are very sensitive and can count single photons.

9.6 Heterojunctions

In Chap. 8 we studied the $p{-}n$ junction, made of two semiconductor samples of the same chemical composition but different dopings ("homojunction"). If we have two semiconductors A and B differing in their chemical composition and possibly their doping, but with sufficiently similar crystalline lattices that epitaxy, i.e., a continuity of the crystal lattices of A and B, is possible, we can form a "heterojunction."

We take the example of a junction formed from p-type GaAs and n-type $Al_x Ga_{1-x}$ As. Figure 9.15 shows the band structure when the two semiconductors are apart.

The material with the larger band gap is $Al_x Ga_{1-x}$ As. The vacuum level, i.e., that of a free electron at rest outside the solid, is the same for the two materials, and the affinities are $e\chi_1$ and $e\chi_2$. The difference between the two band gaps $E_{g2} - E_{g1}$ is divided unequally between the valence bands (ΔE_v) and the conduction bands (ΔE_c). In non-degenerate semiconductors these quantities do not depend on the doping. Once the contact is made the Fermi levels are aligned. An internal potential difference $e\phi = E_{F2} - E_{F1}$ appears here too. Figure 9.16 corresponds to a heterojunction made from the materials depicted in Fig. 9.15.

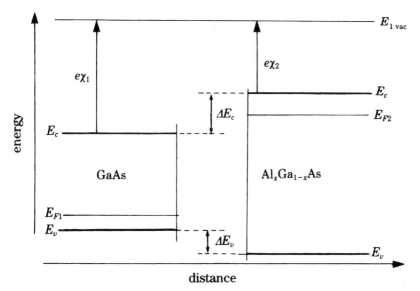

Fig. 9.15. Band profiles of two different semiconductors when far apart from each other: on the left GaAs (p type); on the right $Al_x Ga_{1-x}$ As (n type).

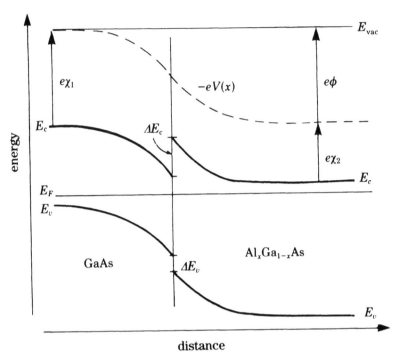

Fig. 9.16. Band profile of a heterojunction formed from the materials of Fig. 9.15. The dashed curve shows the behavior of electrostatic potential energy.

We note from Fig. 9.16 the asymmetry between the electrons and holes for such a band profile: in the conduction band we have a potential well and a barrier, and in the valence band a simple discontinuity, which is increased by ΔE_v with respect to the homojunction. We can guess that this increase of the height of the potential barrier will modify the transport of holes in the *p–n* direction, and that moreover the GaAs electrons will tend to accumulate in the (triangular) conduction well. The width of this well is a fraction of the width of the space charge W, i.e., a few tens of nanometers, depending on the doping. We thus obtain a quasi-two-dimensional electron layer.

Heterostructures are in current use. We noted in Sect. 5.4b that by selective doping of superlattices, which are periodic repetitions of heterojunctions, one can obtain extremely high electron mobilities, the electrons and ionized donors being spatially separated.

10.

The Principles of Some Electronic Devices

In this last chapter we apply the concepts introduced in this book to explain the operation of several electronic devices which make use of junctions: the junction transistor, the field-effect transistor (FET), the junction FET, and the MOSFET. Some problems in Appendices 10.1, 10.2, and 10.3 give a more quantitative description. We shall describe the principle of integration and planar technology, and we shall discuss the concepts and hopes of band structure engineering, which allows the manufacture of semiconductors conceived on paper for particular uses. Miniaturization of circuits and memories has been the great technological revolution of the last 40 years. The physical limitations of this revolution are briefly discussed at the end of this Chapter.

10.1 The Junction Transistor

A junction transistor is a single crystal containing two junctions (in principle monocrystalline) of opposite polarities in series. We can thus have p–n–p or n–p–n transistors. Here we describe the operation of a p–n–p transistor. The results apply without restriction to n–p–n structures if we change the directions of the currents and voltages and permute the symbols p and n. These transistors are also called bipolar transistors, as their operation relies on the existence of two types of carriers.

In Fig. 10.1 three regions are shown: the p^+-doped emitter, the weakly n-doped base, and the moderately p-doped collector.

The base is smaller than the diffusion length in a recombination time. A forward voltage is applied to the emitter-base junction, and reverse voltage to the base-collector junction. We shall confine ourselves to a semi-quantitative description of the working of this junction transistor.

The applied emitter-base voltage is positive and the applied base-collector voltage is positive and large. Thus:

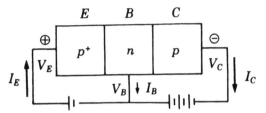

Fig. 10.1. Bias of a *p–n–p* transistor and sign conventions for the currents.

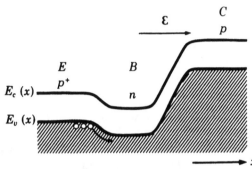

Fig. 10.2. Schematic band diagram for a biased *p–n–p* transistor: E = emitter; B = base; C = collector. (From Dalven, "Introduction to Applied Solid State Physics," Plenum Press, 1980.)

$$V_B - V_C \gg V_E - V_B \geq 0. \tag{10.1}$$

Figure 10.2 shows the band diagram for a transistor biased in this way. The detailed equations governing the working of a junction transistor are given in Appendix 10.1. Consider a flux of particles within the transistor as shown in Fig. 10.3. Since the emitter-base junction is forward biased, there is an injection of holes into the base (flux 1 of the figure). The holes diffuse (flux 2) without recombining if the base is small. However at the base-collector junction there is a strong field that sweeps up the holes reaching the base-collector space-charge region (flux 3). In fact, a few holes diffusing into the base will recombine with electrons, the majority carriers within the base (which is weakly *n*-doped); this is the flux 4 of holes, which do not reach the collector. Electrons must therefore be supplied by the external circuit to replace electrons recombining with the holes. This flux is represented by the dashed arrow (flux 5). Also there are electrons injected from the base to the emitter in the forward biased emitter-base junction, but this injection is small as the base is weakly *n*-doped. This is flux 6.

Overall the relation between the emitter-base current and the voltage differs little from the characteristic equation for a junction, i.e.,

$$I_E = I_s \left[\exp\left(\frac{V_E - V_B}{kT} \right) - 1 \right]. \tag{10.2}$$

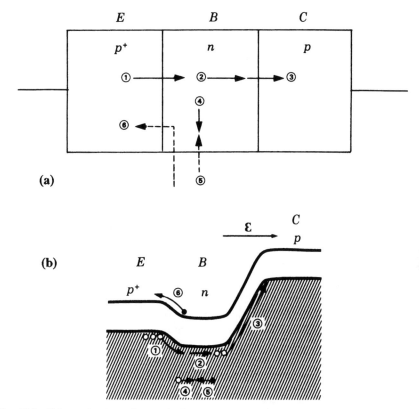

Fig. 10.3. Schematic view of particle fluxes in a p–n–p transistor: (a) across the leads; (b) on the band diagram. The arrows denote the particle currents (thick for electrons and dashed holes). We note that the electric currents corresponding to (5) and (6) add, respectively, to (4) and (1). (After Dalven, "Introduction to Applied Solid State Physics," Plenum Press, 1980.)

On the other hand, we see that all the fluxes are proportional and that I_C is a little smaller than I_E. We set

$$I_C = \beta \, I_B. \tag{10.3}$$

Since

$$I_E = I_C + I_B, \tag{10.4}$$

$$\frac{I_C}{I_E} = \frac{\beta}{1+\beta}. \tag{10.5}$$

The factor β is called the current gain of the transistor. This factor is typically 100 and shows that the current I_C is only slightly less than I_E. A large current gain requires a very thin base, and the diffusion coefficient and recombination time must be large. The effective width of the base depends on the technology used in manufacturing the transistor. But it also depends on the doping of the base. We have seen that if the base is weakly doped the space-charge region will extend further on each side of the base, leaving at the center a narrower neutral zone for the holes to cross. The lifetime of the minority carriers will be longer in a semiconductor with an "indirect gap" where the recombination probability is lower (cf. Appendix 6.2) as the optical absorption is weak. Silicon is a good example.

Application: Transistor Amplifier

Here we shall describe a very simple example of the use of a p–n–p transistor as an amplifier, with the further aim of introducing planar technology, the basis of integration at both small and large scales (cf. Sect. 10.4). The symbols shown in Fig. 10.4 are used to represent p–n–p and n–p–n transistors in circuit diagrams; the base is the connection to the left, and the emitter is represented by the arrow, whose sense is the forward sense of the base-emitter junction. The simplest amplifier then consists of the arrangement in Fig. 10.5. The transistor is suitably biased by the dc voltages V_1 and V_2, and we can use Eqs. (10.2) and (10.3). The generator produces an input signal of small amplitude s which may, for example, be sinusoidal. We can write down Ohm's law for the base loop

$$s + V_2 = (V_B - V_E) - R_E\, I_E. \tag{10.6}$$

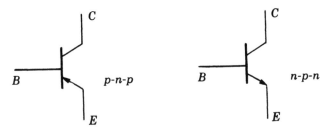

Fig. 10.4. Conventional representation of p–n–p and n–p–n junction transistors. The sense of the arrow is that of the direct electric current across the emitter-base junction.

In this expression $(V_B - V_E)$ is always small compared with the other voltages in the circuit as the junction is forward biased. Thus using Eq. (10.5),

$$I_C = \frac{-\beta}{\beta+1}\frac{(s+V_2)}{R_E} \sim \frac{-(s+V_2)}{R_E}. \tag{10.7}$$

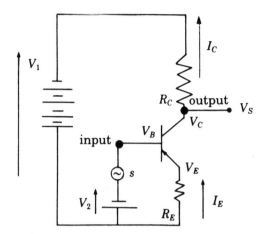

Fig. 10.5. Arrangement of a *p–n–p* transistor as an amplifier.

There is therefore a small sinusoidal current $\sim -s/R_E$ in the collector and thus an output voltage signal $-sR_C/R_E$ at the collector, the output terminal of the amplifier. The voltage gain is therefore

$$|G| = \frac{R_C}{R_E}. \tag{10.8}$$

One can easily obtain gains of the order of 100. An *n–p–n* transistor is (or rather was) made by pulling a large crystal from an *n*-doped bath whose composition is changed by adding acceptors in sufficient numbers, and then again by adding donors in still larger numbers. If necessary this operation can be repeated several times, since as we saw in Chap. 4, what matters is $N_d - N_a$ in the *n* material and $N_a - N_d$ in the *p* material. We thus obtain a cylindrical bar alternately *n* and *p* doped, which we cut off and arrange as in Fig. 10.6.

In the device studied above the intrinsic limit to the operation speed of the transistor is the transit time of the minority carriers in the base, i.e., the time taken for the injected holes to diffuse to the base collector junction. This time is given by

$$\tau = \frac{d^2}{D}, \tag{10.9}$$

where d is the base thickness. With a very thin base of 10^{-6} m this time is about 10^{-9} s for silicon. If we wish to make very high-frequency transistors we must use a material such as gallium arsenide GaAs where the mobility, hence the diffusion coefficient, is higher than in silicon and a thin base.

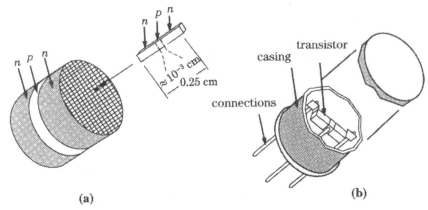

Fig. 10.6. Manufacture (a) and mounting (b) of an n–p–n transistor. (From Marmstadt and Enke, "Digital Electronics for Scientists," Benjamin, 1969.)

10.2 The Field-Effect Transistor

This frequency limitation does not exist in devices whose functioning relies only on the majority carriers, such as field-effect transistors (FETs). In a FET, the current is controlled by an applied voltage: such a transistor is essentially a resistance whose value is controlled by the applied voltage.

10.2a The Junction FET

Figure 10.7(a) shows a rectangular bar of n-type silicon with two metallic contacts at its ends, the source and drain. It acts as a resistor. Suppose now that we create above and below this resistor two p^{+}–n junctions as shown in Fig. 10.7(b). We reverse bias these junctions using a contact. Then the original conductance of the resistor

$$G = \frac{1}{R} = \frac{\sigma W l}{L} \tag{10.10}$$

is reduced as the width of the neutral region where the electrons can flow is decreased by the width of the space charge. This in turn is controlled by the voltage applied at the gate G. We thus obtain a resistor controlled by the voltage. The problems of Appendix 10.2 analyze the junction FET in greater detail.

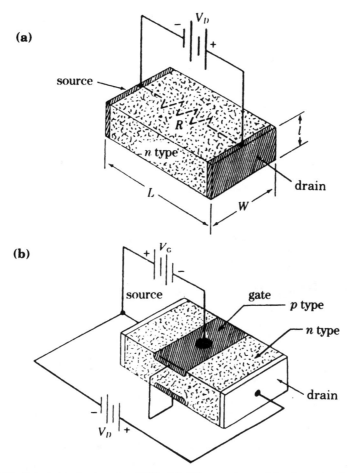

Fig. 10.7. Principle of a junction FET. (a) Silicon resistor; (b) schematic picture of a junction FET. (After Marmstadt and Enke, "Digital Electronics for Scientists," Benjamin, 1969.)

10.2b The MOSFET

The MOSFET (Metal-Oxide–Semiconductor Field-Effect Transistor) is a unipolar device where the resistance of a semiconducting channel is controlled electrostatically. The idea is to use a capacitor, one plate of which is metallic and the other being the semiconductor. The structure is shown in Fig. 10.8(a).

For a silicon structure the insulator is a layer of silica (SiO_2) obtained simply by oxidation of the semiconductor. When a voltage is applied across such a system a charge appears on the two plates of the capacitor. The car-

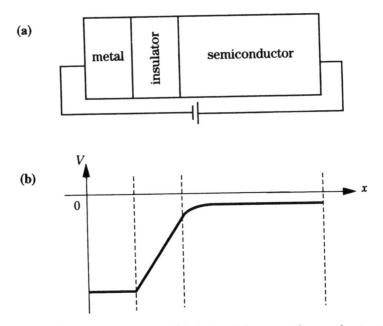

Fig. 10.8. (a) Structure of a MOSFET; (b) behavior of the potential across the structure.

riers distribute themselves so as to create a total charge opposed to that at
the metal surface. There thus appears a space charge near the semiconductor surface. We shall see that depending on the sign and size of the charge,
we may have accumulation or depletion of free carriers, or even inversion,
i.e., the appearance of a layer of minority carriers at the surface. In this
system the semiconductor carriers are in thermodynamic equilibrium, since
there is no current in the semiconductor. This is possible, even though the
Fermi levels in the metal and the semiconductor differ, because the presence
of the insulator prevents charge transport perpendicular to the surface of
the device.

The equations governing the charge distribution are thus the Poisson
equation (Eqs. (8.2) and (8.3)) (8.8) and the law $np = n_i^2$. The problem is
the same as that of the equilibrium junction. We confine ourselves here to
a qualitative description. Appendix 10.3 gives a quantitative treatment.

Figure 10.8(b) shows the behavior of the potential in this structure
for the case where the metal is negative with respect to the semiconductor
$(V_m - V_s < 0)$. In this case the negative potential of the metal attracts holes.
The energy levels of the semiconductor increase near to the insulator. The
band scheme is shown again in Fig. 10.9(a) for p-type material. We see that
the Fermi level, which is constant since we are in equilibrium, is closer to the
valence band at the surface and there will be accumulation or enhancement
of hole concentration at the surface.

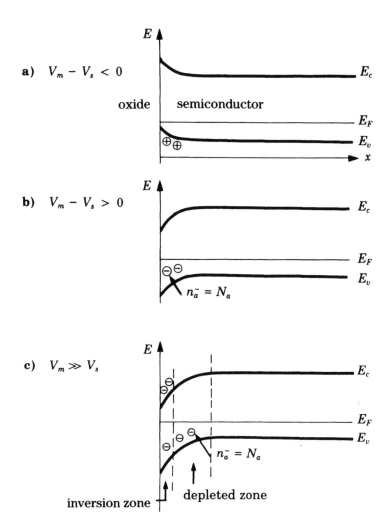

Fig. 10.9. Energy levels in the semiconductor of a MOSFET as a function of distance from the insulator: (a) negative voltage applied to the metal; (b) small positive voltage; (c) large positive voltage.

By contrast in the case where $(V_m - V_s)$ is positive, the holes are repelled and the band scheme is shown in Fig. 10.9(b). At the surface the Fermi level is further from the band edges, and there is depopulation, or depletion, of mobile carriers. Thus there remain only ionized acceptors near the surface with concentration N_a.

Above a certain threshold, for large positive voltages applied to the metal we obtain the band scheme shown in Fig. 10.9(c): there are three zones. At the surface the Fermi level is close to the conduction band, and electrons appear. This is called an inversion layer. The electron number in this inversion layer depends directly on the voltage applied to the metal. Deeper in the semiconductor we encounter a depleted region where the only charges are ionized acceptors, before reaching the neutral semiconductor.

The operation of a MOSFET makes use of these phenomena. The structure of a MOSFET is shown in Fig. 10.10(a). Two contacts of types opposite from the chosen substrate (here p type) constitute the source and the drain. These contacts bound the active region of the MOSFET which is situated below the gate. The gate, oxide, and substrate constitute the structure (metal, oxide, semiconductor) discussed above. The source is linked to the substrate. In the absence of sufficient positive polarization at the gate, the substrate remains p type everywhere and there is no conduction between the source and the drain, as the source-substrate and substrate-drain junctions are back-to-back.

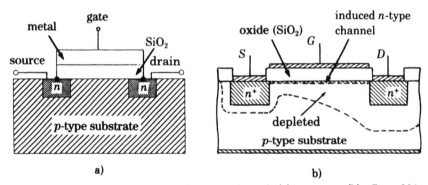

Fig. 10.10. MOSFET transistor with n-type channel: (a) structure; (b) effect of bias. (After Leturcq and Rey, "Physique des Composants Actifs à Semi-Conducteurs," Dunod, 1978.)

In contrast, if a sufficient positive voltage is applied to the gate, one can create a surface inversion layer, called the induced channel—see Fig. 10.10(b). There is then continuity of n-type conduction between source and drain and a source-drain current can flow. The resistance between source and drain is therefore controlled by the value of the applied voltage between gate and source. We can thus construct a device whose current is controlled by a voltage and not a current as in the junction transistor. This constitutes the n-MOS, since the induced channel is n type; the device operates in the enhancement mode, since applying $V_G > 0$ makes electrons appear.

Exercise: Can an n-MOS enhancement transistor work as an n-MOS depletion transistor under certain circumstances?

Present electronic logic provides the logical functions NO, AND, OR by use of inverters, which are just MOS transistors or combinations of

MOS, conducting or not, depending on the voltage applied to the gate. The advantage of these circuits over bipolar transistors is that, because of the oxide layer, no current flows between the gate and the conducting channel.

The complementary MOS inverter, called CMOS (see Appendix 10.3, end of the solutions), is the basic element of logic circuits. It consists of an electron-channel MOS transistor (n-MOS) and a complementary hole-channel transistor (p-MOS) which behaves exactly opposite to an n-MOS transistor under voltages applied to its gate. Whatever the applied voltage, one of the transistors is always blocked: in principle no electric current flows, implying very small dissipation.

10.3 An Application of the MOSFET: The Charge-Coupled Device (CCD)

We have just seen that in a MOSFET made of a p-type semiconductor we can create a potential well for the electrons (Fig. 10.9(c)) in the inversion layer if we apply a strong positive voltage to the metal. The accumulation and storage of minority carriers in a surface potential well are the basic principle of CCD cameras which are now the most common image converter in video systems.

The structure involves a p-type semiconductor covered by an insulating layer and a series of very close metal gates. When we apply to the system of Fig. 10.11(a) positive potentials $V_1 = V_3 = V_4, V_2 > V_1$, a potential well for the electrons appears under electrode 2.

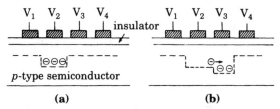

Fig. 10.11. Schematic figure of a CCD. The metallic gates are hatched. The potential profile of the electrons is shown dashed. The potentials applied in (a) are such that $V_2 > V_1 = V_3 = V_4$, and the potential well is under electrode 2. In (b), $V_3 > V_2 > V_1 = V_4$, and the electrons transfer from electrode 2 to 3.

Let us assume that some electrons are introduced under electrode 2, where they are stored: as there are no holes in the inversion layer there is no recombination. If the potentials are changed so that $V_1 = V_4$, $V_1 < V_2 < V_3$, the charges move towards the deeper potential well, ending up under electrode 3 (Fig. 10.11(b)).

By applying an appropriate sequence of voltages on the various gates, we can transfer the charges from one region to another of the surface, and scan it point by point and line by line. We can identify which charges come from which region by analyzing the resulting sequence of electrical signals.

The minority carriers (electrons in this figure) can be produced by luminous excitation. Each surface element or pixel provides a charge, and we thus have the conversion of an image consisting of a set of light or dark points into an electro-optical signal. This provides an electrical signal whose amplitude varies with the light intensity at various points of the original image.

Charge-coupled (CCD) devices are very sensitive light detectors. They have now replaced Vidicon tubes in cameras and kinescopes, and they are used in spectroscopy and astronomical detection, where, for example, arrays of 400×1200 pixels with a quantum efficiency of 0.5 electrons per photon are currently used.

10.4. Concepts of Integration and Planar Technology

The recent and spectacular development of microelectronics is related to the use of planar techniques and the consequent idea of an integrated circuit. Consider the amplifier whose circuit diagram is given in Fig. 10.12. It consists of a p–n–p transistor, resistors, and capacitors. We have seen that a homogeneous semiconductor can behave as a resistor and that a reverse biased junction behaves like a capacitor for ac current. Thus the circuit (Fig. 10.12(a)) can be constructed from fused elements, but all made from silicon as shown in Fig. 10.12(b). The passage to Fig. 10.12(c) is obvious as it is not necessary to link semiconductors of the same type by metallic wires. We are led to a circuit of three elements. Deforming these elements we can combine them into a single semiconducting layer containing n, p, and intrinsic regions, the latter being insulating (Fig. 10.12(d)). It is these same ideas that are currently used in the manufacture of electronic devices from audio sets to computer components: microprocessors, memories, etc.

These circuits are constructed on a substrate: a thin slice, or "wafer," of monocrystalline material of adequate resistivity. The wafers are several cm (typically 10) in diameter and a thickness of several 10^{-1} mm. Their manufacture involves the operations of oxidation, masking, diffusion, metal deposition *in vacuo*, and possibly epitaxy, i.e., the growth of a monocrystalline layer, of adjustable doping on a monocrystalline substrate.

An important step involves the ability to create by "lithography" an oxide layer of a given design on the silicon wafer. This layer acts either as an insulator or as protection for the substrate in a later operation, e.g., diffusion.

The first operation is to (a) oxidize the substrate uniformly (Fig. 10.13). Then a photoresist resin which polymerizes in light is deposited on the

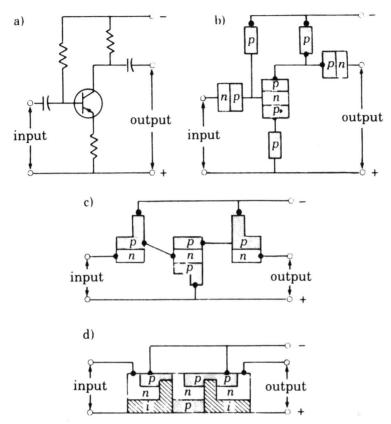

Fig. 10.12. (a) Schematic view of a p–n–p transistor used as an amplifier; (b) circuit made of fused elements; (c) suppression of connecting wires; (d) planar technology. (After Brophy, "Semiconductor Devices," George Allen and Unwin, 1964.)

oxide (b). The system is then illuminated through a mask (c) and the resin polymerizes at the places that receive light. The non-polymerized resin is then dissolved (d) and the silica attacked by an acid that does not affect the polymerized resin (e). Finally the resin is dissolved in an appropriate solvent (f).

The manufacture of an n–p–n transistor is illustrated in Fig. 10.14. An epitaxial n layer is grown on an n^+ substrate. (a) After oxidation, a window is removed by photoetching. Acceptors (e.g., boron) are then diffused through this window. (b) After reoxidizing, a new smaller window is removed, through which phosphorus is diffused to make the n^+ layer, the emitter (c). It remains to make the metallic contacts. The device is oxidized and windows cut over the active regions. A metallic layer is deposited over the device from which we eliminate superfluous zones by photoetching (d).

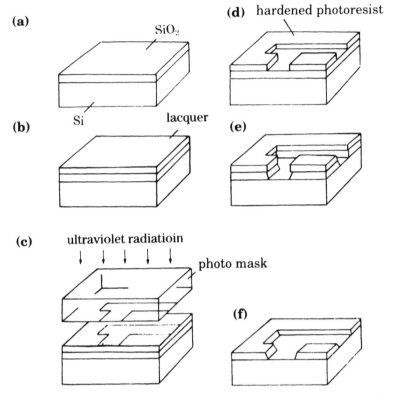

Fig. 10.13. The photolithography process (see the text). (After Tom Forrester, ed. "The Microelectronics Revolution," Basil Blackwell, 1980.)

Note that the exact shape of the base volume is not really limiting, because diffusion currents always follow the concentration gradients, whatever their geometrical distribution.

These techniques allow us to produce from a limited set of operations a considerable variety of logic circuits with complex functions, known colloquially as "chips." One can manufacture connected sets of transistors, resistors, and capacitors, but integration allows us to conceive devices without discrete equivalents, e.g., transistors with multiple emitters. This last technique is called I^2L (Integrated Injection Logic).

Two parameters govern progress in this domain: the size of an element and the speed of operation. The minimum size of an element is determined by the wavelength of the electromagnetic radiation used for the photolithography. Diffraction effects limit the definition of the image to around half a wavelength. The size of an element is currently of the order of 2 μm or about ten times the wavelength of the ultraviolet light used for the photolithography. A transistor and its contacts occupy between 10^{-4} and 10^{-3} mm^2. The use of light of shorter wavelength (UV or X-rays from synchrotron ra-

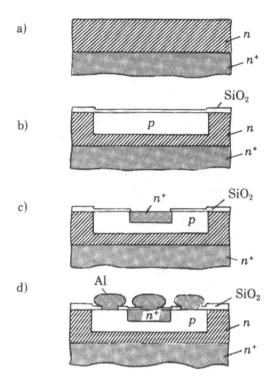

Fig. 10.14. Manufacture of an n–p–n transistor. (After Leturcq and Rey, "Physique des Composants Actifs à Semi-conducteurs," Dunod, 1978.)

diation) should allow the size to be reduced. The precision of positioning the masks in the successive operations will have to conform to this size reduction. A 0.3 μm scale technology is expected to be operative in the late nineties.

Another important parameter is the minimum time for an elementary operation. This is the time taken, for example, by the carriers to cross the channel of a MOSFET. If L is the length of the channel and v_d the drift speed

$$\tau = \frac{L}{v_d} = \frac{L}{\mu E} = \frac{L^2}{\mu \, V_{DS}}.$$

For $L = 10$ μm, $\mu = 0.1$ m$^2 \cdot$ V$^{-1} \cdot$ s^{-1}, and $V_{DS} = 2$ V, this time is of the order of 10^{-9} s. (Compare with (10.9).) The time taken by the whole system to complete a logical operation is not τ but proportional to τ. We see that the reduction in size corresponds not only to a greater integration density but also to increased speed. However, in strong electric fields of order 10^6 V/m the speed v_d saturates at around 10^5 m/s for electrons of silicon and thus τ will in future vary as the size L and not as its square. There are however difficulties in increasing the number of components per chip, related to the fact that a single defect can prevent the functioning of the entire device. These defects can come from the substrate (scratches) but also from dust that may contaminate the surfaces at each stage of the process—the work must therefore be performed in dust-free environments ("clean rooms"). Also circuits that are too complex cannot be tested in all their states. It is now becoming necessary to design chips in which a part of the circuit is devoted to testing the subcircuits of the same chip.

10.5 Band Gap Engineering

The recent ability to manufacture semiconductors layer by atomic layer through molecular beam epitaxy allows one to produce structures that are sequences of heterojunctions. Depending on how successive layers are doped we obtain either a band profile with gaps (intrinsic or weakly doped semiconductors) or of the type shown in Fig. 9.16. This gives a new technique, "band gap engineering," so-called since the band profile can be modeled at will. It takes advantage of several effects: the possibility of confining and controlling the wave function culminates, for example, in the manufacture of a FET in which the conducting channel does not contain any dopant, increasing the mobility (cf. Sect. 5.4). The asymmetry of the electrons and holes in heterojunctions allows the design of circuits where only the majority carriers are active. The ability to stack circuits of several nanometers thickness leads to a miniaturization no longer limited by lithography.

Semiconducting heterostructures that make use of quantum effects have been used commercially for several years: for example, adjusting the energy of the transition between the conduction and the valence ground states through the choice of the well width (cf. Appendix 3.2), together with the large value of the absorption coefficient (cf. Appendix 6.3), allow the construction of efficient laser diodes, in which the rate of conversion of electrical energy into light is large. These systems operate in the red or near infrared range and are currently used in fiber optics communication.

Many other applications of band gap engineering are in progress. As an example we describe the principle of an infrared detector in the 5–10 μm range based on a dissymetrical quantum well structure (Fig. 10.15).

In such a system, the average position of an electron in the quantized level E^2 is shifted in space with respect to that in the quantized level E^1. This charge displacement is associated with a very large transition matrix element between E^1 and E^2. This ensures a very strong coupling of the well with the electromagnetic field. In a sense the well behaves as a giant molecule with a size of the order of nanometers.

If the well is n doped, in the dark electrons occupy the level E^1. By illumination with infrared photons, these electrons are promoted to E^2 (bound-

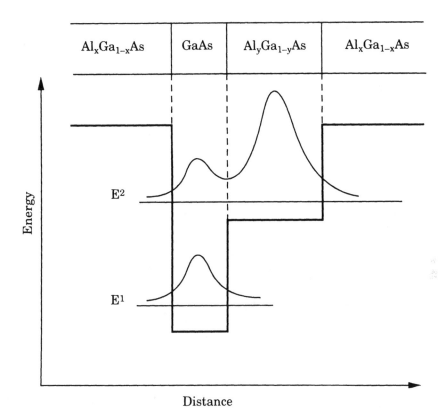

Fig. 10.15. Conduction band profile of a dissymmetrical quantum well structure.

to-bound transition) or to the delocalized levels above the $Al_x\,Ga_{1-x}$ As barrier (bound-to-free transition). If the structure is submitted to a transverse electric field, the energy profile is as in Fig. 10.16.

When an infrared photon induces a bound-to-free (a) transition, like in the figure, i.e., if its energy is larger than the distance between E_1 and the barrier, the photoexcited electron is swept by the electric field, and produces a photocurrent. In the case of a photon exciting the E^1-E^2 bound-to-bound (b) transition, the excited electron can tunnel from the level E^2 through the barrier in presence of the electric field (Fig. 10.16). The barrier being triangular, it is easier to cross by tunneling when the field is strong. In both cases the absorption of an infrared photon leads to a photocurrent.

Moreover in presence of the electric field, the positions of E^1 and E^2 are slightly modified by the linear Stark effect (Fig. 10.17). Then the energy of the photons which are absorbed by the structure can be slightly tuned through this electric field.

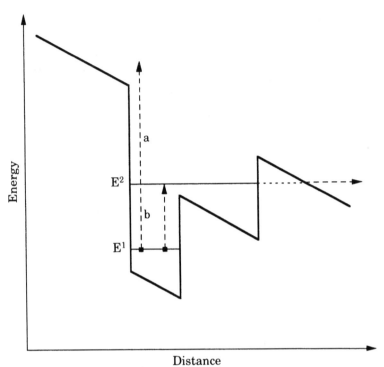

Fig. 10.16. Modification of the band profile of Fig. 10.15 by application of a transverse electric field.

10.6 Physical Limits in Digital Electronics

Since 1958, the beginning of the integrated circuit era, the size of the smallest circuits has decreased by an average of 13% per year. At this rate it will be around 0.1 μm by the year 2000. Assuming that technological progress allows the manufacture of submicron circuits, it is essential to understand the fundamental limits on miniaturization and integration.

We summarize below some of the physical limitations to the electronic treatment of information by solid-state electronics. We shall see that one of the main limitations arises from the need to treat numerical data digitally.

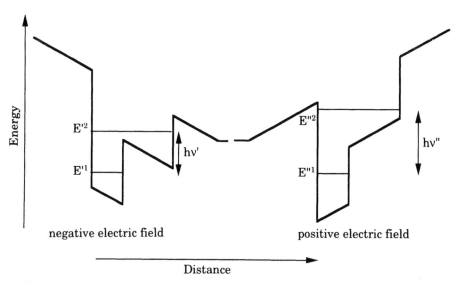

Fig. 10.17. Linear Stark effect on the band profile of Fig. 10.15. The modifications of the energy level positions and of the optical transitions depend on the sign and amplitude of the electric field.

The Need for a Digital Treatment

In electronic information processing or calculation, all data are manipulated in a large number of successive operations. A small error introduced into each operation can cause the complete loss of the original information. A random error of 1% at each step becomes an error of 100% after 10^4 operations. For this reason one uses digital, particularly binary, representation of the input data, which allows one to prevent the buildup of random errors by "restandardizing" each "digit" at each stage. Consider the simplest binary calculation, which associates an output $S = E + 0$ with each input E. If $E = 1$ V it may be that because of an imperfection or perturbation we have $S = 1.01$ V (a 1% error). At the end of this elementary process we recognize that the actual output value (1.01) is close to 1 and thus corresponds *with certainty* to the value 1 and not the value 0. We can thus assign it the value 1. Digital representation thus allows one to obtain arbitrarily high precision by increasing the number of binary digits of the representation.

Minimum Energy of a Logical Operation

We may also ask what is the limiting energy required for a logical operation (Landauer, 1961). The simplest physical system representing a binary state has the potential energy shown in Fig. 10.18.

This system is bistable in the sense that it can exist in either the state 0 or the state 1. Consider the operation of putting the system in state 0 whatever the initial state. If we wish to be certain to find the system in

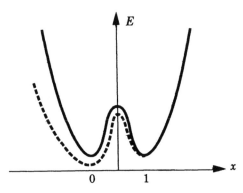

Fig. 10.18. Potential energy of a binary system. Dashed, the energy obtained by application of an external force.

state 0 we have to apply an external force so as to lower the energy of the left-hand potential well (dashed curve in Fig. 10.18). We thus increase the probability of finding the system on the left. For this probability to get very close to 1 we must lower the potential by several times kT. In other words, the system must dissipate an energy several times kT, i.e., $4 \cdot 10^{-21}$ J, in each operation of resetting to zero. In fact, for the best currently available complete circuits, the product of the power P times the duration τ of an operation is of the order of 10^{-10} J. It requires, for example, 10^{-1} W to perform an operation in 10^{-9} s, and we are thus very far above the fundamental limit described here.

The Quantum Limit

If we wish to localize a physical system between two distinct states after a delay τ, these states must differ by an energy $\Delta E \geq \hbar \tau^{-1}$. The dissipated power is thus $\hbar \tau^{-2}$. Here again this limit is remote from practical performances. For $\Delta E = 10^{-10}$ J, $\Delta t \sim 10^{-23}$ s while the best circuits presently work at $\Delta t \sim 10^{-9}$ s. It is therefore not the fundamental limits which determine the present performance of computers. We should nevertheless gather from this discussion that the order of magnitude of the energy necessary to run a logical system must be several times kT.

The "Natural" Voltage Scale for Semiconductor Electronics

The operation of restandardizing requires by definition the use of a non-linear device. Now, the basic non-linear relation available is Shockley's law, which gives the characteristic $J(V)$ of a p–n junction (Chap. 8). For an electronic device built using this law to be strongly non-linear we have to apply voltages of a few times $kT/e = 25 \times 10^{-3}$ V. This is why the "natural" voltage scale used in digital electronics is of the order of a volt.

Transmission of Data

To move from one stage to the next the voltage must be carried by conductors with an impedance Z. The power required is thus at least of the order of $(kT/e)^2 Z^{-1}$. It is very difficult to produce lines where Z is very different from the impedance Z_0 of vacuum, which is 300 Ω. We find in this case a power $\sim 2 \times 10^{-6}$ W. This power corresponds to an energy $10^6 kT$ per nanosecond, showing why the energy dissipated per operation is much higher than the thermodynamic limit kT.

In practice the voltages used are 10 or 100 times larger than kT/e and this is the main source of dissipation in logic circuits. Given this power, one has to efficiently cool the most rapid circuits to prevent their deterioration. Heat removal at present sets the practical limit for the most powerful machines.

Miniaturization Limit

There are other limits which are caused by the nature of the manufacturing process itself (photolithography, chemical etching, and so forth). We mention the striking limit coming from the inherent fluctuation of small objects. A cube of semiconductor of size 0.2 μm contains on average only 80 impurities if the semiconductor is doped at a concentration of 10^{22} m^{-3}. The role of statistical concentration fluctuations becomes very important when we envisage several million circuits per chip as we do now, near the end of the 20th century.

Appendix 10.1

Problems on the n–p–n Transistor

The aim of these problems is to give a simple quantitative description of the operation of the bipolar transistor, which we have already discussed qualitatively. We consider the device shown schematically in Fig. 10.19, consisting of a semiconducting single crystal with three successive regions of different dopings. These regions are conventionally called emitter (here of type n^+, strongly doped), base (weakly p doped), and collector (normal n doping), and we shall denote physical quantities in these regions by indices E, B, and C. For example, for the emitter n_E and p_E are the equilibrium electron and hole concentrations; D_E, τ_E, and L_E the diffusion coefficient, recombination time, and diffusion length of the minority carriers (here the holes since the emitter is n doped).

The crystal has the form of a rectangular parallelepiped with its long side along Ox, and cross section S (Fig. 10.19): the emitter-base and base-collector junctions are in planes perpendicular to Ox, so the currents have non-zero components only along Ox. We assume further the lengths of the emitter and collector sections are infinite. The thickness of the base is Δ.

Problems

1. Show the band energy profile as a function of x when the system is in equilibrium. Note the appearance of two space-charge regions, bounding three neutral regions.

In the transistor's normal regime there is a forward bias on the base-emitter junction $(B\text{–}E)$, and the base-collector junction is strongly biased in the reverse sense: if we adopt the convention $V_B = 0$, the applied voltages shown in Fig. 10.19 are $V_E < 0, |V_E| \gg (kT/e) = 25$ mV at the emitter and $V_C \gg |V_E|$ at the collector. We assume that the voltage drops in the bulk of the semiconductor are negligible. Show the band scheme as a function of x for these conditions.

2. Write down the evolution equations for the minority carrier density in the three neutral regions taking account of diffusion and recombination, and simplify them in the stationary case.

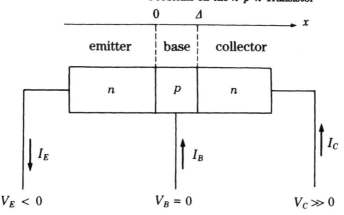

Fig. 10.19. Schematic representation of the structure of an n–p–n transistor with the standard sign conventions for currents.

3. We will assume as for the biased p–n junction that the widths of the space-charge regions are negligible compared with the diffusion lengths and that in the space-charge regions the electrons and holes are in thermal equilibrium (cf. Eq. (8.35)). Deduce the ratio $n(0_+)/n_B$, where $n(0_+)$ is the electron concentration in the base at the edge of the space-charge region. Similarly calculate $p(0_-)/p_E, n(\Delta_-)/n_B$, and $p(\Delta_+)/p_C$.

4. Find the hole concentration in the emitter by solving the differential equation corresponding to **2** with the boundary conditions of **3**. Deduce the current density $J_h(0_-)$ of injected holes at the base-emitter junction. Assume that the current densities are conserved across the junctions, and that the points $0_+, 0_-$ have the same abscissa, 0, in the algebraic expressions obtained.

5. The same question for the holes in the collector. Deduce $J_h(\Delta_+)$, the hole current density at the base-collector junction.

6. The same question for the electrons in the base. Because of boundary conditions at $x \simeq 0$ and at $x \simeq \Delta$ it is convenient to introduce hyperbolic functions $\sinh(x/L_B)$ and $\sinh[(x - \Delta)/L_B]$ rather than exponentials. Deduce the current densities $J_e(0_+)$ and $J_e(\Delta_-)$ at the two junctions. For ease of writing set $eV_E/kT = v_E, eV_C/kT = v_C$.

7. Deduce from questions **4** and **6** the total intensities I_E crossing the emitter and I_C crossing the collector. Take the positive senses as in Figs. 10.19 and 10.20.

8. Using the assumptions $V_E < 0, V_C >> 0$ show that if we also have $\Delta << L_B$, then I_C and I_E are of the same order, and thus $I_B << I_C$. Interpret this physically.

9. In the limit $V_C >> 0$ the currents I_E and I_C depend only on V_E; calculate the differential current gain $\beta = dI_C/dI_B$ of the transistor. How should the transistor be made so that β is as large as possible? Calculate

β for $n_E = 10^{25}$ m^{-3}, $p_B = 3 \cdot 10^{22}$ m^{-3} $n_C = 5 \cdot 10^{21}$ m^{-3}, $\Delta = 3$ μm, $D_E = 10^{-4}$ m^2/s, $D_B = 3.5 \times 10^{-3}$ m^2/s, $L_E = 1$ μm, $L_B = 100$ μm.

10. Draw the characteristics of the transistor, i.e., the set of curves $I_C = f(V_C - V_E)$ for different values of V_E.

11. Why are users of transistors advised not to interchange the collector and emitter leads?

Remark: As seen in this problem, the diffusion of the carriers across the base (here the electrons) determines the operation of the transistor. The geometry of the emitter or collector has no influence. This justifies the choice of simple parallelepiped geometry in this problem.

Solutions

1. In equilibrium the Fermi level is constant throughout the system. We know that each p–n or n–p junction has a space-charge region whose width is determined by the doping as represented in Fig. 10.20(a). Figure 10.20(b) shows the band profile of a polarized transistor.

2. The carrier conservation equation for holes in the emitter is

$$\frac{\partial p}{\partial t} = \frac{1}{e}\frac{\partial J_h}{\partial x} - \frac{p - p_E}{\tau_E}, \tag{10.11}$$

where J_h is the hole electric current.

For the minority carriers, the current is essentially the diffusion term, so for holes in the emitter we have

$$J_h \simeq -e\, D_E \frac{\partial p}{\partial x}. \tag{10.12}$$

Combining these equations we get

$$\frac{\partial p}{\partial t} = D_E \frac{\partial^2 p}{\partial x^2} - \frac{p - p_E}{\tau_E} = 0 \text{ in the steady state.} \tag{10.13}$$

Similarly, in the steady state we write for the base electrons:

$$D_B \frac{\partial^2 n}{\partial x^2} - \frac{n - n_B}{\tau_B} = 0 \tag{10.14}$$

and for the collector holes:

$$D_C \frac{\partial^2 p}{\partial x^2} - \frac{p - p_C}{\tau_C} = 0. \tag{10.15}$$

3. The equations above involve the diffusion lengths

$$L_E = \sqrt{D_E\, \tau_E}, \quad L_B = \sqrt{D_B\, \tau_B}, \quad \text{and} \quad L_C = \sqrt{D_C\, \tau_C}. \tag{10.16}$$

At the edge of the space charge of the emitter-base junction minority carriers are injected into the base since the junction is forward biased. For

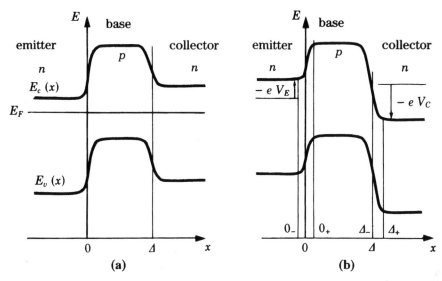

Fig. 10.20. Band profile in an n–p–n transistor: (a) zero bias; (b) profile when biased ($V_E < 0, V_C > 0$). The arrows show the energy shifts, giving the algebraic values on the energy axis.

example, electrons are injected from the emitter to the base. The assumption of thermal equilibrium for the carriers in the space-charge region allows us to write

$$n\,(0_+) = n_B \exp(-e\,V_E/kT), \tag{10.17}$$

$$p\,(0_-) = p_E \exp(-e\,V_E/kT). \tag{10.18}$$

As $V_E < kT/e$, we have

$$n\,(0_+) \gg n_B, \quad p\,(0_-) \gg p_E. \tag{10.19}$$

There are thus more minority carriers than in equilibrium.

The base-collector junction is reverse biased, resulting in extraction of minority carriers, so that

$$n\,(\Delta_-) = n_B \exp(-e\,V_C/kT), \tag{10.20}$$

$$p\,(\Delta_+) = p_C \exp(-e\,V_C/kT). \tag{10.21}$$

As $V_C \gg 0$ we have

$$n\,(\Delta_-) \ll n_B, \quad p\,(\Delta_+) \ll p_C. \tag{10.22}$$

Remark: We can use the language of quasi-Fermi levels introduced in Eqs. (8.48) and (8.49) for the polarized junction. As we are in the low injection regime, the quasi-Fermi levels of the majority carriers in the neutral regions are very close to those in equilibrium. The widths of the space-charge regions are small compared with the diffusion lengths, so the quasi-Fermi levels are little changed in these regions. The position of the quasi-Fermi levels E_{Fe}, E_{Fh} for electrons and holes is shown in Fig. 10.21 for the band profile of Fig. 10.20.

4. The hole concentration in the emitter is

$$p = p_E + \lambda_1 \exp\left(\frac{x}{L_E}\right) + \lambda_2 \exp\left(-\frac{x}{L_E}\right), \tag{10.23}$$

where λ_1, λ_2 are constants determined by the boundary conditions

for $x \to -\infty$ $p \to p_E$ so that $\lambda_2 = 0$,

for $x \to 0_-$ $p = p_E \exp(-e\, V_E/kT)$. \hfill (10.24)

Hence, for $x < 0$,

$$p = p_E \left[\exp\left(-\frac{e\, V_E}{kT}\right) - 1\right] \exp\left(\frac{x}{L_E}\right) + p_E. \tag{10.25}$$

The corresponding current is

$$J_h(x) = -e\, D_E \frac{\partial p}{\partial x}, \tag{10.26}$$

$$J_h(x) = -e\frac{D_E\, p_E}{L_E}\left[\exp\left(-\frac{e\, V_E}{kT}\right) - 1\right]\exp\left(\frac{x}{L_E}\right) \tag{10.27}$$

The current of holes injected at $x = 0_-$ from the base is

$$J_h(0_-) = -\frac{e\, D_E\, p_E}{L_E}\left[\exp\left(-\frac{e\, V_E}{kT}\right) - 1\right]. \tag{10.28}$$

5. Following the method of **4**, we get

$$J_h(x = \Delta_+) = \frac{e\, D_C\, p_C}{L_C}\left[\exp\left(-\frac{e\, V_C}{kT}\right) - 1\right]. \tag{10.29}$$

6. The electron concentration in the base is found from

$$n - n_B = \lambda_1'' \sinh\frac{x}{L_B} + \lambda_2'' \sinh\frac{x - \Delta}{L_B}, \tag{10.30}$$

at $x = 0_+$ $n_B \exp\left(-\frac{e\, V_E}{kT}\right) = n_B - \lambda_2'' \sinh\left(\frac{\Delta}{L_B}\right),$ \hfill (10.31)

at $x = \Delta_-$ $n_B \exp\left(-\frac{e\, V_C}{kT}\right) = n_B + \lambda_1'' \sinh\left(\frac{\Delta}{L_B}\right).$ \hfill (10.32)

For ease of writing we set

$$e\, V_E/kT = v_E, \quad e\, V_C/kT = v_C. \tag{10.33}$$

We deduce

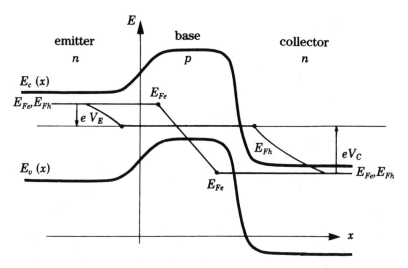

Fig. 10.21. Quasi-Fermi levels E_{Fe}, E_{Fh} in a polarized n–p–n transistor ($V_E < 0$, $V_C > 0$).

$$n - n_B = n_B \left\{ \sinh \frac{\Delta}{L_B} + \left[\sinh \frac{x}{L_B} \right] [\exp(-v_C) - 1] \right.$$
$$\left. - \left[\sinh \frac{x - \Delta}{L_B} \right] [\exp(-v_E) - 1] \right\} / \sinh \frac{\Delta}{L_B}, \qquad (10.34)$$

$$J_e(x) = e \, D_B \frac{\partial n}{\partial x}$$
$$= \frac{e \, D_B \, n_B}{L_B \sinh(\Delta/L_B)} \left\{ \left[\cosh \frac{x}{L_B} \right] [\exp(-v_C) - 1] \right.$$
$$\left. - \left[\cosh \frac{x - \Delta}{L_B} \right] [\exp(-v_E) - 1] \right\}, \qquad (10.35)$$

so that

$$J_e(0_+) = \frac{eD_B\, n_B}{L_B} \times$$

$$\left\{ \frac{1}{\sinh(\Delta/L_B)}[\exp(-v_C) - 1] - \left[\coth\frac{\Delta}{L_B}\right][\exp(-v_E) - 1] \right\}, \tag{10.36}$$

$$J_e(\Delta_-) = \frac{eD_B\, n_B}{L_B} \times$$

$$\left\{ \left[\coth\frac{\Delta}{L_B}\right][\exp(-v_C) - 1] - \frac{1}{\sinh(\Delta/L_B)}[\exp(-v_E) - 1] \right\}. \tag{10.37}$$

7. In the space-charge region of the emitter-base junction the total current is conserved. As in the text, we use the expression for the electron and hole currents in the region where the current is purely diffusive. The currents I_E, I_C flow towards $x < 0$; also the current is the flux of the current density across the section S, so that

$$I_E = -S\, [J_e\, (0_+) + J_h\, (0_-)], \tag{10.38}$$

$$I_C = -S\, [J_e\, (\Delta_-) + J_h\, (\Delta_+)], \tag{10.39}$$

$$\frac{I_E}{eS} = \frac{D_B\, n_B}{L_B}$$

$$\left\{ -\frac{1}{\sinh(\Delta/L_B)}[\exp(-v_C) - 1] + [\coth\frac{\Delta}{L_B}(-v_E) - 1] \right\} + \frac{D_E\, p_E}{L_E}[\exp(-v_E) - 1], \tag{10.40}$$

$$\frac{I_C}{eS} = \frac{D_B\, n_B}{L_B}\left\{ -\coth\frac{\Delta}{L_B}[\exp(-v_C) - 1] + \frac{1}{\sinh(\Delta/L_B)}[\exp(-v_E) - 1] \right\} - \frac{D_C\, p_C}{L_C}[\exp(-v_C) - 1]. \tag{10.41}$$

8. The assumptions about V_E, V_C imply that $\exp(-eV_E/kT) \gg 1$ and $\exp(-eV_C/kT) \ll 1$. Further, if the base is narrow enough that $\Delta/L_B \ll 1$, we can write $\sinh(\Delta/L_B) \simeq \Delta/L_B$, $\coth(\Delta/L_B) \simeq L_B/\Delta \gg 1$ and

$$I_E \simeq eS\left(\frac{D_B\, n_B}{\Delta} + \frac{D_E\, p_E}{L_E}\right)\exp(-v_E), \tag{10.42}$$

$$I_C \simeq \frac{eS\, D_B\, n_B}{\Delta}\exp(-v_E). \tag{10.43}$$

Then the term in $(1/\Delta)$ dominates the terms in $1/L_E$ or $1/L_C$, and I_E, I_C effectively reduce to their first term, and are thus of the same order. As $I_E = I_B + I_C$, this implies $I_B \ll I_E$.

Physically, if the base is narrower than the diffusion length, and if it is rather weakly doped, so that $(p_E/n_B) \cdot (\Delta/L_E) \ll 1$, almost all the electrons injected from the emitter towards the base manage to cross it without recombining, and reach the collector.

9. The differential current gain β is defined by

$$\beta = \frac{d\,I_C}{d\,I_B} = \frac{d\,I_C/d\,V_E}{d\,(I_E - I_C)/d\,V_E}. \tag{10.44}$$

Taking the exact expressions (10.40) and (10.41) for I_E and I_C we get

$$I_E - I_C = eS\left[D_B \frac{n_B}{L_B}\frac{\cosh(\Delta/L_B) - 1}{\sinh(\Delta/L_B)} + \frac{D_E\,p_E}{L_E}\right][\exp(-v_E) - 1] +$$
$$eS\left[\frac{D_B\,n_B}{L_B}\frac{\cosh(\Delta/L_B) - 1}{\sinh(\Delta/L_B)} + \frac{D_C\,p_C}{L_C}\right] \times$$
$$[\exp(-v_C) - 1]. \tag{10.45}$$

For $\Delta \ll L_B$,

$$d\,I_C/d\,V_E \simeq \frac{-e^2\,S\,D_B\,n_B}{kT\,\Delta}\exp(-v_E), \tag{10.46}$$

$$d\,(I_E - I_C)/d\,V_E \simeq \frac{-e^2 S}{kT}\left(\frac{D_B\,n_B}{L_B}\frac{\Delta}{2\,L_B} + \frac{D_E\,p_E}{L_E}\right)\exp(-v_E), \tag{10.47}$$

$$\beta \simeq \frac{1}{(\Delta/L_B)\,[(\Delta/2)\,L_B + (p_E/n_B)\,(D_E/D_B)\,(L_B/L_E)]}. \tag{10.48}$$

To increase β, one adjusts the doping to have $p_E \ll n_B$, or equivalently $n_E \gg p_B$ (since $n_E \cdot p_E = n_B \cdot p_B = n_i^2$). The emitter will be strongly doped, the base weakly doped. An indirect consequence of these different dopings is an increase in the lifetime in the base, and hence L_B.

For the values given in the question β is about 1500.

10. The characteristics (Fig. 10.22) are horizontal straight lines ("saturated" transistor) except when $V_C - V_E$ is small. I_C then decreases abruptly. For $V_E = 0$ and any V_C, the current I_C is also very small ("blocked" transistor).

11. The symmetry of the n–p–n transistor is only superficial. The emitter is strongly doped and the collector is not, so reversing these connections leads to poor performance (for example, a lower β). Moreover, because of the strong doping of the emitter, the base-emitter junction would not in general stand large voltages and would "break down."

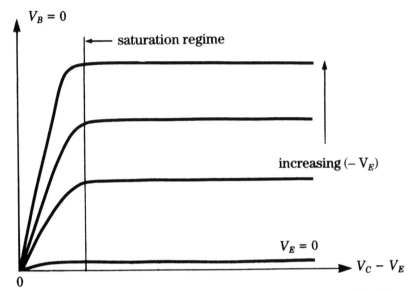

Fig. 10.22. Characteristics $I_C = f(V_C - V_E)$ with the base at zero potential, for different values of the emitter potential.

Appendix 10.2

Problems on the Junction Field-Effect Transistor (JFET)

In 1952 W. Shockley conceived an active device consisting of a slice of semiconductor between two reverse-biased p–n junctions. The principle is to modify the size of the conducting region called the channel by varying the reverse bias of the junctions. The device is shown in Fig. 10.23. We shall study the simplified geometry shown in Fig. 10.24, concentrating on the active part between the junctions.

The p^+ regions (gates) are negatively biased with respect to the contact S, called the source, and taken as the origin of the potentials. The contact D is called the drain. We apply to the drain a voltage positive with respect to the source. The thickness of the n region is $2a$, its length is L, and its width b. The set of characteristics measured for such a device (curves of drain current I_D as a function of the drain voltage V_D for various gate voltages V_G) has the form shown in Fig. 10.25. The aim of these problems is to understand this figure. The various notations are introduced below.

Problems

1. We are first of all interested in the case of infinitesimal drain potentials. In this limit the channel can be regarded as an equipotential, and because of the inverse polarization applied to the gates a depleted zone appears on each side, with uniform width d. Express d as a function of the internal potential ϕ of the junction, the applied gate voltage V_G, the dielectric constant $\epsilon_0 \epsilon_r$ of the semiconductor, and the donor concentration N_d. (Consider the case where $N_d \ll N_a$, the acceptor concentration in the p^+ region, and neglect N_d/N_a compared with 1.)

2. If the electron mobility is μ_e, give the channel conductance G as a function of the voltage applied to the gate. Express the result as a function of $G_0 = 2e\mu_e N_d ab/L$, the channel conductance in the absence of any space charge.

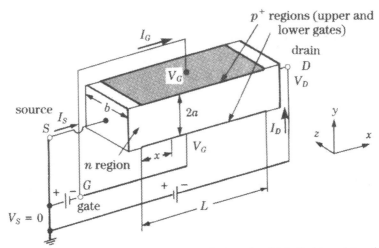

Fig. 10.23. Perspective view of the active part of a junction field-effect transistor. (After Millman and Halkias, "Integrated Electronics: Analog and Digital Circuits and Systems," McGraw–Hill, 1972.)

3. Give the minimum thickness $2a_m$ of the n-doped slice required for the conductance to be non-zero in the absence of a potential. Give this explicitly in the case $N_d = 10^{21}$ m^{-3}; $\epsilon_0\epsilon_r = 10^{-10}$ F · m^{-1}: take $\phi = 0.8$ V.

4. In the same regime where V_D is very small, consider the case where there is non-zero conduction, i.e., $a > a_m$. The device behaves as a variable resistance controlled by the gate. Show that the conductance vanishes for a particular value V_P of the gate voltage, which we call the pinch voltage, and give the expression for it. Find this voltage for $a = 2.65 \cdot 10^{-6}$ m.

5. The gate current does not appear in the results. Why?

6. In the same approximation (V_D small), write the equations of the characteristics $I_D = f(V_D)$ as functions of the parameters $G_0, (\phi - V_G)$, and $\psi = (\phi - V_P)$. Compare briefly with Fig. 10.25.

7. Now discard the assumption that V_D is infinitesimal. The larger V_D is, the less can the channel be regarded as equipotential. For a positive drain voltage the junction is more strongly reverse biased near the drain than near the source, and the width $d(x)$ of the depopulated zone on each side varies with x as shown in Fig. 10.26.

We assume $L \gg 2a$. Then the current lines are effectively parallel to Ox and $J_y \ll J_x$, implying $\mathcal{E}_y \ll \mathcal{E}_x$ (Fig. 10.27). The equipotentials in the conducting channel are to a first approximation planes orthogonal to Ox and $\mathcal{E}_x$ is then independent of y.

We count the current positive when it flows from the drain to the source. Using the functions $d(x), V(x)$, the width and voltage in the depleted region, find the current at x in the useful section.

8. The current $I(x)$ is independent of x. Why?

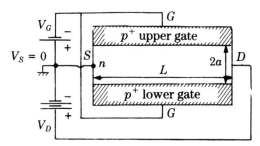

Fig. 10.24. Schematic view of the same transistor showing the doping of various regions. (After Leturcq and Rey, "Physique des Composants Actifs à Semi-conducteurs," Dunod, 1978.)

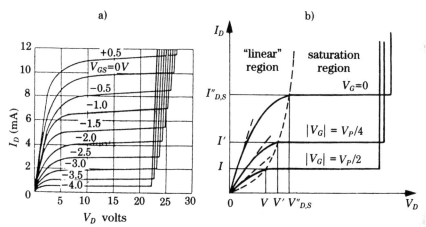

Fig. 10.25. Characteristics of the 2N3278 transistor (Fairchild Semiconductor Company). (a) Experimental curves; (b) simplified version of the same curves: I, I', I'', V, V', V'' denote saturation drain currents and voltages. The pinch voltage V_P is defined in problem 4.

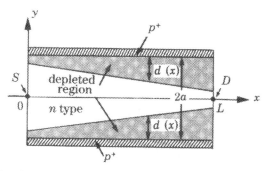

Fig. 10.26. Depleted regions and the conducting channel in a polarized JFET.

9. Deduce a differential equation relating $V(x)$ and $dV(x)/dx$, in which the drain current I_D is a parameter. For convenience use G_0 and ψ.

10. Integrate this differential equation between $x = 0$ and $x = L$ so as to obtain I_D as a function of G_0, V_D, ϕ, ψ, and V_G. Assume the channel is nowhere pinched.

11. What are the conditions on the applied voltages for this equation to hold? For what value of $V_D = V_{D,\text{Saturation}} = V_{D,S}$ is the channel pinched at $x = L$?

12. Under these conditions express the saturation current $I_{D,S}$ as a function of G_0, ψ and $V_{D,S}$. This relation bounds in Fig. 10.25(b) a region called "linear," to the left of the dashed curve. Show that this curve is a parabola with the equation

$$I_{D,S} = \frac{G_0}{4} \frac{(V_{D,S})^2}{\psi} \tag{10.49}$$

for $V_{D,S} \ll \psi$.

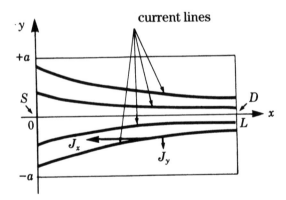

Fig. 10.27. Current lines in the channel of a polarized JFET. (After Leturcq and Rey, "Physique des Composants Actifs à Semi-conducteurs," Dunod, 1978.)

13. One might imagine that for $V_D > V_{D,S}$ the conducting region is closed off at the pinch and the current I_D drops. But experimentally one sees (Fig. 10.25(a)) that this does not happen. The current is stable above $V_{D,S}$ (in fact it rises very slightly). We investigate this regime here.

The problem is as follows: in the pinch region the electric field is very large. But in the presence of such a field the usual transport equations, derived for situations close to thermodynamic equilibrium, are no longer valid. In particular Ohm's law fails. Show first that in the above description the electric field grows in modulus and tends to infinity for $x = L$ when $V_D \rightarrow V_{D,S}$.

14. We thus have to consider the properties of a semiconductor in a very strong field and we may wonder if the electron velocity tends to infinity as we increase the field. It does not. In the presence of a field, the electrons gain between collisions an energy which is no longer negligible compared with the thermal energy kT, where T is the lattice temperature. If the electron collisions with the lattice were strictly elastic the mean energy of the electron gas would increase. But the collisions are not elastic: the electrons lose energy at each collision and an equilibrium is established. We call this a "hot" electron gas, which we describe by an effective temperature T_e which differs from that of the lattice (T), and by a distribution of electron energies E obeying

$$f = A \exp\left(-\frac{E}{kT_e}\right). \tag{10.50}$$

The effective temperature of the electrons is an increasing function of the electric field intensity. By substituting this distribution function in the expression for the mobility deduced from the Boltzmann equation, show that for collisions with crystal vibrations (cf. Sect. 5.4a) the mobility is proportional to $T^{-1}T_e^{-1/2}$, and is thus a decreasing function of the electric field.

15. In fact one observes that in silicon the mobility follows a law which can be modeled as

$$\mu = \frac{\mu_e}{1 + \mu_e \left|dV/dx\right|/v_l}. \tag{10.51}$$

The velocity v_l appearing in this equation is a limiting velocity, which is 10^5 m·s^{-1} in silicon. The behavior of the drift velocity as a function of electric field is shown in Fig. 10.28.

The description of the fields and currents in this case is more complex. In particular the approximation of neglecting the y dependence of the component $\mathcal{E}_x$ (and the $\mathcal{E}_y$ component) no longer holds throughout the device.

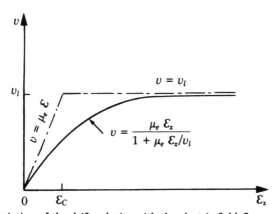

Fig. 10.28. Variation of the drift velocity with the electric field $\mathcal{E}_x$.

We retain the approximation however. Show from the equation obtained in
7 that for a given current the channel is no longer completely pinched in
the strong electric field region, the width of the conducting region remaining approximately constant in this region. Give an order of magnitude of a
"very strong" field.

16. To understand this saturation region we can schematically represent
the problem by decomposing the crystal into two regions: one with a weak
field and constant mobility, of length L_1, and a region with very strong field
and constant velocity with length L_2, with $L = L_1 + L_2$. We can assume
that $V(x = L_1) = V_{D,S}$ and thus that $V(x = L) - V(x = L_1) = V_D - V_{D,S}$.
Give a maximum value for L_2. Evaluate this in the case $V_D - V_{D,S} = 10$ V.

17. Assume that $L = 10^{-4}$ m. Summarize the behavior of the characteristics in the saturation region.

18. What causes the abrupt rise of I_D for very large V_D?

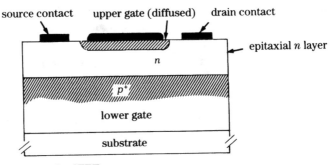

Fig. 10.29. Section of a JFET.

19. A real device, manufactured using planar techniques, is shown in
section form in Fig. 10.29. What other effects, neglected here, can modify
its behavior?

Solutions

1. From Eq. (8.60),

$$d = [2 \, \epsilon_0 \epsilon_r \, (\phi - V_G)/eN_d]^{1/2}. \tag{10.52}$$

2. We have for a conducting parallelepiped of cross section $s = 2(a-d)b$:

$$G = \frac{1}{R} = \sigma \frac{s}{L} = N_d \, e \, \mu_0 \frac{2 \, b}{L}(a - d) \tag{10.53}$$

$$= G_0 \left(1 - \frac{d}{a} \right) \tag{10.54}$$

$$= G_0 \left\{ 1 - \left[\frac{2 \, \epsilon_r \epsilon_0 \, (\phi - V_G)}{e \, N_d \, a^2} \right]^{1/2} \right\}. \tag{10.55}$$

3. For $d = a_m$ the conductance is zero, because the depleted region blocks all the conducting space. From Eq. (10.52), with $V_G = 0$:

$$a_m = \left(\frac{2\epsilon_0\epsilon_r}{e} \frac{\phi}{N_d} \right)^{1/2} \tag{10.56}$$

$$a_m = \left(\frac{2 \cdot 10^{-10}}{1.6 \cdot 10^{-19}} \cdot \frac{0.8}{10^{21}} \right)^{1/2} = 1\mu m \tag{10.57}$$

4. From Eq. (10.54), $G = 0$ when

$$\frac{2\epsilon_r\epsilon_0}{e\, N_d\, a^2}(\phi - V_P) = 1, \tag{10.58}$$

or

$$V_P = \phi - \frac{e\, N_d\, a^2}{2\epsilon_r\epsilon_0}, \tag{10.59}$$

$$V_P = -4.8 \text{ V}. \tag{10.60}$$

5. The gate current is that of a reverse biased junction and is negligible.

6. Using Eq. (10.56) in the form $\psi = eN_d a^2/2\epsilon_r\epsilon_0$, we get

$$I_D = G_0 \left[1 - \left(\frac{\phi - V_G}{\psi} \right)^{1/2} \right] V_D. \tag{10.61}$$

This gives a family of straight lines through the origin. The slopes increase when the magnitude of V_G rises from V_P.

7. The current at x is the flux of the current vector $J(x)$ in the depleted section, with modulus

$$I(x) = \int_{-a+d(x)}^{a-d(x)} b|J(x)|dy. \tag{10.62}$$

$J(x) = N_d e \mu_0 (dV/dx)$ is independent of y in the present approximation, so

$$I(x) = 2[a - d(x)]b\, e\, N_d\, \mu_e \frac{dV}{dx}. \tag{10.63}$$

8. If no current flows in the gate, all the electrons supplied by the source must reach the drain.

9. Equation (10.52), evaluated at the point x where the applied voltage at the junction is $V_G - V(x)$, can be written

$$d(x) = \left[\frac{2\epsilon_0\epsilon_r[\phi - V_G + V(x)]}{eN_d} \right]^{1/2} \tag{10.64}$$

or using Eq. (10.59),

$$d(x) = a \left(\frac{\phi - V_G + V(x)}{\psi} \right)^{1/2}. \tag{10.65}$$

Substituting in Eq. (10.59),

$$I(x) = I_D = 2abeN_d\,\mu_e \left[1 - \left(\frac{\phi - V_G + V(x)}{\psi}\right)^{1/2}\right]\frac{dV}{dx} \qquad (10.66)$$

$$I_D = G_0\,L\left[1 - \left(\frac{\phi - V_G + V(x)}{\psi}\right)^{1/2}\right]\frac{dV}{dx}. \qquad (10.67)$$

10. From Eq. (10.67),

$$\int_0^L I_D\,dx = I_D\,L = G_0\,L\int_0^{V_D}\left[1 - \left(\frac{\phi - V_G + V}{\psi}\right)^{1/2}\right]dV, \qquad (10.68)$$

$$I_D = G_0\left[V_D - \frac{2}{3}\frac{(\phi - V_G + V_D)^{3/2} - (\phi - V_G)^{3/2}}{\psi^{1/2}}\right]. \qquad (10.69)$$

11. For Eq. (10.69) to hold we must have reverse bias, the gate current must be negligible, and the channel must be open everywhere, or

$$V_G < 0 \text{ and } |V_G| < V_P,\ V_D - V_G > 0,\ V_D - V_G < -V_P. \qquad (10.70)$$

The channel is pinched at $x = L$ when $V_{D,S} = V_G - V_P$.

12. The current $I_{D,S}$ is found by substituting $V_{D,S}$ into Eq. (10.69),

$$I_{D,S} = G_0\,\psi\left[\frac{V_{D,S}}{\psi} - \frac{2}{3} + \frac{2}{3}\left(\frac{\psi - V_{D,S}}{\psi}\right)^{3/2}\right]. \qquad (10.71)$$

For $V_{D,S}/\psi << 1$,

$$I_{D,S} \sim G_0\psi\left[\frac{V_{D,S}}{\psi} - \frac{2}{3} + \frac{2}{3}\left(1 - \frac{3}{2}\frac{V_{D,S}}{\psi} + \frac{3}{8}\left(\frac{V_{D,S}}{\psi}\right)^2 + \cdots\right)\right]$$

$$\sim \frac{G_0\,(V_{D,S})^2}{4}\frac{}{\psi}. \qquad (10.72)$$

13. From Eq. (10.67),

$$\left(\frac{dV}{dx}\right)_{x=L} = \frac{I_D}{G_0\,L}\frac{1}{1 - [(\phi - V_G + V_D)/\psi]^{1/2}} \qquad (10.73)$$

is infinite for $V_D = V_{D,S} = V_G - V_P$.

We can also say that the area crossed by the constant current vanishes. This requires the electric field to be infinite.

14. We showed in Chap. 5 that

$$\mu = e\frac{<\tau>}{m}\quad\text{with} \qquad (10.74)$$

$$<\tau> = \frac{\displaystyle\int_0^\infty \tau(E)\,E^{3/2}\,f(E)\,dE}{\displaystyle\int_0^\infty E^{3/2}\,f(E)\,dE}. \qquad (10.75)$$

Replacing $f(E)$ by $A \exp(-E/kT_e)$ and τ by $E^{-1/2}T^{-1}$ for collisions with crystal vibrations for a lattice temperature T (see Sect. 5.4a), we get

$$< \tau > \sim T^{-1} \, T_e^{-1/2}. \tag{10.76}$$

15. Replacing μ in Eq. (10.63) we have

$$I_D = 2[a - d(x)]b \; e \; N_d \; \mu_e \frac{dV/dx}{1 + \mu_e/v_l \, |dV/dx|}. \tag{10.77}$$

For $\dfrac{dV}{dx} >> \dfrac{v_l}{\mu_e}$,

$$I_D = 2[a - d(x)]be \; N_d \; v_l. \tag{10.78}$$

For a very strong field, d becomes independent of x since I_D is conserved. As this constant value is less than a ($I_D > 0$) the channel is no longer pinched. The critical value of the field is $\mathcal{E}_C = v_l/\mu_e$. For $v_l \sim 10^5$ m $\cdot$ s^{-1}, $\mu_e \sim 0.1$ m$^2 \cdot$ V$^{-1} \cdot$ s^{-1}, a "strong field" will be greater than $\mathcal{E}_C$, which is 10^6 V $\cdot$ m^{-1}.

16. In the strong-field region of length L_2, $\dfrac{dV}{dx} > \dfrac{v_l}{\mu_e}$, and

$$V_D - V_{D,S} = \int_{L_1}^{L} \frac{dV}{dx} dx > \frac{v_l}{\mu_e} L_2, \tag{10.79}$$

$$L_2 < \frac{(V_D - V_{D,S}) \, \mu_e}{v_l} \sim 10^{-5} \text{ m}. \tag{10.80}$$

17. In the saturation regime the conductance of the low-field region is given by Eq. (10.71), where we replace L in G_0 by L_1, and thus G_0 by $G_0 L/(L - L_2)$. As L_2 is much smaller than L the current is almost constant (in reality it increases slightly), and the strong-field region does not contribute significantly to the total conductance.

18. This is the reverse breakdown of the gate–drain junction. A current therefore flows in the gate. We note that the more negative V_G is, the lower the breakdown voltage at the drain: the important quantity here is the difference between the gate and drain potentials, which must be kept below the breakdown voltage.

19. We have neglected the y dependence of the electric field and the edge effects, among others arising from the finite width of the device. The coefficients appearing in the calculation are slightly modified without changing the behavior. The main effect is the existence of non-zero resistances between the source contact and the active region and between the active region and the drain contact.

Appendix 10.3

Problems on MOS (Metal-Oxide-Semiconductor) Structure

We study here the MIS (metal–insulator–semiconductor) and MOS (metal–oxide–semiconductor) structures, which are the building blocks of MOSFET transistors, just as p–n junctions are for bipolar transistors. To this end, we consider the plane capacitor shown in Fig. 10.30(a), formed of a semiconducting substrate covered with an insulating oxide layer of thickness d and a metallic layer. The insulating layer prevents a current from flowing between the metal and the semiconductor, and thus allows control of the electric field in the semiconductor by the voltage applied to the metal electrode without there being injection of carriers, unlike the case of a p–n junction.

Figure 10.30(b) gives the energy diagram of the MOS structure when the two ends of the structure are not electrically connected. Conditions are assumed uniform in planes parallel to the interfaces, and thus depend only on the variable x orthogonal to the interfaces, as in the notation of Fig. 10.30.

The energy origin is arbitrary, but a convenient way of relating the semiconductor and metal energies in the present case is to refer the energies of the two media to the energy E_0 of the bottom of the oxide conduction band. The Fermi levels in the metal and semiconductor are E_{Fm}, E_{Fsc}. We denote by ϕ_m, ϕ_{sc} the quantities $E_0 - E_{Fm}, E_0 - E_{Fsc}$. In general the Fermi levels do not coincide since the metal and the semiconductor are isolated from each other and cannot exchange electrons. We set $\phi_m - \phi_{sc} = \Delta\phi$. The case $\Delta\phi > 0$ is shown schematically in Fig. 10.30(b). This figure also shows the bottom E_c of the conduction band, the top of the valence band E_v ($E_c - E_v = E_g$, the semiconductor band gap). We call the interface affinity χ the difference between the energy of the oxide conduction band and the energy of the conduction band of the semiconductor at the oxide–semiconductor interface:

$$\chi = (E_0 - E_c)_{x=0}.$$

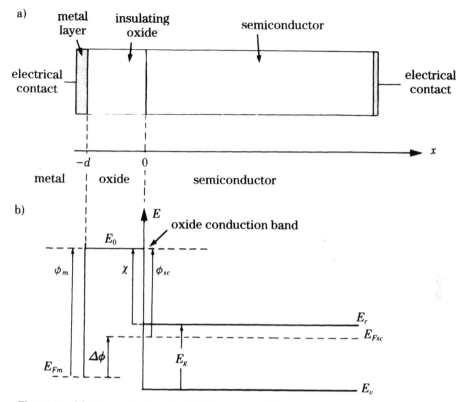

Fig. 10.30. (a) Schematic view of a MOS structure; (b) energy diagram of this structure when the end contacts are not electrically connected.

MOS Capacitor without Applied Voltage

In Part 1 of this discussion we connect the two electric contacts by an external lead without any applied voltage (short circuit). Figure 10.31 shows the resulting energy diagram for this structure.

1. Explain how the charges move to establish equilibrium, defined by the equality of Fermi levels in the metal and semiconductor. Justify the new energy diagram of Fig. 10.31 quantitatively, in particular the bending of the bands at the surface. Can you explain briefly why the interface affinity has not changed?

2. Deduce that there is now an electric field between the two plates of the capacitor formed by the MOS structure. Assume that there are no charges in the oxide and thus that the field is constant over the width of the oxide. Give the sign of the electric field as a function of the sign of $\Delta\phi$.

3. Under the conditions of Fig. 10.31 the energy of an electron depends on x, since there is an electrostatic potential $V(x)$ in the semiconductor and in the oxide caused by the charge distribution induced by the electrical contact between the metal and the semiconductor. The energy of an electron

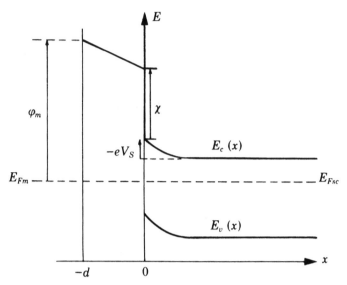

Fig. 10.31. Energy diagram of the structure shown in Fig. 10.30 when the metal and semiconductor are short-circuited by an external lead.

at the bottom of the semiconductor conduction band is $E_c(x) = E_c(+\infty) - eV(x)$, where $-e$ is the electron charge, and we assume that the induced electrostatic potential vanishes far from the interface, i.e., for x large and positive.

Write down the continuity of the electric displacement vector $\boldsymbol{D} = \epsilon\boldsymbol{\mathcal{E}}$, where ϵ is the dielectric constant of the medium ($\epsilon = \epsilon_0\epsilon_r$) and $\boldsymbol{\mathcal{E}}$ the electric field. Deduce a relation between the electrostatic potential $V_S = V(x = 0)$ at the oxide–semiconductor interface and its derivative at $x = 0$ for $x > 0$, $(\partial V/\partial x)_{0_+}$ involving $\Delta\phi, e, d$ and the dielectric constants $\epsilon_{ox}, \epsilon_{sc}$ of the oxide and semiconductor. Sketch the form of $V(x)$ in the case $\Delta\phi > 0$ for $-d \leq x < +\infty$.

4. From now on we study an n-type semiconductor, uniformly doped with donor concentration N_d. We assume further that the Fermi level remains deep enough in the band gap for all the donors to be ionized. Express the densities $n(x), p(x), \rho(x)$ of electrons, holes, and charges as functions of $eV(x)/kT$, N_d, and the intrinsic density n_i. Write down the differential equation obeyed by $V(x)$.

MOS Capacitor with Applied Voltage

To simplify this study (and particularly the notation), we henceforth consider an "ideal" MOS structure, where $\Delta\phi = 0$. The energy level scheme when the metal and semiconductor are short-circuited by an external lead, is given in Fig. 10.32.

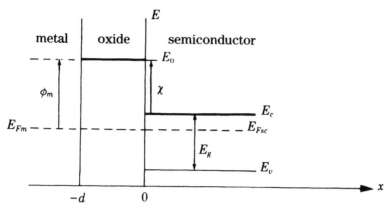

Fig. 10.32. Energy diagram of an "ideal" MOS structure. The figure should be compared with Fig. 10.31, which shows a "non-ideal" structure under the same conditions.

We now consider the MOS capacitor with a voltage V_G applied at the metal layer called the gate, the right-hand region of the n-type semiconductor being now at zero potential (Fig. 10.33).

5. The oxide is a perfect insulator, so, apart from a transient interval during the establishment of the voltage V_G, no current flows in the capacitor. What can we deduce about the Fermi level in the semiconductor and the relative positions of the Fermi levels in the metal and the semiconductor?

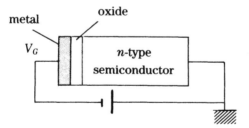

Fig. 10.33. MOS capacitor biased by the voltage V_G.

6. For the moment we limit ourselves to the case where V_G is negative and not too large in modulus ($|V_G| <$ about $E_g/2e$).

(a) Integrate the equation giving $V(x)$, assuming that the donors are ionized and using for $\rho(x)$ an approximation copied from that in the abrupt p-n junction (cf. Chap. 8). State this approximation and show that

$$V(x) = V_S \left(1 - \frac{x}{W}\right)^2 \quad \text{for } 0 < x < W, \tag{10.81}$$

$$V(x) = 0 \quad \text{for } x > W, \tag{10.82}$$

$$V(x) = V_S + (V_S - V_G)\frac{x}{d} \quad \text{for } -d < x < 0, \tag{10.83}$$

where the width W of the space charge depends on V_S. Give an expression for V_S as a function of N_d and W.

(b) Writing the continuity of the electric displacement vector $\epsilon\mathcal{E}$ across the oxide–semiconductor interface, give a relation between V_G, V_S, and W (V_G). Show that

$$W\ (V_G) = d\frac{\epsilon_{sc}}{\epsilon_{ox}}\left[\sqrt{1 - \frac{2V_G\ \epsilon_{ox}^2}{d^2\ e\ N_d\ \epsilon_{sc}}} - 1\right]. \tag{10.84}$$

Applications:

Calculate W for $V_G = -1$ V; $e = 1.6 \cdot 10^{-19}$ C; $\epsilon_{ox} = 3.2 \cdot 10^{-11}$ F $\cdot$ m^{-1}; $\epsilon_{sc} = 10^{-10}$ F $\cdot$ m^{-1}; $d = 150$ nm; $N_d = 10^{22}$ m^{-3}.

The maximum field that can be supported by silicon without breakdown is 10^9 V/m. What is the maximum value of W for n-type silicon ($N_d = 10^{22}$ m^{-3})?

7. Under the preceding approximations, give an expression for the total charge appearing in the semiconductor when we apply a voltage V_G to the gate:

$$Q_S\ (V_G) = \int_0^\infty \rho\ (x)\ dx.$$

A charge $Q_G = -Q_S$ appears at the metal–oxide interface, and we define the differential capacitance per unit area of the MOS structure by the relation $C = dQ_G/dV_G$.

Show that C can be regarded as the resultant capacitance when the capacitance $C_{ox} = \epsilon_{ox}/d$ of the oxide layer and the capacitance C_{sc} of the space-charge region are placed in series, where

$$C_{sc} = -\frac{dQ_S}{dV_S} = \frac{\epsilon_{sc}}{W(V_G)}. \tag{10.85}$$

Calculate C_{ox} and C_{sc} for the data given in **6**.

8. Sketch the forms of the energy levels in the following cases:
(a) $V_G \le 0$; $|V_G| \lesssim E_g/2e$;
(b) $V_G << -E_g/e$;
(c) $V_G > 0$.

Explain briefly why the cases (a), (b), (c) are known, respectively, as depletion, inversion, and enhancement. Sketch the curves giving $V(x)$ for $-d \le x \le +\infty$ in the various cases.

9. Find the value $V_G = V_{\text{threshold}}$ for which there is complete depletion of a region of the semiconductor of thickness $l = 200$ nm doped with $N_d = 10^{23}$ m^{-3}, when the oxide layer has thickness $d = 100$ nm.

For what value of V_G do we reach inversion? Take the intrinsic concentration as $n_i = 1.6 \times 10^{16}$ m^{-3}.

Applications of the MOSFET

In practice the semiconductor is used as a resistor whose value can be controlled by use of the applied gate voltage V_G. We thus consider a square of semiconductor in the y–z plane, with total thickness l in the x direction. On one of its faces is an oxide layer of thickness d and a metal layer. On the other face is a metallic contact, through which we can apply a voltage V_G to the MOS structure as described in Part 2. We denote as the transverse resistance $R(V_G)$ the resistance measured for a current flowing in the z direction while a voltage V_G is applied across the two sides of the structure (Fig. 10.34).

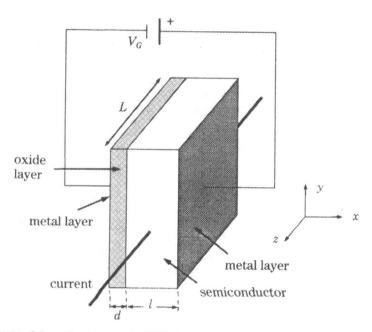

Fig. 10.34. Schematic view of a MOSFET.

10. Show that $R(V_G)$ does not depend on the surface area of the square and express $R(V_G)$ as a function of $R(V_G = 0) = R_0$, l, and $W(V_G)$ in the depletion regime (a) of **8**.

The interesting quantity for applications is the differential conductance $G = d(1/R)/dV_G$. Show that $G(V_G)$ is proportional to $C(V_G)$ and that the constant of proportionality is the mobility of the charge carriers in the semiconductor.

11. We observe that when $|V_G|$ is increased with $V_G < 0$, the resistance R first grows monotonically, and for sufficiently small l it reaches effectively infinite values. On the other hand if l is large enough there is a value of V_G above which R decreases abruptly. Can you explain this?

Solutions

1. The alignment of the Fermi levels required for equilibrium is achieved by a flow of electrons from the semiconductor to the metal, since $E_{Fsc} - F_{Fm} = \Delta\phi > 0$, thus charging the semiconductor positively and the metal negatively. The upward bending of the conduction band results from an electron depletion near the interface. The bands stay parallel because the width of the band gap is an intrinsic property of the semiconductor; similar considerations hold for $\chi = (E_0 - E_c)_{x=0}$, the energy required to transfer an electron from a quantum level in the interior of the semiconductor to a well-defined level at the exterior, near the interface.

2. As charges have now appeared on both sides of the oxide layer, there is an electric field from the semiconductor to the metal if $\Delta\phi > 0$ and in the opposite direction if $\Delta\phi < 0$.

3.

$$\epsilon_{ox}\mathcal{E}_{ox} = \epsilon_{sc}\,\mathcal{E}_{sc}; \quad \mathcal{E}_{ox} = -\frac{\epsilon_{sc}}{\epsilon_{ox}}\left(\frac{dV}{dx}\right)_{0+}. \tag{10.86}$$

The quantity $\Delta\phi$ is now the energy shift between the two regions with flat bands in the metal and the semiconductor. This shift is split between the oxide and the region where the bands bend in the semiconductor. The energy jump χ at the semiconductor surface remains the same. The electrostatic potential $V(x)$ shown in Fig. 10.35 varies continuously.

We have

$$\Delta\phi = [-eV(-d)] - [-eV(\infty)]$$
$$= -e[(V(-d) - V(0)) + (V(0) - V(\infty))]$$
$$= -e\,(d\,\mathcal{E}_{ox} + V_S),$$

so that

$$\frac{\Delta\phi}{e} = \frac{\epsilon_{sc}}{\epsilon_{ox}}d\left(\frac{dV}{dx}\right)_{0+} - V_S. \tag{10.87}$$

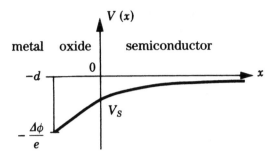

Fig. 10.35. Electrostatic potential across a MOS capacitance in the case $\Delta\phi > 0$.

4.

$$n\,(x) = N_c \exp\left(-\frac{E_c(x) - E_F}{kT}\right)$$

$$= N_c \exp\left(-\frac{[E_c(\infty) - eV(x)] - E_F}{kT}\right),$$

$$n\,(x) = N_c \exp\left(-\frac{E_c(\infty) - E_F}{kT}\right) \exp\frac{eV(x)}{kT}$$

$$= N_d \exp\frac{eV(x)}{kT}, \tag{10.88}$$

$$p\,(x) = \frac{n_i^2}{n(x)} = \frac{n_i^2}{N_d} \exp\left(-\frac{eV(x)}{kT}\right). \tag{10.89}$$

The potential $V(x)$ obeys the Poisson equation

$$\frac{\partial^2 V}{\partial x^2} = -\frac{\rho\,(x)}{\epsilon_{sc}} \quad \text{with} \quad \rho\,(x) = e\,(N_d - n + p) \tag{10.90}$$

or

$$\frac{\partial^2 V(x)}{\partial x^2} = -\frac{eN_d}{\epsilon_{sc}}\left[1 - \exp\frac{eV(x)}{kT} + \frac{n_i^2}{N_d^2} \exp\left(-\frac{eV\,(x)}{kT}\right)\right]. \tag{10.91}$$

5. There is no current, so E_F is constant in the semiconductor.

$$E_{Fm} - E_{Fsc} = -eV_G. \tag{10.92}$$

6. In this case the Fermi level is always within the gap. We assume, as for the p–n junction (Sect. 8.2), that the electron density $n(x) = N_d \exp(eV(x)/kT)$ varies rapidly enough so that we can define a depletion length $W(V_G)$ such that

$$\rho\,(x) = N_d\,e \quad \text{for } 0 < x < W(V_G),$$
$$\rho\,(x) = 0 \quad \text{for } x > W(V_G),$$
$$\rho\,(x) = 0 \quad \text{for } -d < x < 0. \tag{10.93}$$

The distance over which n changes from N_d to nearly zero is much smaller than $W(V_G)$ and we thus take $\rho(x)$ as varying discontinuously, as in Chapt. 8, so that

$$\frac{\partial^2 V}{\partial x^2} = -\frac{eN_d}{\epsilon_{sc}} \quad \text{for } 0 < x < W. \tag{10.94}$$

(a) For $0 < x \leq W$, $V(x) = V_S\left(1 - \frac{x}{W}\right)^2$, \hfill (10.95)

with $V_S = -\dfrac{e\,N_d\,W^2}{2\,\epsilon_{sc}}$ \hfill (10.96)

for $x > W$, $V(x) = 0$, \hfill (10.97)

for $-d < x < 0$, $\dfrac{\partial^2 V}{\partial x^2} = 0$ \hfill (10.98)

with boundary conditions $V(-d) = V_G; V(0) = V_S$, so that

$$V(x) = V_S + (V_S - V_G)\frac{x}{d}. \tag{10.99}$$

(b) We have continuity of the displacement vector at $x = 0$:

$$\epsilon_{ox}\,\mathcal{E}_{ox}(0) = \epsilon_{sc}\,\mathcal{E}_{sc}(0) \tag{10.100}$$

or, using the expressions for $\mathcal{E} = -dV/dx$ in the two materials,

$$-\epsilon_{ox}\frac{V_S - V_G}{d} = \frac{2\,\epsilon_{sc}}{W}V_S \tag{10.101}$$

giving

$$V_G = V_S\left(\frac{\epsilon_{sc}}{\epsilon_{ox}}\frac{2d}{W} + 1\right). \tag{10.102}$$

Substituting expression (10.96) for V_S in Eq. (10.102):

$$V_G = -\frac{e\,N_d}{2\,\epsilon_{sc}}W^2\left(\frac{2\,\epsilon_{sc}}{\epsilon_{ox}}\frac{d}{W} + 1\right), \tag{10.103}$$

$$W^2 + 2d\frac{\epsilon_{sc}}{\epsilon_{ox}}W + \frac{2\epsilon_{sc}}{eN_D}V_G = 0 \tag{10.104}$$

$$W = \frac{\epsilon_{sc}}{\epsilon_{ox}}d\left[\sqrt{1 - \frac{2\,V_G\,\epsilon_{ox}^2}{d^2\,e\,N_d\,\epsilon_{sc}}} - 1\right]. \tag{10.105}$$

Numerical applications: for the given values $W(V_G) \simeq 120$ nm. The maximum field in the oxide is $(\epsilon_{sc}/\epsilon_{ox})\mathcal{E}_{max}$, where $\mathcal{E}_{max}$ is the field in the semiconductor. Integrating $dE/dx = \rho/\epsilon_{sc}$ we get $eN_dW_{max}/\epsilon_{sc} = \mathcal{E}_{max}$, implying $W_{max} = 20\ \mu m$.

7. The charge Q_S appears in the space-charge region, of width $W(V_G)$:

$$Q_S\,(V_G) = \int_0^W e\,N_d\,dx = e\,N_d\,W(V_G),\qquad(10.106)$$

$$Q_G = -Q_S = -e\,N_d\,W,\qquad(10.107)$$

$$dQ_G = -e\,N_d\,dW.$$

Differentiating Eq. (10.98) with respect to $W(V_G)$ we get

$$dV_G = -e\,N_d\left(\frac{d}{\epsilon_{ox}} + \frac{W}{\epsilon_{sc}}\right)dW\qquad(10.108)$$

giving the differential capacitance per unit area C:

$$\frac{dV_G}{dQ_G} = \frac{1}{C} = \frac{d}{\epsilon_{ox}} + \frac{W}{\epsilon_{sc}} = \frac{1}{C_{ox}} + \frac{1}{C_{sc}}\qquad(10.109)$$

with

$$C_{ox} = \epsilon_{ox}/d,\ \ C_{sc} = \epsilon_{sc}/W.\qquad(10.110)$$

For the values given $C_{ox} = 2.1 \times 10^{-4}$ F/m^2; $C_{sc} = 8.3 \times 10^{-4}$ F/m^2.

8. (a) $E_g/2e < V_G < 0$.
We are in the depletion region since $n(0) < N_d$. (See Fig. 10.36.)

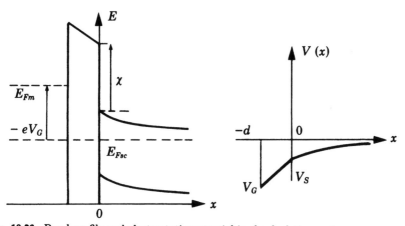

Fig. 10.36. Band profile and electrostatic potential in the depletion regime.

(b) $V_G \ll -E_g/e$.
We are in the inversion region as $p(0) > n(0)$: the semiconductor has become p type (hole conduction) near the interface. (See Fig. 10.37.)
(c) $V_G > 0$.
We now have enhancement as $n(0) > N_d$. Electrons accumulate at the interface, as seen on Fig. 10.38.

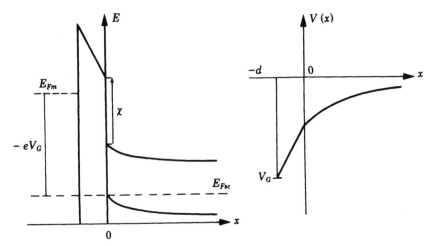

Fig. 10.37. Band profile and electrostatic potential in the inversion regime.

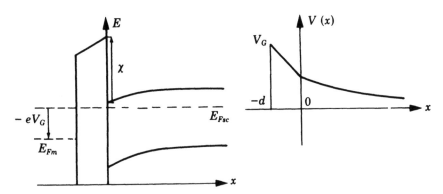

Fig. 10.38. Band profile and electrostatic potential in the enhancement regime.

9. $V_{\text{threshold}}$ is such that $W(V_G) = 200$ nm. From the expression for $W(V_G)$ we get

$$V_{\text{threshold}} = -13.2 \text{ V}.$$

There is inversion when

$$p(0) \approx n(\infty) \approx N_d \tag{10.111}$$

or

$$p\,(0) = \frac{n_i^2}{N_d}\,\exp\left(\frac{eV_S}{kT}\right) \qquad (10.112)$$

so that $\dfrac{V_S}{kT} = -\dfrac{2}{e}\,\log\dfrac{N_d}{n_i} \sim -0.78$ V.

From the expression for V_S:

$$W(V_G) = \sqrt{-\frac{2\,\epsilon_{sc}\,V_S}{eN_D}} \approx 0.1 \;\mu\text{m}.$$

At inversion $W(V_G)$ is less than the channel width. The channel is inverted before it is fully depleted.

From the expression for V_G as a function of V_S: $V_G = -5.65$ V.

10.

$$\frac{1}{R} = \frac{\int_0^\infty n(x)\,e\mu_e\,dx \cdot L}{L} = \int_W^l N_d\,e\mu_e\,dx$$

$$= N_d\,e\mu_e\,l\left(1 - \frac{W}{l}\right), \qquad (10.113)$$

$$R\,(V_G) = R_0\left(1 - \frac{W(V_G)}{l}\right)^{-1}, \qquad (10.114)$$

where L is the length of a side of the square, l the semiconductor thickness, and μ_e the electron mobility in the semiconductor:

$$G\,(V_G) = \frac{d\,(1/R)}{dV_G} = -\mu_e\,N_D\,e\frac{dW(V_G)}{dV_G}$$

$$= \mu_e\frac{dQ_G}{dV_G} = \mu_e\,C(V_G). \qquad (10.115)$$

11. When we increase $|V_G|$ with $V_G < 0$, W grows and can become equal to l before we reach the inversion regime. There are then practically no more mobile carriers in the semiconductor and its resistance is extremely high. In contrast, if l is large enough we reach the inversion regime before completely depleting the semiconductor. The conductance then increases abruptly because it arises from the mobile holes which appear in the inversion region at the oxide–semiconductor interface.

Supplement: Integrated Circuits with MOS Transistors

These are of two types: normally on and normally off, depending on whether the structure at rest $(V_G = 0)$ conducts or not. These structures are also sometimes called depletion-mode or enhancement-mode, respectively. This means that when a voltage is applied the structure becomes non-conducting ("normally on") or conducting ("normally off"). Depending on the material of the channel (n or p) this change of state occurs for a positive or a negative voltage.

Figure 10.10 shows an enhancement-mode MOSFET with an n channel: when a voltage is applied between the source and drain electrodes (n type), one of the junctions between source or drain and substrate (p type) is reverse biased and does not allow current to flow.

When a sufficiently large *positive voltage* is applied to the gate, there will be inversion near the oxide, i.e., the appearance of significant numbers of conduction electrons; hence the name "n channel." Analogous reasoning shows that a structure with source-drain electrodes of p type and a substrate of n type (as in the problems) will conduct if a *negative voltage* is applied at the gate. This would be a p-channel enhancement-mode MOSFET.

The depletion structures are entirely analogous. We start from a situation of *inversion at rest* because of the presence of *static charges in the oxide*. These charges induce charges of opposite sign in the semiconductor. As we can easily introduce positive ions into the oxide (Na^+ ions) the usual case involves the introduction of an n channel with p-type source and drain contacts. Depletion-mode MOSFETs are n-channel transistors with n-type source and drain contacts. Applying a negative voltage to the gate then empties the channel of electrons. The channel becomes p type and one of the contacts blocks the current.

In each case it is clear that the channel conductance is a linear function of the applied voltage in a certain range. There is then a linear amplification effect: linear control of the source-drain current I_D by the gate voltage. The parameter $g_m = \partial I_D / \partial V_G$ is called the transconductance of the transistor and is one of the essential parameters for the user.

Microlithography techniques allow one to manufacture n-channel and p-channel MOSFETs simultaneously on the same substrate. In this way one produces complementary structures called CMOS (Complementary MOS) as in Fig. 10.39(a). The equivalent electrical scheme is given in Fig. 10.39(b). These structures have the property of having two stable states in which the current dissipation is infinitesimal. These are used in binary logic applications in which the signal is either 0 or 0.5 V.

One applies voltages $V_{high} = 5$ V at S_2 and $V_{low} = 0$ V at S_1 throughout. Assume that the input voltage is 5 V (Fig. 10.39(c)). The gates G_1, G_2 are positively polarized and electrons are attracted. We thus create an n channel

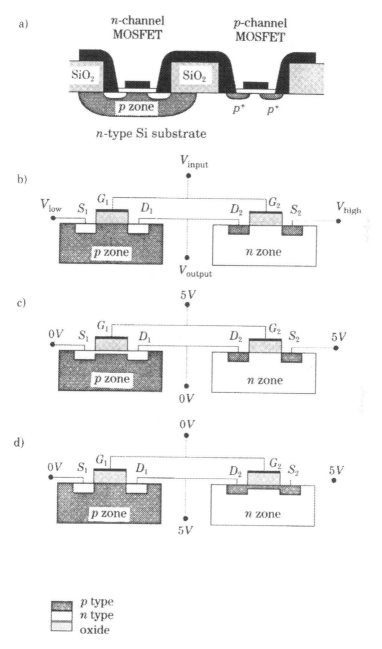

Fig. 10.39. (a) CMOS structure, with two complementary MOSFETs. The p region of the n-channel MOSFET is obtained by diffusion; (b) equivalent electrical scheme; operating as an inverter; (c) $V_{input} = 5$ V; (d) $V_{input} = 0$ V.

in the MOS on the left which becomes conducting. By contrast the MOS on the right is blocked since it contains two head–tail junctions in series. In these conditions the MOS behaves as a short circuit and $V_{\text{output}} \sim 0$ V.

On the other hand if $V_{\text{input}} = 0$ the left-hand MOS is blocked if we chose it "normally off," while that on the right conducts if it is "normally on." The output voltage is then 5 V. In both cases one of the two MOS in series is blocked and the current flowing in the circuit is very small. This device is an inverter of very low power dissipation.

MOSFETs constitute the most rapidly developing sector of integrated circuit production (annual growth rates of between 30%–60%). The reason for this success comes from the ease of manufacture (fewer masking levels than in bipolar circuits) and their low dissipation. The degree of integration is also higher because of the reduced size of the transistors, which are the building blocks of each device; indeed, 64-megabit memory chips are currently being developed and will presumably reach the market shortly. On the other hand MOSFETs are slower than bipolar transistors because of the large gate capacitances, caused by the presence of an insulating layer.

Values of the Important Physical Constants

Avogadro's number	N =	6.02×10^{23}
electron charge	e =	-1.6×10^{-19} coulomb
electron rest-mass	m =	0.91×10^{-30} kg
Planck's constant	h =	6.624×10^{-34} J·s
velocity of light	c =	2.9979×10^{8} m·s^{-1}
electron specific charge	e/m =	1.76×10^{11} C·kg^{-1}
electron radius	$r_0 = e^2/4\pi\varepsilon_0 mc^2$ =	2.82×10^{-15} m
first Bohr radius	a_0 =	5.29×10^{-11} m
proton–electron mass ratio	=	$1{,}836.1$
Boltzmann's constant	k =	$1.38 \cdot 10^{-23}$ J·K^{-1}
Bohr magneton	μ_B =	-9.27×10^{-24} J·T^{-1}
wavelength associated with 1 eV	=	1.239 μm
frequency associated with 1 eV	=	2.418×10^{14} Hz
energy associated with 1 K	=	8.616×10^{-5} eV
temperature associated with 1 eV	=	$11{,}605$ K

Some Physical Properties of Semiconductors (286 K)

		E_g (eV)	μ_e (cm²/V.s)	μ_h (cm²/V.s)	$\varepsilon = n^2$	a angstr.	d (g.cm⁻³)	F.T. (°C)
IV	C	5.4	1 800	1 200	5.5	3.567	3.51	3 550
	Si	1.15	1 900	480	11.8	5.42	2.42	1 412
	Ge	0.65	3 800	1 800	16.0	5.646	5.36	958
	Sn	0.08	2 500	2 400		6.47	6.0	232
VI	Se	1.6		0.6	8.5	4.35 / 4.95	4.8	220
	Te	0.33	1 100	560	5.0	4.447 / 5.915	6.24	452
III V	BP	6		100	11.6	4.537	2.97	3 000
	Al P	2.5	3 500		11.6	5.43	2.85	1 500
	Al As	2.3	1 200	200		5.63	3.81	1 600
	Al Sb	1.52	400	150	10.3	6.13	4.22	1 060
	Ga P	2.25	80	17	8.4	5.44	4.13	1 350
	Ga As	1.42	8 500	400	13.5	5.65	5.31	1 280
	Ga Sb	0.69	4 000	650	15.2	6.095	5.62	728
	In P	1.27	4 600	700	10.6	5.869	4.78	1 055
	In As	0.35	30 000	240	11.5	6.058	5.66	942
	In Sb	0.17	70 000	1 000	16.8	6.48	5.775	525
IV IV	Si C	3.0	60	8	10.2	4.35	3.21	2 700
IV VI	Pb S	0.37	800	1 000	17.9	7.5	7.61	1 114
	Pb Se	0.26	1 500	1 500		6.14	8.15	1 062
	Pb Te	0.25	1 620	750		6.45	8.16	904
V VI	Bi₂Te₃	0.15	1 250	515		10.48	7.7	580
II V	Cd₃As₂	0.13	15 000			8,76	6.21	721
	Cd Sb	0.48	300	300		6,471	6.66	456
II VI	Zn O	3.2	190		8.5	5.18	5.60	1 975
	Zn S	3.65	100		8.3	5.423	4.80	
	Zn Se	2.6	100	16	5.75	5.667	5.42	1 515
	Zn Te	2.15		50	18.6	6.101	5.54	1 239
	Cd S	2.4	200		5.9	5.83	4.82	685
	Cd Se	1.74	500		4.30	6.05	5.81	1 350
	Cd Te	1.50	650	45	11.0	6.48	6.20	1 098
	Hg S	2.5			5.86	5.852	7.67	583
	Hg Se	0.3	18 500		14	6.08	8.5	798
	Hg Te	0.2	22 000	160		6.429	8.42	670

The table gives the values of the band gap energy, electron and hole mobilities, high-frequency dielectric constant, crystal lattice constant, density and fusion temperature.

Bibliography

Ashcroft, N. W. and Mermin, M. D. 1976.
Solid State Physics. New York: Holt, Rinehart and Winston. A general solid state physics textbook.

Handbook on Semiconductors. 1981. Amsterdam: North-Holland Publishing Co. A comprehensive presentation of semiconductor physics.

Kittel, C. 1986.
Introduction to Solid State Physics (6th ed.). New York: John Wiley and Sons, Inc. A general solid state physics textbook.

Sze, S. M. ed., 1991.
Semiconductor Devices: Pioneering Papers. Singapore: World Scientific.

Willardson, R. K., Beer, A. C., and Weber, E. R., eds.
Semiconductors and Semimetals (37 volumes.) Boston: Academic Press. Contains reviews on the various aspects of semiconductor research.

Index